Conversion Factors for Commonly Used Quantities in Heat Trans

Quantity	SI → English	
Specific heat	1 J/(kg · K) = 2.3886×10^{-4} Btu/(lb_m · °F)	1 Btu/(lb_m
Surface tension	1 N/m = 0.06852 lb_f/ft	1 lb_f/ft = 14.594 N/m 1 dyne/cm = 1×10^{-3} N/m
Temperature	T(K) = T(°C) + 273.15 = T(°R)/1.8 = [T(°F) + 459.67]/1.8 T(°C) = [T(°F) − 32]/1.8	T(°R) = 1.8T(K) = T(°F) + 459.67 T(°F) = 1.8T(°C) + 32 = 1.8[T(K) − 273.15] + 32
Temperature difference	1 K = 1°C = 1.8°R = 1.8°F	1°R = 1°F = (5/9)K = (5/9)°C
Thermal conductivity	1 W/(m · K) = 0.57782 Btu/(h · ft · °F)	1 Btu/(h · ft · °F) = 1.731 W/m · K 1 kcal/(h · m · °C) = 1.163 W/m · K
Thermal diffusivity	1 m^2/s = 10.7639 ft^2/s	1 ft^2/s = 0.0929 m^2/s 1 ft^2/h = 2.581×10^{-5} m^2/s
Thermal resistance	1 K/W = 0.5275°F · h/Btu	1°F · h/Btu = 1.896 K/W
Velocity	1 m/s = 3.2808 ft/s	1 ft/s = 0.3048 m/s
Viscosity (dynamic)	1 N · s/m^2 = 0.672 lb_m/(ft · s) = 2419.1 lb_m/(ft · h) = 5.8016×10^{-6} lb_f · h/ft^2	1 lb_m/(ft · s) = 1.488 N · s/m^2 1 lb_m/(ft · h) = 4.133×10^{-4} N · s/m^2 1 centipoise = 0.001 N · s/m^2
Viscosity (kinematic)	1 m^2/s = 10.7639 ft^2/s	1 ft^2/s = 0.0929 m^2/s 1 ft^2/h = 2.581×10^{-5} m^2/s
Volume	1m^3 = 35.3134 ft^3	1 ft^3 = 0.02832 m^3 1 in^3 = 1.6387×10^{-5} m^3 1 gal (U.S. liq.) = 0.003785 m^3
Volume flow rate	1 m^3/s = 35.3134 ft^3/s = 1.2713×10^5 ft^3/h	1 ft^3/h = 7.8658×10^{-6} m^3/s 1 ft^3/s = 2.8317×10^{-2} m^3/s

*Some units in this column belong to the cgs and mks metric systems.

†Definitions of the units of energy which are based on thermal phenomena:

1 Btu = energy required to raise 1 lb_m of water 1°F at 68°F

1 cal = energy required to raise 1 g of water 1°C at 20°C

Seventh Edition

Principles of HEAT TRANSFER

Seventh Edition

Principles of
HEAT TRANSFER

Frank Kreith
Professor Emeritus, University of Colorado at Boulder, Boulder, Colorado

Raj M. Manglik
Professor, University of Cincinnati, Cincinnati, Ohio

Mark S. Bohn
Former Vice President, Engineering Rentech, Inc., Denver, Colorado

Australia • Brazil • Japan • Korea • Mexico • Singapore • Spain • United Kingdom • United States

Principles of Heat Transfer, Seventh Edition
Authors Frank Kreith, Raj M. Manglik, Mark S. Bohn

Publisher, Global Engineering: Christopher M. Shortt

Senior Developmental Editor: Hilda Gowans

Editorial Assistant: Tanya Altieri

Team Assistant: Carly Rizzo

Marketing Manager: Lauren Betsos

Media Editor: Chris Valentine

Director, Content and Media Production: Barbara Fuller-Jacobsen

Content Project Manager: Cliff Kallemeyn

Production Service: RPK Editorial Services, Inc.

Copyeditor: Fred Dahl

Proofreader: Martha McMaster/Erin Wagner

Indexer: Shelly Gerger-Knechtl

Compositor: Integra

Senior Art Director: Michelle Kunkler

Cover Designer: Andrew Adams

Cover Image: Abengoa Solar; SkyTrough™ © Shirley Speer/SkyFuel, Inc. 2009

Internal Designer: Jennifer Lambert/jen2design

Text and Image Permissions Researcher: Kristiina Paul

First Print Buyer: Arethea Thomas

Library of Congress Control Number: 2010922630

ISBN-13: 978-0-495-66770-4

ISBN-10: 0-495-66770-6

Cengage Learning
200 First Stamford Place, Suite 400
Stamford, CT 06902
USA

Cengage Learning is a leading provider of customized learning solutions with office locations around the globe, including Singapore, the United Kingdom, Australia, Mexico, Brazil, and Japan. Locate your local office at: **international.cengage.com/region**.

Cengage Learning products are represented in Canada by Nelson Education Ltd.

For your course and learning solutions, visit **www.cengage.com/engineering**.

Purchase any of our products at your local college store or at our preferred online store **www.CengageBrain.com**.

Printed in the United States of America
1 2 3 4 5 6 7 13 12 11 10

To our students
all over the world

PREFACE

When a textbook that has been used by more than a million students all over the world reaches its seventh edition, it is natural to ask, "What has prompted the authors to revise the book?" The basic outline of how to teach the subject of heat transfer, which was pioneered by the senior author in its first edition, published 60 years ago, has now been universally accepted by virtually all subsequent authors of heat transfer texts. Thus, the organization of this book has essentially remained the same over the years, but newer experimental data and, in particular the advent of computer technology, have necessitated reorganization, additions, and integration of numerical and computer methods of solution into the text.

The need for a new edition was prompted primarily by the following factors: 1) When a student begins to read a chapter in a textbook covering material that is new to him or her, it is useful to outline the kind of issues that will be important. We have, therefore, introduced at the beginning of each chapter a summary of the key issues to be covered so that the student can recognize those issues when they come up in the chapter. We hope that this pedagogic technique will help the students in their learning of an intricate topic such as heat transfer. 2) An important aspect of learning engineering science is to connect with practical applications, and the appropriate modeling of associated systems or devices. Newer applications, illustrative modeling examples, and more current state-of-the art predictive correlations have, therefore, been added in several chapters in this edition. 3) The sixth edition used MathCAD as the computer method for solving real engineering problems. During the ten years since the sixth edition was published, the teaching and utilization of MathCAD has been supplanted by the use of MATLAB. Therefore, the MathCAD approach has been replaced by MATLAB in the chapter on numerical analysis as well as for the illustrative problems in the real world applications of heat transfer in other chapters. 4) Again, from a pedagogic perspective of assessing student learning performance, it was deemed important to prepare general problems that test the students' ability to absorb the main concepts in a chapter. We have, therefore, provided a set of Concept Review Questions that ask a student to demonstrate his or her ability to understand the new concepts related to a specific area of heat transfer. These review questions are available on the book website in the Student Companion Site at www.cengage.com/engineering. Solutions to the Concepts Review Questions are available for Instructors on the same website. 5) Furthermore, even though the sixth edition had many homework problems for the students, we have introduced some additional problems that deal directly with topics of current interest such as the space program and renewable energy.

The book is designed for a one-semester course in heat transfer at the junior or senior level. However, we have provided some flexibility. Sections marked with asterisks can be omitted without breaking the continuity of the presentation. If all the sections marked with an asterisk are omitted, the material in the book can be covered in a single quarter. For a full semester course, the instructor can select five or six of these sections and thus emphasize his or her own areas of interest and expertise.

The senior author would also like to express his appreciation to Professor Raj M. Manglik, who assisted in the task of updating and refreshing the sixth edition to bring it up to speed for students in the twenty-first century. In turn, Raj Manglik is profoundly grateful for the opportunity to join in the authorship of this revised edition, which should continue to provide students worldwide an engaging learning experience in heat transfer. Although Dr. Mark Bohn decided not to participate in the seventh edition, we wish to express our appreciation for his previous contribution. In addition, the authors would like to acknowledge the contributions by the reviewers of the sixth edition who have provided input and suggestions for the update leading to the new edition of the book: B. Rabi Baliga, McGill University; F.C. Lai, University of Oklahoma; S. Mostafa Ghiaasiaan, Georgia Tech; Michael Pate, Iowa State University; and Forman A. Williams, University of California, San Diego. The authors would also like to thank Hilda Gowans, the Senior Developmental Editor for Engineering at Cengage Learning, who has provided support and encouragement throughout the preparation of the new edition. On a more personal level, Frank Kreith would like to express his appreciation to his assistant, Bev Weiler, who has supported his work in many tangible and intangible ways, and to his wife, Marion Kreith, whose forbearance with the time taken in writing books has been of invaluable help. Raj Manglik would like to thank his graduate students Prashant Patel, Rohit Gupta, and Deepak S. Kalaikadal for the computational solutions and algorithms in the book. Also, he would like to express his fond gratitude to his wife, Vandana Manglik, for her patient encouragement during the long hours needed in this endeavor, and to his children, Aditi and Animaesh, for their affection and willingness to forego some of our shared time.

CONTENTS

Chapter 4 Analysis of Convection Heat Transfer 230

Chapter 5 Natural Convection 296

NOMENCLATURE

Symbol	Quantity	International System of Units	English System of Units
a	velocity of sound	m/s	ft/s
a	acceleration	m/s^2	ft/s^2
A	area; A_c cross-sectional area; A_p, projected area of a body normal to the direction of flow; A_q, area through which rate of heat flow is q; A_s, surface area; A_o, outside surface area; A_i, inside surface area	m^2	ft^2
b	breadth or width	m	ft
c	specific heat; c_p, specific heat at constant pressure; c_v, specific heat at constant volume	J/kg K	Btu/lb_m °F
C	constant		
C	thermal capacity	J/K	Btu/°F
C	hourly heat capacity rate in Chapter 8; C_c, hourly heat capacity rate of colder fluid in a heat exchanger; C_h, hourly heat capacity rate of warmer fluid in a heat exchanger	W/K	Btu/h °F
C_D	total drag coefficient		
C_f	skin friction coefficient; C_{fx}, local value of C_f at distance x from leading edge; $\bar{C}_f$, average value of C_f defined by Eq. (4.31)		
d, D	diameter; D_H, hydraulic diameter; D_o, outside diameter; D_i, inside diameter	m	ft
e	base of natural or Napierian logarithm		
e	internal energy per unit mass	J/kg	Btu/lb_m
E	internal energy	J	Btu
E	emissive power of a radiating body; E_b, emissive power of blackbody	W/m^2	Btu/h ft^2

(Continued)

Symbol	Quantity	International System of Units	English System of Units
E_λ	monochromatic emissive power per micron at wavelength λ	$W/m^2\ \mu m$	Btu/h ft^2 micron
$\mathscr{E}$	heat exchanger effectiveness defined by Eq. (8.22)		
f	Darcy friction factor for flow through a pipe or a duct, defined by Eq. (6.13)		
f	friction coefficient for flow over banks of tubes defined by Eq. (7.37)		
F	force	N	lb_f
F_T	temperature factor defined by Eq. (9.119)		
F_{1-2}	geometric shape factor for radiation from one blackbody to another		
$\mathscr{F}_{1-2}$	geometric shape and emissivity factor for radiation from one graybody to another		
g	acceleration due to gravity	m/s^2	ft/s^2
g_c	dimensional conversion factor	1.0 kg m/N s^2	32.2 ft $lb_m/lb_f\ s^2$
G	mass flow rate per unit area ($G = \rho U_\infty$)	$kg/m^2\ s$	$lb_m/h\ ft^2$
G	irradiation incident on unit surface in unit time	W/m^2	Btu/h ft^2
h	enthalpy per unit mass	J/kg	Btu/lb_m
h_c	local convection heat transfer coefficient	W/m^2 K	Btu/h ft^2°F
$\bar{h}$	combined heat transfer coefficient $\bar{h} = \bar{h}_c + \bar{h}_r$; h_b, heat transfer coefficient of a boiling liquid, defined by Eq. (10.1); $\bar{h}_c$, average convection heat transfer coefficient; $\bar{h}_r$, average heat transfer coefficient for radiation	W/m^2 K	Btu/h ft^2°F
h_{fg}	latent heat of condensation or evaporation	J/kg	Btu/lb_m
i	angle between sun direction and surface normal	rad	deg
i	electric current	amp	amp
I	intensity of radiation	W/sr	Btu/h sr
I_λ	intensity per unit wavelength	W/sr μm	Btu/h sr micron
J	radiosity	W/m^2	Btu/h ft^2

Symbol	Quantity	International System of Units	English System of Units
k	thermal conductivity; k_s, thermal conductivity of a solid; k_f, thermal conductivity of a fluid	W/m K	Btu/h ft °F
K	thermal conductance; K_k, thermal conductance for conduction heat transfer; K_c, thermal conductance for convection heat transfer; K_r, thermal conductance for radiation heat transfer	W/K	Btu/h °F
l	length, general	m	ft or in.
L	length along a heat flow path or characteristic length of a body	m	ft or in.
L_f	latent heat of solidification	J/kg	Btu/lb_m
$\dot{m}$	mass flow rate	kg/s	lb_m/s or lb_m/h
M	mass	kg	lb_m
$\mathcal{M}$	molecular weight	gm/gm-mole	lb_m/lb-mole
N	number in general; number of tubes, etc.		
p	static pressure; p_c, critical pressure; p_A, partial pressure of component A	N/m^2	psi, lb_f/ft^2, or atm
P	wetted perimeter	m	ft
q	rate of heat flow; q_k, rate of heat flow by conduction; q_r, rate of heat flow by radiation; q_c, rate of heat flow by convection; q_b, rate of heat flow by nucleate boiling	W	Btu/h
$\dot{q}_G$	rate of heat generation per unit volume	W/m^3	Btu/h ft^3
q''	heat flux	W/m^2	Btu/h ft^2
Q	quantity of heat	J	Btu
$\dot{Q}$	volumetric rate of fluid flow	m^3/s	ft^3/h
r	radius; r_H, hydraulic radius; r_i, inner radius; r_o, outer radius	m	ft or in.
R	thermal resistance; R_c, thermal resistance to convection heat transfer; R_k, thermal resistance to conduction heat transfer; R_r, thermal resistance to radiation heat transfer	K/W	h °F/Btu
R_e	electrical resistance	ohm	ohm

(Continued)

Symbol	Quantity	International System of Units	English System of Units
$\mathcal{R}$	perfect gas constant	8.314 J/K kg-mole	1545 ft lb_f/lb-mole °F
S	shape factor for conduction heat flow		
S	spacing	m	ft
S_L	distance between centerlines of tubes in adjacent longitudinal rows	m	ft
S_T	distance between centerlines of tubes in adjacent transverse rows	m	ft
t	thickness	m	ft
T	temperature; T_b, temperature of bulk of fluid; T_f, mean film temperature; T_s, surface temperature; T_∞, temperature of fluid far removed from heat source or sink; T_m, mean bulk temperature of fluid flowing in a duct; T_{sv}, temperature of saturated vapor; T_{sl}, temperature of a saturated liquid; T_{fr}, freezing temperature; T_l, liquid temperature; T_{as}, adiabatic wall temperature	K or °C	R or °F
u	internal energy per unit mass	J/kg	Btu/lb_m
u	time average velocity in x direction; u', instantaneous fluctuating x component of velocity; $\bar{u}$, average velocity	m/s	ft/s or ft/h
U	overall heat transfer coefficient	W/m^2 K	Btu/h ft^2 °F
U_∞	free-stream velocity	m/s	ft/s
v	specific volume	m^3/kg	ft^3/lb_m
v	time average velocity in y direction; v', instantaneous fluctuating y component of velocity	m/s	ft/s or ft/h
V	volume	m^3	ft^3
w	time average velocity in z direction; w', instantaneous fluctuating z component of velocity	m/s	ft/s
w	width	m	ft or in.
$\dot{W}$	rate of work output	W	Btu/h
x	distance from the leading edge; x_c, distance from the leading edge where flow becomes turbulent	m	ft

Symbol	Quantity	International System of Units	English System of Units
x	coordinate	m	ft
x	quality		
y	coordinate	m	ft
y	distance from a solid boundary measured in direction normal to surface	m	ft
z	coordinate	m	ft
Z	ratio of hourly heat capacity rates in heat exchangers		
	Greek Letters		
α	absorptivity for radiation; α_λ, monochromatic absorptivity at wavelength λ		
α	thermal diffusivity $= k/\rho c$	m^2/s	ft^2/s
β	temperature coefficient of volume expansion	1/K	1/R
β_k	temperature coefficient of thermal conductivity	1/K	1/R
γ	specific heat ratio, c_p/c_v		
Γ	body force per unit mass	N/kg	lb_f/lb_m
Γ_c	mass rate of flow of condensate per unit breadth for a vertical tube	kg/s m	lb_m/h ft
δ	boundary-layer thickness; δ_h, hydrodynamic boundary-layer thickness; δ_{th}, thermal boundary-layer thickness	m	ft
Δ	difference between values		
ε	packed bed void fraction		
ε	emissivity for radiation; ε_λ, monochromatic emissivity at wavelength λ; ε_ϕ, emissivity in direction of ϕ		
ε_H	thermal eddy diffusivity	m^2/s	ft^2/s
ε_M	momentum eddy diffusivity	m^2/s	ft^2/s
ζ	ratio of thermal to hydrodynamic boundary-layer thickness, δ_{th}/δ_h		

(Continued)

Symbol	Quantity	International System of Units	English System of Units
η_f	fin efficiency		
θ	time	s	h or s
λ	wavelength; λ_{max}, wavelength at which monochromatic emissive power $E_{b\lambda}$ is a maximum	μm	micron
λ	latent heat of vaporization	J/kg	Btu/lb_m
μ	absolute viscosity	N s/m^2	lb_m/ft s
ν	kinematic viscosity, μ/ρ	m^2/s	ft^2/s
ν_r	frequency of radiation	1/s	1/s
ρ	mass density, $1/v$; ρ_l, density of liquid; ρ_v, density of vapor	kg/m^3	lb_m/ft^3
ρ	reflectivity for radiation		
τ	shearing stress; τ_s, shearing stress at surface; τ_w, shear at wall of a tube or a duct	N/m^2	lb_f/ft^2
τ	transmissivity for radiation		
σ	Stefan–Boltzmann constant	W/m^2 K^4	Btu/h ft^2 R^4
σ	surface tension	N/m	lb_f/ft
ϕ	angle	rad	rad
ω	angular velocity	rad/s	rad/s
ω	solid angle	sr	steradian
	Dimensionless Numbers		
Bi	Biot number $= \bar{h}L/k_s$ or $\bar{h}r_o/k_s$		
Fo	Fourier modulus $= a\theta/L^2$ or $a\theta/r_o^2$		
Gz	Graetz number $= (\pi/4)\text{RePr}(D/L)$		
Gr	Grashof number $= \beta_g L^3\ \Delta T/\nu^2$		
Ja	Jakob number $= (T_\infty - T_{sat})c_{pl}/h_{fg}$		
M	Mach number $= U_\infty/a$		
Nu_x	local Nusselt number at a distance x from leading edge, $h_c x/k_f$		
$\overline{Nu}_L$	average Nusselt number for blot plate, $\bar{h}_c L/k_f$		
$\overline{Nu}_D$	average Nusselt number for cylinder, $\bar{h}_c D/k_f$		

Symbol	Quantity
Pe	Peclet number = RePr
Pr	Prandtl number = $c_p\mu/k$ or ν/α
Ra	Rayleigh number = GrPr
Re_L	Reynolds number = $U_\infty \rho L/\mu$;
$\mathrm{Re}_x = U_\infty \rho x/\mu$	Local value of Re at a distance x from leading edge
$\mathrm{Re}_D = U_\infty \rho D/\mu$	Diameter Reynolds number
$\mathrm{Re}_b = D_b G_b/\mu_l$	Bubble Reynolds number
θ	Boundary Fourier modulus = $\bar{h}^2 a\theta/k_s^2$
St	Stanton number = $\bar{h}_c/\rho U_\infty c_p$ or $\overline{\mathrm{Nu}}$/RePr
	Miscellaneous
$a > b$	a greater than b
$a < b$	a smaller than b
$\propto$	proportional sign
$\simeq$	approximately equal sign
∞	infinity sign
Σ	summation sign

Seventh Edition

Principles of

HEAT TRANSFER

CHAPTER 1

Basic Modes of Heat Transfer

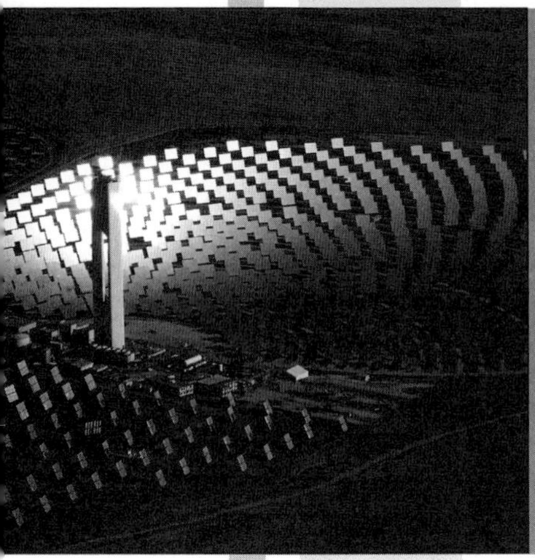

A typical solar power station with its arrays or field of heliostats and the solar power tower in the foreground; such a system involves all modes of heat transfer–radiation, conduction, and convection, including boiling and condensation.

Source: Photo courtesy of Abengoa Solar.

Concepts and Analyses to Be Learned

Heat is fundamentally transported, or "moved," by a temperature gradient; it *flows* or is *transferred* from a high temperature region to a low temperature one. An understanding of this process and its different mechanisms requires you to connect principles of thermodynamics and fluid flow with those of heat transfer. The latter has its own set of concepts and definitions, and the foundational principles among these are introduced in this chapter along with their mathematical descriptions and some typical engineering applications. A study of this chapter will teach you:

- How to apply the basic relationship between thermodynamics and heat transfer.
- How to model the concepts of different modes or mechanisms of heat transfer for practical engineering applications.
- How to use the analogy between heat and electric current flow, as well as thermal and electrical resistance, in engineering analysis.
- How to identify the difference between steady state and transient modes of heat transfer.

1.1 The Relation of Heat Transfer to Thermodynamics

Whenever a temperature gradient exists within a system, or whenever two systems at different temperatures are brought into contact. energy is transferred. The process by which the energy transport taltes place is known as *heat transfer*. The thing in transit, called heat, cannot be observed or measured directly. However, its effects can be identified and quantified through measurements and analysis. The flow of heat, like the performance of work, is a process by which the initial energy of a system is changed.

The branch of science that deals with the relation between heat and other forms of energy, including mechanical work in particular, is called *thermodynamics*. Its principles, like all laws of nature, are based on observations and have been generalized into laws that are believed to hold for all processes occurring in nature because no exceptions have ever been found. For example, the first law of thermodynamics states that energy can be neither created nor destroyed but only changed from one form to another. It governs all energy transformations quantitatively, but places no restrictions on the direction of the transformation. It is known, however, from experience that no process is possible whose sole result is the net transfer of heat from a region of lower temperature to a region of higher temperature. This statement of experimental truth is known as the second law of thermodynamics.

All heat transfer processes involve the exchange and/or conversion of energy. They must, therefore, obey the first as well as the second law of thermodynamics. At first glance, one might therefore be tempted to assume that the principles of heat transfer can be derived from the basic laws of thermodynamics. This conclusion, however, would be erroneous, because classical thermodynamics is restricted primarily to the study of equilibrium states including mechanical, chemical, and thermal equilibriums, and is therefore, by itself, of little help in determining quantitavely the transformations that occur from a lack of equilibrium in engineering processes. Since heat flow is the result of temperature nonequilibriuin, its quantitative treatment must be based on other branches of science. The same reasoning applies to other types of transport processes such as mass transfer and diffusion.

Limitations of Classical Thermodynamics Classical thermodynamics deals with the states of systems from a macroscopic view and makes no hypotheses about the structure of matter. To perform a thermodynamic analysis it is necessary to describe the state of a system in terms of gross characteristics, such as pressure, volume, and temperature, that can be measured directly and involve no special assumptions regarding the structure of matter. These variables (or thermodynamic properties) are of significance for the system as a whole only when they are uniform throughout it, that is, when the system is in equilibrium. Thus, classical thermodynamics is not concerned with the details of a process but rather with equilibrium states and the relations among them. The processes employed in a thermodynamic analysis are idealized processes devised to give information concerning equilibrium states.

The schematic example of an automobile engine in Fig. 1.1 is illustrative of the distinctions between thermodynamic and heat transfer analysis. While the basic law of energy conservation is applicable in both, from a thermodynamic viewpoint, the amount of heat transferred during a process simply equals the difference between the energy change of the system and the work done. It is evident that this type of analysis considers neither the mechanism of heat flow nor the time required to transfer the heat. It simply prescribes how much heat to supply to or reject from a system during a process between specified end states without considering whether, or how, this could be accomplished. The question of how long it would take to transfer a specified amount of heat, via different mechanisms or modes of heat transfer and their processes (both in terms of space and time) by which they occur, although of great practical importance, does not usually enter into the thermodynamic analysis.

Engineering Heat Transfer From an engineering viewpoint, the key problem is the determination of the *rate of heat transfer at a specified temperature difference.*

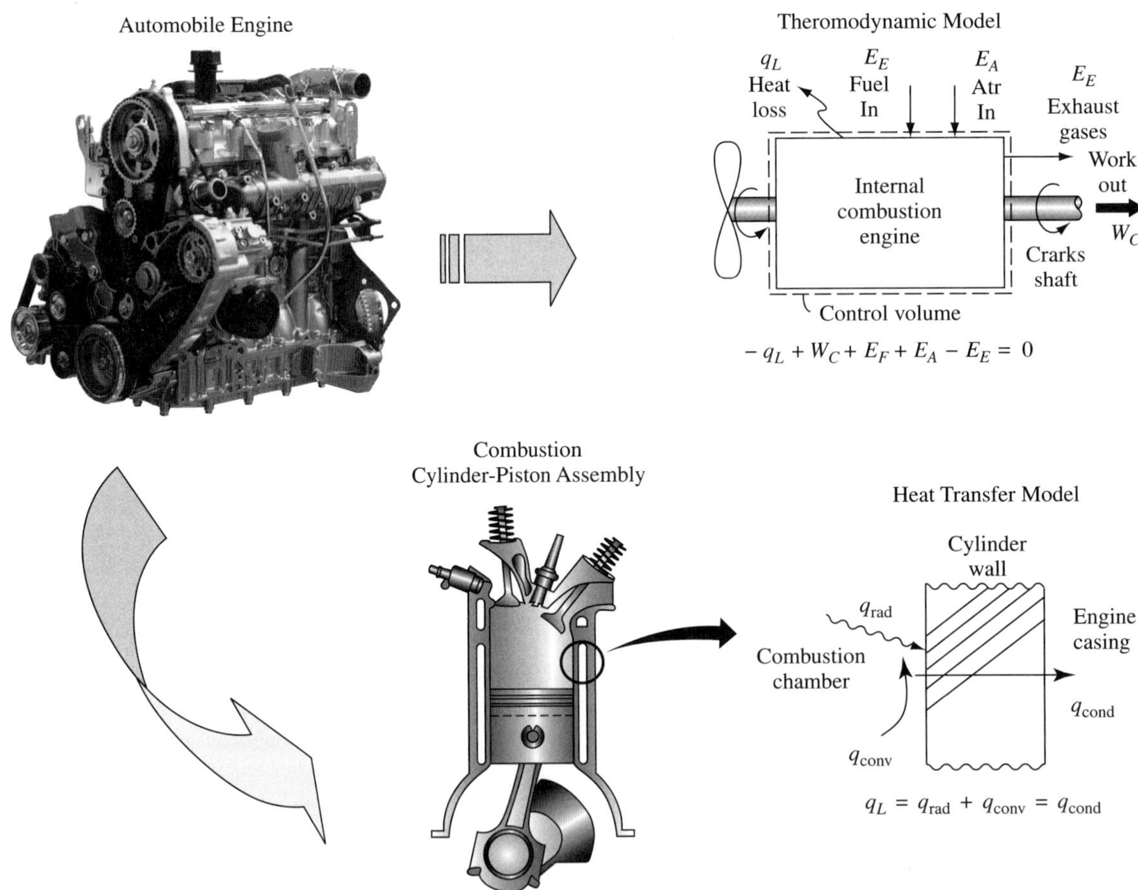

FIGURE 1.1 A classical thermodynamics model and a heat transfer model of a typical automobile (spark-ignition internal combustion) engine.

Source: Photo of automobile engine courtesy of Ajancso/shutterstock.

To estimate the cost, the feasibility, and the size of equipment necessary to transfer a specified amount of heat in a given time, a detailed heat transfer analysis must be made. The dimensions of boilers, heaters, refrigerators, and heat exchangers depends not only on the amount of heat to be transmitted but also on the rate at which the heat is to be transferred under given conditions. The successful operation of equipment components such as turbine blades, or the walls of combustion chambers, depends on the possibility of cooling certain metal parts by continuously removing heat from a surface at a rapid rate. A heat transfer analysis must also be made in the design of electric machines, transformers, and bearings to avoid conditions that will cause overheating and damage the equipment. The listing in Table 1.1, which by no means is comprehensive, gives an indication of the extensive significance of heat transfer and its different practical applications. These examples show that almost every branch of engineering encounters heat transfer problems, which shows that they are not capable of solution by thermodynamic reasoning alone but require an analysis based on the science of heat transfer.

In heat transfer, as in other branches of engineering, the successful solution of a problem requires assumptions and idealizations. It is almost impossible to describe physical phenomena exactly, and in order to express a problem in the form of an equation that can be solved, it is necessary to make some approximations. In electrical circuit calculations, for example, it is usually assumed that the values of the resistances, capacitances, and inductances are independent of the current flowing through them. This assumption simplifies the analysis but *may* in certain cases severely limit the accuracy of the results.

TABLE 1.1 Significance and diverse practical applications of heat transfer

Chemical, petrochemical, and process industry: Heat exchangers, reactors, reboilers, etc.

Power generation and distribution: Boilers, condensers, cooling towers, feed heaters, transformer cooling, transmission cable cooling, etc.

Aviation and space exploration: Gas turbine blade cooling, vehicle heat shields, rocket engine/nozzle cooling, space suits, space power generation, etc.

Electrical machines and electronic equipment: Cooling of motors, generators, computers and microelectronic devices, etc.

Manufacturing and material processing: Metal processing, heat treating, composite material processing, crystal growth, micromachining, laser machining, etc.

Transportation: Engine cooling, automobile radiators, climate control, mobile food storage, etc.

Fire and combustion

Health care and biomedical applications: Blood warmers, organ and tissue storage, hypothermia, etc.

Comfort heating, ventilation, and air-conditioning: Air conditioners, water heaters, furnaces, chillers, refrigerators, etc.

Weather and environmental changes

Renewable Energy System: Flat plate collectors, thermal energy storage, PV module cooling, etc.

It is important to keep in mind the assumptions, idealizations, and approximations made in the course of an analysis when the final results are interpreted. Sometimes insufficient information on physical properties make it necessary to use engineering approximations to solve a problem. For example, in the design of machine parts for operation at elevated temperatures, it may be necessary to estimate the propotional limit or the fatigue strength of the material from low-temperature data. To assure satisfactory operation of a particular part, the designer should apply a factor of safety to the results obtained from the analysis. Similar approximations are also necessary in heat transfer problems. Physical properties such as thermal conductivity or viscosity change with temperature, but if suitable average values are selected, the calculations can be considerably simplified without introducing an appreciable error in the final result. When heat is transferred from a fluid to a wall, as in a boiler, a scale forms under continued operation and reduces the rate of heat flow. To assure satisfactory operation over a long period of time, a factor of safety must be applied to provide for this contingency.

When it becomes necessary to make an assumption or approximation in the solution of a problem, the engineer must rely on ingenuity and past experience. There are no simple guides to new and unexplored problems, and an assumption valid for one problem may be misleading in another. Experience has shown, however, that the first requirement for making sound engineering assumptions or approximations is a complete and thorough physical understanding of the problem at hand. In the field of heat transfer, this means having familiarity not only with the laws and physical mechanisms of heat flow but also with those of fluid mechanics, physics, and mathematics.

Heat transfer can be defined as the transmission of energy from one region to another as a result of a temperature difference between them. Since differences in temperatures exist all over the universe, the phenomenn of heat flow are as universal as those associated with gravitational attractions. Unlike gravity, however, heat flow is governed not by a unique relationship but rather by a combination of various independent laws of physics.

Mechanisms of Heat Transfer The literature of heat transfer generally recognizes three distinct modes of heat transmission: conduction, radiation, and convection. Strictly speaking, only conduction and radiation should be classified as heat transfer processes, because only these two mechanisms depend for their operation on the mere existence of a temperature difference. The last of the three, convection, does not strictly comply with the definition of heat transfer because its operation also depends on mechanical mass transport. But since convection also accomplishes transmission of energy from regions of higher temperature to regions of lower temperature, the term "heat transfer by convection" has become generally accepted.

In Sections 1.3–1.5, we will survey the basic equations governing each of the three modes of heat transfer. Our initial aim is to obtain a broad perspective of the field without becoming involved in details. We shall therefore consider only simple cases. Yet it should be emphasized that in most natural situations heat is transferred not by one but by several mechanisms operating simultaneously. Hence, in Section 1.6 we will show how to combine the simple relations in situations when several heat transfer modes occur simultaneously and in Section 1.7 we will show how to reduce heat flow by insulation. And finally, in Section 1.8, we will illustrate how to use the laws of thermodynamics in heat transfer analyses.

1.2 Dimensions and Units

Before proceeding with the development of the concepts and principles governing the transmission or flow of heat, it is instructive to review the primary dimensions and units by which its descriptive variables are quantified. It is important not to confuse the meaning of the terms ***units*** and ***dimensions***. ***Dimensions*** are our basic concepts of measurements such as length, time, and temperature. For example, the distance between two points is a dimension called length. ***Units*** are the means of expressing dimensions numerically, for instance, meter or foot for length; second or hour for time. Before numerical calculations can be made, dimensions must be quantified by units.

Several different systems of units are in use throughout the world. The SI system (Systeme international d'unites) has been adopted by the International Organization for Standardization and is recommended by most U.S. national standard organizations. Therefore we will primarily use the SI system of units in this book. In the United States, however, the English system of units is still widely used. It is therefore important to be able to change from one set of units to another. To be able to communicate with engineers who are still in the habit of using the English system, several examples and exercise problems in the book will use the English system.

The basic SI units are those for length, mass, time, and temperature. The unit of force, the newton, is obtained from Newton's second law of motion, which states that force is proportional to the time rate of change of momentum. For a given mass, Newton's law can be written in the form

$$F = \frac{1}{g_c} ma \tag{1.1}$$

where F is the force, m is the mass, a is the acceleration, and g_c is a constant whose numerical value and units depend on those selected for F, m, and a.

In the SI system the unit of force, the newton, is defined as

$$1 \text{ newton} = \frac{1}{g_c} \times 1 \text{ kg} \times 1 \text{ m/s}^2$$

Thus, we see that

$$g_c = 1 \text{ kg m/newton s}^2$$

In the English system we have the relation

$$1 \text{ lb}_f = \frac{1}{g_c} \times 1 \text{ lb} \times g \text{ ft/s}^2$$

The numerical value of the conversion constant g_c is determined by the acceleration imparted to a 1-lb mass by a 1-lb force, or

$$g_c = 32.174 \text{ ft lb}_m/\text{lb}_f \text{ s}^2$$

The weight of a body, W, is defined as the force exerted on the body by gravity. Thus

$$W = \frac{g}{g_c} m$$

where g is the local acceleration due to gravity. Weight has the dimensions of a force and a 1-kg_{mass} will weigh 9.8 N at sea level.

It should be noted that g and g_c are not similar quantities. The gravitational acceleration g depends on the location and the altitude, whereas g_c is a constant whose value depends on the system of units. One of the great conveniences of the SI system is that g_c is numerically equal to one and therefore need not be shown specifically. In the English system, on the other hand, the omission of g_c will affect the numerical answer, and it is therefore imperative that it be included and clearly displayed in analysis, especially in numerical calculations.

With the fundamental units of meter, kilogram, second, and kelvin, the units for both force and energy or heat are derived units. For quantifying heat, rate of heat transfer, its flux, and its temperature, the units employed as per the international convention are given in Table 1.2. Also listed are their counterparts in English units, along with the respective conversion factors, in cognizance of the fact that such units are still prevalent in practice in the United States. The joule (newton meter) is the only energy unit in the SI system, and the watt (joule per second) is the corresponding unit of power. In the engineering system of units, on the other hand, the Btu (British thermal unit) is the unit for heat or energy. It is defined as the energy required to raise the temperature of 1 lb of water by 1°F at 60°F and one atmosphere pressure.

The SI unit of temperature is the kelvin, but use of the Celsius temperature scale is widespread and generally considered permissible. The kelvin is based on the thermodynamic scale, while zero on the Celsius scale (0°C) corresponds to the freezing temperature of water and is equivalent to 273.15 K on the thermodynamic scale. Note, however, that temperature differences are numerically equivalent in K and °C, since 1 K is equal to 1°C.

In the English system of units, the temperature is usually expressed in degrees Fahrenheit (°F) or, on the thermodynamic temperature scale, in degrees Rankine (°R). Here, 1 K is equal to 1.8°R and conversions for other temperature scales are given

$$^{\circ}\mathrm{C} = \frac{^{\circ}\mathrm{F} - 32}{1.8}$$

TABLE 1.2 Dimensions and units of heat and temperature

Quantity	SI units	English units	Conversion
Q, quantity of heat	J	Btu	$1\ \mathrm{J} = 9.4787 \times 10^{-4}$ Btu
q, rate of heat transfer	J/s or W	Btu/h	1 W = 3.4123 Btu/h
q'', heat flux	$\mathrm{W/m^2}$	$\mathrm{Btu/h \cdot ft^2}$	$1\ \mathrm{W/m^2} = 0.3171\ \mathrm{Btu/h \cdot ft^2}$
T, temperature	K	°R or °F	T°C = (T°F−32)/1.8
	[K] = [°C] + 273.15	[R] = [°F] + 459.67	T K = T°R/1.8

EXAMPLE 1.1 A masonry brick wall of a house has an inside surface temperature of 55°F and an average outside surface temperature of 45°F. The wall is 1.0 ft thick, and because of the temperature difference, the heat loss through the wall per square foot is $3.4\ \mathrm{Btu/h \cdot ft^2}$. Express the heat loss in SI units. Also, calculate the value of this heat

loss for a 100-ft^2 surface over a 24-h period if the house is heated by an electric resistance heater and the cost of electricity is 10 ¢/kWh.

SOLUTION The rate of heat loss per unit surface area in SI units is

$$q'' = 3.4\left(\frac{\text{Btu}}{\text{ft}^2\text{h}}\right) \times 0.2931\left(\frac{\text{W}}{\text{Btu/h}}\right) \times \frac{1}{0.0929}\left(\frac{\text{ft}^2}{\text{m}^2}\right) = 10.72[\text{W/m}^2]$$

The total heat loss to the environment over the specified surface area of the house wall in 24 hours is

$$Q = 3.4\left(\frac{\text{Btu}}{\text{ft}^2\text{h}}\right) \times 100(\text{ft}^2) \times 24(\text{h}) = 8160\ [\text{Btu}]$$

This can be expressed in SI units as

$$Q = 8160 \times 0.2931 \times 10^{-3}\left(\frac{\text{kWh}}{\text{Btu}}\right) = 2.392\ [\text{kWh}]$$

And at 10 ¢/kW·h, this amounts to ≈ 24 ¢ as the cost of heat loss in 24 h.

1.3 Heat Conduction

Whenever a temperature gradient exists in a solid medium, heat will flow from the higher-temperature to the lower-temperature region. The rate at which heat is transferred by conduction, q_k, is proportional to the temperature gradient dT/dx times the area A through which heat is transferred:

$$q_k \propto A\frac{dT}{dx}$$

In this relation, $T(x)$ is the local temperature and x is the distance in the direction of the heat flow. The actual rate of heat flow depends on the thermal conductivity k, which is a physical property of the medium. For conduction through a homogeneous medium, the rate of heat transfer is then

$$q_k = -kA\frac{dT}{dx} \tag{1.2}$$

The minus sign is a consequence of the second law of thermodynamics, which requires that heat *must* flow in the direction from higher to lower temperature. As illustrated in Fig. 1.2 on the next page, the temperature gradient will be negative if the temperature decreases with increasing values of x. Therefore, if heat transferred in the positive x direction is to be a positive quantity, a negative sign must be inserted on the right side of Eq. (1.2).

Equation (1.2) defines the thermal conductivity. It is called Fourier's law of conduction in honor of the French scientist J. B. J. Fourier, who proposed it in 1822.

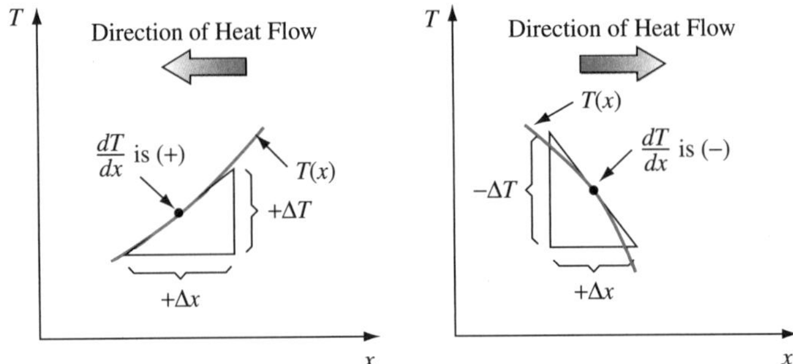

FIGURE 1.2 The sign convention for conduction heat flow.

The thermal conductivity in Eq. (1.2) is a material property that indicates the amount of heat that will flow per unit time across a unit area when the temperature gradient is unity. In the SI system, as reviewed in Section 1.2, the area is in square meters (m^2), the temperature in kelvins (K), x in meters (m), and the rate of heat flow in watts (W). The thermal conductivity therefore has the units of watts per meter per kelvin (W/m K). In the English system, which is still widely used by engineers in the United States, the area is expressed in square feet (ft^2), x in feet (ft), the temperature in degrees Fahrenheit (°F), and the rate of heat flow in Btu/h. Thus, k, has the units Btu/h ft °F. The conversion constant for k between the SI and English systems is

$$1 \text{ W/m K} = 0.578 \text{ Btu/h ft °F}$$

Orders of magnitude of the thermal conductivity of various types of materials are presented in Table 1.3. Although, in general, the thermal conductivity varies with temperature, in many engineering problems the variation is sufficiently small to be neglected.

TABLE 1.3 Thermal conductivities of some metals, nonmetallic solids, liquids, and gases

	Thermal Conductivity at 300 K (540 °R)	
Material	**W/m K**	**Btu/h ft °F**
Copper	399	231
Aluminum	237	137
Carbon steel, 1% C	43	25
Glass	0.81	0.47
Plastics	0.2–0.3	0.12–0.17
Water	0.6	0.35
Ethylene glycol	0.26	0.15
Engine oil	0.15	0.09
Freon (liquid)	0.07	0.04
Hydrogen	0.18	0.10
Air	0.026	0.02

1.3.1 Plane Walls

For the simple case of steady-state one-dimensional heat flow through a plane wall, the temperature gradient and the heat flow do not vary with time and the cross-sectional area along the heat flow path is uniform. The variables in Eq. (1.1) can then be separated, and the resulting equation is

$$\frac{q_k}{A}\int_0^L dx = -\int_{T_{\text{hot}}}^{T_{\text{cold}}} k\,dT = -\int_{T_1}^{T_2} k\,dT$$

The limits of integration can be checked by inspection of Fig. 1.3, where the temperature at the left face ($x = 0$) is uniform at T_{hot} and the temperature at the right face ($x = L$) is uniform at T_{cold}.

If k is independent of T, we obtain, after integration, the following expression for the rate of heat conduction through the wall:

$$q_k = \frac{Ak}{L}(T_{\text{hot}} - T_{\text{cold}}) = \frac{\Delta T}{L/Ak} \tag{1.3}$$

In this equation AT, the difference between the higher temperature T_{hot} and the lower temperature T_{cold} is the driving potential that causes the flow of heat. The quantity L/Ak is equivalent to a thermal resistance R_k that the wall offers to the flow of heat by conduction:

$$R_k = \frac{L}{Ak} \tag{1.4}$$

There is an analogy between heat flow systems and DC electric circuits. As shown in Fig. 1.3 the flow of electric current, i, is equal to the voltage potential, $E_1 - E_2$, divided by the electrical resistance, R_e, while the flow rate of heat, q_k, is equal to the temperature potential $T_1 - T_2$, divided by the thermal resistance R_k. This analogy is a convenient tool, especially for visualizing more complex situations, to be discussed

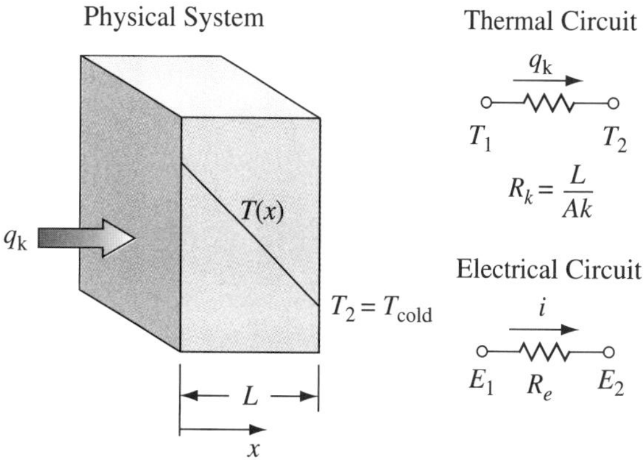

FIGURE 1.3 Temperature distribution for steady-state conduction through a plane wall and the analogy between thermal and electrical circuits.

in later chapters. The reciprocal of the thermal resistance is referred to as the thermal conductance K_k, defined by

$$K_k = \frac{Ak}{L} \tag{1.5}$$

> The French mathematician and physicist Jean Baptiste Joseph Fourier (1768–1830) and the younger German physicist Georg Ohm (1789–1854, the discoverer of Ohm's law that is the fundamental basis of electrical circuit theory) were contemporaries of sorts. It is believed that Ohm's mathematical treatment, published in *Die Galvanische Kette, Mathematisch Bearbeitet* (The Galvanic Circuit Investigated Mathematically) in 1827, was inspired by and based on the work of Fourier, who had developed the rate equation to describe heat flow in a conducting medium. Thus, the analogous treatment of the flow of heat and electricity, in terms of a thermal circuit with a thermal resistance between a temperature difference, is not surprising.

The ratio k/L in Eq. (1.5), the thermal conductance per unit area, is called the *unit thermal conductance* for conduction *heat flow*, while the reciprocal, L/k, is called the *unit thermal resistance*. The subscript k indicates that the transfer mechanism is conduction. The thermal conductance has the units of watts per kelvin temperature difference (Btu/h °F in the English system), and the thermal resistance has the units kelvin per watt (h °F/Btu in the engineering system). The concepts of resistance and conductance are helpful in the analysis of thermal systems where several modes of heat transfer occur simultaneously.

For many materials, the thermal conductivity can be approximated as a linear function of temperature over limited temperature ranges:

$$k(T) = k_0(1 + \beta_k T) \tag{1.6}$$

where β_k is an empirical constant and k_0 is the value of the conductivity at a reference temperature. In such cases, integration of Eq. (1.2) gives

$$q_k \frac{k_0 A}{L}\left[(T_1 - T_2) + \frac{\beta_k}{2}(T_1^2 - T_2^2)\right] \tag{1.7}$$

or

$$q_k = \frac{k_{av} A}{L}(T_1 - T_2) \tag{1.8}$$

where k_{av} is the value of k at the average temperature (T, $+ T_2)/2$.

The temperature distribution for a constant thermal ($\beta_k = 0$) and for thermal conductivities that increase ($\beta_k > 0$) and decrease ($\beta_k < 0$) with temperature are shown in Fig. 1.4.

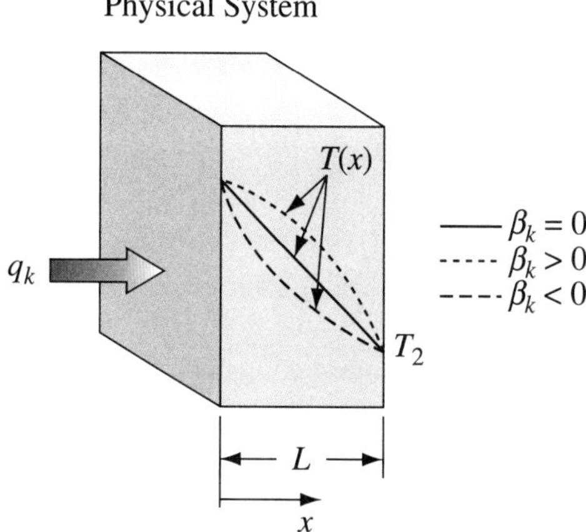

FIGURE 1.4 Temperature distribution in conduction through a plane wall with constant and variable thermal conductivity.

EXAMPLE 1.2 Calculate the thermal resistance and the rate of heat transfer through a pane of window glass (k = 0.81 W/m K) 1 m high, 0.5 m wide, and 0.5 cm thick, if the outer-surface temperature is 24°C and the inner-surface temperature is 24.5°C.

SOLUTION A schematic diagram of the system is shown in Fig. 1.5. Assume that steady state exists and that the temperature is uniform over the inner and outer surfaces. The thermal resistance to conduction R_k is from Eq. (1.4)

$$R_k = \frac{L}{kA} = \frac{0.005\text{ m}}{0.81\text{ W/m K} \times 1\text{ m} \times 0.5\text{ m}} = 0.0123\text{ K/W}$$

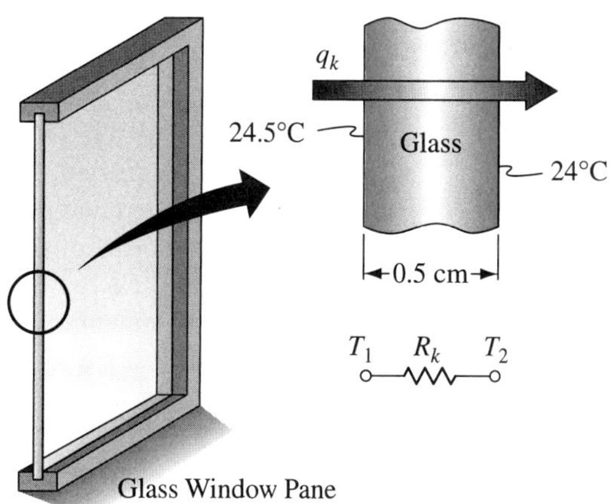

FIGURE 1.5 Heat transfer by conduction through a window pane.

The rate of heat loss from the interior to the exterior surface is obtained from Eq. (1.3):

$$q_k = \frac{T_1 - T_2}{R_k} = \frac{(24.5 - 24.0)°\text{C}}{0.0123 \text{ K/W}} = 40 \text{ W}$$

Note that a temperature difference of 1°C is equal to a temperature difference of 1 K. Therefore, °C and K can be used interchangeably when temperature differences are indicated. If a temperature level is involved, however, it must be remembered that zero on the Celsius scale (0°C) is equivalent to 273.15 K on the thermodynamic or absolute temperature scale and

$$T(\text{K}) = T(°\text{C}) + 273.15$$

1.3.2 Thermal Conductivity

According to Fourier's law, Eq. (1.2), the thermal conductivity is defined as

$$k \equiv \frac{q_k/A}{|dT/dx|}$$

For engineering calculations we generally use experimentally measured values of thermal conductivity, although for gases at moderate temperatures the kinetic theory of gases can be used to predict the experimental values accurately. Theories have also been proposed to calculate thermal conductivities for other materials, but in the case of liquids and solids, theories are not adequate to predict thermal conductivity with satisfactory accuracy [1, 2].

Table 1.3 lists values of thermal conductivity for several materials. Note that the best conductors are pure metals and the poorest ones are gases. In between lie alloys, nonmetallic solids, and liquids.

The mechanism of thermal conduction in a gas can be explained on a molecular level from basic concepts of the kinetic theory of gases. The kinetic energy of a molecule is related to its temperature. Molecules in a high-temperature region have higher velocities than those in a lower-temperature region. But molecules are in continuous random motion, and as they collide with one another they exchange energy as well as momentum. When a molecule moves from a higher-temperature region to a lower-temperature region, it transports kinetic energy from the higher- to the lower-temperature part of the system. Upon collision with slower molecules, it gives up some of this energy and increases the energy of molecules with a lower energy content. In this manner, thermal energy is transferred from higher- to lower-temperature regions in a gas by molecular action.

In accordance with the above simplified description, the faster molecules move, the faster they will transport energy. Consequently, the transport property that we have called thermal conductivity should depend on the temperature of the gas. A somewhat simplified analytical treatment (for example, see [3]) indicates that the thermal conductivity of a gas is proportional to the square root of the absolute temperature. At moderate pressures the space between molecules is large compared to the size of a molecule; thermal conductivity of gases is therefore essentially independent of pressure. The curves in Fig. 1.6(a) show how the thermal conductivities of some typical gases vary with temperature.

The basic mechanism of energy conduction in liquids is qualitatively similar to that in gases. However, molecular conditions in liquids are more difficult to describe and the details of the conduction mechanisms in liquids are not as well understood. The curves in Fig. 1.6(b) show the thermal conductivity of some nonmetallic liquids as a function of temperature. For most liquids, the thermal conductivity decreases with increasing temperature, but water is a notable exception. The thermal conductivity of liquids is insensitive to pressure except near the critical point. As a general rule, the thermal conductivity of liquids decreases with increasing molecular weight. For engineering purposes, values of the thermal conductivity of liquids are taken from tables as a function of temperature in the saturated state. Appendix 2 presents such data for several common liquids. Metallic liquids have much higher conductivities than nonmetallic liquids and their properties are listed separately in Tables 25 through 27 in Appendix 2.

According to current theories, solid materials consist of free electrons and atoms in a periodic lattice arrangement. Thermal energy can thus be conducted by two mechanisms: migration of free electrons and lattice vibration. These two effects are additive, but in general, the transport due to electrons is more effective than the transport

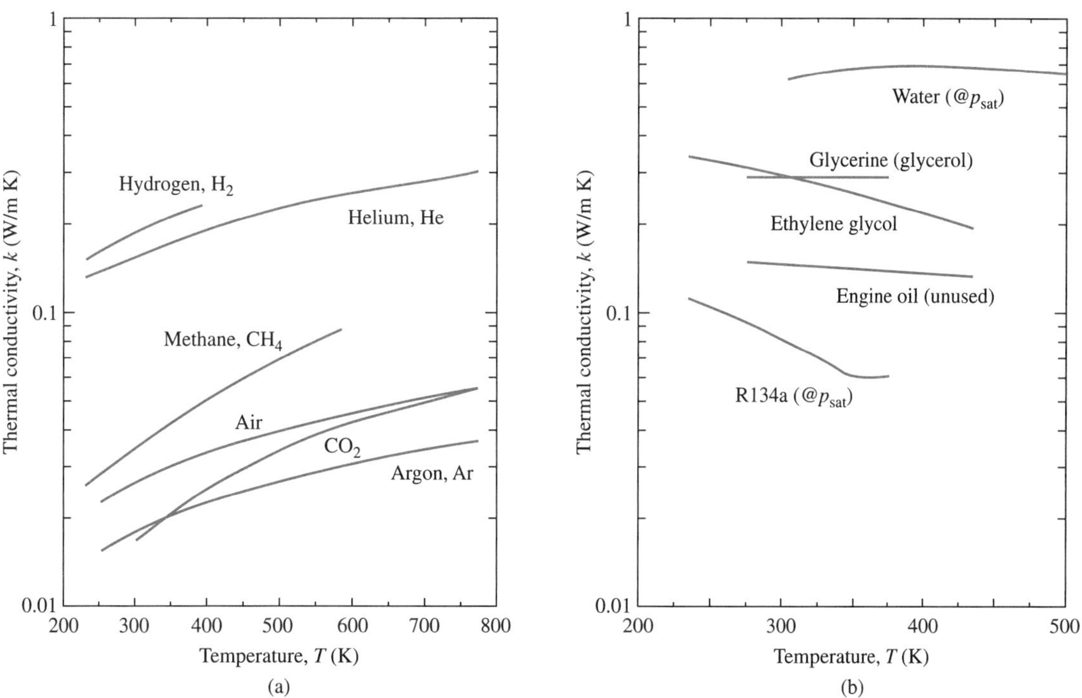

FIGURE 1.6. Variation of thermal conductivity with temperature of typical fluids: (a) gases and (b) liquids.

Sources for property data: *ASHRAE Handbook* 2005, Union Carbide (ethylene glycol) and Dow Chemicals (glycerine or glycerol).

due to vibrational energy in the lattice structure. Since electrons transport electric charge in a manner similar to the way in which they carry thermal energy from a higher- to a lower-temperature region, good electrical conductors are usually also good heat conductors, whereas good electrical insulators are poor heat conductors. In non-metallic solids, there is little or no electronic transport and the conductivity is therefore determined primarily by lattice vibration. Thus these materials have a lower thermal conductivity than metals. Thermal conductivities of some typical metals and alloys are shown in Fig. 1.7.

Thermal insulators [4] are an important group of solid materials for heat transfer design. These materials are solids, but their structure contains air spaces that are sufficiently small to suppress gaseous motion and thus take advantage of the low

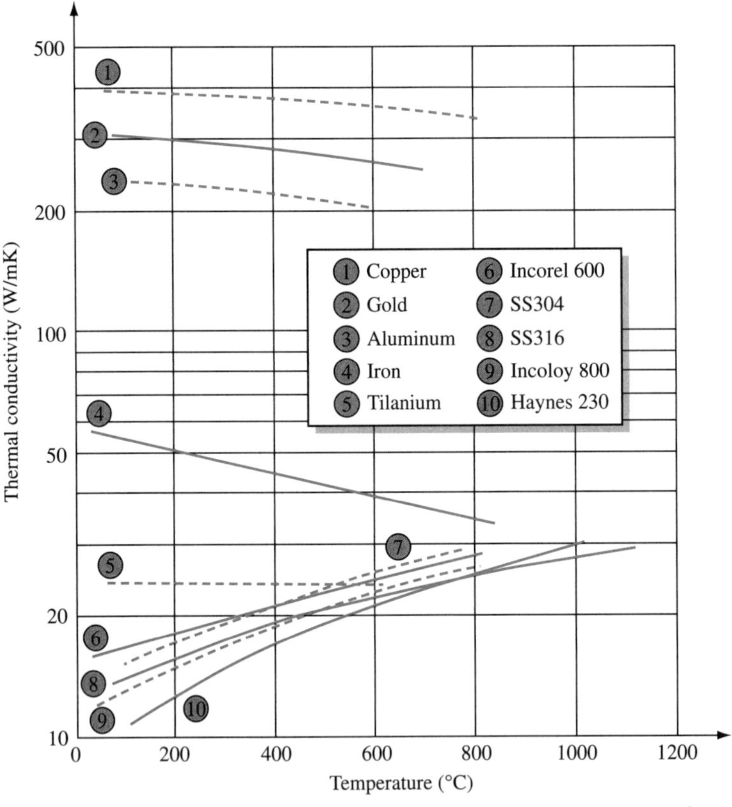

FIGURE 1.7 Variation of thermal conductivity with temperature for typical metallic elements and alloys.

Sources: Aluminum, Copper, Gold, Iron, and Titanium - Y. S. Touloukian, R. W. Powell, C. Y. Ho, and P. G. Klemens, Thermophysical Properties of Matter, Vol. 1, Thermal Conductivity Metallic Elements and Alloys, IFI/Plenum, New York, 1970. Stainless Steel 304 and 316 - D. Pecjner and I. M. Bernstein, Handbook of Stainless Steels, McGraw-Hill, New York, 1977. Inconel 600 and Incoloy 800 - Huntington Alloys, Huntington Alloys Handbook, Fifth Ed. 1970. Haynes 230 - Haynes International, Haynes Alloy No. 230 (Inconel and Incoloy are registered trademarks of Huntington Alloys, Inc. Haynes is a registered trademark of Haynes International.)

thermal conductivity of gases in reducing heat transfer. Although we usually speak of a thermal conductivity for thermal insulators, in reality, the transport through an insulator is comprised of conduction as well as radiation across the interstices filled with gas. Thermal insulation will be discussed further in Section 1.7. Table 11 in Appendix 2 lists typical values of the effective conductivity for several insulating materials.

1.4 Convection

The convection mode of heat transfer actually consists of two mechanisms operating simultaneously. The first is the energy transfer due to molecular motion, that is, the conductive mode. Superimposed upon this mode is energy transfer by the macroscopic motion of fluid parcels. The fluid motion is a result of parcels of fluid, each consisting of a large number of molecules, moving by virtue of an external force. This extraneous force may be due to a density gradient, as in natural convection, or due to a pressure difference generated by a pump or a fan, or possibly to a combination of the two.

Figure 1.8 shows a plate at surface temperature T_s and a fluid at temperature T_∞ flowing parallel to the plate. As a result of viscous forces the velocity of the fluid will be zero at the wall and will increase to U_∞ as shown. Since the fluid is not moving at the interface, heat is transferred at that location only by conduction. If we knew the temperature gradient and the thermal conductivity at this interface, we could calculate the rate of heat transfer from Eq. (1.2):

$$q_c = -k_{\text{fluid}} A \left. \frac{\partial T}{\partial y} \right|_{\text{at } y=0} \tag{1.9}$$

But the temperature gradient at the interface depends on the rate at which the macroscopic as well as the microscopic motion of the fluid carries the heat away from the interface. Consequently, the temperature gradient at the fluid-plate interface depends on the nature of the flow field, particularly the free-stream velocity U_∞.

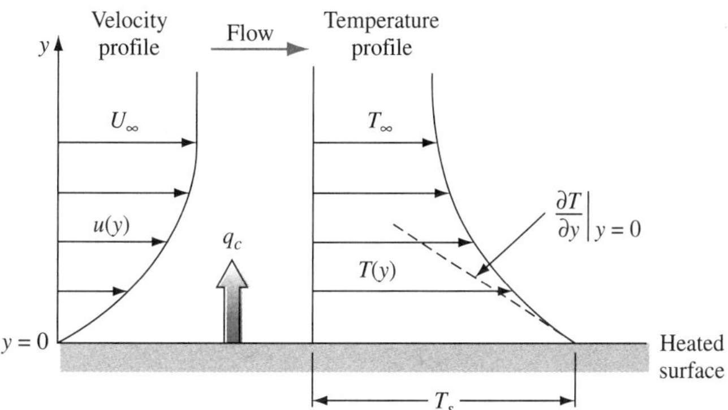

FIGURE 1.8 Velocity and temperature profile for convection heat transfer from a heated plate with flow over its surface.

The situation is quite similar in natural convection. The principal difference is that in forced convection the velocity far from the surface approaches the free-stream value imposed by an external force, whereas in natural convection the velocity at first increases with increasing distance from the heat transfer surface and then decreases, as shown in Fig. 1.9. The reason for this behavior is that the action of viscosity diminishes rather rapidly with distance from the surface, while the density difference decreases more slowly. Eventually, however, the buoyant force also decreases as the fluid density approaches the value of the unheated surrounding fluid. This interaction of forces will cause the velocity to reach a maximum and then approach zero far from the heated surface. The temperature fields in natural and forced convection have similar shapes, and in both cases the heat transfer mechanism at the fluid-solid interface is conduction.

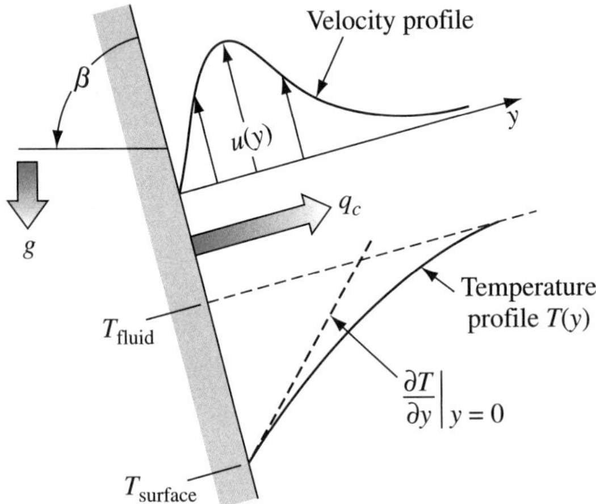

FIGURE 1.9 Velocity and temperature distribution for natural convection over a heated flat plate inclined at angle β from the horizontal.

The preceding discussion indicates that convection heat transfer depends on the density, viscosity, and velocity of the fluid as well as on its thermal properties (thermal conductivity and specific heat). Whereas in forced convection the velocity is usually imposed on the system by a pump or a fan and can be directly specified, in natural convection the velocity depends on the temperature difference between the surface and the fluid, the coefficient of thermal expansion of the fluid (which determines the density change per unit temperature difference), and the body force field, which in systems located on the earth is simply the gravitational force.

In later chapters we will develop methods for relating the temperature gradient at the interface to the external flow conditions, but for the time being we shall use a simpler approach to calculate the rate of convection heat transfer, as shown below.

Irrespective of the details of the mechanism, the rate of heat transfer by convection between a surface and a fluid can be calculated from the relation

$$q_c = \bar{h}_c A \Delta T \tag{1.10}$$

where q_c = rate of heat transfer by convection, W (Btu/h)
A = heat transfer area, m^2 (ft^2)

ΔT = difference between the surface temperature T_s and a temperature of the fluid T_∞ at some specified location (usually far way from the surface), K (°F)

$\bar{h}_c$ = average convection heat transfer coefficient over the area A (often called the surface coefficient of heat transfer or the convection heat transfer coefficient), W/m^2 K (Btu/h ft^2 °F)

The relation expressed by Eq. (1.10) was originally proposed by the British scientist Isaac Newton in 1701. Engineers have used this equation for many years, even though it is a definition of $\bar{h}_c$ rather than a phenomenological law of convection. Evaluation of the convection heat transfer coefficient is difficult because convection is a very complex phenomenon. The methods and techniques available for a quantitative evaluation of $\bar{h}_c$ will be presented in later chapters. At this point it is sufficient to note that the numerical value of $\bar{h}_c$ in a system depends on the geometry of the surface, on the velocity as well as the physical properties of the fluid, and often even on the temperature difference ΔT. In view of the fact that these quantities are not necessarily constant over a surface, the convection heat transfer coefficient may also vary from point to point. For this reason, we must distinguish between a local and an average convection heat transfer coefficient. The local coefficient h_c is defined by

$$dq_c = h_c\, dA(T_s - T_\infty) \tag{1.11}$$

while the average coefficient $\bar{h}_c$ can be defined in terms of the local value by

$$\bar{h}_c = \frac{1}{A}\iint_A h_c\, dA \tag{1.12}$$

For most engineering applications, we are interested in average values. Typical values of the order of magnitude of average convection heat transfer coefficients seen in engineering practice are given in Table 1.4.

Using Eq. (1.10), we can define the *thermal conductance for convection heat transfer* K_c as

$$K_c = \bar{h}_c A \qquad \text{(W/K)} \tag{1.13}$$

TABLE 1.4 Order of magnitude of convection heat transfer coefficients $\bar{h}_c$

Fluid	Convection Heat Transfer Coefficient W/m^2 K	Btu/h ft^2 °F
Air, free convection	6–30	1–5
Superheated steam or air, forced convection	30–300	5–50
Oil, forced convection	60–1,800	10–300
Water, forced convection	300–18,000	50–3,000
Water, boiling	3,000–60,000	500–10,000
Steam, condensing	6,000–120,000	1,000–20,000

and the *thermal resistance to convection heat transfer* R_c, which is equal to the reciprocal of the conductance, as

$$R_c = \frac{1}{\bar{h}_c A} \quad \text{(K/W)} \tag{1.14}$$

EXAMPLE 1.3 Calculate the rate of heat transfer by natural convection between a shed roof of area 20 m × 20 m and ambient air, if the roof surface temperature is 27°C, the air-temperature −3°C, and the average convection heat transfer coefficient 10 W/m^2 K (see Fig. 1.10).

SOLUTION Assume that steady state exists and the direction of heat flow is from the air to the roof. The rate of heat transfer by convection from the air to the roof is then given by Eq. (1.10):

$$\begin{aligned} q_c &= \bar{h}_c A_{\text{roof}}(T_{\text{air}} - T_{\text{roof}}) \\ &= 10\ (\text{W/m}^2\ \text{K}) \times 400\ \text{m}^2(-3 - 27)°\text{C} \\ &= -120{,}000\ \text{W} \end{aligned}$$

Note that in using Eq. (1.10), we initially assumed that the heat transfer would be from the air to the roof. But since the heat flow under this assumption turns out to be a negative quantity, the *direction of heat flow is actually from the roof to the air*. We could, of course, have deduced this at the outset by applying the second law of thermodynamics, which tells us that heat will always flow from a higher to a lower temperature if there is no external intervention. But as we shall see in a later section, thermodynamic arguments cannot always he used at the outset in heat transfer problems because in many real situations the surface temperature is not known.

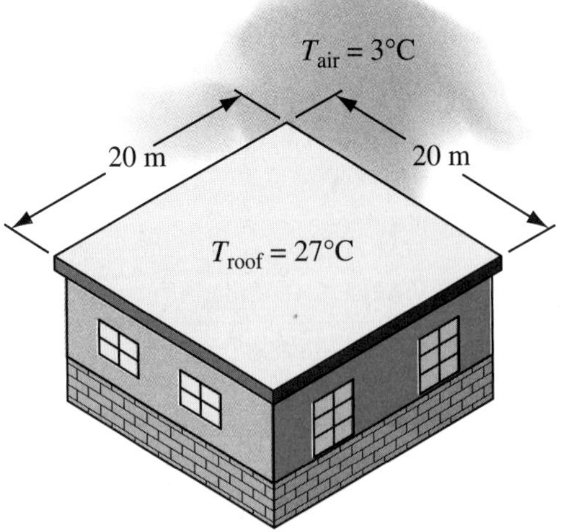

FIGURE 1.10 Schematic sketch of shed for analysis of roof temperature in Example 1.3.

1.5 Radiation

The quantity of energy leaving a surface as radiant heat depends on the absolute temperature and the nature of the surface. A perfect radiator, which is referred to as a *blackbody*,* emits radiant energy from its surface at a rate as given by

$$q_r = \sigma A_1 T_1^4 \tag{1.15}$$

The heat flow rate q_r will be in watts if the surface area A, is in square meters and the surface temperature T_1 is in kelvin; σ is a dimensional constant with a value of 5.67×10^{-8} (W/m^2 K^4). In the English system, the heat flow rate will be in Btu's per hour if the surface area is in square feet, the surface temperature is in degrees Rankine (°R), and σ is 0.1714×10^{-8} (Btu/h ft^2 °R^4). The constant σ is the Stefan-Boltzmann constant; it is named after two Austrian scientists, J. Stefan, who in 1879 discovered Eq. (1.15) experimentally, and L. Boltzmann, who in 1884 derived it theoretically.

Inspection of Eq. (1.15) shows that any blackbody surface above a temperature of absolute zero radiates heat at a rate proportional to the fourth power of the absolute temperature. While the rate of radiant heat emission is independent of the conditions of the surroundings, a *net* transfer of radiant heat requires a difference in the surface temperature of any two bodies between which the exchange is taking place. If the blackbody radiates to an enclosure (see Fig. 1.11) that is also black, (that is, absorbs all the radiant energy incident upon it) the net rate of radiant heat transfer is given by

$$q_r = A_1 \sigma (T_1^4 - T_2^4) \tag{1.16}$$

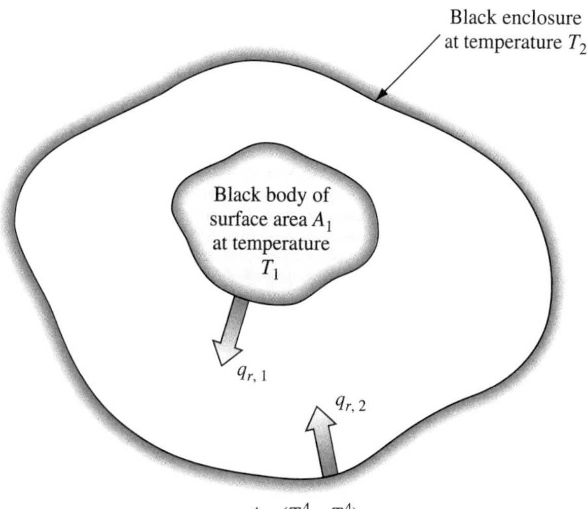

FIGURE 1.11 Schematic diagram of radiation between body 1 and enclosure 2.

* A detailed discussion of the meaning of these terms is presented in Chapter 9.

where T_2 is the surface temperature of the enclosure in kelvin.

Real bodies do not meet the specifications of an ideal radiator but emit radiation at a lower rate than blackbodies. If they emit, at a temperature equal to that of a blackbody (a constant fraction of blackbody emission at each wavelength) they are called gray bodies. A gray body A_1 at T_1 emits radiation at the rate $\varepsilon_1 \sigma A_1 T_1^4$, and the rate of heat transfer between a gray body at a temperature T_1 and a surrounding black enclosure at T_2 is

$$q_r = A_1 \varepsilon_1 \sigma (T_1^4 - T_2^4) \tag{1.17}$$

where ε_1 is the emittance of the gray surface and is equal to the ratio of the emission from the gray surface to the emission from a perfect radiator at the same temperature.

If neither of two bodies is a perfect radiator and if the two bodies have a given geometric relationship to each other, the net heat transfer by radiation between them is given by

$$q_r = A_1 \mathscr{F}_{1-2} \sigma (T_1^4 - T_2^4) \tag{1.18}$$

where $\mathscr{F}_{1-2}$ is a dimensionless modulus that modifies the equation for perfect radiators to account for the emittances and relative geometries of the actual bodies. Methods for calculating $\mathscr{F}_{1-2}$ will be taken up in Chapter 9.

In many engineering problems, radiation is combined with other modes of heat transfer. The solution of such problems can often be simplified by using a thermal conductance K_r, or a thermal resistance R_r, for radiation. The definition of K_r is similar to that of K_k, the thermal conductance for conduction. If the heat transfer by radiation is written

$$q_r = K_r(T_1 - T_2') \tag{1.19}$$

the radiation conductance, by comparison with Eq. (1.12), is given by

$$K_r = \frac{A_1 \mathscr{F}_{1-2} \sigma (T_1^4 - T_2^4)}{T_1 - T_2'} \quad \text{W/K (Btu/h °F)} \tag{1.20}$$

The unit thermal radiation conductance, or *radiation heat transfer coefficients*, $\bar{h}_r$, is then

$$\bar{h}_r = \frac{K_r}{A_1} = \frac{\mathscr{F}_{1-2} \sigma (T_1^4 - T_2^4)}{T_1 - T_2'} \quad \text{W/m}^2\text{ K (Btu/h ft}^2\text{ °F)} \tag{1.21}$$

where T_2' is any convenient reference temperature, whose choice is often dictated by the convection equation, which will be discussed next. Similarly, the unit *thermal resistance for radiation* is

$$R_r = \frac{T_1 - T_2'}{A_1 \mathscr{F}_{1-2} \sigma (T_1^4 - T_2^4)} \quad \text{K/W (°F h/Btu)} \tag{1.22}$$

EXAMPLE 1.4 A long, cylindrical electrically heated rod, 2 cm in diameter, is installed in a vacuum furnace as shown in Fig. 1.12. The surface of the heating rod has an emissivity of 0.9 and is maintained at 1000 K, while the interior walls of the furnace are black and are at 800 K. Calculate the net rate at which heat is lost from the rod per unit length and the radiation heat transfer coefficient.

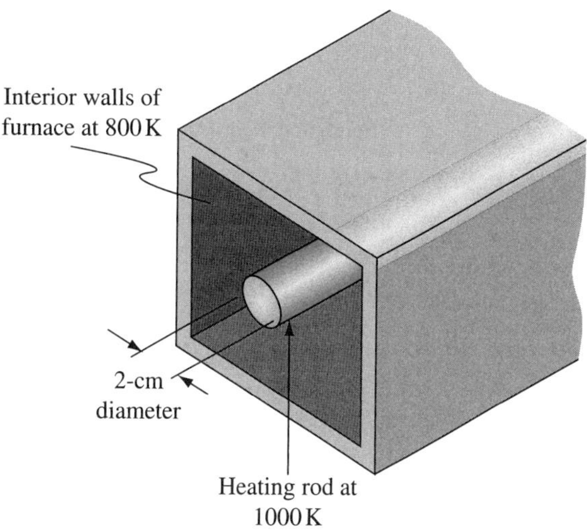

FIGURE 1.12 Schematic diagram of vacuum furnace with heating rod for Example 1.4.

SOLUTION Assume that steady state has been reached. Moreover, note that since the walls of the furnace completely enclose the heating rod, all the radiant energy emitted by the surface of the rod is intercepted by the furnace walls. Thus, for a black enclosure, Eq. (1.17) applies and the net heat loss from the rod of surface A_1 is

$$q_r = A\varepsilon\sigma(T_1^4 - T_2^4) = \pi D_1 L\varepsilon\sigma(T_1^4 - T_2^4)$$

$$= \pi(0.02\text{ m})(1.0\text{ m})(0.9)\left(5.67 \times 10^{-8}\,\frac{\text{W}}{\text{m}^2\text{K}^4}\right)(1000^4 - 800^4)(\text{K}^4)$$

$$= 1893\text{ W}$$

Note that in order for steady state to exist, the heating rod must dissipate electrical energy at the rate of 1893 W and the rate of heat loss through the furnace walls must equal the rate of electric input to the system, that is, to the rod.

From Eq. (1.17), $\mathscr{F}_{1-2} = \varepsilon_1$, and therefore the radiation heat transfer coefficient, according to its definition in Eq. (1.21), is

$$h_r = \frac{\varepsilon_1\sigma(T_1^4 - T_2^4)}{T_1 - T_2} = 151\text{ W/m}^2\text{ K}$$

Here, we have used T_2 as the reference temperature T_2'.

1.6 Combined Heat Transfer Systems

In the preceding sections the three basic mechanisms of heat transfer have been treated separately. In practice, however, heat is usually transferred by several of the basic mechanisms occurring simultaneously. For example, in the winter, heat is transferred from the roof of a house to the colder ambient environment not only by convection but also by radiation, while the heat transfer through the roof from the interior to the exterior surface is by conduction. Heat transfer between the panes of a double-glazed window occurs by convection and radiation acting in parallel, while the transfer through the panes of glass is by conduction with some radiation passing directly through the entire window system. In this section, we will examine combined heat transfer problems. We will set up and solve these problems by dividing the heat transfer path into sections that can be connected in series, just like an electrical circuit, with heat being transferred in each section by one or more mechanisms acting in parallel. Table 1.5 summarizes the basic relations for the rate equation of each of the three basic heat transfer mechanisms to aid in setting up the thermal circuits for solving combined heat transfer problems.

1.6.1 Plane Walls in Series and Parallel

If heat is conducted through several plane walls in good thermal contact, as through a multilayer wall of a building, the rate of heat conduction is the same through all sections. However, as shown in Fig. 1.13 for a three-layer system, the temperature

TABLE 1.5 The three modes of heat transfer

One dimensional conduction heat transfer through a stationary medium

$$q_k = \frac{kA}{L}(T_1 - T_2) = \frac{T_1 - T_2}{R_k}$$

$$R_k = \frac{L}{kA}$$

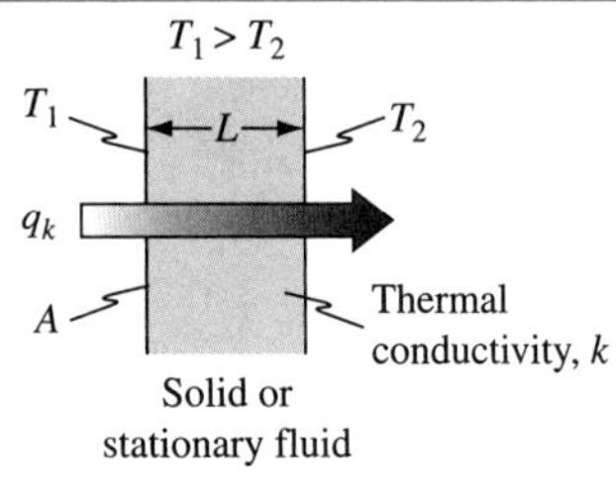

Convection heat transfer from a surface to a moving fluid

$$q_c = \bar{h}_c A(T_s - T_\infty) = \frac{T_s - T_\infty}{R_c}$$

$$R_c = \frac{1}{\bar{h}_c A}$$

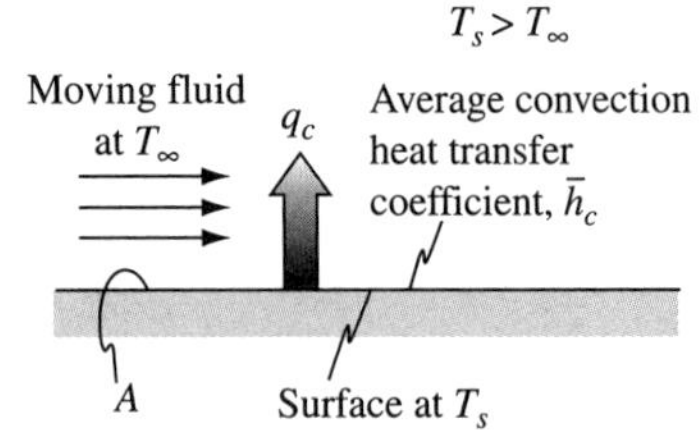

Net radiation heat transfer from surface 1 to surface 2

$$q_r = A_1 \mathscr{F}_{1-2}\sigma(T_1^4 - T_2^4) = \frac{T_2 - T_2}{R_r}$$

$$R_r = \frac{T_1 - T_2}{A_1 \mathscr{F}_{1-2}\sigma(T_1^4 - T_2^4)}$$

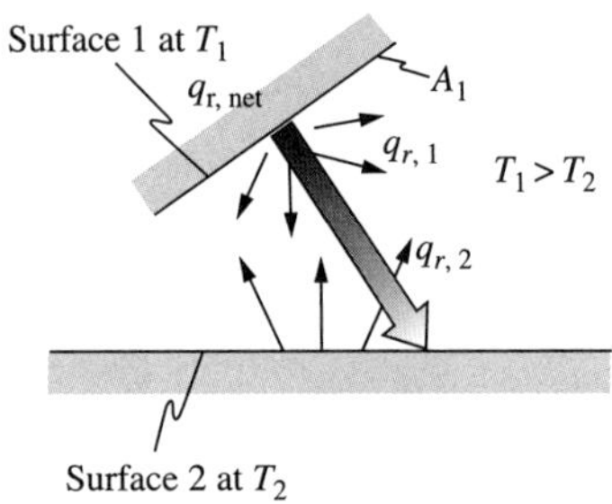

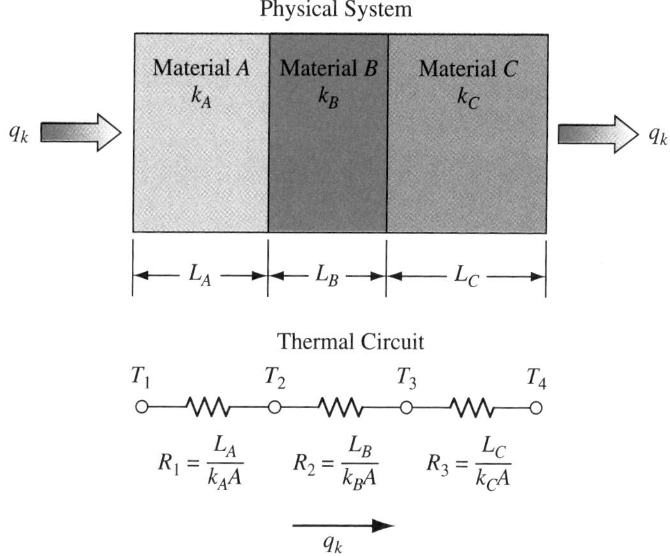

FIGURE 1.13 Conduction through a three-layer system in series.

gradients in the layers are different. The rate of heat conduction through each layer is q_k, and from Eq. (1.2) we get

$$q_k = \left(\frac{kA}{L}\right)_A (T_1 - T_2) = \left(\frac{kA}{L}\right)_B (T_2 - T_3) = \left(\frac{kA}{L}\right)_C (T_3 - T_4) \tag{1.23}$$

Eliminating the intermediate temperatures T_2 and T_3 in Eq. (1.23), q_k can be expressed in the form

$$q_k = \frac{T_1 - T_4}{(L/kA)_A + (L/kA)_B + (L/kA)_C}$$

Similarly, for N layers in series we have

$$q_k = \frac{T_1 - T_{N+1}}{\sum_{n=1}^{n=N} (L/kA)_n} \tag{1.24}$$

where T_1 is the outer-surface temperature of layer 1 and T_{N+1} is the outer-surface temperature of layer N. Using the definition of thermal resistance from Eq. (1.4), Eq. (1.24) becomes

$$q_k = \frac{T_1 - T_{N+1}}{\sum_{n=1}^{n=N} R_{k,n}} = \frac{\Delta T}{\sum_{n=1}^{n=N} R_{k,n}} \tag{1.25}$$

where ΔT is the overall temperature difference, often called the temperature potential. The rate of heat flow is proportional to the temperature potential.

EXAMPLE 1.5 Calculate the rate of heat loss from a furnace wall per unit area. The wall is constructed from an inner layer of 0.5-cm-thick steel ($k = 40$ W/m K) and an outer layer of 10-cm zirconium brick ($k = 2.5$ W/m K) as shown in Fig. 1.14. The inner-surface temperature is 900 K and the outside surface temperature is 460 K. What is the temperature at the interface?

SOLUTION Assume that steady state exists, neglect effects at the corners and edges of the wall, and assume that the surface temperatures are uniform. The physical system and the corresponding thermal circuit are similar to those in Fig. 1.13, but only two layers or walls are present. The rate of heat loss per unit area can be calculated from Eq. (1.24):

$$\frac{q_k}{A} = \frac{(900 - 460)\text{ K}}{(0.005\text{ m})/(40\text{ W/m K}) + (0.1\text{ m})/(2.5\text{ W/m K})}$$

$$= \frac{440\text{ K}}{(0.000125 + 0.04)(\text{m}^2\text{ K/W})} = 10{,}965\text{ W/m}^2$$

The interface temperature T_2 is obtained from

$$\frac{q_k}{A} = \frac{T_1 - T_2}{R_1}$$

Solving for T_2 gives

$$T_2 = T_1 - \frac{q_k}{A_1}\frac{L_1}{k_1}$$

$$= 900\text{ K} - \left(10{,}965\frac{\text{W}}{\text{m}^2}\right)\left(0.00125\frac{\text{m}^2\text{ K}}{\text{W}}\right)$$

$$= 898.6\text{ K}$$

Note that the temperature drop across the steel interior wall is only 1.4 K because the thermal resistance of the wall is small compared to the resistance of the brick, across which the temperature drop is many times larger.

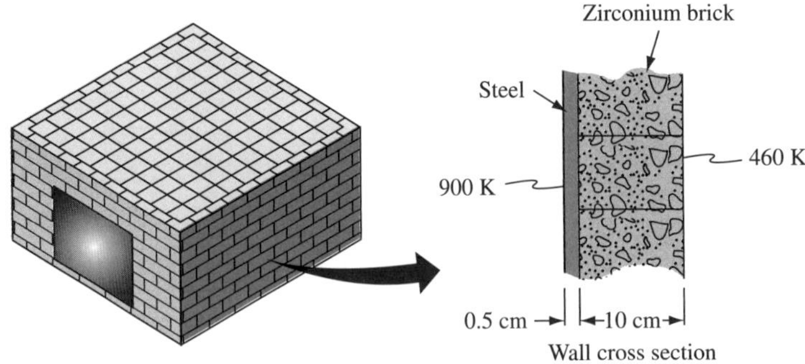

FIGURE 1.14 Schematic diagram of furnace wall for Example 1.5.

The analogy between heat flow systems and electrical circuits has been previously discussed. A contact or interface resistance can be integrated into the thermal circuit approach. The following example illustrates the procedure.

EXAMPLE 1.6 Two large aluminum plates ($k = 240$ W/m K), each 1-cm thick, with 10-μm surface roughness are placed in contact under 10^5 N/m^2 pressure in air as shown in Fig. 1.15. The temperatures at the outside surfaces are 395°C and 405°C. Calculate (a) the heat flux and (b) the temperature drop due to the contact resistance.

SOLUTION (a) The rate of heat flow per unit area, q'', through the sandwich wall is

$$q = \frac{T_{s1} - T_{s3}}{R_1 + R_2 + R_3} = \frac{\Delta T}{(L/k)_1 + R_i + (L/k)_2}$$

From Table 1.6 the contact resistance R_i is 2.75×10^{-4} m^2 K/W while each of the other two resistances is equal to

$$(L/k) = (0.01 \text{ m})/(240 \text{ W/m K}) = 4.17 \times 10^{-5} \text{ m}^2 \text{ K/W}$$

Hence, the heat flux is

$$q'' = \frac{(405 - 395)°\text{C}}{(4.17 \times 10^{-5} + 2.75 \times 10^{-4} + 4.17 \times 10^{-5})\text{m}^2 \text{ K/W}}$$
$$= 2.79 \times 10^4 \text{ W/m}^2 \text{ K}$$

(b) The temperature drop in each section of this one-dimensional system is proportional to the resistance. The fraction of the contact resistance is

$$R_i \Big/ \sum_{n=1}^{3} R_n = 2.75/3.584 = 0.767$$

Hence 7.67°C of the total temperature drop of 10°C is the result of the contact resistance.

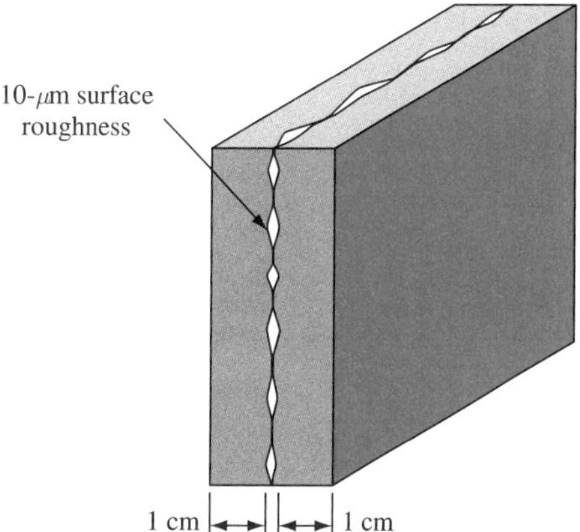

FIGURE 1.15 Schematic diagram of interface between plates for Example 1.6.

Conduction can occur in a section with two different materials in parallel. For example, Fig. 1.16 shows the cross-section of a slab with two different materials of areas A_A and A_B in parallel. If the temperatures over the left and right faces are uniform at T_1 and T_2, we can analyze the problem in terms of the thermal circuit shown to the right of the physical systems. Since heat is conducted through the two materials along separate paths between the same potential, the total rate of heat flow is the sum of the flows through A_A and A_B:

$$\begin{aligned} q_k &= q_1 + q_2 \\ &= \frac{T_1 - T_2}{(L/kA)_A} + \frac{T_1 - T_2}{(L/kA)_B} = \frac{T_1 - T_2}{R_1R_2/(R_1 + R_2)} \end{aligned} \tag{1.26}$$

Note that the total heat transfer area is the sum of A_A and A_B and that the total resistance equals the product of the individual resistances divided by their sum, as in any parallel circuit.

A more complex application of the thermal network approach is illustrated in Fig. 1.17, where heat is transferred through a composite structure involving thermal resistances in series and in parallel. For this system the resistance of the middle layer, R_2 becomes

$$R_2 = \frac{R_BR_C}{R_B + R_C}$$

and the rate of heat flow is

$$q_k = \frac{\Delta T_{\text{overall}}}{\sum_{n=1}^{n=3} R_n} \tag{1.27}$$

where N = number of layers in series (three)
R_n = thermal resistance of nth layer
$\Delta T_{\text{overall}}$ = temperature difference across two outer surfaces

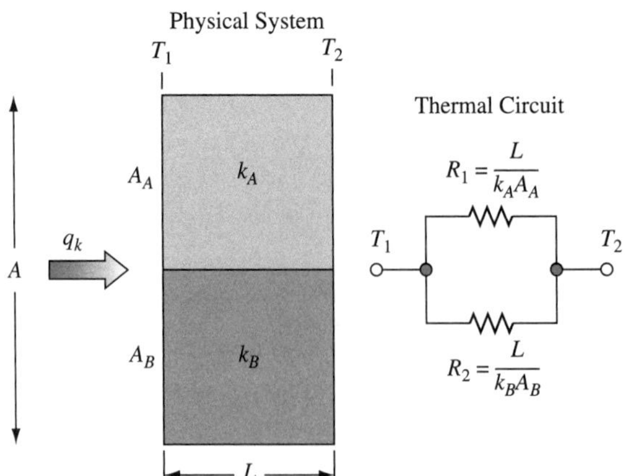

FIGURE 1.16 Heat conduction through a wall section with two paths in parallel.

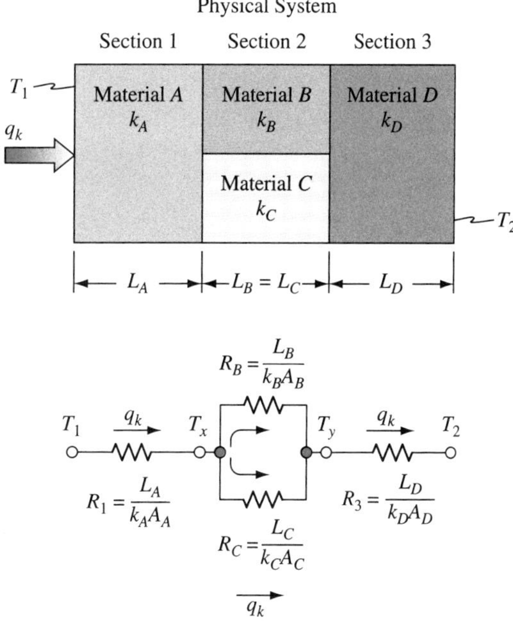

FIGURE 1.17 Conduction through a wall consisting of series and parallel thermal paths.

By analogy to Eqs. (1.4) and (1.5), Eq. (1.27) can also be used to obtain an overall conductance between the two outer surfaces:

$$K_k = \left(\sum_{n=1}^{n=N} R_n \right)^{-1} \quad (1.28)$$

EXAMPLE 1.7 A layer of 2-in.-thick firebrick ($k_b = 1.0$ Btu/h ft °F) is placed between two 1/4-in.-thick steel plates ($k_s = 30$ Btu/h ft °F). The faces of the brick adjacent to the plates are rough, having solid-to-solid contact over only 30 percent of the total area, with the average height of asperities being 1/32 in. If the surface temperatures of the steel plates are 200° and 800°F, respectively, determine the rate of heat flow per unit area.

SOLUTION The real system is first idealized by assuming that the asperities of the surface are distributed as shown in Fig. 1.18 on the next page. We note that the composite wall is symmetrical with respect to the center plane and therefore consider only half of the system. The overall unit conductance for half the composite wall is then, from Eq. (1.28),

$$K_k = \frac{1}{R_1 + [R_4 R_5/(R_4 + R_5)] + R_3}$$

from an inspection of the thermal circuit.

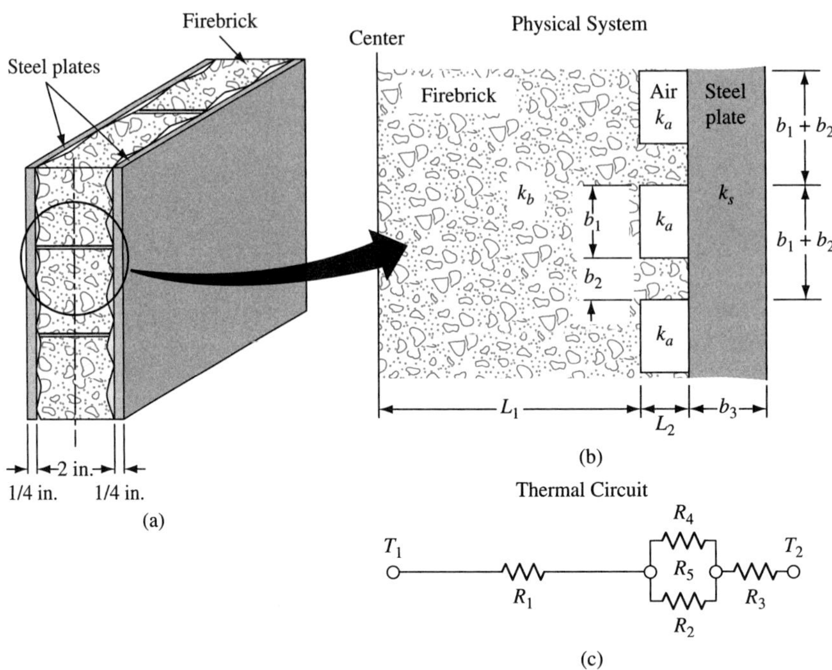

FIGURE 1.18 Thermal circuit for the parallel-series composite wall in Example 1.7. L_1 = 1 in.; L_2 = 1/32 in.; L_3 = 1/4 in.; T_1 is at the center.

The thermal resistance of the steel plate R_3 is, on the basis of a unit wall area, equal to

$$R_3 = \frac{L_3}{k_s} = \frac{(1/4 \text{ in.})}{(12 \text{ in./ft})(30 \text{ Btu/h °F ft})} = 0.694 \times 10^{-3} \simeq 0.69 \times 10^{-3} (\text{Btu/h ft}^2 \text{ °F})^{-1}$$

The thermal resistance of the brick asperities R_4 is, on the basis of a unit wall area, equal to

$$R_4 = \frac{L_2}{0.3k_b} = \frac{(1/32 \text{ in.})}{(12 \text{ in./ft})(0.3)(1 \text{ Btu/h °F ft})} = 8.68 \times 10^{-3} (\text{Btu/h ft}^2 \text{ °F})^{-1}$$

Since the air is trapped in very small compartments, the effects of convection are small and it will be assumed that heat flows through the air by conduction. At a temperature of 300°F, the conductivity of air k_a is about 0.02 Btu/h ft°F. Then R_5, the thermal resistance of the air trapped between the asperities, is, on the basis of a unit area, equal to

$$R_5 = \frac{L_2}{0.7k_a} = \frac{(1/32 \text{ in.})}{(12 \text{ in./ft})(0.02 \text{ Btu/h °F ft})} = 186 \times 10^{-3} (\text{Btu/h ft}^2 \text{ °F})^{-1}$$

The factors 0.3 and 0.7 in R_4 and R_5, respectively, represent the percent of the total area of the two separately heat flow paths.

The total thermal resistance for the two paths, R_4 and R_5 in parallel, is

$$R_2 = \frac{R_4 R_5}{R_4 + R_5} = \frac{(8.7)(187) \times 10^{-6}}{(8.7 + 187) \times 10^{-3}} = 8.29 \times 10^{-3}\ (\text{Btu/h ft}^2\ ^\circ\text{F})^{-1}$$

The thermal resistance of half of the solid brick, R_1, is

$$R_1 = \frac{L_1}{k_b} = \frac{(1\ \text{in.})}{(12\ \text{in./ft})(1\ \text{Btu/h}\ ^\circ\text{F ft})} = 83.3 \times 10^{-3}\ (\text{Btu/h ft}^2\ ^\circ\text{F})^{-1}$$

and the overall unit conductance is

$$K_k = \frac{1/2 \times 10^3}{83.3 + 8.3 + 0.69} = 5.4\ \text{Btu/h ft}^2\ ^\circ\text{F}$$

Inspection of the values for the various thermal resistances shows that the steel offers a negligible resistance, while the contact section although only 1/32 in. thick, contributes 10% to the total resistance. From Eq. (1.27), the rate of heat flow per unit area is

$$\frac{q}{A} = K_k \Delta T = \left(5.4\ \frac{\text{Btu}}{\text{h ft}^2\ ^\circ\text{F}}\right)(800 - 200)(^\circ\text{F}) = 3240\ \text{Btu/h ft}^2$$

1.6.2 Contact Resistance

In many practical applications, when two different conducting surfaces are placed in contact as shown in Fig. 1.19 on the next page, a thermal resistance is present at the interface of the solids. Mounting heat sinks onto microelectronic or IC chip modules and attaching fins to tubular surfaces in evaporators and condensers for air-conditioning systems are some examples where this situation is of significance. The interface resistance, frequently called the *thermal contact resistance*, develops when two materials will not fit tightly together and a thin layer of fluid is trapped between them. Examination of an enlarged view of the contact between the two surfaces shows that the solids touch only at peaks in the surface and that the valleys in the mating surfaces are occupied by a fluid (possibly air), a liquid, or a vacuum.

The interface resistance is primarily a function of surface roughness, the pressure holding the two surfaces in contact, the interface fluid, and the interface temperature. At the interface, the mechanism of heat transfer is complex. Conduction takes place through the contact points of the solid, while heat is transferred by convection and radiation across the trapped interfacial fluid.

If the heat flux through two solid surfaces in contact is q/A and the temperature difference across the fluid gap separating the two solids is ΔT_i, the interface resistance R_i is defined by

$$R_i = \frac{\Delta T_i}{q/A} \tag{1.29}$$

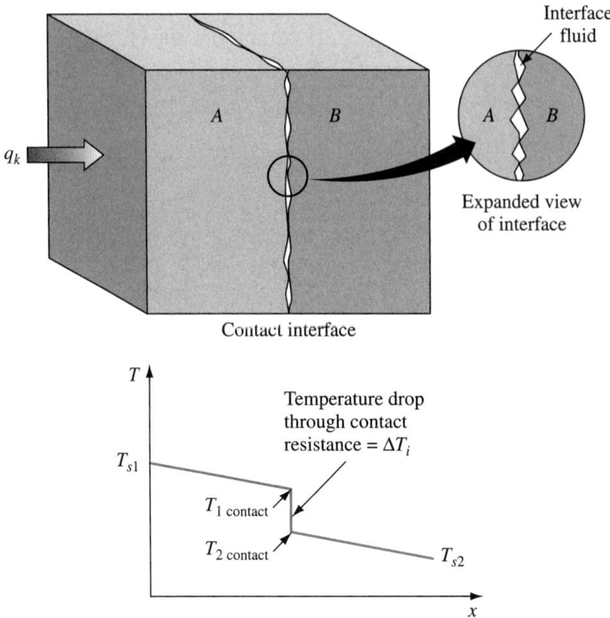

FIGURE 1.19 Schematic diagram showing physical contact between two solid slabs *A* and *B* and the temperature profile through the solids and the contact interface.

When two surfaces are in perfect thermal contact, the interface resistance approaches zero, and there is no temperature difference across the interface. For imperfect thermal contact, a temperature difference occurs at the interface, as shown in Fig. 1.19.

Table 1.6 shows the influence of contact pressure on the thermal contact resistance between metal surfaces under vacuum conditions. It is apparent that an increase in the pressure can reduce the contact resistance appreciably. As shown in Table 1.7, the interfacial fluid also affects the thermal resistance. Putting a viscous liquid such as glycerin on the interface reduces the contact resistance between two aluminum surfaces by a factor of 10 at a given pressure.

Numerous measurements have been made of the contact resistance at the interface between dissimilar metallic surfaces, but no satisfactory correlations have been

TABLE 1.6 Approximate range of thermal contact resistance for metallic interfaces under vacuum conditions [5]

	Resistance, R_i(m^2 K/W $\times$ 10^4)	
Interface Material	**Contact Pressure 100 kN/m^2**	**Contact Pressure 10,000 kN/m^2**
Stainless steel	6–25	0.7–4.0
Copper	1–10	0.1–0.5
Magnesium	1.5–3.5	0.2–0.4
Aluminum	1.5–5.0	0.2–0.4

TABLE 1.7 Thermal contact resistance for aluminum–aluminum interface[a] with different interfacial fluids [5]

Interfacial Fluid	Resistance, R_i(m^2 K/W)
Air	2.75×10^{-4}
Helium	1.05×10^{-4}
Hydrogen	0.720×10^{-4}
Silicone oil	0.525×10^{-4}
Glycerin	0.265×10^{-4}

[a] 10-μm surface roughness under 10^5 N/m^2 contact pressure.

found. Each situation must be treated separately. The results of many different conditions and materials have been summarized by Fletcher [6]. In Fig. 1.20 some experimental results for the contact resistance between dissimilar base metal surfaces at atmospheric pressure are plotted as a function of contact pressure.

Efforts have been made to reduce the contact resistance by placing a soft metallic foil, a grease, or a viscous liquid at the interface between the contacting materials.

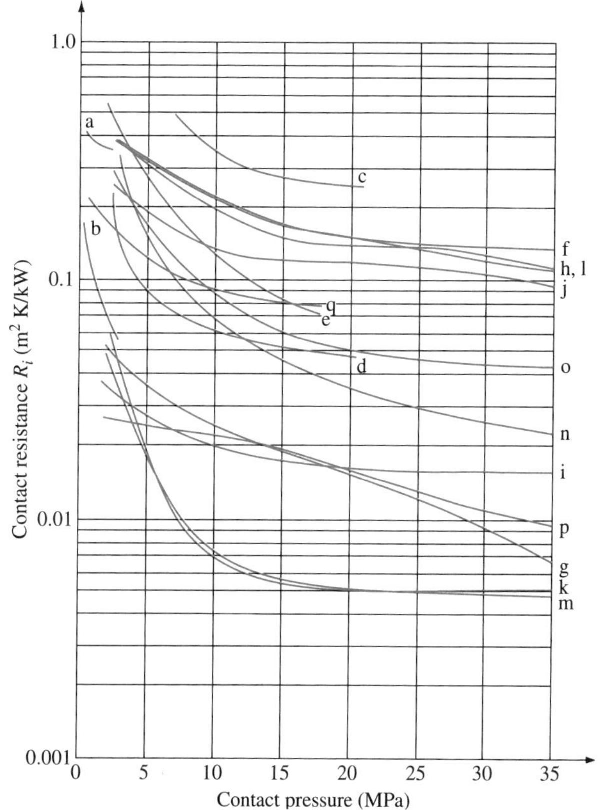

FIGURE 1.20 Contact resistances between dissimilar bare metal surfaces. Solid metal blocks in air at 1 atmosphere absolute pressure (see legend, on the next page).

Legend for Fig. 1.20

Curve in Fig. 1.20	Material	Finish	Roughness rms (μm)	Temp. (°C)	Condition	Scatter of Data
a	416 Stainless 7075(75S)T6 Al	Ground	0.76–1.65	93	Heat flow from stainless to aluminum	±26%
b	7075(75S)T6 Al to Stainless	Ground	1.65–0.76	93–204	Heat flow from aluminum to stainless	±30%
c	Stainless Aluminum		19.94–29.97	20	Clean	
d	Stainless Aluminum		1.02–2.03	20	Clean	
e	Bessemer Steel Foundry Brass	Ground	3.00–3.00	20	Clean	
f	Steel Ct-30	Milled	7.24–5.13	20	Clean	
g	Steel Ct-30	Ground	1.98–1.52	20	Clean	
h	Steel Ct-30 Aluminum	Milled	7.24–4.47	20	Clean	
i	Steel Ct-30 Aluminum	Ground	1.98–1.35	20	Clean	
j	Steel Ct-30 Copper	Milled	7.24–4.42	20	Clean	
k	Steel Ct-30 Copper	Ground	1.98–1.42	20	Clean	
l	Brass Aluminum	Milled	5.13–4.47	20	Clean	
m	Brass Aluminum	Ground	1.52–1.35	20	Clean	
n	Brass Copper	Milled	5.13–4.42	20	Clean	
o	Aluminum Copper	Milled	4.47–4.42	20	Clean	
p	Aluminum Copper	Ground	1.35–1.42	20	Clean	
q	Uranium Aluminum	Ground		20	Clean	

Source: Abstracted from the Heat Transfer and Fluid Flow Data Books, F. Kreith ed., Genium Pub., Comp., Schenectady, NY, 1991, With permission.

This procedure can reduce the contact resistance as shown in Table 1.7, but there is no way to predict the effect quantitatively. High-conductivity pastes are often used to mount electronic components to heat sinks. These pastes fill in the interstices and reduce the thermal resistance at the component/heat sink interface.

EXAMPLE 1.8 An instrument used to study the ozone depletion near the poles is placed on a large 2-cm-thick duralumin plate. To simplify this analysis the instrument can be thought of as a stainless steel plate 1 cm tall with a 10-cm × 10-cm square base, as shown in Fig. 1.21. The interface roughness of the steel and the duralumin is between 20 and 30 rms (μm). Four screws at the corners provide fastening that exerts an average pressure of 1000 psi. The top and sides of the instruments are thermally insulated. An integrated circuit placed between the insulation and the upper surface of the stainless steel plate generates heat. If this heat is to be transferred to the lower surface of the duralumin, estimated to be at a temperature of 0°C, determine the maximum allowable dissipation rate from the circuit if its temperature is not to exceed 40°C.

SOLUTION Since the top and sides of the instrument are insulated, all the heat generated by the circuit must flow downward. The thermal circuit will have three resistances—the stainless steel, the contact, and the duralumin. Using thermal conductivities from Table 10 in Appendix 2, the thermal resistances of the metal plates are calculated from Eq. 1.4:

Stainless:

$$R_k = \frac{L_{ss}}{Ak_{ss}} = \frac{0.01 \text{ m}}{0.01 \text{ m}^2 \times 144 \text{ W/m K}} = 0.07 \frac{\text{K}}{\text{W}}$$

Duralumin:

$$R_k = \frac{L_{\text{Al}}}{Ak_{\text{Al}}} = \frac{0.02 \text{ m}}{0.01 \text{ m}^2 \times 164 \text{ W/m K}} = 0.012 \frac{K}{\text{W}}$$

The contact resistance is obtained from Fig. 1.20. The contact pressure of 1000 psi equals about 7×10^6 N/m^2 or 7 MPa. For that pressure the unit contact resistance given by line c in Fig. 1.20 is 0.5 m^2 K/kW. Hence,

$$R_i = 0.5 \frac{\text{m}^2\text{K}}{\text{kW}} \times 10^{-3} \frac{\text{kW}}{\text{W}} \times \frac{1}{0.01 \text{ m}^2} = 0.05 \frac{\text{K}}{\text{W}}$$

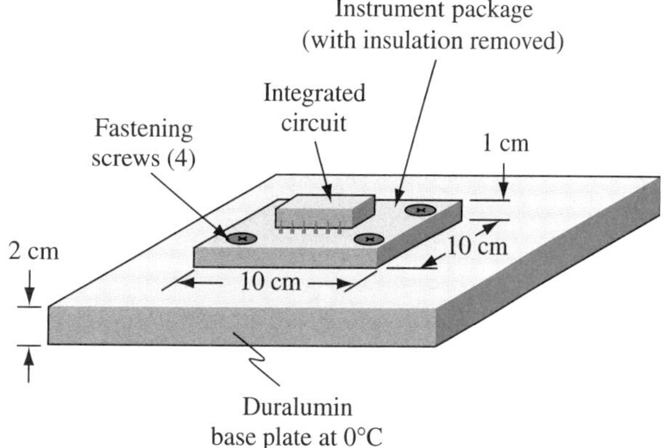

FIGURE 1.21 Schematic sketch of instrument for ozone measurement

The thermal circuit is

$R_k = 0.07$ K/W $R_i = 0.05$ K/W $R_k = 0.012$ K/W

Insulation Heat source 40°C | Stainless plate | Contact resistance | Duralumin plate | Heat sink 0°C

The total resistance is 0.132 K/W, and the maximum allowable rate of heat dissipation is therefore

$$q_{\max} = \frac{\Delta T}{R_{\text{total}}} = \frac{40 \text{ K}}{0.132 \text{ K/W}} = 303 \text{ W}$$

Hence, the maximum allowable heat dissipation rate is about 300 W. Note that if the surfaces were smooth ($1-2\mu$m rms), the contact resistance according to curve *a* in Fig. 1.20 would be only about 0.03 K/W and the heat dissipation could be increased to 357 W without exceeding the upper temperature limit.

Most of the problems at the end of the chapter do not consider interface resistance, even though it exists to some extent whenever solid surfaces are mechanically joined. We should therefore always be aware of the existence of the interface resistance and the resulting temperature difference across the interface. Particularly with rough surfaces and low bonding pressures, the temperature drop across the interface can be significant and cannot be ignored. The subject of interface resistance is complex, and no single theory or set of empirical data accurately describes the interface resistance for surfaces of engineering importance. The reader should consult References 6–9 for more detailed discussions of this subject.

1.6.3 Convection and Conduction in Series

In the preceding section we treated conduction through composite walls when the surface temperatures on both sides are specified. The more common problem encountered in engineering practice, however, is heat being transferred between two fluids of specified temperatures separated by a wall. In such a situation the surface temperatures are not known, but they can be calculated if the convection heat transfer coefficients on both sides of the wall are known.

Convection heat transfer can easily be integrated into a thermal network. From Eq. (1.14), the thermal resistance for convection heat transfer is

$$R_c = \frac{1}{\bar{h}_c A}$$

Figure 1.22 shows a situation in which heat is transferred between two fluids separated by a wall. According to the thermal network shown below the physical

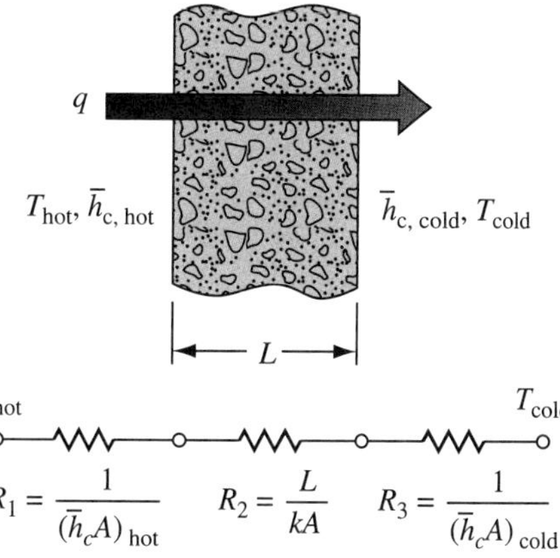

FIGURE 1.22 Thermal circuit with conduction and convection in series.

system, the rate of heat transfer from the hot fluid at temperature T_{hot} to the cold fluid at temperature T_{cold} is

$$q = \frac{T_{hot} - T_{cold}}{\sum_{n=1}^{n=3} R_i} = \frac{\Delta T}{R_1 + R_2 + R_3} \tag{1.30}$$

where
$$R_1 = \frac{1}{(\bar{h}_c A)_{hot}}$$
$$R_2 = \frac{L}{kA}$$
$$R_3 = \frac{1}{(\bar{h}_c A)_{cold}}$$

EXAMPLE 1.9 A 0.1-m-thick brick wall (k = 0.7 W/m K) is exposed to a cold wind at 270 K through a convection heat transfer coefficient of 40 W/m^2 K. On the other side is calm air at 330 K, with a natural-convection heat transfer coefficient of 10 W/m^2 K. Calculate the rate of heat transfer per unit area (i.e., the heat flux).

SOLUTION The three resistances are

$$R_1 = \frac{1}{\bar{h}_{c,hot}A} = \frac{1}{(10\ \text{W/m}^2\ \text{K})(1\ \text{m}^2)} = 0.10\ \text{K/W}$$
$$R_2 = \frac{L}{kA} = \frac{(0.1\ \text{m})}{(0.7\ \text{W/m K})(1\ \text{m}^2)} = 0.143\ \text{K/W}$$
$$R_3 = \frac{1}{\bar{h}_{c,cold}A} = \frac{1}{(40\ \text{W/m}^2\ \text{K})(1\ \text{m}^2)} = 0.025\ \text{K/W}$$

and from Eq. (1.30) the rate of heat transfer per unit area is

$$\frac{q}{A} = \frac{\Delta T}{R_1 + R_2 + R_3} = \frac{(330 - 270)\ \text{K}}{(0.10 + 0.143 + 0.025)\ \text{K/W}} = 223.9\ \text{W}$$

The approach used in Example 1.9 can also be used for composite walls, and Fig. 1.23 shows the structure, temperature distribution, and equivalent network for a wall with three layers and convection on both surfaces.

1.6.4 Convection and Radiation in Parallel

In many engineering problems a surface loses or receives thermal energy by convection and radiation simultaneously. For example, the roof of a house heated from the interior is at a higher temperature than the ambient air and thus loses heat by convection as well as radiation. Since both heat flows emanate from the same potential, that is, the roof, they act in parallel. Similarly, the gases in a combustion chamber contain species that emit and absorb radiation. Consequently, the wall of the combustion chamber receives heat by convection as well as radiation. Figure 1.24 illustrates the cocurrent heat transfer from a

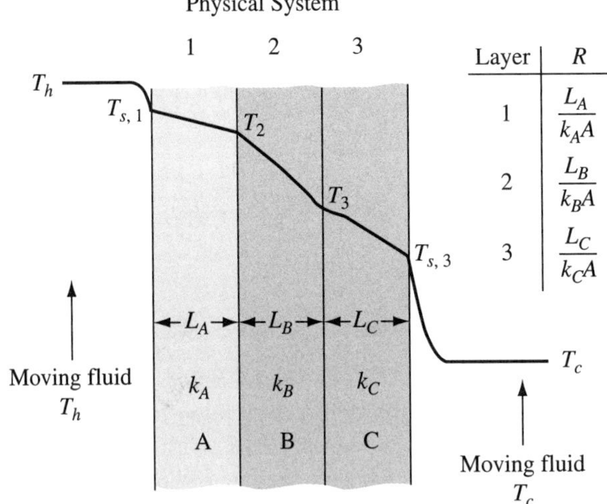

FIGURE 1.23 Schematic diagram and thermal circuit for composite three-layer wall with convection over both exterior surfaces.

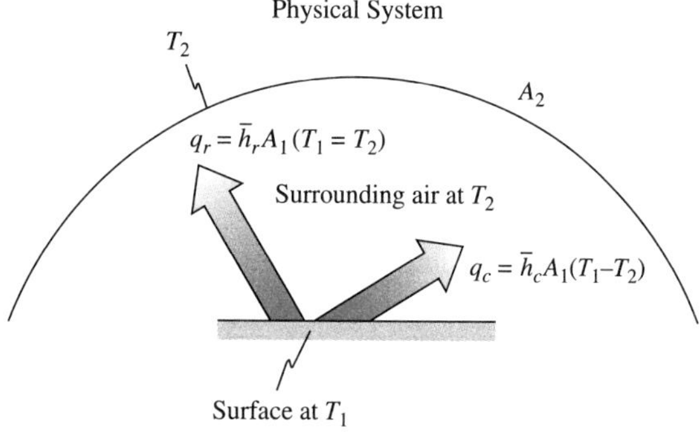

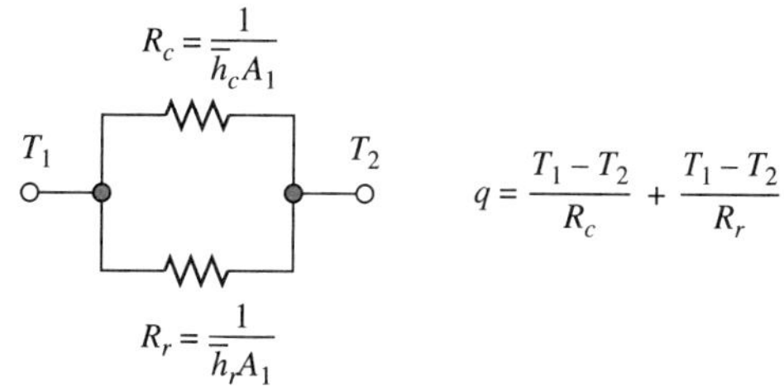

Simplified Circuit

T_1 T_2

$R = \frac{R_c R_r}{R_c + R_r}$

$q = \frac{T_1 - T_2}{R} = \bar{h}A(T_1 - T_2)$

FIGURE 1.24 Thermal circuit with convection and radiation acting in parallel.

surface to its surroundings by convection and radiation. The total rate of heat transfer is the sum of the rates of heat flow by convection and radiation, or

$$\begin{aligned} q &= q_c + q_r \\ &= \bar{h}_c A(T_1 - T_2) + \bar{h}_r A(T_1 - T_2) \\ &= (\bar{h}_c + \bar{h}_r) A(T_1 - T_2) \end{aligned} \tag{1.31}$$

where $\bar{h}_c$ is the average convection heat transfer coefficient between area A_1 and the ambient air at T_2, and, as shown previously, the radiation heat transfer coefficient between A_1 and the surroundings at T_2 is

$$\bar{h}_r = \frac{\varepsilon_1 \sigma (T_1^4 - T_2^4)}{T_1 - T_2} \tag{1.32}$$

The analysis of combined heat transfer, especially at boundaries of a complicated geometry or in unsteady-state conduction, can often be simplified by using an effective heat transfer coefficient that combines convection and radiation. The combined heat transfer coefficient (or heat transfer coefficient for short) is defined by

$$\bar{h} = \bar{h}_c + \bar{h}_r \tag{1.33}$$

The combined heat transfer coefficient specifies the average total rate of heat flow between a surface and an adjacent fluid and the surroundings per unit surface area and unit temperature difference between the surface and the fluid. Its units are W/m^2 K.

EXAMPLE 1.10 A 0.5-m-diameter pipe ($\varepsilon = 0.9$) carrying steam has a surface temperature of 500 K (see Fig. 1.25). The pipe is located in a room at 300 K, and the convection heat transfer coefficient between the pipe surface and the air in the room is 20 W/m^2 K. Calculate the combined heat transfer coefficient and the rate of heat loss per meter of pipe length.

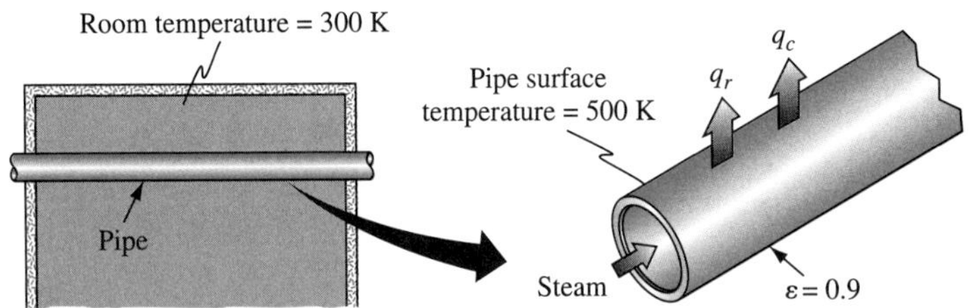

FIGURE 1.25 Schematic diagram of steam pipe for Example 1.10.

SOLUTION This problem may be idealized as a small object (the pipe) inside a large black enclosure (the room). Noting that

$$\frac{T_1^4 - T_2^4}{T_1 - T_2} = (T_1^2 + T_2^2)(T_1 + T_2)$$

the radiation heat transfer coefficient is, from Eq. (1.33),

$$\bar{h}_r = \sigma\varepsilon(T_1^2 + T_2^2)(T_1 + T_2) = 13.9 \text{ W/m}^2\text{ K}$$

The combined heat transfer coefficient is, from Eq. (1.33),

$$\bar{h} = \bar{h}_c + \bar{h}_r = 20 + 13.9 = 33.9 \text{ W/m}^2\text{ K}$$

and the rate of heat loss per meter is

$$q = \pi D L \bar{h}(T_{\text{pipe}} - T_{\text{air}}) = (\pi)(0.5 \text{ m})(1 \text{ m})(33.9 \text{ W/m}^2\text{ K})(200 \text{ K}) = 10{,}650 \text{ W}$$

1.6.4 Overall Heat Transfer Coefficient

We noted previously that a common heat transfer problem is to determine the rate of heat flow between two fluids, gaseous or liquid, separated by a wall (see Fig. 1.26.). If the wall is plane and heat is transferred only by convection on both

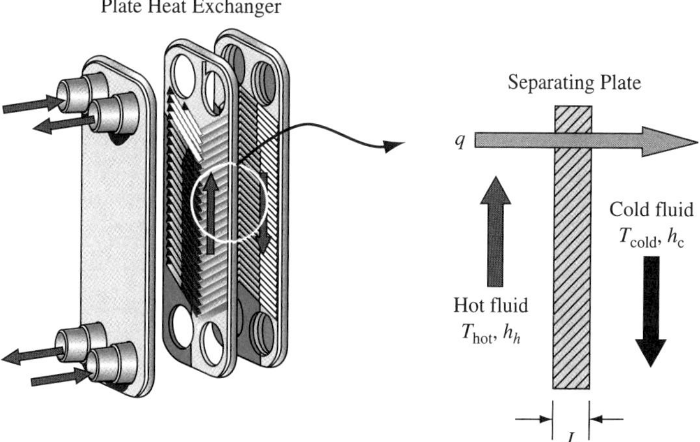

FIGURE 1.26 Heat transfer by convection between two fluid streams in a plate heat exchanger.

sides, the rate of heat transfer in terms of the two fluid temperatures is given by Eq. (1.30):

$$q = \frac{T_{\text{het}} - T_{cold}}{(1/\bar{h}_c A)_{\text{hot}} + (L/kA) + (1/\bar{h}_c A)_{\text{cold}}} = \frac{\Delta T}{R_1 + R_2 + R_3}$$

In Eq. (1.30) the rate of heat flow is expressed only in terms of an overall temperature potential and the heat transfer characteristics of individual sections in the heat flow path. From these relations it is possible to evaluate quantitatively the importance of each individual thermal resistance in the path. Inspection of the orders of magnitude of the individual terms in the denominator often reveals a means of simplifying a problem. When one term dominates quantitatively, it is sometimes permissible to neglect the rest. As we gain facility in the techniques of determining individual thermal resistances and conductances, there will be numerous examples of such approximations. There are, however, certain types of problems, notably in the design of heat exchangers, where it is convenient to simplify the writing of Eq. (1.30) by combining the individual resistances or conductances of the thermal system into one quantity called the overall unit conductance, the overall transmittance, or the overall coefficient of heat transfer U. The use of an overall coefficient is a convenience in notation, and it is important not to lose sight of the significance of the individual factors that determine the numerical value of U.

Writing Eq. (1.30) in terms of an overall coefficient gives

$$q = UA\Delta T_{\text{total}} \tag{1.34}$$

where

$$UA = \frac{1}{R_1 + R_2 + R_3} = \frac{1}{R_{\text{total}}} \tag{1.35}$$

The overall coefficient U can be based on any chosen area. The area selected becomes particularly important in heat transfer through the walls of tubes in a heat exchanger, and to avoid misunderstandings the area basis of an overall coefficient should always be stated. Additional information about the overall heat transfer coefficient U will be presented in later chapters, particularly in Chapter 8.

An overall heat transfer coefficient can also be obtained in terms of individual resistances in the thermal circuit when convection and radiation transfer heat to and/or from one or both surfaces of the wall. In general, radiation will not be of any significance when the fluid is a liquid, but it can play an important role in convection to or from a gas when the temperatures are high or the convection heat transfer coefficient is small, for instance, in natural convection. The integration of radiation into an overall heat transfer coefficient will be illustrated below.

The schematic diagram in Fig. 1.27 shows the heat transfer from hot products of combustion in the chamber of a rocket motor through a wall that is liquid-cooled on the outside by convection. In the first section of this system, heat is transferred by convection and radiation in parallel. Hence, the rate of heat flow to the interior surface of the wall is the sum of the two heat flows

$$\begin{aligned} q &= q_c + q_r \\ &= \bar{h}_c A(T_g - T_{sg}) + \bar{h}_r A(T_g - T_{sg}) \\ &= (\bar{h}_{c1} + \bar{h}_{r1})A(T_g - T_{sg}) = \frac{T_g - T_{sg}}{R_1} \end{aligned} \tag{1.36}$$

where T_g = temperature of the hot gas in the interior
T_{sg} = temperature of the hot wall surface

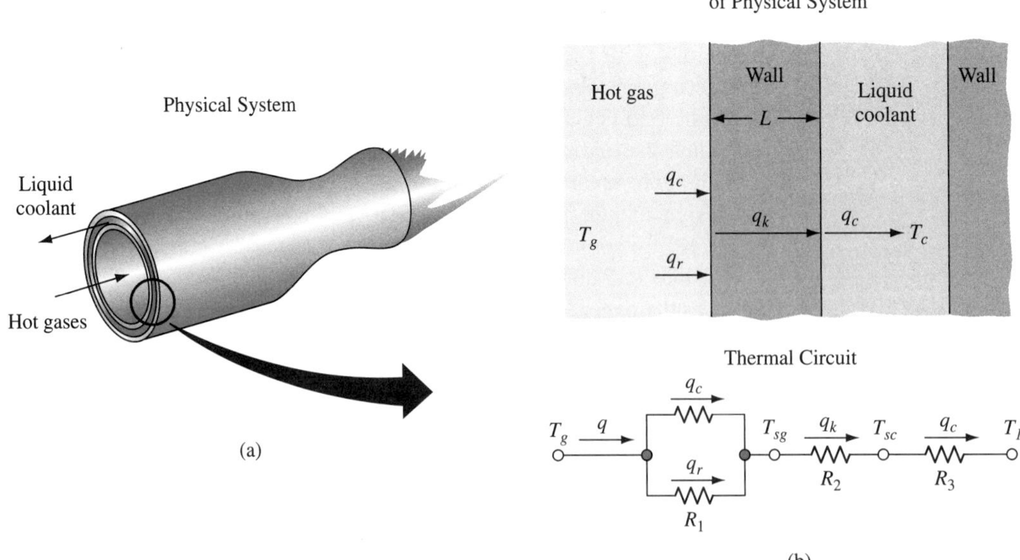

FIGURE 1.27 Heat transfer from combustion gases to a liquid coolant in a rocket motor.

$$\bar{h}_{r1} = \frac{\sigma(T_g^4 - T_{sg}^4)}{T_g - T_{sg}}$$ = the radiation heat transfer coefficient in the first section (ε is assumed unity)

$\bar{h}_{c1}$ = convection heat transfer coefficient from gas to wall

$$R_1 = \frac{1}{(h_r + \bar{h}_{c1})A}$$ = combined thermal resistance of first section

In the steady state, heat is conducted through the shell, the second section of the system, at the same rate as to the surface and

$$q = q_k = \frac{kA}{L}(T_{sg} - T_{sc})$$
$$= \frac{T_{sg} - T_{sc}}{R_2} \tag{1.37}$$

where T_{sc} = surface temperature at wall on coolant side
R_2 = thermal resistance of second section

After passing through the wall, the heat flows through the third section of the system by convection to the coolant. The rate of heat flow in the third and last step is

$$q = q_c = \bar{h}_{c3}A(T_{sc} - T_l)$$
$$= \frac{T_{sc} - T_l}{R_3} \tag{1.38}$$

where T_l = temperature of liquid coolant
R_3 = thermal resistance in third section of system

It should be noted that the symbol $\bar{h}_c$ stands for average convection heat transfer coefficient in general, but the numerical values of the convection coefficients in the first, $\bar{h}_{c1}$, and third, $\bar{h}_{c3}$, sections of the system depend on many factors and will, in general, be different. Also note that the areas of the three-heat-flow sections are not equal, but since the wall is very thin, the change in the heat-flow area is so small that it can be neglected in this system.

In practice, often only the temperatures of the hot gas and the coolant are known. If intermediate temperatures are eliminated by algebraic addition of Eqs. (1.36), (1.37), and (1.38), the rate of heat flow is

$$q = \frac{T_g - T_l}{R_1 + R_2 + R_3} = \frac{\Delta T_{\text{total}}}{R_1 + R_2 + R_3} \tag{1.39}$$

where the thermal resistance of the three series-connected sections or heat flow steps in the system are defined in Eqs. (1.36), (1.37), and (1.38).

EXAMPLE 1.11 In the design of a heat exchanger for aircraft application (Fig. 1.28 on the next page), the maximum wall temperature in steady state is not to exceed 800 K. For the conditions tabulated below, determine the maximum permissible unit thermal resistance per square meter of the metal wall that separates the hot gas from the cold gas.

Schematic of Aircraft Heat Exchanger Section

Hot exhaust gas

Coolant gas

(a)

Physical System

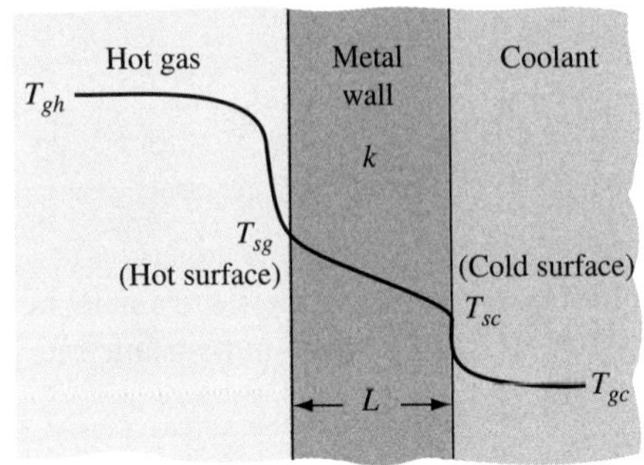

Detailed Thermal Circuit

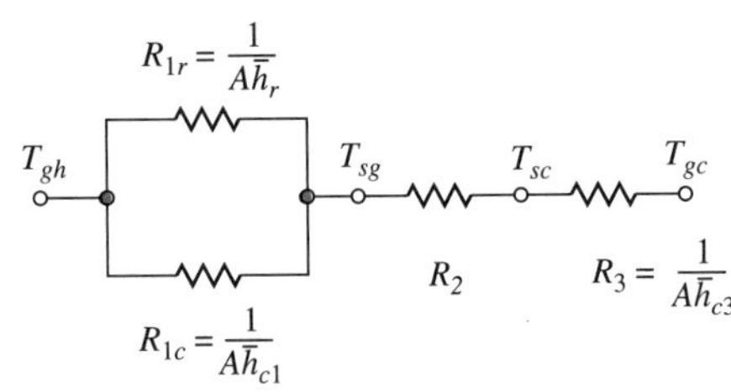

Simplified Circuit

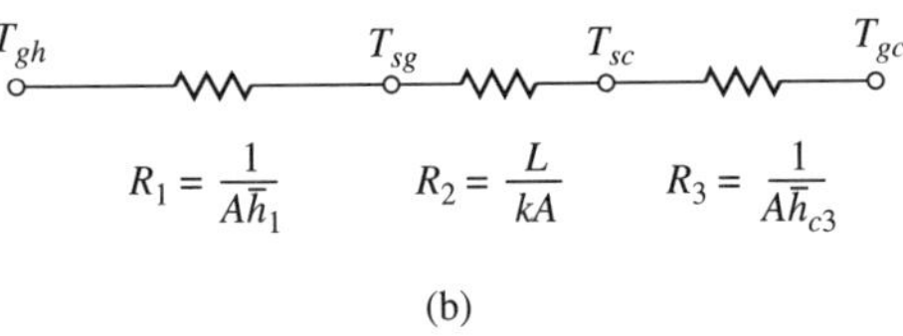

(b)

FIGURE 1.28 Physical system and thermal circuit for Example 1.11.

Hot-gas temperature $= T_{gh} = 1300$ K
Heat transfer coefficient on hot side $= \bar{h}_1 = 200$ W/m^2 K
Heat transfer coefficient on cold side $= \bar{h}_3 = \bar{h}_{c3} = 400$ W/m^2 K
Coolant temperature $= T_{gc} = 300$ K

SOLUTION In the steady state we can write

$$\frac{q}{A} \text{ from hot gas to hot side of wall} = \frac{q}{A} \text{ from hot gas through wall to cold gas}$$

Using the nomenclature in Fig. 1.28, we get

$$\frac{q}{A} = \frac{T_{gh} - T_{sg}}{R_1} = \frac{T_{gh} - T_{gc}}{R_1 + R_2 + R_3}$$

where T_{sg} is the hot-surface temperature. Substituting numerical values for the unit thermal resistances and temperatures yields

$$\frac{300 - 800}{1/200} = \frac{1300 - 300}{1/200 + R_2 + 1/400}$$

$$\frac{1300 - 800}{0.005} = \frac{1300 - 300}{R_2 + 0.0075}$$

Solving for R_2 gives

$$R_2 = 06025 \text{ m}^2 \text{ K/W}$$

Thus, a unit thermal resistance larger than 06025 m^2 K/W for the wall would raise the inner-wall temperature above 800 K. This value can place an upper limit on the wall thickness.

1.7 Thermal Insulation

There are many situations in engineering design when the objective is to reduce the flow of heat. Examples of such cases include the insulation of buildings to minimize heat loss in the winter, a thermos bottle to keep tea or coffee hot, and a ski jacket to prevent excessive heat loss from a skier. All of these examples require the use of thermal insulation.

Thermal insulation materials must have a low thermal conductivity. In most cases, this is achieved by trapping air or some other gas inside small cavities in a solid, but sometimes the same effect can be produced by filling the space across which heat flow is to be reduced with small solid particles and trapping air between the particles. These types of thermal insulation materials use the inherently low conductivity of a gas to inhibit heat flow. However, since gases are fluids, heat can also be transferred by natural convection inside the gas pockets and by radiation between the solid enclosure walls. The conductivity of insulting materials is therefore not really a material property but rather the result of a combination of heat flow mechanisms. The thermal conductivity of insulation is an effective value, k_{eff}, that changes not only with temperature, but also with pressure and environmental conditions, e.g., moisture. The change of k_{eff} with temperature can be quite pronounced, especially at elevated temperatures when radiation plays a significant role in the overall heat transport process.

The many different types of insulation materials can essentially be classified in the following three broad categories:

1. *Fibrous*. Fibrous materials consist of small-diameter particles of filaments of low density that can be poured into a gap as "loose fill" or formed into boards, batts, or blankets. Fibrous materials have very high porosity (–90%). Mineral wool is a common fibrous material for applications at temperatures below 700°C, and fiberglass is often used for temperatures below 200°C. For thermal protection at temperatures between 700°C to 1700°C one can use refractory fibers such as alumina (Al_2O_3) or silica (SiO_2).

2. *Cellular*. Cellular insulations are closed- or open-cell materials that are usually in the form of extended flexible or rigid boards. They can, however, also be foamed or sprayed in place to achieve desired geometrical shapes. Cellular insulation has the advantage of low density, low-heat capacity, and relatively good-compressive strength. Examples are polyurethane and expanded polystyrene foam.
3. *Granular*. Granular insulation consists of small flakes or particles of inorganic materials bonded into preformed shapes or used as powders. Examples are perlite powder, diatomaceous silica, and vermiculite.

For use at cryogenic temperatures, the gases in cellular materials can be condensed or frozen to create a partial vacuum, which improves the effectiveness of the insulation. Fibrous and granular insulation can be evacuated to eliminate convection and conduction, thus decreasing the effective conductivity appreciably. Figure 1.29 shows the ranges of effective thermal conductivity for evacuated and nonevacuated insulation as well as the product of thermal conductivity and bulk density, which is sometimes important in design.

In addition to these three types of thermal insulating materials, insulation can also be achieved by the use of reflective sheets. In this approach two or more thin

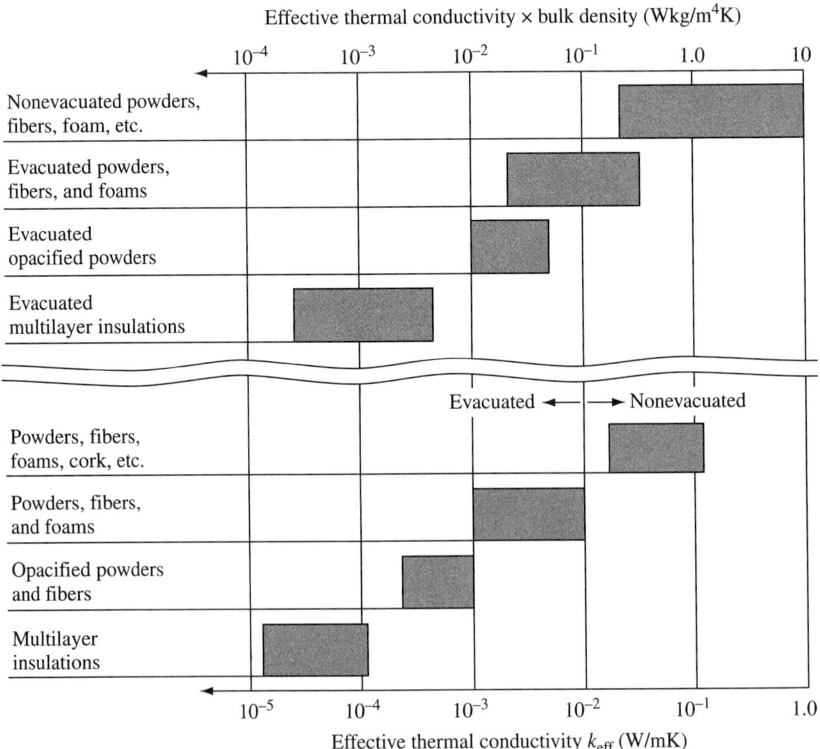

FIGURE 1.29 Ranges of thermal conductivities of thermal insulators and products of thermal conductivity and bulk density.

sheets of metal with low emittance are placed parallel to each other to reflect radiation back to its source. An example is the thermos bottle, in which the space between the reflective surfaces is evacuated to suppress conduction and convection, leaving radiation as the sole transfer mechanism. Reflective insulation will be treated in Chapter 9.

The most important property to consider in selecting an insulation material is the effective thermal conductivity, but the density, the upper limit of temperature, the structural rigidity, degradation, chemical stability, and, of course, the cost are also important factors. Physical properties of insulating materials are usually supplied by the product manufacturer or can be obtained from handbooks. Unfortunately, the data are often quite limited, especially at elevated temperatures. In such cases, it is necessary to extrapolate available information and then use a safety factor in the final design.

Ranges of effective thermal conductivities for several common low-temperature fibrous and cellular-insulation materials are shown in Fig. 1.30. The lower value is for low temperatures, the upper value for temperatures at the upper limit of allowable use. All of the values are for new materials. Polyurethane and polystyrene generally lose between 20% and 50% of their insulation quality during the first year of use. Some other materials experience increases in their effective thermal conductivity as a result of moisture uptake in a high-humidity environment

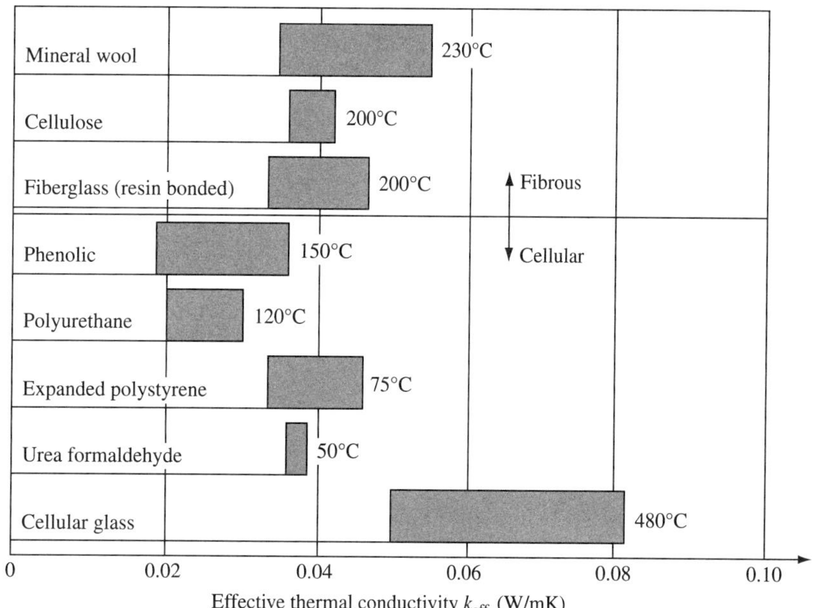

FIGURE 1.30 Effective thermal conductivity ranges for typical fibrous and cellular insulations. Approximate maximum-use temperatures are listed to the right of the insulations.

Source: Adapted from *Handbook of Applied Thermal Design*, E. C. Guyer, ed., McGraw-Hill, 1989.

or loss of vacuum. Note that except for cellular glass, cellular insulating materials are plastics that are inexpensive and lightweight, i.e., they have densities on the order of 30 kg/m^3. All cellular materials are rigid and can be obtained in practically any desired shape.

For high-temperature applications refractory materials are used. They come in the form of bricks and can withstand temperatures up to 1700°C. The effective conductivities range from 1.5 W/m K for fire clay to about 2.5 W/m K for zirconium. Loose-fill types of insulation have much lower thermal conductivities, as shown in Fig. 1.31, but most of them can only be used below about 900°C. Loose-fill materials also trend to "settle," causing potential problems in places that are difficult to access.

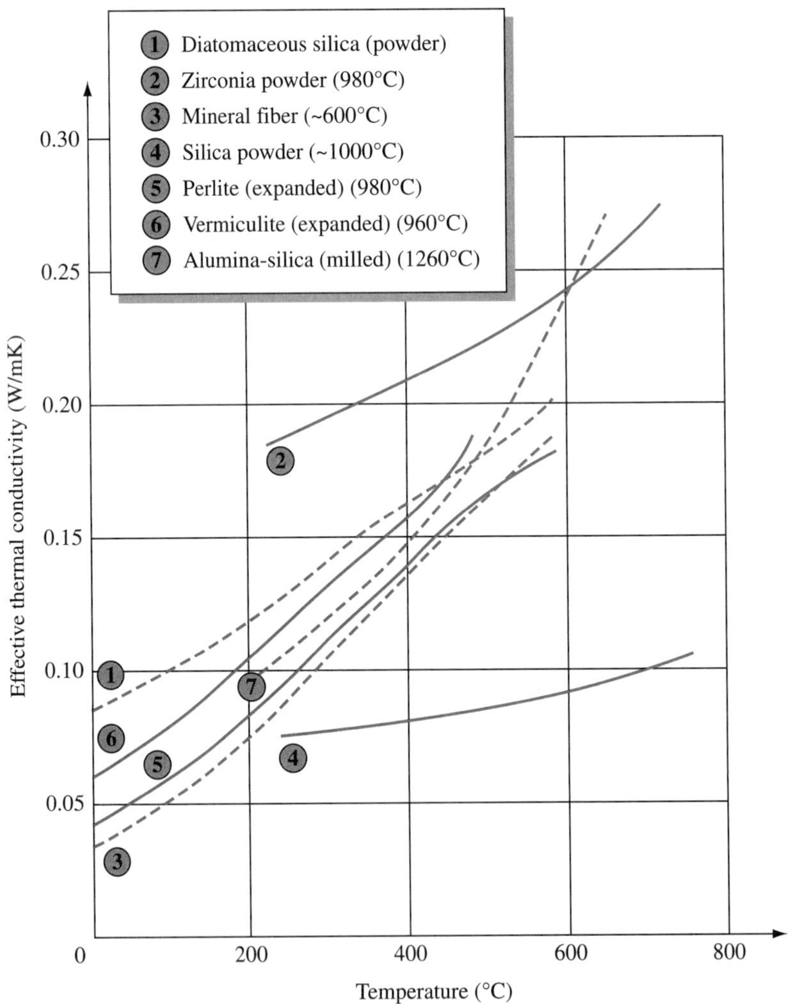

FIGURE 1.31 Effective thermal conductivity vs. temperature for some high-temperature insulations. The maximum useful temperature is given in parentheses.

EXAMPLE 1.12 The door for an industrial gas furnace is 2 m × 4 m in surface area and is to be insulated to reduce heat loss to no more than 1200 W/m^2. The door is shown schematically in Fig. 1.32. The interior surface is a 3/8-in.-thick Inconel 600 sheet, and the outer surface is 1/4-in.-thick sheet of stainless steel 316. Between these metal sheets a suitable thickness of insulation material is to be placed. The effective gas temperature inside the furnace is 1200°C, and the overall heat transfer coefficient between the gas and the door is $U_i = 20$ W/m^2 K. The heat transfer coefficient between the outer surface of the door and the surroundings at 20°C is $\bar{h} = 5$ W/m^2 K. Select a suitable insulation material and size its thickness.

SOLUTION From Fig. 1.7 we estimate the thermal conductivity of the metal sheets to be approximately 43 W/m K. The thermal resistances of the two metal sheets are approximately

$$R = L/k \sim \frac{0.625 \text{ in.}}{43 \text{ W/m K}} \times \frac{1 \text{ m}}{39.4 \text{ in.}} \sim 3.7 \times 10^{-4} \text{ m}^2 \text{ K/W}$$

These resistances are negligible compared to the other three resistances shown in the simplified thermal circuit below:

20°C — Air — Insulation — Gas — 1200°C

$R_a = \frac{1}{\bar{h}}$ $\quad R_{ins}$ $\quad R_g = \frac{1}{\bar{U}_i}$

The temperature drop between the gas and the interior surface of the door at the specified heat flux is:

$$\Delta T = \frac{q/A}{U} = \frac{1200 \text{ W/m}^2}{20 \text{ W/m}^2 \text{ K}} = 60 \text{ K}$$

Hence, the temperature of the Inconel will be about 1140°C. This is acceptable since no appreciable structural load is applied.

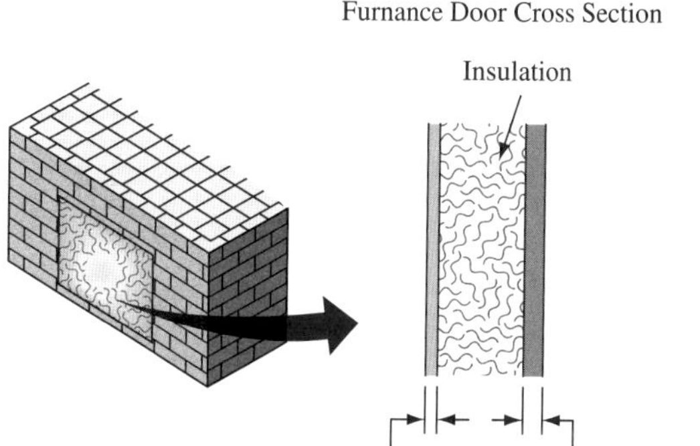

FIGURE 1.32 Cross section of composite wall of gas furnace door for Example 1.12.

From Fig. 1.31 we see that only milled alumina-silica chips can withstand the maximum temperature in the door. Thermal conductivity data are available only between 300 and 650°C. The trend of the data suggests that at higher temperatures when radiation becomes the dominant mechanism, the increase of k_{eff} with T will become more pronounced. We shall select the value at 650°C (0.27 W/mK) and then apply a safety factor to the insulation thickness.
The temperature drop at the outer surface is

$$\Delta T = \frac{1200\ \text{W/m}^2}{5\ \text{W/m}^2\ \text{K}} = 240°\text{C}$$

Hence, ΔT across the insulation is 1180°C − (240 + 60)°C = 880 K. The insulation thickness for k = 0.27 W/m K is:

$$L = \frac{k\Delta T}{q/A} = \frac{0.27\ \text{W/m K} \times 880\ \text{K}}{1200\ \text{W/m}^2} = 0.2\ \text{m}$$

In view of the uncertainty, in the value of k_{eff}, and the possibility that the insulation may become more compact with use, a prudent design would double the value of insulation thickness. Additional insulation would also reduce the temperature of the outer surface of the door for safety, comfort, and ease of operation.

In engineering practice, especially for building materials, insulation is often characterized by a term called *R-value*. The temperature difference divided by the *R*-value gives the rate of heat transfer per unit area. For a large sheet or slab of material:

$$R\text{-value} = \frac{\text{thickness}}{\text{effective average thermal conductivity}}$$

The *R*-value is generally given in the English units of h ft^2 °F/Btu. For example, the *R*-value of a 3.5-in.-thick sheet of fiberglass (k_{eff} = 0.035 Btu/h ft °F from Table 11 in Appendix 2) equals

$$\frac{(3.5\ \text{in.})\ \text{h ft °F}}{0.035\,\text{Btu}} \times \frac{\text{ft}}{12\ \text{in.}} = 8.3\,\frac{\text{h °F ft}^2}{\text{Btu}}$$

R-values can also be assigned to composite structures such as double-glazed windows or walls constructed of wood with insulation between the struts.

In some cases the *R*-value is given on a "per inch" basis. Then its units are h ft^2 °F/Btu in. In the above example, the *R*-value per inch of the fiberglass is $8.3/3.5 \sim 2.4$ h ft^2 °F/Btu in. Note that the *R*-value per inch is equal to $1/12k$ when the thermal conductivity is given in units of Btu/h ft °F. Care should be exercised when using manufacturers' literature for *R*-values because the per-inch value may be given even though the property may be called simply the *R*-value. By examining the units given for the property it should be clear which *R*-value is given.

1.8 Heat Transfer and the Law of Energy Conservation

In addition to the heat transfer rate equations we shall also often use the first law of thermodynamics, or the fundamental law of conservation of energy, in analyzing a system. Although, as mentioned previously, a thermodynamic analysis alone cannot predict the rate at which the transfer will occur in terms of the degree of thermal non-equilibrium, the basic laws of thermodynamics (both first and second) must be obeyed. Thus, any physical law that must be satisfied by a process or a system provides an equation that can be used for analysis. We have already used the second law of thermodynamics to indicate the direction of heat flow. We will now demonstrate how the first law of thermodynamics can be applied in the analysis of heat transfer problems.

1.8.1 First Law of Thermodynamics

The first law of thermodynamics states that energy cannot be created or destroyed but can be transformed from one form to another or transferred as heat or work. To apply the law of conservation of energy, we first need to identify a control volume. A *control volume* is a fixed region in space bounded by a *control surface* through which heat, work, and mass can pass. The conservation of energy requirement for an open system in a fonn useful for heat transfer analysis is:

> *The rate at which thermal and mechanical energies enter a control volume plus the rate at which energy is generated within that volume minus the rate at which thermal and mechanical energies leave the control volume must equal the rate at which energy is stored inside this volume.*

If the sum of the energy inflow and the generation exceeds the outflow, there will be an increase in the amount of energy stored in the control volume, whereas when the outflow exceeds the inflow and generation there will be a decrease in energy storage. But when there is no generation and the rate of energy inflow is equal to the rate of outflow, steady state exists and there is no change in the energy stored in the control volume.

Referring to Fig. 1.33 on the next page, the energy conservation requirements can be expressed in the form

$$(e\dot{m})_{in} + q + \dot{q}_G - (e\dot{m})_{out} - W_{out} = \frac{\partial E}{\partial t} \tag{1.39}$$

where $(e\dot{m})_{in}$ is the rate of energy inflow, $(e\dot{m})_{out}$ is the rate of energy outflow, q is the *net* rate of heat transfer into the control volume $(q_{in} - q_{out})$, W_{out} is the net rate of work output, $\dot{q}_G$ is the rate of energy generation within the control volume, and $\partial E/\partial t$ is the rate of energy storage inside the control volume.

The specific energy carried by the mass flow, e, across the surface may contain potential and kinetic as well as thermal (internal) forms, but for most heat transfer problems the potential and kinetic energy terms are negligible. The inflow and

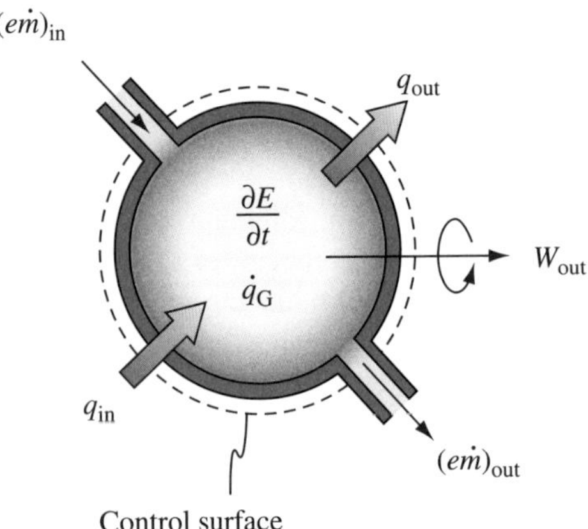

FIGURE 1.33 Control volume for first law of thermodynamics or conservation of energy.

outflow energy terms may also include work interactions, but these phenomena are of significance only in extremely high speed flow processes.

Observe that the inflow and outflow rate terms are surface phenomena and are therefore proportional to the surface area. The internal energy generation term $\dot{q}_G$ is encountered when another form of energy (such as chemical, electrical, or nuclear energy) is converted to thermal energy within the control volume. The generation term is therefore a volumetric phenomenon, and its rate is proportional to the volume within the control surface. Energy storage is also a volumetric phenomenon associated with the internal energy of the mass in the control volume, but the process of energy generation is quite different from that of energy storage, even though both contribute to the rate of energy storage.

Equation (1.39) can be simplified when there is no transport of mass across the boundary. Such a system is called a *closed system*, and for such conditions Eq. (1.39) becomes

$$q + \dot{q}_G - W_{out} = \frac{\partial E}{\partial t} \tag{1.39}$$

where the right side represents the rate of energy storage or the rate of increase in internal energy. Note that E is the total internal energy stored in the system, and it equals the product of the specific internal energy and the mass of the system.

1.8.2 Conservation of Energy Applied to Heat Transfer Analysis

The following two examples demonstrate the use of the energy conservation law in heat transfer analysis. The first example is a steady-state problem in which the storage term is zero, while the second example demonstrates the analytic procedure for a problem in which internal energy storage occurs. The latter is called *transient heat transfer*, and a more detailed analysis of such cases will be presented in the next chapter.

EXAMPLE 1.13 A house has a black tar, flat, horizontal roof. The lower surface of the roof is well insulated, while the upper surface is exposed to ambient air at 300 K through a convective heat transfer coefficient of 10 W/m^2 K. Calculate the roof equilibrium temperature for the following conditions: (a) a clear sunny day with an incident solar radiation flux of 500 W/m^2 and the ambient sky at an effective temperature of 50 K and (b) a clear night with an ambient sky temperature of 50 K.

SOLUTION A schematic sketch of the system is shown in Fig. 1.34. The control volume is the roof. Assume that there are no obstructions between the roof, called surface 1, and the sky, called surface 2, and that both surfaces are black. The sky behaves as a blackbody because it absorbs all the radiation emitted by the roof and reflects none.

Heat is transferred by convection between the ambient air and the roof and by radiation between the sun and roof and between the roof and the sky. This is a closed system in thermal equilibrium. Since there is no generation, storage, or work output, we can express the energy conservation requirement by the conceptual relation

rate of solar radiation heat transfer *to* roof	+	rate of convection heat transfer *to* roof	=	net rate of radiation heat transfer *from* roof to ambient sky

Analytically, this relation can be cast in the form

$$A_1 q_{r,\text{sun}\to\text{roof}} + \bar{h}_c A_1 (T_{\text{air}} - T_{\text{roof}}) = A_1 q_{r,\text{roof}\to\text{ambient sky}}$$

Canceling the roof area A_1 and substituting the Stefan-Boltzmann relation [Eq. (1.17)] for the net radiation from the roof to the ambient sky gives

$$q_{r,\text{sun}\to 1} + \bar{h}_c(300 - T_{\text{roof}}) = \sigma(T_{\text{roof}}^4 - T_{\text{sky}}^4)$$

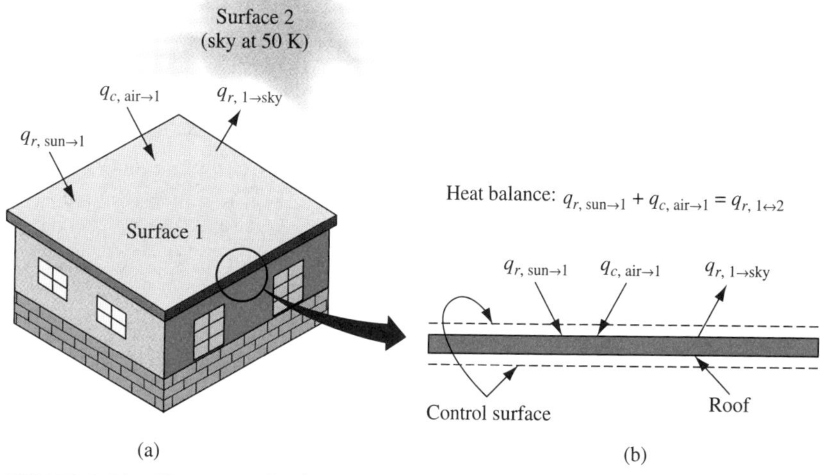

FIGURE 1.34 Heat transfer by convection and radiation for roof in Example 1.13.

(a) When the solar radiation to the roof, $q_{r,\text{sun}\rightarrow 1}$, is 500 W/m^2 and T_{sky} is 50 K, we get

$$500 \text{ W/m}^2 + (10 \text{ W/m}^2 \text{ K})(300 - T_{\text{roof}})(\text{K}) = (5.67 \times 10^{-8} \text{ W/m}^2 \text{ K}^4)(T_{\text{roof}}^4 - 50^4)(\text{K}^4)$$

Solving by trial and error for the roof temperature, we get

$$T_{\text{roof}} = 303 \text{ K} = 30°\text{C}$$

Note that the convection term is negative because the sun heats the roof to a temperature above the ambient air, so that the roof is not heated but is cooled by convection to the air.

(b) At night the term $Q_{r,\text{sun}\rightarrow 1} = 0$ and we get, upon substituting the numerical data in the conservation of energy relation,

$$\bar{h}_c(T_{\text{air}} - T_{\text{roof}}) = \sigma(T_{\text{roof}}^4 - T_{\text{sky}}^4)$$

or

$$(10 \text{ W/m}^2 \text{ K})(300 - T_{\text{roof}})(\text{K}) = (5.67 \times 10^{-8} \text{ W/m}^2 \text{ K}^4)(T_{\text{roof}}^4 - 50^4)(\text{K}^4)$$

Solving this equation for T_{roof} gives

$$T_{\text{roof}} = 270 \text{ K} = -3°\text{C}$$

At night the roof is cooler than the ambient air and convection occurs from the air to the roof, which is heated in the process. Observe also that the conditions at night and during the day are assumed to be steady and that the change from one steady condition to the other requires a period of transition in which the energy stored in the roof changes and the roof temperature also changes. The energy stored in the roof increases during the morning hours and decreases during the evening after the sun has set, but these periods are not considered in this example.

EXAMPLE 1.14 A long, thin copper wire of diameter D and length L has an electrical resistance of ρ_e per unit length. The wire is initially at steady state in a room at temperature T_{air}. At time $t = 0$, an electric current i is passed through the wire. The wire temperature begins to increase due to internal electrical heat generation, but at the same time heat is lost from the wire by convection through a convection coefficient $\bar{h}_c$ to the ambient air.

Set up an equation to determine the change in temperature with time in the wire, assuming that the wire temperature is uniform. This is a good assumption because the thermal conductivity of copper is very large and the wire is thin. We will learn in Chapter 2 how to calculate the transient radial temperature distribution if the conductivity is small.

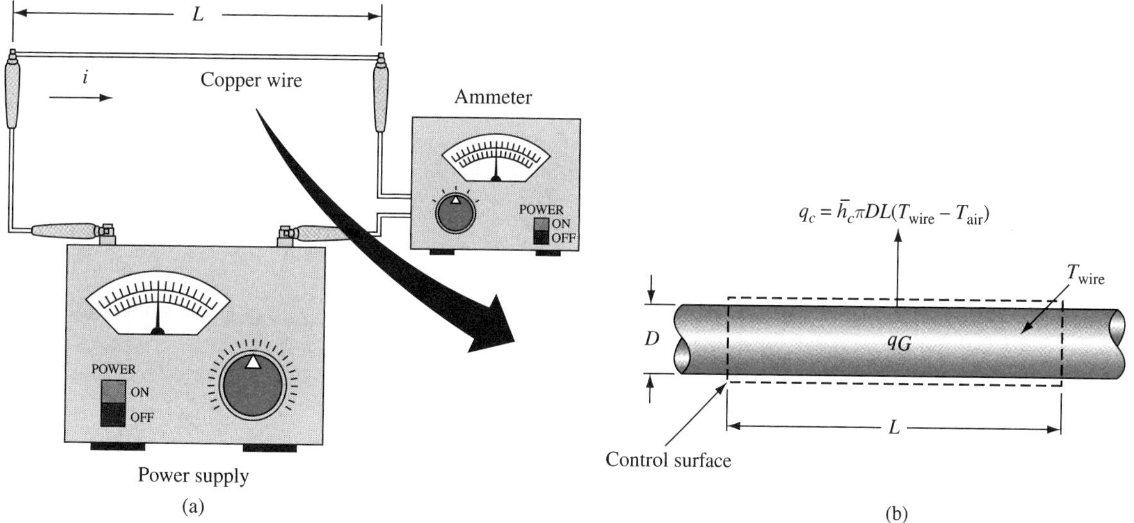

FIGURE 1.35 Schematic diagram for electric heat generation system of Example 1.14.

SOLUTION The sketch in Fig. 1.35 shows the wire and the control volume. We shall assume that radiation losses are negligible so that the net rate of convection heat flow q_c is equal to the rate of heat loss from the wire, q_{out}:

$$q_{out} = \bar{h}_c A_{surf}(T_{wire} - T_{air}) = \bar{h}_c \pi DL(T_{wire} - T_{air})$$

The rate of energy generation (or electrical dissipation) in the wire control volume is

$$\dot{q}_G = i^2 R_e = i^2 \rho_e L$$

where $R_e = \rho_e L$, the electrical resistance.

The rate of internal energy storage in the control volume is

$$\frac{\partial E}{\partial t} = \frac{d[(\pi D^2/4)Lc\rho T_{wire}(t)]}{dt}$$

where c is the specific heat and ρ is the density of the wire material.

Applying the conservation of energy relation for a closed system [Eq. (1.39)] to the problem at hand gives

$$\dot{q}_G - q_{out} = \frac{\partial E}{\partial t}$$

since there is no work output and q_{in} is zero.

Substituting the appropriate relations for the three energy terms in the conservation of energy law gives the different equation

$$i^2 \rho_e L - (\bar{h}_c \pi DL)(T_{wire} - T_{air}) = \left(\frac{\pi D^2}{4} Lc\rho\right) \frac{dT_{wire}(t)}{dt}$$

If the specific heat and density are constant, the solution to this equation for the wire temperature as a function of time, $T(t)$, becomes

$$T_{\text{wire}}(t) - T_{\text{air}} = C_1(1 - e^{-C_2 t})$$

where
$$C_1 = \frac{i^2 \rho_e}{\bar{h}_c \pi D}$$
$$C_2 = \frac{4\bar{h}_c}{c\rho D}$$

Note that as $t \rightarrow \infty$, the second term on the right-hand side approaches C_1 and $dT_{\text{wire}}/dt \rightarrow 0$. This means physically that *the wire temperature has reached a new equilibrium* value that can be evaluated from the steady-state conservation relation $q_{\text{out}} = \dot{q}_G$ or

$$(T_{\text{wire}} - T_{\text{air}})\bar{h}_c \pi DL = i^2 \rho_e L$$

Thermodynamics alone, that is, the law of energy conservation, could predict the differences in the internal energy stored in the control volume between the two equilibrium states at $t = 0$ and $t \rightarrow \infty$, but it could not predict the rate at which the change occurs. For that calculation it is necessary to use the heat transfer rate analysis shown above.

1.8.3 Boundary Conditions

There are many situations in which the conservation of energy requirement is applied at the surface of a system. In these cases, the control surface contains no mass and the volume it encompasses approaches zero, as shown in Fig. 1.36.

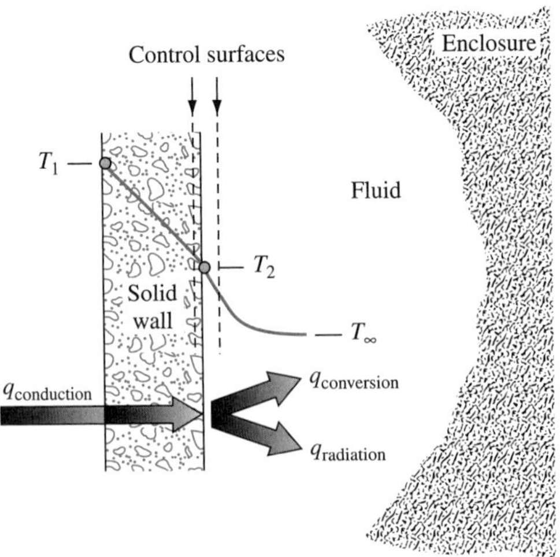

FIGURE 1.36 Application of conservation of energy law at the surface of a system.

Consequently, there can be no storage or generation of energy, and the conservation requirement reduces to

$$q_{net} = q_{in} - q_{out} = 0 \tag{1.40}$$

It is important to note that in this form the conservation law holds for steady-state as well as transient conditions and that the heat inflow and outflow may occur by several heat transfer mechanisms in parallel. Applications of Eq. (1.40) to many different physical situations will be illustrated later.

1. Carefully read the problem and ask yourself *in your own words* what is known about the system, what information can be obtained from sources such as tables of properties, handbooks, or appendices, and what are the unknowns for which an answer must be found.
2. Draw a schematic diagram of the system, including the boundaries to be used in the application of conservation laws. Identify the relevant heat transfer processes, and sketch a thermal circuit for the system. Figures 1.18 and 1.27, for example, are good representations of this procedure.
3. State all the simplifying assumptions that you feel are appropriate for the solution of the problem, and flag those that will need to be verified after an answer has been obtained. Pay particular attention to whether the system is in the steady or unsteady state. Also, compile the physical properties necessary for analyzing the system and cite the sources from which they were obtained.
4. Analyze the problem by means of the appropriate conservation laws and rate equations, using, wherever possible, insight into the processes and intuition. As you develop more insights, refer back to the thermal circuit and modify it, if appropriate. Perform the numerical calculations in a step-by-step manner so that you can easily check your results by an order-of-magnitude analysis.
5. Comment on the results you have obtained and discuss any questionable points, in particular as they apply to the original assumptions. Then summarize the key conclusions at the end.

This method of analysis has been amply demonstrated in the example problems in the previous sections (particularly 1.11–1.13) and their review in the context of the five steps listed above would be instructive. Furthermore, as you progress in your studies of heat transfer in subsequent chapters of the book, the procedure outlined above will become more meaningful and you may wish to refer to it as you begin to analyze and design more complex thermal systems.

Finally, bear in mind that the subject of heat transfer is in a constant state of evolution, and an engineer is well advised to follow the current literature on the subject (often, authoritative reviews are useful) in order to keep up to date. The most important serial publications that present new findings in heat transfer are listed in Appendix 5. In addition to serial publications, the engineer will find it useful to refer from time to time to handbooks and monographs that periodically summarize the current state of knowledge.

References

1. P. G. Klemens, "Theory of the Thermal Conductivity of Solids," in *Thermal Conductivity*, R. P. Tye, ed., vol. 1, Academic Press, London, 1969.
2. E. McLaughlin, "Theory of the Thermal Conductivity of Fluids," in *Thermal Conductivity*, R. P. Tye, ed., vol. 2. Academic Press, London, 1969.
3. W. G. Vincenti and C. H. Kruger Jr., *Introduction to Physical Gas Dynamics*, Wiley, New York. 1965.
4. J. F. Mallory, *Thermal Insulation*, Reinhold, New York. 1969.
5. E. Fried, "Thermal Conduction Contribution to Heat Transfer at Contacts," *Thermal Conductivity*, R. *P*. Tye, ed., vol. 2, Academic press, London, 1969.
6. L. S. Fletcher, "Imperfect Metal-to-Metal Contact," sec. 502.5 in *Heat Transfer and Fluid Flow Data Books*, F. Kreith, ed., Genium, Schenectady, NY, 1991.

Problems

The problems for this chapter are organized by subject matter as shown below.

Topic	Problem Number
Conduction	1.1–1.11
Convection	1.12–1.21
Radiation	1.22–1.29
Conduction in series and parallel	1.30–1.35
Convection and conduction in series and parallel	1.36–1.43
Convection and radiation in parallel	1.44–1.53
Conduction, convection, and radiation combinations	1.54–1.56
Heat transfer and energy conservation	1.57–1.58
Dimensions and units	1.59–1.65
Heat transfer modes	1.66–1.72

1.1 The outer surface of a 0.2-m-thick concrete wall is kept at a temperature of −5°C, while the inner surface is kept at 20°C. The thermal conductivity of the concrete is 1.2 W/m K. Determine the heat loss through a wall 10 m long and 3 m high.

0.2 m
Concrete
−5°C
20°C
$q = ?$

1.2 The weight of the insulation in a spacecraft may be more important than the space required. Show analytically that the lightest insulation for a plane wall with a specified thermal resistance is the insulation that has the smallest product of density times thermal conductivity.

1.3 A furnace wall is to be constructed of brick having standard dimensions of 9 by 4.5 in. × 3 in. Two kinds of material are available. One has a maximum usable temperature of 1900°F and a thermal conductivity of 1 Btu/h ft °F, and the other has a maximum temperature limit of 1600°F and a thermal conductivity of 0.5 Btu/h ft °F. The bricks have the same cost and can be laid in any manner, but we wish to design the most economical wall for a furnace with a temperature of 1900°F on the hot side and 400°F on the cold side. If the maximum amount of heat transfer permissible is 300 Btu/h for each square foot of area, determine the most economical arrangement using the available bricks.

1.4 To measure thermal conductivity, two similar 1-cm-thick specimens are placed in the apparatus shown in the accompanying sketch. Electric current is supplied to the 6-cm × 6-cm guard heater, and a wattmeter shows that the power dissipation is 10 W. Thermocouples attached to the warmer and to the cooler surfaces show temperatures of 322 and 300 K, respectively. Calculate the thermal conductivity of the material at the mean temperature in Btu/h ft °F and W/m K.

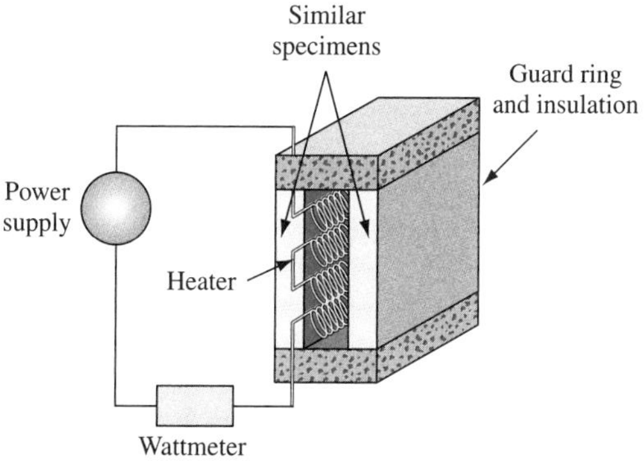

1.5 To determine the thermal conductivity of a structural material, a large 6-in.-thick slab of the material was subjected to a uniform heat flux of 800 Btu/h ft^2, while thermocouples embedded in the wall at 2-in. intervals were read over a period of time. After the system had reached equilibrium, an operator recorded the thermocouple readings shown below for two different environmental conditions:

Distance from the Surface (in.)	Temperature (°F)
Test 1	
0	100
2	150
4	206
6	270
Test 2	
0	200
2	265
4	335
6	406

From these data, determine an approximate expression for the thermal conductivity as a function of temperature between 100 and 400°F.

1.6 A square silicone chip 7 mm × 7 mm in size and 0.5 mm thick is mounted on a plastic substrate as shown in the sketch below. The top surface of the chip is cooled by a synthetic liquid flowing over it. Electronic circuits on the bottom of the chip generate heat at a rate of 5 W that must be transferred through the chip. Estimate the steady-state temperature difference between the front and back surfaces of the chip. The thermal conductivity of silicone is 150 W/m K.

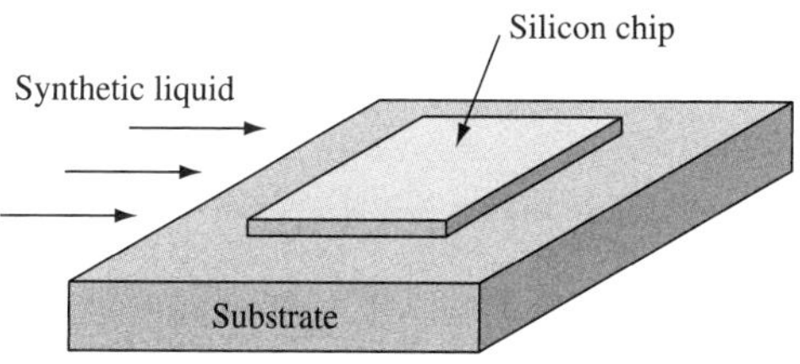

1.7 A warehouse is to be designed for keeping perishable foods cool prior to transportation to grocery stores. The warehouse has an effective surface area of 20,000 ft^2 exposed to an ambient air temperature of 90°F. The warehouse wall insulation (k = 0.1 Btu/h ft °F) is 3 in. thick. Determine the rate at which heat must be removed (Btu/h) from the warehouse to maintain the food at 40°F.

1.8 With increasing emphasis on energy conservation, the heat loss from buildings has become a major concern. The typical exterior surface areas and R-factors (area × thermal resistance) for a small tract house are listed below:

Element	Area (m^2)	R-Factors (m^2 K/W)
Walls	150	2.0
Ceiling	120	2.8
Floor	120	2.0
Windows	20	0.1
Doors	5	0.5

(a) Calculate the rate of heat loss from the house when the interior temperature is 22°C and the exterior is −5°C. (b) Suggest ways and means to reduce the heat loss, and show quantitatively the effect of doubling the wall insulation and substituting double-glazed windows (thermal resistance = 0.2 m^2 K/W) for the single-glazed type in the table above.

1.9 Heat is transferred at a rate of 0.1 kW through glass wool insulation (density = 100 kg/m^3) of 5-cm thickness and 2-m^2 area. If the hot surface is at 70°C, determine the temperature of the cooler surface.

1.10 A heat flux meter at the outer (cold) wall of a concrete building indicates that the heat loss through a wall of 10 cm thickness is 20 W/m^2. If a thermocouple at the inner surface of the wall indicates a temperature of 22°C while another at the outer surface shows 6°C, calculate the thermal conductivity of the concrete and compare your result with the value in Appendix 2, Table 11.

1.11 Calculate the heat loss through a 1 m × 3 m glass window 7 mm thick if the inner surface temperature is 20°C and the outer surface temperature is 17°C. Comment on the possible effect of radiation on your answer.

1.12 If the outer air temperature in Problem 1.11 is −2°C, calculate the convection heat transfer coefficient between the outer surface of the window and the air, assuming radiation is negligible.

1.13 Using Table 1.4 as a guide, prepare a similar table showing the orders of magnitude of the thermal resistances of a unit area for convection between a surface and various fluids.

1.14 A thermocouple (0.8-mm-diameter wire) used to measure the temperature of the quiescent gas in a furnace gives a reading of 165°C. It is known, however, that the rate of radiant heat flow per meter length from the hotter furnace walls to the thermocouple wire is 1.1 W/m and the convection heat transfer coefficient between the wire and the gas is 6.8 W/m^2 K. With this information, estimate the true gas temperature. State your assumptions and indicate the equations used.

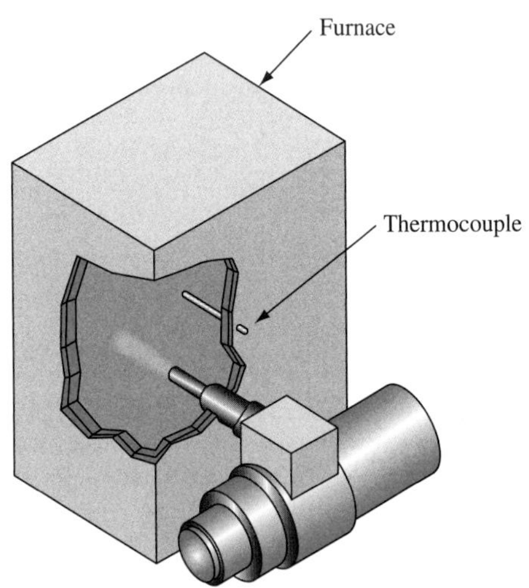

1.15 Water at a temperature of 77°C is to be evaporated slowly in a vessel. The water is in a low-pressure container surrounded by steam as shown in the sketch below. The steam is condensing at 107°C. The overall heat transfer coefficient between the water and the steam is 1100 W/m^2 K. Calculate the surface area of the container that would be required to evaporate water at a rate of 0.01 kg/s.

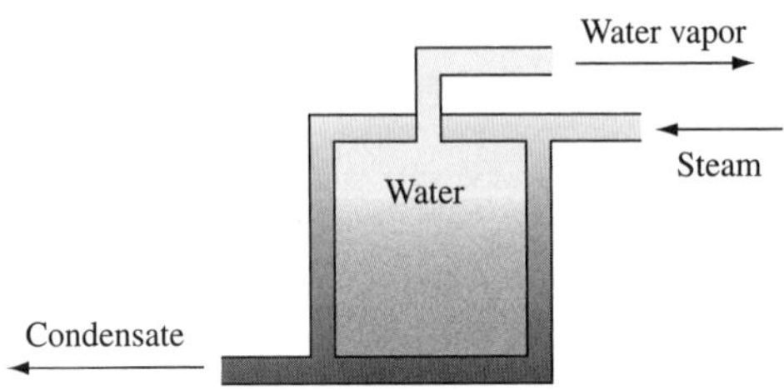

1.16 The heat transfer rate from hot air by convection at 100°C flowing over one side of a flat plate with dimensions 0.1 m by 0.5 m is determined to be 125 W when the surface of the plate is kept at 30°C. What is the average convection heat transfer coefficient between the plate and the air?

1.17 The heat transfer coefficient for a gas flowing over a thin flat plate 3 m long and 0.3 m wide varies with distance from the leading edge according to

$$h_c(x) = 10x^{-1/4} \frac{\text{W}}{\text{m}^2\,\text{K}}$$

If the plate temperature is 170°C and the gas temperature is 30°C, calculate (a) the average heat transfer coefficient, (b) the rate of heat transfer between the plate and the gas, and (c) the local heat flux 2 m from the leading edge.

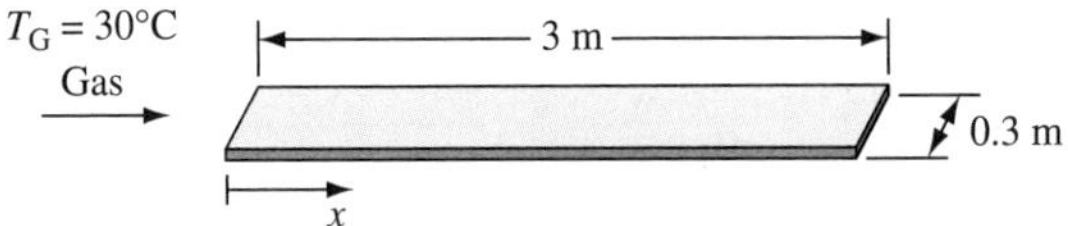

1.18 A cryogenic fluid is stored in a 0.3-m-diameter spherical container in still air. If the convection heat transfer coefficient between the outer surface of the container and the air is 6.8 W/m^2 K, the temperature of the air is 27°C, and the temperature of the surface of the sphere is −183°C, determine the rate of heat transfer by convection.

1.19 A high-speed computer is located in a temperature-controlled room at 26°C. When the machine is operating, its internal heat generation rate is estimated to be 800 W. The external surface temperature of the computer is to be maintained below 85°C. The heat transfer coefficient for the surface of the computer is estimated to be 10 W/m^2 K. What surface area would be necessary to assure safe operation of this machine? Comment on ways to reduce this area.

1.20 In order to prevent frostbite to skiers on chair lifts, the weather report at most ski areas gives both an air temperature and the wind-chill temperature. The air temperature is measured with a thermometer that is not affected by the wind. However, the rate of heat loss from the skier increases with wind velocity, and the wind-chill temperature is the temperature that would result in the same rate of heat loss in still air as occurs at the measured air temperature with the existing wind.

Suppose that the inner temperature of a 3-mm-thick layer of skin with a thermal conductivity of 0.35 W/m K is 35°C and the air temperature is −20°C. Under calm ambient conditions the heat transfer coefficient at the outer skin surface is about 20 W/m^2 K (see Table 1.4), but in a 40-mph wind it increases to 75 W/m^2 K. (a) If frostbite can occur when the skin temperature drops to about 10°C, would you advise the skier to wear a face mask? (b) What is the skin temperature drop due to the wind?

1.21 Using the information in Problem 1.20, estimate the ambient air temperature that could cause frostbite on a calm day on the ski slopes.

1.22 Two large parallel plates with surface conditions approximating those of a blackbody are maintained at 1500°F and 500°F, respectively. Determine the rate of heat transfer by radiation between the plates in Btu/h ft^2 and the radiative heat transfer coefficient in Btu/h ft^2 °F and in W/m^2 K.

1.23 A spherical vessel, 0.3 m in diameter, is located in a large room whose walls are at 27°C (see sketch). If the vessel is used to store liquid oxygen at −183°C and both the surface of the storage vessel and the walls of the room are black, calculate the rate of heat transfer by radiation to the liquid oxygen in watts and in Btu/h.

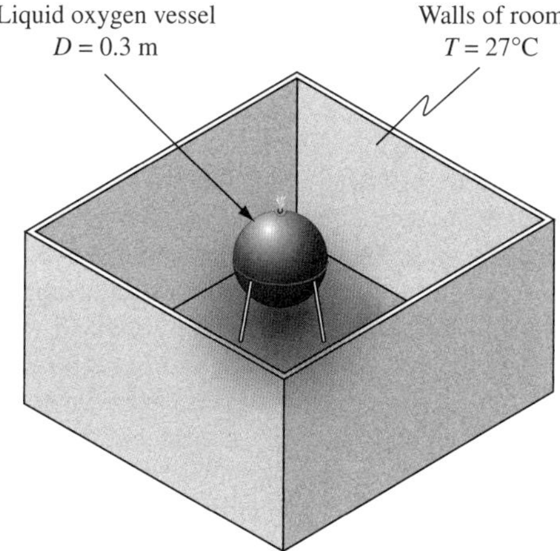

1.24 Repeat Problem 1.23 but assume that the surface of the storage vessel has an absorbance (equal to the emittance) of 0.1. Then determine the rate of evaporation of the liquid oxygen in kilograms per second and pounds per hour, assuming that convection can be neglected. The heat of vaporization of oxygen at −183°C is 213.3 kJ/kg.

1.25 Determine the rate of radiant heat emission in watts per square meter from a blackbody at (a) 150°C, (b) 600°C, (c) 5700°C.

1.26 The sun has a radius of 7×10^8 m and approximates a blackbody with a surface temperature of about 5800 K. Calculate the total rate of radiation from the sun and the emitted radiation flux per square meter of surface area.

1.27 A small gray sphere having an emissivity of 0.5 and a surface temperature of 1000°F is located in a blackbody enclosure having a temperature of 100°F. Calculate for this system (a) the net rate of heat transfer by radiation per unit of surface area of the sphere, (b) the radiative thermal conductance in Btu/h °F if the surface area of the sphere is 0.1 ft^2, (c) the thermal resistance for radiation between the sphere and its surroundings, (d) the ratio of thermal resistance for radiation to thermal resistance for convection if the convection heat transfer coefficient between the sphere and its surroundings is 2.0 Btu/h ft^2 °F, (e) the total rate of heat transfer from the sphere to the surroundings, and (f) the combined heat transfer coefficient for the sphere.

1.28 A spherical communications satellite, 2 m in diameter, is placed in orbit around the earth. The satellite generates 1000 W of internal power from a small nuclear generator. If the surface of the satellite has an emittance of 0.3, and is shaded from solar radiation by the earth, estimate its surface temperature. What would the temperature be if the satellite with an absorptivity of 0.2 were in an orbit in which it would be exposed to solar radiation? Assume the sum is a blackbody at 6,700 K and state your assumptions.

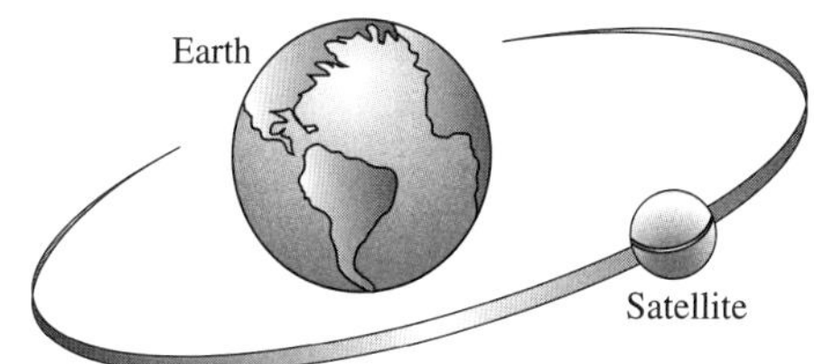

1.29 A long wire 0.03 in. in diameter with an emissivity of 0.9 is placed in a large quiescent air space at 20°F. If the wire is at 1000°F, calculate the net rate of heat loss. Discuss your assumptions.

1.30 Wearing layers of clothing in cold weather is often recommended because dead-air spaces between the layers keep the body warm. The explanation for this is that the heat loss from the body is less. Compare the rate of heat loss for a single 3/4-in.-thick layer of wool (k = 0.020 Btu/h ft °F) with three 1/4-in. layers separated by 1/16-in. air gaps. The thermal conductivity of air is 0.014 Btu/h ft °F.

1.31 A section of a composite wall with the dimensions shown below has uniform temperatures of 200°C and 50°C over the left and right surfaces, respectively. If the thermal conductivities of the wall materials are: k_A = 70 W/m K, k_B = 60 W/m K, k_C = 40 W/m K, and k_D = 20 W/m K, determine the rate of heat transfer through this section of the wall and the temperatures at the interfaces.

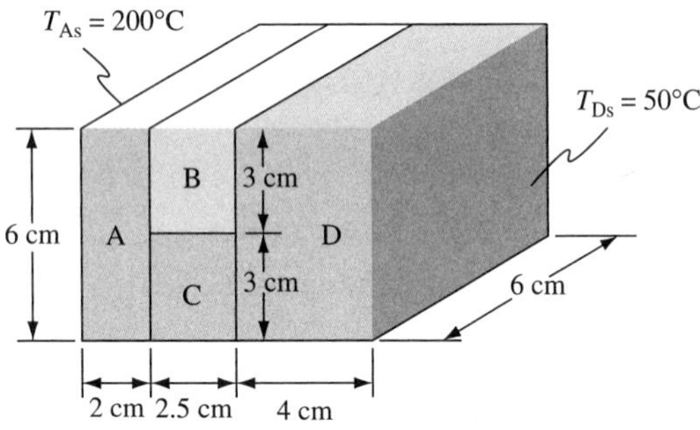

1.32 Repeat Problem 1.31, including a contact resistance of 0.1 K/W at each of the interfaces.

1.33 Repeat Problem 1.32 but assume that instead of surface temperatures, the given temperatures are those of the air on the left and right sides of the wall and that the convection heat transfer coefficients on the left and right surfaces are 6 and 10 W/m² K, respectively.

1.34 Mild steel nails were driven through a solid wood wall consisting of two layers, each 2.5 cm thick, for reinforcement. If the total cross-sectional area of the nails is 0.5% of the wall area, determine the unit thermal conductance of the composite wall and the percent of the total heat flow that passes through the nails when the temperature difference across the wall is 25°C. Neglect contact resistance between the wood layers.

1.35 Calculate the rate of heat transfer through the composite wall in Problem 1.34 if the temperature difference is 25°C and the contact resistance between the sheets of wood is 0.005 m² K/W.

1.36 Heat is transferred through a plane wall from the inside of a room at 22°C to the outside air at −2°C. The convective heat transfer coefficients at the inside and outside surfaces are 12 and 28 W/m² K, respectively. The thermal resistance of a unit area of the wall is 0.5 m² K/W. Determine the temperature at the outer surface of the wall and the rate of heat flow through the wall per unit area.

1.37 How much fiberglass insulation (k = 0.035 W/m K) is needed to guarantee that the outside temperature of a kitchen oven will not exceed 43°C? The maximum oven temperature to be maintained by the conventional type of thermostatic control is 290°C, the kitchen temperature may vary from 15°C to 33°C, and the average heat transfer coefficient between the oven surface and the kitchen is 12 W/m² K.

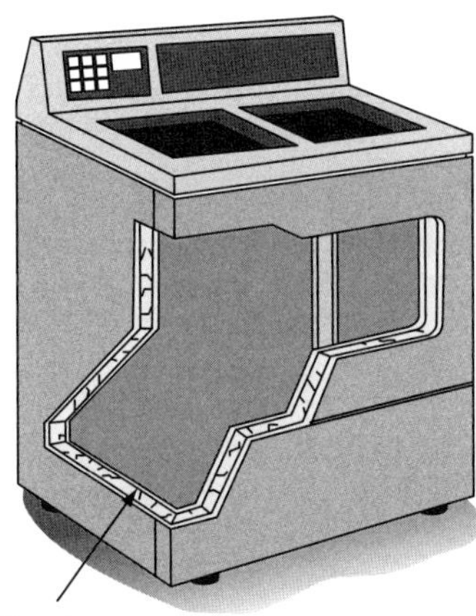

1.38 A heat exchanger wall consists of a copper plate 3/8 in. thick. The heat transfer coefficients on the two sides of the plate are 480 and 1250 Btu/h ft² °F, corresponding to fluid temperatures of 200 and 90°F, respectively. Assuming that the thermal conductivity of the wall is 220 Btu/h ft °F, (a) compute the surface temperatures in °F and (b) calculate the heat flux in Btu/h ft².

1.39 A submarine is to be designed to provide a comfortable temperature of no less than 70°F for the crew. The submarine can be idealized by a cylinder 30 ft in diameter and 200 ft in length, as shown. The combined heat transfer coefficient on the interior is about 2.5 Btu/h ft² °F, while on the outside the heat transfer coefficient is estimated to vary from about 10 Btu/h ft^2 °F (not moving) to 150 Btu/h ft^2 °F (top speed). For the following wall constructions, determine the minimum size (in kilowatts) of the heating unit required if the seawater temperature varies from 34°F to 55°F during operation. The walls of the submarine are (a) 1/2-in. aluminum, (b) 3/4-in. stainless steel with a 1-in.-thick layer of fiberglass insulation on the inside, and (c) of sandwich construction with a 3/4-in.-thick layer of stainless steel, a 1-in.-thick layer of fiberglass insulation, and a 1/4-in.-thick layer of aluminum on the inside. What conclusions can you draw?

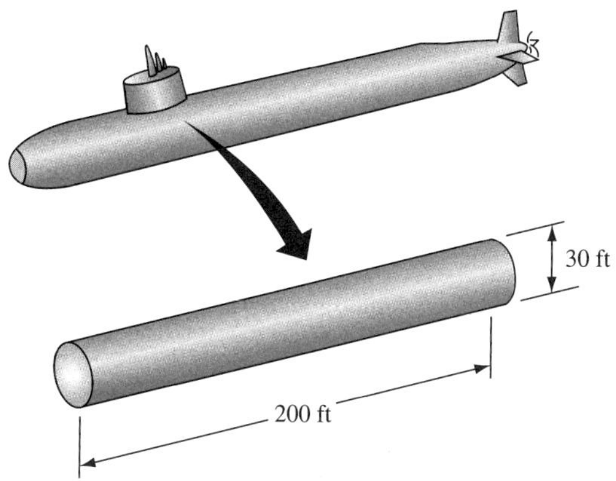

1.40 A simple solar heater consists of a flat plate of glass below which is located a shallow pan filled with water, so that the water is in contact with the glass plate above it. Solar radiation passes through the glass at the rate of 156 Btu/h ft^2. The water is at 200°F and the surrounding air is at 80°F. If the heat transfer coefficients between the water and the glass, and between the glass and the air are 5 Btu/h ft^2 °F and 1.2 Btu/h ft^2 °F, respectively, determine the time required to transfer 100 Btu per square foot of surface to the water in the pan. The lower surface of the pan can be assumed to be insulated.

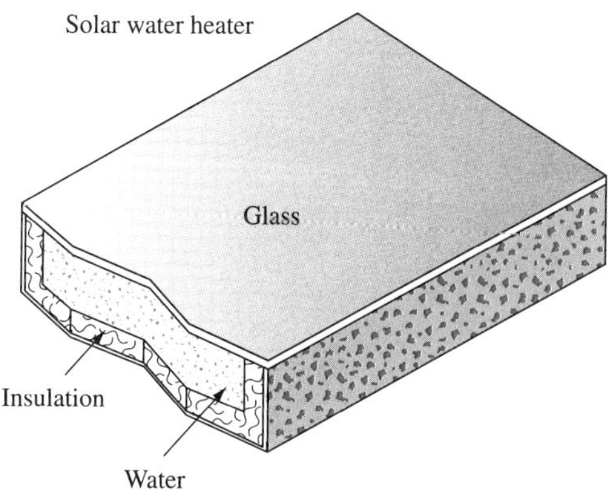

1.41 A composite refrigerator wall is composed of 2 in. of corkboard sandwiched between a 1/2-in.-thick layer of oak and a 1/32-in.-thick layer of aluminum lining on the inner surface. The average convection heat transfer coefficients at the interior and exterior wall are 2 and 1.5 Btu/h ft^2 °F, respectively. (a) Draw the thermal circuit. (b) Calculate the individual resistances of the components of this composite wall and the resistances at the surfaces. (c) Calculate the overall heat transfer coefficient through the wall. (d) For an air temperature of 30°F inside the refrigerator and 90°F outside, calculate the rate of heat transfer per unit area through the wall.

1.42 An electronic device that internally generates 600 mW of heat has a maximum permissible operating temperature of 70°C. It is to be cooled in 25°C air by attaching aluminum fins with a total surface area of 12 cm^2. The convection heat transfer coefficient between the fins and the air is 20 W/m^2 K. Estimate the operating temperature when the fins are attached in such a way that (a) there is a contact resistance of approximately 50 K/W between the surface of the device and the fin array and (b) there is no contact resistance (in this case, the construction of the device is more expensive). Comment on the design options.

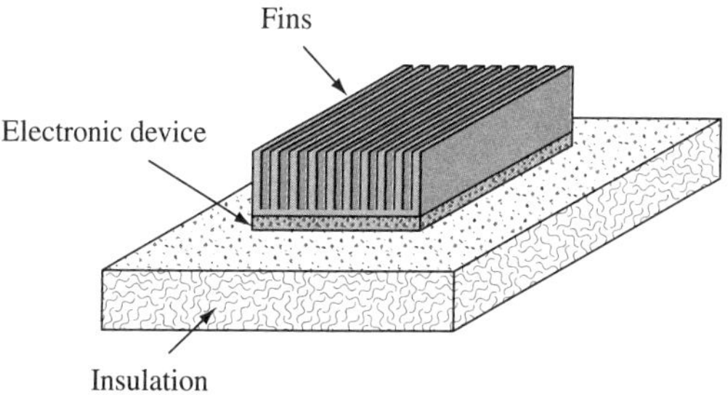

1.43 To reduce home heating requirements, modern building codes in many parts of the country require the use of double-glazed or double-pane windows, i.e., windows with two panes of glass. Some of these so-called thermopane windows have an evacuated space between the two glass panes while others trap stagnant air in the space. (a) Consider a double-pane window with the dimensions shown in the following sketch. If this window has stagnant air trapped between the two panes and the convection heat transfer coefficients on the inside and outside surfaces are 4 W/m^2 K and 15 W/m^2 K, respectively, calculate the overall heat transfer coefficient for the system. (b) If the inside air temperature is 22°C and the outside air temperature is −5°C, compare the heat loss through a 4-m^2 double-pane window with the heat loss through a single-pane window. Comment on the effect of the window frame on this result. (c) If the total window area of a home heated by electric resistance heaters at a cost of $0.10/k Wh is 80 m^2. How much more cost can you justify for the double-pane windows if the average temperature difference during the six winter months when heating is required is about 15°C?

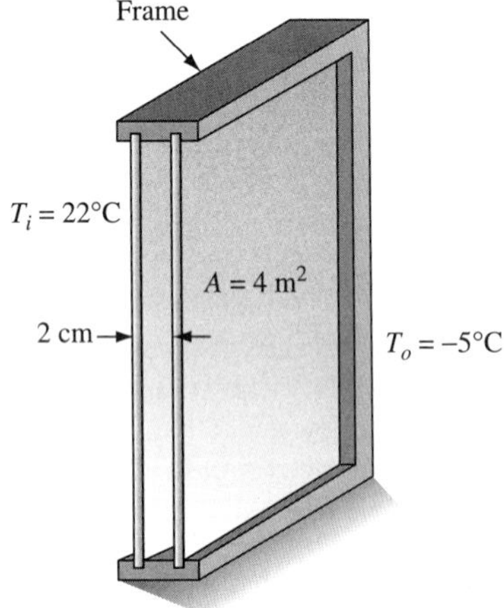

1.44 A flat roof can be modeled as a flat plate insulated on the bottom and placed in the sunlight. If the radiant heat that the roof receives from the sun is 600 W/m^2, the convection heat transfer coefficient between the roof and the air is 12 W/m^2 K, and the air temperature is 27°C, determine the roof temperature for the following two cases: (a) Radiative heat loss to space is negligible. (b) The roof is black (ε = 1.0) and radiates to space, which is assumed to be a blackbody at 0 K.

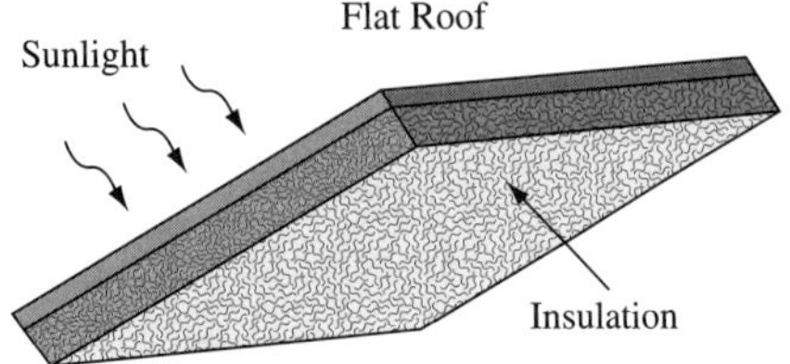

1.45 A horizontal, 3-mm-thick flat-copper plate, 1 m long and 0.5 m wide, is exposed in air at 27°C to radiation from the sun. If the total rate of solar radiation absorbed is 300 W and the combined radiation and convection heat transfer coefficients on the upper and lower surfaces are 20 and 15 W/m^2 K, respectively, determine the equilibrium temperature of the plate.

1.46 A small oven with a surface area of 3 ft^2 is located in a room in which the walls and the air are at a temperature of 80°F. The exterior surface of the oven is at 300°F, and the next heat transfer by radiation between the oven's surface and the surroundings is 2000 Btu/h. If the average convection heat transfer coefficient between the oven and the surrounding air is 2.0 Btu/h ft^2 °F, calculate (a) the net heat transfer between the oven and the surroundings in Btu/h, (b) the thermal resistance at the surface for radiation and convection, respectively, in h °F/Btu, and (c) the combined heat transfer coefficient in Btu/h ft^2 °F.

1.47 A steam pipe 200 mm in diameter passes through a large basement room. The temperature of the pipe wall is 500°C, while that of the ambient air in the room is 20°C. Determine the heat transfer rate by convection and radiation per unit length of steam pipe if the emissivity of the pipe surface is 0.8 and the natural convection heat transfer coefficient has been determined to be 10 W/m^2 K.

1.48 The inner wall of a rocket motor combustion chamber receives 50,000 Btu/h ft^2 by radiation from a gas at 5000°F. The convection heat transfer coefficient between the gas and the wall is 20 Btu/h ft^2 °F. If the inner wall

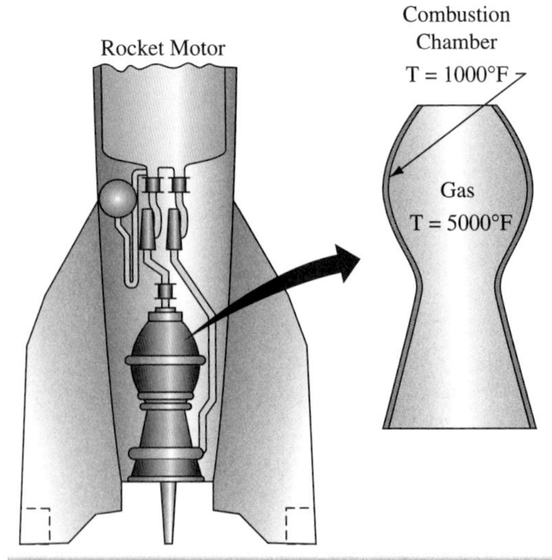

of the combustion chamber is at a temperature of 1000°F, determine (a) the total thermal resistance of a unit area of the wall in h ft^2 °F/Btu and (b) the heat flux. Also draw the thermal circuit.

1.49 A flat roof of a house absorbs a solar radiation flux of 600 W/m^2. The backside of the roof is well insulated, while the outside loses heat by radiation and convection to ambient air at 20°C. If the emittance of the roof is 0.80 and the convection heat transfer coefficient between the roof and the air is 12 W/m^2 K, calculate (a) the equilibrium surface temperature of the roof and (b) the ratio of convection to radiation heat loss. Can one or the other of these be neglected? Explain your answer.

1.50 Determine the power requirement of a soldering iron in which the tip is maintained at 400°C. The tip is a cylinder 3 mm in diameter and 10 mm long. The surrounding air temperature is 20°C, and the average convection heat transfer coefficient over the tip is 20 W/m^2 K. The tip is highly polished initially, giving it a very low emittance.

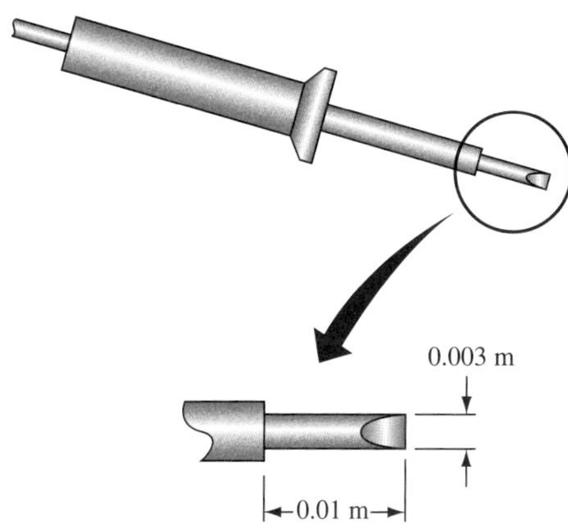

1.51 The soldering iron tip in Problem 1.50 becomes oxidized with age and its gray-body emittance increases to 0.8. Assuming that the surroundings are at 20°C, determine the power requirement for the soldering iron.

1.52 Some automobile manufacturers are currently working on a ceramic engine block that could operate without a cooling system. Idealize such an engine as a rectangular solid, 45 cm × 30 cm × 30 cm. Suppose that under maximum power output the engine consumes 5.7 L of fuel per hour, the heat released by the fuel is 9.29/k Whr per liter, and the net engine efficiency (useful work output divided by the total heat input) is 0.33. If the engine block is aluminum with a graybody emissivity of 0.9, the engine compartment operates at 150°C, and the convection heat transfer coefficient is 30 W/m^2 K, determine the average surface temperature of the engine block. Comment on the practicality of the concept.

1.53 A pipe carrying superheated steam in a basement at 10°C has a surface temperature of 150°C. Heat loss from the pipe occurs by radiation (ε = 0.6) and natural convection ($\bar{h}_c$ = 25 W/m^2 K). Determine the percentage of the total heat loss by these two mechanisms.

1.54 For a furnace wall, draw the thermal circuit, determine the rate of heat flow per unit area, and estimate the exterior surface temperature if (a) the convection heat transfer coefficient at the interior surface is 15 W/m^2 K (b) the rate of heat flow by radiation from hot gases and soot particles at 2000°C to the interior wall surface is 45,000 W/m^2 (c) the unit thermal conductance of the wall (interior surface temperature is about 850°C) is 250 W/m^2 K and (d) there is convection from the outer surface.

1.55 Draw the thermal circuit for heat transfer through a double-glazed window. Identify each of the circuit elements. Include solar radiation to the window and interior space.

1.56 The ceiling of a tract house is constructed of wooden studs with fiberglass insulation between them. On the interior of the ceiling is plaster and on the exterior is a thin layer of sheet metal. A cross section of the ceiling with dimensions is shown below.

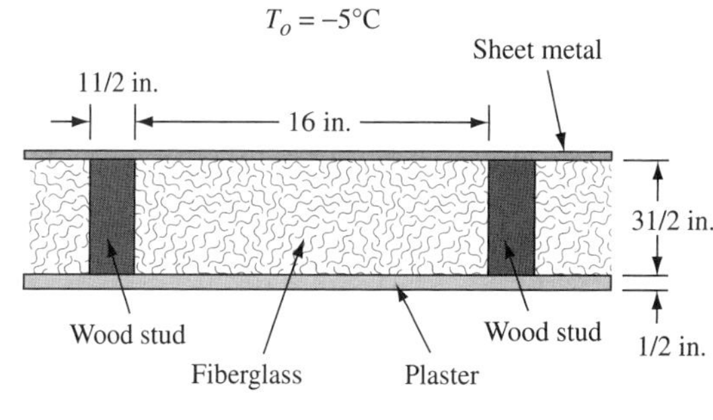

(a) The R-factor describes the thermal resistance of insulation and is defined by

$$R - factor = L/k_{\text{eff}} = \Delta T/(q/A)$$

Calculate the R-factor for this type of ceiling and compare the value of this R-factor with that for a similar thickness of fiberglass. Why are the two different? (b) Estimate the rate of heat transfer per square meter through the ceiling if the interior temperature is 22°C and the exterior temperature is −5°C.

1.57 A homeowner wants to replace an electric hot-water heater. There are two models in the store. The inexpensive model costs \$280 and has no insulation between the inner and outer walls. Due to natural convection, the space between the inner and outer walls has an effective conductivity three times that of air. The more expensive model costs \$310 and has fiberglass insulation in the gap between the walls. Both models are 3.0 m tall and have a cylindrical shape with an inner-wall diameter of 0.60 m and a 5-cm gap. The surrounding air is at 25°C, and the convection heat transfer coefficient on the outside is 15 W/m^2 K. The hot water inside the tank results in an inside wall temperature of 60°C.

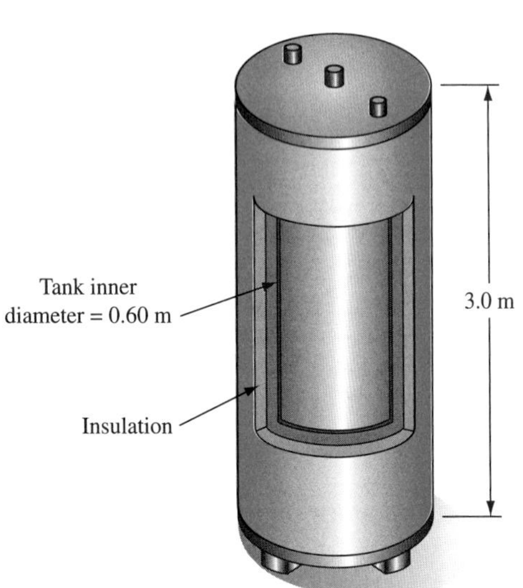

If energy costs 6 ¢/k Wh, estimate how long it will take to pay back the extra investment in the more expensive hot-water heater. State your assumptions.

1.58 Liquid oxygen (LOX) for the space shuttle can be stored at 90 K prior to launch in a spherical container 4 m in diameter. To reduce the loss of oxygen, the sphere is insulated with superinsulation developed at the U.S. National Institute of Standards and Technology's Cryogenic Division; the superinsulation has an effective thermal conductivity of 0.00012 W/m K. If the outside temperature is 20°C on the average and the LOX has a heat of vaporization of 213 J/g, calculate the thickness of insulation required to keep the LOX evaporation rate below 200 g/h.

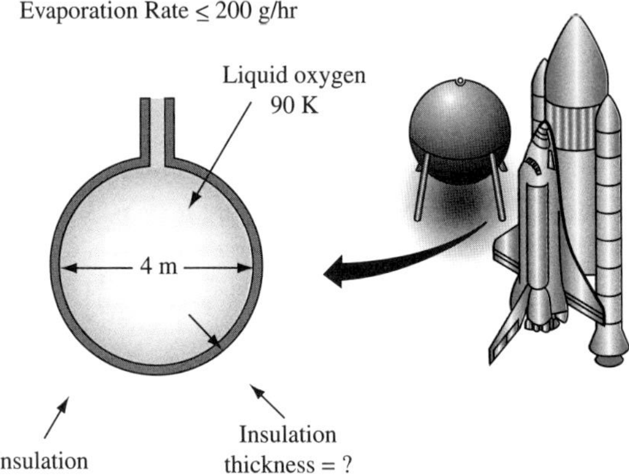

1.59 The heat transfer coefficient between a surface and a liquid is 10 Btu/h ft^2 °F. How many watts per square meter will be transferred in this system if the temperature difference is 10°C?

1.60 The thermal conductivity of fiberglass insulation at 68°F is 0.02 Btu/h ft °F. What is its value in SI units?

1.61 The thermal conductivity of silver at 212°F is 238 Btu/h ft °F. What is the conductivity in SI units?

1.62 An ice chest (see sketch) is to be constructed from styrofoam (k = 0.033 W/m K). If the wall of the chest is 5-cm thick, calculate its R-value in h ft^2 °F/Btu in.

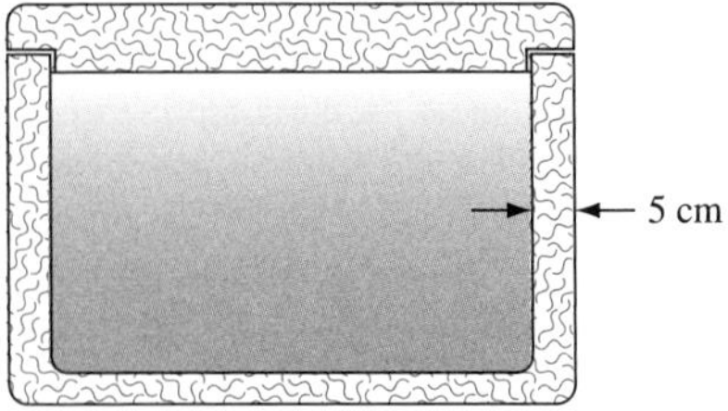

1.63 Estimate the R-values for a 2-inch-thick fiberglass board and a 1-inch-thick polyurethane foam layer. Then, compare their respective conductivity-times-density products if the density for fiberglass is 50 kg/m^3 and the density of polyurethane is 30 kg/m^3. Use the units given in Figure 1.30.

1.64 A manufacturer in the United States wants to sell a refrigeration system to a customer in Germany. The standard measure of refrigeration capacity used in the United States is the ton (T); a 1 T capacity means that the unit is capable of making about 1 T of ice per day or has a heat removal rate of 12,000 Btu/h. The capacity of the American system is to be guaranteed at 3 T. What would this guarantee be in SI units?

1.65 Referring to Problem 1.64, how many kilograms of ice can a 3-ton refrigeration unit produce in a 24-h period? The heat of fusion of water is 330 kJ/kg.

1.66 Explain a fundamental characteristic that differentiates conduction from convection and radiation.

1.67 Explain in your own words (a) what is the mode of heat transfer through a large steel plate that has its surfaces at specified temperatures? (b) What are the modes when the temperature on one surface of the steel plate is not specified, but the surface is exposed to a fluid at a specified temperature?

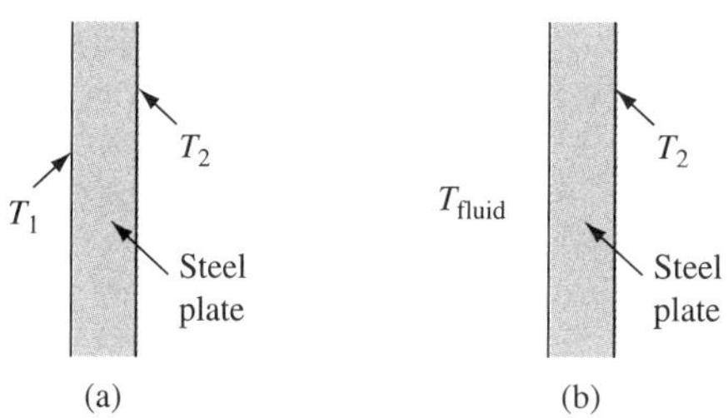

1.68 What are the important modes of heat transfer for a person sitting quietly in a room? What if the person is sitting near a roaring fireplace?

1.69 Consider the cooling of (a) a personal computer with a separate CPU and (b) a laptop computer. The reliable functioning of these machines depends on their effective cooling. Identify and briefly explain all modes of heat transfer involved in the cooling process.

1.70 Describe and compare the modes of heat loss through the single-pane and double-pane window assemblies shown in the sketch below.

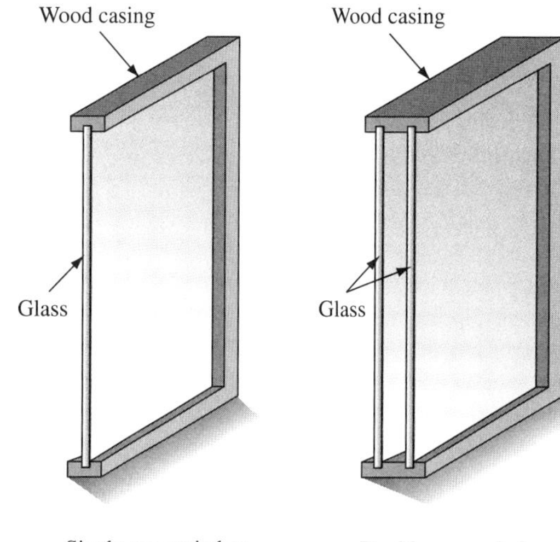

1.71 A person wearing a heavy parka is standing in a cold wind. Describe the modes of heat transfer determining heat loss from the person's body.

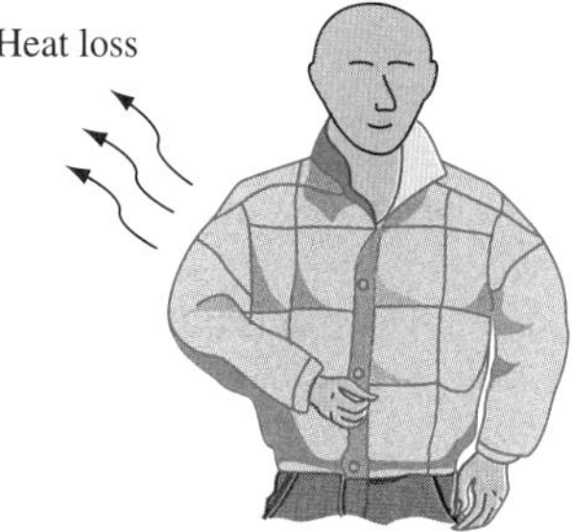

1.72 Discuss the modes of heat transfer that determine the equilibrium temperature of the space shuttle *Endeavour* when it is in orbit. What happens when it reenters the earth's atmosphere?

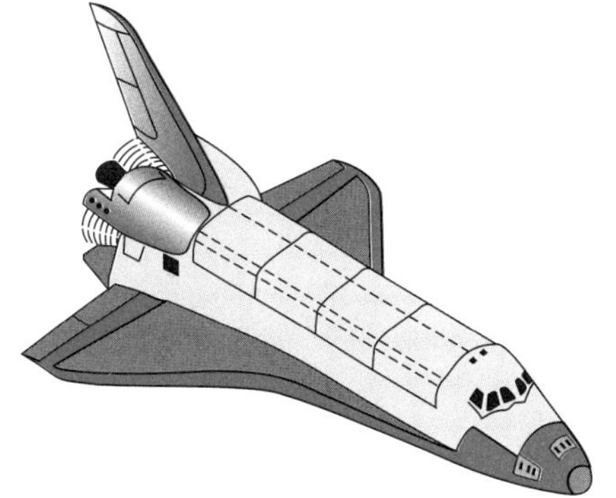

Design Problems

1.1 **Optimum Boiler Insulation Package** (Chapter 1) To insulate high-temperature surfaces it is economical to use two layers of insulation. The first layer is placed next to the hot surface and is suitable for high temperature. It is costly and is usually a relatively poor insulator. The second layer is placed outside the first layer and is cheaper and a good insulator, but will not withstand high temperatures. Essentially, the first layer protects the second layer by providing just enough insulating capability so that the second layer is only exposed to moderate temperatures. Given commercially available insulating materials, design the optimum combination of two such materials to insulate a flat 1000°C surface from ambient air at 20°C. Your goal is to reduce the rate of heat transfer to 0.1% of that without any insulation, to achieve an outer surface temperature that is safe to personnel, and to minimize cost of the insulating package.

1.2 **Thermocouple Radiator Error** (Chapters 1 and 9) Design a thermocouple installation to measure the temperature of air flowing at a velocity of 15 m/s in a 1-m-diameter duct. The air is at approximately 1000°C and the duct walls are at 200°C. Select a type of thermocouple that could be used, and then determine how accurately the thermocouple will measure the air temperature. Prepare a plot of the measurement error vs. air temperature and discuss the result. Use Table 1.4 to estimate the convection heat transfer coefficients.

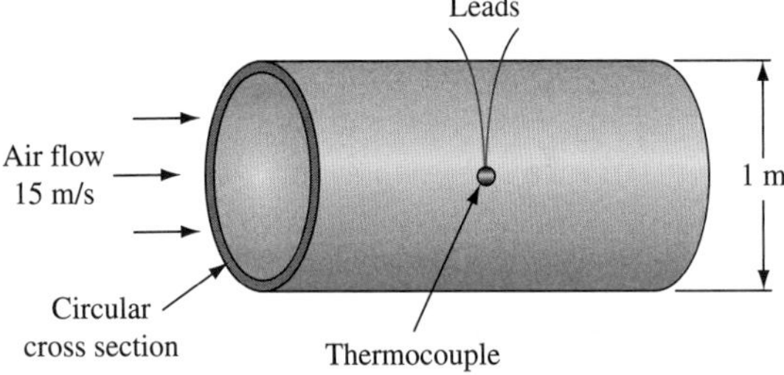

This is a multistep problem; after you have studied convection and radiation, you will improve this design to reduce the measurement error by orienting the thermocouple and its leads differently and using a radiation shield.

1.3 **Heating Load on Factory** (Chapters 1, 4 and 5) Design a heating system for a small factory in Denver, Colorado. This is a multistep problem that will be continued in subsequent chapters. In the first step, you are to determine the heating load on the building, i.e., the rate at which the building loses heat in the winter, if the inside temperature is to be maintained at 20°C. In order to compensate for this heat loss, you will subsequently be asked to design a suitable heater that can provide a rate of heat transfer equal to the heat load from the building. A schematic diagram of the building and construction details for the walls and ceilings are shown in the figure. Additional information may be obtained from the ASHRAE *Handbook of Fundamentals*.

For the purpose of this analysis, it may be assumed that the ambient temperature in Denver is equal to or greater than −10°C 97% of the time. Furthermore, air infiltration through windows and doors may be assumed to be approximately 0.2 times the volume of the building per hour. For the initial estimate of the heat load, you may use average values for the convective heat transfer coefficients over the inside and outside surfaces from Table 1.4. Note that for this design, the outside temperature assumes the worst possible conditions and, if the heater is able to maintain the temperature under these conditions, it will be able to meet less-severe conditions as well.

After you have completed the initial design, examine the results and prepare a report for the architect and the owner of the building, pointing out how the thermal design could be improved. Note especially any areas where excessive heat losses may occur. After you have studied Chapters 4 and 5 you will be asked to repeat the heat loss calculations, but calculate the heat transfer coefficient from information presented in these chapters.

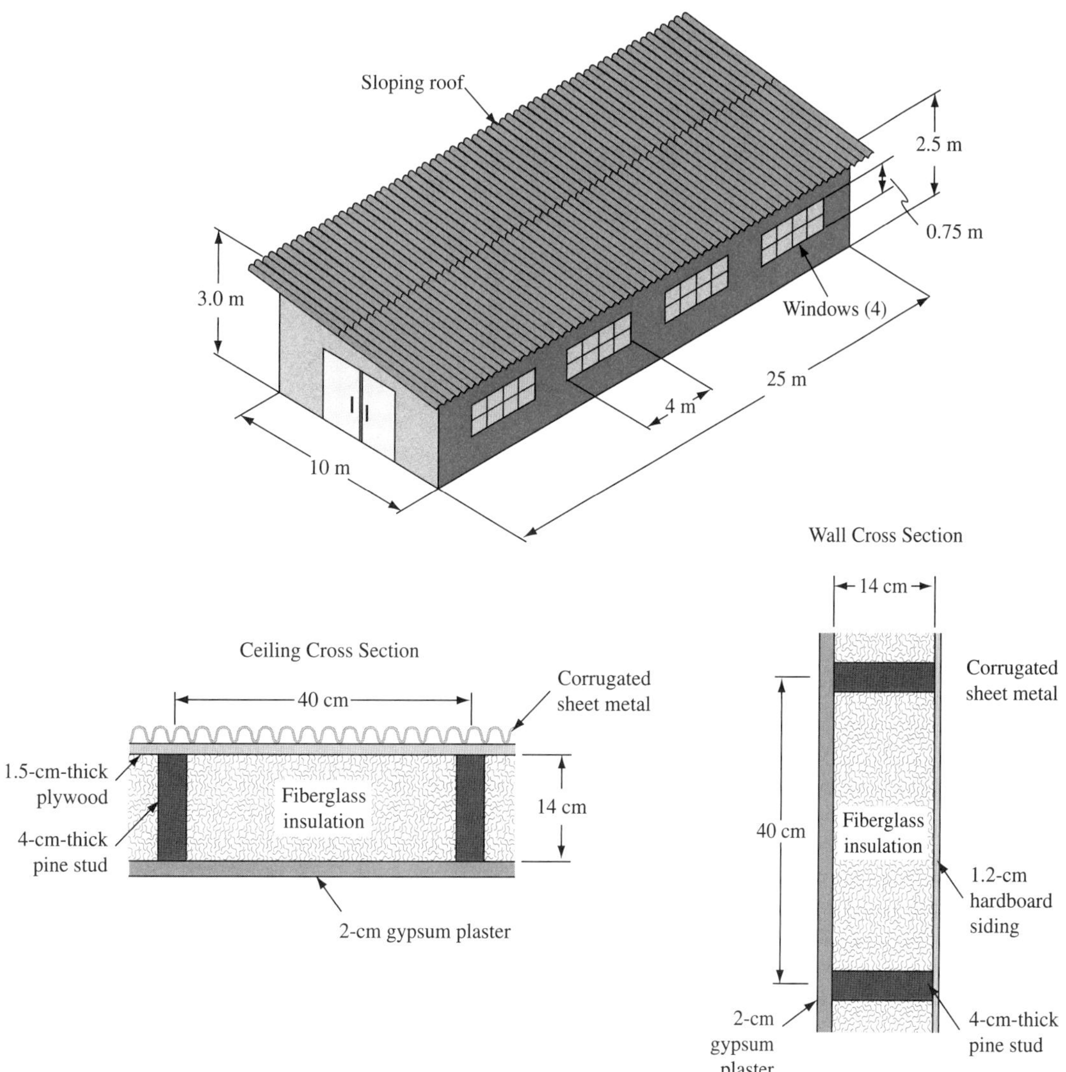
Sloping roof
2.5 m
0.75 m
3.0 m
Windows (4)
25 m
4 m
10 m
Wall Cross Section
14 cm
Ceiling Cross Section
40 cm
Corrugated sheet metal
1.5-cm-thick plywood
4-cm-thick pine stud
Fiberglass insulation
14 cm
2-cm gypsum plaster
Corrugated sheet metal
40 cm
Fiberglass insulation
1.2-cm hardboard siding
2-cm gypsum plaster
4-cm-thick pine stud

Design Problem 1.3

CHAPTER 2

A typical arrangement of rectangular pin-fin heat sinks mounted on a computer/microprocessor hardware for electronic cooling.

Source: Courtesy of Hardware Canucks.

Heat Conduction

Concepts and Analyses to Be Learned

Heat transfer by conduction is a *diffusion* process, whereby thermal energy is transferred from a hot end of a medium (usually solid) to its colder end via an intermolecular energy exchange. Modeling the heat conduction process requires you to apply thermodynamics of energy conservation along with Fourier's law of heat conduction. The consequent mathematical descriptions are usually in the form of ordinary as well as partial differential equations. By considering different engineering applications that represent situations for steady as well as time-dependent (or transient) conduction, a study of this chapter will teach you:

- How to derive the conduction equation in different coordinate systems for both steady-state and transient conditions.
- How to obtain steady-state temperature distributions in simple conducting geometries without and with heat generation.
- How to develop the mathematical formulation of boundary conditions with insulation, constant heat flux, surface convection, and specified changes in surface temperature.
- How to apply the concept of lumped capacitance (conditions under which internal resistance in a conducting body can be neglected) in transient heat transfer.
- How to use charts for transient heat conduction to obtain temperature distribution as a function of time in simple geometries.
- How to obtain temperature distribution and rate of heat loss or gain from extended surfaces, also called *fins*, and use them in typical applications.

2.1 Introduction

Heat flows through a solid by a process that is called *thermal diffusion*, or simply *diffusion* or *conduction.* In this mode, heat is transferred through a complex submicroscopic mechanism in which atoms interact by elastic and inelastic collisions to propagate the energy from regions of higher to regions of lower temperature. From an engineering point of view there is no need to delve into the complexities of the molecular mechanisms, because the rate of heat propagation can be predicted by Fourier's law, which incorporates the mechanistic features of the process into a physical property known as the *thermal conductivity*.

Although conduction also occurs in liquids and gases, it is rarely the predominant transport mechanism in fluids—once heat begins to flow in a fluid, even if no external force is applied, density gradients are set up and convective currents are set in motion. In convection, thermal energy is thus transported on a macroscopic scale as well as on a microscopic scale, and convection currents are generally more effective in transporting heat than conduction alone, where the motion is limited to submicroscopic transport of energy.

Conduction heat transfer can readily be modeled and described mathematically. The associated governing physical relations are partial differential equations, which are susceptible to solution by classical methods [1]. Famous mathematicians, including Laplace and Fourier, spent part of their lives seeking and tabulating useful solutions to heat conduction problems. However, the analytic approach to conduction is limited to relatively simple geometric shapes and to boundary conditions that can only approximate the situation in realistic engineering problems. With the advent of the high-speed computer, the situation changed dramatically and a revolution occurred in the field of conduction heat transfer. The computer made it possible to solve, with relative ease, complex problems that closely approximate real conditions. As a result, the analytic approach has nearly disappeared from the engineering scene. The analytic approach is nevertheless important as background for the next chapter, in which we will show how to solve conduction problems by numerical methods.

2.2 The Conduction Equation

In this section the general conduction equation is derived. A solution of this equation, subject to given initial and boundary conditions, yields the temperature distribution in a solid system. Once the temperature distribution is known, the heat transfer rate in the conduction mode can be evaluated by applying Fourier's law [Eq. (1.2)].

The conduction equation is a mathematical expression of the conservation of energy in a solid substance. To derive this equation we perform an energy balance on an elemental volume of material in which heat is being transferred only by conduction. Heat transfer by radiation occurs in a solid only if the material is transparent or translucent.

The energy balance includes the possibility of heat generation in the material. Heat generation in a solid can result from chemical reactions, electric currents passing through the material, or nuclear reactions. Typical examples are illustrated in Fig. 2.1, which include (a) an element of a planar solid-oxide fuel cell (SOFC) that has a chemical reaction at the electrolyte-electrode interface, (b) a current-carrying electrical cable, and (c) a spherical nuclear fuel element for a pebble-bed nuclear reactor. The general form of the conduction equation also accounts for storage of internal energy. Thermodynamic considerations show that when the internal energy of a material increases, its temperature also increases. A solid material therefore experiences a net increase in stored energy when its temperature increases with time. If the temperature of the material remains constant, no energy is stored and steady-state conditions are said to prevail.

Heat transfer problems are classified according to the variables that influence the temperature. If the temperature is a function of time, the problem is classified as

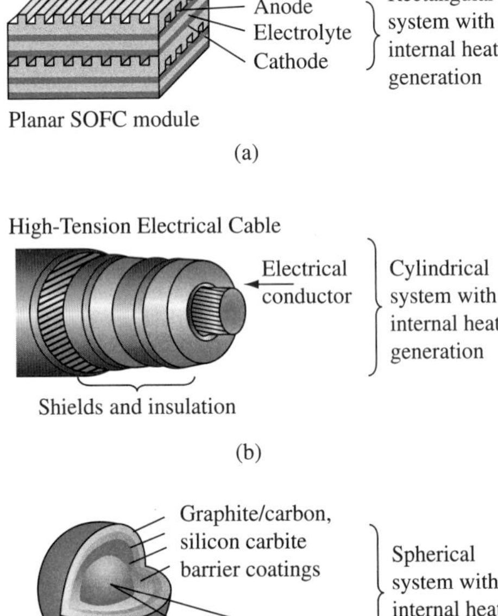

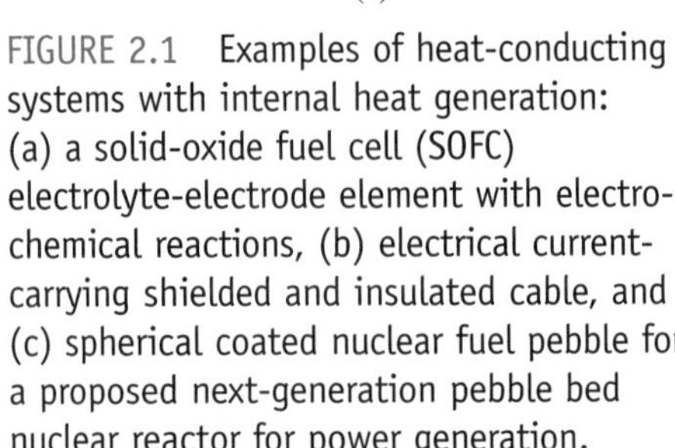

FIGURE 2.1 Examples of heat-conducting systems with internal heat generation: (a) a solid-oxide fuel cell (SOFC) electrolyte-electrode element with electrochemical reactions, (b) electrical current-carrying shielded and insulated cable, and (c) spherical coated nuclear fuel pebble for a proposed next-generation pebble bed nuclear reactor for power generation.

unsteady or *transient*. If the temperature is independent of time, the problem is called a *steady-state* problem. If the temperature is a function of a single space coordinate, the problem is said to be *one-dimensional*. If it is a function of two or three coordinate dimensions, the problem is *two-* or *three-dimensional*, respectively. If the temperature is a function of time and only one space coordinate, the problem is classified as *one-dimensional and transient*.

2.2.1 Rectangular Coordinates

To illustrate the analytic approach, we will first derive the conduction equation for a one-dimensional, rectangular coordinate system as shown in Fig. 2.2. We will assume that the temperature in the material is a function only of the x coordinate and time; that is, $T = T(x,t)$, and the conductivity k, density ρ, and specific heat c of the solid are all constant.

The principle of conservation of energy for the control volume, surface area A, and thickness Δx, of Fig. 2.2 can be stated as follows:

$$\begin{array}{c} \text{rate of heat conduction} \\ \text{into control volume} \\ + \\ \text{rate of heat generation} \\ \text{inside control volume} \end{array} = \begin{array}{c} \text{rate of heat conduction} \\ \text{out of control volume} \\ + \\ \text{rate of energy storage} \\ \text{inside control volume} \end{array} \tag{2.1}$$

We will use Fourier's law to express the two conduction terms and define the symbol $\dot{q}_G$ as the rate of energy generation per unit volume inside the control volume. Then the word equation (Eq. 2.1) can be expressed in mathematical form:

$$-kA\frac{\partial T}{\partial x}\bigg|_x + \dot{q}_G A\,\Delta x = -kA\frac{\partial T}{\partial x}\bigg|_{x+\Delta x} + \rho A\,\Delta x c\,\frac{\partial T(x+\Delta x/2,\,t)}{\partial t} \tag{2.2}$$

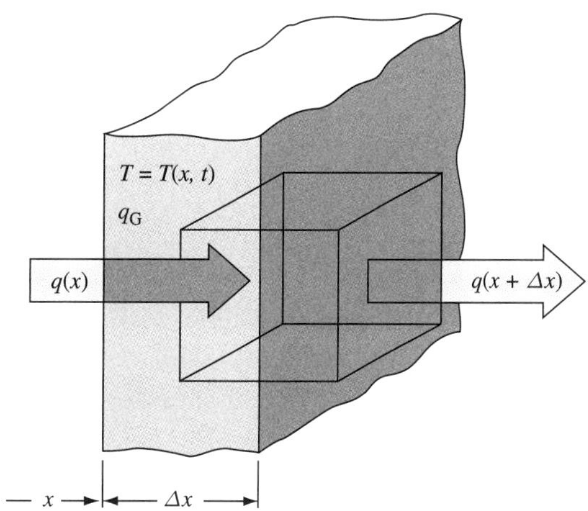

FIGURE 2.2 Control volume for one-dimensional conduction in rectangular coordinates.

Dividing Eq. (2.2) by the control volume $A\ \Delta x$ and rearranging, we obtain

$$k\frac{(\partial T/\partial x)_{x+\Delta x} - (\partial T/\partial x)_x}{\Delta x} + \dot{q}_G = \rho c\frac{\partial T(x + \Delta x/2, t)}{\partial t} \quad (2.3)$$

In the limit as $\Delta x \rightarrow 0$, the first term on the left side of Eq. (2.3) can be expressed in the form

$$\left.\frac{\partial T}{\partial x}\right|_{x+dx} = \left.\frac{\partial T}{\partial x}\right|_x + \frac{\partial}{\partial x}\left(\left.\frac{\partial T}{\partial x}\right|_x\right)dx = \left.\frac{\partial T}{\partial x}\right|_x + \left.\frac{\partial^2 T}{\partial x^2}\right|_x dx \quad (2.4)$$

The right side of Eq. (2.3) can be expanded in a Taylor series as

$$\frac{\partial T}{\partial t}\left[\left(x + \frac{\Delta x}{2}\right), t\right] = \left.\frac{\partial T}{\partial t}\right|_x + \left.\frac{\partial^2 T}{\partial x\, \partial T}\right|_x \frac{\Delta x}{2} + \ldots$$

Equation (2.2) then becomes, to the order of Δx,

$$k\frac{\partial^2 T}{\partial x^2} + \dot{q}_G = \rho c\frac{\partial T}{\partial t} \quad (2.5)$$

Physically, the first term on the left side represents the *net rate of heat conduction* into the control volume per unit volume. The second term on the left side is the *rate of energy generation per unit volume* inside the control volume. The right side represents the *rate of increase in internal energy* inside the control volume per unit volume. Each term has dimensions of energy per unit time and volume with the units (W/m^3) in the SI system and (Btu/h ft^3) in the English system.

Equation (2.5) applies only to unidimensional heat flow because it was derived on the assumption that the temperature distribution is one-dimensional. If this restriction is now removed and the temperature is assumed to be a function of all three coordinates as well as time, that is, $T = T(x, y, z, t)$, terms similar to the first one in Eq. (2.5) but representing the net rate of conduction per unit volume in the y and z directions will appear. The three-dimensional form of the conduction equation then becomes (see Fig. 2.3)

$$\frac{\partial^2 T}{\partial x^2} + \frac{\partial^2 T}{\partial y^2} + \frac{\partial^2 T}{\partial z^2} + \frac{\dot{q}_G}{k} = \frac{1}{\alpha}\frac{\partial T}{\partial t} \quad (2.6)$$

where α is the *thermal diffusivity*, a group of material properties defined as

$$\alpha = \frac{k}{\rho c} \quad (2.7)$$

The thermal diffusivity has units of (m^2/s) in the SI system and (ft^2/s) in the English system. Numerical values of the thermal conductivity, density, specific heat, and thermal diffusivity for several engineeering materials are listed in Appendix 2.

Solutions to the general conduction equation in the form of Eq. (2.6) can be obtained only for simple geometric shapes and easily specified boundary conditions. However, as shown in the next chapter, solutions by numerical methods can be obtained

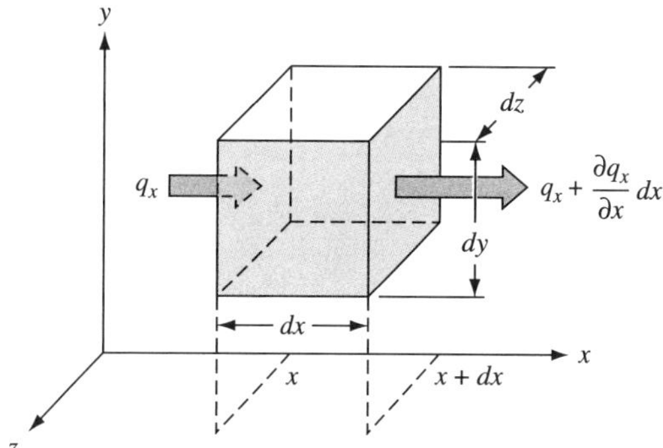

FIGURE 2.3 Differential control volume for three-dimensional conduction in rectangular coordinates.

quite easily for complex shapes and realistic boundary conditions; this procedure is used in engineering practice today for the majority of conduction problems. Nevertheless, a basic understanding of analytic solutions is important in writing computer programs, and in the rest of this chapter we will examine problems for which simplifying assumptions can eliminate some terms from Eq. (2.6) and reduce the complexity of the solution.

If the temperature of a material is not a function of time, the system is in the steady state and does not store any energy. The steady-state form of a three-dimensional conduction equation in rectangular coordinates is

$$\frac{\partial^2 T}{\partial x^2} + \frac{\partial^2 T}{\partial y^2} + \frac{\partial^2 T}{\partial z^2} + \frac{\dot{q}_G}{k} = 0 \tag{2.8}$$

If the system is in the steady state and no heat is generated internally, the conduction equation further simplifies to

$$\frac{\partial^2 T}{\partial x^2} + \frac{\partial^2 T}{\partial y^2} + \frac{\partial^2 T}{\partial z^2} = 0 \tag{2.9}$$

Equation (2.9) is known as the *Laplace equation*, in honor of the French mathematician Pierre Laplace. It occurs in a number of areas in addition to heat transfer, for instance, in diffusion of mass or in electromagnetic fields. The operation of taking the second derivatives of the potential in a field has therefore been given a shorthand symbol, ∇^2, called the *Laplacian operator*. For the rectangular coordinate system Eq. (2.9) becomes

$$\frac{\partial^2 T}{\partial x^2} + \frac{\partial^2 T}{\partial y^2} + \frac{\partial^2 T}{\partial z^2} = \nabla^2 T = 0 \tag{2.10}$$

Since the operator ∇^2 is independent of the coordinate system, the above form will be particularly useful when we want to study conduction in cylindrical and spherical coordinates.

2.2.2 Dimensionless Form

The conduction equation in the form of Eq. (2.6) is dimensional. It is often more convenient to express this equation in a form where each term is dimensionless. In the development of such an equation we will identify dimensionless groups that govern the heat conduction process. We begin by defining a dimensionless temperature as the ratio

$$\theta = \frac{T}{T_r} \tag{2.11}$$

a dimensionless x coordinate as the ratio

$$\xi = \frac{x}{L_r} \tag{2.12}$$

and a dimensionless time as the ratio

$$\tau = \frac{t}{t_r} \tag{2.13}$$

where the symbols T_r, L_r, and t_r represent a reference temperature, a reference length, and a reference time, respectively. Although the choice of reference quantities is somewhat arbitrary, the values selected should be physically significant. The choice of dimensionless groups varies from problem to problem, but the form of the dimensionless groups should be structured so that they limit the dimensionless variables between convenient extremes, such as zero and one. The value for L_r should therefore be selected as the maximum x dimension of the system for which the temperature distribution is sought. Similarly, a dimensionless ratio of temperature differences that varies between zero and unity is often preferable to a ratio of absolute temperatures.

If the definitions of the dimensionless temperature, x coordinate, and time are substituted into Eq. (2.5), we obtain the conduction equation in the nondimensional form

$$\frac{\partial^2 \theta}{\partial \xi^2} + \frac{\dot{q}_G L_r^2}{k T_r} = \frac{L_r^2}{\alpha t_r} \frac{\partial \theta}{\partial \tau} \tag{2.14}$$

The reciprocal of the dimensionless group $(L_r^2/\alpha t_r)$ is called the *Fourier number,* designated by the symbol Fo:

$$\mathrm{Fo} = \frac{\alpha t_r}{L_r^2} = \frac{(k/L_r)}{(\rho c L_r/t_r)} \tag{2.15}$$

In a more fundamental and physical sense, the Fourier number, named after the French mathematician and physicist Jean Baptiste Joseph Fourier (1768–1830), is the ratio of the rate of heat transfer by conduction to the rate of energy storage in the system. This is evident from the expanded second right-hand side of Eq. (2.15). It is an important dimensionless group in transient conduction problems and will be encountered frequently. The choice of reference time and length in the Fourier number depends on the specific problem, but the basic form is always a thermal diffusivity multiplied by time and divided by the square of a characteristic length.

The other dimensionless group appearing in Eq. (2.14) is a ratio of internal heat generation per unit time to heat conduction through the volume per unit time. We will use the symbol $\dot{Q}_G$ to represent this dimensionless heat generation number:

$$\dot{Q}_G = \frac{\dot{q}_G L_r^2}{kT_r} \tag{2.16}$$

The one-dimensional form of the conduction equation expressed in dimensionless form now becomes

$$\frac{\partial^2 \theta}{\partial \xi^2} + \dot{Q}_G = \frac{1}{\text{Fo}} \frac{\partial \theta}{\partial \tau} \tag{2.17}$$

If steady state prevails, the right side of Eq. (2.17) becomes zero.

2.2.3 Cylindrical and Spherical Coordinates

Equation (2.6) was derived for a rectangular coordinate system. Although the generation and energy storage terms are independent of the coordinate system, the heat conduction terms depend on geometry and therefore on the coordinate system. The dependence on the coordinate system used to formulate the problem can be removed by replacing the heat conduction terms with the Laplacian operator.

$$\nabla^2 T + \frac{\dot{q}_G}{k} = \frac{1}{\alpha} \frac{\partial T}{\partial t} \tag{2.18}$$

The differential form of the Laplacian is different for each coordinate system.

For a general transient three-dimensional problem in the cylindrical coordinates shown in Fig. 2.4, $T = T(r, \phi, z, t)$ and $\dot{q}_G = \dot{q}_G(r, \phi, z, t)$. If the Laplacian is substituted into Eq. (2.18), the general form of the conduction equation in cylindrical coordinates becomes

$$\frac{1}{r} \frac{\partial}{\partial r}\left(r \frac{\partial T}{\partial r} \right) + \frac{1}{r^2} \frac{\partial^2 T}{\partial \phi^2} + \frac{\partial^2 T}{\partial z^2} + \frac{\dot{q}_G}{k} = \frac{1}{\alpha} \frac{\partial T}{\partial t} \tag{2.19}$$

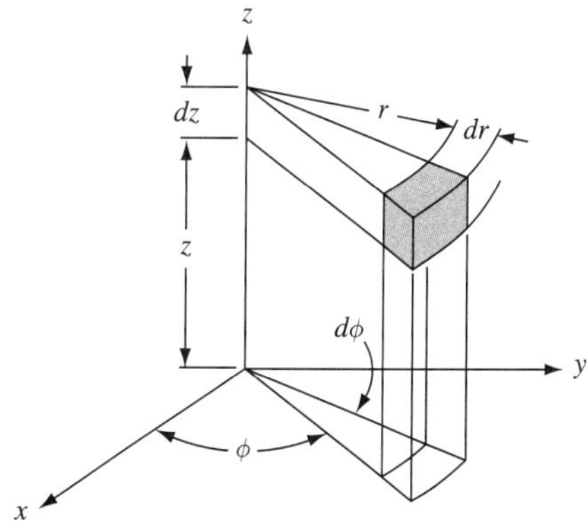

FIGURE 2.4 Cylindrical coordinate system for the general conduction equation.

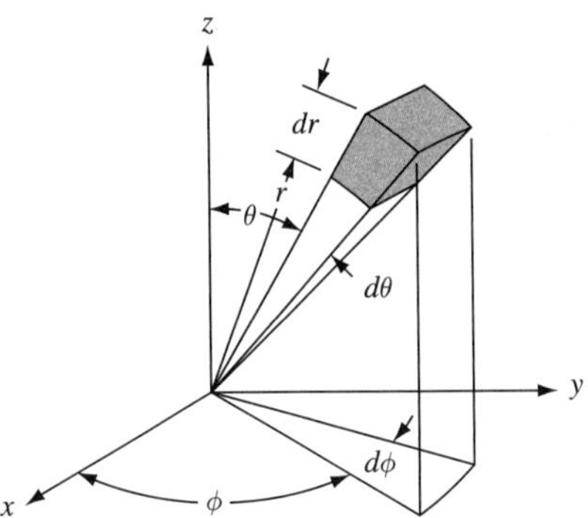

FIGURE 2.5 Spherical coordinate system for the general conduction equation.

If the heat flow in a cylindrical shape is only in the radial direction, $T = T(r,t)$, the conduction equation reduces to

$$\frac{1}{r}\frac{\partial}{\partial r}\left(r\frac{\partial T}{\partial r}\right) + \frac{\dot{q}_G}{k} = \frac{1}{\alpha}\frac{\partial T}{\partial t} \tag{2.20}$$

Furthermore, if the temperature distribution does not vary with time, the conduction equation becomes

$$\frac{1}{r}\frac{d}{dr}\left(r\frac{dT}{dr}\right) + \frac{\dot{q}_G}{k} = 0 \tag{2.21}$$

In this case the equation for the temperature contains only a single variable r and is therefore an ordinary differential equation.

When there is no internal energy generation and the temperature is a function of the radius only, the steady-state conduction equation for cylindrical coordinates is

$$\frac{d}{dr}\left(r\frac{dT}{dr}\right) = 0 \tag{2.22}$$

For spherical coordinates, as shown in Fig. 2.5, the temperature is a function of the three space coordinates r, θ, ϕ and time t, that is, $T = T(r, \theta, \phi, t)$. The general form of the conduction equation in spherical coordinates is then

$$\frac{1}{r^2}\frac{\partial}{\partial r}\left(r^2\frac{\partial T}{\partial r}\right) + \frac{1}{r^2\sin^2\theta}\frac{\partial}{\partial \theta}\left(\sin\theta\frac{\partial T}{\partial \theta}\right) + \frac{1}{r^2\sin\theta}\frac{\partial^2 T}{\partial \phi^2} + \frac{\dot{q}_G}{k} = \frac{1}{\alpha}\frac{\partial T}{\partial t} \tag{2.23}$$

2.3 Steady Heat Conduction in Simple Geometries

In this section we will demonstrate how to obtain solutions to the conduction equations derived in the preceding section for relatively simple geometric configurations with and without internal heat generation.

2.3.1 Plane Wall with and without Heat Generation

In the first chapter we saw that the temperature distribution for one-dimensional, steady conduction through a wall is linear. We can verify this result by simplifying the more general case expressed by Eq. (2.6). For steady state $\partial T/\partial t = 0$, and since T is only a function of x, $\partial T/\partial y = 0$ and $\partial T/\partial z = 0$. Furthermore, if there is no internal generation, $\dot{q}_G = 0$, Eq. (2.6) reduces to

$$\frac{d^2T}{dx^2} = 0 \tag{2.24}$$

Integrating this ordinary differential equation twice yields the temperature distribution

$$T(x) = C_1x + C_2 \tag{2.25}$$

For a wall with $T(x = 0) = T_1$ and $T(x = L) = T_2$ we get

$$T(x) = \frac{T_2 - T_1}{L}x + T_1 \tag{2.26}$$

The above relation agrees with the linear temperature distribution deduced by integrating Fourier's law, $q_k = -kA(dT/dx)$.

Next consider a similar problem, but with heat generation throughout the system, as shown in Fig. 2.6. If the thermal conductivity is constant and the heat generation is uniform, Eq. (2.5) reduces to

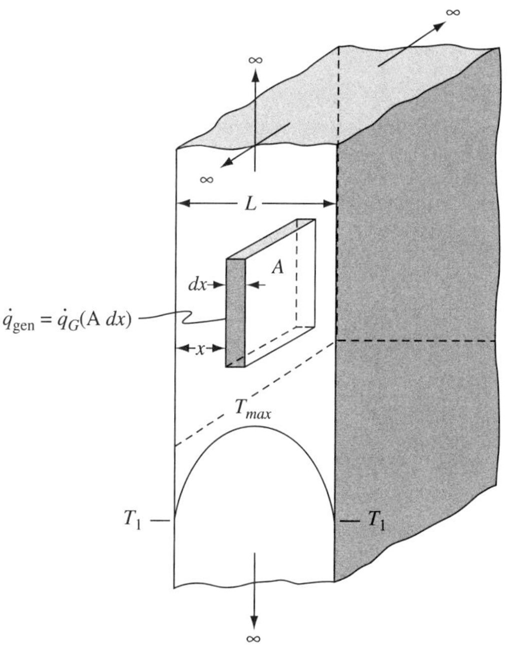

FIGURE 2.6 Conduction in a plane wall with uniform heat generation. Temperature distribution is for the case $T_1 = T_2$ (see Eq. 2.33).

$$k\frac{d^2T(x)}{dx^2} = -\dot{q}_G \tag{2.27}$$

Integrating this equation once gives

$$\frac{dT(x)}{dx} = -\frac{\dot{q}_G}{k}x + C_1 \tag{2.28}$$

and a second integration yields

$$T(x) = -\frac{\dot{q}_G}{2k}x^2 + C_1x + C_2 \tag{2.29}$$

where C_1 and C_2 are constants of integration whose values are determined by the boundary conditions. The specified conditions require that the temperature at $x = 0$ be T_1 and at $x = L$ be T_2. Substituting these conditions successively into the conduction equation gives

$$T_1 = C_2 \; (x = 0) \tag{2.30}$$

and

$$T_2 = -\frac{\dot{q}_G}{2k}L^2 + C_1L + T_1 \qquad (x = L) \tag{2.31}$$

Solving for C_1 and substituting into Eq. (2.29) gives the temperature distribution

$$T(x) = -\frac{\dot{q}_G}{2k}x^2 + \frac{T_2 - T_1}{L}x + \frac{\dot{q}_G L}{2k}x + T_1 \tag{2.32}$$

Observe that Eq. (2.26) is now modified by two terms containing the heat generation and that the temperature distribution is no longer linear.

If the two surface temperatures are equal, $T_1 = T_2$, the temperature distribution becomes

$$T(x) = \frac{\dot{q}_G L^2}{2k}\left[\frac{x}{L} - \left(\frac{x}{L}\right)^2\right] + T_1 \tag{2.33}$$

This temperature distribution is parabolic and symmetric about the center plane with a maximum T_{max} at $x = L/2$. At the centerline $dT/dx = 0$, which corresponds to an insulated surface at $x = L/2$. The maximum temperature is

$$T_{max} = T_1 + \frac{\dot{q}_G L^2}{8k} \tag{2.34}$$

For the symmetric boundary conditions the temperature in dimensionless form is

$$\frac{T(x) - T_1}{T_{max} - T_1} = 4(\xi - \xi^2)$$

where $\xi = x/L$.

EXAMPLE 2.1 A long electrical heating element made of iron has a cross section of 10 cm × 1.0 cm. It is immersed in a heat transfer oil at 80°C as shown in Fig. 2.7. If heat is generated uniformly at a rate of 1,000,000 W/m^3 by an electric current, determine the heat transfer coefficient necessary to keep the temperature of the heater below 200°C. The thermal conductivity for iron at 200°C is 64 W/m K by interpolation from Table 12 in Appendix 2.

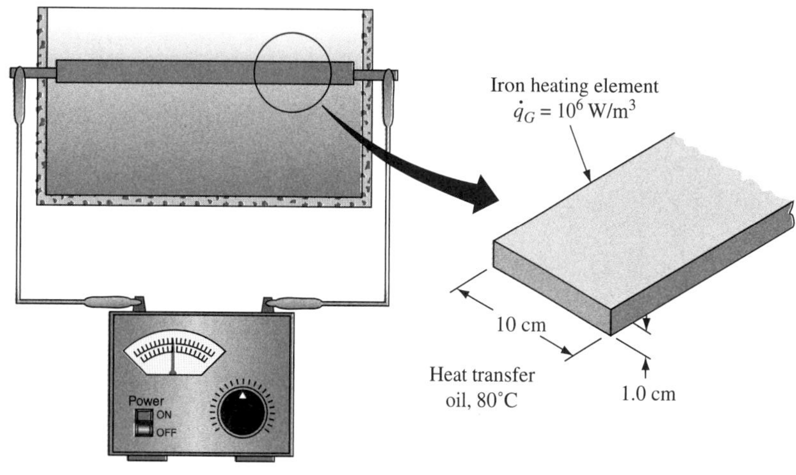

FIGURE 2.7 Electrical heating element for Example 2.1.

SOLUTION If we disregard the heat dissipated from the edges, a reasonable assumption since the heater has a width 10 times greater than its thickness, Eq. (2.34) can be used to calculate the temperature difference between the center and the surface:

$$T_{\max} - T_1 = \frac{\dot{q}_G L^2}{8k} = \frac{(1{,}000{,}000 \text{ W/m}^3)(0.01 \text{ m})^2}{(8)(64 \text{ W/m k})} = 0.2°\text{C}$$

The temperature drop from the center to the surface of the heater is small because the heater material is made of iron, which is a good conductor. We can neglect this temperature drop and calculate the minimum heat transfer coefficient from a heat balance:

$$\dot{q}_G \frac{L}{2} = \bar{h}_c(T_1 - T_\infty)$$

Solving for $\bar{h}_c$:

$$\bar{h}_c = \frac{\dot{q}_G(L/2)}{(T_1 - T_\infty)} = \frac{(10^6 \text{ W/m}^3)(0.005 \text{ m})}{120 \text{ K}} = 42 \text{ W/m}^2 \text{ K}$$

Thus, the heat transfer coefficient required to keep the temperature in the heater from exceeding the set limit must be larger than 42 W/m^2 K.

2.3.2 Cylindrical and Spherical Shapes without Heat Generation

In this section we will obtain solutions to some problems in cylindrical and spherical systems that are often encountered in practice. Probably the most common case is that of heat transfer through a pipe with a fluid flowing inside. This system can be idealized, as shown in Fig. 2.8, by radial heat flow through a cylindrical shell. Our problem is then to determine the temperature distribution and the heat transfer rate in a long hollow cylinder of length L if the inner- and outer-surface temperatures are T_i and T_o, respectively, and there is no internal heat generation. Since the temperatures at the boundaries are constant, the temperature distribution is not a function of time and the appropriate form of the conduction equation is

$$\frac{d}{dr}\left(r\frac{dT}{dr}\right) = 0 \tag{2.35}$$

Integrating once with respect to radius gives

$$r\frac{dT}{dr} = C_1 \text{ or } \frac{dT}{dr} = \frac{C_1}{r}$$

A second integration gives $T = C_1 \ln r + C_2$. The constants of integration can be determined from the boundary conditions:

$$T_i = C_1 \ln r_i + C_2 \qquad \text{at } r = r_i$$

Thus, $C_2 = T_i - C_1 \ln r_i$. Similarly, for T_o,

$$T_o = C_1 \ln r_o + T_i - C_1 \ln r_i \qquad \text{at } r = r_o$$

Thus, $C_1 = (T_o - T_i)/\ln(r_o/r_i)$.

The temperature distribution, written in dimensionless form, is therefore

$$\frac{T(r) - T_i}{T_o - T_i} = \frac{\ln(r/r_i)}{\ln(r_o/r_i)} \tag{2.36}$$

The rate of heat transfer by conduction through the cylinder of length L is, from Eq. (1.1),

$$q_k = -kA\frac{dT}{dr} = -k(2\pi rL)\frac{C_1}{r} = 2\pi Lk\frac{T_i - T_o}{\ln(r_o/r_i)} \tag{2.37}$$

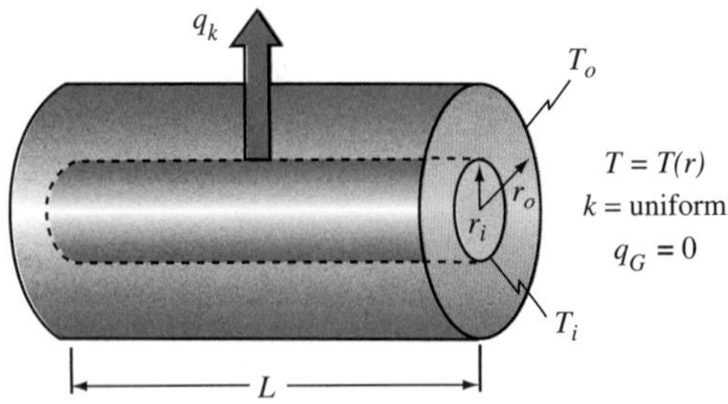

FIGURE 2.8 Radial heat conduction through a cylindrical shell.

In terms of a thermal resistance we can write

$$q_k = \frac{T_i - T_o}{R_{\text{th}}} \tag{2.38}$$

where the resistance to heat flow by conduction through a cylinder of length L, inner radius r_i, and outer radius r_o is

$$R_{\text{th}} = \frac{\ln(r_o/r_i)}{2\pi Lk} \tag{2.39}$$

The principles developed for a plane wall with conduction and convection in series can also be applied to a long hollow cylinder such as a pipe or a tube. For example, as shown in Fig. 2.9, suppose that a hot fluid flows through a tube that is covered by an insulating material. The system loses heat to the surrounding air through an average heat transfer coefficient $\bar{h}_{c,o}$.

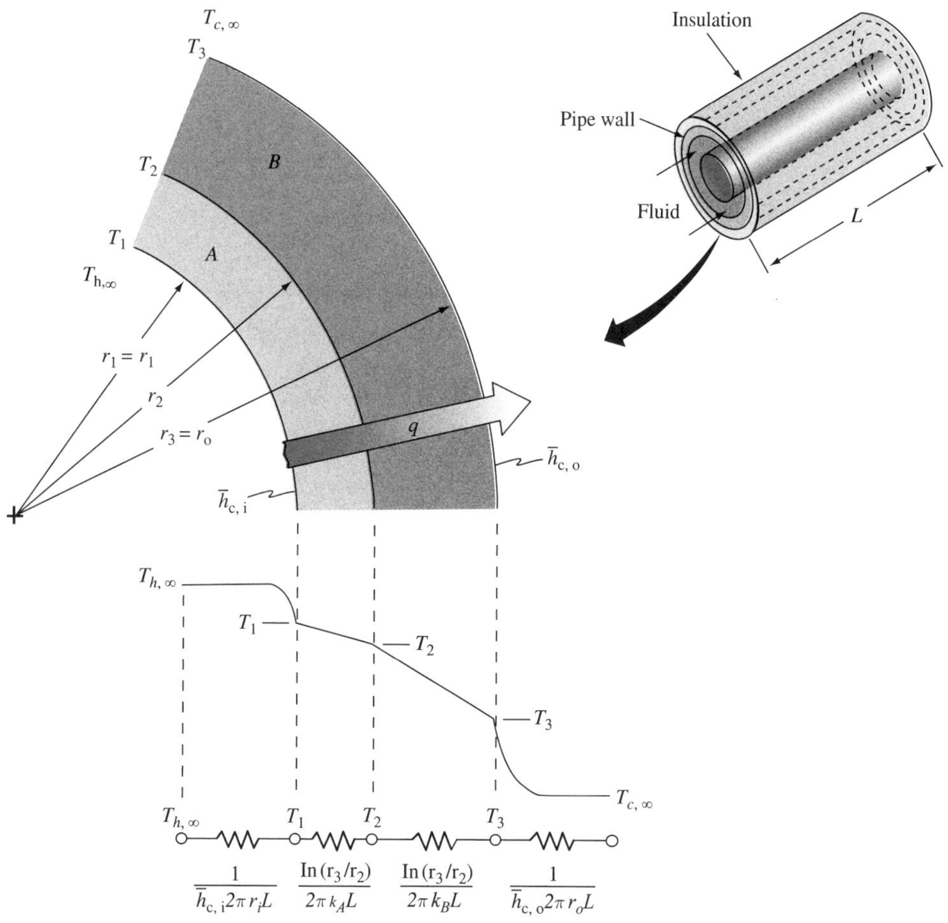

FIGURE 2.9 Temperature distribution for a composite cylindrical wall with convection at the interior and exterior surfaces.

Using Eq. (2.38) for the thermal resistance of the two cylinders and Eq. (1.14) for the thermal resistance at the inside of the tube and the outside of the insulation gives the thermal network shown below the physical system in Fig. 2.9. Denoting the hot-fluid temperature by $T_{h,\infty}$ and the environmental air temperature by $T_{c,\infty}$, the rate of heat flow is

$$q = \frac{\Delta T}{\sum_{1}^{4} R_{\text{th}}} = \frac{T_{h,\infty} - T_{c,\infty}}{\dfrac{1}{\bar{h}_{c,i}2\pi r_1 L} + \dfrac{\ln(r_2/r_1)}{2\pi k_A L} + \dfrac{\ln(r_3/r_2)}{2\pi k_B L} + \dfrac{1}{\bar{h}_{c,o}2\pi r_3 L}} \tag{2.40}$$

EXAMPLE 2.2 Compare the heat loss from an insulated and an uninsulated copper pipe under the following conditions. The pipe (k = 400 W/m K) has an internal diameter of 10 cm and an external diameter of 12 cm. Saturated steam flows inside the pipe at 110°C. The pipe is located in a space at 30°C and the heat transfer coefficient on its outer surface is estimated to be 15 W/m^2 K. The insulation available to reduce heat losses is 5 cm thick and its conductivity is 0.20 W/m K.

SOLUTION The uninsulated pipe is depicted by the system in Fig. 2.10. The heat loss per unit length is therefore

$$\frac{q}{L} = \frac{T_s - T_\infty}{R_1 + R_2 + R_3}$$

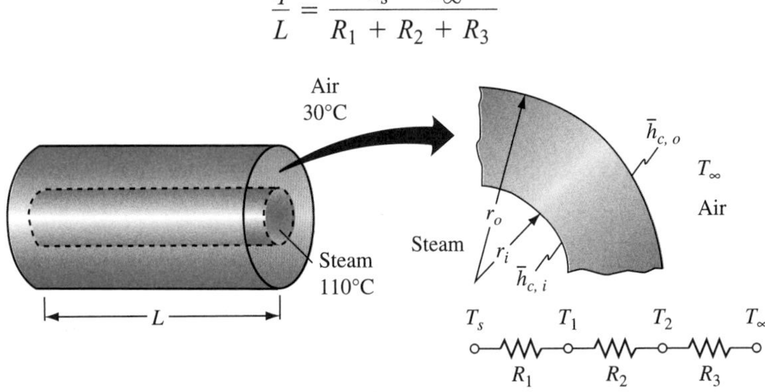

FIGURE 2.10 Schematic diagram and thermal circuit for a hollow cylinder with convection surface conditions (Example 2.2).

For the interior surface resistance we can use Table 1.3 to estimate $\bar{h}_{c,i}$. For saturated steam condensing, $\bar{h}_{c,i}$ = 10,000 W/m^2 K. Hence we get

$$R_1 = R_i = \frac{1}{2\pi r_i \bar{h}_{c,i}} \simeq \frac{1}{(2\pi)(0.05\text{ m})(10{,}000\text{ W/m}^2\text{ K})} = 0.000318\text{ m K/W}$$

$$R_2 = \frac{\ln(r_o/r_i)}{2\pi k_{\text{pipe}}} = \frac{0.182}{(2\pi)(400\text{ W/m K})} = 0.00007\text{ m K/W}$$

$$R_3 = R_o = \frac{1}{2\pi r_o \bar{h}_{c,o}} = \frac{1}{(2\pi)(0.06\text{ m})(15\text{ W/m}^2\text{ K})} = 0.177\text{ m K/W}$$

Since R_1 and R_2 are negligibly small compared to R_3, $q/L = 80/0.177 = 452$ W/m for the uninsulated pipe.

For the insulated pipe the system corresponds to that shown in Fig. 2.9; hence, we must add a fourth resistance between r_1 and r_3.

$$R_4 = \frac{\ln(11/6)}{(2\pi)(0.2 \text{ W/m K})} = 0.482 \text{ m K/W}$$

Also, the outer convection resistance changes to

$$R_o = \frac{1}{(2\pi)(0.11 \text{ m})(15 \text{ W/m}^2 \text{ K})} = 0.096 \text{ m K/W}$$

The total thermal resistance per meter length is therefore 0.578 m K/W and the heat loss is $80/0.578 = 138$ W/m. Adding insulation will reduce the heat loss from the steam by 70%.

Critical Radius of Insulation In the context of Example 2.2, while heat loss from an insulated *cylindrical system* to an external convective environment can generally be minimized by increasing the thickness of insulation, the problem is somewhat different in small-diameter systems. A case of some practical interest is the insulation or sheathing of electrical wires, electrical resistors, and other cylindrical electronic devices through which current flows. Consider an electrical resistor (or wire) with an insulating sleeve of conductivity k, which has an electrical resistivity R_e and carries a current i, as shown in Fig. 2.11, along with its thermal-resistance circuit, where the heat generated in the wire is transferred to the ambient via conduction through the insulation and convection at the outer insulation surface.

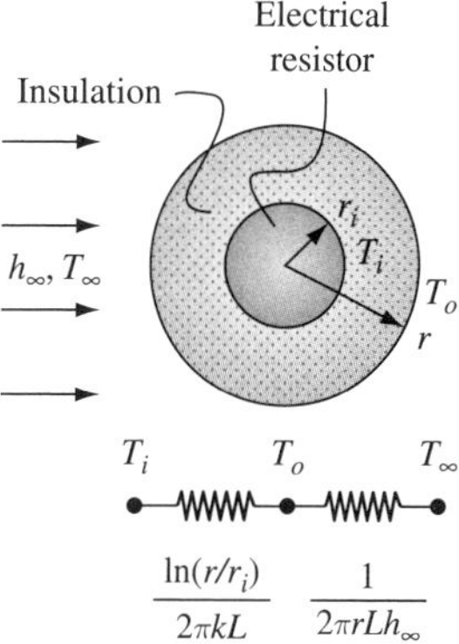

FIGURE 2.11 Current-carrying electrical resistor or wire with an insulating sheath and its thermal-resistance circuit.

Here the electrical-resistance heat dissipated in the wire is transferred (or lost) to the ambient, and the heat transfer rate is given by

$$q = i^2 R_e = \frac{T_i - T_\infty}{R_{\text{total}}} \tag{2.41}$$

where the total thermal resistance R_{total} is the sum of the resistances for conduction through the insulation and external convection, or

$$R_{\text{total}} = R_{\text{cond}} + R_{\text{conv}} = \frac{\ln(r/r_i)}{2\pi kL} + \frac{1}{2\pi rLh_\infty} \tag{2.42}$$

From Eq. (2.42) it is evident that as the outer insulation radius r increases, R_{cond} also increases whereas R_{conv} decreases because of the increasing outer surface area. A relatively larger decrease in the latter would suggest that there is an optimum value of r, or a *critical radius* r_{cr} of insulation, for which R_{total} is minimum and the heat loss q is maximum. This can be readily obtained by differentiating R_{total} in Eq. (2.42) with respect to r and setting the derivative equal to zero as follows:

$$\frac{dR_{\text{total}}}{dr} = \frac{1}{2\pi krL} - \frac{1}{2\pi r^2 Lh_\infty} = 0$$

or

$$r = r_{\text{cr}} = \frac{k}{h_\infty} \tag{2.43}$$

That r_{cr} yields a minimum total resistance can be confirmed by establishing a positive value for the second derivative of Eq. (2.42) with $r = r_{\text{cr}}$, and the student can readily show this as a home exercise.

The graph in Fig. 2.12 depicts the variations in R_{total}, given by Eq. (2.42) for a typical electrical resistor or current-carrying wire, and that the competing changes in

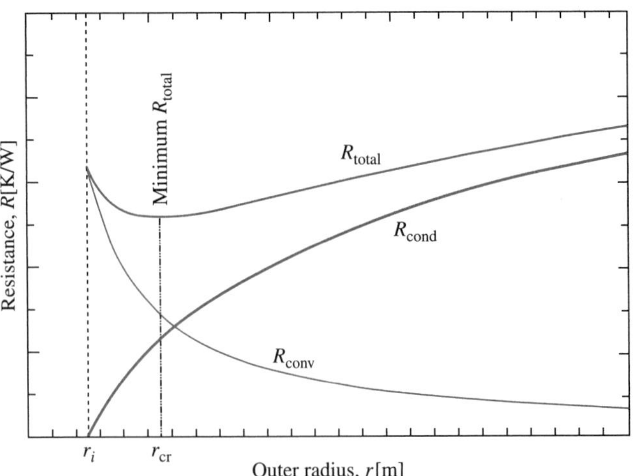

FIGURE 2.12 Variation of thermal resistance with radius of insulation on a cylindrical system and existence of a *critical radius* for minimum total resistance.

R_{cond} and R_{conv} with r result in a minimum value of R_{total} is self-evident. This feature is often employed in *cooling* cylindrical electrical and electronic systems (wires, cables, resistors, etc.) where the design provides effective electrical insulation and at the same time promotes optimum heat loss (*reduces thermal insulation effect*) so as to prevent overheating.

This condition is also encountered in a *spherical system* (see the subsequent treatment of the conduction equation in spherical coordinates), where, based on a similar mathematical treatment, the corresponding critical radius can be determined to be $r_{cr} = (2k/h_\infty)$. The derivation of this result, following the preceding method, is left for the student to carry out as a homework exercise.

Furthermore, it is important to note that the practicality of critical radius is somewhat limited to small-diameter systems in very low convective coefficient environments; in essence, the radius of the cylindrical system, which would need insulation for a "cooling" effect or where the thermal insulation might be ineffective, should be less than (k/h_∞). This can be seen from the numerical extension of Example 2.2, where for the given values of k and h_∞ (or $h_{c,o}$) the critical radius of insulation is $r_{cr} = 1.33$ cm, which is much smaller than 10-cm inner diameter of the steam pipe.

Overall Heat Transfer Coefficient As shown in Chapter 1, Section 1.6.4, for the case of plane walls with convection resistances at the surfaces, it is often convenient to define an overall heat transfer coefficient by the equation

$$q = UA\,\Delta T_{total} = UA(T_{hot} - T_{cold}) \tag{2.44}$$

Comparing Eqs. (2.40) and (2.44) we see that

$$UA = \frac{1}{\sum_1^4 R_{th}} = \frac{1}{\dfrac{1}{\bar{h}_{c,i}A_i} + \dfrac{\text{ln}(r_2/r_1)}{2\pi k_A L} + \dfrac{\text{ln}(r_3/r_2)}{2\pi k_B L} + \dfrac{1}{\bar{h}_{c,o}A_o}} \tag{2.45}$$

For plane walls the areas of all sections in the heat flow path are the same, but for cylindrical and spherical systems the area varies with radial distance and the overall heat transfer coefficient can be based on any area in the heat flow path. Thus, the numerical value of U will depend on the area selected. Since the outermost diameter is the easiest to measure in practice, $A_o = 2\pi r_3 L$ is usually chosen as the base area. The rate of heat flow is then

$$q = (UA)_o\,(T_{hot} - T_{cold}) \tag{2.46}$$

and the overall coefficient becomes

$$U_o = \frac{1}{\dfrac{r_3}{r_1\bar{h}_{c,1}} + \dfrac{r_3\text{ln}(r_2/r_1)}{k_A} + \dfrac{r_3\text{ln}(r_3/r_2)}{k_B} + \dfrac{1}{\bar{h}_{c,o}}} \tag{2.47}$$

EXAMPLE 2.3 A hot fluid at an average temperature of 200°C flows through a plastic pipe of 4 cm OD and 3 cm ID. The thermal conductivity of the plastic is 0.5 W/m^2 K, and the convection heat transfer coefficient at the inside is 300 W/m^2 K. The pipe is located in a room at 30°C, and the heat transfer coefficient at the outer surface is 10 W/m^2 K. Calculate the overall heat transfer coefficient and the heat loss per unit length of pipe.

SOLUTION A sketch of the physical system and the corresponding thermal circuit is shown in Fig. 2.10. The overall heat transfer coefficient from Eq. (2.47) is

$$U_o = \frac{1}{\dfrac{r_o}{r_i \bar{h}_{c,1}} + \dfrac{r_o \text{In}(r_o/r_1)}{k} + \dfrac{1}{\bar{h}_{c,o}}}$$

$$= \frac{1}{\dfrac{0.02}{0.015 \times 300} + \dfrac{0.02 \ln(2/1.5)}{0.5} + \dfrac{1}{10}} = 8.62 \text{ W/m}^2\text{ K}$$

where U_o is based on the outside area of the pipe. The heat loss per unit length is, from Eq. 2.46,

$$\frac{q}{L} = (UA)_o(T_{\text{hot}} - T_{\text{cold}}) = (8.62 \text{ W/m}^2\text{ K})(\pi)(0.04 \text{ m})(200 - 30)(\text{ K})$$
$$= 184 \text{ W/m}$$

Spherical Coordinate System For a hollow sphere with uniform temperatures at the inner and outer surfaces (see Fig. 2.13), the temperature distribution without heat

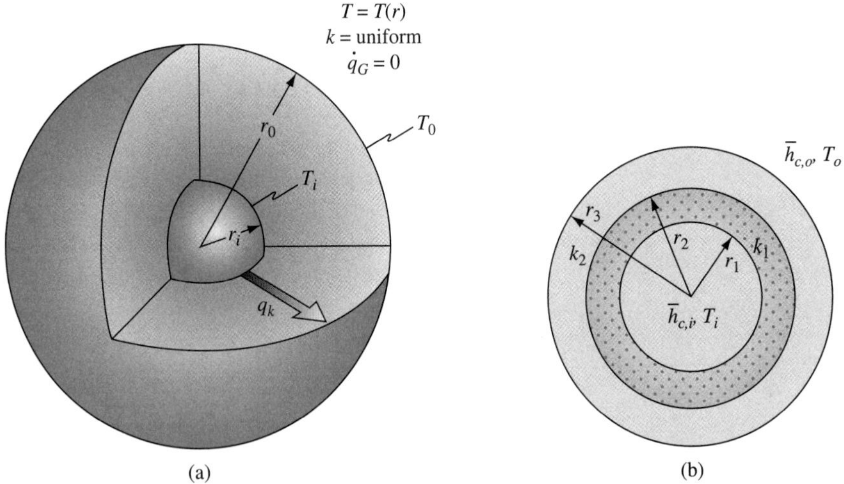

FIGURE 2.13 (a) Hollow sphere with uniform surface temperature and without heat generation; (b) hollow multilayered sphere with convection on inside and outside surfaces.

generation in the steady state can be obtained by simplifying Eq. (2.23). Under these boundary conditions the temperature is only a function of the radius r, and the conduction equation in spherical coordinates is

$$\frac{1}{r^2}\frac{d}{dr}\left(r^2\frac{dT}{dr}\right) = \frac{1}{r}\frac{d^2(rT)}{dr^2} = 0 \tag{2.48}$$

If the temperature at r_i is uniform and equal to T_i and at r_o equal to T_o, the temperature distribution is

$$T(r) - T_i = (T_o - T_i)\frac{r_o}{r_o - r_i}\left(1 - \frac{r_i}{r}\right) \tag{2.49}$$

The rate of heat transfer through the spherical shell is

$$q_k = -4\pi r^2 k\frac{\partial T}{\partial r} = \frac{T_i - T_o}{(r_o - r_i)/4\pi k r_o r_i} \tag{2.50}$$

The thermal resistance for a spherical shell is then

$$R_{\text{th}} = \frac{r_o - r_i}{4\pi k r_o r_i} \tag{2.51}$$

Furthermore, as in the case of a cylindrical system, the *overall heat transfer coefficient* for the multilayered spherical system shown in Fig. 2.13(b) can be expressed as

$$UA = \frac{1}{\sum_1^4 R_{\text{th}}} = \frac{1}{\dfrac{1}{\bar{h}_{c,i}A_i} + \dfrac{r_2 - r_1}{4\pi k_1 r_1 r_2} + \dfrac{r_3 - r_2}{4\pi k_2 r_2 r_3} + \dfrac{1}{\bar{h}_{c,o}A_o}} \tag{2.52}$$

Here the inner and outer surface areas are, respectively, $A_i = 4\pi r_1^2$ and $A_o = 4\pi r_3^2$. The total heat transfer rate is again given by the equation

$$q = (UA)\Delta T_{\text{total}} = \frac{(T_o - T_i)}{\sum_1^4 R_{\text{th}}} \tag{2.53}$$

EXAMPLE 2.4 The spherical, thin-walled metallic container shown in Fig. 2.14 is used to store liquid nitrogen at 77 K. The container has a diameter of 0.5 m and is covered with an evacuated insulation system composed of silica powder (k = 0.0017 W/m K). The insulation is 25 mm thick, and its outer surface is exposed to ambient air at 300 K. The latent heat of vaporization h_{fg} of liquid nitrogen is 2×10^5 J/kg. If the

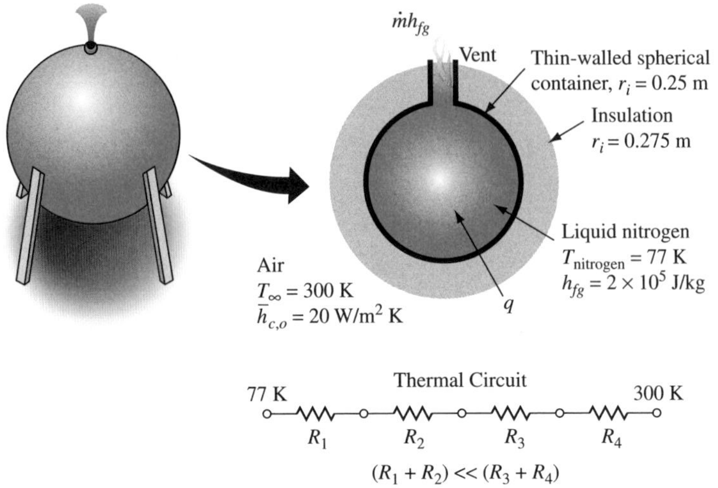

FIGURE 2.14 Schematic diagram of spherical container for Example 2.4.

convection coefficient is 20 W/m^2 K over the outer surface, determine the rate of liquid boil-off of nitrogen per hour.

SOLUTION The rate of heat transfer from the ambient air to the nitrogen in the container can be obtained from the thermal circuit shown in Fig. 2.14. We can neglect the thermal resistances of the metal wall and between the boiling nitrogen and the inner wall because the heat transfer coefficient (see Table 1.3) is large, i.e., neglect R_1 and R_2 from the thermal resistances shown in Fig. 2.14. Hence,

$$q = \frac{T_{\infty,\,\text{air}} - T_{\text{nitrogen}}}{R_3 + R_4} = \frac{(300 - 77)\text{ K}}{\dfrac{1}{\bar{h}_{c,o} 4\pi r_o^2} + \dfrac{r_o - r_i}{4\pi k r_o r_i}}$$

$$= \frac{223\text{ K}}{\dfrac{1}{(20\text{ W/m}^2\text{ K})(4\pi)(0.275\text{ m})^2} + \dfrac{(0.275 - 0.250)\text{ m}}{4\pi(0.0017\text{ W/m K})(0.275\text{ m})(0.250\text{ m})}}$$

$$= \frac{223\text{ K}}{(0.053 + 17.02)\text{ K/W}} = 13.06\text{ W}$$

Observe that almost the entire thermal resistance is in the insulation. To determine the rate of boil-off we perform an energy balance:

$$\begin{matrix}\text{rate of boil-off} \\ \text{of liquid nitrogen}\end{matrix} \times \begin{matrix}\text{nitrogen heat} \\ \text{of vaporization}\end{matrix} = \begin{matrix}\text{rate of heat transfer} \\ \text{to liquid nitrogen}\end{matrix}$$

or

$$\dot{m} h_{fg} = q$$

Solving for $\dot{m}$ gives

$$\dot{m} = \frac{q}{h_{fg}} = \frac{(13.06 \text{ J/s})(3600 \text{ s/h})}{2 \times 10^5 \text{ J/kg}} = 0.235 \text{ kg/h}$$

2.3.3 Long Solid Cylinder with Heat Generation

A long, solid circular cylinder with internal heat generation can be thought of as an idealization of a real system, for example, an electric coil in which heat is generated as a result of the electric current in the wire [see Fig. 2.1(b) for an example], or a cylindrical fuel element of uranium 235, in which heat is generated by nuclear fission. (An example is considered in the ensuing problem, Example 2.5, which is typically used in conventional nuclear reactors and is different from the spherical element shown in Fig. 2.1c.). The energy equation for an annular element (Fig. 2.15) formed between a fictitious inner cylinder of radius r and a fictitious outer cylinder of radius $r + dr$ is

$$-kA_r \frac{dT}{dr}\bigg|_r + \dot{q}_G L 2\pi r\, dr = -kA_{r+dr} \frac{dT}{dr}\bigg|_{r+dr}$$

where $A_r = 2\pi r L$ and $A_{r+dr} = 2\pi(r + dr)L$. Relating the temperature gradient at $r + dr$ to the temperature gradient at r, we obtain, after simplification,

$$r\dot{q}_G = -k\left(\frac{dT}{dr} + r\frac{d^2T}{dr^2}\right) \tag{2.54}$$

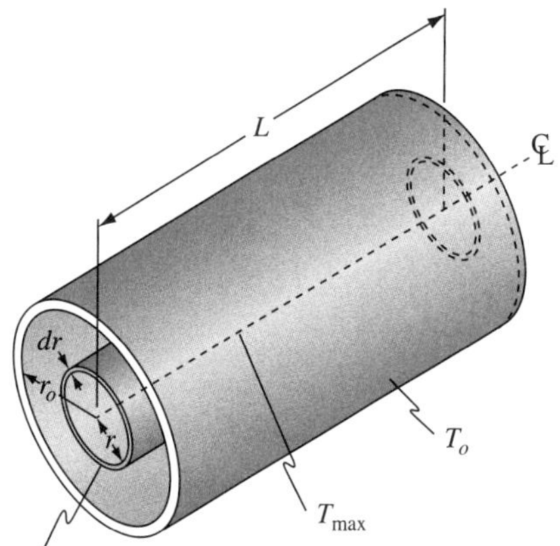

FIGURE 2.15 Nomenclature for heat conduction in a long circular cylinder with internal heat generation.

Integration of Eq. (2.54) can best be accomplished by noting that

$$\frac{d}{dr}\left(r\frac{dT}{dr}\right) = \frac{dT}{dr} + r\frac{d^2T}{dr^2}$$

and rewriting it in the form

$$\dot{q}_G r = -k\frac{d}{dr}\left(r\frac{dT}{dr}\right)$$

This is similar to the result obtained previously by simplifying the general conduction equation [see Eq. (2.21)]. Integration yields

$$\frac{\dot{q}_G r^2}{2} + = -kr\frac{dT}{dr} + C_1$$

from which we deduce that, to satisfy the boundary condition $dT/dr = 0$ at $r = 0$, the constant of integration C_1 must be zero. Another integration yields the temperature distribution

$$T = -\frac{\dot{q}_G r^2}{4k} + C_2$$

To satisfy the condition that the temperature at the outer surface, $r = r_o$, is T_o, $C_2 = (\dot{q}_G r_o^2/4k) + T_o$. The temperature distribution is therefore

$$T = T_o + \frac{\dot{q}_G r_o^2}{4k}\left[1 - \left(\frac{r}{r_o}\right)^2\right] \tag{2.55}$$

The maximum temperature at $r = 0$, $T_{\max}$, is

$$T_{\max} = T_o + \frac{\dot{q}_G r_o^2}{4k} \tag{2.56}$$

In dimensionless form Eq. (2.55) becomes

$$\frac{T(r) - T_o}{T_{\max} - T_o} = 1 - \left(\frac{r}{r_o}\right)^2 \tag{2.57}$$

For a hollow cylinder with uniformly distributed heat sources and specified surface temperatures, the boundary conditions are

$$T = T_i \text{ at } r = r_i \text{ (inside surface)}$$
$$T = T_o \text{ at } r = r_o \text{ (outside surface)}$$

It is left as an exercise to verify that for this case the temperature distribution is given by

$$T(r) = T_o + \frac{\dot{q}_G}{4k}(r_o^2 - r^2) + \frac{\ln(r/r_o)}{\ln(r_o/r_i)}\left[\frac{\dot{q}_G}{4k}(r_o^2 - r_i^2) + T_o - T_i\right] \tag{2.58}$$

If a solid cylinder is immersed in a fluid at a specified temperature T_∞ and the convection heat transfer coefficient at the surface is specified and denoted by $\bar{h}_c$, the surface temperature at r_o is not known a priori. The boundary condition for this case requires that the heat conduction from the cylinder equal the rate of convection at the surface, or

$$-k\frac{dT}{dr}\bigg|_{r=r_o} = \bar{h}_c(T_o - T_\infty)$$

Using this condition to evaluate the constants of integration yields for the dimensionless temperature distribution

$$\frac{T(r) - T_\infty}{T_\infty} = \frac{\dot{q}_G r_o}{4\bar{h}_c T_\infty}\left\{2 + \frac{\bar{h}_c r_o}{k}\left[1 - \left(\frac{r}{r_o}\right)^2\right]\right\} \tag{2.59}$$

and for the dimensionless maximum temperature ratio

$$\frac{T_{\max}}{T_\infty} = 1 + \frac{\dot{q}_G r_o}{4\bar{h}_c T_\infty}\left(2 + \frac{\bar{h}_c r_o}{k}\right) \tag{2.60}$$

In the preceding equations we have two dimensionless parameters of importance in conduction. The first is the heat generation parameter $\dot{q}_G r_o/\bar{h}_c T_\infty$ and the other is the *Biot number*, $\mathrm{Bi} = \bar{h}_c r_o/k$, which appears in problems with simultaneous conduction and convection modes of heat transfer.

Physically, the Biot number is the ratio of a conduction thermal resistance, $R_k = r_o/k$, to a convection resistance, $R_c = 1/\bar{h}_c$. The physical limits on this ratio for the above problem are:

$$\mathrm{Bi} \to 0 \quad \text{when} \quad R_k = \left(\frac{r_o}{k}\right) \to 0 \quad \text{or} \quad R_c = \frac{1}{\bar{h}_c} \to \infty$$

$$\mathrm{Bi} \to \infty \quad \text{when} \quad R_c = \frac{1}{\bar{h}_c} \to 0 \quad \text{or} \quad R_k = \frac{r_o}{k} \to \infty$$

The Biot number approaches zero when the conductivity of the solid or the convection resistance is so large that the solid is practically isothermal and the temperature change is mostly in the fluid at the interface. Conversely, the Biot number approaches infinity when the thermal resistance in the solid predominates and the temperature change occurs mostly in the solid.

EXAMPLE 2.5 Figure 2.16 on the next page shows a graphite-moderated nuclear reactor. Heat is generated uniformly in uranium rods of 0.05 m (1.973 in.) diameter at the rate of 7.5×10^7 W/m^3 (7.24×10^6 Btu/h ft^3). These rods are jacketed by an annulus in which water at an average temperature of 120°C (248°F) is circulated. The water cools the rods and the average convection heat transfer coefficient is estimated to be 55,000 W/m^2 K (9700 Btu/h ft^2 °F). If the thermal conductivity of uranium is 29.5 W/m K (17.04 Btu/h ft °F), determine the center temperature of the uranium fuel rods.

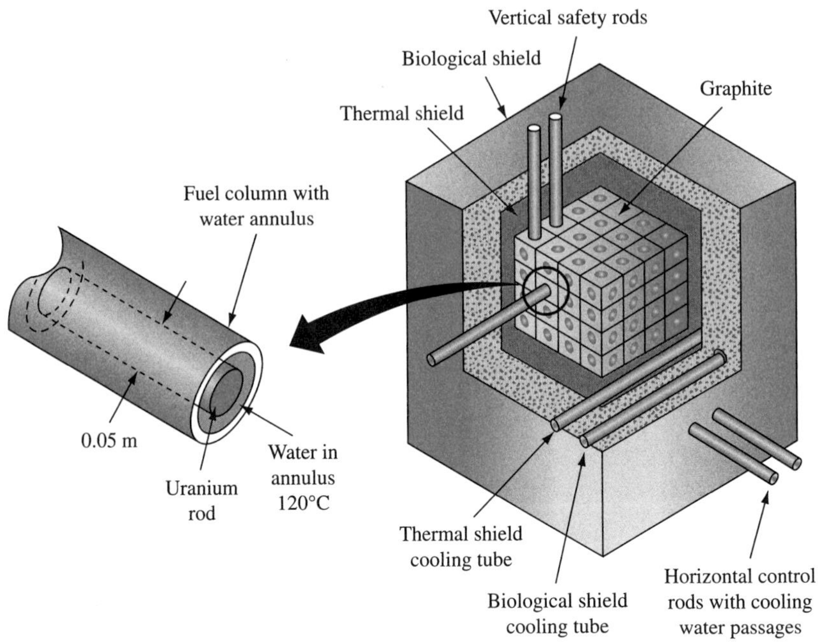

FIGURE 2.16 Nuclear reactor for Example 2.5.
Source: General Electric Review

SOLUTION Assuming that the fuel rods are sufficiently long that end effects can be neglected and that the thermal conductivity of uranium does not change appreciably with temperature, the thermal system can be approximated by the system shown in Fig. 2.16. Then the rate of flow through the surface of the rod equals the rate of internal heat generation:

$$2\pi r_o L\left(-k\frac{dT}{dr}\right)_{r_o} = \dot{q}_G \pi r_o^2 L$$

or

$$-k\frac{dT}{dr}\bigg|_{r_o} = \frac{\dot{q}_G r_o}{2} = \frac{(7.5 \times 10^7 \text{ W/m}^3)(0.025 \text{ m})}{2}$$
$$= 9.375 \times 10^5 \text{ W/m}^2 \ (2.97 \times 10^5 \text{ Btu/h ft}^2)$$

The rate of heat flow by conduction at the outer surface equals the rate of heat flow by convection from the surface to the water:

$$2\pi r_o\left(-k\frac{dT}{dr}\right)\bigg|_{r_o} = 2\pi r_o \bar{h}c,o(T_o - T_{\text{water}})$$

from which

$$T_o = \frac{-k(dT/dr)|_{r_o}}{\bar{h}_{c,o}} + T_{\text{water}}$$

Substituting the numerical data gives for T_o:

$$T_o = \frac{9.375 \times 10^5 \text{ W/m}^2}{5.5 \times 10^4 \text{ W/m}^2 \text{ K}} + 120°\text{C} = 137°\text{C } (279°\text{F})$$

Adding the temperature difference between the center and the surface of the fuel rods to the surface temperature T_o gives the maximum temperature:

$$T_{\max} = T_o + \frac{\dot{q}_G r_o^2}{4k} = 137 + \frac{(7.5 \times 10^7 \text{ W/m}^3)(0.025 \text{ m})^2}{(4)(29.5 \text{ W/m K})}$$
$$= 534°\text{C } (993.6°\text{F})$$

The same result can be obtained from Eq. (2.59). We observe that most of the temperature drop occurs in the solid because the convection resistance is very small (Bi is about 100).

2.4 Extended Surfaces

The problems considered in this section are encountered in practice when a solid of relatively small cross-sectional area protrudes from a large body into a fluid at a different temperature. Such extended surfaces have wide industrial application as fins attached to the walls of heat transfer equipment in order to increase the rate of heating or cooling.

2.4.1 Fins of Uniform Cross Section

As a simple illustration, consider a pin fin having the shape of a rod whose base is attached to a wall at surface temperature T_s (Fig. 2.17). The fin is cooled along its surface by a fluid at temperature T_∞. The fin has a uniform cross-sectional area A and is made of a material having uniform conductivity k; the heat transfer coefficient

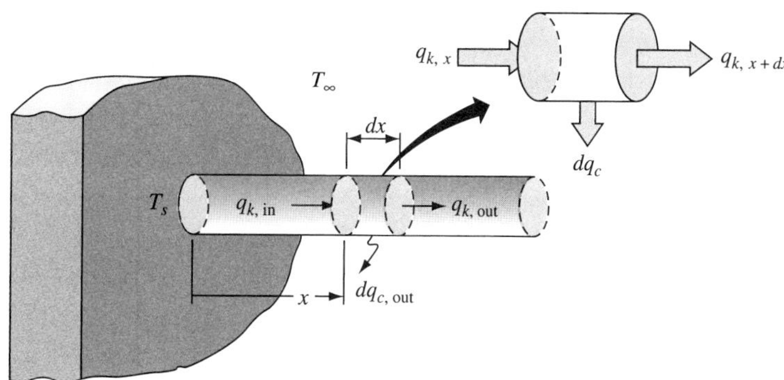

FIGURE 2.17 Schematic diagram of a pin fin protruding from a wall.

between the surface of the fin and the fluid is $\bar{h}_c$. We will assume that transverse temperature gradients are so small that the temperature at any cross section of the rod is uniform, that is, $T = T(x)$ only. As shown in Gardner [2], even in a relatively thick fin the error in a one-dimensional solution is less than 1%.

To derive an equation for temperature distribution, we make a heat balance for a small element of the fin. Heat flows by conduction into the left face of the element, while heat flows out of the element by conduction through the right face and by convection from the surface. Under steady-state conditions,

rate of heat flow by conduction into element at x	=	rate of heat flow by conduction out of element at $x + dx$	+	rate of heat flow by convection from surface between $x + dx$

In symbolic form, this equation becomes

$$q_{k,x} = q_{k,x+dx} + dq_c$$

or

$$-kA\left.\frac{dT}{dx}\right|_x = -kA\left.\frac{dT}{dx}\right|_{x+dx} + \bar{h}_c P\,dx[T(x) - T_\infty] \tag{2.61}$$

where P is the perimeter of the pin and $P\,dx$ is the pin surface area between x and $x + dx$.

If k and $\bar{h}_c$ are uniform, Eq. (2.61) simplifies to the form

$$\frac{d^2T(x)}{dx^2} - \frac{\bar{h}_c P}{kA}[T(x) - T_\infty] = 0 \tag{2.62}$$

It will be convenient to define an excess temperature of the fin above the environment, $\theta(x) = [T(x) - T_\infty]$, and transform Eq. (2.62) into the form

$$\frac{d^2\theta}{dx^2} - m^2\theta = 0 \tag{2.63}$$

where $m^2 = \bar{h}_c P/kA$

Equation (2.63) is a linear, homogeneous, second-order differential equation whose general solution is of the form

$$\theta(x) = C_1 e^{mx} + C_2 e^{-mx} \tag{2.64}$$

To evaluate the constants C_1 and C_2 it is necessary to specify appropriate boundary conditions. One condition is that at the base ($x = 0$) the fin temperature is equal to the wall temperature, or

$$\theta(0) = T_s - T_\infty \equiv \theta_s$$

The other boundary condition depends on the physical condition at the end of the fin. We will treat the following four cases:

1. The fin is very long and the temperature at the end approaches the fluid temperature:

$$\theta = 0 \quad \text{at} \quad x \to \infty$$

2. The end of the fin is insulated:

$$\frac{d\theta}{dx} = 0 \quad \text{at} \quad x = L$$

3. The temperature at the end of the fin is fixed:

$$\theta = \theta_L \quad \text{at} \quad x = L$$

4. The tip loses heat by convection:

$$-k\frac{d\theta}{dx}\bigg|_{x=L} = \bar{h}_{c,L}\theta_L$$

Figure 2.18 illustrates schematically the cases described by these conditions at the tip.

For case 1 the second boundary condition can be satisfied only if C_1 in Eq. (2.64) equals zero, that is,

$$\theta(x) = \theta_s e^{-mx} \tag{2.65}$$

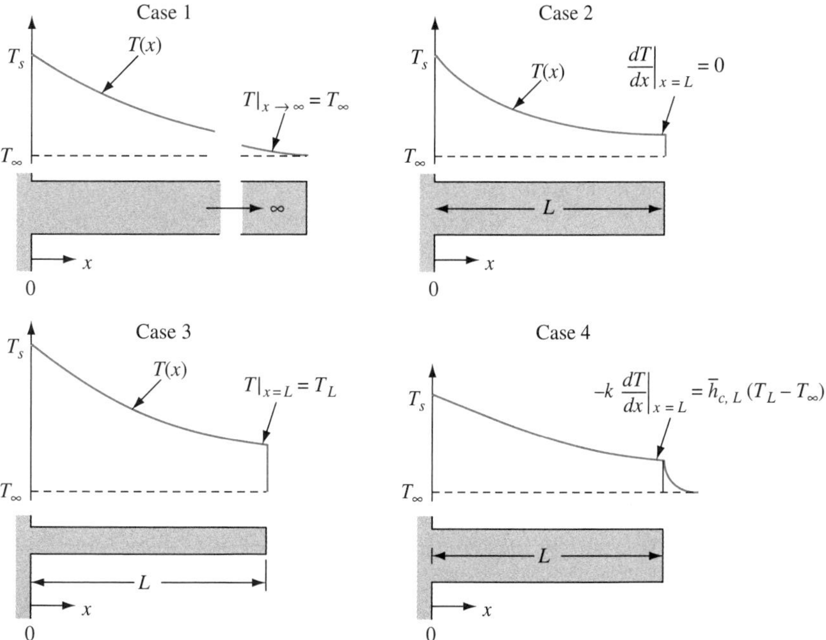

FIGURE 2.18 Schematic representation of four boundary conditions at the tip of a fin.

Usually we are interested not only in the temperature distribution but also in the total rate of heat transfer to or from the fin. The rate of heat flow can be obtained by two different methods. Since the heat conducted across the root of the fin must equal the heat transferred by convection from the surface of the rod to the fluid,

$$q_{\text{fin}} = -kA\frac{dT}{dx}\bigg|_{x=0} = \int_0^\infty \bar{h}_c P[T(x) - T_\infty]\, dx$$
$$= \int_0^\infty \bar{h}_c P\theta(x)\, dx \tag{2.66}$$

Differentiating Eq. (2.65) and substituting the result for $x = 0$ into Eq. (2.66) yields

$$q_{\text{fin}} = -kA[-m\theta(0)e^{(-m)0}] = \sqrt{\bar{h}_c PAk}\,\theta_s \tag{2.67}$$

The same result is obtained by evaluating the convection heat flow from the surface of the rod:

$$q_{\text{fin}} = \int_0^\infty \bar{h}_c P\theta_s e^{-mx}\, dx = \frac{\bar{h}_c P}{m}\theta_s e^{-mx}\bigg|_0^\infty = \sqrt{\bar{h}_c PAk}\,\theta_s$$

Equations (2.65) and (2.67) are reasonable approximations of the temperature distribution and heat flow rate in a finite fin if the square of its length is very large compared to its cross-sectional area. If the rod is of finite length but the heat loss from the end of the rod is neglected, or if the end of the rod is insulated, the second boundary condition requires that the temperature gradient at $x = L$ be zero, that is, $dT/dx = 0$ at $x = L$. These conditions require that

$$\left(\frac{d\theta}{dx}\right)_{x=L} = 0 = mC_1e^{mL} - mC_2e^{-mL}$$

Solving this equation for condition 2 simultaneously with the relation for condition 1, which required that

$$\theta(0) = \theta_s = C_1 + C_2$$

yields

$$C_1 = \frac{\theta_s}{1 + e^{2mL}} \quad C_2 = \frac{\theta_s}{1 + e^{-2mL}}$$

Substituting the above relations for C_1 and C_2 into Eq. (2.64) gives the temperature distribution

$$\theta = \theta_s\left(\frac{e^{mx}}{1 + e^{2mL}} + \frac{e^{-mx}}{1 + e^{-2mL}}\right) = \theta_s\frac{\cosh m(L - x)}{\cosh(mL)} \tag{2.68}^*$$

---

*The derivation of Eq. (2.68) is left as an exercise for the reader. The hyperbolic cosine, cosh for short, is defined by $\cosh x = (e^x + e^{-x})/2$.

TABLE 2.1 Equations for temperature distribution and rate of heat transfer for fins of uniform cross section[a]

Case	Tip Condition ($x = L$)	Temperature Distribution, θ/θ_s	Fin Heat Transfer Rate, q_{fin}
1	Infinite fin ($L \to \infty$): $\theta(L) = 0$	e^{-mx}	M
2	Adiabatic: $\left.\frac{d\theta}{dx}\right\rvert_{x=L} = 0$	$\frac{\cosh m(L-x)}{\cosh mL}$	$M \tanh mL$
3	Fixed temperature: $\theta(L) = \theta_L$	$\frac{(\theta_L/\theta_s)\sinh mx + \sinh m(L-x)}{\sinh mL}$	$M\frac{\cosh mL - (\theta_L/\theta_s)}{\sinh mL}$
4	Convection heat transfer: $\bar{h}_c\theta(L) = -k\left.\frac{d\theta}{dx}\right\rvert_{x=L}$	$\frac{\cosh m(L-x) + (\bar{h}_c/mk)\sinh m(L-x)}{\cosh mL + (\bar{h}_c/mk)\sinh mL}$	$M\frac{\sinh mL + (\bar{h}_c/mk)\cosh mL}{\cosh mL + (\bar{h}_c/mk)\sinh mL}$

[a] $\theta \equiv T - T_\infty$

$\theta_s \equiv \theta(0) = T_s - T_\infty$

$m^2 \equiv \frac{\bar{h}_c P}{kA}$

$M \equiv \sqrt{\bar{h}_c PkA}\ \theta_s$

The heat loss from the fin can be found by substituting the temperature gradient at the root into Eq. (2.66). Noting that $\tanh(mL) = (e^{mL} - e^{-mL})/(e^{mL} + e^{-mL})$, we get

$$q_{\text{fin}} = \sqrt{\bar{h}_c PAk}\,\theta_s \tanh(mL) \tag{2.69}$$

The results for the other two tip conditions can be obtained in a similar manner, but the algebra is more lengthy. For convenience, all four cases are summarized in Table 2.1.

EXAMPLE 2.6 An experimental device that produces excess heat is passively cooled. The addition of pin fins to the casing of this device is being considered to augment the rate of cooling. Consider a copper pin fin 0.25 cm in diameter that protrudes from a wall at 95°C into ambient air at 25°C as shown in Fig. 2.19. The heat transfer is

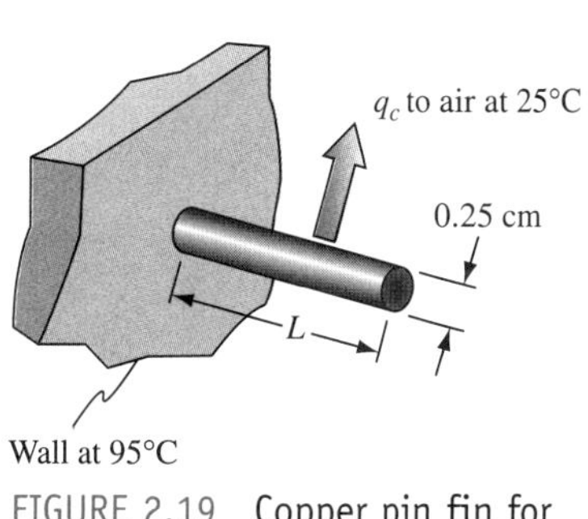

FIGURE 2.19 Copper pin fin for Example 2.6.

mainly by natural convection with a coefficient equal to 10 W/m^2 K. Calculate the heat loss, assuming that (a) the fin is "infinitely long" and (b) the fin is 2.5 cm long and the coefficient at the end is the same as around the circumference. Finally, (c) how long would the fin have to be for the infinitely long solution to be correct within 5%?

SOLUTION Make the following assumptions:

1. Thermal conductivity does not change with temperature.
2. Steady state prevails.
3. Radiation is negligible.
4. The convection heat transfer coefficient is uniform over the surface of the fin.
5. Conduction along the fin is one dimensional.

The thermal conductivity of the copper can be found in Table 12 of Appendix 2. We know that the fin temperature will decrease along its length, but we do not know its value at the tip. As an approximation, choose a temperature of 70°C or 343 K. Interpolating the values in Table 12 gives k = 396 W/m K.

(a) From Eq. (2.67) the heat loss for the "infintely long" fin is

$$\begin{aligned} q_{\text{fin}} &= \sqrt{\bar{h}_c PkA}\,(T_s - T_\infty) \\ &= \left[(10\ \text{W/m}^2\ \text{K})(\pi)(0.0025\ \text{m})(396\ \text{W/m K}) \times \left(\frac{\pi}{4}\right)(0.0025\ \text{m})^2\right]^{1/2} \\ &\quad (95 - 25)°\text{C} \\ &= 0.865\ \text{W} \end{aligned}$$

(b) The equation for the heat loss from the finite fin is case 4 in Table 2.1:

$$\begin{aligned} q_{\text{fin}} &= \sqrt{\bar{h}_c PkA}\,(T_s - T_\infty)\frac{\sinh mL + (\bar{h}_c/mk)\cosh mL}{\cosh mL + (\bar{h}_c/mk)\sinh mL} \\ &= 0.140\ \text{W} \end{aligned}$$

(c) For the two solutions to be within 5%, it is necessary that

$$\frac{\sinh mL + (\bar{h}_c/mk)\cosh mL}{\cosh mL + (\bar{h}_c/mk)\sinh mL} \geq 0.95$$

This condition is satisfied when $mL \geq 1.8$ or $L > 28.3$ cm.

2.4.2 Fin Selection and Design

In the preceding section, we developed equations for the temperature distribution and the rate of heat transfer for extended surfaces and fins. Fins are widely used to increase the rate of heat transfer from a wall. As an illustration of such an application, consider a surface exposed to a fluid at temperature T_∞ flowing over the surface. If the wall is bare and the surface temperature T_s is fixed, the rate of heat transfer per unit area from the plane wall is controlled entirely by the heat transfer coefficient $\bar{h}$. The coefficient at the plane wall may be increased by increasing the fluid velocity, but this also creates a larger pressure drop and requires increased pumping power.

In many cases it is thus preferable to increase the rate of heat transfer from the wall by using fins that extend from the wall into the fluid and increase the contact area between the solid surface and the fluid. If the fin is made of a material with high thermal conductivity, the temperature gradient along the fin from base to tip will be small and the heat transfer characteristics of the wall will be greatly enhanced. Fins come in many shapes and forms, some of which are shown in Fig. 2.20. The selection of fins is made on the basis of thermal performance and cost. The selection of a suitable fin geometry requires a compromise among the cost, the weight, the available space, and the pressure drop of the heat transfer fluid, as well as the heat transfer characteristics of the extended surface. From the point of view of thermal performance, the most desirable size, shape, and length of the fin can be evaluated by an analysis such as that outlined in the following discussion.

The heat transfer effectiveness of a fin is measured by a parameter called the fin efficiency η_f, which is defined as

$$\eta_f = \frac{\text{actual heat transferred by}}{\text{heat that would have been transferred if the entire fin were at the base temperature}}$$

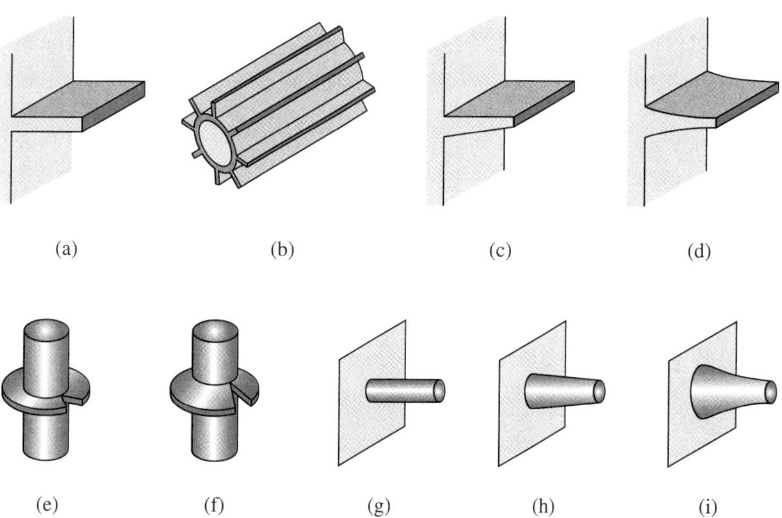

FIGURE 2.20 Schematic diagrams of different types of fins: (a) longitudinal fin of rectangular profile; (b) cylindrical tube with fins of rectangular profile; (c) longitudinal fin of trapezoidal profile; (d) longitudinal fin of parabolic profile; (e) cylindrical tube with radial fin of rectangular profile; (f) cylindrical tube with radial fin of truncated conical profile; (g) cylindrical pin fin; (h) truncated conical spine; (i) parabolic spine.

Using Eq. (2.69), the fin efficiency for a circular pin fin of diameter D and length L with an insulated end is

$$\eta_f = \frac{\tanh\sqrt{4L^2\bar{h}/kD}}{\sqrt{4L^2\bar{h}/kD}} \tag{2.70}$$

whereas for a fin of rectangular cross section (length L and thickness t) and an insulated end the efficiency is

$$\eta_f = \frac{\tanh\sqrt{\bar{h}PL^2/kA}}{\sqrt{\bar{h}PL^2/kA}} \tag{2.71}$$

If a rectangular fin is long, wide, and thin, $P/A \simeq 2/t$, and the heat loss from the end can be taken into account approximately by increasing L by $t/2$ and assuming that the end is insulated. This approximation keeps the surface area from which heat is lost the same as in the real case, and the fin efficiency then becomes

$$\eta_f = \frac{\tanh\sqrt{2\bar{h}L_c^2/kt}}{\sqrt{2\bar{h}L_c^2/kt}} \tag{2.72}$$

where $L_c = (L + t/2)$

The error that results from this approximation will be less than 8% when

$$\left(\frac{\bar{h}t}{2k}\right)^{1/2} \leq \frac{1}{2}$$

It is often convenient to use the profile area of a fin, A_m. For a rectangular shape A_m is Lt, whereas for a triangular cross section A_m is $Lt/2$, where t is the base thickness. In Fig. 2.21 the fin efficiencies for rectangular and triangular fins are compared.

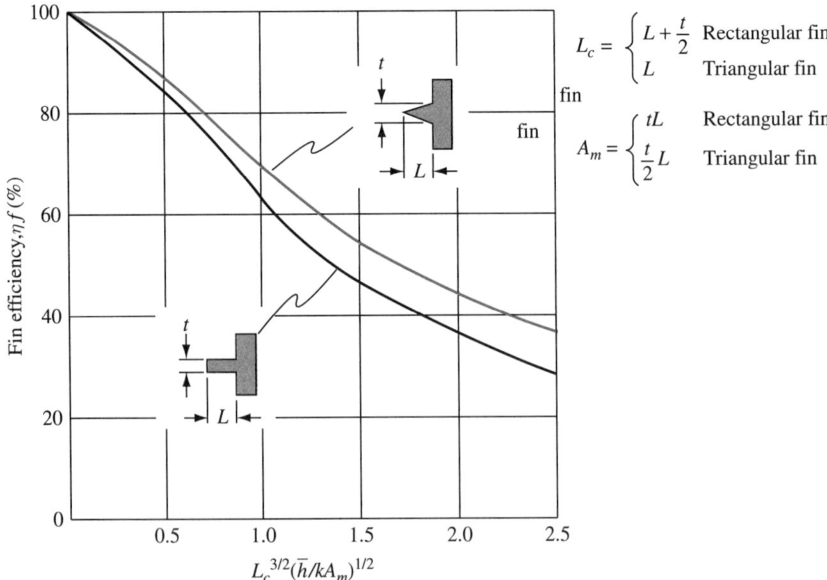

FIGURE 2.21 Efficiency of rectangular and triangular fins.

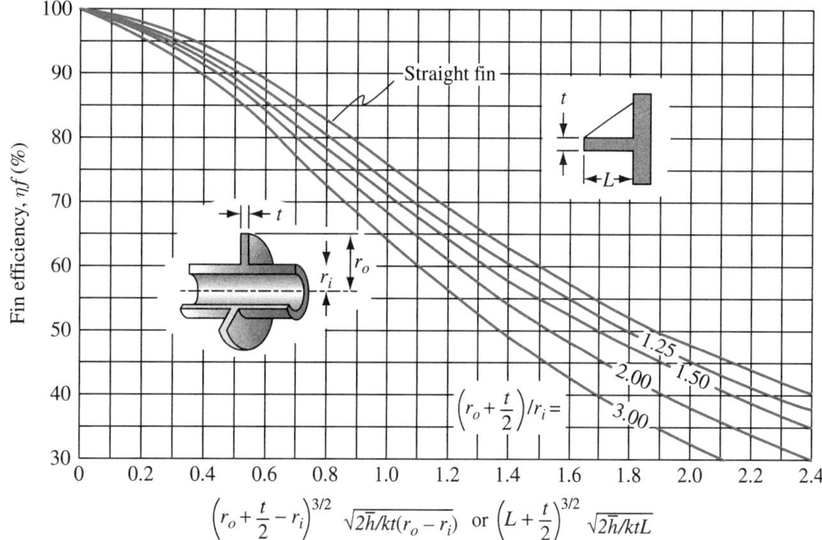

FIGURE 2.22 Efficiency of circumferential rectangular fins.

Figure 2.22 shows the fin efficiency for circumferential fins of rectangular cross section [2, 3].

EXAMPLE 2.7 To increase the heat dissipation from a 2.5-cm-OD tube, circumferential fins made of aluminum (k = 200 W/m K) are soldered to the outer surface. The fins are 0.1 cm thick and have an outer diameter of 5.5 cm as shown in Fig. 2.23. If the tube temperature is 100°C, the environmental temperature is 25°C, and the heat transfer coefficient between the fins and the environment is 65 W/m^2 K, calculate the rate of heat loss from a fin.

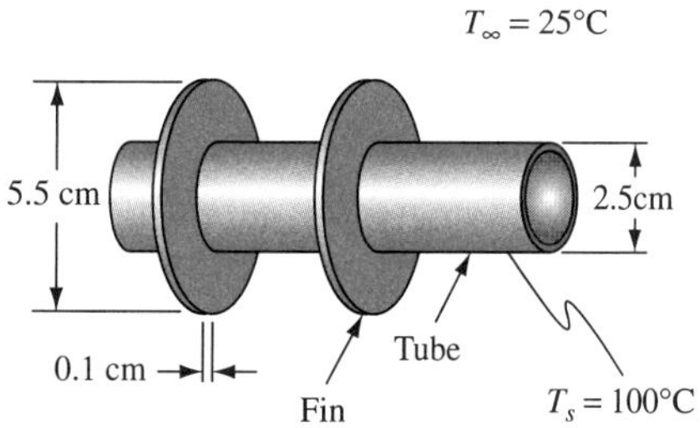

FIGURE 2.23 Circumferential fins for Example 2.7.

SOLUTION The geometry of the fin in this problem corresponds to that in Fig. 2.22, and we can therefore use the fin efficiency curve in Fig. 2.22. The parameters required to obtain the fin efficiency are

$$\left(r_o + \frac{t}{2} - r_i\right)^{3/2} = [(2.75 + 0.05 - 1.25)/100 \text{ m}]^{3/2} = 0.00193 \text{ m}^{3/2}$$

$$[2\bar{h}/kt(r_o - r_i)]^{1/2} = \left[\frac{2(65 \text{ W/m}^2 \text{ K})}{(200 \text{ W/m K})(0.001 \text{ m})(0.0275 - 0.0125)(\text{m})}\right]^{1/2}$$
$$= 208 \text{ m}^{3/2}$$

$$\left(r_o + \frac{t}{2}\right)\Big/ r_i = \frac{(0.0275 + 0.001)(\text{m})}{0.0125 \text{ m}} = 2.24$$

$$\left(r_o + \frac{t}{2} - r_i\right)^{3/2} [2\bar{h}/kt\,(r_o - r_i)]^{1/2} = 0.402$$

From Fig. 2.22 the fin efficiency is found to be 91%. The rate of heat loss from a single fin is

$$q_{\text{fin}} = \eta_{\text{fin}}\bar{h}A_{\text{fin}}(T_s - T_\infty)$$
$$= \eta_{\text{fin}}\bar{h}2\pi\left[\left(r_o + \frac{t}{2}\right)^2 - r_i^2\right](T_s - T_\infty)$$
$$= 0.91(65 \text{ W/m}^2 \text{ K})2\pi(7.84 - 1.56) \times 10^{-4} \text{ m}^2 (75 \text{ K}) = 17.5 \text{ W}$$

For a plane surface of area A, the thermal resistance is $1/\bar{h}A$. Addition of fins increases the surface area, but at the same time it introduces a conduction resistance over that portion of the original surface to which the fins are attached. Addition of fins will therefore not always increase the rate of heat transfer. In practice, addition of fins is hardly ever justified unless $\bar{h}A/Pk$ is considerably less than unity.

It is interesting to note that the fin efficiency reaches its maximum value for the trivial case of $L = 0$, or no fin at all. It is therefore not possible to maximize fin performance with respect to fin length. It is normally more important to maximize the efficiency with respect to the quantity of fin material (mass, volume, or cost), because such an optimization has obvious economic significance.

Using the values of the average heat transfer coefficients in Table 1.3 as a guide, we can easily see that fins effectively increase the heat transfer to or from a gas, are less effective when the medium is a liquid in forced convection, but offer no advantage in heat transfer to boiling liquids or from condensing vapors. For example, for a 0.3175-cm-diameter aluminum pin fin in a typical gas heater, $\bar{h}A/Pk = 0.00045$, whereas in a water heater, for example, $\bar{h}A/Pk = 0.022$. In a gas heater the addition of fins would therefore be much more effective than in a water heater.

It is apparent from these considerations that when fins are used they should be placed on the side of the heat exchange surface where the heat transfer coefficient between the fluid and the surface is lower. Thin, slender, closely spaced fins are superior to fewer and thicker fins from the heat transfer standpoint. Obviously, fins made of materials having a high thermal conductivity are desirable. Fins are sometimes an integral part of the heat transfer surface, but there can be a contact resistance at the base of the fin if the fins are mechanically attached.

To obtain the total efficiency, η_t, of a surface with fins, we combine the unfinned portion of the surface at 100% efficiency with the surface area of the fins at η_f, or

$$A_o\eta_t = (A_o - A_f) + A_f\eta_f \tag{2.73}$$

where A_o = total heat transfer area
A_f = heat transfer area of the fins

In practice, particularly in industrial heat exchangers [4], fins can often be used on either side of the primary heat transfer surface. Thus, for example, the overall heat transfer coefficient U_o, based on the total outer surface area, for heat transfer between two fluids separated by a tubular wall with fins can then be expressed as

$$U_o = \frac{1}{\dfrac{1}{\eta_{to}\bar{h}_o} + R_{k_{\text{wall}}} + \dfrac{A_o}{\eta_{ti}A_i\bar{h}_i}} \tag{2.74}$$

where $R_{k_{\text{wall}}}$ = thermal resistance of the wall to which the fins are attached, m^2 K/W (outside surface)
A_o = total outer surface area, m^2
A_i = total inner surface area, m^2
η_{to} = total efficiency for outer surface
η_{ti} = total efficiency for inner surface
$\bar{h}_o$ = average heat transfer coefficient for outer surface, W/m^2 K
$\bar{h}_i$ = average heat transfer coefficient for inner surface, W/m^2 K

For tubes with fins on the outside only, the more commonly encountered case in practice, η_{ti} is unity and $A_i = \pi D_i L$.

In the analysis presented in this chapter, details of the convection heat flow between the fin surface and the surrounding fluid have been omitted. A complete engineering analysis of heat transfer in heat exchanger systems not only requires an evaluation of the fin performance, but must also take the relation between the fin geometry and the convection heat transfer into account. Problems on the convection heat transfer part of the design will be considered in Chapters 6 and 7, and the application of such analyses in the design of heat exchangers is presented in Chapter 8.

2.5* Multidimensional Steady Conduction

In the preceding part of this chapter we dealt with problems in which the temperature and the heat flow can be treated as functions of a single variable. Many practical problems fall into this category, but when the boundaries of a system are irregular or when the temperature along a boundary is nonuniform, a one-dimensional treatment may no longer be satisfactory. In such cases, the temperature is a function of two or possibly even three coordinates. The heat flow through a corner section where two or three walls meet, the heat conduction through the walls of a short, hollow cylinder, and the heat loss from a buried pipe are typical examples of this class of problem.

We shall now consider some methods for analyzing conduction in two- and three-dimensional systems. The emphasis will be placed on two-dimensional problems because they are less cumbersome to solve, yet they illustrate the basic methods of analysis for three-dimensional systems. Heat conduction in two- and three-dimensional systems can be treated by analytic, graphic, analogic, numerical and computational methods. For some cases, "shape factors" are also available. We will consider in this chapter the analytic, graphic, and shape-factor methods of solution. The numerical approach that requires computational simulation will be taken up in Chapter 3. The analytic treatment in this chapter is limited to an illustrative example, and for more extensive coverage of analytic methods the reader is referred to [1, 4–6]. The analogic method is presented in [7] but is omitted here because it is no longer used in practice.

2.5.1 Analytic Solution

The objective of any heat transfer analysis is to predict the rate of heat flow, the temperature distribution, or both. According to Eq. (2.10), in a two-dimensional system without heat sources the general conduction equation governing the temperature distribution in the steady state is

$$\frac{\partial^2 T}{\partial x^2} + \frac{\partial^2 T}{\partial y^2} = 0 \tag{2.75}$$

if the thermal conductivity is uniform. The solution of Eq. (2.75) will give $T(x, y)$, the temperature as a function of the two space coordinates x and y. The components of the heat flow per unit area or heat flux q'' in the x and y directions (q_x'' and q_y'', respectively) can be obtained from Fourier's law:

$$q_x'' = \left(\frac{q}{A}\right)_x = -k\frac{\partial T}{\partial x}$$

$$q_y'' = \left(\frac{q}{A}\right)_y = -k\frac{\partial T}{\partial y}$$

It should be noted that while the temperature is a scalar, the heat flux depends on the temperature gradient and is therefore a vector. The heat flux q'' at a given point x, y is the resultant of the components q_x'' and q_y'' at that point and is directed perpendicular to the isotherm, as shown in Fig. 2.24. Thus, if the temperature distribution in a system is known, the rate of heat flow can easily be calculated. Therefore, heat transfer analyses usually concentrate on determining the temperature field.

An analytic solution of a heat conduction problem must satisfy the heat conduction equation as well as the boundary conditions specified by the physical conditions of the particular problem. The classical approach to an exact solution of the Fourier equation is the separation-of-variables technique. We shall illustrate this approach by applying it to a relatively simple problem. Consider a thin rectangular plate, free of heat sources and insulated at the top and bottom surfaces (Fig. 2.25). Since $\partial T/\partial z$

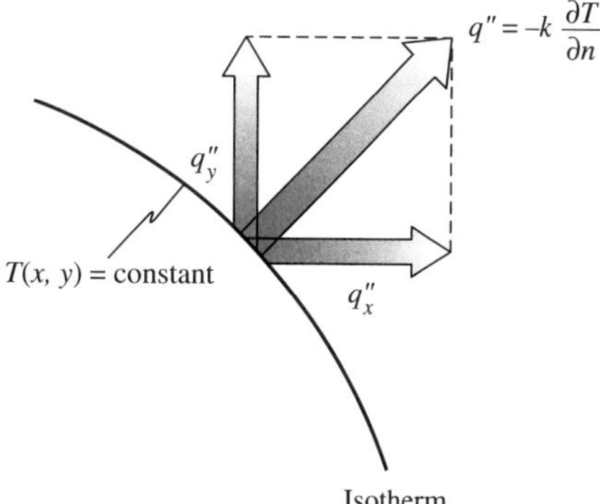

FIGURE 2.24 Sketch showing heat flow in two dimensions.

is assumed to be negligible, the temperature is a function of x and y only. If the thermal conductivity is uniform, the temperature distribution must satisfy Eq. (2.75), a linear and homogeneous partial differential equation that can be integrated by assuming a product solution for $T(x, y)$ of the form

$$T = XY \tag{2.76}$$

where $X = X(x)$, a function of x only, and $Y = Y(y)$, a function of y alone. Substituting Eq. (2.76) into Eq. (2.75) yields

$$-\frac{1}{X}\frac{d^2X}{dx^2} = \frac{1}{Y}\frac{d^2Y}{dy^2} \tag{2.77}$$

The variables are now separated. The left-hand side is a function of x only, while the right-hand side is a function of y alone. Since neither side can change as x and y vary, both must be equal to a constant, say λ^2. We have, therefore, the two ordinary differential equations

$$\frac{d^2X}{dx^2} + \lambda^2 X = 0 \tag{2.78}$$

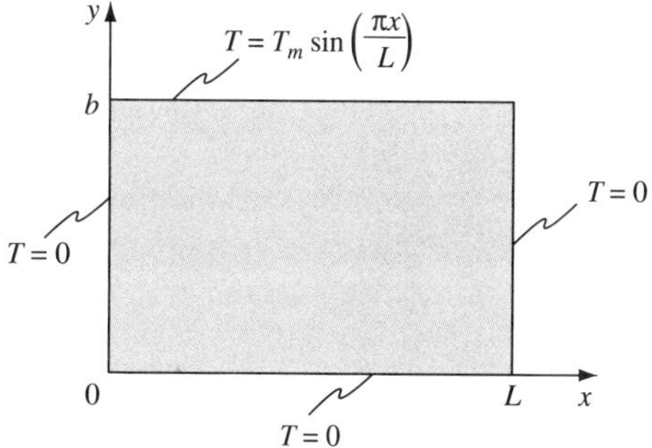

FIGURE 2.25 Rectangular adiabatic plate with sinusoidal temperature distribution on one edge.

and

$$\frac{d^2Y}{dy^2} - \lambda^2 Y = 0 \tag{2.79}$$

The general solution to Eq. (2.78) is

$$X = A \cos \lambda x + B \sin \lambda x$$

and the general solution to Eq. (2.79) is

$$Y = Ce^{-\lambda y} + De^{\lambda y}$$

and therefore, from Eq. (2.76),

$$T = XY = (A \cos \lambda x + B \sin \lambda x)(Ce^{-\lambda y} + De^{\lambda y}) \tag{2.80}$$

where A, B, C, and D are constants to be evaluated from the boundary conditions. As shown in Figure 2.25, the boundary conditions to be satisfied are

$$\begin{aligned} T &= 0 \quad \text{at} \quad y = 0 \\ T &= 0 \quad \text{at} \quad x = 0 \\ T &= 0 \quad \text{at} \quad x = L \\ T &= T_m \sin(\pi x/L) \quad \text{at} \quad y = b \end{aligned}$$

Substituting these conditions into Eq. (2.80) for T, we get from the first condition

$$(A \cos \lambda x + B \sin \lambda x)(C + D) = 0$$

from the second condition

$$A(Ce^{-\lambda y} + De^{\lambda y}) = 0$$

and from the third condition

$$(A \cos \lambda L + B \sin \lambda L)(Ce^{-\lambda y} + De^{\lambda y}) = 0$$

The first condition can be satisfied only if $C = -D$, and the second if $A = 0$. Using these results in the third condition, we obtain

$$2BC \sin \lambda L \sinh \lambda y = 0$$

To satisfy this condition, $\sin \lambda L$ must be zero or $\lambda = n\pi/L$, where $n = 1, 2, 3$, etc.* There exists therefore a different solution for each integer n, and each solution has a separate integration constant C_n. Summing these solutions, we get

$$T = \sum_{n=1}^{\infty} C_n \sin\frac{n\pi x}{L} \sinh\frac{n\pi y}{L} \tag{2.81}$$

* The value $n = 0$ is excluded because it would give the trivial solution $T = 0$.

The last boundary condition demands that, at $y = b$,

$$\sum_{n=1}^{\infty} C_n \sin\frac{n\pi x}{L}\sinh\frac{n\pi b}{L} = T_m \sin\frac{\pi x}{L}$$

so that only the first term in the series solution with $C_1 = T_m/\sinh(\pi b/L)$ is needed. The solution therefore becomes

$$T(x, y) = T_m \frac{\sinh(\pi y/L)}{\sinh(\pi b/L)} \sin\frac{\pi x}{L} \tag{2.82}$$

The corresponding temperature field is shown in Fig. 2.26. The solid lines are isotherms, and the dashed lines are heat flow lines. It should be noted that lines indicating the direction of heat flow are perpendicular to the isotherms.

When the boundary conditions are not as simple as in the illustrative problem, the solution is obtained in the form of an infinite series. For example, if the temperature at the edge $y = b$ is a function of x, say $T(x, b) = F(x)$, then the solution, as shown in [1], is the infinite series

$$T = \frac{2}{L}\sum_{n=1}^{\infty} \frac{\sinh(n\pi/L)y}{\sinh n\pi(b/L)} \sin\frac{\pi n}{L}x \int_0^L F(x')\sin\left(\frac{n\pi}{L}x'\right)dx' \tag{2.83}$$

which is quite laborious to evaluate quantitatively.

The separation-of-variables method can be extended to three-dimensional cases by assuming $T = XYZ$, substituting this expression for T in Eq. (2.9), separating the variables, and integrating the resulting total differential equations subject to the given boundary conditions. Examples of three-dimensional problems are presented in [1, 5, 6, and 17].

2.5.2 Graphic Method and Shape Factors

The graphic method presented in this section can rapidly yield a reasonably good estimate of the temperature distribution and heat flow in geometrically

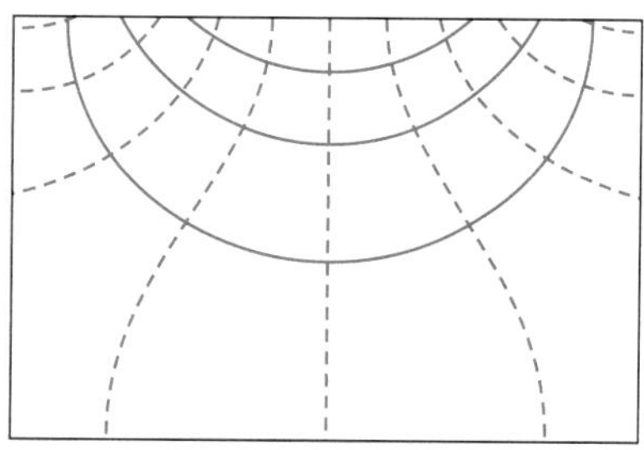

——— Isotherms

- - - - - Heat flow lines

FIGURE 2.26 Isotherms and heat flow lines for the plate shown in Fig. 2.25.

complex two-dimensional systems, but its application is limited to problems with isothermal and insulated boundaries. The object of a graphic solution is to construct a network consisting of isotherms (lines of constant temperature) and constant-flux lines (lines of constant heat flow). The flux lines are analogous to streamlines in a potential fluid flow, that is, they are tangent to the direction of heat flow at any point. Consequently, no heat can flow across the constant-flux lines. The isotherms are analogous to constant-potential lines, and heat flows perpendicular to them. Thus, lines of constant temperature and lines of constant heat flux intersect at right angles. To obtain the temperature distribution one first prepares a scale model and then draws isotherms and flux lines freehand, by trial and error, until they form a network of curvilinear squares. Then a constant amount of heat flows between any two flux lines. The procedure is illustrated in Fig. 2.27 for a corner section of unit depth ($\Delta z = 1$) with faces ABC at temperature T_1, faces FED at temperature T_2, and faces CD and AF insulated. Figure 2.27(a) shows the scale model, and Fig. 2.27(b) shows the curvilinear network of isotherms and flux lines. It should be noted that the flux lines emanating from isothermal boundaries are perpendicular to the boundary, except when they come from a corner. Flux lines leading to or from a corner of an isothermal boundary bisect the angle between the surfaces forming the corner.

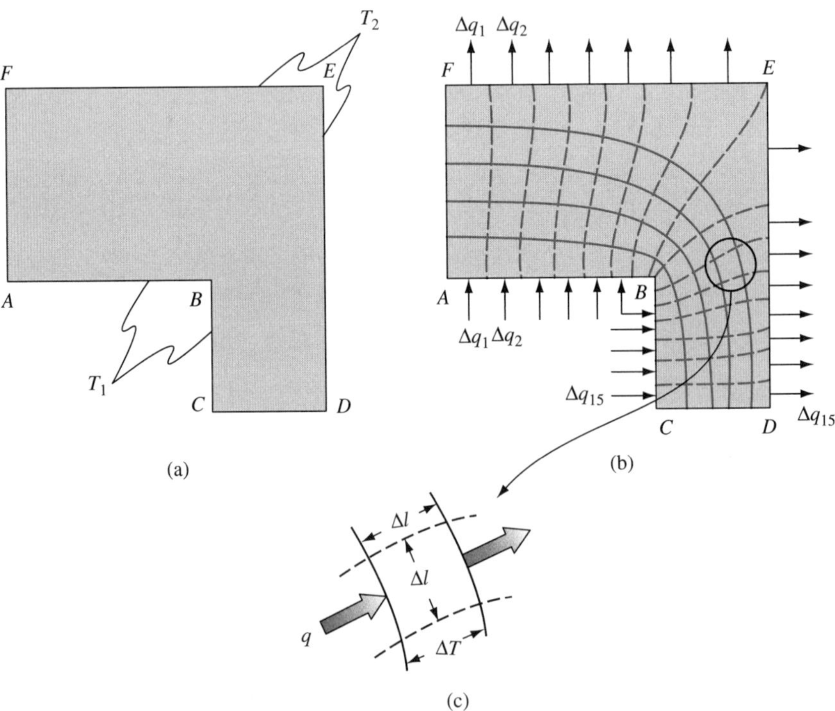

FIGURE 2.27 Construction of a network of curvilinear squares for a corner section: (a) scale model; (b) flux plot; (c) typical curvilinear square.

A graphic solution, like an analytic solution of a heat conduction problem described by the Laplace equation and the associated boundary condition, is unique. Therefore, any curvilinear network, irrespective of the size of the squares, that satisfies the boundary conditions represents the correct solution. For any curvilinear square [for example, see Fig. 2.27(c)] the rate of heat flow is given by Fourier's law:

$$\Delta q = -k(\Delta l \times 1)\frac{\Delta T}{\Delta l} = -k\Delta T$$

This heat flow will remain the same across any square within any one heat flow lane from the boundary at T_1 to the boundary at T_2. The ΔT across any one element in the heat flow lane is therefore

$$\Delta T = \frac{T_2 - T_1}{N}$$

where N is the number of temperature increments between the two boundaries at T_1 and T_2. The total rate of heat flow from the boundary at T_2 to the boundary at T_1 equals the sum of the heat flow through all the lanes. According to the above relations, the heat flow rate is the same through all lanes since it is independent of the size of the squares in a network of curvilinear squares. The total rate of heat transfer can therefore be written

$$q = \sum_{n=1}^{n=M} \Delta q_n = \frac{M}{N} k(T_2 - T_1) = \frac{M}{N} k\, \Delta T_{\text{overall}} \tag{2.84}$$

where Δq_n is the rate of heat flow through the nth lane, and M is the number of heat flow lanes.

Thus, to calculate the rate of heat transfer we need only construct a network of curvilinear squares in the scale model and count the number of temperature increments and heat flow lanes. Although the accuracy of the method depends a good deal on the skill and patience of the person sketching the curvilinear square network, even a crude sketch can give a reasonably good estimate of the temperature distribution; if desired, this type of estimate can be refined by the numerical method described in the next chapter.

In any two-dimensional system in which heat is transferred from one surface at T_1 to another at T_2, the rate of heat transfer per unit depth depends only on the temperature difference $T_1 - T_2 = \Delta T_{\text{overall}}$, the thermal conductivity k, and the ratio M/N. This ratio depends on the shape of the system and is called the *shape factor, S*. The rate of heat transfer can thus be written

$$q = kS\, \Delta T_{\text{overall}} \tag{2.85}$$

when the grid consists of curvilinear squares. Values of S for several shapes of practical significance [7–10] are summarized in Table 2.2.

TABLE 2.2 Conduction shape factor S for various systems $[q_k = Sk(T_1 - T_2)]$

Description of System	Symbolic Sketch	Shape Factor S
Conduction through a homogeneous medium of thermal conductivity k between an isothermal surface and a sphere buried a distance z below the surface	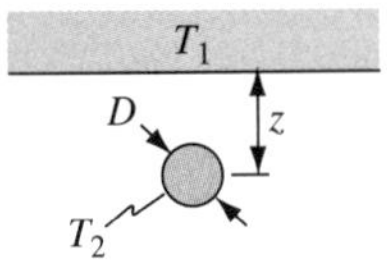	$\dfrac{2\pi D}{1 - D/4z}$
Conduction through a homogeneous medium of thermal conductivity k between an isothermal surface and a horizontal cylinder of length L buried with its axis a distance z below the surface	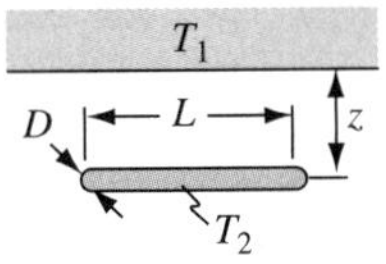	$\dfrac{2\pi L}{\cosh^{-1}(2z/D)}$ if $z/L \ll 1$ and $D/L \ll 1$
Conduction through a homogeneous medium of thermal conductivity k between an isothermal surface and an infinitely long cylinder buried a distance z below (per unit length of cylinder)	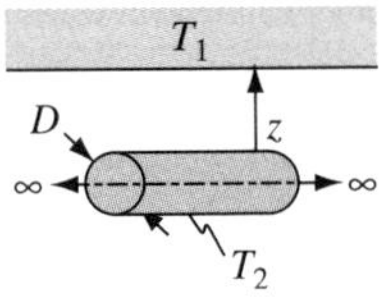	$\dfrac{2\pi}{\cosh^{-1}(2z/D)}$
Conduction through a homogeneous medium of thermal conductivity k between an isothermal surface and a vertical cylinder of length L	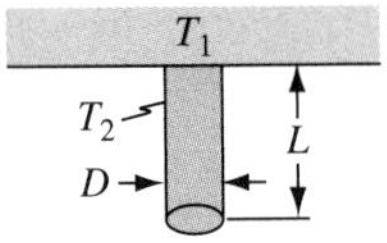	$\dfrac{2\pi L}{\mathrm{In}(4L/D)}$ if $D/L \ll 1$
Horizontal thin circular disk buried far below an isothermal surface in a homogeneous material of thermal conductivity k	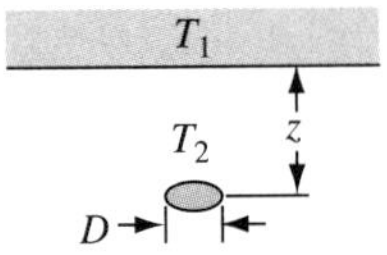	$\dfrac{4.45D}{1 - D/5.67z}$
Conduction through a homogeneous material of thermal conductivity k between two long parallel cylinders a distance l apart (per unit length of cylinders)	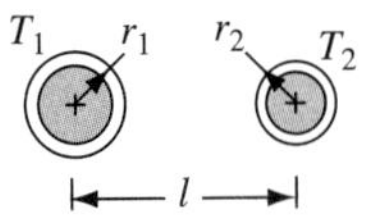	$\dfrac{2\pi}{\cosh^{-1}\left(\dfrac{L^2 - 1 - r^2}{2r}\right)}$ ($r = r_1/r_2$ and $L = l/r_2$)
Conduction through two plane sections and the edge[a] section of two walls of thermal conductivity k, with inner- and outer-surface temperatures uniform	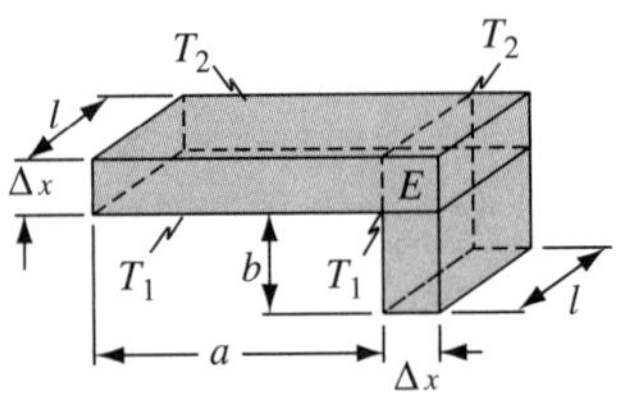	$\dfrac{al}{\Delta x} + \dfrac{bl}{\Delta x} + 0.54l$

(Continued)

TABLE 2.2 (Continued)

Description of System	Symbolic Sketch	Shape Factor S
Conduction through the corner section C of three[a] homogeneous walls of thermal conductivity k, inner- and outer-surface temperatures uniform		$(0.15\ \Delta x)$ if Δx is small compared to the lengths of walls

[a]Sketch illustrating dimensions for use in calculating three-dimensional shape factors:

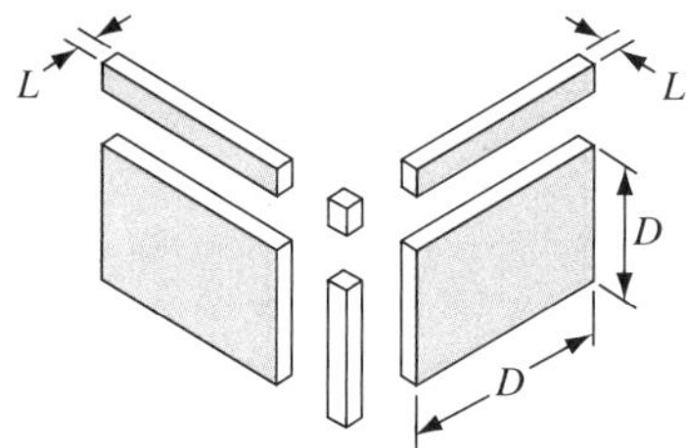

EXAMPLE 2.8 A long, 10-cm-OD pipe is buried with its centerline 60 cm below the surface in soil having a thermal conductivity of 0.4 W/m K, as shown in Fig. 2.28. (a) Prepare a curvilinear square network for this system and calculate the heat loss per meter length if the pipe surface temperature is 100°C and the soil surface is at 20°C. (b) Compare the result from part (a) with that obtained using the appropriate shape factor S.

SOLUTION (a) The curvilinear square network for the system is shown in Fig. 2.29 on the next page. Because of symmetry, only half of this heat flow field needs to be plotted.

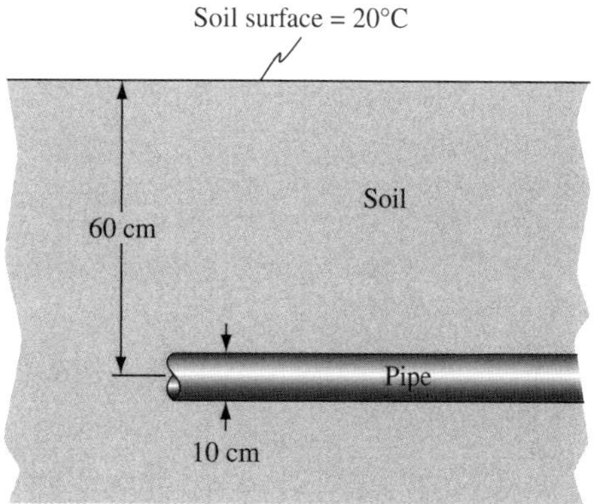

FIGURE 2.28 Heat loss from a buried pipe, Example 2.8.

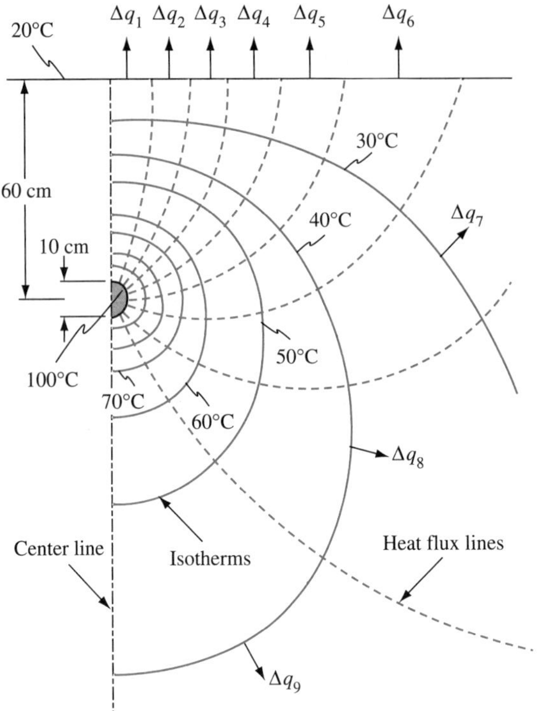

FIGURE 2.29 Potential field for a buried pipe for Example 2.8.

There are 18 heat flow lanes leading from the pipe to the surface, and each lane consists of 8 curvilinear squares. The shape factor is therefore

$$S = \frac{18}{8} = 2.25$$

and the rate of heat flow per meter is, from Eq. (2.85),

$$q = (0.4 \text{ W/m K})(2.25)(100 - 20)(\text{K}) = 72 \text{ W/m}$$

(b) From Table 2.2

$$S = \frac{2\pi(1)}{\cosh^{-1}(120/10)} = \frac{2\pi}{3.18} = 1.98$$

and the rate of heat loss per meter length is

$$q = (0.4)(1.98)(100 - 20) = 63.4 \text{ W/m}$$

The reason for the difference in the calculated heat loss is that the potential field in Fig. 2.29 has as finite number of flux lines and isotherms and is therefore only approximate.

For a three-dimensional wall, as in a furnace, separate shape factors are used to calculate the heat flow through the edge and corner sections. When all the interior dimensions are greater than one-fifth of the wall thickness,

$$S_{\text{wall}} = \frac{A}{L} \quad S_{\text{edge}} = 0.54D \quad S_{\text{corner}} = 0.15L$$

where A = inside area of wall
L = wall thickness
D = length of edge

These dimensions are illustrated in Table 2.2. Note that the shape factor per unit depth is given by the ratio M/N when the curvilinear-squares method is used for calculations.

EXAMPLE 2.9

A small cubic furnace 50 cm × 50 cm on the inside is constructed of fireclay brick ($k = 1.04$ W/m °C) with a wall thickness of 10 cm as shown in Fig. 2.30. The inside of the furnace is maintained at 500°C and the outside at 50°C. Calculate the heat lost through the walls.

SOLUTION

We compute the total shape factor by adding the shape factors for the walls, edges, and corners.

Walls:

$$S = \frac{A}{L} = \frac{(0.5)(0.5)}{0.1} = 2.5 \text{ m}$$

Edges:

$$S = 0.54D = (0.54)(0.5) = 0.27 \text{ m}$$

Corners:

$$S = 0.15L = (0.15)(0.1) = 0.015 \text{ m}$$

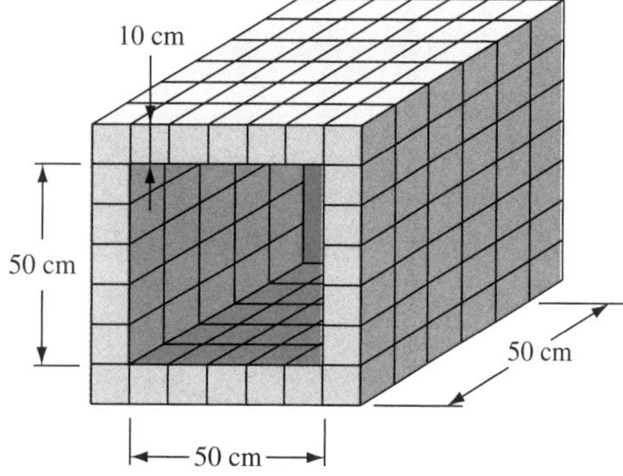

FIGURE 2.30 Cubic furnace for Example 2.9.

(a)

(b)

FIGURE 2.31 Typical kilns and furnaces: (a) a set of brick kilns and (b) heat treating industrial furnaces.

Sources: (a) Courtesy of ©iStockphoto.com/TimMcClean. (b) Courtesy of Kusuma Baja.

There are 6 wall sections, 12 edges, and 8 corners, so that the total shape factor is

$$S = (6)(2.5) + (12)(0.27) + (8)(0.015) = 18.36 \text{ m}$$

and the heat flow is calculated as

$$q = ks\,\Delta T = (1.04 \text{ W/m K})(18.36 \text{ m})(500 - 50)(\text{K}) = 8.59 \text{ kW}$$

2.6 Unsteady or Transient Heat Conduction

So far we have only dealt with steady-state conduction in this chapter, but some time must elapse after the heat transfer process is initiated before steady-state conditions are reached. During this transient period the temperature changes, and the analysis must take into account changes in the internal energy. Example 1.14 in Chapter 1 illustrates this phenomenon for a simple case. In the remainder of this chapter we will deal with methods for analyzing more complex unsteady heat flow problems, because transient heat flow is of great practical importance in industrial heating and cooling.

In addition to unsteady heat flow when the system undergoes a transition from one steady state to another, there are also engineering problems involving periodic variations in heat flow and temperature. Examples of such cases are the periodic heat flow in a building between day and night and the heat flow in an internal combustion engine.

We shall first analyze problems that can be simplified by assuming that the temperature is only a function of time and is uniform throughout the system at any instant. This type of analysis is called the lumped-heat-capacity method. In subsequent sections of this chapter we shall consider methods for solving problems of unsteady heat flow when the temperature not only depends on time but also varies in the interior of the system. Throughout this chapter we shall not be concerned with the mechanisms of heat transfer by convection or radiation. Where these modes of heat transfer affect the boundary conditions of the system, an appropriate value for the heat transfer coefficient will simply be specified.

2.6.1 Systems with Negligible Internal Resistance

Even though no materials in nature have an infinite thermal conductivity, many transient heat flow problems can be readily solved with acceptable accuracy by assuming that the internal conductive resistance of the system is so small that the temperature within the system is substantially uniform at any instant. This simplification is justified when the external thermal resistance between the surface of the system and the surrounding medium is so large compared to the internal thermal resistance of the system that it controls the heat transfer process.

A measure of the relative importance of the thermal resistance within a solid body is the Biot number Bi, which is the ratio of the internal to the external resistance and can be defined by the equation

$$\text{Bi} = \frac{R_{\text{internal}}}{R_{\text{external}}} = \frac{\bar{h}L}{k_s} \tag{2.86}$$

where $\bar{h}$ is the average heat transfer coefficient, L is a significant length dimension obtained by dividing the volume of the body by its surface area, and k_s is the thermal conductivity of the solid body. In bodies whose shape resembles a plate, a cylinder, or a sphere, the error introduced by the assumption that the temperature at any instant is uniform will be less than 5% when the internal resistance is less than 10% of the external surface resistance, that is, when $\bar{h}L/k_s < 0.1$. A transient heat conducting system in which $\text{Bi} < 0.1$ is often referred to as a *lumped capacitance*, and, as shown subsequently, this reflects the fact that its internal resistance is very small or negligible.

As a typical example of this type of transient heat flow, consider the cooling of a small metal casting or a billet in a quenching bath after its removal from a hot furnace. Suppose that the billet is removed from the furnace at a uniform temperature T_0 and is quenched so suddenly that we can approximate the environmental temperature change by a step. Designate the time at which the cooling begins as $t = 0$, and assume that the heat transfer coefficient $\bar{h}$ remains constant during the process and that the bath temperature T_∞ at a distance far removed from the billet does not vary with time. Then, in accordance with the assumption that the temperature within the body is substantially uniform at any instant, an energy balance for the billet over a small time interval dt is

$$\begin{matrix}\text{change in internal energy} \\ \text{of the billet during } dt\end{matrix} = \begin{matrix}\text{net heat flow from the} \\ \text{billet to the bath during } dt\end{matrix}$$

or

$$-c\rho V\,dT = \bar{h}A_s(T - T_\infty)dt \tag{2.87}$$

where c = specific heat of billet, J/kg K
ρ = density of billet, kg/m^3
V = volume of billet, m^3
T = average temperature of billet, K
$\bar{h}$ = average heat transfer coefficient, W/m^2 K
A_s = surface area of billet, m^2
dT = temperature change (K) during time interval dt (s)

The minus sign in Eq. (2.87) indicates that the internal energy decreases when $T > T_\infty$. The variables T and t can be readily separated, and for a differential time interval dt, Eq. (2.87) becomes

$$\frac{dT}{T - T_\infty} = \frac{d(T - T_\infty)}{(T - T_\infty)} = -\frac{\bar{h}A_s}{c\rho V}dt \tag{2.88}$$

where it is noted that $d(T - T_\infty) = dT$, since T_∞ is constant. With an initial temperature of T_0 and a temperature at time t of T as limits, integration of Eq. (2.88) yields

$$\mathrm{In}\frac{T - T_\infty}{T_0 - T_\infty} = -\frac{\bar{h}A_s}{c\rho V}t$$

or

$$\frac{T - T_\infty}{T_0 - T_\infty} = e^{-(\bar{h}\pi A_s/c\rho V)t} \tag{2.89}$$

where the exponent $\bar{h}A_s t/c\rho V$ must be dimensionless. The combination of variables in this exponent can in fact be expressed as the product of two dimensionless groups we encountered previously, as follows:

$$\frac{\bar{h}A_s t}{c\rho V} = \left(\frac{\bar{h}L}{k_s}\right)\left(\frac{\alpha t}{L^2}\right) = \text{Bi Fo} \tag{2.90}$$

where the characteristic length L is the volume of the body V divided by its surface area A_s.

An electrical network analogous to the thermal network for a lumped-single-capacity system is shown in Fig. 2.32. In this network the capacitor is initially "charged" to the potential T_0 by closing the switch S. When the switch is opened, the energy stored in the capacitance is discharged through the resistance $1/\bar{h}A_s$. The analogy between this thermal system and an electrical system is apparent. The thermal resistance is $R = 1/\bar{h}A_s$, and the thermal capacitance is $C = \rho Vc$, while R_e and C_e are the electrical resistance and capacitance, respectively. To construct an electrical system that would behave exactly like the thermal system we would only have to make the ratio $\bar{h}A_s/c\rho V$ equal $1/R_eC_e$. In the thermal system internal energy is stored, while in the electrical system electric charge is stored. The flow of energy in the thermal system is heat, and the flow of charge is electric current. The

Thermal Circuit

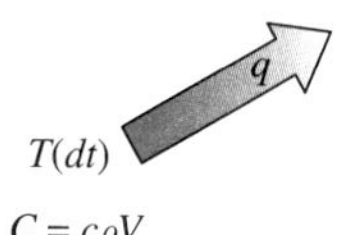

$T(dt)$

$C = c\rho V$

$R = \dfrac{1}{\bar{h}A_s}$

$$q = \frac{T - T_\infty}{R} = -C\frac{dT}{dt}$$

$$\frac{T - T_\infty}{T_0 - T_\infty} = e^{-(1/CR)t}$$

$t = 0$ when billet is immersed in fluid and heat begins to flow.

(a)

Electrical System

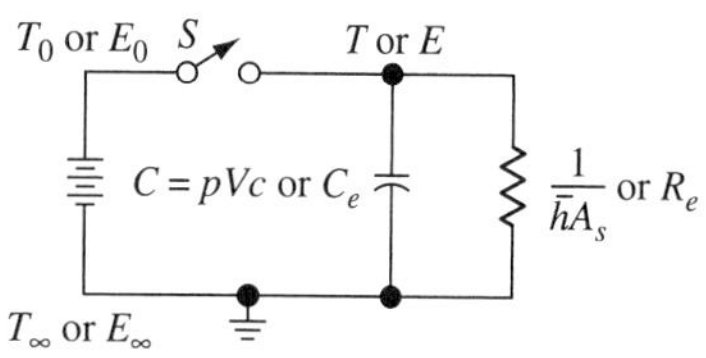

T_0 or E_0 S T or E

$C = \rho Vc$ or C_e

$\dfrac{1}{\bar{h}A_s}$ or R_e

T_∞ or E_∞

$$i = \frac{E - E_\infty}{R_e} = -C_e\frac{dE}{dt}$$

$$\frac{E - E_\infty}{E_0 - E_\infty} = e^{-(1/C_eR_e)t}$$

$t = 0$ when switch S is opened and the condenser begins to discharge.

(b)

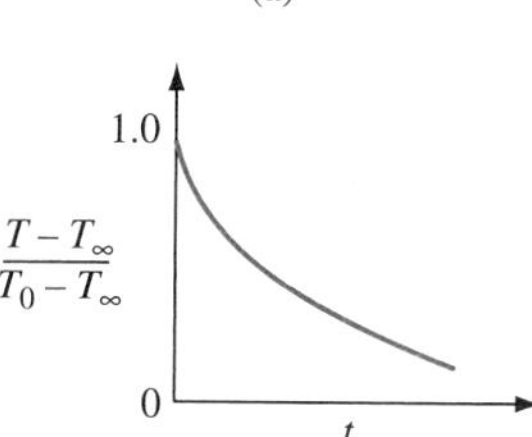

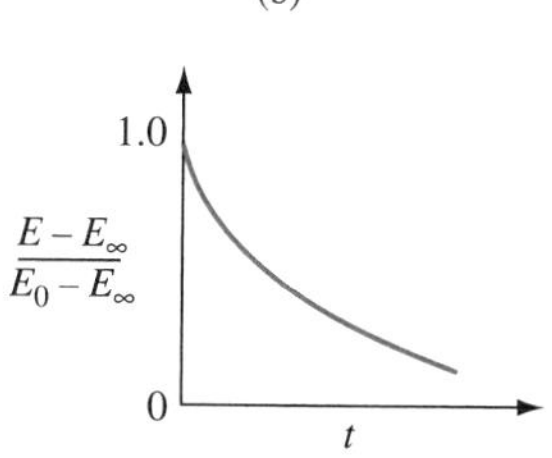

Rate of heat flow q (J/s or W)	Current flow i (amps)
Thermal capacity $C = \rho Vc$ (J/K)	Electrical capacity C_e (farads)
Thermal resistance $R = 1/\bar{h}A_s$ (K/W)	Electrical resistance R_e (ohms)
Thermal potential $(T - T_\infty)$ (K)	Electrical potential $(E - E_\infty)$ (volts)

FIGURE 2.32 Network and schematic of transient lumped-capacity system.

quantity $c\rho V/\bar{h}A$ is called the *time constant* of the system, since it has the dimensions of time. Its value is indicative of the rate of response of a single-capacity system to a sudden change in the environmental temperature. Observe that when the time $t = c\rho V/\bar{h}A_s$ the temperature difference $T - T_\infty$ is equal to 36.8% of the initial difference $T_0 - T_\infty$.

EXAMPLE 2.10 When a thermocouple is moved from one medium to another medium at a different temperature, the thermocouple must be given sufficient time to come to thermal equilibrium with the new conditions before a reading is taken. Consider a 0.10-cm-diameter copper thermocouple wire originally at 150°C. Determine the temperature response when this wire is suddenly immersed in (a) water at 40°C ($\bar{h} = 80$ W/m^2 K) and (b) air at 40°C ($\bar{h} = 10$ W/m^2 K).

SOLUTION From Table 12, Appendix 2, we get

$$k_s = 391 \text{ W/m K}$$
$$c = 383 \text{ J/kg K}$$
$$\rho = 8930 \text{ kg/m}^3$$

The surface area A_s and the volume of the wire per unit length are

$$A_s = \pi D = (\pi)(0.001 \text{ m}) = 3.14 \times 10^{-3} \text{ m}$$
$$V = \frac{\pi D^2}{4} = (\pi)(0.001^2 \text{ m}^2)/4 = 7.85 \times 10^{-7} \text{ m}^2$$

The Biot number in water is

$$\text{Bi} = \frac{\bar{h}D}{4k_s} = \frac{(80 \text{ W/m}^2 \text{ K})(0.001 \text{ m})}{(4)(391 \text{ W/m K})} \ll 1$$

Since the Biot number for air is even smaller, the internal resistance can be neglected for both cases and Eq. (2.89) applies. From Eq. (2.90),

$$\text{Bi Fo} = \frac{\bar{h}A}{c\rho V}t = \frac{4\bar{h}}{c\rho D}t$$

From the property values we obtain:

$$\text{Bi Fo} = \frac{4(80 \text{ J/s m}^2 \text{ K})}{(383 \text{ J/kg K})(8930 \text{ kg/m}^3)(0.001 \text{ m})}$$
$$= 0.0936t \qquad \text{for water}$$
$$\text{Bi Fo} = \frac{4(10 \text{ J/s m}^2 \text{ K})}{(383 \text{ J/kg K})(8930 \text{ kg/m}^3)(0.001 \text{ m})}$$
$$= 0.0117t \qquad \text{for air}$$

The temperature response is given by Eq. (2.84):

$$\frac{T - T_\infty}{T_0 - T_\infty} = e^{-\text{Bi Fo}}$$

The results are plotted in Fig. 2.33. Note that the time required for the temperature of the wire to reach 67°C is more than 2 min in air but only 15 s in water. A thermocouple 0.1 cm in diameter would therefore lag considerably if it were used to measure rapid change in air temperature, and it would be advisable to use wire of smaller diameter to reduce this lag.

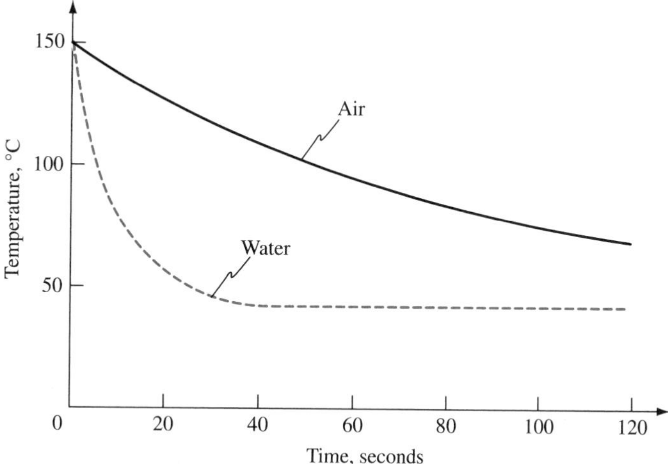

FIGURE 2.33 Temperature response of thermocouple in Example 2.10 after immersion in air and water.

The same general method can also be used to estimate the temperature-time history and internal energy change of a well-stirred fluid in a container suddenly immersed in a medium at a different temperature. If the walls of the container are so thin that their heat capacity is negligible, the temperature-time history of the fluid is given by a relation similar to Eq. (2.89):

$$\frac{T - T_\infty}{T_0 - T_\infty} = e^{-(UA_s/c\rho V)t}$$

where U is the overall heat transfer coefficient between the fluid and the surrounding medium, V is the volume of the fluid in the container, A_s is its surface area, and c and ρ are the specific heat and density of the fluid, respectively.

The lumped-capacity method of analysis can also be applied to composite systems or bodies. For example, if the walls of the container shown in Fig. 2.34 have a substantial thermal capacitance $(c\rho V)_2$, the heat transfer coefficient at A_1, the inner surface of the container, is $\bar{h}_1$, the heat transfer coefficient at A_2, the outer surface of the container, is $\bar{h}_2$, and the thermal capacitance of the fluid in the container is $(c\rho V)_1$, the temperature-time history of the fluid $T_1(t)$ is obtained by solving simultaneously the energy balance equations for the fluid:

$$-(c\rho V)_1 \frac{dT_1}{dt} = \bar{h}_1 A_1 (T_1 - T_2) \tag{2.91a}$$

and for the container:

$$-(c\rho V)_2 \frac{dT_2}{dt} = \bar{h}_2 A_2 (T_2 - T_\infty) - \bar{h}_1 A_1 (T_1 - T_2) \tag{2.91b}$$

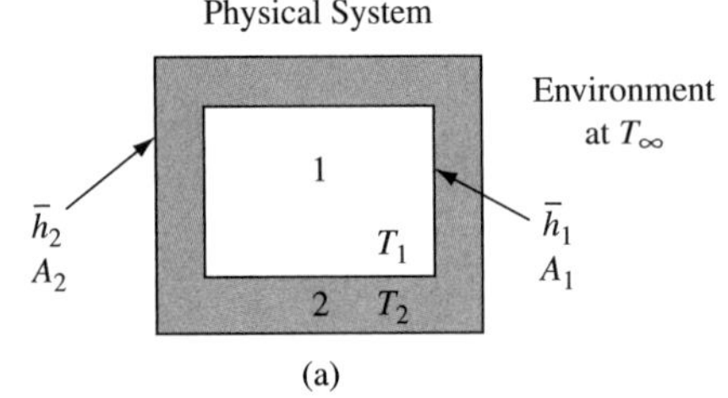

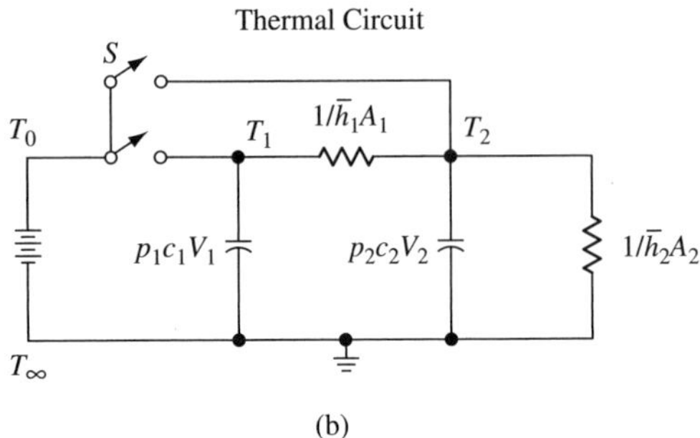

FIGURE 2.34 Schematic diagram and thermal network for a two-lump heat capacity system.

where T_2 is the temperature of the walls of the container. Inherent in this approach is the assumption that both the fluid and the container can be considered isothermal.

The preceding two simultaneous linear differential equations can be solved for the temperature history in each of the bodies. If the fluid and the container are initially at T_0, the initial conditions for the system are

$$T_1 = T_2 = T_0 \quad \text{at} \quad t = 0$$

which implies that at $t = 0$, $dT_1/dt = 0$ from Eq. (2.86a).

Equations (2.91a) and (2.91b) can be rewritten in operator form as

$$\left(D + \frac{\bar{h}_1A_1}{\rho_1c_1V_1}\right)T_1 - \left(\frac{\bar{h}_1A_1}{\rho_1c_1V_1}\right)T_2 = 0$$

$$-\left(\frac{\bar{h}_1A_1}{\rho_2c_2V_2}\right)T_1 + \left(D + \frac{\bar{h}_1A_1 + \bar{h}_2A_2}{\rho_2c_2V_2}\right)T_2 = \frac{\bar{h}_2A_2}{\rho_2c_2V_2}T_\infty$$

where the symbol D denotes differentiation with respect to time. For convenience let

$$K_1 = \frac{\bar{h}_1A_1}{\rho_1c_1V_1} \qquad K_2 = \frac{\bar{h}_1A_1}{\rho_2c_2V_2} \qquad K_3 = \frac{\bar{h}_2A_2}{\rho_2c_2V_2}$$

Then

$$(D + K_1)T_1 - K_1T_2 = 0$$
$$-K_2T_1 + (D + K_2 + K_3)T_2 = K_3T_\infty$$

Solving the equations simultaneously, we get a differential equation involving only T_1:

$$[D^2 + (K_1 + K_2 + K_3)D + K_1K_3]T_1 = K_1K_3T_\infty$$

The general solution of this equation is

$$T = T_\infty + Me^{m_1t} + Ne^{m_2t}$$

where m_1 and m_2 are given by

$$m_1 = \frac{-(K_1 + K_2 + K_3) + [(K_1 + K_2 + K_3)^2 - 4K_1K_3]^{1/2}}{2}$$
$$m_2 = \frac{-(K_1 + K_2 + K_3) + [(K_1 + K_2 + K_3)^2 - 4K_1K_3]^{1/2}}{2}$$

The arbitrary constants M and N can be obtained by applying the initial conditions

$$T_1 = T_0 \quad \text{at} \quad t = 0$$

and

$$\frac{dT_1}{dt} = 0 \quad \text{at} \quad t = 0$$

This leads to the two equations

$$T_0 = T_\infty + M + N$$
$$0 = m_1M + m_2N$$

The final solution for T_1, in dimensionless form, is

$$\frac{T_1 - T_\infty}{T_0 - T_\infty} = \frac{m_2}{m_2 - m_1}e^{m_1t} - \frac{m_1}{m_2 - m_1}e^{m_2t} \tag{2.92}$$

The solution for $T_2(t)$ is obtained by substituting the relation for T_1 from Eq. (2.92) into Eq. (2.91a).

The network analogy for the two-lump system is shown in Fig. 2.34. When the switch S is closed, the two thermal capacitances are charged to the potential T_0. At time zero, the switch is opened and the capacitances discharge through the two thermal resistances shown.

2.6.2* Infinite Slab

In the remainder of this chapter we will consider some transient conduction problems in which the temperature of the system interior is not uniform. An example of such a problem is transient heat flow in an infinite slab, as shown in Fig. 2.35. If the temperatures over the two surfaces are uniform, the problem is one-dimensional and

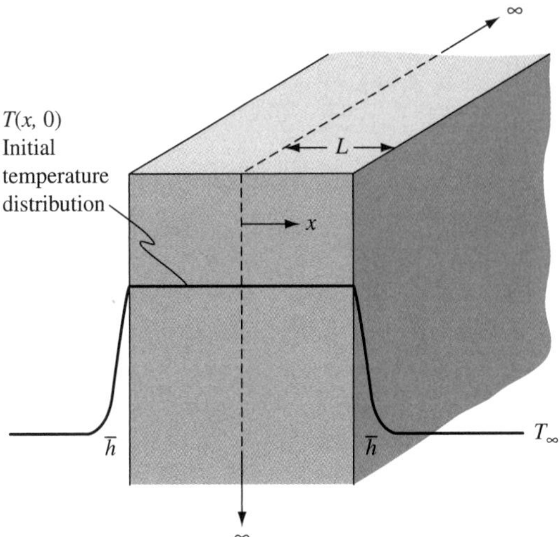

FIGURE 2.35 Nomenclature for analytical solution of a slab, initially at uniform temperature, subjected at time zero to a change in environmental temperature through a unit surface conductance $\bar{h}$.

transient. If, furthermore, there are no internal heat sources and the physical properties of the slab are constant, the general heat conduction equation reduces to the form

$$\frac{1}{\alpha}\frac{\partial T}{\partial t} = \frac{\partial^2 T}{\partial x^2} \qquad 0 \le x \le L \tag{2.93}$$

The thermal diffusivity α, which appears in all unsteady heat conduction problems, is a property of the material, and the time rate of temperature change depends on its numerical value. Qualitatively we observe that, in a material that combines a low thermal conductivity with a large specific heat (small α), the rate of temperature change will be slower than in a material with a large thermal diffusivity.

Since the temperature T must be a function of time t and x, we begin by assuming a product solution

$$T(x, t) = X(x)\Theta(t)$$

Note that

$$\frac{\partial T}{\partial t} = X\frac{\partial \Theta}{\partial t} \quad \text{and} \quad \frac{\partial^2 T}{\partial x^2} = \Theta\frac{\partial^2 X}{\partial x^2}$$

Substituting these partial derivatives into Eq. (2.93) yields

$$\frac{1}{\alpha}X\frac{\partial \Theta}{\partial t} = \Theta\frac{\partial^2 X}{\partial x^2}$$

We can now separate the variables, that is, bring all functions that depend on x to one side of the equation and all functions that depend on t to the other. By dividing both sides by $X\Theta$, we obtain

$$\frac{1}{\alpha\Theta}\frac{\partial\Theta}{\partial t} = \frac{1}{X}\frac{\partial^2 X}{\partial x^2}$$

Now observe that the left-hand side is a function of t only and therefore is independent of x, whereas the right-hand side is a function of x only and will not change as t varies. Since neither side can change as t and x vary, both sides are equal to a constant, which we will call μ. Hence, we have two ordinary and linear differential equations with constant coefficients:

$$\frac{d\Theta(t)}{dt} = \alpha\mu\Theta(t) \tag{2.94}$$

and

$$\frac{d^2X}{dx^2} = \mu X(x) \tag{2.95}$$

The general solution for Eq. (2.94) is

$$\Theta(t) = C_1 e^{\alpha\mu t}$$

If μ were a positive number, the temperature of the slab would become infinitely high as t increased, which is physically impossible. Therefore, we must reject the possibility that $\mu > 0$. If μ were zero, the slab temperature would be a constant. Again, this possibility must be rejected because it would not be consistent with the physical conditions of the problem. We therefore conclude that μ must be a negative number, and for convenience we let $\mu = -\lambda^2$. The time-dependent function then becomes

$$\Theta(t) = C_1 e^{-\alpha\lambda^2 t} \tag{2.96}$$

Next we direct attention to the equation involving x, (Eq. (2.95)). Its general solution can be written in terms of a sinusoidal function. Since this is a second-order equation, there must be two constants of integration in the solution. In convenient form, the solution to the equation

$$\frac{d^2X(x)}{dx^2} = -\lambda^2 X(x)$$

can be written as

$$X(x) = C_2 \cos \lambda x + C_3 \sin \lambda x \tag{2.97}$$

The temperature as a function of distance and time in the slab is given by

$$\begin{aligned} T(x, t) &= C_1 e^{-\alpha\lambda^2 t}(C_2 \cos \lambda x + C_3 \sin \lambda x) \\ &= e^{-\alpha\lambda^2 t}(A \cos \lambda x + B \sin \lambda x) \end{aligned} \tag{2.98}$$

where $A = C_1C_2$ and $B = C_1C_3$ are constants that must be evaluated from the boundary and initial conditions. In addition, we must determine the value of the constant λ in order to complete the solution.

The boundary and initial conditions are:

1. At $x = 0$, $\partial T/\partial x = 0$.
2. At $x = \pm L$, $-(\partial T/\partial x)\,|_{x=\pm L} = (\bar{h}/k_s)(T_{x=\pm L} - T_\infty)$.
3. At $t = 0$, $T = T_i$.

Boundary condition 1 requires that

$$\left.\frac{\partial T}{\partial x}\right|_{x=0} = e^{-\alpha\lambda^2 t}(-A\lambda \sin \lambda x + B\lambda \cos \lambda x)\Big|_{x=0} = 0$$

Now sin 0 = 0, but the second term in the parentheses, involving cos 0, can be zero only if $B = 0$ or $\lambda = 0$. Since $\lambda = 0$ gives a trivial solution, we reject it, and the solution for $T(x, t)$ therefore becomes

$$T(x, t) = e^{-\alpha\lambda^2 t} A cos \lambda x$$

To satisfy the second boundary condition, namely, that the heat flow by conduction at the interface must be equal to the heat flow by convection, the equality

$$-\left.\frac{\partial T}{\partial x}\right|_{x=L} = e^{-\alpha\lambda^2 t} A\lambda \sin \lambda L = \frac{\bar{h}}{k_s}(T_{x=L} - 0) = \frac{\bar{h}}{k_s} e^{-\alpha\lambda^2 t} A \cos \lambda L$$

must hold for all values of t, which gives

$$\frac{\bar{h}}{k_s} \cos \lambda L = \lambda \sin \lambda L \quad \text{or}$$

$$\cot \lambda L = \frac{k_s}{\bar{h}L} \lambda L = \frac{\lambda L}{\text{Bi}} \tag{2.99}$$

Equation (2.99) is *transcendental*, and there are an infinite number of values of λ, called characteristic values, that will satisfy it. The simplest way to determine the numerical values of λ is to plot cot λL and λL/Bi against λL. The values of λ at the points of intersection of these curves are the characteristic values and satisfy the second boundary condition. Figure 2.36 is a plot of these curves, and if $L = 1$ we can read off the first few characteristic values as $\lambda_1 = 0.86$ Bi, $\lambda_2 = 3.43$ Bi, $\lambda_3 = 6.44$ Bi, etc. The value $\lambda = 0$ is disregarded because it leads to the trivial solution $T = 0$. A particular solution of Eq. (2.99) corresponds to each value of λ. Therefore, we shall adopt a subscript notation to identify the correspondence between A and λ. For instance, A_1 corresponds to λ_1 or, in general, A_n to λ_n. The complete solution is formed as the sum of the solutions corresponding to each characteristic value:

$$T(x, t) = \sum_{n=1}^{\infty} e^{-\alpha\lambda_n^2 t} A_n \cos \lambda_n x \tag{2.100}$$

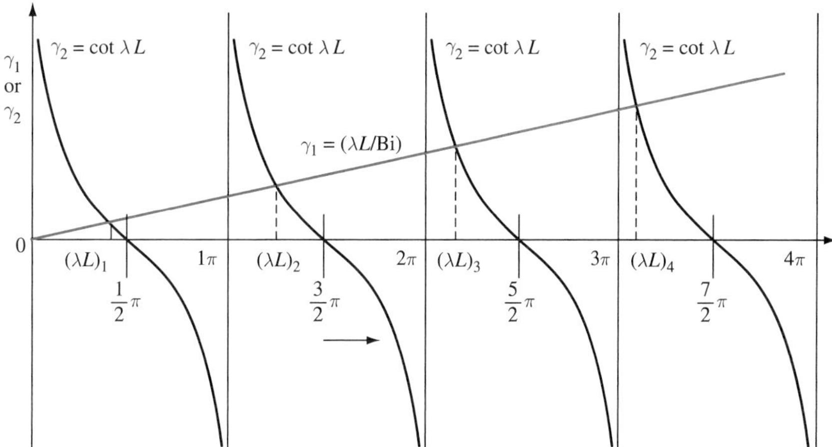

FIGURE 2.36 Graphic solution of transcendental equation.

Each term of this infinite series contains a constant. These constants are evaluated by substituting the initial condition into Eq. (2.100):

$$T(x, 0) = T_i = \sum_{n=1}^{\infty} A_n \cos \lambda_n x \tag{2.101}$$

It can be shown that the characteristic functions $\cos \lambda_n x$ are orthogonal between $x = 0$ and $x = L$ and therefore*

$$\int_0^L \cos \lambda_n x \cos \lambda_m x \, dx \begin{cases} = 0 & \text{if } m \neq n \\ \neq 0 & \text{if } m = n \end{cases} \tag{2.102}$$

where λ_m may be any characteristic value of λ. To obtain a particular value of A_n, we multiply both sides of Eq. (2.96) by $\cos \lambda_m x$ and integrate between 0 and L.

* This can be verified by performing the integration, which yields

$$\int_0^L \cos \lambda_n x \cos \lambda_m x \, dx = \frac{\lambda_n \sin L\lambda_n \cos L\lambda_m - \lambda_m \sin L\lambda_m \cos L\lambda_n}{2L(\lambda_m^2 - \lambda_n^2)}$$

when $m \neq n$. However, from Eq. (2.99) we have

$$\frac{\cot \lambda_m L}{\lambda_m} = \frac{k_s}{\bar{h}} = \frac{\cot \lambda_n L}{\lambda_n}$$

or

$$\lambda_n \cos \lambda_m L \sin \lambda_n L = \lambda_m \cos \lambda_n L \sin \lambda_m L$$

Therefore, the integral is zero when $m \neq n$.

In accordance with Eq. (2.102), all terms on the right-hand side disappear except the one involving the square of the characteristic function, $\cos \lambda_n x$, and we obtain

$$\int_0^L (T_i - T_\infty) \cos \lambda_n x \, dx = A_n \int_0^L \cos^2 \lambda_n x \, dx$$

From standard integral tables (11) we get

$$\int_0^L \cos^2 \lambda_n x \, dx = \frac{1}{2}x + \frac{1}{2\lambda_n} \sin\lambda_n x \cos\lambda_n x \Big|_0^L = \frac{L}{2} + \frac{1}{2\lambda_n} \sin\lambda_n L \cos \lambda_n L$$

and

$$\int_0^L \cos\lambda_n x \, dx = \frac{1}{\lambda_n} \sin\lambda_n L$$

whence the constant A_n is

$$A_n = \frac{2\lambda_n}{L\lambda_n + \sin \lambda_n L \cos\lambda_n L} \frac{(T_i - T_\infty) \sin\lambda_n L}{\lambda_n} = \frac{2(T_i - T_\infty) \sin\lambda_n L}{L\lambda_n + \sin\lambda_n L \cos\lambda_n L} \tag{2.103}$$

As an illustration of the general procedure outlined above, let us determine A_1 when $\bar{h} = 1$, $k_s = 1$, and $L = 1$. From the graph of Fig. 2.36, the value of λ_1 is 0.86 radians or 49.2°. Then we have

$$\begin{aligned} A_1 &= (T_i - T_\infty) \frac{2 \sin 49.2}{(1)(0.86) + \sin 49.2 \cos 49.2} \\ &= (T_i - T_\infty) \frac{(2)(0.757)}{0.86 + (0.757)(0.653)} \\ &= 1.12(T_i - T_\infty) \end{aligned}$$

Similarly, we obtain

$$A_2 = -0.152(T_i - T_\infty) \quad \text{and} \quad A_3 = 0.046(T_i - T_\infty)$$

The series converges rapidly, and for Bi = 1, three terms represent a fairly good approximation for practical purposes.

To express the temperature in the slab in terms of conventional dimensionless moduli, we let $\lambda_n = \delta_n/L$. The final form of the solution, obtained by substituting Eq. (2.103) into Eq. (2.101), is then

$$\frac{T(x, t) - T_\infty}{T_i - T_\infty} = \sum_{n=1}^{\infty} e^{-\delta_n^2 (t\alpha/L^2)} 2 \frac{\sin\delta_n \cos(\delta_n x/L)}{\delta_n + \sin \delta_n \cos \delta_n} \tag{2.104}$$

The time dependence is now contained in the dimensionless Fourier modulus, Fo = $t\alpha/L^2$. Furthermore, if we write the second boundary condition in terms of δ_n, we obtain from Eq. (2.99)

$$\cot \delta_n = \frac{k_s}{\bar{h}L} \delta_n \tag{2.105}$$

or

$$\delta_n \tan \delta_n = \frac{\bar{h}L}{k_s} = \text{Bi}$$

Since δ_n is a function only of the dimensionless Biot number, $\text{Bi} = \bar{h}L/k_s$, the temperature $T(x, t)$ can be expressed in terms of the three dimensionless quantities, $\text{Fo} = t\alpha/L^2$, $\text{Bi} = \bar{h}L/k_s$, and x/L.

The rate of internal energy change of the slab per unit area of the surface of the slab, dQ/dt, is given by

$$\frac{dQ}{dt} = \frac{q}{A} = -k_s \frac{\partial T}{\partial x}\bigg|_{x=L} \tag{2.106}$$

The temperature gradient can be obtained by differentiating Eq. (2.104) with respect to x for a given value of t, or

$$\frac{\partial T}{\partial x}\bigg|_{x=L} = -\frac{2(T_0 - T_\infty)}{L} \sum_{n=1}^{\infty} e^{-\delta_n^2 \text{Fo}} \frac{\delta_n \sin^2 \delta_n}{\delta_n + \sin \delta_n \cos \delta_n} \tag{2.107}$$

Substituting Eq. (2.107) into Eq. (2.106) and integrating between the limits of $t = 0$ and t gives the change in internal energy of the slab during the time t, which is equal to the amount of heat Q absorbed by (or removed from) the slab. After some algebraic simplification, we obtain

$$Q = 2(T_0 - T_\infty)Lc\rho \sum_{n=1}^{\infty} (1 - e^{-\delta_n^2 \text{Fo}}) \frac{\sin^2 \delta_n}{\delta_n^2 + \delta_n \sin \delta_n \cos \delta_n} \tag{2.108}$$

To make Eq. (2.108) dimensionless, we note that $c\rho LT_0$ represents the initial internal energy per unit area of the slab. If we denote $c\rho L(T_0 - T_\infty)$ by Q_0, we get

$$\frac{Q}{Q_0} = \sum_{n=1}^{\infty} \frac{2 \sin^2 \delta_n}{\delta_n^2 + \delta_n \sin \delta_n \cos \delta_n} (1 - e^{-\delta_n^2 \text{Fo}}) \tag{2.109}$$

The temperature distribution and the amount of heat transferred at any time can be determined from Eqs. (2.104) and (2.109), respectively. The final expressions are in the form of infinite series. These series have been evaluated, and the results are available in the form of charts. Use of the charts for the problem treated in this section as well as for other cases of practical interest will be taken up in Section 2.7. A complete understanding of the methods by which the mathematical solutions have been obtained is helpful but is not necessary for using the charts.

2.6.3* Semi-Infinite Solid

Another simple geometric configuration for which analytic solutions are available is the semi-infinite solid. Such a solid extends to infinity in all but one direction and can therefore be characterized by a single surface (Fig. 2.37). A semi-infinite solid approximates many practical problems. It can be used to estimate transient heat transfer effects near the surface of the earth or to approximate the transient response of finite solid, such as a thick slab, during the early portion of a transient when the temperature in the slab interior is not yet influenced by the change in surface conditions.

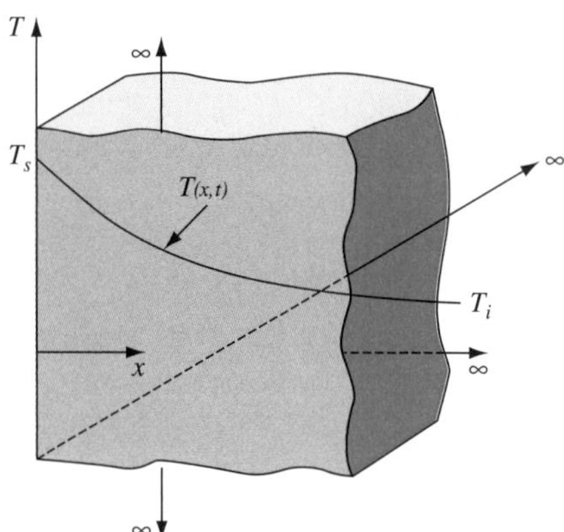

FIGURE 2.37 Schematic diagram and nomenclature for transient conduction in a semi-infinite solid.

An example of the latter is the heat treating (transient heating as well as cooling) of a large rectangular steel slab, seen in Fig. 2.38, where the thickness is substantially smaller than the length and width of the slab.

If a thermal change is suddenly imposed at this surface, a one-dimensional temperature wave will be propagated by conduction within the solid. The appropriate equation for transient conduction in a semi-infinite solid is Eq. (2.93) in the domain $0 \leq x < \infty$. To solve this equation we must specify two boundary conditions and the initial temperature distribution. For the initial condition we shall specify that the

FIGURE 2.38 A large rectangular steel slab as it exits a heat treatment furnace.

Source: Courtesy of SBS, *www.sbs-forge.com*

temperature inside the solid is uniform at T_i, that is, $T(x, 0) = T_i$. For one of the two required boundary conditions we postulate that far from the surface the interior temperature will not be affected by the temperature wave, that is, $T(\infty, t) = T_i$, with the above specifications.

Closed-form solutions have been obtained for three types of changes in surface conditions, instantaneously applied at $t = 0$:

1. A sudden change in surface temperature, $T_s \neq T_i$
2. A sudden application of a specified heat flux q_0'', as, for example, exposing the surface to radiation
3. A sudden exposure of the surface to a fluid at a different temperature through a uniform and constant heat transfer coefficient $\bar{h}$

These three cases are illustrated in Fig. 2.39 and the solutions are summarized below.

Case 1. Change in surface temperature:

$$T(0, t) = T_s$$

$$\frac{T(x, t) - T_s}{T_i - T_s} = \mathrm{erf}\left(\frac{x}{2\sqrt{\alpha t}}\right) \tag{2.110}$$

$$q_s''(t) = -k\frac{\partial T}{\partial x}\bigg|_{x=0} = \frac{k(T_s - T_i)}{\sqrt{\pi \alpha t}}$$

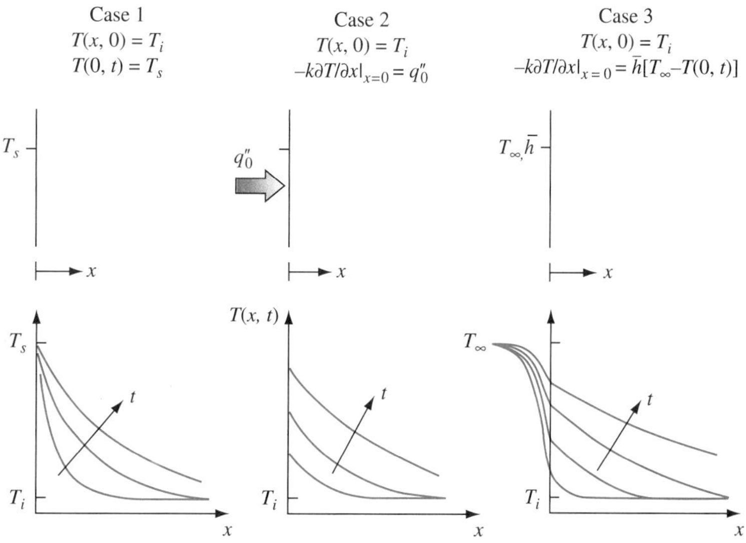

FIGURE 2.39 Transient temperature distributions in a semi-infinite solid for three surface conditions: (1) constant surface temperature, (2) constant surface heat flux, and (3) surface convection.

Case 2. Constant surface heat flux:

$$q_s'' = q_0''$$

$$T(x, t) - T_i = \frac{2q_0''(\alpha t/\pi)^{1/2}}{k_s}\exp\left(\frac{-x^2}{4\alpha t}\right) - \frac{q_0''x}{k_s}\operatorname{erfc}\left(\frac{x}{2\sqrt{\alpha t}}\right) \tag{2.111}$$

Case 3. Surface heat transfer by convection and radiation:

$$-k\frac{\partial T}{\partial x}\bigg|_{x=0} = \bar{h}[T_\infty - T(0, t)]$$

$$\frac{T(x, t) - T_i}{T_\infty - T_i} = \operatorname{erfc}\left(\frac{x}{2\sqrt{\alpha t}}\right) - \exp\left(\frac{\bar{h}x}{k} + \frac{\bar{h}^2\alpha t}{k^2}\right)\operatorname{erfc}\left(\frac{x}{2\sqrt{\alpha t}} + \frac{\bar{h}\sqrt{\alpha t}}{k}\right) \tag{2.112}$$

Note that the quantity $\bar{h}^2\alpha t/k^2$ equals the product of the Biot number squared ($\mathrm{Bi} = \bar{h}x/k$) times the Fourier number ($\mathrm{Fo} = \alpha t/x^2$).

The function erf appearing in Eq. (2.110) is the *Gaussian error function*, which is encountered frequently in engineering and is defined as

$$\operatorname{erf}\left(\frac{x}{2\sqrt{\alpha t}}\right) = \frac{2}{\sqrt{\pi}}\int_0^{x/2\sqrt{\alpha t}} e^{-\eta^2}\,d\eta \tag{2.113}$$

Values of this function are tabulated in Table 43 of the Appendix. The complementary error function, $\operatorname{erfc}(w)$, is defined as

$$\operatorname{erfc}(w) = 1 - \operatorname{erf}(w) \tag{2.114}$$

Temperature histories for the three cases are illustrated qualitatively in Fig. 2.39. For Case 3, the specific temperature histories computed from Eq. (2.112) are plotted in Fig. 2.40. The curve corresponding to $\bar{h} = \infty$ is equivalent to the result that would be

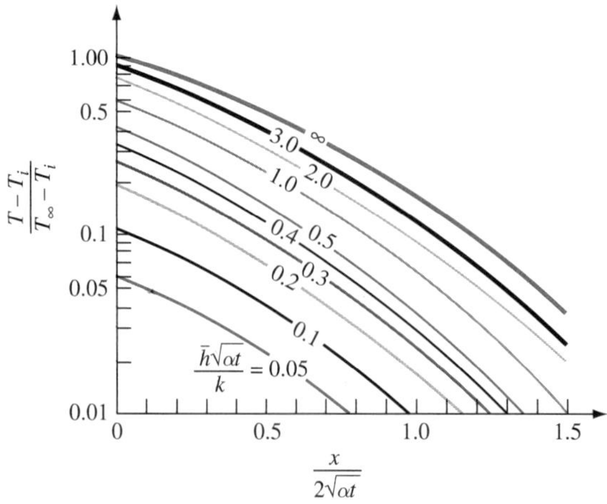

FIGURE 2.40 Dimensionless transient temperatures for a semi-infinite solid with surface heat transfer.

obtained for a sudden change in the surface temperature to $T_s = T(x,0)$ because when $\bar{h} = \infty$ the second term on the right-hand side of Eq. (2.112) is zero, and the result is equivalent to Eq. (2.110) for Case 1.

EXAMPLE 2.11 Estimate the minimum depth x_m at which one must place a water main below the surface to avoid freezing. The soil is initially at a uniform temperature of 20°C. Assume that under the worst conditions anticipated it is subjected to a surface temperature of −15°C for a period of 60 days. Use the following properties for soil (300 K):

$$\rho = 2050 \text{ kg/m}^3 \quad k = 0.52 \text{ W/m K} \quad c = 1840 \text{ J/kg K}$$

$$\alpha = \frac{k}{\rho c} = 0.138 \times 10^{-6} \text{ m}^2/\text{s}$$

A sketch of the system is shown in Fig. 2.41.

SOLUTION To simplify the problem assume that

1. Conduction is one-dimensional.
2. The soil is a semi-infinite medium.
3. The soil has uniform and constant properties.

The prescribed conditions correspond to those of Case 1 of Fig. 2.39, and the transient temperature response of the soil is governed by Eq. (2.112). At the time $t = 60$ days after the change in surface temperature, the temperature distribution in the soil is

$$\frac{T(x_m, t) - T_s}{T_i - T_s} = \text{erf}\left(\frac{x_m}{2\sqrt{\alpha t}}\right)$$

or

$$\frac{0 - (-15°\text{C})}{20°\text{C} - (-15°\text{C})} = 0.43 = \text{erf}\left(\frac{x_m}{2\sqrt{\alpha t}}\right)$$

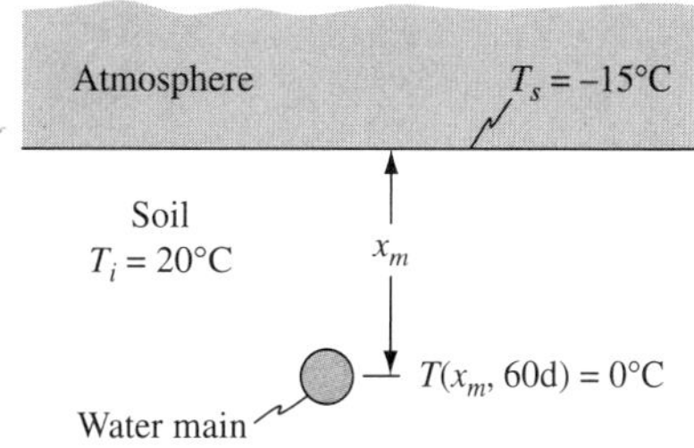

FIGURE 2.41 Schematic diagram for Example 2.11.

From Table 43 we find by interpolation that when $x_m/2\sqrt{\alpha t} = 0.4$, $\text{erf}(0.4) = 0.43$ to satisfy the above relation. Thus

$$\begin{aligned} x_m &= (0.4)(2\sqrt{\alpha t}) \\ &= 0.8[(0.138 \times 10^{-6}\ \text{m}^2/\text{s})(60\ \text{days})(24\ \text{h/day})(3600\ \text{s/h})]^{1/2} = 0.68\ \text{m} \end{aligned}$$

To use Fig. 2.40, first calculate $[T(x, t) - T_s]/(T_\infty - T_s) = (0 - 20)/(-15 - 20) = 0.57$, then enter the curve for $\bar{h}\sqrt{\alpha t}/k = \infty$ and obtain $x/2\sqrt{\alpha t} = 0.4$, the same result as above.

2.7* Charts for Transient Heat Conduction

For transient heat conduction in several simple shapes, subject to boundary conditions of practical importance, the temperature distribution and the heat flow have been calculated and the results are available in the form of charts or tables [5, 6,12–14]. Although most transient conduction problems can be readily computed with ease employing modern tools such as spreadsheets and programmable calculators, the charts and tables presented here are still useful in providing a means for obtaining quick solutions for most engineering problems. In this section we shall illustrate the application of some of these charts to typical problems of transient heat conduction in solids having a Biot number larger than 0.1.

2.7.1 One-Dimensional Solutions

Three simple geometries for which results have been prepared in graphic form are:

1. An infinite plate of width $2L$ (see Fig. 2.42 on pages 135 and 136)
2. An infinitely long cylinder of radius r_0 (see Fig. 2.43 on pages 137 and 138)
3. A sphere of radius r_0 (see Fig. 2.44 on pages 139 and 140)

The boundary conditions and the initial conditions for all three geometries are similar. One boundary condition requires that the temperature gradient at the mid-plane of the plate, the axis of the cylinder, and the center of the sphere be equal to zero. Physically, this corresponds to no heat flow at these locations.

The other boundary condition requires that the heat conducted to or from the surface be transferred by convection to or from a fluid at temperature T_∞ through a uniform and constant convection heat transfer coefficient $\bar{h}_c$, or

$$\bar{h}_c(T_s - T_\infty) = -k\left.\frac{\partial T}{\partial n}\right|_s \tag{2.115}$$

where the subscript s refers to conditions at the surface and n to the coordinate direction normal to the surface. It should be noted that the limiting case of $\text{Bi} \to \infty$ corresponds to a negligible thermal resistance at the surface ($\bar{h}_c \to \infty$) so that the surface temperature is specified as equal to T_∞ for $t > 0$.

The initial conditions for all three chart solutions require that the solid be initially at a uniform temperature T_i and that when the transient begins at time zero ($t = 0$), the entire surface of the body is contacted by fluid at T_∞.

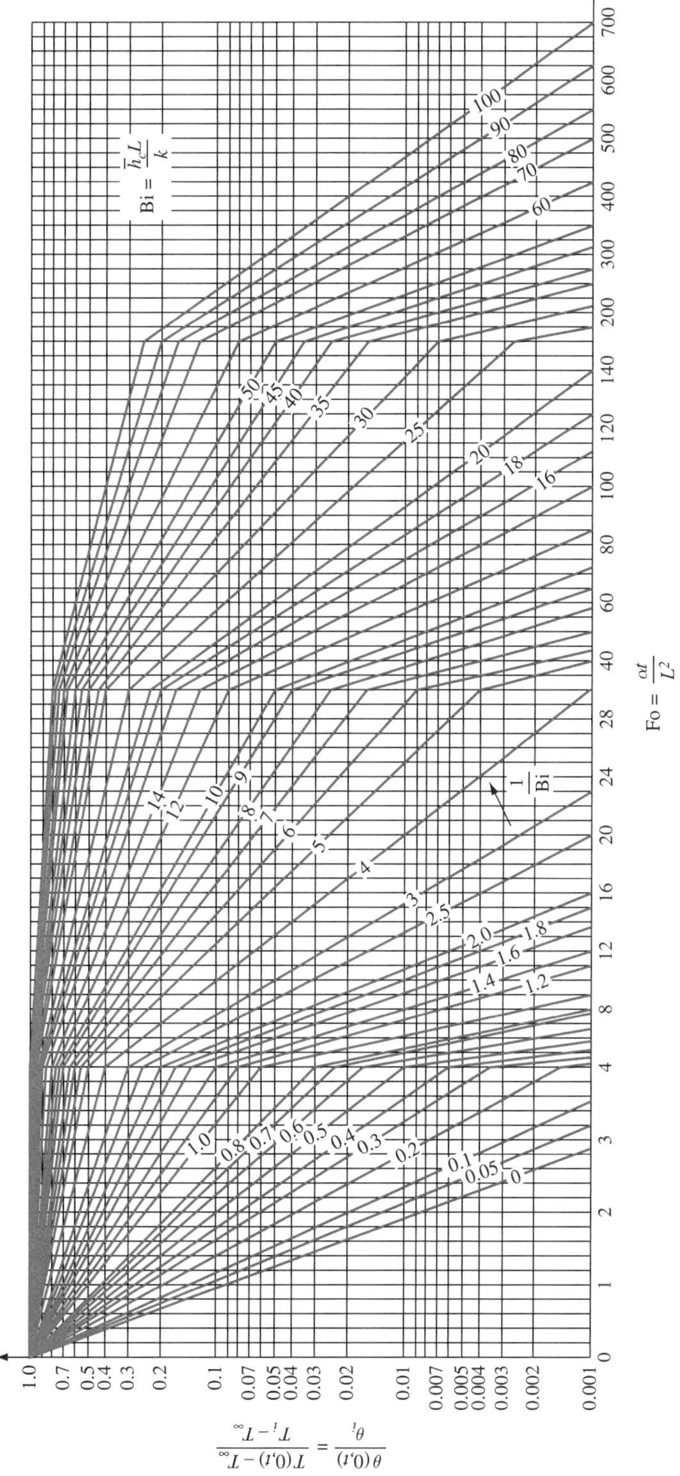

FIGURE 2.42 Dimensionless transient temperatures and heat flow in an infinite plate of width 2*L*.

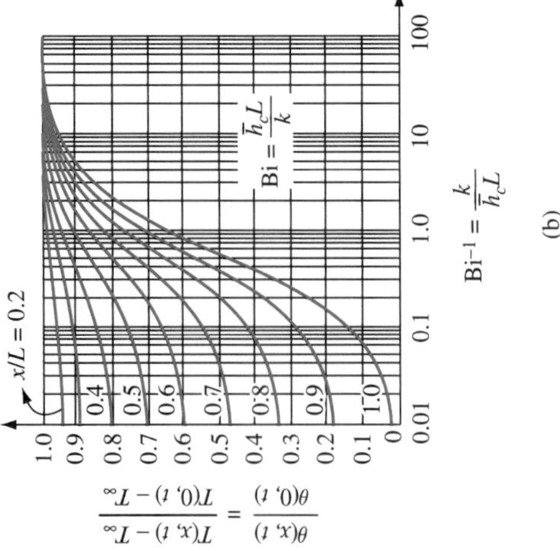

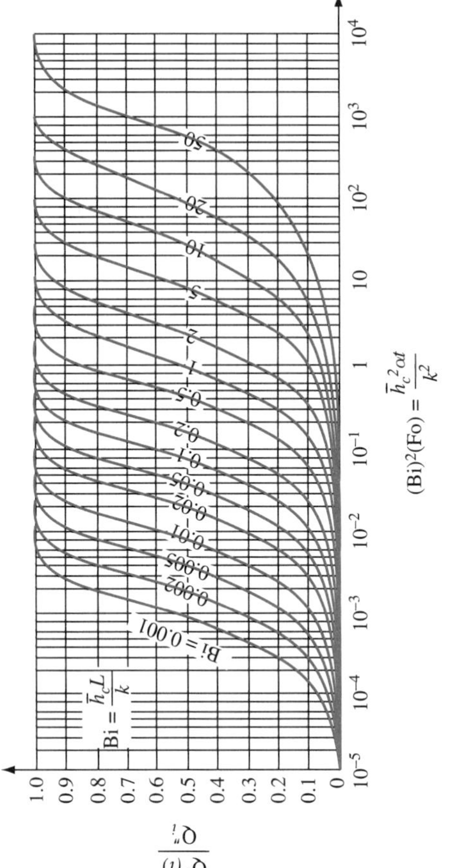

FIGURE 2.42 *(Continued)*

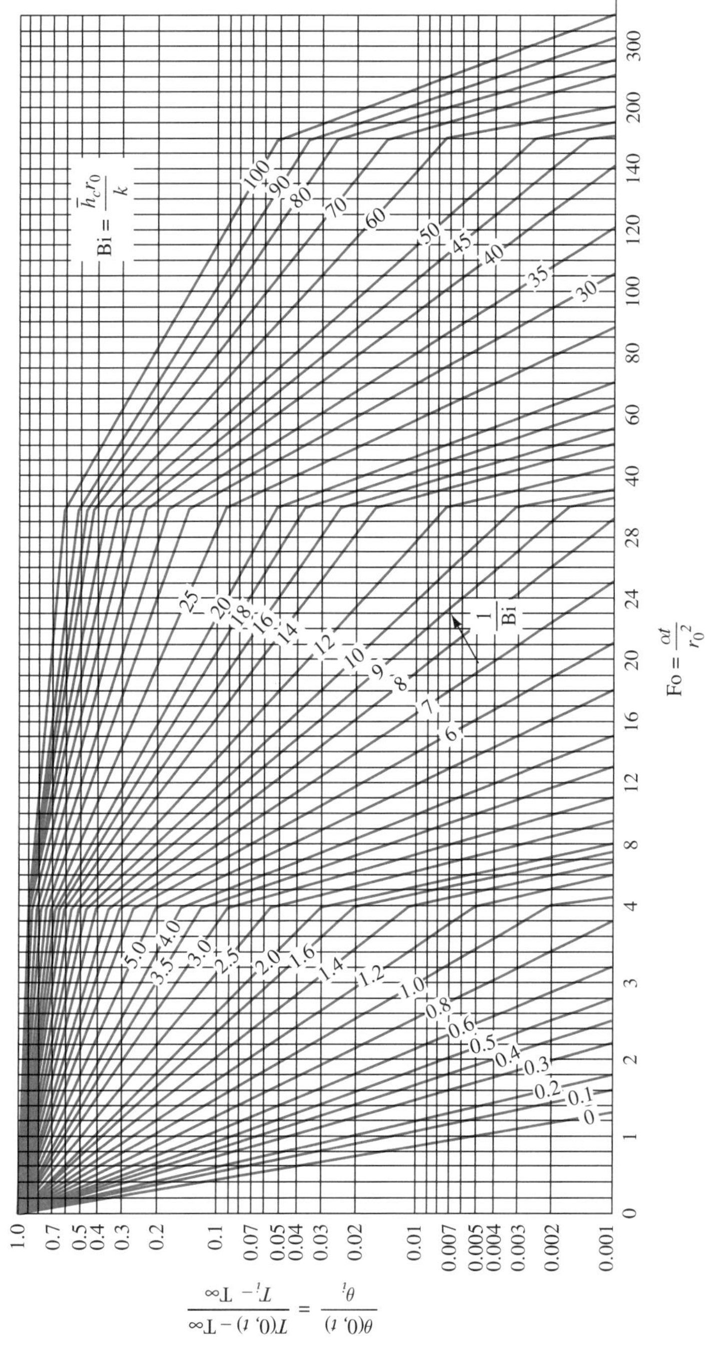

FIGURE 2.43 Dimensionless transient temperatures and heat flow for a long cylinder.

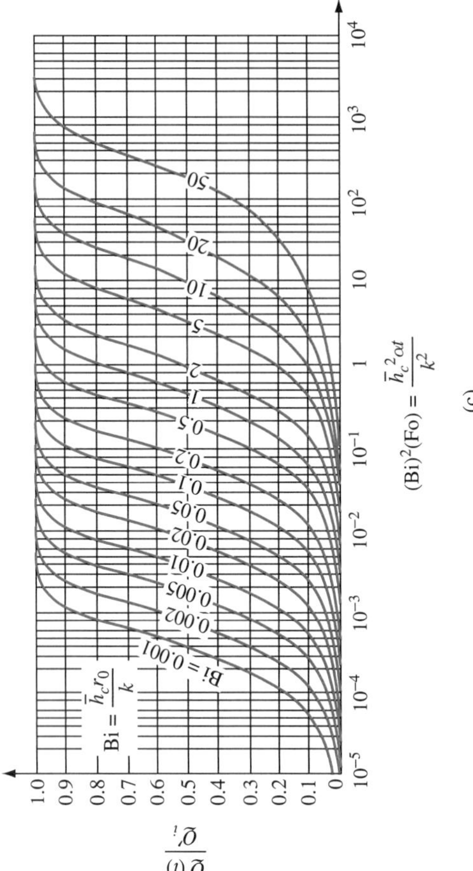

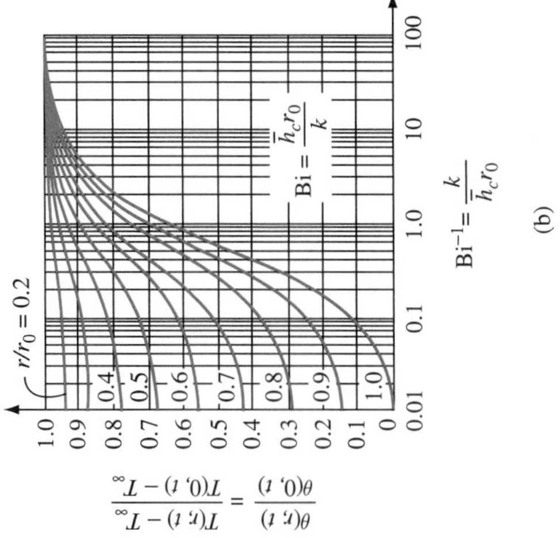

FIGURE 2.43 *(Continued)*

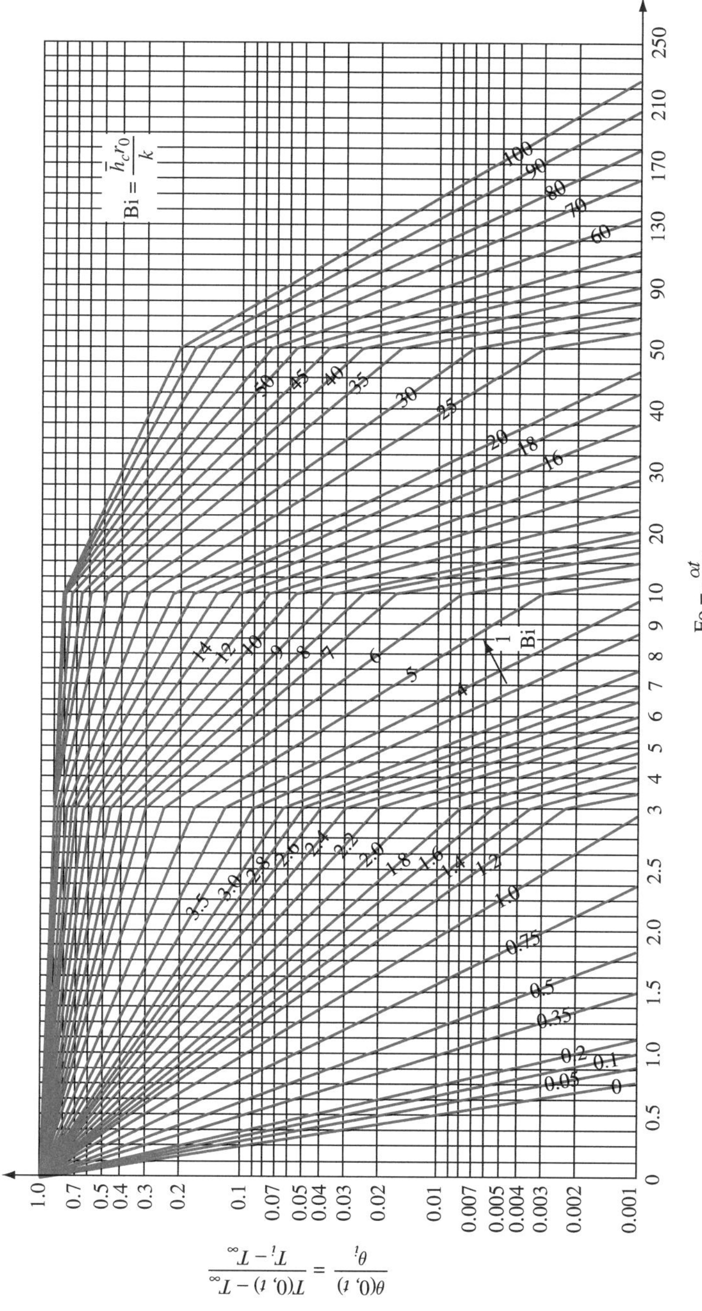

(a)

FIGURE 2.44 Dimensionless transient temperatures and heat flow for a sphere.

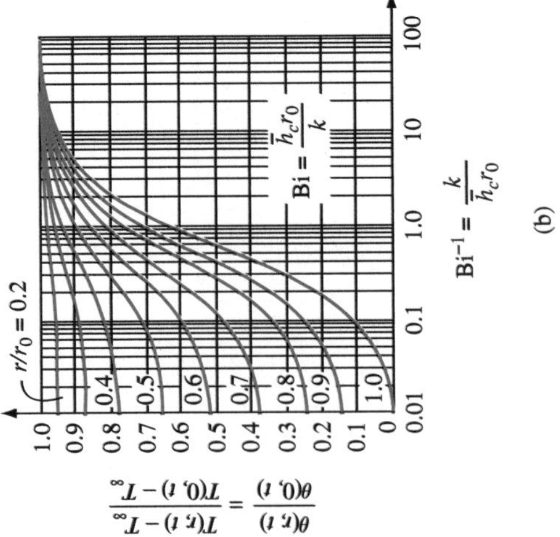

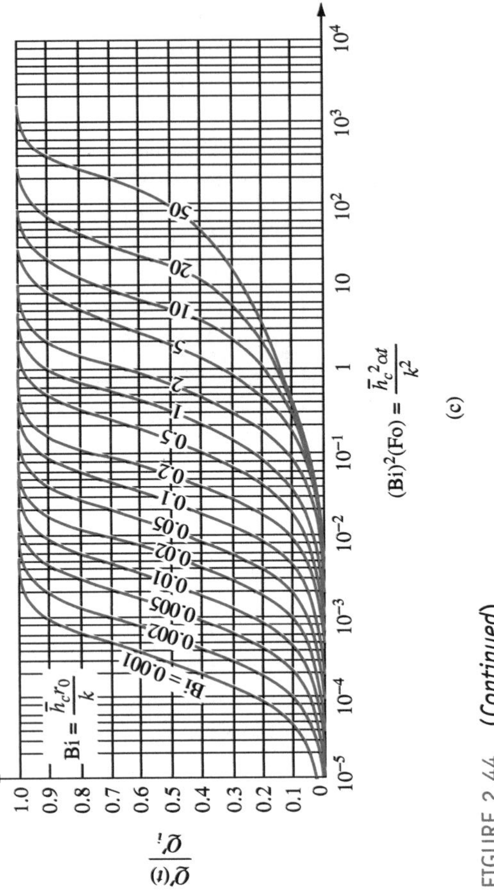

FIGURE 2.44 *(Continued)*

The solutions for all three cases are plotted in terms of dimensionless parameters. The forms of the dimensionless parameters are summarized in Table 2.3. Use of the graphic solutions is discussed below.

For each geometry there are three graphs, the first two for the temperatures and the third for the heat flow. The dimensionless temperatures are presented in the form of two interrelated graphs for each shape. The first set of graphs, Figs. 2.42(a) for the plate, 2.43(a) for the cylinder, and 2.44(a) for the sphere, gives the dimensionless temperature at the center or midpoint as a function of the Fourier number, that is, dimensionless time, with the inverse of the Biot number as the constant parameter. The dimensionless center or midpoint temperature for these graphs is defined as

$$\frac{T(0,t) - T_\infty}{T_i - T_\infty} \equiv \frac{\theta(0,t)}{\theta_i} \tag{2.116}$$

TABLE 2.3 Summary of dimensionless parameters for use with transient heat conduction charts in Figs. 2.42, 2.43, and 2.44

Situation	Infinite Plate, Width $2L$	Infinitely Long Cylinder, Radius r_0	Sphere, Radius r_0
Geometry	Fluid $\bar{h}_c, T_\infty$; k, α; x; Fluid $\bar{h}_c, T_\infty$; $2L$	Fluid $\bar{h}_c, T_\infty$; k, α; r; r_0	Fluid $\bar{h}_c, T_\infty$; k, α; r; r_0
Dimensionless position	$\frac{x}{L}$	$\frac{r}{r_0}$	$\frac{r}{r_0}$
Biot number	$\frac{\bar{h}_c L}{k}$	$\frac{\bar{h}_c r_0}{k}$	$\frac{\bar{h}_c r_0}{k}$
Fourier number	$\frac{\alpha t}{L^2}$	$\frac{\alpha t}{r_0^2}$	$\frac{\alpha t}{r_0^2}$
Dimensionless centerline temperature $\frac{\theta(0,t)}{\theta_i}$	Fig. 2.42(a)	Fig. 2.43(a)	Fig. 2.44(a)
Dimensionless local temperature $\frac{\theta(x,t)}{\theta(0,t)}$ or $\frac{\theta(r,t)}{\theta(0,t)}$	Fig. 2.42(b)	Fig. 2.43(b)	Fig. 2.44(b)
Dimensionless heat transfer $\frac{Q''(t)}{Q_i''}, \frac{Q'(t)}{Q_i'}, \frac{Q(t)}{Q_i}$	Fig. 2.42(c) $Q_i'' = \rho c L(T_i - T_\infty)$	Fig. 2.43(c) $Q_i' = \rho c \pi r_0^2 (T_i - T_\infty)$	Fig. 2.44(c) $Q_i = \rho c \frac{4}{3}\pi r_0^3 (T_i - T_\infty)$

To evaluate the local temperature as a function of time, the second temperature graph must be used. The second set of graphs, Figs. 2.42(b) for a plate, 2.43(b) for a cylinder, and 2.44(b) for a sphere, gives the ratio of the local temperature to the center or midpoint temperature as a function of the inverse of the Biot number for various values of the dimensionless distance parameter, x/L for the slab and r/r_0 for the cylinder and the sphere. For the infinite plate this temperature ratio is

$$\frac{T(x, t) - T_\infty}{T(0, t) - T_\infty} = \frac{\theta(x, t)}{\theta(0, t)} \tag{2.117}$$

For the cylinder and the sphere the expressions are similar, but x is replaced by r.

To determine the local temperature at any time t, form the product

$$\begin{aligned}\frac{T(x, t) - T_\infty}{T_i - T_\infty} &= \left[\frac{T(0, t) - T_\infty}{T_i - T_\infty}\right]\left[\frac{T(x, t) - T_\infty}{T(0, t) - T_\infty}\right] \\ &= \frac{\theta(0, t)\,\theta(x, t)}{\theta_i\quad \theta(0, t)}\end{aligned} \tag{2.118}$$

for the plate and

$$\frac{T(r, t) - T_\infty}{T_i - T_\infty} = \left[\frac{T(0, t) - T_\infty}{T_i - T_\infty}\right]\left[\frac{T(r, t) - T_\infty}{T(0, t) - T_\infty}\right] \tag{2.119}$$

for the cylinder and the sphere.

The instantaneous rate of heat transfer to or from the surface of the solid can be evaluated from Fourier's law once the temperature distribution is known. The change in internal energy between time $t = 0$ and $t = t$ can be obtained by integrating the instantaneous heat transfer rates, as shown for the slab by Eqs. (2.106) and (2.108). Denoting by $Q(t)$ the internal energy relative to the fluid at time t, and by Q_i the initial internal energy relative to the fluid, the ratios $Q(t)/Q_i$ are plotted against $\mathrm{Bi}^2\mathrm{Fo} = \bar{h}^2\alpha t/k^2$ for various values of Bi in Fig. 2.42(c) for the plate, Fig. 2.43(c) for the cylinder, and Fig. 2.44(c) for the sphere.

Each heat transfer value $Q(t)$ is the total amount of heat that is transferred from the surface to the fluid during the time from $t = 0$ to $t = t$. The normalizing factor Q_i is the initial amount of energy in the solid at $t = 0$ when the reference temperature for zero energy is T_∞. The values for Q_i for each of the three geometries are listed in Table 2.3 for convenience. Since the volume of the plate is infinite, the dimensionless heat transfer for this geometry, per unit surface area, is designed by the ratio $Q''(t)/Q_i''$. The volume of an infinitely long cylinder is also infinite, so the dimensionless heat transfer ratio is written, per unit length, as $Q'(t)/Q_i'$. The sphere has a finite volume, so the heat transfer ratio is simply $Q(t)/Q_i$ for that geometry. If the value of $Q(t)$ is positive, heat flows from the solid into the fluid, that is, the body is cooled. If it is negative, the solid is heated by the fluid.

Two general classes of transient problems can be solved by using the charts. One class of problem involves knowing the time, while the local temperature at that time is unknown. In the other type of problem, the local temperature is the known quantity and the time required to reach that temperature is the unknown. The first class of problems can be solved in a straightforward fashion by use of the charts. The second class of problem occasionally involves a trial-and-error procedure. Both types of solutions will be illustrated in the following examples.

EXAMPLE 2.12 In a fabrication process, steel components are formed hot and then quenched in water. Consider a 2.0-m-long, 0.2-m-diameter steel cylinder ($k = 40$ W/m K, $\alpha = 1.0 \times 10^{-5}$ m^2/s), initially at 400°C, that is suddenly quenched in water at 50°C. If the heat transfer coefficient is 200 W/m^2 K, calculate the following 20 min after immersion:

1. The center temperature
2. The surface temperature
3. The heat transferred to the water during the initial 20 min

SOLUTION Since the cylinder has a length 10 times the diameter, we can neglect end effects. To determine whether the internal resistance is negligible, we calculate first the Biot number

$$\text{Bi} = \frac{\bar{h}_c r_0}{k} = \frac{(200\ \text{W/m}^2\text{K})(0.1\ \text{m})}{40\ \text{W/m K}} = 0.5$$

Since the Biot number is larger than 0.1, the internal resistance is significant and we cannot use the lumped-capacitance method. To use the chart solution we calculate the appropriate dimensionless parameters according to Table 2.3:

$$\text{Fo} = \frac{\alpha t}{r_0^2} = \frac{(1 \times 10^{-5}\ \text{m}^2/\text{s})(20\ \text{min})(60\ \text{s/min})}{0.1^2\ \text{m}^2} = 1.2$$

and

$$\text{Bi}^2\,\text{Fo} = (0.5^2)(1.2) = 0.3$$

The initial amount of internal energy stored in the cylinder per unit length is

$$Q_i' = c\rho\pi r_0^2(T_i - T_\infty) = \left(\frac{k}{\alpha}\right)\pi r_0^2(T_i - T_\infty)$$

$$= \frac{40\ \text{W/m K}}{1 \times 10^{-5}\ \text{m}^2/\text{s}}(\pi)(0.1^2\ \text{m}^2)(350\ \text{K}) = 4.4 \times 10^7\ \text{W s/m}$$

The dimensionless centerline temperature for $1/\text{Bi} = 2.0$ and $\text{Fo} = 1.2$ from Fig. 2.43(a) is

$$\frac{T(0, t) - T_\infty}{T_i - T_\infty} = 0.35$$

Since $T_i - T_\infty$ is specified as 350°C and $T_\infty = 50$°C, $T(0, t) = (0.35)(350) + 50 = 172.5$°C.

The surface temperature at $r/r_0 = 1.0$ and $t = 1200$ s is obtained from Fig. 2.43(b) in terms of the centerline temperature:

$$\frac{T(r_0, t) - T_\infty}{T(0, t) - T_\infty} = 0.8$$

The surface temperature ratio is thus

$$\frac{T(r_0, t) - T_\infty}{(T_i - T_\infty)} = 0.8\frac{T(0, t) - T_\infty}{T_i - T_\infty} = (0.8)(0.35) = 0.28$$

and the surface temperature after 20 min is

$$T(r_0, t) = (0.28)(350) + 50 = 148°\text{C}$$

Then the amount of heat transferred from the steel rod to the water can be obtained from Fig. 2.43(c). Since $Q'(t)/Q'_i = 0.61$,

$$Q(t) = (0.61)\frac{(2\text{ m})(4.4 \times 10^7\text{ W s/m})}{3600\text{ s/h}} = 14.9\text{ kWh}$$

EXAMPLE 2.13 A large concrete wall 50 cm thick is initially at 60°C. One side of the wall is insulated. The other side is suddenly exposed to hot combustion gases at 900°C through a heat transfer coefficient of 25 W/m^2 K. Determine (a) the time required for the insulated surface to reach 600°C, (b) the temperature distribution in the wall at that instant, and (c) the heat transferred during the process. The following average physical properties are given:

$$k_s = 1.25\text{ W/m K}$$

$$c = 837\text{ J/kg K}$$

$$\rho = 500\text{ kg/m}^3$$

$$\alpha = 0.30 \times 10^{-5}\text{ m}^2\text{/s}$$

SOLUTION Note that the wall thickness is equal to L since the insulated surface corresponds to the center plane of a slab of thickness $2L$ when both surfaces experience a thermal change. The temperature ratio $(T_s - T_\infty)/(T_i - T_\infty)$ for the insulated face at the time sought is

$$\left.\frac{T_s - T_\infty}{T_i - T_\infty}\right|_{x=0} = \frac{600 - 900}{60 - 900} = 0.357$$

and the reciprocal of the Biot number is

$$\frac{k_s}{\bar{h}L} = \frac{1.25\text{ W/m K}}{(25\text{ W/m}^2\text{ K})(0.5\text{ m})} = 0.10$$

From Fig. 2.42(a) we find that for the above conditions the Fourier number $\alpha t/L^2 =$ 0.70 at the midplane. Therefore,

$$t = \frac{(0.7)(0.5^2\text{ m}^2)}{0.3 \times 10^{-5}\text{ m}^2\text{/s}}$$

$$= 58{,}333\text{ s} = 16.2\text{ h}$$

The temperature distribution in the wall 16 h after the transient was initiated can be obtained from Fig. 2.42(b) for various values of x/L, as shown below:

$\frac{x}{L}$	**1.0**	**0.8**	**0.6**	**0.4**	**0.2**
$\frac{T\left(\frac{x}{L}\right) - T_\infty}{T(0) - T_\infty}$	0.13	0.41	0.64	0.83	0.96

From the above dimensionless data we can obtain the temperature distribution as a function of distance from the insulated surface:

***x*,m**	**0.5**	**0.4**	**0.3**	**0.2**	**0.1**	**0**
$T_\infty - T(x)$, °C	39	123	192	249	288	300
$T(x)$, °C	861	777	708	651	612	600

The heat transferred to the wall per square meter of surface area during the transient can be obtained from Fig. 2.42(c). For Bi = 10, $Q''(t)/Q_i''$ at Bi^2 Fo = 70 is 0.70. Thus we get

$$Q''(t) = c\rho L(T_i - T_\infty) = (837 \text{ J/kg K})(500 \text{ kg/m}^3)(0.5 \text{ m})(-840 \text{ K})$$
$$= -1.758 \times 10^8 \text{ J/m}^2$$

The minus sign indicates that the heat was transferred into the wall and the internal energy increased during the process.

2.7.2* Multidimensional Systems†

The use of the one-dimensional transient charts can be extended to two- and three-dimensional problems [15]. The method involves using the product of multiple values from the one-dimensional charts, Figs. 2.40, 2.42, and 2.43. The basis for obtaining two- and three-dimensional solutions from one-dimensional charts is the manner in which partial differential equations can be separated into the product of two or three ordinary differential equations. A proof of the method can be found in Arpaci ([16], Section 5-2). Again, it should be recognized that although computational techniques (discussed in Chapter 3) are now increasingly used to solve most multidimensional transient conduction, the use of charts provides a quick estimate tool in most cases before one carries out a more detailed analysis.

The product solution method can best be illustrated by an example. Suppose we wish to determine the transient temperature at point P in a cylinder of finite length, as shown in Fig. 2.45. The point P is located by the two coordinates (x, r), where x is the axial location measured from the center of the cylinder and r is the radial position. The initial condition and boundary conditions are the same as those that apply to the

†F. Kreith and W.Z. Black, *Basic Heat Transfer*, Harper & Row, New York, 1980.

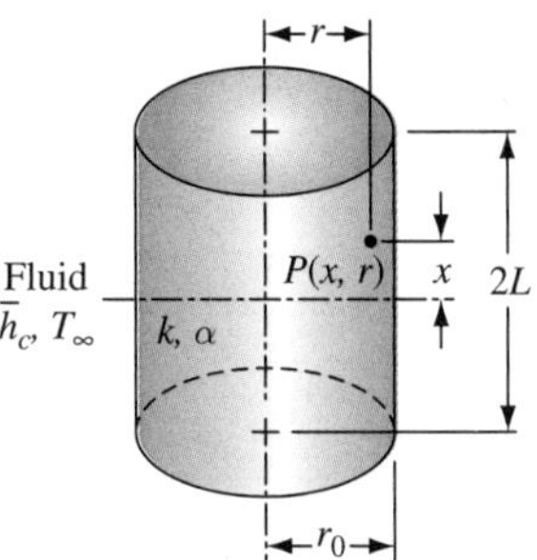

FIGURE 2.45 Geometry for a short-cylinder product solution.

transient one-dimensional charts. The cylinder is initially at a uniform temperature T_i. At time $t = 0$ the entire surface is subjected to a fluid with constant ambient temperature T_∞, and the convection heat transfer coefficient between the cylinder surface area and fluid is a uniform and constant value $\bar{h}_c$.

The radial temperature distribution for an infintely long cylinder is given in Fig. 2.43. For a cylinder with finite length, the radial and axial temperature distribution is given by the product solution of an infinitely long cylinder and infinite plate

$$\frac{\theta_p(r, x)}{\theta_i} = C(r)P(x)$$

where the symbols $C(r)$ and $P(x)$ are the dimensionless temperatures of the infinite cylinder and infinite plate, respectively:

$$C(r) = \frac{\theta(r, t)}{\theta_i}$$

$$P(x) = \frac{\theta(x, t)}{\theta_t}$$

The solution for $C(r)$ is obtained from Figs. 2.43(a) and (b), while the value for $P(x)$ is obtained from Figs. 2.42(a) and (b).

Solutions for other two- and three-dimensional geometries can be obtained using a procedure similar to the one illustrated for the finite cylinder. Three-dimensional problems involve the product of three solutions, while two-dimensional problems can be solved by taking the product of two solutions.

Two-dimensional geometries that have chart solutions are summarized in Table 2.4. Three-dimensional solutions are outlined in Table 2.5. The symbols used in the two tables represent the following solutions:

$S(x) = \dfrac{\theta(x, t)}{\theta_i}$ for a semi-infinite solid. The ordinate in Fig. 2.40 gives $1 - S(x)$.

$P(x) = \dfrac{\theta(x, t)}{\theta_i}$ for an infinite plate, Figs. 2.42(a) and (b)

$C(r) = \dfrac{\theta(r, t)}{\theta_i}$ for a long cylinder, Figs. 2.43(a) and (b)

TABLE 2.4 Schematic diagrams and nomenclature for product solutions to transient conduction problems with Figs. 2.40, 2.42, and 2.43 for two-dimensional systems

	Geometry	Dimensionless Temperature at Point *P*
Semi-infinite plate	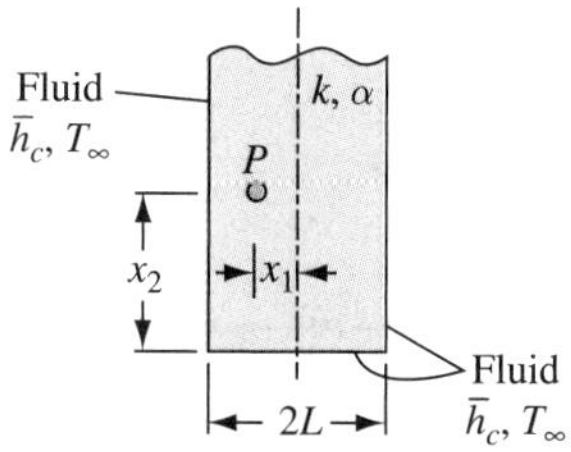	$\dfrac{\theta_p(x_1, x_2)}{\theta_i} = P(x_1)S(x_2)$
Infinite rectangular bar	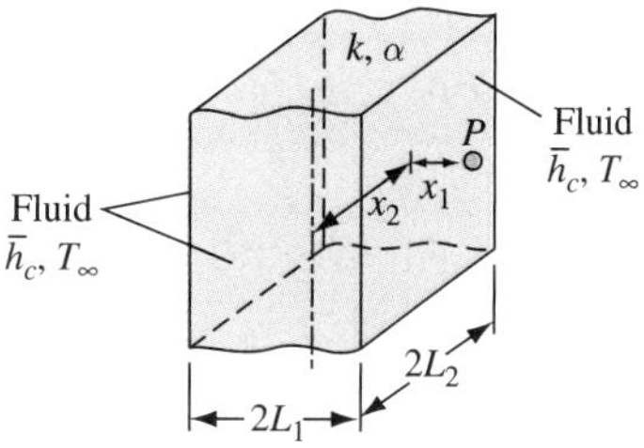	$\dfrac{\theta_p(x_1, x_2)}{\theta_i} = P(x_1)P(x_2)$
One-quarter infinite solid	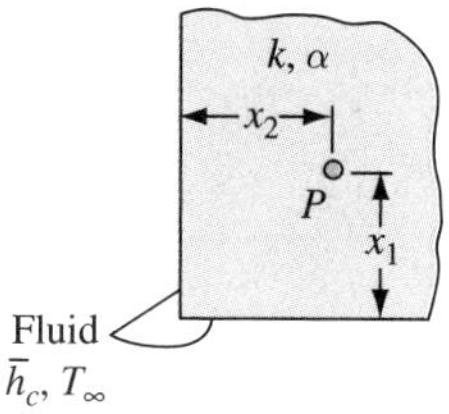	$\dfrac{\theta_p(x_1, x_2)}{\theta_i} = S(x_1)S(x_2)$
Semi-infinite cylinder	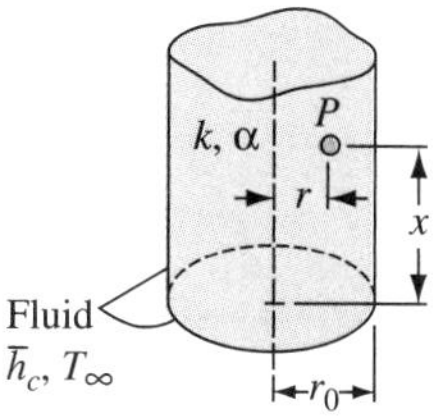	$\dfrac{\theta_p(x, r)}{\theta_i} = S(x)C(r)$
Finite cylinder	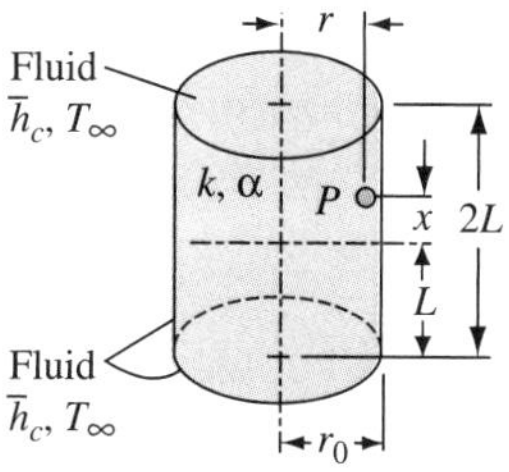	$\dfrac{\theta_p(x, r)}{\theta_i} = P(x)C(r)$

TABLE 2.5 Schematic diagrams and nomenclature for product solutions to transient conduction problems with Figs. 2.40, 2.42, and 2.43 for three-dimensional systems

	Geometry	Dimensionless Temperature at Point P
Semi-infinite rectangular bar	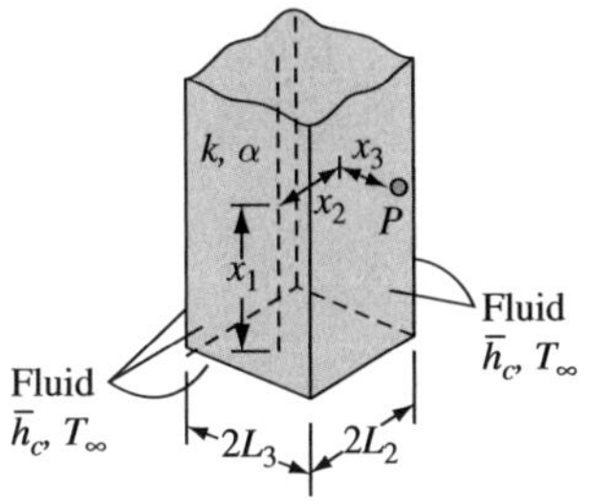	$\dfrac{\theta_p(x_1, x_2, x_3)}{\theta_i} = S(x_1)P(x_2)P(x_3)$
Rectangular parallelepiped	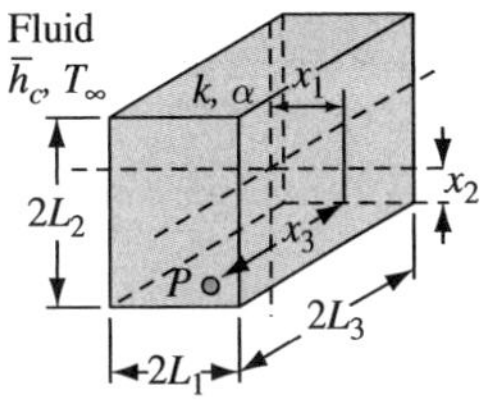	$\dfrac{\theta_p(x_1, x_2, x_3)}{\theta_i} = P(x_1)P(x_2)P(x_3)$
One-quarter infinite plate	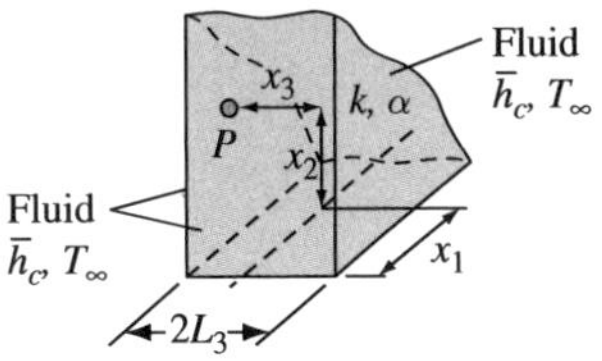	$\dfrac{\theta_p(x_1, x_2, x_3)}{\theta_i} = S(x_1)S(x_2)P(x_3)$
One-eighth infinite plate	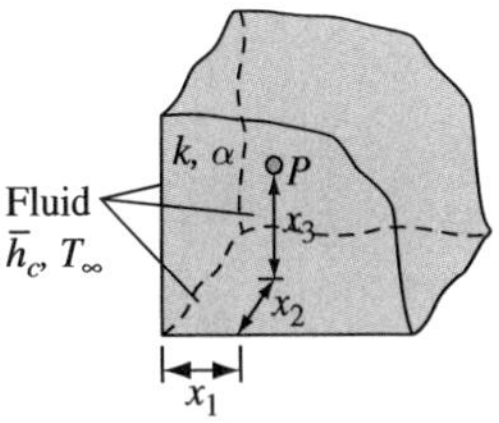	$\dfrac{\theta_p(x_1, x_2, x_3)}{\theta_i} = S(x_1)S(x_2)S(x_3)$

The extension of the one-dimensional charts to two- and three-dimensional geometries allows us to solve a large variety of transient conduction problems.

EXAMPLE 2.14 A 10-cm-diameter, 16-cm-long cylinder with properties $k = 0.5$ W/m K and $\alpha = 5 \times 10^{-7}$ m^2/s is initially at a uniform temperature of 20°C. The cylinder is placed in an oven where the ambient air temperature is 500°C and $\bar{h}_c = 30$ W/m^2 K. Determine the minimum and maximum temperatures in the cylinder 30 min after it has been placed in the oven.

SOLUTION The Biot number based on the cylinder radius is

$$\text{Bi} = \frac{\bar{h}_c r_0}{k} = \frac{(30\ \text{W/m}^2\ \text{K})(0.05\ \text{m})}{(0.5\ \text{W/m K})} = 3.0$$

The problem cannot be solved by using the simplified approach assuming negligible internal resistance; a chart solution is necessary.

Table 2.4 indicates that the temperature distribution in a cylinder of finite length can be determined by the product of the solution for an infinite plate and an infinite cylinder. At any time, the minimum temperature is at the geometric center of the cylinder and the maximum temperature is at the outer circumference at each end of the cylinder. Using the coordinates for the finite cylinder shown in Fig. 2.45, we have

$$\text{minimum temperature at:} \quad x = 0 \quad r = 0$$
$$\text{maximum temperature at:} \quad x = L \quad r = r_0$$

The calculations are summarized in the following tables.

Infinite Plate

$\text{Fo} = \dfrac{\alpha t}{L^2}$	$\text{Bi}^{-1} = \dfrac{k}{\bar{h}_c L}$	$P(0) = \dfrac{\theta(0, t)}{\theta_i}$ [Fig. 2.42(a)]	$P(L) = \dfrac{\theta(L, t)}{\theta_i}$ [Figs. 2.42(a) and (b)]
$\dfrac{(5 \times 10^{-7})(1800)}{(0.08)^2} = 0.14$	$\dfrac{0.5}{(30)(0.08)} = 0.21$	0.90	$(0.90)(0.27) = 0.243$

Infinite Cylinder

$\text{Fo} = \dfrac{\alpha t}{r_0^2}$	$\text{Bi}^{-1} = \dfrac{k}{\bar{h}_c r_0}$	$C(0) = \dfrac{\theta(0, t)}{\theta_i}$ [Fig. 2.43(a)]	$C(r_0) = \dfrac{\theta(r_0, t)}{\theta_i}$ [Figs. 2.43(a) and (b)]
$\dfrac{(5 \times 10^{-7})(1800)}{(0.05)^2} = 0.36$	$\dfrac{0.5}{(30)(0.05)} = 0.33$	0.47	$(0.47)(0.33) = 0.155$

The minimum cylinder temperature is

$$\frac{\theta_{\min}}{\theta_i} = P(0)C(0) = (0.90)(0.47) = 0.423$$
$$T_{\min} = 0.423(20 - 500) + 500 = 297°\text{C}$$

The maximum cylinder temperature is

$$\frac{\theta_{\max}}{\theta_i} = P(L)C(r_0) = (0.243)(0.155) = 0.038$$
$$T_{\max} = 0.038(20 - 500) + 500 = 482°\text{C}$$

2.8 Closing Remarks

In this chapter, we have considered methods of analyzing heat conduction problems in the steady and unsteady states. Problems in the steady state are divided into one-dimensional and multidimensional geometries. For one-dimensional problems, solutions are available in the form of simple equations that can incorporate various boundary conditions by using thermal circuits. For problems of heat conduction in more than one dimension, solutions can be obtained by analytic, graphic, and numerical means. The analytic approach is recommended only for situations involving systems with a simple geometry and simple boundary conditions. It is accurate and lends itself readily to parameterization, but when the boundary conditions are complex, the analytic approach usually becomes too involved to be practical, and for complex geometries it is impossible to obtain a closed-form solution.

Systems of complex geometries but having isothermal and insulated boundaries are readily amenable to graphic solutions. The graphic method, however, becomes unwieldy when the boundary conditions involve heat transfer through a surface conductance. For such cases the numerical approach to be considered in the next chapter is recommended because it can easily be adapted to all kinds of boundary conditions and geometric shapes.

Conduction problems in the unsteady state can be subdivided into those that can be handled by the lumped-capacity method and those in which the temperature is a function not only of time, but also of one or more spatial coordinates. In the lumped-capacity method, which is a good approximation for conditions in which the Biot number is less than 0.1, it is assumed that internal conduction is sufficiently large that the temperature throughout the system can be considered uniform at any instant in time. When this approximation is not permissible, it is necessary to set up and solve partial differential equations, which generally require series solutions that are attainable only for simple geometric shapes. However, for spheres, cylinders, slabs, plates, and other simple geometric shapes, the results of analytic solutions have been presented in the form of charts that are relatively easy and straightforward to use. As in the case of steady-state conduction problems, when the geometries are complex and when the boundary conditions vary with time or have other complex features, it is necessary to obtain the solution by numerical means, as discussed in the next chapter.

References

1. H. S. Carslaw and J. C. Jaeger, *Conduction of Heat in Solids*, 2d ed., Oxford University Press, London, 1986.
2. K. A. Gardner, "Efficiency of Extended Surfaces," *Trans. ASME*, vol. 67, pp. 621–631, 1945.
3. W. P. Harper and D. R. Brown, "Mathematical Equation for Heat Conduction in the Fins of Air-Cooled Engines," NACA Rep. 158, 1922.
4. R. M. Manglik, "Heat Transfer Enhancement," *Heat Transfer Handbook*, A. Bejan and A. D. Kraus, eds., Wiley, Hoboken, NJ, 2003, Ch. 14.
5. P. J. Schneider, *Conduction Heat Transfer*, Addison-Wesley, Cambridge, Mass., 1955.
6. M. N. Ozisik, *Boundary Value Problems of Heat Conduction*, International Textbook Co., Scranton, Pa., 1968.

7. L. M. K. Boelter, V. H. Cherry, and H. A. Johnson, *Heat Transfer Notes*, 3d ed., University of California Press, Berkeley, 1942.
8. C. F. Kayan, "An Electrical Geometrical Analogue for Complex Heat Flow," *Trans. ASME*, vol. 67, pp. 713–716, 1945.
9. I. Langmuir, E. Q. Adams, and F. S. Meikle, "Flow of Heat through Furnace Walls," *Trans. Am. Electrochem. Soc.*, vol. 24, pp. 53–58, 1913.
10. O. Rüdenberg, "Die Ausbreitung der Luft und Erdfelder um Hochspannungsleitungen besonders bei Erd-und Kurzschlüssen," *Electrotech. Z*, vol. 46, pp. 1342–1346, 1925.
11. B. O. Pierce, *A Short Table of Integrals*, Ginn, Boston, 1929.
12. M. P. Heisler, "Temperature Charts for Induction and Constant Temperature Heating," *Trans. ASME*, vol. 69, pp. 227–236, 1947.
13. H. Gröber, S. Erk, and U. Grigull, *Fundamentals of Heat Transfer*, McGraw-Hill, New York, 1961.
14. P. J. Schneider, *Temperature Response Charts*, Wiley, New York, 1963.
15. F. Kreith and W. Z. Black, *Basic Heat Transfer*, Harper & Row, New York, 1980.
16. V. Arpaci, *Heat Transfer*, Prentice Hall, Upper Saddle River, NJ, 2000.
17. S. Kakaç and Y. Yener, *Heat Conduction*, 2d ed., Hemisphere, Washington, D.C., 1988.

Problems

The problems for this chapter are organized by subject matter as shown below.

Topic	Problem Number
Conduction equation	2.1–2.2
Steady-state conduction in simple geometries	2.3–2.30
Extended surfaces	2.31–2.42
Multidimensional steady-state conduction	2.43–2.57
Transient conduction (analytical solutions)	2.58–2.69
Transient conduction (chart solutions)	2.70–2.87

2.1 The heat conduction equation in cylindrical coordinates is

$$\rho c \frac{\partial T}{\partial t} = k\left(\frac{\partial^2 T}{\partial r^2} + \frac{1}{r}\frac{\partial T}{\partial r} + \frac{1}{r^2}\frac{\partial^2 T}{\partial \phi^2} + \frac{\partial^2 T}{\partial z^2}\right) + \dot{q}_G$$

(a) Simplify this equation by eliminating terms equal to zero for the case of steady-state heat flow without sources or sinks around a right-angle corner such as the one in the accompanying sketch. It may be assumed that the corner extends to infinity in the direction perpendicular to the page. (b) Solve the resulting equation for the temperature distribution by substituting the boundary condition. (c) Determine the rate of heat flow from T_1 to T_2. Assume $k = 1$ W/m K and unit depth.

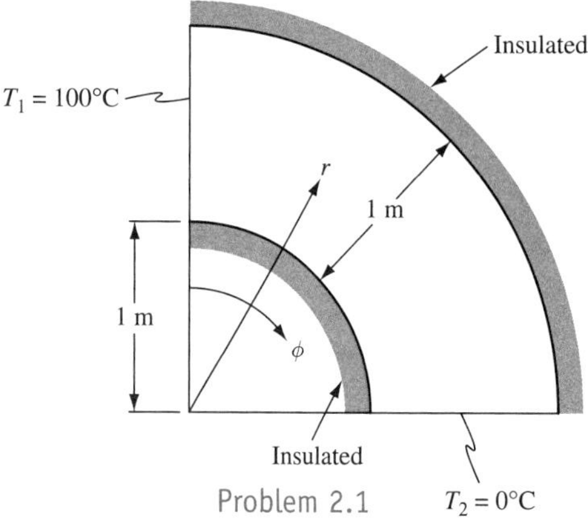

Problem 2.1

2.2 Write Eq. (2.20) in a dimensionless form similar to Eq. (2.17).

2.3 Calculate the rate of heat loss per foot and the thermal resistance for a 6-in. schedule 40 steel pipe covered with a 3-in.-thick layer of 85% magnesia. Superheated steam at 300°F flows inside the pipe ($\bar{h}_c = 30$ Btu/h ft^2 °F), and still air at 60°F is on the outside ($\bar{h}_c = 5$ Btu/h ft^2 °F).

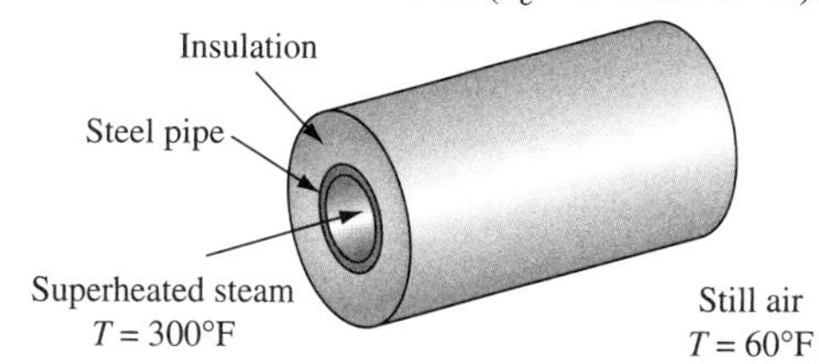

Problem 2.3

2.4 Suppose that a pipe carrying a hot fluid with an external temperature of T_i and outer radius r_i is to be insulated with an insulation material of thermal conductivity k and outer radius r_o. Show that if the convection heat transfer coefficient on the outside of the insulation is $\bar{h}$ and the environmental temperature is T_∞, the addition of insulation can actually increase the rate of heat loss if $r_o < k/\bar{h}$ and the maximum heat loss occurs when $r_o = k/\bar{h}$. This radius, r_c, is often called the critical radius.

2.5 A solution with a boiling point of 180°F boils on the outside of a 1-in. tube with a No. 14 BWG gauge wall. On the inside of the tube flows saturated steam at 60 psia. The convection heat transfer coefficients are 1500 Btu/h ft^2 °F on the steam side and 1100 Btu/h ft^2 °F on the exterior surface. Calculate the increase in the rate of heat transfer if a copper tube is used instead of a steel tube.

2.6 Steam having a quality of 98% at a pressure of 1.37×10^5 N/m^2 is flowing at a velocity of 1 m/s through a steel pipe of 2.7-cm OD and 2.1-cm ID. The heat transfer coefficient at the inner surface, where condensation occurs, is 567 W/m^2 K. A dirt film at the inner surface adds a unit thermal resistance of 0.18 m^2 K/W. Estimate the rate of heat loss per meter length of pipe if (a) the pipe is bare, (b) the pipe is covered with a 5-cm layer of 85% magnesia insulation. For both cases assume that the convection heat transfer coefficient at the outer surface is 11 W/m^2 K and that the environmental temperature is 21°C. Also estimate the quality of the steam after a 3-m length of pipe in both cases.

2.7 Estimate the rate of heat loss per unit length from a 2-in.ID, $2\frac{3}{8}$ in. OD steel pipe covered with high temperature insulation having a thermal conductivity of 0.065 Btu/h ft and a thickness of 0.5 in. Steam flows in the pipe. It has a quality of 99% and is at 300°F. The unit thermal resistance at the inner wall is 0.015 h ft^2 °F/Btu, the heat transfer coefficient at the outer surface is 3.0 Btu/h ft^2 °F, and the ambient temperature is 60°F.

2.8 The rate of heat flow per unit length q/L through a hollow cylinder of inside radius r_i and outside radius r_o is

$$q/L = (\bar{A}\, k\, \Delta T)/(r_o - r_i)$$

where $\bar{A} = 2\pi(r_o - r_i)/\ln(r_o/r_i)$. Determine the percent error in the rate of heat flow if the arithmetic mean area $\pi(r_o + r_i)$ is used instead of the logarithmic mean area $\bar{A}$ for ratios of outside-to-inside diameters (D_o/D_i) of 1.5, 2.0, and 3.0. Plot the results.

2.9 A 2.5-cm-OD, 2-cm-ID copper pipe carries liquid oxygen to the storage site of a space shuttle at −183°C and 0.04 m^3/min. The ambient air is at 21°C and has a dew point of 10°C. How much insulation with a thermal conductivity of 0.02 W/m K is needed to prevent condensation on the exterior of the insulation if $h_c + h = 17$ W/m^2 K on the outside?

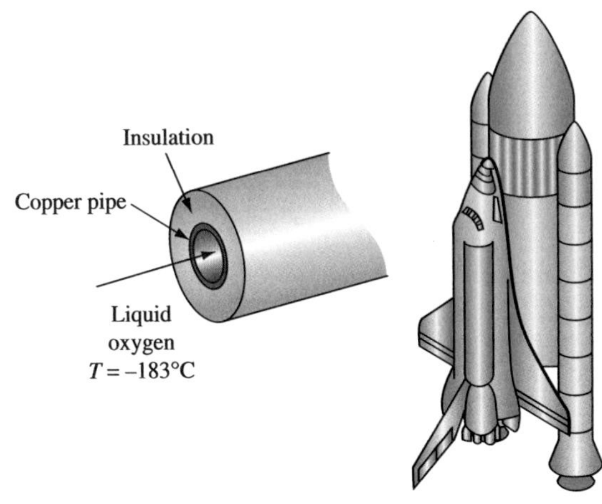

Problem 2.9

2.10 A salesperson for insulation material claims that insulating exposed steam pipes in the basement of a large hotel will be cost-effective. Suppose saturated steam at 5.7 bars flows through a 30-cm-OD steel pipe with a 3-cm wall thickness. The pipe is surrounded by air at 20°C. The convection heat transfer coefficient on the outer surface of the pipe is estimated to be 25 W/m^2 K. The cost of generating steam is estimated to be \$5 per 10^9 J, and the salesman offers to install a 5-cm-thick layer of 85% magnesia insulation on the pipes for \$200/m or a 10-cm-thick layer for \$300/m. Estimate the payback time for these two alternatives, assuming that the steam line operates all year long, and make a recommendation to the hotel owner. Assume that the surface of the pipe as well as the insulation have a low emissivity and radiative heat transfer is negligible.

2.11 A hollow sphere with inner and outer radii of R_1 and R_2, respectively, is covered with a layer of insulation having an outer radius of R_3. Derive an expression for the rate of heat transfer through the insulated sphere in terms of the radii, the thermal conductivities, the heat transfer coefficients, and the temperatures of the interior and the surrounding medium of the sphere.

2.12 The thermal conductivity of a material can be determined in the following manner. Saturated steam at 2.41×10^5 N/m^2 is condensed at the rate of 0.68 kg/h inside a hollow iron sphere that is 1.3 cm thick and has an internal diameter of 51 cm. The sphere is coated with the material whose thermal conductivity is to be evaluated. The thickness of the material to be tested is 10 cm, and there are two thermocouples embedded in it, one 1.3 cm from the surface of the iron sphere and one 1.3 cm from the exterior surface of the system. If the inner thermocouple indicates a temperature of

110°C and the outer thermocouple a temperature of 57°C, calculate (a) the thermal conductivity of the material surrounding the metal sphere, (b) the temperatures at the interior and exterior surfaces of the test material, and (c) the overall heat transfer coefficient based on the interior surface of the iron sphere, assuming the thermal resistances at the surfaces, as well as at the interface between the two spherical shells, are negligible.

2.13 A cylindrical liquid oxygen (LOX) tank has a diameter of 4 ft, a length of 20 ft, and hemispherical ends. The boiling point of LOX is −297°F. An insulation is sought that will reduce the boil-off rate in the steady state to no more than 25 lb/h. The heat of vaporization of LOX is 92 Btu/lb. If the thickness of this insulation is to be no more than 3 in., what would the value of its thermal conductivity have to be?

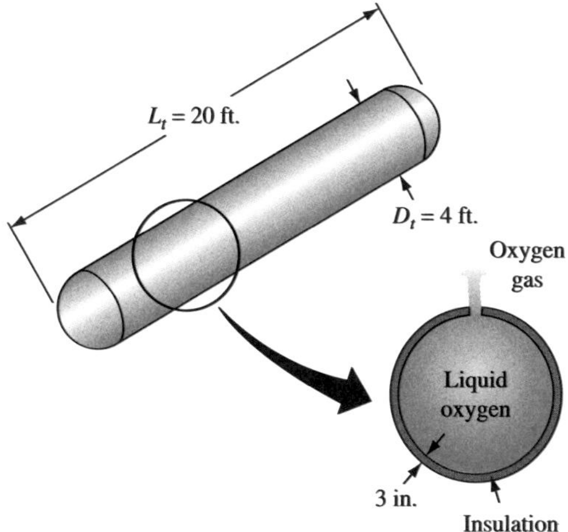

Problem 2.13

2.14 The addition of insulation to a cylindrical surface such as a wire, may increase the rate of heat dissipation to the surroundings (see Problem 2.4). (a) For a No. 10 wire (0.26 cm in diameter), what is the thickness of rubber insulation ($k = 0.16$ W/m K) that will maximize the rate of heat loss if the heat transfer coefficient is 10 W/m^2 K? (b) If the current-carrying capacity of this wire is considered to be limited by the insulation temperature, what percent increase in capacity is realized by addition of the insulation? State your assumptions.

2.15 For the system outlined in Problem 2.11, determine an expression for the critical radius of the insulation in terms of the thermal conductivity of the insulation and the surface coefficient between the exterior surface of the insulation and the surrounding fluid. Assume that the temperature difference, R_1, R_2, the heat transfer coefficient on the interior, and the thermal conductivity of the material of the sphere between R_1 and R_2 are constant.

2.16 A standard 4-in. steel pipe (ID = 4.026 in., OD = 4.500 in.) carries superheated steam at 1200°F in an enclosed space where a fire hazard exists, limiting the outer surface temperature to 100°F. To minimize the insulation cost, two materials are to be used: first a high-temperature (relatively expensive) insulation is to be applied to the pipe, and then magnesia (a less expensive material) will be applied on the outside. The maximum temperature of the magnesia is to be 600°F. The following constants are known:

steam-side coefficient	$h = 100$ Btu/h ft^2 °F
high-temperature insulation conductivity	$k = 0.06$ Btu/h ft °F
magnesia conductivity	$k = 0.045$ Btu/h ft °F
outside heat transfer coefficient	$h = 2.0$ Btu/h ft^2 °F
steel conductivity	$k = 25$ Btu/h ft °F
ambient temperature	$T_a = 70$°F

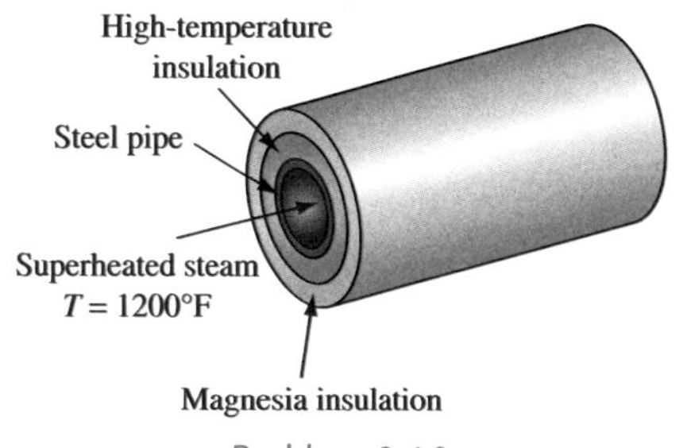

Problem 2.16

(a) Specify the thickness for each insulating material. (b) Calculate the overall heat transfer coefficient based on the pipe OD. (c) What fraction of the total resistance is due to (1) steam-side resistance, (2) steel pipe resistance, (3) insulation (the combination of the two), and (4) outside resistance? (d) How much heat is transferred per hour per foot length of pipe?

2.17 Show that the rate of heat conduction per unit length through a long, hollow cylinder of inner radius r_i and outer radius r_o, made of a material whose thermal conductivity varies linearly with temperature, is given by

$$\frac{q_k}{L} = \frac{T_i - T_o}{(r_o - r_i)/k_m\bar{A}}$$

where
T_i = temperature at the inner surface
T_o = temperature at the outer surface
$\bar{A} = 2\pi(r_o - r_i)/\ln(r_o/r_i)$
$k_m = k_o[1 + \beta_k(T_i + T_o)/2]$
L = length of cylinder

2.18 A long, hollow cylinder is constructed from a material whose thermal conductivity is a function of temperature according to $k = 0.060 + 0.00060T$, where T is in °F and k

is in Btu/h ft °F. The inner and outer radii of the cylinder are 5 and 10 in., respectively. Under steady-state conditions, the temperature at the interior surface of the cylinder is 800°F and the temperature at the exterior surface is 200°F. (a) Calculate the rate of heat transfer per foot length, taking into account the variation in thermal conductivity with temperature. (b) If the heat transfer coefficient on the exterior surface of the cylinder is 3 Btu/h ft^2 °F, calculate the temperature of the air on the outside of the cylinder.

2.19 A plane wall 15-cm thick has a thermal conductivity given by the relation

$$k = 2.0 + 0.0005T \text{ W/m K}$$

where T is in kelvin. If one surface of this wall is maintained at 150°C and the other at 50°C, determine the rate of heat transfer per square meter. Sketch the temperature distribution through the wall.

2.20 A plane wall, 7.5 cm thick, generates heat internally at the rate of 10^5 W/m^3. One side of the wall is insulated, and the other side is exposed to an environment at 90°C. The convection heat transfer coefficient between the wall and the environment is 500 W/m^2 K. If the thermal conductivity of the wall is 12 W/m K, calculate the maximum temperature in the wall.

2.21 A small dam, which can be idealized by a large slab 1.2-m thick, is to be completely poured in a short period of time. The hydration of the concrete results in the equivalent of a distributed source of constant strength of 100 W/m^3. If both dam surfaces are at 16°C, determine the maximum temperature to which the concrete will be subjected, assuming steady-state conditions. The thermal conductivity of the wet concrete can be taken as 0.84 W/m K.

2.22 Two large steel plates at temperatures of 90°C and 70°C are separated by a steel rod 0.3 m long and 2.5 cm in diameter. The rod is welded to each plate. The space between the plates is filled with insulation that also insulates the circumference of the rod. Because of a voltage difference between the two plates, current flows through the rod, dissipating electrical energy at a rate of 12 W. Determine the maximum temperature in the rod and the heat flow rate at each end. Check your results by comparing the net heat flow rate at the two ends with the total rate of heat generation.

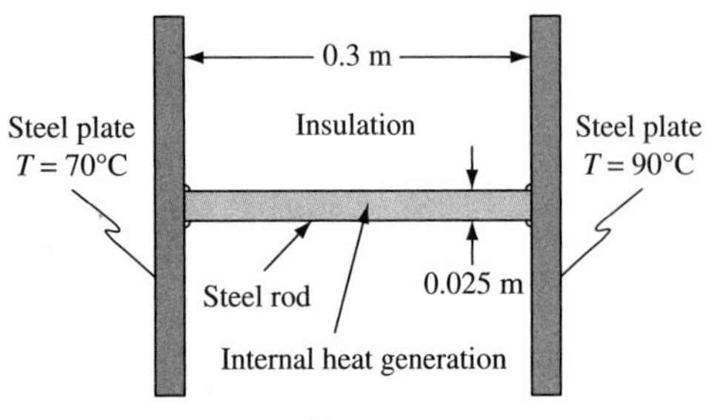

Problem 2.22

2.23 The shield of a nuclear reactor can be idealized by a large 10-in.-thick flat plate having a thermal conductivity of 2 Btu/h ft °F. Radiation from the interior of the reactor penetrates the shield and there produces heat generation that decreases exponentially from a value of 10 Btu/h in.^{3} at the inner surface to a value of 1.0 Btu/h in.^{3} at a distance of 5 in. from the interior surface. If the exterior surface is kept at 100°F by forced convection, determine the temperature at the inner surface of the field. *Hint*: First set up the differential equation for a system in which the heat generation rate varies according to $\dot{q}(x) = \dot{q}(0)e^{-Cx}$.

2.24 Derive an expression for the temperature distribution in an infinitely long rod of uniform cross section within which there is uniform heat generation at the rate of 1 W/m. Assume that the rod is attached to a surface at T_s and is exposed through a convection heat transfer coefficient h to a fluid at T_f.

2.25 Derive an expression for the temperature distribution in a plane wall in which there are uniformly distributed heat sources that vary according to the linear relation

$$\dot{q}_G = \dot{q}_w[1 - \beta(T - T_w)]$$

where $\dot{q}_w$ is a constant equal to the heat generation per unit volume at the wall temperature T_w. Both sides of the plate are maintained at T_w and the plate thickness is $2L$.

2.26 A plane wall of thickness $2L$ has internal heat sources whose strength varies according to

$$\dot{q}_G = \dot{q}_o \cos(ax)$$

where $\dot{q}_o$ is the heat generated per unit volume at the center of the wall ($x = 0$) and a is a constant. If both sides of the wall are maintained at a constant temperature of T_w, derive an expression for the total heat loss from the wall per unit surface area.

2.27 Heat is generated uniformly in the fuel rod of a nuclear reactor. The rod has a long, hollow cylindrical shape with its inner and outer surfaces at temperatures of T_i and T_o, respectively. Derive an expression for the temperature distribution.

2.28 Show that the temperature distribution in a sphere of radius r_o, made of a homogeneous material in which energy is released at a uniform rate per unit volume $\dot{q}_G$, is

$$T(r) = T_o + \frac{\dot{q}_G r_o^2}{6k}\left[1 - \left(\frac{r}{r_o}\right)^2\right]$$

2.29 In a cylindrical fuel rod of a nuclear reactor, heat is generated internally according to the equation

$$\dot{q}_G = \dot{q}_1\left[1 - \left(\frac{r}{r_o}\right)^2\right]$$

where $\dot{q}_G$ = local rate of heat generation per unit volume at r

r_o = outside radius

$\dot{q}_1$ = rate of heat generation per unit volume at the centerline

Calculate the temperature drop from the centerline to the surface for a 1-in.-diameter rod having a thermal conductivity of 15 Btu/h ft °F if the rate of heat removal from its surface is 500,000 Btu/h ft^2.

2.30 An electrical heater capable of generating 10,000 W is to be designed. The heating element is to be a stainless steel wire having an electrical resistivity of 80×10^{-6} ohm-centimeter. The operating temperature of the stainless steel is to be no more than 1260°C. The heat transfer coefficient at the outer surface is expected to be no less than 1720 W/m^2 K in a medium whose maximum temperature is 93°C. A transformer capable of delivering current at 9 and 12 V is available. Determine a suitable size for the wire, the current required, and discuss what effect a reduction in the heat transfer coefficient would have. (*Hint*: Demonstrate *first* that the temperature drop between the center and the surface of the wire is independent of the wire diameter, and determine its value.)

2.31 The addition of aluminum fins has been suggested to increase the rate of heat dissipation from one side of an electronic device 1 m wide and 1 m tall. The fins are to be rectangular in cross section, 2.5 cm long and 0.25 cm thick, as shown in the figure. There are to be 100 fins per meter. The convection heat transfer coefficient, both for the wall and the fins, is estimated to be 35 W/m^2 K. With this information determine the percent increase in the rate of heat transfer of the finned wall compared to the bare wall.

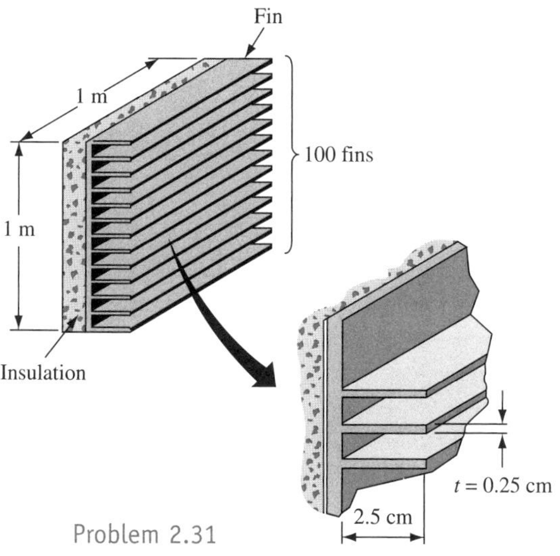

Problem 2.31

2.32 The tip of a soldering iron consists of a 0.6-cm-diameter copper rod, 7.6 cm long. If the tip must be 204°C, what are the required minimum temperature of the base and the heat flow, in Btu's per hour and in watts, into the base? Assume that $\bar{h} = 22.7$ W/m^2 K and $T_{air} = 21$°C.

2.33 One end of a 0.3-m-long steel rod is connected to a wall at 204°C. The other end is connected to a wall that is maintained at 93°C. Air is blown across the rod so that a heat transfer coefficient of 17 W/m^2 K is maintained over the entire surface. If the diameter of the rod is 5 cm and the temperature of the air is 38°C, what is the net rate of heat loss to the air?

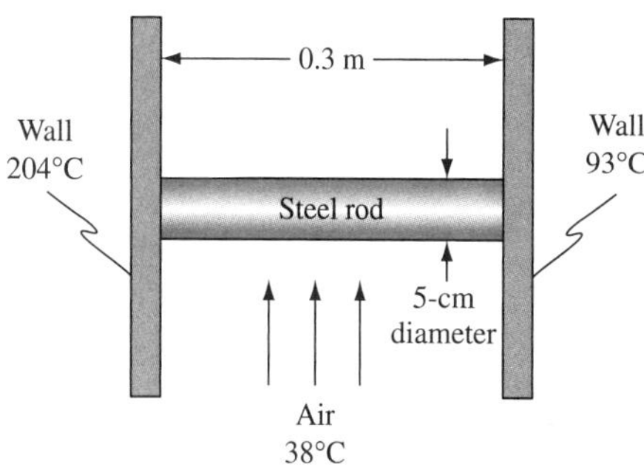

Problem 2.33

2.34 Both ends of a 0.6-cm copper U-shaped rod are rigidly affixed to a vertical wall as shown in the accompanying sketch. The temperature of the wall is maintained at 93°C. The developed length of the rod is 0.6 m, and it is exposed to air at 38°C. The combined radiation and convection heat transfer coefficient for this system is 34 W/m^2 K. (a) Calculate the temperature of the midpoint of the rod. (b) What will the rate of heat transfer from the rod be?

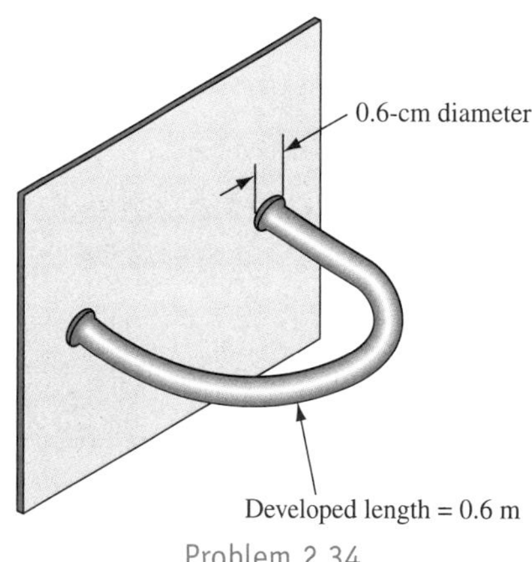

Problem 2.34

2.35 A circumferential fin of rectangular cross section, 3.7-cm OD and 0.3 cm thick, surrounds a 2.5-cm-diameter tube as shown below. The fin is constructed of mild steel. Air blowing over the fin produces a heat transfer coefficient of 28.4 W/m^2 K. If the temperatures of the base of the fin and the air are 260°C and 38°C, respectively, calculate the heat transfer rate from the fin.

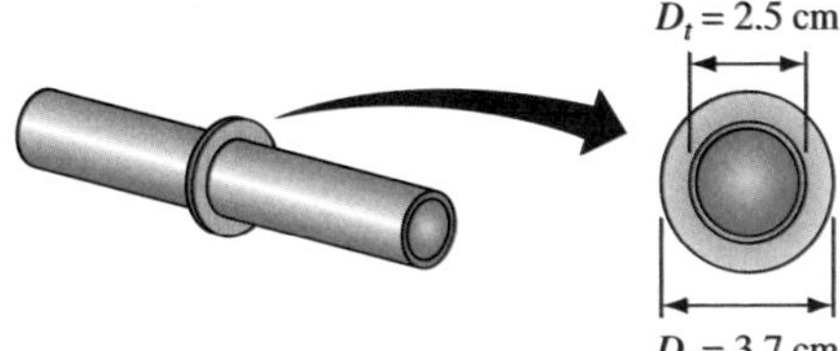

Problem 2.35

2.36 A turbine blade 6.3 cm long, with cross-sectional area $A = 4.6 \times 10^{-4}$ m^2 and perimeter $P = 0.12$ m, is made of stainless steel ($k = 18$ W/m K). The temperature of the root, T_s, is 482°C. The blade is exposed to a hot gas at 871°C, and the heat transfer coefficient $\bar{h}$ is 454 W/m^2 K. Determine the temperature of the blade tip and the rate of heat flow at the root of the blade. Assume that the tip is insulated.

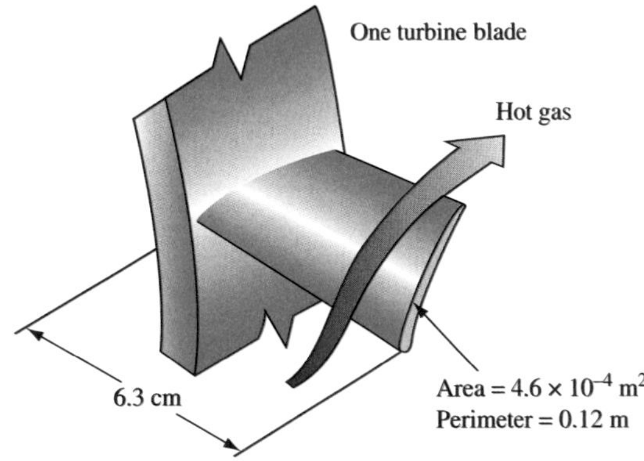

Problem 2.36

2.37 To determine the thermal conductivity of a long, solid 2.5-cm-diameter rod, one half of the rod was inserted into a furnace while the other half was projecting into air at 27°C. After steady state had been reached, the temperatures at two points 7.6 cm apart were measured and found to be 126°C and 91°C, respectively. The heat transfer coefficient over the surface of the rod exposed to the air was estimated to be 22.7 W/m^2 K. What is thermal conductivity of the rod?

2.38 Heat is transferred from water to air through a brass wall ($k = 54$ W/m K). The addition of rectangular brass fins, 0.08 cm thick and 2.5 cm long, spaced 1.25 cm apart, is contemplated. Assuming a water-side heat transfer coefficient of 170 W/m^2 K and an air-side heat transfer coefficient of 17 W/m^2 K, compare the gain in heat transfer rate achieved by adding fins to (a) the water side, (b) the air side, and (c) both sides. (Neglect temperature drop through the wall.)

2.39 The wall of a liquid-to-gas heat exchanger has a surface area on the liquid side of 1.8 m^2 (0.6 m × 3.0 m) with a heat transfer coefficient of 255 W/m^2 K. On the other side of the heat exchanger wall flows a gas, and the wall has 96 thin rectangular steel fins 0.5 cm thick and 1.25 cm high ($k = 3$ W/m K) as shown in the accompanying sketch. The fins are 3 m long and the heat transfer coefficient on the gas side is 57 W/m^2 K. Assuming that the thermal resistance of the wall is negligible, determine the rate of heat transfer if the overall temperature difference is 38°C.

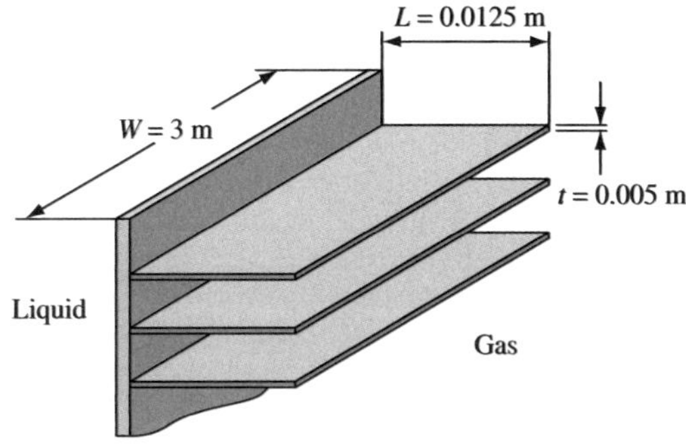

Problem 2.39

2.40 The top of a 12-in. I-beam is maintained at a temperature of 500°F, while the bottom is at 200°F. The thickness of the web is 1/2 in. Air at 500°F is blowing along the side of the beam so that $\bar{h} = 7$ Btu/h ft^2 °F. The thermal conductivity of the steel may be assumed constant and equal to 25 Btu/h ft °F. Find the temperature distribution along the web from top to bottom and plot the results.

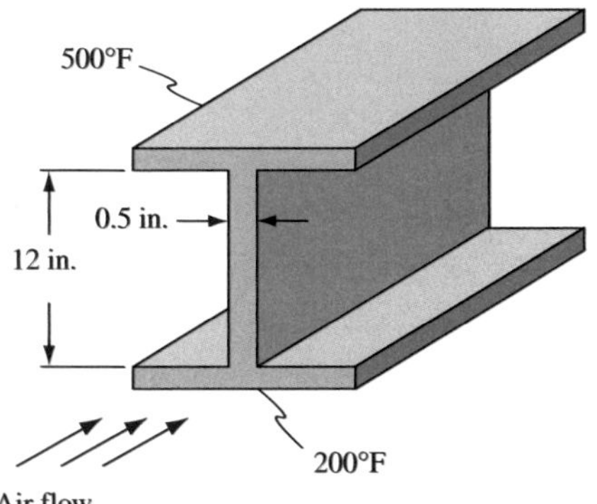

Problem 2.40

2.41 The handle of a ladle used for pouring molten lead is 30 cm long. Originally the handle was made of 1.9 cm × 1.25 cm mild steel bar stock. To reduce the grip temperature, it is proposed to form the handle of tubing 0.15 cm thick to the same rectangular shape. If the average heat transfer coefficient over the handle surface is 14 W/m^2 K, estimate the reduction of the temperature at the grip in air at 21°C.

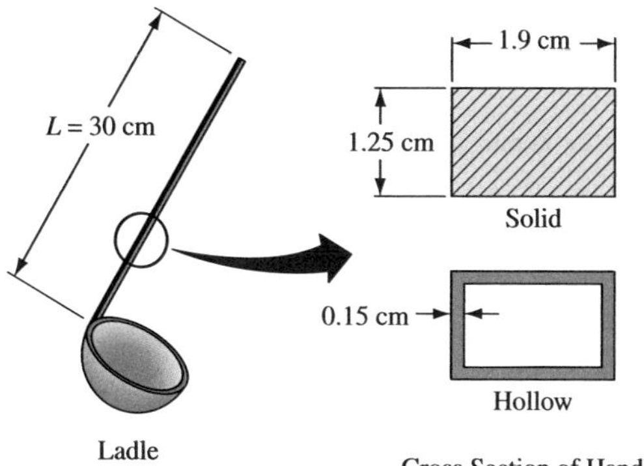

Problem 2.41

2.42 A 0.3-cm-thick aluminum plate has rectangular fins 0.16 cm × 0.6 cm, on one side, spaced 0.6 cm apart. The finned side is in contact with low pressure air at 38°C, and the average heat transfer coefficient is 28.4 W/m^2 K. On the unfinned side, water flows at 93°C and the heat transfer coefficient is 284 W/m^2 K. (a) Calculate the efficiency of the fins, (b) calculate the rate of heat transfer per unit area of wall, and (c) comment on the design if the water and air were interchanged.

2.43 Compare the rate of heat flow from the bottom to the top of the aluminum structure shown in the sketch below with the rate of heat flow through a solid slab. The top is at −10°C, the bottom at 0°C. The holes are filled with insulation that does not conduct heat appreciably.

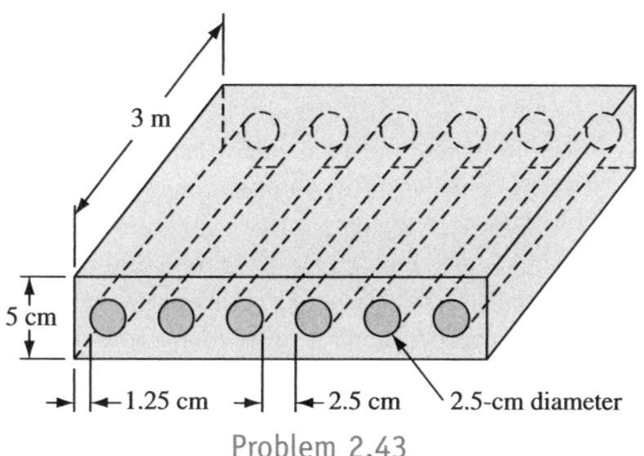

Problem 2.43

2.44 Determine by means of a flux plot the temperatures and heat flow per unit depth in the ribbed insulation shown in the accompanying sketch.

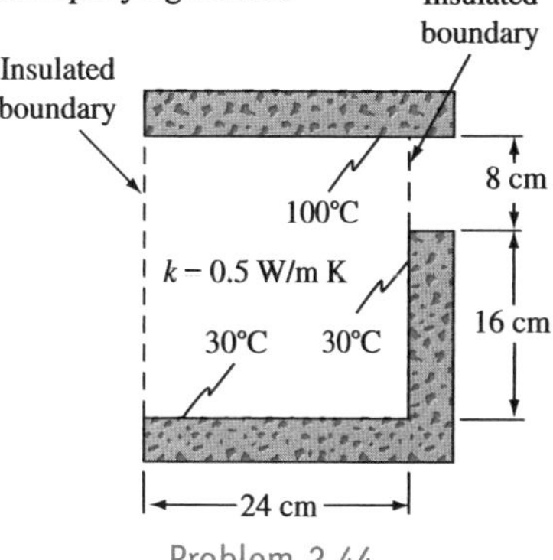

Problem 2.44

2.45 Use a flux plot to estimate the rate of heat flow through the object shown in the sketch. The thermal conductivity of the material is 15 W/m K. Assume no heat is lost from the sides.

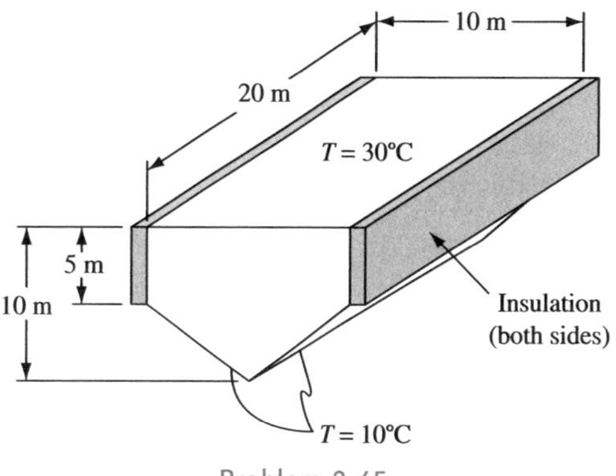

Problem 2.45

2.46 Determine the rate of heat transfer per meter length from a 5-cm-OD pipe at 150°C placed eccentrically within a larger cylinder of 85% magnesia wool as shown in the

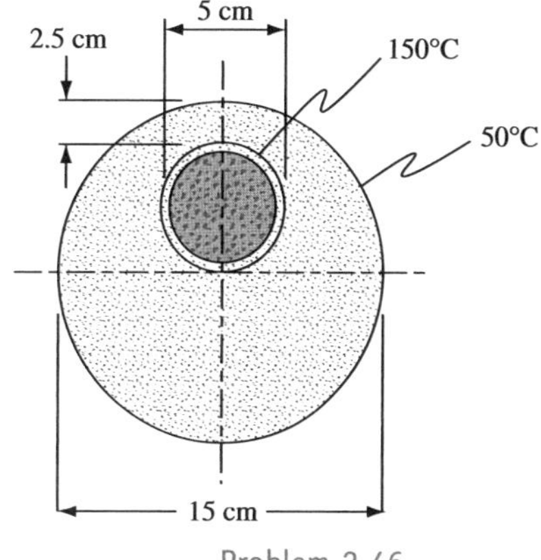

Problem 2.46

sketch. The outside diameter of the larger cylinder is 15 cm and the surface temperature is 50°C.

2.47 Determine the rate of heat flow per foot length from the inner to the outer surface of the molded insulation in the accompanying sketch. Use $k = 0.1$ Btu/h ft°F.

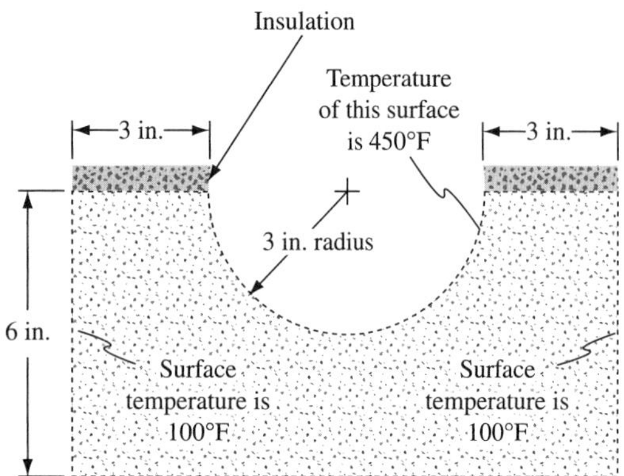

Problem 2.47

2.48 A long, 1-cm-diameter electric copper cable is embedded in the center of a 25-cm-square concrete block. If the outside temperature of the concrete is 25°C and the rate of electrical energy dissipation in the cable is 150 W per meter length, determine temperatures at the outer surface and at the center of the cable.

2.49 A large number of 1.5-in.-OD pipes carrying hot and cold liquids are embedded in concrete in an equilateral staggered arrangement with centerlines 4.5 in. apart as shown in the sketch. If the pipes in rows A and C are at 60°F while the pipes in rows B and D are at 150°F, determine the rate of heat transfer per foot length from pipe X in row B.

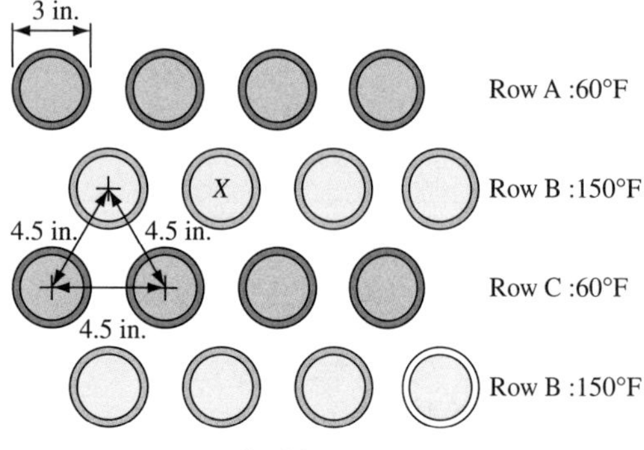

Problem 2.49

2.50 A long, 1-cm-diameter electric cable is embedded in a concrete wall ($k = 0.13$ W/m K) that is 1 m by 1 m, as shown in the sketch below. If the lower surface is insulated, the surface of the cable is 100°C, and the exposed surface of the concrete is 25°C, estimate the rate of energy dissipation per meter of cable.

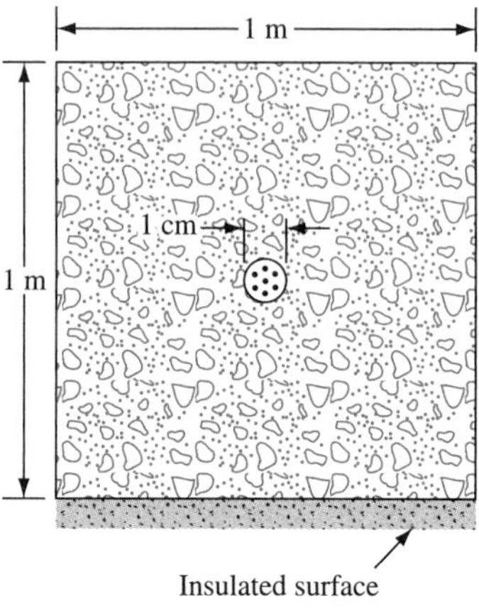

Problem 2.50

2.51 Determine the temperature distribution and heat flow rate per meter length in a long concrete block having the shape shown below. The cross-sectional area of the block is square and the hole is centered.

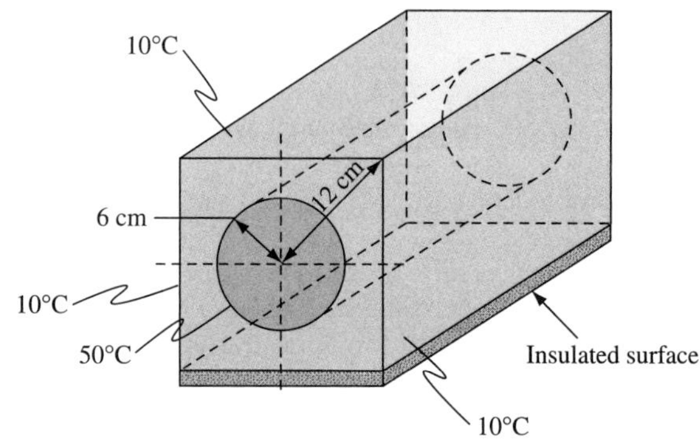

Problem 2.51

2.52 A 30-cm-OD pipe with a surface temperature of 90°C carries steam over a distance of 100m. The pipe is buried with its centerline at a depth of 1 m, the ground surface is −6°C, and the mean thermal conductivity of the soil is 0.7 W/m K. Calculate the heat loss per day, and the cost of this loss if steam heat is worth \$3.00 per 10^6 kJ. Also estimate the thickness of 85% magnesia insulation necessary to achieve the same insulation as provided by the soil with a total heat transfer coefficient of 23 W/m^2 K on the outside of the pipe.

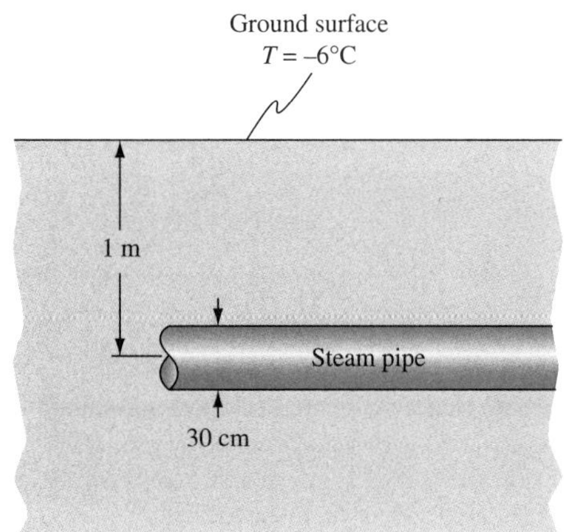

Problem 2.52

2.53 Two long pipes, one having a 10-cm OD and a surface temperature of 300°C, the other having a 5-cm OD and a surface temperature of 100°C, are buried deeply in dry sand with their centerlines 15 cm apart. Determine the rate of heat flow from the larger to the smaller pipe per meter length.

2.54 A radioactive sample is to be stored in a protective box with 4-cm-thick walls and interior dimensions of 4 cm × 4 cm × 12 cm. The radiation emitted by the sample is completely absorbed at the inner surface of the box, which is made of concrete. If the outside temperature of the box is 25°C but the inside temperature is not to exceed 50°C, determine the maximum permissible radiation rate from the sample, in watts.

2.55 A 6-in.-OD pipe is buried with its centerline 50 in. below the surface of the ground (k of soil is 0.20 Btu/h ft °F). An oil having a density of 6.7 lb/gal and a specific heat of 0.5 Btu/lb °F flows in the pipe at 100 gpm. Assuming a ground surface temperature of 40°F and a pipe wall temperature of 200°F, estimate the length of pipe in which the oil temperature decreases by 10°F.

2.56 A 2.5-cm-OD hot steam line at 100°C runs parallel to a 5.0-cm-OD cold water line at 15°C. The pipes are 5 cm apart (center to center) and deeply buried in concrete with a thermal conductivity of 0.87 W/m K. What is the heat transfer per meter of pipe between the two pipes?

2.57 Calculate the rate of heat transfer between a 15-cm-OD pipe at 120°C and a 10-cm-OD pipe at 40°C. The two pipes are 330 m long and are buried in sand ($k = 0.33$ W/m K) 12 m below the surface ($T_s = 25°C$). The pipes are parallel and are separated by 23 cm (center to center).

2.58 A 0.6-cm-diameter mild steel rod at 38°C is suddenly immersed in a liquid at 93°C with $\bar{h}_c = 110$ W/m^2 K. Determine the time required for the rod to warm to 88°C.

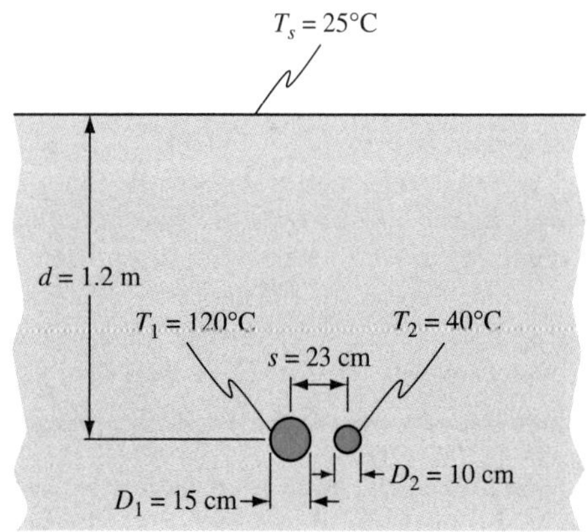

Problem 2.57

2.59 A spherical shell satellite (3-m-OD, 1.25-cm-thick stainless steel walls) reenters the atmosphere from outer space. If its original temperature is 38°C, the effective average temperature of the atmosphere is 1093°C, and the effective heat transfer coefficient is 115 W/m^2 °C, estimate the temperature of the shell after reentry, assuming the time of reentry is 10 min and the interior of the shell is evacuated.

2.60 A thin-wall cylindrical vessel (1 m in diameter) is filled to a depth of 1.2 m with water at an initial temperature of 15°C. The water is well stirred by a mechanical agitator. Estimate the time required to heat the water to 50°C if the tank is suddenly immersed in oil at 105°C. The overall heat transfer coefficient between the oil and the water is 284 W/m^2 K, and the effective heat transfer surface is 4.2 m^2.

2.61 A thin-wall jacketed tank heated by condensing steam at one atmosphere contains 91 kg of agitated water. The

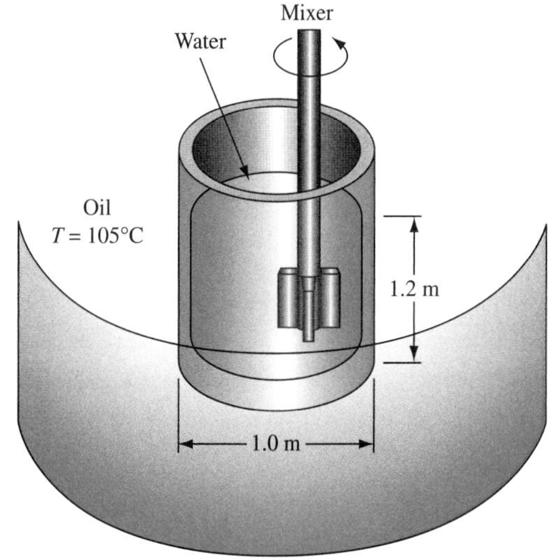

Problem 2.60

heat transfer area of the jacket is 0.9 m^2 and the overall heat transfer coefficient $U = 227$ W/m^2 K based on that area. Determine the heating time required for an increase in temperature from 16°C to 60°C.

2.62 The heat transfer coefficients for the flow of 26.6°C air over a sphere of 1.25 cm in diameter are measured by observing the temperature-time history of a copper ball the same dimension. The temperature of the copper ball ($c = 376$ J/kg K, $\rho = 8928$ kg/m^3) was measured by two thermocouples, one located in the center and the other near the surface. The two thermocouples registered, within the accuracy of the recording instruments, the same temperature at any given instant. In one test run, the initial temperature of the ball was 66°C, and the temperature decreased by 7°C in 1.15 min. Calculate the heat transfer coefficient for this case.

2.63 A spherical stainless steel vessel at 93°C contains 45 kg of water initially at the same temperature. If the entire system is suddenly immersed in ice water, determine (a) the time required for the water in the vessel to cool to 16°C, and (b) the temperature of the walls of the vessel at that time. Assume that the heat transfer coefficient at the inner surface is 17 W/m^2 K, the heat transfer coefficient at the outer surface is 22.7 W/m^2 K, and the wall of the vessel is 2.5 cm thick.

2.64 A copper wire, 1/32-in. OD, 2 in. long, is placed in an air stream whose temperature rises at a rate given by $T_{air} = (50 + 25t)$°F, where t is the time in seconds. If the initial temperature of the wire is 50°F, determine its temperature after 2 s, 10 s, and 1 min. The heat transfer coefficient between the air and the wire is 7 Btu/h ft^2 °F.

2.65 A large, 2.54-cm.-thick copper plate is placed between two air streams. The heat transfer coefficient on one side is 28 W/m^2 K and on the other side is 57 W/m^2 K. If the temperature of both streams is suddenly changed from 38°C to 93°C, determine how long it will take for the copper plate to reach a temperature of 82°C.

2.66 A 1.4-kg aluminum household iron has a 500-W heating element. The surface area is 0.046 m^2. The ambient temperature is 21°C, and the surface heat transfer coefficient is 11 W/m^2 K. How long after the iron is plugged in will its temperature reach 104°C?

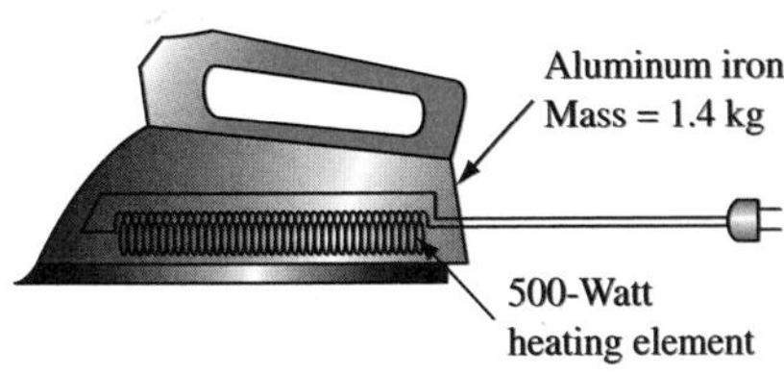

Problem 2.66

2.67 Estimate the depth in moist soil at which the annual temperature variation will be 10% of that at the surface.

2.68 A small aluminum sphere of diameter D, initially at a uniform temperature T_o, is immersed in a liquid whose temperature, T_∞, varies sinusoidally according to

$$T_\infty - T_m = A \sin(\omega t)$$

where T_m = time-averaged temperature of the liquid
A = amplitude of the temperature fluctuation
ω = frequency of the fluctuations

If the heat transfer coefficient between the fluid and the sphere, $\bar{h}_o$, is constant and the system can be treated as a "lumped capacity," derive an expression for the sphere temperature as a function of time.

2.69 A wire of perimeter P and cross-sectional area A emerges from a die at a temperature T (above the ambient temperature) and with a velocity U. Determine the temperature distribution along the wire in the steady state if the exposed length downstream from the die is quite long. State clearly and try to justify all assumptions.

2.70 Ball bearings are to be hardened by quenching them in a water bath at a temperature of 37°C. You are asked to devise a continuous process in which the balls could roll from a soaking oven at a uniform temperature of 870°C into the water, where they are carried away by a rubber conveyor belt. The rubber conveyor belt, however, would not be satisfactory if the surface temperature of the balls leaving the water is above 90°C. If the surface coefficient of heat transfer between the balls and the water can be assumed to be equal to 590 W/m^2 K, (a) find an approximate relation giving the minimum allowable cooling time in the water as a function of the ball radius for balls up to 1.0 cm in diameter, (b) calculate the cooling time, in seconds, required for a ball having a 2.5-cm diameter, and (c) calculate the total amount of heat in watts that would have to be removed from the water bath in order to maintain a uniform temperature if 100,000 balls of 2.5-cm diameter are to be quenched per hour.

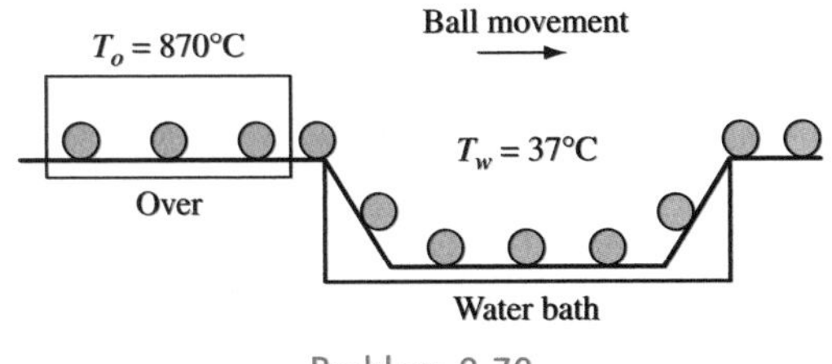

Problem 2.70

2.71 Estimate the time required to heat the center of a 1.5-kg roast in a 163°C oven to 77°C. State your assumptions carefully and compare your results with cooking instructions in a standard cookbook.

2.72 A stainless steel cylindrical billet ($k = 14.4$ W/m K, $\alpha = 3.9 \times 10^{-6}$ m²/s) is heated to 593°C preparatory to a forming process. If the minimum temperature permissible for forming is 482°C, how long can the billet be exposed to air at 38°C if the average heat transfer coefficient is 85 W/m² K? The shape of the billet is shown in the sketch.

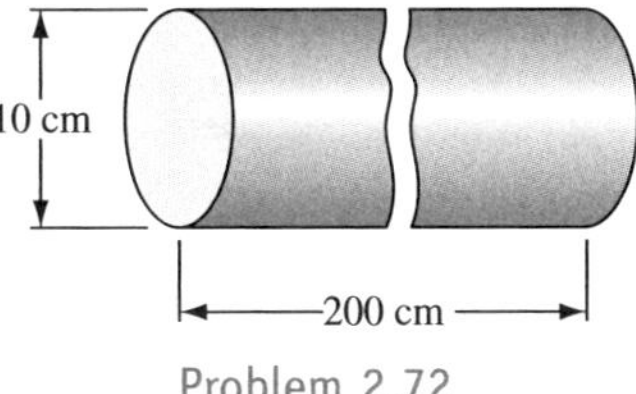

Problem 2.72

2.73 In the vulcanization of tires, the carcass is placed into a jig and steam at 149°C is admitted suddenly to both sides. If the tire thickness is 2.5 cm, the initial temperature is 21°C, the heat transfer coefficient between the tire and the steam is 150 W/m² K, and the specific heat of the rubber is 1650 J/kg K, estimate the time required for the center of the rubber to reach 132°C.

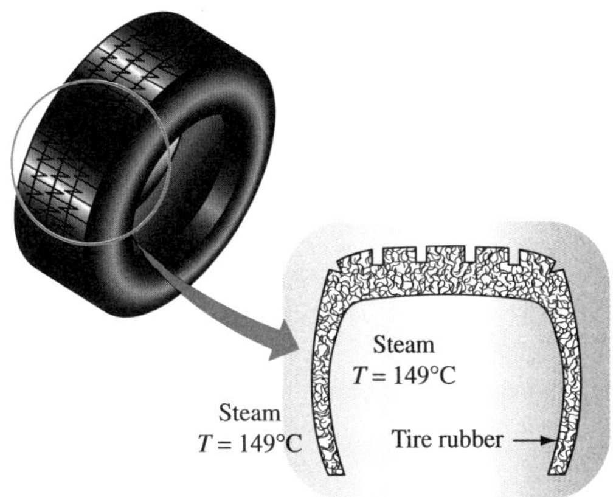

Problem 2.73

2.74 A long copper cylinder 0.6 m in diameter and initially at a uniform temperature of 38°C is placed in a water bath at 93°C. Assuming that the heat transfer coefficient between the copper and the water is 1248 W/m² K, calculate the time required to heat the center of the cylinder to 66°C. As a first approximation, neglect the temperature gradient within the cylinder; then repeat your calculation without this simplifying assumption and compare your results.

2.75 A steel sphere with a diameter of 7.6 cm is to be hardened by first heating it to a uniform temperature of 870°C and then quenching it in a large bath of water at a temperature of 38°C. The following data apply:

surface heat transfer coefficient $\bar{h}_c = 590$ W/m² K

thermal conductivity of steel = 43 W/m K

specific heat of steel = 628 J/kg K

density of steel = 7840 kg/m³

Calculate (a) the time elapsed in cooling the surface of the sphere to 204°C and (b) the time elapsed in cooling the center of the sphere to 204°C.

2.76 A 2.5-cm-thick sheet of plastic initially at 21°C is placed between two heated steel plates that are maintained at 138°C. The plastic is to be heated just long enough for its midplane temperature to reach 132°C. If the thermal conductivity of the plastic is 1.1×10^{-3} W/m K, the thermal diffusivity is 2.7×10^{-6} m/s, and the thermal resistance at the interface between the plates and the plastic is negligible, calculate (a) the required heating time, (b) the temperature at a plane 0.6 cm from the steel plate at the moment the heating is discontinued, and (c) the time required for the plastic to reach a temperature of 132°C 0.6 cm from the steel plate.

2.77 A monster turnip (assumed spherical) weighing in at 0.45 kg is dropped into a cauldron of water boiling at atmospheric pressure. If the initial temperature of the turnip is 17°C, how long does it take to reach 92°C at the center? Assume that

$$\bar{h}_c = 1700 \text{ W/m}^2\text{ K} \qquad c_p = 3900 \text{ J/kg K}$$
$$k = 0.52 \text{ W/m K} \qquad \rho = 1040 \text{ kg/m}^3$$

2.78 An egg, which for the purposes of this problem can be assumed to be a 5-cm-diameter sphere having the thermal properties of water, is initially at a temperature of 4°C. It is immersed in boiling water at 100°C for 15 min. The heat transfer coefficient from the water to the egg can be assumed to be 1700 W/m² K. What is the temperature of the egg center at the end of the cooking period?

2.79 A long wooden rod at 38°C with a 2.5-cm-OD is placed into an airstream at 600°C. The heat transfer coefficient between the rod and air is 28.4 W/m² K. If the ignition temperature of the wood is 427°C, $\rho = 800$ kg/m³, $k = 0.173$ W/m K, and $c = 2500$ J/kg K, determine the time between initial exposure and ignition of the wood.

2.80 In the inspection of a sample of meat intended for human consumption, it was found that certain undesirable organisms were present. To make the meat safe for consumption, it is ordered that the meat be kept at a temperature of at least 121°C for a period of at least 20 min during the preparation. Assume that a 2.5-cm-thick slab of this meat is originally at a uniform temperature of 27°C, that it is to be heated from both sides in a constant temperature oven, and that the maximum temperature meat can withstand is 154°C. Assume furthermore that the surface coefficient of heat transfer remains constant and is 10 W/m^2 K. The following data can be assumed for the sample of meat: specific heat = 4184 J/kg K; density = 1280 kg/m^3; thermal conductivity = 0.48 W/m K. Calculate the oven temperature and the minimum total time of heating required to fulfill the safety regulation.

2.81 A frozen-food company freezes its spinach by first compressing it into large slabs and then exposing the slab of spinach to a low-temperature cooling medium. The large slab of compressed spinach is initially at a uniform temperature of 21°C; it must be reduced to an average temperature over the entire slab of −34°C. The temperature at any part of the slab, however, must never drop below −51°C. The cooling medium that passes across both sides of the slab is at a constant temperature of −90°C. The following data can be used for the spinach: density = 80 kg/m^3; thermal conductivity = 0.87 W/m K; specific heat = 2100 J/kg K. Present a detailed analysis outlining a method to estimate the maximum slab thickness that can be safely cooled in 60 min.

2.82 In the experimental determination of the heat transfer coefficient between a heated steel ball and crushed mineral solids, a series of 1.5% carbon steel balls were heated to a temperature of 700°C and the center temperature-time history of each was measured with a thermocouple as it cooled in a bed of crushed iron ore that was placed in a steel drum rotating horizontally at about 30 rpm. For a 5-cm-diameter ball, the time required for the temperature difference between the ball center and the surrounding ore to decrease from an initial 500°C to 250°C was found to be 64, 67, and 72 s, respectively, in three different test runs. Determine the average heat transfer coefficient between the ball and the ore. Compare the results obtained by assuming the thermal conductivity to be infinite with those obtained by taking the internal thermal resistance of the ball into the account.

2.83 A mild-steel cylindrical billet 25 cm in diameter is to be raised to a minimum temperature of 760°C by passing it through a 6-m long strip-type furnace. If the furnace gases are at 1538°C and the overall heat transfer coefficient on the outside of the billet is 68 W/m^2 K, determine the maximum speed at which a continuous billet entering at 204°C can travel through the furnace.

2.84 A solid lead cylinder 0.6 m in diameter and 0.6 m long, initially at a uniform temperature of 121°C, is dropped into a 21°C liquid bath in which the heat transfer coefficient $\bar{h}_c$ is 1135 W/m^2 K. Plot the temperature-time history of the center of this cylinder and compare it with the time histories of a 0.6-m diameter, infinitely long lead cylinder and a lead slab 0.6 m thick.

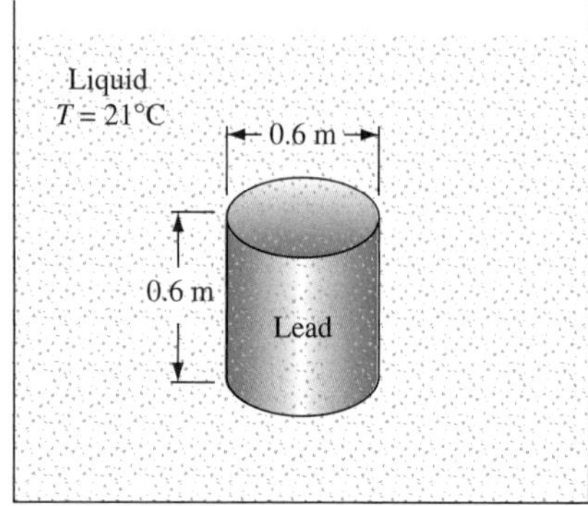

Problem 2.84

2.85 A long, 0.6-m-OD 347 stainless steel (k = 14 W/m K) cylindrical billet at 16°C room temperature is placed in an oven where the temperature is 260°C. If the average heat transfer coefficient is 170 W/m^2 K, (a) estimate the time required for the center temperature to increase to 232°C by using the appropriate chart and (b) determine the instantaneous surface heat flux when the center temperature is 232°C.

2.86 Repeat Problem 2.85(a), but assume that the billet is only 1.2 m long with the average heat transfer coefficient at both ends equal to 136 W/m^2 K.

2.87 A large billet of steel initially at 260°C is placed in a radiant furnace where the surface temperature is held at 1200°C. Assuming the billet to be infinite in extent, compute the temperature at point P (see the accompanying sketch) after 25 min have elapsed. The average properties of steel are: k = 28 W/m K, ρ = 7360 kg/m^3, and c = 500 J/kg K.

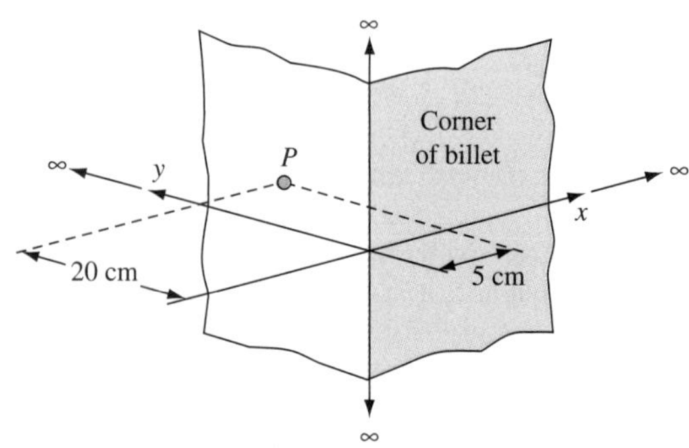

Problem 2.87

Design Problems

2.1 **Fins for Heat Recovery** (Chapters 2 and 5)

An inventor wants to increase the efficiency of wood-burning stoves by reducing the energy lost through the exhaust stack. He proposes to accomplish this by attaching fins to the outer surface of the chimney, as shown schematically below. The fins are attached circumferentially to the stack, having a base of 0.5 cm, 2 cm long perpendicular to the surface, and 6 cm long in the vertical direction. The surface temperature of the stack is 500°C and the surrounding temperature is 20°C. For this initial thermal design, assume that each fin loses heat by natural convection with a convection heat transfer coefficient of 10 W/m^2 K. Select a suitable material for the fin and discuss the manner of attachment, as well as the effect of contact resistance. In Chapter 5 you will be asked to reconsider this design and calculate the natural convection heat transfer coefficient from information presented in that chapter. For further information on this concept, you may consult U.S. Patent 4,236,578, F. Kreith and R. C. Corneliusen, *Heat Exchange Enhancement Structure*, Washington, D.C., December 2, 1980.

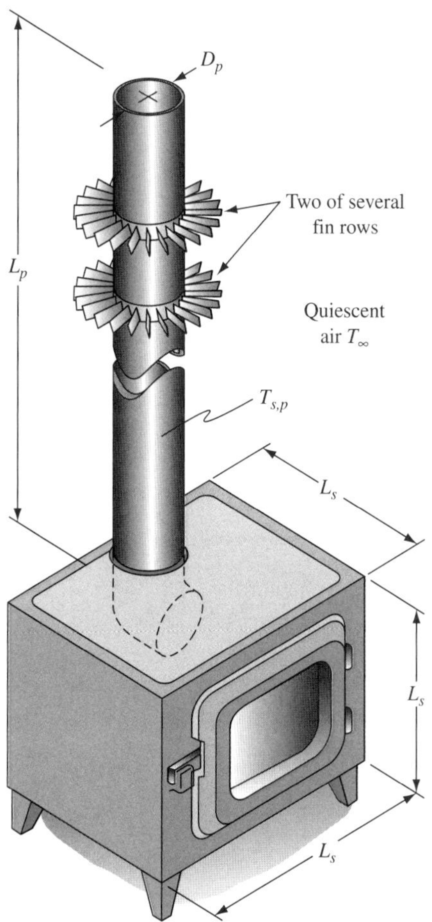

Design Problem 2.1

2.2 **Camping Cooler** (Chapter 2)

Design a cooler that can be used on camping trips. Primary considerations in the design are weight, capacity, and how long the cooler can keep items cold. Investigate commercially available insulation materials and advanced insulation concepts to determine an optimum design. The internal volume of the cooler should be nominally 2 ft^3 and it should be able to maintain an internal temperature of 40°F when the outside temperature is 90°F.

2.3 **Pressure Vessel** (Chapter 2)

Design a pressure vessel that can hold 100 lb of saturated steam at 400 psia for a chemical process. The shape of the vessel is to be a cylinder with hemispherical ends. The vessel is to have sufficient insulation to maintain equilibrium with a maximum internal heat input of 3000 MW. For the initial phase of this design, determine the thickness of insulation necessary if heat loss were to occur only by conduction with an outside temperature of 70°F. For this design, examine Section VIII, Division I of the ASME Boiler and Pressure Vessel Code to determine allowable strength and shell thickness. After completing the initial design, repeat your calculations, assuming that the heat transfer from the steam to the inside of the vessel is by condensation with an average heat transfer coefficient from Table 1.4. On the outside, heat transfer is by nature convection with a heat transfer coefficient of 15 W/m^2 K. Select an appropriate steel for the vessel to guarantee a lifetime of at least 12 years.

2.4 **Waste Heat Recovery** (Chapter 2)

Suppose that waste heat from a refinery is available for a chemical plant located one mile away. The waste stream from the refinery consists of 2000 standard cubic feet per minute of corrosive gas at 300°F and 500 psi. The refinery is located on one side of a highway with the chemical plant on the other side. To bring the waste heat to the chemical process, a pipe has to be laid underground and buried in the soil. The pipe is to be made of a material that can withstand corrosion. Select an appropriate material for the pipe and its insulation, and then estimate the heat loss from the gas between the source and the place of utilization as a function of insulating thickness for two different insulating materials.

The velocity of the gas stream inside the pipe is of the order of 5 m/s and has a heat transfer coefficient of 100 W/m^2 K. As part of the assignment, outline any safety problems that need to be addressed with an insurance company in order to protect against a claim in case of an accident.

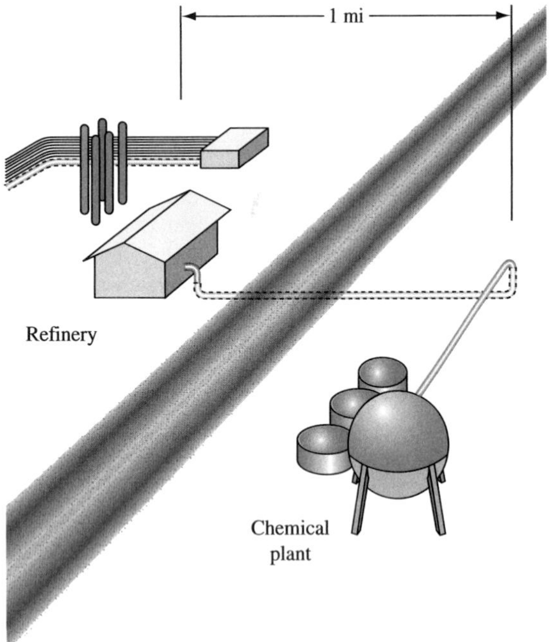

Design Problem 2.4

2.5 **Hydrogen Fuel System** (Chapter 2)
There is worldwide concern that the availability of oil will diminish within 20 or 30 years. (See, for example, Frank Kreith et al., *Ground Transportation for the 21st Century*, ASME Press, 2000.) In an effort to maintain the availability of a convenient fuel while at the same time reducing adverse environmental impact, some have suggested that in the future there will be a paradigm shift from oil to hydrogen as the primary fuel. Hydrogen, however, it not available in nature as is oil. Consequently, it must be produced by splitting water electrically or by producing it from a hydrogen-rich fuel. Moreover, to transport hydrogen, it has to be liquified and transported through pipelines to the location where it is needed. Prepare a preliminary assessment for the feasibility of a hydrogen fuel supply system.

As a first step, it is necessary to split water into hydrogen and oxygen. To do this, wind turbines will be used to generate electricity for the electrolytic separation of hydrogen and oxygen. This can be accomplished at a cost of \$0.06/kWh in parts of the country that have adequate wind resources, such as, for example, North Dakota. Begin your thermal analysis by calculating the energy required to cool the gaseous hydrogen from a temperature of 30°C to a temperature at which it will become a liquid. Assume, for this estimate, that refrigeration can be achieved with a COP of approximately 50% of Carnot efficiency between appropriate temperature limits. Now that hydrogen is available as a liquid, estimate the heat loss from a pipe laid at a reasonable distance underground and insulated with cryogenic insulation in transporting the hydrogen from North Dakota to Chicago. Also estimate the pumping requirements of moving the hydrogen, assuming that suitable pumps with an overall efficiency of 65% are available. Once the liquified hydrogen has reached its destination, it must be stored in a suitable spherical container. Estimate the size of the container sufficient to supply approximately 100 MW of electric power in Chicago by means of a fuel cell that has an efficiency of 60%. The cost of the fuel cell is estimated to be about \$5,000/kW. After having completed these estimates, prepare a brief analysis on whether or not a hydrogen economy appears to be technically and economically feasible.

For some additional background on this problem, refer also to P. Sharpe, "Fueling the Cells," *Mech. Eng.*, Dec. 1999, pp. 46–49.

2.6 **Refrigerated Truck** (Chapters 2 and 4)
Prepare the thermal design for a refrigerated container truck to carry frozen meat from Butte, Montana, to Phoenix, Arizona. The refrigerated shipping container has dimensions of 20 ft × 10 ft × 8 ft and will use dry ice as the refrigerant. For this design it is necessary to select suitable insulation type and thickness. Also estimate the size of the dry ice compartment sufficient to maintain the temperature inside the container at 32°F when the average outside surface temperature of the container during the trip may rise up to 100°F. Dry ice currently costs \$0.6/kg and the shipping company would like to know the amount of dry ice required for one trip and its cost. Assuming that the insulation on the truck will last for 10 years, prepare a cost comparison of insulation thickness for the container vs. the amount of dry ice necessary to maintain the refrigeration temperature during the trip. Clearly state all of your assumptions.

2.7 **Electrical Resistance Heater** (Chapters 2, 3, 6, and 10)
Electrical resistance heaters are usually made from coils of nichrome wire. The coiled wire can be supported between insulators and backed with a reflector, for example, as in a supplemental room heater. In other applications, however, it is often necessary to protect the nichrome wire from its environment. An example of such an application is a process heater where a flowing fluid is to be heated. In such a case,

the nichrome wire is embedded in an electrical insulator and covered by a metal sheath. Sketch (a) shows the construction details. Since the sheathed heater is often used to heat a fluid flowing over its outside surface, it may be necessary to increase the surface area of the heater sheath. A proposed design for such an application is shown in sketch (b).

The preliminary design of a fast-response hot-water heater using this proposed heater element design is shown in sketch (c). The heating element is located inside a pipe carrying the water to be heated. The heating element dissipates 4800 watts per meter length and has a maximum temperature limit of 200°C. Water is to be heated to 65°C by the device and the surface of the heating element should not exceed 100°C to avoid boiling.

For simplicity, assume that the heat dissipated by the nichrome wire is distributed uniformly over the cross section of the heating element shown in sketch (b) and that the thermal conductivity of the MgO insulation is 2 W/m K. You may also assume that the metal sheath is very thin.

First, perform an order-of-magnitude analysis to estimate the required convective heat transfer coefficient and to determine whether the temperature constraints given above can be met. Next, use analytical tools developed in this chapter to refine your answer. In Chapters 3, 6, and 10, you will be asked to refine these estimates further.

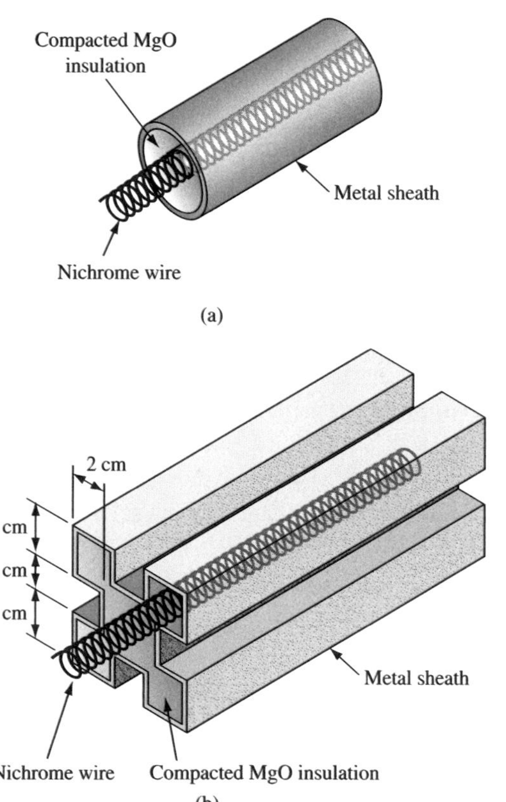

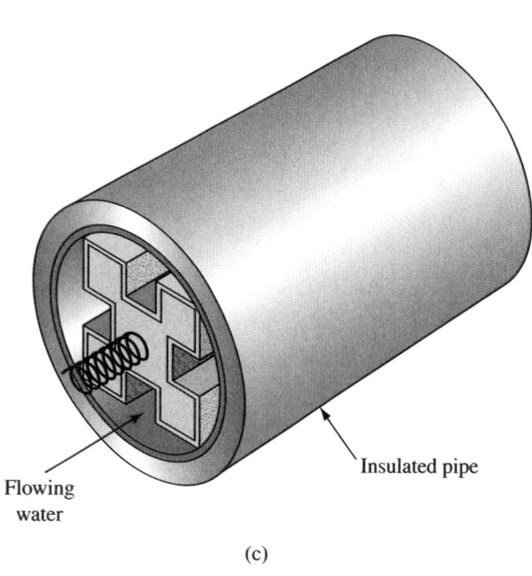

Design Problem 2.7

CHAPTER 3

Numerical Analysis of Heat Conduction

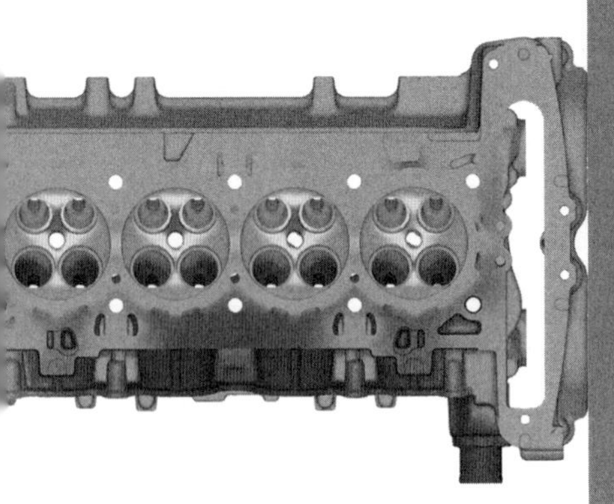

Temperature distribution in the structural side of a coolant jacket of an automobile engine obtained from a computational simulation (numerical analysis) of the heat transfer.

Source: Courtesy of General Motors and CD-adapco.

Concepts and Analyses to Be Learned

Practical problems of heat transfer by conduction are often quite intricate and cannot be solved by analytical methods. Their mathematical models may include nonlinear differential equations with complex boundary conditions. In such cases, the only recourse is to obtain approximate solutions by employing discrete numerical techniques. Such computational techniques provide an effective way not only for resolving such problems, but also for simulating intricate multidimensional models for a variety of applications. A study of this chapter will help you understand the mechanisms of control-volume-based finite difference methods for solving differential equations and will teach you:

- How to solve one-dimensional heat conduction equations for steady-state and unsteady (or transient) conditions with different boundary conditions.
- How to perform numerical analysis of steady and unsteady-state heat conduction equations with different boundary conditions.
- How to obtain numerical solutions for problems in cylindrical coordinates as well as those having irregular boundaries.

3.1 Introduction

Mathematical models and their governing equations that describe the transfer of heat by conduction were developed in Chapter 2, and analytical solutions for several conduction problems for typical engineering applications were presented. It should be clear from the types of problems addressed in Chapter 2 that analytical solutions are usually possible only for relatively simple cases. Nevertheless, these solutions play an important role in heat transfer analysis because they provide insight into complex engineering problems that can be simplified using certain assumptions.

Many practical problems, however, involve complex geometries, complex boundary conditions, and variable thermophysical properties and cannot be solved analytically. However, these problems can be solved by numerical or computational methods that include, among others, finite-difference, finite-element, and boundary element methods. In addition to providing a solution method for these more complex problems, numerical analysis is often more efficient in terms of the total time required to find the solution. Another advantage is that changes in problem parameters can be made more easily allowing an engineer to determine the behavior of a thermal system or perhaps optimize a thermal system much more easily.

Analytical solution methods such as those described in Chapter 2 solve the governing differential equations and can provide a solution at every point in space and time within the problem boundaries. In contrast, numerical methods provide the solution only at discrete points within the problem boundaries and give only an approximation to the exact solution. However, by dealing with the solution at only a finite number of discrete points, we simplify the solution method to one of solving a system of simultaneous algebraic equations as opposed to solving the differential equation. The solution of a system of simultaneous equations is a task ideally suited to computers.

In addition to replacing the differential equation with a system of algebraic equations, a process called *discretization*, there are several other important considerations for a complete numerical solution. First, boundary conditions or initial conditions that have been specified for the problem must be discretized. Second, we need to be aware that as an approximation to the exact solution, the numerical method introduces errors into the solution. We need to know how to estimate and minimize these errors. Finally, under some conditions, the numerical method may give a solution that oscillates in time or space. We need to know how to avoid these *stability* problems.

Several methods are available for discretizing the differential equations of heat conduction. Among these methods are the finite difference, finite element, and control volume approaches. There are advantages and disadvantages to each method. In this chapter, we have chosen to use the control volume approach, which is relatively more prevalent in scientific usage as well as employed in several commercial codes and software [1]. The control volume approach considers the energy balance on a small but finite volume within the boundaries of the problem. This approach was used in Chapter 2 to develop the *differential* equation for one-dimensional unsteady conduction. There, the energy balance equation for a slab Δx thick was written by determining the heat conduction into the left face, the heat conduction out of the right face, and the energy stored in the slab. The final step was to mathematically decrease the size of the control volume so that the energy balance equation became, in the limit of infinitesimal Δx, a differential equation [Eq. (2.5)]. In this chapter, we follow the same procedure except that we will skip the last step, leaving the energy balance in the form of a *difference* equation. The control volume method determines the difference equation directly from energy conservation considerations.

One advantage of this method is that we already know how to determine an energy balance on a control volume. We only need to add the boundary conditions and implement a method to solve the resulting system of difference equations. Another advantage is that energy is conserved regardless of the size of the control volume. Thus a problem can be solved quickly on a fairly coarse grid to develop the numerical technique and then on a finer grid to find the final, more accurate solution. Finally, the control volume method minimizes complex mathematics and therefore promotes a better physical feel for the problem.

In this chapter we introduce the control volume approach to solving conduction heat transfer problems. First, we will develop the numerical analysis of the one-dimensional steady conduction problem. Complexity will then be increased by examining one-dimensional unsteady conduction, two-dimensional steady and unsteady conduction, conduction in a cylindrical geometry, and finally, irregular boundaries. In each case, the appropriate difference equation and boundary conditions will be derived from energy balance considerations. Methods for solving the resulting set of difference equations are discussed for each type of problem.

3.2 One-Dimensional Steady Conduction

3.2.1 The Difference Equation

We will first considersteady conduction with heat generation in a semi-infinite slab (i.e., thickness of slab L is orders of magnitude smaller than its length or height and width). Thus, the one-dimensional, steady-state domain of interest is $0 \leq x \leq L$ as depicted in the schematic of Fig. 3.1. For this geometry and as described in Chapter 2, the general heat conduction equation, given by Eq. (2.8), reduces to that given by Eq. (2.27) that can be rewritten as follows:

$$\frac{d^2T}{dx^2} + \frac{\dot{q}_G}{k} = 0 \tag{3.1}$$

In order to discretize this equation and to apply the control volume method, we first divide the domain into $N - 1$ equal segments of width $\Delta x = L/(N - 1)$ as shown in Fig. 3.1. With this arrangement, we can identify the boundaries of each segment with

$$x_i = (i - 1)\Delta x, \quad i = 1, 2, \ldots, N$$

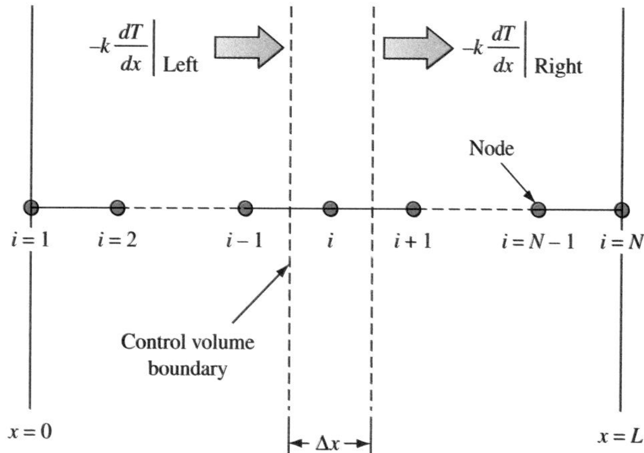

FIGURE 3.1 Control volume for one-dimensional conduction.

The x_i locations are called *nodes*, and nodes 1 and N are called the boundary nodes. We can then identify the temperature at each node as $T(x_i)$ or, for short, T_i. Now, center a slab of thickness Δx over one of the interior nodes (see the shaded portion of Fig. 3.1). Since we are considering one-dimensional conduction we can take a unit length in the y and z directions for this slab. Then this slab has dimensions Δx by 1 by 1 and becomes our control volume.

Consider an energy balance on this control volume as we developed in Chapter 2 and expressed by Eq. (2.1) as follows:

$$\begin{matrix}\text{rate of heat conduction} \\ \text{into control volume}\end{matrix} + \begin{matrix}\text{rate of heat generation} \\ \text{inside control volume}\end{matrix} = \begin{matrix}\text{rate of heat conduction} \\ \text{out of control volume}\end{matrix} \tag{2.1}$$

The energy storage term has been dropped from Eq. (2.1) because here we are concerned only with steady-state behavior. The first term on the left side of Eq. (2.1) can be written according to Eq. (1.1) (see Chapter 1) as

$$\text{rate of heat conduction into control volume} = -k\frac{dT}{dx}\bigg|_{\text{left}}$$

where the temperature gradient is to be evaluated at the left face of the control volume. Our ultimate goal is to determine the values T_i at all node points. We are not especially concerned about the temperature distribution between the nodes; therefore it is reasonable to assume that the temperature varies linearly between the nodes. With this assumption, the temperature gradient at the left face of the control volume is exactly

$$\frac{dT}{dx}\bigg|_{\text{left}} = \frac{T_i - T_{i-1}}{\Delta x}$$

If we are given the volumetric rate of heat generation, $\dot{q}_G(x)$, then the second term on the left side of Eq. (2.1) is just $\dot{q}_G(x_i)\Delta x$ or, for short, $\dot{q}_{G,i}\Delta x$. Here, we are assuming that the heat generation rate is constant over the entire control volume. Finally, the term on the right side of Eq. (2.1) is

$$\text{rate of heat conduction out of control volume} = -k\frac{dT}{dx}\bigg|_{\text{right}}$$

By arguments similar to those used to find $\frac{dT}{dx}\big|_{\text{left}}$, we can write

$$\frac{dT}{dx}\bigg|_{\text{right}} = \frac{T_{i+1} - T_i}{\Delta x}$$

In terms of the nodal temperatures, we can now write the control volume energy balance as

$$-k\frac{T_i - T_{i-1}}{\Delta x} + \dot{q}_{G,i}\Delta x = -k\frac{T_{i+1} - T_i}{\Delta x}$$

Rearranging, we have

$$\frac{T_{i+1} - 2T_i + T_{i-1}}{\Delta x^2} - \frac{\dot{q}_{G,i}}{k} = 0 \tag{3.2}$$

By comparing the above expression with Eq. (3.2) we can readily see how it is a discretized version of the differential equation, where the second-order derivative of temperature with respect to x is now expressed in terms of discrete values of T in the domain $i = 1, 2, \ldots N$, or $<= x <= L$.

In the above treatment, the heat conducted *into* the left face is on the left side of the energy balance equation, while the heat conducted *out* of the right face is on the right side of the energy balance equation. This convention was followed to be consistent with Eq. (2.1). Actually, the choice of direction of heat flow at the control volume boundaries is arbitrary as long as it is correctly accounted for in the energy balance equation. For the term "rate of heat conduction out of control volume" in Eq. (2.1) we could have written

$$\begin{pmatrix}\text{rate of heat conduction}\\ \textit{into}\text{ control volume}\end{pmatrix} = k\frac{dT}{dx}\bigg|_{\text{right}} = -\begin{pmatrix}\text{rate of heat conduction}\\ \textit{out}\text{ of control volume}\end{pmatrix}$$

The control volume energy balance would then be

$$k\frac{T_{i-1} - T_i}{\Delta x} + k\frac{T_{i+1} - T_i}{\Delta x} + \dot{q}_{G,i}\Delta x = 0$$

which is equivalent to Eq. (3.2). This formulation may be easier to remember because we can think of all conduction terms being positive when heat flow is *into* the control volume. The conduction terms will then always be on the same side of the equation. In addition, they will be proportional to the node temperature T_i *subtracted from* the temperature of the node just outside the surface in question.

Equation (3.2) is called the *finite difference equation*, and it represents the energy balance on a finite control volume of width Δx. In contrast, Eq. (2.27) is the *differential equation*, and it represents an energy balance on a control volume of infinitesimal width dx. It can be shown that in the limit as $\Delta x \rightarrow 0$, Eq. (3.2) and Eq. (2.27) are identical.

In the absence of heat generation, Eq. (3.2) becomes

$$T_{i+1} - 2T_i + T_{i-1} = 0 \tag{3.3}$$

Therefore, the temperature at each node is just the average of its neighbors if there is no heat generation. Equation (3.3), it may again be noted, is the discretized form of Eq. (2.24) or the heat conduction equation for a semi-infinite slab without internal heat generation.

If the thermal conductivity k varies with temperature and therefore with x, for example, $k = k[T(x)]$, we need to modify the evaluation of the terms in Eq. (2.1) by a method suggested by Patankar [2]. The conductivity appropriate to the heat flux at the left face of the control volume is

$$k_{\text{left}} = \frac{2k_i k_{i-1}}{k_i + k_{i-1}}$$

Similarly, at the right face we use

$$k_{\text{right}} = \frac{2k_i k_{i+1}}{k_i + k_{i+1}}$$

In Section 3.2.3 we will discuss how to use this method to solve a problem with variable thermal conductivity.

How do we choose the size of the control volume Δx? Generally, a smaller value of Δx will give a more accurate solution but will increase the computer time required to find the solution. Essentially, our pointwise temperature distribution can more accurately represent a nonlinear temperature distribution when we reduce Δx. Some trial and error may be needed to determine a desirable accuracy for a reasonable computation time. Usually a series of computations is performed for smaller and smaller values of Δx. At some point, further reduction in Δx will produce no significant change in the solution. It is not necessary to reduce Δx beyond this value.

In some situations it is beneficial to allow the node spacing, Δx, to vary throughout the spatial domain of the problem. One example of such a situation occurs when a high heat flux is imposed at a boundary and a large temperature gradient is expected near that boundary. Near the surface, small values of Δx would be used so that the large temperature gradient can be accurately represented. Farther away from this boundary, where the temperature gradient is small, Δx could be made larger because the small temperature gradient can be accurately represented by larger Δx. This technique allows one to use the minimum number of nodes to achieve a desired accuracy without using excessive computation time or computer memory. Details of the *variable node spacing method*, or what is often also referred to as *nonuniform grid* or *nonuniform mesh method*, are given by Patankar and others [1–3].

It was mentioned previously that one advantage of the control volume approach is that energy is conserved regardless of the size of the control volume. This feature makes it convenient to start with a fairly coarse grid, i.e., relatively few control volumes, to develop the numerical solution. In this way the computer runs required to debug the program execute quickly and do not consume much memory. When the program is debugged, a finer grid can then be used to determine the solution to the desired accuracy.

A final consideration is round-off error. Because the computer deals with only a given number of digits, each mathematical operation results in some rounding off of the solution. As the number of mathematical operations needed to produce a numerical solution increases, these round-off errors can accumulate and, under some circumstances, adversely affect the solution.

The method used in this section to develop the difference equation will be used throughout this chapter. Regardless of whether the problem under consideration is steady, unsteady, one-dimensional, two-dimensional, cartesian, or cylindrical, we will first determine the appropriate control volume shape. Then we will determine all heat flows into and out of all the control volume boundaries and write the energy balance equation. For steady problems, the sum of all heat flows into the control volume plus heat generated inside the control volume must equal the sum of all heat flows out of the control volume. For unsteady problems, the difference between the heat flow in and out of the control volume plus heat generated inside the control volume must equal the rate at which energy is stored in the control volume.

3.2.2 Boundary Conditions

Recall that the solution of a differential equation requires the application of boundary conditions. So also is the case in numerical analysis, and hence to complete the problem statement, we must incorporate boundary conditions into our control volume method. The following three boundary conditions were discussed in Chapter 2: (i) specified surface temperature, (ii) specified surface heat flux, and (iii) specified surface convection. The techniques to incorporate each of these into the control volume method are described below.

The simplest of these three boundary conditions is the *specified surface temperature* for which

$$T(x_1) = T_1 \qquad T(x_N) = T_N \tag{3.4}$$

where T_1 and T_N are the specified surface temperatures at the left and right boundaries, respectively. The specified surface temperature boundary condition is illustrated in Fig. 3.2(a). This boundary condition is very simple to implement because we just assign the given surface temperatures to the boundary nodes. We do not need to write an energy balance at a surface node where the temperature is prescribed in order to solve the problem. However, in problems where the surface temperature is prescribed, we often need to determine the heat flow at that boundary, and in this situation an energy balance, as described below, is needed.

If the boundary condition consists of a *specified heat flux* into the boundary, q_1'', we can calculate the boundary temperature in terms of the flux by considering an energy balance over the control volume extending from $x = 0$ to $x = \Delta x/2$, as shown in Fig. 3.2(b). Note that this boundary control volume has a length half that of the internal control volumes. Using Eq. (2.1) again we have

$$q_1'' + \dot{q}_{G,1}\frac{\Delta x}{2} = -k\frac{T_2 - T_1}{\Delta x} \tag{3.5}$$

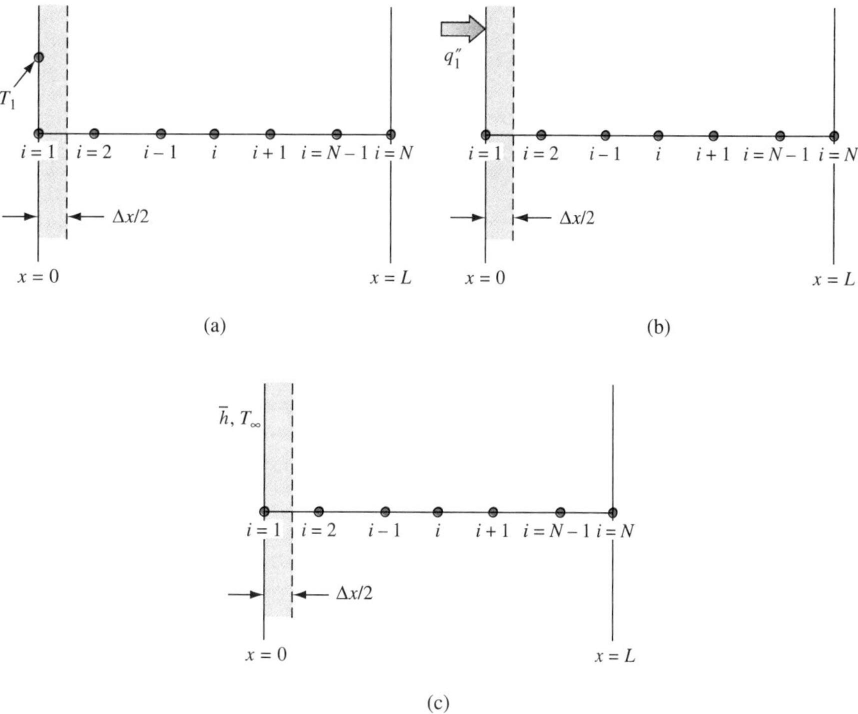

FIGURE 3.2 Boundary control volume for one-dimensional conduction, (a) specified temperature boundary condition, (b) specified heat flux boundary condition, (c) specified surface convection boundary condition.

Solving for T_1 yields

$$T_1 = T_2 + \frac{\Delta x}{k}\left(q_1'' + \dot{q}_{G,1}\frac{\Delta x}{2}\right)$$

Eq. (3.5) can also be used to solve for the surface heat flux in problems in which the boundary temperature is specified. In this case, the temperatures T_1 and T_2 as well as the heat generation term are known and the heat flux can be calculated.

For an *insulated surface* boundary condition, or $q_1'' = 0$, Eq. (3.5) yields

$$T_1 = T_2 + \dot{q}_{G,1}\frac{\Delta x^2}{2k}$$

Finally, if a *surface convection* is specified at the left face, applying Eq. (2.1) to the control volume shown in Fig. 3.2(c) gives

$$\bar{h}(T_\infty - T_1) + \dot{q}_{G,1}\frac{\Delta x}{2} = -k\frac{T_2 - T_1}{\Delta x} \tag{3.6}$$

where T_∞ is the temperature of the ambient fluid in contact with the left face and $\bar{h}$ is the convection heat transfer coefficient. Solving Eq. (3.5) for T_1 gives

$$T_1 = \frac{T_2 + \dfrac{\Delta x}{k}\left(\bar{h}T_\infty + \dot{q}_{G,1}\dfrac{\Delta x}{2}\right)}{1 + \dfrac{\bar{h}\Delta x}{k}} \tag{3.7}$$

Note that if the heat transfer coefficient is very large, T_1 approaches T_∞ as expected. If the heat transfer coefficient is very small, again as expected, we get the insulated-surface boundary condition.

A variation of this type of boundary condition is when the *surface radiation* is specified instead of *surface convection*. In such a case, the convective heat transfer coefficient in Eqs. (3.6) and (3.7) can be replaced by the radiation heat transfer coefficient given by Eq. (1.21) (Chapter 1; also see Eq. (9.118) in Chapter 9). Numerical treatment of radiation heat transfer coefficient, however, is somewhat complex because it is a function of the surface temperature, not an independent variable.

For all three types of boundary conditions, the surface temperature can be expressed in terms of the known heat flux or known convection conditions ($\bar{h}$ and T_∞) and the nodal temperature, T_2. That is, we could write all three boundary conditions as

$$a_1 T_1 = b_1 T_2 + d_1 \tag{3.8}$$

For the *specified surface temperature* boundary condition,

$$a_1 = 1 \qquad b_1 = 0 \qquad d_1 = T_1$$

For the *specified heat flux* boundary condition

$$a_1 = 1 \qquad b_1 = 1 \qquad d_1 = \frac{\Delta x}{k}\left(q_1'' + \dot{q}_{G,1}\frac{\Delta x}{2}\right)$$

For the *specified surface convection* boundary condition

$$a_1 = 1 \qquad b_1 = \frac{1}{1 + \dfrac{\bar{h}\Delta x}{k}} \qquad d_1 = \frac{\Delta x}{k}\,\frac{\left(\bar{h}T_\infty + \dot{q}_{G,1}\dfrac{\Delta x}{2}\right)}{1 + \dfrac{\bar{h}\Delta x}{k}}$$

Similarly, conditions at the right boundary could be written as

$$a_N T_N = c_N T_{N-1} + d_N \tag{3.9}$$

The coefficients a_N, c_N, and d_N are given in Table 3.1. Derivation of these coefficients is left as an exercise.

TABLE 3.1 Matrix coefficients for one-dimensional steady conduction [Eq. (3.11)]

	a_i	b_i	c_i	d_i
$i = 1$, specified surface temperature	1	0	0	T_i
$i = 1$, specified heat flux	1	1	0	$\frac{\Delta x}{k}\left(q_1'' + \dot{q}_{G,1}\frac{\Delta x}{2}\right)$
$i = 1$, specified surface convection	1	$\left(1 + \frac{\bar{h}_1 \Delta x}{k}\right)^{-1}$	0	$\dfrac{\frac{\Delta x}{k}\left(\bar{h}_1 T_{\infty,1} + \dot{q}_{G,1}\frac{\Delta x}{2}\right)}{1 + \frac{\bar{h}_1 \Delta x}{k}}$
$1 < i < N$	2	1	1	$\frac{\Delta x^2}{k}\dot{q}_{G,i}$
$i = N$, specified surface temperature	1	0	0	T_N
$i = N$, specified heat flux	1	0	1	$\frac{\Delta x}{k}\left(q_N'' + \dot{q}_{G,N}\frac{\Delta x}{2}\right)$
$i = N$, specified surface convection	1	0	$\left(1 + \frac{\bar{h}_N \Delta x}{k}\right)^{-1}$	$\frac{\Delta x}{k}\left(\dfrac{\bar{h}_N T_{\infty,N} + \dot{q}_{G,N}\frac{\Delta x}{2}}{1 + \frac{\bar{h}_N \Delta x}{k}}\right)$

Note: q_A'' is the heat flux *into* surface A.

3.2.3 Solution Methods

The difference equation, Eq. (3.2), can be written using the notation used above in the boundary condition equations:

$$a_i T_i = b_i T_{i+1} + c_i T_{1-1} + d_i, \quad 1 < i < N \tag{3.10}$$

where

$$a_i = 2 \qquad b_i = 1 \qquad c_i = 1 \qquad d_i = \frac{\Delta x^2}{k}\dot{q}_{G,i}$$

Since $c_1 = b_N = 0$, Eq. (3.10) represents the difference equation for all nodes, including the boundary nodes.

The entire set of simultaneous difference equations can thus be expressed in matrix notation as follows:

$$\begin{bmatrix} a_1 & -b_1 & & & \\ -c_2 & a_2 & -b_2 & & \\ & & \vdots & & \\ & & -c_{n-1} & a_{N-1} & -b_{N-1} \\ & & & -c_N & -a_N \end{bmatrix} \begin{bmatrix} T_1 \\ T_2 \\ \vdots \\ T_{N-1} \\ T_N \end{bmatrix} = \begin{bmatrix} d_1 \\ d_2 \\ \vdots \\ d_{N-1} \\ d_N \end{bmatrix} \tag{3.11}$$

Blank spaces in the matrix represent zeros. We can now write Eq. (3.11) as

$$\mathbf{AT} = \mathbf{D}$$

and by inverting the matrix **A**, the solution for the temperature vector **T** is

$$\mathbf{T} = \mathbf{A}^{-1}\mathbf{D}$$

Since all the matrix coefficients a_i, b_i, c_i and d_i are known, the problem has been reduced to one of finding the inverse of a matrix with known coefficients, a task that is easily handled by a computer. For example, most spreadsheet programs for personal computers incorporate *matrix inversion* and *matrix multiplication*, and for many problems this will be satisfactory. Coefficients for the matrix **A** and the vector **D** in Eq. (3.11) are summarized in Table 3.1 for all three boundary conditions and for the interior nodes.

For a problem with a large number of nodes, using a spreadsheet may not be practical or efficient. In such cases, we can take advantage of a special characteristic of the matrix **A**. As can be seen in Eq. (3.11), each row of the matrix has at most three nonzero elements, and for this reason **A** is called a *tridiagonal matrix*. Special methods that are very efficient have been developed for solving tridiagonal systems. Computer pragrams that implement a popular tridiagonal matrix algorithm (TDMA) are given in Appendix 3. These programs are written for MATLAB as well as in C^{++} as a user-defined function or subroutine so that they can be easily adapted to a wide range of problems and computer codes. Also included is an older FORTRAN version of the subroutine; many currently popular commercial codes and software developed in the 1970s and 1980s are in this language, and a listing of some of these software is given in Appendix 4.

An alternative solution method called *iteration* can be used if software for matrix inversion is not available. In this method we start with an initial guess of the entire temperature distribution for the problem. Denote this initial guess of the temperature distribution by superscript zero, i.e., $T_i^{(0)}$. This temperature distribution is used in the right sides of Eqs. (3.8, 3.9, 3.10). The left side of each of these equations will then give a revised estimate of the temperature distribution. Equation (3.8) gives the revised temperature at the left boundary, T_1. Equation (3.9) gives the revised temperature at the right boundary, T_N. Equation (3.10) gives the revised temperature for all the interior nodes. Call this temperature distribution $T_i^{(1)}$ since it is the first revision to our initial guess. This completes the first iteration. The revised temperature distribution is then inserted into the right side of the same equations to produce the next revision, $T_i^{(2)}$. This procedure is repeated until the temperature distribution ceases to change significantly between iterations. Figure 3.3 shows the procedure in the form of a flowchart.

The iterative method shown in Fig. 3.3 is called *Jacobi iteration*. Close inspection of the procedure shows that after the first temperature $T_1^{(1)}$ is calculated, we have an updated nodal temperature that can be used in place of $T_1^{(0)}$ in the righthand side of Eq. (3.9) as we calculate $T_2^{(1)}$:

$$T_2^{(1)} = (b_2T_3^{(0)} + c_2T_1^{(1)} + d_2)/a_2$$

The equation for $T_3^{(1)}$ can now use the updated values $T_1^{(1)}$ and $T_2^{(1)}$ instead of $T_1^{(0)}$ and $T_2^{(0)}$. This observation can be generalized for any iteration p: the equation for $T_i^{(p)}$ can use $T_j^{(p)}$ for $j < i$ and $T_j^{(p-1)}$ for $j > i$. Because we are using updated nodal

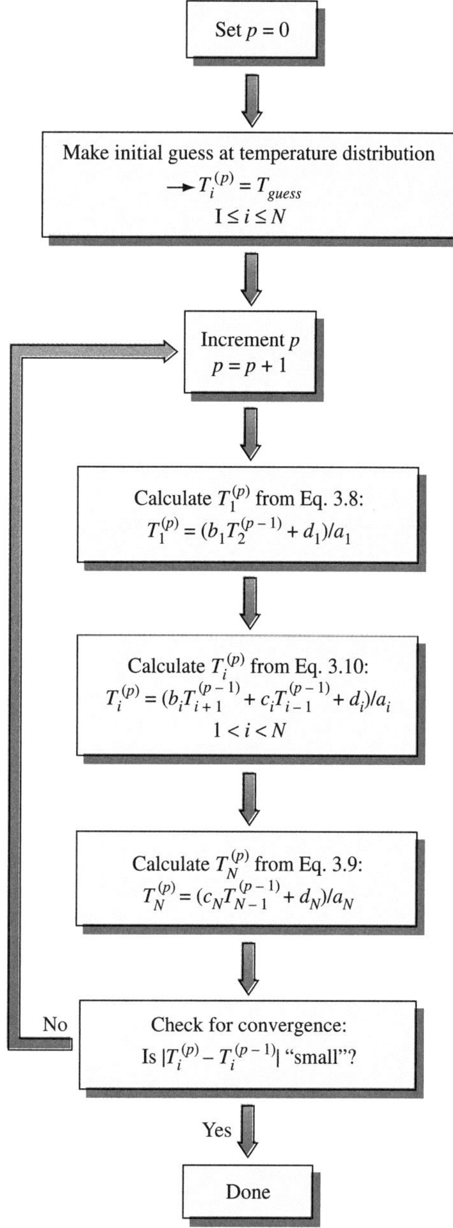

FIGURE 3.3 Flowchart for the node-by-node iterative solution of a one-dimensional steady conduction problem.

temperatures as soon as they become available, convergence is more rapid. This improved version of Jacobi iteration is called *Gauss-Seidel iteration*; both are, however, iterative schemes for point-by-point or node-by-node calculations. For very large matrices there are other faster variants of iterative schemes, also known as line-by-line or block iterative methods [1–3].

It should be obvious that the better the first guess, $T_i^{(0)}$, the more quickly the solution will converge. We can usually make a reasonably good first guess based on the boundary conditions.

When either iterative method is used, the temperature distribution will converge to the correct solution if one condition is met—either we must specify the temperature for at least one boundary node or we must specify a convection-type boundary condition with given ambient fluid temperature over at least one boundary node. Remaining boundaries can then have any type of boundary condition. This constraint is reasonable since the difference equations cannot, by themselves, establish an absolute temperature at any node; they can only establish relative temperature differences among the nodes. By specifying at least one boundary temperature or an ambient fluid temperature for the convection boundary condition, we can tie down the absolute temperature for the problem.

The method for handling variable thermal conductivity described in Section 3.2.1 will result in coefficients d_i that depend on the temperature at that node and surrounding nodes. Thus an iterative solution procedure *must* be used for this type of problem. An initial guess at the temperature distribution, T_i, must be made to allow the d_i to be determined. An updated temperature distribution can then be determined by the method described in the previous paragraphs. This updated temperature distribution is used to revise the d_i, and the procedure is repeated until the temperature distribution ceases to change.

EXAMPLE 3.1 Use a control volume approach to verify the results of Example 2.1. Use 5 nodes ($N = 5$).

Recall that Example 2.1 involves a long electrical heating element made of iron. It has a cross section of 10 cm × 1.0 cm and is immersed in a heat transfer oil at 80°C. We were to determine the convection heat transfer coefficient necessary to keep the temperature of the heater below 200°C when heat was generated uniformly at a rate of 10^6 W/m^3 by an electrical current. The thermal conductivity for iron at 200°C (64 W/m K) was taken from Table 12 in Appendix 2 by interpolation.

SOLUTION Because of symmetry we need to consider only half of the thickness of the heating element, as shown in Fig. 3.4. Define the nodes as

$$x_i = (i - 1)\Delta x \qquad \text{where} \qquad i = 1, 2, \ldots, N$$

$$\text{and} \quad \Delta x = \frac{L}{N-1}, \; L = 0.005\,\text{m, and } N = 5$$

Choose the top face, $x_N = L$, to correspond to the plane of symmetry. Since no heat flows across this plane it corresponds to a zero-heat flux boundary condition. Applying Eq. (2.1) to a control volume extending from $x = L - \Delta x/2$ to L we have

$$\dot{q}_G \frac{\Delta x}{2} = k \frac{T_N - T_{N-1}}{\Delta x} \quad \text{or} \quad T_N = T_{N-1} + \dot{q}_G \frac{\Delta x^2}{2k}$$

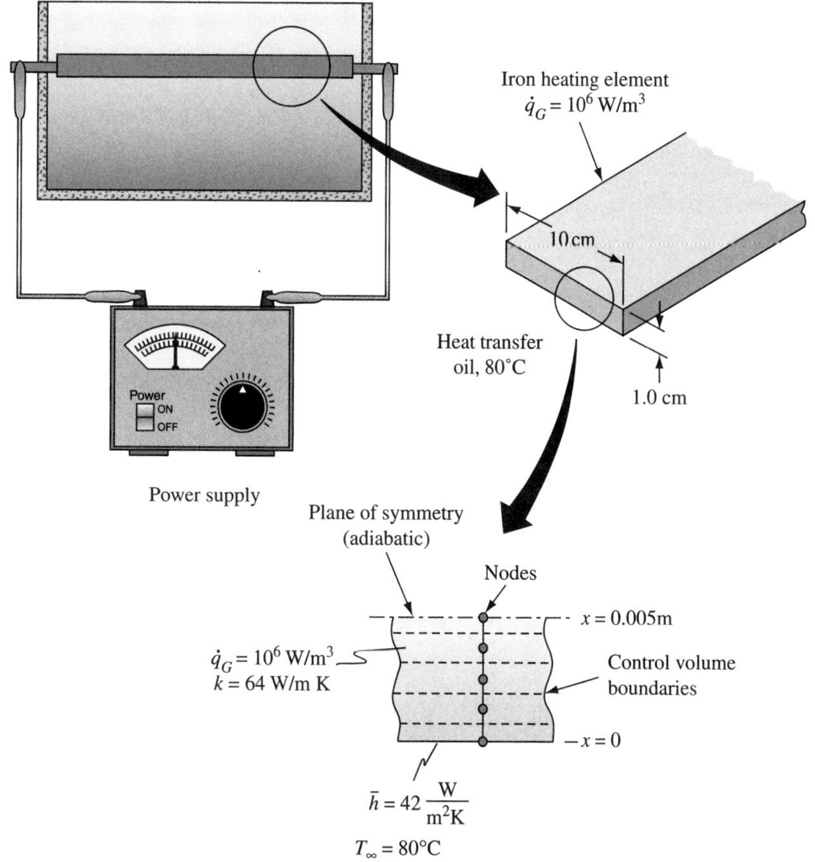

FIGURE 3.4 Heating element for Example 3.1.

At the left face, $x_1 = 0$, we have a surface convection boundary condition to which Eq. (3.7) can be applied. We can now determine all the matrix coefficients in Eq. (3.11):

$$a_1 = 1 \qquad b_1 = \frac{1}{1 + \dfrac{\bar{h}\,\Delta x}{k}} \qquad c_1 = 0 \qquad d_1 = \frac{\Delta x}{k} \frac{\left(\bar{h}T_\infty + \dot{q}_G \dfrac{\Delta x}{2}\right)}{1 + \dfrac{\bar{h}\,\Delta x}{k}}$$

$$a_i = 2 \qquad b_i = 1 \qquad c_i = 1 \qquad d_i = \frac{\Delta x^2}{k} \dot{q}_G \qquad 1 < i < N$$

$$a_N = 1 \qquad b_N = 0 \qquad c_N = 1 \qquad d_N = \dot{q}_G \frac{\Delta x^2}{2k}$$

Note that these coefficients can be taken directly from Table 3.1.

The following parameters are given in the problem statement:

$$\bar{h} = 42 \frac{\text{W}}{\text{m}^2\text{K}}$$

$$T_\infty = 80°\text{C}$$

$$k = 64 \frac{\text{W}}{\text{m K}}$$

$$\dot{q}_G = 10^6 \frac{\text{W}}{\text{m}^3}$$

Using the computer program in Appendix 3 for solving the tridiagonal system, we find the temperature distribution given next:

i	T_i (°C)
1	199.0556
2	199.1410
3	199.2021
4	199.2387
5	199.2509

The results show that a heat transfer coefficient of 42 W/m^2 K does allow the heater to operate below 200°C. Note that $T_5 - T_1 = 0.1953$°C, while the exact value from Example 2.1 is 0.2°C. Since we have an exact solution for comparison, we can be reasonably confident of our numerical solution. If we did not have the exact solution available, it would probably be worthwhile to repeat the numerical solution for smaller Δx (larger N) to assess its accuracy, as discussed in Section 3.2.1.

3.3 One-Dimensional Unsteady Conduction

3.3.1 The Difference Equation

To develop a difference equation for unsteady conduction problems we need to consider the energy storage term in Eq. (2.1). First, we define a discrete time step Δt analogous to the discrete spatial step Δx introduced in the previous section:

$$t_m = m\,\Delta t \qquad m = 0,1,\ldots$$

Nodal temperatures now depend on two indices, i and m, which correspond to the spatial and time dependencies, respectively:

$$T_{i,m} \equiv T(x_i, t_m)$$

Since $\Delta y = \Delta z = 1$, the energy storage term in Eq. (2.1) can be written

$$\text{rate of energy storage inside control volume} = \rho c \, \Delta x \frac{T_{i,m+1} - T_{i,m}}{\Delta t} \tag{3.12}$$

This term represents the energy stored from time t_m to t_{m+1} in a slab of thickness Δx divided by the time step $\Delta t = t_{m+1} - t_m$. Just as we allowed the temperature to vary linearly between spatial nodes in Section 3.2, here we allow the temperature to vary linearly between time steps.

We add this energy storage term to the right side of Eq. (2.1) because terms on the left side of the equation represent energy that flows into the control volume and tends to increase the nodal temperature with time. This gives, after some algebraic rearrangement, the following expression:

$$-k\frac{T_{i,m} - T_{i-1,m}}{\Delta x} + \dot{q}_{G,i,m}\,\Delta x = -k\frac{T_{i+1,m} - T_{i,m}}{\Delta x} + \rho c\,\Delta x\frac{T_{i,m+1} - T_{i,m}}{\Delta t} \tag{3.13}$$

Equation (3.13) is the discretized form of Eq. (2.5), and the heat generation rate and all but one temperature in this expression are evaluated at time t_m. One temperature in Eq. (3.13) is evaluated at time $t_{m+1} = t_m + \Delta t$. Solving for this temperature gives

$$T_{i,m+1} = T_{i,m} + \frac{\Delta t}{\rho c \, \Delta x}\left\{\frac{k}{\Delta x}(T_{i+1,m} - 2T_{i,m} + T_{i-1,m}) + \dot{q}_{G,\,i,\,m}\,\Delta x\right\} \tag{3.14}$$

Equation (3.14) is called the *explicit difference equation* because the temperature distribution at the new time t_{m+1} can be determined if the complete temperature distribution at time t_m is known. Since any unsteady problem statement must include an initial condition, T_i,0 is given for all i. Equation (3.14) can then be used to calculate T_i,1, T_i, 2, and so forth for all required time steps. This procedure is called *marching* because the solution is essentially marched forward from one time step to another.

Thus, solution of the explicit equation is very straightforward. There is, however, a limitation on the size of the time step Δt. We require

$$\Delta t < \frac{\Delta x^2}{2\alpha} \tag{3.15}$$

Based on the definition of the Fourier number given in Chapter 2, we also can write Eq. (3.15) in terms of this dimensionless parameter:

$$\text{Fo} < \frac{1}{2} \qquad \text{where} \qquad \text{Fo} \equiv \frac{\alpha \Delta t}{\Delta x^2}$$

If a time step larger than that prescribed by Eq. (3.15) is used, the solution will begin to exhibit growing oscillations. In this situation the solution is said to be *unstable*. This behavior can be explained both mathematically and physically. First, rearrange Eq. (3.14):

$$T_{i,m+1} = T_{i,m}\left\{1 - \frac{\Delta t}{(\Delta x^2/2\alpha)}\right\} + \frac{\Delta t}{\rho c\,\Delta x}\left\{\frac{k}{\Delta x}(T_{i+1,m} + T_{i-1,m}) + \dot{q}_{G,i,m}\,\Delta x\right\}$$

Note that if the condition expressed by Eq. (3.15) is violated, the coefficient on $T_{i,m}$ in this equation is negative. This will lead to oscillations in the solution because the temperature at node i for the new time step $m + 1$ will have a negative dependence on the value at that node at time step m. Physically, the right side of Eq. (3.15) can be thought of as the time required for a temperature field to diffuse through the control volume Δx. If our solution method uses time steps larger than this diffusion time, unrealistic results are likely to appear after a few time steps.

We can eliminate the oscillations by making the time steps sufficiently small. However, very small time steps are undesirable because they require more computation time for a given total elapsed time. Also note that if we wish to increase solution accuracy by using smaller values of Δx, we are forced by Eq. (3.15) to use smaller time steps. This also increases the computation time.

In practice, the time step Δt will be set to a value somewhat smaller than that prescribed by Eq. (3.15). The actual value of Δt to be used in the final numerical solution can be determined by trial and error, just as the appropriate value of Δx was determined for steady-state problems (see Section 3.2.1). A series of solutions are found for decreasing values of Δt until further reductions in Δt cease to affect the temperature distribution. Round-off error is a consideration here just as it was for steady-state problems because as Δt is decreased, more mathematical operations are required to produce the solution.

It should be pointed out that the difference equations for the boundary control volumes also impose stability restrictions on the size of the time step. This will be discussed in Sections 3.3.2 and 3.3.4.

EXAMPLE 3.2 Consider the problem given in Example 2.11—determining the minimum depth x_m at which a water main must be buried to avoid freezing (see Fig. 3.5). For the parameters given in that problem, a depth of 0.68 m was required. Determine x_m by solving the explicit difference equation.

In Example 2.11 the soil was initially at a uniform temperature of 20°C and we assumed that under the worst conditions anticipated it would be subjected to a surface temperature of −15°C for a period of 60 days. We used the following properties for soil (at 300 K):

$$\rho = 2050 \text{ kg/m}^3 \qquad k = 0.52\,\text{W/m K} \qquad c = 1840\,\text{J/kg K}$$

$$\alpha = 0.138 \times 10^{-6}\,\text{m}^2\text{/s}$$

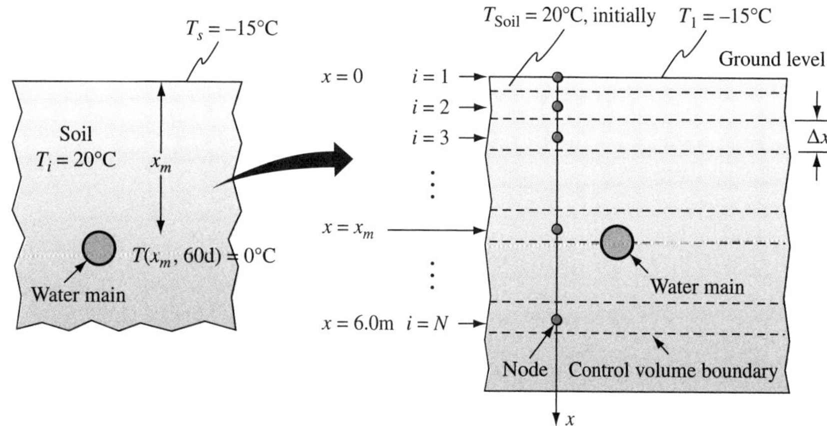

FIGURE 3.5 Water main depth schematic, Example 3.2.

SOLUTION We need to implement Eq. (3.14) in a computer program. Figure 3.5 shows the relationship between depth and nodes. Since there is no heat generation ($\dot{q}_G = 0$), the equation simplifies to

$$T_{i,m+1} = T_{i,m} + \text{Fo}(T_{i+1,m} - 2T_{i,m} + T_{i-1,m})$$

where Fo is the Fourier number, $\text{Fo} = \Delta t\alpha/\Delta x^2$, as discussed earlier.

Let us create a one-dimensional array to contain the known temperatures at time step m and a similar array for the unknown temperature at time step $m + 1$:

$$T_{i,\text{old}} \equiv T_{i,m}$$
$$T_{i,\text{new}} \equiv T_{i,m+1}$$

We now have an equation of the form

$$T_{i,\text{new}} = T_{i,\text{old}} + \text{Fo}(T_{i+1,\text{old}} - 2T_{i,\text{old}} + T_{i-1,\text{old}})$$

which is valid for all nodes except the boundary nodes. For the boundary nodes we have $T_{1,\text{old}} = T_{1,\text{new}} = -15°\text{C}$ and $T_{N,\text{old}} = T_{N,\text{new}} = 20°\text{C}$.

We must now decide to what depth the calculation should extend and the size of the control volumes. At a very large depth we would expect to see no temperature change for the entire 60 days. Let us select a maximum depth of 6 m and check our solution to ensure that no temperature change is seen at 6 m depth. If the solution shows that the temperature has changed significantly at 6 m, we will need to allow the calculation to extend to a greater depth.

We are free to select the control volume size, Δx, but remember that accuracy improves as Δx decreases. As far as the time step is concerned, we must follow the constraint given in Eq. (3.15). As the time step decreases, we expect greater accuracy. A further constraint is that we want the calculation to end at exactly 60 days, so we want 60 days to be an integer multiple of Δt.

The solution procedure is shown in the flow chart of Fig. 3.6 on the next page.

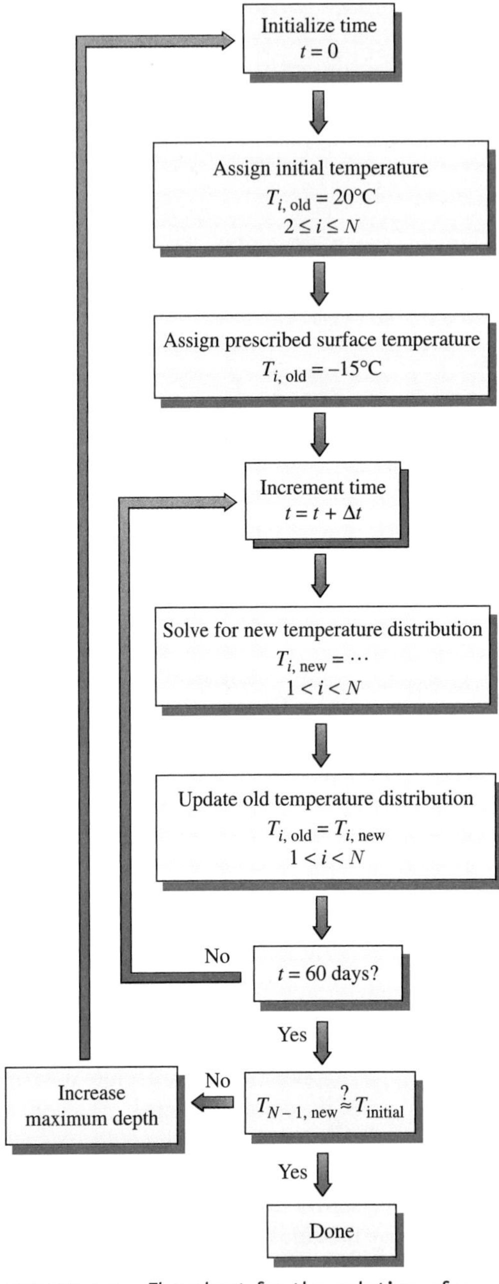

FIGURE 3.6 Flowchart for the solution of a one-dimensional transient conduction problem, Example 3.2.

First, let us choose $N = 6$, giving $\Delta x = 1.2$ m. With the given value of α, the maximum allowable value of Δt from Eq. (3.15) is 60.43 days. Selecting $\Delta t = 30$ days, we find the following temperature distribution at $t = 60$ days:

i	T_i(°C)
1	−15.000
2	11.306
3	20.000
4	20.000
5	20.000
6	20.000

By linear interpolation between $i = 1$ where $T_{x=0} = -15$°C and $i = 2$ where $T_{x=1,2m} = 11.306$°C, x_m, the depth at which $T = 0$°C, is found to be 0.684 m.

Next, we need to determine whether we have used a sufficiently small time step for accuracy. We successively halve Δt, finding x_m for each Δt. This gives

Δt (days)	x_m(m)
30	0.684
15	0.724
7.5	0.743
3.75	0.753
1.875	0.757

Reducing the time step to less than 1.875 days will give no further significant change in x_m.

Now we need to determine whether the control volume is sufficiently small. Fix Δt at 1.875 days. We then successively halve Δx, finding x_m for each value of Δx. This gives

Δx (m)	x_m(m)
1.2	0.757
0.6	0.680
0.3	0.676

Reducing the node spacing to less than 0.3 m will give no further significant change in x_m. (Note that if we want to try $\Delta x = 0.15$ m we would need to reduce Δt to less than 1.875 days.)

To demonstrate what happens if Δt is too large, two additional runs were made. We used $\Delta x = 0.1$ m, which according to Eq. (3.15), requires a time step less than $\Delta t = 0.4197$ days. We used $\Delta t_1 = 1.1 \times \Delta t = 0.4616$ days for one run and

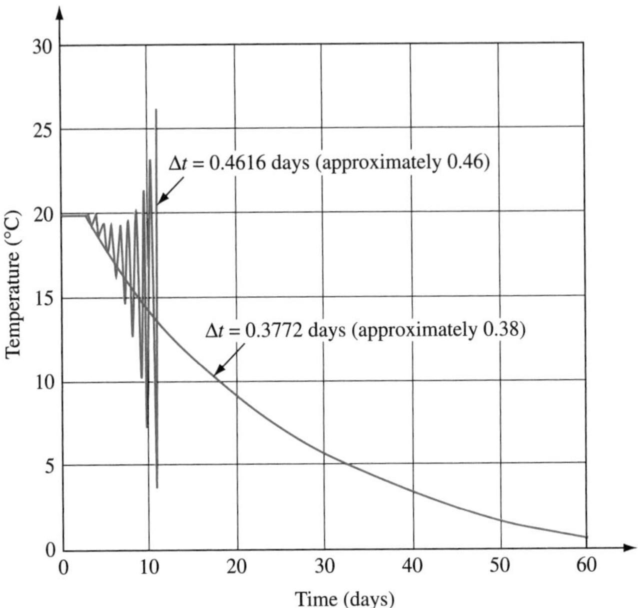

FIGURE 3.7 Temperature as a function of time for Example 3.2 using two values of the time step Δt.

$\Delta t_2 = 0.9 \times \Delta t = 0.3772$ days for the other run. Figure 3.7 presents the temperature at one point in the soil ($x = x_m - \Delta x$) as a function of time for both runs. The solution for the larger time step is unstable as seen by the growing oscillations with time.

Example 3.2 demonstrates the main disadvantages of the explicit method—small time steps and long computation times are required to ensure accuracy and stability. We can circumvent this problem by writing the difference equation in its *implicit* form. To do this, we evaluate the term in braces in Eq. (3.14) at time step $m + 1$ rather than at m:

$$T_{i,m+1} = T_{i,m} + \frac{\Delta t}{\rho c\,\Delta x} \times \left\{ \frac{k}{\Delta x}(T_{i+1,m+1} - 2T_{i,m+1} + T_{i-1,m+1}) + \dot{q}_{G,i,m+1}\,\Delta x \right\} \tag{3.16}$$

Eq. (3.16) is called implicit because the temperatures at time step $m + 1$ must be determined simultaneously. This requirement complicates the solution over that for the explicit method but the restriction on the time step expressed by Eq. (3.15) is eliminated. Hence, the solution of Eq. (3.16) is stable for any time step Δt. Large time steps sacrifice accuracy, just as before, but at least we need not be concerned with stability. Solution of implicit difference equations will be discussed in Section 3.3.3.

3.3.2 Boundary Conditions

As in the steady conduction problem, specified surface temperature for the unsteady problem is simple to implement. Specified surface flux or surface convection boundary conditions are more complex because we must account for energy storage in the control volumes nearest the surfaces. Consider the *specified flux boundary condition* first. For steady conduction we derived an energy balance over a control volume extending from $x = 0$ to $x = \Delta x/2$, Eq. (3.5). Over the time step Δt, the rate at which the control volume stores energy is

$$\rho c \frac{\Delta x}{2} \frac{T_{i,m+1} - T_{i,m}}{\Delta t}$$

This term must be added to the right side of Eq. (3.5), giving

$$q''_{1,m} + \dot{q}_{G,1,m} \frac{\Delta x}{2} = -k \frac{T_{2,m} - T_{1,m}}{\Delta x} + \rho c \frac{\Delta x}{2} \frac{T_{1,m+1} - T_{1,m}}{\Delta t}$$

Solving for $T_{1,m+1}$ gives

$$T_{1,m+1} = T_{1,m} + \frac{2\,\Delta t}{\rho c\,\Delta x} \left\{ q''_{1,m} + \frac{\Delta x}{2} \dot{q}_{G,1,m} + \frac{k}{\Delta x} (T_{2,m} - T_{1,m}) \right\} \tag{3.17}$$

Equation (3.17) can be rearranged to show that the coefficient on $T_{1,m}$ is

$$\left(1 - \frac{2\alpha\,\Delta t}{\Delta x^2}\right)$$

Thus the stability criterion imposed by this equation is identical to that for the interior control volumes as expressed by Eq. (3.15).

This expression for the boundary condition is explicit since the new boundary temperature can be determined once the temperature distribution at time m is known. In an implicit equation we would evaluate the terms in braces in Eq. (3.17) at time step $m + 1$.

For the *surface convection boundary condition* we add the storage term to the right side of Eq. (3.6), giving

$$\bar{h}(T_\infty - T_{1,m}) + \dot{q}_{G,1,m} \frac{\Delta x}{2} = -k \frac{T_{2,m} - T_{1,m}}{\Delta x} + \rho c \frac{\Delta x}{2} \frac{T_{1,m+1} - T_{1,m}}{\Delta t}$$

Solving for $T_{1,m+1}$ yields

$$T'_{1,m+1} = T_{1,m} + \frac{2\,\Delta t}{\rho c\,\Delta x} \left\{ \bar{h}(T_\infty - T_{1,m}) + \dot{q}_{G,i,m} \frac{\Delta x}{2} + k \frac{T_{2,m} - T_{1,m}}{\Delta x} \right\} \tag{3.18}$$

Rearranging this equation shows that the coefficient on $T_{1,m}$ is

$$1 - \frac{2\alpha\,\Delta t}{\Delta x^2}\left(1 + \frac{\bar{h}\Delta x}{k}\right)$$

This stability constraint is more restrictive than that for the interior nodes [Eq. (3.15)]. (If $\Delta x/k$ is small compared to 1, then the constraints are similar.) The numerical solutions for problems with very high convection coefficients can become unstable unless sufficiently small steps are chosen so that the expected large temperature gradients near the boundary can be accurately represented.

Similar expressions can be developed for specified conditions at $x = L$. For a *specified heat flux* at $x = L$,

$$T_{N,m+1} = T_{N,m} + \frac{2\,\Delta t}{\rho c\,\Delta x}\left\{q''_{N,m} + \frac{\Delta x}{2}\dot{q}_{G,N,m} + \frac{k}{\Delta x}(T_{N-1,m} - T_{N,m})\right\} \tag{3.19}$$

where $q''_{N,m}$ is the specified flux *into* the right boundary. For a *specified surface convection* at $x = L$,

$$T_{N,m+1} = T_{N,m} + \frac{2\Delta t}{\rho c\,\Delta x}\left\{\bar{h}(T_\infty - T_{N,m}) + \frac{\Delta x}{2}\dot{q}_{G,N,m} + \frac{k}{\Delta x}(T_{N-1,m} - T_{N,m})\right\} \tag{3.20}$$

EXAMPLE 3.3 A large sheet of stainless steel is removed from an annealing bath and allowed to cool in air (see Fig. 3.8). If the initial temperature of the sheet is 500°C, determine how long it will take for the center of the sheet to cool to 250°C. The sheet is 2 cm thick and has a density of 8500 kg/m^3, a specific heat of 460 J/kg K, and a thermal conductivity of 20 W/m K. The heat transfer coefficient to the air is 80 W/m^2 K, and the ambient air temperature is 20°C. Use an explicit method and compare your results with those from a chart solution.

SOLUTION The transient cooling of stainless steel sheet can be modeled as a semi-infinite slab (one dimension in space) problem, because the thickness of the sheet is much

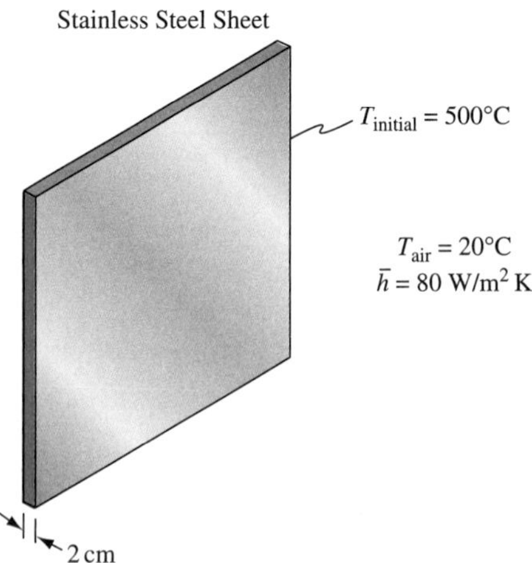

FIGURE 3.8 Stainless steel sheet, Example 3.3.

smaller than its width and length. Also, before carrying out the numerical analysis it is instructive first to find the chart solution. The Biot number is

$$\text{Bi} = \frac{\bar{h}L}{k} = \frac{\left(80\,\frac{\text{W}}{\text{m}^2\text{K}}\right)(0.01\,\text{m})}{\left(20\,\frac{\text{W}}{\text{m K}}\right)} = 0.04$$

or $1/\text{Bi} = 25$. Note that $\text{Bi} < 0.1$ and hence the sheet can be treated as a lumped capacitance. For the ordinate in Fig. 2.42(a) we have

$$\frac{T(0, t) - T_\infty}{T(x, 0) - T_\infty} = \frac{250 - 20}{500 - 20} = 0.479$$

which gives

$$\frac{\alpha t}{L^2} = 19$$

from which

$$t = \frac{(19)(0.01\,\text{m}^2)}{\left(20\,\frac{\text{W}}{\text{m K}}\right)}\left(8500\,\frac{\text{kg}}{\text{m}^3}\right)\left(460\,\frac{\text{W s}}{\text{kg K}}\right) = 371\ \text{s}$$

To implement the explicit numerical method, consider half the sheet thickness, with $x = 0$ being the exposed left face and $x = L = 0.01\,\text{m}$ being the sheet centerline. Eq. (3.14) can be used for all interior nodes:

$$T_{i,m+1} = T_{i,m} + \frac{\Delta t\alpha}{\Delta x^2}(T_{i+1,m} - 2T_{i,m} + T_{i-1,m})$$

For the left face Eq. (3.18) simplifies to

$$T_{1,m+1} = T_{1,m} + \frac{2\,\Delta t}{\rho c\,\Delta x}\left(\bar{h}(T_\infty - T_{1,m}) + k\frac{T_{2,m} - T_{1,m}}{\Delta x}\right)$$

Since the right boundary corresponds to the plane of symmetry, there can be no heat flow across the $x = L$ plane, and Eq. (3.19) therefore simplifies to

$$T_{N,m+1} = T_{N,m} + \frac{2\Delta t\alpha}{\Delta x^2}(T_{N-1,m} - T_{N,m})$$

To simplify the notation, let us define

$$C_1 = \frac{\alpha\Delta t}{\Delta x^2} \qquad C_2 = \frac{2\bar{h}\Delta t}{\rho c\,\Delta x} \qquad C_3 = \frac{2\alpha\Delta t}{\Delta x^2}$$

Note that $C_1 = C_3/2 = \text{Fo}$.

As before, let $T_{i,\text{old}} \equiv T_{i,m}$ and $T_{i,\text{new}} \equiv T_{i,m+1}$. Then our equations become

$$T_{1,\text{new}} = T_{1,\text{old}} + C_2(T_\infty - T_{1,\text{old}}) + C_3(T_{2,\text{old}} - T_{1,\text{old}}) \qquad \text{(a)}$$

$$T_{i,\text{new}} = T_{i,\text{old}} + C_1(T_{i+1,\text{old}} - 2T_{i,\text{old}} + T_{i-1,\text{old}}) \qquad i = 2, 3, \ldots, N-1 \text{ (b)}$$

$$T_{N,\text{new}} = T_{N,\text{old}} + C_3(T_{N-1,\text{old}} - T_{N,\text{old}}) \qquad \text{(c)}$$

The initial condition is

$$T_{i,m=0} = 500°\text{C} \qquad \text{for all } i$$

The solution procedure is shown in Fig. 3.9. Using 20 control volumes ($N = 21$) gives $\Delta x = L/20 = 0.01/20 = 0.0005$ m. Since we are using an explicit scheme, we must have

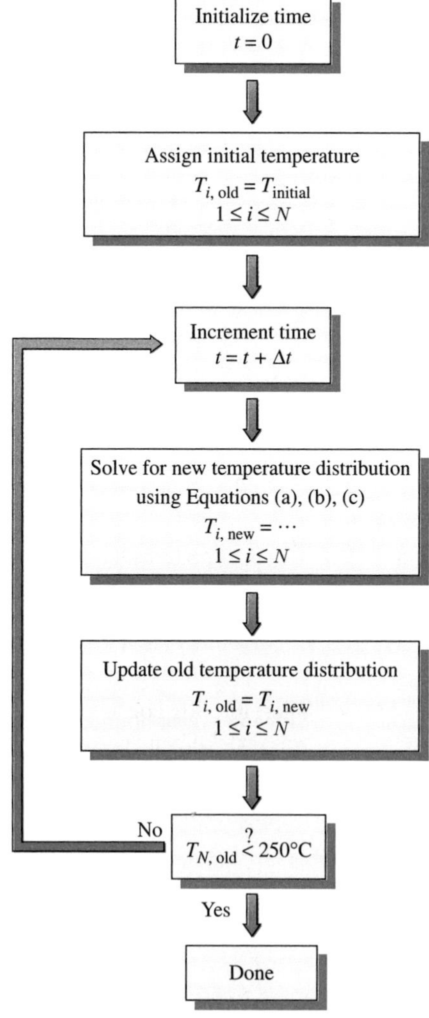

FIGURE 3.9 Flowchart for the explicit solution of a one-dimensional transient conduction problem, Example 3.3.

$$\Delta t < \frac{\Delta x^2}{2\alpha} = \frac{(0.0005\,\text{m})^2}{2\left(20\,\dfrac{\text{W}}{\text{m K}}\right)}\left(8500\,\frac{\text{kg}}{\text{m}^3}\right)\left(460\,\frac{\text{W s}}{\text{kg K}}\right) = 0.0244\,\text{s}$$

Since $\bar{h}\,\Delta x/k \ll 1$, the stability criterion for the boundary control volume will give a similar value for Δt. To provide a margin, we will use $\Delta t = 0.02\,\text{s}$.

A computer program that carries out the procedure just described gives a result of $t = 367.5\,\text{s}$, which is about 1.5% less than the chart solution. Considering that the charts cannot be read precisely, this agreement is quite good. Running the program with half the number of control volumes, ($N = 11$) and a time step of 0.08 s gives $t = 367.6\,\text{s}$, indicating that the solution for $\Delta t = 0.02$ s is sufficiently accurate.

3.3.3 Solution Methods

As demonstrated in Examples 3.2 and 3.3, the solution of explicit difference equations is straightforward. It is simply a matter of carrying forward the solution from the previous time step to the new time step. In this section we will focus on the solution of the implicit difference equation, Eq. (3.16), with associated boundary conditions. The interrelationships of the temperatures at time $m + 1$ in Eq. (3.16) are similar to those in Eq. (3.2), suggesting a means of solving the implicit equation. First rearrange Eq. (3.16) to give

$$(1 + 2\,\text{Fo})T_{i,m+1} = \text{Fo}\,T_{i+1,m+1} + \text{Fo}\,T_{i-1,m+1} + T_{i,m} + \frac{\Delta t}{\rho c}\dot{q}_{G,i,m+1}$$

which can be written in a form like Eq. (3.10):

$$a_i T_{i,m+1} = b_i T_{i+1,m+1} + c_i T_{i-1,m+1} + d_i \tag{3.21}$$

Comparing Eq. (3.21) with the rearranged form of Eq. (3.16), we see that

$$a_i = 1 + 2\,\text{Fo}, \qquad b_i = c_i = \text{Fo}, \qquad d_i = T_{i,m} + \frac{\Delta t}{\rho c}\dot{q}_{G,i,m+1}$$

Therefore the coefficients a_i, b_i, and c_i ($1 < i < N$) are known. The coefficients d_i are also known, either from the given initial condition, $T_{i,0}$, or because we have solved for $T_{i,m}$ in the previous time step.

If we could express the boundary conditions in a form like Eqs. (3.8) and (3.9), then we could solve the implicit transient difference equation by inverting a matrix, just as we did to solve the steady difference equation. At the left boundary we seek an equation like

$$a_1 T_{1,m+1} = b_1 T_{2,m+1} + d_1$$

For a specified surface temperature we therefore have

$$a_1 = 1 \qquad b_1 = 0 \qquad d_1 = T_1$$

Inspection of the implicit forms of Eq. (3.17) and (3.18) shows that for a specified surface heat flux boundary condition we have

$$a_1 = 1 + \frac{2\,\Delta t\alpha}{\Delta x^2} \qquad b_1 = \frac{2\,\Delta t\alpha}{\Delta x^2} \qquad d_1 = T_{1,m} + \frac{2\,\Delta t}{\rho c\,\Delta x}\left(q''_{1,m+1} + \frac{\Delta x}{2}\dot{q}_{G,1,m+1}\right)$$

For the surface convection boundary condition, the implicit form of Eq. (3.18) gives

$$a_1 = 1 + \frac{2\,\Delta t\alpha}{\Delta x^2} + \frac{2\,\Delta t\bar{h}}{\rho c\,\Delta x} \qquad b_1 = \frac{2\,\Delta t\alpha}{\Delta x^2}$$

$$d_1 = T_{1,m} + \frac{2\,\Delta t}{\rho c\,\Delta x}\left(\bar{h}T_\infty + \frac{\Delta x}{2}\dot{q}_{G,1,m+1}\right)$$

At the right boundary we seek an equation like

$$a_N T_{N,m+1} = c_N T_{N-1,m+1} + d_N$$

For specified surface temperature we have

$$a_N = 1 \qquad c_N = 0 \qquad d_N = T_N$$

For a specified surface heat flux

$$a_N = 1 + \frac{2\,\Delta t\alpha}{\Delta x^2} \qquad c_N = \frac{2\,\Delta t\alpha}{\Delta x^2}$$

$$d_N = T_{N,m} + \frac{2\,\Delta t}{\rho c\,\Delta x}\left(q''_{N,m+1} + \frac{\Delta x}{2}\dot{q}_{G,N,m+1}\right)$$

For the surface convection boundary condition

$$a_N = 1 + \frac{2\,\Delta t\alpha}{\Delta x^2} + \frac{2\,\Delta t\bar{h}}{\rho c\,\Delta x} \qquad c_N = \frac{2\,\Delta t\alpha}{\Delta x^2}$$

$$d_N = T_{N,m} + \frac{2\,\Delta t}{\rho c\,\Delta x}\left(\bar{h}T_\infty + \frac{\Delta x}{2}\dot{q}_{G,N,m+1}\right)$$

Coefficients for Eq. (3.21) are summarized in Table 3.2 for all three boundary conditions.

The solution using matrix inversion closely follows the procedure for the steady difference equation. The only difference between the two is that for the unsteady problem, we must update the constants d_i between time steps because the d_i depend on the temperature at node i, which depends on the time step. After updating d_i, we apply matrix inversion on our tridiagonal matrix solution technique to get the new temperature distribution, $T_{i,m+1}$. We can then update the coefficients d_i again and repeat the process. This can be illustrated by applying the procedure to Example 3.3.

TABLE 3.2 Matrix coefficients for one-dimensional unsteady conduction [Eq. (3.21)], implicit solution

	a_i	b_i	c_i	d_i
$i = 1$, specified surface temperature	1	0	0	T_1
$i = 1$, specified heat flux	$1 + 2\text{ Fo}$	2 Fo	0	$T_{1,m} + \dfrac{2\,\Delta t}{\rho c\,\Delta x}\left(q''_{1,m+1} + \dfrac{\Delta x}{2}\dot{q}_{G,1,m+1}\right)$
$i = 1$, specified surface convection	$1 + 2\text{ Fo} + \dfrac{2\bar{h}\Delta t_1}{\rho c\,\Delta x}$	2 Fo	0	$T_{1,m} + \dfrac{2\,\Delta t}{\rho c\,\Delta x}\left(\bar{h}_1 T_{\infty,1} + \dfrac{\Delta x}{2}\dot{q}_{G,1,m+1}\right)$
$1 < i < N$	$1 + 2\text{ Fo}$	Fo	Fo	$T_{i,m} + r_s\dot{q}_{G,i,m+1}$
$i = N$, specified surface temperature	1	0	0	T_N
$i = N$, specified heat flux	$1 + 2\text{ Fo}$	0	2 Fo	$T_{N,m} + \dfrac{2\,\Delta t}{\rho c\,\Delta x}\left(q''_{N,m+1} + \dfrac{\Delta x}{2}\dot{q}_{G,N,m+1}\right)$
$i = N$, specified surface convection	$1 + 2\text{ Fo} + \dfrac{2\bar{h}\Delta t_N}{\rho c\,\Delta x}$	0	2 Fo	$T_{N,m} + \dfrac{2\,\Delta t}{\rho c\,\Delta x}\left(\bar{h}_N T_{\infty,N} + \dfrac{\Delta x}{2}\dot{q}_{G,N,m+1}\right)$

Note: $\text{Fo} = \dfrac{\alpha\,\Delta t}{\Delta x^2}$

$q''_{A,m+1}$ is the heat flux *into* surface A.

In Example 3.3 we had a convection boundary condition at $x = 0$ and a specified flux boundary condition ($q'' = 0$) at $x = L$. Since there is no heat generation, the coefficients of Eq. (3.21) are

$$a_1 = 1 + 2\text{ Fo} + \frac{2\bar{h}\Delta t}{\rho c\,\Delta x} \qquad b_1 = 2\text{Fo} \qquad d_1 = T_{1,m} + \frac{2\bar{h}\Delta t T_\infty}{\rho c\,\Delta x}$$

$$a_i = 1 + 2\text{ Fo} \qquad b_i = c_i = \text{Fo} \qquad d_i = T_{i,m} \qquad 1 < i < N$$

$$a_N = 1 + 2\text{ Fo} \qquad c_N = 2 \qquad \text{Fo} \qquad d_N = T_{N,m}$$

The matrix representation of the problem is

$$\begin{bmatrix} a_1 & -b_1 & & & \\ -c_2 & a_2 & -b_2 & & \\ & & \vdots & & \\ & & -c_{N-1} & a_{N-1} & -b_{N-1} \\ & & & -c_N & a_N \end{bmatrix} \begin{bmatrix} T_{1,\text{new}} \\ T_{2,\text{new}} \\ \vdots \\ T_{N-1,\text{new}} \\ T_{N,\text{new}} \end{bmatrix} = \begin{bmatrix} T_{1,\text{old}} + 2\bar{h}\Delta t T_\infty/\rho c\,\Delta x \\ T_{2,\text{old}} \\ \vdots \\ T_{N-1,\text{old}} \\ T_{N,\text{old}} \end{bmatrix}$$

Write this as

$$\mathbf{AT}_{\text{new}} = \mathbf{D}$$

Then for each time step we can find the $\mathbf{T}_{\text{new}}$ vector by solving this tridiagonal system. The solution procedure is shown in Fig. 3.10.

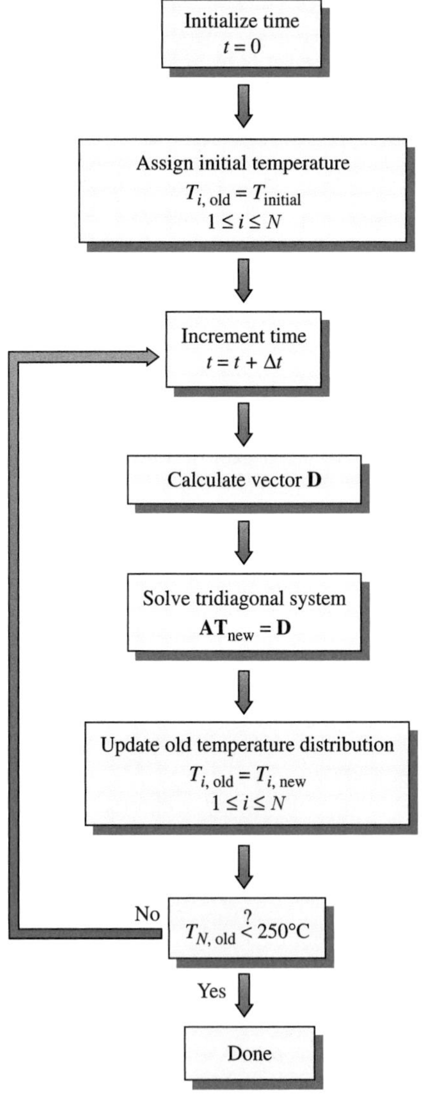

FIGURE 3.10 Flowchart for Example 3.3, using the implicit solution of a one-dimensional transient conduction problem.

3.4* Two-Dimensional Steady and Unsteady Conduction

3.4.1 The Difference Equation

With the control-volume method the extension to two- or three-dimensional systems is straightforward. Consider Fig. 3.11, which shows a control volume for a two-dimensional problem. Since the problem is two-dimensional, let $\Delta z = 1$. As in Section 3.2, the x nodes are identified by

$$x_i = (i - 1)\Delta x \qquad i = 1, 2, \ldots, M$$

In a similar manner the y nodes are identified by

$$y_i = (j - 1)\Delta y \qquad j = 1, 2, \ldots, N$$

The control volume size is Δx by Δy, and it is centered about the node, i, j. Now, applying Eq. (2.1), we have

$$\text{rate of conduction into the control volume} = -k\frac{\partial T}{\partial x}\bigg|_{\text{left}}\Delta y - k\frac{\partial T}{\partial y}\bigg|_{\text{bottom}}\Delta x$$

and

$$\text{rate of conduction out of the control volume} = -k\frac{\partial T}{\partial x}\bigg|_{\text{right}}\Delta y - k\frac{\partial T}{\partial y}\bigg|_{\text{top}}\Delta x$$

where “left,” “right,” “top,” and “bottom” refer to the control volume faces shown in Fig. 3.11. Note that the surface area of the control volume face normal to the temperature gradient has been accounted for by Δy in the left and right terms and by Δx in the bottom and top terms.

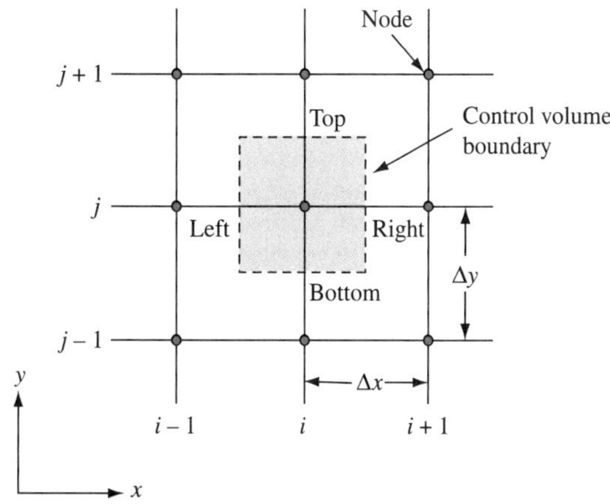

FIGURE 3.11 Control volume for two-dimensional conduction.

The temperature gradients at the control volume surfaces are evaluated as in Section 3.2,

$$\left.\frac{\partial T}{\partial x}\right|_{\text{left}} = \frac{T_{i,j,m} - T_{i-1,j,m}}{\Delta x} \qquad \left.\frac{\partial T}{\partial x}\right|_{\text{right}} = \frac{T_{i+1,j,m} - T_{i,j,m}}{\Delta x}$$

$$\left.\frac{\partial T}{\partial x}\right|_{\text{bottom}} = \frac{T_{i,j,m} - T_{i,j-1,m}}{\Delta y} \qquad \left.\frac{\partial T}{\partial y}\right|_{\text{top}} = \frac{T_{i,j+1,m} - T_{i,j,m}}{\Delta y}$$

If the rate of volumetric heat generation is $\dot{q}_G(x, y, t)$, then

$$\text{rate of heat generation is inside the control volume} = \dot{q}_{G,i,j,m}\,\Delta x\,\Delta y$$

Finally,

$$\text{rate of energy storage inside the control volume} = \rho c\,\Delta x\,\Delta y\,\frac{T_{i,j,m+1} - T_{i,j,m}}{\Delta t}$$

The overall heat balance on the control volume is therefore

$$-k\left(\frac{T_{i,j,m} - T_{i-1,j,m}}{\Delta x}\Delta y + \frac{T_{i,j,m} - T_{i,j-1,m}}{\Delta y}\Delta x\right) + \dot{q}_{G,i,j,m}\Delta x\,\Delta y$$
$$= -k\left(\frac{T_{i+1,j,m} - T_{i,j,m}}{\Delta x}\Delta y + \frac{T_{i,j+1,m} - T_{i,j,m}}{\Delta y}\Delta x\right) \tag{3.22}$$
$$+ \rho c\,\Delta x \Delta y - \frac{T_{i,j,m+1} - T_{i,j,m}}{\Delta t}$$

Dividing by $k\,\Delta x\,\Delta y$ we get the desired difference equation

$$\frac{T_{i+1,j,m} - 2T_{i,j,m} + T_{i-1,j,m}}{\Delta x^2} + \frac{T_{i,j+1,m} - 2T_{i,j,m} + T_{i,j-1,m}}{\Delta y^2} + \frac{\dot{q}_{G,i,j,m}}{k}$$
$$= \frac{\rho c}{k}\frac{T_{i,j,m+1} - T_{i,j,m}}{\Delta t} \tag{3.23}$$

For steady two-dimensional conduction without heat generation, Eq. (3.23) becomes

$$\frac{T_{i+1,j} - 2T_{i,j} + T_{i-1,j}}{\Delta x^2} + \frac{T_{i,j+1} - 2T_{i,j} + T_{i,j-1}}{\Delta y^2} = 0 \tag{3.24}$$

Solving for $T_{i,j}$

$$T_{i,j} = \frac{\Delta y^2(T_{i+1,j} + T_{i-1,j}) + \Delta x^2(T_{i,j+1} + T_{i,j-1})}{2(\Delta x^2 + \Delta y^2)}$$

If $\Delta x = \Delta y$, we have

$$T_{i,j} = \frac{1}{4}(T_{i+1,j} + T_{i-1,j} + T_{i,j+1} + T_{i,j-1})$$

As in one-dimensional steady conduction with no heat generation, the temperature at any interior node i, j is the average of the temperatures at the neighboring nodes.

Referring to Eq. (3.23), we see that the temperature distribution at time step $m + 1$ is easily determined from the distribution at time step m. In other words, Eq. (3.23) is in the explicit form. It is stable only if

$$\Delta t < \frac{1}{2\alpha}\left(\frac{1}{\Delta x^2} + \frac{1}{\Delta y^2}\right)^{-1}$$

The implicit form of the difference equation is

$$\frac{T_{i+1,j,m+1} - 2T_{i,j,m+1} + T_{i-1,j,m+1}}{\Delta x^2} + \frac{T_{i,j+1,m+1} - 2T_{i,j,m+1} + T_{i,j-1,m+1}}{\Delta y^2} + \frac{\dot{q}_{G,i,j,m+1}}{k} = \frac{\rho c}{k}\frac{T_{i,j,m+1} - T_{i,j,m}}{\Delta t} \tag{3.25}$$

Variable thermal conductivity can be handled as in Section 3.2.1. The thermal conductivity appropriate for determining the flux at the left and right faces of the control volume in Fig. 3.11 can be calculated from

$$k_{\text{left}} = \frac{2k_{i,j}k_{i-1,j}}{k_{i,j}k_{i-1,j}} \quad \text{and} \quad k_{\text{right}} = \frac{2k_{i,j}k_{i+1,j}}{k_{i,j} + k_{i+1,j}}$$

and for the bottom and top faces of the control volume

$$k_{\text{bottom}} = \frac{2k_{i,j}k_{i,j-1}}{k_{i,j} + k_{i,j-1}} \quad \text{and} \quad k_{\text{top}} = \frac{2k_{i,j}k_{i,j+1}}{k_{i,j} + k_{i,j+1}}$$

3.4.2 Boundary Conditions

Developing difference equations for the boundary nodes in multiple dimensions is similar to developing them in one-dimension, as described in Section 3.2.2. We first define the control volume that contains the boundary node. Then we define all the energy flows into and out of the control volume boundaries and the volumetric terms, including the heat generation and energy storage terms.

Consider the vertical edge of the two-dimensional geometric shape shown in Fig. 3.12. A control volume of width $\Delta x/2$ and height Δy is shown by the shaded area. The temperature at the boundary node is $T_{i,j,m}$. The size and shape of this control volume was selected so that if interior control volumes are as shown in Fig. 3.11 and remaining edge control volumes are as shown in Fig. 3.12 on the next page, the entire volume of the problem domain (with the exception of the corners, to be discussed next) will be covered.

Consider a heat balance on the control volume in Fig. 3.12:

$$\text{heat flow into the control volume left face via conduction} = k\frac{T_{i-1,j,m} - T_{i,j,m}}{\Delta x}\Delta y$$

$$\text{heat flow out of the right face} = q''_{x,i,j,m}\Delta y$$

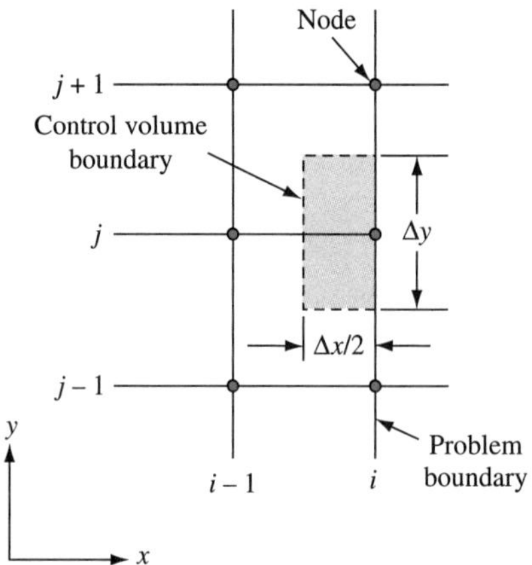

FIGURE 3.12 Boundary control volume for two-dimensional conduction—vertical edge.

where $q''_{x,i,j,m}$ is a specified heat flux in the $+x$ direction at the location i, j on the edge at time m.

$$\begin{aligned}
&\text{heat flow into the control volume bottom face by conduction} = k\frac{T_{i,j-1,m} - T_{i,j,m}}{\Delta y}\frac{\Delta x}{2} \\
&\text{heat flow out of the control volume top face by conduction} = k\frac{T_{i,j,m} - T_{i,j,+1,m}}{\Delta y}\frac{\Delta y}{2}
\end{aligned}$$

If the rate of volumetric heat generation is $\dot{q}_{G,i,j,m}$, then the rate of heat generation in the control volume is

$$\dot{q}_{G,i,j,m}\frac{\Delta x\,\Delta y}{2}.$$

The rate at which energy is stored in the control volume over a time step Δt is

$$\rho c\frac{\Delta x\,\Delta y}{2}\frac{T_{i,j,m+1} - T_{i,j,m}}{\Delta t}$$

and the control volume energy balance is:

$$\begin{aligned}
&k\frac{T_{i-1,j,m} - T_{i,j,m}}{\Delta x}\Delta y + k\frac{T_{i,j-1,m} - T_{i,j,m}}{\Delta y}\frac{\Delta x}{2} + \dot{q}_{G,i,j,m}\frac{\Delta x\,\Delta y}{2} \\
&\quad = q''_{x,i,j,m}\Delta y + k\frac{T_{i,j,m} - T_{i,j+1,m}}{\Delta y}\frac{\Delta x}{2} + \rho c\frac{\Delta x \Delta y}{2}\frac{T_{i,j,m+1} - T_{i,j,m}}{\Delta t}
\end{aligned}$$

which can be rearranged to give an explicit equation for the boundary temperature at the next time step, $T_{i,j,m+1}$:

$$T_{i,j,m+1} = T_{i,j,m}\left[1 - 2\alpha\,\Delta t\left(\frac{1}{\Delta x^2} + \frac{1}{\Delta y^2}\right)\right] + T_{i-1,j,m}\left(\frac{2\alpha\,\Delta t}{\Delta x^2}\right) + (T_{i,j-1,m} + T_{i,j+1,m})\left(\frac{\alpha\,\Delta t}{\Delta y^2}\right) + \dot{q}_{G,i,j,m}\left(\frac{\alpha\,\Delta t}{k}\right) - q''_{x,i,j,m}\left(\frac{2\alpha\,\Delta t}{k\,\Delta x}\right) \tag{3.26}$$

Following this same procedure for an outside corner, Fig. 3.13, we find

$$T_{i,j,m+1} = T_{i,j,m}\left[1 - 2\alpha\,\Delta t\left(\frac{1}{\Delta x^2} + \frac{1}{\Delta y^2}\right)\right] + T_{i-1,j,m}\left(\frac{2\alpha\,\Delta t}{\Delta x^2}\right) + T_{i,j-1,m}\left(\frac{2\alpha\,\Delta t}{\Delta y^2}\right) + \dot{q}_G\left(\frac{\alpha\,\Delta t}{k}\right) - \frac{2\alpha\,\Delta t}{k}\left(q''_{x,i,j,m}\frac{1}{\Delta x} + q''_{y,i,j,m}\frac{1}{\Delta y}\right) \tag{3.27}$$

where $q''_{y,i,j,m}$ is a specified heat flux in the $+y$ direction on the upper surface of the control volume at the node i, j at time m.

Finally, for an inside corner as in Fig. 3.14

$$T_{i,j,m+1} = T_{i,j,m}\left[1 - 2\alpha\,\Delta t\left(\frac{1}{\Delta x^2} + \frac{1}{\Delta y^2}\right)\right] + T_{i-1,j,m}\left(\frac{4}{3}\,\frac{\alpha\,\Delta t}{\Delta x^2}\right) + T_{i+1,j,m}\left(\frac{2}{3}\,\frac{\alpha\,\Delta t}{\Delta x^2}\right) + T_{i,j+1,m}\left(\frac{4}{3}\,\frac{\alpha\,\Delta t}{\Delta y^2}\right) + T_{i,j-1,m}\left(\frac{2}{3}\,\frac{\alpha\,\Delta t}{\Delta y^2}\right) + \dot{q}_{G,i,j,m}\left(\frac{\alpha\,\Delta t}{k}\right) + \frac{2}{3}\,\frac{\alpha\,\Delta t}{k}\left(-\frac{q''_{x,i,j,m}}{\Delta x} + \frac{q''_{y,i,j,m}}{\Delta y}\right) \tag{3.28}$$

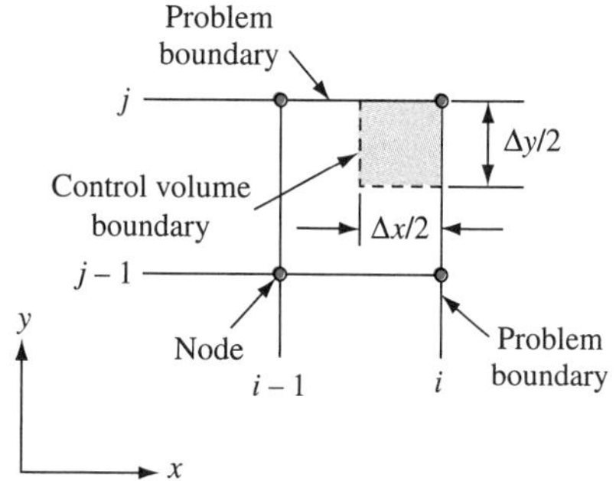

FIGURE 3.13 Boundary control volume for two-dimensional conduction—outside corner.

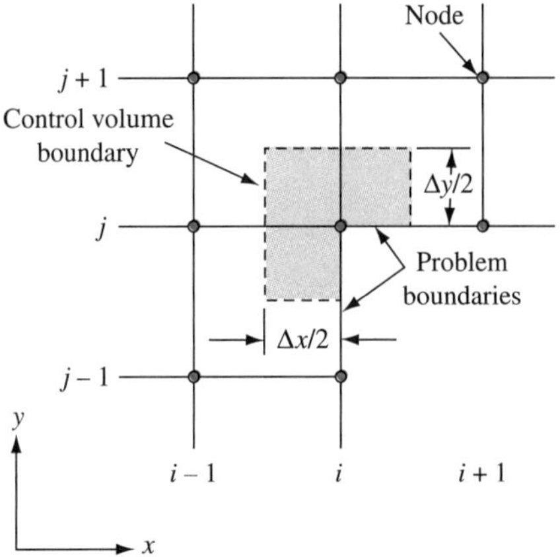

FIGURE 3.14 Boundary control volume for two-dimensional conduction—inside corner.

Note that with our sign convention on the specified boundary heat fluxes, a positive value of $q''_{y,i,j,m}$ increases the temperature at the node i, j, while a positive value of $q''_{x,i,j,m}$ decreases the temperature at the node i, j.

In Eqs. (3.26) through (3.28) the indices i, j of the boundary control volume depend on the location of the boundary control volume within the problem geometry. For example, for a rectangular geometry, the corner and edge control volumes on the right side of the rectangle would have $i = M$, the top corner would have $j = N$, the bottom corner would have $j = 1$, and so forth.

The energy balance equations for these boundary control volumes each impose their own stability criteria as expressed by the coefficient of $T_{i,j,m}$. Because we have chosen the specified heat flux boundary condition here, these criteria are identical to the criteria for the interior control volumes. As suggested by the one-dimensional boundary control volume energy equations for surface convection boundary conditions, we would expect that the surface convection boundary condition could impose more restrictive stability criteria for two-dimensional problems.

Boundary conditions expressed by Eq. (3.26) to (3.28) are explicit. Implicit versions can be derived as follows. Replace the subscript m with $m + 1$ in all but the $T_{i,j,m}$ terms on the right side of the equations. For the $T_{i,j,m}$ term, retain the subscript m for the term 1 inside the bracket, and use $m + 1$ for the second term inside the bracket. For the implicit equation, we are determining the temperature change $T_{i,j,m+1} - T_{i,j,m}$ from temperatures at new time step $m + 1$ as opposed to determining the temperature change from the temperatures at time step m for the explicit equation.

For steady-state boundary equations, the term representing the rate of energy storage

$$\rho c \frac{T_{i,j,m+1} - T_{i,j,m}}{\Delta t}$$

is, by definition, zero. This can be most easily accommodated in Eqs. (3.26) to (3.28) by dividing by Δt and letting $\Delta t \rightarrow \infty$. Then the two terms in each equation that are not proportional to Δt drop out. For example, Eq. (3.26) becomes

$$T_{i,j} = \frac{\dfrac{2\alpha}{\Delta x^2} T_{i-1,j} + \dfrac{\alpha}{\Delta y^2}(T_{i,j+1} + T_{i,j-1}) + \dfrac{\alpha}{k}\dot{q}_{G,i,j} - q''_{x,i,j}\dfrac{2\alpha}{k\,\Delta x}}{2\alpha\left(\dfrac{1}{\Delta x^2} + \dfrac{1}{\Delta y^2}\right)}$$

3.4.3 Solution Methods

Solution Methods for Steady State For steady, two-dimensional conduction with heat generation, Eq. (3.23) becomes

$$\frac{T_{i+1,j} - 2T_{i,j} + T_{i-1,j}}{\Delta x^2} + \frac{T_{i,j+1} - 2T_{i,j} + T_{i,j-1}}{\Delta y^2} + \frac{\dot{q}_{G,i,j}}{k} = 0 \quad (3.29)$$

Note that the temperature of each interior node $T_{i,j}$ depends on its four neighbors. Solving for $T_{i,j}$ yields

$$T_{i,j} = \frac{\Delta y^2(T_{i+1,j} + T_{i-1,j}) + \Delta x^2(T_{i,j+1} + T_{i,j-1}) + \dfrac{\Delta x^2 \Delta y^2}{k}\dot{q}_{G,i,j}}{2\,\Delta x^2 + 2\,\Delta y^2}$$

$$1 < i < M, \qquad 1 < j < N \quad (3.30)$$

The indices i, j are restricted to the indicated ranges because special difference equations like those developed in Section 3.4.2 are required at the boundaries.

The iterative method introduced in Section 3.2 is the most straightforward method for solving Eq. (3.30). To apply Jacobi iteration, we start with a guess of the temperature distribution for the problem. Call this temperature distribution guess

$$T_{i,j}^{(0)} \quad 1 \le i \le M, \qquad 1 \le j \le N$$

If we use this distribution in the right side of Eq. (3.30) and in the appropriate boundary equations, we can calculate a new value for $T_{i,j}$ at each node. Denote this new temperature distribution by the superscript 1, or

$$T_{i,j}^{(1)} \quad 1 \le i \le M, \qquad 1 \le j \le N$$

since it is the first revision to our initial guess, $T_{i,j}^{(0)}$. The new temperature distribution $T_{i,j}^{(1)}$ is now used in the right side of Eq. (3.30) to give a new temperature distribution, $T_{i,j}^{(2)}$. If we continue this iterative procedure, the temperature distribution will converge to the correct solution provided we meet the same criteria specified in Section 3.2: we must either specify the temperature for at least one boundary node or we must specify a convection-type boundary condition with given ambient fluid temperature at at least one boundary node. Remaining boundaries can then have any type of boundary condition.

EXAMPLE 3.4 A long rod of 1-in. × 1-in. cross section is to undergo a thermal stress test. Two opposing sides of the rod are held at 0°C while the other two sides are held at 50°C and 100°C (see Fig. 3.15). Using a node spacing of $\frac{1}{3}$ in. determine the steady-state temperature in the rod cross section.

SOLUTION We wish to find the temperature distribution in the square domain $0 \leq x \leq 1$, $0 \leq y \leq 1$, as shown in Fig. 3.15. Since $\Delta x = \Delta y = \frac{1}{3}$ in. $M = N = 4$. Due to the simple boundary conditions, we need to consider difference equations only for the four interior nodes. Since $\dot{q}_G = 0$ and $\Delta x = \Delta y$, Eq. (3.30) gives

$$T_{2,2} = \frac{1}{4}(50 + 0 + T_{2,3} + T_{3,2})$$

$$T_{2,3} = \frac{1}{4}(50 + 0 + T_{2,2} + T_{3,3})$$

$$T_{3,2} = \frac{1}{4}(100 + 0 + T_{2,2} + T_{3,3})$$

$$T_{3,3} = \frac{1}{4}(100 + 0 + T_{2,3} + T_{3,2})$$

Let's begin with a guess of $T_{i,j} = 0$ for all four interior nodes. This is not an especially good first guess, but it will illustrate the procedure. (A better first guess may be to roughly interpolate the interior node temperatures from the boundary temperatures.) The table on p. 203 shows the results of each iteration.

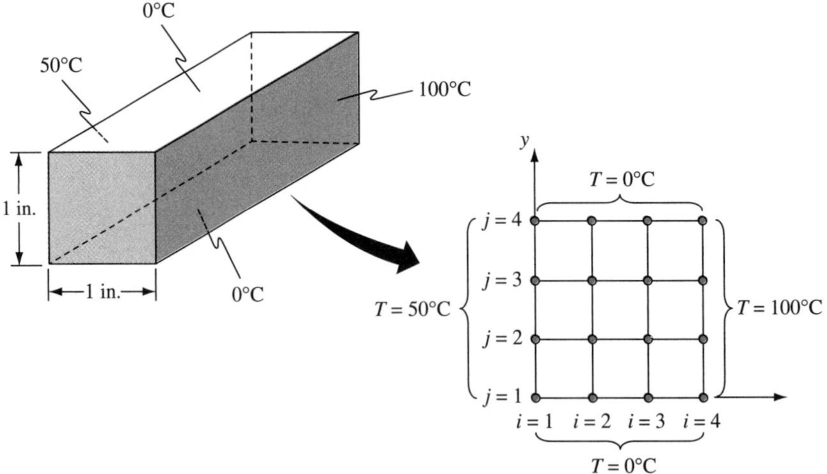

FIGURE 3.15 Sketch for Example 3.4.

Iteration Number	$T_{2.2}$ (°C)	$T_{2.3}$ (°C)	$T_{3.2}$ (°C)	$T_{3.3}$ (°C)
0	0	0	0	0
1	12.5	12.5	25	25
2	21.875	21.875	34.375	34.375
3	26.563	26.563	39.063	39.063
4	28.907	28.907	41.407	41.407
5	30.079	30.079	42.579	42.579
6	30.665	30.665	43.165	43.165
7	30.957	30.957	43.457	43.457
8	31.104	31.104	43.604	43.604
9	31.177	31.177	43.677	43.677
10	31.214	31.214	43.714	43.714
⋮	⋮	⋮	⋮	⋮
20	31.25	31.25	43.75	43.75

After 20 iterations, the solution ceases to change significantly.

The procedure illustrated by Example 3.4 can be accelerated somewhat if Gauss-Seidel iteration is used. In Example 3.4, the solution will be reached after 11 iterations using Gauss-Seidel iteration instead of the 20 iterations required by Jacobi iteration, a significant improvement.

EXAMPLE 3.5 The alloy bus bar (k = 20 W/m K) shown in Fig. 3.16 carries sufficient electrical current to have a heat generation rate of 10^6 W/m^3. The bus bar is 10 cm high by 5 cm wide by 1 cm thick, with current flowing in the direction of the long dimension between two water-cooled electrodes. The electrodes maintain the left end of the bus bar at 40°C and the right end at 10°C. Both of the large faces and one long edge is insulated. The other is cooled by natural convection with a heat transfer coefficient of 75 W/m^2 K and an ambient air temperature of 0°C. Determine the temperature distribution along both edges, the maximum temperature in the bus bar, and the distribution of heat loss along the edge cooled by natural convection.

SOLUTION Figure 3.16 shows the nodal network. We have chosen $\Delta x = \Delta y = 1$, giving $M =$ 11 nodes in the x direction and $N = 6$ nodes in the y direction. Since we expect no gradients in the z direction, let $\Delta z = 1$ cm. Also note that the volumetric heat generation is constant and uniform.

For all interior nodes, $1 < i < M$, $1 < j < N$, Eq. (3.30) applies directly. For $\Delta x = \Delta y$, Eq. (3.30) simplifies to

$$T_{i,j} = \frac{T_{i+1,j} + T_{i-1,j} + T_{i,j+1} + T_{i,j-1} + \left(\dfrac{\Delta x^2}{k}\right)\dot{q}_G}{4}$$

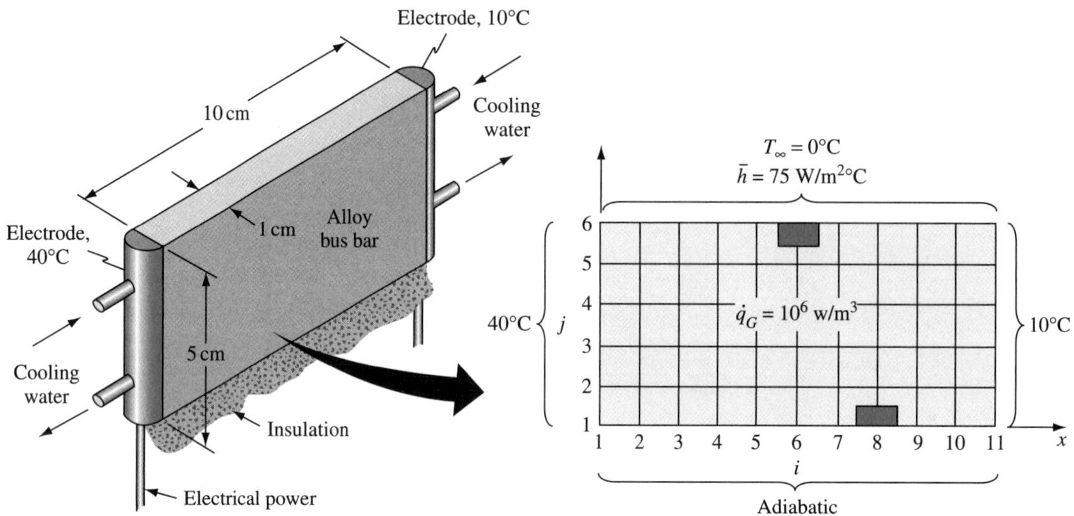

FIGURE 3.16 Alloy bus bar, Examples 3.5 and 3.6 (insulation over the large faces not shown for clarity).

For nodes on the short edges we have

$$T_{1,j} = 40°\text{C}, \qquad T_{M,j} = 10°\text{C} \qquad 1 \leq j \leq N$$

For nodes along the top edge, $1 < i < M, j = N$, we need to derive a difference equation similar to the steady form of Eq. (3.26). The shaded area on the upper edge of Fig. 3.16 represents a typical control volume for the top edge. An energy balance on the control volume gives the difference equation we need:

$$k\left[\left(\frac{T_{i-1,j} - T_{i,j}}{\Delta x} + \frac{T_{i+1,j} - T_{i,j}}{\Delta x}\right)\frac{\Delta y}{2} + \left(\frac{T_{i,j-1} - T_{i,j}}{\Delta y}\right)\Delta x\right] + \dot{q}_G\frac{\Delta x\,\Delta y}{2}$$
$$= \bar{h}\Delta x(T_{i,j} - T_\infty) \qquad 1 < i < M, \qquad j = N$$

Solving for $T_{i,N}$ and using $\Delta x = \Delta y$,

$$T_{i,N} = \frac{\bar{h}T_\infty + \dot{q}_G\dfrac{\Delta x}{2} + \dfrac{k}{\Delta x}\left(\dfrac{1}{2}(T_{i-1,N} + T_{i+1,N}) + T_{i,N-1}\right)}{\bar{h} + 2\dfrac{k}{\Delta x}} \qquad 1 < i < M$$

For nodes along the adiabatic edge consider an energy balance on the shaded area on the lower edge of Fig. 3.16:

$$k\left[\left(\frac{T_{i-1,j} - T_{i,j}}{\Delta x} + \frac{T_{i+1,j} - T_{i,j}}{\Delta x}\right)\frac{\Delta y}{2} + \left(\frac{T_{i,j+1} - T_{i,j}}{\Delta y}\right)\Delta x\right] + \dot{q}_G\frac{\Delta x\Delta y}{2} = 0$$
$$1 < i < M, \qquad j = 1$$

Solving for $T_{i,1}$ and using $\Delta x = \Delta y$, we obtain

$$T_{i,1} = \frac{1}{4}(T_{i+1,1} + T_{i-1,1}) + \frac{1}{2}T_{i,2} + \dot{q}_G \frac{\Delta x^2}{4k} \qquad 1 < i < M$$

The solution procedure for the set of difference equations is shown in Fig. 3.17 on the next page. After 198 iterations, the results for the top and bottom edges are:

	Node				
	1	**2**	**3**	**4**	**5**
Top edge, °C	40.000	55.138	66.643	74.143	77.553
Bottom edge, °C	40.000	58.089	71.333	79.859	83.756

	Node					
	6	**7**	**8**	**9**	**10**	**11**
Top edge, °C	76.847	71.998	62.960	49.668	32.035	10.000
Bottom edge, °C	83.070	77.811	67.950	53.426	34.148	10.000

The maximum temperature occurs on the insulated edge at the fifth node from the left edge ($i = 5, j = 1, x(5 - 1)\Delta x = 0.04\,\text{m}, y = 0.0\,\text{m}$), where $T_{5,1} \approx 83.8°\text{C}$.

The heat loss along the top edge for each node i is

$$q_i = \bar{h}A_i(T_i - T_\infty)$$

where $A_i = \Delta x\, \Delta z$ for $i = 2, 3, \ldots, N - 1$ and
$A_i = \Delta x\, \Delta z/2$ for $i = 1$ and $i = N$

Using $\bar{h} = 75\,\text{W/m}^2\,\text{K} = 0.0075\,\text{W/cm}^2\,\text{K}$, $T_\infty = 0°\text{C}$, $\Delta x = \Delta z = 1\,\text{cm}$ and the values for T_i along the top edge given in the table above, we find the following distribution of heat loss on the top edge:

	Node				
	1	**2**	**3**	**4**	**5**
Heat loss (watts)	0.15	0.414	0.500	0.556	0.582

	Node					
	6	**7**	**8**	**9**	**10**	**11**
Heat loss (watts)	0.576	0.540	0.472	0.373	0.240	0.0375

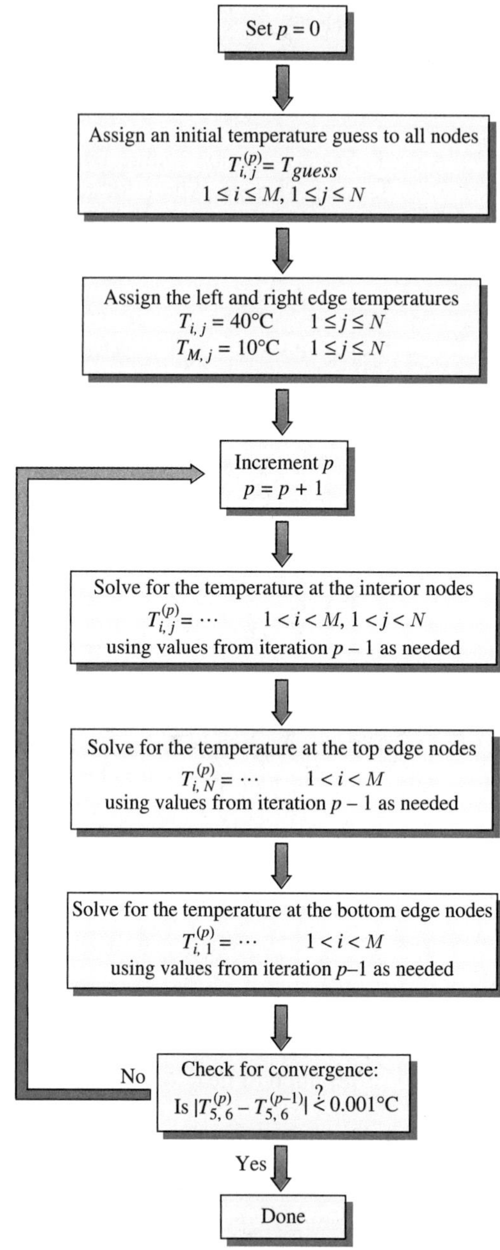

FIGURE 3.17 Flowchart for the iterative solution of a two-dimensional steady conduction problem, Example 3.5.

It may be recalled that numerical solutions tend to be dependent on the size of the nodal network (or grid or mesh) that describes the spatial domain of the problem. The preceding solution was obtained with 11 nodes in the x direction and 6 nodes in the y direction (or an 11×6 grid). It is instructive to see how the solution for the

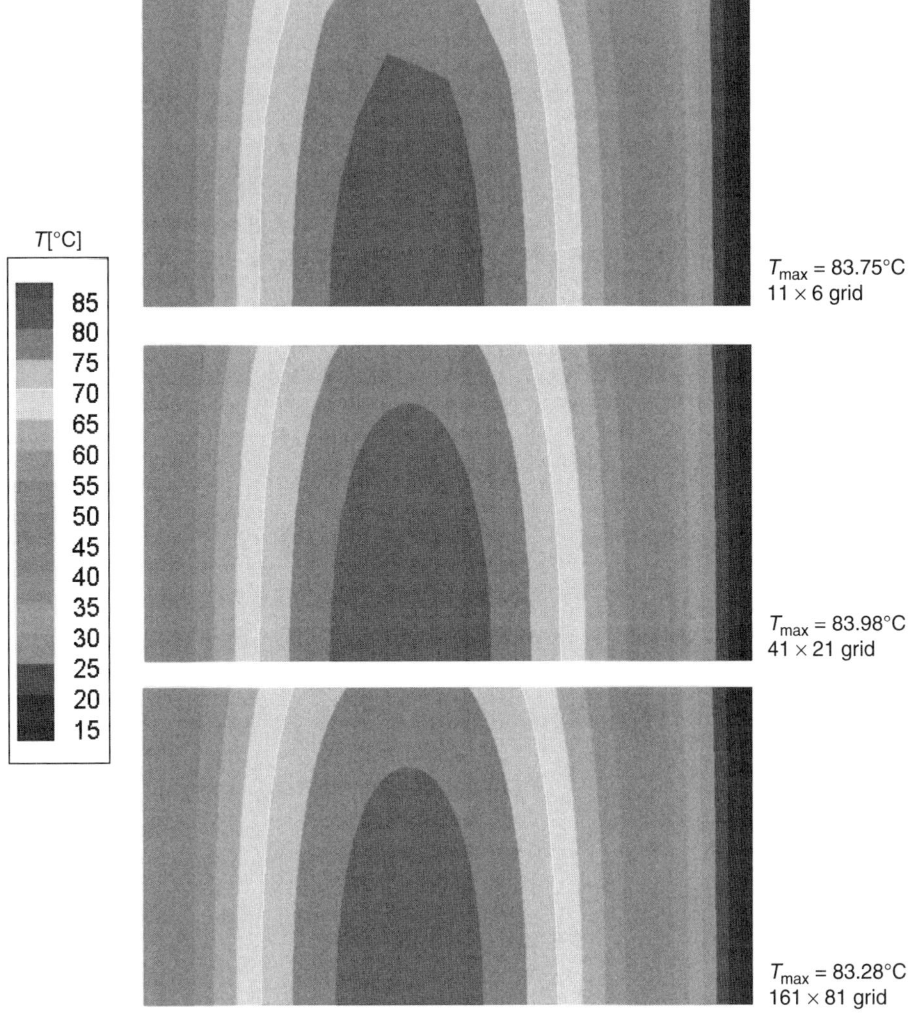

FIGURE 3.18 Effect of grid or mesh refinement (increasing number of node points) on the temperature distribution in the alloy bus bar.

Source: Courtesy of Prof. Raj M. Manglik, Thermal-Fluids & Thermal Processing Laboratory, University of Cincinnati.

temperature distribution in the bus bar changes when we choose a larger nodal network. This is seen in the graphical representation of the temperature distribution with an increasing grid size in Fig. 3.18, where a substantial change in the resolution of the temperature distribution from that in our earlier solution is seen as the grid is increased from 11 $\times$ 6 to 41 $\times$ 21, and then to 161 $\times$ 81. Most significantly the maximum temperature at the insulated edge (including its spatial region within the bus bar) changes from 83.75°C to 83.98°C to 83.28°C, respectively, with successive refinement in the grid size. A nodal network larger than 161 $\times$ 81 did not result in any significant change in the temperature distribution, and for most practical cases this solution can be considered as what is often termed a *grid independent* solution in the numerical analysis literature. This example clearly highlights the approximate nature of numerical solution and the problems that may be encountered with calculations based on a rather *coarse* grid or nodal network. Of course, the larger the number of nodes or more refined the mesh, the greater the computational time, especially in more complex three-dimensional problems.

Because of the change in the temperature distribution with grid refinement, the values for heat loss from the top edge of the bus bar will also change. The revised values can be readily calculated by following the methodology outlined in the example, and this is left for the student as a home exercise.

Unsteady-State Solution Methods If the difference equation for the interior nodes and boundary nodes is written in the explicit form, e.g., Eq. (3.23), the solution can be found by the same method used to solve explicit one-dimensional unsteady difference equations. The initial temperature distribution, $T_{i,j,m=0}$, is inserted into Eq. (3.23) to give the temperature distribution at time step $m = 1$, i.e., $T_{i,j,m=1}$. The new distribution $T_{i,j,m=1}$ is then inserted into Eq. (3.23) to give $T_{i,j,m=2}$. This procedure is repeated for as many time steps as are needed to reach the final condition.

If the implicit form has been chosen, the iterative method can be applied fairly simply. Consider Eq. (3.25), for example. The temperature at the new time step, $T_{i,j,m+1}$, is given in terms of the temperature at that node at the previous time step, $T_{i,j,m}$, which is known, and the temperatures at the nodes surrounding the node i, j at the new time step, which are not known. An estimate of the complete temperature distribution $T_{i,j,m+1}$ can be had by using the most recently calculated temperature distribution for all temperatures in Eq. (3.25), except for $T_{i,j,m}$ which is known. If this estimate is determined for the temperature at all nodes, we have completed one iteration. The correct temperature distribution for all nodes at the new time step, $m + 1$, is determined when subsequent iterations converge, i.e., when the temperature distribution ceases to change significantly from one iteration to the next. We are then ready to go to the next time step.

EXAMPLE 3.6 Consider the transient behavior of the bus bar in Example 3.5. Let the thermal diffusivity of the alloy be $\alpha = 8 \times 10^{-6}\ \text{m}^2/\text{s}$. Initially, the bus bar temperature is uniform at 20°C. At $t = 0$, water flow is started through the electrodes, cooling air flow is applied to the top edge, and electrical power is applied across the electrodes. How much time is needed to reach steady state? Use both explicit and implicit solution methods.

SOLUTION (a) *Explicit method.* We will utilize the same node network as in Fig. 3.16. Let us first use the explicit form of the difference equation. The largest permissible time step is

$$\Delta t_{\max} = \frac{1}{2\alpha\left(\dfrac{1}{\Delta x^2} + \dfrac{1}{\Delta y^2}\right)}$$

Using $\Delta x = \Delta y = 0.01\,\text{m}$ as before, we find $\Delta t_{\max} = 3.13\,\text{s}$. Since $h\,\Delta x/k = 0.03 \ll 1$, the stability criterion for the boundary control volume is very close to that for the interior control volumes. We will use $\Delta t = 3.0$ s.

Let us define steady state to mean that the temperature change in the interior of the bus bar per time step is less than 0.0001°C.

Solving the explicit difference equation for $T_{i,j,m+1}$, Eq. (3.23) gives

$$T_{i,j,m+1} = T_{i,j,m} + \alpha\,\Delta t\left\{\frac{T_{i+1,j,m} - 2T_{i,j,m} + T_{i-1,j,m}}{\Delta x^2} + \frac{T_{i,j+1,m} - 2T_{i,j,m} + T_{i,j-1,m}}{\Delta y^2} + \frac{\dot{q}_G}{k}\right\}$$

As before, the left and right boundary conditions are

$$T_{1,j,m} = 40°\text{C} \qquad \text{T}_{M,j,m} = 10°\text{C}$$

For the nodes along the top edge, the energy balance is similar to that in Example 3.5, except that we must consider the rate at which energy is stored in the shaded control volume shown on the top edge of Fig. 3.16:

$$k\left[\left(\frac{T_{i-1,j,m} - T_{i,j,m}}{\Delta x} + \frac{T_{i+1,j,m} - T_{i,j,m}}{\Delta x}\right)\frac{\Delta y}{2} + \left(\frac{T_{i,j-1,m} - T_{i,j,m}}{\Delta y}\right)\Delta x\right] + \dot{q}_G\frac{\Delta x\Delta y}{2}$$
$$= \bar{h}\Delta x(T_{i,j,m} - T_\infty) + \rho c\frac{\Delta x\Delta y}{2}\left(\frac{T_{i,j,m+1} - T_{i,j,m}}{\Delta t}\right) \qquad 1 < i < M, \qquad j = N$$

Solving to find $T_{i,j,m+1}$, using $\Delta x = \Delta y$, and setting $j = N$, we find

$$T_{i,N,m+1} = T_{i,N,m} + \frac{2\,\Delta t}{\Delta x^2\rho c}\left\{k\left(\frac{T_{i-1,N,m} + T_{i+1,N,m}}{2} + T_{i,N-1,m} - 2T_{i,N,m}\right) + \dot{q}_G\frac{\Delta x^2}{2} - \bar{h}\Delta x(T_{i,N,m} - T_\infty)\right\} \qquad 1 < i < M$$

Along the adiabatic edge an energy balance gives

$$T_{i,1,m+1} = T_{i,1,m} + \frac{2\,\Delta t}{\Delta x^2\rho c} \times \left\{k\left(\frac{T_{i-1,1,m} + T_{i+1,1m}}{2} + T_{i,2,m} - 2T_{i,1,m}\right) + \dot{q}_G\frac{\Delta x^2}{2}\right\} \qquad 1 < i < M$$

The solution procedure is shown in Fig. 3.19.

The results of the procedure show that steady state is achieved in 1131 s or 18.85 min. Similar results are found for smaller values of Δt, as long as the definition of steady state is expressed as a rate of change, for example,

$$\frac{T_{5,6,m+1} - T_{5,6,m}}{\Delta t} = 0.0001/3 = 3.3 \times 10^{-5}\,\frac{°\text{C}}{\text{s}}$$

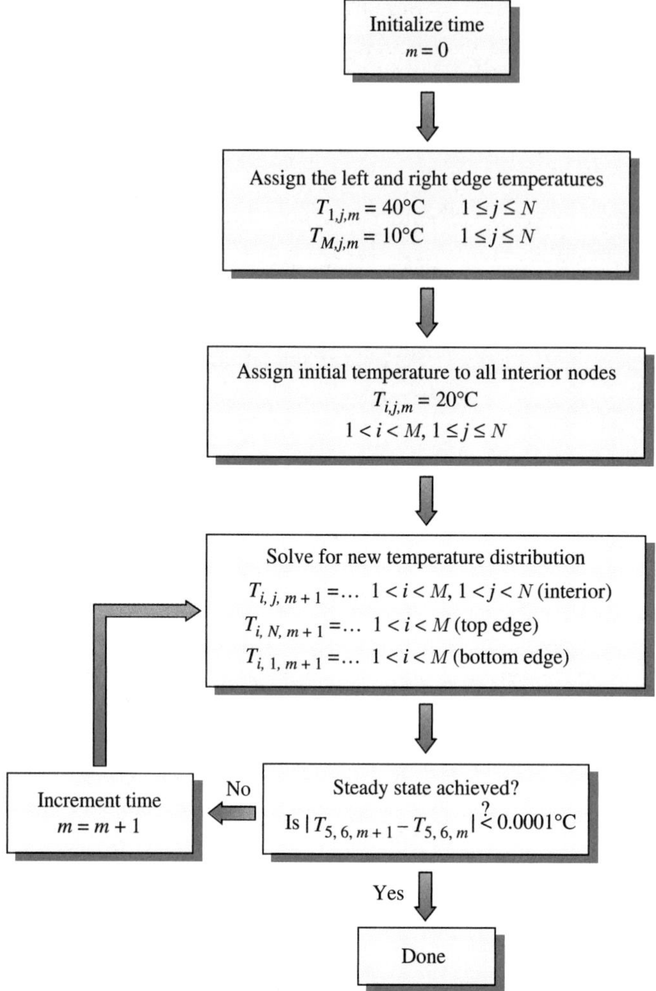

FIGURE 3.19 Flowchart for the explicit solution of a two-dimensional transient conduction problem, Example 3.6.

Using $\Delta t = 0.3$ s, the time required to reach steady state is found to be 1140 s, and for $\Delta t = 0.1$ s, it is 1141 s. In all cases the resulting temperature distribution is within 0.002°C of that determined by the steady-state calculation in Example 3.5.

(b) *Implicit method.* The difference equations given above are easily converted to their implicit form. For the interior nodes we have

$$T_{i,j,m+1} = T_{i,j,m} + \alpha \Delta t \left\{ \frac{T_{i+1,j,m+1} - 2T_{i,j,m+1} + T_{i-1,j,m+1}}{\Delta x^2} + \frac{T_{i,j+1,m+1} - 2T_{i,j,m+1} + T_{i,j-1,m+1}}{\Delta y^2} + \frac{\dot{q}_G}{k} \right\}$$

Solving for $T_{i,j,m+1}$ gives

$$T_{i,j,m+1} = \frac{T_{i,j,m} + \alpha\,\Delta t\left\{\dfrac{T_{i+1,j,m+1} + T_{i-1,j,m+1}}{\Delta x^2} + \dfrac{T_{i,j+1,m+1} + T_{i,j-1,m+1}}{\Delta y^2} + \dfrac{\dot{q}_G}{k}\right\}}{1 + 2\alpha\,\Delta t\left(\dfrac{1}{\Delta x^2} + \dfrac{1}{\Delta y^2}\right)}$$

$$1 < i < M, \qquad 1 < j < N \tag{a}$$

At the top edge

$$T_{i,N,m+1} = T_{i,N,m} + \frac{2\,\Delta t}{\Delta x^2 \rho c} \times \left\{k\left(\frac{T_{i+1,N,m+1} + T_{i-1,N,m+1}}{2} + T_{i,N-1,m+1} - 2T_{i,N,m+1}\right)\right\} + \dot{q}_G\frac{\Delta x^2}{2} - \bar{h}\Delta x(T_{i,N,m+1} - T_\infty)\Big\}$$

Solving for $T_{i,N,m+1}$ yields

$$T_{i,N,m+1} = \frac{T_{i,N,m} + \dfrac{2\Delta t}{\Delta x^2 \rho c}\left\{k\left(\dfrac{T_{i+1,N,m+1} + T_{i-1,N,m+1}}{2} + T_{i,N-1,m+1}\right) + \dot{q}_G\dfrac{\Delta x^2}{2} + \bar{h}\Delta x T_\infty\right\}}{1 + \dfrac{2\,\Delta t}{\Delta x^2 \rho c}(2k + \bar{h}\Delta x)} \tag{b}$$

where $1 < i < M$

Similarly, along the adiabatic edge the implicit difference equation is

$$T_{i,1,m+1} = T_{i,1,m} + \frac{2\,\Delta t}{\Delta x^2 \rho c}\left\{k\left(\frac{T_{i+1,1,m+1} + T_{i-1,1,m+1}}{2} + T_{i,2,m+1} - 2T_{i,1,m+1}\right) + \dot{q}_G\frac{\Delta x^2}{2}\right\}$$

So,

$$T_{i,1,m+1} = \frac{T_{i,1,m} + \dfrac{2\,\Delta t}{\Delta x^2 \rho c}\left\{k\left(\dfrac{T_{i+1,1,m+1} + T_{i-1,1,m+1}}{2} + T_{i,2,\,m+1}\right) + \dot{q}_G\dfrac{\Delta x^2}{2}\right\}}{1 + \dfrac{2\,\Delta t}{\Delta x^2 \rho c}(2k)} \tag{c}$$

The right sides of Eqs. (a), (b), and (c) have terms evaluated at time step $m + 1$ and one term evaluated at time step m. We can express these equations as

$$T_{i,j,m+1} = f(T_{i,j,m}, T_{i,j,m+1})$$

The term evaluated at time step m is known from our previous time step. To find the $m + 1$ terms we will use iteration. For our first guess at the $m + 1$ terms we will use the values from the previous time step:

$$T^{(0)}_{i,j,m+1} = T_{i,j,m}$$

Again we have used the superscript notation $T^{(0)}$ to denote the first guess.

We insert this guess in the right sides of Eqs. (a), (b), and (c) to get a revised value for the temperature at time step $m + 1$:

$$T^{(1)}_{i,j,m+1} = f(T_{i,j,m}, T^{(0)}_{i,j,m+1})$$

These updated values $T^{(1)}_{i,j,m+1}$ can then go into the right side of Eqs. (a), (b), and (c) to provide the next update:

$$T^{(2)}_{i,j,m+1} = f(T_{i,j,m}, T^{(1)}_{i,j,m+1})$$

This process is repeated until the change between iterations is small, say

$$|T^{(p)}_{5,6,m+1} - T^{(p-1)}_{5,6,m+1}| < \delta T$$

where δT is some small specified temperature difference and the superscript p denotes the pth iteration. When this convergence criterion is met, we have the solution for time step $m + 1$. We can then compare the change *per time step* to see if we have achieved steady state:

$$|T^{(p)}_{5,6,m+1} - T_{5,6,m}| < \frac{dT}{dt}$$

where the right side of this equation is some specified rate of temperature change per unit time.

The solution procedure is shown in Fig. 3.20.

Using the same time step as used for the explicit solution, 3 s, the time required to reach steady state is determined to be 1143 s, and the resulting temperature distribution is essentially identical to that for the explicit solution.

The variation in the spatial (x-y plane of the bus bar) temperature distribution with time, at three different intermediate time steps ($t = 50$, 150, and 300 s) between the initial condition ($t = 0$) and steady state, based on calculation with a grid composed of 161×81 nodes, is presented in Fig. 3.21. The corresponding steady state results, attained after 499 s in this case (which is significantly less than that with a coarser grid), may be recalled from Fig. 3.18. Replicating this exercise with a different x-y grid, altering the time step appropriately, and recalculating the time required for attaining steady state conditions is left for the student to carry out as a home exercise.

Recall that the main reason for using the implicit method is that if we need to reduce the node spacing, Δx, then the explicit method forces us to reduce the time

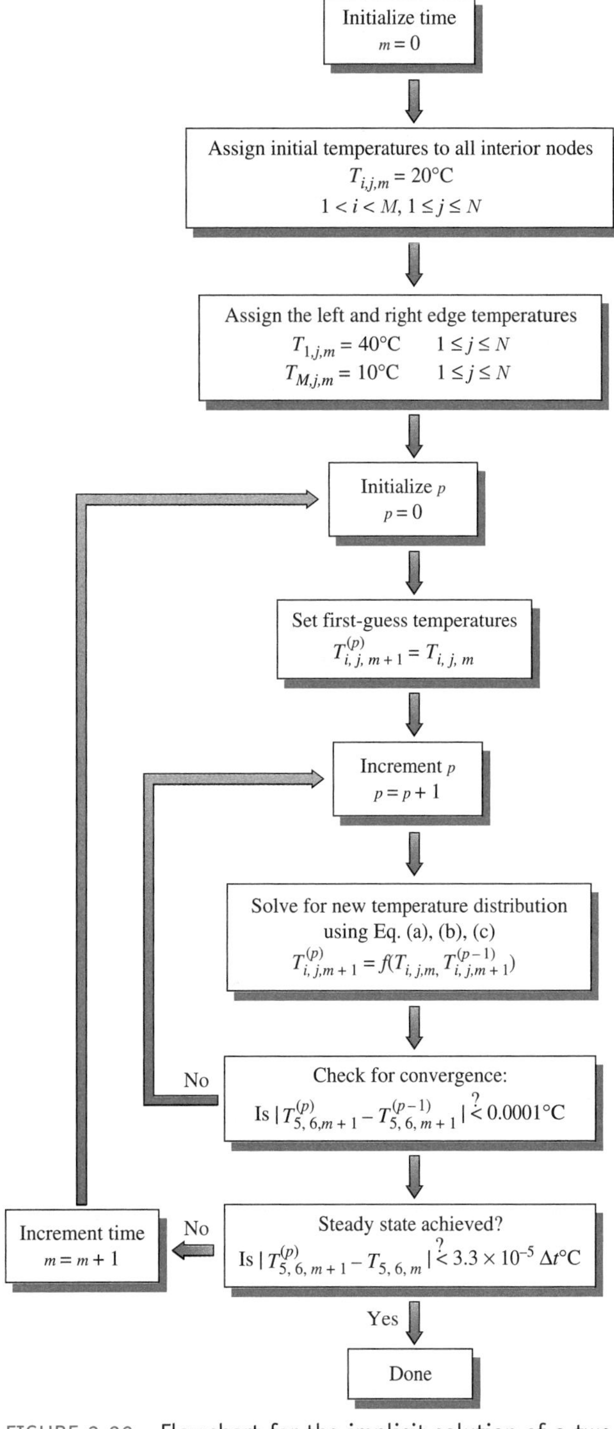

FIGURE 3.20 Flowchart for the implicit solution of a two-dimensional transient conduction problem, Example 3.6.

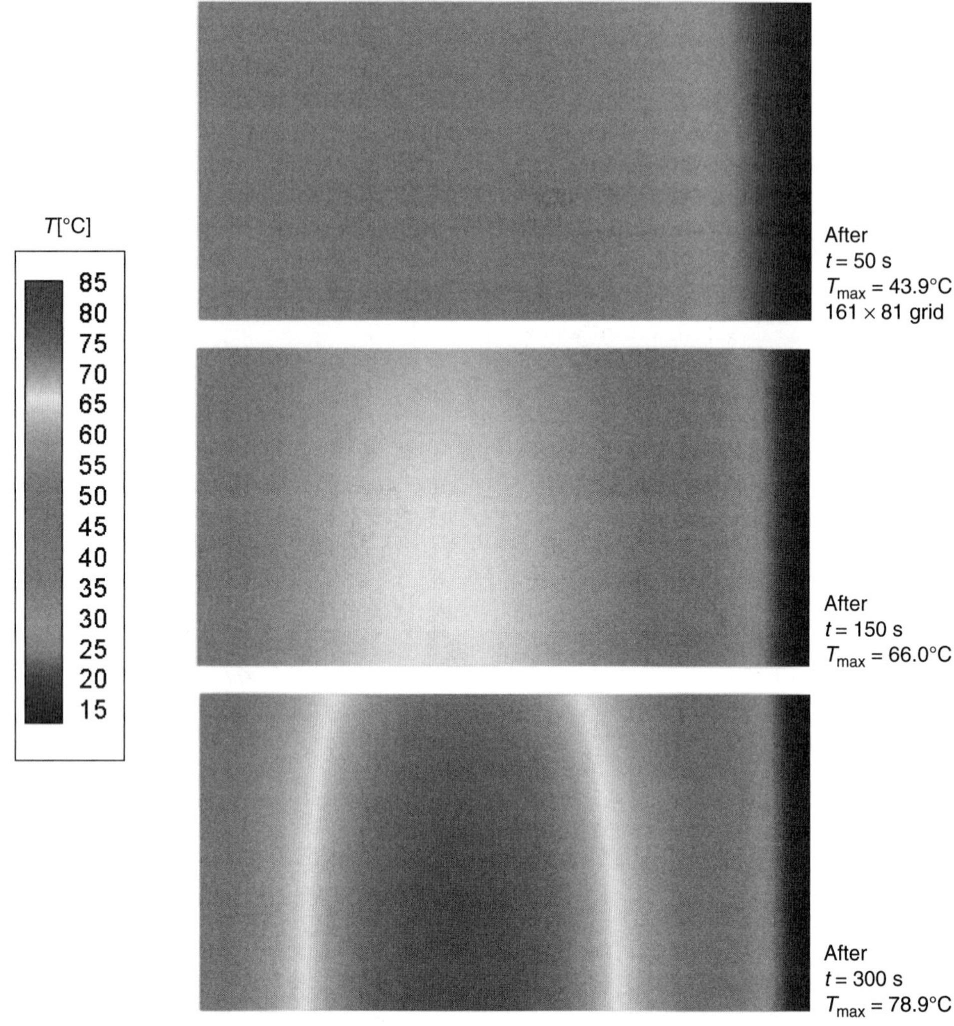

FIGURE 3.21 Variation with time ($t > 0 \rightarrow 50$, 150, and 300 s) in the temperature distribution in the *x-y* cross section of the alloy bus bar; note that $T_{max} = 83.3°C$ at steady state ($t \rightarrow \infty$).

Source: Courtesy of Prof. Raj M. Manglik, Thermal-Fluids & Thermal Processing Laboratory, University of Cincinnati.

step in proportion to Δx^2 while the implicit method does not have this constraint. A further advantage of the iterative method used to solve the implicit problem is that nonlinear behavior can be accommodated. For example, if the heat generation term in Example 3.5 depended on temperature (as it would if the bus bar electrical resistivity depended strongly on temperature), we would need to use the iterative method to ensure convergence. In this case the $\dot{q}_G$ term would be recalculated at each iteration from the most recently calculated temperature.

3.5* Cylindrical Coordinates

The cylindrical system is an important coordinate system because it has a number of applications including heat loss from pipes, wires, heat exchanger shells, reactors, and so forth. Therefore, it is worthwhile to develop the control volume equations for cylindrical coordinates.

Consider a two-dimensional system with coordinates r and θ and the shaded control volume as shown in Fig. 3.22. Let the radius r be determined by the index i

$$r_i = (i - 1)\Delta r \quad i = 1, 2, \ldots, N$$

and the angle θ be determined by the index j

$$\theta_j = (j - 1)\Delta\theta \quad j = 1, 2, \ldots, M$$

For unsteady problems, use the index m to indicate time

$$t = m\,\Delta t \quad m = 0, 1, \ldots$$

Now, for a unit length in the z direction, the area of the control volume normal to the radial direction is $(r - \Delta r/2)\Delta\theta$ at the inner surface and $(r + \Delta r/2)\Delta\theta$ at the outer surface. The area normal to the circumferential direction is Δr. The distance between nodes is

$$(i, j) \text{ to } (i \pm 1, j){:}\,\Delta r$$

$$(i, j) \text{ to } (i, j \pm 1){:}\,r\,\Delta\theta$$

The volume of the control volume per unit width is

$$\frac{1}{2}\Delta\theta\left[\left(r + \frac{\Delta r}{2}\right)^2 - \left(r - \frac{\Delta r}{2}\right)^2\right] = r\,\Delta\theta\,\Delta r$$

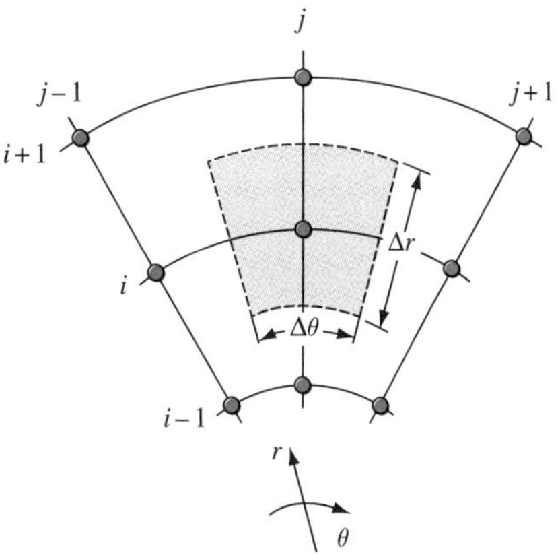

FIGURE 3.22 Control volume for cylindrical geometry.

With this information the difference equation can be determined from the energy balance:

$$\text{heat conducted into the face between } (i, j) \text{ and } (i, j - 1) = k\Delta r \frac{T_{i,j-1,m} - T_{i,j,m}}{r\,\Delta\theta}$$

$$\text{heat conducted into the face between } (i, j) \text{ and } (i, j + 1) = k\Delta r \frac{T_{i,j+1,m} - T_{i,j,m}}{r\,\Delta\theta}$$

$$\begin{array}{l}\text{heat conducted into the face}\\ \text{between } (i, j) \text{ and } (i, - 1, j)\end{array} = k\left(r - \frac{\Delta r}{2}\right)\Delta\theta \frac{T_{i-1,j,m} - T_{i,j,m}}{\Delta r}$$

$$\begin{array}{l}\text{heat conducted into the face}\\ \text{between } (i, j) \text{ and } (i + 1, j)\end{array} = k\left(r + \frac{\Delta r}{2}\right)\Delta\theta \frac{T_{i+1,j,m} - T_{i,j,m}}{\Delta r}$$

The rate at which energy is stored in the control volume is

$$\rho c r\,\Delta\theta\,\Delta r \frac{T_{i,j,m+1} - T_{i,j,m}}{\Delta t}$$

If the heat generation rate is nonzero ($\dot{q}_G \neq 0$), then the rate of heat generation inside the control volume is

$$\dot{q}_{G,i,j,m} r\,\Delta\theta\,\Delta r$$

The resulting energy balance on the control volume according to Eq. (2.1) is

$$k\left\{\frac{\Delta r}{r\,\Delta\theta}(T_{i,j+1,m} - 2T_{i,j,m} + T_{i,j-1,m}) + \frac{r\,\Delta\theta}{\Delta r}(T_{i+1,j,m} - 2T_{i,j,m} + T_{i-1,j,m}) + \frac{\Delta\theta}{2}(T_{i+1,j,m} - T_{i-1,j,m})\right\} + \dot{q}_{G,i,j,m} r\,\Delta\theta\,\Delta r = \rho c r\,\Delta\theta\,\Delta r \frac{T_{i,j,m+1} - T_{i,j,m}}{\Delta t} \tag{3.31}$$

In the implicit form of Eq. (3.31), all subscripts m on the left side of the equation would be replaced with $m + 1$.

By comparing Eq. (3.31) with Eq. (3.22), we see that the forms of the difference equations for two-dimensional unsteady heat conduction with heat generation are identical for a Cartesian and a cylindrical system. The only difference is in the coefficients multiplying the nodal temperatures and the heat generation term. Because of this, the solution techniques for cylindrical geometries are identical to those described for the Cartesian system in Section 3.4.2.

Control volume energy balances for the boundary nodes in a cylindrical geometry are developed just as they were in a Cartesian geometry. Heat transferred into or out of the control volume by conduction or by convection must be considered along with the volumetric generation and energy storage terms. The main difference in the cylindrical geometry is that the volume and surface areas of the control volume are slightly more complicated to calculate since they depend on radius. This aspect of the method for cylindrical geometries is left as an exercise (see Problems 3.15 and 3.38).

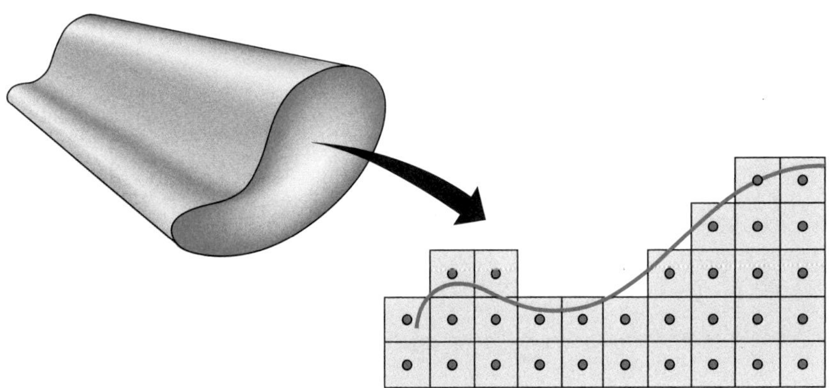

FIGURE 3.23 Control volumes near an irregular boundary.

3.6* Irregular Boundaries

In Section 3.4.1 we worked with rectangular control volumes, and in Section 3.5 we worked with control volumes that were sections of a circle (see Figs. 3.11 and 3.22). The choice of control volume shape depends on the shape of the overall geometry of the system under consideration. If the system geometry is rectangular, it makes sense to work in a Cartesian coordinate system and to use rectangular control volumes. We can then easily fill the entire geometry with control volumes. In Section 3.5 we used control volumes that were shaped like sections of a circle so that we could fill circular geometries easily. For geometries that cannot be classified as strictly rectangular or circular, neither of these control volume shapes will completely fill the system geometry. We say that these systems have an *irregular boundary*.

A method for solving problems with irregular boundaries is suggested by Fig. 3.23, where we have a long cylinder of noncircular cross section. If the cylinder is long enough, we can consider the system to be two-dimensional. We *approximate* the curved boundary with rectangular control volumes as shown in Fig. 3.23. Even though we are only approximating the curved boundary, this approach is often satisfactory. The only additional complexity introduced by this approach is that we need to develop energy balance equations for full-sized control volumes that have exposed surfaces.

EXAMPLE 3.7 A long cylinder of circular cross section has been heated uniformly to 500°C and is to be cooled by sudden immersion in a coolant bath at 0°C. The cylinder diameter is 10 cm, and it has a thermal conductivity of 20 W/m K and a thermal diffusivity of 10^{-5} m^2/s. The heat transfer coefficient at the cylinder surface is 200 W/m^2 K. Use chart methods to determine how long it will take to cool the center of the cylinder to 100°C, and compare your results with those from an explicit numerical solution using square control volumes 1 cm × 1 cm.

SOLUTION First, let's find the chart solution. The Biot number is

$$\text{Bi} = \frac{\bar{h}R_0}{k} = \frac{\left(200\,\dfrac{\text{W}}{\text{m}^2\,\text{K}}\right)(0.05\,\text{m})}{\left(20\,\dfrac{\text{W}}{\text{m K}}\right)} = 0.5$$

and

$$\frac{T(r = 0, t) - T_\infty}{T(r = 0, t = 0) - T_\infty} = \frac{100 - 0}{500 - 0} = 0.2$$

Figure 2.43 (a) then gives

$$\frac{\alpha t}{R_0^2} = 1.8$$

from which we find $t = 450$ s.

Figure 3.24 shows the arrangement of control volumes and nodes for the numerical solution. The placement of control volumes is a matter of judgment, but the goal is to represent the curved boundary as well as possible. Because of symmetry we need to consider only one quarter of the circular cross section. The vertical and horizontal radii are then adiabatic surfaces. We have nine different types of control

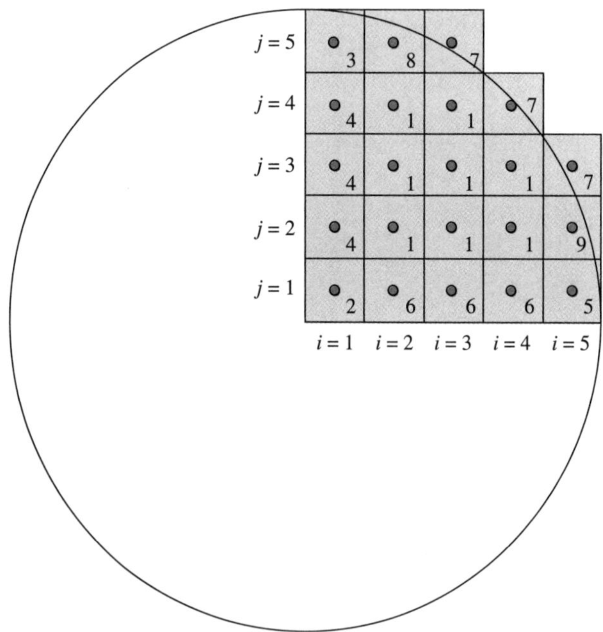

FIGURE 3.24 Arrangement of control volumes and nodes for Example 3.7.

volumes identified by the number in the lower right corner of each control volume. To simplify the notation, we define

$$\begin{aligned}
T_l &\equiv T_{i-1,j,m} \quad \text{(left)} \\
T_r &\equiv T_{i+1,j,m} \quad \text{(right)} \\
T_u &\equiv T_{i,j+1,m} \quad \text{(up)} \\
T_d &\equiv T_{i,j-1,m} \quad \text{(down)} \\
T &\equiv T_{i,j,m} \\
T_{\text{new}} &\equiv T_{i,j,m+1} \\
P_1 &\equiv \frac{\Delta x^2}{\alpha\,\Delta t} \\
P_2 &\equiv \frac{\bar{h}\,\Delta x}{k}
\end{aligned}$$

Control volume type 1 is an interior control volume. The energy balance equation for this control volume type is

$$T_l + T_r + T_u + T_d - 4T = P_1(T_{\text{new}} - T)$$

Control volume type 2 has adiabatic surfaces at the left and at the bottom. The energy balance equation for this control volume type is

$$T_r + T_u - 2T = P_1(T_{\text{new}} - T)$$

Control volume type 3 has an adiabatic surface at its left face, and its upper face is exposed to ambient conditions. The energy balance equation for this control volume type is

$$T_r + T_d - 2T + P_2(T_\infty - T) = P_1(T_{\text{new}} - T)$$

Control volume type 4 has an adiabatic surface at its left face. The energy balance equation for this control volume type is

$$T_u + T_r + T_d - 3T = P_1(T_{\text{new}} - T)$$

Control volume type 5 has an adiabatic lower surface and a right surface exposed to ambient conditions. The energy balance equation for this control volume type is

$$T_l + T_u - 2T + P_2(T_\infty - T) = P_1(T_{\text{new}} - T)$$

Control volume type 6 has an adiabitic lower surface. The energy balance equation for this control volume type is

$$T_l + T_u + T_r - 3T = P_1(T_{\text{new}} - T)$$

Control volume type 7 has its upper and right surfaces exposed to ambient conditions. The energy balance equation for this control volume type is

$$T_l + T_d - 2T + 2P_2(T_\infty - T) = P_1(T_{\text{new}} - T)$$

Control volume type 8 has its upper surface exposed to ambient conditions. The energy balance equation for this control volume type is

$$T_l + T_r + T_d - 3T + P_2(T_\infty - T) = P_1(T_{\text{new}} - T)$$

Control volume type 9 has its right surface exposed to ambient conditions. The energy balance equation for this control volume type is

$$T_l + T_u + T_d - 3T + P_2(T_\infty - T) = P_1(T_{\text{new}} - T)$$

Each of the control volume energy balance equations can be solved for T_{new}. The explicit solution can then be found as before. We anticipate that the stability criterion will come from control type 7 because it has two surfaces exposed to ambient conditions. The coefficient on T for control volume type 7 is

$$\frac{\Delta x^2}{\alpha\,\Delta t} - 2 - 2\frac{\bar{h}\,\Delta x}{k}$$

For this coefficient to remain positive, the maximum time step is

$$\Delta t_{\max} = \frac{\Delta x^2}{2\alpha\left(1 + \dfrac{\bar{h}\,\Delta x}{k}\right)}$$

so the value of Δt we use in the numerical solution must be smaller than this maximum value. The calculation is continued until the temperature for the control volume nearest the cylinder axis is less than 100°C, that is,

$$T_{1,1,m_{\text{final}}} < 100°\text{C}$$

The value of m_{final} then gives the desired time from

$$t_{\text{final}} = m_{\text{final}}\,\Delta t$$

The results of the numerical calculation give $t_{\text{final}} = 431$ s, about 4% less than the chart solution of 450 s.

Accuracy can be improved by using smaller control volumes. We can use smaller control volumes throughout the cylinder, or we can use variable control volume sizing as discussed earlier in the chapter. In the former case, the computation time will be increased. In the latter case, we will need to develop control volume energy equations for both sizes of control volumes and for special control volumes where the two control volume sizes meet.

In Examples 3.6 and 3.7 we saw that even a relatively simple two-dimensional problem can become quite involved because we end up with one type of control volume energy balance equation for the interior control volumes and one for each type of boundary control volume. Clearly, if a problem involves a large number of different boundary control volumes, a good deal of work will be required to set up all control volume equations.

3.7 Closing Remarks

This chapter can be considered an extension of Chapter 2. Here we have developed numerical methods for the analysis of conduction problems that cannot be easily solved by analytical, graphical, or chart methods. Problems that fall into this category include those with complex geometries, complex boundary conditions, or variable properties. Recent advances in both computer hardware and software now make it practical for an engineer to efficiently solve many conduction problems with numerical methods.

We presented the control volume method in this chapter because its implementation is the same for all problems, including one-dimensional, multidimensional, steady and unsteady problems. The basis of the method involves dividing the problem domain into discrete control volumes and writing an energy balance for each control volume. The result is a set of algebraic equations involving the temperatures at the center of each control volume. The resulting set of equations can be solved by methods such as matrix inversion, tridiagonal matrix solvers, marching, iteration, or combinations of these methods. For one-dimensional steady and unsteady problems, Tables 3.1 and 3.2, respectively, can be used to determine the matrix coefficients for three types of boundary conditions. In setting up the control volumes and choosing the solution method for the problem, the issues of solution accuracy and stability must be considered.

While the energy balance equation for the interior control volumes can be generalized, the equations for the boundary control volumes may be specialized for each problem, especially for multidimensional problems with complex boundaries. In some cases the effort required to develop these specialized equations for boundary control volumes may be excessive. For such problems commercial software packages that solve conduction problems are worth serious consideration. A listing of several commercial software including some open-source or public, domain software that are currently in vogue in engineering practice is given in Appendix 4.

References

1. P. Majumdar, *Computational Methods for Heat and Mass Transfer*, Taylor & Francis, New York, 2005.
2. S. V. Patankar, *Numerical Heat Transfer and Fluid Flow*, Hemisphere Publishing Corp., Washington, D.C., 1980.
3. J. C. Tannehill, D. A. Anderson, and R. H. Pletcher, *Computational Fluid Mechanics and Heat Transfer*, 2nd ed., Taylor & Frances, Washington, DC, 1997.

Problems

The problems for this chapter are organized by subject matter as shown below.

Topic	Problem Number
One-dimensional, steady conduction	3.1–3.16
One-dimensional, transient conduction	3.17–3.26
Two-dimensional steady conduction, rectangular coordinates	3.27–3.37
Two-dimensional steady conduction, cylindrical coordinates	3.38–3.40
Two-dimensional transient conduction, rectangular coordinates	3.41–3.46
Two-dimensional transient conduction, cylindrical coordinates	3.47–3.48
Irregular boundaries	3.49
Three-dimensional transient conduction	3.50–3.52

3.1 Show that in the limit as $\Delta x \to 0$, the difference equation for one-dimensional steady conduction with heat generation, Eq. (3.2), is equivalent to the differential equation, Eq. (2.27).

3.2 What is the physical significance of the statement that the temperature of each node is just the average of its neighbors if there is no heat generation [with reference to Eq. (3.3)]?

3.3 Give an example of a practical problem in which the variation of thermal conductivity with temperature is significant and for which a numerical solution is therefore the only viable solution method.

3.4 Discuss the advantages and disadvantages of using a large control volume.

3.5 For one-dimensional conduction, why are the boundary control volumes half the size of the interior control volumes?

3.6 Discuss the advantages and disadvantages of two methods for solving one-dimensional steady conduction problems.

3.7 Solve the system of equations:

$$2T_1 + T_2 - T_3 = 30$$

$$T_1 - T_2 + 7T_3 = 270$$

$$T_1 + 6T_2 - T_3 = 160$$

by Jacobi and Gauss-Seidel iteration. Use as a convergence criterion $|T_2^{(p)} - T_2^{(p-1)}| < 0.001$. Compare the rate of convergence for the two methods.

3.8 Develop the control volume difference equation for one-dimensional steady conduction in a fin with variable cross-sectional area $A(x)$ and perimeter $P(x)$. The heat transfer coefficient from the fin to ambient is a constant $\bar{h}_0$ and the fin tip is adiabatic. See sketch for Problem 3.9.

3.9 Using your results from Problem 3.8, find the heat flow at the base of the fin for the following conditions.

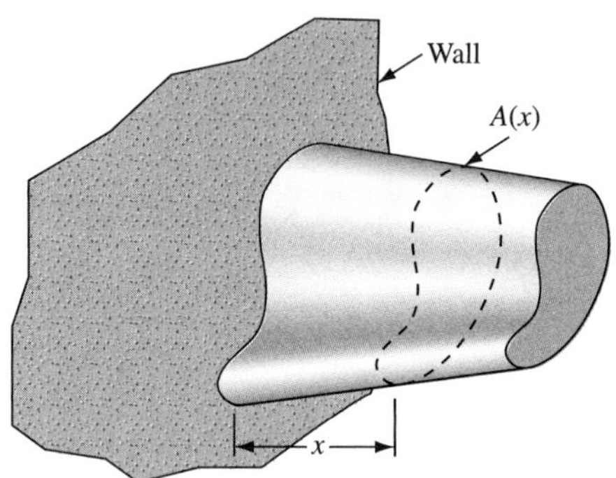

Problem 3.9

$$k = 20\,\text{Btu/h ft °F}$$

$$L = 2\text{in}$$

$$A(x) = 0.5\left(1 - \frac{1}{3}\sinh\left(\frac{x}{L}\right)\right)\text{in.}^2$$

$$P(x) = [A(x)]^{1/2}$$

$$\bar{h}_0 = 20\,\text{Btu/h ft}^{2\circ}\text{F}$$

$$\text{T}_0 = 200°\text{F}$$

$$T_\infty = 80°\text{F}$$

Use a grid spacing of 0.2 in.

3.10 Consider a pin fin with variable conductivity $k(T)$, constant cross-sectional area A_c and constant perimeter, P. Develop the difference equations for steady one-dimensional conduction in the fin, and suggest a method for solving the equations. The fin is exposed to ambient temperature T_a through a heat transfer coefficient h. The fin tip is insulated and the fin root is at temperature T_0.

3.11 How would you treat a radiation heat transfer boundary condition for a one-dimensional steady problem? Develop the difference equation for a control volume near the boundary, and explain how to solve the entire system of difference equations. Assume that the heat flux at the surface is $q = \epsilon\sigma(T_s^4 - T_e^4)$, where T_s is the surface temperature and T_e is the temperature of an enclosure surrounding the surface.

3.12 How should the control volume method be implemented at an interface between two materials with different thermal conductivities? Illustrate with a steady, one-dimensional example. Neglect contact resistance.

3.13 How would you include contact resistance between the two materials in Problem 3.12? Derive the appropriate difference equations.

3.14 A turbine blade 5 cm long with a cross-sectional area $A = 4.5\ \text{cm}^2$ and a perimeter $P = 12$ cm is made of a high-alloy steel ($k = 25$ W/m K). The temperature of the blade attachment point is 500°C, and the blade is exposed to combustion gases at 900°C. The heat transfer coefficient between the blade surface and the combustion gases is 500 W/m^2 K. Using the nodal network shown in the accompanying sketch, (a) determine the temperature distribution in the blade, the rate of heat transfer to the blade, and the fin efficiency of the blade and, (b) compare the fin efficiency calculated numerically with that calculated by the exact method.

Font View of Blade

5 cm

x

1 2 3 4 5 6

500°C

Cross-Sectional View of Blade

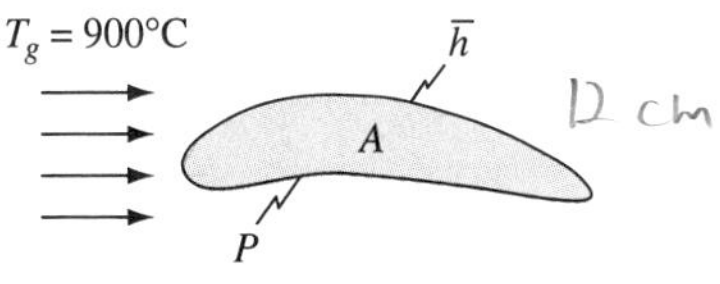

Problem 3.14

3.15 Determine the difference equations applicable at the centerline and at the surface of an axisymmetric cylindrical geometry with volumetric heat generation and convection boundary condition. Assume steady-state conditions.

3.16 Determine the appropriate difference equations for an axisymmetric, steady, spherical geometry with volumetric heat generation. Explain how to solve the equations.

3.17 Show that in the limit as $\Delta x \rightarrow 0$ and $\Delta t \rightarrow 0$, the difference equation, Eq. (3.13), is equivalent to the differential equation, Eq. (2.5).

3.18 Determine the largest permissible time step for a one-dimensional transient conduction problem to be solved by an explicit method if the node spacing is 1 mm and the material is (a) carbon steel 1C, (b) window glass. Explain the difference in the two results.

3.19 Consider one-dimensional transient conduction with a convection boundary condition in which the ambient temperature near the surface is a function of time. Determine the energy balance equation for the boundary control volume. How would the solution method need to be modified to accommodate this complexity?

3.20 What are the advantages and disadvantages of using explicit and implicit difference equations?

3.21 Equation (3.16) is often called the *fully-implicit* form of the one-dimensional transient conduction difference equation because all quantities in the equation, except for the temperatures in the energy storage term, are evaluated at the new time step, $m + 1$. In an alternative form called the Crank-Nicholson form, these quantities are evaluated at both time step m and time step $m + 1$ and then averaged. This averaging significantly improves the accuracy of the numerical solution relative to the fully-implicit form without increasing the complexity of the solution method. Derive the one-dimensional transient conduction difference equation in the Crank-Nicholson form.

3.22 A 3-m-long steel rod ($k = 43$ W/m K, $\alpha = 1.17) \times 10^{-5}\ \text{m}^2/\text{s}$) is initially at 20°C and is insulated completely except for its end faces. One end is suddenly exposed to the flow of combustion gases at 1000°C through a heat transfer coefficient of 250 W/m^2 K and the other end is held at 20°C. How long will it take for the exposed end to reach 700°C? How much energy will the rod have absorbed if it is circular in cross section and has a diameter of 3 cm?

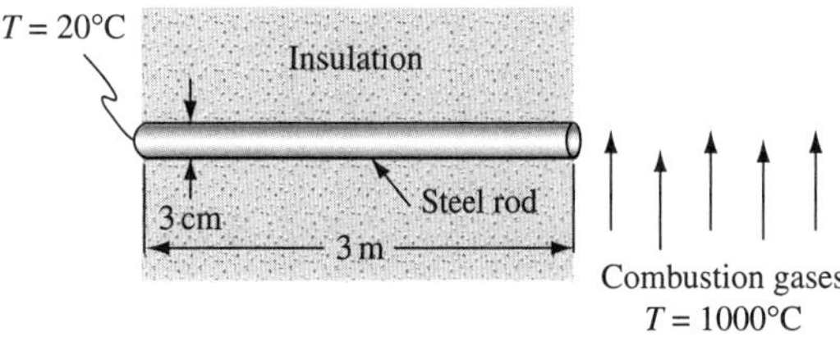

Problem 3.22

3.23 A Trombe wall is a masonry wall often used in passive solar homes to store solar energy. Suppose that such a wall, fabricated from 20-cm-thick solid concrete blocks ($k = 0.13$ W/m K, $\alpha = 5 \times 10^{-7}\text{m}^2/\text{s}$), is initially at 15°C in equilibrium with the room in which it is located. It is suddenly exposed to sunlight and absorbs 500 W/m^2 on the exposed face. The exposed face loses heat by radiation and convection to the outside ambient temperature of −15°C through a combined heat transfer coefficient of 10 W/m^2 K. The other face of the wall is exposed to room air through a heat transfer coefficient of 10 W/m^2 K. Assuming that the room air temperature does not change, determine (a) the maximum temperature in the wall after 4 h of exposure and (b) the net heat transferred to the room (see figure).

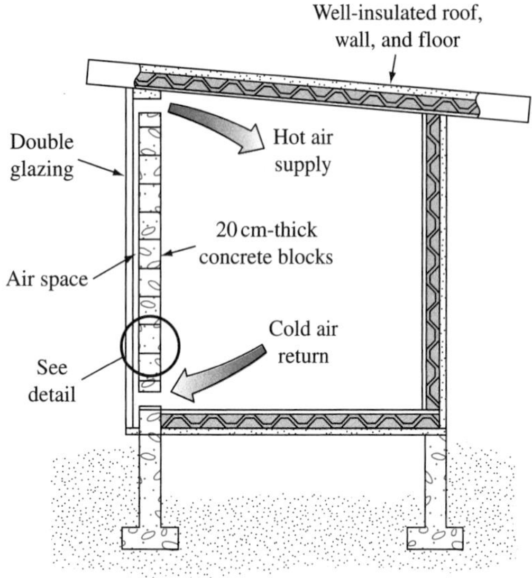

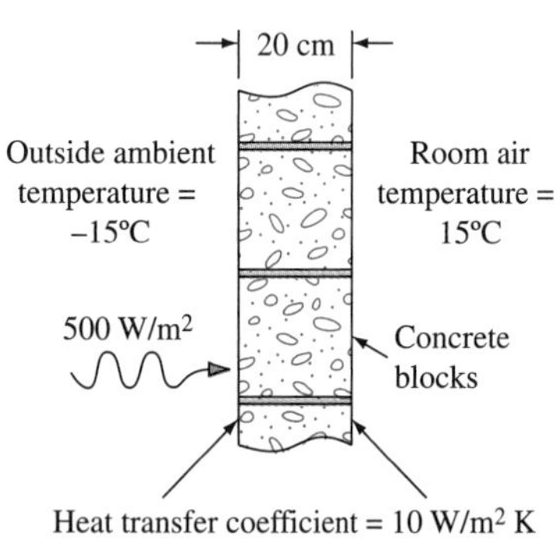

Problem 3.23

3.24 To more accurately model the energy input from the sun, suppose the absorbed flux in Problem 3.23 is given by

$$q_{abs}(t) = t(375 - 46.875t)$$

where t is in hours and q_{abs} is in W/m^2. (This time variation of q_{abs} gives the same total heat input to the wall as in Problem 3.23, i.e., 2000 W h/m^2). Repeat Problem 3.23 with the above equation for q_{abs} in place of the constant values of 500 W/m^2. Explain your results.

3.25 An interior wall of a cold furnace, initially at 0°C, is suddenly exposed to a radiant flux of 15 kW/m^2 when the furnace is brought on line. The outer surface of the wall is exposed to ambient air at 20°C through a heat transfer coefficient of 10 W/m^2 K. The wall is 20 cm thick and is made of expanded perlite ($k = 0.10$ W/m K, $\alpha = 3 \times 10^{-7}$ m^2/s) sandwiched between two sheets of oxidized steel (see the following sketch). Determine how long after start-up the inner (hot) sheet metal surface will get hot enough so that reradiation becomes significant.

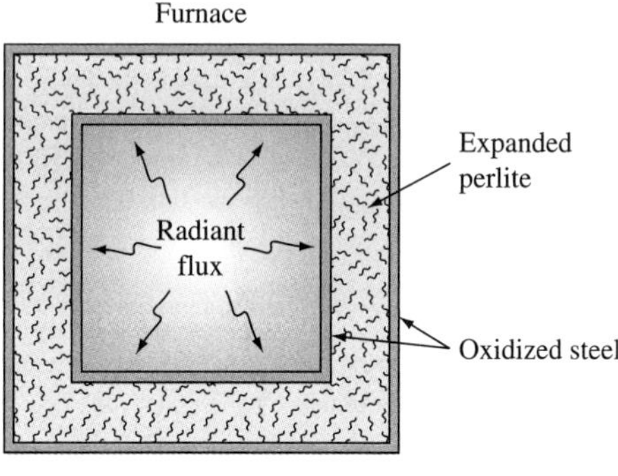

Problem 3.25

3.26 A long cylindrical rod, 8 cm in diameter, is initially at a uniform temperature of 20°C. At time $t = 0$, the rod is exposed to an ambient temperature of 400°C through a heat transfer coefficient of 20 W/m^2 K. The thermal conductivity of the rod is 0.8 W/m K, and the thermal diffusivity is 3×10^{-6} m^2/s. Determine how much time will be required for the temperature *change* at the centerline of the rod to reach 93.68% of its maximum value. Use an explicit difference equation and compare your numerical results with a chart solution from Chapter 2.

3.27 Develop a reasonable layout of nodes and control volumes for the geometry shown in the sketch below. Provide a scale drawing showing the problem geometry overlaid with the nodes and control volumes.

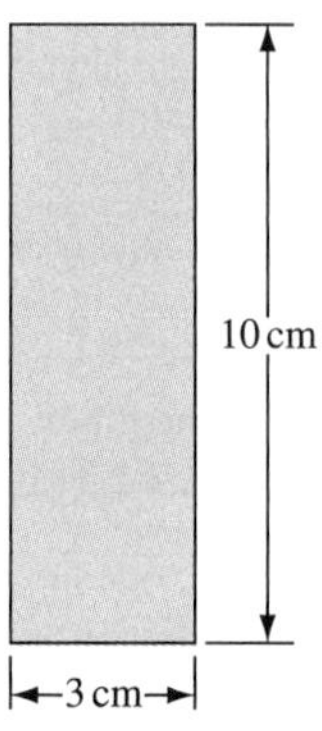

Problem 3.27

3.28 Develop a reasonable layout of nodes and control volumes for the geometry shown in the sketch below. Provide a scale drawing showing the problem geometry overlaid with the nodes and control volumes. Identify each type of control volume used.

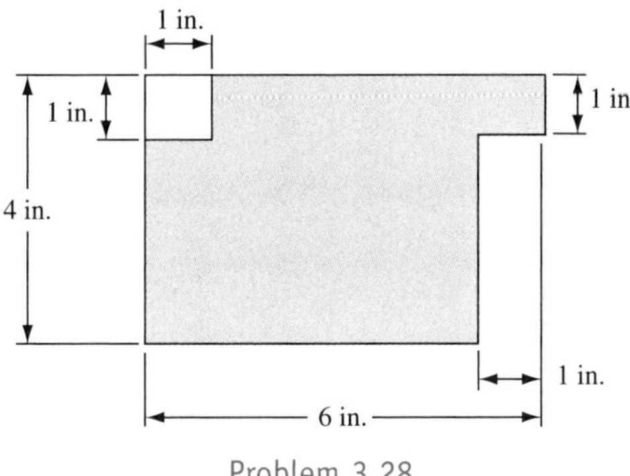

Problem 3.28

3.29 Determine the temperature at the four nodes shown in the sketch. Assume steady conditions and two-dimensional heat conduction. The four faces of the square shape are at different temperatures as shown.

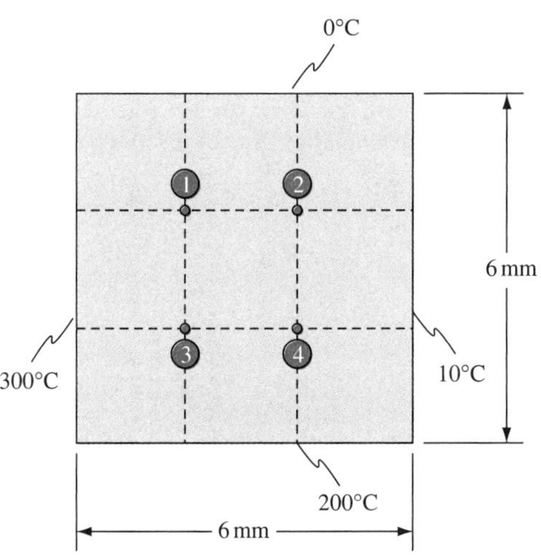

Problem 3.29

3.30 The horizontal cross section of an industrial chimney is shown in the accompanying sketch. Flue gases maintain the interior surface of the chimney at 300°C, and the outside is exposed to an ambient temperature of 0°C through a heat transfer coefficient of 5 W/m^2 K. The thermal conductivity of the chimney is $k = 0.5$ W/m K. For a grid spacing of 0.2 m, determine the temperature distribution in the chimney and the rate of heat loss from the flue gases per unit length of the chimney.

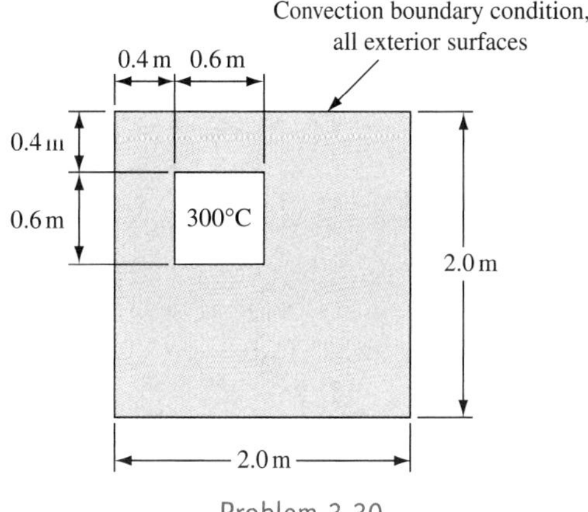

Problem 3.30

3.31 In the long, 30-cm-square bar shown in the accompanying sketch, the left face is maintained at 40°C and the top face is maintained at 250°C. The right face is in contact with a fluid at 40°C through a heat transfer coefficient of 60 W/m^2 K, and the bottom face is in contact with a fluid at 250°C through a heat transfer coefficient of 100 W/m^2 K. If the thermal conductivity of the bar is 20 W/m K, calculate the temperature at the nine nodes shown in the sketch.

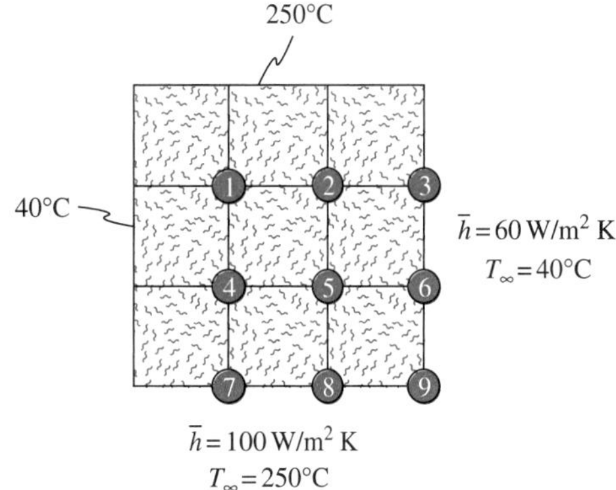

Problem 3.31

3.32 Repeat Problem 3.31 if the temperature distribution on the top surface of the bar varies sinusoidally from 40°C at the left edge to a maximum of 250°C in the center and back to 40°C at the right edge.

3.33 A 1-cm-thick, 1-m-square steel plate is exposed to sunlight and absorbs a solar flux of 800 W/m^2. The bottom of the plate is insulated, the edges are maintained at 20°C by water-cooled clamps, and the exposed face is cooled by a convection coefficient of 10 W/m^2 K to an ambient temperature of 10°C. The plate is polished to minimize reradiation. Determine the temperature distribution in the plate using a node spacing of 20 cm. The thermal conductivity of the steel is 40 W/m K.

3.34 The plate in Problem 3.33 gradually oxidizes over time so that the surface emissivity increases to 0.5. Calculate the resulting temperature in the plate, including radiation heat transfer to the surroundings at the same temperature as the ambient temperature.

3.35 Determine (a) the temperature at the 16 equally spaced points shown in the accompanying sketch to an accuracy of three significant figures and (b) the rate of heat flow per meter thickness. Assume two-dimensional heat flow and $k = 1$ W/m K.

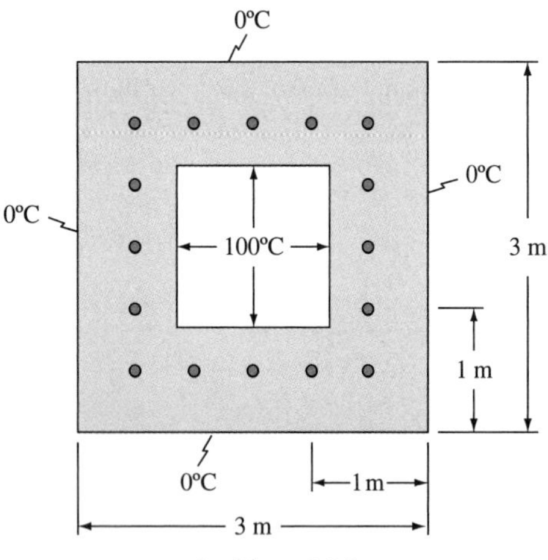

Problem 3.35

3.36 A long, steel beam with a rectangular cross section of 40 cm × 60 cm is mounted on an insulating wall as shown in the following sketch. The rod is heated by radiant heaters that maintain the top and bottom surfaces at 300°C. A stream of air at 130°C cools the exposed face through a heat transfer coefficient of 20 W/m^2 K. Using a node spacing of 1 cm, determine the temperature distribution in the rod and the rate of heat input to the rod. The thermal conductivity of the steel is 40 W/m K.

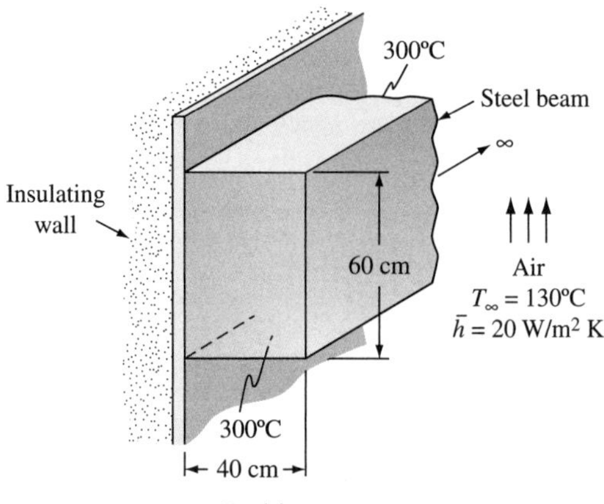

Problem 3.36

3.37 Consider a band saw blade that is to cut steel bar stock. The blade thickness is 2 mm, its height is 20 mm, and it has penetrated the steel workpiece to a depth of 5 mm (see the accompanying sketch). Exposed surfaces of the blade are cooled by an ambient temperature of 20°C through a convection coefficient of 40 W/m^2 K. The thermal conductivity of the blade steel is 30 W/m K. Energy dissipated by the cutting process supplies a heat flux of 10^4 W/m^2 to the surfaces of the blade that are in contact with the workpiece. Assuming two-dimensional, steady conduction, determine the maximum and minimum temperatures in the blade cross section. Use a node spacing of 0.5 mm horizontally and 2 mm vertically.

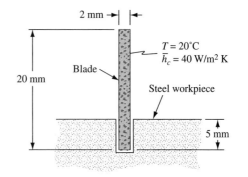

2 mm
$T = 20$°C
$\bar{h}_c = 40$ W/m^2 K
Blade
20 mm
Steel workpiece
5 mm

Problem 3.37

3.38 How would the results of Problem 3.15 be modified if the problem were not axisymmetric?

3.39 For the geometry shown in the sketch below, determine the layout of nodes and control volumes. Provide a scale drawing showing the problem geometry overlaid with the nodes and control volumes. Explain how to derive the energy balance equation for all the boundary control volumes.

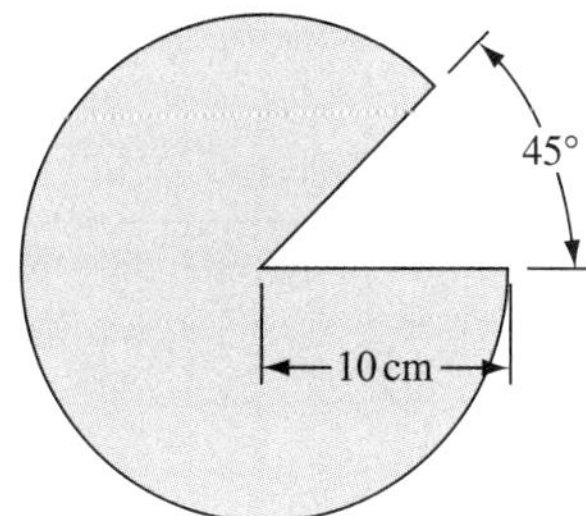

Problem 3.39

3.40 Hot flue gases from a combustion furnace flow through a chimney, which is 7 m tall and has a hollow cylindrical cross section with inner diameter $d_i = 30$ cm and outer diameter $d_o = 50$ cm. The flue gases flow with an average temperature of $T_g = 300°C$ and convective heat transfer coefficient of $h_g = 75\,W/m^2K$. The chimney is made of concrete, which has a thermal conductivity of $k = 1.4\,W/mK$. It is exposed to outside air that has an average temperature of $T_a = 25°C$ and convective heat transfer coefficient of $15\,W/m^2K$. For steady-state conditions, (a) determine the inner and outer wall temperatures, (b) plot the temperature distribution along the thickness of the chimney wall, and (c) determine the rate of heat loss to outside air from the chimney. Solve the problem by numerical analysis using a nodal network with $\Delta r = 2$ cm and $\Delta\theta = 10°$.

3.41 Show that in the limit as $\Delta x \to 0$, $\Delta y \to 0$ and $\Delta t \to 0$, the difference equation, Eq. (3.23), is equivalent to the two-dimensional version of the differential equation, Eq. (2.6).

3.42 Derive the stability criterion for the explicit solution of two-dimensional transient conduction.

3.43 Derive Eq. (3.28).

3.44 Derive the stability criterion for an inside-corner boundary control volume for two-dimensional steady conduction when a convection boundary condition exists.

3.45 A long concrete beam is to undergo a thermal test to determine its loss of strength in the event of a building fire. The beam cross section is triangular as shown in the sketch. Initially, the beam is at a uniform temperature of 20°C. At the start of the test, one of the short faces and the long face are exposed to hot gases at 400°C through a

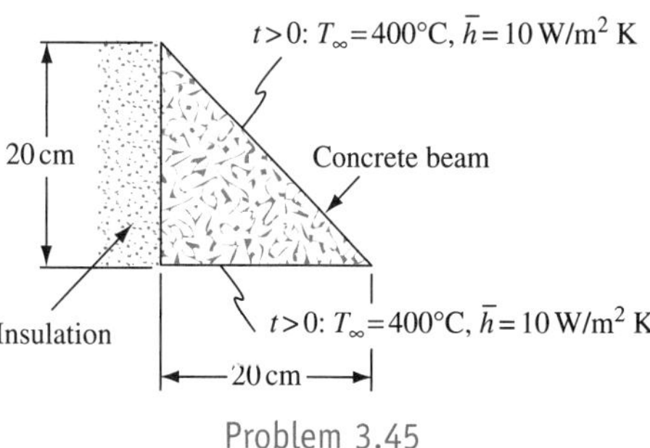

Problem 3.45

heat transfer coefficient of $10\,W/m^2K$, and the other short face is insulated. Produce a graph showing the highest and lowest temperatures in the beam as a function of time for the first hour of exposure. For the concrete properties, use $k = 0.5\,W/m\,K$ and $\alpha = 5 \times 10^{-7} m^2/s$. Use a node spacing of 4 cm and use an explicit difference scheme.

3.46 A steel billet is to be heat treated by immersion in a molten salt bath. The billet is 5 cm square and 1 m long. Prior to immersion in the bath, the billet is at a uniform temperature of 20°C, The bath is 600°C, and the heat transfer coefficient at the billet surface is 20 W/m^2K. Plot the temperature at the center of the billet as a function of time. How much time is needed to heat the billet center to 500°C? Use an implicit difference scheme with a node spacing of 1 cm. The thermal conductivity of the steel is 40 W/mK, and the thermal diffusivity is $1 \times 10^{-5} m^2/s$.

3.47 It has been proposed that a highly concentrating solar collector such as the one that follows can be used to process materials economically when it is desirable to heat the material surface rapidly without significantly heating the bulk. In one such process for case hardening low-cost carbon steel, the surface of a thin disk is to be exposed to concentrated solar flux. The distribution of absorbed solar flux on the disk is given by

$$q''(r) = q''_{max}\,(1 - 0.09(r/R_0)^2)$$

where r is the distance from the disk axis and q''_{max} and R_0 are parameters that describe the flux distribution. The disk diameter is $2R_s$, its thickness is Z_s, its thermal conductivity is k, and its thermal diffusivity is α. The disk is initially at temperature T_{init}, and at time $t = 0$ it is suddenly exposed to the concentrated flux. Derive the set of explicit difference equations needed to predict how the disk temperature distribution evolves with time. The edge and bottom surface of the disk are insulated, and reradiation from the disk can be neglected.

Problem 3.47 Solar furnace used to generate highly concentrated solar energy.
Iconotec / Alamy

3.48 Solve the set of difference equations derived in Problem 3.47, given the following values of the problem parameters:

$$k = 40.0\,\text{W/m K, disk thermal conductivity}$$
$$\alpha = 1 \times 10^{-5}\,\text{m}^2/\text{s, disk thermal diffusivity}$$
$$R_s = 25\,\text{mm, disk radius}$$
$$Z_s = 5\,\text{mm, disk thickness}$$
$$q''_{\text{max}} = 3 \times 10^6\,\text{W/m}^2\text{, peak absorbed flux}$$
$$R_0 = 50\,\text{mm, parameter in flux distribution}$$
$$T_{\text{init}} = 20°\text{C, disk initial temperature}$$

Determine the temperature distribution in the disk when the maximum temperature is 300°C.

3.49 Consider two-dimensional steady conduction near a curved boundary. Determine the difference equation for an appropriate control volume near the node (i, j). The boundary experiences convection heat transfer through a coefficient h to ambient temperature T_a. The surface of the boundary is given by $y_s = f(x)$.

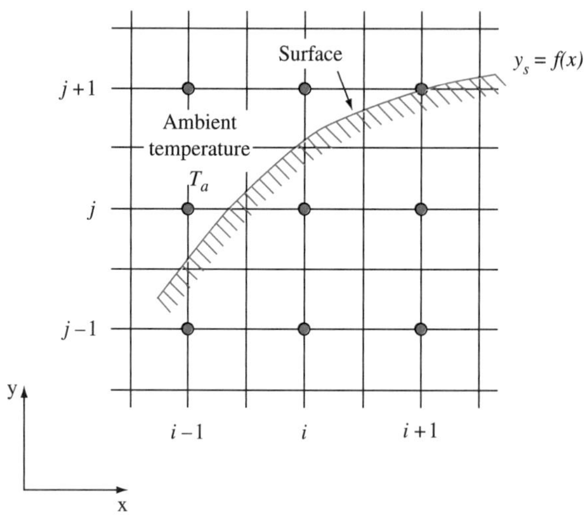

Problem 3.49

3.50 Derive the control volume energy balance equation for three-dimensional transient conduction with heat generation in a rectangular coordinate system.

3.51 Derive the energy balance equation for a corner control volume in a three-dimensional steady conduction problem with heat generation in a rectangular coordinate system. Assume an adiabatic boundary condition and equal node spacing in all three dimensions.

3.52 Determine the stability criterion for an explicit solution of three-dimensional transient conduction in a rectangular geometry.

Design Problems

3.1 **Cooling Analysis of Aluminum Extrusion** (Chapters 3 and 7)
This is the first step in a two-step problem. Assume that a long aluminum extrusion having dimensions shown in the following figure has been heated uniformly to 150°C. It is necessary to determine the time required for this extrusion to cool to a maximum temperature of 40°C in ambient still air with a temperature of 20°C. For the initial estimate, use an

average heat transfer coefficient from Table 1.3. After you have studied convection heat transfer, you will be asked to reconsider this problem with convection heat transfer coefficients calculated from information learned in Chapter 7.

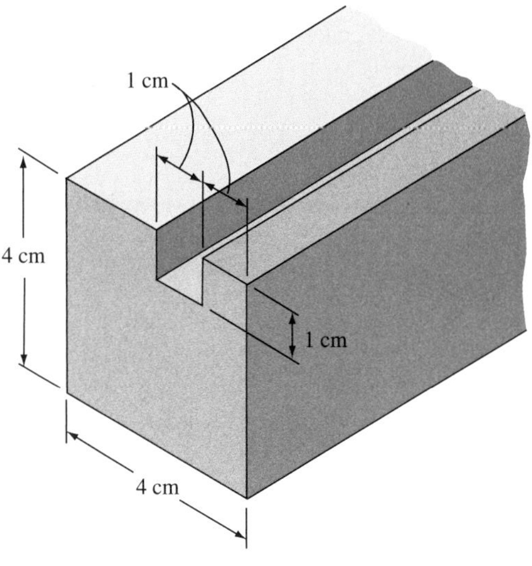

Design Problem 3.1

3.2 **Fast-Response Hot-Water Heater** (Chapters 3, 6 and 10) The preliminary design of a fast-response hot-water heater has been completed in Design Problem 2.7. In the sketch shown below, a cross section of the heater is shown. An extruded heating element is located inside the pipe carrying water to be heated. (The cross-sectional shape of the heater element provides more surface area than a square or round cross-sectional heating element.) The heating element dissipates 4800 watt per meter length, has a thermal conductivity of 20 W/m K, and has a maximum operating temperature of 110°C. Water is to be heated up to 65°C by the device. The surface of the heating element should not exceed 100°C to avoid boiling. Determine the required convective heat transfer coefficient, assuming it to be uniform over the surface of the heating element.

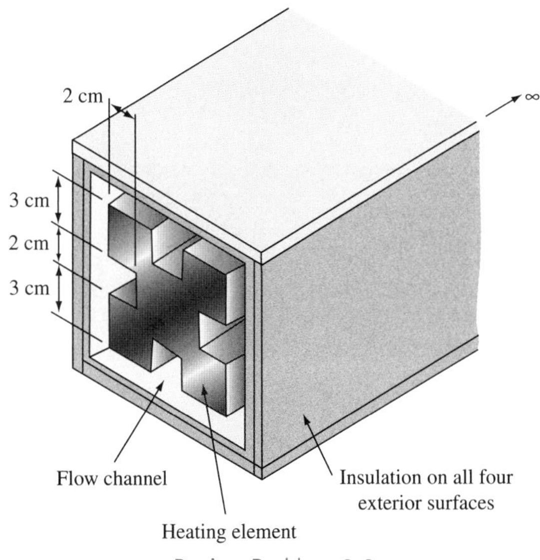

Design Problem 3.2

3.3 **Gas Preheater** (Chapter 3)
A cross-sectional view of a 1-m-long gas preheater is shown in the sketch below. The device provides 14 g/min preheated gas at 150°C to a chemical reactor. The heater rods dissipate 700 watts each. To prevent chemical breakdown of the gas, it is important that the wall of the gas flow channel not exceed 300°C. The aluminum body containing the gas flow channel and the heater elements distributes heat transferred from the heater elements to the gas but cannot exceed 325°C. Assuming a heat transfer coefficient of 500 W/m^2 K between the gas and the surfaces of the rectangular flow channel, determine whether the aluminum body can operate within the required maximum temperature. Estimate the heat transfer coefficients for the exterior surfaces from Table 1.3 and compare your results with an assumption that the surfaces are insulated.

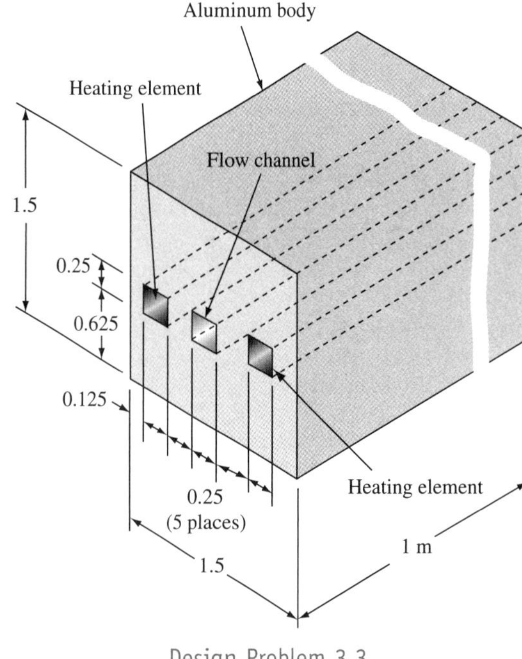

Design Problem 3.3

CHAPTER 4

Analysis of Convection Heat Transfer

High-speed flow over a generic missile body depicting an oblique shock wave at the tip of the body, expansion fan at the shoulder, a dead air region due to shockwave-laminar boundary layer interaction, and high speed base flow features at the back.

Source: Printed under the permission of Erdem E., Yang L., and Kontis K. Presented as "Drag Reduction by Energy Deposition in Hypersonic Flows" in 16th AIAA/DLR/DGLR International Space Planes and Hypersonic Systems and Technologies Conference Bremen, Germany 2009. Affiliation: The School of MACE, The University of Manchester.

Concepts and Analyses to Be Learned

Heat transfer by convection involves two simultaneously occurring mechanisms, diffusion or conduction, accompanied with macroscopic transport of heat to (or from) a moving or flowing fluid. This mode of heat transfer is encountered in a wide spectrum of applications that include, among many others, wind chill effects in winter times, cooling of rocket nozzles, cooling of microelectronic chips, flue-gas heat recovery in a heat exchanger, cooling of gas turbine blades, water heating in a solar collector, and cooling water-glycol mixture with air in an automobile radiator. Thus, a good understanding of the concepts and mathematical expressions that describe convection heat transfer is necessary for the engineering design of such systems and devices. These are developed in this chapter, and its study will teach you:

- How to model a boundary layer in convection heat transfer.
- How to derive the mathematical equations for conservation of mass, momentum, and thermal energy.
- How to perform dimensionless analysis and develop correlations for convection heat transfer with different fluids in laminar and turbulent flow.
- How to obtain analytical solutions for typical laminar flow boundary layer equations.
- How to apply the analogy between momentum and heat transfer to solve turbulent flow convection problems.

4.1 Introduction

In the preceding chapters, convection has been considered only to the extent that it provides the boundary condition when the surface of a body is in contact with a fluid at a different temperature. However, from the illustrative problems, it has probably already become apparent that there are hardly any practical problems that can be treated without a knowledge of the mechanism by which heat is transferred between the surface of a body and the surrounding medium. In this chapter, we shall therefore extend our treatment of convection to gain a better understanding of the mechanism and some of the key parameters that influence it.

4.2 Convection Heat Transfer

Before attempting to calculate a heat transfer coefficient, we shall examine the convection process in some detail and relate the transfer of heat to or from the flowing fluid. Figure 4.1 shows a heated flat plate cooled by a stream of air flowing over it. Also shown are the velocity and temperature distributions that represent this convection situation. The first point to note is that the velocity decreases in the direction toward the surface as a result of viscous forces acting in the fluid. Since the velocity of the fluid layer adjacent to the wall is zero, the heat transfer between the surface and this fluid layer must be by conduction:

$$q_c'' = -k_f \frac{\partial T}{\partial y}\bigg|_{y=0} = h_c(T_s - T_\infty) \tag{4.1}$$

Although this equation suggests that the process can be viewed as conduction, the temperature gradient at the surface $(\partial T/\partial y)_{y=0}$ is determined by the rate at which the fluid farther from the wall can transport the energy into the mainstream. Thus, the temperature gradient at the wall depends on the flow field, with higher

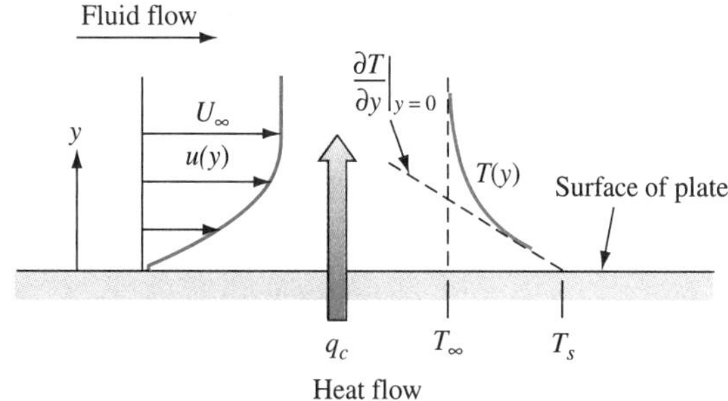

FIGURE 4.1 Temperature and velocity distributions in laminar forced convection flow over a heated flat plate at temperature T_s.

velocities being able to produce larger temperature gradients and higher rates of heat transfer; as explained later, convection heat transfer in higher-velocity turbulent flow is generally greater than in lower-velocity laminar flow of fluids. At the same time, however, the thermal conductivity of the fluid plays a role. For example, the value of k_f for water is an order of magnitude larger than that for air; thus, as shown in Table 1.3, the convection heat transfer coefficient for water is larger than for air.

EXAMPLE 4.1 Air at 20°C is flowing over a flat plate whose surface temperature is 100°C. At a certain location, the temperature is measured as a function of distance from the surface of the plate; the results are plotted in Fig. 4.2. From these data, determine the convection heat transfer coefficient at this location.

SOLUTION From Eq. (4.1), the heat transfer coefficient can be expressed in the form

$$h_c = \frac{-k_f(\partial T/\partial y)_{y=0}}{T_s - T_\infty}$$

Table 28 in Appendix 2 shows that the thermal conductivity of air at the average temperature between the plate and the fluid stream (60°C) is 0.028 W/mK. The temperature gradient $\partial T/\partial y$ at the surface is obtained graphically by drawing the tangent to the measured temperature data shown in Fig. 4.2. We thus obtain $(\partial T/\partial y)_{y=0} \approx -66.7$ K/mm. Substituting this value for the gradient at the heated surface of the plate in Eq. (4.1) yields

$$h_c = \frac{-(0.028 \text{ W/m K})(-66.7 \text{ K/mm})}{(100 - 20) \text{ K}} \times 10^3 \text{ mm/m}$$
$$= 23.3 \text{ W/m}^2 \text{K}$$

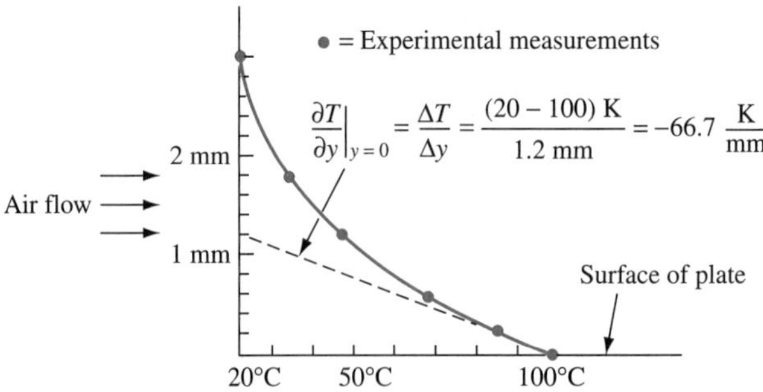

FIGURE 4.2 Experimental data on temperature distribution for Example 4.1.

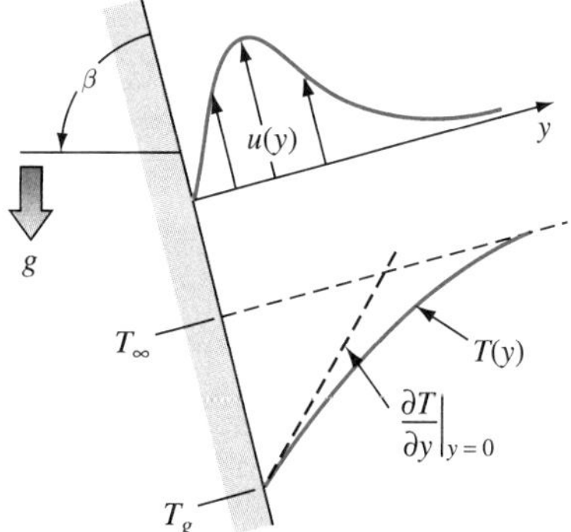

FIGURE 4.3 Temperature and velocity distributions in natural convection over a heated plate inclined at an angle β from the horizontal.

The situation is quite similar in natural convection, as shown in Fig. 4.3. It should be recalled that in natural or free convection, the fluid motion is caused by buoyancy effects due to fluid density inversion (which requires a temperature difference), whereas in forced convection, an imposed pressure gradient drives the fluid flow. The principal difference in their respective velocity distribution is that in forced convection the velocity approaches the free-stream value imposed by an external force, whereas in natural convection, the velocity at first increases with increasing distance from the plate because the effect of viscosity diminishes rather rapidly while the density difference decreases more slowly. Eventually, however, the buoyant force decreases as the fluid density approaches the value of the surrounding fluid, causing the velocity to first reach a maximum and then approach zero far away from the heated surface. The temperature fields in natural and forced convection have similar shapes, and in both cases, the heat transfer mechanism at the fluid/solid interface is conduction.

The preceding discussion indicates that the convection heat transfer coefficient depends on the density, viscosity, and velocity of the fluid as well as on its thermal properties. In forced convection, the velocity is usually imposed on the system by a pump or a fan and can be directly specified. In natural convection, the velocity depends on the buoyancy effect due to the temperature difference between the surface and the fluid, the coefficient of thermal expansion of the fluid, which determines the density change per unit temperature difference, and the force field, which in systems located on the earth is simply the gravitational force.

4.3 Boundary Layer Fundamentals

To gain an understanding of the parameters that are significant in forced convection, we shall examine the flow field in more detail. Figure 4.4 shows the velocity distribution at various distances from the leading edge of a plate. From the

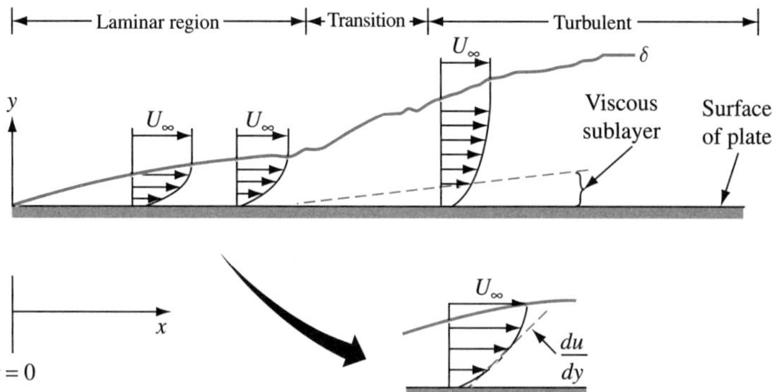

FIGURE 4.4 Velocity profiles in laminar, transition, and turbulent boundary layers in flow over a flat plate.

leading edge onward, a region develops in the flow in which viscous forces cause the fluid to slow down. These viscous forces depend on the shear stress τ. In flow over a flat plate, the fluid velocity parallel to the plate can be used to define this stress as

$$\tau = \mu \frac{du}{dy} \tag{4.2}$$

where du/dy is the velocity gradient and the constant of proportionality μ is called the dynamic viscosity. If the shear stress is expressed in newtons per square meter and the velocity gradient in $(\text{seconds})^{-1}$, then μ has the units newton-seconds per square meter (N s/m^2).*

The flow region near the plate where the velocity of the fluid is decreased by viscous forces is called the *boundary layer*. The distance from the plate at which the velocity reaches 99% of the free-stream velocity is arbitrarily designated as the boundary layer thickness, and the region beyond this point is called the undisturbed free stream or potential flow regime.

Initially, the flow in the boundary layer is completely laminar. The boundary layer thickness grows with increasing distance from the leading edge, and at some critical distance x_c, the inertial effects become sufficiently large compared to the viscous damping action that small disturbances in the flow begin to grow. As these disturbances become amplified, the regularity of the viscous flow is disturbed and a transition from laminar to turbulent flow takes place. In the turbulent-flow region, macroscopic chunks of fluid move across streamlines and vigorously transport thermal energy as well as momentum. As shown in books on fluid mechanics [1], the dimensionless parameter that quantitatively relates the viscous and inertial forces

*Note that μ can also be expressed in the units kg/m s. With these units it is necessary to use the conversion constant g_c for dimensional consistency, and Eq. 4.2 would then read $\tau = (\mu/g_c)\, du/dy$.

and whose value determines the transition from laminar to turbulent flow is the Reynolds number* Re_x, which is defined as

$$\mathrm{Re}_x = \frac{\rho U_\infty x}{\mu} = \frac{U_\infty x}{\nu} \tag{4.3}$$

where U_∞ = free-stream velocity
x = distance from the leading edge
$\nu = \mu/\rho$ = kinematic viscosity of the fluid
ρ = density of the fluid

Approximate shapes of the velocity profiles in laminar and turbulent flow are sketched in Fig. 4.4. In the laminar range, the boundary layer velocity profile is approximately parabolic. In the turbulent range, there is a thin layer near the surface (the viscous sublayer), across which the velocity profile is nearly linear. Outside this layer the velocity profile is flat compared to the laminar profile.

The critical value of Re_x at which transition occurs, $\mathrm{Re}_{c,x}$, depends on the surface roughness and the level of turbulent activity—the turbulence level—in the mainstream. When large disturbances are present in the main flow, transition begins when $\mathrm{Re}_x = 10^5$, but in less disturbed flow fields, it will not start until $\mathrm{Re}_x = 2 \times 10^5$ [1, 2]. If the flow is very free from disturbances, transition may not start until $\mathrm{Re} = 10^6$. For example, consider the flow of air at 1 m/s and 20°C parallel to a flat plate. Since $\nu = 15.7 \times 10^{-6}\ \mathrm{m^2/s}$, the distance from the leading edge where transition occurs is given by

$$x_c = \frac{\mathrm{Re}_{c,x}\nu}{U_\infty} = \frac{(10^5)\left(15.7 \times 10^{-6}\,\dfrac{\mathrm{m}^2}{\mathrm{s}}\right)}{\left(1\dfrac{\mathrm{m}}{\mathrm{s}}\right)} = 1.57\,\mathrm{m}$$

The transition regime extends to a Reynolds number about twice the value at which transition began, and beyond this point the boundary layer is turbulent.

4.4 Conservation Equations of Mass, Momentum, and Energy for Laminar Flow Over a Flat Plate

In the classical approach to convection, one derives differential equations for the momentum and energy balances in the boundary layer and then solves these equations for the temperature gradient in the fluid at the fluid/wall interface to evaluate

*The Reynolds number, which describes the dimensionless dynamic similarity of fluid flows given by the ratio of inertia to viscous forces, is named after the British mathematician and professor of engineering, Osborne Reynolds (1842–1912), who was born in Ireland; studied at Cambridge University; worked at Owens College in Manchester, England; and conducted experiments that developed the similitude basis for determining the transition from laminar to turbulent flows.

the convection heat transfer coefficient. A somewhat simpler but more useful approach is to derive integral instead of differential equations and use an approximate analysis to obtain a solution. In this section, the differential equations governing the flow of a fluid over a flat plate will be derived to illustrate the similarity between heat and momentum transfer and to introduce appropriate dimensionless parameters relating the processes. Then the integral equations for flow over a flat surface will be developed and solved to illustrate an analytical approach that will also be used to obtain the heat transfer boundary layer coefficients in turbulent flow.

To derive the *conservation of mass* or continuity equation, consider a control volume within the boundary layer, as shown in Fig. 4.5, and assume that steady-state conditions prevail. There are no gradients in the z direction (perpendicular to the plane of the sketch), and the fluid is incompressible. Then the rates of mass flow into and out of the control volume, respectively, in the x direction are

$$\rho u \, dy \quad \text{and} \quad \rho\left(u + \frac{\partial u}{\partial x} dx\right) dy$$

Thus, the net mass flow into the element in the x direction is

$$-\rho \frac{\partial u}{\partial x} dx \, dy$$

Similarly, the net mass flow into the control volume in the y direction is

$$-\rho \frac{\partial v}{\partial y} dx \, dy$$

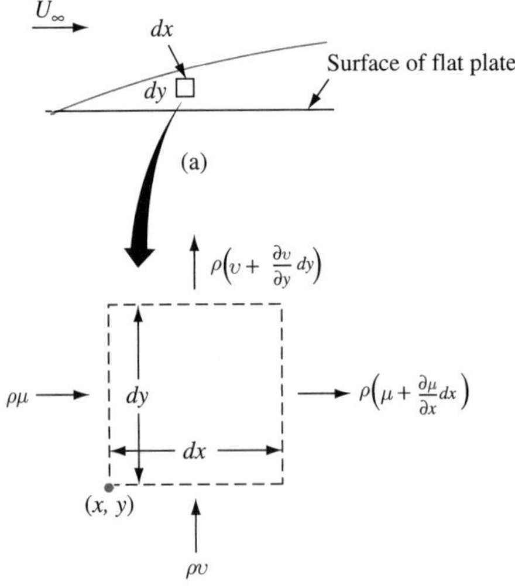

FIGURE 4.5 Control volume ($dx\, dy \cdot 1$) for conservation of mass in an incompressible boundary layer in flow over a flat plate.

Since the net mass flow rate out of the control volume must be zero, we obtain

$$-\rho\left(\frac{\partial u}{\partial x} + \frac{\partial v}{\partial y}\right)dx\,dy = 0$$

from which it follows that in two-dimensional steady flow, conservation of mass requires that

$$\frac{\partial u}{\partial x} + \frac{\partial v}{\partial y} = 0 \tag{4.4}$$

The *conservation of momentum* equation is obtained from application of Newton's second law of motion to the element. Assuming that the flow is Newtonian, that there are no pressure gradients in the y direction, and that viscous shear in the y direction is negligible, the rates of momentum flow in the x direction for the fluid flowing across the left- and right-hand vertical faces (see Fig. 4.6) are $\rho u^2\,dy$ and $\rho[u + (\partial u/\partial x)\,dx]^2\,dy$, respectively. It should be noted, however, that flow across the horizontal faces will also contribute to the momentum balance in the x direction. The x-momentum flow entering through the bottom face is $\rho uv\,dx$, and the momentum flow per unit width leaving through the upper face is

$$\rho\left(v + \frac{\partial v}{\partial y}dy\right)\left(u + \frac{\partial u}{\partial y}dy\right)dx$$

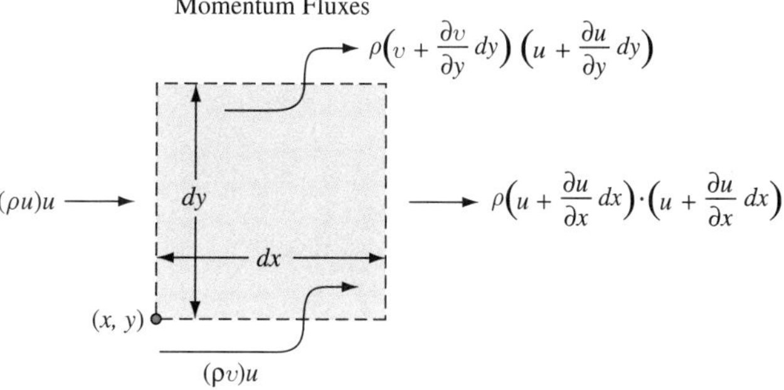

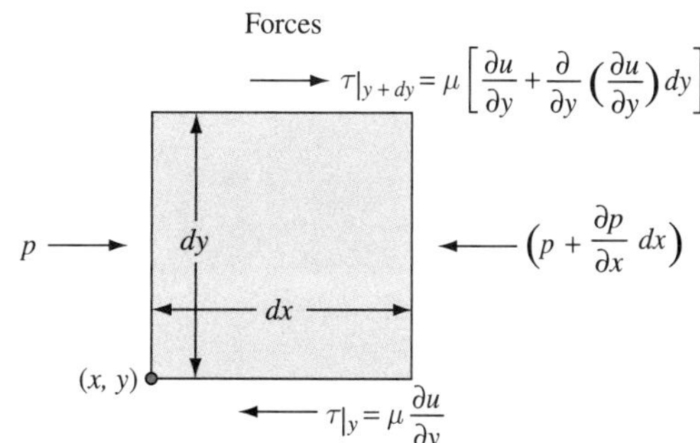

FIGURE 4.6 Differential control volume for conservation of momentum in a two-dimensional incompressible boundary layer.

The viscous shear force at the bottom face is $\tau|_y = -\mu(\partial u/\partial y)\,dx$ and over the top face is

$$|\tau|_{y+dy} = \mu\,dx\left[\frac{\partial u}{\partial y} + \frac{\partial}{\partial y}\left(\frac{\partial u}{\partial y}\right)dy\right]$$

Thus, the net viscous shear in the x direction is $\mu\,dx(\partial^2 u/\partial y^2)\,dy$.

The pressure force over the left face is $p\,dy$ and over the right face is $-[p + (\partial p/\partial x)\,dx]\,dy$. Thus the net pressure force in the direction of motion is $-(\partial p/\partial x)\,dx\,dy$. Equating the sum of the forces to the momentum flow rate out of the control volume in the x direction gives

$$\rho\left(u + \frac{\partial u}{\partial x}dx\right)^2 dy - \rho u^2\,dy + \rho\left(v + \frac{\partial v}{\partial y}dy\right)\left(u + \frac{\partial u}{\partial y}dy\right)dx - \rho v u\,dx = \mu\frac{\partial^2 u}{\partial y^2}dx\,dy - \frac{\partial p}{\partial x}dx\,dy$$

Neglecting second-order differentials and using the conservation of mass equation, the conservation of momentum equation reduces to

$$\rho\left(u\frac{\partial u}{\partial x} + v\frac{\partial u}{\partial y}\right) = \mu\frac{\partial^2 u}{\partial y^2} - \frac{\partial p}{\partial x} \tag{4.5}$$

Figure 4.7 shows the rate at which energy is conducted and convected into and out of the control volume. There are four convective terms in addition to the conductive terms derived in Chapter 2. An energy balance requires that the net rate of conduction and convection be zero. This yields

$$k\,dx\,dy\left(\frac{\partial^2 T}{\partial x^2} + \frac{\partial^2 T}{\partial y^2}\right) - \left[\rho c_p\left(u\frac{\partial T}{\partial x} + \frac{\partial u}{\partial x}T + \frac{\partial u}{\partial x}\frac{\partial T}{\partial x}dx\right)\right]dx\,dy - \left[\rho c_p\left(v\frac{\partial T}{\partial y} + \frac{\partial v}{\partial y}T + \frac{\partial v}{\partial y}\frac{\partial T}{\partial y}dy\right)\right]dx\,dy = 0$$

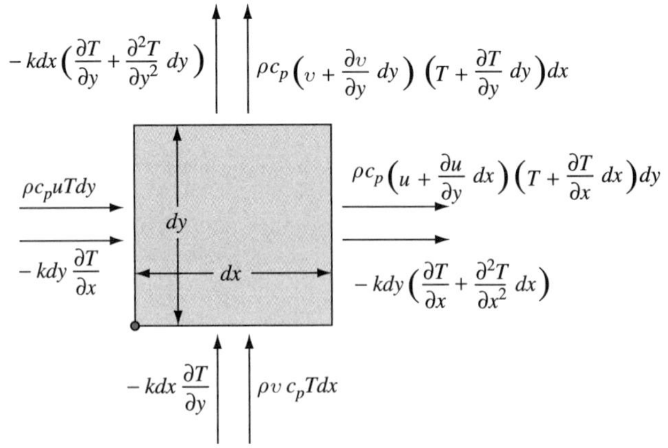

$-kdx\left(\frac{\partial T}{\partial y} + \frac{\partial^2 T}{\partial y^2}dy\right)$ $\rho c_p\left(v + \frac{\partial v}{\partial y}dy\right)\left(T + \frac{\partial T}{\partial y}dy\right)dx$

$\rho c_p u T dy$ $\rho c_p\left(u + \frac{\partial u}{\partial y}dx\right)\left(T + \frac{\partial T}{\partial x}dx\right)dy$

dy

$-kdy\frac{\partial T}{\partial x}$ dx $-kdy\left(\frac{\partial T}{\partial x} + \frac{\partial^2 T}{\partial x^2}dx\right)$

$-kdx\frac{\partial T}{\partial y}$ $\rho v c_p T dx$

FIGURE 4.7 Differential control volume for conservation of energy.

The *conservation of energy* equation has been derived on the assumption that all physical properties are temperature-independent and that the flow velocity is sufficiently small that viscous dissipation can be neglected. When viscous dissipation cannot be neglected, mechanical energy is irreversibly converted into thermal energy, giving rise to an additional term on the left-hand side of the equation. The term, called the *viscous dissipation* Φ, is given by

$$\Phi = \mu\left\{\left(\frac{\partial u}{\partial y} + \frac{\partial v}{\partial x}\right)^2 + 2\left[\left(\frac{\partial u}{\partial x}\right)^2 + \left(\frac{\partial v}{\partial y}\right)^2\right] - \frac{2}{3}\left(\frac{\partial u}{\partial x} + \frac{\partial v}{\partial y}\right)^2\right\}$$

for constant properties. The effect of viscous dissipation can be significant if the fluid is very viscous, as in journal bearings, or if the fluid shear rate is extremely high [3, 4].

Using the conservation of mass equation and neglecting second-order terms, as we did in the derivation of the conservation of momentum equation, gives the following expression for the energy equation without dissipation:

$$u\frac{\partial T}{\partial x} + v\frac{\partial T}{\partial y} = \alpha\left(\frac{\partial^2 T}{\partial x^2} + \frac{\partial^2 T}{\partial y^2}\right) \tag{4.6}$$

Since a boundary layer is quite thin, under normal conditions $\partial T/\partial y \gg \partial T/\partial x$. Also, the pressure term in the momentum equation is zero for flow over a flat plate since $(\partial U_\infty/\partial x) = 0$. Then the similarity between the momentum and energy equations becomes apparent:

$$u\frac{\partial u}{\partial x} + v\frac{\partial u}{\partial y} = \nu\left(\frac{\partial^2 u}{\partial y^2}\right) \tag{4.7a}$$

$$u\frac{\partial T}{\partial x} + v\frac{\partial T}{\partial y} = \alpha\left(\frac{\partial^2 T}{\partial y^2}\right) \tag{4.7b}$$

In the preceding relations, ν is the kinematic viscosity. It is equal to μ/ρ and is often called momentum diffusivity. The ratio ν/α is equal to $(\mu/\rho)/(k/\rho c_p)$, which is the Prandtl number, Pr:

$$\mathrm{Pr} = \frac{c_p \mu}{k} = \frac{\nu}{\alpha} \tag{4.8}$$

If ν equals α, then Pr is 1 and the momentum and energy equations are identical. For this condition, nondimensional solutions of $u(y)$ and $T(y)$ are identical if the boundary conditions are similar. Thus, it is apparent that the Prandtl number, which is the ratio of fluid properties (or more rigorously the ratio of momentum and thermal diffusivities), controls the relation between the velocity and temperature distributions.

4.5 Dimensionless Boundary Layer Equations and Similarity Parameters

Solutions to Eqs. (4.7a–b), the so-called laminar boundary layer equations for low-speed forced convection, will yield the velocity and temperature profiles. In general, these solutions are quite complicated, and the reader is referred to Schlichting [1],

Van Driest [2], and Langhaar [5] for a treatment of the mathematical procedures, which are beyond the scope of this text. However, considerable additional insight into the physical aspects of boundary layer flow as well as the form of similarity parameters governing the transport processes can be gained by nondimensionalizing the governing equations, even without solving them.

Figure 4.8 shows the development of the velocity and thermal boundary layers in flow over a flat surface of arbitrary shape. To express the boundary layer equations in dimensionless form, we define the following dimensionless variables, which are similar to those defined in Section 2.2:

$$x^* = \frac{x}{L} \qquad v^* = \frac{v}{U_\infty}$$

$$y^* = \frac{y}{L} \qquad p^* = \frac{p}{\rho_\infty U_\infty^2}$$

$$u^* = \frac{u}{U_\infty} \qquad T^* = \frac{T - T_s}{T_\infty - T_s}$$

where L is a characteristic length dimension such as the length of a plate, U_∞ is the free-stream velocity, T_s is the surface temperature, T_∞ is the free-stream temperature, and ρ_∞ is the free-stream density.

Substituting the above dimensionless variables into the dimensional Eqs. (4.4), (4.5), and (4.7b) yields the corresponding boundary layer equations:

$$\frac{\partial u^*}{\partial x^*} + \frac{\partial v^*}{\partial y^*} = 0 \tag{4.9a}$$

$$u^* \frac{\partial u^*}{\partial x^*} + v^* \frac{\partial u^*}{\partial y^*} = -\frac{\partial p^*}{\partial x^*} + \frac{1}{\mathrm{Re}_L} \frac{\partial^2 u^*}{\partial y^{*2}} \tag{4.9b}$$

$$u^* \frac{\partial T^*}{\partial x^*} + v^* \frac{\partial T^*}{\partial y^*} = \frac{1}{\mathrm{Re}_L \mathrm{Pr}} \frac{\partial^2 T^*}{\partial y^{*2}} \tag{4.9c}$$

Observe that by nondimensionalizing the boundary layer equations we have cast them into a form in which the dimensionless similarity parameters Re_L and Pr appear. These similarity parameters permit us to apply solutions from one system to

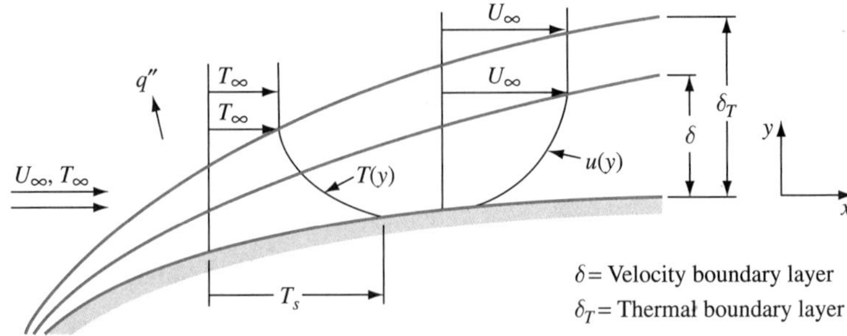

FIGURE 4.8 Development of the velocity and thermal boundary layers in flow over a flat surface of arbitrary shape.

another geometrically similar system provided the similarity parameters have the same value in both. For example, if the Reynolds number is the same, the dimensionless velocity distributions for air, water, and glycerin flowing over a flat plate will be the same at given values of x^*.

Inspection of Eq. (4.9a) shows that v^* is related to u^*, y^*, and x^*:

$$v^* = f_1(u^*, y^*, x^*) \tag{4.10}$$

and that from Eq. (4.9b) the solution for u^*, accordingly, can be expressed in the form

$$u^* = f_2\left(x^*, y^*, \mathrm{Re}_L, \frac{\partial p^*}{\partial x^*}\right) \tag{4.11}$$

The pressure distribution over the surface of a body is determined by its shape. From the y-momentum equation it can be shown that $\partial p^*/\partial y^* = 0$ and p^* is only a function of x^*. Hence, dp^*/dx^* can be obtained independently. It represents the influence of the shape on the velocity distribution in the free stream just outside the boundary layer.

4.5.1 Friction Coefficient

From Eq. (4.2), the surface shear stress τ_s is given by

$$\tau_s = \mu \left.\frac{\partial u}{\partial y}\right|_{y=0} = \frac{\mu U_\infty}{L} \left.\frac{\partial u^*}{\partial y^*}\right|_{y^*=0} \tag{4.12}$$

Defining the local frictional drag coefficient C_f as

$$C_{fx} = \frac{\tau_s}{\rho U_\infty^2/2} \tag{4.13}$$

and substituting Eq. (4.12) for τ_s gives

$$C_{fx} = \frac{2}{\mathrm{Re}_L} \left.\frac{\partial u^*}{\partial y^*}\right|_{y^*=0} \tag{4.14}$$

From Eq. (4.11), it is apparent that the dimensionless velocity gradient $\partial u^*/\partial y^*$ at the surface ($y^* = 0$) depends only on x^*, Re_L, and dp^*/dx^*. But since dp^*/dx^* is entirely determined by the geometric shape of a body, Eq. (4.14) reduces to the form

$$C_{fx} = \frac{2}{\mathrm{Re}_L} f_3(x^*, \mathrm{Re}_L) \tag{4.15}$$

for bodies of similar shape. The above relation implies that for flow over bodies of similar shape the local frictional drag coefficient is related to x^* and Re_L by a universal function that is independent of the fluid or the free-stream velocity.

The average frictional drag over a body $\bar{\tau}$ can be determined by integrating the local shear stress τ over the surface of the body. Hence, $\bar{\tau}$ must be independent of x^*, and the average friction coefficient $\bar{C}_f$ depends only on the value of the Reynolds number for flow over geometrically similar bodies:

$$\bar{C}_f = \frac{\bar{\tau}}{\rho U_\infty^2/2} = \frac{2}{\text{Re}_L} f_4(\text{Re}_L) \tag{4.16}$$

EXAMPLE 4.2 For flow over a slightly curved surface, the local shear stress is given by the relation

$$\tau_s(x) = 0.3\left(\frac{\rho\mu}{x}\right)^{0.5} U_\infty^{1.5}$$

From this dimensional equation, obtain nondimensional relations for the local and average friction coefficients.

SOLUTION From Eq. (4.13), the local friction coefficient is

$$C_{fx} = \frac{\tau_s(x)}{\frac{1}{2}\rho U_\infty^2} = 0.6\left(\frac{\rho\mu}{x}\right)^{0.5} \frac{U_\infty^{1.5}}{\rho U_\infty^2}$$

$$= 0.6\left(\frac{\mu}{\rho U_\infty x}\right)^{0.5} = \frac{0.6}{\text{Re}_x^{0.5}} = \frac{0.6}{(\text{Re}_L x^*)^{0.5}}$$

Integrating the local value and dividing by the area per unit width ($L \times 1$) gives the average shear $\bar{\tau}$:

$$\bar{\tau} = \frac{1}{L}\int_0^L 0.3\left(\frac{\rho\mu}{x}\right)^{0.5} U_\infty^{1.5}\, dx = 0.6\left(\frac{\rho\mu}{L}\right)^{0.5} U_\infty^{1.5}$$

and the average friction coefficient is therefore

$$\bar{C}_f = \frac{\bar{\tau}}{\rho U_\infty^2/2} = \frac{1.2}{\text{Re}_L^{0.5}}$$

4.5.2 Nusselt Number

In convection heat transfer, the key unknown is the heat transfer coefficient. From Eq. (4.1), we obtain the following equation in terms of the dimensionless parameters:

$$h_c = -\frac{k_f}{L}\left[\frac{(T_\infty - T_s)}{(T_s - T_\infty)}\right]\frac{\partial T^*}{\partial y^*}\bigg|_{y^*=0} = +\frac{k_f}{L}\frac{\partial T^*}{\partial y^*}\bigg|_{y^*=0} \tag{4.17}$$

Inspection of this equation suggests that the appropriate dimensionless form of the heat transfer coefficient is the so-called Nusselt number, Nu, defined by

$$\mathrm{Nu} = \frac{h_c L}{k_f} \equiv \left.\frac{\partial T^*}{\partial y^*}\right|_{y^*=0} \tag{4.18}$$

From Eqs. (4.9a) and (4.9c), it is apparent that for a prescribed geometry the local Nusselt number depends only on x^*, Re_L, and Pr:

$$\mathrm{Nu} = f_5(x^*, \mathrm{Re}_L, \mathrm{Pr}) \tag{4.19}$$

Once this functional relation is known, either from an analysis or from experiments with a particular fluid, it can be used to obtain the value of Nu for other fluids and for any values of U_∞ and L. Moreover, from the local value of Nu, we can first obtain the local value of h_c and then an average value of the heat transfer coefficient $\bar{h}_c$ and an average Nusselt number $\overline{\mathrm{Nu}}_L$. Since the average heat transfer coefficient is obtained by integrating over the heat transfer surface of a body, it is independent of x^*, and the average Nusselt number is a function of only Re_L and Pr:

$$\overline{\mathrm{Nu}}_L = \frac{\bar{h}_c L}{k_f} = f_6(\mathrm{Re}_L, \mathrm{Pr}) \tag{4.20}$$

4.6 Evaluation of Convection Heat Transfer Coefficients

Five general methods are available for the evaluation of convection heat transfer coefficients:

1. Dimensional analysis combined with experiments
2. Exact mathematical solutions of the boundary layer equations
3. Approximate analyses of the boundary layer equations by integral methods
4. The analogy between heat and momentum transfer
5. Numerical analysis, or modeling with computational fluid dynamics (CFD) methods

All five of these techniques have contributed to our understanding of convection heat transfer. Yet no single method can solve all the problems, because each one has limitations that restrict its scope of application.

Dimensional analysis is mathematically simple and has found a wide range of application [5, 6]. The chief limitation of this method is that the results obtained are incomplete and quite useless without experimental data. Dimensional analysis contributes little to our understanding of the transfer process but facilitates the interpretation and extends the range of experimental data by correlating them in terms of dimensionless groups.

There are two different methods for determining dimensionless groups suitable for correlating experimental data. The first of these methods, discussed in the following section, requires only listing of the variables pertinent to a phenomenon. This technique is simple to use, but if a pertinent variable is omitted, erroneous results ensue. In the second method, the dimensionless groups and similarity conditions are deduced from the differential equations describing the phenomenon. This method is preferable when the phenomenon can be described mathematically, but the solution of the resulting equations is often too involved to be practical. This technique was presented in Section 4.5.

Exact mathematical analyses require simultaneous solution of the equations describing the fluid motion and the transfer of energy in the moving fluid [7]. The method presupposes that the physical mechanisms are sufficiently well understood to be described in mathematical language. This preliminary requirement limits the scope of exact solutions because complete mathematical equations describing the fluid flow and the heat transfer mechanisms can be written only for laminar flow. Even for laminar flow, the equations are quite complicated, but solutions have been obtained for a number of simple systems such as flow over a flat plate, an airfoil, or a circular cylinder [7].

Exact solutions are important because the assumptions made in the course of the analysis can be specified accurately and their validity can be checked by experiment. They also serve as a basis of comparison and as a check on simpler approximate methods. Furthermore, the development of high-speed computers has increased the range of problems amenable to mathematical solution, and results of computations for different systems are continually being published in the literature.

Approximate analysis of the boundary layer avoids the detailed mathematical description of the flow in the boundary layer. Instead, a plausible but simple equation is used to describe the velocity and temperature distributions in the boundary layer. The problem is then analyzed on a macroscopic basis by applying the equation of motion and the energy equation to the aggregate of the fluid particles contained within the boundary layer. This method is relatively simple; moreover, it yields solutions to problems that cannot be treated by an exact mathematical analysis. In instances where other solutions are available, they agree within engineering accuracy with the solutions obtained by this approximate method. The technique is not limited to laminar flow but also can be applied to turbulent flow.

The analogy between heat and momentum transfer is a useful tool for analyzing turbulent transfer processes. Our knowledge of turbulent-exchange mechanisms is insufficient for us to write mathematical equations describing the temperature distribution directly, but the transfer mechanism can be described in terms of a simplified model. According to one such model that has been widely accepted, a mixing motion in a direction perpendicular to the mean flow accounts for the transfer of momentum as well as energy. The mixing motion can be described on a statistical basis by a method similar to that used to picture the motion of gas molecules in the kinetic theory. There is by no means general agreement that this model corresponds to conditions actually existing in nature, but for practical purposes, its use can be justified by the fact that experimental results are substantially in agreement with analytical predictions based on the hypothetical model.

Numerical methods can solve in an approximate form the exact equations of motion [8, 9]. The approximation results from the need to express the field variables (temperature, velocity, and pressure) at discrete points in time and space rather than continuously. However, the solution can be made sufficiently accurate if care is taken in discretizing the exact equations. One of the most important advantages of numerical methods is that once the solution procedure has been programmed, solutions for different boundary conditions, property variables, and so on can be easily computed. Generally, numerical methods can handle complex boundary conditions easily. Numerical methods for solving convection problems are discussed in [9] and are an extension of the methods presented in Chapter 3 for conduction problems.

4.7 Dimensional Analysis

Dimensional analysis differs from other approaches in that it does not yield equations that can be solved. Instead, it combines several variables into dimensionless groups, such as the Nusselt number, which facilitate the interpretation and extend the range of application of experimental data. In practice, convection heat transfer coefficients are generally calculated from empirical equations obtained by correlating experimental data with the aid of dimensional analysis.

The most serious limitation of dimensional analysis is that it gives no information about the nature of a phenomenon. In fact, to apply dimensional analysis it is necessary to know beforehand what variables influence the phenomenon, and the success or failure of the method depends on the proper selection of these variables. It is therefore important to have at least a preliminary theory or a thorough physical understanding of a phenomenon before a dimensional analysis can be performed. However, once the pertinent variables are known, dimensional analysis can be applied to most problems by a routine procedure that is outlined below.*

4.7.1 Primary Dimensions and Dimensional Formulas

The first step is to select a system of *primary dimensions*. The choice of the primary dimensions is arbitrary, but the dimensional formulas of all pertinent variables must be expressible in terms of them. In the SI system, the primary dimensions of length L, time t, temperature T, and mass M are used.

The dimensional formula of a physical quantity follows from definitions or physical laws. For instance, the dimensional formula for the length of a bar is $[L]$ by definition.† The average velocity of a fluid particle is equal to a distance divided by the time interval taken to traverse it. The dimensional formula of velocity is therefore $[L/t]$ or $[Lt^{-1}]$ (i.e., a distance or length divided by a time). The units of velocity could be expressed in meters per second, feet per second, or miles per hour, since they all are a length divided by a time. The dimensional formulas and the symbols of physical quantities occurring frequently in heat transfer problems are given in Table 4.1.

---

*The algebraic theory of dimensional analysis will not be developed here. For a rigorous and comprehensive treatment of the mathematical background, Chapters 3 and 4 of Langhaar [5] are recommended.

†Square brackets indicate that the quantity has the dimensional formula stated within the brackets.

TABLE 4.1 Important heat and mass transfer physical quantities and their dimensions

Quantity	Symbol	Dimensions in *MLtT* System
Length	L, x	L
Time	t	t
Mass	M	M
Force	F	ML/t^2
Temperature	T	T
Heat	Q	ML^2/t^2
Velocity	u, v, U_∞	L/t
Acceleration	a, g	L/t^2
Work	W	ML^2/t^2
Pressure	p	M/t^2L
Density	ρ	M/L^3
Internal energy	e	L^2/t^2
Enthalpy	i	L^2/t^2
Specific heat	c	L^2/t^2T
Absolute viscosity	μ	M/Lt
Kinematic viscosity	$\nu = \mu/\rho$	L^2/t
Thermal conductivity	k	ML/t^3T
Thermal diffusivity	α	L^2/t
Thermal resistance	R	Tt^3/ML^2
Coefficient of expansion	β	$1/T$
Surface tension	σ	M/t^2
Shear stress	τ	M/Lt^2
Heat transfer coefficient	h	M/t^3T
Mass flow rate	$\dot{m}$	M/t

4.7.2 Buckingham π Theorem

To determine the number of independent dimensionless groups required to obtain a relation describing a physical phenomenon, the Buckingham π theorem may be used.‡ According to this rule, the required number of independent dimensionless groups that can be formed by combining the physical variables pertinent to a problem is equal to the total number of these physical quantities n (e.g., density, viscosity, heat transfer coefficient) minus the number of primary dimensions m required to express the dimensional formulas of the n physical quantities. If we call these groups π_1, π_2, and so forth, the equation expressing the relationship among the variables has a solution of the form

$$F(\pi_1, \pi_2, \pi_3, \ldots) = 0 \tag{4.21}$$

‡A more rigorous rule proposed by Van Driest [26] shows that the π theorem holds as long as the set of simultaneous equations formed by equating the exponents of each primary dimension to zero is linearly independent. If one equation in the set is a linear combination of one or more of the other equations (i.e., if the equations are linearly dependent), the number of dimensionless groups is equal to the total number of variables n minus the number of independent equations.

In a problem involving five physical quantities and three primary dimensions, $n - m$ is equal to two and the solution either has the form

$$F(\pi_1, \pi_2) = 0 \tag{4.22}$$

or the form

$$\pi_1 = f(\pi_2) \tag{4.23}$$

Experimental data for such a case can be presented conveniently by plotting π_1 against π_2. The resulting empirical curve reveals the functional relationship between π_1 and π_2, which cannot be deduced from dimensional analysis.

For a phenomenon that can be described in terms of three dimensionless groups (i.e., if $n - m = 3$), Eq. (4.21) has the form

$$F(\pi_1, \pi_2, \pi_3) = 0 \tag{4.24}$$

but can also be written as

$$\pi_1 = f(\pi_2, \pi_3)$$

For such a case, experimental data can be correlated by plotting π_1 against π_2 for various values of π_3. Sometimes it is possible to combine two of the π's in some manner and to plot this parameter against the remaining π on a single curve.

4.7.3 Determination of Dimensionless Groups

A simple method for determining dimensionless groups will now be illustrated by applying it to the problem of correlating experimental convection heat transfer data for a fluid flowing across a heated tube. Exactly the same approach could be used for heat transfer in flow through a tube or over a plate.

From the description of the convection heat transfer process, it is reasonable to expect that the physical quantities listed in Table 4.2 are pertinent to the problem.

There are seven physical quantities and four primary dimensions. We therefore expect that three dimensionless groups will be required to correlate the data. To find these dimensionless groups, we write π as a product of the variables, each raised to an unknown power:

$$\pi = D^a k^b U_\infty^c \rho^d \mu^e c_p^f \bar{h}_c^g \tag{4.25}$$

TABLE 4.2 Pertinent physical quantities in convection heat transfer

Variable	Symbol	Dimensions
Tube diameter	D	$[L]$
Thermal conductivity of fluid	k	$[ML/t^3T]$
Free-stream velocity of fluid	U_∞	$[L/t]$
Density of fluid	ρ	$[M/L^3]$
Viscosity of fluid	μ	$[M/Lt]$
Specific heat at constant pressure	c_p	$[L^2/t^2T]$
Heat transfer coefficient	$\bar{h}_c$	$[M/t^3T]$

and substitute the dimensional formulas

$$\pi = [L]^a \left[\frac{ML}{t^3T}\right]^b \left[\frac{L}{t}\right]^c \left[\frac{M}{L^3}\right]^d \left[\frac{M}{Lt}\right]^e \left[\frac{L^2}{t^2T}\right]^f \left[\frac{M}{t^3T}\right]^g \tag{4.26}$$

For π to be dimensionless, the exponents of each primary dimension must separately add up to zero. Equating the sum of the exponents of each primary dimension to zero, we obtain the set of equations

$$\begin{aligned} b + d + e + g &= 0 \quad \text{for} \quad M \\ a + b + c - 3d - e + 2f &= 0 \quad \text{for} \quad L \\ -3b - c - e - 2f - 3g &= 0 \quad \text{for} \quad t \\ -b - f - g &= 0 \quad \text{for} \quad T \end{aligned}$$

Evidently, any set of values of a, b, c, d, and e that simultaneously satisfies this set of equations will make π dimensionless. There are seven unknowns, but only four equations. We can therefore choose values for three of the exponents in each of the dimensionless groups. The only restriction on the choice of the exponents is that each of the selected exponents be independent of the others. An exponent is independent if the determinant formed with the coefficients of the remaining terms does not vanish (i.e., is not equal to zero).

Since $\bar{h}_c$, the convection heat transfer coefficient, is the variable we eventually want to evaluate, it is convenient to set its exponent g equal to unity. At the same time, we let $c = d = 0$ to simplify the algebraic manipulations. Solving the equations simultaneously, we obtain $a = 1$, $b = -1$, $e = f = 0$. The first dimensionless group is then

$$\pi_1 = \frac{\bar{h}_c D}{k}$$

which we recognize as the *Nusselt number*, $\overline{\mathrm{Nu}}_D$.

For π_2, we select g equal to zero, so that $\bar{h}_c$ will not appear again, and let $a = 1$ and $f = 0$. Simultaneous solution of the equations with these choices yields $b = 0$, $c = d = 1$, $e = -1$, and

$$\pi_2 = \frac{U_\infty D \rho}{\mu}$$

This dimensionless group is a *Reynolds number*, Re_D, with the tube diameter as the length parameter.

If we let $e = 1$ and $c = g = 0$, we obtain the third dimensionless group,

$$\pi_3 = \frac{c_p \mu}{k}$$

which is the *Prandtl number*, Pr.

We observe that although the convection heat transfer coefficient is a function of six variables, with the aid of dimensional analysis the seven original variables have been combined into three dimensionless groups. According to Eq. (4.24), the functional relationship can be written

$$\overline{\mathrm{Nu}}_D = f(\mathrm{Re}_D, \mathrm{Pr})$$

and experimental data now can be correlated in terms of three variables instead of the original seven. The importance of this reduction in the number of variables becomes apparent when we attempt to plan experiments and correlate experimental data.

4.7.4 Correlation of Experimental Data

Suppose that in a series of tests with air flowing over a 25-mm-OD pipe, the heat transfer coefficient has been measured experimentally at velocities ranging from 0.15 to 30 m/s. This range of velocities corresponds to Reynolds numbers based on the diameter $D\rho U_\infty/\mu$ ranging from 250 to 50,000. Since the velocity was the only variable in these tests, the results are correlated in Fig. 4.9(a) by plotting the heat

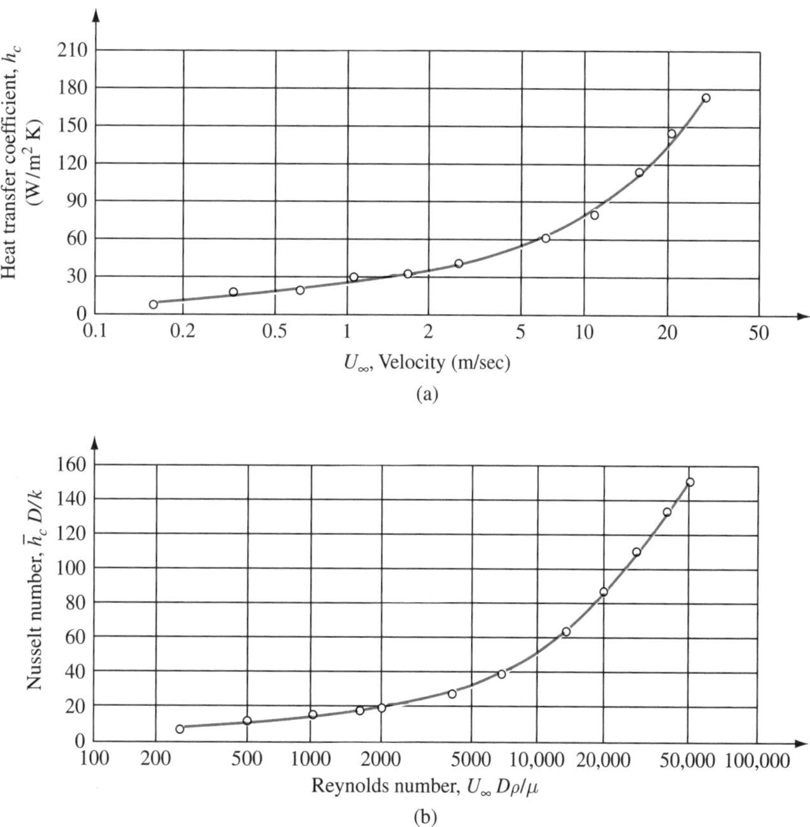

FIGURE 4.9 Variation of Nusselt number with Reynolds number for cross-flow of air over a pipe or a long cylinder (a) dimensional plot, (b) dimensionless plot.

transfer coefficient $\bar{h}_c$ against the velocity U_∞. The resulting curve permits a direct determination of $\bar{h}_c$ at any velocity for the system used in the tests, but it cannot be used to determine the heat transfer coefficients for cylinders that are larger or smaller than the one used in the tests. Nor could the heat transfer coefficient be evaluated if the air were under pressure and its density were different from that used in the tests. Unless experimental data could be correlated more effectively, it would be necessary to perform separate experiments for every cylinder diameter, every density, etc. The effort would be enormous.

With the aid of dimensional analysis, however, the results of one series of tests can be applied to a variety of other problems, as illustrated in Fig. 4.9(b), where the data of Fig. 4.9(a) are replotted in terms of pertinent dimensionless groups. The abscissa in Fig. 4.9(b) is the Reynolds number $U_\infty D\rho/\mu$, and the ordinate is the Nusselt number $\bar{h}_c D/k$. This correlation of the data permits the evaluation of the convection heat transfer coefficient for air flowing over any size of pipe or wire as long as the Reynolds number of the system falls within the range covered in the experiment and the systems are geometrically similar.

Experimental data obtained with air alone do not reveal the dependence of the Nusselt number on the Prandtl number, since the Prandtl number is a combination of physical properties whose value does not vary appreciably for gases. To determine the influence of the Prandtl number, it is necessary to use different fluids. According to the preceding analysis, experimental data with several fluids whose physical properties yield a wide range of Prandtl numbers are necessary to complete the correlation.

In Fig. 4.10, the experimental results of several independent investigations for heat transfer between air, water, and oils in cross-flow over a tube or a wire are plotted for a wide range of temperatures, cylinder sizes, and velocities. The ordinate

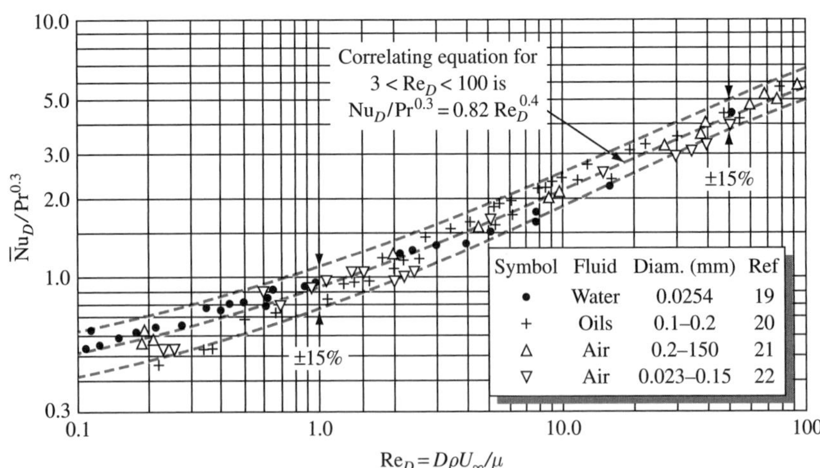

FIGURE 4.10 Correlation of experimental heat transfer data for various fluids in cross-flow over pipes, wires, and circular cylinders.

in Fig. 4.10 is the dimensionless quantity* $\overline{\mathrm{Nu}}_D/\mathrm{Pr}^{0.3}$ and the abscissa is Re_D. An inspection of the results shows that all of the data follow a single line reasonably well, and thus they can be correlated empirically.

It should be noted that the experimental data in Figs. 4.9 and 4.10 cover different ranges of Reynolds numbers: between 0.1 and 100 in Fig. 4.10 and between about 200 and 50,000 in Fig. 4.9. Extrapolation of the correlation equation in Fig. 4.10 into a Reynolds number range much above 200 or 300 would lead to serious error. For example, for air (Pr = 0.71) flowing over a cylinder at $\mathrm{Re}_D = 20{,}000$, the correlation equation in Fig. 4.10 would predict

$$\mathrm{Nu}_D = \mathrm{Pr}^{0.3} \times 0.82 \times \mathrm{Re}_D^{0.4} = 0.71^{0.3} \times 0.82 \times 20{,}000^{0.4} = 39$$

while the experimental data in Fig. 4.9 give $\mathrm{Nu}_D = 85$, a substantial difference. When no data are available in a range encountered in a design, some extrapolation may be necessary, but as the above example shows, extrapolation of data well beyond the range of the dimensionless parameters covered in experiments should be avoided if possible. If there is no opportunity to conduct appropriate experiments, the results of extrapolation must be treated with caution.

4.7.5 Principle of Similarity

The remarkable result of Fig. 4.10 can be explained by the principle of similarity. According to this principle, often called the model law, the behavior of two systems will be similar if the ratios of their linear dimensions, forces, velocities, and so forth are the same. Under conditions of forced convection in geometrically similar systems, the velocity fields will be similar provided the ratio of inertial forces to viscous forces is the same in both fluids. The Reynolds number is the ratio of these forces, and consequently, we expect similar flow conditions in forced convection for a given value of the Reynolds number. The Prandtl number is the ratio of two molecular transport properties, the kinematic viscosity $\nu = \mu/\rho$, which affects the velocity distribution, and the thermal diffusivity $k/\rho c_p$, which affects the temperature profile. In other words, it is a dimensionless group that relates the temperature distribution to the velocity distribution. Hence, in geometrically similar systems having the same Prandtl and Reynolds numbers, the temperature distributions will be similar. The Nusselt number is equal to the ratio of the temperature gradient at a fluid-to-surface interface to a reference temperature gradient. We therefore expect that, in systems having similar geometries and similar temperature fields, the numerical values of the Nusselt numbers will be equal. This prediction is borne out by the experimental results in Fig. 4.10.

Dimensional analyses have been performed for numerous heat transfer systems, and Table 4.3 summarizes the most important dimensionless groups used in design.

---

*Combining the Nusselt number with the Prandtl number for plotting the data is simply a matter of convenience. As mentioned previously, any combination of dimensionless parameters is satisfactory. The selection of the most convenient parameter is usually made on the basis of experience by trial and error with the aid of experimental results, although sometimes the characteristic groups are suggested by the results of analytic solutions.

TABLE 4.3 Dimensionless groups of importance for heat transfer and fluid flow

Group	Definition	Interpretation
Biot number (Bi)	$\dfrac{\bar{h}L}{k_s}$	Ratio of internal thermal resistance of a solid body to its surface thermal resistance
Drag coefficient (C_f)	$\dfrac{\tau_s}{\rho U_\infty^2/2}$	Ratio of surface shear stress to free-stream kinetic energy
Eckert number (Ec)	$\dfrac{U_\infty^2}{c_p(T_s - T_\infty)}$	Kinetic energy of flow relative to boundary layer enthalpy difference
Fourier number (Fo)	$\dfrac{\alpha t}{L^2}$	Dimensionless time; ratio of rate of heat conduction to rate of internal energy storage in a solid
Friction factor (f)	$\dfrac{\Delta p}{(L/D)(\rho U_m^2/2)}$	Dimensionless pressure drop for internal flow through ducts
Grashof number (Gr_L)	$\dfrac{g\beta(T_s - T_\infty)L^3}{\nu^2}$	Ratio of buoyancy to viscous forces
Colburn j factor (j_H)	$StPr^{2/3}$	Dimensionless heat transfer coefficient
Nusselt number (Nu_L)	$\dfrac{\bar{h}_c L}{k_f}$	Dimensionless heat transfer coefficient; ratio of convection heat transfer to conduction in a fluid layer of thickness L
Peclet number (Pe_L)	$Re_L Pr$	Product of Reynolds and Prandtl numbers
Prandtl number (Pr)	$\dfrac{c_p\mu}{k} = \dfrac{\nu}{\alpha}$	Ratio of molecular momentum diffusivity to thermal diffusivity
Rayleigh number (Ra)	$Gr_L Pr$	Product of Grashof and Prandtl numbers
Reynolds number (Re_L)	$\dfrac{U_\infty L}{\nu}$	Ratio of inertia to viscous forces
Stanton number (St)	$\dfrac{\bar{h}_c}{\rho U_\infty c_p} = \dfrac{Nu_L}{Re_L Pr}$	Dimensionless heat transfer coefficient

4.8* Analytic Solution for Laminar Boundary Layer Flow Over a Flat Plate†

In the preceding section, we determined dimensionless groups for correlating experimental data for heat transfer by forced convection. We found that the Nusselt number depends on the Reynolds number and the Prandtl number, i.e.,

$$\text{Nu} = \phi(\text{Re})\psi(\text{Pr}) \tag{4.27}$$

†In the remainder of this chapter, the mathematical details can be omitted in an introductory course without breaking the continuity of the presentation.

In the next few sections, we shall consider analytical methods for determining the functional relations in Eq. (4.27) for low-speed flow over a flat plate. This system has been selected primarily because it is the simplest to analyze, but the results have many practical applications, for instance, they are good approximations for flow over the surfaces of streamlined bodies such as airplane wings or turbine blades.

In view of the differences in the flow characteristics, the frictional forces as well as the heat transfer are governed by different relations for laminar and turbulent flow. We will first treat the laminar boundary layer, which is amenable to an exact and an approximate method of solution. The turbulent boundary layer is taken up in Section 4.10.

To determine the forced-convection heat transfer coefficient and the friction coefficient for incompressible flow over a flat surface, we must satisfy the continuity, momentum, and energy equations simultaneously. These relations were derived in Section 4.4 and are repeated below for convenience.

Continuity:

$$\frac{\partial u}{\partial x} + \frac{\partial v}{\partial y} = 0 \tag{4.4}$$

Momentum:

$$\rho\left(u\frac{\partial u}{\partial x} + v\frac{\partial u}{\partial y}\right) = \mu\frac{\partial^2 u}{\partial y^2} - \frac{\partial p}{\partial x} \tag{4.5}$$

Energy:

$$u\frac{\partial T}{\partial x} + v\frac{\partial T}{\partial y} = \alpha\frac{\partial^2 T}{\partial y^2} \tag{4.7b}$$

4.8.1 Boundary Layer Thickness and Skin Friction

Equation (4.5) must be solved simultaneously with the continuity equation, Eq. (4.4), in order to determine the velocity distribution, boundary layer thickness, and friction force at the wall. These equations are solved by first defining a stream function $\psi(x,y)$ that automatically satisfies the continuity equation, that is,

$$u = \frac{\partial \psi}{\partial y} \quad \text{and} \quad v = -\frac{\partial \psi}{\partial x}$$

Introducing the new variable

$$\eta = y\sqrt{\frac{U_\infty}{\nu x}}$$

we can let

$$\psi = \sqrt{\nu x U_\infty}\, f(\eta)$$

where $f(\eta)$ denotes a dimensionless stream function. In terms of $f(\eta)$, the velocity components are

$$u = \frac{\partial \psi}{\partial y} = \frac{\partial \psi}{\partial \eta}\frac{\partial \eta}{\partial y} = U_\infty \frac{d[f(\eta)]}{d\eta}$$

and

$$v = -\frac{\partial \psi}{\partial x} = \frac{1}{2}\sqrt{\frac{\nu U_\infty}{x}}\left\{\frac{d[f(\eta)]}{d\eta}\eta - f(\eta)\right\}$$

Expressing $\partial u/\partial x$, $\partial u/\partial y$, and $\partial^2 u/\partial y^2$ in terms of η and inserting the resulting expressions in the momentum equation yields the ordinary, nonlinear, third-order differential equation

$$f(\eta)\frac{d^2[f(\eta)]}{d\eta^2} + 2\frac{d^3[f(\eta)]}{d\eta^3} = 0$$

which is to be solved subject to the three boundary conditions

$$f(\eta) = 0 \qquad \text{and} \qquad \frac{d[f(\eta)]}{d\eta} = 0 \qquad \text{at } \eta = 0$$

and

$$\frac{d[f(\eta)]}{d\eta} = 1 \qquad \text{at } \eta = \infty$$

The solution to this differential equation was obtained numerically by Blasius in 1908 [10]. The significant results are shown in Figs. 4.11 and 4.12.

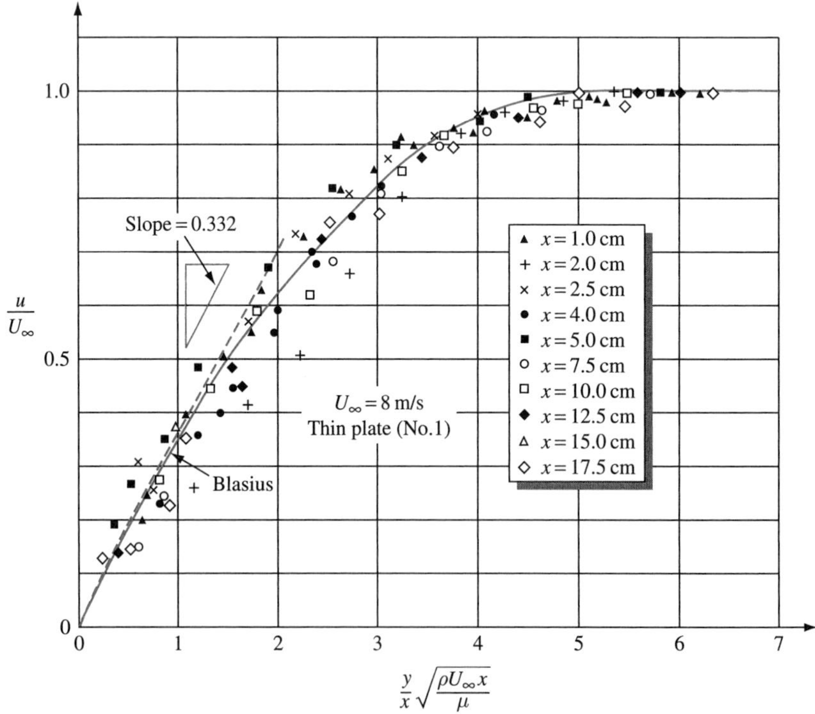

FIGURE 4.11 Velocity profile in a laminar boundary layer according to Blasius, with experimental data of Hansen [11].

Source: Courtesy of the National Advisory Committee for Aeronautics, NACA TM 585.

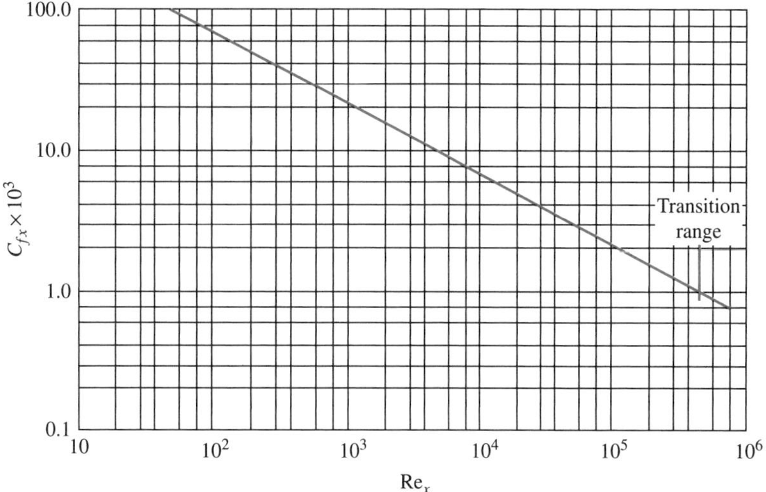

FIGURE 4.12 Local friction coefficient versus Reynolds number based on distance from leading edge for laminar flow over a flat plate.

In Fig. 4.11, the Blasius velocity profiles in the laminar boundary layer on a flat plate are plotted in dimensionless form together with experimental data obtained by Hansen [11]. The ordinate is the local velocity in the x direction u divided by the free-stream velocity U_∞, and the abscissa is a dimensionless distance parameter $(y/x)\sqrt{(\rho U_\infty x)/\mu}$. We note that a single curve is sufficient to correlate the velocity distributions at all stations along the plate. The velocity u reaches 99% of the free-stream value U_∞ at $(y/x)\sqrt{(\rho U_\infty x)/\mu} = 5.0$. If we define the hydrodynamic boundary layer thickness as that distance from the surface at which the local velocity u reaches 99% of the free-stream value U_∞, the boundary layer thickness δ becomes

$$\delta = \frac{5x}{\sqrt{\mathrm{Re}_x}} \tag{4.28}$$

where $\mathrm{Re}_x = \rho U_\infty x/\mu$, which is the local Reynolds number. Equation (4.28) satisfies the qualitative description of the boundary layer growth, δ being zero at the leading edge ($x = 0$) and increasing with x along the plate. At any station, that is, at any given value of x, the thickness of the boundary layer is inversely proportional to the square root of the local Reynolds number. Hence, an increase in velocity results in a decrease of the boundary layer thickness.

The shear force at the wall can be obtained from the velocity gradient at $y = 0$ in Fig. 4.11. We see that

$$\left.\frac{\partial(u/U_\infty)}{\partial(y/x)\sqrt{\mathrm{Re}_x}}\right|_{y=0} = 0.332$$

and thus at any specified value of x the velocity gradient at the surface is

$$\left.\frac{\partial u}{\partial y}\right|_{y=0} = 0.332\,\frac{U_\infty}{x}\sqrt{\mathrm{Re}_x}$$

Substituting this velocity gradient in the equation for the shear, the wall shear per unit area τ_s becomes

$$\tau_s = \mu \frac{\partial u}{\partial y}\bigg|_{y=0} = 0.332\mu \frac{U_\infty}{x} \sqrt{\text{Re}_x} \tag{4.29}$$

We note that the wall shear near the leading edge is very large and decreases with increasing distance from the leading edge.

For a graphic presentation, it is more convenient to use dimensionless coordinates. Dividing both sides of Eq. (4.29) by the velocity pressure of the free stream $\rho U_\infty^2/2$, we obtain

$$C_{fx} = \frac{\tau_s}{\rho U_\infty^2/2} = \frac{0.664}{\sqrt{\text{Re}_x}} \tag{4.30}$$

where C_{fx} is the dimensionless local drag or friction coefficient. Figure 4.12 is a plot of C_{fx} against Re_x and shows the variation of the local friction coefficient graphically. The average friction coefficient is obtained by integrating Eq. (4.30) between the leading edge $x = 0$ and $x = L$:

$$\bar{C}_f = \frac{1}{L}\int_0^L C_{fx}\,dx = 1.33\sqrt{\frac{\mu}{U_\infty \rho L}} \tag{4.31}$$

Thus, for laminar flow over a flat plate, the average friction coefficient $\bar{C}_f$ is equal to twice the value of the local friction coefficient at $x = L$.

4.8.2 Convection Heat Transfer

The energy conservation equation for a laminar boundary layer is

$$u\frac{\partial T}{\partial x} + v\frac{\partial T}{\partial y} = \alpha \frac{\partial^2 T}{\partial y^2} \tag{4.7b}$$

The velocities in the energy conservation equation, u and v, have the same values at any point x, y as in the fluid dynamic equation, Eq. (4.5). For the case of the flat plate, Pohlhausen [12] used the velocities calculated previously by Blasius [10] to obtain the solution of the heat transfer problem. Without considering the details of this mathematical solution, we can obtain significant results by comparing Eq. (4.7b) with Eq. (4.5), the momentum equation. The two equations are similar; in fact $u(x, y)$ is also a solution for the temperature distribution $T(x, y)$ if $\nu = \alpha$ and if the temperature of the plate T_s is constant. We can easily verify this by replacing the symbol T in Eq. (4.7b) by the symbol u and noting that the boundary conditions for both T and u are identical. If we use the surface temperature as our datum and let the variable in Eq. (4.7b) be $(T - T_s)/(T_\infty - T_s)$, then the boundary conditions are

$$\frac{T - T_s}{T_\infty - T_s} = 0 \quad \text{and} \quad \frac{u}{U_\infty} = 0 \quad \text{at } y = 0$$

$$\frac{T - T_s}{T_\infty - T_s} = 1 \quad \text{and} \quad \frac{u}{U_\infty} = 1 \quad \text{at } y \to \infty$$

where T_∞ is the free-stream temperature.

The condition that $\nu = \alpha$ corresponds to a Prandtl number of unity since

$$\text{Pr} = \frac{c_p \mu}{k} = \frac{\nu}{\alpha}$$

For Pr = 1, the velocity distribution is therefore identical to the temperature distribution. An interpretation in terms of physical processes is that the transfer of momentum is analogous to the transfer of heat when Pr = 1. The physical properties of most gases are such that they have Prandtl numbers ranging from 0.6 to 1.0, and the analogy is therefore satisfactory. Liquids, on the other hand, have Prandtl numbers considerably different from unity, and the preceding analysis cannot be applied directly [13].

Using the analytical results of Pohlhausen's work, the temperature distribution in the laminar boundary layer for Pr = 1 can be modified empirically to include fluids having Prandtl numbers different from unity. In Fig. 4.13, theoretically calculated temperature profiles in the boundary layer are shown for values of Pr of 0.6, 0.8, 1.0, 3.0, 7.0, 15, and 50. We now define a thermal boundary layer thickness δ_{th} as the distance from the surface at which the temperature difference between the wall and the fluid reaches 99% of the free-stream value. Inspection of the temperature profiles shows that the thermal boundary layer is larger than the hydrodynamic boundary layer for fluids having Pr less than unity, but smaller when Pr is greater

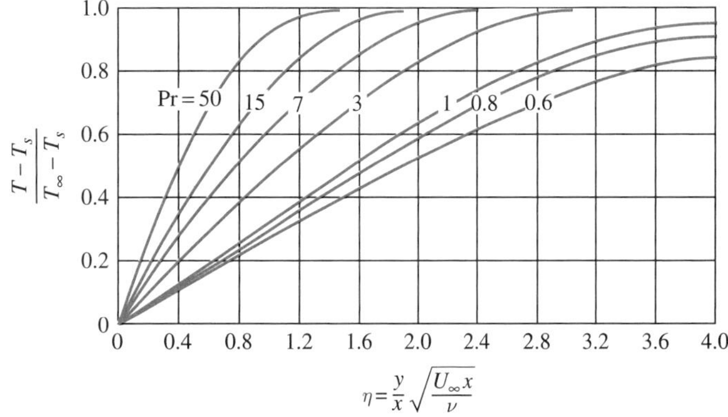

FIGURE 4.13 Dimensionless temperature distributions in a fluid flowing over a heated plate for various Prandtl numbers.

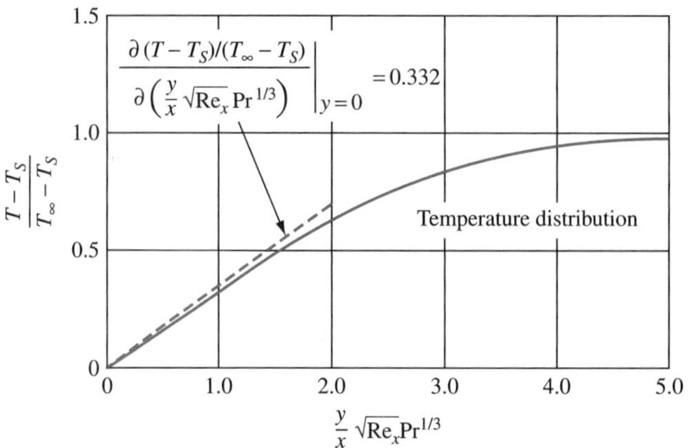

FIGURE 4.14 Dimensionless temperature distribution for laminar flow over a heated plate at uniform temperature.

than unity. According to Pohlhausen's calculations, the relationship between the thermal and hydrodynamic boundary layer is approximately

$$\delta/\delta_{th} = \mathrm{Pr}^{1/3} \tag{4.32}$$

Using the same correction factor, $\mathrm{Pr}^{1/3}$, at any distance from the surface, the curves of Fig. 4.13 are replotted in Fig. 4.14. The new abscissa is $\mathrm{Pr}^{1/3}(y/x)\sqrt{\mathrm{Re}_x}$ and the ordinate is the dimensionless temperature $(T - T_s)/(T_\infty - T_s)$, where T is the local temperature of the fluid, T_s the surface temperature of the plate, and T_∞ the free-stream temperature. This modification of the ordinate brings the temperature profiles for a wide range of Prandtl numbers together on a single line, which is the curve for $\mathrm{Pr} = 1$.

4.8.3 Evaluation of the Convection Heat Transfer Coefficient

The rate of heat transfer by convection and the convection heat transfer coefficient can now be determined. The dimensionless temperature gradient at the surface (at $y = 0$) is

$$\left.\frac{\partial[(T - T_s)/(T_\infty - T_s)]}{\partial[(y/x)\sqrt{\mathrm{Re}_x}\,\mathrm{Pr}^{1/3}]}\right|_{y=0} = 0.332$$

Therefore, at any specified value of x,

$$\left.\frac{\partial T}{\partial y}\right|_{y=0} = 0.332\,\frac{\mathrm{Re}_x^{1/2}\mathrm{Pr}^{1/3}}{x}(T_\infty - T_s) \tag{4.33}$$

and the local rate of heat transfer by convection per unit area becomes, on substituting $\partial T/\partial y$ from Eq. (4.33),

$$q_c'' = -k\left.\frac{\partial T}{\partial y}\right|_{y=0} = -0.332k\,\frac{\mathrm{Re}_x^{1/2}\mathrm{Pr}^{1/3}}{x}(T_\infty - T_s) \tag{4.34}$$

The total rate of heat transfer from a plate of width b and length L, obtained by integrating q_c'' from Eq. (4.34) between $x = 0$ and $x = L$, is

$$q = 0.664k\,\mathrm{Re}_L^{1/2}\mathrm{Pr}^{1/3}b(T_s - T_\infty) \tag{4.35}$$

The local convection heat transfer coefficient is

$$h_{cx} = \frac{q_c''}{(T_s - T_\infty)} = 0.332\frac{k}{x}\,\mathrm{Re}_x^{1/2}\mathrm{Pr}^{1/3} \tag{4.36}$$

and the corresponding local Nusselt number is

$$\mathrm{Nu}_x = \frac{h_{cx}x}{k} = 0.332\,\mathrm{Re}_x^{1/2}\mathrm{Pr}^{1/3} \tag{4.37*}$$

The average Nusselt number $\bar{h}_c L/k$ is obtained by integrating the right-hand side of Eq. (4.36) between $x = 0$ and $x = L$ and dividing the result by L to obtain $\bar{h}_c$, which is the average value of h_{cx}; multiplying $\bar{h}_c$ by L/k gives

$$\overline{\mathrm{Nu}_L} = 0.664\,\mathrm{Re}_L^{1/2}\mathrm{Pr}^{1/3} \tag{4.38}$$

The average value of the Nusselt number over a length L of the plate is therefore twice the local value of Nu_x at $x = L$. It easily can be verified that the same relation between the average and the local value holds also for the heat transfer coefficient, that is,

$$\bar{h}_c = 2h_{c(x=L)} \tag{4.39}$$

In practice, the physical properties in Eqs. (4.32) to (4.38) vary with temperature, while for the purpose of analysis it was assumed that the physical properties are constant. Experimental data have been found to agree satisfactorily with the results predicted analytically if the properties are evaluated at a mean temperature halfway between that of the surface and the free-stream temperature; this mean temperature is called the *film temperature*.

EXAMPLE 4.3 A flat-plate solar collector is placed horizontally on a roof, as shown in Fig. 4.15. To determine its efficiency, it is necessary to calculate the heat loss from its surface to the environment. The collector is a long strip 1 ft wide. The surface temperature of the collector is 140°F. If a wind at 60°F is blowing over the collector at a velocity of 10 ft/s, calculate the following quantities at $x = 1$ ft and $x = x_c$ in English and SI units:

(a) boundary layer thickness
(b) local friction coefficient
(c) average friction coefficient
(d) local drag or shearing stress due to friction

*Note that Eq. (4.37) was derived under the assumption that $\mathrm{Pr} \geq 1$. It is therefore not valid for small values of Pr, i.e., liquid metals. An empirical equation for the local Nusselt numbers valid for liquid metals ($\mathrm{Pr} < 0.1$) is given in Table 4.5 on page 283.

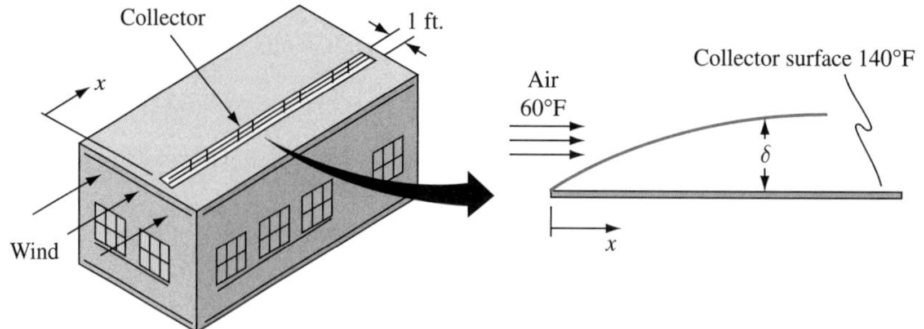

FIGURE 4.15 Flat-plate solar collector for Example 4.3.

(e) thickness of thermal boundary layer
(f) local convection heat transfer coefficient
(g) average convection heat transfer coefficient
(h) rate of heat transfer by convection

SOLUTION The relevant properties of air at 100°F in English units are

$$\begin{aligned} \rho &= 0.071\,\text{lb}_\text{m}/\text{ft}^3 \\ c_p &= 0.240\,\text{Btu/lb}_\text{m}\,^\circ\text{F} \\ \mu &= 1.285 \times 10^{-5}\,\text{lb}_\text{m}/\text{ft s} \\ k &= 0.0154\,\text{Btu/h ft}\,^\circ\text{F} \\ \text{Pr} &= 0.72 \end{aligned}$$

The local Reynolds number at $x = 1$ ft (0.305 m) is

$$\text{Re}_{x=1} = \frac{U_\infty \rho x}{\mu} = \frac{(10\,\text{ft/s})(0.071\,\text{lb}_\text{m}/\text{ft}^3)(1\,\text{ft})}{1.285 \times 10^{-5}\,\text{lb}_\text{m}/\text{ft s}} = 55{,}200$$

and at $x = 9$ ft is

$$\text{Re}_{x=9} = 5 \times 10^5$$

Assuming that the critical Reynolds number is 5×10^5, the critical distance is

$$x_c = \frac{\text{Re}_c \mu}{U_\infty \rho} = \frac{(5 \times 10^5)(1.285 \times 10^{-5}\,\text{lb}_\text{m}/\text{ft s})}{(10\,\text{ft/s})(0.071\,\text{lb}_\text{m}/\text{ft}^3)} = 9.0\,\text{ft}\ (2.8\,\text{m})$$

The desired quantities are determined by substituting appropriate values of the variable into the pertinent equations. The results of the calculations are shown in Table 4.4, and it is suggested that the reader verify them.

TABLE 4.4 Results for Example 4.3

Part	Symbol	Equation Used	English Units	Result ($x = 1$ ft)	Result ($x = 9$ ft)	SI Units	Result ($x = 0.305$ m)	Result ($x = 2.8$ m)
(a)	δ	(4.28)	ft	0.0213	0.0638	m	0.00648	0.0195
(b)	C_{fx}	(4.30)	—	0.00283	0.000942	—	0.00283	0.000942
(c)	$\bar{C}_f$	(4.31)	—	0.00566	0.00189	—	0.00566	0.00189
(d)	τ_s	(4.29)	lb_f/ft^2	3.12×10^{-4}	1.04×10^{-4}	N/m^2	0.0149	0.00497
(e)	δ_{th}	(4.32)	ft	0.0237	0.0712	m	0.00723	0.0217
(f)	h_{cx}	(4.36)	Btu/h ft^2 °F	1.08	0.359	W/m^2 K	6.12	2.04
(g)	$\bar{h}_c$	(4.39)	Btu/h ft^2 °F	2.18	0.718	W/m^2 K	12.23	4.08
(h)	q	(4.35)	Btu/h	172	517	W	50.5	152

A useful relation between the local Nusselt number, Nu_x, and the corresponding friction coefficient, C_{fx}, is obtained by dividing Eq. (4.37) by $Re_x Pr^{1/3}$:

$$\left(\frac{Nu_x}{Re_x Pr}\right) Pr^{2/3} = \frac{0.332}{Re_x^{1/2}} = \frac{C_{fx}}{2} \tag{4.40}$$

The dimensionless ratio $Nu_x/Re_x Pr$ is known as the *Stanton number*, St_x. According to Eq. (4.40), the Stanton number times the Prandtl number raised to the two-thirds power is equal to one-half the value of the friction coefficient. This relation between heat transfer and fluid friction was proposed by Colburn [14] and illustrates the interrelationship of the two processes.

4.9* Approximate Integral Boundary Layer Analysis

To circumvent the problems involved in solving the partial differential equations of the boundary layer, an integral approach can be used. For that purpose, let us consider an elemental control volume that extends from the wall to beyond the limit of the boundary layer in the y direction, is dx thick in the x direction, and has a unit width in the z direction, as shown in Fig. 4.16. To obtain a relationship for the net

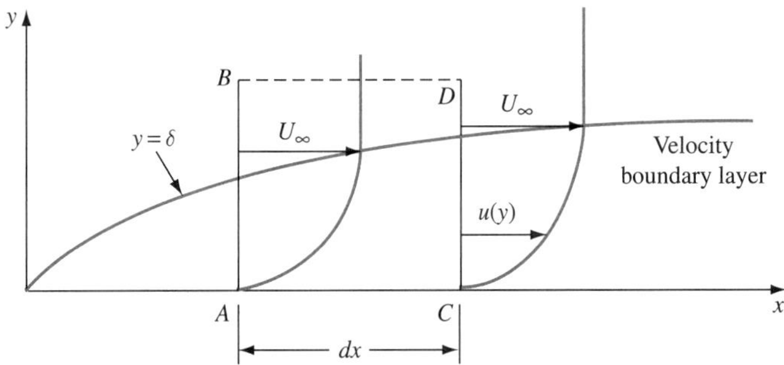

FIGURE 4.16 Control volume for integral conservation of momentum analysis.

momentum inflow and the net energy transport, we proceed in a manner similar to that used to derive the boundary layer equations in the preceding section.

The momentum flow across face AB in Fig. 4.16 will be

$$\int_0^\delta \rho u^2\, dy$$

Similarly, the momentum flow across face CD will be

$$\int_0^\delta \rho u^2\, dy + \frac{d}{dx}\int_0^\delta \rho u^2\, dy\, dx$$

Fluid also enters the control volume across face BD at the rate

$$\frac{d}{dx}\int_0^\delta \rho u\, dy\, dx$$

This quantity is the difference between the rate of flow leaving across face CD and that entering across face AB. Since the fluid entering across BD has a velocity component in the x direction equal to the free-stream velocity U_∞, the flow of x momentum into the control volume across the upper face is

$$U_\infty \frac{d}{dx}\int_0^\delta \rho u\, dy\, dx$$

Adding up the x-momentum components gives

$$\frac{d}{dx}\int_0^\delta \rho u^2\, dy\, dx - U_\infty \frac{d}{dx}\int_0^\delta \rho u\, dy\, dx = -\frac{d}{dx}\int_0^\delta \rho u(U_\infty - u)\, dy$$

There will be no shear across face BD, since this face is outside the boundary layer, where du/dy is equal to zero. There is, however, a shear force τ_w acting at the fluid-solid interface, and there will be pressure forces acting on faces AB and CD. Writing out the net forces acting on the control volume and adding them yields the relation

$$p\delta - \left(p + \frac{dp}{dx}\, dx\right)\delta - \tau_w\, dx = -\delta \frac{dp}{dx}\, dx - \tau_w\, dx \tag{4.41}$$

For flow over a flat plate, the pressure gradient in the x direction can be neglected, and the momentum equation then can be written in the form

$$\frac{d}{dx}\int_0^\delta \rho u(U_\infty - u) dy = \tau_w \tag{4.42}$$

The integral energy equation can be derived in a similar fashion. In this case, however, a control volume extending beyond the limits of both the temperature and the velocity boundary layers must be used in the derivation (see Fig. 4.17). The first

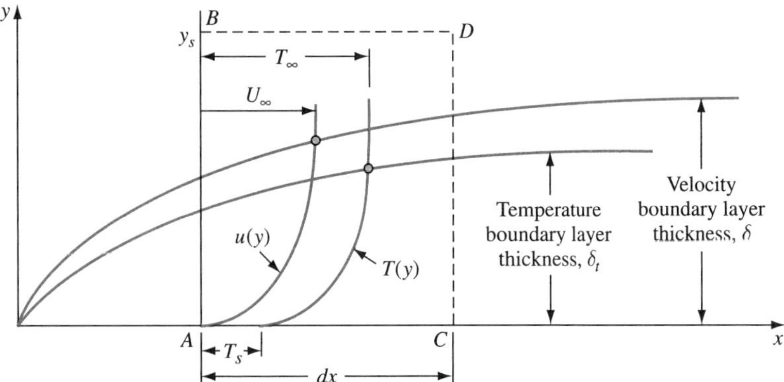

FIGURE 4.17 Control volume for integral conservation of energy analysis.

law of thermodynamics demands that energy in the form of enthalpy, kinetic energy, and heat, as well as shear work, should be considered. For low velocities, however, the kinetic energy terms and the shear work are small compared to the other quantities and can be neglected; then the rate at which enthalpy enters across face *AB* is given by

$$\int_0^{y_s} c_p \rho u T \, dy$$

whereas the rate of enthalpy outflow across face *CD* is

$$\int_0^{y_s} c_p \rho u T \, dy + \frac{d}{dx}\int_0^{y_s} c_p \rho u T dy \, dx$$

The enthalpy carried into the control volume across the upper face is given by

$$c_p T_s \frac{d}{dx}\int_0^{y_s} \rho u \, dy \, dx$$

Finally, heat will be conducted across the interface between the fluid and the solid surface at the rate

$$-k\,dx\left(\frac{\partial T}{\partial y}\right)_{y=0}$$

Adding up all the energy quantities yields the integral equation for the conservation of energy in the form

$$c_p T_\infty \frac{d}{dx}\int_0^{y_s} \rho u \, dy \, dx - \frac{d}{dx}\int_0^{y_s} \rho c_p T u \, dy \, dx - k\,dx\left(\frac{\partial T}{\partial y}\right)_{y=0} = 0 \tag{4.43}$$

It should be noted, however, that outside the limit of the temperature boundary layer, the temperature equals the free-stream temperature, T_∞, so that integration need only be taken up to $y = \delta_t$. Equation (4.43) therefore can be simplified to the form

$$\frac{d}{dx}\int_0^{\delta_t}(T_\infty - T)u\,dy - \alpha\left(\frac{\partial T}{\partial y}\right)_{y=0} = 0 \tag{4.44}$$

which is usually known as the integral energy equation of the laminar boundary layer for low-speed flow.

4.9.1 Evaluation of Heat Transfer and Friction Coefficients in Laminar Flow

In the approximate integral method, the first step is to assume velocity and temperature contours in the form of polynomials. Then, the coefficients of the polynomial are evaluated to satisfy the boundary conditions. Assuming a four-term polynomial for the velocity distribution [15]

$$u(y) = a + by + cy^2 + dy^3 \tag{4.45}$$

the constants are evaluated by applying the boundary conditions

$$\text{at } y = 0\text{:}\quad u = 0 \quad \text{and therefore} \quad a = 0$$

$$u = v = 0 \quad \text{and therefore} \quad \frac{\partial^2 u}{\partial y^2} = 0$$

$$y = \delta\text{:}\quad u = U_\infty \quad \text{and} \quad \frac{\partial u}{\partial y} = 0$$

These conditions provide four equations for the evaluation of the four unknown coefficients in terms of the free-stream velocity and the boundary layer thickness. It easily can be verified that the coefficients that satisfy these boundary conditions are

$$a = 0 \qquad b = \frac{3}{2}\frac{U_\infty}{\delta} \qquad c = 0 \qquad d = -\frac{U_\infty}{2\delta^3}$$

Substituting these coefficients in Eq. (4.45) and dividing through by the free-stream velocity U_∞ to nondimensionalize the result yields

$$\frac{u}{U_\infty} = \frac{3}{2}\frac{y}{\delta} - \frac{1}{2}\left(\frac{y}{\delta}\right)^3 \tag{4.46}$$

Substituting Eq. (4.46) for the velocity distribution in the integral momentum equation [Eq. (4.42)] yields

$$\frac{d}{dx}\int_0^{\delta}\rho U_\infty^2\left[\frac{3}{2}\frac{y}{\delta} - \frac{1}{2}\left(\frac{y}{\delta}\right)^3\right]\cdot\left[1 - \frac{3}{2}\frac{y}{\delta} + \frac{1}{2}\left(\frac{y}{\delta}\right)^3\right]dy = \tau_w = \mu\left(\frac{du}{dy}\right)_{y=0} \tag{4.47}$$

The wall shear stress τ_w can be obtained by evaluating the velocity gradient from Eq. (4.46) at $y = 0$. Substituting for τ_w and performing the integration in Eq. (4.47) yields

$$\frac{d}{dx}\left(\rho U_\infty^2 \frac{39\delta}{280}\right) = \frac{3}{2}\mu\frac{U_\infty}{\delta} \tag{4.48}$$

Equation (4.48) can be rearranged and integrated to obtain the boundary layer thickness in terms of viscosity, distance from the leading edge, and free-stream velocity distribution:

$$\frac{\delta^2}{2} = \frac{140\nu x}{13U_\infty} + C \tag{4.49}$$

Since $\delta = 0$ at the leading edge (i.e., $x = 0$), the coefficient C in the preceding relation must equal 0 and

$$\delta^2 = \frac{280\nu x}{13U_\infty}$$

or

$$\frac{\delta}{x} = \frac{4.64}{\mathrm{Re}_x^{1/2}} \tag{4.50}$$

To evaluate the friction coefficient, substitute Eq. (4.46) into Eq. (4.47):

$$\tau_w = \mu\frac{du}{dy}\bigg|_{y=0} = \mu\frac{3}{2}\frac{U_\infty}{\delta}$$

Substituting for δ from Eq. (4.50) gives

$$\tau_w = \frac{3}{9.28}\frac{\mu U_\infty}{x}\mathrm{Re}_x^{1/2}$$

and the friction coefficient C_{fx} is

$$C_{fx} = \frac{\tau_w}{\frac{1}{2}\rho U_\infty^2} = \frac{0.647}{\mathrm{Re}_x^{1/2}} \tag{4.51}$$

We next turn to the energy equation and propose a temperature distribution in the boundary layer of the same form as the velocity distribution:

$$T(y) = e + fy + gy^2 + hy^3 \tag{4.52}$$

The boundary conditions for the temperature distribution are that at $y = 0$, $T = T_s$; at $y = \delta_t$ (the thickness of the temperature boundary layer), $T = T_\infty$ and $dT/dy = 0$. Also, from Eq. (4.7b), d^2T/dy^2 at $y = 0$ must be zero because both u and v are zero at the interface. From these conditions, it follows that the constants are

$$e = T_s \qquad f = \frac{3}{2}\frac{(T_\infty - T_s)}{\delta_t} \qquad g = 0 \qquad h = \frac{(T_\infty - T_s)}{2\delta_t^3}$$

If the variable in the energy equation is taken as the temperature in the fluid minus the wall temperature, the temperature distribution can be written in the dimensionless form

$$\frac{T - T_s}{T_\infty - T_s} = \frac{3}{2}\frac{y}{\delta_t} - \frac{1}{2}\left(\frac{y}{\delta_t}\right)^3 \tag{4.53}$$

Using Eqs. (4.53) and (4.46) for $T - T_s$ and u, respectively, the integral in Eq. (4.44) can be written as

$$\int_0^{\delta_t} (T_\infty - T)u\,dy = \int_0^{\delta_t} [(T_\infty - T_s) - (T - T_s)]u\,dy$$

$$= (T_\infty - T_s)U_\infty \int_0^{\delta_t} \left[1 - \frac{3}{2}\frac{y}{\delta_t} + \frac{1}{2}\left(\frac{y}{\delta_t}\right)^3\right]\left[\frac{3}{2}\frac{y}{\delta} - \frac{1}{2}\left(\frac{y}{\delta}\right)^3\right]dy$$

Performing the multiplication under the integral sign, we obtain

$$(T_\infty - T_s)U_\infty \int_0^{\delta_t} \left(\frac{3}{2\delta}y - \frac{9}{4\delta\delta_t}y^2 + \frac{3}{4\delta\delta_t^3}y^4 - \frac{1}{2\delta^3}y^3 + \frac{3}{4\delta_t\delta^3}y^4 - \frac{1}{4\delta_t^3\delta^3}y^6\right)dy$$

which yields, after integration,

$$(T_\infty - T_s)U_\infty\left(\frac{3}{4}\frac{\delta_t^2}{\delta} - \frac{3}{4}\frac{\delta_t^2}{\delta} + \frac{3}{20}\frac{\delta_t^2}{\delta} - \frac{1}{8}\frac{\delta_t^4}{\delta^3} + \frac{3}{20}\frac{\delta_t^4}{\delta^3} - \frac{1}{28}\frac{\delta_t^4}{\delta^3}\right)$$

If we let $\zeta = \delta_t/\delta$, the above expression can be written as

$$(T_\infty - T_s)U_\infty\delta\left(\frac{3}{20}\zeta^2 - \frac{3}{280}\zeta^4\right)$$

For fluids having a Prandtl number equal to or larger than unity, ζ is equal to or less than unity and the second term in the parentheses can be neglected compared to the first.* Substituting this approximate form for the integral into Eq. (4.44), we obtain

$$\frac{3}{20}U_\infty(T_\infty - T_s)\zeta^2\frac{\partial\delta}{\partial x} = \alpha\frac{\partial T}{\partial y}\bigg|_{y=0} = \frac{3}{2}\alpha\frac{T_\infty - T_s}{\delta\zeta}$$

or

$$\frac{1}{10}U_\infty\zeta^3\delta\frac{\partial\delta}{\partial x} = \alpha$$

*For liquid metals, which have Pr $\ll 1$, $\zeta > 1$ and the second term cannot be neglected.

From Eq. (4.50), we obtain

$$\delta \frac{\partial \delta}{\partial x} = 10.75 \frac{\nu}{U_\infty}$$

and with this expression we get

$$\zeta^3 = \frac{10}{10.75} \frac{\alpha}{\nu}$$

or

$$\delta_t = 0.976 \delta \, \mathrm{Pr}^{-1/3} \tag{4.54}$$

Except for the numerical constant (0.976 compared with 1.0), the foregoing result is in agreement with the exact calculation of Pohlhausen [12].

The rate of heat flow by convection from the plate per unit area is, from Eqs. (4.1) and (4.53),

$$q_c'' = -k \frac{\partial T}{\partial y}\bigg|_{y=0} = -\frac{3}{2} \frac{k}{\delta_t} (T_\infty - T_s)$$

Substituting Eqs. (4.50) and (4.54) for δ and δ_t yields

$$q'' = -\frac{3}{2} \frac{k}{x} \frac{\mathrm{Pr}^{1/3} \mathrm{Re}_x^{1/2}}{(0.976)(4.64)} (T_\infty - T_s) = 0.33 \frac{k}{x} \mathrm{Re}_x^{1/2} \mathrm{Pr}^{1/3} (T_s - T_\infty) \tag{4.55}$$

and the local Nusselt number, Nu_x, is

$$\mathrm{Nu}_x = \frac{h_{cx} x}{k} = \frac{q_c''}{(T_s - T_\infty)} \frac{x}{k} = 0.33 \mathrm{Re}_x^{1/2} \mathrm{Pr}^{1/3} \tag{4.56}$$

This result is in excellent agreement with Eq. (4.37), the result of an exact analysis of Pohlhausen [12].

The foregoing example illustrates the usefulness of the approximate boundary layer analysis. Guided by a little physical insight and intuition, this technique yields satisfactory results without the mathematical complications inherent in the exact boundary layer equations. The approximate method has been applied to many other problems, and the results are available in the literature.

4.10* Analogy Between Momentum and Heat Transfer in Turbulent Flow Over a Flat Surface

In a majority of practical applications, the flow in the boundary layer is turbulent rather than laminar. Qualitatively, the exchange mechanism in turbulent flow can be pictured as a magnification of the molecular exchange in laminar flow. In steady laminar flow, fluid particles follow well-defined streamlines. Heat and momentum are transferred across streamlines only by molecular diffusion, and the cross flow is so small that when a colored dye is injected into the fluid at some point, it follows a streamline without appreciable mixing. In turbulent flow, on the other hand, the color will be distributed over a wide area a short distance downstream from the point of injection. The mixing mechanism consists of rapidly fluctuating eddies that transport fluid particles in an irregular manner. Groups of particles collide with each other at random, establish cross flow on a macroscopic scale, and effectively mix the fluid. Since the mixing in turbulent flow is on a macroscopic scale with groups of particles transported in a zigzag path through the fluid, the exchange mechanism is many times more effective than in laminar flow. As a result, the rates of heat and momentum transfer in turbulent flow and the associated friction and heat transfer coefficients are many times larger than in laminar flow.

If turbulent flow at a point is averaged over a long period of time (compared with the period of a single fluctuation), the time-mean properties and the velocity of the fluid are constant if the average flow remains steady. It is therefore possible to describe each fluid property and the velocity in turbulent flow in terms of a *mean value* that does not vary with time and a *fluctuating component* that is a function of time. To simplify the problem, consider a two-dimensional flow (Fig. 4.18) in which the mean value of the velocity is parallel to the x direction. The instantaneous velocity components u and v then can be expressed in the form

$$u = \bar{u} + u' \qquad (4.57)$$

$$v = v'$$

where the bar over a symbol denotes the temporal mean value, and the prime denotes the instantaneous deviation from the mean value. According to the model used to describe the flow,

$$\bar{u} = \frac{1}{t^*}\int_0^{t^*} u\,dt \qquad (4.58)$$

where t^* is a time interval large compared with the period of the fluctuations. Figure 4.19 shows qualitatively the time variation of u and u'. From Eq. (4.58) or from an inspection of the graph, it is apparent that the time average of u' is zero (i.e., $\overline{u'} = 0$). A similar argument shows that $\overline{v'}$ and $\overline{(\rho v')}$ are also zero.

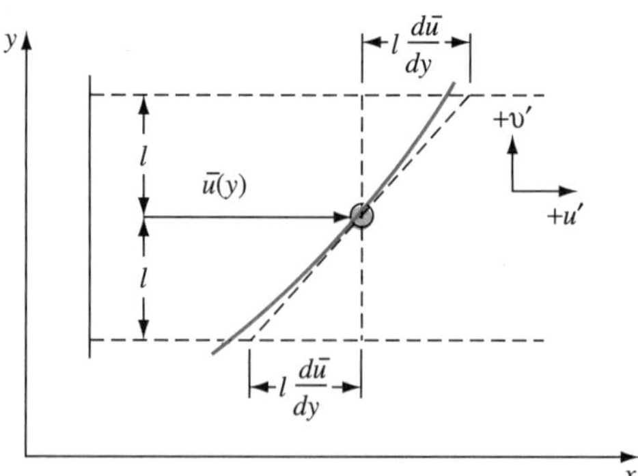

FIGURE 4.18 Mixing length for momentum transfer in turbulent flow.

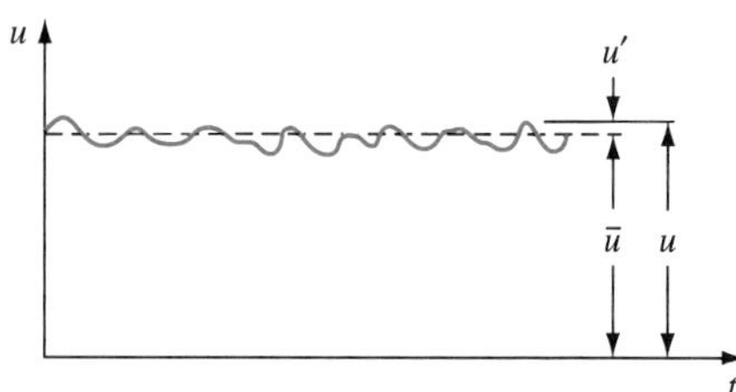

FIGURE 4.19 Time variation of instantaneous velocity in turbulent flow.

The fluctuating velocity components continuously transport mass, and consequently momentum, across a plane normal to the y direction. The instantaneous rate of transfer in the y direction of x momentum per unit area at any point is

$$-(\rho v)'(\bar{u} + u')$$

where the minus sign, as will be shown later, takes account of the statistical correlation between u' and v'.

The time average of the x-momentum transfer gives rise to an *apparent* turbulent shear or Reynolds stress τ_t, defined by

$$\tau_t = -\frac{1}{t^*}\int_0^{t^*} (\rho v)'(\bar{u} + u')\, dt \tag{4.59}$$

Breaking this term up into parts, the time average of the first is

$$\frac{1}{t^*}\int_0^{t^*} (\rho v)'\bar{u}\, dt = 0$$

since $\bar{u}$ is a constant and the time average of $(\rho v)'$ is zero. Integrating the second term, Eq. (4.59) becomes

$$\tau_t = -\frac{1}{t^*}\int_0^{t^*} (\rho v)'u'\, dt = -\overline{(\rho v)'u'} \tag{4.60}$$

or, if ρ is constant,

$$\tau_t = -\rho\overline{(v'u')} \tag{4.61}$$

where $\overline{(v'u')}$ is the time average of the product of u' and v'.

It is not difficult to visualize that the time averages of the mixed products of velocity fluctuations, such as $\overline{v'u'}$, differ from zero. From Fig. 4.18, we can see that the particles that travel upward ($v' > 0$) arrive at a layer in the fluid in which the mean velocity $\bar{u}$ is larger than in the layer from which they come. Assuming that the fluid particles preserve, on the average, their original velocity $\bar{u}$ during their migration, they will tend to slow down other fluid particles after they have reached their destination and thereby give rise to a negative component u'. Conversely, if v' is negative, the observed value of u' at the new destination will be positive. On the average, therefore, a positive v' is associated with a negative u', and vice versa. The time average of $\overline{u'v'}$ is therefore on the average not zero but a negative quantity. The turbulent shearing stress defined by Eq. (4.61) is thus positive and has the same sign as the corresponding laminar shearing stress,

$$\tau_l = \mu \frac{d\bar{u}}{dy} = \rho \nu \frac{d\bar{u}}{dy}$$

It should be noted, however, that the laminar shearing stress is a true stress, whereas the apparent turbulent shearing stress is simply a concept introduced to account for the effects of the momentum transfer by turbulent fluctuations. This concept allows us to express the total shear stress in turbulent flow as

$$\tau = \frac{\text{viscous force}}{\text{area}} + \text{turbulent momentum flux} \tag{4.62}$$

To relate the turbulent momentum flux to the time-average velocity gradient $d\bar{u}/dy$, we postulate that fluctuations of macroscopic fluid particles in turbulent flow are, on the average, similar to the motion of molecules in a gas [i.e., they travel, on the average, a distance l perpendicular to $\bar{u}$ (Fig. 4.18) before coming to rest in another y plane]. This distance l is known as *Prandtl's mixing length* [16, 17] and corresponds qualitatively to the mean-free path of a gas molecule. Assuming that the fluid particles retain their identity and physical properties during the cross motion and that the turbulent fluctuation arises chiefly from the difference in the time-mean properties between y planes spaced a distance l apart, if a fluid particle travels from a layer y to a layer $y + l$,

$$u' \simeq l \frac{d\bar{u}}{dy} \tag{4.63}$$

With this model, the turbulent shearing stress τ_t in a form analogous to the laminar shearing stress is

$$\tau_t = -\rho \overline{v'u'} = \rho \varepsilon_M \frac{d\bar{u}}{dy} \tag{4.64}$$

where the symbol ε_M is called the *eddy viscosity* or the turbulent exchange coefficient for momentum. The eddy viscosity ε_M is formally analogous to the kinematic viscosity ν, but whereas ν is a physical property, ε_M depends on the dynamics of the flow. Combining Eqs. (4.63) and (4.64) shows that $\varepsilon_M = -v'l$, and Eq. (4.62) gives the total shearing stress in the form

$$\tau = \rho(\nu + \varepsilon_M) \frac{d\bar{u}}{dy} \tag{4.65}$$

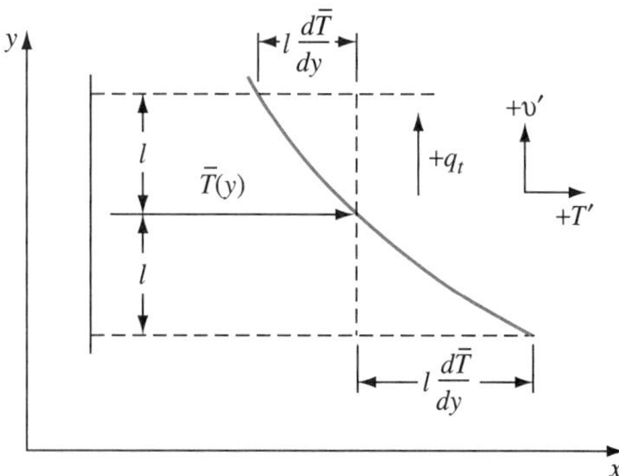

FIGURE 4.20 Mixing length for energy transfer in turbulent flow.

In turbulent flow, ε_M is much larger than ν and the viscous term can therefore be neglected.

The transfer of energy as heat in a turbulent flow can be pictured in an analogous fashion. Let us consider a two-dimensional time-mean temperature distribution as shown in Fig. 4.20. The fluctuating velocity components continuously transport fluid particles and the energy stored in them across a plane normal to the y direction. The instantaneous rate of energy transfer per unit area at any point in the y direction is

$$(\rho v')(c_p T) \tag{4.66}$$

where $T = \bar{T} + T'$. Following the same line of reasoning that led to Eq. (4.61), the time average of energy transfer due to the fluctuations, called the *turbulent rate of heat transfer, q_t,* is

$$q_t = A\rho c_p \overline{v'T'} \tag{4.67}$$

Using Prandtl's concept of mixing length, we can relate the temperature fluctuation to the time-mean temperature gradient by the equation

$$T' \simeq l\frac{d\bar{T}}{dy} \tag{4.68}$$

This means physically that when a fluid particle migrates from layer y to another layer a distance l above or below, the resulting temperature fluctuation is caused chiefly by the difference between the time-mean temperatures in the layers. Assuming that the transport mechanisms of temperature (or energy) and velocity are similar, the mixing lengths in Eqs. (4.63) and (4.68) are equal. The product $\overline{v'T'}$, however, is positive on the average because a positive v' is accompanied by a positive T', and vice versa.

Combining Eqs. (4.67) and (4.68), the turbulent rate of heat transfer per unit area becomes

$$q_t'' = \frac{q_t}{A} = c_p\rho\overline{v'T'} = -\,c_p\rho\overline{v'l}\,\frac{d\bar{T}}{dy} \tag{4.69}$$

where the minus sign is a consequence of the second law of thermodynamics (see Chapter 1). To express the turbulent heat flux in a form analogous to the Fourier conduction equation, we define ε_H, a quantity called the *turbulent exchange coefficient for temperature* or *eddy diffusivity of heat* by the equation $\varepsilon_H = \overline{v'l}$. Substituting ε_H for $\overline{v'l}$ in Eq. (4.69) gives

$$q_t'' = -c_p \rho \varepsilon_H \frac{d\bar{T}}{dy} \tag{4.70}$$

The total rate of heat transfer per unit area normal to the mean stream velocity can then be written as

$$q'' = \frac{q}{A} = \frac{\text{molecular conduction}}{\text{unit area}} + \frac{\text{turbulent transfer}}{\text{unit area}}$$

or in symbolic form as

$$q'' = -c_p \rho (\alpha + \varepsilon_H) \frac{d\bar{T}}{dy} \tag{4.71}$$

where $\alpha = k/c_p\rho$, which is the molecular diffusivity of heat. The contribution to the heat transfer by molecular conduction is proportional to α, and the turbulent contribution is proportional to ε_H. For all fluids except liquid metals, ε_H is much larger than α in turbulent flow. The ratio of the molecular kinematic viscosity to the molecular diffusivity of heat, ν/α, has previously been named the Prandtl number. Similarly, the ratio of the turbulent eddy viscosity to the eddy diffusivity $\varepsilon_M/\varepsilon_H$ could be considered a turbulent Prandtl number Pr_t. According to the Prandtl mixing-length theory, the turbulent Prandtl number is unity, since $\varepsilon_M = \varepsilon_H = \overline{v'l}$.

Although this treatment of turbulent flow is oversimplified, experimental results indicate it is at least qualitatively correct. Isakoff and Drew [18] found that Pr_t for the heating of mercury in turbulent flow inside a tube may vary from 1.0 to 1.6, and Forstall and Shapiro [19] found that Pr_t is about 0.7 for gases. The latter investigators also showed that Pr_t is substantially independent of the value of the laminar Prandtl number as well as of the type of experiment. Assuming that Pr_t is unity, the turbulent heat flux can be related to the turbulent shear stress by combining Eqs. (4.64) and (4.70):

$$q_t'' = -\tau_t c_p \frac{d\bar{T}}{d\bar{u}} \tag{4.72}$$

This relation was originally derived in 1874 by the British scientist Osborn Reynolds and is called the *Reynolds analogy*. It is a good approximation whenever the flow is turbulent and can be applied to turbulent boundary layers as well as to turbulent flow in pipes or ducts. However, the Reynolds analogy does not hold in the viscous sublayer [20]. Since this layer offers a large thermal resistance to the flow of heat, Eq. (4.72) does not, in general, suffice for a quantitative solution. Only for fluids having a Prandtl number of unity can it be used directly to calculate the rate of heat transfer. This special case will now be considered.

4.11 Reynolds Analogy for Turbulent Flow Over Plane Surfaces

To derive a relation between the heat transfer and the skin friction in flow over a plane surface for a Prandtl number of unity, we recall that the laminar shearing stress τ is

$$\tau = \mu \frac{du}{dy}$$

and the rate of heat flow per unit area across any plane perpendicular to the y direction is

$$q'' = -k \frac{dT}{dy}$$

Combining these equations yields

$$q'' = -\tau \frac{k}{\mu} \frac{dT}{du} \tag{4.73}$$

An inspection of Eqs. (4.72) and (4.73) shows that if $c_p = k/\mu$ (i.e., for Pr $= 1$), the same equation of heat flow applies in the laminar and turbulent layers.

To determine the rate of heat transfer from a flat plate to a fluid with Pr $= 1$ flowing over it in turbulent flow, we replace k/μ by c_p and separate the variables in Eq. (4.73). Assuming that q'' and τ are constant, we get

$$\frac{q_s''}{\tau_s c_p} du = -dT \tag{4.74}$$

where the subscript s is used to indicate that both q'' and τ are taken at the surface of the plate. Integrating Eq. (4.74) between the limits $u = 0$ when $T = T_s$ and $u = U_\infty$ when $T = T_\infty$ yields

$$\frac{q_s''}{\tau_s c_p} U_\infty = (T_s - T_\infty) \tag{4.75}$$

But since by definition the local heat-transfer and friction coefficients are

$$h_{cx} = \frac{q_s''}{(T_s - T_\infty)} \quad \text{and} \quad \tau_{sx} = C_{fx} \frac{\rho U_\infty^2}{2}$$

Eq. (4.75) can be written

$$\frac{h_{cx}}{c_p \rho U_\infty} = \frac{\text{Nu}_x}{\text{Re}_x \text{Pr}} = \frac{C_{fx}}{2} \tag{4.76}$$

Equation (4.76) is satisfactory for gases in which Pr is approximately unity. Colburn [14] has shown that Eq. (4.76) also can be used for fluids having Prandtl numbers ranging from 0.6 to about 50 if it is modified in accordance with experimental results to read

$$\frac{\mathrm{Nu}_x}{\mathrm{Re}_x\mathrm{Pr}}\mathrm{Pr}^{2/3} = \mathrm{St}_x\mathrm{Pr}^{2/3} = \frac{C_{fx}}{2} \tag{4.77}$$

where the subscript x denotes the distance from the leading edge of the plate.

To apply the analogy between heat transfer and momentum transfer in practice, it is necessary to know the friction coefficient C_{fx}. For turbulent flow over a plane surface, the empirical equation for the local friction coefficient

$$C_{fx} = 0.0576\left(\frac{U_\infty x}{\nu}\right)^{-1/5} \tag{4.78a}$$

is in good agreement with experimental results in the Reynolds number rangeing between 5×10^5 and 10^7 as long as no separation occurs. Assuming that the turbulent boundary layer starts at the leading edge, the average friction coefficient over a plane surface of length L can be obtained by integrating Eq. (4.78a):

$$\bar{C}_f = \frac{1}{L}\int_0^L C_{fx}\,dx = 0.072\left(\frac{U_\infty L}{\nu}\right)^{-1/5} \tag{4.78b}$$

Using the integral method in Section 4.9 combined with experimental data for the friction coefficient, one can also derive an expression for the boundary layer thickness in turbulent flow. According to Schlichting [1], the hydrodynamic boundary layer thickness can be approximated by the relation

$$(\delta/x) = 0.37/\mathrm{Re}_x^{0.2} \tag{4.79}$$

Comparing Eqs. (4.78a) and (4.79) with the results for laminar flow, Eqs. (4.30) and (4.28), it is apparent that the decay of the friction coefficient with distance is more gradual in turbulent than in laminar flow ($x^{-0.2}$ versus $x^{-0.5}$), while the boundary layer thickness increases more rapidly in turbulent than in laminar flow ($\delta_t \propto x^{0.8}$ versus $\delta \propto x^{0.5}$).

In turbulent flow, the boundary layer growth is influenced more by random fluctuations in the fluid than by molecular diffusion. Hence, the Prandtl number does not play a major part in the process, and for $\mathrm{Pr} > 0.5$, Eq. (4.79) is also a reasonable approximation for the thermal boundary layer thickness, that is, $\delta \approx \delta_t$ in turbulent flow [21].

4.12 Mixed Boundary Layer

In reality, a laminar boundary layer precedes the turbulent boundary layer between $x = 0$ and $x = x_c$. Since the local frictional drag of a laminar boundary layer is less than the local frictional drag of a turbulent boundary layer at the same Reynolds number, the average drag calculated from Eq. (4.78b) without correcting for the laminar portion of the boundary layer is too large. The actual drag can be closely estimated, however, by assuming that behind the point of transition, the turbulent boundary layer behaves as though it had started at the leading edge.

Adding the laminar friction drag between $x = 0$ and $x = x_c$ to the turbulent drag between $x = x_c$ and $x = L$ gives, per unit width,

$$\bar{C}_f = \frac{[0.072\mathrm{Re}_L^{-1/5}L - 0.072\mathrm{Re}_{x_c}^{-1/5}x_c + 1.33\mathrm{Re}_{x_c}^{-1/2}x_c]}{L}$$

For a critical Reynolds number of 5×10^5, this yields

$$\bar{C}_f = 0.072\left(\mathrm{Re}_L^{-1/5} - \frac{0.0464x_c}{L}\right) \tag{4.80}$$

Substituting Eq. (4.78a) for C_{fx} in Eq. (4.77) yields the local Nusselt number at any value of x larger than x_c:

$$\mathrm{Nu}_x = \frac{h_{cx}x}{k} = 0.0288\mathrm{Pr}^{1/3}\left(\frac{U_\infty x}{\nu}\right)^{0.8} \tag{4.81}$$

We observe that the local heat transfer coefficient h_{cx} for heat transfer by convection through a turbulent boundary layer decreases with the distance x as $h_{cx} \propto 1/x^{0.2}$. Equation (4.81) shows that in comparison with laminar flow, where $h_{cx} \propto 1/x^{1/2}$, the heat transfer coefficient in turbulent flow decreases less rapidly with x and that the turbulent heat transfer coefficient is much larger than the laminar heat transfer coefficient at a given value of the Reynolds number.

The average heat transfer coefficient in turbulent flow over a plane surface of length L can be calculated to a first approximation by integrating Eq. (4.81) between $x = 0$ and $x = L$:

$$\bar{h}_c = \frac{1}{L}\int_0^L h_{cx}\,dx$$

In dimensionless form, we get

$$\overline{\mathrm{Nu}}_L = \frac{\bar{h}_c L}{k} = 0.036\mathrm{Pr}^{1/3}\mathrm{Re}_L^{0.8} \tag{4.82}$$

Equation (4.82) neglects the existence of the laminar boundary layer and is therefore valid only when $L \gg x_c$. The laminar boundary layer can be included in the analysis if Eq. (4.56) is used between $x = 0$ and $x = x_c$ and if Eq. (4.81) is between $x = x_c$ and $x = L$ for the integration of h_{cx}. This yields, with $\mathrm{Re}_c = 5 \times 10^5$,

$$\overline{\mathrm{Nu}}_L = 0.036\mathrm{Pr}^{1/3}(\mathrm{Re}_L^{0.8} - 23{,}200) \tag{4.83}$$

EXAMPLE 4.4 The crankcase of an automobile is approximately 0.6 m long, 0.2 m wide, and 0.1 m deep (see Fig. 4.21). Assuming that the surface temperature of the crankcase is 350 K, estimate the rate of heat flow from the crankcase to atmospheric air at

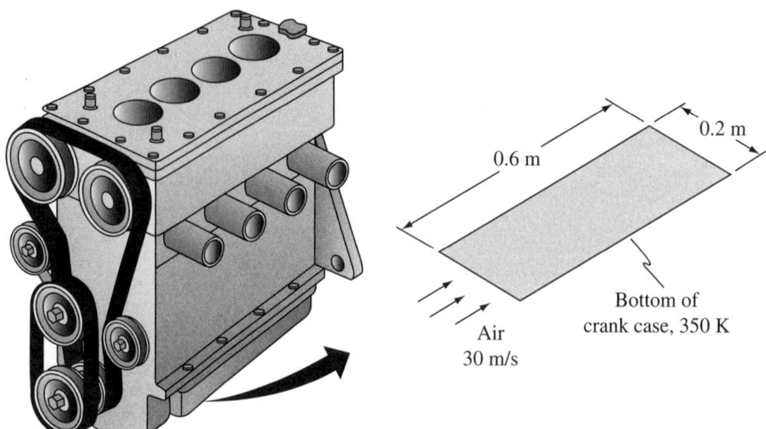

FIGURE 4.21 Automobile crankcase for Example 4.4.

276 K at a road speed of 30 m/s. Assume that the vibration of the engine and the chassis induce the transition from laminar to turbulent flow so near to the leading edge that, for practical purposes, the boundary layer is turbulent over the entire surface. Neglect radiation and use for the front and rear surfaces the same average convection heat transfer coefficient as for the bottom and sides.

SOLUTION Using the properties of air at 313 K, the Reynolds number is

$$\mathrm{Re}_L = \frac{\rho U_\infty L}{\mu} = \frac{1.092 \times 30 \times 0.6}{19.123 \times 10^{-6}} = 1.03 \times 10^6$$

From Eq. (4.82), the average Nusselt number is

$$\begin{aligned} \overline{\mathrm{Nu}}_L &= 0.036\mathrm{Pr}^{1/3}\mathrm{Re}_L^{0.8} \\ &= 0.036(0.71)^{1/3}(1.03 \times 10^6)^{0.8} \\ &= 2075 \end{aligned}$$

and the average convection heat transfer coefficient becomes

$$\bar{h}_c = \frac{\overline{\mathrm{Nu}}_L k}{L} = \frac{2075 \times (0.0265\,\mathrm{W/mK})}{(0.6\,\mathrm{m})} = 91.6\,\mathrm{W/m^2K}$$

The surface area that dissipates heat is 0.28 m^2, and the rate of heat loss from the crankcase is therefore

$$\begin{aligned} q_c &= \bar{h}_c A(T_s - T_\infty) = (91.6\,\mathrm{W/m^2K})(0.28\,\mathrm{m^2})(350 - 276)(\mathrm{K}) \\ &= 1898\,\mathrm{W} \end{aligned}$$

4.13* Special Boundary Conditions and High-Speed Flow

In the preceding sections of this chapter, the boundary condition at the surface assumed a constant and uniform temperature. There are, however, other situations of practical interest, for example, a surface with constant heat flux and a flat plate with an unheated starting length followed by a uniform surface temperature above ambient.

The boundary conditions for a flat plate with flow over its surface and an unheated starting length is illustrated in Fig. 4.22. The hydrodynamic boundary layer begins at $x = 0$, while the thermal boundary layer begins at $x = \zeta$, where the surface temperature increases from T_∞ to T_s. No heat transfer occurs between $0 \le x < \zeta$. For this case, assuming laminar flow, the integral method gives

$$\mathrm{Nu}_x/\mathrm{Nu}_{x,\zeta=0} = \{1 - (\zeta/x)^{3/4}\}^{-0.33} \tag{4.84}$$

where $\mathrm{Nu}_{x,\zeta=0}$ is given by Eq. (4.56).

For turbulent flow with an unheated starting length,

$$\mathrm{Nu}_x/\mathrm{Nu}_{x,\zeta=0} = \{1 - (\zeta/x)^{0.9}\}^{-0.9} \tag{4.85}$$

where $\mathrm{Nu}_{x,\zeta=0}$ is given by Eq. (4.81).

When flow occurs over a flat plate with a constant heat flux imposed (for example, by electric heating), the Nusselt number in laminar flow is given by

$$\mathrm{Nu}_x = 0.453\mathrm{Re}_x^{0.5}\mathrm{Pr}^{0.33} \tag{4.86}$$

while for turbulent flow

$$\mathrm{Nu}_x = 0.0308\mathrm{Re}_x^{0.8}\mathrm{Pr}^{0.33} \tag{4.87}$$

For an unheated starting length with constant heat flux at $x > \zeta$, Eqs. (4.86) and (4.87) can be modified by using correction factors equal to the right-hand sides of Eqs. (4.84) and (4.85), respectively.

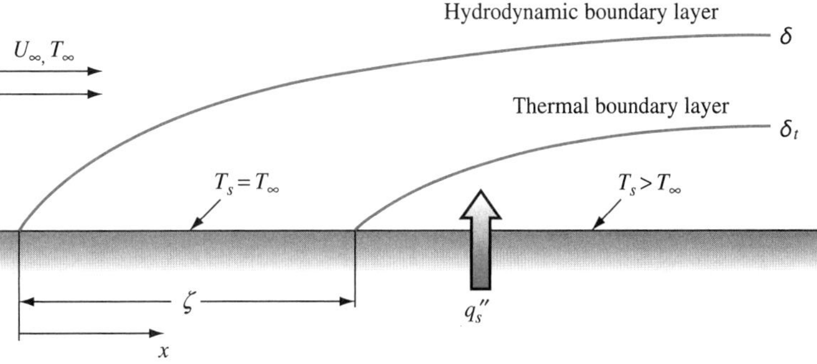

FIGURE 4.22 Flat plate in parallel flow with unheated starting length.

With constant heat flux, the surface temperature is not uniform, but once the heat transfer coefficient is known, the surface temperature can be obtained from

$$T_s(x) = T_\infty + (q_s''/h_x) \tag{4.88}$$

at any given location.

Another special surface condition occurs in high-speed flow. Convection heat transfer in high-speed flow is important for aircraft and missiles when the velocity approaches or exceeds the velocity of sound. For a perfect gas the sound velocity a can be obtained from

$$a = \sqrt{\frac{\gamma \mathcal{R}_u T}{\mathcal{M}}} \tag{4.89}$$

where γ = specific heat ratio, c_p/c_v (about 1.4 for air)
$\mathcal{R}_u$ = universal gas constant
T = absolute temperature
$\mathcal{M}$ = molecular weight of the gas

When the velocity of a gas flowing over a heated or cooled surface is of the order of the velocity of sound or larger, the flow field can no longer be described solely in terms of the Reynolds number; instead, the ratio of the gas-flow velocity to the accoustical velocity—the Mach number, $M = U_\infty/a_\infty$—also must be considered. When the gas velocity in a flow system reaches a value of about half the speed of sound, the effects of viscous dissipation in the boundary layer become inportant. Under such conditions, the temperature of a surface over which a gas is flowing can actually exceed the free-stream temperature. For flow over an adiabatic surface, such as a perfectly insulated wall, Fig. 4.23 shows the velocity and temperature distributions schematically. The high temperature at the surface is the combined result of the heating due to viscous dissipation and the temperature rise of the fluid as the kinetic energy of the flow is converted to internal energy, while the flow decelerates through the boundary layer. The actual shape of the temperature profile depends on the relation between the rate at which viscous shear work increases the internal energy of the fluid and the rate at which heat is conducted toward the free stream.

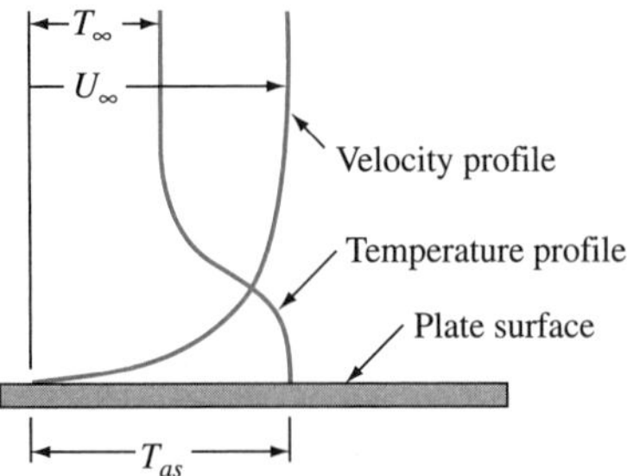

FIGURE 4.23 Velocity and temperature distribution in high-speed flow over an insulated plate.

Although the processes in a high-speed boundary layer are not adiabatic, it is general practice to relate them to adiabatic processes. The conversion of kinetic energy in a gas being slowed down adiabatically to zero velocity is described by

$$i_0 = i_\infty + \frac{U_\infty^2}{2g_c} \tag{4.90}$$

where i_0 is the stagnation enthalpy and i_∞ is the enthalpy of the gas in the free stream. For an ideal gas, Eq. (4.90) becomes

$$T_0 = T_\infty + \frac{U_\infty^2}{2g_c c_p} \tag{4.91}$$

or, in terms of the Mach number,

$$\frac{T_0}{T_\infty} = 1 + \frac{\gamma - 1}{2} M_\infty^2 \tag{4.92}$$

where T_0 is the stagnation temperature and T_∞ is the free-stream temperature.

In a real boundary layer, the fluid is not brought to rest reversibly because the viscous shearing process is thermodynamically irreversible. To account for the irreversibility in a boundary layer flow, we define a recovery factor r as

$$r = \frac{T_{as} - T_\infty}{T_0 - T_\infty} \tag{4.93}$$

where T_{as} is the adiabatic surface temperature.

Experiments [22] have shown that in laminar flow:

$$r = \mathrm{Pr}^{1/2} \tag{4.94}$$

whereas in turbulent flow:

$$r = \mathrm{Pr}^{1/3} \tag{4.95}$$

When a surface is not insulated, the rate of heat transfer by convection between a high-speed gas and that surface is governed by the relation

$$q_c'' = -k \left. \frac{\partial T}{\partial y} \right|_{y=0}$$

The influence of heat transfer to and from the surface on the temperature distribution is illustrated in Fig. 4.24. We observe that in high-speed flow heat can be transferred to the surface even when the surface temperature is above the free-stream temperature. This phenomenon is the result of viscous shear and is often called aerodynamic heating. The heat transfer rate in high-speed flow over a flat surface can be predicted [1] from the boundary layer energy equation:

$$u\frac{\partial T}{\partial x} + v\frac{\partial T}{\partial y} = \alpha\frac{\partial^2 T}{\partial y^2} + \frac{\mu}{\rho c_p}\left(\frac{\partial u}{\partial x}\right)^2$$

where the last term accounts for the viscous dissipation. However, for most practical purposes, the rate of heat transfer can be calculated with the same relations used

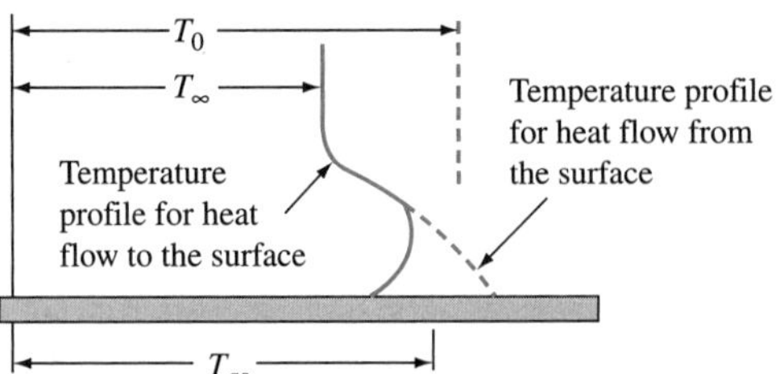

FIGURE 4.24 Temperature profiles in a high-speed boundary layer for heating and cooling.

for low-speed flow if the average convection heat transfer coefficient is redefined by the relation

$$q_c'' = \bar{h}_c(T_s - T_{as}) \tag{4.96}$$

which will yield zero heat flow when the surface temperature T_s equals the adiabatic surface temperature.

Since in high-speed flow the temperature gradients in a boundary layer are large, variations in the physical properties of the fluid also will be substantial. Eckert [23] has shown, however, that the constant-property heat transfer equations still can be used if all the properties are evaluated at a reference temperature T^* given by

$$T^* = T_\infty + 0.5(T_s - T_\infty) + 0.22(T_{as} - T_\infty) \tag{4.97}$$

The local values of the heat transfer coefficient, defined by

$$h_{cx} = \frac{q_c''}{T_s - T_{as}} \tag{4.98}$$

can be obtained from the following equations.

Laminar Boundary Layer ($\text{Re}_x^* < 10^5$):

$$\text{St}_x^* = \left(\frac{h_{cx}}{c_p \rho U_\infty}\right)^* = 0.332(\text{Re}_x^*)^{-1/2}(\text{Pr}^*)^{-2/3} \tag{4.99}$$

Turbulent Boundary Layer ($10^5 < \text{Re}_x^* < 10^7$):

$$\text{St}_x^* = \left(\frac{h_{cx}}{c_p \rho U_\infty}\right)^* = 0.0288(\text{Re}_x^*)^{-1/5}(\text{Pr}^*)^{-2/3} \tag{4.100}$$

Turbulent Boundary Layer ($10^7 < \text{Re}_x^* < 10^9$):

$$\text{St}_x^* = \left(\frac{h_{cx}}{c_p \rho U_\infty}\right)^* = \frac{2.46}{(\ln \text{Re}_x^*)^{2.584}}(\text{Pr}^*)^{-2/3} \tag{4.101}$$

Equation (4.101) is based on experimental data for local friction coefficients in high-speed gas flow [23] in the Reynolds number range 10^7 to 10^9 that are correlated by

$$C_{fx} = \frac{4.92}{(\ln \mathrm{Re}_x^*)^{2.584}} \tag{4.102}$$

If the average value of the heat transfer coefficients is to be determined, the expressions above must be integrated between $x = 0$ and $x = L$. However, the integration may have to be done numerically in most practical cases because the reference temperature T^* is not the same for the laminar and turbulent portions of the boundary layer, as shown by Eqs. (4.94) and (4.95).

When the speed of a gas is exceedingly high, the boundary layer may become so hot that the gas begins to dissociate. In such situations, Eckert [23] recommends that the heat transfer coefficient be based on the enthalpy difference between the surface and the adiabatic state, $(i_s - i_{as})$, and be defined by

$$q_c'' = h_{ci}(i_s - i_{as}) \tag{4.103}$$

If an enthalpy recovery factor is defined by

$$r_i = \frac{i_{as} - i_\infty}{i_0 - i_\infty}$$

the same relation used previously to calculate the reference temperature can be used to calculate a reference enthalpy:

$$i^* = i_\infty + 0.5(i_s - i_\infty) + 0.22(i_{as} - i_\infty) \tag{4.104}$$

The local Stanton number is then redefined as

$$\mathrm{St}_{x,t}^* = \frac{h_{c,i}}{\rho^* u_\infty} \tag{4.105}$$

and used in Eqs. (4.99), (4.100), and (4.101) to calculate the heat transfer coefficient. It should be noted that the enthalpies in the above relations are the total values, which include the chemical energy of dissociation as well as the internal energy. As shown in [23], this method of calculation is in excellent agreement with experimental data.

In some situations, for instance, at extremely high altitudes, the fluid density may be so small that the distance between gas molecules becomes of the same order of magnitude as the boundary layer. In such cases, the fluid cannot be treated as a continuum, and it is necessary to subdivide the flow processes into regimes. These flow regimes are characterized by the ratio of the molecular free path to a significant physical scale of the system; this ratio is called the Knudsen number, Kn. Continuum flow corresponds to small values of Kn, while at larger values of Kn, molecular collisions occur primarily at the surface and in the main stream. Since energy transport is by free motion of molecules between the surface and the main stream, this regime is called the "free-molecule" regime. Between the

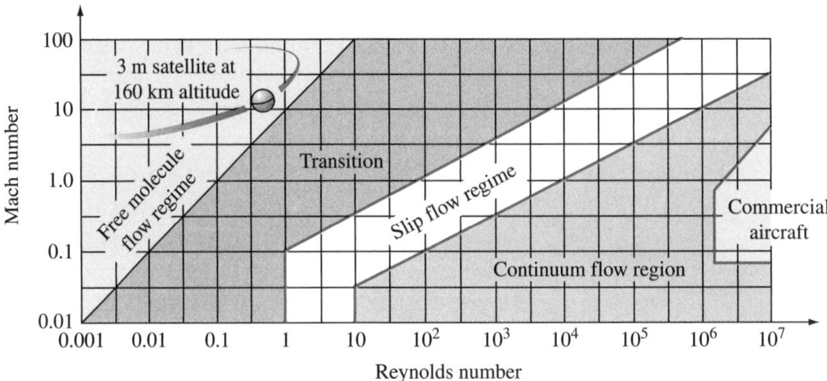

FIGURE 4.25 Flow regimes in high-speed flow.

free-molecule and the continuum regimes is a transition range, called the "slip-flow" regime because it is treated by assuming temperature and velocity "slip" at fluid-solid interfaces. Figure 4.25 shows a map of these flow regimes. For a treatment of heat transfer and friction in these specialized flow systems the reader is referred to [24–27].

4.14 Closing Remarks

In this chapter, we have studied the principles of heat transfer by forced convection. We have seen that the transfer of heat by convection is intimately related to the mechanics of the fluid flow, particularly to the flow in the vicinity of the transfer surface. We have also observed that the nature of heat transfer, as well as flow phenomena, depends greatly on whether the fluid far away from the surface is in laminar or in turbulent flow.

To become familiar with the basic principles of boundary layer theory and forced-convection heat transfer, we have considered the problem of convection in flow over a flat plate in some detail. This system is geometrically simple, but it illustrates the most important features of forced convection. In subsequent chapters, we shall treat heat transfer by convection in geometrically more complicated systems. In the next chapter, we shall examine natural-convection phenomena. In Chapter 6, heat transfer by convection to and from fluids flowing inside pipes and ducts will be taken up. In Chapter 7, forced convection in flow over the exterior surfaces of bodies such as cylinders, spheres, tubes, and tube bundles will be considered. The application of the principles of forced-convection heat transfer to the selection and design of heat transfer equipment will be taken up in Chapter 8.

For the convenience of the reader, a summary of the equations used to calculate the heat transfer and friction coefficients in low-speed flow of gases and liquids over flat or only slightly curved plane surfaces is presented in Table 4.5. For additional information and correlations or equations for other conditions and geometries, the reader is referred to [3, 18, 21, and 28].

TABLE 4.5 Summary of useful empirical equations for calculating friction factors and heat transfer coefficients in flow over flat or slightly curved surfaces at zero angle of attack[a]

Coefficient	Equation	Conditions
	Laminar Flow	
Local friction coefficient	$C_{fx} = 0.664\text{Re}_x^{-0.5}$	$\text{Re}_x < 5 \times 10^5$
Local Nusselt number at distance x from leading edge	$\text{Nu}_x = 0.332\text{Re}_x^{0.5}\text{Pr}^{0.33}$	$\text{Pr} > 0.5$, $\text{Re}_x < 5 \times 10^5$
	$\text{Nu}_x = 0.565(\text{Re}_x\text{Pr})^{0.5}$	$\text{Pr} < 0.1$, $\text{Re}_x < 5 \times 10^5$
Average friction coefficient	$\bar{C}_f = 1.33\text{Re}_L^{-0.5}$	$\text{Re}_L < 5 \times 10^5$
Average Nusselt number between $x = 0$ and $x = L$	$\overline{\text{Nu}}_L = 0.664\text{Re}_L^{0.5}\text{Pr}^{0.33}$	$\text{Pr} > 0.5$, $\text{Re}_L < 5 \times 10^5$
	Turbulent Flow	
Local friction coefficient	$C_{fx} = 0.0576\text{Re}_x^{-0.2}$	$\text{Re}_x > 5 \times 10^5$, $\text{Pr} > 0.5$
Local Nusselt number at distance x from leading edge	$\text{Nu}_x = 0.0288\text{Re}_x^{0.8}\text{Pr}^{0.33}$	
Average friction coefficient	$\bar{C}_f = 0.072[\text{Re}_L^{-0.2} - 0.0464(x_c/L)]$	$\text{Re}_L > 5 \times 10^5$, $\text{Pr} > 0.5$
Average Nusselt number between $x = 0$ and $x = L$ with transition at $\text{Re}_{x,c} = 5 \times 10^5$	$\overline{\text{Nu}}_L = 0.036\text{Pr}^{0.33}[\text{Re}_L^{0.8} - 23{,}200]$	

[a]Applicable to low-speed flow (Mach number <0.5) of gases and liquids with all physical properties evaluated at the mean film temperature, $T_f = (T_s + T_\infty)/2$.

$$C_{fx} = \tau_s/(\rho U_\infty^2/2g_c) \qquad \bar{C}_f = (1/L)\int_0^L C_{fx}\,dx \qquad \text{Pr} = c_p\mu/k$$

$$\text{Nu}_x = h_c x/k \qquad \overline{\text{Nu}} = \bar{h}_c L/k \qquad \bar{h}_c = (1/L)\int_0^L h_c(x)\,dx$$

$$\text{Re}_x = \rho U_\infty x/\mu \qquad \text{Re}_L = \rho U_\infty L/\mu$$

References

1. H. Schlichting, *Boundary Layer Theory*, 6th ed., J. Kestin, transl., McGraw-Hill, New York, 1968.
2. E. R. Van Driest, "Calculation of the Stability of the Laminar Boundary Layer in a Compressible Fluid on a Flat Plate with Heat Transfer," *J. Aero. Sci.*, vol. 19, pp. 801–813, 1952.
3. S. Kakaç, R. K. Shah, and W. Aung, *Handbook of Single-Phase Convective Heat Transfer*, Wiley, New York, 1987.
4. L. D. Landau and E. M. Lifshitz, *Fluid Mechanics*, Pergamon Press, New York, 1959.
5. H. L. Langhaar, *Dimensional Analysis and Theory of Models*, Wiley, New York, 1951.
6. E. R. Van Driest, "On Dimensional Analysis and the Presentation of Data in Fluid Flow Problems," *J. Appl. Mech.*, vol. 13, p. A-34, 1940.
7. *Handbook of Heat Transfer*, W. M. Rohsenow, J. P. Hartnett, and Y. I. Cho, eds., McGraw-Hill, New York, 1998.
8. S. V. Patankar and D. B. Spalding, *Heat and Mass Transfer in Boundary Layers*, 2d ed., International Textbook Co., London, 1970.
9. P. Majumdar, *Computational Methods for Heat and Mass Transfer*, Taylor & Francis, New York, NY, 2005.
10. M. Blasius, "Grenzschichten in Flüssigkeiten mit Kleiner Reibung," *Z. Math. Phys.*, vol. 56, no. 1, 1908.

11. M. Hansen, "Velocity Distribution in the Boundary Layer of a Submerged Plate," NACA TM 582, 1930.
12. E. Pohlhausen, "Der Wärmeaustausch zwischen festen Körpern und Flüssigkeiten mit kleiner Reibung und kleiner Wärmeleitung," *Z. Angew. Math. Mech.*, vol. 1, p. 115, 1921.
13. B. Gebhart, *Heat Transfer*, 2d ed., McGraw-Hill, New York, 1971.
14. A. P. Colburn, "A Method of Correlating Forced Convection Heat Transfer Data and a Comparison with Fluid Friction," *Trans. AIChE*, vol. 29, pp. 174–210, 1993.
15. E. R. G. Eckert and R. M. Drake, *Heat and Mass Transfer*, 2d ed., McGraw-Hill, New York, 1959.
16. L. Prandtl, "Bemerkungen über den Wärmeübergang irn Rohr," *Phys. Zeit.*, vol. 29, p. 487, 1928.
17. L. Prandtl, "Eine Beziehung zwischen Wärmeaustauch und Ströhmungswiederstand der Flüssigkeiten," *Phys. Zeit.*, vol. 10, p. 1072, 1910.
18. S. E. Isakoff and T. B. Drew, "Heat and Momentum Transfer in Turbulent Flow of Mercury," *Institute of Mechanical Engineers and ASME, Proceedings, General Discussion on Heat Transfer*, pp. 405–409, 1951.
19. W. Forstall, Jr., and A. H. Shapiro, "Momentum and Mass Transfer in Co-axial Gas Jets," *J. Appl. Mech.*, vol. 17, p. 399, 1950.
20. R. C. Martinelli, "Heat Transfer to Molten Metals," *Trans. ASME*, vol. 69, pp. 947–959, 1947.
21. W. Kays, M. Crawford, and B. Weigand, *Convective Heat and Mass Transfer*, 4th ed., McGraw-Hill, New York, NY, 2005.
22. J. Kaye, "Survey of Friction Coefficients, Recovery Factors, and Heat Transfer Coefficients for Supersonic Flow," *J. Aeronaut. Sci.*, vol. 21, no. 2, pp. 117–229, 1954.
23. E. R. G. Eckert, "Engineering Relations for Heat Transfer and Friction in High-Velocity Laminar and Turbulent. Boundary Layer Flow over Surfaces with Constant Pressure and Temperature," *Trans. ASME*, vol. 78, pp. 1273–1284, 1956.
24. E. R. Van Driest, "Turbulent Boundary Layer in Compressible Fluids," *J. Aeronaut. Sci.*, vol. 18, no. 3, pp. 145–161, 1951.
25. A. K. Oppenheim, "Generalized Theory of Convective Heat Transfer in a Free-Molecule Flow," *J. Aeronaut. Sci.*, vol. 20, pp. 49–57, 1953.
26. W. D. Hayes and R. F. Probstein, *Hypersonic Flow Theory*, Academic Press, New York, 1959.
27. F. M. White, *Viscous Fluid Flow*, 2nd ed., McGraw-Hill, New York, 1991.
28. R. B. Bird, W. E. Stewart, and E. N. Lightfoot, *Transport Phenomena*, 2nd ed., Wiley, New York, NY, 2007.

Problems

The problems for this chapter are organized by subject matter as shown below.

Topic	Problem Number
Dimensionless numbers	4.1–4.6
Dimensional analysis	4.7–4.19
Boundary layers	4.20–4.28
Flow over a flat plate	4.29–4.40
Analogy between heat and momentum transfer	4.41–4.44
Viscous dissipation	4.45–4.48
Design problems	4.49–4.59
High-speed flow	4.60–4.66

4.1 Evaluate the Reynolds number for flow over a tube from the following data: $D = 6$ cm, $U_\infty = 1.0$ m/s, $\rho = 300$ kg/m^3, $\mu = 0.04$ N s/m^2.

4.2 Evaluate the Prandtl number from the following data: $c_p = 0.5$ Btu/lb$_m$ °F, $k = 2$ Btu/ft h °F, $\mu = 0.3$ lb$_m$/ft s.

4.3 Evaluate the Nusselt number for flow over a sphere for the following conditions: $D = 6$ in., $k = 0.2$ W/m K, $\bar{h}_c = 18$ Btu/h ft^2 °F.

4.4 Evaluate the Stanton number for flow over a tube from the following data: $D = 10$ cm, $U_\infty = 4$ m/s, $\rho = 13{,}000$ kg/m^3, $\mu = 1 \times 10^{-3}$ N s/m^2, $c_p = 140$ J/kg K, $\bar{h}_c = 1000$ W/m^2 K.

4.5 Evaluate the dimensionless groups $\bar{h}_c D/k$, $U_\infty D\rho/\mu$, and $c_p\mu/k$ for water, *n*-butyl alcohol, mercury, hydrogen, air, and saturated steam at a temperature of 100°C. Let $D = 1$ m, $U_\infty = 1$ m/sec, and $\bar{h}_c = 1$ W/m^2 K.

4.6 A fluid flows at 5 m/s over a wide, flat plate 15 cm long. For each from the following list, calculate the Reynolds number at the downstream end of the plate. Indicate whether the flow at that point is laminar, transition, or turbulent. Assume all fluids are at 40°C. (a) air, (b) CO_2, (c) water, (d) engine oil.

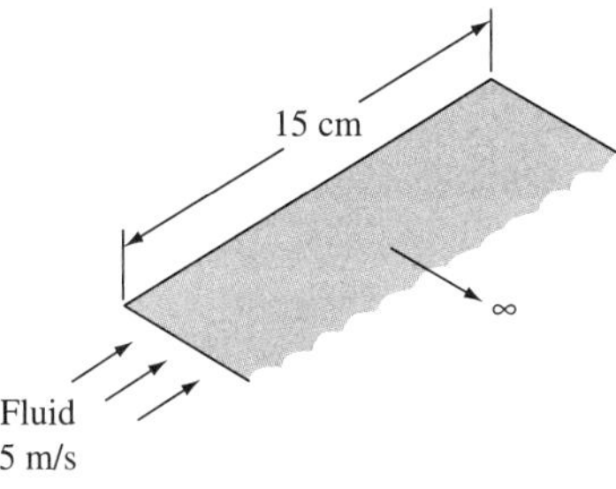

Problem 4.6

4.7 Replot the data points of Fig. 4.9(b) on log-log paper and find an equation approximating the best correlation line. Compare your results with Fig. 4.10. Then suppose that steam at 1 atm and 100°C is flowing across a 5-cm-OD pipe at a velocity of 1 m/s. Using the data in Fig. 4.10, estimate the Nusselt number, the heat transfer coefficient, and the rate of heat transfer per meter length of pipe if the pipe is at 200°C; compare these results with predictions from your correlation equation.

4.8 The average Reynolds number for air passing in turbulent flow over a 2-m-long, flat plate is 2.4×10^6. Under these conditions, the average Nusselt number was found to be equal to 4150. Determine the average heat transfer coefficient for an oil having thermal properties similar to those in Appendix 2, Table 18, at 30°C at the same Reynolds number and flowing over the same plate.

4.9 The dimensionless ratio $U_\infty/\sqrt{Lg}$, called the Froude number, is a measure of similarity between an ocean-going ship and a scale model of the ship to be tested in a laboratory water channel. A 500-ft-long cargo ship is designed to run at 20 knots, and a 5-ft geometrically similar model is towed in a water channel to study wave resistance. What should be the towing speed in ms^{-1}?

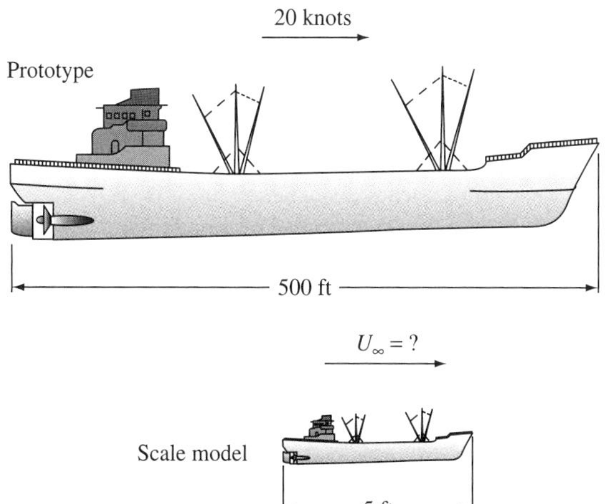

Problem 4.9

4.10 The torque due to the frictional resistance of the oil film between a rotating shaft and its bearing is found to be dependent on the force F normal to the shaft, the speed of rotation N of the shaft, the dynamic viscosity μ of the oil, and the shaft diameter D. Establish a correlation among these variables by using dimensional analysis.

4.11 When a sphere falls freely through a homogeneous fluid, it reaches a terminal velocity at which the weight of the sphere is balanced by the buoyant force and the frictional resistance of the fluid. Make a dimensional analysis of this problem and indicate how experimental data for this problem could be correlated. Neglect compressibility effects and the influence of surface roughness.

4.12 Experiments have been performed on the temperature distribution in a homogeneous long cylinder (0.1 m diameter, thermal conductivity of 0.2 W/m K) with uniform internal heat generation. By dimensional analysis, determine the relation between the steady-state temperature at the center of the cylinder T_c, the diameter, the thermal conductivity, and the rate of heat generation. Take the temperature at the surface as your datum. What is the equation for the center temperature if the difference between center and surface temperature is 30°C when the heat generation is 3000 W/m^3?

4.13 The convection equations relating the Nusselt, Reynolds, and Prandtl numbers can be rearranged to show that for gases the heat transfer coefficient h_c depends on the absolute temperature T and the group $\sqrt{U_\infty/x}$. This formulation is of the form $h_{c,x} = CT^n\sqrt{U_\infty/x}$, where n and C are constants. Indicate clearly how such a relationship could be obtained for the laminar flow case from $\text{Nu}_x = 0.332\,\text{Re}_x^{0.5}\text{Pr}^{0.333}$ for the condition $0.5 < \text{Pr} < 5.0$. State restrictions as necessary.

4.14 A series of tests in which water was heated while flowing through a 38.6-in-long electrically heated tube of 0.527-in. ID yielded the experimental pressure-drop data shown next.

Mass Flow Rate $\dot{m}$ (lb/sec)	Fluid Bulk Temp T_b (°F)	Tube Surface Temp T_s (°F)	Pressure Drop with Heat Transfer, Δp_{ht} (psi)
3.04	90	126	9.56
2.16	114	202	4.74
1.82	97	219	3.22
3.06	99	248	8.34
2.15	107	283	4.45

Isothermal pressure-drop data for the same tube are given below in terms of the dimensionless friction factor $f = (\Delta p/\rho \bar{u}^2)(2D/L)g_c$ and Reynolds number based on the pipe diameter, $\mathrm{Re}_D = 4\dot{m}/\pi D\mu$. The symbol $\bar{u}$ denotes the average pipe velocity.

Re_D	f
1.71×10^5	0.0189
1.05×10^5	0.0205
1.9×10^5	0.0185
2.41×10^5	0.0178

By comparing the isothermal with the nonisothermal friction coefficients at similar bulk Reynolds numbers, derive a dimensionless equation for the nonisothermal friction coefficients in the form

$$f = \text{constant} \times \mathrm{Re}_D^n(\mu_s/\mu_b)^m$$

where μ_s = viscosity at surface temperature, μ_b = viscosity at bulk temperature, and n and m = empirical constants.

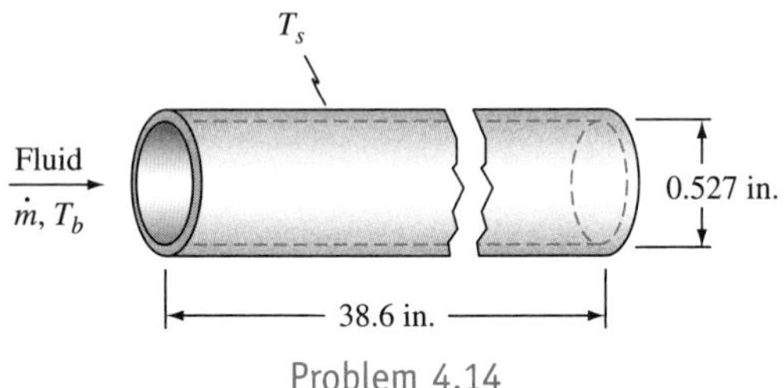

Problem 4.14

4.15 The experimental data shown tabulated were obtained by passing n-butyl alcohol at a bulk temperature of 15°C over a heated flat plate 0.3-m long, 0.9-m wide, and with a surface temperature of 60°C. Correlate the experimental data using appropriate dimensionless numbers and compare the line that best fits the data with Eq. (4.38).

Velocity (m/s)	Average Heat Transfer Coefficient (W/m^2°C)
0.089	121
0.305	218
0.488	282
1.14	425

4.16 The test data tabulated below were reduced from measurements made to determine the heat transfer coefficient inside tubes at Reynolds numbers only slightly above transition and at relatively high Prandtl numbers (as associated with oils). Tests were made in a double-tube exchanger with a counterflow of water to provide the cooling. The pipe used to carry the oils was 5/8-in. OD, 18 BWG, and 121 in. long. Correlate the data in terms of appropriate dimensionless parameters.

Test No.	Fluid	$\bar{h}_c$	ρu	c_p	k_f	μ_b	μ_f
11	10C oil	87.0	1,072,000	0.471	0.0779	13.7	19.5
19	10C oil	128.2	1,504,000	0.472	0.0779	13.3	19.1
21	10C oil	264.8	2,460,000	0.486	0.0776	9.60	14.0
23	10C oil	143.8	1,071,000	0.495	0.0773	7.42	9.95
24	10C oil	166.5	2,950,000	0.453	0.0784	23.9	27.3
25	10C oil	136.3	1,037,000	0.496	0.0773	7.27	11.7
36	1488 pyranol	140.7	1,795,000	0.260	0.0736	12.1	16.9
39	1488 pyranol	133.8	2,840,000	0.260	0.0740	23.0	29.2
45	1488 pyranol	181.4	1,985,000	0.260	0.0735	10.3	12.9
48	1488 pyranol	126.4	3,835,000	0.260	0.0743	40.2	53.5
49	1488 pyranol	105.8	3,235,000	0.260	0.0743	39.7	45.7

where $\bar{h}_c$ = mean surface heat transfer coefficient based on the mean temperature difference, Btu/h ft^2°F
ρu = mass velocity, lb$_m$/h ft^2
c_p = specific heat, Btu/lb$_m$ °F
k_f = thermal conductivity, Btu/h ft °F (based on film temperature)
μ_b = viscosity, based on average bulk (mixed mean) temperature, lb$_m$/h ft
μ_f = viscosity, based on average film temperature, lb$_m$/h ft

Hint: Start by correlating Nu and Re_D irrespective of the Prandtl numbers, since the influence of the Prandtl number on the Nusselt number is expected to be relatively small. By plotting Nu versus Re on log-log paper, one can guess the nature of the correlation equation, Nu $= f_1$(Re). A plot of Nu/f_1(Re) versus Pr will then reveal the dependence upon Pr. For the final equation, the influence of the viscosity variation should also be considered.

4.17 A turbine blade with a characteristic length of 1 m is cooled in an atmospheric pressure wind tunnel by air at 40°C with a velocity of 100 m/s. For a surface temperature of 500 K, the cooling rate is found to be 10,000 watts. Use these results to estimate the cooling rate from another turbine blade of similar shape but with a characteristic length of 0.5 m and operating with a surface temperature of 600 K in air at 40°C with a velocity of 200 m/s.

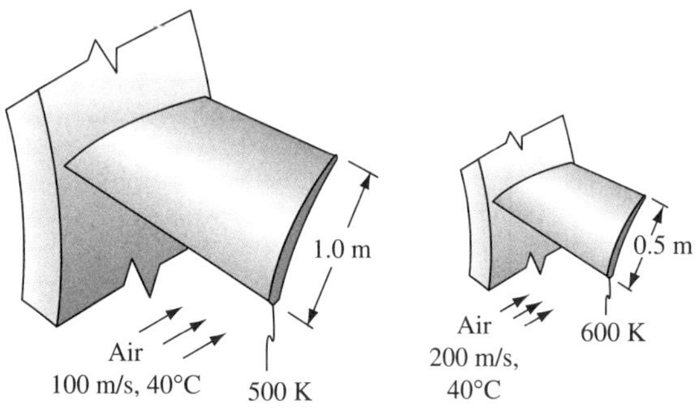

Problem 4.17

4.18 The drag on an airplane wing in flight is known to be a function of the density of air (ρ), the viscosity of air (μ), the free-stream velocity (U_∞), a characteristic dimension of the wing (s), and the shear stress on the surface of the wing (τ_s).

Show that the dimensionless drag, $\frac{\tau_s}{\rho U_\infty^2}$, can be expressed as a function of the Reynolds number, $\frac{\rho U_\infty s}{\mu}$.

4.19 Suppose that the graph below shows measured values of h_c for air in forced convection over a cylinder of diameter D, plotted on a logarithmic graph of Nu_D as a function of Re_DPr.

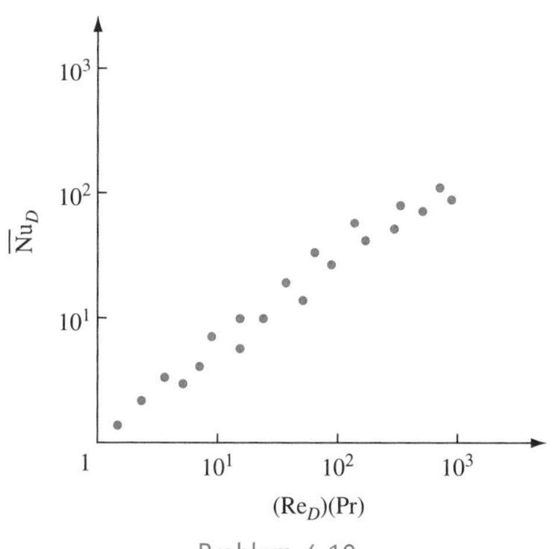

Problem 4.19

Write an appropriate dimensionless correlation for the average Nusselt number for these data and state any limitations to your equation.

4.20 Engine oil at 100°C flows over and parallel to a flat surface at a velocity of 3 m/s. Calculate the thickness of the hydrodynamic boundary layer at a distance 0.3 m from the leading edge of the surface.

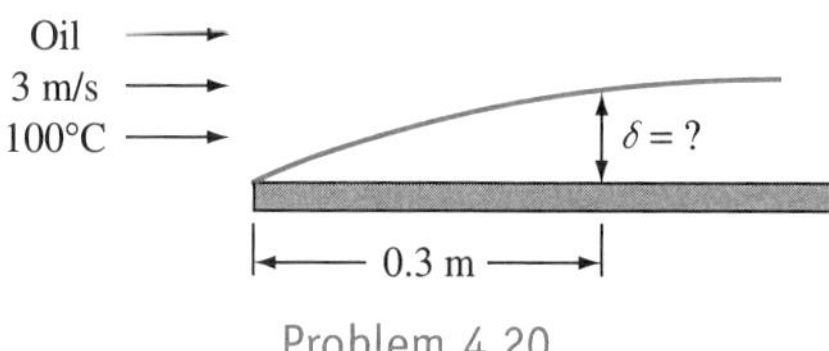

Problem 4.20

4.21 Assuming a linear velocity distribution and a linear temperature distribution in the boundary layer over a flat plate, derive a relation between the thermal and hydrodynamic boundary layer thicknesses and the Prandtl number.

4.22 Air at 20°C flows at 1 m/s between two parallel flat plates spaced 5 cm apart. Estimate the distance from the entrance to the point at which the hydrodynamic boundary layers meet.

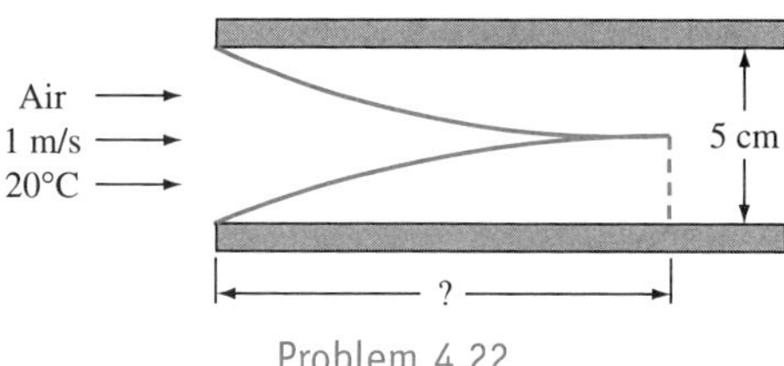

Problem 4.22

4.23 A fluid at temperature T_∞ is flowing at a velocity U_∞ over a flat plate that is at the same temperature as the fluid for a distance x_0 from the leading edge but at a higher temperature T_s beyond this point.

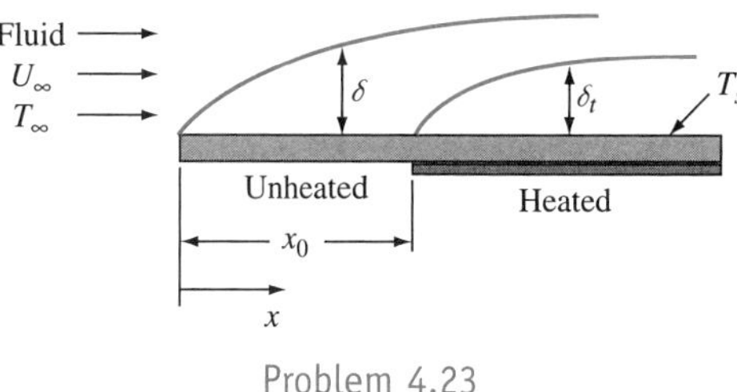

Problem 4.23

Show by means of the integral boundary layer equations that ζ, the ratio of the thermal boundary layer thickness to the hydrodynamic boundary layer thickness, over the heated portion of the plate is approximately

$$\zeta \approx \Pr^{-1/3}\left[1 - \left(\frac{x_0}{x}\right)^{3/4}\right]^{1/3}$$

if the flow is laminar.

4.24 Air at 1000°C flows at an inlet velocity of 2 m/s between two parallel flat plates spaced 1 cm apart. Estimate the distance from the entrance to the point where the boundary layers meet.

4.25 Experimental measurements of the temperature distribution during the flow of air at atmospheric pressure over the wing of an airplane indicate that the temperature distribution near the surface can be approximated by a linear equation:

$$(T - T_s) = ay(T_\infty - T_s)$$

where a = a constant = 2 m^{-1}, T_s = surface temperature, K, T_∞ = free-stream temperature, K, and y = perpendicular distance from surface (mm). (a) Estimate the convection heat transfer coefficient if $T_s = 50°C$ and $T_\infty = -50°C$. (b) Calculate the heat flux in W/m^2.

4.26 For flow over a slightly curved isothermal surface, the temperature distribution inside the boundary layer δ_t can be approximated by the polynomial $T(y) = a + by + cy^2 + dy^3 (y < \delta_t)$, where y is the distance normal to the surface. (a) By applying appropriate boundary conditions, evaluate the constants a, b, c, and d.

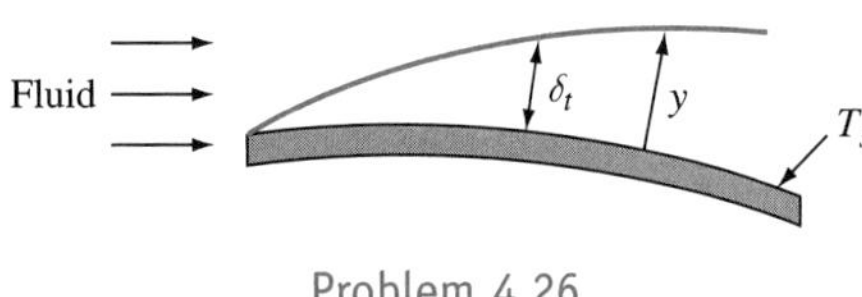

Problem 4.26

(b) Then obtain a dimensionless relation for the temperature distribution in the boundary layer.

4.27 The integral method also can be applied to turbulent-flow conditions if experimental data for the wall shear stress are available. In one of the earliest attempts to analyze turbulent flow over a flat plate, Ludwig Prandtl proposed in 1921 the following relations for the dimensionless velocity and temperature distributions:

$$u/U_\infty = (y/\delta)^{1/7}$$

$$(T - T_\infty)/(T_s - T_\infty) = 1 - (y/\delta_t)^{1/7} \; (T_s > T > T_\infty)$$

From experimental data, an empirical relation relating the shear stress at the wall to boundary layer thickness is:

$$\tau_s = 0.023\rho U_\infty^2/\mathrm{Re}_\delta^{0.25} \quad \text{where} \quad \mathrm{Re}_\delta = U_\infty \delta/\nu$$

Following the approach outlined in Section 4.9.1 for laminar conditions, substitute the relations in the boundary layer momentum and energy integral equations and derive equations for: (a) the boundary layer thickness, (b) the local friction coefficient, and (c) the local Nusselt number. Assume $\delta = \delta_t$ and discuss the limitations of your results.

4.28 For liquid metals with Prandtl numbers much less than unity, the hydrodynamic boundary layer is much thinner than the thermal boundary layer. As a result, one may assume that the velocity in the boundary layer is uniform [$u = U_\infty$ and $v = 0$]. Starting with Eq. (4.7b), show that the energy equation and its boundary condition are analogous to those for a semi-infinite slab with a sudden change in surface temperature [see Eq. (2.93)]. Then show that the local Nusselt number is given by

$$\mathrm{Nu}_x = 0.56\,(\mathrm{Re}_x \Pr)^{0.5}$$

Compare this equation with the appropriate relation in Table 4.5.

4.29 Hydrogen at 15°C and a pressure of 1 atm is flowing along a flat plate at a velocity of 3 m/s. If the plate is 0.3-m wide and at 71°C, calculate the following quantities at $x = 0.3$ m and at the distance corresponding to the transition point, i.e., $\mathrm{Re}_x = 5 \times 10^5$ (use properties at 43°C): (a) hydrodynamic boundary layer thickness, in cm, (b) thickness of thermal boundary layer, in cm, (c) local friction coefficient, dimensionless, (d) average friction coefficient, dimensionless, (e) drag force, in N, (f) local convection heat transfer coefficient, in W/m^2°C, (g) average convection heat transfer coefficient, in W/m^2°C, and (h) rate of heat transfer, in W.

4.30 Repeat Problem 4.29, parts (d), (e), (g), and (h) for $x = 4.0$ m and $U_\infty = 80$ m/s, (a) taking the laminar boundary layer into account and (b) assuming that the turbulent boundary layer starts at the leading edge.

4.31 Determine the rate of heat loss in Btu/h from the wall of a building resulting from a 10-mph wind blowing parallel to its surface. The wall is 80 ft long and 20 ft high, its surface temperature is 80°F, and the temperature of the ambient air is 40°F.

4.32 A flat-plate heat exchanger will operate in a nitrogen atmosphere at a pressure of about 10^4 N/m^2 and 38°C. The flat-plate heat exchanger was originally designed to operate in air at 1 atm and 38°C turbulent flow. Estimate the ratio of the heat transfer coefficient in air to that in nitrogen, assuming forced circulation cooling of the flat-plate surface at the same velocity in both cases.

4.33 A heat exchanger for heating liquid mercury is under development. The exchanger can be visualized as a 6-in.-long and 1-ft-wide flat plate. The plate is maintained at 160°F, and the mercury flows parallel to the short side at 60°F with a velocity of 1 ft/sec.(a) Find the local friction coefficient at the midpoint of the plate and the total drag force on the plate. (b) Determine the temperature of the mercury at a point 4 in. from the leading edge and 0.05 in. from the surface of the plate. (c) Calculate the Nusselt number at the end of the plate.

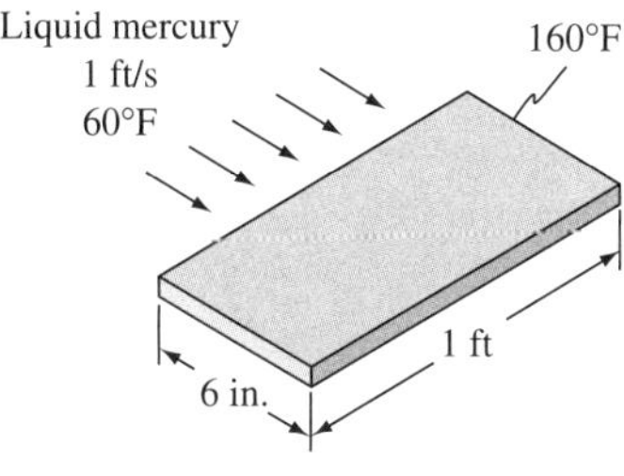

Problem 4.33

4.34 Water at a velocity of 2.5 m/s flows parallel to a 1-m-long horizontal, smooth, and thin, flat plate. Determine the local thermal and hydrodynamic boundary layer thicknesses and the local friction coefficient at the midpoint of the plate. What is the rate of heat transfer from one side of the plate to the water per unit width of the plate if the surface temperature is kept uniformly at 150°C and the temperature of the main water stream is 15°C?

4.35 A thin, flat plate is placed in an atmospheric pressure air stream that flows parallel to it at a velocity of 5 m/s. The temperature at the surface of the plate is maintained uniformly at 200°C, and that of the main air stream is 30°C. Calculate the temperature and horizontal velocity at a point 30 cm from the leading edge and 4 mm above the surface of the plate.

4.36 The surface temperature of a thin, flat plate located parallel to an air stream is 90°C. The free-stream velocity is 60 m/s, and the temperature of the air is 0°C. The plate is 60 cm wide and 45 cm long in the direction of the air stream. Neglecting the end effect of the plate and assuming that the flow in the boundary layer changes abruptly from laminar to turbulent at a transition Reynolds number of $Re_{tr} = 4 \times 10^5$, find: (a) the average heat transfer coefficient in the laminar and turbulent regions, (b) the rate of heat transfer for the entire plate, considering both sides, (c) the average friction coefficient in the laminar and turbulent regions, and (d) the total drag force. Also plot the heat transfer coefficient and local friction coefficient as a function of the distance from the leading edge of the plate.

4.37 The wing of an airplane has a polished aluminum skin. At a 1500 m altitude, it absorbs 100 W/m^2 by solar radiation. Assuming that the interior surface of the wing's skin is well insulated and the wing has a chord of 6-m length (i.e., $L = 6$ m), estimate the equilibrium temperature of the wing at a flight speed of 150 m/s at distances of 0.1 m, 1 m, and 5 m from the leading edge. Discuss the effect of a temperature gradient along the chord.

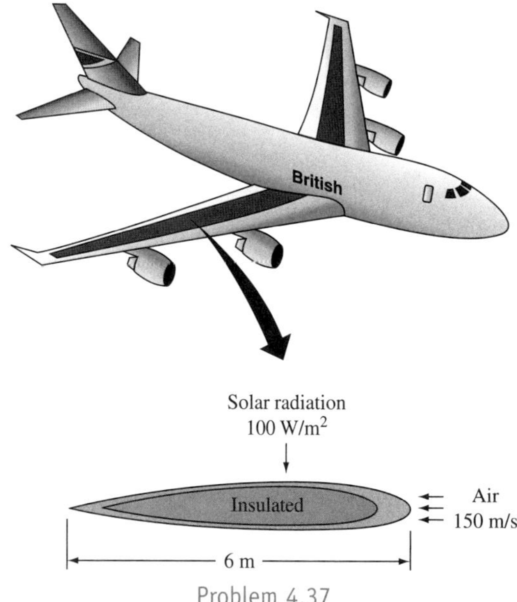

Problem 4.37

4.38 An aluminum cooling fin for a heat exchanger is situated parallel to an atmospheric pressure air stream. The fin, as shown in the sketch, is 0.075-m high, 0.005-m thick and 0.45-m in the flow direction. Its base temperature is 88°C, and the air is at 10°C. The velocity of the air is 27 m/s. Determine the total drag force and the total rate of heat transfer from the fin to the air.

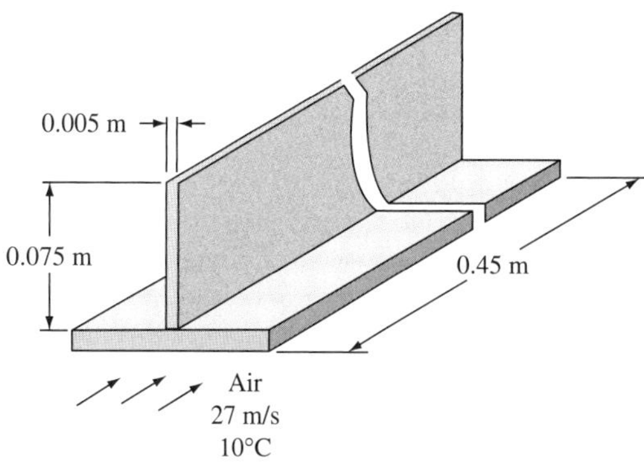

Problem 4.38

4.39 Air at 320 K with a free-stream velocity of 10 m/s is used to cool small electronic devices mounted on a printed circuit board as shown in the sketch below. Each device is 5 mm × 5 mm square in planform and dissipates 60 mW. A turbulator is located at the leading edge to trip the boundary layer so that it will become turbulent. Assuming that the lower surfaces of the electronic devices are insulated, estimate the surface temperature at the center of the fifth device on the circuit board (see next page).

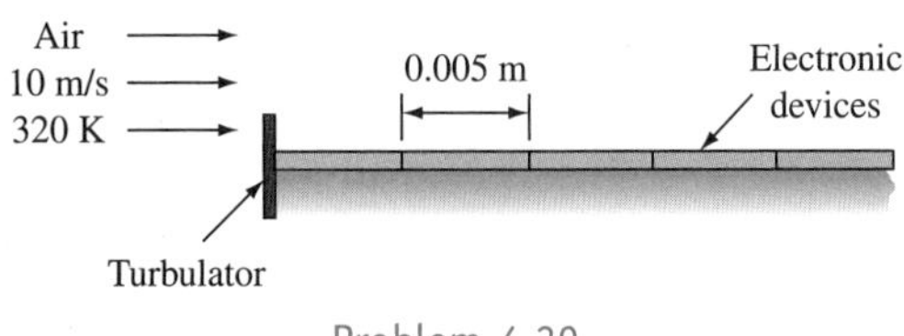

Problem 4.39

4.40 The average friction coefficient for flow over a 0.6-m long plate is 0.01. What is the value of the drag force in N per m width of the plate for the following fluids: (a) air at 15°C, (b) steam at 100°C and atmospheric pressure, (c) water at 40°C, (d) mercury at 100°C, and (e) *n*-butyl alcohol at 100°C?

4.41 A thin, flat plate 6-in. square is tested for drag in a wind tunnel with air at 100 ft/s, 14.7 psia, and 60°F flowing across and parallel to the top and bottom surfaces. The observed total drag force is 0.0135 lb. Using the definition of friction coefficient [Eq. (4.13)] and the Reynolds analogy, calculate the rate of heat transfer from this plate when the surface temperature is maintained at 250°F.

4.42 A thin, flatplate 15-cm square is suspended from a balance in a uniformly flowing stream of engine oil in such a way that the oil flows parallel to and along both surfaces of the plate. The total drag on the plate is measured and found to be 55.5 N. If the oil flows at the rate of 15 m/s with a temperature of 45°C, calculate the heat transfer coefficient using the Reynolds analogy.

4.43 For a study on global warming, an electronic instrument has to be designed to map the CO_2 absorption characteristics of the Pacific Ocean. The instrument package resembles a flat plate with a total (upper and lower) surface area of 2 m^2. For safe operation, its surface temperature must not exceed the ocean temperature by more than 2°C. To monitor the temperature of the instrument package, which is towed by a ship moving at 20 m/s, the tension in the towing cable is measured. If the tension is 400 N, calculate the maximum permissible heat generation rate from the instrument package.

4.44 For flow of gas over a flat surface that has been artificially roughened by sandblasting, the local heat transfer by convection can be correlated by the dimensionless relation $\mathrm{Nu}_x = 0.05\mathrm{Re}_x^{0.9}$. (a) Derive a relationship for the average heat transfer coefficient in flow over a flat plate of length L. (b) Assuming the analogy between heat and momentum transfer to be valid, derive a relationship for the local friction coefficient. (c) If the gas is air at a temperature of 400 K flowing at a velocity of 50 m/s, estimate the heat flux 1 m from the leading edge for a plate surface temperature of 300 K.

4.45 When viscous dissipation is appreciable, conservation of energy [Eq. (4.6)] must take into account the rate at which mechanical energy is irreversibly converted to thermal energy due to viscous effects in the fluid. This gives rise to an additional term, ϕ, the viscous dissipation, on the right-hand side, where:

$$\frac{\phi}{\mu} = \left(\frac{\partial u}{\partial y} + \frac{\partial v}{\partial x}\right)^2 + 2\left[\left(\frac{\partial u}{\partial x}\right)^2 + \left(\frac{\partial v}{\partial y}\right)^2\right] - \frac{2}{3}\left(\frac{\partial u}{\partial x} + \frac{\partial v}{\partial y}\right)^2$$

Apply the resulting equation to laminar flow between two infinite parallel plates, with the upper plate moving at a velocity U. Assuming constant physical properties (ρ, c_p, k, and μ), obtain expressions for the velocity and temperature distributions. Compare the solution with the dissipation term included with the results when dissipation is neglected. Find the plate velocity required to produce a 1-K temperature rise in nominally 40°C air relative to the case where dissipation is neglected.

4.46 A journal bearing can be idealized as a stationary flat plate and a moving flat plate that moves parallel to it. The space between the two plates is filled by an incompressible fluid. Consider such a bearing in which the stationary and moving plates are at 10°C and 20°C, respectively, the distance between them is 3 mm, the speed of the moving plate is 5 m/s, and there is engine oil between the plates. (a) Calculate the heat flux to the upper and lower plates. (b) Determine the maximum temperature of the oil.

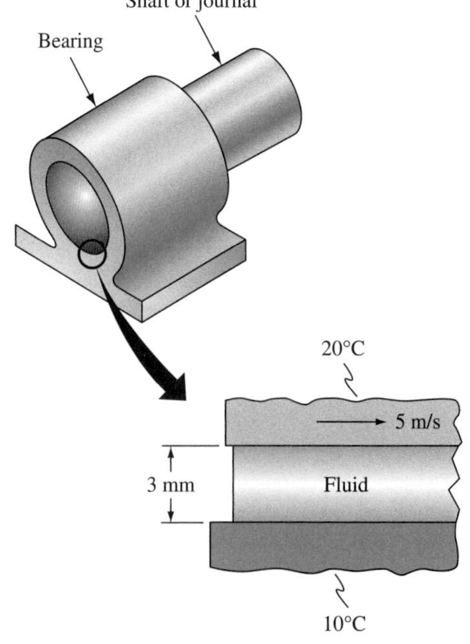

Problem 4.46

4.47 A journal bearing has a clearance of 0.5 mm. The journal has a diameter of 100 mm and rotates at 3600 rpm within the bearing. It is lubricated by an oil having a density of 800 kg/m^3, a viscosity of 0.01 kg/ms, and a thermal conductivity of 0.14 W/m K. If the bearing surface is at 60°C, determine the temperature distribution in the oil film, assuming that the journal surface is insulated.

4.48 A journal bearing has a clearance of 0.5 mm. The journal has a diameter of 100 mm and rotates at 3600 rpm within the bearing. It is lubricated by an oil having a density of 800 kg/m^3, a viscosity of 0.01 kg/ms, and a thermal conductivity of 0.14 W/m K. If both the journal and the bearing temperatures are maintained at 60°C, calculate the rate of heat transfer from the bearing and the power required for rotation per unit length.

4.49 A refrigeration truck is traveling at 80 mph on a desert highway where the air temperature is 50°C. The body of the truck can be idealized as a rectangular box 3 m wide, 2.1 m high, and 6 m long, at a surface temperature 10°C. Assume that (1) the heat transfer from the front and back of the truck can be neglected, (2) the stream does not separate from the surface, and (3) the boundary layer is turbulent over the whole surface. If a one-ton refrigerating capacity is required for every 3600 W of heat loss, calculate the required tonnage of the refrigeration unit.

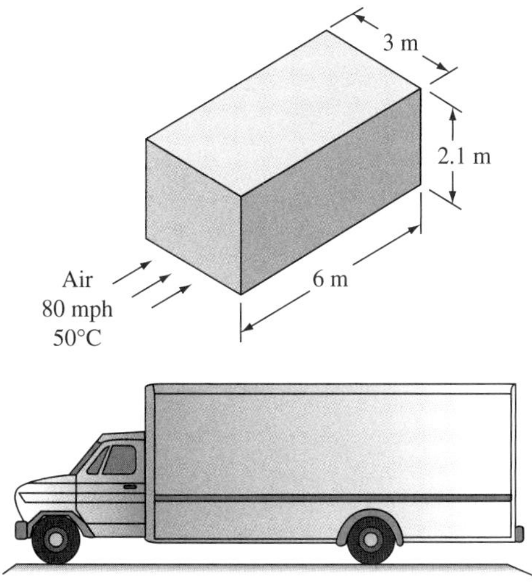

Problem 4.49

4.50 At the equator, where the sun is approximately overhead at noon, a near optimum orientation for a flat-plate solar hot-water heater is in the horizontal position. Suppose a 4-m × 4-m solar collector for domestic hot water use is mounted on a horizontal roof as shown in the accompanying sketch. The surface temperature of the glass cover is estimated to be 40°C, and air at 20°C is blowing over the roof at a velocity of 15 mph. Estimate the heat loss by convection from the collector to the air when the collector is mounted (a) at the leading edge of the roof ($L_c = 0$) and (b) at a distance of 10 m from the leading edge.

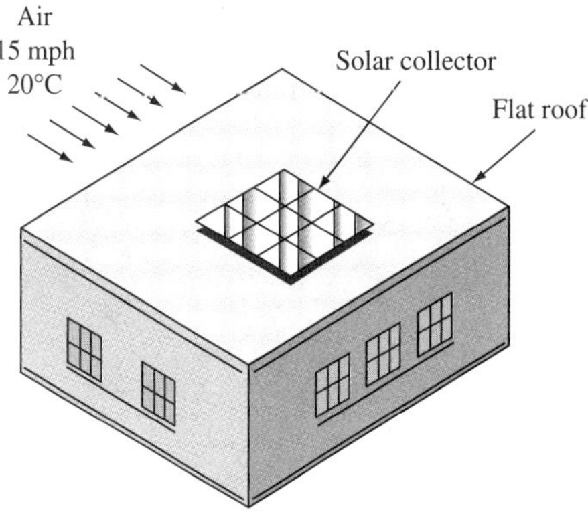

Problem 4.50

4.51 An electronic device is to be cooled by air flowing over aluminum fins attached to lower surface as shown:

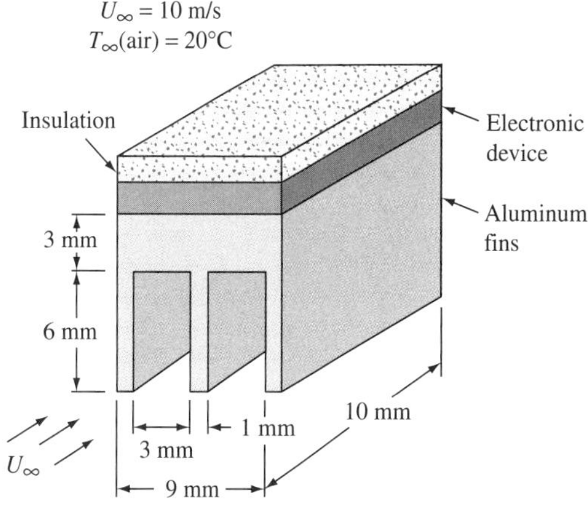

Problem 4.51

The device dissipates 5 W, and the thermal contact resistance between the lower surface of the device and the upper surface of the cooling fin assembly is 0.1 cm^2 K/W. If the device is at a uniform temperature and is insulated at the top, estimate that temperature under steady state.

4.52 An array of 16 silicon chips arranged in 2 rows is insulated at the bottom and cooled by air flowing in forced convection over the top. The array can be located either with its long side or its short side facing the cooling air. If each chip is 10 mm × 10 mm in surface area and dissipates the same power, calculate the rate of maximum power dissipation permissible for both possible arrangements if the maximum permissible surface temperature of the chips is 100°C. What would be the effect of a turbulator on the leading edge to trip the boundary layer into turbulent flow? The air temperature is 30°C and its velocity is 25 m/s.

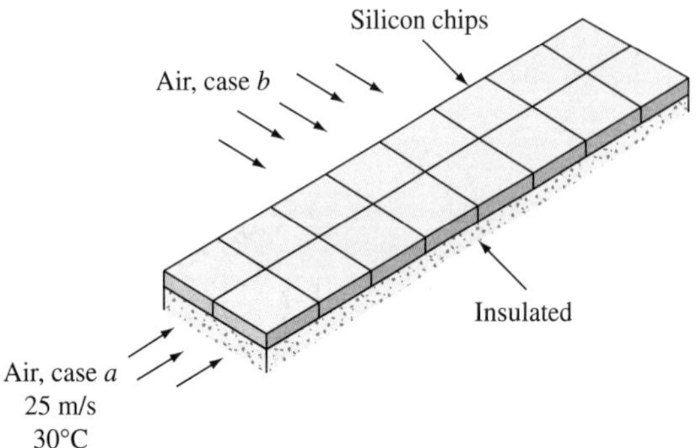

Problem 4.52

4.53 The air-conditioning system in a Chevrolet van for use in desert climates is to be sized. The system is to maintain an interior temperature of 20°C when the van travels at 100 km/h through dry air at 30°C at night. If the top of the van can be idealized as a flat plate 6 m long and 2 m wide and the sides as flat plates 3 m tall and 6 m long, estimate the rate at which heat must be removed from the interior to maintain the specified comfort conditions. Assume the heat transfer coefficient on the inside of the van wall is 10 W/m^2 K.

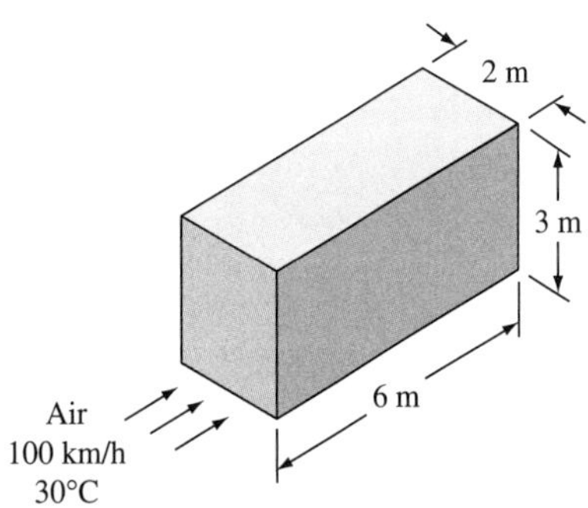

Problem 4.53

4.54 The six identical aluminum fins shown in the sketch below are attached to an electronic device for cooling. Cooling air is available at a velocity of 5 m/s from a fan at 20°C. If the average temperature at the base of a fin is not to exceed 100°C, estimate the maximum permissible power dissipation for the device.

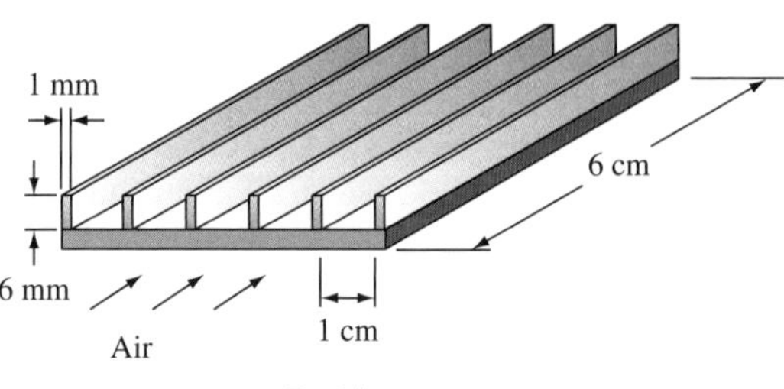

Problem 4.54

4.55 Twenty-five square computer chips, each 10 mm × 10 mm in size and 1 mm thick, are mounted 1 mm apart on an insulating plastic substrate as shown below. The chips are to be cooled by nitrogen flowing along the length of the row at −40°C and atmospheric pressure to prevent their temperature from exceeding 30°C. The design is to provide for a dissipation rate of 30 mW per chip. Estimate the minimum free-stream velocity required to provide safe operating conditions for every chip in the array.

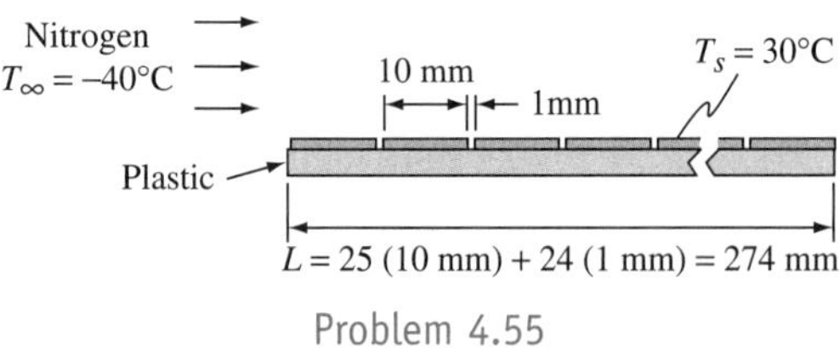

Problem 4.55

4.56 It has been proposed to tow icebergs from the polar region to the Middle East in order to supply potable water to arid regions there. A typical iceberg suitable for towing should be relatively broad and flat. Consider an iceberg 0.25 km thick and 1 km square. This iceberg is to be towed at 1 km/h over a distance of 6000 km through water whose average temperature during the trip is 8°C. Assuming that the interaction of the iceberg with its surroundings can be approximated by the heat transfer and friction at its bottom surface, calculate the following parameters: (a) the average rate at which ice will melt at the bottom surface, and (b) the power required to tow the iceberg at the designated speed. (c) If towing energy costs are approximately 50 ¢/kW · h hour of power and the cost of delivering water at the destination can be approximated by the same figure, calculate the cost of fresh water. The latent heat of fusion of the ice is 334 kJ/kg, and its density is 900 kg/m^3.

4.57 In a manufacturing operation, a long strip of sheet metal is transported on a conveyor at a velocity of 2 m/s while a coating on its top surface is to be cured by radiant heating. Suppose that infrared lamps mounted above the conveyor provide a radiant flux of 2500 W/m^2 on the coating. The coating absorbs 50% of the incident radiant flux, has an emissivity of 0.5, and radiates to the surroundings at a temperature of 25°C. In addition, the coating loses heat by convection through a heat transfer coefficient between both the upper and lower surface and the ambient air, which can be assumed to be at the same temperature as the environment. Estimate the temperature of the coating under steady-state conditions.

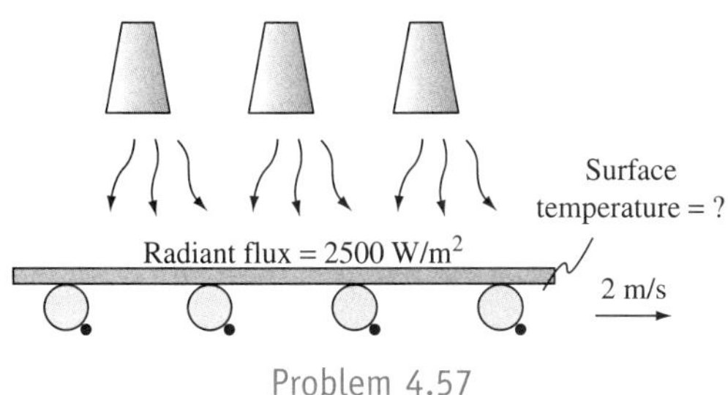

Problem 4.57

4.58 A 4-m^2 square flat-plate solar collector for domestic hot-water heating is shown schematically. Solar radiation at a rate of 750 W/m^2 is incident on the glass cover, which transmits 90% of the incident flux. Water flows through the tubes soldered to the backside of the absorber plate, entering with a temperature of 25°C. The glass cover has a temperature of 27°C in the steady state and radiates heat with an emissivity of 0.92 to the sky at −50°C. In addition, the glass cover loses heat by convection to air at 20°C flowing over its surface at 20 mph.

(a) Calculate the rate at which heat is collected by the working fluid, i.e., the water in the tubes, per unit area of the absorber plate. (b) Calculate the collector efficiency η_c, which is defined as the ratio of useful energy transferred to the water in the tubes to the solar energy incident on the collector cover plate. (c) Calculate the outlet temperature of the water if its flow rate through the collector is 0.02 kg/s. The specific heat of the water is 4179 J/kg K.

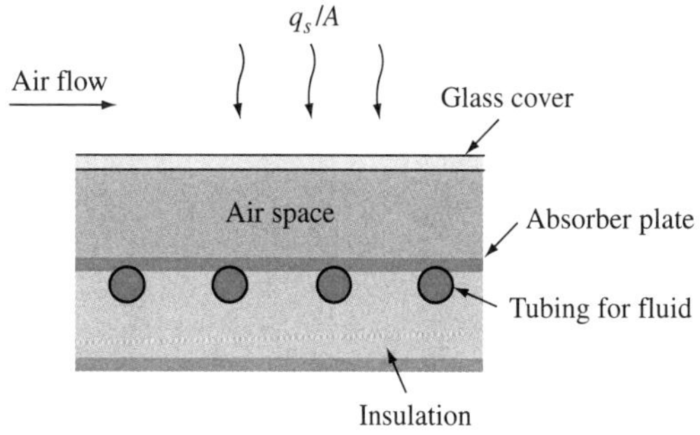

Problem 4.58

4.59 A 1-in.-diameter, 6-in.-long transit rod ($k = 0.56$, Btu/h ft °F, $\rho = 100$ lb/cu ft, $c = 0.20$ Btu/lb °F) on the end of a 1-in.-diameter wood rod at a uniform temperature of 212°F is suddenly placed into a 60°F, 100 ft/sec air stream flowing parallel to the axis of the rod. Estimate the average centerline temperature of the transite rod 8 min after cooling starts. Assume radial heat conduction, but include radiation losses based on an emissivity of 0.90 to black surroundings at the air temperature.

4.60 A highly polished chromium flat plate is placed in a high-speed wind tunnel to simulate flow over the fuselage of a supersonic aircraft. The air flowing in the wind tunnel is at a temperature of 0°C and a pressure of 3500 N/m^2 and has a velocity parallel to the plate of 800 m/s. What is the adiabatic wall temperature in the laminar region, and how long is the laminar boundary layer?

4.61 Air at a static temperature of 70°F and a static pressure of 0.1 psia flows at zero angle of attack over a thin, electrically heated flat plate at a velocity of 200 fps. If the plate is 4 in. long in the direction of flow and 24 in. in the direction normal to the flow, determine the rate of electrical heat dissipation necessary to maintain the plate at an average temperature of 130°F.

4.62 Heat rejection from high-speed racing automobiles is a problem because the required heat exchangers generally create additional drag. Integration of heat rejection into the skin of the vehicle has been proposed for a car to be tested at the Bonneville Salt Flats. Preliminary tests are to be performed in a wind tunnel on a flat plate without heat rejection. Atmospheric air in the tunnel is at 10°C and flows at 250 m/s^{-1} over the 3-m-long thermally nonconducting flat plate. What is the plate temperature 1 m downstream from the leading edge? How much does this temperature differ from the temperature 0.005 m from the leading edge?

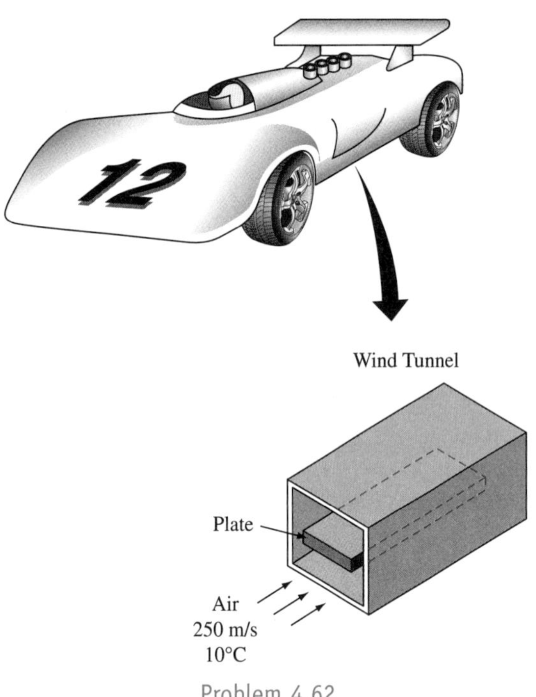

Problem 4.62

4.63 Air at 15°C and 0.01 atmosphere pressure flows at a velocity of 250 m/s over a thin flat strip of metal that is 0.1 m long in the direction of flow. Determine (a) the surface temperature of the plate at equilibrium and (b) the rate of heat removal required per meter width if the surface temperature is to be maintained at 30°C.

4.64 A flat plate is placed in a supersonic wind tunnel with air flowing over it at a Mach number of 2.0, a pressure of 25,000 N/m^2, and an ambient temperature of -15°C. If the plate is 30 cm long in the direction of flow, calculate the cooling rate per unit area that is required to maintain the plate temperature below 120°C.

4.65 A satellite reenters the earth's atmosphere at a velocity of 2700 m/s. Estimate the maximum temperature the heat shield would reach if the shield material is not allowed to ablate and radiation effects are neglected. The temperature of the upper layer of the atmosphere is $\cong -50$°C.

4.66 A scale model of an airplane wing section is tested in a wind tunnel at a Mach number of 1.5. The air pressure and temperature in the test section are 20,000 N/m^2 and -30°C, respectively. If the wing section is to be kept at an average temperature of 60°C, determine the rate of cooling required. The wing model can be approximated by a flat plate of 0.3 m length in the flow direction.

Design Problems

4.1 **Heating Load on Factory** (Continuation of Design Problem 1.3)
In Design Problem 1.3, you calculated the heat loss from a small industrial building in the winter. In the initial calculations, you estimated the convection heat transfer coefficient from Table 1.3. Repeat now the heat loss calculations, but calculate the external heat transfer coefficient from material presented in this chapter. To estimate the forced convection conditions on the building, assume that winds in Denver, Colorado, can reach as high as 70 mph, but under normal conditions will not exceed 20 mph. Discuss the effect of lighting on the heat load and estimate what the effect of electric lights would be if 20 150-W incandescent bulbs are needed to provide adequate lighting for this industrial building. Can you suggest an improved lighting method?

4.2 **Refrigerated Truck** (Continuation of Design Problem 2.10)
Refine the thermal design for the refrigerated shipping container from Design Problem 2.10 by calculating the convection heat transfer coefficient over the two sides and the top at 65 mph from information presented in this chapter. Heat loss from the front and the back of the truck can be estimated from information in Table 1.3. Heat loss from the bottom of the container will be relatively small because it sits close to the road on the bed of the truck.

4.3 **Solar-Heated Pool**
Estimate the heat loss from the surface of a swimming pool with a 3-m × 10-m surface area and an average depth of 1.5 m and then design a solar-heating system for it. Previously, the pool had been heated by an electric heater that can be used as a backup when there is insufficient solar radiation to maintain the pool's temperature, but which can be shut down when there is enough radiation for the collectors to heat the pool. The collector panels are

Retrofit Solar Swimming Pool Heating System

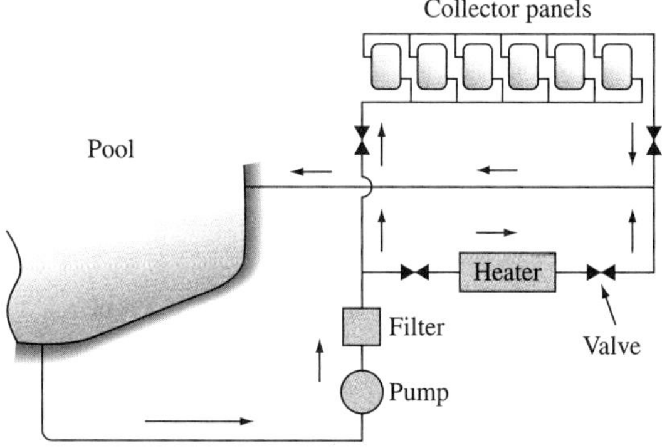

Solar Panel Detail

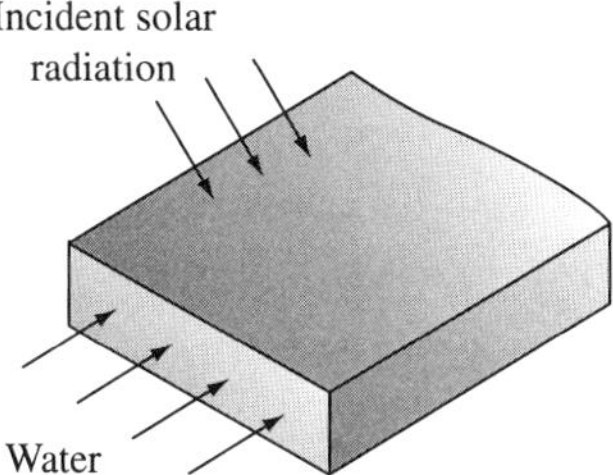

Design Problem 4.3

oriented toward the south at a slope equal to the latitude minus ten degrees, as recommended by solar experts (see, e.g., J. F. Kreider, C. J. Hoogendoorn, and F. Kreith, *Solar Design: Component, Systems, Economics*, Hemisphere Publishing, 1989). The pool is located in San Diego, California, where wind speeds average 10 mph, and is to be maintained at a temperature of 28°C year-round. The solar-heating system is a closed water loop with a piping arrangement that permits flow to the existing heater alone, flow through the electric heater and solar panels, and to the solar panels only as shown in the schematic diagram. The solar collector is to be made of a black plastic extrusion without a cover to minimize cost. The heat transfer coefficient for the water in the rectangular-flow passages of the collector can be estimated from Table 1.3. Suggest ways to reduce the heat loss from the pool at night and estimate the cost-effectiveness of the system.

CHAPTER 5

Natural Convection

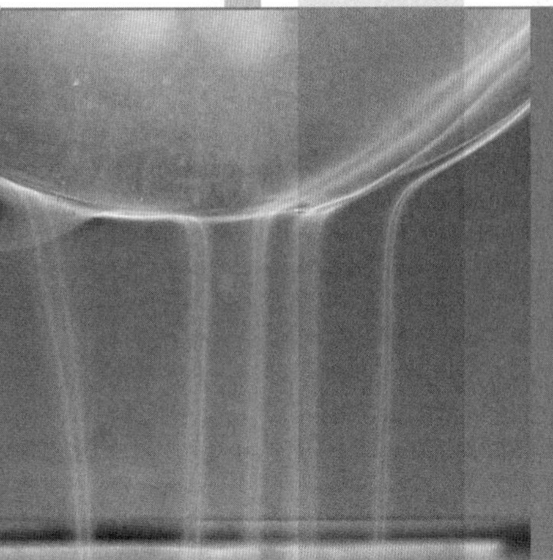

Boundary layer in natural convection around a blunt object depicted by rising warm smoke streaklines impinging on the cooler bluff body.

Source: Courtesy of Sanjeev Sharma and Prof. Jean Hertzberg, Department of Mechanical Engineering, University of Colorado, Boulder, Co.

Concepts and Analyses to Be Learned

Natural-convection heat transfer, also referred to as *free convection* or *buoyancy-induced flow with heat transfer*, is the result of fluid motion produced by density inversion. For example, air in contact with a hot surface gets heated, its density decreases, and in the presence of gravity, it rises upward due to buoyancy. Cold air then moves in from the surroundings to fill this void, and an upward airflow current sets in. The converse process occurs when air comes in contact with a colder surface. It sinks, or moves downward, and a reverse current sets up. This, in essence, describes natural convection, and it is the mode of heat transfer that is observed when, for example, a cup of coffee cools on a table. A study of this chapter will teach you:

- How to mathematically model natural-convection heat transfer for steady-state conditions.
- How to obtain similarity or dimensionless scaling parameters.
- How to apply different correlations to determine natural-convection heat transfer for different geometries and orientation of surfaces, bluff bodies, and enclosed spaces.
- How to determine the influence of natural convection on forced convection.
- How to calculate heat transfer due to natural convection from fins and finned surfaces.

5.1 Introduction

Natural-convection heat transfer occurs whenever a body is placed in a fluid at a higher or a lower temperature than that of the body. As a result of the temperature difference, heat flows between the fluid and the body and causes a change in the density of the fluid in the vicinity of the surface. The difference in density leads to downward flow of the heavier fluid and upward flow of the lighter one. If the motion of the fluid is caused solely by differences in density resulting from temperature gradients and is not aided by a pump or a fan, the associated heat transfer mechanism is called *natural convection.* Natural-convection currents transfer internal energy stored in the fluid in essentially the same manner as forced-convection currents. However, the intensity of the mixing motion is generally less in natural convection, and consequently, the heat transfer coefficients are lower than in forced convection.

Although natural-convection heat transfer coefficients are relatively small, many devices depend largely on this mode of heat transfer for cooling. In the electrical engineering field, transmission lines, transformers, rectifiers, electronic devices, and electrically heated wires such as the heating elements of an electric furnace are cooled in part by natural convection. The temperatures of these bodies rise above that of the surroundings as a result of the heat generated internally. As the temperature difference increases, the rate of heat flow also increases until a state of equilibrium in which the rate of heat generation equal to the rate of heat dissipation is reached.

Natural convection is the dominant heat flow mechanism from steam radiators, the walls of the buildings, or the stationary human body in a quiescent atmosphere. The determination of the heat load on heating and air-conditioning equipment and computers requires, therefore, a knowledge of natural-convection heat transfer coefficients. Natural convection is also responsible for heat losses from pipes carrying steam or other heated fluids. Natural convection has been proposed in nuclear power applications for cooling the surfaces of bodies in which heat is generated by fission [1]. The importance of natural-convection heat transfer has led to the publication of a textbook devoted entirely to the subject [2].

In all of the aforementioned examples, gravitational attraction is the body force responsible for the convection currents. Gravity, however, is not the only body force that can produce natural convection. In certain aircraft applications, there are components such as the blades of gas turbines and helicopter ramjets that rotate at high speeds. Associated with these rotative speeds are large centrifugal forces whose magnitudes, like the gravitational force, are also proportional to the fluid density and hence can generate strong natural-convection currents. Cooling of rotating components by natural convection is therefore feasible even at high heat fluxes.

The fluid velocities in natural-convection currents, especially those generated by gravity, are generally low, but the characteristics of the flow in the vicinity of the heat transfer surface are similar to those in forced convection. A boundary layer forms near the surface, and the fluid velocity at the interface is zero. Figure 5.1 shows the velocity and temperature distributions near a heated flat plate placed in a vertical position in

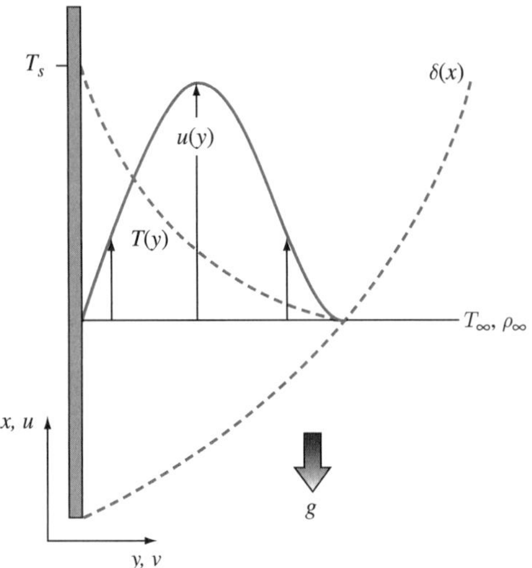

FIGURE 5.1 Temperature and velocity distributions in the vicinity of a heated flat plate placed vertically in still air.
Source: After E. Schmidt and W. Beckmann [3].

air [3]. At a given distance from the bottom of the plate, the local upward velocity increases with increasing distance from the surface to reach a maximum value close to the surface, then decreases and approaches zero again, as shown in Fig. 5.1. Although the velocity profile is different from that observed in forced convection over a flat plate, where the velocity approaches the free-stream velocity asymptotically, in the vicinity of the surface the characteristics of both types of boundary layers are similar. In natural convection, as in forced convection, the flow may be laminar or turbulent, depending on the distance from the leading edge, the fluid properties, the body force, and the temperature difference between the surface and the fluid.

The temperature field in natural convection (Fig. 5.1) is similar to that observed in forced convection. Hence, the physical interpretation of the Nusselt number presented in Section 4.5 applies. For practical applications, however, Newton's equation, Eq. (1.11),

$$dq = h_c \, dA(T_s - T_\infty)$$

is generally used. The equation is written for a differential area dA, because in natural convection, the heat transfer coefficient h_c is not uniform over a surface. As in forced convection over a flat plate, we shall therefore distinguish between a local value of h_c and an average value $\bar{h}_c$ obtained by averaging h_c over the entire surface. The temperature T_∞ refers to a point in the fluid sufficiently removed from the body that the temperature of the fluid is not affected by the presence of a heating (or cooling) source in the body.

Exact evaluation of the heat transfer coefficient for natural convection from the boundary layer is very difficult. The problem has been solved only for simple geometries such as a vertical flat plate and a horizontal cylinder [3, 4]. We shall not discuss these specialized solutions here. Instead, we shall set up the differential equations for natural convection from a vertical flat plate by using only fundamental physical principles. From these equations, without actually solving them, we shall determine the similarity conditions and associated dimensionless parameters that correlate experimental data. In Section 5.3, pertinent experimental data for various shapes of practical interest will be presented in terms of these dimensionless parameters, and their physical significance will be discussed. Section 5.4 treats natural convection from rotating objects, in which the body force due to centrifugal acceleration may be more important than the gravitational body force. Section 5.5 deals with problems in which natural convection and forced convection act at the same time—that is, mixed convection. Section 5.6 treats heat transfer to and from finned surfaces in natural convection.

5.2 Similarity Parameters for Natural Convection

In the analysis of natural convection, we shall make use of a phenomenon that is commonly referred to as buoyancy and is often phrased somewhat as follows: A body immersed in a fluid experiences a buoyant or lifting force equal to the mass of the displaced fluid. Hence, a submerged body rises when its density is less than that of the surrounding fluid and sinks when its density is greater. This in essence is the buoyant effect, and it is the driving force in natural convection.

For the purposes of our analysis, consider a domestic heating panel. The panel can be idealized as a vertical flat plate, very long and very wide in the plane perpendicular to the floor, so that the flow is two dimensional (Fig. 5.2).

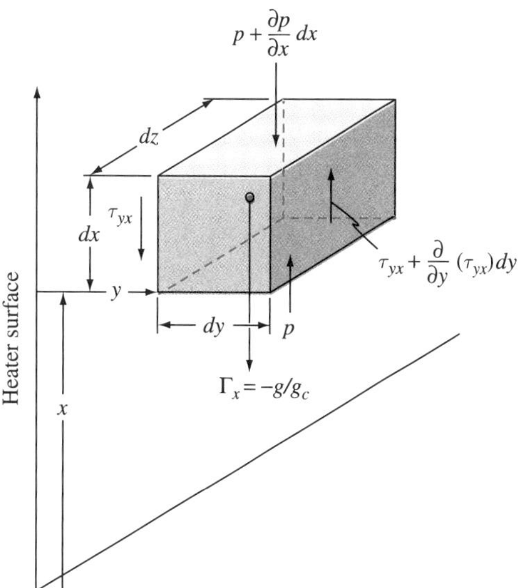

FIGURE 5.2 Forces acting on a fluid element in natural convection flow.

When the heater is turned off, the panel is at the same temperature as the surrounding air. The gravitational or body force acting on each fluid element is in equilibrium with the hydrostatic pressure gradient, and the air is motionless. When the heater is turned on, the fluid in the vicinity of the panel will be heated, and its density will decrease. Hence, the body force (defined as the force per unit mass) on a unit volume in the heated portion of the fluid is less than in the unheated fluid. This imbalance causes the heated fluid to rise—a phenomenon that is well known from experience. In addition to the buoyant force, there are pressure forces and also frictional forces that act when the air is in motion. Once steady-state conditions have been established, the total force on a volume element $dx\,dy\,dz$ in the positive x direction perpendicular to the floor consists of the following:

1. The force due to the pressure gradient

$$p\,dy\,dz - \left(p + \frac{\partial p}{\partial x}dx\right)dy\,dz = -\frac{\partial p}{\partial x}(dx\,dy\,dz)$$

2. The body force $\Gamma_x \rho(dx\,dy\,dz)$, where $\Gamma_x = -g/g_c$, since gravity alone is active*
3. The frictional shearing forces due to the velocity gradient

$$(-\tau_{yx})\,dx\,dz + \left(\tau_{yx} + \frac{\partial \tau_{yx}}{\partial y}dy\right)dx\,dz$$

Since $\tau_{yx} = \mu(\partial u/\partial y)/g_c$ in laminar flow, the net frictional force is

$$\left(\frac{\mu}{g_c}\frac{\partial^2 u}{\partial y^2}\right)dx\,dy\,dz$$

Forces due to the deformation of the fluid element will be neglected in view of the low velocity. Ostrach [1] has shown that the effects of compression work and frictional heat may be important in natural-convection problems when very large temperature differences exist, very large length scales are involved, or very high body forces occur, such as in high-speed rotating machinery.

The rate of change of momentum of the fluid element is $\rho\,dx\,dy\,dz \times [u(\partial u/\partial x) + v(\partial u/\partial y)]$, as shown in Section 4.4. Applying Newton's second law to the elemental volume yields

$$\rho\left(u\frac{\partial u}{\partial x} + v\frac{\partial u}{\partial y}\right) = -g_c\frac{\partial p}{\partial x} - \rho g + \mu\frac{\partial^2 u}{\partial y^2} \tag{5.1}$$

after canceling $dx\,dy\,dz$. The unheated fluid far removed from the plate is in hydrostatic equilibrium, or $g_c(\partial p_e/\partial x) = -\rho_e g$, where the subscript e denotes equilibrium

*g_c is the gravitational constant, equal to 1 kg m/N s^2 in the SI system.

conditions. At any elevation, the pressure is uniform and therefore $\partial p/\partial x = \partial p_e/\partial x$. Substituting $\rho_e g$ for $-(\partial p/\partial x)$ in Eq. (5.1) gives

$$\rho\left(u\frac{\partial u}{\partial x} + v\frac{\partial u}{\partial y}\right) = (\rho_e - \rho)g + \mu\frac{\partial^2 u}{\partial y^2} \tag{5.2}$$

A further simplification can be made by assuming that the density ρ depends only on the temperature and not on the pressure. For an incompressible fluid, this is self-evident, but for a gas, it implies that the vertical dimension of the body is small enough that the hydrostatic density ρ_e is constant. This simplification is referred to as the *Boussinesq approximation.* With these assumptions, the buoyant term can be written

$$g(\rho_e - \rho) = g(\rho_\infty - \rho) = -g\rho\beta(T_\infty - T) \tag{5.3}$$

where β is the coefficient of thermal expansion, defined as

$$\beta = -\frac{1}{\rho}\frac{\partial \rho}{\partial T}\bigg|_p \cong \frac{\rho_\infty - \rho}{\rho(T - T_\infty)} \tag{5.4}$$

For an ideal gas (i.e., $\rho = p/RT$), the coefficient of thermal expansion is

$$\beta = \frac{1}{T_\infty} \tag{5.5}$$

where the temperature T_∞ is the absolute temperature far from the plate.

The equation of motion for natural convection is obtained by substituting the buoyant term, Eq. (5.3), into Eq. (5.2), yielding

$$u\frac{\partial u}{\partial x} + v\frac{\partial u}{\partial y} = g\beta(T - T_\infty) + \nu\frac{\partial^2 u}{\partial y^2} \tag{5.6}$$

In deriving the conservation of energy equation for the flow near the plate, we follow the same steps used in Chapter 4 to derive the conservation of energy equation for the forced flow near a flat plate. This leads to Eq. (4.7b), which also describes the temperature field for the natural-convection problem:

$$u\frac{\partial T}{\partial x} + v\frac{\partial T}{\partial y} = \alpha\frac{\partial^2 T}{\partial y^2} \tag{4.7b}$$

The dimensionless parameters can be determined from the Buckingham π theorem, Section 4.7. We have seven physical quantities:

U_∞ = characteristic velocity

L = characteristic length

g = acceleration due to gravity

β = coefficient of expansion

$(T - T_\infty)$ = temperature difference

ν = kinematic viscosity

α = thermal diffusivity

that can be expressed in four primary dimensions: mass, length, time, and temperature. We should therefore be able to express the dimensionless heat transfer coefficient (Nusselt number) in terms of 7 − 4 = 3 dimensionless groups:

$$\mathrm{Nu} = \mathrm{Nu}(\pi_1, \pi_2, \pi_3) \tag{5.7}$$

Using the method described in Section 4.7, we find

$$\begin{aligned} \pi_1 &= \frac{U_\infty L}{\nu} \\ \pi_2 &= \frac{\nu}{\alpha} \\ \pi_3 &= \frac{g\beta(T - T_\infty)L^3}{\nu^2} \end{aligned} \tag{5.8}$$

We recognize π_1 as the Reynolds number and π_2 as the Prandtl number. The third dimensionless group is called the *Grashof number*, Gr, and represents the ratio of buoyant forces to viscous forces. Consistent units are:

α,ν (m^2/s)	g (m/s^2)
L (m)	β (1/K)
U_∞ (m/s)	$(T - T_\infty)$ (K)

Since the flow velocity is determined by the temperature field, π_1 is not an independent parameter. Therefore, we eliminate the dependence of Nusselt number on π_1. Experimental results for natural-convection heat transfer can therefore be correlated by an equation of the type

$$\mathrm{Nu} = \phi(\mathrm{Gr})\psi(\mathrm{Pr}) \tag{5.9}$$

The Grashof number and the Prandtl number are often grouped together as a product GrPr, which is called the *Rayleigh number*, Ra. Then the Nusselt number relation becomes

$$\mathrm{Nu} = \phi(\mathrm{Ra}) \tag{5.10}$$

Using an equation of this type, experimental data from various sources for natural convection from horizontal wires and tubes of diameter D are correlated in Fig. 5.3 by plotting $\bar{h}_c D/k$, the average Nusselt number, against $c_p\rho^2 g\beta\Delta T D^3/\mu k$, which is the Rayleigh number. The physical properties are evaluated at the film temperature. We observe that data for fluids as different as air, glycerin, and water are well correlated over a range of Rayleigh numbers from 10^{-5} to 10^9 for cylinders ranging from small wires to large pipes.

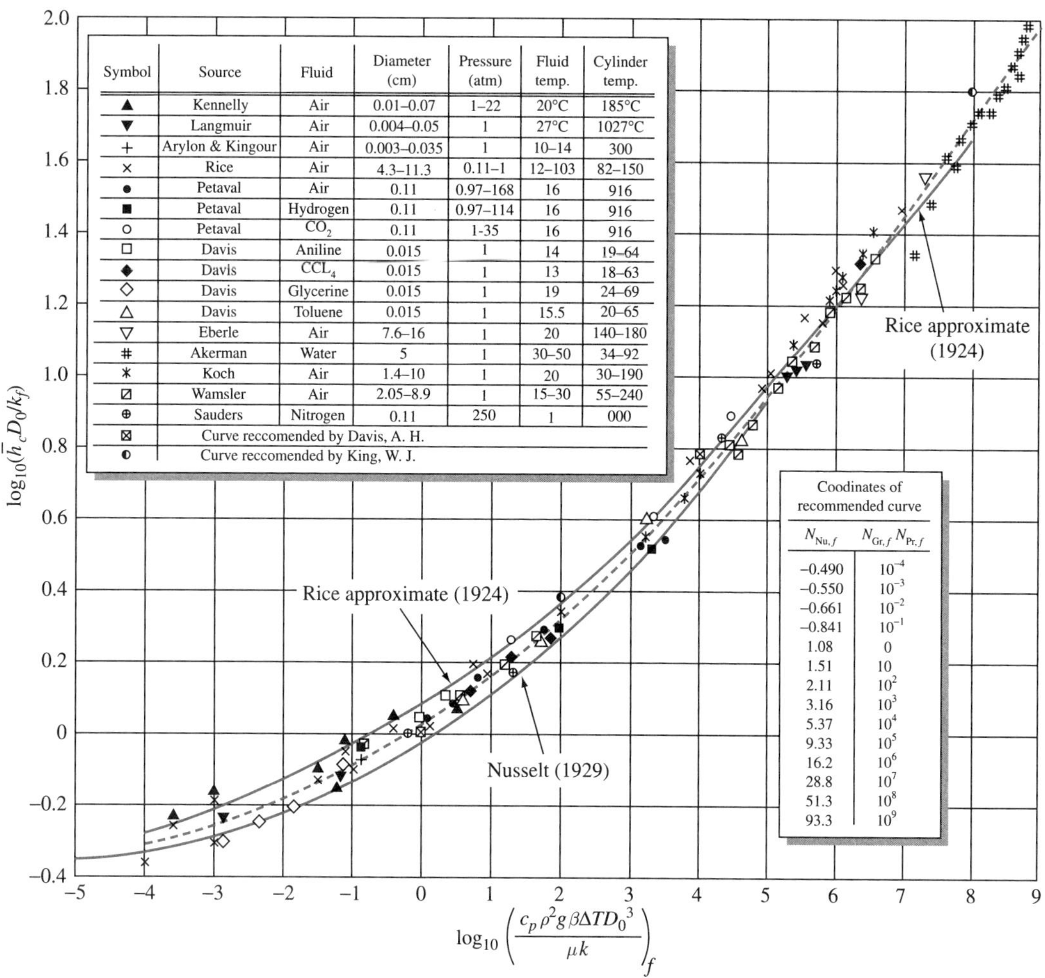

FIGURE 5.3 Correlation of data for natural-convection heat transfer from horizontal cylinders in gases and liquids.

EXAMPLE 5.1 An electrical room heater consists of a horizontal coil of electrical resistance wire, as shown in Fig. 5.4. Such a coil is to be tested at a low power that will result in a wire temperature of 127°C. Calculate the rate of convection heat loss per unit length from the wire, which is 1 mm in diameter. For the purposes of this calculation, the wire can be approximated as being straight and horizontal. Room air is at 27°C. Repeat the calculation for a test performed in a carbon dioxide atmosphere, also at 27°C.

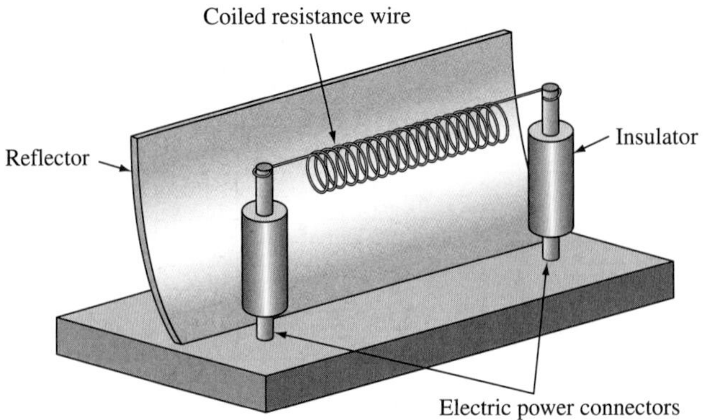

FIGURE 5.4 Schematic diagram of electrical heater for Example 5.1.

SOLUTION Using the film temperature of 77°C to calculate properties from Appendix 2, Table 28, the Rayleigh number is

$$\mathrm{Ra}_D = \frac{g\beta\Delta T D^3}{\nu^2}\mathrm{Pr}$$

$$= \frac{(9.8\ \mathrm{m/s^2})(350\,\mathrm{K})^{-1}(100\ \mathrm{K})(0.001\ \mathrm{m})^3}{(2.12\times 10^{-5}\ \mathrm{m^2/s})^2}(0.71) = 4.43$$

$$\log_{10}\mathrm{Ra}_D = 0.646$$

From Fig. 5.3, $\log_{10}\mathrm{Nu}_D = 0.12$, $\mathrm{Nu}_D = 1.32$, and

$$\bar{h}_c = \frac{(1.32)(0.0291\ \mathrm{W/m\,K})}{0.001\ \mathrm{m}}$$

$$= 38.4\ \mathrm{W/m^2\,K}$$

The rate of heat loss per meter length in air is

$$q = (38.4\ \mathrm{W/m^2\,K})(100\ \mathrm{K})\pi(0.001\ \mathrm{m^2/m})$$

$$= 12.1\ \mathrm{W/m}$$

Using Table 29 in Appendix 2 for properties of carbon dioxide yields

$$\mathrm{Ra}_D = 16.90$$

$$\log_{10}\mathrm{Ra}_D = 1.23$$

$$\log_{10}\mathrm{Nu}_D = 0.21$$

$$\mathrm{Nu}_D = 1.62$$

$$\bar{h}_c = 33.2\ \mathrm{W/m^2K}$$

$$q = 10.4\ \mathrm{W/m}$$

It has been claimed [5] that the correlation in Fig. 5.3 also gives approximate results for three-dimensional shapes such as short cylinders and blocks if the characteristic length dimension is determined by

$$\frac{1}{L} = \frac{1}{L_{\text{hor}}} + \frac{1}{L_{\text{vert}}}$$

where L_{vert} is the height and L_{hor} the average horizontal dimension of the body. Sparrow and Ansari [6], however, have shown that the characteristic length given by this equation may lead to large errors in predicting $\overline{\text{Nu}}_L$ for some three-dimensional bodies. Their data suggest, in fact, that it is likely that no single characteristic length will collapse data for a wide range of geometric shapes and that a separate correlation equation may be required for each shape.

A correlation for natural convection from vertical plates and vertical cylinders is shown in Fig. 5.5.* The ordinate is $\bar{h}_c L/K$, the average Nusselt number based on the height of the body, and the abscissa is $c_p \rho^2 \beta g \Delta T L^3/\mu k$, the Rayleigh number. We note that there is a change in the slope of the line correlating the experimental

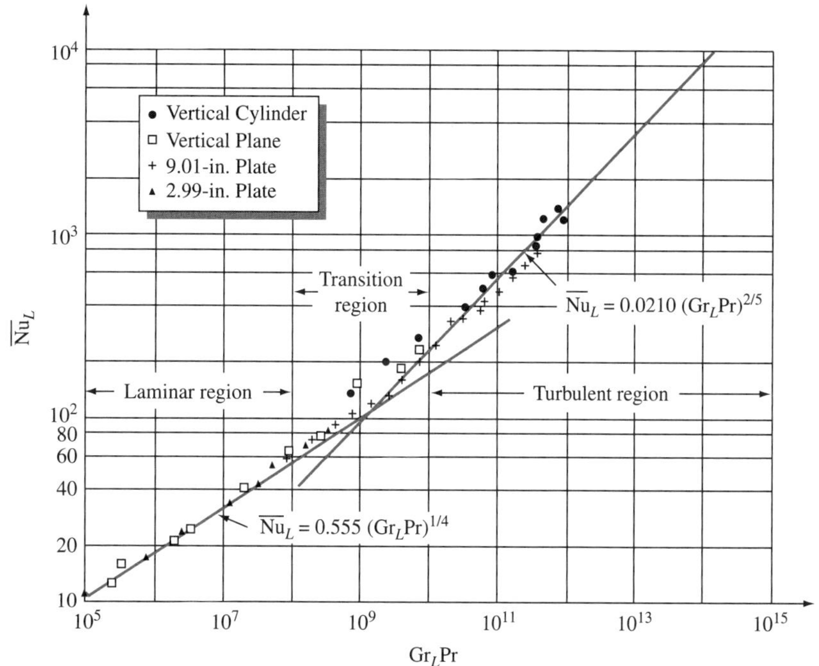

FIGURE 5.5 Correlation of data for natural-convection heat transfer from vertical plates and cylinders [10].

*According to Gebhart [9], a vertical cylinder of diameter D can be treated as a flat plate of height L when $D/L > 35\text{Gr}_L^{-1/4}$.

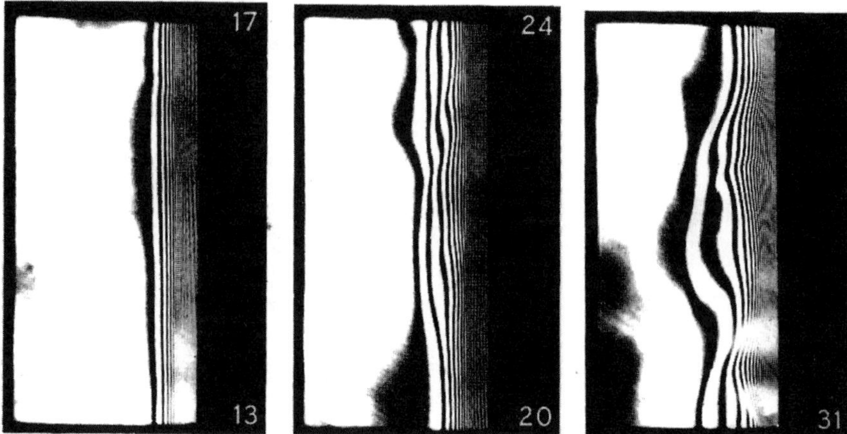

FIGURE 5.6 Interference photograph illustrating laminar and turbulent natural-convection flow of air along a vertical flat plate. Numbers in the photograph indicate height from the bottom edge in inches.

Source: Courtesy of Professor E. R. C. Eckert.

data at a Rayleigh number of 10^9. The reason for the change in slope is that the flow is laminar up to a Rayleigh number of about 10^8, passes through a transition regime between 10^8 and 10^{10}, and becomes fully turbulent at Rayleigh numbers above 10^{10}. These changes are illustrated in the photographs in Fig. 5.6. These pictures show lines of constant density in natural convection from a vertical flat plate to air at atmospheric pressure and were obtained with a Mach-Zehnder optical interferometer [7, 8]. This instrument produces interference fringes that are recorded by a camera. The fringes are the result of density gradients caused by temperature gradients in gases. The spacing of the fringes is a direct measure of the density distribution, which is related to the temperature distribution. Figure 5.6 shows the fringe pattern observed in air near a heated vertical flat plate 0.91 m high and 0.46 m wide. The flow is laminar for about 51 cm from the bottom of the plate. Transition to turbulent flow begins at 53 cm, corresponding to a critical Rayleigh number about 4×10^8. Near the top of the plate, turbulent flow is approached. This type of behavior is typical of natural convection on vertical surfaces, and under normal conditions, the critical value of the Rayleigh number is usually taken as 10^9 for air. Extensive treatments of transition and stability in natural-convection systems are presented in [2] and [9].

When the physical properties of the fluid vary considerably with temperature and the temperature difference between the body surface T_s and the surrounding medium T_∞ is large, satisfactory results can be obtained by evaluating the physical properties in Eq. (5.10) at the mean temperature $(T_s + T_\infty)/2$. However, when the surface temperature is not known, a value must be assumed initially. It then can be used to calculate the heat transfer coefficient to a first approximation. The surface temperature is then recalculated with this value of the heat transfer coefficient, and if there is a significant discrepancy between the assumed and the calculated values of T_s, the latter is used to recalculate the heat transfer coefficient for the second approximation. Correlations that specifically include the effect of variable properties are given by Clausing [11].

EXAMPLE 5.2 The rating for the small vertical-plate resistance heater shown in Fig. 5.7 is to be determined. Estimate the electrical power required to maintain the vertical heater surface at 130°C in ambient air at 20°C. The plate is 15 cm high and 10 cm wide. Compare with results for a plate 450 cm high. The heat transfer coefficient for radiation $\bar{h}_r$ is 8.5 W/m^2 K for the specified surface temperature.

SOLUTION The film temperature is 75°C, and the corresponding value of Gr_L is found to be $65L^3(T_s - T_\infty)$, where L is in centimeters and T is in K, from the last column in Table 28, Appendix 2 by interpolation. For the specified conditions, we get

$$\mathrm{Gr}_L = (65\ \mathrm{cm}^{-3}\ \mathrm{K}^{-1})(15\ \mathrm{cm})^3(110\ \mathrm{K}) = 2.41 \times 10^7$$

for the smaller plate. Since the Grashof number is less than 10^9, the flow is laminar. For air at 75°C, the Prandtl number is 0.71, and GrPr is therefore 1.17×10^7. From Fig. 5.5, the average Nusselt number is 35.7 at GrPr $= 1.17 \times 10^7$, and therefore

$$\bar{h}_c = 35.7\frac{k}{L} = (35.7)\frac{(2.9 \times 10^{-2}\ \mathrm{W/m\,K})}{(0.15\,\mathrm{m})} = 6.90\,\mathrm{W/m^2\,K}$$

Combining the effects of convection and radiation as shown in Chapter 1, the total dissipation rate from both sides of the plate is therefore

$$\begin{aligned} q &= A(\bar{h}_c + \bar{h}_r)(T_s - T_\infty) \\ &= [(2)(0.15)(0.10)\ \mathrm{m}^2][(6.9 + 8.5)\,\mathrm{W/m^2\,K}](110\,\mathrm{K}) = 50.8\,\mathrm{W} \end{aligned}$$

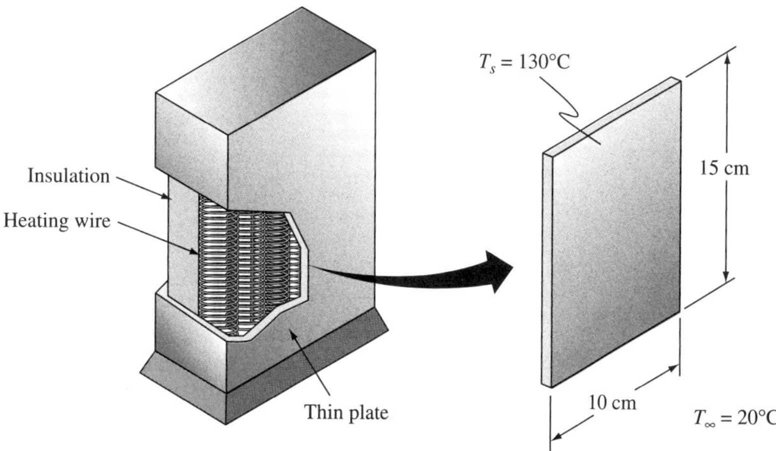

FIGURE 5.7 Schematic diagram of vertical-plate resistance heater for Example 5.2.

For the large plate, the Rayleigh number is $(450/15)^3$ times larger or $\mathrm{Ra} = 4.62 \times 10^{11}$, indicating that the flow is turbulent. From Fig. 5.5, the average Nusselt number is 973 and $\bar{h}_c = 6.3\,\mathrm{W/m^2\,K}$. The total heat dissipation rate from both sides of the plate is therefore

$$q = [(2)(4.5)(0.10)\,\mathrm{m}^2][(6.3 + 8.5)\,\mathrm{W/m^2\,K}](110\,\mathrm{K}) = 1465\,\mathrm{W}$$

5.3 Empirical Correlation for Various Shapes

After experimental data have been correlated by dimensional analysis, it is general practice to determine an equation for the line that best fits the data. It is also useful to compare the experimental results with those obtained by analytic means, if they are available. This comparison allows one to determine whether the analytic method adequately describes the experimental results. If the two agree, one can describe with confidence the physical mechanisms that are important for the problem.

The section presents the results of some experimental studies on natural convection for a number of geometric shapes of practical interest. Each shape is associated with a characteristic dimension, such as its distance from the leading edge x, length L, diameter D, and so on. The characteristic dimension is attached as a subscript to the dimensionless parameters Nu and Gr. Average values of the Nusselt number for a given surface are identified by a bar, that is, $\overline{\mathrm{Nu}}$; local values are shown without a bar. All physical properties are to be evaluated at the arithmetic mean between the surface temperature T_s and the temperature of the undisturbed fluid T_∞. The temperature difference in the Grashof number, ΔT, represents the absolute value of the difference between the temperatures T_s and T_∞. The accuracy with which the heat transfer coefficient can be predicted from any of the equations in practice is generally no better than 20% because most experimental data scatter by as much as $\pm 15\%$ or more, and in most engineering applications, stray currents due to some interaction with surfaces other than the one transferring the heat are unavoidable.

In the following subsections, we present correlation equations for several important geometries. That information is also shown in condensed form in the Closing Remarks, Section 5.7, where brief descriptions and simple illustrations of the geometries are given along with the appropriate correlation equations.

5.3.1 Vertical Plates and Cylinders

For a flat vertical surface, it is possible to find analytical and approximate solutions to the momentum and energy equations, Eqs. (5.6) and (4.7b), using the integral boundary layer analysis introduced in Section 4.9. Details of the method for natural convection can be found in [2]. The results indicate that the local value of the heat transfer coefficient for laminar natural convection from an isothermal vertical plate or cylinder at a distance x from the leading edge is

$$h_{cx} = 0.508\mathrm{Pr}^{1/2}\frac{\mathrm{Gr}_x^{1/4}}{(0.952 + \mathrm{Pr})^{1/4}}\frac{k}{x} \tag{5.11a}$$

and the boundary layer thickness is given by

$$\delta(x) = 4.3x\left[\frac{\mathrm{Pr} + 0.56}{\mathrm{Pr}^2\mathrm{Gr}_x}\right]^{1/4} \tag{5.11b}$$

Since $\mathrm{Gr}_x \sim x^3$, Eq. (5.11a) shows that the heat transfer coefficient decreases with the distance from the leading edge to the 1/4 power, while Eq. (5.11b) shows that the boundary layer thickness increases with $x^{1/4}$. The leading edge is the lower edge for a heated surface and the upper edge for a surface cooler than the surrounding fluid. The average value of the heat transfer coefficient for a height L is obtained by integrating Eq. (5.11a) and dividing by L:

$$\bar{h}_c = \frac{1}{L}\int_0^L h_{cx}\,dx = 0.68\mathrm{Pr}^{1/2}\frac{\mathrm{Gr}_L^{1/4}}{(0.952 + \mathrm{Pr})^{1/4}}\frac{k}{L} \tag{5.12a}$$

In dimensionless form, the average Nusselt number is

$$\overline{\mathrm{Nu}}_L\frac{\bar{h}_c L}{k} = 0.68\mathrm{Pr}^{1/2}\frac{\mathrm{Gr}_L^{1/4}}{(0.952 + \mathrm{Pr})^{1/4}} \tag{5.12b}$$

Gryzagoridis [12] has shown experimentally that Eq. (5.12b) adequately represents the data in the regime $10 < \mathrm{Gr}_L\mathrm{Pr} < 10^8$.

For a vertical plane submerged in a liquid metal ($\mathrm{Pr} < 0.03$), the average Nusselt number in laminar flow is [13]

$$\overline{\mathrm{Nu}}_L\frac{\bar{h}_c L}{k} = 0.68(\mathrm{Gr}_L\mathrm{Pr}^2)^{1/3} \tag{5.12c}$$

For natural convection over a vertical flat plate or vertical cylinder in the turbulent region, the value of h_{cx}, the local heat transfer coefficient, is nearly constant over the surface. In fact, McAdams [5] recommends for $\mathrm{Gr}_L > 10^9$ the equation

$$\overline{\mathrm{Nu}}_L\frac{\bar{h}_c L}{k} = 0.13(\mathrm{Gr}_L\mathrm{Pr})^{1/3} \tag{5.13}$$

According to this equation, the heat transfer coefficient is independent of the length L.

In addition to problems in which the body has a uniform surface temperature, there are sometimes situations, e.g., electric heating, in which a uniform surface heat flux is specified. Since in this case the temperature difference is not known a priori, one must either assume a value and iterate or follow a procedure proposed by Sparrow and Gregg [14], who solved the uniform heat flux problem for a vertical flat plate with various Prandtl numbers in laminar flow. However, experimental data from Dotson [15] indicate that the equations for laminar natural convection from a vertical flat plate apply to a constant surface temperature as well as to a uniform heat flux over the surface (in the latter case, the surface temperature T_s is taken at half of the total height of the plate). An experimental study by Yan and Lin [16] has shown that for constant heat flux the relation for vertical flat plates also can be applied to natural convection for fluids inside vertical pipes at high Rayleigh numbers. Other types of correlations for constant heat flux are presented in [17] and [18].

For a long vertical plate or a long plate tilted at an angle θ from the vertical with the heated surface facing downward (Fig. 5.8a) (or cooled surface facing upward (Fig. 5.8b), Fujii and Imura [19] found that the equation

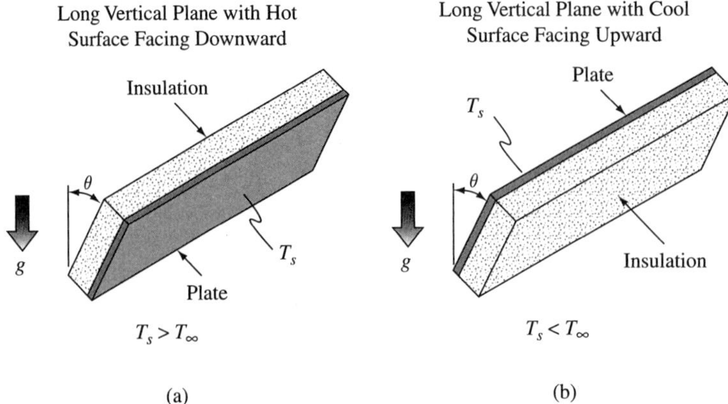

FIGURE 5.8 (a) Long vertical plate with heated surface facing downward, (b) Long vertical plate with cooled surface facing upward.

$$\overline{\mathrm{Nu}}_L = 0.56(\mathrm{Gr}_L \mathrm{Pr}\cos\theta)^{1/4} \tag{5.14}$$

applies in the range

$$10^5 < \mathrm{Gr}_L \mathrm{Pr}\cos\theta < 10^{11} \quad \text{and} \quad 0 \le \theta \le 89°$$

In Eq. (5.14), L is the plate length—the dimension that rotates in a vertical plane as θ increases. If the heated surface is facing upward (or cooled surface facing downward), Eq. (5.13) is recommended.

5.3.2 Horizontal Plates

For two-dimensional horizontal plates as shown in Figs. 5.9 and 5.10, the following equations correlate experimental data [5, 20].

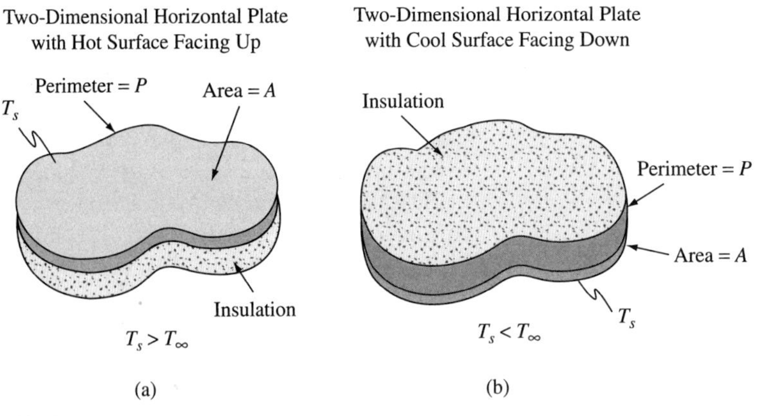

FIGURE 5.9

Two-Dimensional Horizontal Plate with Cool Surface Facing Up

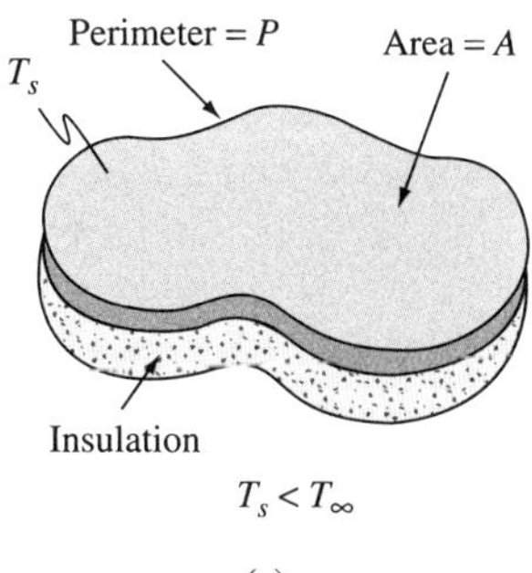

(a)

Two-Dimensional Horizontal Plate with Hot Surface Facing Down

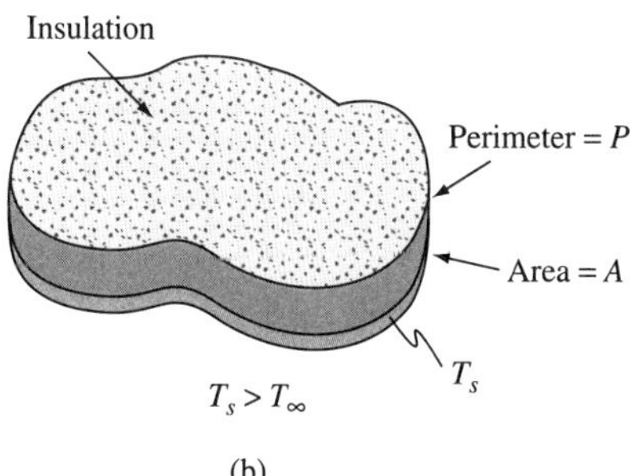

(b)

FIGURE 5.10

Upper surface hot or lower surface cool [Figs. 5.9(a) and (b)]:

$$\overline{\text{Nu}}_L = 0.54\text{Ra}_L^{1/4} \quad (10^5 \lesssim \text{Ra}_L \lesssim 10^7) \tag{5.15}$$

$$\overline{\text{Nu}}_L = 0.15\text{Ra}_L^{1/3} \quad (10^7 \lesssim \text{Ra}_L \lesssim 10^{10}) \tag{5.16}$$

Lower surface hot or upper surface cool [Figs. 5.10(a) and (b)]:

$$\overline{\text{Nu}}_L = 0.27\text{Ra}_L^{1/4} \quad (10^5 \lesssim \text{Ra}_L \lesssim 10^{10}) \tag{5.17}$$

where $L = \dfrac{\text{surface area}}{\text{perimeter}}$

Experimental data for a cooled circular horizontal plate facing down in a liquid metal are correlated by the relation [21]

$$\overline{\text{Nu}}_D = \frac{\bar{h}_c D}{k} = 0.26(\text{Gr}_D\text{Pr}^2)^{0.35} \tag{5.18}$$

EXAMPLE 5.3 Calculate the rate of convection heat loss from the top and bottom of a flat, 1-m square, horizontal restaurant grill heated to 227 °C in ambient air at 27 °C (see Fig. 5.11).

SOLUTION The appropriate length dimension for a square plate is $L^2/4L = 0.25\,\text{m}$. Using properties of air at the mean temperature we find

$$\text{Ra}_L = \frac{(9.8\,\text{m/s}^2)(200\,\text{K})(0.25\,\text{m})^3 0.71}{(396\,\text{K})(2.7 \times 10^{-5}\,\text{m}^2/\text{s})^2} = 7.55 \times 10^7$$

From Eq. (5.17), the Nusselt number for heat transfer from the bottom of the plate is

$$\overline{\text{Nu}}_L = 0.27(7.55 \times 10^7)^{0.25} = 25.2$$

and from Eq. (5.16), the Nusselt number from the top surface is

$$\overline{\text{Nu}}_L = 0.15(7.55 \times 10^7)^{0.33} = 63.4$$

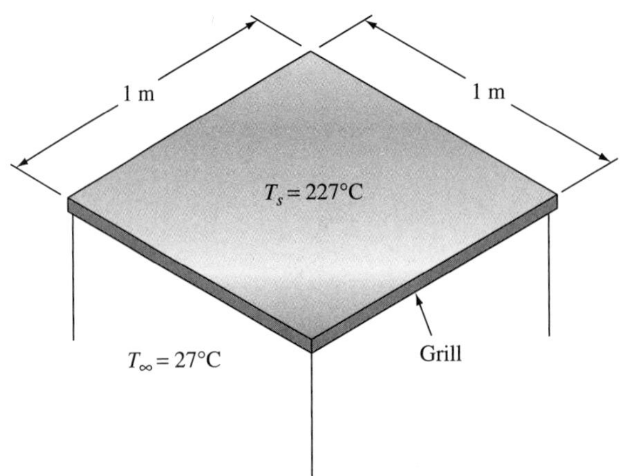

FIGURE 5.11 Schematic diagram of grill for Example 5.3.

The corresponding heat transfer coefficients are

Bottom: $\bar{h}_c = (25.2)(0.032\,\text{W/m K})/(0.25\,\text{m}) = 3.23\,\text{W/m}^2\text{K}$

Top: $\bar{h}_c = (63.4)(0.032\,\text{W/m K})/(0.25\,\text{m}) = 8.11\,\text{W/m}^2\text{K}$

Hence, the total convection heat loss is

$$q = (1\,\text{m}^2)(3.23 + 8.11)(\text{W/m}^2\,\text{K})(200\,\text{K}) = 2268\,\text{W}$$

Note that the heat dissipated by the upward-facing surface is nearly 72% of the total.

5.3.3 Cylinders, Spheres, Cones, and Three-Dimensional Bodies

The temperature field around a horizontal cylinder heated in air is illustrated in Fig. 5.12, which shows interference fringes photographed by Eckert and Soehnghen [8]. The flow is laminar over the entire surface. The closer spacing of the interference fringes over the lower portion of the cylinder indicates a steeper temperature gradient and consequently a larger local heat transfer coefficient than over the top portion. The variation of the heat transfer coefficient with angular position α is shown in Fig. 5.13 for two Grashof numbers. The experimental results do not differ appreciably from the theoretical calculations of Hermann [4], who derived the equation

$$\text{Nu}_{D\alpha} = 0.604\,\text{Gr}_D^{1/4}\phi(\alpha) \tag{5.19}$$

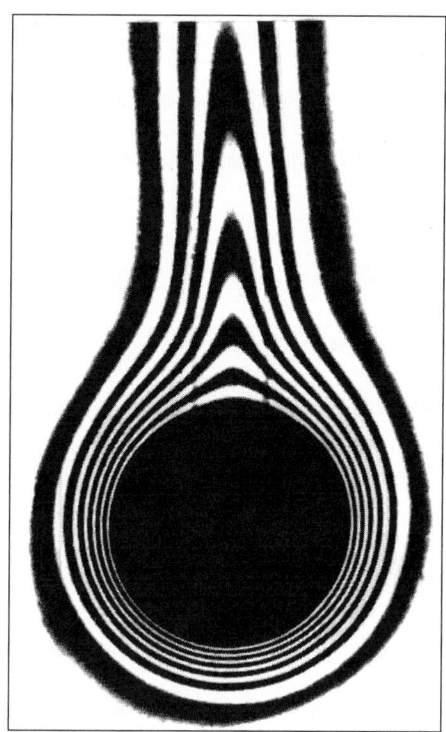

FIGURE 5.12 Interference photograph illustrating temperature field around a horizontal cylinder in laminar flow.

Source: Courtesy of Professor E. R. G. Eckert.

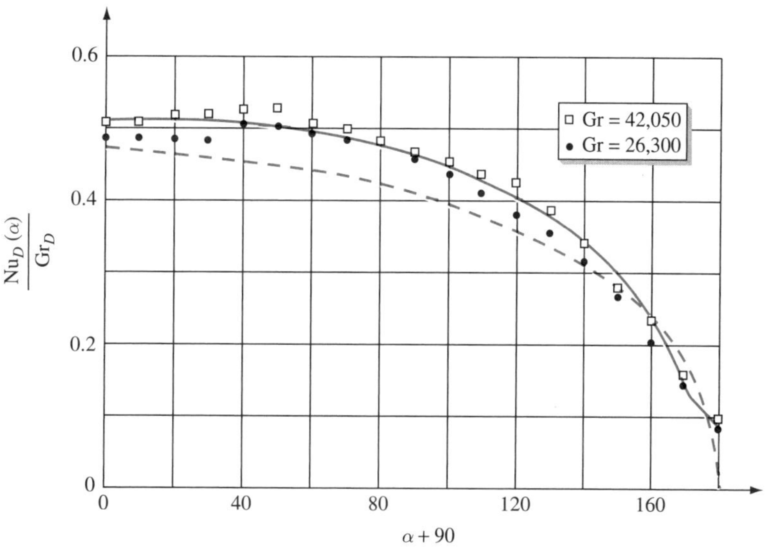

FIGURE 5.13 Local dimensionless heat transfer coefficient along the circumference of a horizontal cylinder in laminar natural convection.

Source: E. R. G. Eckert and E. E. Soehnghen, "Studies on Heat Transfer in Laminar Free Convection with the Zehnder-Mach Interferometer," USAF Technical Report 5747, December 1948; dashed line according to Hermann [4].

for air, that is, Pr = 0.71. The angle α is measured from the horizontal position and numerical values of the function $\phi(\alpha)$ are as follows:

	Bottom Half			Top Half				
α	−90	−60	−30	0	30	60	75	90
$\phi(\alpha)$	0.76	0.75	0.72	0.66	0.58	0.46	0.36	0

An equation for the average heat transfer coefficient from single horizontal wires or pipes in natural convection, based on the experimental data in Fig. 5.3, is

$$\overline{\mathrm{Nu}_D} = 0.53(\mathrm{Gr}_D \mathrm{Pr})^{1/4} \tag{5.20}$$

This equation is valid for Prandtl numbers larger than 0.5 and Grashof numbers ranging from 10^3 to 10^9. For very small diameters, Langmuir showed that the rate of heat dissipation per unit length is nearly independent of the wire diameter, a phenomenon he applied in his invention of the coiled filaments in gas-filled incandescent lamps. The average Nusselt number for Gr_D less than 10^3 is most conveniently evaluated from the dashed line drawn through the experimental points in Fig. 5.3 in the low-Grashof-number range.

In turbulent flow, it has been observed that the heat flux can be increased substantially without a corresponding increase in the surface temperature. It appears that in natural convection the turbulent-exchange mechanism increases in intensity as the rate of heat flow is increased and thereby reduces the thermal resistance.

EXAMPLE 5.4 At what temperature will a long, heated, horizontal steel pipe 1 m in diameter produce turbulent flow in air at 27 °C? Repeat for the case where the pipe is placed in a water bath at 27 °C. Use property values at 27 °C.

SOLUTION The criterion for transition is $\mathrm{Ra}_D = 10^9$. For air at 27 °C, this gives

$$\mathrm{Ra}_D = \frac{(9.8\,\mathrm{m/s^2})(300\,\mathrm{K})^{-1}(\Delta T)(1\,\mathrm{m})^3(0.71)}{(1.64 \times 10^{-5}\,\mathrm{m^2/s})^2} = 10^9$$

Therefore,

$$\Delta T = 12°\mathrm{C}$$
$$T_{\mathrm{pipe}} = 12 + 27 = 39°\mathrm{C}$$

For water (Table 13, Appendix 2), we get

$$\mathrm{Ra}_D = \frac{(9.8\,\mathrm{m/s^2})(2.73 \times 10^{-4}\,\mathrm{K^{-1}})(\Delta T)(1\,\mathrm{m})^3(5.9)}{(0.861 \times 10^{-6}\,\mathrm{m^2/s})^2} = 10^9$$

Solving for ΔT, we find $\Delta T = 0.05°\mathrm{C}$. Note that in water even a small temperature difference will induce turbulence.

For liquid metals in laminar flow, the equation

$$\overline{\mathrm{Nu}}_D = 0.53(\mathrm{Gr}_D\mathrm{Pr}^2)^{1/4} \tag{5.21}$$

correlates the available data [22] for horizontal cylinders.

Al-Arabi and Khamis [23] have correlated heat transfer data for cylinders of various lengths, diameters, and angles of inclination from the vertical, as shown in Fig. 5.14. Their results are of the form $\overline{\mathrm{Nu}}_L = m(\mathrm{Gr}_L\mathrm{Pr})^n$, where m and n are functions of the cylinder diameter and angle of inclination from the vertical, θ. Transition to turbulent flow occurred near

$$(\mathrm{Gr}_L\mathrm{Pr})_{\mathrm{cr}} = 2.6 \times 10^9 + 1.1 \times 10^9 \tan\theta \tag{5.22}$$

In the laminar regime, $9.88 \times 10^7 \le \mathrm{Gr}_L\mathrm{Pr} \le (\mathrm{Gr}_L\mathrm{Pr})_{\mathrm{cr}}$, they found

$$\overline{\mathrm{Nu}}_L = [2.9 - 2.32(\sin\theta)^{0.8}](\mathrm{Gr}_D)^{-1/12}(\mathrm{Gr}_L\mathrm{Pr})^{[1/4+(1/12)(\sin\theta)1.2]} \tag{5.23}$$

and in the turbulent regime, $(\mathrm{Gr}_L\mathrm{Pr})_{\mathrm{cr}} \le \mathrm{Gr}_L\mathrm{Pr} \le 2.95 \times 10^{10}$, they found

$$\overline{\mathrm{Nu}}_L = [0.47 + 0.11(\sin\theta)^{0.8}](\mathrm{Gr}_D)^{-1/12}(\mathrm{Gr}_L\mathrm{Pr})^{1/3} \tag{5.24}$$

In both regimes, the Grashof number based on the cylinder diameter is restricted to the range $1.08 \times 10^4 \le \mathrm{Gr}_D \le 6.9 \times 10^5$.

Sparrow and Stretton [24] have correlated natural-convection data for cubes, spheres, and short vertical cylinders over a Rayleigh-number range from about 200 to 1.5×10^9 by the empirical relation:

$$\mathrm{Nu}_{L^+} = 5.75 + 0.75[\mathrm{Ra}_{L^+}/F(\mathrm{Pr})]^{0.252} \tag{5.25}$$

where

$$F(\mathrm{Pr}) = [1 + (0.49/\mathrm{Pr})^{9/16}]^{16/9}$$

In Eq. (5.25), the length dimension in Nu_{L^+} and Ra_{L^+} is defined by the relation

$$L^+ = A/(4A_{\mathrm{horiz}}/\pi)^{1/2}$$

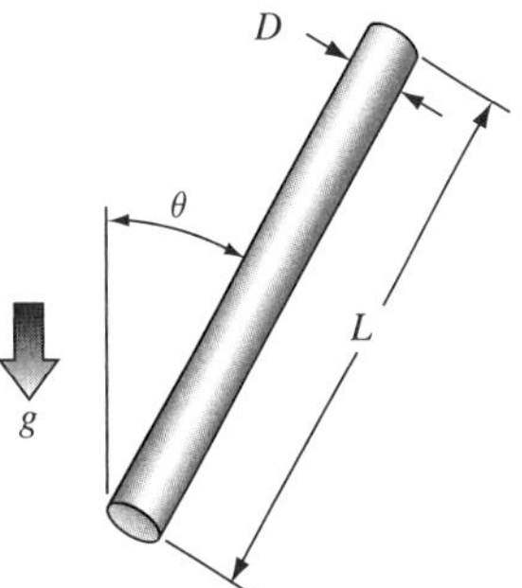

FIGURE 5.14 Nomenclature for heated or cooled finite cylinder of length L and diameter D inclined from vertical.

where A = the surface area of the body

A_{horiz} = the area of horizontal projection of the body

For example, for a cube with sides of length S and one surface horizontal,

$$L^+ = \frac{6S^2}{\sqrt{\dfrac{4S^2}{\pi}}} = 5.32S$$

while for a sphere of diameter D,

$$L^+ = \frac{\pi D^2}{\sqrt{\dfrac{4}{\pi}\dfrac{\pi D^2}{4}}} = \pi D$$

For natural convection to or from small spheres of diameter D, the empirical equation

$$\overline{\text{Nu}}_D = 2 + 0.392(\text{Gr}_D)^{1/4} \quad \text{for } 1 < \text{Gr}_D < 10^5 \tag{5.26}$$

is recommended [25]. For very small spheres, as the Grashof number approaches zero, the Nusselt number approaches a value of 2, that is, $\bar{h}_c D/k \rightarrow 2$. This condition corresponds to pure conduction through a stagnant layer of fluid surrounding the sphere.

Experimental data for natural convection from vertical cones pointing downward with vertex angles between 3° and 12° have been correlated [26] by

$$\overline{\text{Nu}}_L = 0.63(1 + 0.72\varepsilon)\text{Gr}_L^{1/4} \tag{5.27}$$

where $3° < \phi < 12°$, $7.5 < \log \text{Gr}_L < 8.7$, $0.2 \leq \varepsilon \leq 0.8$

$$\varepsilon = \frac{2}{\text{Gr}_L^{1/4} \tan(\phi/2)}$$

ϕ = vertex angle

L = slant height of the cone

5.3.4 Enclosed Spaces

Natural-convection heat transfer across enclosures such as shown schematically in Fig. 5.15 is important for determining heat loss through double-glazed windows, from flat-plate solar collectors, through building walls, and in many other applications. If the enclosure consists of two isothermal parallel surfaces at temperatures T_1 and T_2 spaced a distance δ apart and of height L and the top and bottom of the enclosure are insulated, the Grashof number is defined by

$$\text{Gr}_\delta = \frac{g\beta(T_1 - T_2)\delta^3}{\nu^2}$$

and the parameter L/δ is called the *aspect ratio.* A temperature difference will produce flow in the enclosure. In vertical cavities ($\tau = 90°$), Hollands and

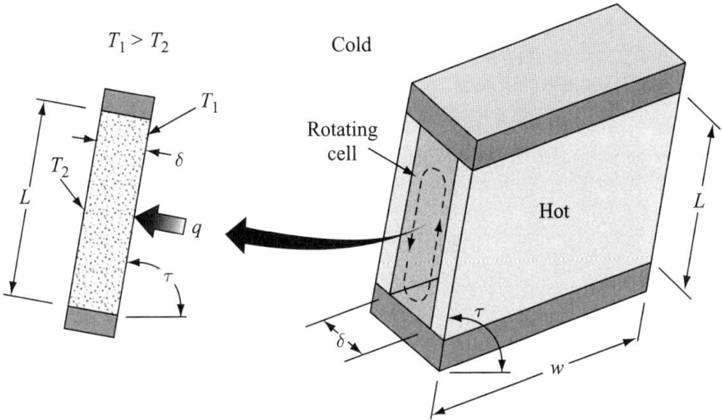

FIGURE 5.15 Nomenclature for natural convection in inclined enclosed spaces.

Konicek [27] found that for $\mathrm{Gr}_\delta \gtrsim 8000$ the flow consists of one large cell rotating in the enclosure. The heat transfer mechanism is essentially by conduction across the enclosure for $\mathrm{Gr}_\delta < 8000$. As the Grashof number is increased beyond this value, the flow becomes more of a boundary layer type with fluid rising in a layer near the heated surface, turning the corner at the top, and flowing downward in a layer near the cooled surface. The boundary layer thickness decreases with $\mathrm{Gr}_\delta^{1/4}$, and the core region is more or less inactive and thermally stratified.

For the geometry in Fig. 5.15 with $\tau = 90°$, Catton [28] recommends the correlations of Berkovsky and Polevikov:

$$\overline{\mathrm{Nu}}_\delta = 0.22\left(\frac{L}{\delta}\right)^{-1/4}\left(\frac{\mathrm{Pr}}{0.2 + \mathrm{Pr}}\mathrm{Ra}_\delta\right)^{0.28} \tag{5.28a}$$

in the range

$$2 < L/\delta < 10, \quad \mathrm{Pr} < 10, \quad \text{and} \quad \mathrm{Ra}_\delta < 10^{10}$$

and

$$\overline{\mathrm{Nu}}_\delta = 0.18\left(\frac{\mathrm{Pr}}{0.2 + \mathrm{Pr}}\mathrm{Ra}_\delta\right)^{0.29} \tag{5.28b}$$

in the range

$$1 < L/\delta < 2, \quad 10^{-3} < \mathrm{Pr} < 10^5, \quad \text{and} \quad 10^3 < \frac{\mathrm{Ra}_\delta \mathrm{Pr}}{0.2 + \mathrm{Pr}}$$

For larger aspect ratios and $\tau = 90°$, the following relation is recommended [29]:

$$\overline{\mathrm{Nu}}_\delta = 0.42\mathrm{Ra}_\delta^{0.25}\mathrm{Pr}^{0.012}/(L/\delta)^{0.3} \tag{5.29a}$$

in the range $10 < L/\delta < 40$, $1 < \mathrm{Pr} < 2 \times 10^4$, and $10^4 < \mathrm{Ra}_\delta < 10^7$.

For larger Rayleigh numbers in the range $10^6 < \mathrm{Ra}_\delta < 10^9$, aspect ratios in the range $1 < L/\delta < 40$ and $1 < \mathrm{Pr} < 20$, the relation

$$\overline{\mathrm{Nu}}_\delta = 0.046\,\mathrm{Ra}_\delta^{0.33} \tag{5.29b}$$

is recommended [29]. All properties in Eqs. (5.28) and (5.29) are to be evaluated at the mean temperature $(T_1 + T_2)/2$.

Data are lacking for aspect ratios less than one. Imberger [30] found that as $\mathrm{Ra}_\delta \rightarrow \infty$, $\mathrm{Nu}_\delta \rightarrow (L/\delta)\mathrm{Ra}_\delta^{1/4}$ for $L/\delta = 0.01$ and 0.02. Bejan et al. [31] found that $\mathrm{Nu}_\delta = 0.014\mathrm{Ra}_\delta^{0.38}$ for $L/\delta = 0.0625$ and $2 \times 10^8 < \mathrm{Ra}_\delta < 2 \times 10^9$. Nansteel and Greif [32] found $\mathrm{Nu}_\delta = 0.748\mathrm{Ra}_\delta^{0.226}$ for $L/\delta = 0.5$, $2 \times 10^{10} < \mathrm{Ra}_\delta \leq 10^{11}$, and $3.0 \leq \mathrm{Pr} \leq 4.3$.

In a horizontal fluid layer with heating from above, heat transfer is by conduction only. Heating from below results in conduction heat transfer only if $\mathrm{Ra}_\delta < 1700$, where the length scale is the spacing enclosing layer. Above this value of Ra_δ, the fluid motion is in the form of multiple cells rotating with a horizontal axis, which are known as Benard cells. The flow begins to become turbulent for $\mathrm{Ra}_\delta \sim 5500$ for $\mathrm{Pr} = 0.7$ and for $\mathrm{Ra}_\delta \sim 55{,}000$ for $\mathrm{Pr} = 8500$ [34] and becomes fully turbulent for $\mathrm{Ra}_\delta \sim 10^6$.

Hollands et al. [34] correlated data for horizontal air layers contained between two flat plates and heated from below (see Fig. 5.16) over a very wide range of Rayleigh numbers with

$$\overline{\mathrm{Nu}}_\delta = 1 + 1.44\left[1 - \frac{1708}{\mathrm{Ra}_\delta}\right]^{\cdot} + \left[\left(\frac{\mathrm{Ra}_\delta}{5830}\right)^{1/3} - 1\right]^{\cdot} \tag{5.30a}$$

where the notation $[\ ]^{\cdot}$ indicates that if the quantity inside the bracket is negative the quantity is to be taken as zero. This equation closely represented data for air from the critical Rayleigh number ($\mathrm{Ra}_\delta = 1700$) to $\mathrm{Ra}_\delta = 10^8$. To closely match data for water, it was necessary to add a term to the previous equation:

$$\overline{\mathrm{Nu}}_\delta = 1 + 1.44\left[1 - \frac{1708}{\mathrm{Ra}_\delta}\right]^{\cdot} + \left[\left(\frac{\mathrm{Ra}_\delta}{5830}\right)^{1/3} - 1\right]^{\cdot} + 2.0\left[\frac{\mathrm{Ra}_\delta^{1/3}}{140}\right]^{[1 - \ln(\mathrm{Ra}_\delta^{1/3}/140)]} \tag{5.30b}$$

which is then valid from the critical Rayleigh number ($\sim$1700) to $\mathrm{Ra}_\delta = 3.5 \times 10^9$. These two correlation equations are shown with experimental data in Figs. 5.17 and 5.18.

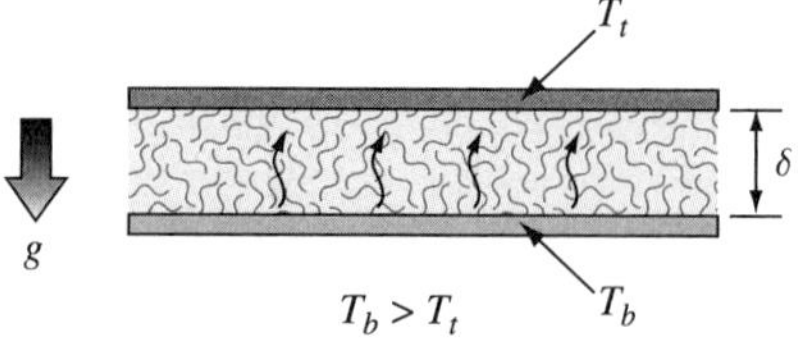

FIGURE 5.16 Horizontal air layer heated from below.

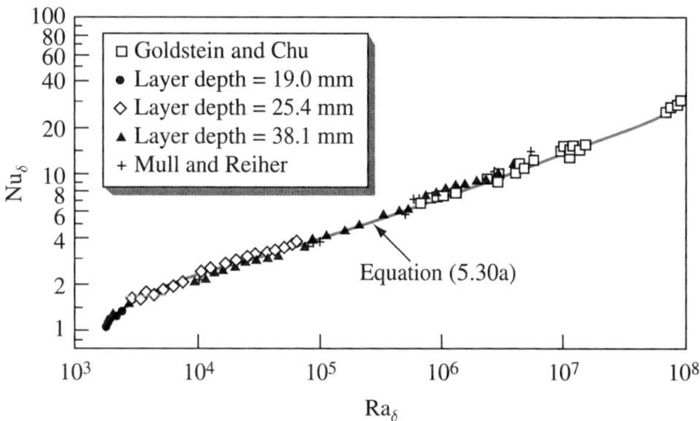

FIGURE 5.17 Correlation of data for natural-convection heat transfer across a horizontal layer of air contained between two flat plates and heated from below.

Source: Reprinted from Krisnamurti [34] with permission from Pergamon Press, Ltd.

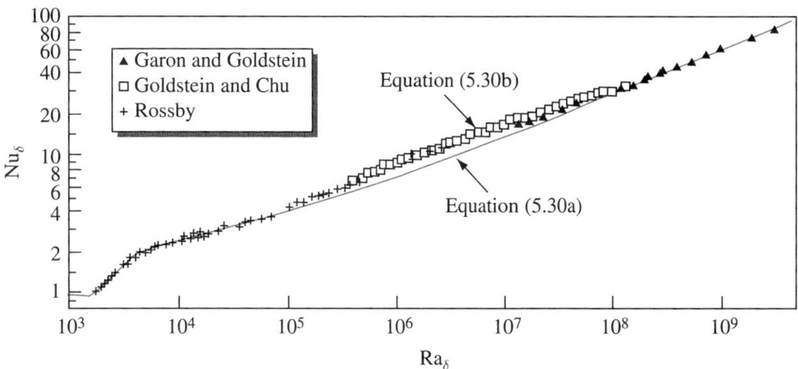

FIGURE 5.18 Correlation of data for natural-convection heat transfer across a horizontal layer of water heated from below.

Source: Reprinted from Hollands et al. [34] with permission from Pergamon Press, Ltd.

EXAMPLE 5.5 A covered pan of water 8 cm deep is placed on a stove-top burner as shown in Fig. 5.19. The burner element is thermostatically controlled and maintains the bottom of the pan at 100 °C. Assuming the top surface of the water is initially at room temperature, 20 °C, what is the initial rate of heat transfer from the burner to the water? The pan is circular and is 15 cm in diameter.

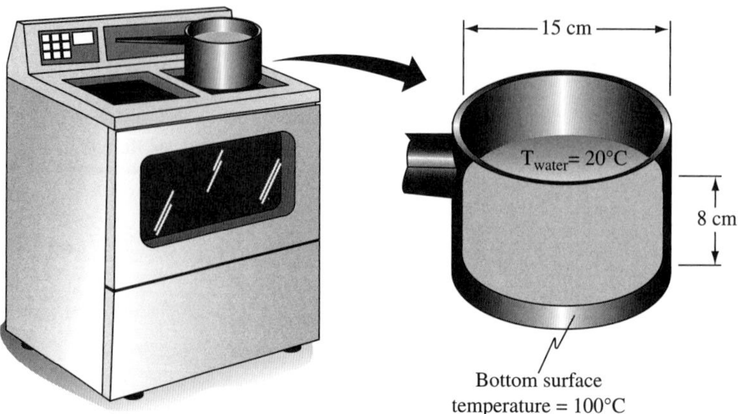

FIGURE 5.19 Schematic diagram for Example 5.5.

SOLUTION For the properties of water at 60 °C, we have

$$\mathrm{Ra}_\delta = \frac{(9.8\ \mathrm{m/s^2})(5.18 \times 10^{-4}\ \mathrm{K^{-1}})(80\ \mathrm{K})(0.08\ \mathrm{m})^3(3.02)}{(0.478 \times 10^{-6}\ \mathrm{m^2/s})^2}$$

$$= 2.75 \times 10^9$$

From Eq. (5.30b), we find

$$\overline{\mathrm{Nu}}_\delta = 1 + 1.44 + 76.8 + 0.1 = 79.3$$

$$\bar{h}_c = \overline{\mathrm{Nu}}_\delta \frac{k}{\delta} = \frac{(79.3)(0.657\ \mathrm{W/m\ K})}{0.08\ \mathrm{m}} = 651\ \mathrm{W/m^2K}$$

The initial rate of heat transfer is therefore

$$q = (651\,\mathrm{W/m^2\,K})\left(\frac{\pi\, 0.15^2\,\mathrm{m^2}}{4}\right)(80\,\mathrm{K})$$

$$= 920\,\mathrm{W}$$

Natural convection in a cavity formed between two inclined plates (see Fig. 5.15) is encountered in flat-plate solar collectors and in double-glazed windows ($\tau = 90°$). This configuration has been investigated for large aspect ratios ($L/\delta > 12$) by Hollands et al. [35]. They found that the following equation correlated experimental data at tilt angles, τ, less than 70°:

$$\overline{\mathrm{Nu}}_L = 1 + 1.44\left[1 - \frac{1708}{\mathrm{Ra}_L \cos\tau}\right]^{\cdot}\left[1 - \frac{1708(\sin 1.8\tau)^{1.6}}{\mathrm{Ra}_L \cos\tau}\right]$$

$$+ \left[\left(\frac{\mathrm{Ra}_L \cos\tau}{5830}\right)^{1/3} - 1\right]^{\cdot}$$

Again, the notation [ ] implies that, if the quantity in brackets is negative, it must be set equal to zero. The implication is that, if the Rayleigh number is less than a critical value $\text{Ra}_{L,c} = 1708/\cos\tau$, there is no flow within the cavity.

For tilt angles between 70° and 90°, Catton [28] recommends that the Nusselt number for a vertical enclosure ($\tau = 90°$) be multiplied by $(\sin\tau)^{1/4}$, i.e., $\overline{\text{Nu}}_L(\tau) = \overline{\text{Nu}}_L(\tau = 90°)\sin\tau^{1/4}$. Catton also gives correlations for aspect ratios less than 12.

For natural convection inside spherical cavities of diameter D, the relation

$$\frac{D\bar{h}_c}{k} = C(\text{Gr}_D\text{Pr})^n \tag{5.32}$$

is recommended [36] with the constants C and n selected from the tabulation below:

Gr_DPr	C	n
10^4–10^9	0.59	1/4
10^9–10^{12}	0.13	1/3

For natural-convection heat transfer across the gap between two horizontal concentric cylinders as shown in Fig. 5.20, Raithby and Hollands [37] suggest the correlation equation

$$\frac{k_{\text{eff}}}{k} = 0.386\left[\frac{\ln(D_o/D_i)}{b^{3/4}(1/D_i^{3/5} + 1/D_o^{3/5})^{5/4}}\right]\left(\frac{\text{Pr}}{0.861 + \text{Pr}}\right)^{1/4}\text{Ra}_b^{1/4} \tag{5.33}$$

Here, D_o is the diameter of the outer cylinder, D_i is the diameter of the inner cylinder, $2b = D_o - D_i$, and the Rayleigh number Ra_b is based on the temperature difference across the gap. The effective thermal conductivity k_{eff} is the thermal conductivity that a motionless fluid (with conductivity k) in the gap must have to transfer the same amount of heat as the moving fluid.

The correlation equation, Eq. (5.33), is valid over the following range of parameters:

$$0.70 \le \text{Pr} \le 6000$$

$$10 \le \left[\frac{\ln(D_o/D_i)}{b^{3/4}(1/D_i^{3/5} + 1/D_o^{3/5})^{5/4}}\right]^4 \text{Ra}_b \le 10^7$$

FIGURE 5.20 Nomenclature in natural convection between two horizontal concentric cylinders.

For concentric spheres, Raithby and Hollands [37] recommend

$$\frac{k_{\text{eff}}}{k} = 0.74\left[\frac{b^{1/4}}{D_oD_i(D_i^{-7/5} + D_o^{-7/5})^{5/4}}\right]\text{Ra}_b^{1/4}\left(\frac{\text{Pr}}{0.861 + \text{Pr}}\right)^{1/4} \tag{5.34}$$

Eq. (5.34) is valid for

$$0.70 \le \text{Pr} \le 4200$$

and

$$10 \le \left[\frac{b}{(D_oD_i)^4(D_i^{-7/5} + D_o^{-7/5})^5}\right]\text{Ra}_b \le 10^7$$

where $2b = D_o - D_i$.

5.4* Rotating Cylinders, Disks, and Spheres

Heat transfer by convection between a rotating body and a surrounding fluid is of importance in the thermal analysis of shafting, flywheels, turbine rotors, and other rotating components of various machines. Convection from a heated rotating horizontal cylinder to ambient air has been studied by Anderson and Saunders [38].

With heat transfer, a critical velocity is reached when the circumferential speed of the cylinder surface becomes approximately equal to the upward natural-convection velocity at the side of a heated stationary cylinder. Below the critical velocity, simple natural convection, characterized by the conventional Grashof number $\beta g(T_s - T_\infty)D^3/\nu^2$, controls the rate of heat transfer. At speeds greater than critical ($\text{Re}_\omega > 8000$ in air), the peripheral-speed Reynolds number $\pi D^2\omega/\nu$ becomes the controlling parameter. The combined effects of the Reynolds, Prandtl, and Grashof numbers on the average Nusselt number for a horizontal cylinder rotating in air above the critical velocity (see Fig. 5.21) can be expressed by the empirical equation [39]

$$\overline{\text{Nu}}_D = \frac{\bar{h}_c D}{k} = 0.11(0.5\text{Re}_\omega^2 + \text{Gr}_D\text{Pr})^{0.35} \tag{5.35}$$

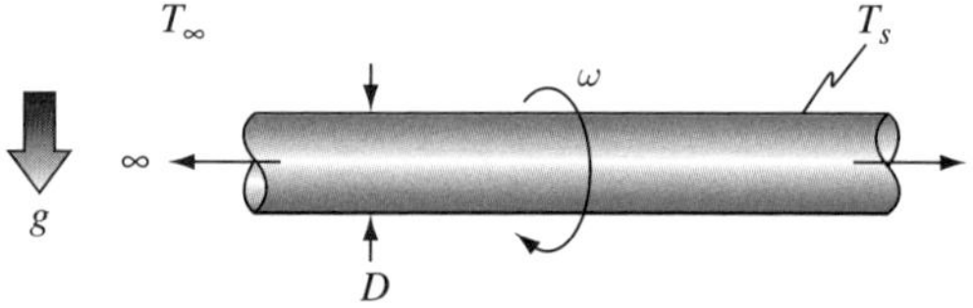

$\text{Gr} = \rho^2\beta g\,(T_s - T_\infty)D^3/\mu^2$
$\text{Re}_\omega = \rho\pi D^3\omega/\mu$

FIGURE 5.21 Horizontal cylinder rotating in air.

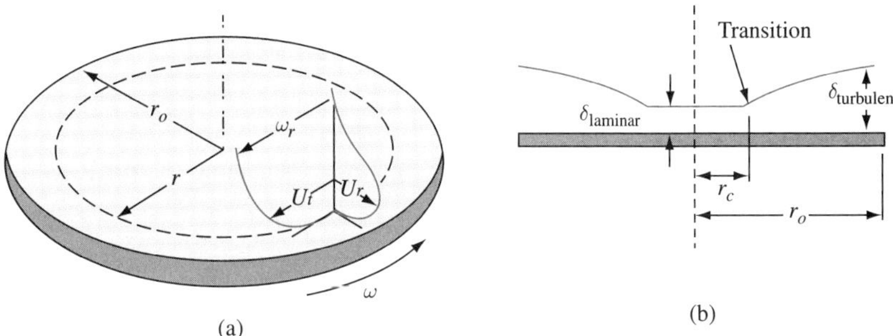

FIGURE 5.22 Velocity and boundary layer profiles for a disk rotating in an infinite environment.

Heat transfer from a rotating disk has been investigated experimentally by Cobb and Saunders [40] and theoretically by Millsap and Pohlhausen [41] and Kreith and Taylor [42], among others. The boundary layer on the disk is laminar and of uniform thickness at rotational Reynolds numbers $\omega D^2/\nu$ below about 10^6. At higher Reynolds numbers, the flow becomes turbulent near the outer edge, and as Re_ω is increased, the transition point moves radially inward. The boundary layer thickens with increasing radius (see Fig. 5.22). For the laminar regime, the average Nusselt number for a disk rotating in air is [40, 43]

$$\overline{\text{Nu}}_D = \frac{\bar{h}D}{k} = 0.36\left(\frac{\omega D^2}{\nu}\right)^{1/2} \tag{5.36}$$

for $\omega D^2/\nu < 10^6$

In the turbulent-flow regime of a disk rotating in air [40], the local value of the Nusselt number at a radius r is given approximately by

$$\text{Nu}_r = \frac{h_c r}{k} = 0.0195\left(\frac{\omega r^2}{\nu}\right)^{0.8} \tag{5.37}$$

The average value of the Nusselt number for laminar flow between $r = 0$ and r_c and turbulent flow in the outer ring between $r = r_c$ and r_o is

$$\overline{\text{Nu}}_{r_o} = \frac{\bar{h}_c r_o}{k} = 0.36\left(\frac{\omega r_o^2}{\nu}\right)^{1/2}\left(\frac{r_c}{r_o}\right)^2 + 0.015\left(\frac{\omega r_o^2}{\nu}\right)^{0.8}\left(1 - \left(\frac{r_c}{r_o}\right)^{2.6}\right) \tag{5.38}$$

for $r_c < r_o$.

EXAMPLE 5.6 A 20-cm-diameter steel shaft is heated to 400°C for heat treating. The shaft is then allowed to cool in air (at 20°C) while rotating about its own (horizontal) axis at 3 rpm. Compute the rate of convection heat transfer from the shaft when it has cooled to 100°C.

SOLUTION The rotation speed of the shaft is

$$\omega = \frac{3\,\text{rev/min}(2\pi\,\text{rad/rev})}{(60\,\text{s/min})} = 0.31\,\text{rad/s}$$

From the properties of air at 60 °C, the Reynolds number is

$$\text{Re}_\omega = \frac{\pi(0.2\,\text{m})^2(0.31\,\text{s}^{-1})}{1.94 \times 10^{-5}\,\text{m}^2/\text{s}} = 2008$$

and the Rayleigh number is

$$\text{Ra} = \frac{(9.8\,\text{m/s}^2)(333\,\text{K})^{-1}(80\,\text{K})(0.2\,\text{m})^3(0.71)}{(1.94 \times 10^{-5}\,\text{m}^2/\text{s})^2} = 3.55 \times 10^7$$

From Eq. (5.35),

$$\overline{\text{Nu}}_D = 0.11[0.5(2008)^2 + 3.55 \times 10^7]^{0.35} = 49.2$$

$$\bar{h}_c = \frac{(49.2)(0.0279\,\text{W/m K})}{0.20\,\text{m}} = 6.86\,\text{W/m}^2\,\text{K}$$

and

$$q = (6.86\,\text{W/m}^2\,\text{K})[\pi(0.2)(1)\,\text{m}^2](80\,\text{K}) = 345\,\text{W/m}$$

Note that the effect of gravity-induced natural convection is large relative to that induced by the rotation of the shaft.

For a disk rotating in a fluid having a Prandtl number larger than unity, the local Nusselt number can be obtained, according to [44], from the equation

$$\text{Nu}_r = \frac{\text{Re}_r\text{Pr}\sqrt{(C_{Dr}/2)}}{5\text{Pr} + 5\ln(5\text{Pr} + 1) + \sqrt{(2/C_{Dr})} - 14} \tag{5.39}$$

where C_{Dr} is the local drag coefficient at radius r, which, according to [45], is given by

$$\frac{1}{\sqrt{C_{Dr}}} = -2.05 + 4.07\log_{10}\text{Re}_r\sqrt{C_{Dr}} \tag{5.40}$$

For a sphere of diameter D rotating in an infinite environment with $\text{Pr} > 0.7$ in the laminar-flow regime ($\text{Re}_\omega = \omega D^2/\nu < 5 \times 10^4$), the average Nusselt number ($\bar{h}_c D/k$) can be obtained from

$$\overline{\text{Nu}}_D = 0.43\text{Re}_\omega^{0.5}\text{Pr}^{0.4} \tag{5.41}$$

while in the Reynolds-number range between 5×10^4 and 7×10^5, the equation

$$\overline{\text{Nu}}_D = 0.066\text{Re}_\omega^{0.67}\text{Pr}^{0.4} \tag{5.42}$$

correlates the available experimental data [46].

5.5 Combined Forced and Natural Convection

In Chapter 4, forced convection in flow over a flat surface was treated, and the preceding sections of this chapter dealt with heat transfer in natural-convection systems. In this section, the interaction between natural- and forced-convection processes will be considered.

In any heat transfer process, density gradients occur, and in the presence of a force field natural-convection, currents arise. If the forced-convection effects are very large, the influence of natural-convection currents may be negligible, and similarly, when the natural-convection forces are very strong, the forced-convection effects may be negligible. The questions we wish to consider now are, under what circumstances can either forced or natural convection be neglected, and what are the conditions when both effects are of the same order of magnitude?

To obtain an indication of the relative magnitudes of natural- and forced-convection effects, we consider the differential equation describing the uniform flow over a vertical flat plate with the buoyancy effect and the free-stream velocity U_∞ in the same direction. This would be the case when the plate is heated and the forced flow is upward or when the plate is cooled and the forced flow is downward. Taking the flow direction as x and assuming that the physical properties are uniform except for the temperature effect on the density, the Navier-Stokes boundary layer equation including natural-convection forces is

$$u\frac{\partial u}{\partial x} + v\frac{\partial u}{\partial y} = -\frac{1}{\rho}\frac{\partial p}{\partial x} + \frac{\mu}{\rho}\frac{\partial^2 u}{\partial y^2} + g\beta(T - T_\infty) \tag{5.43}$$

This equation can be generalized as follows. Substituting X for x/L, Y for y/L, θ for $(T - T_\infty)/(T_0 - T_\infty)$, P for $(p - p_\infty)/(\rho U_\infty^2/2g_c)$, U for u/U_∞, and V for v/U_∞ in Eq. (5.43) gives

$$\begin{aligned} U\frac{\partial U}{\partial X} + V\frac{\partial U}{\partial Y} &= -\frac{1}{2}\frac{\partial P}{\partial X} + \left(\frac{\mu}{\rho U_\infty L}\right)\frac{\partial^2 U}{\partial Y^2} \\ &\quad + \left[\frac{g\beta L^3(T_0 - T_\infty)}{\nu^2}\right]\frac{\nu^2}{U_\infty^2 L^2}\theta \end{aligned} \tag{5.44}$$

In the region near the surface—that is, in the boundary layer—$\partial U/\partial X$ and U are of the order of unity. Since U changes from 1 at $x = 0$ to a very small value at $x = 1$, and since u is of the same order of magnitude as U_∞, the left-hand side of Eq. (5.44) is of the order of unity. Similar reasoning indicates that the first two terms on the right-hand side as well as θ are of the order of unity. Consequently, the buoyancy effect will influence the velocity distribution, on which, in turn, the temperature distribution depends, if the coefficient of θ is of the order of 1 or larger; that is, if

$$\frac{[g\beta L^3(T_0 - T_\infty)]/\nu^2}{(U_\infty L/\nu)^2} = \frac{\mathrm{Gr}_L}{\mathrm{Re}_L^2} \cong 1 \tag{5.45}$$

In other words, the ratio Gr/Re^2 gives a qualitative indication of the influence of buoyancy on forced convection. When the Grashof number is of the same order of magnitude as or larger than the square of the Reynolds number, natural-convection effects cannot be ignored, compared to forced convection. Similarly, in a natural-convection process, the influence of forced convection becomes significant when the square of the Reynolds number is of the same order of magnitude as the Grashof number.

Several special cases have been treated in the literature [47–49]. For example, for laminar forced convection over a vertical flat plate, Sparrow and Gregg [47] showed that for Prandtl numbers between 0.01 and 10 the effect of buoyancy on the local heat transfer coefficient for pure forced convection will be less than 10% if

$$Gr_x \leq 0.150 Re_x^2 \tag{5.46}$$

Eckert and Diaguila [49] studied mixed convection in a vertical tube with air, primarily in the turbulent regime. When the buoyancy-induced flow was in the same direction as the forced flow, they found that the local heat transfer coefficient differed from that for pure natural-convection behavior by less than 10% if

$$Gr_x > 0.007 Re_x^{2.5} \tag{5.47a}$$

and from pure forced-convection behavior by less than 10% if

$$Gr_x < 0.0016 Re_x^{2.5} \tag{5.47b}$$

In Fig. 5.23, Eqs. (5.46) and (5.47) are plotted to delineate the regimes of pure natural convection in boundary layer flow, mixed-convection boundary layer flow, and

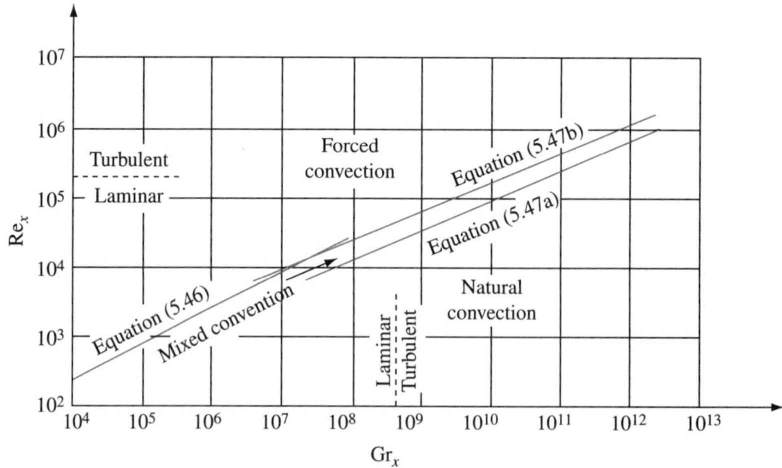

FIGURE 5.23 Regimes of convection for flow and buoyancy effects parallel; boundary layer processes.

Source: Courtesy of B. Gebhart, *Heat Transfer,* 2nd ed., McGraw-Hill, New York, 1971.

pure forced convection in boundary layer flow for geometries in which the buoyancy and forced-flow effects are parallel.

Siebers et al. [50] measured mixed-convection heat transfer from a large (3.03 m high × 2.95 m wide) vertical flat plate. The plate was heated electrically to produce natural-convection flow up the plate, and it was located in a wind tunnel to simultaneously expose it to a horizontal forced flow parallel to the plate. Therefore, this study was concerned with vertical buoyancy flow and horizontal forced flow. They based the magnitude of the natural convection on Gr_H, where H is the plate height, and the magnitude of the forced convection on Re_L, where L is the plate width. Their results indicate that if $\mathrm{Gr}_H/\mathrm{Re}_L^2 < 0.7$ then the heat transfer is essentially due to forced convection and if $\mathrm{Gr}_H/\mathrm{Re}_L^2 > 10$ then natural convection dominates. For intermediate values, that is, for mixed convection, they provide correlating equations for local heat transfer coefficients. Figure 5.24 shows the different flow regimes for this geometry for laminar flow (Fig. 5.24a) and turbulent flow (Fig. 5.24b).

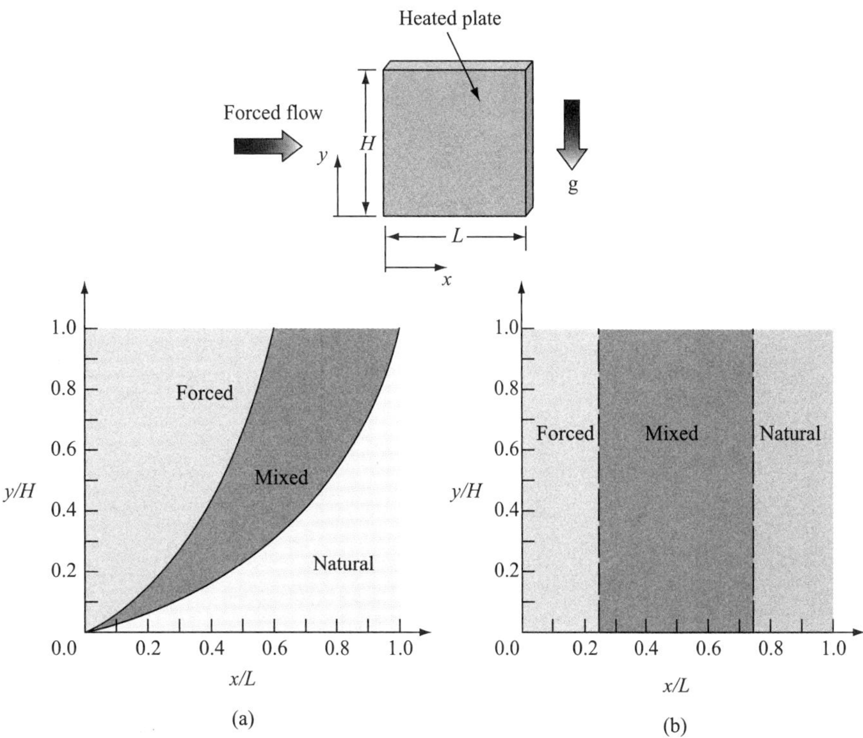

FIGURE 5.24 (a) Laminar zones of forced, mixed, and natural convection for a heated vertical plate in horizontal forced flow [50]; (b) turbulent zones of forced, mixed, and natural convection for a heated vertical plate in horizontal forced flow [50].

An empirical method for estimating the Nusselt number for combined, forced, and natural convection has been proposed [20]

$$(\overline{\text{Nu}})^n_{\text{combined}} = (\overline{\text{Nu}})^n_{\text{forced}} \pm (\overline{\text{Nu}})^n_{\text{natural}} \tag{5.48}$$

where $n = 3$ for vertical plates, the $+$ sign applies when the flows are in the same direction, and the $-$ sign when they are in opposite directions.

The influence of natural convection on forced flow in tubes and ducts is discussed in Chapter 6, Section 6.3.

5.6* Finned Surfaces

Finned or extended surfaces are commonly used to increase the surface area in a heat exchanger or heat sink to promote enhanced heat transfer [57]. The design of natural convection fins and fin arrays, however, can be difficult because of the coupling between the flow and temperature fields and the limited experimental data available. Although the fin relations developed in Chapter 2 are applicable, the evaluation of the appropriate heat transfer coefficient for the physical design geometry can present difficulties. In this section, the results of experiments and some correlations for common natural convection fin geometries are summarized.

5.6.1 Fins on Horizontal Tubes

Many types of heat exchangers (e.g., baseboard heaters or devices for cooling electronic equipment) use circular or square fins attached to a tube as shown in Fig. 5.25. For square fins, data are available only without a tube in air in the range $0.2 < \text{Ra}_s < 4 \times 10^4$ from experiments by Elenbaas [52] and Bahrani and Sparrow [53]. Raithby and Hollands [54] recommended the relation

$$\text{Nu}_s = \left\{\left(\frac{\text{Ra}_s^{0.89}}{18}\right)^{2.7} + (0.62\text{Ra}_s^{1/4})^{2.7}\right\}^{0.37} \tag{5.49}$$

The nomenclature is defined in Fig. 5.25.

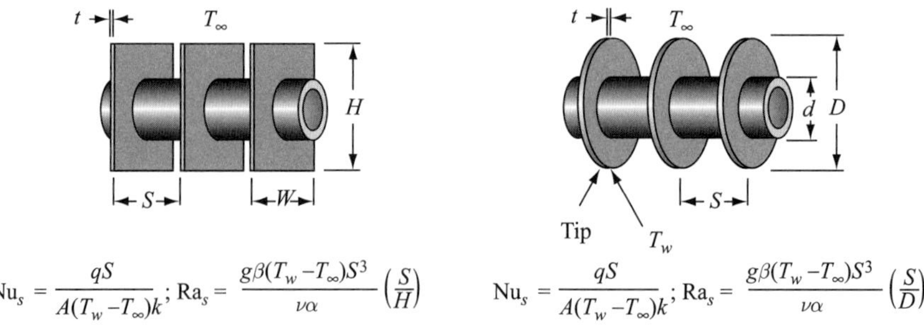

FIGURE 5.25 Plate and circular fins attached to a horizontal tube.

For circular fins attached to horizontal tubes, Tsubouchi and Masuda [55] did experiments in air in which they measured separately the heat transfer from the circumferential tips of the fins and the tubes plus the vertical fin surfaces. Using the nomenclature in Fig. 5.25, the correlation proposed for the heat transfer from the tips is

$$\mathrm{Nu}_s = C\,\mathrm{Ra}_s^b \tag{5.50}$$

where $b = 0.9$, $C = (0.44 + 0.12\xi)$, and $\xi = (D/d)$. Data were obtained for $2 < \mathrm{Ra}_s < 10^4$ and $1.36 < \xi < 3.73$ with properties evaluated at the film temperature.

The heat transfer from the lateral fin surfaces together with the supporting cylinder were correlated for long fins ($1.67 < \xi$) by

$$\mathrm{Nu}_s = \frac{\mathrm{Ra}_s}{12\pi}\left\{2 - \exp\left[-\left(\frac{C}{\mathrm{Ra}_s}\right)^{3/4}\right] - \exp\left[-\beta\left(\frac{C}{\mathrm{Ra}_s}\right)^{3/4}\right]\right\} \tag{5.51a}$$

where

$$\beta = (0.17/\xi) + e^{-(4.8/\xi)} \quad \text{and} \quad C = \left\{\frac{23.7 - 1.1[1 + (152/\xi^2)]^{1/2}}{1 + \beta}\right\}^{4/3}$$

For short fins ($1.0 < \xi < 1.67$), the correlation replacing Eq. (5.51a) is

$$\mathrm{Nu}_s = C_0\,\mathrm{Ra}_0^P\left\{1 - \exp\left[-\left(\frac{C_1}{\mathrm{Ra}_0}\right)^{C_2}\right]\right\}^{C_3} \tag{5.51b}$$

where

$$C_0 = -0.15 + (0.3/\xi) + 0.32\xi^{-16} \qquad C_1 = -180 + (480/\xi) - 1.4\xi^8$$

$$C_2 = 0.04 + (0.9/\xi) \qquad C_3 = 1.3(1 - \xi^{-1}) + 0.0017\xi^{12}$$

$$P = 0.25 + C_2C_3 \qquad \mathrm{Ra}_0 = \mathrm{Ra}_s\xi$$

with properties evaluated at the wall temperature, T_w.

The Nusselt number relations from the lateral surfaces of circular and square fins are equivalent if $D = 1.23H$. Hence, the above equations also can be used to estimate the combined heat transfer from square plate fins on a tube or cylinder as shown in Fig. 5.25.

Edwards and Chaddock [56] correlated experimental data for heat transfer from the entire surface of circular fins, including the tip, for $(D/d) = 1.94$, in the range $5 < \mathrm{Ra}_s < 10^4$, by

$$\mathrm{Nu}_s = 0.125\,\mathrm{Ra}_s^{0.55}\left[1 - \exp\left(-\frac{137}{\mathrm{Ra}_s}\right)\right]^{0.294} \tag{5.52}$$

with properties evaluated at $[T_\infty + 0.62(T_w + T_\infty)]$. Subsequent measurements of Jones and Nwizu [57] fall slightly below values calculated using Eq. (5.52).

5.6.2 Horizontal Triangular Fins

A heated, horizontal array of triangular corrugations with slant height L, as shown in Fig. 5.26, can be treated as a surface with triangular fins. Al-Arabi and El-Rafaee [58] measured heat transfer from such a surface in air with $W \gg L$ over a range

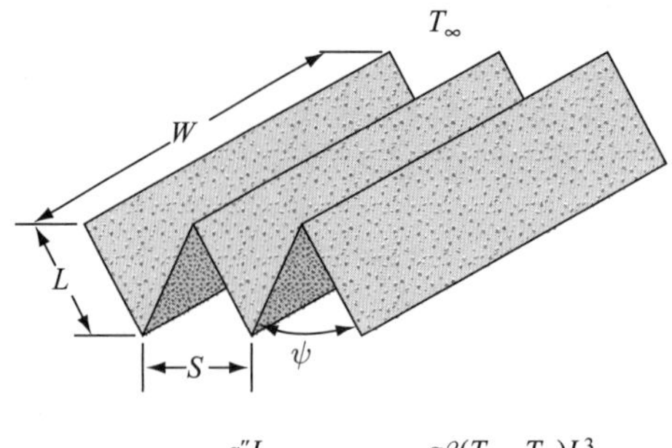

$$\text{Nu}_L = \frac{q''L}{(T_w - T_\infty)k}\;;\; \text{Ra}_L = \frac{g\beta(T_w - T_\infty)L^3}{\nu\alpha}$$

FIGURE 5.26 Nomenclature for triangular fins.

$1.8 \times 10^4 < \text{Ra}_L < 1.4 \times 10^7$ and correlated the data using the following expressions:

$$\text{Nu}_L = \left[\frac{0.46}{\sin\left(\dfrac{\psi}{2}\right)} - 0.32\right]\text{Ra}_L^m \tag{5.53a}$$

for $1.8 \times 10^4 < \text{Ra}_L < \text{Ra}_c$ and

$$\text{Nu}_L = \left[0.090 + \frac{0.054}{\sin\left(\dfrac{\psi}{2}\right)}\right]\text{Ra}_L^{1/3} \tag{5.53b}$$

for $\text{Ra}_c < \text{Ra}_L < 1.4 \times 10^7$

where ψ = the apex angle, as shown in Fig. 5.26

$\text{Ra}_c = [15.8 - 14.0 \sin(\psi/2)] \times 10^5$

$m = 0.148 \sin(\psi/2) + 0.187$

5.6.3 Rectangular Fins on Horizontal Surfaces

Heat transfer data to or from horizontal surfaces with rectangular thin fins, as shown in Fig. 5.27 (upward-facing for $T_w > T_\infty$ or downward-facing for $T_w > T_\infty$), has been correlated by Jones and Smith [59] to within about $\pm 25\%$ over the range $2 \times 10^2 < \text{Ra}_s < 6 \times 10^5$, $\text{Pr} = 0.71$, $0.026 < H/W < 0.19$, and $0.0160 < S/W < 0.20$ by the equation

$$\text{Nu}_s = \left[\left(\frac{1500}{\text{Ra}_s}\right)^2 + (0.081\text{Ra}_s^{0.39})^{-2}\right]^{-1/2} \tag{5.54}$$

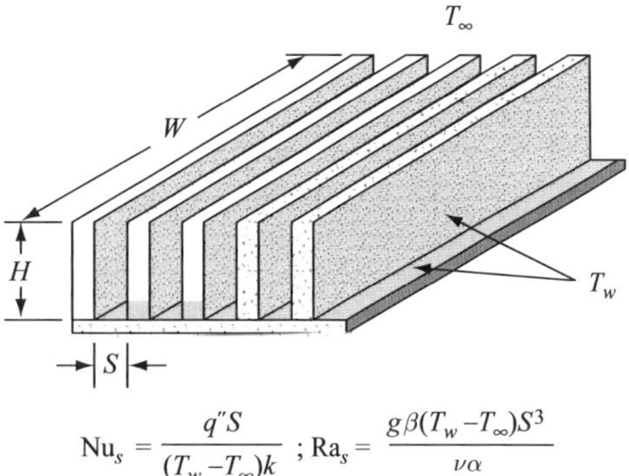

$$\mathrm{Nu}_s = \frac{q''S}{(T_w - T_\infty)k}\ ;\ \mathrm{Ra}_s = \frac{g\beta(T_w - T_\infty)S^3}{\nu\alpha}$$

FIGURE 5.27 Rectangular fins on a horizontal surface.

This relation ignores the effect of the geometrical parameters H/S and H/W. While H/S does not appear to play a strong role, H/W is known to have a significant effect. When H/W is large, horizontal inflow through the open ends of the fins results in higher heat transfer coefficients. For smaller H/W, the cooling fluid along much of the fin length is drawn downward from above by thermosyphon action, reducing the heat transfer coefficients.

5.6.4 Rectangular Fins on Vertical Surfaces

Vertical parallel-plate fins resemble two-dimensional channels formed by parallel plates. This configuration is frequently encountered in natural-convection cooling of electrical equipment ranging from transformers to mainframe computers and from transistors to power supplies.

In relatively short channels, individual boundary layers develop along each surface, and conditions approach those of isolated plates in infinite media, as discussed previously. For longer channels, the boundary layers merge, and then the fluid temperature at a given height is not explicitly known. The heat transfer coefficient is therefore based on the ambient or inlet temperature, and this convention will be followed here.

Bar-Cohen and Rohsenow [60] have compiled a tabulation of the Nusselt-number relations recommended for vertical parallel-plate fins under various thermal boundary conditions encountered in practice (Table 5.1). Figure 5.28 shows a typical array of printed circuit cards in a computer with the geometric definitions necessary for using Table 5.1. The Nusselt number, Nu_0, for all cases is based on the spacing between adjacent fins, S, as the characteristic length and the average heat flux from the fin, q_0''. Two Rayleigh numbers are used in Table 5.1, one for the isothermal case when the temperature difference, θ_0, is specified explicitly as the difference between the fin surface and the ambient inlet temperatures; the other is used for the isoflux condition when the heat flux is specified

TABLE 5.1 Composite Nu_0 relations for parallel-plate fins [60]

Boundary Conditions	Composite Relations
Symmetrically isothermal fins	$Nu_0 = \{576/(Ra')^2 + 2.873/\sqrt{Ra'}\}^{-1/2}$
Asymmetrically isothermal fins (one side insulated)	$Nu_0 = \{144/(Ra')^2 + 2.873/\sqrt{Ra'}\}^{-1/2}$
Symmetrically isoflux fins (θ_0 at $L/2$)	$Nu_{0,L/2} = \{12/Ra'' + 1.88/(Ra'')^{2/5}\}^{-1/2}$
Asymmetrically isoflux fins (one side insulated, θ_0 at $L/2$)	$Nu_{0,L/2} = \{6/Ra'' + 1.88/(Ra'')^{2/5}\}^{-1/2}$

where $Nu_0 \equiv q_0''S/k\theta_0$

$Ra' \equiv \bar{\rho}^2 g\beta c_p S^4\theta_0/\mu kL$

$Ra'' \equiv \bar{\rho}^2 g\beta c_p S^5 q_0''/\mu k^2 L$

S = card spacing (m)
c_p = specific heat (J/kg K)
g = gravitational acceleration (m/s^2)
K = thermal conductivity (W/m K)
L = channel height (m)
q_0'' = heat flux (W/m^2)
β = volumetric expansion coefficient (K^{-1})
$\bar{\rho}$ = density (kg/m^3)
θ_0 = temperature difference (k)
μ = dynamic viscosity (kg/ms)
Ra' = channel Rayleigh number (dimensionless)
Ra'' = modified-channel Rayleigh number (dimensionless)

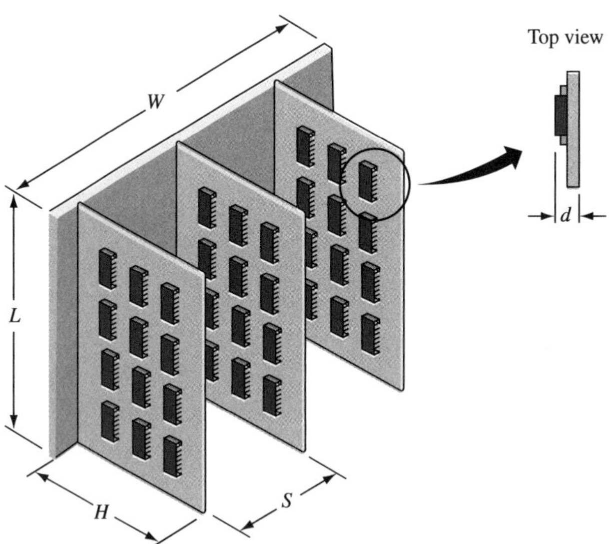

FIGURE 5.28 Natural convection cooled array of printed circuit cards—geometric definition. *Note: d* is the effective card thickness, including the card and the circuits mounted on it.

and the surface temperature is not known explicitly. For the latter case, the heat transfer coefficient is based on the temperature difference between the surface at midheight, $L/2$, and the inlet to the array.

In addition to compiling relations for the Nusselt number, Bar-Cohen and Rohsenow [60] also determined the spacing between adjacent fins that will maximize the volumetric rate of heat dissipation. This "optimum" spacing, S_{opt}, depends on the card or plate thickness and a parameter P defined

$$P = c_p \rho^2 g \beta \Delta T / \mu k L \tag{5.55}$$

For plates of negligible thickness, the optimum spacing with isothermal fins is

$$S_{\text{opt}} = 2.7/P^{0.25} \tag{5.56a}$$

while for the asymmetric isothermal conditions (one side at constant temperature and the other side insulated),

$$S_{\text{opt}} = 2.15/P^{0.25} \tag{5.56b}$$

For uniform heat-flux conditions, S_{opt} is defined as the spacing that yields the maximum volumetric (or prime area) heat dissipation rate per unit temperature difference (based on midheight minus inlet temperature). For both surfaces at constant heat flux, q'',

$$S_{\text{opt}} = 1.5/R^{0.2} \tag{5.56c}$$

while for asymmetric conditions,

$$S_{\text{opt}} = 1.2/R^{0.2} \tag{5.56d}$$

where

$$R = c_p \rho^2 g \beta q'' / \mu L k^2 \tag{5.57}$$

in both cases.

A discussion of three-dimensional flow and geometric effects is presented in [60]. When natural convection cannot adequately cool an electronic device, it becomes necessary to resort to forced convection.

5.7 Closing Remarks

For the convenience of the reader, useful correlation equations for the determination of the average value of the natural-convection heat transfer coefficients for several important geometries are presented in Table 5.2.

TABLE 5-2 Natural-convection heat transfer correlations

Geometry	Correlation Equation	Restrictions
Long vertical or tilted plate with hot surface facing downward	$\overline{\text{Nu}}_L = 0.56(\text{Gr}_L \text{ Pr} \cos\theta)^{1/4}$	$10^5 < \text{Gr}_L\text{Pr} \cos\theta < 10^{11}$ $0 \le \theta \le 89°$
Long horizontal plate with hot surface facing upward or cool surface facing downward	$\overline{\text{Nu}}_L = 0.54\,\text{Ra}_L^{1/4}$ $\overline{\text{Nu}}_L = 0.15\,\text{Ra}_L^{1/3}$ $L = A/P$	$10^5 \lesssim \text{Ra}_L \lesssim 10^7$ $10^7 \lesssim \text{Ra}_L \lesssim 10^{10}$
Horizontal plate with hot surface facing downward or cool surface facing upward	$\overline{\text{Nu}}_L = 0.27\,\text{Ra}_L^{1/4}$ $L = A/P$	$10^5 \lesssim \text{Ra}_L \lesssim 10^{10}$

Geometry	Correlation	Range of validity
Single long horizontal cylinder 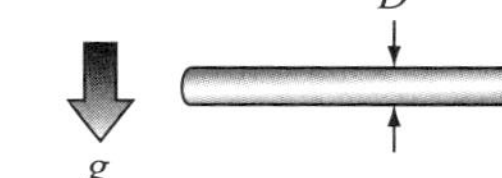	$\overline{\mathrm{Nu}}_D = 0.53(\mathrm{Gr}_D\mathrm{Pr})^{1/4}$ $\overline{\mathrm{Nu}}_D = 0.53(\mathrm{Gr}_D\mathrm{Pr}^2)^{1/4}$	$Pr > 0.5;\ 10^3 < Gr_D < 10^9$ Liquid metals, Laminar flow
Inclined cylinder, length L 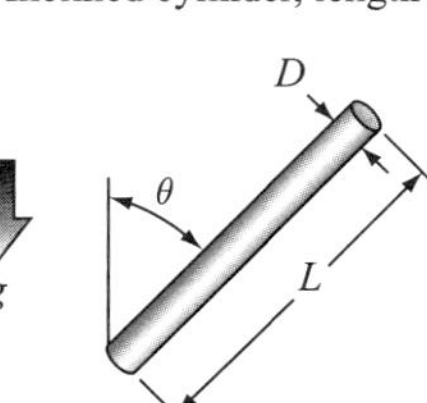	$\overline{\mathrm{Nu}}_L = [2.9 - 2.32(\sin\theta)^{0.8}] \times (\mathrm{Gr}_D)^{-1/12}[\mathrm{Gr}_L\mathrm{Pr}]^{(1/4+1/12(\sin\theta)1.2)}$ $\overline{\mathrm{Nu}}_L = [0.47 + 0.11(\sin\theta)^{0.8}](\mathrm{Gr}_D)^{-1/12}(\mathrm{Gr}_L\mathrm{Pr})^{1/3}$	Laminar: $9.88 \times 10^7 \le \mathrm{Gr}_L\mathrm{Pr} \le (\mathrm{Gr}_L\mathrm{Pr})_{cr}$ $1.08 \times 10^4 \le \mathrm{Gr}_D \le 6.9 \times 10^5$ Turbulent: $(\mathrm{Gr}_L\mathrm{Pr})_{cr} \le \mathrm{Gr}_L\mathrm{Pr} \le 2.95 \times 10^{10}$ $1.08 \times 10^4 \le \mathrm{Gr}_D \le 6.9 \times 10^5$ where $(\mathrm{Gr}_L\mathrm{Pr})_{cr} = 2.6 \times 10^9 + 1.1 \times 10^9 \tan\theta$
Sphere 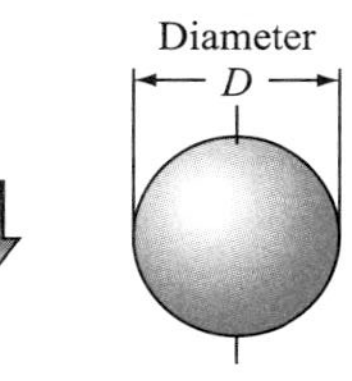	$\overline{\mathrm{Nu}}_D = 2 + 0.392(\mathrm{Gr}_D)^{1/4}$	$1 < \mathrm{Gr}_D < 10^5$
Vertical cone 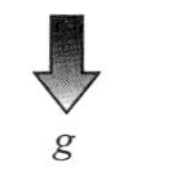	$\overline{\mathrm{Nu}}_L = 0.63(1 + 0.72\varepsilon)\mathrm{Gr}_L^{1/4})$	$3^\circ < \phi < 12^\circ$ $7.5 < \log \mathrm{Gr}_L < 8.7$ $0.2 \le \varepsilon < 0.8$ where $\varepsilon = 2/[\mathrm{Gr}_L^{1/4}\tan(\phi/z)]$

(Continued)

TABLE 5.2 (*Continued*)

Geometry	Correlation Equation	Restrictions
Space enclosed between two vertical plates heated from one side	$\overline{\text{Nu}}_\delta = 0.22\left(\dfrac{L}{\delta}\right)^{-1/4}\left(\dfrac{\text{Pr}}{0.2+\text{Pr}}\text{Ra}_\delta\right)^{0.28}$	$\begin{cases} 2 < \dfrac{L}{\delta} < 10,\ \text{Pr} < 10 \\ \text{Ra}_\delta < 10^{10} \end{cases}$
	$\overline{\text{Nu}}_\delta = 0.18\left(\dfrac{\text{Pr}}{0.2+\text{Pr}}\text{Ra}_\delta\right)^{0.29}$	$\begin{cases} 1 < \dfrac{L}{\delta} < 2,\ 10^{-3} < \text{Pr} < 10^{5} \\ 10^{3} < \dfrac{\text{Ra}_\delta \text{Pr}}{0.2+\text{Pr}} \end{cases}$
Space enclosed between two horizontal plates heated from below	$\overline{\text{Nu}}_\delta = 1 + 1.44\left[1 - \dfrac{1708}{\text{Ra}_\delta}\right]^{\cdot} + \left[\left(\dfrac{\text{Ra}_\delta}{5830}\right)^{1/3} - 1\right]^{\cdot}$	Air, $1700 < \text{Ra}_\delta < 10^{8}$
	$\overline{\text{Nu}}_\delta = 1 + 1.44\left[1 - \dfrac{1708}{\text{Ra}_\delta}\right]^{\cdot} + \left[\left(\dfrac{\text{Ra}_\delta}{5830}\right)^{1/3} - 1\right]^{\cdot} + 20\left[\dfrac{\text{Ra}_\delta^{1/3}}{140}\right]^{(1-\ln(\text{Ra}_\delta^{1/3}/140))}$	Water, $1700 < \text{Ra}_\delta < 3.5 \times 10^{9}$
Spherical cavity interior 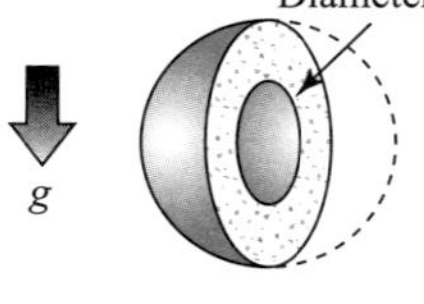	$\overline{\text{Nu}}_D = C(\text{Gr}_D \text{Pr})^{n}$	See table following Eq. (5.32)
Long concentric cylinders	$\dfrac{k_{\text{eff}}}{k} = 0.386\left[\dfrac{\ln(D_o/D_i)}{b^{3/4}(1/D_i^{3/5} + 1/D_o^{3/5})^{5/4}}\right] \times \left(\dfrac{\text{Pr}}{0.861+\text{Pr}}\right)^{1/4}\text{Ra}_b^{1/4}$	$0.70 \le \text{Pr} \le 6000$; $10 \le \left[\dfrac{\ln(D_o/D_i)}{b^{3/4}(1/D_i^{3/5} + 1/D_o^{3/5})^{5/4}}\right]\text{Ra}_b \le 10^{7}$

Geometry	Correlation	Range
Concentric spheres $2b = D_o - D_i$ 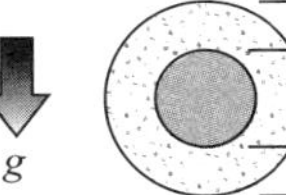	$\frac{k_{eff}}{k} = 0.74\left[\frac{b^{1/4}}{D_oD_i(D_i^{-7/5} + D_o^{-7/5})^{5/4}}\right] \times Ra_b^{1/4}\left(\frac{Pr}{0.861 + Pr}\right)^{1/4}$	$0.70 \le Pr \le 4200$; $10 \le \left[\frac{b}{(D_oD_i)^4(D_i^{-7/5} + D_o^{-7/5})^5}\right]Ra_b \le 10^7$
Long rotating cylinder (ω, D, $\to \infty$)	$\overline{Nu}_D = \frac{\bar{h}_cD}{k} = 0.11(0.5Re_\omega^2 + Gr_DPr)^{0.35}$	$Re_\omega = \frac{\pi D^2\omega}{\nu} > 8000$
Rotating disk (Diameter D, ω)	$\overline{Nu}_D = \frac{\bar{h}_cD}{k} = 0.36(Re_\omega)^{1/2}$	$Re_\omega = \frac{\omega D^2}{\nu} < 10^6$
Rotating sphere 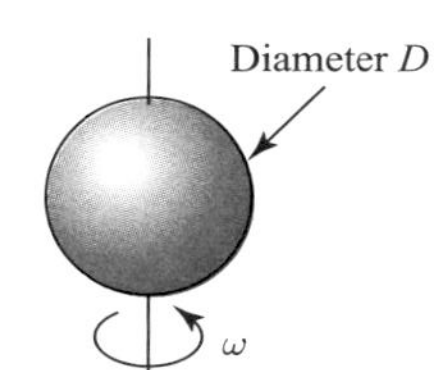	$\overline{Nu}_D = 0.43Re_\omega^{0.5}Pr^{0.4}$	$Re_\omega = \frac{\omega D^2}{\nu} < 5 \times 10^4$; $Pr > 0.7$
	$\overline{Nu}_D = 0.066Re_\omega^{0.67}Pr^{0.4}$	$5 \times 10^4 < Re_\omega < 7 \times 10^5$

References

1. S. Ostrach, "New Aspects of Natural-Convection Heat Transfer," *Trans. ASME*, vol. 75, pp. 1287–1290, 1953.
2. Y. Jaluria, *Natural Convection Heat and Mass Transfer*, Pergamon, New York, 1980.
3. E. Schmidt and W. Beckmann, "Das Temperatur und Geschwindigkeitsfeld vor einer wärmeabgebenden senkrechten Platte bei natürlicher Konvection," *Tech. Mech. Thermodyn.*, vol. 1, no. 10, pp. 341–349, October 1930; cont., vol. 1, no. 11, pp. 391–406, November 1930.
4. R. Hermann, "Wärmeübergang bei freir Ströhmung am wagrechten Zylinder in zweiatomic Gasen," *VDI Forschungsh.*, no. 379, 1936; translated in NACA Tech. Memo. 1366, November 1954.
5. W. H. McAdams, *Heat Transmission*, 3rd ed., McGraw-Hill, New York, 1954.
6. E. M. Sparrow and M. A. Ansari, "A Refutation of King's Rule for Multi-Dimensional External Natural Convection," *Int. J. Heat Mass Transfer*, vol. 26, pp. 1357–1364, 1983.
7. E. R. G. Eckert and E. Soehnghen, "Interferometric Studies on the Stability and Transition to Turbulence of a Free-Convection Boundary Layer," Proceedings of the General Discussion on Heat Transfer, pp. 321–323, ASME-IME, London, 1951.
8. E. R. G. Eckert and E. Soehnghen, "Studies on Heat Transfer in Laminar Free Convection with the Zehnder-Mach Interferometer," USAF Tech. Rept. 5747, December 1948.
9. B. Gebhart, *Heat Transfer*, 2nd ed., chap. 8, McGraw-Hill, New York, 1970.
10. E. R. G. Eckert and T. W. Jackson, "Analysis of Turbulent Free Convection Boundary Layer on Flat Plate," NACA Rept. 1015, July 1950.
11. A. M. Clausing, "Natural Convection Correlations for Vertical Surfaces Including Influences of Variable Properties," *J. Heat Transfer*, vol. 105, no. 1, pp. 138–143, 1983.
12. J. Gryzagoridis, "Natural Convection from a Vertical Flat Plate in the Low Grashof Number Range," *Int. J. Heat Mass Transfer*, vol. 14, pp. 162–164, 1971.
13. O. E. Dwyer, "Liquid-Metal Heat Transfer," chapter 5 in Sodium and NaK Supplement to *Liquid Metals Handbook*, Atomic Energy Commission, Washington, D.C., 1970.
14. E. M. Sparrow and J. L. Gregg, "Laminar Free Convection from a Vertical Flat Plate," *Trans. ASME*, vol. 78, pp. 435–440, 1956.
15. J. P. Dotson, *Heat Transfer from a Vertical Flat Plate to Free Convection*, M.S. thesis, Purdue University, May 1954.
16. W. M. Yan and T. F. Lin, "Theoretical and Experimental Study of Natural Convection Pipe Flows at High Rayleigh Numbers," *Int. J. Heat Mass Transfer*, vol. 34, pp. 291–302, 1991.
17. G. C. Vliet, "Natural Convection Local Heat Transfer on Constant Heat Flux Inclined Surfaces," *Trans. ASME, Ser. C, J. Heat Transfer*, vol. 91, pp. 511–516, 1969.
18. G. C. Vliet and C. K. Liu, "An Experimental Study of Turbulent Natural Convection Boundary Layers," *Trans. ASME, Ser. C, J. Heat Transfer*, vol. 91, pp. 517–531, 1969.
19. T. Fujii and H. Imura, "Natural Convection Heat Transfer from a Plate with Arbitrary Inclination," *Int. J. Heat Mass Transfer*, vol. 15, pp. 755–767, 1972.
20. F. P. Incropera and D. P. DeWitt, *Introduction to Heat Transfer*, 2nd ed., Wiley, New York, 1990.
21. J. S. McDonald and T. J. Connally, "Investigation of Natural Convection Heat Transfer in Liquid Sodium," *Nucl. Sci. Eng.*, vol. 8, pp. 369–377, 1960.
22. S. C. Hyman, C. F. Bonilla, and S. W. Erhlich, "Heat Transfer to Liquid Metals and Non-Metals at Horizontal Cylinders," in *AIChE Symposium on Heat Transfer*, Atlantic City, pp. 21–23, 1953.
23. M. Al-Arabi and M. Khamis, "Natural Convection Heat Transfer from Inclined Cylinders," *Int. J. Heat Mass Transfer*, vol. 25, pp. 3–15, 1982.
24. E. M. Sparrow and A. J. Stretton, "Natural Convection from Variously Oriented Cubes and from Other Bodies of Unity Aspect Ratios," *Int. J. Heat Mass Transfer*, vol. 28, no. 4, pp. 741–752, 1985.
25. T. Yuge, "Experiments on Heat Transfer from Spheres Including Combined Natural and Forced Convection," *Trans. ASME, Ser. C, J. Heat Transfer*, vol. 82, pp. 214–220, 1960.
26. P. H. Oosthuizen and E. Donaldson, "Free Convection Heat Transfer from Vertical Cones," *Trans. ASME, Ser. C, J. Heat Transfer*, vol. 94, pp. 330–331, 1972.
27. K. G. T. Hollands and L. Konicek, "Experimental Study of the Stability of Differentially Heated Inclined Air Layers," *Int. J. Heat Mass Transfer*, vol. 16, pp. 1467–1476, 1973.
28. I. Catton, "Natural Convection in Enclosures," in *Proceedings, Sixth International Heat Transfer Conference, Toronto*, vol. 6, pp. 13–31, Hemisphere, Washington, D.C., 1978.

29. R. K. MacGregor and A. P. Emery, "Free Convection through Vertical Plane Layers: Moderate and High Prandtl Number Fluid," *J. Heat Transfer*, vol. 91, p. 391, 1969.
30. J. Imberger, "Natural Convection in a Shallow Cavity with Differentially Heated End Walls, Part 3. Experimental Results," *J. Fluid Mech.*, vol. 65, pp. 247–260, 1974.
31. A. Bejan, A. A. Al-Homoud, and J. Imberger, "Experimental Study of High Rayleigh Number Convection in a Horizontal Cavity with Different End Temperatures," *J. Heat Transfer*, vol. 109, pp. 283–299, 1981.
32. M. Nansteel and R. Greif, "Natural Convection in Undivided and Partially Divided Rectangular Enclosures," *J. Heat Transfer*, vol. 103, pp. 623–629, 1981.
33. R. Krisnamurti, "On the Transition to Turbulent Convection, Part 2, The Transition to Time-Dependent Flow," *J. Fluid Mech.*, vol. 42, pp. 309–320, 1970.
34. K. G. T. Hollands, G. D. Raithby, and L. Konicek, "Correlation Equations for Free Convection Heat Transfer in Horizontal Layers of Air and Water," *Int. J. Heat Mass Transfer*, vol. 18, pp. 879–884, 1975.
35. K. G. T. Hollands, T. E. Unny, G. D. Raithby, and L. Konicek, "Free Convection Heat Transfer Across Inclined Air Layers," *J. Heat Transfer*, vol. 98, pp. 189–193, 1976.
36. F. Kreith, "Thermal Design of High Altitude Balloons and Instrument Packages," *J. Heat Transfer*, vol. 92, pp. 307–332, 1970.
37. G. D. Raithby and K. G. T. Hollands, "A General Method of Obtaining Approximate Solutions to Laminar and Turbulent Free Convection Problems," in *Advances in Heat Transfer*, Academic Press, New York, 1974.
38. J. T. Anderson and O. A. Saunders, "Convection from an Isolated Heated Horizontal Cylinder Rotating about Its Axis," *Proc. R. Soc. London Ser. A*, vol. 217, pp. 555–562, 1953.
39. W. M. Kays and I. S. Bjorklund, "Heat Transfer from a Rotating Cylinder with and without Cross Flow," *Trans. ASME, Ser. C*, vol. 80, pp. 70–78, 1958.
40. E. C. Cobb and O. A. Saunders, "Heat Transfer from a Rotating Disk," *Proc. R. Soc. London Ser. A*, vol. 220, pp. 343–351, 1956.
41. K. Millsap and K. Pohlhausen, "Heat Transfer by Laminar Flow from a Rotating Plate," *J. Aero-sp. Sci.*, vol. 19, pp. 120–126, 1952.
42. F. Kreith and J. H. Taylor, Jr., "Heat Transfer from a Rotating Disk in Turbulent Flow," ASME paper 56-A-146, 1956.
43. C. Wagner, "Heat Transfer from a Rotating Disk to Ambient Air," *J. Appl. Phys.*, vol. 19, pp. 837–841, 1948.
44. F. Kreith, J. H. Taylor, and J. P. Chang, "Heat and Mass Transfer from a Rotating Disk," *Trans. ASME, Ser. C*, vol. 81, pp. 95–105, 1959.
45. T. Theodorsen and A. Regier, "Experiments on Drag of Revolving Disks, Cylinders, and Streamlined Rods at High Speeds," NACA Rept. 793, Washington, D.C., 1944.
46. F. Kreith, L. G. Roberts, J. A. Sullivan, and S. N. Sinha, "Convection Heat Transfer and Flow Phenomena of Rotating Spheres," *Int. J. Heat Mass Transfer*, vol. 6, pp. 881–895, 1963.
47. E. M. Sparrow and J. L. Gregg, "Buoyancy Effects in Forced Convection Flow and Heat Transfer," *Trans. ASME, J. Appl. Mech.*, sect. E, vol. 81, pp. 133–135, 1959.
48. Y. Mori, "Buoyancy Effects in Forced Laminar Convection Flow over a Horizontal Flat Plate," *Trans. ASME, J. Heat Transfer*, sect. C, vol. 83, pp. 479–482, 1961.
49. E. Eckert and A. J. Diaguila, "Convective Heat Transfer for Mixed, Free, and Forced Flow through Tubes," *Trans. ASME*, vol. 76, pp. 497–504, 1954.
50. D. L. Siebers, R. G. Schwind, and R. J. Moffat, "Experimental Mixed Convection Heat Transfer from a Large, Vertical Surface in Horizontal Flow," Sandia Rept. SAND 83–8225, Sandia National Laboratories, Albuquerque, N.M., 1983.
51. R. M. Manglik, "Heat Transfer Enhancement," A. Bejan and A. D. Kraus, eds., *Heat Transfer Handbook*, ch. 14, Wiley, Hoboken, NJ, 2003.
52. W. Elenbaas, "Heat Dissipation of Parallel Plates by Free Convection," *Physica* IX, no. 1, pp. 2–28, January 1942.
53. P. A. Bahrani and E. M. Sparrow, "Experiments on Natural Convection from Vertical Parallel Plates with Either Open or Closed Edges," ASME *J. Heat Transfer*, vol. 102, pp. 221–227, 1980.
54. G. D. Raithby and K. G. T. Hollands, "Natural Convection," in *CRC Handbook of Mechanical Engineering*, F. Kreith, ed., CRC Press, Boca Raton, FL, 1998.
55. T. Tsubouchi and M. Masuda, "Natural Convection Heat Transfer from Horizontal Cylinders with Circular Fins," *Proc. 6th Int. Heat Transfer Conf.*, Paper NC 1.10, Paris, 1970.
56. J. A. Edwards and J. B. Chaddock, "Free Convection and Radiation Heat Transfer from Fin-on-Tube Heat Exchangers," ASME Paper No. 62-WA-205, 1962; see also *Trans. of ASHRAE*, vol. 69, pp. 313–322, 1963.
57. C. D. Jones and E. I. Nwizu, "Optimum Spacing of Circular Fins on Horizontal Tubes for Natural Convection Heat Transfer," ASHRAE *Symp. Bull.* DV-69-3, pp. 11–15, 1969.
58. M. Al-Arabi and M. M. El-Rafaee, "Heat Transfer by Natural Convection from Corrugated Plates to Air," *Int. J. Heat Mass Transfer*, vol. 21, pp. 357–359, 1978.

59. C. D. Jones and L. F. Smith, "Optimum Arrangement of Rectangular Fins on Horizontal Surfaces for Free Convection Heat Transfer," *J. Heat Transfer*, vol. 92, pp. 6–10, 1970.

60. A. Bar-Cohen and W. M. Rohsenow, "Thermally Optimum Arrays of Cards and Fins in Natural Convection," *IEEE Trans. on Components, Hybrids, and Mfg. Tech.*, vol. CHMT-6, June 1983.

Problems

The problems for this chapter are organized by subject matter as shown below.

Topic	Problem Number
Fundamentals	5.1–5.6
Vertical plates and cylinders	5.7–5.14
Horizontal and inclined plates	5.15–5.22
Cylinders, spheres, and 3-D bodies	5.23–5.32
Enclosed spaces	5.33–5.41
Rotating surfaces	5.42–5.44
Vertical fins	5.45
Combined forced and natural convection	5.46–5.49
Design problems	5.50–5.53

5.1 Show that the coefficient of thermal expansion for an ideal gas is $1/T$, where T is the absolute temperature.

5.2 From its definition and from the property values in Appendix 2, Table 13, calculate the coefficient of thermal expansion, β, for saturated water at 403 K. Then compare your results with the value in the table.

5.3 Using standard steam tables, calculate the coefficient of thermal expansion, β, from its definition for steam at 450°C and pressures of 0.1 atm and 10 atm. Then compare your results with the value obtained by assuming that steam is a perfect gas, and explain the difference.

5.4 A long cylinder 0.1 m in diameter has a surface temperature of 400 K. If it is immersed in a fluid at 350 K, natural convection will occur as a result of the temperature difference. Calculate the Grashof and Rayleigh numbers that will determine the Nusselt number if the fluid is (a) nitrogen, (b) air, (c) water, (d) oil, (e) mercury.

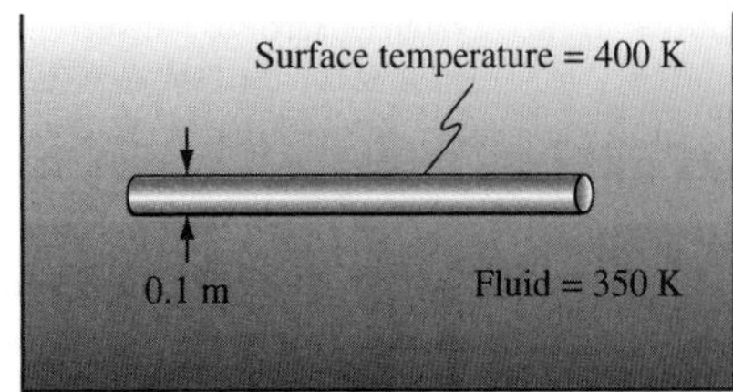

Problem 5.4

5.5 Use Fig. 5.3 to determine the Nusselt number and the heat transfer coefficient for the conditions given in Problem 5.4.

5.6 The following equation has been proposed for the heat transfer coefficient in natural convection from long vertical cylinders to air at atmospheric pressure:

$$\bar{h}_c = \frac{536.5(T_s - T_\infty)^{0.33}}{T}$$

where T = the film temperature = $(T_s + T_\infty)/2$ and T is in the range 0 to 200 °C. The corresponding equation in dimensionless form is

$$\bar{h}_c L/k = C(\mathrm{GrPr})^m$$

Compare the two equations to determine the values of C and m such that the second equation will give the same results as the first equation.

5.7 Solar One, located near Barstow, CA, was the first large-scale (10-MW electric) solar-thermal electric-power-generating plant in the United States. A schematic diagram of the plant is shown on page 341. The receiver can be treated as a cylinder 7 m in diameter and 13.5 m tall. At the design operating conditions, the average outer surface temperature of the receiver is about 675°C and ambient air temperature is about 40°C. Estimate the rate of heat loss, in MW, from the receiver *via natural convection only* for the temperatures given. What are other mechanisms by which heat may be lost from the receiver?

5.8 Compare the rate of heat loss from a human body with the typical energy intake from consumption of food (1033 kcal/day). Model the body as a vertical cylinder 30 cm in diameter and 1.8 m high in still air. Assume the skin

Problem 5.7

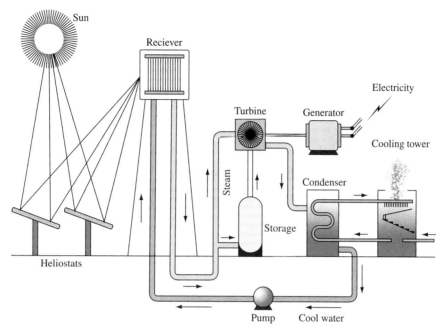

temperature is 2°C below normal body temperature. Neglect radiation, transpiration cooling (sweating), and the effects of clothing.

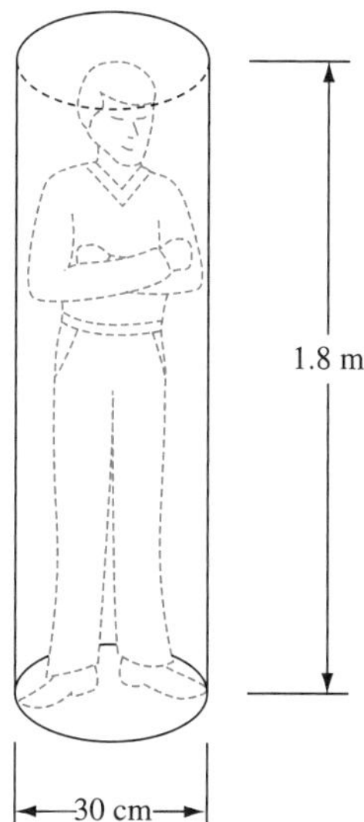

Problem 5.8

5.9 An electric room heater has been designed in the shape of a vertical cylinder 2 m tall and 30 cm in diameter. For safety, the heater surface cannot exceed 35°C. If the room air is at 20°C, find the power rating of the heater in watts.

5.10 Consider a design for a nuclear reactor using natural-convection heating of liquid bismuth, as shown at top of following page. The reactor is to be constructed of parallel vertical plates 6 ft tall and 4 ft wide in which heat is generated uniformly. Estimate the maximum possible heat dissipation rate from each plate if the average surface temperature of the plate is not to exceed 1600°F and the lowest allowable bismuth temperature is 600°F.

5.11 A mercury bath at 60°C is to be heated by immersing cylindrical electric heating rods, each 20 cm tall and 2 cm in diameter. Calculate the maximum electric power rating of a typical rod if its maximum surface temperature is 140°C.

5.12 An electric heating blanket is subjected to an acceptance test. It is to dissipate 400 W on the high setting when hanging in air at 20°C. (a) If the blanket is 1.3-m wide, what is the length required If its average temperature at the high setting is to be 40°C? (b) If the average temperature at the low setting is to be 30°C, what rate of dissipation would be possible?

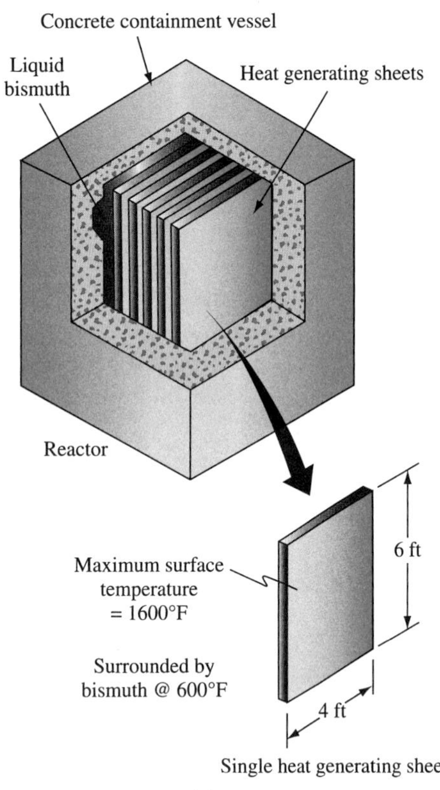

Problem 5.10

5.13 An aluminum sheet 0.4 m tall, 1 m long, and 0.002 m thick is to be cooled from an initial temperature of 150°C to 50°C by immersing it suddenly in water at 20°C. The sheet is suspended from two wires at the upper corners as shown in the sketch. (a) Determine the initial and the final rate of heat transfer from the plate. (b) Estimate the time required. (*Hint:* Note that in laminar natural convection, $h = \Delta T^{0.25}$.)

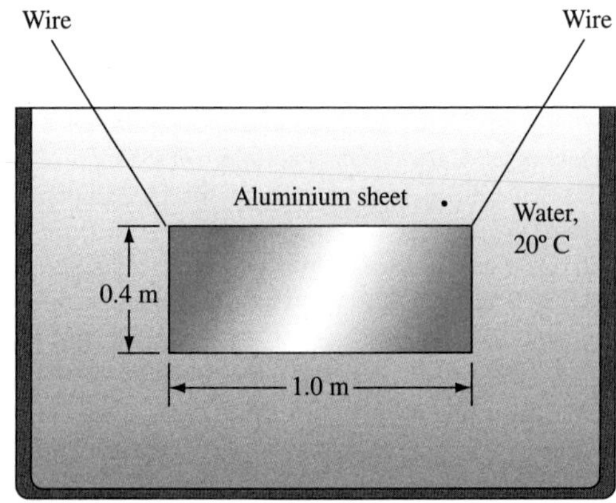

Problem 5.13

5.14 A 0.1-cm-thick square, flat copper plate, 2.5 m × 2.5 m, is to be cooled in a vertical position. The initial temperature of the plate is 90°C with the ambient fluid at 30°C. The fluid medium is either atmospheric air or water. For both fluids, (a) calculate the Grashof number, (b) determine the initial heat transfer coefficient, (c) calculate the initial rate of heat transfer by convection, and (d) estimate the initial rate of temperature change for the plate.

5.15 A laboratory apparatus is used to maintain a horizontal slab of ice at 28°F so that specimens can be prepared on the surface of the ice and kept close to 32°F. If the ice is 4 in. × 1.5 in. and the laboratory is kept at 60°F, find the cooling rate in watts that the apparatus must provide to the ice.

5.16 An electronic circuit board the shape of a flat plate is 0.3 m × 0.3 m in planform and dissipates 15 W. It will be placed in operation on an insulated surface either in a horizontal position or at an angle of 45° to horizontal; in both cases, it will be in still air at 25°C. If the circuit would fail above 60°C, determine whether the two proposed installations are safe.

5.17 Cooled air is flowing through a long, sheet metal air conditioning duct 0.2 m high and 0.3 m wide. If the duct temperature is 10°C and passes through a crawl space under a house at 30°C, estimate (a) the heat transfer rate to the cooled air per meter length of duct and (b) the additional air conditioning load if the duct is 20 m long. (c) Discuss qualitatively the energy conservation that would result if the duct were insulated with glass wool.

5.18 Solar radiation at 600 W/m^2 is absorbed by a black roof inclined at 30° as shown. If the underside of the roof is well insulated, estimate the maximum roof temperature in 20°C air.

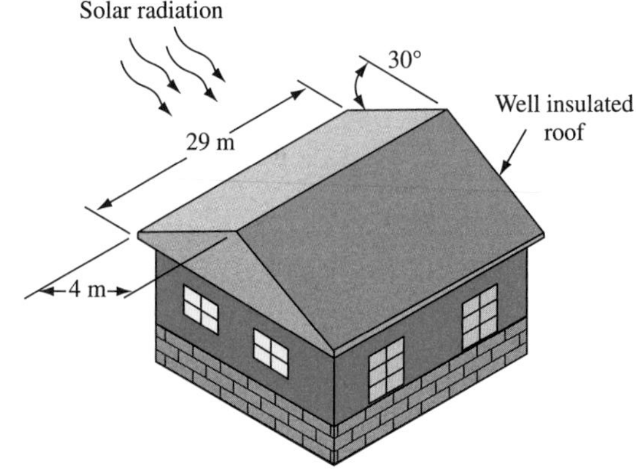

Problem 5.18

5.19 A 1-m-square copper plate is placed horizontally on 2-m-high legs. The plate has been coated with a material that provides a solar absorptivity of 0.9 and an infrared emissivity of 0.25. If the air temperature is 30°C, determine the equilibrium temperature on an average clear day in which the solar radiation incident on a horizontal surface is 850 W/m^2.

5.20 A 2.5-m × 2.5-m steel sheet 1.5 mm thick is removed from an annealing oven at a uniform temperature of 425°C and placed in a large room at 20°C in a horizontal position. (a) Calculate the rate of heat transfer from the steel sheet immediately after its removal from the furnace, considering both radiation and convection. (b) Determine the time required for the steel sheet to cool to a temperature of 60°C. (*Hint:* This will require numerical integration.)

5.21 A thin electronic circuit board, 0.1 m × 0.1 m in size, is to be cooled in air at 25°C, as shown in the sketch. The board is placed in a vertical position, and the back side is well insulated. If the heat dissipation is uniform at 200 W/m^2, determine the average temperature of the surface of the board cover.

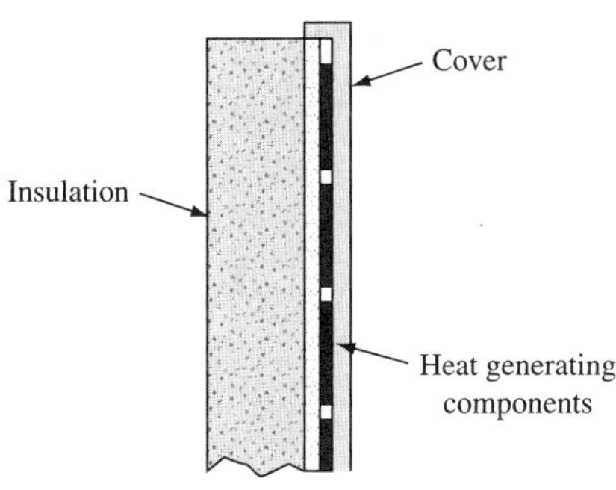

Problem 5.21

5.22 A pot of coffee has been allowed to cool to 17°C. If the electric coffee maker is turned back on, the hot plate on which the pot rests is brought up to 70°C immediately and held at that temperature by a thermostat. Consider the pot to be a vertical cylinder 130 mm in diameter and the depth of coffee in the pot to be 100 mm. Neglect heat losses from the sides and top of the pot. How long will it take before the coffee is drinkable (50°C)? How much did it cost to heat the coffee if electricity costs \$0.05/kW h?

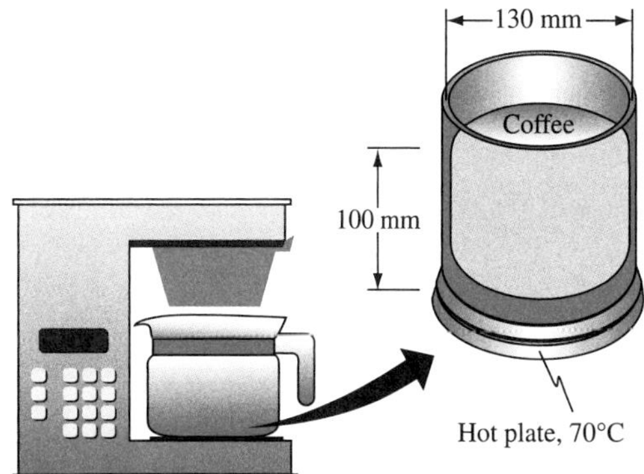

Problem 5.22

5.23 A laboratory experiment has been performed to determine the natural-convection heat transfer correlation for a horizontal cylinder of elliptical cross section in air. The cylinder is 1 m long, has a hydraulic diameter of 1 cm, a surface area of 0.0314 m^2, and is heated internally by electrical resistance heating. Recorded data include power dissipation, cylinder surface temperature, and ambient air temperature. The power dissipation has been corrected for radiation effects:

$T_s - T_\infty$(°C)	q (W)
15.2	4.60
40.7	15.76
75.8	34.29
92.1	43.74
127.4	65.62

Assume that all air properties can be evaluated at 27°C, and determine the constants in the correlation equation: Nu = $C(\text{GrPr})^m$.

5.24 A long, 2-cm-diameter horizontal copper pipe carries dry saturated steam at 1.2 atm absolute pressure. The pipe is contained within an environmental testing chamber in which the ambient air pressure can be adjusted from 0.5 to 2.0 atm absolute, while the ambient air temperature is held constant at 20°C. What is the effect of this pressure change on the rate of condensate flow per meter length of pipe? Assume that the pressure change does not affect the absolute viscosity, thermal conductivity, or specific heat of the air.

5.25 Compare the rate of condensate flow from the pipe in Problem 5.24 (air pressure = 2.0 atm) with that for a 3.89-cm-OD pipe and 2.0 atm air pressure. What is the rate of condensate flow if the 2 cm pipe is submerged in a 20°C constant-temperature water bath?

5.26 A thermocouple (0.8-mm OD) is located horizontally in a large enclosure whose walls are at 37°C. The enclosure is filled with a transparent, quiescent gas that has the same properties as air. The electromotive force (emf) of the thermocouple indicates a temperature of 230°C. Estimate the true gas temperature if the emissivity of the thermocouple is 0.8.

5.27 Only 10% of the energy dissipated by the tungsten filament of an incandescent lamp is in the form of useful visible light. Consider a 100-W lamp with a 10-cm spherical glass bulb, as shown in the sketch. Assuming an emissivity of 0.85 for the glass and an ambient air temperature of 20°C, what is the temperature of the glass bulb?

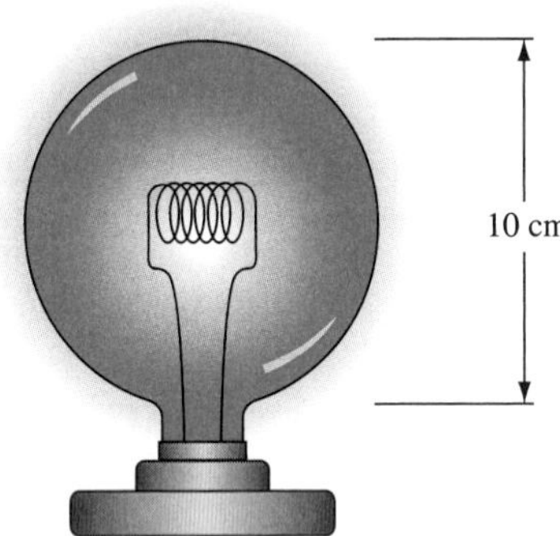

Problem 5.27

5.28 A sphere 20 cm in diameter containing liquid air (−140°C) is covered with 5-cm-thick glass wool (50 kg/m^3 density) with an emissivity of 0.8. Estimate the rate of heat transfer to the liquid air from the surrounding air at 20°C by convection and radiation. How would you reduce the heat transfer?

5.29 A 2-cm-diameter bare aluminum electric power transmission line with an emissivity of 0.07 carries 500 A at 400 kV. The wire has an electrical resistivity of 1.72 $\mu\Omega$ cm^2/cm at 20°C and is supended horizontally between two towers separated by 1 km. Determine the surface temperature of the transmission line if the air temperature is 20°C. What fraction of the dissipated power is due to radiation heat transfer?

5.30 An 8-in.-diameter horizontal steam pipe carries 220 lbm/h of dry, pressurized, saturated steam at 250°F. If the ambient air temperature is 70°F, determine the rate of condensate flow at the end of 10 ft of pipe. Use an emissivity of 0.85 for the pipe surface. If heat losses are to be kept below 1% of the rate of energy transport by the steam, what thickness of fiberglass insulation is required? The rate of energy transport by the steam is the heat of condensation of the steam flow. The heat of vaporization of the steam is 950 Btu/lb.

5.31 A long steel rod (2 cm in diameter, 2 m long) has been heat treated and quenched to a temperature of 100°C in an oil bath. To cool the rod further, it is necessary to remove it from the bath and expose it to room air. Will cool-down be faster by cooling the cylinder in the vertical or horizontal position? How long will the two methods require to allow the rod to cool to 40°C in 20°C air?

5.32 In petroleum processing plants, it is often necessary to pump highly viscous liquids such as asphalt through pipes. To keep pumping costs within reason, the pipelines are electrically heated to reduce the viscosity of the asphalt. Consider a 15-cm-OD uninsulated pipe and an ambient temperatures of 20°C. How much power per meter of pipe length is necessary to maintain the pipe at 50°C? If the pipe is insulated with 5 cm of fiberglass insulation, what is the power requirement?

5.33 Estimate the rate of convection heat transfer across a 1-m-tall double-pane window assembly in which the outside pane is at 0°C and the inside pane is at 20°C. The panes are spaced 2.5 cm apart. What is the thermal resistance (R-value) of the window if the rate of radiative heat flux is 84 W/m^2?

5.34 An architect is asked to determine the heat loss through a wall of a building constructed as shown in the sketch. The space between the walls is 10 cm and contains air. If the inner surface is at 20°C, and the

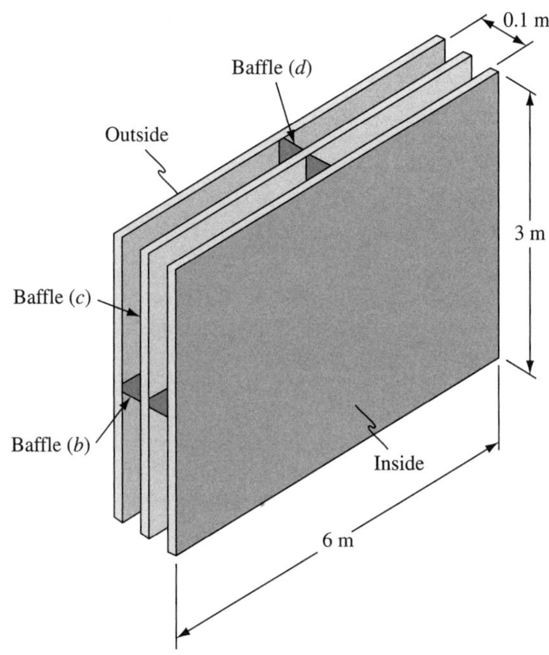

Problem 5.34

outer surface is at −8°C, (a) estimate the heat loss by natural convection. Then determine the effect of placing a baffle (b) horizontally at the midheight of the vertical section, (c) vertically at the center of the horizontal section, and (d) vertically halfway between the two surfaces.

5.35 A flat-plate solar collector of 3 m × 5 m area has an absorber plate that is to operate at a temperature of 70°C. To reduce heat losses, a glass cover is placed 0.05 m from the absorber. Its operating temperature is estimated to be 35°C. Determine the rate of heat loss from the absorber if the 3 m edge is tilted at angles of inclination from the horizontal of 0°, 30°, and 60°.

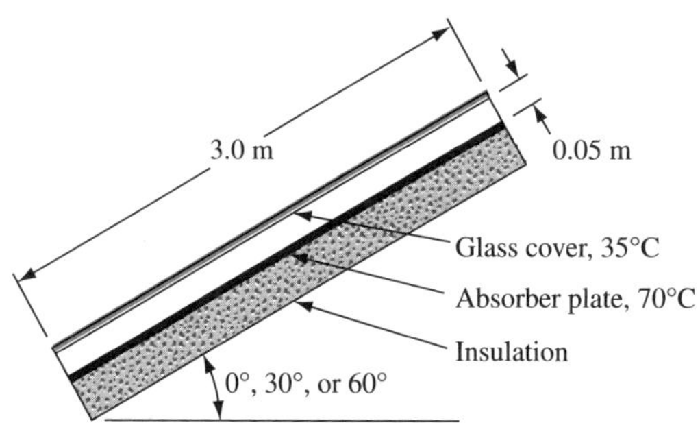

Problem 5.35

5.36 Determine the rate of heat loss through a doublepane window shown in the sketch if the inside room temperature is 65°C and the average outside air is 0°C during December. Neglect the effect of the window frame. If the house is electrically heated at a cost of $.06/kW h, estimate the savings achieved during December by using a double-pane compared to a single-pane window.

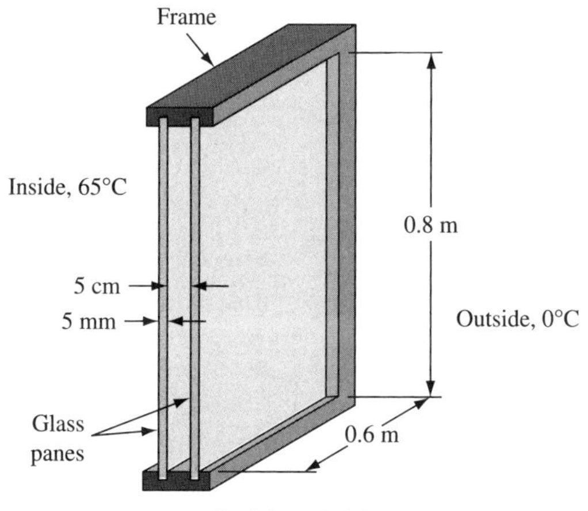

Problem 5.36

5.37 Calculate the rate of heat transfer between a pair of concentric horizontal cylinders 20 mm and 126 mm in diameter. The inner cylinder is maintained at 37°C, and the outer cylinder is maintained at 17°C.

5.38 Two long, concentric, horizontal aluminum tubes of 0.2 m and 0.25 m diameter are maintained at 300 K and 400 K, respectively. The space between the tubes is filled with nitrogen. If the surfaces of the tubes are polished to prevent radiation, estimate the rate of heat transfer for gas pressures in the annulus of (a) 10 atm and (b) 0.1 atm.

5.39 A solar collector design consists of several parallel tubes, each enclosed concentrically in an outer tube that is transparent to solar radiation. The tubes are thin walled with inner and outer cylinder diameters of 0.10 and 0.15 m, respectively. The annual space between the tubes is filled with air at atmospheric pressure. Under operating conditions, the inner and outer tube surface temperatures are 70°C and 30°C, respectively. (a) What is the convection heat loss per meter of the tube length? (b) If the emissivity of the outer surface of the inner tube is 0.2 and the outer cylinder behaves as though it were a blackbody, estimate the radiation loss. (c) Discuss design options for reducing the total heat loss.

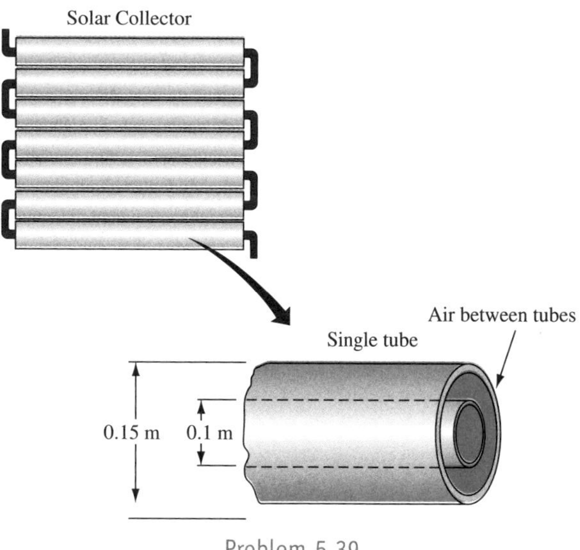

Problem 5.39

5.40 Liquid oxygen at −183°C is stored in a thin-walled spherical container with an outside diameter of 2 m. This container is surrounded by another sphere of 2.5-m inside diameter to reduce heat loss. The inner spherical surface has an emissivity of 0.05, and the outer sphere is black. Under normal operation the space between the spheres is evacuated, but an accident resulted in a leak in the outer sphere, so the space is now filled with air at one atm. If the outer sphere is at 25°C, compare the heat losses before and after the accident.

5.41 The surfaces of two concentric spheres with radii of 75 and 100 mm are maintained at 325 K and 275 K, respectively. (a) If the space between the spheres is filled with nitrogen at 5 atm, estimate the convection heat transfer rate. (b) If both sphere surfaces are black, estimate the total rate of heat transfer between them. (c) Suggest ways to reduce the heat transfer.

5.42 Estimate the rate of heat transfer from one side of a 2-m-diameter disk with a surface temperature of 50°C rotating at 600 rev/min in 20°C air.

5.43 A sphere 0.1 m in diameter is rotating at 20 rpm in a large container of CO_2 at atmospheric pressure. If the sphere is at 60°C and the CO_2 at 20°C, estimate the rate of heat transfer.

5.44 A mild steel (1% carbon), 2-cm-OD shaft rotating in 20°C air at 20,000 rev/min is attached to two bearings 0.7 m apart, as shown below. If the temperature at the bearings is 90°C, determine the temperature distribution along the shaft. (*Hint:* Show that for the high rotational speeds Eq. (5.35) approaches $\overline{\text{Nu}}_D = 0.086(\pi D^2\omega/\nu)^{0.7}$.)

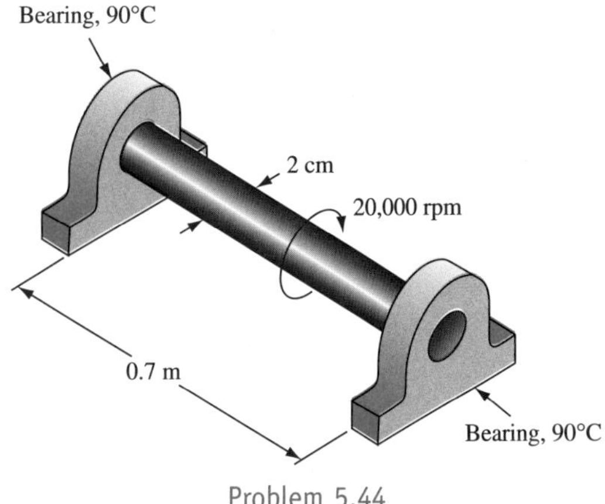

Problem 5.44

5.45 An electronic device is to be cooled in air at 20°C by an array of equally spaced vertical rectangular fins, as shown in the following sketch. The fins are made of aluminum and their average temperature, T_s, is 100°C. Estimate (a) the optimum spacing, *S*, (b) the number of fins, (c) the rate of heat transfer from one fin, and (d) the total rate of heat dissipation. (e) Is the assumption of a uniform fin temperature justified?

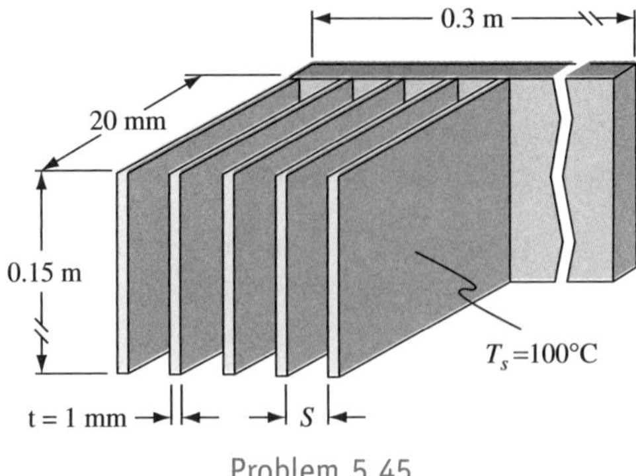

Problem 5.45

5.46 Consider a vertical, 20-cm-tall, flat plate at 120°C, suspended in a fluid at 100°C. If the fluid is being forced past the plate from above, estimate the fluid velocity for which natural convection becomes negligible (less than 10%) in (a) mercury, (b) air, (c) water.

5.47 Suppose a thin, vertical, flat plate 60 cm high and 40 cm wide is immersed in a fluid flowing parallel to its surface. If the plate is at 40°C and the fluid at 10°C, estimate the Reynolds number at which buoyancy effects are essentially negligible for heat transfer from the plate if the fluid is (a) mercury, (b) air, (c) water. Then calculate the corresponding fluid velocity for the three fluids.

5.48 A vertical, isothermal plate 30 cm high is suspended in an atmospheric airstream flowing at 2 m/s in a vertical direction. If the air is at 16°C, estimate the plate temperature for which the natural-convection effect on the heat transfer coefficient will be less than 10%.

5.49 A horizontal disk 1 m in diameter rotates in air at 25°C. If the disk is at 100°C, estimate the number of revolutions per minute at which natural convection for a stationary disk becomes less than 10% of the heat transfer for a rotating disk.

5.50 The refrigeration system for an indoor ice rink is to be sized by an HVAC contractor. The refrigeration system has a COP (coefficient of performance) of 0.5. The ice surface is estimated to be −2°C, and the ambient air is 24°C. Determine the size of the refrigeration system (in kW) required for a 110-m-diameter circular ice surface.

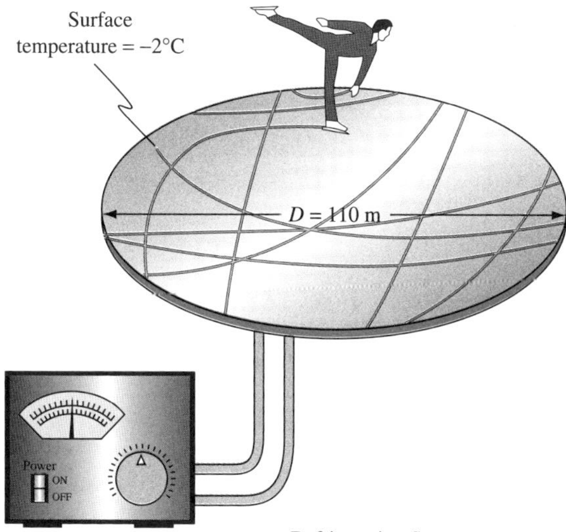

Problem 5.50

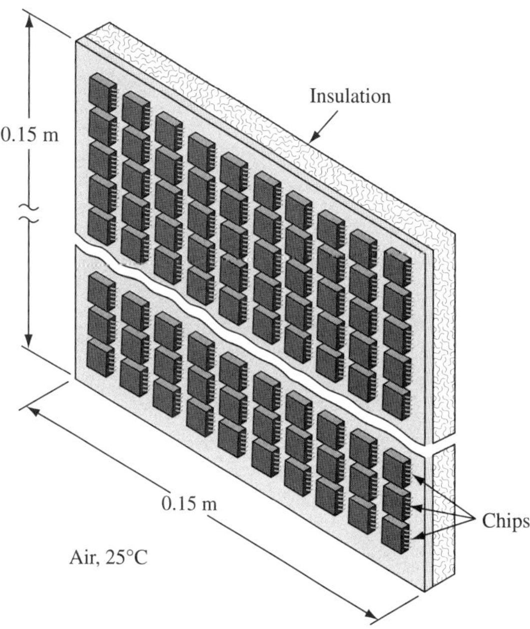

Problem 5.51

5.51 A 0.15-m-square circuit board is to be cooled in a vertical position, as shown in the sketch. The board is insulated on one side, while on the other, 100 closely spaced square chips are mounted. Each chip dissipates 0.06 W of heat. The board is exposed to air at 25°C, and the maximum allowable chip temperature is 60°C. Investigate the following cooling options: (a) natural convection, (b) air cooling with upward flow at a velocity of 0.5 m/s, (c) air cooling with downward flow at a velocity of 0.5 m/s.

5.52 A gas-fired industrial furnace is used to generate steam. The furnace is a 3-m cubic structure, and the interior surfaces are completely covered with boiler tubes transporting pressurized wet steam at 150°C. It is desired to keep the furnace losses to 1% of the total heat input of 1 MW. The outside of the furnace can be insulated with a blanket-type mineral wool insulation (k = 0.13W/m °C), which is protected by a polished metal-sheet outer shell. Assume the floor of the furnace is insulated. What is the temperature of the metal-shell sides? What thickness of insulation is required?

5.53 An electronic device is to be cooled by natural convection in atmospheric air at 20°C. The device generates 50 W internally, and only one of its external surfaces is suitable for attaching fins. The surface available for attaching cooling fins is 0.15 m tall and 0.4 m wide. The maximum length of a fin perpendicular to the surface is limited to 0.02 m, and the temperature at the base of the fin is not to exceed 70°C in one design and 100°C in another. Design an array of fins spaced at a distance S from each other so that the boundary layers will not interfere with each other appreciably and the maximum rate of heat dissipation is approached. For the evaluation of this spacing, assume that the fins are at a uniform temperature. Then select a thickness t that will provide good fin efficiency and ascertain which base temperature is feasible. (For a complete thermal analysis see ASME *J. Heat Transfer*, 1977, p. 369; *J. Heat Transfer*, 1979, p. 569; and *J. Heat Transfer*, 1984, p. 116.)

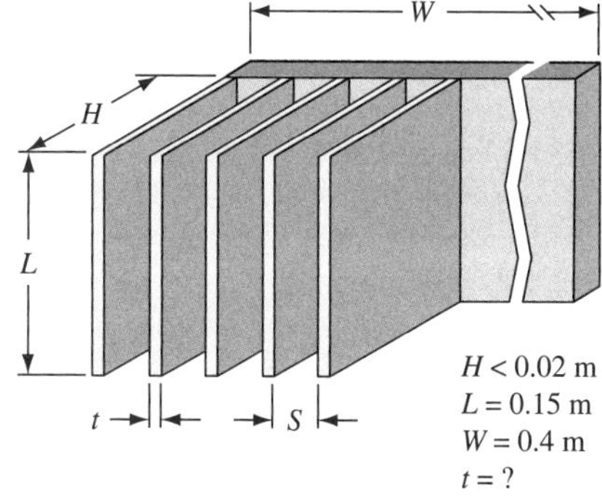

Problem 5.53

Design Problems

5.1 **Residential Baseboard Heater Design Improvement** (Chapter 5)
Baseboard heaters used in residential applications have not changed in over 30 years. In either electrical or hot-water heating systems, the baseboard heater is a horizontal tube with closely spaced slip-on vertical aluminum fins. An extruded sheet metal shroud directs natural-convection flow of cool air near the floor over the finned tube. Consider alternative designs for this heat transfer device with the objective of reducing the purchase price per unit heat transfer. To that end, you should consider materials selection, manufacturability, and heat transfer performance. Clearly, any design that increase operating costs (e.g., a requirement for periodic cleaning), is to be avoided.

5.2 **Heater Design** (Chapter 5)
In Design Problems 1.5 and 4.5, you have calculated the heat load on a small industrial building in Denver, Colorado. Repeat the heat load estimate, but calculate the convection heat transfer with equations presented in this chapter. In order to maintain the temperature at 20°C, it is necessary to provide a heating system; two options are available. One would be an electric baseboard heater and the other would be a hot-water system that circulates water through a thin tube on the inside of the inside of the building. The water can be heated by natural-convection combustion from a temperature of 10 to 80°C with 80% efficiency. Natural gas in Colorado costs about \$4 per 1000 ft^3. Electric power for an industrial organization in Colorado costs about 5 ¢/kW h. Recommend the preferred heating system on the basis of an economic analysis.

5.3 **Skin Temperature Probe** (Chapter 5)
Physicians can use the local temperature of the skin as an indicator of underlying inflammation. U.S. Patent 3,570,312, "*Skin Temperature Sensing Device*" by F. Kreith, March 16, 1971, describes such a device. It utilizes a small thin-walled tube with a thermocouple or thermistor at the end. In order to obtain reproducible results, it is necessary to exert the same pressure on the skin with repeated measurements.

Design a skin-temperature sensing device that is no larger than a pencil and can be stored in a pocket alongside a pen. Select an appropriate thermocouple or a thermistor from available literature and devise a means of exerting repeatable constant pressure with the device. Also estimate the possible error that could ensue because heat is lost from the outside of the cylinder after an equilibrium temperature between skin and sensing device has been established. For experience with this device, you can consult by F. Kreith and D. Gudagni, "Skin Temperature Sensing Device," *Journal of Physics*, E. Sci, Inst., vol. 5, pp. 869–876, 1971.

5.4 **Fin Design** (Chapter 5)
Reconsider the fin design from Design Problem 2.1, but calculate the natural-convection heat transfer coefficient from information presented in this chapter. As shown in the schematic diagram, the invention envisions several flat fins of 6-cm height in a staged

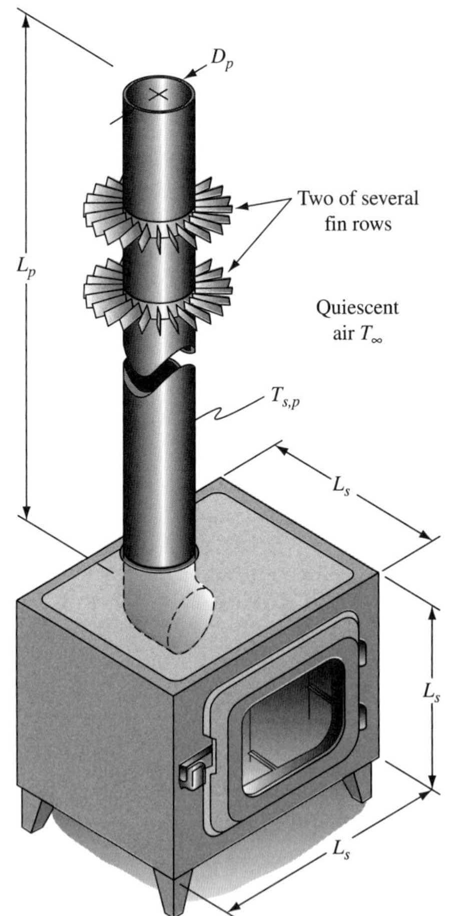

Design Problem 5.4

arrangement attached circumferentially along the exhaust stack. Explain why the inventor did not provide a continuous fin from the top of the stove to the ceiling. Also calculate the amount of heat that will be recouped by the attachment of the fins, assuming that the stove operates eight hours per day. Then calculate the cost of the material that you have selected for the fin and, assuming that the manufacturing cost is approximately the same as the cost of material, estimate the value in dollars per kilowatt hour of the heat recovered from this enhancement structure. Taking into account the cost of manufacturing the circumferential fins, would a simple flat fin that could be stamped from sheet metal be cost-effective?

CHAPTER 6

Forced Convection Inside Tubes and Ducts

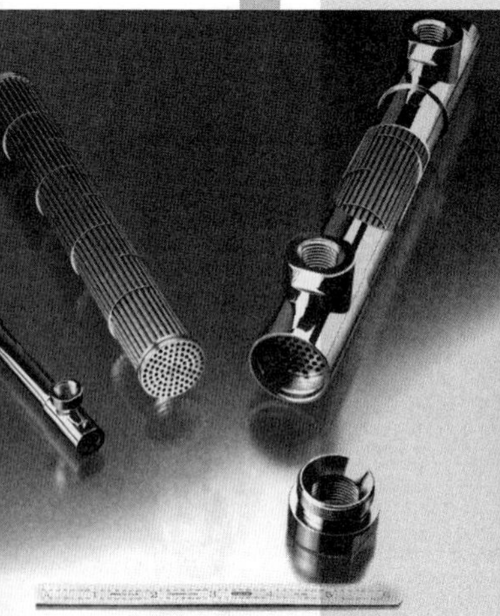

Typical tube bundle of multiple circular tubes and cutaway section of a mini shell-and-tube heat exchanger.

Source: Courtesy of Exergy, LLC.

Concepts and Analyses to Be Learned

The process of transferring heat by convection when the fluid flow is driven by an applied pressure gradient is referred to as *forced convection*. When this flow is confined in a tube or a duct of any arbitrary geometrical cross section, the growth and development of boundary layers are also confined. In such flows, the hydraulic diameter of the duct, rather than its length, is the characteristic length for scaling the boundary layer as well as for dimensionless representation of flow-friction loss and the heat transfer coefficient. Convective heat transfer inside tubes and ducts is encountered in numerous applications where heat exchangers, made up of circular tubes as well as a variety of noncircular cross-sectional geometries, are employed. A study of this chapter will teach you:

- How to express the dimensionless form of the heat transfer coefficient in a duct, and its dependence on flow properties and tube geometry.
- How to mathematically model forced-convection heat transfer in a long circular tube for laminar fluid flow.
- How to determine the heat transfer coefficient in ducts of different geometries from different theoretical and/or empirical correlations in both laminar and turbulent flows.
- How to model and employ the analogy between heat and momentum transfer in turbulent flow.
- How to evaluate heat transfer coefficients in some examples where enhancement techniques, such as coiled tubes, finned tubes, and twisted-tape inserts, are employed.

6.1 Introduction

Heating and cooling of fluids flowing inside conduits are among the most important heat transfer processes in engineering. The design and analysis of heat exchangers require a knowledge of the heat transfer coefficient between the wall of the conduit and the fluid flowing inside it. The sizes of boilers, economizers, superheaters, and preheaters depend largely on the heat transfer coefficient between the inner surface of the tubes and the fluid. Also, in the design of air-conditioning and refrigeration equipment, it is necessary to evaluate heat transfer coefficients for fluids flowing inside ducts. Once the heat transfer coefficient for a given geometry and specified flow conditions is known, the rate of heat transfer at the prevailing temperature difference can be calculated from the equation

$$q_c = \bar{h}_c A(T_{\text{surface}} - T_{\text{fluid}}) \tag{6.1}$$

The same relation also can be used to determine the area required to transfer heat at a specified rate for a given temperature potential. But when heat is transferred to a fluid inside a conduit, the fluid temperature varies along the conduit and at any cross section. The fluid temperature for flow inside a duct must therefore be defined with care and precision.

The heat transfer coefficient $\bar{h}_c$ can be calculated from the Nusselt number $\bar{h}_c D_H/k$, as shown in Section 4.5. For flow in long tubes or conduits (Fig. 6.1a), the significant length in the Nusselt number is the *hydraulic diameter*, D_H, defined as

$$D_H = 4\,\frac{\text{flow cross-sectional area}}{\text{wetted perimeter}} \tag{6.2}$$

For a circular tube or a pipe, the flow cross-sectional area is $\pi D^2/4$, the wetted perimeter is πD, and therefore, the inside diameter of the tube equals the hydraulic

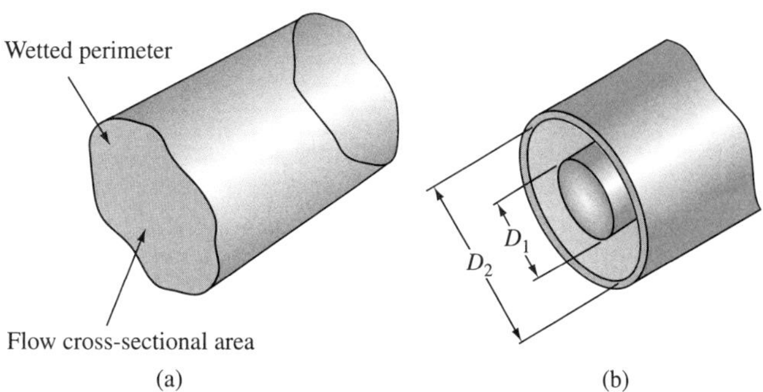

FIGURE 6.1 Hydraulic diameter for (a) irregular cross section and (b) annulus.

diameter. For an annulus formed between two concentric tubes (Fig. 6.1b), we have

$$D_H = 4\frac{(\pi/4)(D_2^2 - D_1^2)}{\pi(D_1 + D_2)} = D_2 - D_1 \tag{6.3}$$

In engineering practice, the Nusselt number for flow in conduits is usually evaluated from empirical equations based on experimental results. The only exception is laminar flow inside circular tubes, selected noncircular cross-sectional ducts, and a few other conduits for which analytical and theoretical solutions are available [13]. Some simple examples of laminar-flow heat transfer in circular tubes are dealt with in Section 6.2. From a dimensional analysis, as shown in Section 4.5, the experimental results obtained in forced-convection heat transfer experiments in long ducts and conduits can be correlated by an equation of the form

$$\text{Nu} = \phi(\text{Re})\psi(\text{Pr}) \tag{6.4}$$

where the symbols ϕ and ψ denote functions of the Reynolds number and Prandtl number, respectively. For short ducts, particularly in laminar flow, the right-hand side of Eq. (6.4) must be modified by including the aspect ratio x/D_H:

$$\text{Nu} = \phi(\text{Re})\psi(\text{Pr})f\left(\frac{x}{D_H}\right)$$

where $f(x/D_H)$ denotes the functional dependence on the aspect ratio.

6.1.1 Reference Fluid Temperature

The convection heat transfer coefficient used to build the Nusselt number for heat transfer to a fluid flowing in a conduit is defined by Eq. (6.1). The numerical value of $\bar{h}_c$, as mentioned previously, depends on the choice of the reference temperature in the fluid. For flow over a plane surface, the temperature of the fluid far away from the heat source is generally uniform, and its value is a natural choice for the fluid temperature in Eq. (6.1). In heat transfer to or from a fluid flowing in a conduit, the temperature of the fluid does not level out but varies both along the direction of mass flow and in the direction of heat flow. At a given cross section of the conduit, the temperature of the fluid at the center could be selected as the reference temperature in Eq. (6.1). However, the center temperature is difficult to measure in practice; furthermore, it is not a measure of the change in internal energy of all the fluid flowing in the conduit. It is therefore a common practice, and one we shall follow here, to use the *average fluid bulk temperature*, T_b, as the reference fluid temperature in Eq. (6.1). The average fluid temperature at a station of the conduit is often called the *mixing-cup temperature* because it is the temperature which the fluid passing a cross-sectional area of the conduit during a given time internal would assume if the fluid were collected and mixed in a cup.

Use of the fluid bulk temperature as the reference temperature in Eq. (6.1) allows us to make heat balances readily, because in the steady state, the difference in average bulk temperature between two sections of a conduit is a direct measure of the rate of heat transfer:

$$q_c = \dot{m} c_p \Delta T_b \tag{6.5}$$

where q_c = rate of heat transfer to fluid, W
$\dot{m}$ = flow rate, kg/s
c_p = specific heat at constant pressure, kJ/kg K
ΔT_b = difference in average fluid bulk temperature between cross sections in question, K or °C

The problems associated with variations of the bulk temperature in the direction of flow will be considered in detail in Chapter 8, where the analysis of heat exchangers is taken up. For preliminary calculations, it is common practice to use the bulk temperature halfway between the inlet and the outlet section of a duct as the reference temperature in Eq. (6.1). This procedure is satisfactory when the wall heat flux of the duct is constant but may require some modification when the heat is transferrred between two fluids separated by a wall, as, for example, in a heat exchanger where one fluid flows inside a pipe while another passes over the outside of the pipe. Although this type of problem is of considerable practical importance, it will not concern us in this chapter, where the emphasis is placed on the evaluation of convection heat transfer coefficients, which can be determined in a given flow system when the pertinent bulk and wall temperatures are specified.

6.1.2 Effect of Reynolds Number on Heat Transfer and Pressure Drop in Fully Established Flow

For a given fluid, the Nusselt number depends primarily on the flow conditions, which can be characterized by the Reynolds number, Re. For flow in long conduits, the characteristic length in the Reynolds number, as in the Nusselt number, is the hydraulic diameter, and the velocity to be used is the average over the flow cross-sectional area, $\bar{U}$, or

$$\mathrm{Re}_{D_H} = \frac{\bar{U} D_H \rho}{\mu} = \frac{\bar{U} D_H}{\nu} \tag{6.6}$$

In long ducts, where the entrance effects are not important, the flow is laminar when the Reynolds number is below about 2100. In the range of Reynolds numbers between 2100 and 10,000, a transition from laminar to turbulent flow takes place. The flow in this regime is called transitional. At a Reynolds number of about 10,000, the flow becomes fully turbulent.

In laminar flow through a duct, just as in laminar flow over a plate, there is no mixing of warmer and colder fluid particles by eddy motion, and the heat transfer takes place solely by conduction. Since all fluids with the exception of liquid metals have small thermal conductivities, the heat transfer coefficients in laminar flow are relatively small. In transitional flow, a certain amount of mixing occurs through eddies that carry warmer fluid into cooler regions and vice versa. Since the mixing

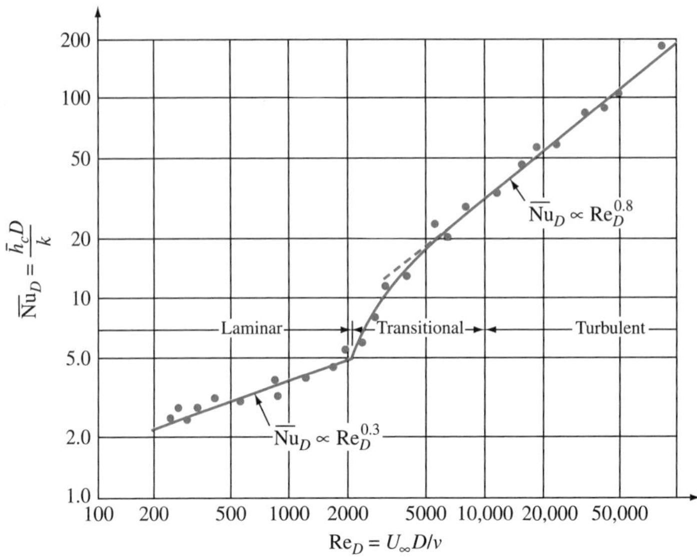

FIGURE 6.2 Nusselt number versus Reynolds number for air flowing in a long heated pipe at uniform wall temperature.

motion, even if it is only on a small scale, accelerates the transfer of heat considerably, a marked increase in the heat transfer coefficient occurs above $Re_{D_H} = 2100$ (it should be noted, however, that this change, or *transition*, can generally occur over a range of Reynolds number, $2000 < Re_{D_H} < 5000$). This change can be seen in Fig. 6.2, where experimentally measured values of the average Nusselt number for atmospheric air flowing through a long heated tube are plotted as a function of the Reynolds number. Since the Prandtl number for air does not vary appreciably, Eq. (6.4) reduces to $Nu = \phi(Re_{D_H})$, and the curve drawn through the experimental points shows the dependence of Nu on the flow conditions. We note that in the laminar regime, the Nusselt number remains small, increasing from about 3.5 at $Re_{D_H} = 300$ to 5.0 at $Re_{D_H} = 2100$. Above a Reynolds number of 2100, the Nusselt number begins to increase rapidly until the Reynolds number reaches about 8000. As the Reynolds number is further increased, the Nusselt number continues to increase, but at a slower rate.

A qualitative explanation for this behavior can be given by observing the fluid flow field shown schematically in Fig. 6.3. At Reynolds numbers above 8000, the flow inside the conduit is fully turbulent except for a very thin layer of fluid adjacent to the wall. In this layer, turbulent eddies are damped out as a result of the viscous forces that predominate near the surface, and therefore heat flows through this layer mainly by conduction.* The edge of this sublayer is indicated by a dashed line

*Although some studies [1] have shown that turbulent transport also exists to some extent near the wall, especially when the Prandtl number is larger than 5, the layer near the wall is commonly referred to as the "viscous sublayer."

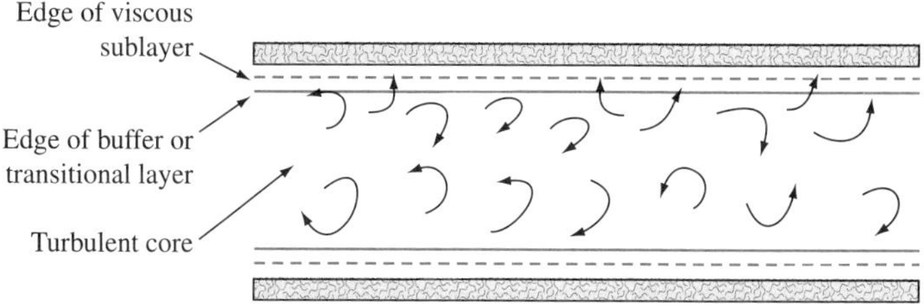

FIGURE 6.3 Flow structure for a fluid in turbulent flow through a pipe.

in Fig. 6.3. The flow beyond it is turbulent, and the circular arrows in the turbulent-flow regime represent the eddies that sweep the edge of the layer, probably penetrate it, and carry along with them fluid at the temperature prevailing there. The eddies mix the warmer and cooler fluids so effectively that heat is transferred very rapidly between the edge of the viscous sublayer and the turbulent bulk of the fluid. It is thus apparent that except for fluids of high thermal conductivity (e.g., liquid metals), the thermal resistance of the sublayer controls the rate of heat transfer, and most of the temperature drop between the bulk of the fluid and the surface of the conduit occurs in this layer. The turbulent portion of the flow field, on the other hand, offers little resistance to the flow of heat. The only effective method of increasing the heat transfer coefficient is therefore to decrease the thermal resistance of the sublayer. This can be accomplished by increasing the turbulence in the main stream so that the turbulent eddies can penetrate deeper into the layer. An increase in turbulence, however, is accompanied by large energy losses that increase the frictional pressure drop in the conduit. In the design and selection of industrial heat exchangers, where not only the initial cost but also the operating expenses must be considered, the pressure drop is an important factor. An increase in the flow velocity yields higher heat transfer coefficients, which, in accordance with Eq. (6.1), decrease the size and consequently the initial cost of the equipment for a specified heat transfer rate. At the same time, however, the pumping cost increases. The optimum design therefore requires a compromise between the initial and operating costs. In practice, it has been found that increases in pumping costs and operating expenses often outweigh the saving in the initial cost of heat transfer equipment under continuous operating conditions. As a result, the velocities used in a majority of commercial heat exchange equipment are relatively low, corresponding to Reynolds numbers of no more than 50,000. Laminar flow is usually avoided in heat exchange equipment because of the low heat transfer coefficients obtained. However, in the chemical industry, where very viscous liquids must frequently be handled, laminar flow sometimes cannot be avoided without producing undesirably large pressure losses.

It was shown in Section 4.12 that, for turbulent flow of liquids and gases over a flat plate, the Nusselt number is proportional to the Reynolds number raised to the 0.8 power. Since in turbulent forced convection the viscous sublayer generally controls the rate of heat flow irrespective of the geometry of the system, it is not surprising that for turbulent forced convection in conduits the Nusselt number is related to the Reynolds number by the same type of power law. For the case of air flowing in a pipe, this relation is illustrated in the graph of Fig. 6.2.

6.1.3 Effect of Prandtl Number

The Prandtl number Pr is a function of the fluid properties alone. It has been defined as the ratio of the kinematic viscosity of the fluid to the thermal diffusivity of the fluid:

$$\mathrm{Pr} = \frac{\nu}{\alpha} = \frac{c_p\mu}{k}$$

The kinematic viscosity ν, or μ/ρ, is often referred to as the molecular diffusivity of momentum because it is a measure of the rate of momentum transfer between the molecules. The thermal diffusivity of a fluid, $k/c_p\rho$, is often called the molecular diffusivity of heat. It is a measure of the ratio of the heat transmission and energy storage capacities of the molecules.

The Prandtl number relates the temperature distribution to the velocity distribution, as shown in Section 4.5 for flow over a flat plate. For flow in a pipe, just as over a flat plate, the velocity and temperature profiles are similar for fluids having a Prandtl number of unity. When the Prandtl number is smaller, the temperature gradient near a surface is less steep than the velocity gradient, and for fluids whose Prandtl number is larger than one, the temperature gradient is steeper than the velocity gradient. The effect of Prandtl number on the temperature gradient in turbulent flow at a given Reynolds number in tubes is illustrated schematically in Fig. 6.4, where temperature profiles at different Prandtl numbers are shown at $\mathrm{Re}_D = 10{,}000$. These curves reveal that, at a specified Reynolds number, the temperature gradient at the wall is steeper in a fluid having a large Prandtl number than in a fluid having a small Prandtl number. Consequently, at a given Reynolds number, fluids with larger Prandtl numbers have larger Nusselt numbers.

Liquid metals generally have a high thermal conductivity and a small specific heat; their Prandtl numbers are therefore small, ranging from 0.005 to 0.01. The Prandtl numbers of gases range from 0.6 to 1.0. Most oils, on the other hand, have large Prandtl numbers, some up to 5000 or more, because their viscosity is large at low temperatures and their thermal conductivity is small.

6.1.4 Entrance Effects

In addition to the Reynolds number and the Prandtl number, several other factors can influence heat transfer by forced convection in a duct. For example, when the conduit is short, entrance effects are important. As a fluid enters a duct with a uniform velocity, the fluid immediately adjacent to the tube wall is brought to rest. For a short distance from the entrance, a laminar boundary layer is formed along the tube wall. If the turbulence in the entering fluid stream is high, the boundary layer will quickly become turbulent. Irrespective of whether the boundary layer remains laminar or becomes turbulent, it will increase in thickness until it fills the entire duct. From this point on, the velocity profile across the duct remains essentially unchanged.

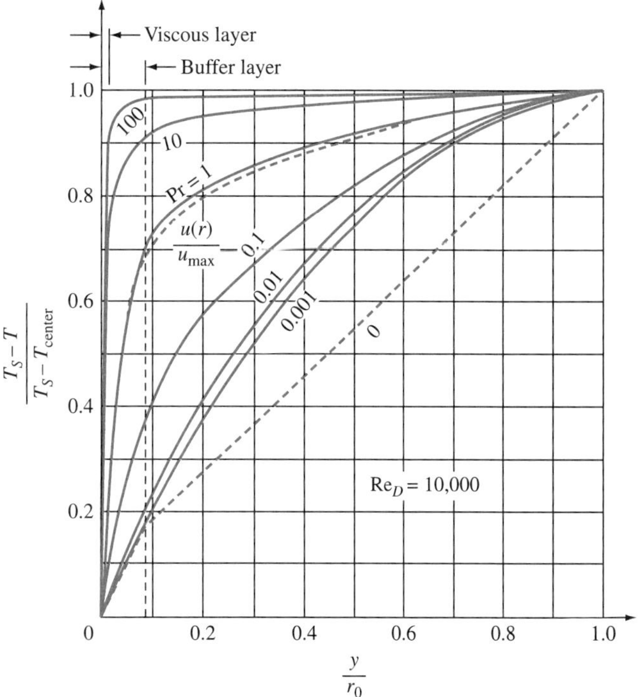

FIGURE 6.4 Effect of Prandtl number on temperature profile for turbulent flow in a long pipe (y is the distance from the tube wall and r_0 is the inner pipe radius).

Source: Courtesy of R. C. Martinelli, "Heat Transfer to Molten Metals", Trans. ASME, Vol. 69, 1947, p. 947. Reprinted by permission of The American Society of Mechanical Engineers International.

The development of the thermal boundary layer in a fluid that is heated or cooled in a duct is qualitatively similar to that of the hydrodynamic boundary layer. At the entrance, the temperature is generally uniform transversely, but as the fluid flows along the duct, the heated or cooled layer increases in thickness until heat is transferred to or from the fluid in the center of the duct. Beyond this point, the temperature profile remains essentially constant if the velocity profile is fully established.

The final shapes of the velocity and temperature profiles depend on whether the fully developed flow is laminar or turbulent. Figures 6.5 on the next page and Figure 6.6 on page 359 qualitatively illustrate the growth of the boundary layers as well as the variations in the local convection heat transfer coefficient near the entrance of a tube for laminar and turbulent conditions, respectively. Inspection of these figures shows that the convection heat transfer coefficient varies considerably near the entrance. If the entrance is square-edged, as in most heat exchangers, the initial development of the hydrodynamic and thermal boundary layers along the walls of the tube is quite similar to that along a flat plane. Consequently,

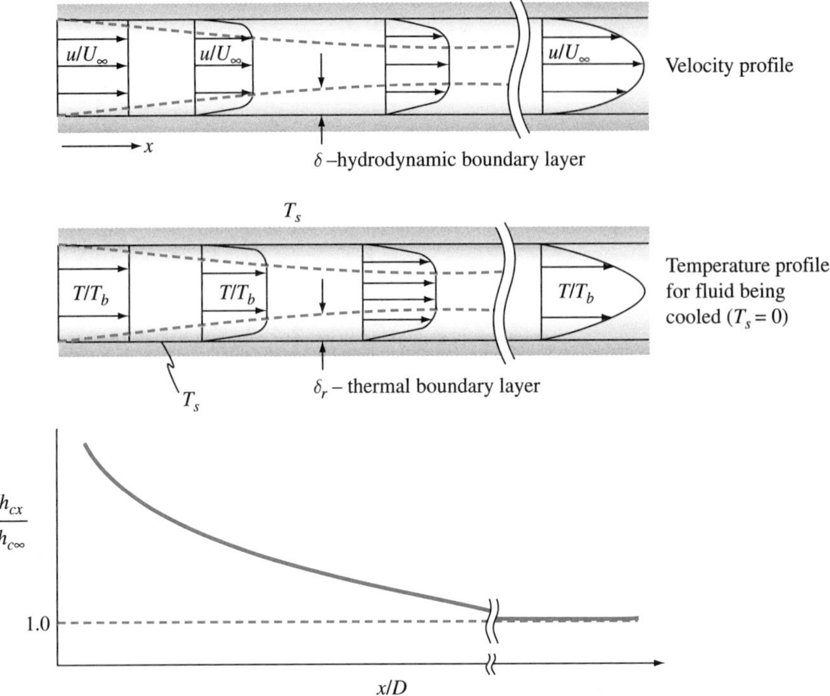

FIGURE 6.5 Velocity distribution, temperature profiles, and variation of the local heat transfer coefficient near the inlet of a tube for air being cooled in laminar flow (surface temperature T_s uniform).

the heat transfer coefficient is largest near the entrance and decreases along the duct until both the velocity and the temperature profiles for the fully developed flow have been established. If the pipe Reynolds number for the fully developed flow $\bar{U}D\rho/\mu$ is below 2100, the entrance effects may be appreciable for a length as much as 100 hydraulic diameters from the entrance. For laminar flow in a circular tube, the hydraulic entry length at which the velocity profile approaches its fully developed shape can be obtained from the relation [3]

$$\left(\frac{x_{\text{fully developed}}}{D}\right)_{\text{lam}} = 0.05\text{Re}_D \tag{6.7}$$

whereas the distance from the inlet at which the temperature profile approaches its fully developed shape is given by the relation [4]

$$\left(\frac{x_{\text{fully developed}}}{D}\right)_{\text{lam},T} = 0.05\text{Re}_D\,\text{Pr} \tag{6.8}$$

In turbulent flow, conditions are essentially independent of Prandtl numbers, and for average pipe velocities corresponding to turbulent-flow Reynolds numbers, entrance effects disappear about 10 or 20 diameters from the inlet.

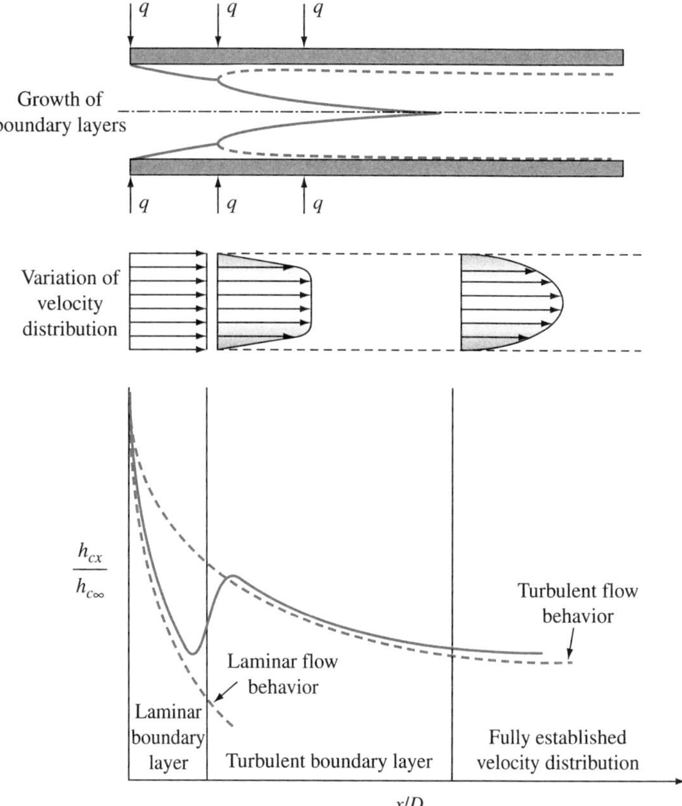

FIGURE 6.6 Velocity distribution and variation of local heat transfer coefficient near the entrance of a uniformly heated tube for a fluid in turbulent flow.

6.1.5 Variation of Physical Properties

Another factor that can influence the heat transfer and friction considerably is the variation of physical properties with temperature. When a fluid flowing in a duct is heated or cooled, its temperature and consequently its physical properties vary along the duct as well as over any given cross section. For liquids, only the temperature dependence of the viscosity is of major importance. For gases, on the other hand, the temperature effect on the physical properties is more complicated than for liquids because the thermal conductivity and the density, in addition to the viscosity, vary significantly with temperature. In either case, the numerical value of the Reynolds number depends on the location at which the properties are evaluated. It is believed that the Reynolds number based on the average bulk temperature is the significant parameter to describe the flow conditions. However, considerable success in the empirical correlation of experimental heat transfer data has been achieved by evaluating the viscosity at an *average film temperature*, defined as a temperature approximately halfway between the wall and the average bulk temperatures. Another method of taking account of the variation of physical properties with temperature is to evaluate all properties at the average bulk temperature and to correct for the thermal effects by multiplying the right-hand side of Eq. (6.4) by a function proportional to the ratio of bulk to wall temperatures or bulk to wall viscosities.

6.1.6 Thermal Boundary Conditions and Compressibility Effects

For fluids having a Prandtl number of unity or less, the heat transfer coefficient also depends on the thermal boundary condition. For example, in geometrically similar liquid metal or gas heat transfer systems, a uniform wall temperature yields smaller convection heat transfer coefficients than a uniform heat input at the same Reynolds and Prandtl numbers [5–7]. When heat is transferred to or from gases flowing at very high velocities, compressibility effects influence the flow and the heat transfer. Problems associated with heat transfer to or from fluids at high Mach numbers are referenced in [8–10].

6.1.7 Limits of Accuracy in Predicted Values of Convection Heat Transfer Coefficients

In the application of any empirical equation for forced convection to practical problems, it is important to bear in mind that the predicted values of the heat transfer coefficient are not exact. The results obtained by various experimenters, even under carefully controlled conditions, differ appreciably. In turbulent flow, the accuracy of a heat transfer coefficient predicted from any available equation or graph is no better than $\pm 20\%$, whereas in laminar flow, the accuracy may be of the order of $\pm 30\%$. In the transition region, where experimental data are scant, the accuracy of the Nusselt number predicted from available information may be even lower. Hence, the number of significant figures obtained from calculations should be consistent with these accuracy limits.

6.2* Analysis of Laminar Forced Convection in a Long Tube

To illustrate some of the most important concepts in forced convection, we will analyze a simple case and calculate the heat transfer coefficient for laminar flow through a tube under fully developed conditions with a constant heat flux at the wall. We begin by deriving the velocity distribution. Consider a fluid element as shown in Fig. 6.7. The pressure is uniform over the cross section, and the pressure forces are balanced by the viscous shear forces acting over the surface:

$$\pi r^2[p - (p + dp)] = \tau 2\pi r\, dx = -\left(\mu \frac{du}{dr}\right) 2\pi r\, dx$$

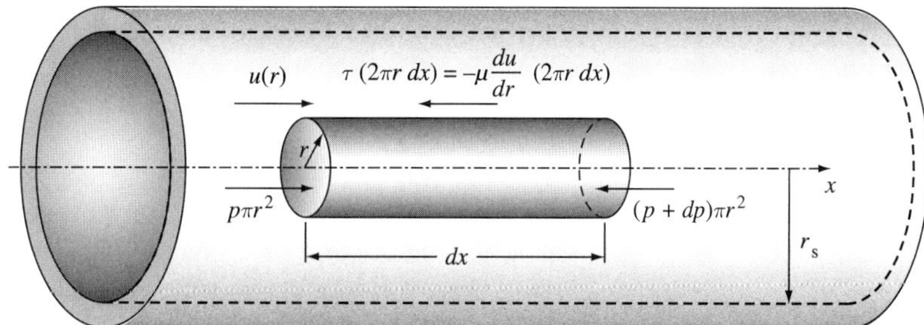

FIGURE 6.7 Force balance on a cylindrical fluid element inside a tube of radius r_s.

From this relation, we obtain

$$du = \frac{1}{2\mu}\left(\frac{dp}{dx}\right) r\,dr$$

where dp/dx is the axial pressure gradient. The radial distribution of the axial velocity is then

$$u(r) = \frac{1}{4\mu}\left(\frac{dp}{dx}\right) r^2 + C$$

where C is a constant of integration whose value is determined by the boundary condition that $u = 0$ at $r = r_s$. Using this condition to evaluate C gives the velocity distribution

$$u(r) = \frac{r^2 - r_s^2}{4\mu}\frac{dp}{dx} \tag{6.9}$$

The maximum velocity u_{max} at the center ($r = 0$) is

$$u_{max} = -\frac{r_s^2}{4\mu}\frac{dp}{dx} \tag{6.10}$$

so that the velocity distribution can be written in dimensionless form as

$$\frac{u}{u_{max}} = 1 - \left(\frac{r}{r_s}\right)^2 \tag{6.11}$$

The above relation shows that the velocity distribution in fully developed laminar flow is parabolic.

In addition to the heat transfer characteristics, engineering design requires consideration of the pressure loss and pumping power required to sustain the convection flow through the conduit. The pressure loss in a tube of length L is obtained from a force balance on the fluid element inside the tube between $x = 0$ and $x = L$ (see Fig. 6.7):

$$\Delta p \pi r_s^2 = 2\pi r_s \tau_s L \tag{6.12}$$

where $\Delta p = p_1 - p_2$ = pressure drop in length $L(\Delta_p = -(dp/dx)L)$ and
τ_s = wall shear stress ($\tau_s = -\mu(du/dr)|_{r=r_s}$)

The pressure drop also can be related to a so-called *Darcy friction factor* f according to

$$\Delta p = f\frac{L}{D}\frac{\rho\bar{U}^2}{2g_c} \tag{6.13}$$

where $\bar{U}$ is the average velocity in the tube.

It is important to note that f, the friction factor in Eq. (6.13), is not the same quantity as the friction coefficient C_f, which was defined in Chapter 4 as

$$C_f = \frac{\tau_s}{\rho\bar{U}^2/2g_c} \tag{6.14}$$

C_f is often referred to as the *Fanning friction coefficient*. Since $\tau_s = -\mu(du/dr)_{r=r}$ it is apparent from Eqs. (6.12), (6.13), and (6.14) that

$$C_f = \frac{f}{4}$$

For flow through a pipe the mass flow rate is obtained from Eq. (6.9)

$$\dot{m} = \rho\int_0^{r_s} u2\pi r\,dr = \frac{\Delta p\pi\rho}{2L\mu}\int_0^{r_x}(r^2 - r_s^2)r\,dr = -\frac{\Delta p\pi r_s^4\rho}{8L\mu} \tag{6.15}$$

and the average velocity $\bar{U}$ is

$$\bar{U} = \frac{\dot{m}}{\rho\pi r_s^2} = -\frac{\Delta p r_s^2}{8L\mu} \tag{6.16}$$

equal to one-half of the maximum velocity in the center. Equation (6.13) can be rearranged into the form

$$p_1 - p_2 = \Delta p = \frac{64L\mu}{\rho\bar{U}^2 D}\frac{\bar{U}^2}{2} = \frac{64}{\text{Re}_D}\frac{L}{D}\frac{\rho\bar{U}^2}{2g_c} \tag{6.17}$$

Comparing Eq. (6.17) with Eq. (6.13), we see that for fully developed laminar flow in a tube the friction factor in a pipe is a simple function of Reynolds number

$$f = \frac{64}{\text{Re}_D} \tag{6.18}$$

The pumping power, P_p, is equal to the product of the pressure drop and the volumetric flow rate of the fluid, Q, divided by the pump efficiency, η_p, or

$$P_p = \Delta p\dot{Q}/\eta_p \tag{6.19}$$

The analysis above is limited to laminar flow with a parabolic velocity distribution in pipes or circular tubes, known as Poiseuille flow, but the approach taken to derive this relation is more general. If we know the shear stress as a function of the velocity and its derivative, the friction factor also could be obtained for turbulent flow. However, for turbulent flow, the relationship between the shear and the average velocity is not well understood. Moreover, while in laminar flow, the friction factor is independent of surface roughness; in turbulent flow, the quality of the pipe surface influences the pressure loss. Therefore, friction factors for turbulent flow cannot be derived analytically but must be measured and correlated empirically.

6.2.1 Uniform Heat Flux

For the energy analysis, consider the control volume shown in Fig. 6.8. In laminar flow, heat is transferred by conduction into and out of the element in a radial direction, whereas in the axial direction, the energy transport is by convection. Thus, the rate of heat conduction into the element is

$$dq_{k,r} = -k2\pi r\,dx\,\frac{\partial T}{\partial r}$$

while the rate of heat conduction out of the element is

$$dq_{k,r+dr} = -k2\pi(r + dr)dx\left[\frac{\partial T}{\partial r} + \frac{\partial^2 T}{\partial r^2}dr\right]$$

The net rate of convection out of the element is

$$dq_c = 2\pi r\,dr\,\rho c_p u(r)\,\frac{\partial T}{\partial x}\,dx$$

Writing a net energy balance in the form

$$\begin{matrix}\text{net rate of conduction} \\ \text{into the element}\end{matrix} = \begin{matrix}\text{net rate of convection} \\ \text{out of the element}\end{matrix}$$

we get, neglecting second-order terms,

$$k\left(\frac{\partial T}{\partial r} + r\frac{\partial^2 T}{\partial r^2}\right)dx\,dr = r\rho c_p u\,\frac{\partial T}{\partial x}\,dx\,dr$$

which can be recast in the form

$$\frac{1}{ur}\frac{\partial}{\partial r}\left(r\frac{\partial T}{\partial r}\right) = \frac{\rho c_p}{k}\frac{\partial T}{\partial x} \tag{6.20}$$

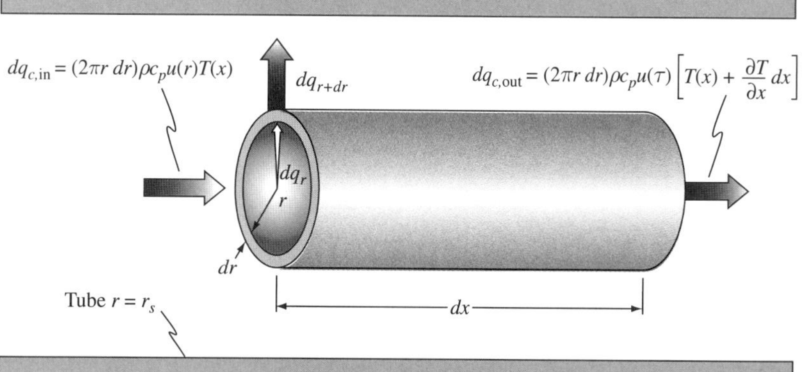

FIGURE 6.8 Schematic sketch of control volume for energy analysis in flow through a pipe.

The fluid temperature must increase linearly with distance x since the heat flux over the surface is specified to be uniform, so

$$\frac{\partial T}{\partial x} = \text{constant} \tag{6.21}$$

When the axial temperature gradient $\partial T/\partial x$ is constant, Eq. (6.20) reduces from a partial to an ordinary differential equation with r as the only space coordinate.

The symmetry and boundary conditions for the temperature distribution in Eq. (6.20) are

$$\frac{\partial T}{\partial r} = 0 \qquad \text{at } r = 0$$

$$\left| k\frac{\partial T}{dr} \right|_{r=r_s} = q_s'' = \text{constant} \quad \text{at } r = r_s$$

To solve Eq. (6.20), we substitute the velocity distribution from Eq. (6.11). Assuming that the temperature gradient does not affect the velocity profile, that is, the properties do not change with temperature, we get

$$\frac{\partial}{\partial r}\left(r\frac{\partial T}{\partial r}\right) = \frac{1}{\alpha}\frac{\partial T}{\partial x}u_{\text{max}}\left(1 - \frac{r^2}{r_s^2}\right)r \tag{6.22}$$

The first integration with respect to r gives

$$r\frac{\partial T}{\partial r} = \frac{1}{\alpha}\frac{\partial T}{\partial x}\frac{u_{\text{max}}r^2}{2}\left(1 - \frac{r^2}{2r_s^2}\right) + C_1 \tag{6.23}$$

A second integration with respect to r gives

$$T(r, x) = \frac{1}{\alpha}\frac{\partial T}{\partial x}\frac{u_{\text{max}}}{4}r^2\left(1 - \frac{r^2}{4r_s^2}\right) + C_1 \ln r + C_2 \tag{6.24}$$

But note that $C_1 = 0$ since $(\partial T/\partial r)_{r=0} = 0$ and that the second boundary condition is satisfied by the requirement that the axial temperature gradient $\partial T/\partial x$ is constant. If we let the temperature at the center ($r = 0$) be T_c, then $C_2 = T_c$ and the temperature distribution becomes

$$T - T_c = \frac{1}{\alpha}\frac{\partial T}{\partial x}\frac{u_{\text{max}}r_s^2}{4}\left[\left(\frac{r}{r_s}\right)^2 - \frac{1}{4}\left(\frac{r}{r_s}\right)^4\right] \tag{6.25}$$

The average bulk temperature T_b that was used in defining the heat transfer coefficient can be calculated from

$$T_b = \frac{\int_0^{r_s}(\rho u c_p T)(2\pi r\,dr)}{\int_0^{r_s}(\rho u c_p)2\pi r\,dr} = \frac{\int_0^{r_s}(\rho u c_p T)2\pi r\,dr}{c_p\dot{m}} \tag{6.26}$$

Since the heat flux from the tube wall is uniform, the enthalpy of the fluid in the tube must increase linearly with x, and thus $\partial T_b/\partial x$ = constant. We can calculate the bulk temperature by substituting Eqs. (6.25) and (6.11) for T and u, respectively, in Eq. (6.26). This yields

$$T_b - T_c = \frac{7}{96}\frac{u_{\max} r_s^2}{\alpha}\frac{\partial T}{\partial x} \tag{6.27}$$

while the wall temperature is

$$T_s - T_c = \frac{3}{16}\frac{u_{\max} r_s^2}{\alpha}\frac{\partial T}{\partial x} \tag{6.28}$$

In deriving the temperature distributions, we used a parabolic velocity distribution, which exists in fully developed flow in a long tube. Hence, with $\partial T/\partial x$ equal to a constant, the average heat transfer coefficient is

$$\bar{h}_c = \frac{q_c}{A(T_s - T_b)} = \frac{k(\partial T/\partial r)_{r=r_s}}{T_s - T_b} \tag{6.29}$$

Evaluating the radial temperature gradient at $r = r_s$ from Eq. (6.23) and substituting it with Eqs. (6.27) and (6.28) in the above definition yields

$$\bar{h}_c = \frac{24k}{11r_s} = \frac{48k}{11D} \tag{6.30}$$

or

$$\overline{\mathrm{Nu}}_D = \frac{\bar{h}_c D}{k} = 4.364 \quad \text{for } q_s'' = \text{constant} \tag{6.31}$$

EXAMPLE 6.1 Water entering at 10°C is to be heated to 40°C in a tube of 0.02-m-ID at a mass flow rate of 0.01 kg/s. The outside of the tube is wrapped with an insulated electric-heating element (see Fig. 6.9) that produces a uniform flux of 15,000 W/m^2 over the surface. Neglecting any entrance effects, determine

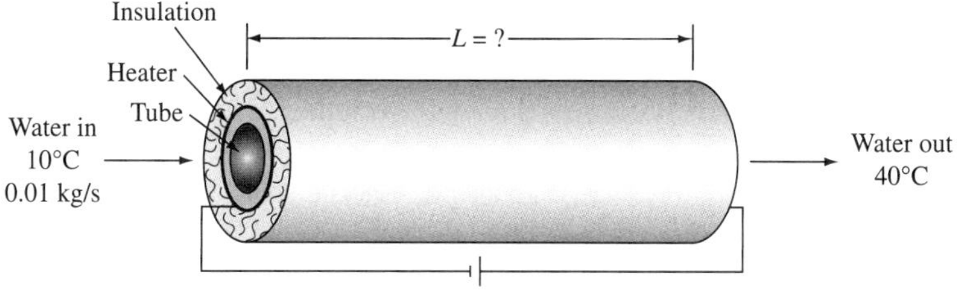

FIGURE 6.9 Schematic diagram of water flowing through electrically heated tube, Example 6.1.

(a) the Reynolds number
(b) the heat transfer coefficient
(c) the length of pipe needed for a 30°C increase in average temperature
(d) the inner tube surface temperature at the outlet
(e) the friction factor
(f) the pressure drop in the pipe
(g) the pumping power required if the pump is 50% efficient.

SOLUTION From Table 13 in Appendix 2, the appropriate properties of water at an average temperature between inlet and outlet of 25°C are obtained by interpolation:

$$\rho = 997 \text{ kg/m}^3$$

$$c_p = 4180 \text{ J/kg K}$$

$$k = 0.608 \text{ W/m K}$$

$$\mu = 910 \times 10^{-6} \text{ N s/m}^2$$

(a) The Reynolds number is

$$\text{Re}_D = \frac{\rho \bar{U} D}{\mu} = \frac{4\dot{m}}{\pi D \mu} = \frac{(4)(0.01 \text{ kg/s})}{(\pi)(0.02 \text{ m})(910 \times 10^{-6} \text{ N s/m}^2)} = 699$$

This establishes that the flow is laminar.

(b) Since the thermal-boundary condition is one of uniform heat flux, $\text{Nu}_D = 4.36$ from Eq. (6.31) and

$$\bar{h}_c = 4.36 \frac{k}{D} = 4.36 \frac{0.608 \text{ W/m K}}{0.02 \text{ m}} = 132 \text{ W/m}^2 \text{ K}$$

(c) The length of pipe needed for a 30°C temperature rise is obtained from a heat balance

$$q'' \pi D L = \dot{m} c_p (T_{\text{out}} - T_{\text{in}})$$

Solving for L when $T_{\text{out}} - T_{\text{in}} = 30$ K gives

$$L = \frac{\dot{m} c_p \Delta T}{\pi D q''} = \frac{(0.01 \text{ kg/s})(4180 \text{ J/kg K})(30 \text{ K})}{(\pi)(0.02 \text{ m})(15{,}000 \text{ W/m}^2)} = 1.33 \text{ m}$$

Since $L/D = 66.5$ and $0.05\,\text{Re}_D = 33.5$, entrance effects are negligible according to Eq. (6.7). Note that if L/D had been significantly less than 33.5, the calculations would have to be repeated with entrance effects taken into account, using relations to be presented.

(d) From Eq. (6.1)

$$q'' = \frac{q_c}{A} = \bar{h}_c (T_s - T_b)$$

and

$$T_s = \frac{q_c}{A \bar{h}_c} + T_b = \frac{15{,}000 \text{ W/m}^2}{132 \text{ W/m}^2 \,^\circ\text{C}} + 40^\circ\text{C} = 154^\circ\text{C}$$

(e) The friction factor is found from Eq. (6.18):

$$f = \frac{64}{\mathrm{Re}_D} = \frac{64}{699} = 0.0915$$

(f) The pressure drop in the pipe is, from Eq. (6.17),

$$p_1 - p_2 = \Delta p = f\left(\frac{L}{D}\right)\left(\frac{\rho \bar{U}^2}{2g_c}\right)$$

Since

$$\bar{U} = \frac{4\dot{m}}{\rho \pi D^2} = \frac{4\left(0.01\ \frac{\mathrm{kg}}{\mathrm{s}}\right)}{\left(997\ \frac{\mathrm{kg}}{\mathrm{m}^3}\right)(\pi)(0.02\ \mathrm{m})^2} = 0.032\ \frac{\mathrm{m}}{\mathrm{s}}$$

we have

$$\Delta p = (0.0915)(66.5)\frac{\left(997\ \frac{\mathrm{kg}}{\mathrm{m}^3}\right)(0.032\ \frac{\mathrm{m}}{2})^2}{2\left(1\ \frac{\mathrm{kg\ m}}{\mathrm{N\ s}^2}\right)} = 3.1\ \frac{\mathrm{N}}{\mathrm{m}^2}$$

(g) The pumping power P_p is obtained from Eq. 6.19 or

$$P_p = \dot{m}\frac{\Delta p}{\rho \eta_p} = \frac{(0.01\ \mathrm{kg/s})(3.1\ \mathrm{N/m}^2)}{(997\ \mathrm{kg/m}^3)(0.5)} = 6.2 \times 10^{-5}\ \mathrm{W}$$

6.2.2* Uniform Surface Temperature

When the tube surface temperature rather than the heat flux is uniform, the analysis is more complicated because the temperature difference between the wall and bulk varies along the tube, that is, $\partial T_b/\partial x = f(x)$. Equation (6.20) can be solved subject to the second boundary condition that at $r = r_s$, $T(x, r_s) =$ constant, but an iterative procedure is necessary. The result is not a simple algebraic expression, but the Nusselt number is found (for example, see Kays and Perkins [11]) to be a constant:

$$\overline{\mathrm{Nu}}_D = \frac{\bar{h}_c D}{k} = 3.66 \quad (T_s = \text{constant}) \tag{6.32}$$

In addition to the value of the Nusselt number, the constant-temperature boundary condition also requires a different temperature to evaluate the rate of heat transfer to or from a fluid flowing through a duct. Except for the entrance region, in which the boundary layer develops and the heat transfer coefficient decreases, the temperature difference between the surface of the duct and the bulk remains constant along the duct when the heat flux is uniform. This is apparent

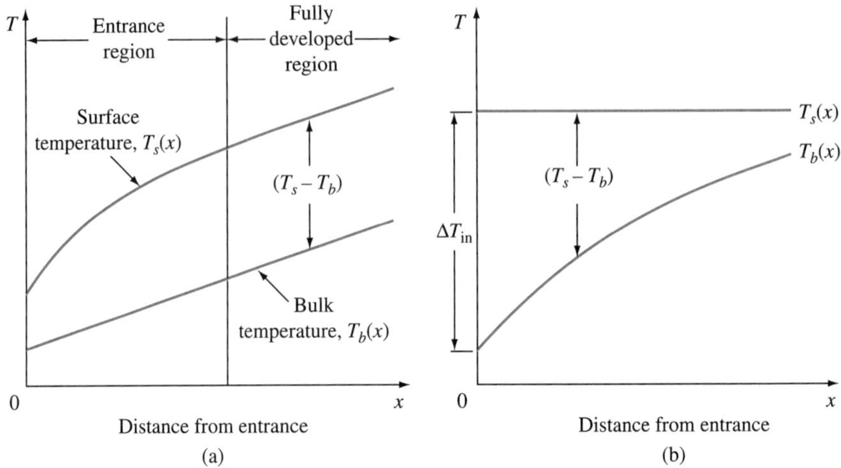

FIGURE 6.10 Variation of average bulk temperature with constant heat flux and constant wall temperature: (a) constant heat flux, $q_s(x)$ = constant; (b) constant surface temperature, $T_s(x)$ = constant.

from an examination of Eq. (6.20) and is illustrated graphically in Fig. 6.10. For a constant wall temperature, on the other hand, only the bulk temperature increases along the duct and the temperature potential decreases (see Fig. 6.10). We first write the heat balance equation

$$dq_c = \dot{m}c_p\, dT_b = q_s'' P\, dx$$

where P is the perimeter of the duct and q_s'' is the surface heat flux. From the preceding we can obtain a relation for the bulk temperature gradient in the x-direction

$$\frac{dT_b}{dx} = \frac{q_s'' P}{\dot{m}c_p} = \frac{P}{\dot{m}c_p} h_c(T_s - T_b) \tag{6.33}$$

Since $dT_b/dx = d(T_b - T_s)/dx$ for a constant surface temperature, after separating variables, we have

$$\int_{\Delta T_{in}}^{\Delta T_{out}} \frac{d(\Delta T)}{\Delta T} = -\frac{P}{\dot{m}c_p} \int_0^L h_c\, dx \tag{6.34}$$

where $\Delta T = T_s - T_b$ and the subscripts "in" and "out" denote conditions at the inlet ($x = 0$) and the outlet ($x = L$) of the duct, respectively. Integrating Eq. (6.34) yields

$$\ln\left(\frac{\Delta T_{out}}{\Delta T_{in}}\right) = -\frac{PL}{\dot{m}c_p} \bar{h}_c \tag{6.35}$$

where

$$\bar{h}_c = \frac{1}{L} \int_0^L h_c\, dx$$

Rearranging Eq. (6.35) gives

$$\frac{\Delta T_{\text{out}}}{\Delta T_{\text{in}}} = \exp\left(\frac{-\bar{h}_c PL}{\dot{m}c_p}\right) \tag{6.36}$$

The rate of heat transfer by convection to or from a fluid flowing through a duct with T_s = constant can be expressed in the form

$$q_c = \dot{m}c_p[(T_s - T_{b,\text{in}}) - (T_s - T_{b,\text{out}})] = \dot{m}c_p(\Delta T_{\text{in}} - \Delta T_{\text{out}})$$

and substituting $\dot{m}c_p$ from Eq. (6.35), we get

$$q_c = \bar{h}_c A_s\left[\frac{\Delta T_{\text{out}} - \Delta T_{\text{in}}}{\ln(\Delta T_{\text{out}}/\Delta T_{\text{in}})}\right] \tag{6.37}$$

The expression in the square bracket is called the *log mean temperature difference* (*LMTD*).

EXAMPLE 6.2 Used engine oil can be recycled by a patented reprocessing system. Suppose that such a system includes a process during which engine oil flows through a 1-cm-ID, 0.02-cm-wall copper tube at the rate of 0.05 kg/s. The oil enters at 35°C and is to be heated to 45°C by atmospheric-pressure steam condensing on the outside, as shown in Fig. 6.11. Calculate the length of the tube required.

SOLUTION We shall assume that the tube is long and that its temperature is uniform at 100°C. The first approximation must be checked; the second assumption is an engineering approximation justified by the high thermal conductivity of copper and the large heat transfer coefficient for a condensing vapor (see Table 1.4). From Table 16 in Appendix 2, we get the following properties for oil at 40°C:

$$c_p = 1964 \text{ J/kg K}$$
$$\rho = 876 \text{ kg/m}^3$$
$$k = 0.144 \text{ W/m K}$$
$$\mu = 0.210 \text{ N s/m}^2$$
$$\text{Pr} = 2870$$

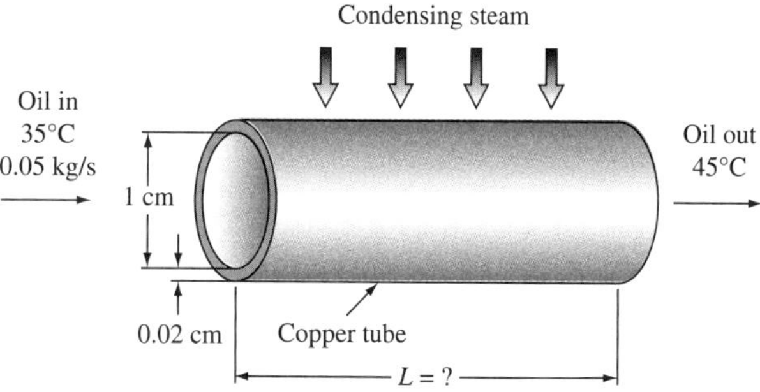

FIGURE 6.11 Schematic diagram for Example 6.2.

The Reynolds number is

$$\mathrm{Re}_D = \frac{4\dot{m}}{\mu\pi D} = \frac{(4)(0.05\ \mathrm{kg/s})}{(\pi)(0.210\ \mathrm{N\ s/m^2})(0.01\ \mathrm{m})} = 30.3$$

The flow is therefore laminar, and the Nusselt number for a constant surface temperature is 3.66. The average heat transfer coefficient is

$$\bar{h}_c = \overline{\mathrm{Nu}}_D \frac{k}{D} = 3.66 \frac{0.144\ \mathrm{W/m\ K}}{0.01\ \mathrm{m}} = 52.7\ \mathrm{W/m^2\ K}$$

The rate of heat transfer is

$$\begin{aligned} q_c &= c_p \dot{m}(T_{b,\mathrm{out}} - T_{b,\mathrm{in}}) \\ &= (1964\ \mathrm{J/kg\ K})(0.05\ \mathrm{kg/s})(45 - 35)\ \mathrm{K} = 982\ \mathrm{W} \end{aligned}$$

Recalling that $\ln(1/x) = -\ln x$, we find the LMTD is

$$\mathrm{LMTD} = \frac{\Delta T_{\mathrm{out}} - \Delta T_{\mathrm{in}}}{\ln(\Delta T_{\mathrm{out}}/\Delta T_{\mathrm{in}})} = \frac{55 - 65}{\ln(55/65)} = \frac{10}{0.167} = 59.9\ \mathrm{K}$$

Substituting the preceding information in Eq. (6.37), where $A_s = L\pi D_i$, gives

$$L = \frac{q_c}{\pi D_i \bar{h}_c \mathrm{LMTD}} = \frac{982\ \mathrm{W}}{(\pi)(0.01\ \mathrm{m})(52.7\ \mathrm{W/m^2 K})(59.9\ \mathrm{K})} = 9.91\ \mathrm{m}$$

Checking our first assumption, we find $L/D \sim 1000$, justifying neglect of entrance effects. Note also that LMTD is very nearly equal to the difference between the surface temperature and the average bulk fluid temperature halfway between the inlet and outlet. The required length is not suitable for a practical design with a straight pipe. To achieve the desired thermal performance in a more convenient shape, one could route the tube back and forth several times or use a coiled tube. The first approach will be discussed in Chapter 8 on heat exchanger design, and the coiled-tube design is illustrated in an example in the next section.

6.3 Correlations for Laminar Forced Convection

This section presents empirical correlations and analytic results that can be used in thermal design of heat transfer systems composed of tubes and ducts containing gaseous or liquid fluids in laminar flow. Although heat transfer coefficients in laminar flow are considerably smaller than in turbulent flow, in the design of heat exchange equipment for viscous liquids, it is often necessary to accept a smaller heat transfer coefficient in order to reduce the pumping power requirements. Laminar gas flow is encountered in high-temperature, compact heat exchangers, where tube diameters are very small and gas densities low. Other applications of laminar-flow forced convection occur in chemical processes and in the food industry, in electronic cooling as well as in solar and nuclear power plants, where liquid metals are used as heat transfer media. Since liquid metals have a high thermal conductivity, their heat transfer coefficients are relatively large, even in laminar flow.

6.3.1 Short Circular and Rectangular Ducts

The details of the mathematical solutions for laminar flow in short ducts with entrance effects are beyond the scope of this text. References listed at the end of this chapter, especially [4] and [11], contain the mathematical background for the engineering equations and graphs that are presented and discussed in this section.

For engineering applications, it is most convenient to present the results of analytic and experimental investigations in terms of a Nusselt number defined in the conventional manner as $h_c D/k$. However, the heat transfer coefficient h_c can vary along the tube, and for practical applications, the average value of the heat transfer coefficient is most important. Consequently, for the equations and charts presented in this section, we shall use a mean Nusselt number, $\overline{\text{Nu}}_D = \bar{h}_c D/k$, averaged with respect to the circumference and length of the duct L:

$$\overline{\text{Nu}}_D = \frac{1}{L}\int_0^L \frac{D}{k} h_{c(x)} dx = \frac{\bar{h}_c D}{k}$$

where the subscript x refers to local conditions at x. This Nusselt number is often called the *log mean Nusselt number,* because it can be used directly in the log mean rate equations presented in the preceding section and can be applied to heat exchangers (see Chapter 8).

Mean Nusselt numbers for laminar flow in tubes at a uniform wall temperature have been calculated analytically by various investigators. Their results are shown in Fig. 6.12

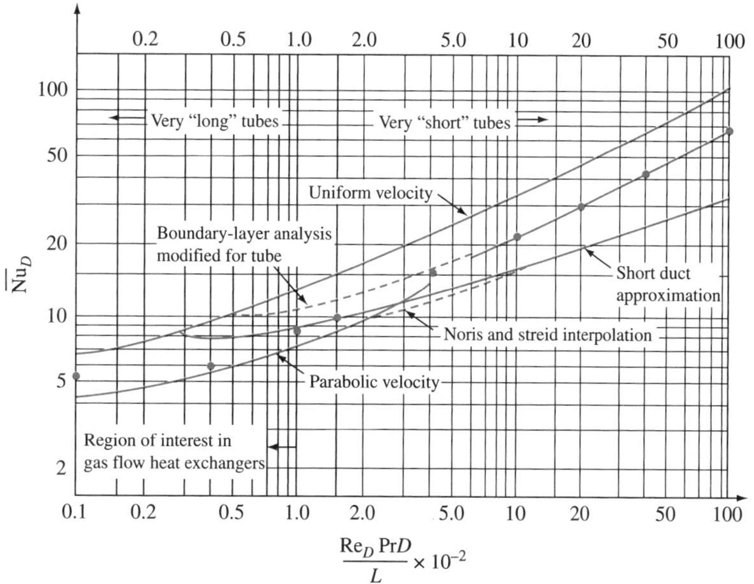

FIGURE 6.12 Analytic solutions and empirical correlations for heat transfer in laminar flow through circular tubes at constant wall temperature, $\overline{\text{Nu}}_D$ versus $\text{Re}_D\text{Pr}D/L$. The dots represent Eq. (6.38).

Source: Courtesy of W. M. Kays, "Numerical Solution for Laminar Flow Heat Transfer in Circular Tubes," Trans. ASME, vol. 77, pp. 1265–1274, 1955.

for several velocity distributions. All of these solutions are based on the idealizations of a constant tube-wall temperature and a uniform temperature distribution at the tube inlet, and they apply strictly only when the physical properties are independent of temperature. The abscissa is the dimensionless quantity $\mathrm{Re}_D\mathrm{Pr}D/L$.* To determine the mean value of the Nusselt number for a given tube of length L and diameter D, one evaluates the Reynolds number, Re_D, and the Prandtl number, Pr, forms the dimensionless parameter $\mathrm{Re}_D\mathrm{Pr}D/L$, and enters the appropriate curve from Fig. 6.12. The selection of the curve representing the conditions that most nearly correspond to the physical conditions depends on the nature of the fluid and the geometry of the system. For high Prandtl number fluids such as oils, the velocity profile is established much more rapidly than the temperature profile. Consequently, application of the curve labeled "parabolic velocity" does not lead to a serious error in long tubes when $\mathrm{Re}_D\mathrm{Pr}D/L$ is less than 100. For very long tubes, the Nusselt number approaches a limiting minimum value of 3.66 when the tube temperature is uniform. When the heat transfer rate instead of the tube temperature is uniform, the limiting value of $\overline{\mathrm{Nu}}_D$ is 4.36.

For very short tubes or rectangular ducts with initially uniform velocity and temperature distribution, the flow conditions along the wall approximate those along a flat plate, and the boundary Layer analysis presented in Chapter 4 is expected to yield satisfactory results for liquids having Prandtl numbers between 0.7 and 15.0. The boundary layer solution applies [14, 15] when L/D is less than $0.0048\mathrm{Re}_D$ for tubes and when L/D_H is less than $0.0021\mathrm{Re}_{D_H}$ for flat ducts of rectangular cross section. For these conditions, the equation for flow of liquids and gases over a flat plate can be converted to the coordinates of Figs. 6.12, leading to

$$\overline{\mathrm{Nu}}_{D_H} = \frac{\mathrm{Re}_{D_H}\mathrm{Pr}D_H}{4L}\ln\left[\frac{1}{1-(2.654/\mathrm{Pr}^{0.167})(\mathrm{Re}_{D_H}\mathrm{Pr}D_H/L)^{-0.5}}\right] \tag{6.38}$$

An analysis for longer tubes is presented in [12], and the results are shown in Fig. 6.12 for $\mathrm{Pr} = 0.73$ in the $\mathrm{Re}_D\mathrm{Pr}D/L$ range of 10 to 1500, where this approximation is applicable.

For laminar flows in circular tubes, whether in the thermal entrance region or for fully developed conditions, a convenient set of correlations [13] for determining the mean Nusselt number, and hence the heat transfer coefficient for both uniform heat flux and uniform surface temperature conditions, are given below.

For tube wall with $q_s'' = $ constant,

$$\overline{\mathrm{Nu}}_D = \begin{cases} 1.953[L/(D\mathrm{Re}_D\mathrm{Pr})]^{1/3} & \text{for } [L/(D\mathrm{Re}_D\mathrm{Pr})] \le 0.03 \\ 4.364 + (0.0722(D\mathrm{Re}_D\mathrm{Pr})]/L & \text{for } [L/(D\mathrm{Re}_D\mathrm{Pr})] \le 0.03 \end{cases} \tag{6.39}$$

For tube wall with $T_s = $ constant,

$$\overline{\mathrm{Nu}}_D = \begin{cases} 1.615[L/(D\mathrm{Re}_D\mathrm{Pr})]^{-1/3} - 0.7 & \text{for } [L/(D\mathrm{Re}_D\mathrm{Pr})] \le 0.005 \\ 1.615[L/(D\mathrm{Re}_D\mathrm{Pr})]^{-1/3} - 0.2 & \text{for } 0.005 < [L/(D\mathrm{Re}_D\mathrm{Pr})] < 0.03 \\ 3.657 + (0.0499(D\mathrm{Re}_D\mathrm{Pr})/L) & \text{for } [L/(D\mathrm{Re}_D\mathrm{Pr})] \ge 0.03 \end{cases} \tag{6.40}$$

*Instead of the dimensionless ratio $\mathrm{Re}_D\mathrm{Pr}D/L$, some authors use the Graetz number, Gz, which is $\pi/4$ times this ratio [13].

Note that when L is very large ($\rightarrow \infty$), the values of $\overline{\text{Nu}}_D$ are obtained as 4.364 and 3.657, respectively, for the mean Nusselt number with the two boundary conditions from Eqs. (6.39) and (6.40).

6.3.2 Ducts of Noncircular Cross Section

Heat transfer and friction in fully developed laminar flow through ducts with a variety of cross sections have been treated analytically [13]. The results are summarized in Table 6.1 on the next page, using the following nomenclature:

$\overline{\text{Nu}}_{H1}$ = average Nusselt number for uniform heat flux in flow direction and uniform wall temperature at any cross section

$\overline{\text{Nu}}_{H2}$ = average Nusselt mumber for uniform heat flux both axially and circumferentially

$\overline{\text{Nu}}_{T}$ = average Nusselt number for uniform wall temperature

$f\text{Re}_{D_H}$ = product of firction factor and Reynolds number

A duct geometry encountered quite often is the concentric tube annulus shown schematically in Fig. 6.1(b). Heat transfer to or from the fluid flowing through the space formed between the two concentric tubes may occur at the inner surface, the outer surface, or both surfaces simultaneously. Moreover, the heat transfer surface may be at constant temperature or constant heat flux. An extensive treatment of this topic has been presented by Kays and Perkins [11], and includes entrance effects and the impact of eccentricity. Here we shall consider only the most commonly encountered case of an annulus in which one side is insulated and the other is at constant temperature.

Denoting the inner surface by the subscript i and the outer surface by o, the rate of heat transfer and the corresponding Nusselt numbers are

$$q_{c,i} = \bar{h}_{c,i}\pi D_i L(T_{s,i} - T_b)$$

$$q_{c,o} = \bar{h}_{c,o}\pi D_o L(T_{s,o} - T_b)$$

$$\overline{\text{Nu}}_i = \frac{\bar{h}_{c,i} D_H}{k}$$

$$\overline{\text{Nu}}_o = \frac{\bar{h}_{c,o} D_H}{k}$$

where $D_H = D_o - D_i$.

The Nusselt numbers for heat flow at the inner surface only with the outer surface insulated, $\overline{\text{Nu}}_i$, and the heat flow at the outer surface with the inner surface insulated, $\overline{\text{Nu}}_o$, as well as the product of the friction factor and the Reynolds number for fully developed laminar flow are presented in Table 6.2 on page 375. For other conditions, such as constant heat flux and short annuli, the reader is referred to [13].

TABLE 6.1 Nusselt number and friction factor for fully developed laminar flow of a Newtonian fluid through specific ducts[a]

Geometry $\left(\frac{L}{D_H} > 100\right)$		$\overline{\text{Nu}}_{H1}$	$\overline{\text{Nu}}_{H2}$	$\overline{\text{Nu}}_T$	$f\,\text{Re}_{D_H}$	$\frac{\overline{\text{Nu}}_{H1}}{\overline{\text{Nu}}_T}$
Triangle, 60°, 2a, 2b	$\frac{2b}{2a} = \frac{\sqrt{3}}{2}$	3.111	1.892	2.47	53.33	1.26
Square, 2b, 2a	$\frac{2b}{2a} = 1$	3.608	3.091	2.976	56.91	1.21
Hexagon, sides a		4.002	3.862	3.34[b]	60.22	1.20
Rectangle, 2b, 2a	$\frac{2b}{2a} = \frac{1}{2}$	4.123	3.017	3.391	62.19	1.22
Circle		4.364	4.364	3.657	64.00	1.19
Rectangle, 2b, 2a	$\frac{2b}{2a} = \frac{1}{4}$	5.331	2.930	4.439	72.93	1.20
Insulation, 2b, 2a	$\frac{2b}{2a} = \frac{1}{4}$	6.279[b]	—	5.464[b]	72.93	1.15
Ellipse, 2b, 2a	$\frac{2b}{2a} = 0.9$	5.099	4.35[b]	3.66	74.80	1.39
Rectangle, 2b, 2a	$\frac{2b}{2a} = \frac{1}{8}$	6.490	2.904	5.597	82.34	1.16
Parallel plates	$\frac{2b}{2a} = 0$	8.235	8.235	7.541	96.00	1.09
Parallel plates, 2a, 2b, Insulation	$\frac{2b}{2a} = 0$	5.385	—	4.861	96.00	1.11

[a] Source: Abstracted from Shah and London [13].

[b] Interpolated values.

TABLE 6.2 Nusselt number and friction factor for fully developed laminar flow in an annulus[a]

$\frac{D_i}{D_o}$	$\overline{\text{Nu}}_i$	$\overline{\text{Nu}}_o$	$f\,\text{Re}_{D_H}$
0.00	—	3.66	64.00
0.05	17.46	4.06	86.24
0.10	11.56	4.11	89.36
0.25	7.37	4.23	93.08
0.50	5.74	4.43	95.12
1.00	4.86	4.86	96.00

[a]One surface at constant temperature and the other insulated [13].

EXAMPLE 6.3 Calculate the average heat transfer coefficient and the friction factor for flow of n-butyl alcohol at a bulk temperature of 293 K through a 0.1-m × 0.1-m-square duct, 5 m long, with walls at 300 K, and an average velocity of 0.03 m/s (see Fig. 6.13).

SOLUTION The hydraulic diameter is

$$D_H = 4\left(\frac{0.1 \times 0.1}{4 \times 0.1}\right) = 0.1 \text{ m}$$

Physical properties at 293 K from Table 19 in Appendix 2 are

$$\rho = 810 \text{ kg/m}^3$$

$$c_p = 2366 \text{ J/kg K}$$

$$\mu = 29.5 \times 10^{-4} \text{ N s/m}^2$$

$$\nu = 3.64 \times 10^{-6} \text{ m}^2/\text{s}$$

$$k = 0.167 \text{ W/m K}$$

$$\text{Pr} = 50.8$$

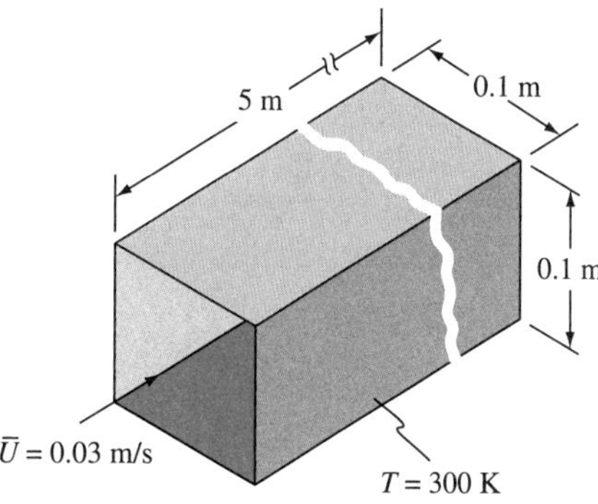

FIGURE 6.13 Schematic diagram of heating duct for Example 6.3.

The Reynolds number is

$$\mathrm{Re}_{D_H} = \frac{\bar{U} D_H \rho}{\mu} = \frac{(0.03\ \mathrm{m/s})(0.1\ \mathrm{m})(810\ \mathrm{kg/m^3})}{29.5 \times 10^{-4}\ \mathrm{N\ s/m^2}} = 824$$

Hence, the flow is laminar. Assuming fully developed flow, we get the Nusselt number for a uniform wall temperature from Table 6.1:

$$\overline{\mathrm{Nu}}_{D_H} = \frac{\bar{h}_c D_H}{k} = 2.98$$

This yields for the average heat transfer coefficient

$$\bar{h}_c = 2.98 \frac{0.167\ \mathrm{W/m\ K}}{0.1\ \mathrm{m}} = 4.98\ \mathrm{W/m^2 K}$$

Similarly, from Table 6.1, the product $\mathrm{Re}_{D_H} f = 56.91$ and

$$f = \frac{56.91}{824} = 0.0691$$

Recall that for a fully developed velocity profile the duct length must be at least $0.05\mathrm{Re} \times D_H = 4.1$ m, but for a fully developed temperature profile, the duct must be 172 m long. Thus, *fully developed flow will not exist.*

If we use Fig. 6.12 with $\mathrm{Re}_{D_H}\mathrm{Pr}D/L = (824)(50.8)(0.1/5) = 837$, the average Nusselt number is about 15, and $\bar{h}_c = (15)(0.167\ \mathrm{W/m\ K})/0.1\ \mathrm{m} = 25\ \mathrm{W/m^2\ K}$. This value is five times larger than that for fully developed flow.

Note that for this problem the difference between bulk and wall temperature is small. Hence, property variations are not significant in this case.

6.3.3 Effect of Property Variations

Since the microscopic heat-flow mechanism in laminar flow is conduction, the rate of heat flow between the walls of a conduit and the fluid flowing in it can be obtained analytically by solving the equations of motion and of conduction heat flow simultaneously, as shown in Section 6.2. But to obtain a solution, it is necessary to know or assume the velocity distribution in the duct. In fully developed laminar flow through a tube without heat transfer, the velocity distribution at any cross section is parabolic. But when appreciable heat transfer occurs, temperature differences are present, and the fluid properties of the wall and the bulk may be quite different. These property variations distort the velocity profile.

In liquids, the viscosity decreases with increasing temperature, while in gases the reverse trend is observed. When a liquid is heated, the fluid near the wall is less viscous than the fluid in the center. Consequently, the velocity of the heated fluid is larger than that of an unheated fluid near the wall, but less than that of the unheated fluid in the center. The distortion of the parabolic velocity profile for heated or cooled liquids is shown in Fig. 6.14. For gases, the conditions are reversed, but the variation of density with temperature introduces additional complications.

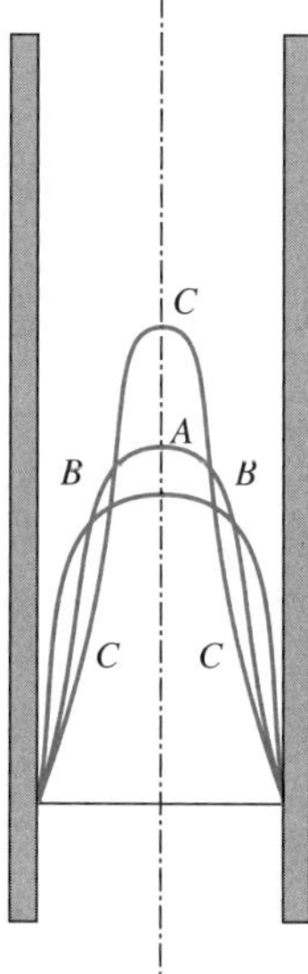

FIGURE 6.14 Effect of heat transfer on velocity profiles in fully developed laminar flow through a pipe. Curve *A*, isothermal flow; curve *B*, heating of liquid or cooling of gas; curve *C*, cooling of liquid or heating of gas.

Empirical viscosity correction factors are merely approximate rules, and recent data indicate that they may not be satisfactory when very large temperature gradients exist. As an approximation in the absence of a more satisfactory method, it is suggesed [16] that for liquids, the Nusselt number obtained from the analytic solutions presented in Fig. 6.12 be multiplied by the ratio of the viscosity at the bulk temperature μ_b to the viscosity at the surface temperature μ_s, raised to the 0.14 power, that is, $(\mu_b/\mu_s)^{0.14}$, to correct for the variation of properties due to the temperature gradients. For gases, Kays and London [17] suggest that the Nusselt number be multiplied by the temperature correction factor shown below. If all fluid properties are evaluated at the average bulk temperature, the corrected Nusselt number is

$$\overline{\mathrm{Nu}}_D = \overline{\mathrm{Nu}}_{D,\mathrm{Fig\ 6.12}}\left(\frac{T_b}{T_s}\right)^n$$

where $n = 0.25$ for a gas heating in a tube and 0.08 for a gas cooling in a tube. Hausen [18] recommended the following relation for the average convection coefficient in laminar flow through ducts with uniform surface temperature:

$$\overline{\mathrm{Nu}}_{D_H} = 3.66 + \frac{0.668\mathrm{Re}_{D_H}\mathrm{Pr}D/L}{1 + 0.045(\mathrm{Re}_{D_H}\mathrm{Pr}D/L)^{0.66}}\left(\frac{\mu_b}{\mu_s}\right)^{0.14} \tag{6.41}$$

where $100 < \mathrm{Re}_{D_H}\mathrm{Pr}D/L < 1500$.

A relatively simple empirical equation suggested by Sieder and Tate [16] has been widely used to correlate experimental results for liquids in tubes and can be written in the form

$$\overline{\mathrm{Nu}}_{D_H} = 1.86\left(\frac{\mathrm{Re}_{D_H}\mathrm{Pr}D_H}{L}\right)^{0.33}\left(\frac{\mu_b}{\mu_s}\right)^{0.14} \tag{6.42}$$

where all the properties in Eqs. (6.41) and (6.42) are based on the bulk temperature and the empirical correction factor $(\mu_b/\mu_s)^{0.14}$ is introduced to account for the effect of the temperature variation on the physical properties. Equation (6.42) can be applied when the surface temperature is uniform in the range $0.48 < \mathrm{Pr} < 16{,}700$ and $0.0044 < (\mu_b/\mu_s) < 9.75$. Whitaker [19] recommends use of Eq. (6.42) only when $(\mathrm{Re}_D \mathrm{Pr} D/L)^{0.33}(\mu_b/\mu_s)^{0.14}$ is larger than 2.

For laminar flow of gases between two parallel, uniformly heated plates a distance $2y_0$ apart, Swearingen and McEligot [20] showed that gas property variations can be taken into account by the relation

$$\overline{\mathrm{Nu}} = \mathrm{Nu}_{\text{constant properties}} + 0.024Q^{+0.3}Gz_b^{0.75} \tag{6.43}$$

where
$$Q^+ = q_s''y_0/(KT)_{\text{entrance}}$$
$$q_s'' = \text{surface heat flux at the walls}$$
$$\mathrm{Gz}_b = (\mathrm{Re}_{D_H}\mathrm{Pr}D_H/L)_b$$

and the subscript b denotes that the physical properties are to be evaluated at T_b.

The variation in physical properties also affects the friction factor. To evaluate the friction factor of fluids being heated or cooled, it is suggested that for liquids the isothermal friction factor be modified by

$$f_{\text{heat transfer}} = f_{\text{isothermal}}\left(\frac{\mu_s}{\mu_b}\right)^{0.14} \tag{6.44}$$

and for gases by

$$f_{\text{heat transfer}} = f_{\text{isothermal}}\left(\frac{T_s}{T_b}\right)^{0.14} \tag{6.45}$$

EXAMPLE 6.4 An electronic device is cooled by water flowing through capillary holes drilled in the casing as shown in Fig. 6.15. The temperature of the device casing is constant at 353 K. The capillary holes are 0.3 m long and 2.54×10^{-3} m in diameter. If water enters at a temperature of 333 K and flows at a velocity of 0.2 m/s, calculate the outlet temperature of the water.

SOLUTION The properties of water at 333 K, from Table 13 in Appendix 2, are

$$\rho = 983 \text{ kg/m}^3$$
$$c_p = 4181 \text{ J/kg K}$$
$$\mu = 4.72 \times 10^{-4} \text{ N s/m}^2$$
$$k = 0.658 \text{ W/m K}$$
$$\mathrm{Pr} = 3.00$$

To ascertain whether the flow is laminar, evaluate the Reynolds number at the inlet bulk temperature,

$$\mathrm{Re}_D = \frac{\rho\bar{U}D}{\mu} = \frac{(983 \text{ kg/m}^3)(0.2 \text{ m/s})(0.00254 \text{ m})}{4.72 \times 10^{-4} \text{ kg/ms}} = 1058$$

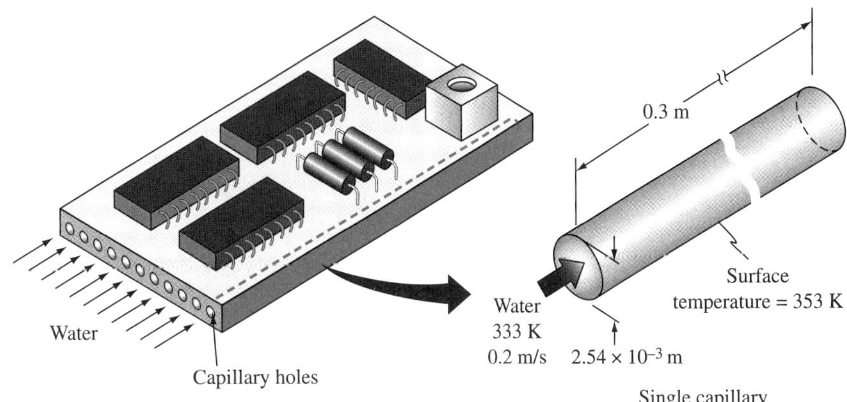

FIGURE 6.15 Schematic diagram for Example 6.4.

The flow is laminar and because

$$\mathrm{Re}_D \mathrm{Pr} \frac{D}{L} = \frac{(10.58)(3.00)(0.00254 \text{ m})}{0.3 \text{ m}} = 26.9 > 10$$

Eq. (6.42) can be used to evaluate the heat transfer coefficient. But since the mean bulk temperature is not known, we shall evaluate all the properties first at the inlet bulk temperature T_{b1}, then determine an exit bulk temperature, and then make a second iteration to obtain a more precise value. Designating inlet and outlet condition with the subscripts 1 and 2, respectively, the energy balance becomes

$$q_c = \bar{h}_c \pi D L\left(T_s - \frac{T_{b1} + T_{b2}}{2}\right) = \dot{m} c_p (T_{b2} - T_{b1}) \qquad \text{(a)}$$

At the wall temperature of 353 K, $\mu_s = 3.52 \times 10^{-4}$ N s/m^2 from Table 13 in Appendix 2. From Eq. (6.42), we can calculate the average Nusselt number.

$$\overline{\mathrm{Nu}}_D = 1.86\left[\frac{(1058)(3.00)(0.00254 \text{ m})}{0.3 \text{ m}}\right]^{0.33}\left(\frac{4.72}{3.52}\right)^{0.14} = 5.74$$

and thus

$$\bar{h}_c = \frac{k\overline{\mathrm{Nu}}_D}{D} = \frac{(0.658 \text{ W/m K})(5.74)}{0.00254 \text{ m}} = 1487 \text{ W/m}^2 \text{ K}$$

The mass flow rate is

$$\dot{m} = \rho \frac{\pi D^2}{4} \bar{U} = \frac{(983 \text{ kg/m}^3)\pi(0.00254 \text{ m})^2(0.2 \text{ m/s})}{4} = 0.996 \times 10^{-3} \text{ kg/s}$$

Inserting the calculated values for $\bar{h}_c$ and $\dot{m}$ into Eq. (a), along with $T_{b1} = 333$ K and $T_s = 353$ K, gives

$$(1487 \text{ W/m}^2 \text{ K})\pi(0.00254 \text{ m})(0.3 \text{ m})\left(353 - \frac{333 + T_{b2}}{2}\right)(\text{K})$$
$$= (0.996 \times 10^{-3} \text{ kg/s})(4181 \text{ J/kg K})(T_{b2} - 333)(\text{K}) \qquad \text{(b)}$$

Solving for T_{b2} gives

$$T_{b2} = 345 \text{ K}$$

For the second iteration, we shall evaluate all properties at the new average bulk temperature

$$\bar{T}_b = \frac{345 + 333}{2} = 339 \text{ K}$$

At this temperature, we get from Table 13 in Appendix 2:

$$\begin{aligned} \rho &= 980 \text{ kg/m}^2 \\ c_p &= 4185 \text{ J/kg K} \\ \mu &= 4.36 \times 10^{-4} \text{ N s/m}^2 \\ k &= 0.662 \text{ W/m K} \\ \text{Pr} &= 2.78 \end{aligned}$$

Recalculating the Reynolds number with properties based on the new mean bulk temperature gives

$$\text{Re}_D = \frac{\rho \bar{U} D}{\mu} = \frac{(980 \text{ kg/m}^3)(0.2 \text{ m/s})(2.54 \times 10^{-3} \text{ m})}{4.36 \times 10^{-4} \text{ kg/ms}} = 1142$$

With this value of Re_D, the heat transfer coefficient can now be calculated. One obtains on the second iteration $\text{Re}_D\text{Pr}(D/L) = 26.9$, $\overline{\text{Nu}}_D = 5.67$, and $\bar{h}_c = 1479 \text{ W/m}^2 \text{ K}$. Substituting the new value of $\bar{h}_c$ in Eq. (b) gives $T_{b2} = 345$ K. Further iterations will not affect the results appreciably in this example because of the small difference between bulk and wall temperature. In cases where the temperature difference is large, a second iteration may be necessary.

It is recommended that the reader verify the results using the LMTD method with Eq. (6.37).

6.3.4 Effect of Natural Convection

An additional complication in the determination of a heat transfer coefficient in laminar flow arises when the buoyancy forces are of the same order of magnitude as the external forces due to the forced circulation. Such a condition may arise in oil coolers when low-flow velocities are employed. Also, in the cooling of rotating parts, such as the rotor blades of gas turbines and ramjets attached to the propellers of helicopters, the natural-convection forces may be so large that their effect on the velocity pattern cannot be neglected even in high-velocity flow. When the buoyancy forces are in the same direction as the external forces, such as the gravitational forces superimposed on upward flow, they increase the rate of heat transfer. When the external and buoyancy forces act in opposite directions, the heat transfer is reduced. Eckert, et al. [14, 15] studied heat transfer in mixed flow, and their results are shown qualitatively in Fig. 6.16(a) and (b). In the darkly shaded area, the contribution of natural convection to the total heat transfer is less

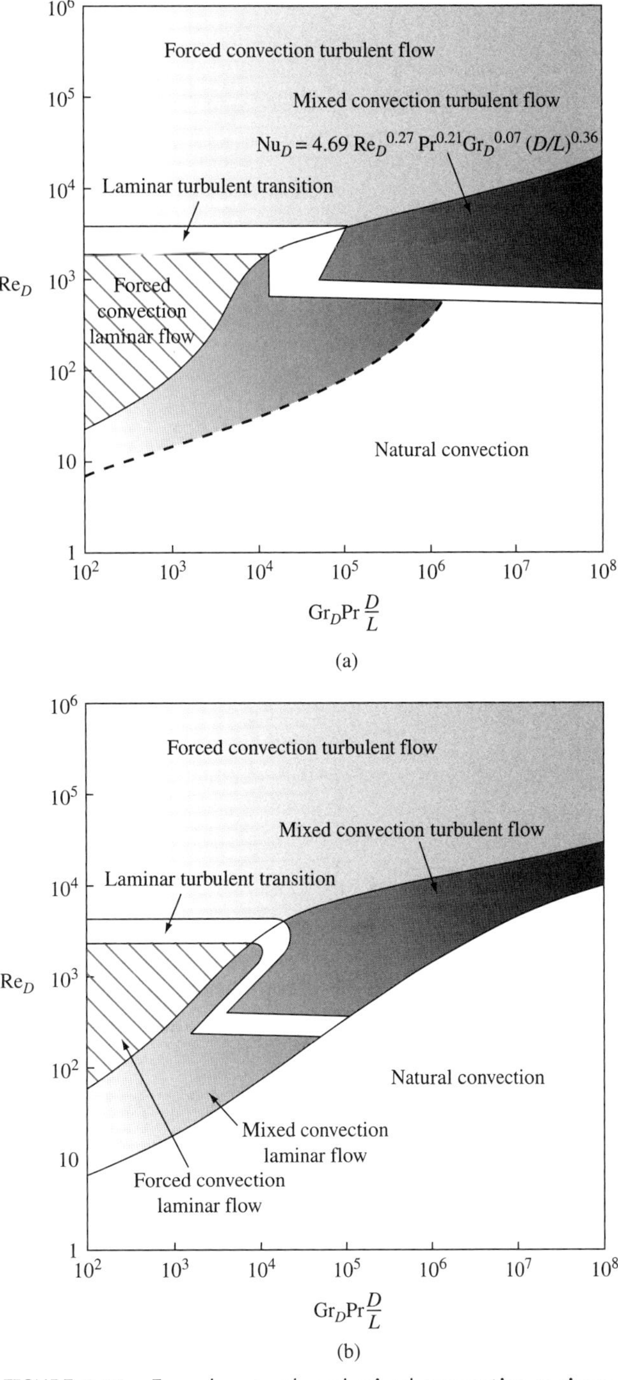

FIGURE 6.16 Forced, natural, and mixed convection regimes for (a) horizontal pipe flow and (b) vertical pipe flow.

Source: Courtesy of B. Metais and E. R. G. Eckert, "Forced, Free, and Mixed Convection Regimes," Trans. ASME. Ser. C. J. Heat Transfer, Vol. 86, pp. 295–298, 1964.

than 10%, whereas in the lightly shaded area, forced-convection effects are less than 10% and natural convection predominates. In the unshaded area, natural and forced convection are of the same order of magnitude. In practice, natural-convection effects are hardly ever significant in turbulent flow [21]. In cases where it is doubtful whether forced- or natural-convection flow applies, the heat transfer coefficient is generally calculated by using forced- and natural-convection relations separately, and the larger one is used [22]. The accuracy of this rule is estimated to be about 25%.

The influence of natural convection on the heat transfer to fluids in horizontal isothermal tubes has been investigated by Depew and August [23]. They found that their own data for $L/D = 28.4$ as well as previously available data for tubes with $L/D > 50$ could be correlated by the equation

$$\overline{\mathrm{Nu}}_D = 1.75\left(\frac{\mu_b}{\mu_s}\right)^{0.14}[\mathrm{Gz} + 0.12(\mathrm{Gz}\mathrm{Gr}_D^{1/3}\,\mathrm{Pr}^{0.36})^{0.88}]^{1/3} \tag{6.46}$$

In Eq. (6.46), Gz is the Graetz number, defined by

$$\mathrm{Gz} = \left(\frac{\pi}{4}\right)\mathrm{Re}_D\mathrm{Pr}\left(\frac{D}{L}\right)$$

The Grashof number, Gr_D, is defined by Eq. (5.8). Equation (6.46) was developed from experimental data with dimensionless parameters in the range $25 < \mathrm{Gz} < 700$, $5 < \mathrm{Pr} < 400$, and $250 < \mathrm{Gr}_D < 10^5$. Physical properties, except for μ_s, are to be evaluated at the average bulk temperature.

Correlations for vertical tubes and ducts are considerably more complicated because they depend on the relative direction of the heat flow and the natural convection. A summary of available information is given in Metais and Eckert [24] and Rohsenow, et al. [25].

6.4* Analogy Between Momentum and Heat Transfer in Turbulent Flow

To illustrate the most important physical variables affecting heat transfer by turbulent forced convection to or from fluids flowing in a long tube or duct, we shall now develop the so-called Reynolds analogy between heat and momentum transfer [26]. The assumptions necessary for the simple analogy are valid only for fluids having a Prandtl number of unity, but the fundamental relation between heat transfer and fluid friction for flow in ducts can be illustrated for this case without introducing mathematical difficulties. The results of the simple analysis can also be extended to other fluids by means of empirical correction factors.

The rate of heat flow per unit area in a fluid can be related to the temperature gradient by the equation developed previously:

$$\frac{q_c}{A\rho c_p} = -\left(\frac{k}{\rho c_p} + \varepsilon_H\right)\frac{dT}{dy} \tag{6.47}$$

Similarly, the shearing stress caused by the combined action of the viscous forces and the turbulent momentum transfer is given by

$$\frac{\tau}{\rho} = \left(\frac{\mu}{\rho} + \varepsilon_M\right)\frac{du}{dy} \tag{6.48}$$

According to the Reynolds analogy, heat and momentum are transferred by analogous processes in turbulent flow. Consequently, both q and τ vary with y, the distance from the surface, in the same manner. For fully developed turbulent flow in a pipe, the local shearing stress increases linearly with the radial distance r. Hence, we can write

$$\frac{\tau}{\tau_s} = \frac{r}{r_s} = 1 - \frac{y}{r_s} \tag{6.49}$$

and

$$\frac{q_c/A}{(q_c/A)_s} = \frac{r}{r_s} = 1 - \frac{y}{r_s} \tag{6.50}$$

where the subscript s denotes conditions at the inner surface of the pipe. Introducing Eqs. (6.49) and (6.50) into Eqs. (6.47) and (6.48), respectively, yields

$$\frac{\tau_s}{\rho}\left(1 - \frac{y}{r_s}\right) = \left(\frac{\mu}{\rho} + \varepsilon_M\right)\frac{du}{dy} \tag{6.51}$$

and

$$\frac{q_{c,s}}{A_s\rho c_p}\left(1 - \frac{y}{r_s}\right) = -\left(\frac{k}{\rho c_p} + \varepsilon_H\right)\frac{dT}{dy} \tag{6.52}$$

If $\varepsilon_H = \varepsilon_M$, the expressions in parentheses on the right-hand sides of Eqs. (6.51) and (6.52) are equal, provided the molecular diffusivity of momentum μ/ρ equals the molecular diffusivity of heat $k/\rho c_p$, that is, the Prandtl number is unity. Dividing Eq. (6.52) by Eq. (6.51) yields, under these restrictions,

$$\frac{q_{c,s}}{A_s c_p \tau_s}\, du = -dT \tag{6.53}$$

Integration of Eq. (6.53) between the wall, where $u = 0$ and $T = T_s$, and the bulk of the fluid, where $u = \bar{U}$ and $T = T_b$, yields

$$\frac{q_s \bar{U}}{A_s c_p \tau_s} = T_s - T_b$$

which can also be written in the form

$$\frac{\tau_s}{\rho \bar{U}^2} = \frac{q_s}{A_s(T_s - T_b)}\,\frac{1}{c_p \rho \bar{U}} = \frac{\bar{h}_c}{c_p \rho \bar{U}} \tag{6.54}$$

since $\bar{h}_c$ is by definition equal to $q_s/A_s(T_s - T_b)$. Multiplying the numerator and the denominator of the right-hand side by $D_H \mu k$ and regrouping yields

$$\frac{\bar{h}_c}{c_p\rho\bar{U}}\frac{D_H\mu k}{D_H\mu k} = \left(\frac{\bar{h}_c D_H}{k}\right)\left(\frac{k}{c_p\mu}\right)\left(\frac{\mu}{\bar{U}D_H\rho}\right) = \frac{\overline{\text{Nu}}}{\text{RePr}} = \overline{\text{St}}$$

where $\overline{\text{St}}$ is the *Stanton number*.

To bring the left-hand side of Eq. (6.54) into a more convenient form, we use Eqs. (6.13) and (6.14):

$$\tau_s = f\frac{\rho\bar{U}^2}{8}$$

Substituting Eq. (6.14) for τ_s in Eq. (6.54) finally yields a relation between the Stanton number $\overline{\text{St}}$ and the friction factor

$$\overline{\text{St}} = \frac{\overline{\text{Nu}}}{\text{RePr}} = \frac{f}{8} \tag{6.55}$$

known as the *Reynolds analogy* for flow in a tube. It agrees fairly well with experimental data for heat transfer in gases whose Prandtl number is nearly unity.

According to experimental data for fluids flowing in smooth tubes in the range of Reynolds numbers from 10,000 to 1,000,000, the friction factor is given by the empirical relation [17]

$$f = 0.184\text{Re}_D^{-0.2} \tag{6.56}$$

Using this relation, Eq. (6.55) can be written as

$$\overline{\text{St}} = \frac{\overline{\text{Nu}}}{\text{RePr}} = 0.023\text{Re}_D^{-0.2} \tag{6.57}$$

Since Pr was assumed unity,

$$\overline{\text{Nu}} = 0.023\text{Re}_D^{0.8} \tag{6.58}$$

or

$$\bar{h}_c = 0.023\bar{U}^{0.8}D^{-0.2}k\left(\frac{\mu}{\rho}\right)^{-0.8} \tag{6.59}$$

Note that in fully established turbulent flow, the heat transfer coefficient is directly proportional to the velocity raised to the 0.8 power, but inversely proportional to the tube diameter raised to the 0.2 power. For a given flow rate, an increase in the tube diameter reduces the velocity and thereby causes a decrease in $\bar{h}_c$ proportional to $1/D^{1.8}$. The use of small tubes and high velocities is therefore conducive to large heat transfer coefficients, but at the same time, the power required to overcome the frictional resistance is increased. In the design of heat exchange equipment, it is therefore necessary to strike a balance between the gain in heat transfer rates achieved by the use of ducts having small cross-sectional areas and the accompanying increase in pumping requirements.

Figure 6.17 shows the effect of surface roughness on the friction coefficient. We observe that the friction coefficient increases appreciably with the relative roughness, defined as ratio of the average asperity height ε to the diameter D. According to Eq. (6.55), one would expect that roughening the surface, which

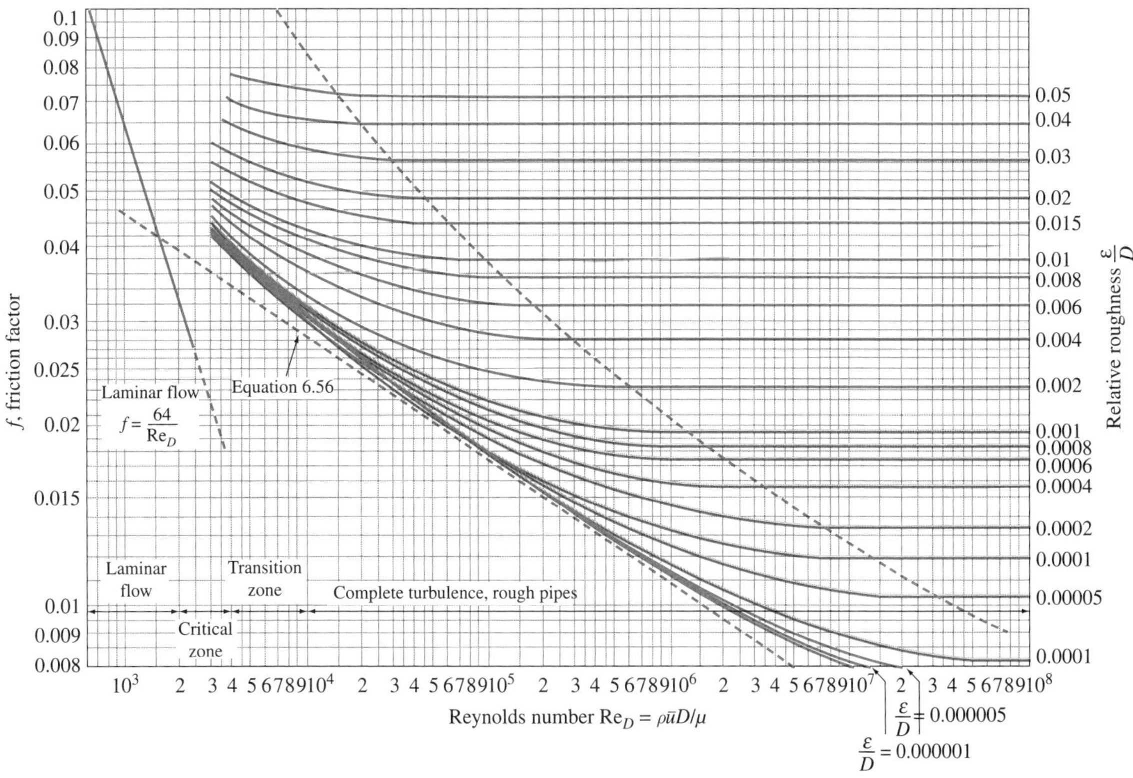

FIGURE 6.17 Friction factor versus Reynolds number for laminar and turbulent flow in tubes with various surface roughnesses.

Source: Courtesy of L. F. Moody, "Friction Factor for Pipe Flow," Trans. ASME, vol. 66, 1944.

increases the friction coefficient, also increases the convection conductance. Experiments performed by Cope [28] and Nunner [29] are qualitatively in agreement with this prediction, but a considerable increase in surface roughness is required to improve the rate of heat transfer appreciably. Since an increase in the surface roughness causes a substantial increase in the frictional resistance, for the same pressure drop, the rate of heat transfer obtained from a smooth tube is larger than from a rough one in turbulent flow.

Measurements by Dipprey and Sabersky [30] in tubes artificially roughened with sand grains are summarized in Fig. 6.18 on the next page. Where the Stanton number is plotted against the Reynolds number for various values of the roughness ratio ε/D. The lower straight line is for smooth tubes. At small Reynolds numbers, St has the same value for rough and smooth tube surfaces. The larger the value ε/D, the smaller the value of Re at which the heat transfer begins to improve with increase in Reynolds number. But for each value of ε/D, the Stanton number reaches a maximum and, with a further increase in Reynolds number, begins to decrease.

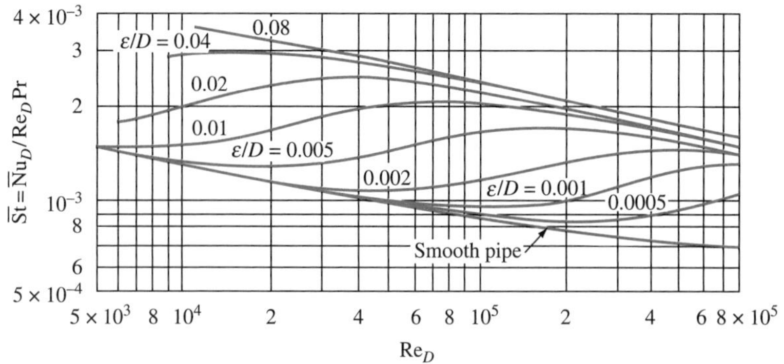

FIGURE 6.18 Heat transfer in artificially roughened tubes, $\overline{\mathrm{St}}$ versus Re for various values of ε/D according to Dipprey and Sabersky [30].

Source: Courtesy of T. von. Karman, "The Analogy between Fluid Friction and Heat Transfer," Trans. ASME, vol. 61, p. 705, 1939.

6.5 Empirical Correlations for Turbulent Forced Convection

The Reynolds analogy presented in the preceding section was extended semi-analytically to fluids with Prandtl numbers larger than unity in [31–34] and to liquid metals with very small Prandtl numbers in [31], but the phenomena of turbulent forced convection are so complex that empirical correlations are used in practice for engineering design.

6.5.1 Ducts and Tubes

The Dittus-Boelter equation [35] extends the Reynolds analogy to fluids with Prandtl numbers between 0.7 and 160 by multiplying the right-hand side of Eq. (6.58) by a correction factor of the form Pr^n:

$$\overline{\mathrm{Nu}}_D = \frac{\bar{h}_c D}{k} = 0.023\mathrm{Re}_D^{0.8}\,\mathrm{Pr}^{\,n} \tag{6.60}$$

where

$$n = \begin{cases} 0.4 & \text{for heating } (T_s > T_b) \\ 0.3 & \text{for cooling } (T_s < T_b) \end{cases}$$

With all properties in this correlation evaluated at the bulk temperature T_b, Eq. (6.60) has been confirmed experimentally to within $\pm 25\%$ for uniform wall temperature as well as uniform heat-flux conditions within the following ranges of parameters:

$$0.5 < \mathrm{Pr} < 120$$

$$6000 < \mathrm{Re}_D < 10^7$$

$$60 < (L/D)$$

Since this correlation does not take into account variations in physical properties due to the temperature gradient at a given cross section, it should be used only for situations with moderate temperature differences ($T_s - T_b$).

For situations in which significant property variations due to a large temperature difference ($T_s - T_b$) exist, a correlation developed by Sieder and Tate [16] is recommended:

$$\overline{\mathrm{Nu}}_D = 0.027\mathrm{Re}_D^{0.8}\,\mathrm{Pr}^{1/3}\left(\frac{\mu_b}{\mu_s}\right)^{0.14} \tag{6.61}$$

In Eq. (6.61), all properties except μ_s are evaluated at the bulk temperature. The viscosity μ_s is evaluated at the surface temperature. Equation (6.61) is appropriate for uniform wall temperature and uniform heat flux in the following range of conditions:

$$0.7 < \mathrm{Pr} < 10{,}000$$
$$6000 < \mathrm{Re}_D < 10^7$$
$$60 < (L/D)$$

To account for the variation in physical properties due to the temperature gradient in the flow direction, the surface and bulk temperatures should be the values halfway between the inlet and the outlet of the duct. For ducts of other than circular cross-sectional shapes, Eqs. (6.60) and (6.61) can be used if the diameter D is replaced by the hydraulic diameter D_H.

A correlation similar to Eq. (6.61) but restricted to gases was proposed by Kays and London [17] for long ducts:

$$\overline{\mathrm{Nu}}_{D_H} = C\mathrm{Re}_{D_H}^{0.8}\,\mathrm{Pr}^{0.3}\left(\frac{T_b}{T_s}\right)^n \tag{6.62}$$

where all properties are based on the bulk temperature T_b. The constant C and the exponent n are:

$$C = \begin{cases} 0.020 & \text{for uniform surface temperature } T_s \\ 0.020 & \text{for uniform heat flux } q_s'' \end{cases}$$

$$n = \begin{cases} 0.020 & \text{for heating} \\ 0.150 & \text{for cooling} \end{cases}$$

More complex empirical correlations have been proposed by Petukhov and Popov [38] and by Sleicher and Rouse [37]. Their results are shown in Table 6.3 on the next page, which presents four empirical correlation equations widely used by engineers to predict the heat transfer coefficient for turbulent forced convection in long, smooth, circular tubes. A careful experimental study with water heated in smooth tubes at Prandtl numbers of 6.0 and 11.6 showed that the Petukhov-Popov and the Sleicher-Rouse correlations argeed with the data over a Reynolds number range between 10,000 and 100,000 to within ±5%, while the Dittus-Boelter and Sieder-Tate correlations, popular with heat transfer engineers, underpredicted the data by 5 to 15% [38]. Figure 6.19 on the next page shows a comparison of these equations with experimental data at Pr = 6.0 (water at 26.7°C). The following example illustrates the use of some of these empirical correlations.

TABLE 6.3 Heat transfer correlations for liquids and gases in incompressible flow through tubes and pipes

Name (reference)	Formula[a]	Conditions	Equation
Dittus-Boelter [35]	$\overline{\mathrm{Nu}}_D = 0.23\mathrm{Re}_D^{0.8}\mathrm{Pr}^n$ $n\begin{cases} = 0.4 \text{ for heating} \\ = 0.3 \text{ for cooling}\end{cases}$	$0.5 < \mathrm{Pr} < 120$ $6000 < \mathrm{Re}_D < 10^7$	(6.60)
Sieder-Tate [16]	$\overline{\mathrm{Nu}}_D = 0.027\mathrm{Re}_D^{0.8}\mathrm{Pr}^{0.3}\left(\frac{\mu_b}{\mu_s}\right)^{0.14}$	$6000 < \mathrm{Re}_D < 10^7$ $0.7 < \mathrm{Pr} < 10^4$	(6.61)
Petukhov-Popov [36]	$\overline{\mathrm{Nu}}_D = \frac{(f/8)\mathrm{Re}_D\mathrm{Pr}}{K_1 + K_2(f/8)^{1/2}(\mathrm{Pr}^{2/3} - 1)}$ where $f = (1.82 \log_{10}\mathrm{Re}_D - 1.64)^{-2}$ $K_1 = 1 + 3.4f$ $K_2 = 11.7 + \frac{1.8}{\mathrm{Pr}^{1/3}}$	$0.5 < \mathrm{Pr} < 2000$ $10^4 < \mathrm{Re}_D < 5 \times 10^6$	(6.63)
Sleicher-Rouse [37]	$\overline{\mathrm{Nu}}_D = 5 + 0.015\mathrm{Re}_D^a\mathrm{Pr}_s^b$ where $a = 0.88 - \frac{0.24}{4 + \mathrm{Pr}_s}$ $b = 1/3 + 0.5e^{-0.6\mathrm{Pr}_s}$	$0.1 < \mathrm{Pr} < 10^5$ $10^4 < \mathrm{Re}_D < 10^6$	(6.64)

[a] All properties are evaluated at the bulk fluid temperature except where noted. Subscripts b and s indicate bulk and surface temperatures, respectively.

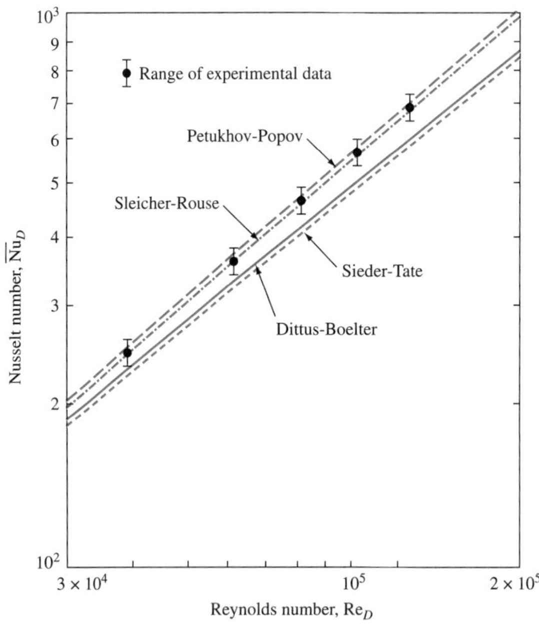

FIGURE 6.19 Comparison of predicted and measured Nusselt number for turbulent flow of water in a tube (26.7°C; Pr = 6.0).

EXAMPLE 6.5 Determine the Nusselt number for water flowing at an average velocity of 10 ft/s in an annulus formed between a 1-in.-OD tube and a 1.5-in.-ID tube as shown in Fig. 6.20. The water is at 180°F and is being cooled. The temperature of the inner wall is 100°F, and the outer wall of the annulus is insulated. Neglect entrance effects and compare the results obtained from all four equations in Table 6.3. The properties of water are given below in engineering units.

T (°F)	*m* (lb_m/h ft)	*k* (Btu/h ft °F)	*r* (lb_m/ft^3)	*c* (Btu/lb_m °F)
100	1.67	0.36	62.0	1.0
140	1.14	0.38	61.3	1.0
180	0.75	0.39	60.8	1.0

SOLUTION The hydraulic diameter D_H for this geometry is 0.5 in. The Reynolds number based on the hydraulic diameter and the bulk temperature properties is

$$\mathrm{Re}_{D_H} = \frac{\rho \bar{U} D_H}{\mu} = \frac{(10\,\mathrm{ft/s})(0.5/12\,\mathrm{ft})(60.8\,\mathrm{lb_m/ft^3})(3600\,\mathrm{s/h})}{0.75\,\mathrm{lb_m/h\,ft}}$$
$$= 125{,}000$$

The Prandtl number is

$$\mathrm{Pr} = \frac{c_p \mu}{k} = \frac{(1.0\,\mathrm{Btu/lb_m\,°F})(0.75\,\mathrm{lb_m/h\,ft})}{(0.39\,\mathrm{Btu/h\,ft\,°F})} = 1.92$$

The Nusselt number according to the Dittus-Boelter correlation [Eq. (6.60)] is

$$\overline{\mathrm{Nu}}_{D_H} = 0.023\,\mathrm{Re}_{D_H}^{0.8}\,\mathrm{Pr}^{0.3} = (0.023)(11{,}954)(1.22) = 334$$

Using the Sieder-Tate correlation [Eq. (6.61)], we get

$$\overline{\mathrm{Nu}}_{D_H} = 0.27\mathrm{Re}_{D_H}^{0.8}\,\mathrm{Pr}^{0.3}\left(\frac{\mu_b}{\mu_s}\right)^{0.14}$$
$$= (0.027)(11{,}954)(1.24)\left(\frac{0.75}{1.67}\right)^{0.14} = 358$$

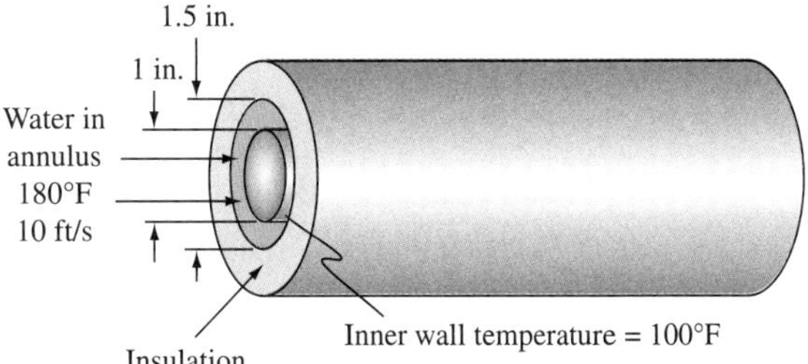

FIGURE 6.20 Schematic diagram of annulus for cooling of water in Example 6.5.

The Petukhov-Popov correlation [Eq. (6.63)] gives

$$f = (1.82 \log_{10}\text{Re}_{D_H} - 1.64)^{-2} = (9.276 - 1.64)^{-2} = 0.01715$$

$$K_1 = 1 + 3.4f = 1.0583$$

$$K_2 = 11.7 + \frac{1.8}{\text{Pr}^{0.33}} = 13.15$$

$$\overline{\text{Nu}}_{D_H} = \frac{f\,\text{Re}_{D_H}\text{Pr}/8}{K_1 + K_2(f/8)^{1/2}(\,\text{Pr}^{\,0.67} - 1)}$$

$$= \frac{(0.01715)(125{,}000)(1.92/8)}{1.0583 + (13.15)(0.01715/8)^{1/2}(0.548)} = 370$$

The Sleicher-Rouse correlation [Eq. (6.64)] yields

$$\overline{\text{Nu}}_{D_H} = 5 + 0.015\text{Re}_D^a\text{Pr}_s^b$$

$$a = 0.88 - \frac{0.24}{4 + 4.64} = 0.88 - 0.0278 = 0.852$$

$$b = \frac{1}{3} + \frac{0.5}{e^{0.6\,\text{Pr}_s}} = 0.333 + \frac{0.5}{16.17} = 0.364$$

$$\text{Re}_D = 82{,}237$$

$$\overline{\text{Nu}}_{D_H} = 5 + (0.015)(82{,}237)^{0.852}(4.64)^{0.364}$$

$$= 5 + (0.015)(15{,}404)(1.748) = 409$$

Assuming that the correct answer is $\overline{\text{Nu}}_{D_H} = 370$, the first two correlations underpredict $\overline{\text{Nu}}_{D_H}$ by about 10% and 3.5%, respectively, while the Sleicher-Rouse method overpredicts by about 10.5%.

It should be noted that in general, the surface and film temperatures are not known and therefore the use of Eq. (6.64) requires iteration for large temperature differences. The main difficulty in applying Eq. (6.63) for conditions with varying properties is that the friction factor f may be affected by heating or cooling to an unknown extent. Thus, to account for variable property effects in the flow cross section due to a significant temperature difference between the tube surface and bulk fluid, a correction factor is commonly employed. This is usually in the form of a bulk-to-surface viscosity ratio or temperature ratio raised to some power, depending on whether the fluid is heated or cooled in the tube; two examples are given in Eqs. (6.61) and (6.62)

For gases and liquids flowing in short circular tubes ($2 < L/D < 60$) with abrupt contraction entrances, the entrance configuration of greatest interest in heat exchanger design, the entrance effect for Reynolds numbers corresponding to turbulent flow becomes important [40]. An extensive theoretical analysis of the heat transfer and the pressure drop in the entrance regions of smooth passages is given in [41], and a complete survey of experimental results for various types of inlet conditions is given in [40].

The most commonly used and widely accepted correlation in current practice for turbulent flows in circular tubes, however, and one that accounts for both variable property and entrance length effects is the Gnielinski correlation [42]. It is a modification of the Petukhov and Popov [36] equation, is valid for the transition flow and fully developed turbulent flow regimes ($2300 \leq \text{Re}_D \leq 5 \times 10^6$) as well as a broad spectrum of fluids ($0.5 < \text{Pr} \leq 200$), and is expressed as follows:

$$\overline{\text{Nu}}_D = \frac{(f/8)(\text{Re}_D - 1000)\,\text{Pr}}{1 + 12.7(f/8)^{1/2}(\text{Pr}^{2/3} - 1)}\left[1 + (D/L)^{2/3}\right]K \tag{6.65}$$

where

$$K = \begin{cases} (\text{Pr}_b/\text{Pr}_s)^{0.11} & \text{for liquids} \\ (T_b/T_s)^{0.45} & \text{for gases} \end{cases}$$

and the friction factor f is calculated from the same expression used in the Petukhov-Popov correlation of Eq. (6.65), as listed in Table 6.3. Note that instead of a viscosity ratio, the ratio of Prandtl number at bulk fluid and tube surface temperatures has been used to account for variable property effects. This same correction factor can be used as a multiplier to calculate f as well.

6.5.2 Ducts of Noncircular Shape

In many heat exchangers, rectangular, oval, trapezoidal, and concentric annular flow passages, among others, are often employed. Some examples include plate-fin, oval-tube-fin, and double-pipe heat exchangers. The generally accepted practice in most such cases, to a fair degree of accuracy as verified with experimental data [43], is to use the circular-tube correlations with all dimensionless variables based on the hydraulic diameter to estimate both the convective heat transfer coefficient and friction factor in turbulent flows. Thus, any of the correlations listed in Table 6.3 could be employed, although the more popular recommendation in many handbooks is for the Gnielinski correlation of Eq. (6.65)

The exception to this rule is the case of turbulent flows in concentric annuli where the curvatures of the inner and outer diameters, or D_i and D_o, tend to have an effect on the convective behavior, particularly when the ratio (D_i/D_o) is small [44, 45]. Based on experimental data and an extended analysis [44], the following correlation has been proposed:

$$\overline{\text{Nu}}_{D_H} = \overline{\text{Nu}}_c\left[1 + \{0.8(D_i/D_o)^{-0.16}\}^{15}\right]^{1/15} \tag{6.66}$$

where $\overline{\text{Nu}}_c$ is calculated from Eq. (6.65), again by using the hydraulic diameter of the annular cross section, $D_H = (D_o - D_i)$, as the length scale. The duct-wall curvature effect, represented by the diameter ratio used in Eq. (6.66) is a modified form of the correction factor considered by Petukhov and Roizen [45]. Furthermore, if effects of temperature-dependent fluid property variations in the flow cross section have to be included in the analysis, then the same correction factor K recommended in Eq. (6.65) may be employed for liquids or gases, as the case may be.

6.5.3 Liquid Metals

Liquid metals have been employed as heat transfer media because they have certain advantages over other common liquids used for heat transfer purposes. Liquid metals, such as sodium, mercury, lead, and lead-bismuth alloys, have relatively low melting points and combine high densities with low vapor pressures at high temperatures as well as with large thermal conductivities, which range from 10 to 100 W/m K. These metals can be used over wide ranges of temperatures, they have a large heat capacity per unit volume, and they have large convection heat transfer coefficients. They are especially suitable for use in nuclear power plants, where large amounts of heat are liberated and must be removed in a small volume. Liquid metals pose some safety difficulties in handling and pumping. The development of electromagnetic pumps has eliminated some of these problems.

Even in a highly turbulent stream, the effect of eddying in liquid metals is of secondary importance compared to conduction. The temperature profile is established much more rapidly than the velocity profile. For typical applications, the assumption of a uniform velocity profile (called "slug flow") may give satisfactory results, although experimental evidence is insufficient for a quantitative evaluation of the possible deviation from the analytic solution for slug flow. The empirical equations for gases and liquids therefore do not apply. Several theoretical analyses for the evaluation of the Nusselt number are available, but there are still some unexplained discrepancies between many of the experimental data and the analytic results. Such discrepancies can be seen in Fig. 6.21, where experimentally measured Nusselt numbers for heating of mercury in long tubes are compared with the analysis of Martinelli [2].

Lubarsky and Kaufman [46] found that the relation

$$\overline{\text{Nu}}_D = 0.625(\text{Re}_D\text{Pr})^{0.4} \tag{6.67}$$

empirically correlated most of the data in Fig. 6.21, but the error band was substantial. Those points in Fig. 6.21 that fall far below the average are believed to have been obtained in systems where the liquid metal did not wet the surface. However, no final conclusions regarding the effect of wetting have been reached to date.

According to Skupinski, et al. [47], the Nusselt number for liquid metals flowing in smooth tubes can be obtained from

$$\overline{\text{Nu}}_D = 4.82 + 0.0185(\text{Re}_D\text{Pr})^{0.827} \tag{6.68}$$

if the heat flux is uniform in the range $\text{Re}_D\text{Pr} > 100$ and $L/D > 30$, with all properties evaluated at the bulk temperature.

According to an investigation of the thermal entry region for turbulent flow of a liquid metal in a pipe with uniform heat flux, the Nusselt number depends only on

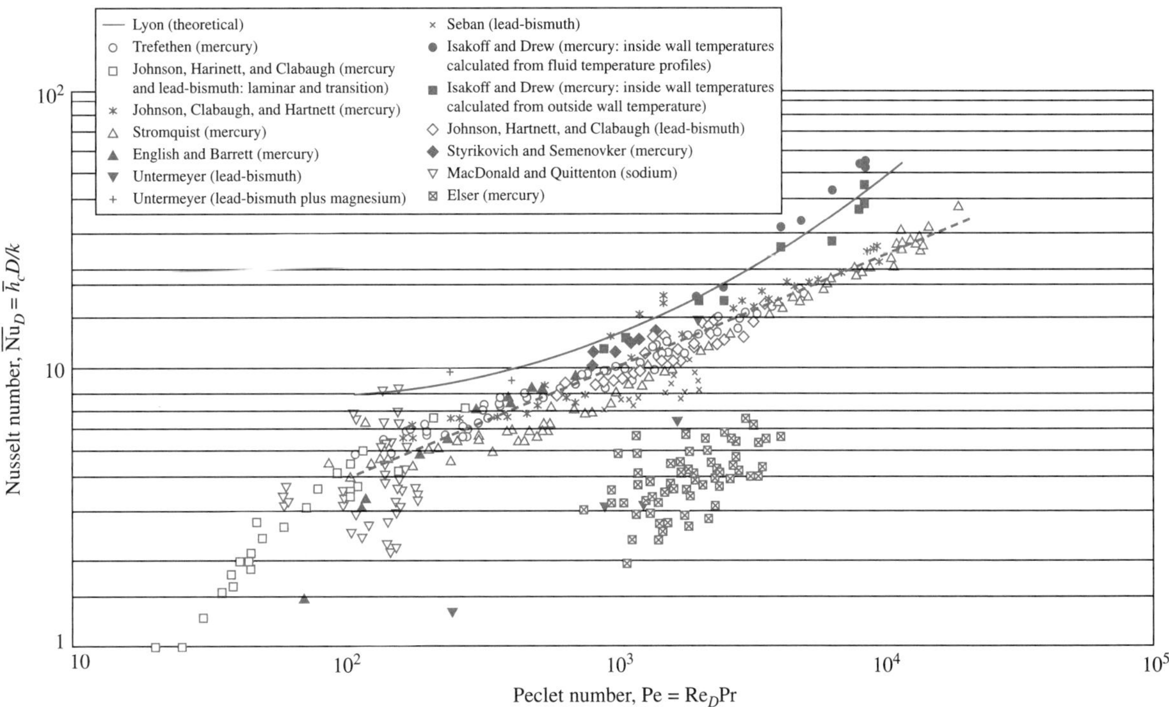

FIGURE 6.21 Comparison of measured and predicted Nusselt numbers for liquid metals heated in long tubes with uniform heat flux.

Source: Courtesy of the National Advisory Committee for Aeronautics, NACA TN 3363.

the Reynolds number when $Re_D Pr < 100$. For these conditions, Lee [48] found that the equation

$$\overline{Nu}_D = 3.0 Re_D^{0.0833} \tag{6.69}$$

fits data and analysis well. Convection in the entrance regions for fluids with small Prandtl numbers has also been investigated analytically by Deissler [41], and experimental data supporting the analysis are summarized in [49] and [50]. In turbulent flow, the thermal entry length $(L/D_H)_{entry}$ is approximately 10 equivalent diameters when the velocity profile is already developed and 30 equivalent diameters when it develops simultaneously with the temperature profile.

For a constant surface temperature the data are correlated, according to Seban and Shimazaki [51], by the equation

$$\overline{Nu}_D = 5.0 + 0.025(Re_D Pr)^{0.8} \tag{6.70}$$

in the range $RePr > 100$, $L/D > 30$.

EXAMPLE 6.6 A liquid metal flows at a mass rate of 3 kg/s through a constant-heat-flux 5-cm-ID tube in a nuclear reactor. The fluid at 473 K is to be heated, and the tube wall is 30 K above the fluid temperature. Determine the length of the tube required for a 1-K rise in bulk fluid temperature, using the following properties:

$$\rho = 7.7 \times 10^3\ \text{kg/m}^3$$
$$\nu = 8.0 \times 10^{-8}\ \text{m}^2\text{/s}$$
$$c_p = 130\ \text{J/kg K}$$
$$k = 12\ \text{W/mK}$$
$$\text{Pr} = 0.011$$

SOLUTION The rate of heat transfer per unit temperature rise is

$$q = \dot{m}c_p\,\Delta T = (3.0\,\text{kg/s})(130\,\text{J/kg K})(1\,\text{K}) = 390\,\text{W}$$

The Reynolds number is

$$\text{Re}_D = \frac{\dot{m}D}{\rho A\nu} = \frac{(3\,\text{kg/s})(0.05\,\text{m})}{(7.7 \times 10^3\,\text{kg/m}^3)[\pi(0.5\,\text{m})^2/4](8.0 \times 10^{-8}\text{m}^2\text{/s})}$$
$$= 1.24 \times 10^5$$

The heat transfer coefficient is obtained from Eq. (6.67):

$$\bar{h}_c = \left(\frac{k}{D}\right)0.625(\text{Re}_D\text{Pr})^{0.4}$$
$$= \left(\frac{12\,\text{W/m K}}{0.05\,\text{m}}\right)0.625[(1.24 \times 10^5)(0.011)]^{0.4}$$
$$= 2692\,\text{W/m}^2\,\text{K}$$

The surface area required is

$$A = \pi DL = \frac{q}{\bar{h}_c(T_s - T_b)}$$
$$= \frac{390}{(2692\,\text{W/m}^2\,\text{K})(30\,\text{K})}$$
$$= 4.83 \times 10^{-3}\,\text{m}^2$$

Finally, the required length is

$$L = \frac{A}{\pi D} = \frac{4.83 \times 10^{-3}\,\text{m}}{(\pi)(0.05\,\text{m})}$$
$$= 0.0307\,\text{m}$$

6.6 Heat Transfer Enhancement and Electronic-Device Cooling

6.6.1 Enhancement of Forced Convection Inside Tubes

The need to increase the heat transfer performance of heat exchangers so as to reduce energy and material consumption, as well as the associated impact on environmental degradation, has led to the development and usage of many heat transfer *enhancement* techniques [52–54]. A variety of methods have been developed, and they are characterized as either *passive* or *active* techniques. The main distinguishing feature between the two is that the former, unlike active methods, does not require additional input of external power other than that needed for fluid motion. Passive techniques generally consist of geometric or material modification of the primary heat transfer surface, and examples include finned surfaces, swirl-flow-producing tube inserts, and coiled tubes, among others [52–54].

The objective of enhancement of forced convection is to increase the heat transfer rate q_c, which is expressed by the following rate equation:

$$q_c = \bar{h}_c A \Delta T$$

Thus, for a fixed temperature difference ΔT, by increasing the surface area A (as is done in the case of finned tubes), or the convective heat transfer coefficient $\bar{h}_c$ by altering the fluid motion (as is produced by swirl-flow inserts in tubes), or both (as is the case with using coiled tubes or helical, serrated, and other types of fins), the heat transfer rate q can be increased. There is, however, an associated pressure-drop penalty due to increased frictional losses; the analogy between heat and momentum transfer discussed in Section 6.4 and some form of interconnected relationship between the two suggest this outcome. The consequent assessment of any effective heat transfer enhancement requires some extended analysis based on different evaluation criteria or figures of merit, and details of such performance evaluation can be found in [52–54].

Finned Tubes In single-phase forced convection applications, tubes with fins on the inner, outer, or both surfaces have long been used in double-pipe and shell-and-tube heat exchangers. Some examples of tubes with fins are shown in Figs. 6.22 and 6.23 on the next page. The focus of discussion in this section is on tubes with fins on their inner surface. While experimental data for several different geometries and flow arrangements have been reported in the literature, their analysis and interpretation to devise correlations for the Nusselt number and friction factor have been rather sparse. Some theoretical studies based on computational simulations of forced convective flows (both laminar and turbulent regimes,) in finned tubes also have been carried out. Issues such as modeling the effects of fin size and thickness, along with its longitudinal geometry (helical or spiral fin, for instance), have been addressed in these studies [53].

For laminar flows inside tubes that have straight or spiral fins, based on experimental data for oil flows and employing the hydraulic diameter D_H length scale,

FIGURE 6.22 Typical examples of tubes with fins that are used in commercial heat exchangers.

Source: F. W. Brökelmann Aluminiumwerk

Watkinson et al. [55] have given the following correlations for the isothermal friction factor, which is common for both straight-fin and spiral-fin tubes:

$$f_{D_H} = \frac{65.6}{\mathrm{Re}_{D_H}} \left(\frac{D_H}{D_o} \right)^{1.4} \tag{6.71}$$

where D_o is the inner diameter of the "bare" tube, i.e., the diameter when all the fins are removed. To calculate the Nusselt number, two different equations have been proposed. For straight-fin tubes, the equation is

$$\mathrm{Nu}_{D_H} = \frac{1.08 \times \log \mathrm{Re}_{D_H}}{N^{0.5}(1 + 0.01\,\mathrm{Gr}_{D_H}^{1/3})} \mathrm{Re}_{D_H}^{0.46}\,\mathrm{Pr}^{1/3} \left(\frac{L}{D_h} \right)^{1/3} \left(\frac{\mu_s}{\mu_b} \right)^{0.14} \tag{6.72}$$

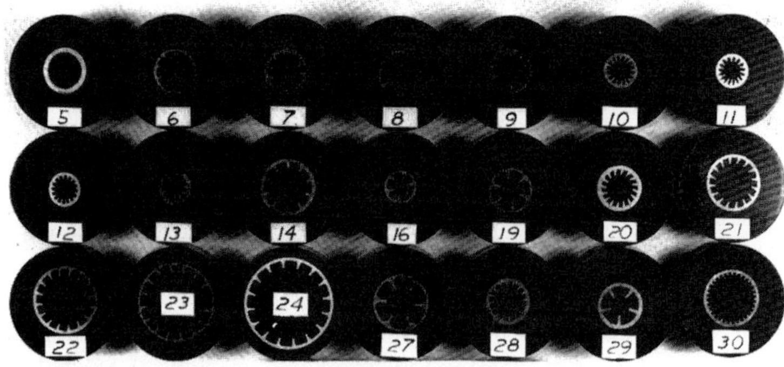

FIGURE 6.23 Profiles of internally finned tubes.

Source: "Cooling Air in Turbulent Flow with Internally Finned Tubes," T. C. Carnavos, Heat Transfer Eng., vol. 1, 1979, reprinted by permission of the publisher, Taylor & Francis Group, *http://www.informaworld.com*.

where N is the number of fins on the tube periphery. For spiral-fin tubes, it is

$$\mathrm{Nu}_{D_H} = \frac{8.533 \times \log \mathrm{Re}_{D_H}}{(1 + 0.01\mathrm{Gr}_{D_H}^{1/3})} \mathrm{Re}_{D_H}^{0.26} \mathrm{Pr}^{1/3} \left(\frac{t}{p}\right)^{0.5} \left(\frac{L}{D_h}\right)^{1/3} \left(\frac{\mu_s}{\mu_b}\right)^{0.14} \tag{6.73}$$

where t is the thickness and p is the spiral pitch of the fin. Note that while temperature-dependent viscosity correction has been included in the expressions for the Nusselt number, it is missing in the friction factor given by Eq. (6.71). Of course, for heating or cooling conditions, f_{D_H} would be different than in isothermal conditions, with lower friction when the fluid is being heated and conversely higher when it is cooled. In such instances, a good engineering approximation can be made by including the correction given by Eqs. (6.44) and (6.45).

Heat transfer performance for the cooling of air in turbulent flow with 21 different tubes having integral internal spiral and longitudinal (or straight) fins has been studied by Carnavos [56]. For the 21 tube profiles shown in Fig. 6.22, the heat transfer data were correlated within $\pm 6\%$ at Reynolds numbers between 10^4 and 10^5 by the equation

$$\overline{\mathrm{Nu}}_{D_H} = 0.023\mathrm{Re}_{D_H}^{0.8} \mathrm{Pr}^{0.4} \left(\frac{A_{fa}}{A_{fc}}\right)^{0.1} \left(\frac{A_n}{A_a}\right)^{0.5} (\sec \alpha)^3 \tag{6.74}$$

The friction factor f_{D_H} was correlated within $\pm 7\%$ for all configurations except 11, 12, and 28 (see Fig. 6.22) by the relation

$$f_{D_H} = \frac{0.184}{\mathrm{Re}_{D_H}^{0.2}} \left(\frac{A_{fa}}{A_{fn}}\right)^{0.5} (\cos \alpha)^{0.5} \tag{6.75}$$

where A_{fa} = actual free-flow cross-sectional area

A_{fc} = open-core flow area inside fins

A_a = actual heat transfer area

A_n = nominal heat transfer area based on tube ID without fins

α = helix angle for spiral fins

A_{fn} = nominal flow area based on tube ID without fins

To apply these correlations, all physical properties should be based on the average bulk temperature.

Twisted-Tape Inserts An effective and widely used device for enhancing a single-phase flow heat transfer coefficient is the twisted-tape insert. It has been shown to increase the heat transfer coefficient substantially with a relatively small pressure-drop penalty [57]. It is often used in a new exchanger design so that, for a specified heat duty, a significant reduction in size can be achieved. It is also employed in the retrofit of existing shell-and-tube heat exchangers so as to upgrade their heat loads. The ease with which multitube bundles can be fitted with twisted-tape inserts and their removal, as depicted in Fig. 6.24 on the next page, makes them very useful in applications where fouling may occur and where frequent tube-side cleaning may be required.

The geometrical features of a twisted tape, as shown in Fig. 6.24(b), are described by the 180° twist pitch H, tape thickness δ, and tape width d (which is usually about

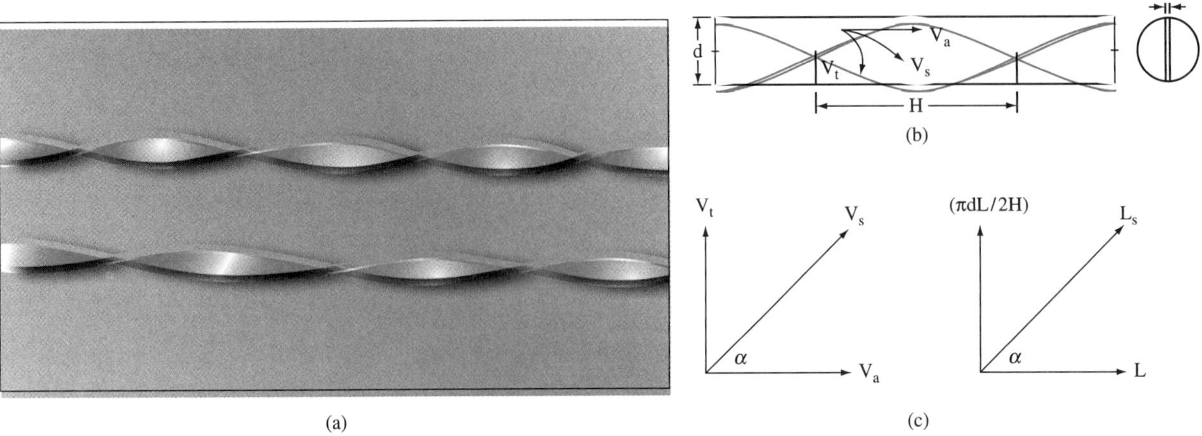

FIGURE 6.24 Twisted-tape inserts: (a) typical application in a shell-and-tube heat exchanger; (b) characteristic geometrical features; and (c) representation of the tape-induced swirl-flow velocity and helical-flow length along with their respective components [53, 57].

the same as the tube inside diameter D in snug- to tight-fitting tapes). The severity of tape twist is given by the dimensionless twist ratio y $(= H/D)$, and depending on the tube diameter and tape material, inserts with a very small twist ratio can be employed. When placed inside a circular tube, the flow field gets altered in several different ways: increased axial velocity and wetted perimeter due to the blockage and partitioning of the flow cross section, longer effective flow length in the helically twisting partitioned duct, and tape's helical-curvature-induced secondary fluid circulation or swirl. However, the most dominant mechanism is swirl generation, which can be scaled in laminar flow conditions by a dimensionless swirl parameter [58] defined as

$$\mathrm{Sw} = \frac{\mathrm{Re}_s}{\sqrt{y}} \tag{6.76}$$

where

$$\mathrm{Re}_s = \rho V_s D/\mu \qquad V_s = (G/\rho)\left[1 + (\pi/2y)^2\right]^{1/2} \qquad G = \dot{m}/(\pi D^2/4) - 2\delta \tag{6.77}$$

Based on this scaling of the swirl behavior in the laminar flow regime, Manglik and Bergles [58] have developed the following correlation for the isothermal Fanning friction factor:

$$C_{f,s} = \frac{15.767}{\mathrm{Re}_s}\left[\frac{\pi + 2 - 2(\delta/D)}{\pi - 4(\delta/D)}\right]^2 (1 + 10^{-6}\mathrm{Sw}^{2.55})^{1/6} \tag{6.78}$$

where $C_{f,s}$ is based on the effective swirl velocity and swirl-flow length [see Fig. 6.24c], or

$$C_{f,s} = \frac{g_c \Delta p D}{2\rho V_s^2 L_s} \qquad L_s = L\left[1 + \left(\frac{\pi}{2y}\right)^2\right]^{1/2} \tag{6.79}$$

This correlation has been found to predict a large set of experimental data for a very wide range of fluids, flow conditions ($0 \le \text{Sw} \le 2000$), and tape geometry ($1.5 \le y \le \infty$, $0.02 \le (\delta/D) \le 0.12$) to within $\pm 10\%$ [59]. For the heat transfer in laminar flows inside circular tubes fitted with a twisted tape and maintained at a uniform or constant wall temperature, Manglik and Bergles [58] have given the following correlation:

$$\overline{\text{Nu}}_D = 4.612\left(\frac{\mu_b}{\mu_s}\right)^{0.14}\Big[\Big\{\left(1 + 0.0951\,\text{Gz}^{0.894}\right)^{2.5} + 6.413 \times 10^{-9}\left(\text{Sw}\cdot \text{Pr}^{\,0.391}\right)^{3.835}\Big\}^2 + 2.132 \times 10^{-14}\left(\text{Re}_D\cdot\text{Ra}\right)^{2.23}\Big]^{0.1} \tag{6.80}$$

Once again, for the more practical conditions of heating or cooling, the friction factor given by Eq. (6.78) requires a correction factor to account for fluid property variations in the flow cross section of the tube, and this can be made as

$$C_{f,\text{heat transfer}} = C_{f,\text{isothermal}} \times \begin{cases} (\mu_b/\mu_w)^m & m = \begin{cases} 0.65 & \text{liquid heating} \\ 0.58 & \text{liquid cooling} \end{cases} \\ (T_b/T_w)^{0.1} & \text{for heating/cooling of gases} \end{cases} \tag{6.81}$$

In the turbulent flow regime, the scaling of swirl flows due to twisted-tape inserts with Sw is found to be inapplicable, and instead Manglik and Bergles [60] have correlated the data for isothermal Fanning friction factor as

$$C_f = \left(\frac{0.0791}{\text{Re}_D^{0.25}}\right)\left(1 + \frac{2.752}{y^{1.29}}\right)\left[\frac{\pi}{\pi - (4\delta/D)}\right]^{1.75}\left[\frac{\pi + 2 - (2\delta/D)}{\pi - (4\delta/D)}\right]^{1.25} \tag{6.82}$$

This equation is able to predict the available experimental data within $\pm 5\%$ [57], and to correct for heating/cooling conditions, the following may be adopted:

$$C_{f,\text{ heat transfer}} = C_{f,\text{ isothermal}} \begin{cases} (\mu_b/\mu_s)^{0.35(d_h/d)} & \text{for liquids} \\ (T_b/T_s)^{0.1} & \text{for gases} \end{cases} \tag{6.83}$$

For turbulent flow heat transfer with $\text{Re}_D \ge 10^4$, the Nusselt number correlation developed by Manglik and Bergles [60] is expressed as

$$\overline{\text{Nu}}_D = 0.023\text{Re}_D^{0.8}\,\text{Pr}^{0.4}\left[1 + \frac{0.769}{y}\right]\left[\frac{\pi + 2 - (2\delta/D)}{\pi - (4\delta/D)}\right]^{0.2} \times \left[\frac{\pi}{\pi - (4\delta/D)}\right]^{0.8}\phi \tag{6.84}$$

where the property-ratio correction factor ϕ is given by

$$\phi = (\mu_b/\mu_s)^n \text{ or } (T_b/T_s)^m$$

$$n = \begin{cases} 0.18 & \text{liquid heating} \\ 0.30 & \text{liquid cooling} \end{cases} \quad \text{and} \quad m = \begin{cases} 0.45 & \text{gas heating} \\ 0.15 & \text{gas cooling} \end{cases}$$

The predictions from this correlation have been found [57, 60] to describe a large set of experimental data for a wide range of tape-twist ratios ($2 \leq y \leq \infty$) to within $\pm 10\%$ for both gas and liquid turbulent flows in circular tubes with twisted-tape inserts.

Coiled Tubes Coiled tubes are used in heat exchange equipment to not only increase the heat transfer surface area per unit volume but to also enhance the heat transfer coefficient of the flow inside the tube. The basic configuration is shown in Fig. 6.25. As a result of the centrifugal forces, a secondary flow pattern consisting of two vortices perpendicular to the axial-flow direction is set up, and heat transport will occur not only by diffusion in the radial direction but also by convection. The contribution of this secondary convective transport dominates the overall process and enhances the rate of heat transfer per unit length of tube compared to a straight tube of equal length.

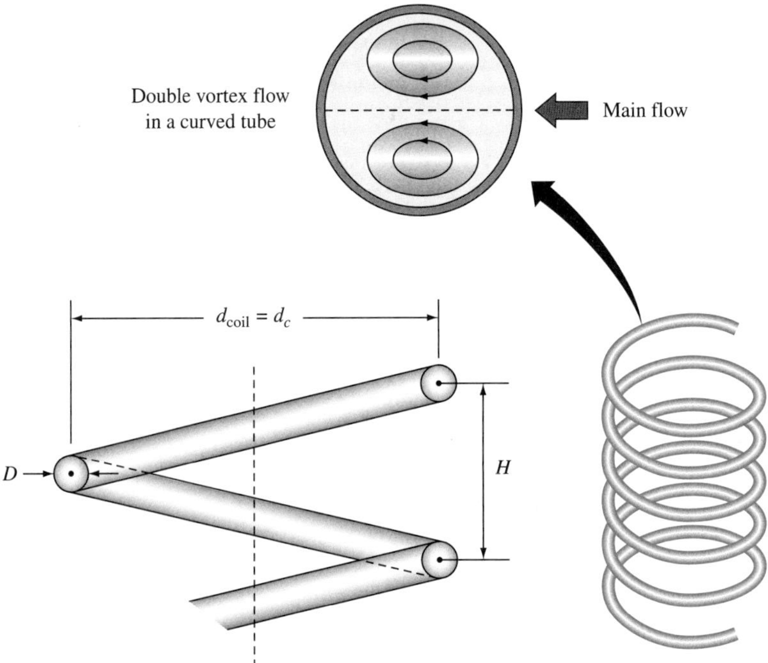

FIGURE 6.25 Schematic diagram illustrating flow and nomenclature for heat transfer in helically coiled tubes.

The flow characterization and the associated convection heat transfer coefficient in coiled tubes are governed by the flow Reynolds number and the ratio of tube diameter to coil diameter, D/d_c. The product of these two dimensionless numbers is called the *Dean number*, $\mathrm{De} \equiv \mathrm{Re}_D(D/d_c)^{1/2}$.

Three regions can be distinguished [61]: the region of small Dean number, $\mathrm{De} < 20$, in which inertia forces due to secondary flow are negligible; the region of intermediate Dean numbers, $20 < \mathrm{De} < 40$, where inertial forces due to secondary flow balance the viscous forces; and the region of large Dean numbers, $\mathrm{De} > 40$, where viscous forces are significant only in the boundary near the tube wall. While several different investigators have reported different correlations [53] for isothermal friction factors in fully developed coiled-tube swirl flows, the following equation given by Manlapaz and Churchill [62] perhaps provides the most generalized predictions for a wide range of coiled tube geometry and operating conditions that cover all three Dean number flow regions:

$$f = \left(\frac{64}{\mathrm{Re}_D}\right)\left[\left(1 - \frac{0.18}{\left\{1 + (35/\mathrm{He})^2\right\}^{0.5}}\right)^m + \left(1 + \frac{D}{d_c}\right)^2\left(\frac{\mathrm{He}}{88.33}\right)\right]^{0.5} \tag{6.85}$$

where

$$m = \begin{cases} 2 & \mathrm{De} < 20 \\ 1 & 20 < \mathrm{De} < 40, \text{ and } \mathrm{He} = \mathrm{De}\left[1 + (H/\pi d_c)^2\right]^{1/2} \\ 0 & \mathrm{De} > 40 \end{cases}$$

It may be noted here that the helical number (He, defined above, which groups the Dean number De, coil diameter d_c, and coil pitch H) reduces to the Dean number when $H = 0$ or $d_c \rightarrow \infty$, i.e., when a simple curved tube is considered.

Manlapaz and Churchill [62] have also given two separate, but similar, expressions for predicting average Nusselt numbers in fully developed laminar swirl flows in circular-tube coils maintained at the two fundamental thermal boundary conditions. For coils with the tube-wall with uniform wall temperature,

$$\overline{\mathrm{Nu}}_D = \left[\left\{3.657 + \frac{4.343}{\left[1 + (957/\mathrm{Pr}\cdot\mathrm{He}^2)\right]^2}\right\}^3 + 1.158\left\{\frac{\mathrm{He}}{[1 + (0.477/\mathrm{Pr})]}\right\}^{3/2}\right]^{1/3} \tag{6.86}$$

and for the uniform heat flux condition at the tube wall,

$$\overline{\mathrm{Nu}}_D = \left[\left\{4.364 + \frac{4.636}{[1 + (1342/\mathrm{Pr}\cdot\mathrm{He})]^2}\right\}^3 + 1.816\left\{\frac{\mathrm{He}}{[1 + (1.15/\mathrm{Pr})]}\right\}^{3/2}\right]^{1/3} \tag{6.87}$$

Predictions from these equations have been shown to agree with a fairly large data set from different experimental investigations [53].

As in the case of swirl flows generated by twisted-tape inserts, it generally has been found that the flow inside coiled tubes remains in the viscous regime for up to a much higher Reynolds number than that in a straight tube [53, 63]. The swirl or helical vortices tend to suppress the onset of turbulence, thereby delaying transition, and to determine the critical Reynolds number for transition, the following correlation given by Srinivasan, et al. [63] is perhaps the more widely cited:

$$\mathrm{Re}_{D,\,\mathrm{transition}} = 2100\left[1 + 12\sqrt{D/d_c}\right]^2, 10 < \left(d_c/D\right) < \infty \tag{6.88}$$

For predicting the isothermal Fanning friction factors for fully developed turbulent flows in coiled tubes, Mishra and Gupta [64] have developed a correlation by the superposition of swirl-flow effects on straight flows that is given as

$$C_f = \frac{0.079}{\mathrm{Re}_D^{0.25}} + 0.0075\left[\frac{D}{d_c\{1 + (H/\pi d_c)^2\}}\right]^{0.5} \tag{6.89}$$

This equation is valid for $\mathrm{Re}_{D,\mathrm{transition}} < \mathrm{Re}_D < 10^5$, $6.7 < (d_c/D) < 346$, and $0 < (H/d_c) < 25.4$ and has been shown to describe the literature database rather well [53]. For the turbulent flow regime, Mori and Nakayama [65] suggest that the Nusselt number can be correlated for gas flows ($\mathrm{Pr} \approx 1$) as

$$\overline{\mathrm{Nu}}_D = \frac{\mathrm{Pr}}{26.2(\mathrm{Pr}^{2/3} - 0.074)}\mathrm{Re}_D^{4/5}\left(\frac{D}{d_c}\right)^{1/10}\left[1 + 0.098\left\{\mathrm{Re}_D\left(\frac{D}{d_c}\right)^2\right\}^{1/5}\right] \tag{6.90}$$

and for liquid flows ($\mathrm{Pr} > 1$) as

$$\overline{\mathrm{Nu}}_D \frac{\mathrm{Pr}^{0.4}}{41.0}\mathrm{Re}_D^{5/6}\left(\frac{D}{d_c}\right)^{1/12}\left[1 + 0.61\left\{\mathrm{Re}_D\left(\frac{D}{d_c}\right)^{2.5}\right\}^{1/6}\right] \tag{6.91}$$

In general, the gains from enhanced heat transfer by coiling a circular tube are less in turbulent flows when compared to that in the laminar regime.

6.6.2 Forced Convection Cooling of Electronic Devices

Recent advances in the design of integrated circuits (ICs) have resulted in ICs that contain the equivalent of millions of transistors in an area roughly 1 cm square. The large number of circuits in an IC allows designers to build ever-increasing functionality in a very small space. However, since each transistor dissipates electrical power in the form of heat, large-scale integration has resulted in a much larger cooling demand to maintain the ICs at their required operating temperature. Because of the need for improved cooling for such devices, there has been recently great interest in the heat transfer literature on electronic cooling. In this section, we briefly discuss some of the recent advances in this field that involve forced convection inside ducts.

A fairly common method for using ICs in an electronic device is to install an array of several ICs on a printed circuit board (PCB), as shown in Fig. 6.26. Signals from the ICs are routed to the edge of the PCB, where a connector is attached. The PCB then can be plugged into a larger circuit board. In this way, the assembly and repair of a device containing many PCBs is greatly simplified. A good example of this type of arrangement is in a personal computer, where PCBs containing circuitry for disk controllers, memory, video, and so forth are plugged into the main circuit board.

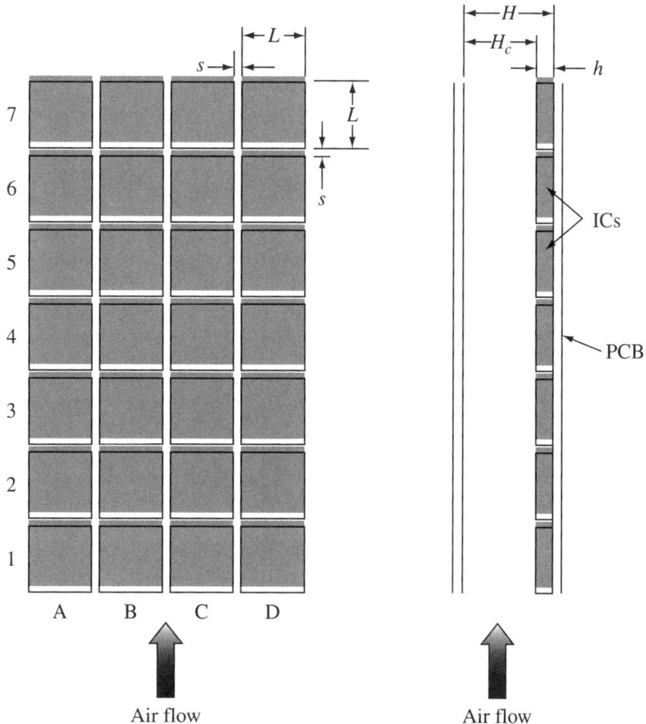

FIGURE 6.26 Fully-populated array of uniform-size modules.

Since the PCBs are mounted in parallel and are fairly close to each other, they form a flow channel through which cool air can be forced. This type of channel flow differs from the channel flow discussed earlier in this chapter in two ways. First, the channel length in the flow direction is fairly small compared to the hydraulic diameter of the flow channel. Thus, entrance effects are important, perhaps more so than in most channel-flow applications. Second, as can be seen in Fig. 6.26, the surface of the PCB is not smooth. One surface of the channel is covered with the ICs that typically are several mm thick and are spaced several mm apart.

Sparrow et al. [66] investigated the forced-convection heat transfer characteristics for this geometry. They studied the heat transfer from an array of 27-mm-square, 10-mm-high ICs mounted on a PCB. The IC array contained 17 ICs in the flow direction and 4 ICs across the flow direction, with 6.7-mm spacing between ICs in the array. Spacing to the adjacent PCB was 17 mm. The experimental results are shown in Fig. 6.27, where the Nusselt number, Nu_L, for each IC is plotted as a function of its row number (location from the entrance of the cooling air flow to the PCB). The length scale in the Nusselt number is the length of the IC, and the Reynolds number is based on spacing, H_c, between the PCBs (see Fig. 6.26). The results clearly show the entrance effect. From the fifth row on, the heat transfer appears to be fully developed. In this fully developed regime, the data were correlated by

$$\mathrm{Nu}_n = C\,\mathrm{Re}_{H_c}^{0.72} \tag{6.92}$$

where $C = 0.093$ in the range $2000 \le \mathrm{Re}_{H_c} \le 7000$
n = row number

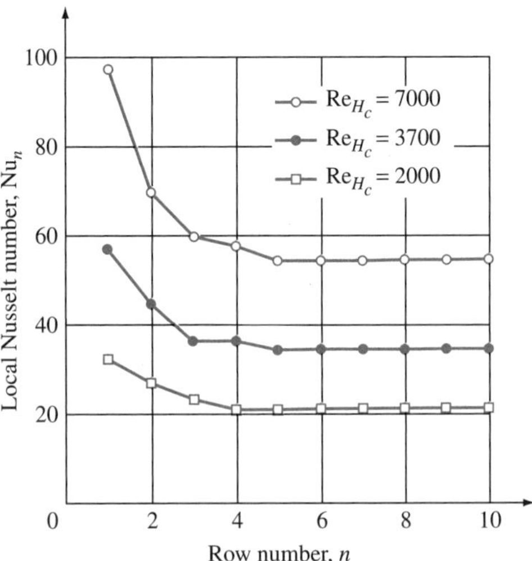

FIGURE 6.27 Local Nusselt number for fully populated array.

Source: Data from Sparrow et al. [66].

In the regime $5000 < \mathrm{Re}_{H_c} < 17{,}000$, the coefficient C in Eq. (6.79) varies with the roughness of the flow channel, expressed by the height of the ICs, h, as shown below [67]:

h (mm)	C
5	0.0571
7.5	0.0503
10	0.0602

In many PCBs, the arrays of ICs are not necessarily made up of identical ICs. They may be of different height, they may be of rectangular shape with various dimensions, and there are likely to be some locations in the array at which no IC is installed. Sparrow et al. [66, 68] examined the effect of a missing IC in an array and the effect of ICs of different height in an irregular array.

Since the purpose of cooling is to ensure that the temperature of an individual IC does not exceed some maximum allowable value, it is important to discuss a complicating factor that affects the individual IC temperatures. Ordinarily in channel flow, we would be able to calculate the local wall temperature according to methods described earlier in the chapter. However, with flow channels comprised of PCBs, some of the cooling flow in the channel can bypass the ICs, resulting in a higher air temperature approaching the ICs than would be predicted from the mean bulk temperature at a given IC row. This effect increases as the PCBs or the ICs on an individual PCB are spaced further apart, because the flow can more easily bypass the ICs. At this time, there are no general correlations that would allow one to predict the correction to the IC temperature, and the designer is advised to use a safety factor to protect the array from overheating.

EXAMPLE 6.7 An array of integrated circuits on a printed circuit board is to be cooled by forced-convection cooling with an airstream at 20°C flowing at a velocity of 1.8 m/s in the channel between adjacent printed circuit boards. The integrated circuits are 27 mm square and 10 mm high, and spacing between the integrated circuits and the adjacent printed circuit board is 17 mm. Determine the heat transfer coefficients for the second and sixth integrated circuits along the flow path.

SOLUTION At 20°C, the properties of air from Table 28, Appendix 2, are $\nu = 15.7 \times 10^{-6}\,\mathrm{m^2/s}$ and $k = 0.0251\,\mathrm{W/m\,K}$. Since the Reynolds number is based on the spacing, H_c, we have

$$\mathrm{Re}_{H_c} = \frac{UH_c}{\nu} = \frac{(1.8\ \mathrm{m/s})(0.017\ \mathrm{m})}{15.7 \times 10^{-6}\ \mathrm{m^2/s}} = 1949$$

From Fig. (6.27), we see that the second integrated circuit is in the inlet region and estimate $\mathrm{Nu}_2 = 29$. This gives

$$h_{c,2} = \frac{\mathrm{Nu}_2 k}{L} = \frac{(29)\left(0.0251\,\dfrac{\mathrm{W}}{\mathrm{m\,K}}\right)}{0.027\,\mathrm{m}} = 27.0\,\frac{\mathrm{W}}{\mathrm{m^2K}}$$

The sixth integrated circuit is in the developed region and from Eq. (6.79)

$$\mathrm{Nu}_6 = 0.093(1949)^{0.72} = 21.7$$

or

$$h_{c,6} = \frac{\mathrm{Nu}_6 k}{L} = \frac{(21.7)(0.0251\,\mathrm{W/m\,K})}{0.027\,\mathrm{m}} = 20.2\,\frac{\mathrm{W}}{\mathrm{m^2K}}$$

6.7 Closing Remarks

In this chapter, we have presented theoretical and empirical correlations that can be used to calculate the Nusselt number, from which the heat transfer coefficient for convection heat transfer to or from a fluid flowing through a duct can be obtained. It cannot be overemphasized that empirical equations derived from experimental data by means of dimensional analysis are applicable only over the range of parameters for which data exist to verify the relation within a specified error band. Serious errors can result if an empirical relation is applied beyond the parameter range over which it has been verified.

When applying an empirical relationship to calculate a convection heat transfer coefficient, the following sequence of steps should be followed:

1. Collect appropriate physical properties for the fluid in the temperature range of interest.
2. Establish the appropriate geometry for the system and the correct significant length for the Reynolds and Nusselt numbers.
3. Determine whether the flow is laminar, turbulent, or transitional by calculating the Reynolds number.
4. Determine whether natural-convection effects may be appreciable by calculating the Grashof number and comparing it with the square of the Reynolds number.
5. Select an appropriate equation that applies to the geometry and flow required. If necessary, iterate initial calculations of dimensionless parameters in accordance with the stipulations of the equation selected.
6. Make an order-of-magnitude estimate of the heat transfer coefficient (see Table 1.4).
7. Calculate the value of the heat transfer coefficient from the equation in step 5 and compare with the estimate in step 6 to spot possible errors in the decimal point or units.

It should be noted that experimental data on which empirical relations are based generally have been obtained under controlled conditions in a laboratory, whereas most practical applications occur under conditions that deviate from laboratory conditions in one way or another. Consequently, the predicted value of a heat transfer coefficient may deviate from the actual value, and since such uncertainties are unavoidable, it is often satisfactory to use a simple correlation, especially for preliminary designs.

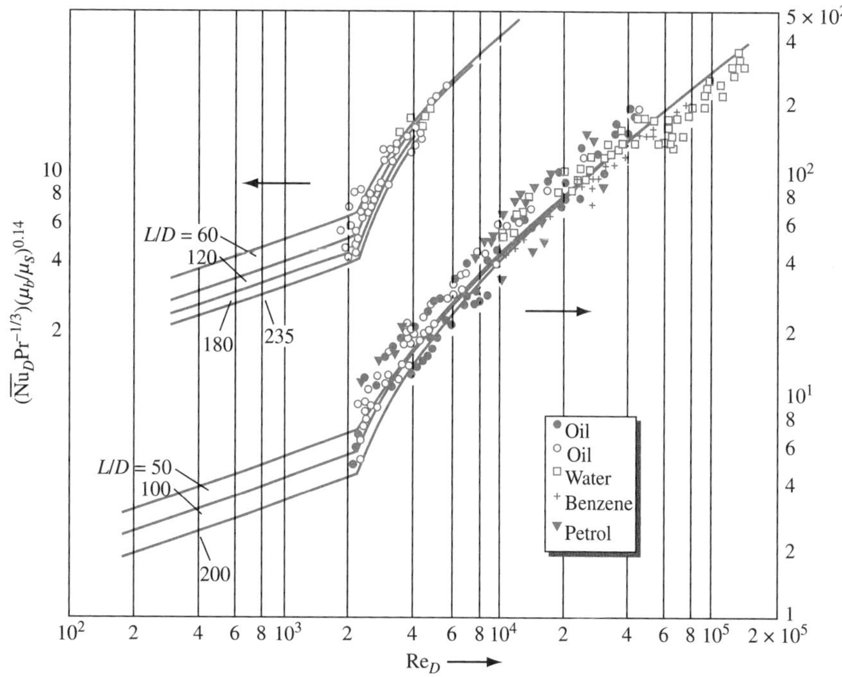

FIGURE 6.28 Recommended correlation curves for heat transfer coefficients in the transition regime.

Source: From E. N. Sieder and C. E. Tate [16], with permission of the copyright owner, the American Chemical Society.

A special note of caution is in order for the transition regime. The mechanisms of heat transfer and fluid flow in the transition region, (Re_D between 2100 and 6000) vary considerably from system to system. In this region, the flow may be unstable, and fluctuations in pressure drop and heat transfer have been observed. There is therefore a large uncertainty in the basic heat transfer and flow-friction performance, and consequently, the designer is advised to design equipment to operate outside this region, if possible; the curves of Fig. 6.28 can be used, but the actual performance may deviate considerably from that predicted on the basis of these curves. Often, instead of estimating the transition Reynolds number, the current practice is to simply use the Gnielinski correlation given by Eq. (6.65) for $\mathrm{Re}_D > 2300$ with the caveat that there always will be some uncertainty in the transition region.

To aid in the rapid selection of an appropriate relation for obtaining the heat transfer coefficient for flow in a duct, some of the most commonly used empirical equations are summarized in Table 6.4. A more complete summary of equations can be found in [25, 68, and 69].

TABLE 6.4 Summary of forced convection correlations for incompressible flow inside tubes and ducts[a,b,c]

System Description	Recommended Correlation	Equation in Text
Friction factor for laminar flow in long tubes and ducts	Liquids: $f = (64/\mathrm{Re}_D)(\mu_s/\mu_b)^{0.14}$ Gases: $f = (64/\mathrm{Re}_D)(T_s/T_b)^{0.14})$	(6.44) (6.45)
Nusselt number for fully developed laminar flow in long tubes with uniform heat flux, $\mathrm{Pr} > 0.6$	$\overline{\mathrm{Nu}}_D = 4.36$	(6.31)
Nusselt number for fully developed laminar flow in long tubes with uniform wall temperature, $\mathrm{Pr} > 0.6$	$\overline{\mathrm{Nu}}_D = 3.36$	(6.32)
Average Nusselt number for laminar flow in tubes and ducts of intermediate length with uniform wall temperature, $(\mathrm{Re}_{D_H}\mathrm{Pr}D_H/L)^{0.33}(\mu_b/\mu_s)^{0.14} > 2$, $0.004 < (\mu_b/\mu_s) < 10$, and $0.5 < \mathrm{Pr} < 16{,}000$	$\overline{\mathrm{Nu}}_{D_H} = 1.86(\mathrm{Re}_{D_H}\mathrm{Pr}D_H/L)^{0.33}(\mu_b/\mu_s)^{0.14}$	(6.42)
Average Nusselt number for laminar flow in short tubes and ducts with uniform wall temperature, $100 < (\mathrm{Re}_{D_H}\mathrm{Pr}D_H/L) < 1500$ and $\mathrm{Pr} < 0.7$	$\overline{\mathrm{Nu}}_{D_H} = 3.66 + \dfrac{0.0668\mathrm{Re}_{D_H}\mathrm{Pr}D/L}{1 + 0.045(\mathrm{Re}_{D_H}\mathrm{Pr}D/L)^{0.66}}\left(\dfrac{\mu_b}{\mu_s}\right)^{0.14}$	(6.41)
Friction factor for fully developed turbulent flow through smooth, long tubes and ducts	$f = 0.184/\mathrm{Re}_{D_H}^{0.2}(10{,}000 < \mathrm{Re}_{D_H} < 10^6)$	(6.56)
Average Nusselt number for fully developed turbulent flow through smooth, long tubes and ducts, $6000 < \mathrm{Re}_{D_H} < 10^7$, $0.7 < \mathrm{Pr} < 10{,}000$, and $L/D_H > 60$	$\overline{\mathrm{Nu}}_{D_H} = 0.027\ \mathrm{Re}_{D_H}^{0.8}\mathrm{Pr}^{1/3}(\mu_b/\mu_s)^{0.14}$ or Table 6.3 or the Gnielinski correlation, Eq. (6.65) for $\mathrm{Re}_D > 2300$	(6.61) (6.63)
Average Nusselt number for liquid metals in turbulent, fully developed flow through smooth tubes with uniform heat flux, $100 < \mathrm{Re}_D\mathrm{Pr} < 10^4$ and $L/D > 30$	$\overline{\mathrm{Nu}}_D = 4.82 + 0.0185\ (\mathrm{Re}_D\mathrm{Pr})^{0.827}$	(6.68)
Same as above, but in thermal entry region when $\mathrm{Re}_D\ \mathrm{Pr} < 100$	$\overline{\mathrm{Nu}}_D = 3.0\mathrm{Re}_D^{0.0833}$	(6.69)
Average Nusselt number for liquid metals in turbulent fully developed flow through smooth tubes with uniform surface temperature, $\mathrm{Re}_D\mathrm{Pr} > 100$ and $L/D > 30$	$\overline{\mathrm{Nu}}_D = 5.0 + 0.025(\mathrm{Re}_D\mathrm{Pr})^{0.8}$	(6.70)

[a]All physical properties in the correlations are evaluated at the bulk temperature T_b except μ_s, which is evaluated at the surface temperature T_s.

[b]$\mathrm{Re}_{D_H} = D_H\bar{U}\rho/\mu$, $D_H = 4A_c/P$, and $\bar{U} = \dot{m}/\rho A_c$.

[c]Incompressible flow correlations apply when average velocity is less than half the speed of sound (Mach number <0.5) to gases and vapors.

References

1. R. H. Notter and C. A. Sleicher, "The Eddy Diffusivity in the Turbulent Boundary Layer near a Wall," *Eng. Sci.*, vol. 26, pp. 161–171, 1971.
2. R. C. Martinelli, "Heat Transfer to Molten Metals," *Trans. ASME*, vol. 69, p. 947, 1947.
3. H. L. Langhaar, "Steady Flow in the Transition Length of a Straight Tube," *J. Appl. Mech.*, vol. 9, pp. 55–58, 1942.
4. W. M. Kays and M. E. Crawford, *Convective Heat and Mass Transfer*, 2d ed., McGraw-Hill, New York, 1980.

5. O. E. Dwyer, "Liquid-Metal Heat Transfer," chapter 5 in Sodium and NaK Supplement to *Liquid Metals Handbook*, Atomic Energy Commission, Washington, D.C., 1970.
6. J. R. Sellars, M. Tribus, and J. S. Klein, "Heat Transfer to Laminar Flow in a Round Tube or Flat Conduit—the Graetz Problem Extended," *Trans. ASME*, vol. 78, pp. 441–448, 1956.
7. C. A. Schleicher and M. Tribus, "Heat Transfer in a Pipe with Turbulent Flow and Arbitrary Wall-Temperature Distribution," *Trans. ASME*, vol. 79, pp. 789–797, 1957.
8. E. R. G. Eckert, "Engineering Relations for Heat Transfer and Friction in High Velocity Laminar and Turbulent Boundary Layer Flow over Surfaces with Constant Pressure and Temperature," *Trans ASME*, vol. 78, pp. 1273–1284, 1956.
9. W. D. Hayes and R. F. Probstein, *Hypersonic Flow Theory*, Academic Press, New York, 1959.
10. F. Kreith, *Principles of Heat Transfer*, 2d ed., chap. 12, International Textbook Co., Scranton, Pa., 1965.
11. W. M. Kays and K. R. Perkins, "Forced Convection, Internal Flow in Ducts," in *Handbook of Heat Transfer Applications*, W. R. Rohsenow, J. P. Hartnett, and E. N. Ganic, eds., vol. 1, chap. 7, McGraw-Hill, New York, 1985.
12. W. M. Kays, "Numerical Solution for Laminar Flow Heat Transfer in Circular Tubes," *Trans. ASME*, vol. 77, pp. 1265–1274, 1955.
13. R. K. Shah and A. L. London, *Laminar Flow Forced Convection in Ducts*, Academic Press, New York, 1978.
14. R. G. Eckert and A. J. Diaguila, "Convective Heat Transfer for Mixed Free and Forced Flow through Tubes," *Trans. ASME*, vol. 76, pp. 497–504, 1954.
15. B. Metais and E. R. G. Eckert, "Forced, Free, and Mixed Convection Regimes," *Trans. ASME. Ser. C. J. Heat Transfer*, vol. 86, pp. 295–296, 1964.
16. E. N. Sieder and C. E. Tate, "Heat Transfer and Pressure Drop of Liquids in Tubes," *Ind. Eng. Chem.*, vol. 28, p. 1429, 1936.
17. W. M. Kays and A. L. London, *Compact Heat Exchangers*, 3rd ed., McGraw-Hill, New York, 1984.
18. H. Hausen, *Heat Transfer in Counter Flow, Parallel Flow and Cross Flow*, McGraw-Hill, New York, 1983.
19. S. Whitaker, "Forced Convection Heat Transfer Correlations for Flow in Pipes, Past Flat Plates, Single Cylinders, and for Flow in Packed Beds and Tube Bundles," *AIChE J.*, vol. 18, pp. 361–371, 1972.
20. T. W. Swearingen and D. M. McEligot, "Internal Laminar Heat Transfer with Gas-Property Variation," *Trans. ASME, Ser. C. J. Heat Transfer*, vol. 93, pp. 432–440, 1971.
21. "Engineering Sciences Data," Heat Transfer Subsciences, Technical Editing and Production Ltd., London, 1970.
22. W. M. McAdams, *Heat Transmission*, 3d ed., McGraw-Hill, New York, 1954.
23. C. A. Depew and S. E. August, "Heat Transfer due to Combined Free and Forced Convection in a Horizontal and Isothermal Tube," *Trans. ASME. Ser. C. J. Heat Transfer*, vol. 93, pp. 380–384, 1971.
24. B. Metais and E. R. G. Eckert, "Forced, Free, and Mixed Convection Regimes," *Trans. ASME, Ser. C. J. Heat Transfer*, vol. 86, pp. 295–296, 1964.
25. W. M. Rohsenow, J. P. Hartnett, and Y. I. Cho, eds., *Handbook of Heat Transfer*, McGraw-Hill, New York, 1998.
26. O. Reynolds, "On the Extent and Action of the Heating Surface for Steam Boilers," *Proc. Manchester Lit. Philos. Soc.*, vol. 8, 1874.
27. L. F. Moody, "Friction Factor for Pipe Flow," *Trans. ASME*, vol. 66, 1944.
28. W. F. Cope, "The Friction and Heat Transmission Coefficients of Rough Pipes," *Proc. Inst. Mech. Eng.*, vol. 145, p. 99, 1941.
29. W. Nunner, "Wärmeübergang and Druckabfall in Rauhen Rohren," *VDI Forschungsh.*, no. 455, 1956.
30. D. F. Dipprey and R. H. Sabersky, "Heat and Momentum Transfer in Smooth and Rough Tubes at Various Prandtl Numbers," *Int. J. Heat Mass Transfer*, vol. 5, pp. 329–353, 1963.
31. L. Prandtl, "Eine Beziehung zwischen Wärmeaustausch und Strömungswiederstand der Flüssigkeiten," *Phys. Z.*, vol. 11, p. 1072, 1910.
32. T. von. Karman, "The Analogy between Fluid Friction and Heat Transfer," *Trans. ASME*, vol. 61, p. 705, 1939.
33. L. M. K. Boelter, R. C. Martinelli, and F. Jonassen, "Remarks on the Analogy between Heat and Momentum Transfer," *Trans. ASME*, vol. 63, pp. 447–455, 1941.
34. R. G. Deissler, "Investigation of Turbulent Flow and Heat Transfer in Smooth Tubes Including the Effect of Variable Properties," *Trans. ASME*, vol. 73, p. 101, 1951.
35. F. W. Dittus and L. M. K. Boelter, *Univ. Calif. Berkeley Publ. Eng.*, vol. 2, p. 433, 1930.
36. B. S. Petukhov, "Heat Transfer and Friction in Turbulent Pipe Flow with Variable Properties," *Adv. Heat Transfer*, vol. 6, Academic Press, New York, pp. 503–564, 1970.
37. C. A. Sleicher and M. W. Rouse, "A Convenient Correlation for Heat Transfer to Constant and Variable Property Fluids in Turbulent Pipe Flow," *Int. J. Heat Mass Transfer*, vol. 18, pp. 677–683, 1975.
38. J. J. Lorentz, D. T. Yung, C. B. Parchal, and G. E. Layton, "An Assessment of Heat Transfer Correlations for

Turbulent Water Flow through a Pipe at Prandtl Numbers of 6.0 and 11.6," ANL/OTEC-PS-11, Argonne Natl. Lab., Argonne, Ill. January 1982.

39. W. M. McAdams, *Heat Transmission*, 3d ed., McGraw-Hill, New York, 1954.
40. J. P. Hartnett, "Experimental Determination of the Thermal Entrance Length for the Flow of Water and of Oil in Circular Pipes," *Trans. ASME*, vol. 77, pp. 1211–1234, 1955.
41. R. G. Deissler, "Turbulent Heat Transfer and Friction in the Entrance Regions of Smooth Passages," *Trans. ASME*, vol. 77, pp. 1221–1234, 1955.
42. V. Gnielinski, "New Equations for Heat and Mass Transfer in Turbulent Pipe and Channel Flow," *International Chemical Engineering*, vol. 16, no. 2, pp. 359–368, 1976; originally appeared in German in *Forschung im Ingenieurwesen*, vol. 41, no. 1, pp. 8–16, 1975.
43. M. S. Bhatti and R. K. Shah, "Turbulent and Transition Flow Convective Heat Transfer in Ducts," in *Handbook of Single-Phase Convective Heat Transfer*, S. Kakaç, R. K. Shah, and W. Aung, eds., Wiley, New York, 1987.
44. R. M. Manglik and A. E. Bergles, "Experimental Investigation of Turbulent Flow Heat Transfer in Horizontal Concentric Annular Ducts," *Experimental Heat Transfer, Fluid Mechanics and Thermodynamics 1997*, M. Giot, F. Mayinger, and G. P. Celata, eds., Edzioni ETS, Pisa, Italy, vol. 3, pp. 1393–1400, 1997.
45. B. S. Petukhov and L. I. Roizen, "Generalized Dependence for Heat Transfer in Tubes of Annular Cross Section," *High Temperature*, vol. 12, pp. 485–489, 1974.
46. B. Lubarsky and S. J. Kaufman, "Review of Experimental Investigations of Liquid-Metal Heat Transfer," NACA TN 3336, 1955.
47. E. Skupinski, J. Tortel, and L. Vautrey, "Determination des Coefficients de Convection d'un Alliage Sodium-Potassium dans un Tube Circulative," *Int. J. Heat Mass Transfer*, vol. 8, pp. 937–951, 1965.
48. S. Lee, "Liquid Metal Heat Transfer in Turbulent Pipe Flow with Uniform Wall Flux," *Int. J. Heat Mass Transfer*, vol. 26, pp. 349–356, 1983.
49. R. P. Stein, "Heat Transfer in Liquid Metals," in *Advances in Heat Transfer*, J. P. Hartnett and T. F. Irvine, eds., vol. 3, Academic Press, New York, 1966.
50. N. Z. Azer, "Thermal Entry Length for Turbulent Flow of Liquid Metals in Pipes with Constant Wall Heat Flux," *Trans. ASME, Ser. C, J. Heat Transfer*, vol. 90, pp. 483–485, 1968.
51. R. A. Seban and T. T. Shimazaki, "Heat Transfer to Fluid Flowing Turbulently in a Smooth Pipe with Walls at Constant Temperature," *Trans. ASME*, vol. 73, pp. 803–807, 1951.
52. A. E. Bergles, "Techniques to Enhance Heat Transfer," in *Handbook of Heat Transfer*, 3rd ed., W. M. Rohsenow, J. P. Hartnett and Y. I. Cho, eds., McGraw-Hill, New York, NY, ch. 11, 1998.
53. R. M. Manglik, "Heat Transfer Enhancement," in *Heat Transfer Handbook*, A. Bejan and A. D. Kraus, eds., Wiley, Hoboken, NJ, 2003.
54. R. L. Webb and N.-H. Kim, *Principles of Enhanced Heat Transfer*, 2nd ed., Taylor & Francis, Boca Raton, FL, 2005.
55. A. P. Watkinson, D. C., Miletti, and G. R., Kubanek, "Heat Transfer and Pressure Drop of Internally Finned Tubes in Laminar Oil Flows," Paper no. 75-HT-41, ASME, New York, 1975.
56. T. C. Carnavos, "Cooling Air in Turbulent Flow with Internally Finned Tubes," *Heat Transfer Eng.*, vol. 1, pp. 43–46, 1979.
57. R. M. Manglik and A. E. Bergles, "Swirl Flow Heat Transfer and Pressure Drop with Twisted-Tape Inserts, *Advances in Heat Transfer*, vol. 36, pp. 183–266, Academic Press, New York, 2002.
58. R. M. Manglik and A. E. Bergles, "Heat Transfer and Pressure Drop Correlations for Twisted-Tape Inserts in Isothermal Tubes: Part I—Laminar Flows," *Journal of Heat Transfer*, vol. 115, no. 4, pp. 881–889, 1993.
59. R. M. Manglik, S. Maramraju, and A. E. Bergles, "The Scaling and Correlation of Low Reynolds Number Swirl Flows and Friction Factors in Circular Tubes with Twisted-Tape Inserts," *Journal of Enhanced Heat Transfer*, vol. 8, no. 6, pp. 383–395, 2001.
60. R. M. Manglik and A. E. Bergles, "Heat Transfer and Pressure Drop Correlations for Twisted-Tape Inserts in Isothermal Tubes: Part II—Transition and Turbulent Flows," *Journal of Heat Transfer*, vol. 115, no. 4, pp. 890–896, 1993.
61. L. A. M. Janssen and C. J. Hoogendoorn, "Laminar Convective Heat Transfer in Helically Coiled Tubes," *Int. J. Heat Mass Transfer*, vol. 21, pp. 1197–1206, 1978.
62. R. L. Manlapaz and S. W. Churchill, "Fully Developed Laminar Flow in a Helically Coiled Tube of Finite Pitch," *Chemical Engineering Communications*, vol. 7, pp. 57–78, 1980.
63. P. S. Srinivasan, S. S., Nandapurkar, and F. A. Holland, "Pressure Drop and Heat Transfer in Coils," *The Chemical Engineer*, no. 218, pp. 113–119, May 1968.
64. P. Mishra and S. N. Gupta, "Momentum Transfer in Curved Pipes, 1. Newtonian Fluids; 2. Non-Newtonian Fluids," *Industrial and Engineering Chemistry, Process Design and Development*, vol. 18, pp. 130–142, 1979.

65. Y. Mori and W. Nakayama, "Study on Forced Convective Heat Transfer in Curved Pipes (3rd Report, Theoretical Analysis Under the Condition of Uniform Wall Temperature and Practical Formulae)," *International Journal of Heat and Mass Transfer*, vol. 10, pp. 681–695, 1967.
66. E. M. Sparrow, J. E. Niethamer, and A. Chaboki, "Heat Transfer and Pressure Drop Characteristics of Arrays of Rectangular Modules in Electronic Equipment," *Int. J. Heat Mass Transfer*, vol. 25, pp. 961–973, 1982.
67. V. W. Antonetti, "Cooling Electronic Equipment," sec. 517 in *Heat Transfer and Fluid Flow Data Books*, F. Kreith, ed., Genium Publ. Co., Schenectady, N.Y., 1992.
68. *International Encyclopedia of Heat and Mass Transfer*, G. F. Hewitt, G. L. Shires, and Y. V. Polezhaev, eds., CRC Press, Boca Raton, FL, 1997.
69. F. Kreith, ed., *CRC Handbook of Thermal Engineering*, CRC Press, Boca Raton, FL, 2000.

Problems

The problems for this chapter are organized by subject matter as shown below.

Topic	Problem Number
Laminar, fully-developed flow	6.1–6.5
Laminar, entrance region	6.6–6.10
Turbulent, fully-developed flow	6.11–6.22
Turbulent, entrance region	6.23–6.28
Mixed convection	6.29–6.30
Liquid metals	6.31–6.34
Combined heat transfer mechanisms	6.35–6.43
Analysis problems	6.44–6.49

6.1 To measure the mass flow rate of a fluid in a laminar flow through a circular pipe, a hot-wire-type velocity meter is placed in the center of the pipe. Assuming that the measuring station is far from the entrance of the pipe, the velocity distribution is parabolic:

$$u(r)/U_{max} = [1 - (2r/D)^2]$$

where U_{max} is the centerline velocity ($r = 0$), r is the radial distance from the pipe centerline, and D is the pipe diameter.

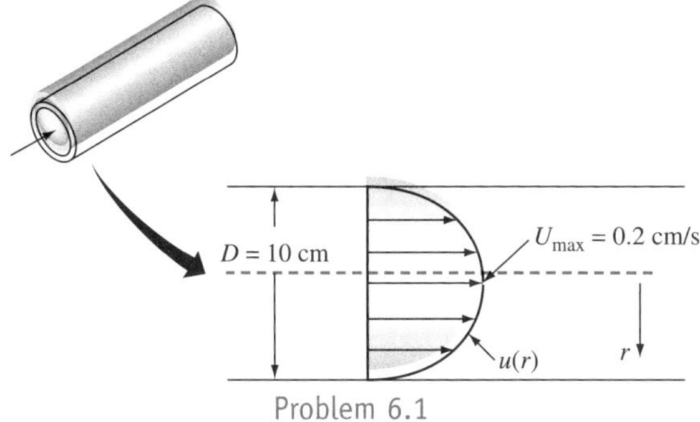

Problem 6.1

(a) Derive an expression for the average fluid velocity at the cross section in terms of U_{max} and D. (b) Obtain an expression for the mass flow rate. (c) If the fluid is mercury at 30°C, D = 10 cm, and the measured value of U_{max} is 0.2 cm/s, calculate the mass flow rate from the measurement.

6.2 Nitrogen at 30°C and atmospheric pressure enters a triangular duct 0.02 m on each side at a rate of 4×10^{-4} kg/s. If the duct temperature is uniform at 200°C, estimate the bulk temperature of the nitrogen 2 m and 5 m from the inlet.

6.3 Air at 30°C enters a rectangular duct 1 m long and 4 mm by 16 mm in cross section at a rate of 0.0004 kg/s. If a uniform heat flux of 500 W/m^2 is imposed on both of the long sides of the duct, calculate (a) the air outlet temperature, (b) the average duct surface temperature, and (c) the pressure drop.

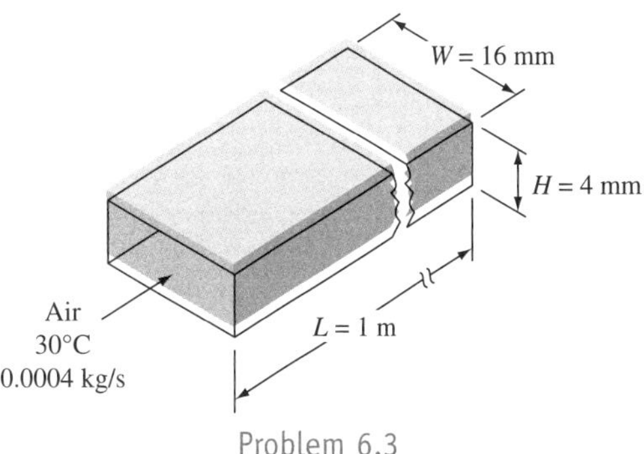

Problem 6.3

6.4 Engine oil flows at a rate of 0.5 kg/s through a 2.5-cm-ID tube. The oil enters at 25°C while the tube wall is at 100°C. (a) If the tube is 4 m long, determine whether the flow is fully developed. (b) Calculate the heat transfer coefficient.

6.5 The equation:

$$\overline{\mathrm{Nu}} = \frac{\bar{h}_c D}{k} = \left[3.65 + \frac{0.668(D/L)\mathrm{Re}\mathrm{Pr}}{1 + 0.04[(D/L)\mathrm{Re}\mathrm{Pr}]^{2/3}}\right]\left(\frac{\mu_b}{\mu_s}\right)^{0.14}$$

was recommended by H. Hausen (*Zeitschr. Ver. Deut. Ing., Beiheft*, No. 4, 1943) for forced-convection heat transfer in fully developed laminar flow through tubes. Compare the values of the Nusselt number predicted by Hausen's equation for Re = 1000, Pr = 1, and $L/D = 2$, 10, and 100 with those obtained from two other appropriate equations or graphs in the text.

6.6 Air at an average temperature of 150°C flows through a short, square duct 10 × 10 × 2.25 cm at a rate of 15 kg/h, as shown in the sketch below. The duct wall temperature is 430°C. Determine the average heat transfer coefficient using the duct equation with appropriate L/D correction. Compare your results with flow-over-flat-plate relations.

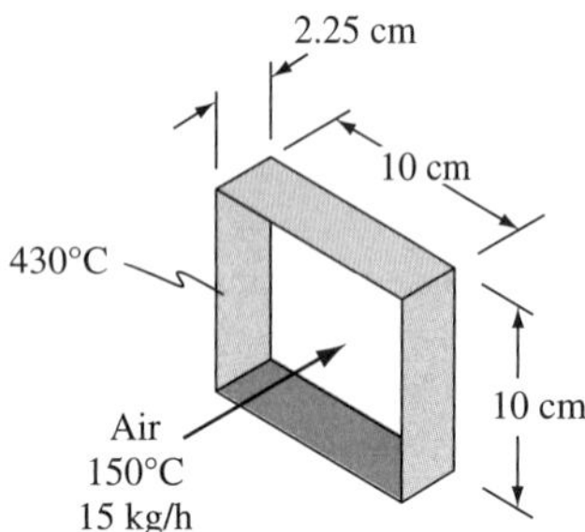

Problem 6.6

6.7 Water enters a double-pipe heat exchanger at 60°C. The water flows on the inside through a copper tube of 2.54-cm-ID at an average velocity of 2 cm/s. Steam flows in the annulus and condenses on the outside of the copper tube at a temperature of 80°C. Calculate the outlet temperature of the water if the heat exchanger is 3 m long.

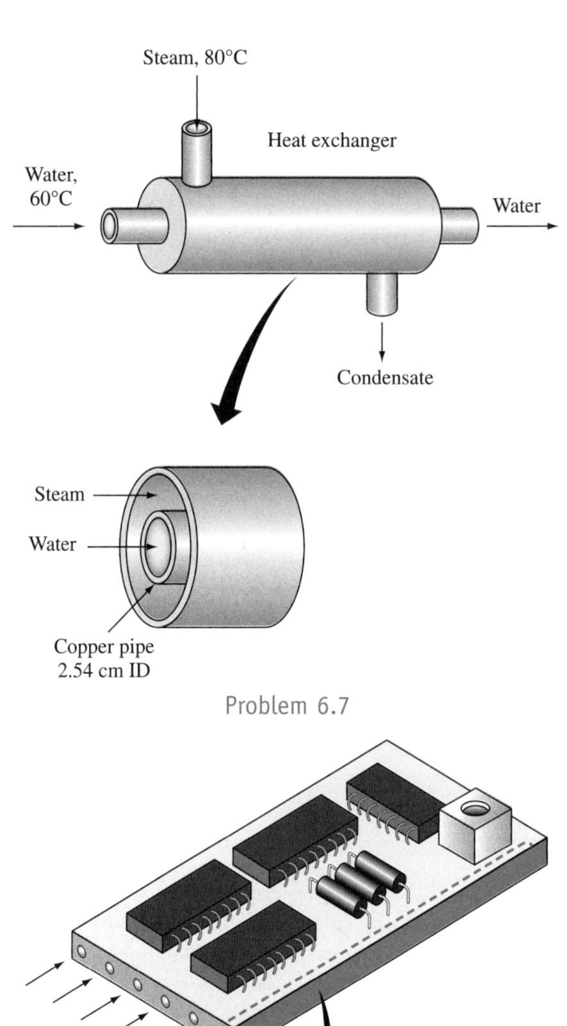

Problem 6.7

6.8 An electronic device is cooled by passing air at 27°C through six small tubular passages drilled through the bottom of the device in parallel as shown. The mass flow rate per tube is 7×10^{-5} kg/s. Heat is generated in the device, resulting in approximately uniform heat flux to the air in the cooling passage. To determine the heat flux, the air-outlet temperature is measured and found to be 77°C. Calculate the rate of heat generation, the average heat transfer coefficient, and the surface temperature of the cooling channel at the center and at the outlet.

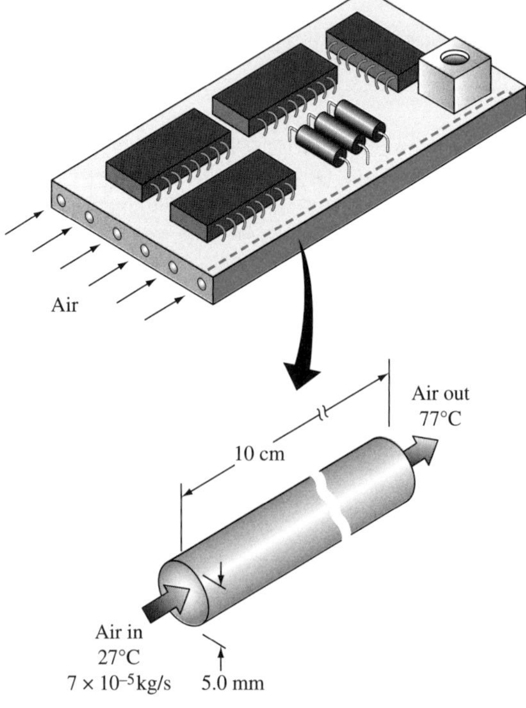

Problem 6.8

6.9 Unused engine oil with a 100°C inlet temperature flows at a rate of 250 g/sec through a 5.1-cm-ID pipe that is enclosed by a jacket containing condensing steam at 150°C. If the pipe is 9 m long, determine the outlet temperature of the oil.

6.10 Determine the rate of heat transfer per foot length to a light oil flowing through a 1-in.-ID, 2-ft-long copper tube at a velocity of 6 fpm. The oil enters the tube at 60°F, and the tube is heated by steam condensing on its outer surface at atmospheric pressure with a heat transfer coefficient of 2000 Btu/h ft^2 °F. The properties of the oil at various temperatures are listed in the following table:

	Temperature, T(°F)				
	60	80	100	150	212
ρ(lb/ft^3)	57	57	56	55	54
c(Btu/lb °F)	0.43	0.44	0.46	0.48	0.51
k(Btu/h ft °F)	0.077	0.077	0.076	0.075	0.074
μ(lb/h ft)	215	100	55	19	8
Pr	1210	577	330	116	55

6.11 Calculate the Nusselt number and the convection heat transfer coefficient by three different methods for water at a bulk temperature of 32°C flowing at a velocity of 1.5 m/s through a 2.54-cm-ID duct with a wall temperature of 43°C. Compare the results.

6.12 Atmospheric pressure air is heated in a long annulus (25-cm-ID, 38-cm-OD) by steam condensing at 149°C on the inner surface. If the velocity of the air is 6 m/s and its bulk temperature is 38°C, calculate the heat transfer coefficient.

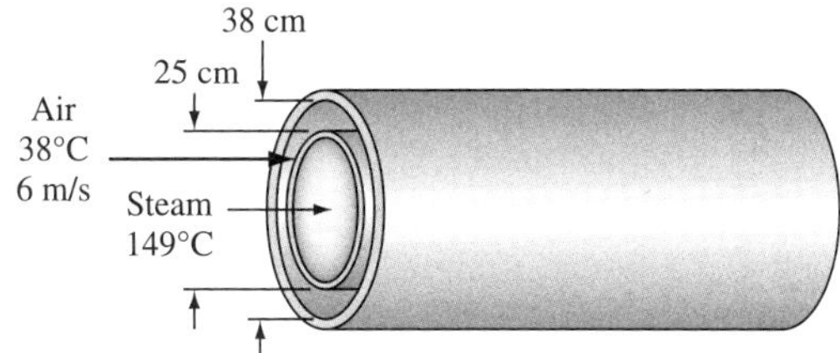

Problem 6.12

6.13 If the total resistance between the steam and the air (including the pipe wall and scale on the steam side) In Problem 6.12 is 0.05 m^2 K/W, calculate the temperature difference between the outer surface of the inner pipe and the air. Show the thermal circuit.

6.14 Atmospheric air at a velocity of 61 m/s and a temperature of 16°C enters a 0.61-m-long square metal duct of 20-cm × 20-cm cross section. If the duct wall is at 149°C, determine the average heat transfer coefficient. Comment briefly on the L/D_h effect.

6.15 Compute the average heat transfer coefficient h_c for 10°C water flowing at 4 m/s in a long, 2.5-cm-ID pipe (surface temperature 40°C) using three different equations. Compare your results. Also determine the pressure drop per meter length of pipe.

6.16 Water at 80°C is flowing through a thin copper tube (15.2-cm-ID) at a velocity of 7.6 m/s. The duct is located in a room at 15°C, and the heat transfer coefficient at the outer surface of the duct is 14.1 W/m^2 K. (a) Determine the heat transfer coefficient at the inner surface. (b) Estimate the length of duct in which the water temperature drops 1°C.

6.17 Mercury at an inlet bulk temperature of 90°C flows through a 1.2-cm-ID tube at a flow rate of 4535 kg/h. This tube is part of a nuclear reactor in which heat can be generated uniformly at any desired rate by adjusting the neutron flux level. Determine the length of tube required to raise the bulk temperature of the mercury to 230°C without generating any mercury vapor, and determine the corresponding heat flux. The boiling point of mercury is 355°C.

6.18 Exhaust gases having properties similar to dry air enter a thin-walled cylindrical exhaust stack at 800 K. The stack is made of steel and is 8 m tall with a 0.5-m inside diameter. If the gas flow rate is 0.5 kg/s and the heat transfer coefficient at the outer surface is 16 W/m^2 K, estimate the outlet temperature of the exhaust gas if the ambient temperature is 280 K.

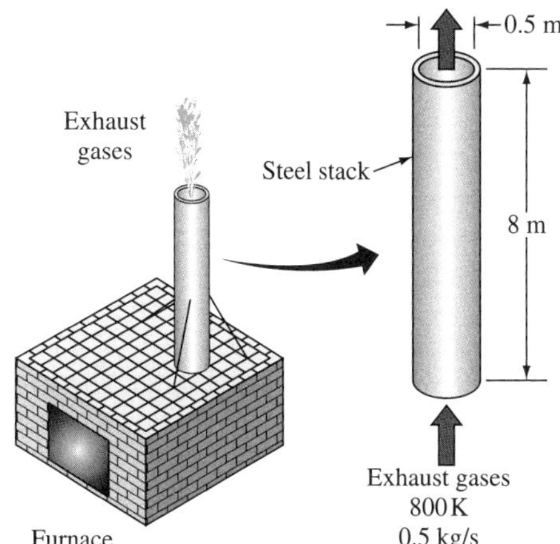

Problem 6.18

6.19 Water at an average temperature of 27°C is flowing through a smooth 5.08-cm-ID pipe at a velocity of 0.91 m/s. If the temperature at the inner surface of the pipe is 49°C, determine (a) the heat transfer coefficient, (b) the rate of heat flow per meter of pipe, (c) the bulk temperature rise per meter, and (d) the pressure drop per meter.

6.20 An aniline-alcohol solution is flowing at a velocity of 10 fps through a long, 1-in.-ID thin-wall tube. Steam is condensing at atmospheric pressure on the outer surface of the tube, and the tube wall temperature is 212°F. The tube is clean, and there is no thermal resistance from scale deposits on the inner surface. Using the physical properties tabulated below, estimate the heat transfer coefficient between the fluid and the pipe using Eqs. (6.60) and (6.61) and compare the results. Assume that the bulk temperature of the aniline solution is 68°F, and neglect entrance effects.

Temperature (°F)	Viscosity (centipoise)	Thermal Conductivity (Btu/h ft °F)	Specific Gravity	Specific Heat (Btu/lb °F)
68	5.1	0.100	1.03	0.50
140	1.4	0.098	0.98	0.53
212	0.6	0.095		0.56

6.21 Brine (10% NaCl by weight) having a viscosity of 0.0016 N s/m^2 and a thermal conductivity of 0.85 W/m K is flowing through a long, 2.5-cm-ID pipe in a refrigeration system at 6.1 m/s. Under these conditions, the heat transfer coefficient was found to be 16,500 W/m^2 K. For a brine temperature of -1°C and a pipe temperature of 18.3°C, determine the temperature rise of the brine per meter length of pipe if the velocity of the brine is doubled. Assume that the specific heat of the brine is 3768 J/kg K and that its density is equal to that of water.

6.22 Derive an equation of the form $h_c = f(T, D, U)$ for the turbulent flow of water through a long tube in the temperature range between 20° and 100°C.

6.23 The intake manifold of an automobile engine can be approximated as a 4-cm-ID tube, 30 cm in length. Air at a bulk temperature of 20°C enters the manifold at a flow rate of 0.01 kg/s. The manifold is a heavy aluminum casting and is at a uniform temperature of 40°C. Determine the temperature of the air at the end of the manifold.

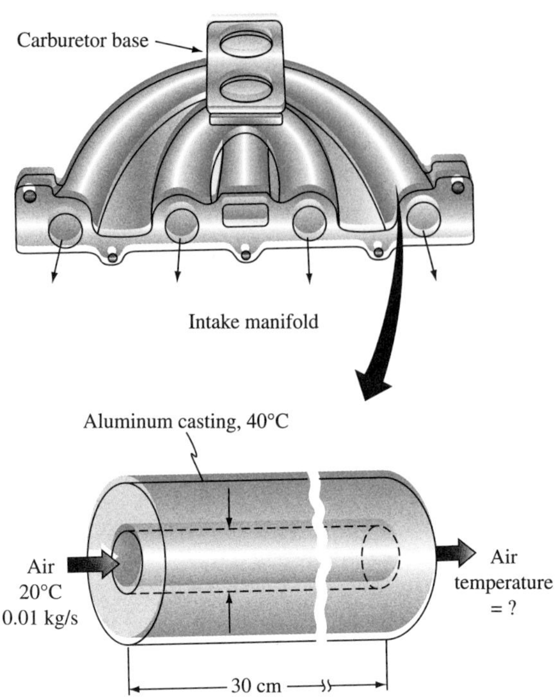

Problem 6.23

6.24 High-pressure water at a bulk inlet temperature of 93°C is flowing with a velocity of 1.5 m/s through a 0.015-m-diameter tube, 0.3 m long. If the tube wall temperature is 204°C, determine the average heat transfer coefficient and estimate the bulk temperature rise of the water.

6.25 Suppose an engineer suggests that air instead of water could flow through the tube in Problem 6.24 and that the velocity of the air could be increased until the heat transfer coefficient with the air equals that obtained with water at 1.5 m/s. Determine the velocity required and comment on the feasibility of the engineer's suggestion. Note that the speed of sound in air at 100°C is 387 m/s.

6.26 Atmospheric air at 10°C enters a 2-m-long smooth, rectangular duct with a 7.5-cm × 15-cm cross section. The mass flow rate of the air is 0.1 kg/s. If the sides are at 150°C, estimate (a) the heat transfer coefficient, (b) the air outlet temperature, (c) the rate of heat transfer, and (d) the pressure drop.

6.27 Air at 16°C and atmospheric pressure enters a 1.25-cm-ID tube at 30 m/s. For an average wall temperature of 100°C, determine the discharge temperature of the air and the pressure drop if the pipe is (a) 10 cm long, (b) 102 cm long.

6.28 The equation

$$\overline{\text{Nu}} = 0.116(\text{Re}^{2/3} - 125)\text{Pr}^{1/3}\left[1 + \left(\frac{D}{L}\right)^{2/3}\right]\left(\frac{\mu_b}{\mu_s}\right)^{0.14}$$

has been proposed by Hausen for the transition range ($2300 < \text{Re} < 8000$) as well as for higher Reynolds numbers. Compare the values of Nu predicted by Hausen's equation for Re = 3000 and Re = 20,000 at $D/L = 0.1$ and 0.01 with those obtained from appropriate equations or charts in the text. Assume the fluid is water at 15°C flowing through a pipe at 100°C.

6.29 Water at 20°C enters a 1.91-cm-ID, 57-cm-long tube at a flow rate of 3 gm/s. The tube wall is maintained at 30°C. Determine the water outlet temperature. What percent error in the water temperature results if natural convection effects are neglected?

6.30 A solar thermal central receiver generates heat by using a field of mirrors to focus sunlight on a bank of tubes through which a coolant flows. Solar energy absorbed by the tubes is transferred to the coolant, which can then deliver useful heat to a load. Consider a receiver fabricated from multiple horizontal tubes in parallel. Each tube is 1-cm-ID and 1 m long. The coolant is molten salt that enters the tubes at 370°C. Under start-up conditions, the salt flow is 10 gm/s in each tube and the net solar flux absorbed by the tubes is 10^4 W/m^2. The tube-wall will tolerate temperatures up to 600°C. Will the tubes survive start-up? What is the salt outlet temperature?

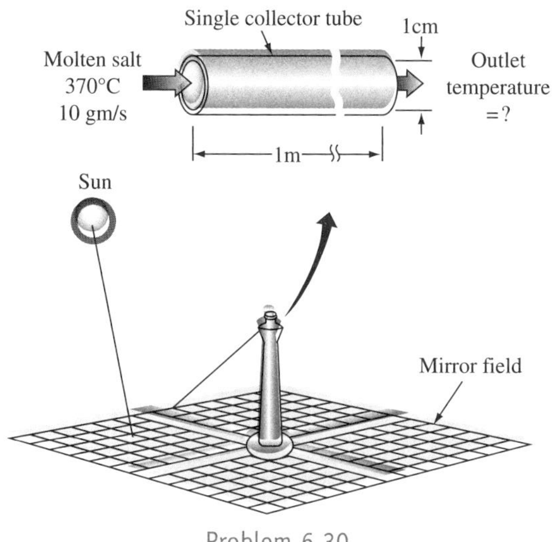

Problem 6.30

6.31 Determine the heat transfer coefficient for liquid bismuth flowing through an annulus (5-cm-ID, 6.1-cm-OD) at a velocity of 4.5 m/s. The wall temperature of the inner surface is 427°C, and the bismuth is at 316°C. It can assumed that heat losses from the outer surface are negligible.

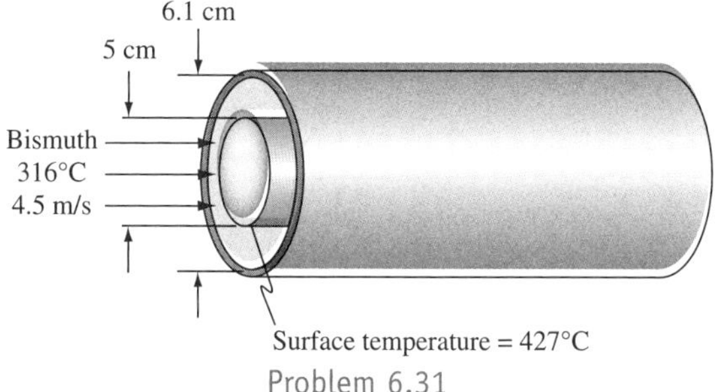

Problem 6.31

6.32 Mercury flows inside a copper tube 9 m long with a 5.1-cm inside diameter at an average velocity of 7 m/s. The temperature at the inside surface of the tube is 38°C uniformly throughout the tube, and the arithmetic mean bulk temperature of the mercury is 66°C. Assuming the velocity and temperature profiles are fully developed, calculate the rate of heat transfer by convection for the 9-m length by considering the mercury as (a) an ordinary liquid and (b) liquid metal. Compare the results.

6.33 A heat exchanger is to be designed to heat a flow of molten bismuth from 377°C to 477°C. The heat exchanger consists of a 50-mm-ID tube with a surface temperature maintained uniformly at 500°C by an electric heater. Find the length of the tube and the power required to heat 4 kg/s and 8 kg/s of bismuth.

6.34 Liquid sodium is to be heated from 500 K to 600 K by passing it at a flow rate of 5.0 kg/s through a 5-cm-ID tube whose surface is maintained at 620 K. What length of tube is required?

6.35 A 2.54-cm-OD, 1.9-cm-ID steel pipe carries dry air at a velocity of 7.6 m/s and a temperature of −7°C. Ambient air is at 21°C and has a dew point of 10°C. How much insulation with a conductivity of 0.18 W/mK is needed to prevent condensation on the exterior of the insulation if $\bar{h} = 2.4\,\text{W/m}^2\,\text{K}$ on the outside?

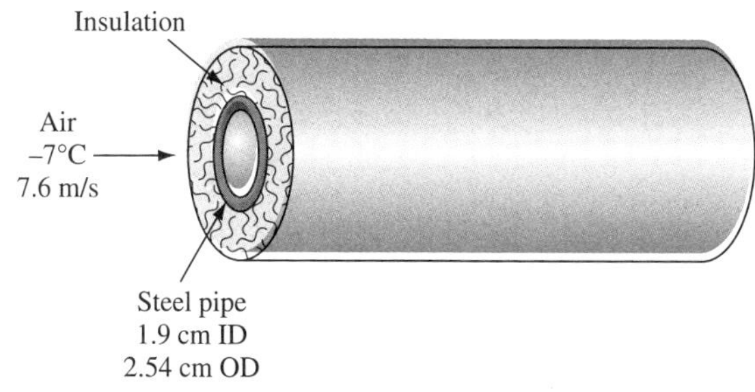

Problem 6.35

6.36 A double-pipe heat exchanger is used to condense steam at 7370 N/m^2. Water at an average bulk temperature of 10°C flows at 3.0 m/s through the inner pipe, which is made of copper and has a 2.54-cm ID and a 3.05-cm OD. Steam at its saturation temperature flows in the annulus formed between the outer surface of the inner pipe and an outer pipe of 5.08-cm-ID. The average heat transfer coefficient of the condensing steam is 5700 W/m^2 K, and the thermal resistance of a surface scale on the outer surface of the copper pipe is 0.000118 m^2 K/W. (a) Determine the overall heat transfer coefficient between the steam and the water based on the outer area of the copper pipe and sketch the thermal circuit. (b) Evaluate the temperature at the inner surface of the pipe. (c) Estimate the length required to condense 45 gm/s of steam. (d) Determine the water inlet and outlet temperatures.

6.37 Assume that the inner cylinder in Problem 6.31 is a heat source consisting of an aluminum-clad rod of uranium with a 5-cm diameter and 2 m long. Estimate the heat flux that will raise the temperature of the bismuth 40°C and the maximum center and surface temperatures necessary to transfer heat at this rate.

6.38 Evalute the rate of heat loss per meter from pressurized water flowing at 200°C through a 10-cm-ID pipe at a velocity of 3 m/s. The pipe is covered with a 5-cm-thick layer of 85% magnesia wool with an emissivity of 0.5. Heat is transferred to the surroundings at 20°C by natural convection and radiation. Draw the thermal circuit and state all assumptions.

6.39 In a pipe-within-a-pipe heat exchanger, water flows in the annulus and an aniline-alcohol solution having the properties listed in Problem 6.20 flows in the central pipe. The inner pipe has a 0.527-in.-ID and a 0.625-in.-OD, and the ID of the outer pipe is 0.750 in. For a water bulk temperature of 80°F and an aniline bulk temperature of 140°F, determine the overall heat transfer coefficient based on the outer diameter of the central pipe and the frictional pressure drop per unit length for water and the aniline for the following volumetric flow rates: (a) water rate 1 gpm, aniline rate 1 gpm; (b) water rate 10 gpm, aniline rate 1 gpm; (c) water rate 1 gpm, aniline rate 10 gpm; and (d) water rate 10 gpm, aniline rate 10 gpm. ($L/D = 400$.) Physical properties of aniline solution:

Temperature (°F)	Viscosity (centipoise)	Thermal Conductivity (Btu/h ft °F)	Specific Gravity	Specific Heat (Btu/lb °F)
68	5.1	0.100	1.03	0.50
140	1.4	0.098	0.98	0.53
212	0.6	0.095		0.56

6.40 A plastic tube of 7.6-cm-ID and 1.27-cm wall thickness has a thermal conductivity of 1.7 W/m K, a density of 2400 kg/m^3, and a specific heat of 1675 J/kg K. It is cooled from an initial temperature of 77°C by passing air at 20°C inside and outside the tube parallel to its axis. The velocities of the two airstreams are such that the coefficients of heat transfer are the same on the interior and exterior surfaces. Measurements show that at the end of 50 min, the temperature difference between the tube surfaces and the air is 10% of the initial temperature difference. A second experiment has been proposed in which a tube of a similar material with an inside diameter of 15 cm and a wall thickness of 2.5 cm will be cooled from the same initial temperature, again using air at 20°C and feeding it to the inside of the tube the same number of kilograms of air per hour that was used in the first experiment. The air-flow rate over the exterior surfaces will be adjusted to give the same heat transfer coefficient on the outside as on the inside of the tube. It can be assumed that the air-flow rate is so high that the temperature rise along the axis of the tube can be neglected. Using the experience gained initially with the 4.5-cm tube, estimate how long it will take to cool the surface of the larger tube to 27°C under the conditions described. Indicate all assumptions and approximations in your solution.

6.41 Exhaust gases having properties similar to dry air enter an exhaust stack at 800 K. The stack is made of steel and is 8 m tall with a 0.5-m-ID. The gas flow rate is 0.5 kg/s, and the ambient temperature is 280 K. The outside of the stack has an emissivity of 0.9. If heat loss from the outside is by radiation and natural convection, calculate the gas outlet temperature.

6.42 A 10-ft-long (3.05 m) vertical cylindrical exhaust duct from a commercial laundry has an ID of 6.0 in. (15.2 cm). Exhaust gases having physical properties approximating those of dry air enter at 600°F (316°C). The duct is insulated with 4 in. (10.2 cm) of rock wool having a thermal conductivity of $k = 0.25 + 0.005\,T$ (where T is in °F and k in Btu/h ft °F). If the gases enter at a velocity of 2 ft/s (0.61 m/s), calculate (a) the rate of heat transfer to quiescent ambient air at 60°F (15.6°C) and (b) the outlet temperature of the exhaust gas. Show your assumptions and approximations.

6.43 A long, 1.2-m-OD pipeline carrying oil is to be installed in Alaska. To prevent the oil from becoming too viscous for pumping, the pipeline is buried 3 m below ground. The oil is also heated periodically at pumping stations, as shown schematically in the figure that follows. The oil pipe is to be covered with insulation having a thickness t and a thermal conductivity of 0.05 W/m K. It is specified by the engineer installing the pumping station, that the temperature drop of the oil in a distance of 100 km should not exceed 5°C when the soil surface temperature $T_s = -40$°C. The temperature of the pipe after each heating is to be 120°C, and the flow rate is 500 kg/s. The properties of the oil being pumped are given as follows.

density (ρ_{oil}) = 900 kg/m^3
thermal conductivity (k_{oil}) = 0.14 W/m K
kinematic viscosity (ν_{oil}) = 8.5×10^{-4} m^2/s
specific heat (c_{oil}) = 2000 J/kg K

The soil under arctic conditions is dry (from Appendix 2 Table 11, k_s = 0.35 W/m K). (a) Estimate the thickness of insulation necessary to meet the specifications of the engineer. (b) Calculate the required rate of heat transfer to the oil at each heating point. (c) Calculate the pumping power required to move the oil between two adjacent heating stations.

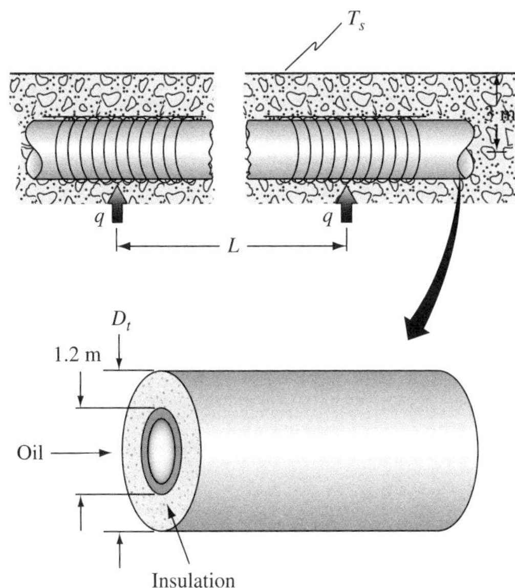

Problem 6.41

6.44 Show that for fully developed laminar flow between two flat plates spaced $2a$ apart, the Nusselt number based on the "bulk mean" temperature and the passage spacing is 4.12 if the temperature of both walls varies linearly with the distance x, i.e., $\partial T/\partial x = C$. The "bulk mean" temperature is defined as

$$T_b = \frac{\int_{-a}^{a} u(y)T(y)dy}{\int_{-a}^{a} u(y)dy}$$

6.45 Repeat Problem 6.44 but assume that one wall is insulated while the temperature of the other wall increases linearly with x.

6.46 For fully turbulent flow in a long tube of diameter D, develop a relation between the ratio $(L/\Delta T)/D$ in terms of flow and heat transfer parameters, where $L/\Delta T$ is the tube length required to raise the bulk temperature of the fluid by ΔT. Use Eq. (6.60) for fluids with Prandtl number of the order of unity or larger and Eq. (6.67) for liquid metals.

6.47 Water in turbulent flow is to be heated in a single-pass tubular heat exchanger by steam condensing on the outside of the tubes. The flow rate of the water, its inlet and outlet temperatures, and the steam pressure are fixed. Assuming that the tube wall temperature remains constant, determine the dependence of the total required heat exchanger area on the inside diameter of the tubes.

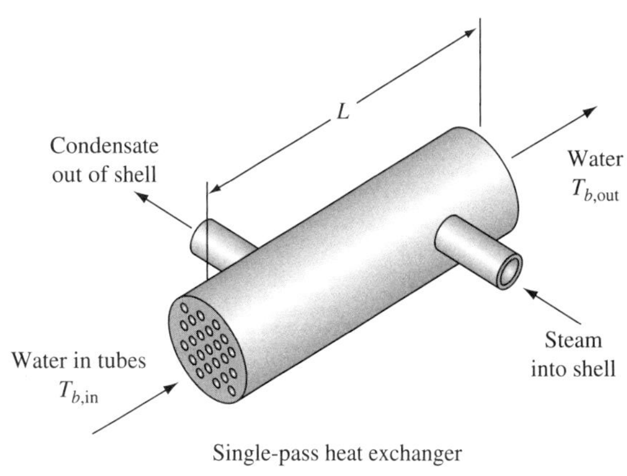

Problem 6.47

6.48 A 50,000-ft^2 condenser is constructed with 1-in.-OD brass tubes that are $23\frac{3}{4}$ long and have a 0.049-in. wall thickness. The following thermal resistance data were obtained at various water velocities inside the tubes (*Trans. ASME*, Vol. 58, p. 672, 1936).

$1/U_0 \times 10^3$ (h ft^2 °F/Btu)	Water Velocity (fps)	$1/U_0 \times 10^3$ (h ft^2 °F/Btu)	Water Velocity (fps)
2.060	6.91	3.076	2.95
2.113	6.35	2.743	4.12
2.212	5.68	2.498	6.76
2.374	4.90	3.356	2.86
3.001	2.93	2.209	6.27
2.081	7.01		

Assuming that the heat transfer coefficient on the steam side is 2000 Btu/h ft^2 °F and the mean bulk water temperature is 50°C, determine the scale resistance.

6.49 A nuclear reactor has rectangular flow channels with a large aspect ratio $(W/H) \gg 1$.

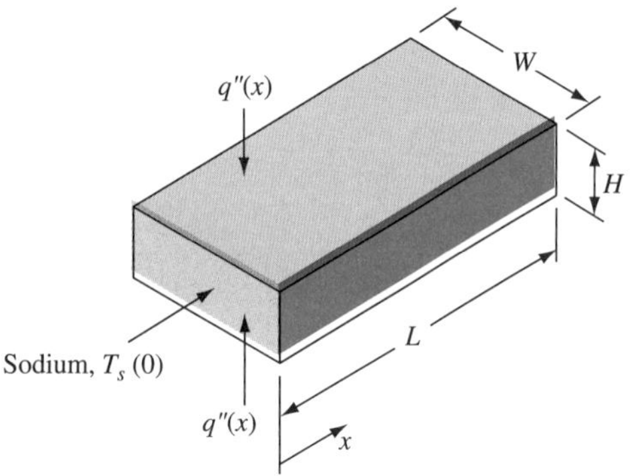

Heat generation from the upper and lower surfaces is equal and uniform at any value of x. However, the rate varies along the flow path of the sodium coolant according to

$$q''(x) = q''_0 \sin(\pi x/L)$$

Assuming that entrance effects are negligible so that the convection heat transfer coefficient is uniform, (a) obtain an expression for the variation of the mean bulk temperature of the sodium, $T_m(x)$, (b) derive a relation for the surface temperature of the upper and lower portion of the channel, $T_s(x)$, and (c) determine the distance x_{max} at which $T_s(x)$ is maximum.

Design Problems

6.1 **Chemical Reactor Cooling System.** (Chapter 6)
Design an internal cooling system for a chemical reactor. The reactor has a cylindrical shape 2 m in diameter is 14 m tall, and is well-insulated externally. The exothermic reaction releases 50 kW/m^3 of reacting medium, and the reacting medium operates at 250°C. It has been experimentally determined that the heat transfer coefficient between the reacting medium and a heat transfer surface inside the reactor is 1700 W/m^2 K. In designing the system, consider (a) capital cost, (b) operating and maintenance cost, (c) how much volume is taken up by the cooling system, inside the reactor and the concomitant reduction in reactor production, (d) availability of the removed heat for use outside the reactor, (e) and choice of cooling medium.

6.2 **Cooling High-Powered Silicone Chips**
Timothy L. Hoopman of the 3M Corporation described a novel method for cooling high-powered-density silicone chips (D. Cho et al., eds., *Microchanneled Structures in DCS, Vol. 19: Microstructures, Sensors, and Actuators*, ASME Winter Annual Meeting, Dallas, Texas, November 1990). This method involves etching microchannels in the back surface of the chip. These microchannels typically have hydraulic diameters of 10 μ to 100 μ with length-to-diameter ratios of 50–1,000. Microchannel center-to-center distances can be as small as 100 μ, depending upon geometry.

Design a suitable microchannel cooling system for a 10-mm × 10-mm chip. The microetched channels are covered with a silicone cap as shown in the schematic diagram. The chip and cap are to be maintained at a temperature of 350 K and the system has to remove a heat flux of 50 W/cm^2. Explain the reason why microchannels, even in laminar flow, produce very high heat transfer coefficients. Also, compare the temperature difference achievable with the microchannel design with a conventional design using water-forced convection cooling in a channel covering the chip surface.

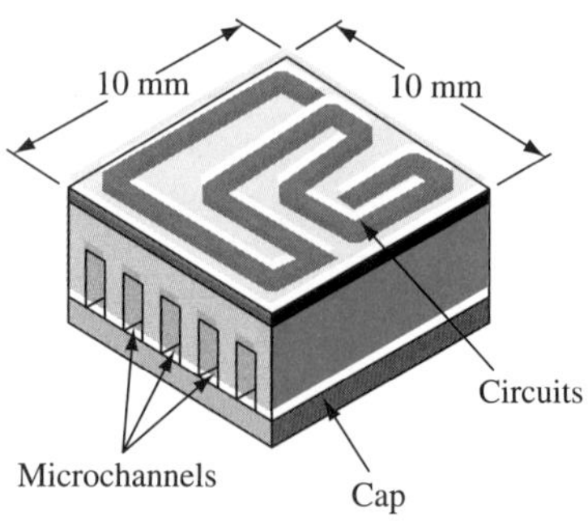

Desgin Problem 6.2

6.3 Electrical Resistance Heater (Chapters 2, 3, 6, and 10)
In Design Problems 2.7 and 3.2, you determined the required heat transfer coefficient for water flowing over the outside surface of a heating element. Determine the pipe length, the required water volumetric flow rate, and the pressure drop if the element is located inside a 15-cm-ID pipe. Give the hot-water delivery rate of a typical domestic hot-water heater, determine how many such pipes would be needed, and how they could be arranged and electrically connected. Make a rough cost estimate, and decide whether the pressure drop is reasonable. Finally, compare the results with those from the design using a simple circular cross-section heater element shown in Fig. 2.7(a).

CHAPTER 7

Forced Convection Over Exterior Surfaces

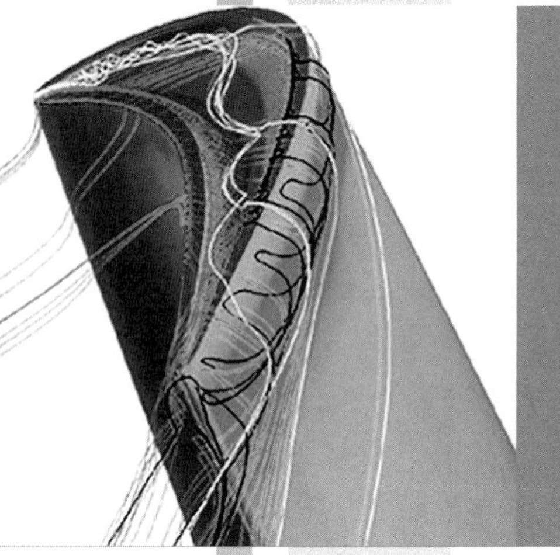

Computational simulation of the vortex flow and heat transfer over the tip of a high-pressure gas turbine rotor blade, where the darkest areas represent undesirable high-heat-flux regions.

Source: Courtesy of NASA.

Concepts and Analyses to Be Learned

In fluid flow and forced-convection heat transfer over exterior surfaces or bluff bodies, the boundary layer growth is not confined, and its spatial development along the surface influences the local heat flow process. In external flows, the length of the surface provides the characteristic length for scaling the boundary layer as well as for dimensionless representation of flow-friction loss and the heat transfer coefficient. A variety of different applications of convective heat transfer over exterior surfaces are encountered in engineering practice. They include flow over tube banks in shell-and-tube heat exchangers, deicing of aircraft wings, metal heat treating, and cooling of electrical and electronics equipment, among others. A study of this chapter will teach you:

- How to characterize the flow behavior over exterior surfaces and bluff bodies and determine the associated fluid drag and convective heat transfer.
- How to calculate the heat transfer coefficient in packed-bed systems and devices.
- How to analyze the forced convection in cross-flow over multitube banks or bundles and predict the frictional loss and heat transfer coefficient.
- How to characterize jet flows as they impinge on bluff surfaces and to determine the heat transfer due to single and multiple jet-impingement systems, as well as submerged jets.

7.1 Flow Over Bluff Bodies

In this chapter, we shall consider heat transfer by forced convection between the exterior surface of bluff bodies, such as spheres, wires, tubes and tube bundles, and fluids flowing perpendicularly and at angles to the axes of these bodies. The heat transfer phenomena for these systems, like those for systems in which a fluid flows inside a duct or along a flat plate, are closely related to the nature of the flow. The most important difference between the flow over a bluff body and the flow over a flat plate or a streamlined body lies in the behavior of the boundary layer. We recall that the boundary layer of a fluid flowing over the surface of a streamlined body will separate when the pressure rise along the surface becomes too large. On a streamlined body, the separation, if it takes place at all, occurs near the rear. On a bluff body, on the other hand, the point of separation often lies not far from the leading edge. Beyond the point of separation of the boundary layer, the fluid in a region near the surface flows in a direction opposite to the main stream, as shown in Fig. 7.1. The local reversal in the flow results in disturbances that produce turbulent eddies. This is illustrated in Fig. 7.2 on the next page, which is a photograph of the flow pattern of a stream flowing at a right angle to a cylinder. We can see that eddies from both sides of the cylinder extend downstream, so that a turbulent wake is formed at the rear of the cylinder.

Large pressure losses are associated with the separation of the flow, since the kinetic energy of the eddies that pass off into the wake cannot be regained. In flow over a streamlined body, the pressure loss is caused mainly by the skin friction drag. For a bluff body, on the other hand, the skin friction drag is small compared to the form drag in the Reynolds number range of commercial interest. The form or pressure drag arises from the separation of the flow, which prevents closing of the streamlines and thereby induces a low-pressure region in the rear of the body. When the pressure over the rear of the body is lower than that over the front, there exists a pressure difference that produces a drag force over and above that of the skin friction. The magnitude of the form drag decreases as the separation moves farther toward the rear.

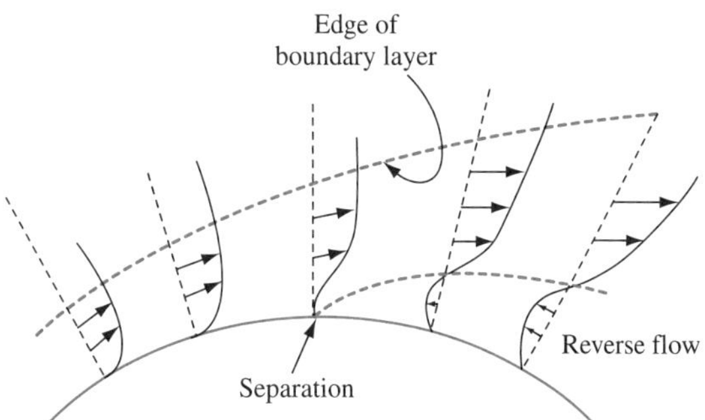

FIGURE 7.1 Schematic sketch of the boundary layer on a circular cylinder near the separation point.

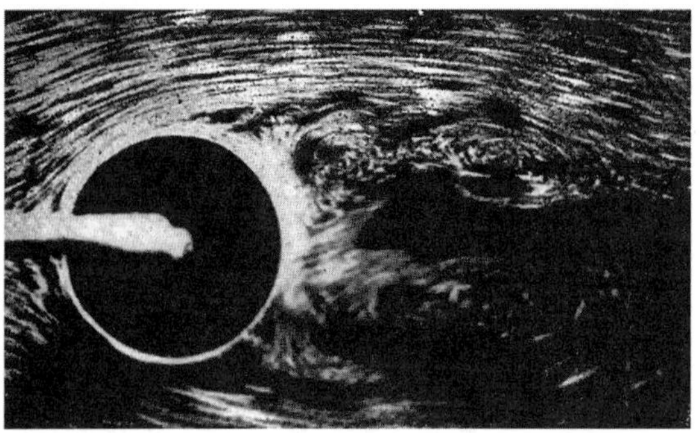

FIGURE 7.2 Flow pattern in cross-flow over a single horizontal cylinder.

Source: Photograph by H. L. Rubach, Mitt. Forschungsarb., vol. 185, 1916.

The geometric shapes that are most important for engineering work are the long circular cylinder and the sphere. The heat transfer phenomena for these two shapes in cross-flow have been studied by a number of investigators, and representative data are summarized in Section 7.2. In addition to the average heat transfer coefficient over a cylinder, the variation of the coefficient around the circumference will be considered. A knowledge of the peripheral variation of the heat transfer associated with flow over a cylinder is important for many practical problems such as heat transfer calculations for airplane wings, whose leading-edge contours are approximately cylindrical. The interrelation between heat transfer and flow phenomena will also be stressed because it can be applied to the measurement of velocity and velocity fluctuations in a turbulent stream using a hot-wire anemometer.

Section 7.3 treats heat transfer in packed beds. These are systems in which heat transfer to or from spherical or other shaped particles is important. Sections 7.4 and 7.5 deal with heat transfer to or from bundles of tubes in cross-flow, a configuration that is widely used in boilers, air-preheaters, and conventional shell-and-tube heat exchangers. Section 7.6 treats heat transfer with jets.

7.2 Cylinders, Spheres, and Other Bluff Shapes

Photographs of typical flow patterns for flow over a single cylinder and a sphere are shown in Figs. 7.2 and 7.3, respectively. The most forward points of these bodies are called stagnation points. Fluid particles striking there are brought to rest, and the pressure at the stagnation point, p_0, rises approximately one velocity head, that is, $(\rho U_\infty^2/2g_c)$, above the pressure in the oncoming free stream, p_∞. The flow divides at the stagnation point of the cylinder, and a boundary layer builds up along the surface. The fluid accelerates when it flows past the surface of the cylinder, as can be seen by the crowding of the streamlines shown in Fig. 7.4. This flow pattern for a nonviscous fluid in irrotational flow, a highly idealized case, is called *potential flow*. The velocity reaches a maximum at both sides of the cylinder, then falls again to zero at the stagnation point in the rear. The pressure distribution corresponding to this idealized flow pattern is shown by the solid line in Fig. 7.5 on page 424.

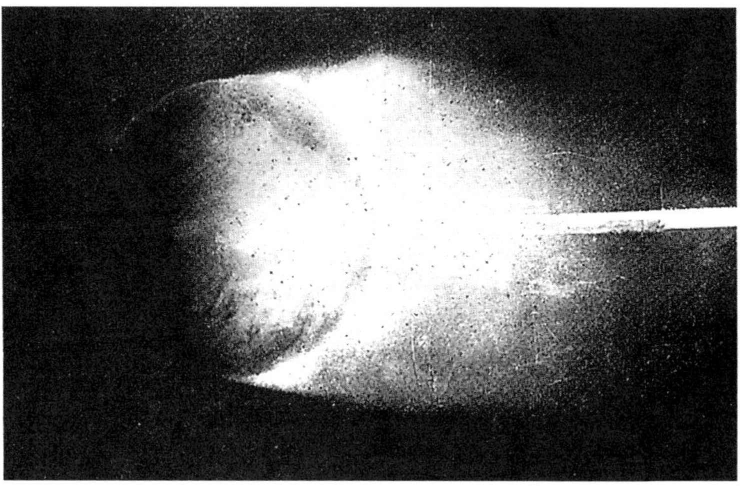

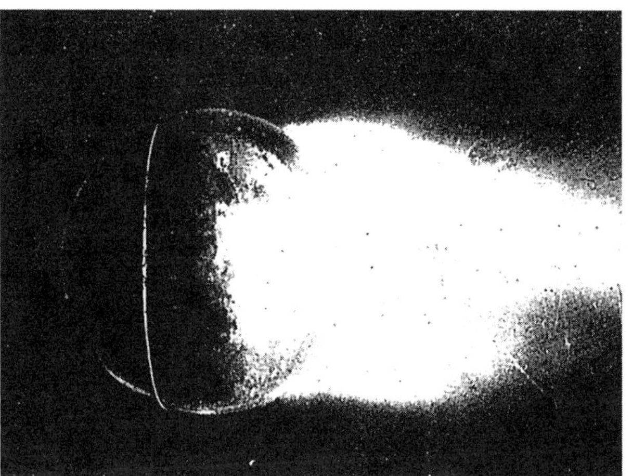

FIGURE 7.3 Photographs of air flowing over a sphere. In the lower picture a "tripping" wire induced early transition and delayed separation.

Source: Courtesy of L. Prandtl and the *Journal of the Royal Aeronautical Society*.

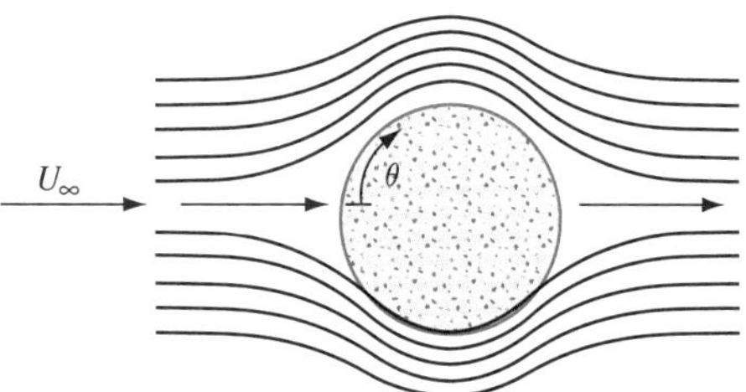

FIGURE 7.4 Streamlines for potential flow over a circular cylinder.

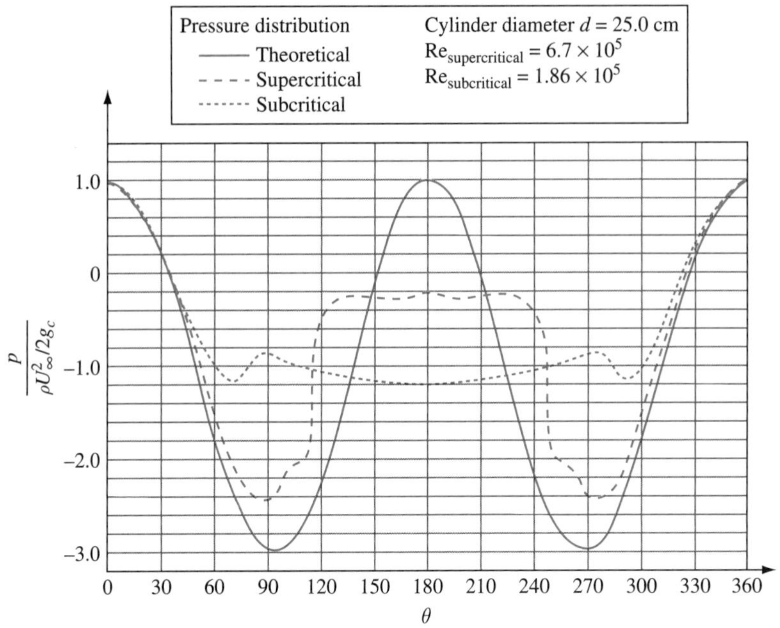

FIGURE 7.5 Pressure distribution around a circular cylinder in crossflow at various Reynolds numbers; p is the local pressure, $\rho U_\infty^2/2g_c$ is the free-stream impact pressure; θ is the angle measured from the stagnation point.

Source: By permission from L. Flachsbart, *Handbuch der Experimental Physik*, Vol. 4, part 2.

Since the pressure distribution is symmetric about the vertical center plane of the cylinder, it is clear that there will be no pressure drag in irrotational flow. However, unless the Reynolds number is very low, a real fluid will not adhere to the entire surface of the cylinder, but as mentioned previously, the boundary layer in which the flow is not irrotational will separate from the sides of the cylinder as a result of the adverse pressure gradient. The separation of the boundary layer and the resultant wake in the rear of the cylinder give rise to pressure distributions that are shown for different Reynolds numbers by the dashed lines in Fig. 7.5. It can be seen that there is fair agreement between the ideal and the actual pressure distribution in the neighborhood of the forward stagnation point. In the rear of the cylinder, however, the actual and ideal distributions differ considerably. The characteristics of the flow pattern and of the boundary layer depend on the Reynolds number, $\rho U_\infty D/\mu$, which for flow over a cylinder or a sphere is based on the velocity of the oncoming free stream U_∞ and the outside diameter of the body D. Properties are evaluated at free-stream conditions. The flow pattern around the cylinder undergoes a series of changes as the Reynolds number is increased, and since the heat transfer depends largely on the flow, we shall consider first the effect of the Reynolds number on the flow and then interpret the heat transfer data in the light of this information.

The sketches in Fig. 7.6 illustrate flow patterns typical of the characteristic ranges of Reynolds numbers. The letters in Fig. 7.6 correspond to the flow regimes indicated in Fig. 7.7, where the total dimensionless drag coefficients of a cylinder and a sphere, C_D, are plotted as a function of the Reynolds number. The force term

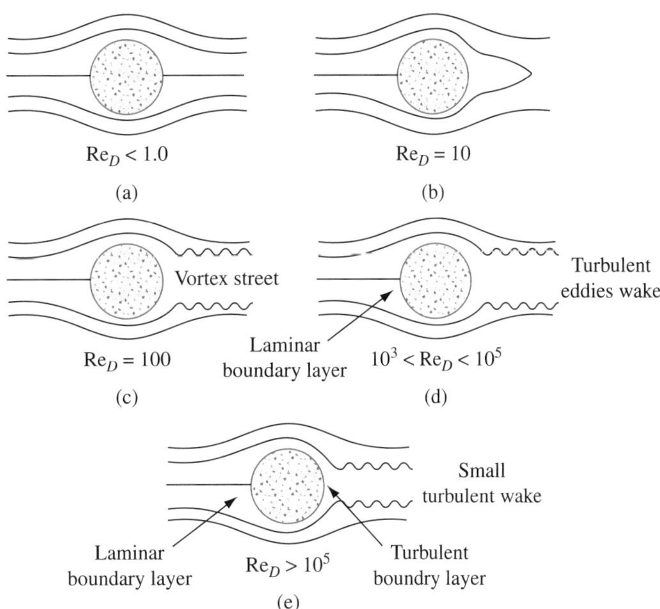

FIGURE 7.6 Flow patterns for cross-flow over a cylinder at various Reynolds numbers.

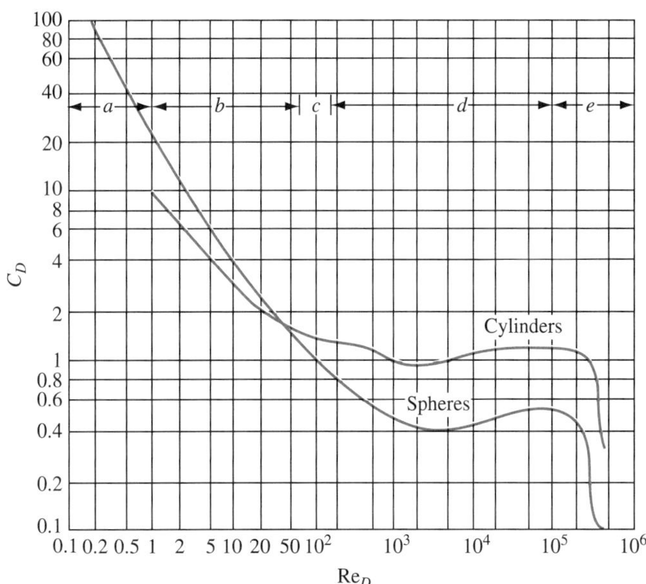

FIGURE 7.7 Drag coefficient versus Reynolds number for long circular cylinders and spheres in cross-flow.

in the total drag coefficient is the sum of the pressure and frictional forces; it is defined by the equation

$$C_D = \frac{\text{drag force}}{A_f(\rho U_\infty^2/2g_c)}$$

where ρ = free-stream density
U_∞ = free-stream velocity
A_f = frontal projected area = πDL (cylinder) or $\pi D^2/4$ (sphere)
D = outside cylinder diameter, or diameter of sphere
L = cylinder length

The following discussion strictly applies only to long cylinders, but it also gives a qualitative picture of the flow past a sphere. The letters (a) to (e) refer to Figs. 7.6 and 7.7.

(a) At Reynolds numbers of the order of unity or less, the flow adheres to the surface and the streamlines follow those predicted from potential-flow theory. The inertia forces are negligibly small, and the drag is caused only by viscous forces, since there is no flow separation. Heat is transferred by conduction alone.
(b) At Reynolds numbers of the order of 10, the inertia forces become appreciable and two weak eddies stand in the rear of the cylinder. The pressure drag accounts now for about half of the total drag.
(c) At a Reynolds number of the order of 100, vortices separate alternately from the two sides of the cylinder and stretch a considerable distance downstream. These vortices are referred to as *von Karman vortex streets* in honor of the scientist Theodore von Karman, who studied the shedding of vortices from bluff objects. The pressure drag now predominates.
(d) In the Reynolds number range between 10^3 and 10^5, the skin friction drag becomes negligible compared to the pressure drag caused by turbulent eddies in the wake. The drag coefficient remains approximately constant because the boundary layer remains laminar from the leading edge to the point of separation, which lies throughout this Reynolds number range at an angular position θ between 80° and 85° measured from the direction of the flow.
(e) At Reynolds numbers larger than about 10^5 (the exact value depends on the turbulence level of the free stream) the kinetic energy of the fluid in the laminar boundary layer over the forward part of the cylinder is sufficient to overcome the unfavorable pressure gradient without separating. The flow in the boundary layer becomes turbulent while it is still attached, and the separation point moves toward the rear. The closing of the streamlines reduces the size of the wake, and the pressure drag is therefore also substantially reduced. Experiments by Fage and Falkner [1, 2] indicate that once the boundary layer has become turbulent, it will not separate before it reaches an angular position corresponding to a θ of about 130°.

Analyses of the boundary layer growth and the variation of the local heat transfer coefficient with angular position around circular cylinders and spheres have been only partially successful. Squire [3] has solved the equations of motion and energy for a cylinder at constant temperature in cross-flow over that portion of the surface to which a laminar boundary layer adheres. He showed that at the stagnation point and in its immediate neighborhood, the convection heat transfer coefficient can be calculated from the equation

$$\mathrm{Nu}_D = \frac{h_c D}{k} = C\sqrt{\frac{\rho U_\infty D}{\mu}} \tag{7.1}$$

where C is a constant whose numerical value at various Prandtl numbers is tabulated below:

Pr	0.7	0.8	1.0	5.0	10.0
C	1.0	1.05	1.14	2.1	1.7

Over the forward portion of the cylinder ($0 < \theta < 80°$), the empirical equation for $h_c(\theta)$, the local value of the heat transfer coefficient at θ

$$\mathrm{Nu}(\theta) = \frac{h_c(\theta)D}{k} = 1.14\left(\frac{\rho U_\infty D}{\mu}\right)^{0.5}\mathrm{Pr}^{0.4}\left[1 - \left(\frac{\theta}{90}\right)^3\right] \tag{7.2}$$

has been found to agree satisfactorily [4] with experimental data.

Giedt [5] has measured the local pressures and the local heat transfer coefficients over the entire circumference of a long, 10.2-cm-OD cylinder in an airstream over a Reynolds number range from 70,000 to 220,000. Giedt's results are shown in Fig. 7.8, and similar data for lower Reynolds numbers are shown in Fig. 7.9 (tooth figures are shown on the next page). If the data shown in Figs. 7.8 and 7.9 are compared at corresponding Reynolds numbers with the flow patterns and the boundary layer characteristics described earlier, some important observations can be made.

At Reynolds numbers below 100,000, separation of the laminar boundary layer occurs at an angular position of about 80°. The heat transfer and the flow characteristics over the forward portion of the cylinder resemble those for laminar flow over a flat plate, which were discussed earlier. The local heat transfer is largest at the stagnation point and decreases with distance along the surface as the boundary layer thickness increases. The heat transfer reaches a minimum on the sides of the cylinder near the separation point. Beyond the separation point, the local heat transfer increases because considerable turbulence exists over the rear portion of the cylinder, where the eddies of the wake sweep the surface. However, the heat transfer coefficient over the rear is no larger than that over the front because the eddies recirculate part of the fluid and, despite their high turbulence, are not as effective as a turbulent boundary layer in mixing the fluid in the vicinity of the surface with the fluid in the main stream.

At Reynolds numbers large enough to permit transition from laminar to turbulent flow in the boundary layer without separation of the laminar boundary layer, the heat transfer coefficient has two minima around the cylinder. The first minimum occurs at the point of transition. As the transition from laminar to turbulent flow progresses, the heat transfer coefficient increases and reaches a maximum approximately at the point where the boundary layer becomes fully turbulent. Then the heat transfer coefficient begins to decrease again and reaches a second minimum at about 130°, the point at which the turbulent boundary layer separates from the cylinder. Over the rear of the cylinder, the heat transfer coefficient increases to another maximum at the rear stagnation point.

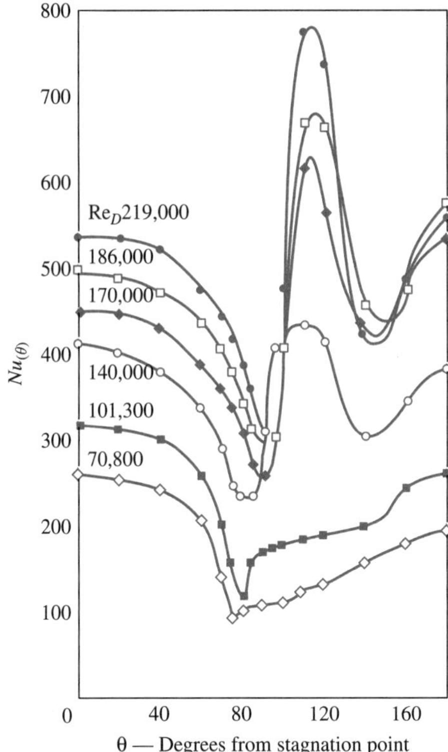

FIGURE 7.8 Circumferential variation of the dimensionless heat transfer coefficient (Nu_θ) at high Reynolds numbers for a circular cylinder in cross-flow.

Source: Courtesy of W. H. Giedt, "Investigation of Variation of Point Unit-Heat-Transfer Coeffient around a Cylinder Normal to an Air Stream", Trans. ASME, Vol. 71, 1949, pp. 375–381. Reprinted by permission of The American Society of Mechanical Engineers International.

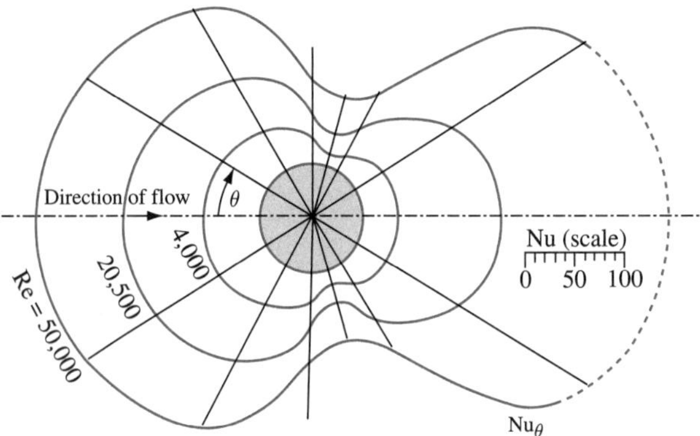

FIGURE 7.9 Circumferential variation of the local Nusselt number $Nu(\theta) = h_c(\theta)D_o/k_f$ at low Reynolds numbers for a circular cylinder in cross-flow.

Source: According to W. Lorisch, from M. ten Bosch, *Die Wärmeübertragung*, 3d ed., Springer Verlag, Berlin, 1936.

EXAMPLE 7.1 To design a heating system for the purpose of preventing ice formation on an aircraft wing, it is necessary to know the heat transfer coefficient over the outer surface of the leading edge. The leading-edge contour can be approximated by a half-cylinder of 30-cm diameter, as shown in Fig. 7.10. The ambient air is at −34°C, and the surface temperature is to be no less than 0°C. The plane is designed to fly at 7500-m altitude at a speed of 150 m/s. Calculate the distribution of the convection heat transfer coefficient over the forward portion of the wing.

SOLUTION At an altitude of 7500 m the standard atmospheric air pressure is 38.9 kPa and the density of the air is 0.566 kg/m^3 (see Table 38 in Appendix 2).

The heat transfer coefficient at the stagnation point ($\theta = 0$) is, according to Eq. (7.2),

$$\begin{aligned} h_c(\theta = 0) &= 1.14\left(\frac{\rho U_\infty D}{\mu}\right)^{0.5} \mathrm{Pr}^{0.4}\frac{k}{D} \\ &= (1.14)\left(\frac{(0.566\,\mathrm{kg/m^3}) \times (150\,\mathrm{m/s}) \times (0.30\,\mathrm{m})}{1.74 \times 10^{-5}\,\mathrm{kg/m\ s}}\right)^{0.5}(0.72)^{0.4}\left(\frac{0.024\,\mathrm{W/m\ K}}{0.30\,\mathrm{m}}\right) \\ &= 96.7\ \mathrm{W/m^2\ °C} \end{aligned}$$

The variation of h_c with θ is obtained by multipling the value of the heat transfer coefficient at the stagnation point by $1 - (\theta/90)^3$. The results are tabulated below.

θ (deg)	0	15	30	45	60	75
$h_c(\theta)$(W/m^2 °C)	96.7	96.3	93.1	84.6	68.0	40.7

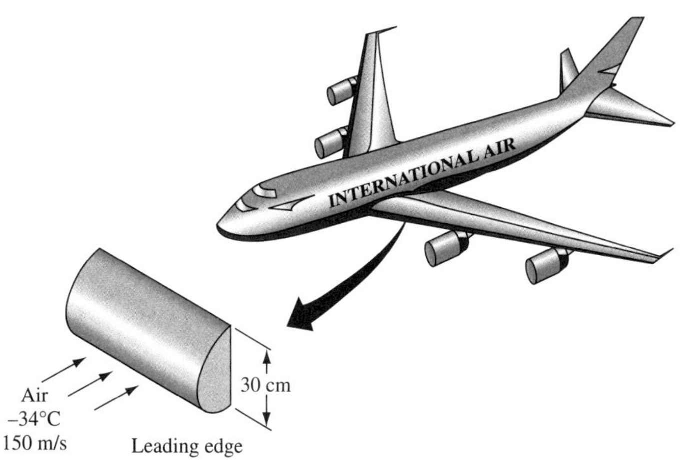

FIGURE 7.10 Approximation of the leading edge of an aircraft wing for Example 7.1.

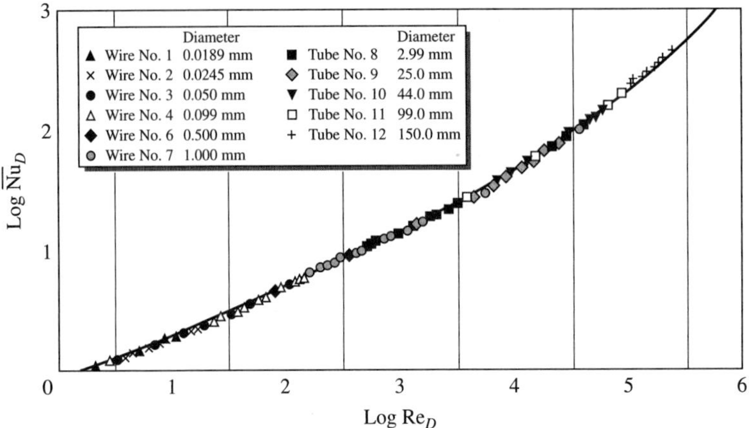

FIGURE 7.11 Average Nusselt number versus Reynolds number for a circular cylinder in cross-flow with air.

Source: After R. Hilpert [6, p. 220].

It is apparent from the foregoing discussion that the variation of the heat transfer coefficient around a cylinder or a sphere is a very complex problem. For many practical applications, it is fortunately not necessary to know the local value $h_{c\theta}$ but is sufficient to evaluate the average value of the heat transfer coefficient around the body. A number of observers have measured mean heat transfer coefficients for flow over single cylinders and spheres. Hilpert [6] accurately measured the average heat transfer coefficients for air flowing over cylinders of diameters ranging from 19 μm to 15 cm. His results are shown in Fig. 7.11, where the average Nusselt $\bar{h}_c D/k$ is plotted as a function of the Reynolds number $U_\infty D/\nu$.

A correlation for a cylinder at uniform temperature T_s in cross-flow of liquids and gases has been proposed by Žukauskas [7]:

$$\overline{\text{Nu}}_D = \frac{\bar{h}_c D}{k} = C\left(\frac{U_\infty D}{\nu}\right)^m \text{Pr}^n \left(\frac{\text{Pr}}{\text{Pr}_s}\right)^{0.25} \tag{7.3}$$

where all fluid properties are evaluated at the free-stream fluid temperature except for Pr_s, which is evaluated at the surface temperature. The constants in Eq. (7.3) are given in Table 7.1. For $\text{Pr} < 10$, $n = 0.37$, and for $\text{Pr} > 10$, $n = 0.36$.

TABLE 7.1 Coefficients for Eq. (7.3)

Re_D	C	m
1–40	0.75	0.4
$40–1 \times 10^3$	0.51	0.5
$1 \times 10^3–2 \times 10^5$	0.26	0.6
$2 \times 10^5–1 \times 10^6$	0.076	0.7

For cylinders that are not normal to the flow, Groehn [8] developed the following correlation

$$\overline{\text{Nu}}_D = 0.206\,\text{Re}_N^{0.63}\text{Pr}^{0.36} \tag{7.4}$$

In Eq. (7.4), the Reynolds number Re_N is based on the component of the flow velocity normal to the cylinder axis:

$$\text{Re}_N - \text{Re}_D \sin\theta$$

and the yaw angle, θ, is the angle between the direction of flow and the cylinder axis, for example, $\theta = 90°$ for cross-flow.

Equation (7.4) is valid from $\text{Re}_N = 2500$ up to the critical Reynolds number, which depends on the yaw angle as follows:

θ	Re_{crit}
15°	2×10^4
30°	8×10^4
45°	2.5×10^5
>45°	$>2.5 \times 10^5$

Groehn also found that, in the range $2 \times 10^5 < \text{Re}_D < 10^6$, the Nusselt number is independent of yaw angle

$$\overline{\text{Nu}}_D = 0.012\,\text{Re}_D^{0.85}\,\text{Pr}^{0.36} \tag{7.5}$$

For cylinders with noncircular cross sections in gases, Jakob [9] compiled data from two sources and presented the coefficients of the correlation equation

$$\overline{\text{Nu}}_D = B\,\text{Re}_D^n \tag{7.6}$$

in Table 7.2 on the next page. In Eq. (7.6), all properties are to be evaluated at the film temperature, which was defined in Chapter 4 as the mean of the surface and free-stream fluid temperatures.

For heat transfer from a cylinder in cross-flow of liquid metals, Ishiguro et al. [10] recommended the correlation equation

$$\overline{\text{Nu}}_D = 1.125(\text{Re}_D\text{Pr})^{0.413} \tag{7.7}$$

in the range $1 \le \text{Re}_D\text{Pr} \le 100$. Equation (7.7) predicts a somewhat lower $\overline{\text{Nu}}_D$ than that of analytic studies for either constant temperature $[\overline{\text{Nu}}_D = 1.015(\text{Re}_D\text{Pr})^{0.5}]$ or constant flux $[\overline{\text{Nu}}_D = 1.145(\text{Re}_D\text{Pr})^{0.5}]$. As pointed out in [10], neither boundary condition was achieved in the experimental effort. The difference between Eq. (7.7) and the correlation equations for the two analytic studies is apparently due to the assumption of inviscid flow in the analytic studies. Such an assumption cannot allow for a separated region at large values of Re_DPr, which is where Eq. (7.7) deviates from the analytic results.

TABLE 7.2 Constants in Eq. (7.6) for forced convection perpendicular to noncircular tubes

Flow Direction and Profile	Re_D From	Re_D To	n	B
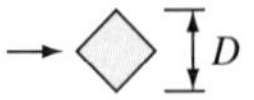 → D	5,000	100,000	0.588	0.222
 → D	2,500	15,000	0.612	0.224
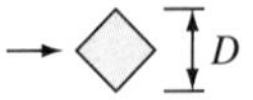 → D	2,500	7,500	0.624	0.261
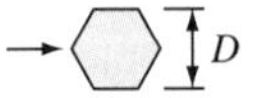 → D	5,000	100,000	0.638	0.138
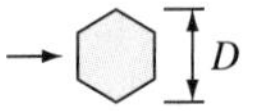 → D	5,000	19,500	0.638	0.144
 → D	5,000	100,000	0.675	0.092
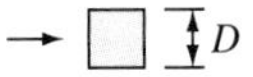 → D	2,500	8,000	0.699	0.160
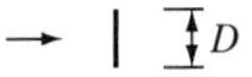 → D	4,000	15,000	0.731	0.205
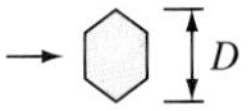 → D	19,500	100,000	0.782	0.035
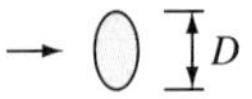 → D	3,000	15,000	0.804	0.085

Quarmby and Al-Fakhri [11] found experimentally that the effect of the tube aspect ratio (length-to-diameter ratio) is negligible for aspect ratio values greater than 4. The forced air flow over the cylinder was essentially that of an infinite cylinder in cross-flow. They examined the effect of heated-length variations, and thus aspect ratio, by independently heating five longitudinal sections of the cylinder. Their data for large aspect ratios compared favorably with the data of Žukauskas [7] for cylinders in cross-flow. For aspect ratios less than 4, they recommend

$$\overline{\text{Nu}}_D = 0.123\ \text{Re}_D^{0.651} + 0.00416\left(\frac{D}{L}\right)^{0.85} \text{Re}_D^{0.792} \tag{7.8}$$

in the range

$$7 \times 10^4 < \text{Re}_D < 2.2 \times 10^5$$

Properties in Eq. (7.8) are to be evaluated at the film temperature. Equation (7.8) agrees well with data of Žukauskas [7] in the limit $L/D \rightarrow \infty$ for this relatively small Reynolds number range.

Several studies have attempted to determine the heat transfer coefficient near the base of a cylinder attached to a wall and exposed to cross-flow or near the tip of a cylinder exposed to cross-flow. The objective of these studies was to more accurately predict the heat transfer coefficient for fins and tube banks and the cooling of electronic components. Sparrow and Samie [12] measured the heat transfer coefficient at the tip of a cylinder and also for a length of the cylindrical portion (equal to 1/4 of the diameter) near the tip. They found that heat transfer coefficients are 50% to 100% greater, depending on the Reynolds number, than those that would be predicted from Eq. (7.3). Sparrow et al. [13] examined the heat transfer near the attached end of a cylinder in cross-flow. They found that in a region approximately one diameter from the attached end, the heat transfer coefficients were about 9% less than those that would be predicted from Eq. (7.3).

Turbulence in the free stream approaching the cylinder can have a relatively strong influence on the average heat transfer. Yardi and Sukhatme [14] experimentally determined an increase of 16% in the average heat transfer coefficient as the free-stream turbulence intensity was increased from 1% to 8% in the Reynolds number range 6000 to 60,000. On the other hand, the length scale of the free-stream turbulence did not affect the average heat transfer coefficient. Their local heat transfer measurements showed that the effect of free-stream turbulence was largest at the front stagnation point and diminished to an insignificant effect at the rear stagnation point. Correlations given in this chapter generally assume that the free-stream turbulence is very low.

7.2.1 Hot-Wire Anemometer

The relationship between the velocity and the rate of heat transfer from a single cylinder in cross-flow is used to measure velocity and velocity fluctuations in turbulent flow and in combustion processes through the use of a hot-wire anemometer. This instrument consists basically of a thin (3- to 30-μm diameter) electrically heated wire stretched across the ends of two prongs. When the wire is exposed to a cooler fluid stream, it loses heat by convection. The temperature of the wire, and consequently its electrical resistance, depends on the temperature and the velocity of the fluid and the heating current. To determine the fluid velocity, either the wire is maintained at a constant temperature by adjusting the current and determining the fluid speed from the measured value of the current, or the wire is heated by a constant current and the speed is deduced from a measurement of the electrical resistance or the voltage drop in the wire. In the first method, the constant-temperature method, the hot wire forms one arm in the circuit of a Wheatstone bridge, as shown in Fig. 7.12(a) on the next page. The resistance of the rheostat arm, R_e, is adjusted to balance the bridge when the temperature, and consequently the resistance, of the wire has reached some desired value. When the fluid velocity increases, the current required to maintain the temperature and resistance of the wire constant must also increase. This change in the current is accomplished by adjusting the rheotstat in series with the voltage supply. When the galvanometer indicates that the bridge is in balance again, the change in current, read on the ammeter, indicates the change in speed. In the other method, the wire current is held constant, and the fluctuations in voltage drop caused by variations in the fluid velocity are impressed across the input of an amplifier, the output of which is connected to an oscilloscope. Figure 7.12(b) schematically shows an arrangement for the voltage measurement. Additional information on the hot-wire method is given in Dryden and Keuthe [15] and Pearson [16]. Although the circuitry required to maintain constant wire temperature is more

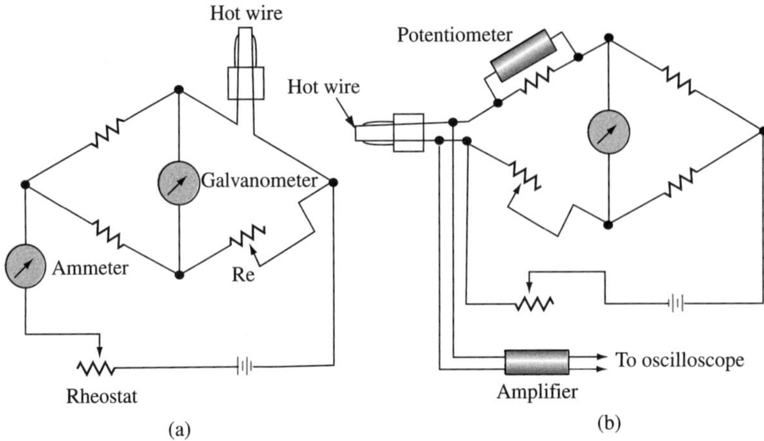

FIGURE 7.12 Schematic circuits for hot-wire probes and associated equipment. (a) constant-temperature method, (b) constant-current method.

complex than that required for constant current operation, it is often preferred since the fluid properties affecting heat transfer from the wire are constant if the wire temperature and free-stream temperatures are constant. This greatly simplifies the determination of velocity from wire current.

EXAMPLE 7.2 A 25-μm-diameter polished-platinum wire 6 mm long is to be used for a hot-wire anemometer to measure the velocity of 20°C air in the range between 2 and 10 m/s (see Fig. 7.13). The wire is to be placed into the circuit of the Wheatstone bridge shown in Fig. 7.12(a). Its temperature is to be maintained at 230°C by adjusting the current using the rheostat. To design the electrical circuit, it is necessary to know the required current as a function of air velocity. The electrical resistivity of platinum at 230°C is 17.1 $\mu\Omega$ cm.

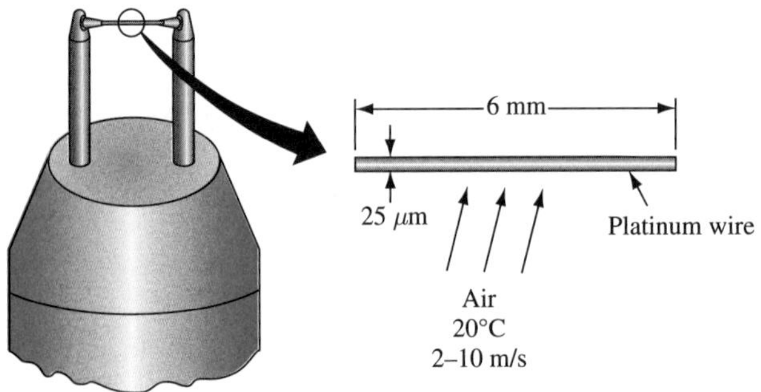

FIGURE 7.13 Sketch of hot-wire anemometer for Example 7.2.

SOLUTION Since the wire is very thin, conduction along it can be neglected; also, the temperature gradient in the wire at any cross section can be disregarded. At the free-stream temperature, the air has a thermal conductivity of 0.0251 W/m °C and a kinematic viscosity of 1.57×10^{-5} m^2/s. At a velocity of 2 m/s, the Reynolds number is

$$\text{Re}_D = \frac{(2\,\text{m/s})(25 \times 10^{-6}\,\text{m})}{1.57 \times 10^{-5}\,\text{m}^2/\text{s}} = 3.18$$

The Reynolds number range of interest is therefore 1 to 40, so the correlation equation from Eq. (7.3) and Table 7.1 is

$$\frac{\bar{h}_c D}{k} = 0.75\,\text{Re}_D^{0.4}\text{Pr}^{0.37}\left(\frac{\text{Pr}}{\text{Pr}_s}\right)^{0.25}$$

Neglecting the small variation in Prandtl number from 20° to 230°C, the average convection heat transfer coefficient as a function of velocity is

$$\begin{aligned}\bar{h}_c &= (0.75)(3.18)^{0.4}\left(\frac{U_\infty}{2}\right)^{0.4}(0.71)^{0.37}\left(\frac{0.0251\,\text{W/m K}}{25 \times 10^{-6}\,\text{m}}\right)\\ &= 799\,U_\infty^{0.4}\,\text{W/m}^2\,°\text{C}\end{aligned}$$

At this point, it is necessary to estimate the heat transfer coefficient for radiant heat flow. According to Eq. (1.21), we have

$$\bar{h}_r = \frac{q_r}{A(T_s - T_\infty)} = \frac{\sigma\epsilon(T_s^4 - T_\infty^4)}{T_s - T_\infty} = \sigma\epsilon(T_s^2 + T_\infty^2)(T_s + T_\infty)$$

or, since

$$(T_s^2 + T_\infty^2)(T_s + T_\infty) \approx 4\left(\frac{T_s + T_\infty}{2}\right)^3$$

we have approximately

$$\bar{h}_r = \sigma\epsilon 4\left(\frac{T_s + T_\infty}{2}\right)^3$$

The emissivity of polished platinum from Appendix 2, Table 7 is about 0.05, so $\bar{h}_r$ is about 0.05 W/m^2 °C. This shows that the amount of heat transferred by radiation is negligible compared to the heat transferred by forced convection.

The rate at which heat is transferred from the wire is therefore

$$\begin{aligned}q_c &= \bar{h}_c A(T_s - T_\infty) = (799U_\infty^{0.4})(\pi)(25 \times 10^{-6})(6 \times 10^{-3})(210)\\ &= 0.0790U_\infty^{0.4}\,\text{W}\end{aligned}$$

which must equal the rate of dissipation of electrical energy to maintain the wire at 230°C. The electrical resistance of the wire, R_e, is

$$R_e = (17.1 \times 10^{-6}\,\text{ohm cm})\frac{0.6\,\text{cm}}{\pi(25 \times 10^{-4}\,\text{cm})^2/4} = 2.09\,\text{ohm}$$

A heat balance with the current i in amperes gives

$$i^2 R_e = 0.0790 U_\infty^{0.4}$$

Solving for the current as a function of velocity, we get

$$i = \left(\frac{0.0790}{2.09}\right)^{1/2} U_\infty^{0.2} = 0.19 U_\infty^{0.20} \text{ amp}$$

7.2.2 Spheres

A knowledge of heat transfer characteristics to or from spherical bodies is important for predicting the thermal performance of systems where clouds of particles are heated or cooled in a stream of fluid. An understanding of the heat transfer from isolated particles is generally needed before attempting to correlate data for packed beds, clouds of particles, or other situations where the particles may interact. When the particles have an irregular shape, the data for spheres will yield satisfactory results if the sphere diameter is replaced by an equivalent diameter, that is, if D is taken as the diameter of a spherical particle having the same surface area as the irregular particle.

The total drag coefficient of a sphere is shown as a function of the free-stream Reynolds number in Fig. 7.7*, and corresponding data for heat transfer between a sphere and air are shown in Fig. 7.14. In the Reynolds number range from about 25 to 100,000, the equation recommended by McAdams [17] for calculating the average heat transfer coefficient for spheres heated or cooled by a gas is

$$\overline{\text{Nu}}_D = \frac{\bar{h}_c D}{k} = 0.37\left(\frac{\rho D U_\infty}{\mu}\right)^{0.6} = 0.37\,\text{Re}_D^{0.6} \tag{7.9}$$

For Reynolds numbers between 1.0 and 25, the equation

$$\bar{h}_c = c_p U_\infty \rho\left(\frac{2.2}{\text{Re}_D} + \frac{0.48}{\text{Re}_D^{0.5}}\right) \tag{7.10}$$

can be used for heat transfer in a gas. For heat transfer in liquids as well as gases, the equation

$$\overline{\text{Nu}}_D = \frac{\bar{h}_c D}{k} = 2 + (0.4\,\text{Re}_D^{0.5} + 0.06\,\text{Re}_D^{0.67})\text{Pr}^{0.4}\left(\frac{\mu}{\mu_s}\right)^{0.25} \tag{7.11}$$

correlates available data in the Reynolds number ranges between 3.5 and 7.6×10^4 and Prandtl numbers between 0.7 and 380 [18].

Achenbach [19] has measured the average heat transfer from a constant-surface-temperature sphere in air for Reynolds numbers beyond the critical

*When the sphere is dragged along by a stream (for example, a liquid droplet in a gas stream), the pertinent velocity for the Reynolds number is the velocity difference between the stream and the body.

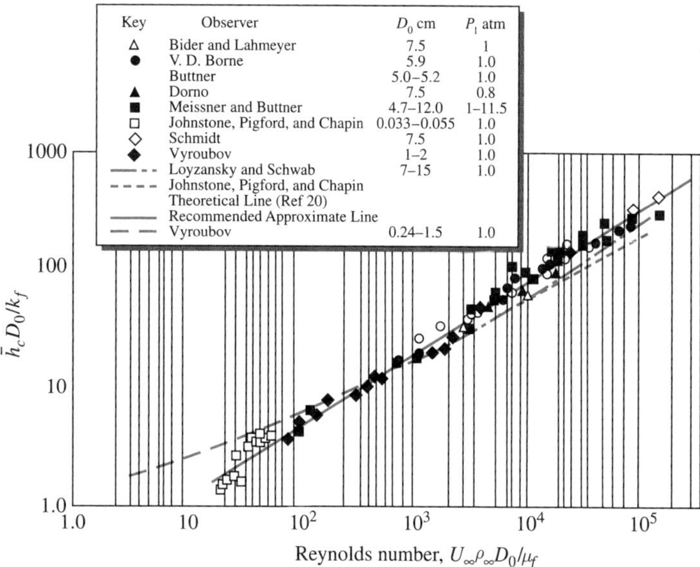

FIGURE 7.14 Correlations of experimental average heat transfer coefficients for flow over a sphere.

value. For Reynolds numbers below the critical value $100 < \text{Re}_D < 2 \times 10^5$, he found

$$\overline{\text{Nu}}_D = 2 + \left(\frac{\text{Re}_D}{4} + 3 \times 10^{-4}\,\text{Re}_D^{1.6}\right)^{1/2} \tag{7.12}$$

which can be compared with the data from several sources presented in Fig. 7.14. In the limiting case when the Reynolds number is less than unity, Johnston et al. [20] have shown from theoretical considerations that the Nusselt number approaches a constant value of 2 for a Prandtl number of unity unless the spheres have diameters of the order of the mean free path of the molecules in the gas. Beyond the critical point, $4 \times 10^5 < \text{Re}_D < 5 \times 10^6$, Achenbach recommended

$$\overline{\text{Nu}}_D = 430 + 5 \times 10^{-3}\,\text{Re}_D + 0.25 \times 10^{-9}\,\text{Re}_D^2 - 3.1 \times 10^{-17}\,\text{Re}_D^3 \tag{7.13}$$

In the case of heat transfer from a sphere to a liquid metal, Witte [21] used a transient measurement technique to determine the correlation equation

$$\overline{\text{Nu}}_D = \frac{\bar{h}_c D}{k} = 2 + 0.386(\text{Re}_D \text{Pr})^{1/2} \tag{7.14}$$

in the range $3.6 \times 10^4 < \text{Re}_D < 2 \times 10^5$. Properties are to be evaluated at the film temperature. The only liquid metal they tested was sodium. The data fell somewhat below those for previous results for air or water, but gave close agreement with previous anlayses that assumed potential flow of liquid sodium around a sphere.

7.2.3 Bluff Objects

Sogin [22] experimentally determined the heat transfer coefficient in the separated wake region behind a flat plate of width D placed perpendicular to the flow and a half-round cylinder of diameter D over Reynolds numbers between 1 and 4×10^5 and found that the following equations correlated the mean heat transfer results in air:

Normal flat plate:

$$\overline{\text{Nu}}_D = \frac{\bar{h}_c D}{k} = 0.20\ \text{Re}_D^{2/3} \tag{7.15}$$

Half-round cylinder with flat rear surface:

$$\overline{\text{Nu}}_D = \frac{\bar{h}_c D}{k} = 0.16\ \text{Re}_D^{2/3} \tag{7.16}$$

Properties are to be evaluated at the film temperature. These results are in agreement with an analysis by Mitchell [23].

Sparrow and Geiger [24] developed the following correlation for heat transfer from the upstream face of a disk oriented with its axis aligned with the free-stream flow:

$$\overline{\text{Nu}}_D = 1.05\ \text{Re}_D^{1/2}\text{Pr}^{0.36} \tag{7.17}$$

which is valid for $5000 < \text{Re}_D < 50{,}000$. Properties are to be evaluated at free-stream conditions.

Tien and Sparrow [25] measured mass transfer coefficients from square plates to air at various angles to a free stream. They studied the range $2 \times 10^4 < \text{Re}_L < 10^5$ for angles of attack and pitch of 25°, 45°, 65°, and 90° and yaw angles of 0°, 22.5°, and 45°. They found the rather unexpected result that all the data could be correlated accurately (±5%) with a single equation

$$(\bar{h}_c / c_p \rho U_\infty)\text{Pr}^{2/3} = 0.930\ \text{Re}_L^{-1/2} \tag{7.18}$$

where the length scale L is the length of the plate edge. Properties are to be evaluated at the free-stream temperature.

The insensitivity to the flow approach angle was attributed to a relocation of the stagnation point as the angle was changed, with the flow adjusting to minimize the drag force on the plate. Because the plate was square, this movement of the stagnation point did not appear to alter the mean flow-path length. For shapes other than squares, this insensitivity to the flow approach angle may not hold.

EXAMPLE 7.3 Determine the rate of convection heat loss from a solar collector panel array attached to a roof and exposed to an air velocity of 0.5 m/s, as shown in Fig. 7.15. The array is 2.5 m square, the surface of the collectors is at 70°C, and the ambient air temperature is 20°C.

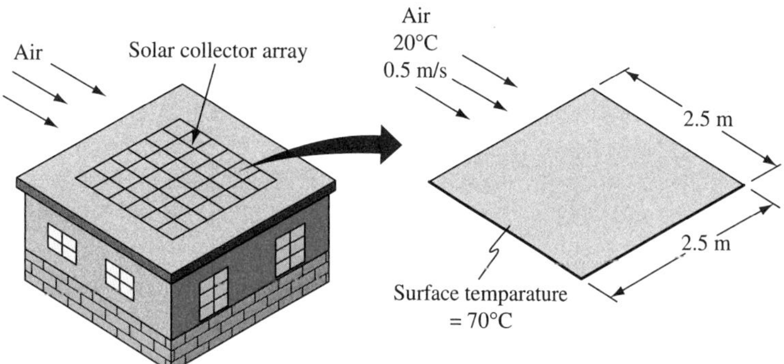

FIGURE 7.15 Sketch for Example 7.3.

SOLUTION At the free-stream temperature of 20°C, the kinematic viscosity of air is 1.57×10^{-5} m^2/s, the density is 1.16 kg/m^3, the specific heat is 1012 W s/kg °C, and Pr = 0.71. The Reynolds number is then

$$\text{Re}_L = \frac{U_\infty L}{\nu} = \frac{(0.5\text{ m/s})(2.5\text{ m})}{(1.57 \times 10^{-5}\text{ m}^2/\text{s})} = 79{,}618$$

Equation (7.18) gives

$$(\bar{h}_c/c_p\rho U_\infty)\text{Pr}^{2/3} = 0.930(79{,}618)^{-1/2} = 0.0033$$

The average heat transfer coefficient is

$$\bar{h}_c = (0.0033)(0.71)^{-2/3}(1.16\text{ kg/m}^3)(1012\text{ W s/kg K})(0.5\text{ m/s}) = 2.43\text{ W/m}^2\text{ °C}$$

and the rate of heat loss from the array is

$$q = (2.43\text{ W/m}^2\text{ K})(70 - 20)(\text{K})(2.5\text{ m})(2.5\text{ m}) = 759\text{ W}$$

Wedekind [26] measured the convection heat transfer from an isothermal disk with its axis aligned perpendicular to the free-stream gas flow. Although not strictly a bluff body, this geometry is important in the field of electronic component cooling. His data are correlated by the relation

$$\overline{\text{Nu}}_D = 0.591\,\text{Re}_D^{0.564}\text{Pr}^{1/3} \tag{7.19}$$

which is valid in the range $9 \times 10^2 < \text{Re}_D < 3 \times 10^4$.

In Eq. (7.19), D is the diameter of the disk. The range of disk thickness-to-diameter ratios tested by Wedekind was 0.06 to 0.16. Property values are to be evaluated at the film temperature. Data were correlated using heat transfer from the entire disk surface area.

7.3* Packed Beds

Many important processes require contact between a gas or a liquid stream and solid particles. These processes include catalytic reactors, grain dryers, beds for storage of solar thermal energy, gas chromatography, regenerators, and desiccant beds. Contact between the fluid and the surface of the particle allows transfer of heat and/or mass between the fluid and the particle. The device may consist of a pipe, vessel, or some other containment for the particle bed through which the gas or liquid flows. Figure 7.16(a) depicts a packed bed that could be used for heat storage of solar energy. The bed would be heated during the charging cycle by pumping hot air or another heated working fluid through the bed. The particles, which comprise the packed bed, heat up to the air temperature, thereby storing heat sensibly. During the discharge cycle, cooler air would be pumped through the bed, cooling the particles and removing the stored heat. The particles, sometimes called the bed packing, may take one of several forms, including rocks, catalyst pellets, or commercially manufactured shapes, as shown in Fig. 7.16(b), depending on the intended use of the packed bed.

Depending on the use of the packed bed, it may be necessary to transfer heat or mass between the particle and the fluid, or it may be necessary to transfer heat through the wall of the containment vessel. For example, in the packed bed in Fig. 7.16(a), one needs to predict the rate of heat transfer between the air and the particles. On the other hand, a catalytic reactor may need to reject the heat of reaction (which occurs on the particle surface) through the walls of the reactor vessel. The presence of the catalyst particles modifies the wall heat transfer to the extent that correlations for flow through an empty tube are not applicable.

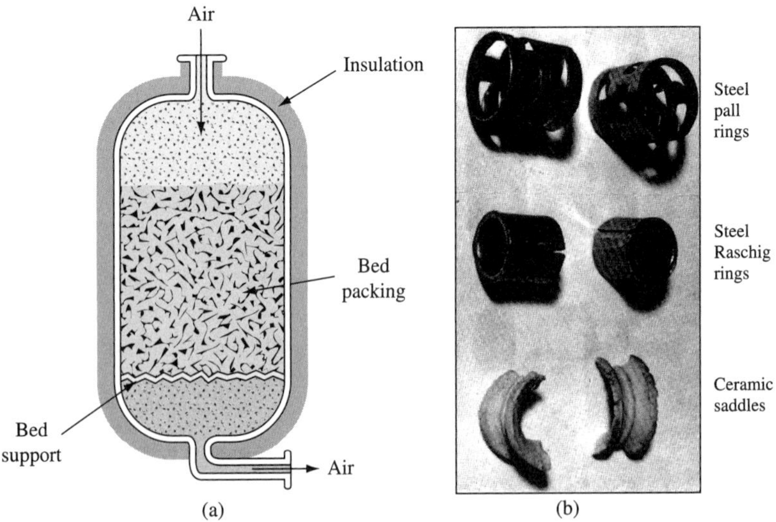

FIGURE 7.16 Packed-bed heat exchanger.

Source: Courtesy of Frank Kreith.

Correlations for heat or mass transfer in packed beds utilize a Reynolds number based on the superficial fluid velocity U_s, that is, the fluid velocity that would exist if the bed were empty. The length scale used in the Reynolds and Nusselt numbers is generally the equivalent diameter of the packing D_p. Since spheres are only one possible type of packing, an equivalent particle diameter that is based in some way on the particle volume and surface area must be defined. Such a definition may vary from one correlation to another, so some care is needed before attempting to apply the correlation. Another important parameter in packed beds is the void fraction ε, which is the fraction of the bed volume that is empty (1 – fraction of bed volume occupied by solid). The void fraction sometimes appears explicitly in correlations and is sometimes used in the Reynolds number. In addition, the Prandtl number may appear explicitly in the correlation even though the original data may have been for gases only. In such a case, the correlation is probably not reliable for liquids.

Whitaker [18] correlated data for heat transfer from gases to different kinds of packing from several sources. The types of packing included cylinders with diameter equal to height, spheres, and several types of commercial packings such as Raschig rings, partition rings, and Berl saddles. The data are correlated with ±25% by the equation

$$\frac{\bar{h}_c D_p}{k} = \frac{1-\varepsilon}{\varepsilon}\left(0.5\,\mathrm{Re}_{D_p}^{1/2} + 0.2\,\mathrm{Re}_{D_p}^{2/3}\right)\mathrm{Pr}^{1/3} \tag{7.20}$$

in the range $20 < \mathrm{Re}_{D_p} < 10^4$, $0.34 < \varepsilon < 0.78$.

The packing diameter D_p is defined as six times the volume of the particle divided by the particle surface area, which for a sphere reduces to the diameter. All fluid properties are to be evaluated at the bulk fluid temperature. If the bulk fluid temperature varies significantly through the heat exchanger, one may use the average of the inlet and outlet values. Whitaker defined the Reynolds number as

$$\mathrm{Re}_{D_p} = \frac{D_p U_s}{\nu(1-\varepsilon)}$$

Equation (7.20) does not correlate data for cubes as well because a significant reduction in surface area can occur when the cubes stack against each other. Also, data for a regular arrangement (body-centered cubic) of spheres lie well above the correlations given by Eq. (7.20).

Upadhyay [27] used the mass transfer analogy to study heat and mass transfer in packed beds at very low Reynolds numbers. Upadhyay recommends the correlation

$$(\bar{h}_c/c_p\rho U_s)\mathrm{Pr}^{2/3} = \frac{1}{\varepsilon}\,1.075\,\mathrm{Re}_{D_p}^{-0.826} \tag{7.21}$$

in the range $0.01 < \mathrm{Re}_{D_p} < 10$ and

$$(\bar{h}_c/c_p\rho U_s)\mathrm{Pr}^{2/3} = \frac{1}{\varepsilon}\,0.455\,\mathrm{Re}_{D_p}^{-0.4} \tag{7.22}$$

in the range $10 < \mathrm{Re}_{D_p} < 200$.

The Reynolds number in Eqs. (7.21) and (7.22) is defined as

$$\text{Re}_{D_p} = \frac{D_p U_s}{\nu}$$

where the partial diameter is

$$D_p = \sqrt{\frac{A_p}{\pi}}$$

and A_p is the particle surface area.

The range of void fraction tested by Upadhyay was fairly narrow, $0.371 < \varepsilon < 0.451$, and data were for cylindrical pellets only. The actual data were for a mass-transfer operation, dissolution of the solid particles in water. Use of this correlation for gases, Pr = 0.71, may be questionable.

For computing heat transfer from the wall of the packed bed to a gas, Beek [28] recommends

$$\frac{\bar{h}_c D_p}{k} = 2.58\ \text{Re}_{D_p}^{1/3}\text{Pr}^{1/3} + 0.094\ \text{Re}_{D_p}^{0.8}\text{Pr}^{0.4} \tag{7.23}$$

for particles like cylinders, which can pack next to the wall, and

$$\frac{\bar{h}_c D_p}{k} = 0.203\ \text{Re}_{D_p}^{1/3}\text{Pr}^{1/3} + 0.220\ \text{Re}_{D_p}^{0.8}\text{Pr}^{0.4} \tag{7.24}$$

for particles like spheres, which contact the wall at one point. In Eqs. (7.23) and (7.24), the Reynolds number is

$$40 < \text{Re}_{D_p} = \frac{U_s D_p}{\nu} < 2000$$

where D_p is defined by Beek as the diameter of the sphere or cylinder. For other types of packings, a definition such as that used by Whitaker should suffice. Properties in Eqs. (7.23) and (7.24) are to be evaluated at the film temperature. Beek also gives a correlation equation for the friction factor

$$f = \frac{D_p}{L}\frac{\Delta p}{\rho U_s^2 g_c} = \frac{1-\varepsilon}{\varepsilon^3}\left(1.75 + 150\frac{1-\varepsilon}{\text{Re}_{D_p}}\right) \tag{7.25}$$

In Eq. (7.25), Δp is the pressure drop over a length L of the packed bed.

EXAMPLE 7.4 Carbon monoxide at atmospheric pressure is to be heated from 50° to 350°C in a packed bed. The bed is a pipe with a 7.62-cm-ID, filled with a random arrangement of solid cylinders 0.93 cm in diameter and 1.17 cm long (see Fig. 7.17). The flow

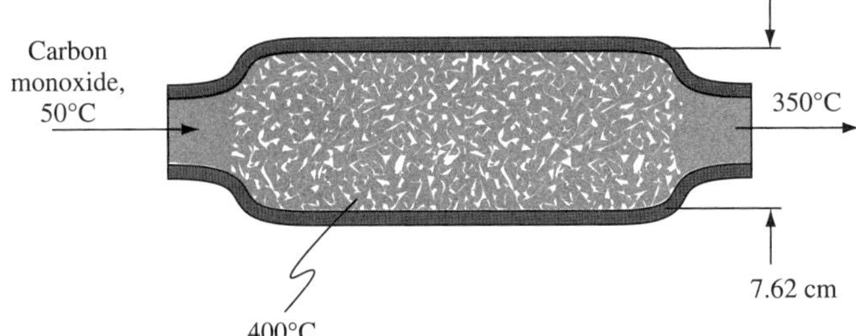

FIGURE 7.17 Schematic sketch of packed bed for Example 7.4.

rate of carbon monoxide is 5 kg/h, and the inside surface of the pipe is held at 400°C. Determine the average heat transfer coefficient at the pipe wall.

SOLUTION The film temperature is 225°C at the preheater inlet and 375°C at the preheater outlet. Evaluating properties of carbon monoxide (Table 30, Appendix 2) at the average of these, or 300°C, we find a kinematic viscosity of 4.82×10^{-5} m^2/s, a thermal conductivity of 0.042 W/m °C, a density of 0.60 kg/m^3, a specific heat of 1081 J/kg °C, and Prandtl number of 0.71. The superficial velocity is

$$U_s = \frac{(5\,\text{kg/h})}{(0.6\,\text{kg/m}^3)(\pi\, 0.0762^2/4)(\text{m}^2)} = 1827\,\text{m/h}$$

The cylindrical packing volume is $[\pi(0.93\text{ cm})^2/4](1.17\text{ cm}) = 0.795\text{ cm}^3$, and the surface area is $(2)[\pi(0.93\text{ cm})^2/4] + \pi(0.93\text{ cm})(1.17\text{ cm}) = 4.78\text{ cm}^2$. Therefore, the equivalent packing diameter is

$$D_p = \frac{(6)(0.795\,\text{cm}^3)}{4.78\,\text{cm}^2} = 1\,\text{cm} = 0.01\,\text{m}$$

giving a Reynolds number of

$$\text{Re}_{D_p} = \frac{(1827\,\text{m/h})/(3600\,\text{s/h})(0.01\,\text{m})}{(4.82 \times 10^{-5}\,\text{m}^2/\text{s})} = 105$$

From Eq. (7.23), we find

$$\frac{\bar{h}_c D_p}{k} = 2.58(105)^{1/3}(0.71)^{1/3} + 0.094(105)^{0.8}(0.71)^{0.4}$$
$$= 14.3$$

or

$$\bar{h}_c = \frac{(14.3)(0.042\,\text{W/m K})}{0.01\,\text{m}} = 60.1\,\text{W/m}^2\,^\circ\text{C}$$

7.4 Tube Bundles in Cross-Flow

The evaluation of the convection heat transfer coefficient between a bank of tubes and a fluid flowing at right angles to the tubes is an important step in the design and performance analysis of many types of commercial heat exchangers. There are, for example, a large number of gas heaters in which a hot fluid inside the tubes heats a gas passing over the outside of the tubes. Figure 7.18 shows several arrangements of tubular air heaters in which the products of combustion, after they leave a boiler, economizer, or superheater, are used to preheat the air going to the steam-generating units. The shells of these gas heaters are usually rectangular, and the shell-side gas flows in the space between the outside of the tubes and the shell. Since the flow cross-sectional area is continuously changing along the path, the shell-side gas speeds up and slows down periodically. A similar situation exists in some unbaffled short-tube liquid-to-liquid heat exchangers in which the shell-side fluid flows over the tubes. In these units, the tube arrangement is similar to that in a gas heater except that the shell cross-sectional area varies where a cylindrical shell is used.

Heat transfer and pressure-drop data for a large number of these heat exchanger cores have been compiled by Kays and London [29]. Their summary includes data on banks of bare tubes as well as tubes with plate fins, strip fins, wavy plate fins, pin fins, and so on.

In this section, we discuss some of the flow and heat transfer characteristics of bare-tube bundles. Rather than concern ourselves with detailed information on a specific heat exchanger core or tube arrangement or a particular type of tube fin, we shall focus on the common element of most heat exchangers, the tube bundle in cross-flow. This information is directly applicable to one of the most common heat exchangers, shell-and-tube, and will provide a basis for understanding the engineering data on specific heat exchangers presented in [29].

The heat transfer in flow over tube bundles depends largely on the flow pattern and the degree of turbulence, which in turn are functions of the velocity of the fluid and the size and arrangement of the tubes. The photographs in Figure 7.19 on page 446 and Figure 7.20 on 447 illustrate the flow patterns for water flowing in the low-turbulent range over tubes arranged *in line* and *staggered*, respectively. The photographs were obtained [30] by sprinkling fine aluminum powder on the surface of water flowing perpendicularly to the axis of vertically placed tubes. We observe that the flow patterns around tubes in the first transverse rows are similar to those for flow around single tubes. Focusing our attention on a tube in the first row of the in-line arrangement, we see that the boundary layer separates from both sides of the tube, and a wake forms behind it. The turbulent wake extends to the tube located in the second transverse row. As a result of the high turbulence in the wakes, the boundary layers around tubes in the second and subsequent rows become progressively thinner. It is therefore not unexpected that in turbulent flow, the heat transfer coefficients of tubes in the first row are smaller than the heat transfer coefficients of tubes in subsequent rows. In laminar flow, on the other hand, the opposite trend has been observed [31] due to a shading effect by the upstream tubes.

For a closely spaced staggered-tube arrangement (Fig. 7.20), the turbulent wake behind each tube is somewhat smaller than for similar in-line arrangements, but there is no appreciable reduction in the overall energy dissipation. Experiments on various types of tube arrangements [7] have shown that, for practical units, the

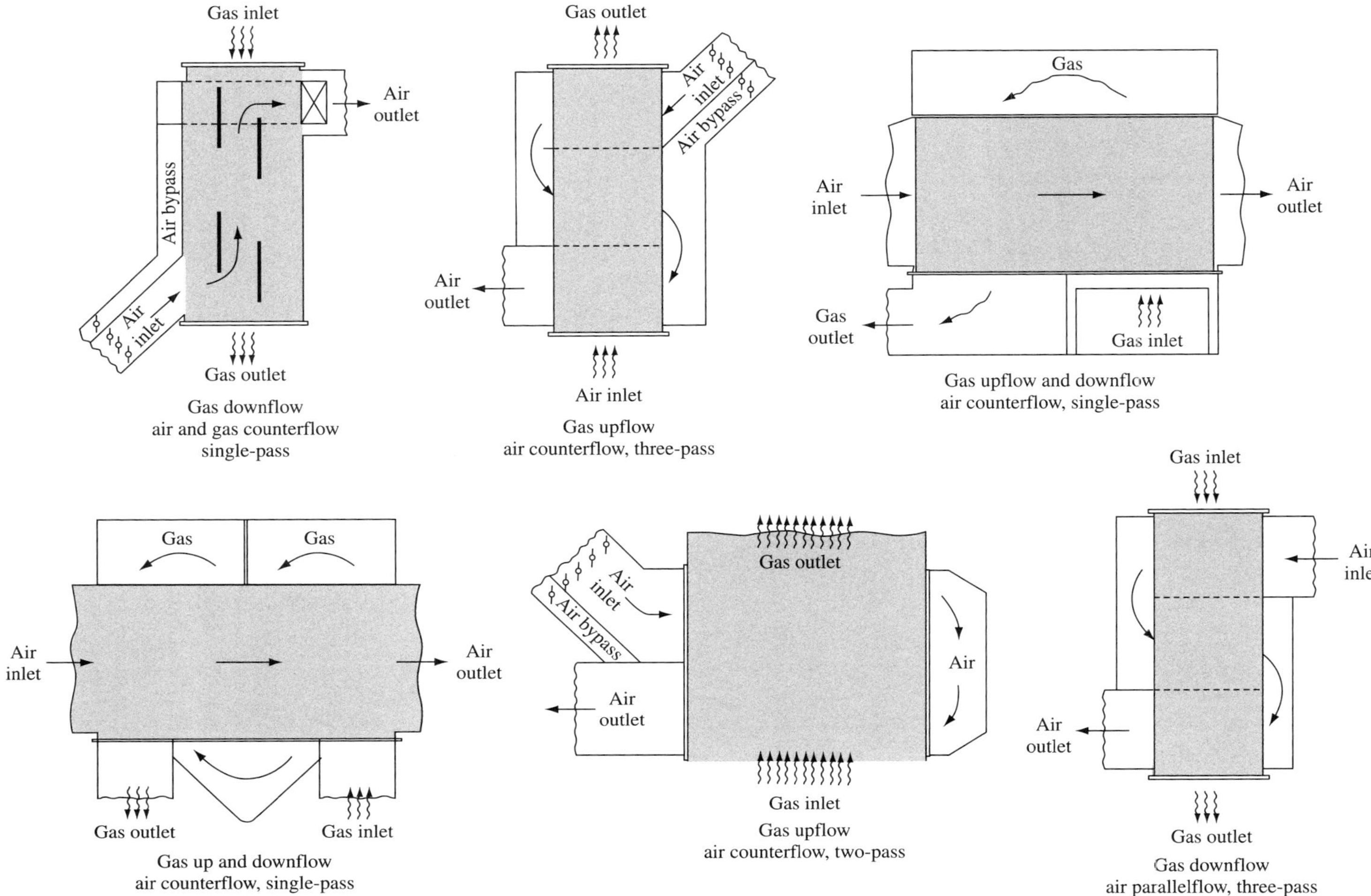

FIGURE 7.18 Some arrangements for tubular air heaters.

Source: Courtesy of the Babcock & Wilcox Company.

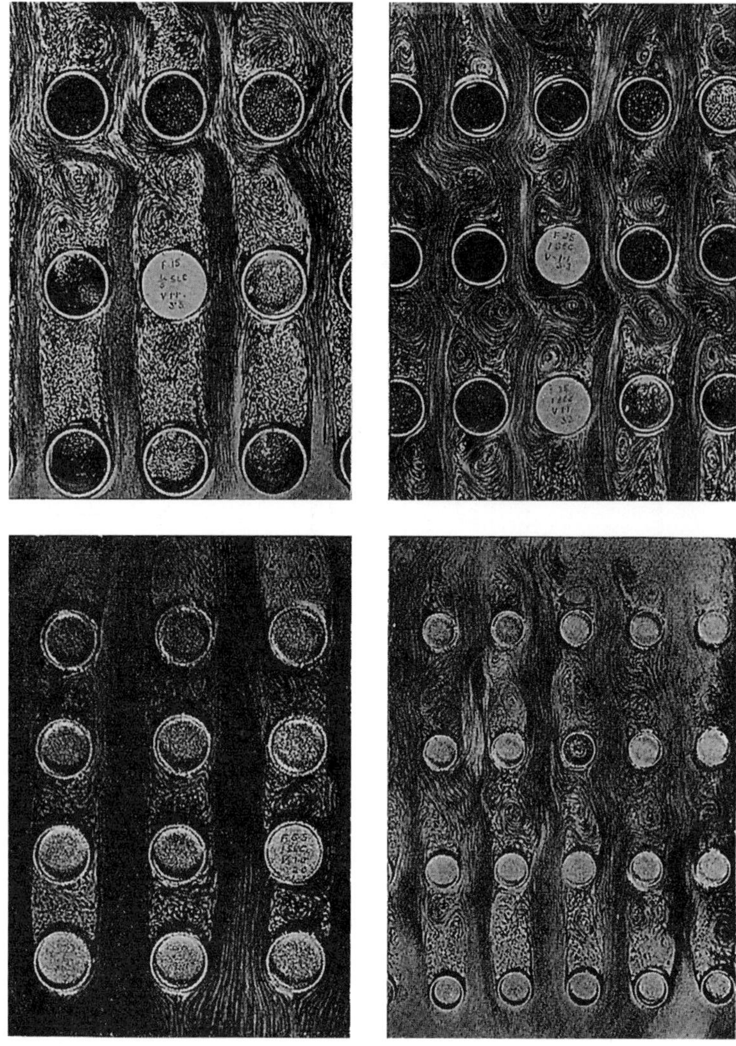

FIGURE 7.19 Flow patterns for in-line tube bundles. Flow in all photographs is upward.

Source: "Photographic Study of Fluid Flow between Banks of Tubes," Pendennis Wallis, Proceedings of the Institution of Mechanical Engineers, Professional Engineering Publishing, ISSN 0020-3483, Volume 142/1939, DOI: 10. 1243/PIME_PROC_1939_142_027_02, pp. 379–387.

relation between heat transfer and energy dissipation depends primarily on the velocity of the fluid, the size of the tubes, and the distance between the tubes. However, in the transition zone, the performance of a closely spaced, staggered tube arrangement is somewhat superior to that of a similar in-line tube arrangement. In the laminar regime, the first row of tubes exhibits lower heat transfer than the downstream rows, just the opposite behavior of the in-line arrangement.

The equations available for the calculation of heat transfer coefficients in flow over tube banks are based entirely on experimental data because the flow pattern is too complex to be treated analytically. Experiments have shown that in flow over staggered-tube banks, the transition from laminar to turbulent flow is more gradual than in flow

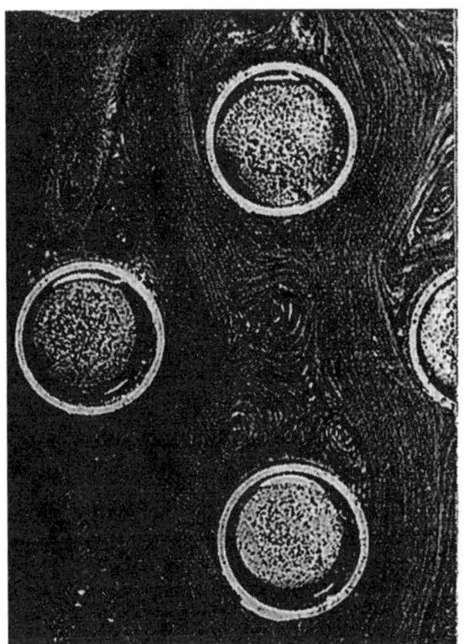

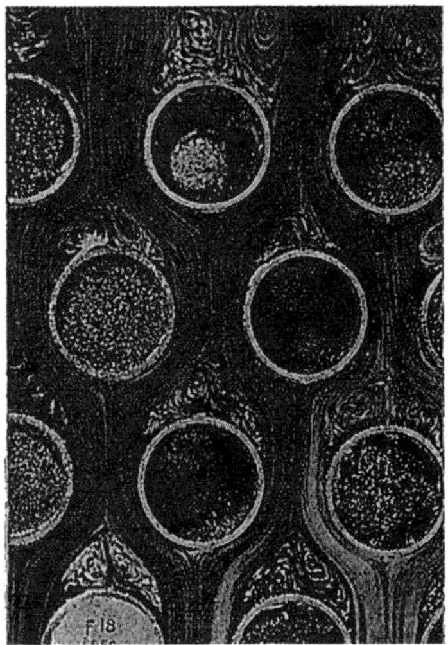

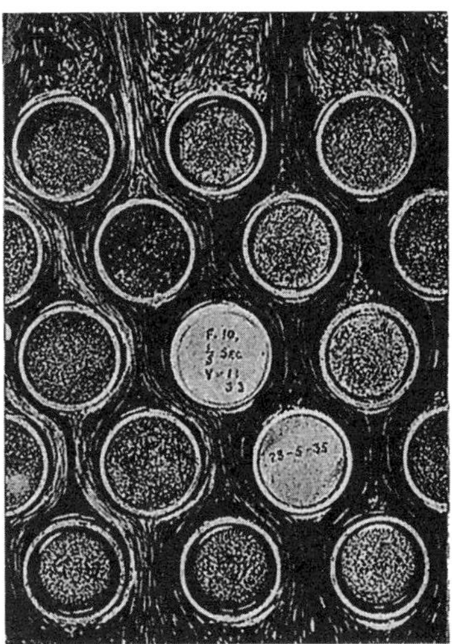

FIGURE 7.20 Flow patterns for staggered tube bundles. Flow in all photographs is upward.

Source: "Photographic Study of Fluid Flow between Banks of Tubes," Pendennis Wallis, Proceedings of the Institution of Mechanical Engineers, Professional Engineering Publishing, ISSN 0020-3483, Volume 142/1939, DOI: 10. 1243/PIME_PROC_1939_142_027_02, pp. 379–387.

through a pipe, whereas for in-line tube bundles the transition phenomena resemble those observed in pipe flow. In either case, the transition from laminar to turbulent flow begins at a Reynolds number based on the velocity at the minimum flow area, about 200, and the flow becomes fully turbulent at a Reynolds number of about 6000.

For engineering calculations, the average heat transfer coefficient for the entire tube bundle is of primary interest. The experimental data for heat transfer in flow over banks of tubes are usually correlated by an equation of the form $\overline{\text{Nu}}_D = \text{const}(\text{Re}_D)^m(\text{Pr})^n$, which was previously used to correlate the data for flow over a single tube. To apply this equation to flow over tube bundles, it is necessary to select a reference velocity, since the speed of the fluid varies along its path. The velocity used to build the Reynolds number for flow over tube bundles is based on

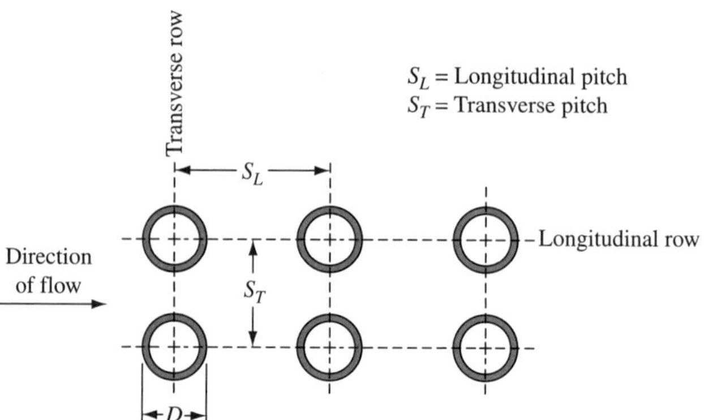

FIGURE 7.21 Nomenclature for in-line tube arrangements.

the *minimum free area* available for fluid flow, regardless of whether the minimum area occurs in the transverse or diagonal openings. For in-line tube arrangements (Fig. 7.21), the minimum free-flow area per unit length of tube A_{min} is always $A_{min} = S_T - D$, where S_T is the distance between the centers of the tubes in adjacent longitudinal rows (measured perpendicularly to the direction of flow), or the *transverse pitch*. Then the maximum velocity is $S_T/(S_T - D)$ times the free-flow velocity based on the shell area without tubes. The symbol S_L denotes the center-to-center distance between adjacent transverse rows of tubes or pipes (measured in the direction of flow) and is called the *longitudinal pitch*.

For staggered arrangements (Fig. 7.22) the minimum free-flow area can occur, as in the previous case, either between adjacent tubes in a row or, if S_L/S_T is so small that $\sqrt{(S_T/2)^2 + S_L^2} < (S_T + D)/2$, between diagonally opposed tubes. In the latter case, the maximum velocity U_{max} is $(S_T/2)/(\sqrt{S_L^2 + (S_T/2)^2} - D)$times the free-flow velocity based on the shell area without tubes.

Having determined the maximum velocity, the Reynolds number is

$$\mathrm{Re}_D = \frac{U_{max} D}{\nu}$$

where D is the tube diameter.

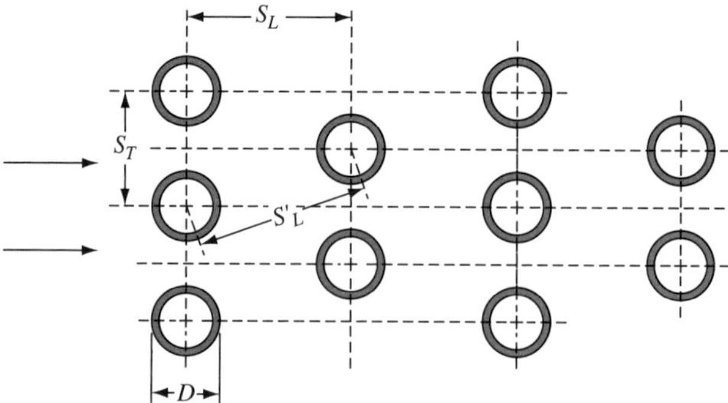

FIGURE 7.22 Sketch illustrating nomenclature for staggered tube arrangements.

Žukauskas [7] has developed correlation equations for predicting the mean heat transfer from tube banks. The equations are primarily for tubes in the inner rows of the tube bank. However, the mean heat transfer coefficients for rows 3, 4, 5, . . . are indistinguishable from one another; the second row exhibits a 10 to 25% lower heat transfer than the internal rows for $\text{Re} < 10^4$ and equal heat transfer for $\text{Re} > 10^4$; the heat transfer of the first row may be 60% to 75% of that of the internal rows, depending on longitudinal pitch. Therefore, the correlation equations will predict tube-bank heat transfer within 6% for 10 or more rows. The correlations are valid for $0.7 < \text{Pr} < 500$.

The correlation equations are of the form

$$\overline{\text{Nu}}_D = C\,\text{Re}_D^m \text{Pr}^{0.36}\left(\frac{\text{Pr}}{\text{Pr}_s}\right)^{0.25} \tag{7.26}$$

where the subscript s means that the fluid property value is to be evaluated at the tube-wall temperature. Other fluid properties are to be evaluated at the bulk fluid temperature.

For in-line tubes in the laminar flow range $10 < \text{Re}_D < 100$,

$$\overline{\text{Nu}}_D = 0.8\,\text{Re}_D^{0.4} \text{Pr}^{0.36}\left(\frac{\text{Pr}}{\text{Pr}_s}\right)^{0.25} \tag{7.27}$$

and for staggered tubes in the laminar flow range $10 < \text{Re}_D < 100$,

$$\overline{\text{Nu}}_D = 0.9\,\text{Re}_D^{0.4} \text{Pr}^{0.36}\left(\frac{\text{Pr}}{\text{Pr}_s}\right)^{0.25} \tag{7.28}$$

Chen and Wung [32] validated Eqs. (7.27) and (7.28) using a numerical solution for $50 < \text{Re}_D < 1000$.

In the transition regime, $10^3 < \text{Re}_D < 2 \times 10^5$, m is the exponent on Re_D and varies from 0.55 to 0.73 for in-line banks, depending on the tube pitch. A mean value of 0.63 is recommended for in-line banks with $S_T/S_L \geq 0.7$:

$$\overline{\text{Nu}}_D = 0.27\,\text{Re}_D^{0.63} \text{Pr}^{0.36}\left(\frac{\text{Pr}}{\text{Pr}_s}\right)^{0.25} \tag{7.29}$$

[For $S_T/S_L < 0.7$, Eq. (7.29) significantly overpredicts $\overline{\text{Nu}}_D$; however, this tube arrangement yields an ineffective heat exchanger.]

For staggered banks with $S_T/S_L < 2$,

$$\overline{\text{Nu}}_D = 0.35\left(\frac{S_T}{S_L}\right)^{0.2} \text{Re}_D^{0.60} \text{Pr}^{0.36}\left(\frac{\text{Pr}}{\text{Pr}_s}\right)^{0.25} \tag{7.30}$$

and for $S_T/S_L \geq 2$,

$$\overline{\text{Nu}}_D = 0.40\,\text{Re}_D^{0.60} \text{Pr}^{0.36}\left(\frac{\text{Pr}}{\text{Pr}_s}\right)^{0.25} \tag{7.31}$$

In the turbulent regime, $\text{Re}_D > 2 \times 10^5$, heat transfer for the inner tubes increases rapidly due to turbulence generated by the upstream tubes. In some cases, the

Reynolds number exponent m exceeds 0.8, which corresponds to the exponent on Reynolds numbers for the turbulent boundary layer on the front of the tube. This means that the heat transfer on the rear portion of the tube must increase even more rapidly. Therefore, the value of m depends on tube arrangement, tube roughness, fluid properties, and free-stream turbulence. An average value $m = 0.84$ is recommended.

For in-line tube banks,

$$\overline{\text{Nu}}_D = 0.021\,\text{Re}_D^{0.84}\text{Pr}^{0.36}\left(\frac{\text{Pr}}{\text{Pr}_s}\right)^{0.25} \tag{7.32}$$

For staggered rows with Pr > 1,

$$\overline{\text{Nu}}_D = 0.022\,\text{Re}_D^{0.84}\text{Pr}^{0.36}\left(\frac{\text{Pr}}{\text{Pr}_s}\right)^{0.25} \tag{7.33}$$

and if Pr = 0.7,

$$\overline{\text{Nu}}_D = 0.019\,\text{Re}_D^{0.84} \tag{7.34}$$

The preceding correlation equations, Eqs. (7.27) to (7.34), are compared with experimental data from several sources in Fig. 7.23 for in-line arrangements and

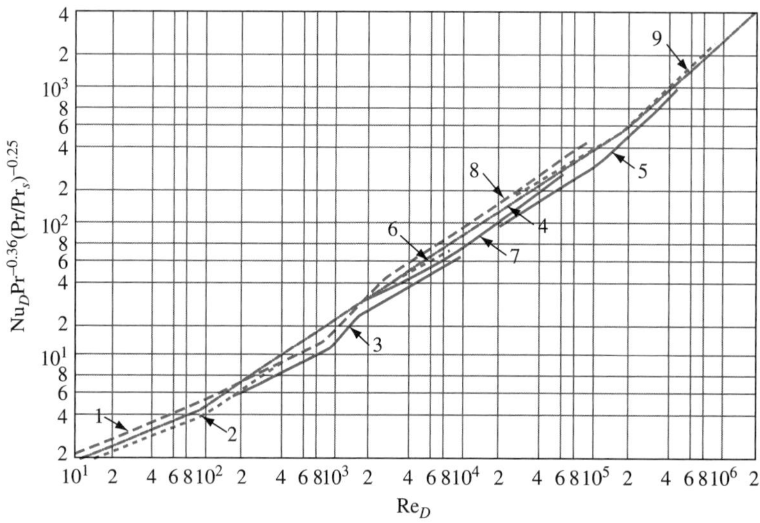

FIGURE 7.23 Comparison of heat transfer of in-line banks. Curve 1, $S_T/D \times S_L/D = 1.25 \times 1.25$, and curve 2, 1.5×1.5 (after Bergelin et al); curve 3, 1.25×1.25 (after Kays and London); curve 4, 1.45×1.45 (after Kuznetsov and Turilin); curve 5, 1.3×1.5 (after Lyapin); curve 6, 2.0×2.0 (after Isachenko); curve 7, 1.9×1.9 (after Grimson); curve 8, 2.4×2.4 (after Kuznetsov and Turilin); curve 9, 2.1×1.4 (after Hammecke et al.).

Source: "Heat Transfer from Tubes in Cross Flow" by A. A. Zukauskas, Advances in Heat Transfer, Vol. 8, 1972, pp. 93–106.

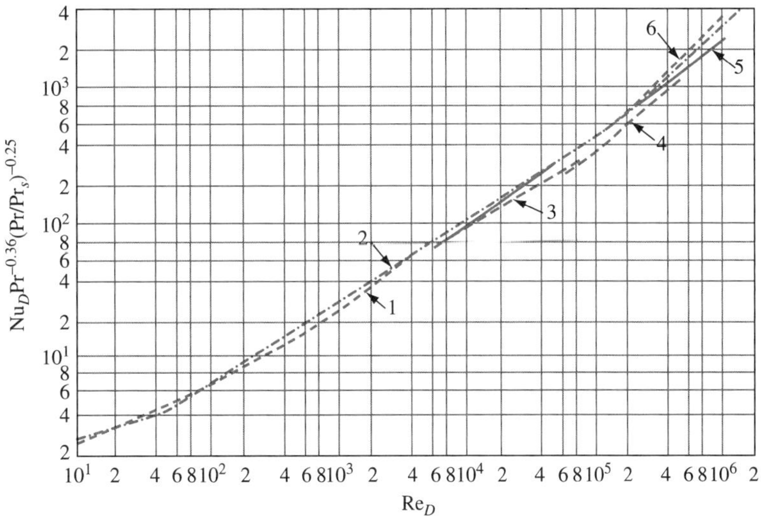

FIGURE 7.24 Comparison of heat transfer of staggered banks. Curve 1, $S_T/D \times S_L/D = 1.5 \times 1.3$ (after Bergelin et al); curve 2, 1.5×1.5 and 2.0×2.0 (after Grimson and Isachenko); curve 3, 2.0×2.0 (after Antuf'yev and Beletsky, Kuznetsov and Turilin, and Kazakevich); curve 4, 1.3×1.5 (after Lyapin); curve 5, 1.6×1.4 (after Dwyer and Sheeman); curve 6, 2.1×1.4 (after Hammecke et al.).

Source: "Heat Transfer from Tubes in Cross Flow" by A. A. Zukauskas, Advances in Heat Transfer, Vol. 8, 1972, pp. 93–106.

in Fig. 7.24 for staggered arrangements. Solid lines in the figures represent the correlation equations.

Achenbach [33] extended the tube-bundle data up to $\text{Re}_D = 7 \times 10^6$ for a staggered arrangement with transverse pitch $S_T/D = 2$ and lateral pitch $S_L/D = 1.4$. His data are correlated by the relation

$$\overline{\text{Nu}}_D = 0.0131\,\text{Re}_D^{0.883}\text{Pr}^{0.36} \tag{7.35}$$

which is valid in the range $4.5 \times 10^5 < \text{Re}_D < 7 \times 10^6$.

Achenbach also investigated the effect of tube roughness on heat transfer and pressure drop in in-line tube bundles in the turbulent regime [34]. He found that the pressure drop through a rough-tube bundle was about 30% less than that for a smooth-tube bundle, while the heat transfer coefficient was about 40% greater than that for the smooth-tube bundle. The maximum effect was seen for a surface roughness of about 0.3% of the tube diameter and was attributed to the early onset of turbulence promoted by the roughness.

For closely spaced in-line banks, it is necessary to base the Reynolds number on the average velocity integrated over the perimeter of the tube so that the results for various spacings will collapse to a single correlation line. Such results, presented in [7], indicate that this procedure correlates data for $2 \times 10^3 < \text{Re}_D < 2 \times 10^5$ and for spacings $1.01 \le S_T/D = S_L/D \le 1.05$. However, Aiba et al. [35] show that for a single row of closely spaced tubes a critical Reynolds number, Re_{Dc}, exists. Below

Re_{Dc}, a stagnant region forms behind the first cylinder, reducing heat transfer to the remaining (three) cylinders below that for a single cylinder. Above Re_{Dc}, the stagnant region rolls up into a vortex and significantly increases the heat transfer from the downstream cylinders.

In the range $1.15 \leq S_L/D \leq 3.4$, Re_{Dc} may be calculated from

$$\mathrm{Re}_{Dc} = 1.14 \times 10^5 \left(\frac{S_L}{D}\right)^{-5.84} \tag{7.36}$$

From data [7] on closely spaced tube banks ($1.01 \leq S_T/D = S_L/D \leq 1.05$), one would conclude that the discontinuous behavior does not occur when the single row of tubes is placed in a bank consisting of several such tube rows.

The pressure drop for a bank of tubes in cross-flow can be calculated from

$$\Delta p = f \frac{\rho U_{\max}^2}{2g_c} N \tag{7.37}$$

where the velocity is that in the minimum free-flow area, N is the number of transverse rows, and the friction coefficient f depends on Re_D (also based on velocity in the minimum free-flow area) according to Fig. 7.25 for in-line banks and Fig. 7.26 for staggered banks [7]. The correlation factor x shown in those figures accounts for nonsquare in-line arrangements and nonequilateral-triangle staggered arrangements.

The variation of the average heat transfer coefficient of a tube bank with the number of transverse rows is shown in Table 7.3 for *turbulent* flow. To calculate the average heat transfer coefficient for tube banks with less than 10 rows, the $\bar{h}_c$ obtained from Eqs. (7.32) to (7.34) should be multiplied by the appropriate ratio $\bar{h}_{cN}/\bar{h}_c$.

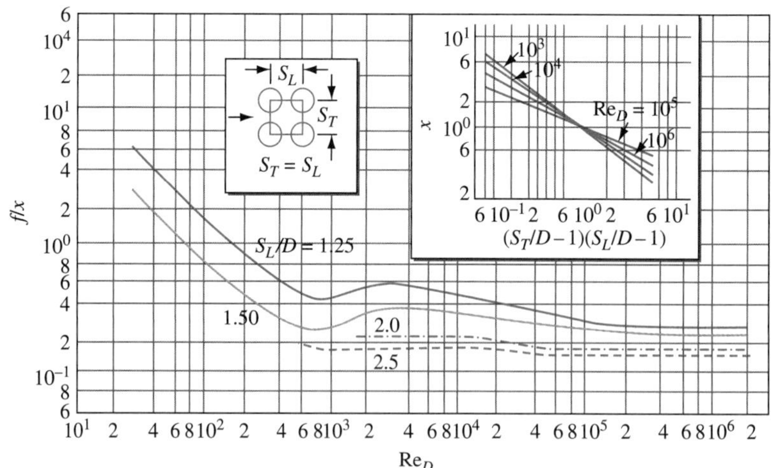

FIGURE 7.25 Pressure-drop coefficients of in-line banks as referred to the relative longitudinal pitch S_L/D.

Source: "Heat Transfer from Tubes in Cross Flow" by A. A. Zukauskas, Advances in Heat Transfer, Vol. 8, 1972, pp. 93–106. Copyright © 1972 by Academic Press. Reprinted by permission of the publisher.

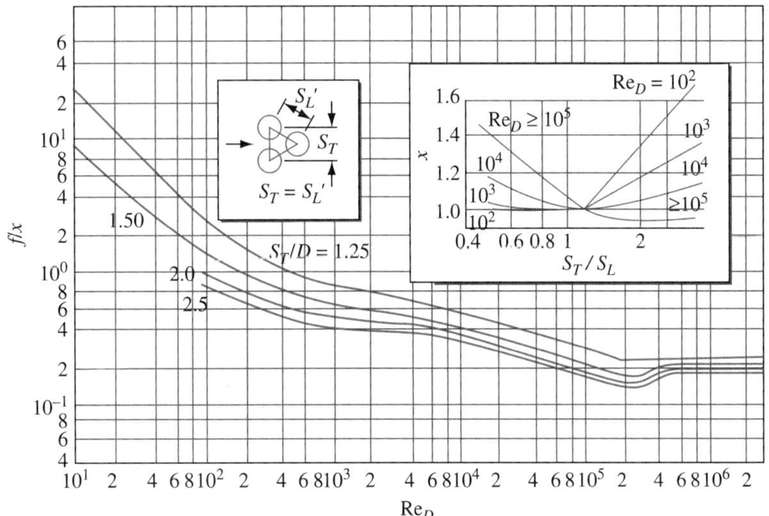

FIGURE 7.26 Pressure-drop coefficients of staggered banks as referred to the relative transverse pitch S_T/D.

Source: "Heat Transfer from Tubes in Cross Flow" by A. A. Zukauskas, Advances in Heat Transfer, Vol. 8, 1972, pp. 93–106. Copyright © 1972 by Academic Press. Reprinted by permission of the publisher.

TABLE 7.3 Ratio of h_c for N transverse rows to $\bar{h}_c$ for 10 transverse rows in turbulent flow[a]

Ratio	N									
$\bar{h}_{cN}/\bar{h}_c$	1	2	3	4	5	6	7	8	9	10
Staggered tubes	0.68	0.75	0.83	0.89	0.92	0.95	0.97	0.98	0.99	1.0
In-line tubes	0.64	0.80	0.87	0.90	0.92	0.94	0.96	0.98	0.99	1.0

[a]From W. M. Kays and R. K. Lo [36].

EXAMPLE 7.5 Atmospheric air at 58°F is to be heated to 86°F by passing it over a bank of brass tubes inside which steam at 212°F is condensing. The heat transfer coefficient on the inside of the tubes is about 1000 Btu/h ft^2 °F. The tubes are 2 ft long, 1/2-in.-OD, BWG No. 18 (0.049-in. wall-thickness). They are to be arranged in-line in a square pattern with a pitch of 3/4-in. inside a rectangular shell 2 ft wide and 15 in. high. The heat exchanger is shown schematically in Fig. 7.27 on the next page. If the total mass rate of flow of the air to be heated is 32,000 lb$_m$/h, estimate (a) the number of transverse rows required and (b) the pressure drop.

SOLUTION (a) The mean bulk temperature of the air T_{air} will be approximately equal to

$$\frac{58 + 86}{2} = 72°\text{F}$$

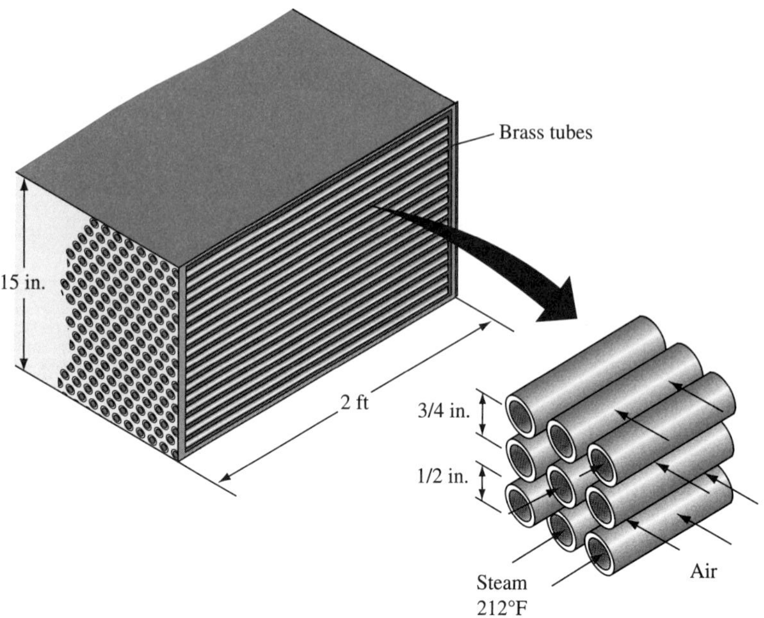

FIGURE 7.27 Sketch of tube bank for Example 7.5.

Appendix 2, Table 28 then gives for the properties of air at this mean bulk temperature: $\rho = 0.072$ lb/ft^3, $k = 0.0146$ Btu/h °F ft, $\mu = 0.0444$ lb/ft h, Pr = 0.71, and $\text{Pr}_s = 0.71$. The mass velocity at the minimum cross-sectional area, which is between adjacent tubes, is calculated next. The shell is 15 in. high and consequently holds 20 longitudinal rows of tubes. The minimum free area is

$$A_{\text{min}} = (20)(2\text{ ft})\left(\frac{0.75 - 0.50}{12}\text{ ft}\right) = 0.833\text{ ft}^2$$

and the maximum mass velocity ρU_{max} is

$$G_{\text{max}} = \frac{(32{,}000\text{ lb/h})}{(0.833\text{ ft}^2)} = 38{,}400\text{ lb}_\text{m}\text{/h ft}^2$$

Hence, the Reynolds number is

$$\text{Re}_{\text{max}} = \frac{G_{\text{max}} D_0}{\mu} = \frac{(38{,}400\text{ lb/h ft}^2)(0.5/12\text{ ft})}{0.0444\text{ lb/h ft}} = 36{,}036$$

Assuming that more than 10 rows will be required, the heat transfer coefficient is calculated from Eq. (7.29). We get

$$\begin{aligned}\bar{h}_c &= \left(\frac{0.0146\text{ Btu/h ft °F}}{0.5/12\text{ ft}}\right)(0.27)(36{,}036)^{0.63}(0.71)^{0.36}\\ &= 62.1\text{ Btu/h ft}^2\text{ °F}\end{aligned}$$

We can now determine the temperature at the outer tube wall. There are three thermal resistances in series between the steam and the air. The resistance at the steam side per tube is approximately

$$R_1 = \frac{1/\bar{h}_i}{\pi D_i L} = \frac{1/1000}{3.14(0.402/12)2} = 0.00474 \text{ h °F/Btu}$$

The resistance of the pipe wall ($k = 60$ Btu/h ft °F) is approximately

$$R_2 = \frac{0.049/k}{\pi[(D_0 + D_i)/2]L} = \frac{0.049/60}{(3.14)(0.451)(2)} = 0.000287 \text{ h °F/Btu}$$

The resistance at the outside of the tube is

$$R_3 = \frac{1/\bar{h}_0}{\pi D_0 L} = \frac{1/62.1}{3.14(0.5/12)2} = 0.0615 \text{ h °F/Btu}$$

The total resistance is then

$$R_1 + R_2 + R_3 = 0.0667 \text{ h °F/Btu}$$

Since the sum of the resistance at the steam side and the resistance of the tube wall is about 8% of the total resistance, about 8% of the total temperature drop occurs between the steam and the outer tube wall. The tube surface temperature can be corrected, and we get

$$T_s = 201 \text{ °F}$$

This will not change the values of the physical properties appreciably, and no adjustment in the previously calculated value of $\bar{h}_c$ is necessary.

The mean temperature difference between the steam and the air now can be calculated. Using the arithmetic average, we get

$$\Delta T_{\text{avg}} = T_{\text{steam}} - T_{\text{air}} = 212 - \left(\frac{58 + 86}{2}\right) = 140\text{°F}$$

The specific heat of air at constant pressure is 0.241 Btu/lb$_{\text{m}}$ °F. Equating the rate of heat flow from the steam to the air to the rate of enthalpy rise of the air gives

$$\frac{20N\,\Delta T_{\text{avg}}}{R_1 + R_2 + R_3} = \dot{m}_{\text{air}} c_p (T_{\text{out}} - T_{\text{in}})_{\text{air}}$$

Solving for N, which is the number of transverse rows, yields

$$N = \frac{(32{,}000 \text{ lb/h})(0.24 \text{ Btu/lb °F})(86 - 58)(\text{°F})(0.0667 \text{ h °F/Btu})}{(20)(140\text{°F})}$$

$$= 5.12, \text{ i.e., 5 rows}$$

Since the number of tubes is less than 10, it is necessary to correct $\bar{h}_c$ in accordance with Table 7.3, or

$$\bar{h}_{c6 \text{ rows}} = 0.92\bar{h}_{c10 \text{ rows}} = (0.92)(62.1) = 57.1 \text{ Btu/h ft}^2 \text{ °F}$$

Repeating the calculations with the corrected values of the average heat transfer coefficient on the air side, we find that six transverse rows are sufficient for heating the air according to the specifications.

(b) The pressure drop is obtained from Eq. (7.37) and Fig. 7.25. Since $S_T = S_L = 1.5D$, we have

$$\left(\frac{S_T}{D} - 1\right)\left(\frac{S_L}{D} - 1\right) = 0.5^2 = 0.25$$

For $\mathrm{Re}_D = 36{,}000$ and $(S_T/D - 1)(S_L/D - 1) = 0.25$, the correction factor is $x = 2.5$, and the friction factor from Fig. 7.24 is

$$f = (2.5)(0.3) = 0.75$$

The velocity is

$$U_{\max} = \frac{G_{\max}}{\rho} = \frac{(38{,}400\,\mathrm{lb_m/h\ ft^2})}{(0.072\,\mathrm{lb_m/ft^3})(3600\ \mathrm{s/h})}$$

$$= 148\ \mathrm{ft/s}$$

with $N = 6$, the pressure drop is therefore

$$\Delta p = 0.75\,\frac{(0.072\,\mathrm{lb_m/ft^3})(148\,\mathrm{ft/s})^2}{2(32.2\,\mathrm{lb_m\ ft/lb_f\,s^2})}\,6 = 110\ \mathrm{lb_f/ft^2}$$

EXAMPLE 7.6 Methane gas at 20°C is to be preheated in a heat exchanger consisting of a staggered arrangement of 4-cm-OD tubes, 5 rows deep, with a longitudinal spacing of 6 cm and a transverse spacing of 8 cm (see Fig. 7.28). Subatmospheric-pressure steam is condensing inside the tubes, maintaining the tube wall temperature at 50°C. Determine (a) the average heat transfer coefficient for the tube bank and (b) the pressure drop through the tube bank. The methane flow velocity is 10 m/s upstream of the tube bank.

SOLUTION For methane at 20°C, Table 36, Appendix 2 gives $\rho = 0.668\ \mathrm{kg/m^3}$, $k = 0.0332$ W/m K, $\nu = 16.27 \times 10^{-6}\ \mathrm{m^2/s}$, and $\mathrm{Pr} = 0.73$. At 50°C, $\mathrm{Pr} = 0.73$.

(a) From the geometry of the tube bundle, we see that the minimum flow area is between adjacent tubes in a row and that this area is half the frontal area of the tube bundle. Thus,

$$U_{\max} = 2\left(10\,\frac{\mathrm{m}}{\mathrm{s}}\right) = 20\,\frac{\mathrm{m}}{\mathrm{s}}$$

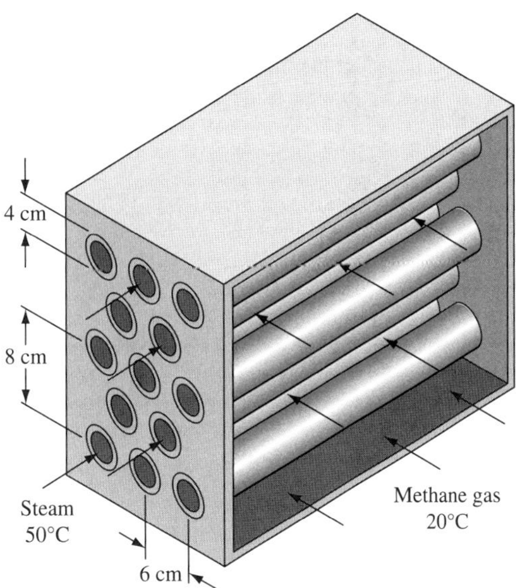

FIGURE 7.28 Sketch of heat exchanger for Example 7.6.

and

$$\mathrm{Re}_D = \frac{U_{\max} D}{\nu} = \frac{\left(20\,\frac{\mathrm{m}}{\mathrm{s}}\right)(0.04\,\mathrm{m})}{\left(16.27 \times 10^{-6}\,\frac{\mathrm{m}^2}{\mathrm{s}}\right)} = 49{,}170$$

which is in the transition regime.

Since $S_T/S_L = 8/6 < 2$, we use Eq. (7.30):

$$\begin{aligned}\overline{\mathrm{Nu}}_D &= 0.35\left(\frac{S_T}{S_L}\right)^{0.2} \mathrm{Re}_D^{0.60}\mathrm{Pr}^{0.36}\left(\frac{\mathrm{Pr}}{\mathrm{Pr}_s}\right)^{0.25} \\ &= (0.35)\left(\frac{8}{6}\right)^{0.2}(49{,}170)^{0.6}(0.73)^{0.36}(1) \\ &= 216\end{aligned}$$

and

$$\bar{h}_c = \frac{\overline{\mathrm{Nu}}\,k}{D} = \frac{(216)\left(0.0332\,\frac{\mathrm{W}}{\mathrm{m\,K}}\right)}{(0.04\,\mathrm{m})} = 179\,\frac{\mathrm{W}}{\mathrm{m}^2\,\mathrm{K}}$$

Since there are fewer than 10 rows, the correlation factor in Table 7.3 gives $\bar{h}_c = (0.92)(179) = 165\ \mathrm{W/m^2\,K}$.

(b) Tube-bundle pressure drop is given by Eq. (7.37). The insert in Fig. (7.26) gives the correction factor x. We have $S_T/S_L = 8/6 = 1.33$ and $\text{Re}_D = 49{,}170$, giving $x = 1.0$. Using the main body of the figure with $S_T/D = 8/4 = 2$, we find that $f/x = 0.25$ or $f = 0.25$. Now the pressure drop can be calculated from Eq. (7.37):

$$\Delta p = (0.25)\frac{\left(0.668\dfrac{\text{kg}}{\text{m}^3}\right)\left(20\dfrac{\text{m}}{\text{s}}\right)^2}{2\left(1.0\dfrac{\text{kg m}}{\text{N s}^2}\right)}(5) = 167\,\frac{\text{N}}{\text{m}^2}$$

7.4.1 Liquid Metals

Experimental data for the heat transfer characteristics of liquid metals in cross-flow over a tube bank have been obtained at Brookhaven National Laboratory [37, 38]. In these tests, mercury (Pr = 0.022 [37]) and NaK (Pr = 0.017 [38]) were heated while flowing normal to a staggered-tube bank consisting of 60 to 70 1.2-cm tubes, 10 rows deep, arranged in an equilateral triangular array with a 1.375 pitch-to-diameter ratio. Both local and average heat transfer coefficients were measured in turbulent flow. The average heat transfer coefficients in the interior of the tube bank are correlated by the equation

$$\text{Nu}_D = 4.03 + 0.228(\text{Re}_D\text{Pr})^{0.67} \tag{7.38}$$

in the Reynolds number range 20,000 to 80,000. Additional data are presented in [39].

The measurements of the distribution of the local heat transfer coefficient around the circumference of a tube indicate that for a liquid metal the turbulent effects in the wake upon heat transfer are small compared to the heat transfer by conduction within the fluid. Whereas with air and water, a marked increase in the local heat transfer coefficient occurs in the wake region of the tube (see Fig. 7.8), and with mercury, the heat transfer coefficient decreases continuously with increasing θ. At a Reynolds number of 83,000, the ratio $h_{c\theta}/\bar{h}_c$ was found to be 1.8 at the stagnation point, 1.0 at $\theta = 90°$, 0.5 at $\theta = 145°$, and 0.3 at $\theta = 180°$.

7.5* Finned Tube Bundles in Cross-Flow

As in the case of flows inside a tube, particularly in gas flows where the heat transfer coefficient is relatively low, numerous applications require the use of enhancement techniques [40, 41] in cross-flow over multitube bundles or tube arrays. The objective, it may be recalled from the discussion in Section 6.6, is to increase the surface area A and/or the convective heat transfer coefficient $\bar{h}_c$, thereby reducing the thermal resistance in flow over tube bundles. This, as is evident from the heat transfer rate equation,

$$q_c = \bar{h}_c A \Delta T$$

results in either increased q_c for a fixed temperature difference ΔT or a reduction in the required ΔT for a fixed heat load q_c. The most widely used method to meet these enhancement objectives is to employ externally finned tubes. A typical example of such tubes for a variety of industrial heat exchangers is shown in Fig. 7.29.

For cross-flow over finned tube banks, a large set of experimental data and correlations for tubes with circular or helical fins have been reviewed by Žukauskas [42]. In calculating the pressure drop and heat transfer, recall that the Reynolds number is based on the maximum flow velocity in the tube bank, and it is given by

$$U_{\max} = U_\infty \times \max\left[\frac{S_T}{S_T - D}, \frac{(S_T/2)}{[S_L^2 + (S_T/2)^2]^{1/2} - D}\right]$$

and

$$\mathrm{Re} = (\rho U_{\max} D/\mu) \tag{7.39}$$

where S_T and S_L are the transverse and longitudinal pitch, respectively, of the tube array. Also, based on the analysis and results of Lokshin and Fomina [43] and Yudin [44], the friction loss is given in terms of the Euler number Eu, and the pressure drop is obtained from

$$\Delta p = \mathrm{Eu}(\rho V_\infty^2 N_L) C_z \tag{7.40}$$

FIGURE 7.29 Typical tube with fins on its outer surfaces that are used in industrial heat exchangers.

where C_z is a correction factor for tube bundles with $N_L < 5$ rows of tubes in the flow direction, and it can be obtained from the following table:

N_L	1	2	3	4	≥5
Aligned	2.25	1.6	1.2	1.05	1.0
Staggered	1.45	1.25	1.1	1.05	1.0

In flows across *inline* (aligned) *tube* banks with *circular* or *helical fins,* where ε is the finned surface extension ratio (ε = ratio of total surface area with fins to the bare tube surface area without fins), the Euler number and the Nusselt number, respectively, are given by the following equations:

$$\mathrm{Eu} = 0.068\varepsilon^{0.5}\left(\frac{S_T - 1}{S_L - 1}\right)^{-0.4} \tag{7.41}$$

for $10^3 \le \mathrm{Re}_D \le 10^5$, $1.9 \le \varepsilon \le 16.3$, $2.38 \le (S_T/D) \le 3.13$, and $1.2 \le (S_L/D) \le 2.35$,

$$\mathrm{Nu}_D = 0.303\varepsilon^{-0.375}\,\mathrm{Re}_D^{0.625}\,\mathrm{Pr}^{0.36}\left(\frac{\mathrm{Pr}}{\mathrm{Pr}_w}\right)^{0.25} \tag{7.42}$$

for $5 \times 10^3 \le \mathrm{Re}_D \le 10^5$, $5 \le \varepsilon \le 12$, $1.72 \le (S_T/D) \le 3.0$, and $1.8 \le (S_L/D) \le 4.0$,

Likewise for cross-flow over *staggered tube* bundles with *circular* or *helical fins*, the recommended correlation for Euler number is

$$\mathrm{Eu} = C_1\,\mathrm{Re}_D^a\varepsilon^{0.5}\,(S_T/D)^{-0.55}\,(S_L/D)^{-0.5} \tag{7.43}$$

where

$C_1 = 67.6$, $a = -0.7$ for $10^2 \le \mathrm{Re}_D < 10^3$, $1.5 \le \varepsilon \le 16$, $1.13 \le S_T/D \le 2.0$, $1.06 \le S_L/D \le 2.0$

$C_1 = 3.2$, $a = -0.25$ for $10^3 \le \mathrm{Re}_D < 10^5$, $1.9 \le \varepsilon \le 16$, $1.6 \le S_T/D \le 4.13$, $1.2 \le S_L/D \le 2.35$

$C_1 = 0.18$, $a = 0$ for $10^5 \le \mathrm{Re}_D < 1.4 \times 10^6$, $1.9 \le \varepsilon \le 16$, $1.6 \le S_T/D \le 4.13$, $1.2 \le S_L/D \le 2.35$

and the Nusselt number is given by

$$\mathrm{Nu} = C_2\,\mathrm{Re}_D^a\,\mathrm{Pr}^b\,(S_T/S_L)^{0.2}\,(p_f/D)^{0.18}\,(h_f/D)^{-0.14}\,(\mathrm{Pr}/\mathrm{Pr}_w)^{0.25} \tag{7.44}$$

where p_f is the fin pitch, h_f is the fin height, and

$$C_2 = 0.192,\ a = 0.65,\ b = 0.36 \text{ for } 10^2 \le \text{Re}_D \le 2 \times 10^4$$
$$C_2 = 0.0507,\ a = 0.8,\ b = 0.4 \text{ for } 2 \times 10^4 \le \text{Re}_D \le 2 \times 10^5$$
$$C_2 = 0.0081,\ a = 0.95,\ b = 0.4 \text{ for } 2 \times 10^5 \le \text{Re}_D \le 1.4 \times 10^6$$

Also, Eq. (7.44) is valid for the general range of the following fin-and-tube pitch parameters:

$$0.06 \le (p_f/D) \le 0.36,\ 0.07 \le h_f/D \le 0.715,\ 1.1 \le (S_T/D) \le 4.2,\ 1.03 \le (S_L/D) \le 2.5$$

In evaluating the Euler number Eu and the Nusselt number Nu given by the correlations in Eqs. (7.41) through (7.44), and hence the pressure drop and heat transfer coefficient in cross-flow over finned tube banks, it would be instructive to compare the results with those for plain or unfinned tubes. To this end, the student should repeat as a home exercise the problems of Examples 7.5 and 7.6 (Section 7.4) by using finned tubes instead of plain tubes.

7.6* Free Jets

One method of expending high convective heat flux from (or to) a surface is with the use of a fluid jet impinging on the surface. The heat transfer coefficient on an area directly under a jet is high. With a properly designed multiple jet on a surface with nonuniform heat flux, a substantially uniform surface temperature can be achieved. The surface on which the jet impinges is termed the target surface.

Confined and Free Jets The jet can be either a confined jet or a free jet. With a confined jet, the fluid flow is affected by a surface parallel to the target surface [Fig. 7.30(a)]. If the parallel surface is sufficiently far away from the target surface, the jet is not affected by it, and we have a free jet [Fig. 7.30(b)].

Heat transfer from the target surface may or may not lead to a change in phase of the fluid. In this section, only free jets without change in phase are considered.

Classification of Free Jets Depending on the cross section of the jet issuing from a nozzle and the number of nozzles, jets are classified as

Single Round or Circular Jet (SRJ)
Single Slot or Rectangular Jet (SSJ)
Array of Round Jets (ARJ)
Array of Slot Jets (ASJ)

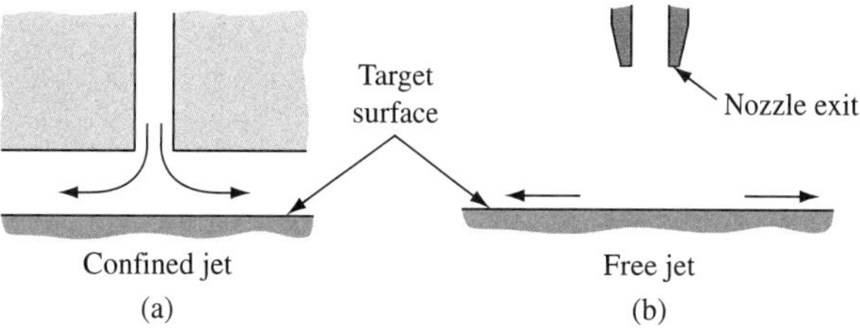

FIGURE 7.30 Confined and free jets.

Free jets are further classified as free-surface or submerged jets. In the case of a free-surface jet, the effect of the surface shear stress on the flow of the jet is negligible. A liquid jet surrounded by a gas is a good example of a free-surface jet. In the case of a submerged jet, the flow is affected by the shear stress at the surface. As a result of the surface shear stress, a significant amount of the surrounding fluid is dragged by the jet. The entrained fluid (that part of the surrounding fluid dragged by the jet) affects the flow and heat transfer characteristics of the jet. A gaseous jet issuing into a gaseous medium (e.g., an air jet issuing into an atmosphere of air) or a liquid jet into a liquid medium are examples of submerged jets. Another difference between the two is that gravity usually plays a part in free-surface jets; the effect of gravity is usually negligible in submerged jets. The two types of jets are illustrated in Fig. 7.31.

In a free-surface round jet, the liquid film thickness along the target surface continuously decreases [Fig. 7.31(a)]. With a slotted free-surface jet, the thickness of the liquid film attains a constant value some distance from the axis of the jet [Fig. 7.31(b)]. With a submerged jet, because of the entrainment of the surrounding fluid, the fluid thickness increases in the direction of flow [Fig. 7.31(c)].

Flow with Single Jets Three distinct regions are identified in single jets (Fig. 7.32). For some distance from the nozzle exit, the jet flow is not significantly affected by the target surface; this region is the free-jet region. In the free-jet region, the velocity component perpendicular to the axis of the jet is negligible compared with the axial component. In the next region, the stagnation region, the jet flow is influenced by the target surface. The magnitude of the axial velocity decreases while the magnitude of the velocity parallel to the surface increases. Following the stagnation region is the wall-jet region where the axial velocity component is negligible compared with the velocity component parallel to the surface.

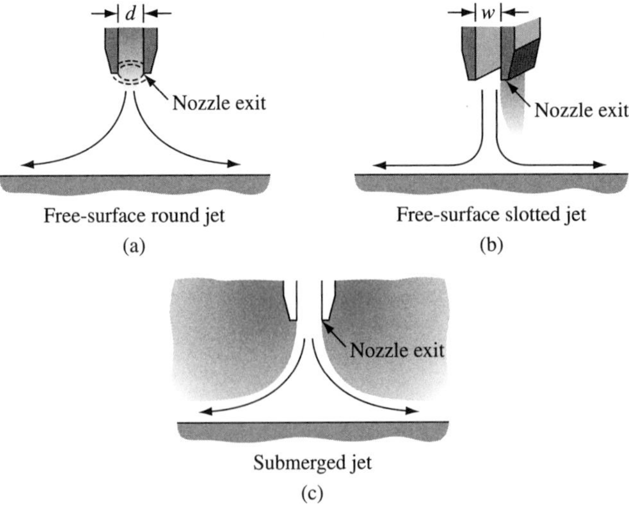

FIGURE 7.31 Free surface and submerged jets.

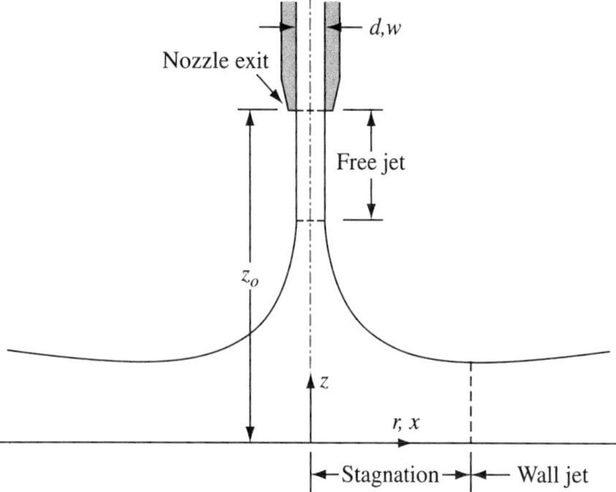

FIGURE 7.32 The three regions in a jet and definition of coordinates.

7.6.1 Free-Surface Jets—Heat Transfer Correlations

Unless the turbulence level in the issuing jet is very high, a laminar boundary layer develops adjacent to the target surface. The laminar boundary layer has four regions, as shown in Fig. 7.33.

The delineation of the four regions for an SRJ with Pr > 0.7 are

Region I	Stagnation layer: The velocity and temperature boundary layer thicknesses are constant, $\delta > \delta_t$
Region II	The velocity and temperature boundary layer thicknesses increase with r but neither has reached the free surface of the fluid film.
Region III	The velocity boundary layer has reached the free surface but the temperature boundary layer has not.
Region IV	Both velocity and temperature boundary layers have reached the free surface.

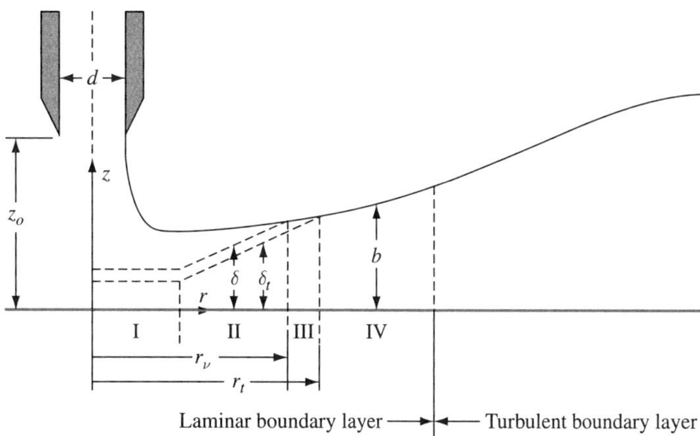

FIGURE 7.33 Definitions of the four regions in the laminar boundary layer.

Heat Transfer Correlations with a Free-Surface SRJ *Uniform Heat Flux* (Liu et al. [45])

Region I: $r < 0.8\ d$

$$\Pr > 3 \qquad \overline{\mathrm{Nu}}_d = 0.797\,\mathrm{Re}_d^{1/2}\Pr^{1/3} \tag{7.45}$$

$$0.15 \le \Pr \le 3 \qquad \overline{\mathrm{Nu}}_d = 0.715\,\mathrm{Re}_d^{1/2}\Pr^{2/5} \tag{7.46}$$

Region II: $0.8 < r/d < r_v/d$

$$\frac{r_v}{d} = 0.1773\,\mathrm{Re}_d^{1/3} \tag{7.47}$$

$$\overline{\mathrm{Nu}}_d = 0.632\,\mathrm{Re}_d^{1/2}\Pr^{1/3}\left(\frac{d}{r}\right)^{1/2} \tag{7.48}$$

The Reynolds number in this section is based on the jet velocity, v_j.

Region III: $r_v < r < r_t$ (from Suryanarayana [46])

$$\frac{r_t}{d} = \left\{-\frac{s}{2} + \left[\left(\frac{s}{2}\right)^2 + \left(\frac{p}{3}\right)^3\right]^{1/2}\right\}^{1/3} + \left\{-\frac{s}{2} + \left[\left(\frac{s}{2}\right)^2 - \left(\frac{p}{3}\right)^3\right]^{1/2}\right\}^{1/3} \tag{7.49}$$

$$p = \frac{-2c}{0.2058\,\Pr - 1}$$

$$s = \frac{0.00686\,\mathrm{Re}_d\Pr}{0.2058\,\Pr - 1}$$

$$c = -5.051 \times 10^{-5}\,\mathrm{Re}_d^{2/3}$$

$$\overline{\mathrm{Nu}}_d = \frac{0.407\,\mathrm{Re}_d^{1/3}\Pr^{1/3}\left(\dfrac{d}{r}\right)^{2/3}}{\left[0.1713\left(\dfrac{d}{r}\right)^2 + \dfrac{5.147}{\mathrm{Re}_d}\left(\dfrac{r}{d}\right)\right]^{2/3}\left[\dfrac{1}{2}\left(\dfrac{r}{d}\right)^2 + c\right]^{1/3}} \tag{7.50}$$

Region IV: $r > r_t$

$$\overline{\mathrm{Nu}}_d = \frac{0.25}{\dfrac{1}{\mathrm{Re}_d\Pr}\left[1 - \left(\dfrac{r_t}{r}\right)^2\right]\left(\dfrac{r}{d}\right)^2 + 0.13\left(\dfrac{b}{d}\right) + 0.0371\left(\dfrac{b_t}{d}\right)} \tag{7.51}$$

where $\dfrac{b}{d} = 0.1713\left(\dfrac{d}{r}\right) + \dfrac{5.147}{\mathrm{Re}_d}\left(\dfrac{r}{d}\right)^2 \qquad b_t = b \quad \text{at } r_t$

Region IV occurs only for $\Pr < 4.86$ and is not valid for $\Pr > 4.86$. Values of r_v/d and r_t/d are given in Table 7.4.

Equations (7.45) through (7.51) are applicable for laminar jets. With a round nozzle, the upper limit of Reynolds number for laminar flow is between 2000 and 4000. In the experiments leading to the correlations, specially designed sharp-edged

TABLE 7.4 Values of r_v/d [Eq. (7.47)] and r_t/d [Eq. (7.49)]

		r_t/d			
Re_d	r_t/d	Pr = 1	Pr = 2	Pr = 3	Pr = 4
1,000	1.773	4.1	5.71	7.55	10.75
4,000	2.81	6.51	9.07	11.98	17.06
10,000	3.82	8.8	12.3	16.3	23.2
20,000	4.82	11.1	15.5	20.5	29.2
30,000	5.5	12.8	17.8	23.5	33.4
40,000	6.1	14.0	19.5	25.8	36.8
50,000	6.5	15.1	21.0	27.8	39.6

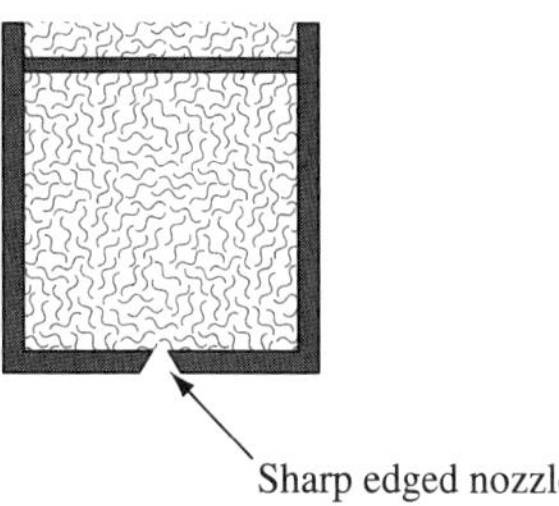

FIGURE 7.34 Sharp-edged orifice.

nozzles (with an inlet momentum break-up plate), as shown in Fig. 7.34, were employed. In those experiments, even with Reynolds numbers as high as 80,000, there was no splattering. Usually, pipe-type nozzles are used, and it is recommended that Eqs. (7.45) through (7.51) be used for laminar flow in pipes. With turbulent flows in pipe nozzles, splattering results. For information on heat transfer with splattering, refer to Lienhard et al. [47].

EXAMPLE 7.7 A jet of water (at 20°C) issues from a 6-mm-diameter (1/4-inch) nozzle at a rate of 0.008 kg/s. The jet impinges on a 4-cm-diameter disk which is subjected to a uniform heat flux of 70,000 W/m^2 (total heat transfer rate of 88 W). Find the surface temperature at radial distances of (a) 3 mm and (b)12 mm from the axis of the jet.

SOLUTION Properties of water (from Appendix 2, Table 13):

$$\mu = 993 \times 10^{-6}\ \mathrm{N\ s/m^2}$$

$$k = 0.597\ \mathrm{W/m\ K}$$

$$\mathrm{Pr} = 7.0$$

$$\mathrm{Re}_d = \frac{4\dot{m}}{\pi d \mu} = \frac{4 \times 0.008}{\pi \times 0.006 \times 993 \times 10^{-6}} = 1709$$

(a) For $r = 3$ mm, $r/d = 0.003/0.006 = 0.5$ (<0.8).

From Eq. (7.45),

$$\overline{\mathrm{Nu}}_d = \frac{\bar{h}_c d}{k} = 0.797 \times 1709^{1/2} \times 7.0^{1/3} = 63.0$$

$$\bar{h}_c = \frac{63.0 \times 0.597}{0.006} = 6269 \text{ W/m}^2\,°\text{C}$$

$$T_s = T_j + \frac{q''}{\bar{h}_c} = 20 + \frac{70{,}000}{6269} = 31.2\,°\text{C}$$

(b) For $r = 12$ mm, $r_v = 0.1773 \times 1709^{1/3} \times 0.006 = 0.013$ m and $r < r_v$. From Eq. (7.48) for Region II,

$$\overline{\mathrm{Nu}}_d = 0.632 \times 1709^{1/2} \times 7.0^{1/3} \times \left(\frac{0.006}{0.012}\right)^{1/2} = 35.3$$

$$\bar{h}_c = \frac{35.3 \times 0.597}{0.006} = 3512 \text{ W/m}^2\,°\text{C} \qquad T_s = 20 + \frac{70{,}000}{3512} = 39.9\,°\text{C}$$

The boundary layer becomes turbulent at some point downstream. Different criteria for the transition to turbulent flow have been suggested. Denoting the radius at which the flow becomes turbulent by r_c, $r_c/d = 1200\,\mathrm{Re}_d^{-0.422}$. The criterion of Liu et al. [45] for the radius r_h at which the flow becomes fully developed turbulent and the heat transfer correlation in that region are given here.

Fully developed turbulent flow:

$$\frac{r_h}{d} = \frac{28{,}600}{\mathrm{Re}_d^{0.68}}$$

$$\mathrm{Nu}_d = \frac{8\,\mathrm{Re}_d \mathrm{Pr} f}{49\left(\dfrac{b}{d}\right) + 28\left(\dfrac{r}{d}\right)^2 f} \tag{7.52}$$

where

$$f = \frac{C_f/2}{1.07 + 12.7(\mathrm{Pr}^{2/3} - 1)\sqrt{C_f/2}} \qquad C_f = 0.073\,\mathrm{Re}_d^{-1/4}\left(\frac{r}{d}\right)^{1/4}$$

$$\frac{b}{d} = \frac{0.02091}{\mathrm{Re}_d^{1/4}}\left(\frac{r}{d}\right)^{5/4} + C\left(\frac{d}{r}\right) \qquad C = 0.1713 + \frac{5.147}{\mathrm{Re}_d}\left(\frac{r_c}{d}\right) - \frac{0.02091}{\mathrm{Re}_d^{1/4}}\left(\frac{r_c}{d}\right)^{1/4}$$

Although the stagnation region is limited to less than $0.8d$ from the axis of the jet, one can take advantage of the high heat transfer coefficient for cooling in regions of high heat fluxes.

Heat Transfer Correlations with a Free-Surface SRJ *Uniform Surface Temperature* (Webb and Ma [48]) $\mathrm{Pr} > 1$.

Region I: $r/d < 1$

$$\overline{\mathrm{Nu}}_d = 0.878\ \mathrm{Re}_d^{1/2}\mathrm{Pr}^{1/3} \tag{7.53}$$

Region II: $\delta < b \qquad r < r_v \qquad \dfrac{r_v}{d} - 0.141\ \mathrm{Re}_d^{1/3} \qquad \hat{r} = \dfrac{r}{d}\dfrac{1}{\mathrm{Re}_d^{1/3}}$

$$\mathrm{Nu}_d = 0.619\ \mathrm{Re}_d^{1/3}\mathrm{Pr}^{1/3}(\hat{r})^{-1/2} \tag{7.54}$$

Region III: $\delta = b \qquad \delta_t < b \quad r_v < r < r_t \qquad \hat{r} = \dfrac{r}{d}\dfrac{1}{\mathrm{Re}_d^{1/3}}$

$$\mathrm{Nu}_d = \frac{2\ \mathrm{Re}_d^{1/3}\mathrm{Pr}^{1/3}}{(6.41\hat{r}^2 + 0.161/\hat{r})[6.55\ \ln(35.9\hat{r}^3 + 0.899) + 0.881]^{1/3}} \tag{7.55}$$

In general, the convective heat transfer coefficients with uniform surface temperature are less than those with uniform surface heat flux.

Heat Transfer Correlations with a Free-Surface SSJ Local convective heat transfer coefficient—*Uniform Heat Flux* (Wolf et al. [49], valid for $17{,}000 < \mathrm{Re}_w < 79{,}000$, $2.8 < \mathrm{Pr} < 5$:

$$\mathrm{Nu}_w = \mathrm{Re}_w^{0.71}\mathrm{Pr}^{0.4}f(x/w) \tag{7.56}$$

For $0 \le \dfrac{x}{w} \le 1.6$, use

$$f(x/w) = 0.116 + \left(\frac{x}{w}\right)^2\left[0.00404\left(\frac{x}{w}\right)^2 - 0.00187\left(\frac{x}{w}\right) - 0.0199\right] \tag{7.57}$$

For $1.6 \le \left(\dfrac{x}{w}\right) \le 6$, use

$$f(x/w) = 0.111 - 0.02\left(\frac{x}{w}\right) + 0.00193\left(\frac{x}{w}\right)^2 \tag{7.58}$$

Figure 7.32 defines x and w.

Turbulent Flow Correlation Equation (7.56) is valid for laminar flows. Transition to turbulence is affected by the free-stream turbulence level. Turbulent flow occurs for Re_x in the range of 4.5×10^6 (low free-stream turbulence of 1.2%) to 1.5×10^6 (high turbulence of 5%). In the turbulent region for the local convective heat transfer coefficient, McMurray et al. [50] proposes

$$\mathrm{Nu}_x = 0.037\ \mathrm{Re}_x^{4/5}\mathrm{Pr}^{1/3} \tag{7.59}$$

where $\mathrm{Nu}_x = (h_c x/k)$ and $\mathrm{Re}_x = \upsilon_J x/\nu$. Equation (7.59) is valid to a local Reynolds number $\mathrm{Re}_x = 2.5 \times 10^6$.

Heat Transfer Correlations with an Array of Jets With single jets, the heat transfer coefficient in the stagnation zone is quite high but decreases rapidly with r/d or x/w. High heat transfer rates from large surfaces can be achieved with multiple jets by taking advantage of the high heat transfer coefficients in the stagnation zone. If the separation distance between two jets is approximately equal to the stagnation zone, one may expect such a high heat transfer coefficient. However, unless the fluid is removed rapidly, the presence of the spent fluid leads to a degradation in heat transfer rate and the average heat transfer coefficient may not reach the high values obtained in the stagnation region with single jets.

The number of variables with an array of jets is quite large, and it is unlikely that a single correlation can be developed to encompass all possible variables. Some of the variables are the spacing between the jets and the target surface, the jet Reynolds number, fluid Prandtl number, the pitch of the jets (distance between the axis of two adjacent jets), and arrangement of the array [square or triangular—see Fig. (7.35)]. In most cases, it is expected that the Reynolds number for each jet has the same value; although with nonuniform heat flux, employing different jet Reynolds numbers may lead to a more uniform surface temperature.

From experimental data with in-line and triangular jets, Pan and Webb [51] suggest the following correlation.

$$\overline{\text{Nu}}_d = 0.225\ \text{Re}_d^{2/3}\text{Pr}^{1/3}e^{-0.095(S/d)} \tag{7.60}$$

Equation (7.60) is valid for

$$2 \le \frac{z_o}{d} \le 5 \qquad 2 \le \frac{S}{d} \le 8 \qquad 5000 \le \text{Re}_d \le 22{,}000$$

For larger values of S/d, based on experimental results, Pan and Webb [51] recommend

$$\overline{\text{Nu}}_d = 2.38\ \text{Re}_d^{2/3}\text{Pr}^{1/3}\left(\frac{d}{S}\right)^{4/3} \tag{7.61}$$

Equation (7.61) is valid for $13.8 < S/d < 330$ and $7100 < \text{Re}_d < 48{,}000$. For other configurations, refer to the review by Webb and Ma [48].

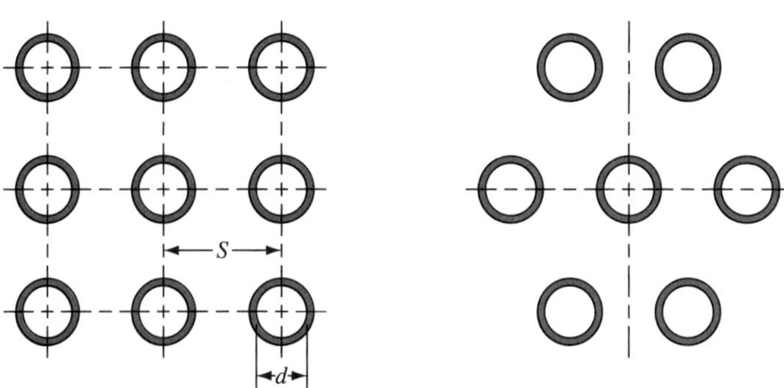

FIGURE 7.35 Definition of in-line and triangular arrangements of jet arrays.

It must be noted that with a vertical nozzle the fluid velocity increases (or decreases) as the fluid issuing from the nozzle approaches the target surface. If such an increase (or decrease) in the jet velocity is significant, the jet velocity and diameter or width used in the computations of the Reynolds number and Nusselt number must reflect the change in the velocity. The modified velocity is $v_m = v_j \pm \sqrt{2gz_o}$, where v_j is the jet velocity at the nozzle exit and z_o is the distance between the nozzle exit and the target surface. The jet velocity is increased if the target surface is below the nozzle and decreased if the surface is above the nozzle. The corresponding diameter and width are $d_j\sqrt{v_j/v_m}$, or $w_j v_j/v_m$ where the subscript j denotes the values at exit of the nozzle.

7.5.2 Submerged Jets—Heat Transfer Correlations

When the jet fluid is surrounded by the same type of fluid (liquid jet in a liquid or gaseous jet in a gas) we have a submerged jet. Most engineering applications of submerged jets involve gaseous jets, usually air jets into air. The surrounding fluid is entrained by the jet both in the free-jet and the wall-jet regions. Because of such entrainment, the thickness of the fluid in motion increases in the direction of flow. With free jets, the thickness is substantially constant for slotted jets and decreases for round jets in the wall-jet region. Consequently, both fluid mechanical and heat transfer characteristics of submerged jets are different from those of free surface jets.

Single Round Jets For local heat transfer with uniform heat flux, Ma and Bergles [52] proposed

$$\mathrm{Nu}_d = \mathrm{Nu}_{d,o}\left[\frac{\tanh(0.88r/d)}{(r/d)}\right]^{1/2} \qquad \frac{r}{d} < 2 \tag{7.62}$$

$$\mathrm{Nu}_d = \frac{1.69\ \mathrm{Nu}_{d,o}}{(r/d)^{1.07}} \qquad \frac{r}{d} > 2 \tag{7.63}$$

where

$$\mathrm{Nu}_{d,o} = 1.29\ \mathrm{Re}_d^{0.5}\mathrm{Pr}^{0.4} \tag{7.64}$$

For liquid jets, replace the exponent of 0.4 for Pr in Eq. (7.64) by 0.33.

A composite equation for both the stagnation and wall jet regions by Sun et al. [53] is

$$\mathrm{Nu}_d = \mathrm{Nu}_{d,o}\left\{\left[\frac{\sqrt{\tanh(0.88r/d)}}{\sqrt{r/d}}\right]^{-17} + \left[\frac{1.69}{(r/d)^{1.07}}\right]^{-17}\right\}^{-1/17} \tag{7.65}$$

where $\mathrm{Nu}_{d,o}$ is given by Eq. (7.64).

A correlation for the average heat transfer coefficient to radius r with uniform surface temperature by Martin [54] is

$$\overline{\mathrm{Nu}}_d = 2\frac{d}{r}\,\frac{1 - 1.1(d/r)}{1 + 0.1\left(\dfrac{z_o}{d} - 6\right)\dfrac{d}{r}}\left[\mathrm{Re}_d\left(1 + \frac{\mathrm{Re}_d^{0.55}}{200}\right)\right]^{0.5}\mathrm{Pr}^{0.42} \tag{7.66}$$

Equation (7.66) is valid for

$$2{,}000 < \mathrm{Re}_d < 400{,}000 \qquad 2.5 \le r/d < 7.5 \qquad 2 \le z_o/d \le 12$$

with properties evaluated at $(T_s + T_j)/2$.

Sitharamayya and Raju [55] proposed

$$\overline{\mathrm{Nu}}_d = [8.1\ \mathrm{Re}_d^{0.523} + 0.133(r/d - 4)\mathrm{Re}_d^{0.828}](d/r)^2\mathrm{Pr}^{0.33} \tag{7.67}$$

Single Slotted Jets For the average heat transfer coefficient up to x with uniform surface temperature, Martin [54] proposed the relation

$$\overline{\mathrm{Nu}}_w = \frac{1.53(2\ \mathrm{Re}_w)^m \mathrm{Pr}^{0.42}}{\dfrac{x}{w} + \dfrac{z_o}{w} + 2.78} \tag{7.68}$$

where $m = 0.695 - 2\left[\dfrac{x}{w} + 0.796\left(\dfrac{z_o}{w}\right)^{1.33} + 6.12\right]^{-1}$ and $\mathrm{Re}_w = \dfrac{v_j w}{\mu}$

Equation (7.68) is valid for $1500 \le \mathrm{Re}_w \le 45{,}000$, $4 \le x/w < 50$, and $4 \le z_o/w \le 20$. Evaluate properties at $(T_s + T_j)/2$.

EXAMPLE 7.8 Air at 20°C issues from a 3-mm-wide, 20-mm-long slotted jet with a velocity of 10 m/s. It impinges on a plate maintained at 60°C. The nozzle exit is at a distance of 10 mm from the plate. Estimate the heat transfer rate from the 4-cm-wide region of the plate directly below the jet.

SOLUTION Properties of air (from Appendix 2, Table 13) at (20 + 60)/2 = 40°C

$$\rho = 1.092\ \mathrm{kg/m^3} \qquad \mu = 1.912 \times 10^{-5}\ \mathrm{Ns/m^2}$$

$$k = 0.0265\ \mathrm{W/m\ K} \qquad \mathrm{Pr} = 0.71$$

$$\mathrm{Re}_w = \frac{1.092 \times 10 \times 0.003}{1.912 \times 10^{-5}} = 1713$$

From Eq. (7.68) with $x = 0.02$ m, $z_o = 0.01$ m, and $w = 0.003$ m,

$$m = 0.695 - 2\left[\frac{0.02}{0.003} + 0.796\left(\frac{0.01}{0.003}\right)^{1.33} + 6.12\right]^{-1} = 0.575$$

$$\overline{\mathrm{Nu}}_w = \frac{1.53 \times (2 \times 1713)^{0.575} \times 0.71^{0.42}}{\dfrac{0.02}{0.003} + \dfrac{0.01}{0.003} + 2.78} = 11.2$$

$$\bar{h}_c = \frac{11.2 \times 0.0265}{0.003} = 98.9\,\text{W/m}^2\,°\text{C}$$

$$q = 98.9 \times 0.04 \times 0.02 \times (60 - 20) = 3.2\text{ W}$$

Array of Round Jets The average heat transfer coefficient with uniform surface temperature for aligned (square) or triangular (hexagonal) arrangement [Fig. (7.35)] (Martin [54]) is

$$\overline{\text{Nu}}_d = K \frac{\sqrt{f}(1 - 2.2\sqrt{f})}{1 + 0.2(z_o/d - 6)\sqrt{f}} \text{Re}_d^{2/3} \text{Pr}^{0.42} \quad (7.69)$$

where
$$K = \left[1 + \left(\frac{z_o/d}{0.6}\sqrt{f}\right)^6\right]^{-1/20}$$

and
$$f = \text{relative nozzle area} = \frac{\pi d^2/4}{\text{area of the square or hexagon}}$$

Equation (7.69) is valid for $2000 \leq \text{Re}_d \leq 100{,}000$, $0.004 \leq f \leq 0.04$, and $2 \leq z_o/d \leq 12$. Evaluate properties at $(T_s + T_j)/2$.

Array of Slotted Jets For the average heat transfer coefficient with uniform surface temperature, Martin [54] proposed

$$\overline{\text{Nu}}_w = \frac{1}{3} f_o^{3/4} \left(\frac{4\,\text{Re}_w}{f/f_o + f_o/f}\right)^{2/3} \text{Pr}^{0.42} \quad (7.70)$$

where $f_o = \left[60 + 4\left(\frac{z_o}{2w} - 2\right)^2\right]^{-1/2}$ and $f = \frac{w}{S}$

Eq. (7.70) is valid in the range

$$750 \leq \text{Re}_w \leq 20{,}000 \qquad 0.008 \leq f \leq 2.5 f_o \qquad 2 \leq x/w \leq 80$$

with properties evaluated at $(T_s + T_j)/2$.

Heat transfer with jets is affected by many factors, such as jet inclination, extended surfaces on the target surface, surface roughness, jet splattering, jet pulsation, hydraulic jump, and rotation of target surface. For a discussion of those effects and more details, refer to Webb and Ma [48] and Lienhard [56]. Martin [54] discusses the optimal spatial arrangement of submerged jets.

7.7 Closing Remarks

For the convenience of the reader, useful correlation equations for determining the average value of the convection heat transfer coefficients in cross-flow over exterior surfaces are tabulated in Table 7.5.

TABLE 7.5 Heat transfer correlations for external flow

Geometry	Correlation Equation	Restrictions
Long circular cylinder normal to gas or liquid flow	$\overline{Nu}_D = C\,Re_D^m Pr^n (Pr/Pr_s)^{1/4}$ (see Table 7.1)	$1 < Re_D < 10^6$
Noncircular cylinder in a gas	$\overline{Nu}_D = B\,Re_D^n$ (see Table 7.2)	$2500 < Re_D < 10^5$
Circular cylinder in a liquid metal	$\overline{Nu}_D = 1.125(Re_D Pr)^{0.413}$	$1 < Re_D Pr < 100$
Short cylinder in a gas	$\overline{Nu}_D = 0.123\,Re_D^{0.651} + 0.00416(D/L)^{0.85}\,Re_D^{0.792}$	$7 \times 10^4 < Re_D < 2.2 \times 10^5$ $L/D < 4$
Sphere in a gas	$\frac{\bar{h}_c}{c_p \rho U_\infty} = (2.2/Re_D + 0.48/Re_D^{0.5})$	$1 < Re_D < 25$
	$\overline{Nu}_D = 0.37\,Re_D^{0.6}$	$25 < Re_D < 10^5$
	$\overline{Nu}_D = 430 + 5 \times 10^{-3}\,Re_D + 0.25 \times 10^{-9}\,Re_D^2 - 3.1 \times 10^{-17}\,Re_D^3$	$4 \times 10^5 < Re_D < 5 \times 10^6$
Sphere in a gas or a liquid	$\overline{Nu}_D = 2 + (0.4\,Re_D^{1/2} + 0.06\,Re_D^{2/3})Pr^{0.4}(\mu/\mu_s)^{1/4}$	$3.5 < Re_D < 7.6 \times 10^4$ $0.7 < Pr < 380$
Sphere in a liquid metal	$\overline{Nu}_D = 2 + 0.386(Re_D Pr)^{1/2}$	$3.6 \times 10^4 < Re_D < 2 \times 10^5$
Long, flat plate, width D, perpendicular to flow in a gas	$\overline{Nu}_D = 0.20\,Re_D^{2/3}$	$1 < Re_D < 4 \times 10^5$
Half-round cylinder with flat rear surface, in a gas	$\overline{Nu}_D = 0.16\,Re_D^{2/3}$	$1 < Re_D < 4 \times 10^5$
Square plate, dimension, L, flow of a gas or a liquid	$(\bar{h}_c/c_p\rho U_\infty)Pr^{2/3} = 0.930\,Re_L^{-1/2}$	$2 \times 10^4 < Re_L < 10^5$ angles of pitch and attack from 25° to 90° yaw angle from 0° to 45°
Upstream face of a disk with axis aligned with flow, gas, or liquid	$\overline{Nu}_D = 1.05\,Re^{1/2}Pr^{0.36}$	$5 \times 10^3 < Re_D < 5 \times 10^4$
Isothermal disk with axis perpendicular to flow, gas, or liquid	$\overline{Nu}_D = 0.591\,Re_D^{0.564}Pr^{1/3}$	$9 \times 10^2 < Re_D < 3 \times 10^4$
Packed bed—heat transfer to or from packing, in a gas	$\overline{Nu}_{D_p} = \frac{1 - \varepsilon}{\varepsilon}(0.5\,Re_{D_p}^{1/2} + 0.2\,Re_{D_p}^{2/3})Pr^{1/3}$	$20 < Re_{D_p} < 10^4$ $0.34 < \varepsilon < 0.78$
(ε = void fraction of bed)	$(\bar{h}_c/c_p\rho U_s)Pr^{2/3} = \frac{1.075}{\varepsilon}\,Re_{D_p}^{-0.826}$	$0.01 < Re_{D_p} < 10$
D_p = equivalent packing diameter (see Eq. 7.20)	$(\bar{h}_c/c_p\rho U_s)Pr^{2/3} = \frac{0.455}{\varepsilon}\,Re_{D_p}^{-0.4}$	$10 < Re_{D_p} < 200$

(Continued)

TABLE 7.5 (Continued)

Geometry	Correlation Equation			Restrictions
Packed bed—heat transfer to or from containment wall, gas	$\overline{Nu}_{D_p} = 2.58\, Re_{D_p}^{1/3} Pr^{1/3} + 0.094\, Re_{D_p}^{0.8} Pr^{0.4}$			$40 < Re_{D_p} < 2000$ cylinderlike packing
	$\overline{Nu}_{D_p} = 0.203\, Re_{D_p}^{1/3} Pr^{1/3} + 0.220\, Re_{D_p}^{0.8} Pr^{0.4}$			$40 < Re_{D_p} < 2000$ spherelike packing
Tube bundle in cross-flow (see Figs. 7.21 and 7.22)	$\overline{Nu}_D Pr^{-0.36} (Pr/Pr_s)^{-0.25} = C (S_T/S_L)^n Re_D^m$			
	C	*m*	*n*	
	0.8	0.4	0	$10 < Re_D < 100$, in-line
	0.9	0.4	0	$10 < Re_D < 100$, staggered
	0.27	0.63	0	$1000 < Re_D < 2 \times 10^5$, in-line $S_T/S_L \geq 0.7$
	0.35	0.60	0.2	$1000 < Re_D < 2 \times 10^5$, staggered $S_T/S_L < 2$
	0.40	0.60	0	$1000 < Re_D < 2 \times 10^5$, staggered $S_T/S_L \geq 2$
	0.021	0.84	0	$Re_D > 2 \times 10^5$, in-line
	0.022	0.84	0	$Re_D > 2 \times 10^5$, staggered $Pr > 1$
	$\overline{Nu}_D = 0.019\, Re_D^{0.84}$			$Re_D > 2 \times 10^5$, staggered $Pr = 0.7$
Flow over staggered tube bundle, gas or liquid ($Pr > 0.5$)	$\overline{Nu}_D = 0.0131\, Re_D^{0.883} Pr^{0.36}$			$4.5 \times 10^5 < Re_D < 7 \times 10^6$ $S_T/D = 2$, $S_L/D = 1.4$
Liquid metals	$\overline{Nu}_D = 4.03 + 0.228 (Re_D Pr)^{2/3}$			$2 \times 10^4 < Re_D < 8 \times 10^4$, staggered

References

1. A. Fage, "The Air Flow around a Circular Cylinder in the Region Where the Boundary Layer Separates from the Surface," Brit. Aero. Res. Comm. R and M 1179, 1929.
2. A. Fage and V. M. Falkner, "The Flow around a Circular Cylinder," Brit. Aero Res. Comm. R and M 1369, 1931.
3. H. B. Squire, *Modern Developments in Fluid Dynamics*, 3d ed., vol. 2, Clarendon, Oxford, 1950.
4. R. C. Martinelli, A. G. Guibert, E. H. Morin, and L. M. K. Boelter, "An Investigation of Aircraft Heaters VIII—a Simplified Method for Calculating the Unit-Surface Conductance over Wings," NACA ARR, March 1943.
5. W. H. Giedt, "Investigation of Variation of Point Unit-Heat-Transfer Coefficient around a Cylinder Normal to an Air Stream," *Trans. ASME*, vol. 71, pp. 375–381, 1949.
6. R. Hilpert, "Wärmeabgabe von geheizten Drähten und Rohren," *Forsch. Geb. Ingenieurwes.*, vol. 4, p. 215, 1933.

7. A. A. Žukauskas, "Heat Transfer from Tubes in Cross Flow," *Advances in Heat Transfer*, Academic Press, vol. 8, pp. 93–106, 1972.
8. H. G. Groehn, "Integral and Local Heat Transfer of a Yawed Single Circular Cylinder up to Supercritical Reynolds Numbers," *Proc. 8th Int. Heat Transfer Conf.*, vol. 3, Hemisphere, Washington, D.C., 1986.
9. M. Jakob, *Heat Transfer*, vol. 1, Wiley, New York, 1949.
10. R. Ishiguro, K. Sugiyama, and T. Kumada, "Heat Transfer around a Circular Cylinder in a Liquid-Sodium Crossflow," *Int. J. Heat Mass Transfer*, vol. 22, pp. 1041–1048, 1979.
11. A. Quarmby and A. A. M. Al-Fakhri, "Effect of Finite Length on Forced Convection Heat Transfer from Cylinders," *Int. J. Heat Mass Transfer*, vol. 23, pp. 463–469, 1980.
12. E. M. Sparrow and F. Samie, "Measured Heat Transfer Coefficients at and Adjacent to the Tip of a Wall-Attached Cylinder in Crossflow—Application to Fins," *J. Heat Transfer*, vol. 103, pp. 778–784, 1981.
13. E. M. Sparrow, T. J. Stahl, and P. Traub, "Heat Transfer Adjacent to the Attached End of a Cylinder in Crossflow," *Int. J. Heat Mass Transfer*, vol. 27, pp. 233–242, 1984.
14. N. R. Yardi and S. P. Sukhatme, "Effects of Turbulence Intensity and Integral Length Scale of a Turbulent Free Stream on Forced Convection Heat Transfer from a Circular Cylinder in Cross Flow," *Proc. 6th Int. Heat Transfer Conf.*, Hemisphere, Washington, D.C., 1978.
15. H. Dryden and A. N. Kuethe, "The Measurement of Fluctuations of Air Speed by the Hot-Wire Anemometer," NACA Rept. 320, 1929.
16. C. E. Pearson, "Measurement of Instantaneous Vector Air Velocity by Hot-Wire Methods," *J. Aerosp. Sci.*, vol. 19, pp. 73–82, 1952.
17. W. H. McAdams, *Heat Transmission*, 3d ed., McGraw-Hill, New York, 1953.
18. S. Whitaker, "Forced Convection Heat Transfer Correlations for Flow in Pipes, Past Flat Plates, Single Cylinders, Single Spheres, and for Flow in Packed Beds and Tube Bundles," *AIChE J.*, vol. 18, pp. 361–371, 1972.
19. E. Achenbach, "Heat Transfer from Spheres up to Re = 6×10^6," *Proc. 6th Int. Heat Transfer Conf.*, vol. 5, Hemisphere, Washington, D.C., 1978.
20. H. F. Johnston, R. L. Pigford, and J. H. Chapin, "Heat Transfer to Clouds of Falling Particles," *Univ. of Ill. Bull.*, vol. 38, no. 43, 1941.
21. L. C. Witte, "An Experimental Study of Forced-Convection Heat Transfer from a Sphere to Liquid Sodium," *J. Heat Transfer*, vol. 90, pp. 9–12, 1968.
22. H. H. Sogin, "A Summary of Experiments on Local Heat Transfer from the Rear of Bluff Obstacles to a Lowspeed Airstream," *Trans. ASME, Ser. C. J. Heat Transfer*, vol. 86, pp. 200–202, 1964.
23. J. W. Mitchell, "Base Heat Transfer in Two-Dimensional Subsonic Fully Separated Flows," *Trans. ASME, Ser. C, J. Heat Transfer*, vol. 93, pp. 342–348, 1971.
24. E. M. Sparrow and G. T. Geiger, "Local and Average Heat Transfer Characteristics for a Disk Situated Perpendicular to a Uniform Flow," *J. Heat Transfer*, vol. 107, pp. 321–326, 1985.
25. K. K. Tien and E. M. Sparrow, "Local Heat Transfer and Fluid Flow Characteristics for Airflow Oblique or Normal to a Square Plate," *Int. J. Heat Mass Transfer*, vol. 22, pp. 349–360, 1979.
26. G. L. Wedekind, "Convective Heat Transfer Measurement Involving Flow Past Stationary Circular Disks," *J. Heat Transfer*, vol. 111, pp. 1098–1100, 1989.
27. S. N. Upadhyay, B. K. D. Agarwal, and D. R. Singh, "On the Low Reynolds Number Mass Transfer in Packed Beds," *J. Chem. Eng. Jpn.*, vol. 8, pp. 413–415, 1975.
28. J. Beek, "Design of Packed Catalytic Reactors," *Adv. Chem. Eng.*, vol. 3, pp. 203–271, 1962.
29. W. M. Kays and A. L. London, *Compact Heat Exchangers*, 2d ed., McGraw-Hill, New York, 1964.
30. R. D. Wallis, "Photographic Study of Fluid Flow between Banks of Tubes," *Engineering*, vol. 148, pp. 423–425, 1934.
31. W. E. Meece, "The Effect of the Number of Tube Rows upon Heat Transfer and Pressure Drop during Viscous Flow across In-Line Tube Banks," M.S. thesis, Univ. of Delaware, 1949.
32. C. J. Chen and T-S. Wung, "Finite Analytic Solution of Convective Heat Transfer for Tube Arrays in Crossflow: Part II—Heat Transfer Analysis," *J. Heat Transfer*, vol. 111, pp. 641–648, 1989.
33. E. Achenbach, "Heat Transfer from a Staggered Tube Bundle in Cross-Flow at High Reynolds Numbers," *Int. J. Heat Mass Transfer*, vol. 32, pp. 271–280, 1989.
34. E. Achenbach, "Heat Transfer from Smooth and Rough In-line Tube Banks at High Reynolds Number," *Int. J. Heat Mass Transfer*, vol. 34, pp. 199–207, 1991.
35. S. Aiba, T. Ota, and H. Tsuchida, "Heat Transfer of Tubes Closely Spaced in an In-Line Bank," *Int. J. Heat Mass Transfer*, vol. 23, pp. 311–319, 1980.
36. W. M. Kays and R. K. Lo, "Basic Heat Transfer and Flow Friction Design Data for Gas Flow Normal to Banks of Staggered Tubes—Use of a Transient Technique," Tech. Rept. 15, Navy Contract N6-ONR-251 T. O. 6, Stanford Univ., 1952.
37. R. J. Hoe, D. Dropkin, and O. E. Dwyer, "Heat Transfer Rates to Crossflowing Mercury in a Staggered Tube Bank—I," *Trans. ASME*, vol. 79, pp. 899–908, 1957.
38. C. L. Richards, O. E. Dwyer, and D. Dropkin, "Heat Transfer Rates to Crossflowing Mercury in a Staggered

Tube Bank—II," *ASME—AIChE Heat Transfer Conf.*, paper 57-HT-11, 1957.

39. S. Kalish and O. E. Dwyer, "Heat Transfer to NaK Flowing through Unbaffled Rod Bundles," *Int. J. Heat Mass Transfer*, vol. 10, pp. 1533–1558, 1967.
40. A. E. Bergles, "Techniques to Enhance Heat Transfer," in *Handbook of Heat Transfer*, 3rd ed., W. M. Rohsenow, J. P. Hartnett, and Y. I. Cho, eds., McGraw-Hill, New York, 1998.
41. R. M. Manglik, "Heat Transfer Enhancement," in *Heat Transfer Handbook*, A. Bejan and A. D. Kraus, eds., Wiley, Hoboken, NJ, 2003.
42. A. Žukauskas, *High-Performance Single-Phase Heat Exchangers*, Hemisphere, New York, 1989.
43. V. A. Lokshin and V. N. Fomina, "Correlation of Experimental Data on Finned Tube Bundles," *Teploenergetika*, Vol. 6, pp. 36–39, 1978.
44. V. F. Yudin, *Teploobmen Poperechnoorebrenykh Trub* [Heat Transfer of Crossfinned Tubes], Mashinostroyeniye Publishing House, Leningrad, Russia, 1982.
45. X. Liu, J. H. Lienhard V, and J. S. Lombara, "Convective Heat Transfer by Impingement of Circular Liquid Jets, *J. Heat Transfer*, vol. 113, pp. 571–582, 1991.
46. N. V. Suryanarayana, "Forced Convection—External Flows," in CRC Handbook of Mechanical Engineering, F. Kreith, ed., CRC Press, 1998.
47. J. H. Lienhard V., X. Liu, and L. A. Gabour, "Splattering and Heat Transfer During Impingement of a Turbulent Liquid Jet," *J. Heat Transfer*, vol. 114, pp. 362–372, 1992.
48. B. W. Webb and C. F. Ma, "Single-phase Liquid Jet Impingement Heat Transfer," in *Advances in Heat Transfer*, J. P. Hartnett and R. F. Irvine, eds., vol. 26, pp. 105–217, Academic Press, New York, 1995.
49. D. H. Wolf, R. Viskanta, and F. P. Incropera, "Local Convective Heat Transfer from a Heated Surface to a Planar Jet of Water with a Non-uniform Velocity Profile," *J. Heat Transfer*, vol. 112, pp. 899–905, 1990.
50. D. C. McMurray, P. S. Meyers, and O. A. Uyehara, "Influence of Impinging Jet Variables on Local Heat Transfer Coefficients along a Flat Surface with Constant Heat Flux," *Proc. 3d Int. Heat Transfer Conference*, vol. 2, pp. 292–299, 1966.
51. Y. Pan and B. W. Webb, "Heat Transfer Characteristics of Arrays of Free-Surface Liquid Jets," *J. Heat Transfer*, vol. 117, pp. 878–886, 1995.
52. C. F. Ma and A. E. Bergles, "Convective Heat Transfer on a Small Vertical Heated Surface in an Impinging Circular Liquid Jet," in *Heat Transfer Science and Technology*, B. X. Wang, ed., pp. 193–200, Hemisphere, New York, 1988.
53. H. Sun, C. F. Ma, and W. Nakayama, "Local Characteristics of Convective Heat Transfer from Simulated Microelectronic Chips to Impinging Submerged Round Jets," *J. Electronic Packaging*, vol. 115, pp. 71–77, 1993.
54. H. Martin, "Impinging Jets," in *Handbook of Heat Exchanger Design*, G. F. Hewitt, ed., Hemisphere, New York, 1990.
55. S. Sitharamayya and K. S. Raju, "Heat Transfer between an Axisymmetric Jet and a Plate Held Normal to the Flow," *Can. J. Chem. Eng.*, vol. 45, pp. 365–369, 1969.
56. J. H. Lienhard V., "Liquid Jet Impingement," in *Annual Review of Heat Transfer*, C. L. Tien, ed., vol. 6, Begell House, New York, 1995.

Problems

The problems for this chapter are organized by subject matter as shown below.

Topic	Problem Number
Cylinders in cross- or yawed-flow	7.1–7.18
Hot-wire anemometer	7.19–7.22
Spheres	7.23–7.31
Bluff bodies	7.32–7.36
Packed beds	7.37–7.39
Tube banks	7.40–7.46

7.1 Determine the heat transfer coefficient at the stagnation point and the average value of the heat transfer coefficient for a single 5-cm-OD, 60-cm-long tube in cross-flow. The temperature of the tube surface is 260°C, the velocity of the fluid flowing perpendicular to the tube axis is 6 m/s, and the temperature of the fluid is 38°C. Consider the following fluids: (a) air, (b) hydrogen, and (c) water.

7.2 A mercury-in-glass thermometer at 100°F (OD = 0.35 in.) is inserted through a duct wall into a 10-ft/s airstream at 150°F. Estimate the heat transfer coefficient between the air and the thermometer.

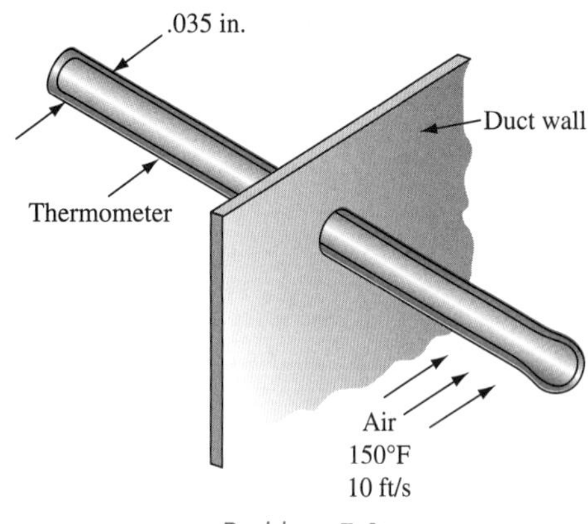

Problem 7.2

7.3 Steam at 1 atm and 100°C is flowing across a 5-cm-OD tube at a velocity of 6 m/s. Estimate the Nusselt number, the heat transfer coefficient, and the rate of heat transfer per meter length of pipe if the pipe is at 200°C.

7.4 An electrical transmission line of 1.2-cm diameter carries a current of 200 amps and has a resistance of 3×10^{-4} ohm per meter of length. If the air around this line is at 16°C, determine the surface temperature on a windy day, assuming a wind blows across the line at 33 km/h.

7.5 Derive an equation in the form $\bar{h}_c = f(T, U, U_\infty)$ for the flow of air over a long, horizontal cylinder for the temperature range 0°C to 100°C. Use Eq. (7.3) as a basis.

7.6 Repeat Problem 7.5 for water in the temperature range 10°C to 40°C.

7.7 The Alaska pipeline carries 2 million barrels of crude oil per day from Prudhoe Bay to Valdez, covering a distance of 800 miles. The pipe diameter is 48 in., and it is insulated with 4 in. of fiberglass covered with steel sheathing. Approximately half of the pipeline length is above ground, running nominally in the north-south direction. The insulation maintains the outer surface of the steel sheathing at approximately 10°C. If the ambient temperature averages 0°C and prevailing winds are 2 m/s from the northeast, estimate the total rate of heat loss from the above-ground portion of the pipeline.

7.8 An engineer is designing a heating system that will consist of multiple tubes placed in a duct carrying the air supply for a building. She decides to perform preliminary tests with a single copper tube of 2-cm OD carrying condensing steam at 100°C. The air velocity in the duct is 5 m/s, and its temperature is 20°C. The tube can be placed normal to the flow, but it may be advantageous to place the tube at an angle to the air flow and thus increase the heat transfer surface area. If the duct width is 1 m, predict the outcome of the planned tests and estimate how the angle θ will affect the rate of heat transfer. Are there limits?

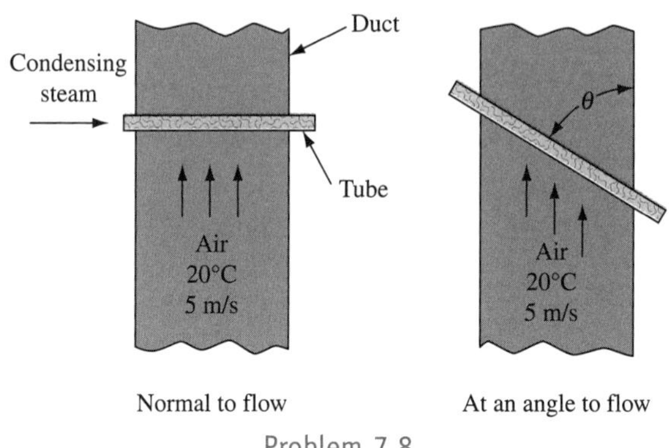

Problem 7.8

7.9 A long, hexagonal copper extrusion is removed from a heat-treatment oven at 400°C and immersed in a 50°C airstream flowing perpendicular to its axis at 10 m/s. The surface of the copper has an emissivity of 0.9 due to oxidation. The rod is 3 cm across opposing flat sides and has a cross-sectional area of 7.79 cm^2 and a perimeter of 10.4 cm. Determine the time required for the center of the copper to cool to 100°C.

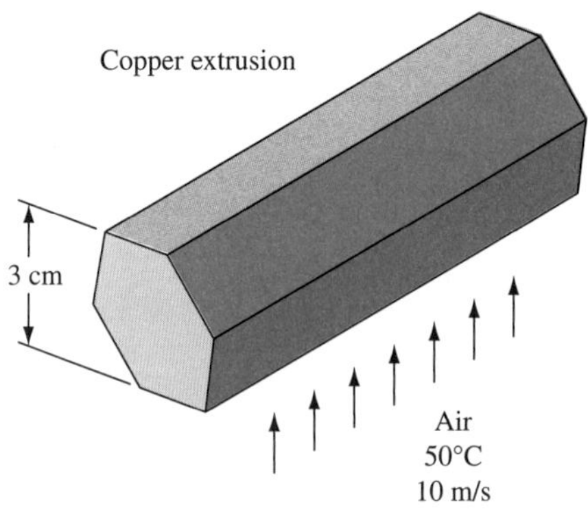

Problem 7.9

7.10 Repeat Problem 7.9 if the extrusion cross-section is elliptical with the major axis normal to the air flow and the same mass per unit length. The major axis of the elliptical cross section is 5.46 cm, and its perimeter is 12.8 cm.

7.11 Calculate the rate of heat loss from a human body at 37°C in an airstream of 5 m/s at 35°C. The body can be modeled as a cylinder 30 cm in diameter and 1.8 m high. Compare your results with those for natural convention from a body (Problem 5.8) and with the typical energy intake from food, 1033 kcal/day.

7.12 A nuclear reactor fuel rod is a circular cylinder 6 cm in diameter. The rod is to be tested by cooling it with a flow of sodium at 205°C with a velocity of 5 cm/s perpendicular to its axis. If the rod surface is not to exceed 300°C, estimate the maximum allowable power dissipation in the rod.

7.13 A stainless steel pin fin 5 cm long and with a 6-mm OD, extends from a flat plate into a 175 m/s airstream, as shown in the sketch at top of next column. Estimate (a) the average heat transfer coefficient between air and the fin, (b) the temperature at the end of the fin, and (c) the rate of heat flow from the fin.

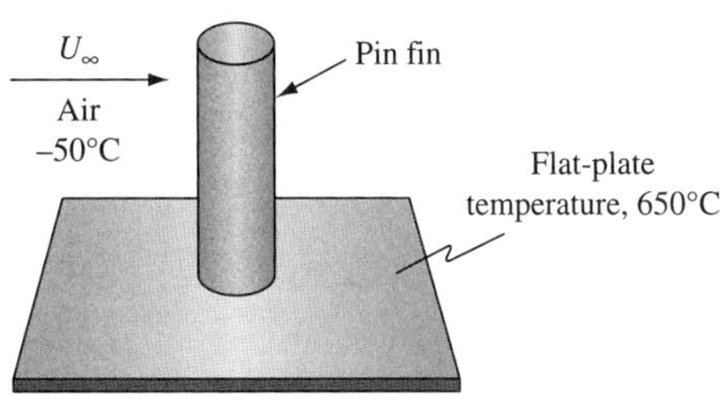

Problem 7.13

7.14 Repeat Problem 7.13 with glycerol at 20°C flowing over the fin at 2 m/s. The plate temperature is 50°C.

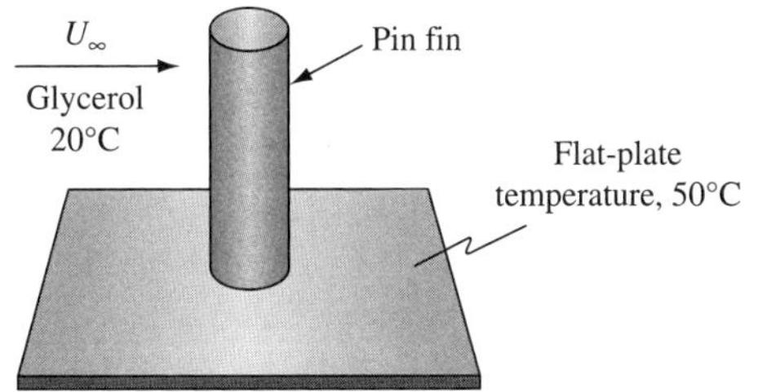

Problem 7.14

7.15 Water at 180°C enters a bare, 15-m-long, 2.5-cm-diameter wrought iron pipe at 3 m/s. If air at 10°C flows perpendicular to the pipe at 12 m/s, determine the outlet temperature of the water. (Note that the temperature difference between the air and the water varies along the pipe.)

7.16 The temperature of air flowing through a 25-cm-diameter duct whose inner walls are at 320°C is to be-measured using a thermocouple soldered in a cylindrical steel well of 1.2-cm OD with an oxidized exterior, as shown in the accompanying sketch. The air flows normal to the cylinder at a mass velocity of 17,600 kg/h m^2. If the temperature indicated by the thermocouple is 200°C, estimate the actual temperature of the air.

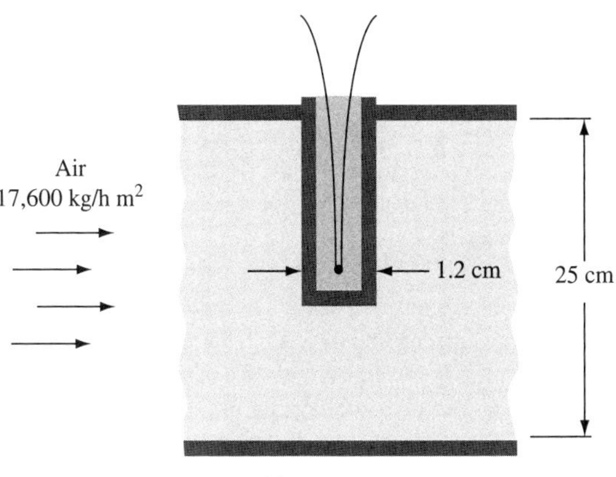

Problem 7.16

7.17 Develop an expression for the ratio of the rate of heat transfer to water at 40°C from a thin flat strip of width $\pi D/2$ and length L at zero angle of attack and from a tube of the same length and diameter D in cross-flow with its axis normal to the water flow in the Reynolds number range between 50 and 1000. Assume both surfaces are at 90°C.

7.18 Repeat Problem 7.17 for air flowing over the same two surfaces in the Reynolds number range between 40,000 and 200,000. Neglect radiation.

7.19 The instruction manual for a hot-wire anemometer states that "roughly speaking, the current varies as the one-fourth power of the average velocity at a fixed wire resistance," Check this statement, using the heat transfer characteristics of a thin wire in air and in water.

7.20 A hot-wire anemometer is used to determine the boundary layer velocity profile in the air flow over a scale model of an automobile. The hot wire is held in a traversing mechanism that moves the wire in a direction normal to the surface of the model. The hot-wire is operated at constant temperature. The boundary layer thickness is to be defined as the distance from the model surface at which the velocity is 90% of the free-stream velocity. If the probe current is I_0 when the hot-wire is held in the free-stream velocity, U_∞, what current will indicate the edge of the boundary layer? Neglect radiation heat transfer from the hot-wire and conduction from the ends of the wire.

7.21 A platinum hot-wire anemometer operated in the constant-temperature mode has been used to measure the velocity of a helium stream. The wire diameter is 20 μm, its length is 5 mm, and it is operated at 90°C. The electronic circuit used to maintain the wire temperature has a maximum power output of 5 W and is unable to accurately control the wire temperature if the voltage applied to the wire is less than 0.5 V. Compare the operation of the wire in the helium stream at 20°C and 10 m/s with its operation in air and water at the same temperature and velocity. The electrical resistance of the platinum at 90°C is 21.6 $\mu\Omega$ cm.

7.22 A hot-wire anemometer consists of a 5-mm-long, 5-μm-diameter platinum wire. The probe is operated at a constant current of 0.03 A. The electrical resistivity of platinum is 17 $\mu\Omega$ cm at 20°C and increases by 0.385% per °C. (a) If the voltage across the wire is 1.75 V, determine the velocity of the air flowing across it and the wire temperature if the free-stream air temperature is 20°C. (b) What are the wire temperature and voltage if the air velocity is 10 m/s? Neglect radiation and conduction heat transfer from the wire.

7.23 A 2.5-cm sphere is to be maintained at 50°C in either an airstream or a water stream, both at 20°C and 2 m/s velocity. Compare the rate of heat transfer and the drag on the sphere for the two fluids.

7.24 Compare the effect of forced convection on heat transfer from an incandescent lamp with that of natural convection (see Problem 5.27). What will the glass temperature be for air velocities of 0.5, 1, 2, and 4 m/s?

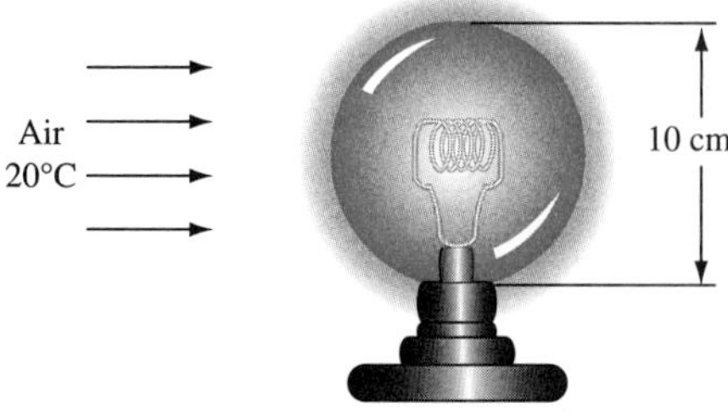

Problem 7.24

7.25 An experiment was conducted in which the heat transfer from a sphere in sodium was measured. The sphere, 0.5 in. in diameter, was pulled through a large sodium bath at a given velocity while an electrical heater inside the sphere maintained the temperature at a set point. The following table gives the results of the experiment. Determine how well the above data are predicted by the appropriate correlation given in the text. Express your results in terms of the percent difference between the experimentally determined Nusselt number and that calculated from the equation.

	Run Number				
	1	2	3	4	5
Velocity (m/s)	3.44	3.14	1.56	3.44	2.16
Sphere surface temp (°C)	478	434	381	350	357
Sodium bath temp (°C)	300	300	300	200	200
Heater temp (°C)	486	439	385	357	371
Heat flux × 10^{-6} W/m^2	14.6	8.94	3.81	11.7	8.15

7.26 A copper sphere initially at a uniform temperature of 132°C is suddenly released at the bottom of a large bath of bismuth at 500°C. The sphere diameter is 1 cm, and it rises through the bath at 1 m/s. How far will the sphere rise before its center temperature is 300°C? What is its surface temperature at that point? (The sphere has a thin nickel plating to protect the copper from the bismuth.)

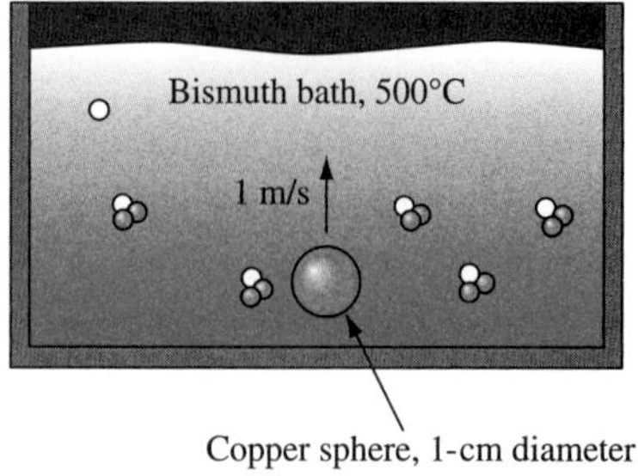

Problem 7.26

7.27 A spherical water droplet of 1.5-mm diameter is freely falling in atmospheric air. Calculate the average convection heat transfer coefficient when the droplet has reached its terminal velocity. Assume that the water is at 50°C and the air is at 20°C. Neglect mass transfer and radiation.

7.28 In a lead-shot tower, spherical 0.95-cm-diameter BB shots are formed by drops of molten lead, which solidify as they descend in cooler air. At the terminal velocity, i.e., when the drag equals the gravitational force, estimate the total heat transfer coefficient if the lead surface is at 171°C, the surface of the lead has an emissivity of 0.63, and the air temperature is 16°C. Assume $C_D = 0.75$ for the first trial calculation.

7.29 A copper sphere 2.5 cm in diameter is suspended by a fine wire in the center of an experimental hollow, cylindrical furnace whose inside wall is maintained uniformly at 430°C. Dry air at a temperature of 90°C and a pressure of 1.2 atm is blown steadily through the furnace at a velocity of 14 m/s. The interior surface of the furnace wall is black. The copper is slightly oxidized, and its emissivity is 0.4. Assuming that the air is completely transparent to radiation, calculate for the steady state: (a) the convection heat transfer coefficient between the copper sphere and the air and (b) the temperature of the sphere.

7.30 A method for measuring the convection heat transfer from spheres has been proposed. A 20-mm-diameter copper sphere with an embedded electrical heater is to be suspended in a wind tunnel. A thermocouple inside the sphere measures the sphere surface temperature. The sphere is supported in the tunnel by a type 304 stainless steel tube with a 5-mm OD, a 3-mm ID, and 20-cm length. The steel tube is attached to the wind tunnel wall in such a way that no heat is transferred through the wall. For this experiment, examine the magnitude of the correction that must be applied to the sphere heater power to account for conduction along the support tube. The air temperature is 20°C, and the desired range of Reynolds numbers is 10^3 to 10^5.

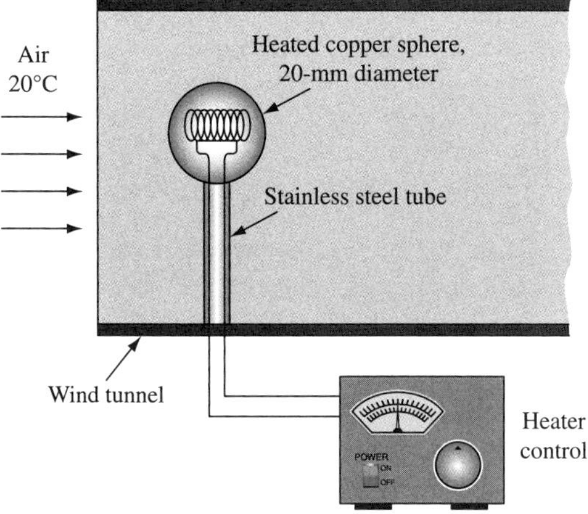

Problem 7.30

7.31 (a) Estimate the heat transfer coefficient for a spherical fuel droplet injected into a diesel engine at 80°C and 90 m/s. The oil droplet is 0.025 mm in diameter, the cylinder pressure is 4800 kPa, and the gas temperature is 944 K. (b) Estimate the time required to heat the droplet to its self-ignition temperature of 600°C.

7.32 Heat transfer from an electronic circuit board is to be determined by placing a model for the board in a wind tunnel. The model is a 15-cm-square plate with embedded electrical heaters. The wind from the tunnel air is delivered at 20°C. Determine the average temperature of the model as a function of power dissipation for an air velocity of 2.5 and 10 m/s. The model is pitched 30° and yawed 10° with respect to the air flow direction as shown below. The surface of the model acts as a blackbody.

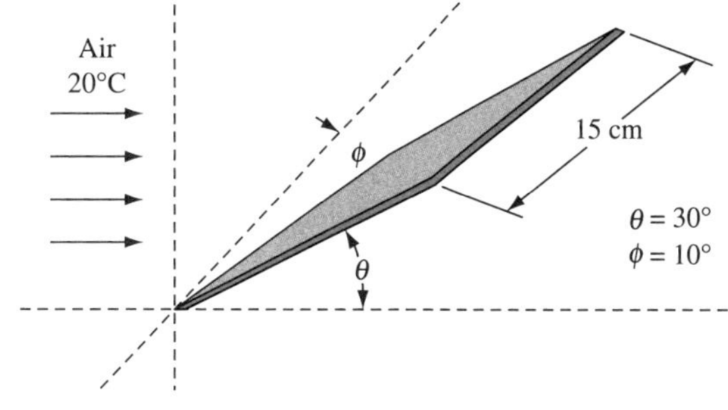

Problem 7.32

7.33 An electronic circuit contains a power resistor that dissipates 1.5 W. The designer wants to modify the circuitry in such a way that it will be necessary for the resistor to dissipate 2.5 W. The resistor is in the shape of a disk 1 cm in diameter and 0.6-mm thick. Its surface is aligned with a cooling air flow at 30°C and 10 m/s velocity. The resistor lifetime becomes unacceptable if its surface temperature exceeds 90°C. Is it necessary to replace the resistor for the new circuit?

7.34 Suppose the resistor in Problem 7.33 is rotated so that its axis is aligned with the flow. What is the maximum permissible power dissipation?

7.35 To decrease the size of personal computer mother boards, designers have turned to a more compact method of mounting memory chips on the board. The single in-line memory modules, as they are called, essentially mount the chips on their edges so that their thin dimension is horizontal, as shown in the sketch. Determine the maximum power dissipation of momory chips operating at 90°C if they are cooled by an airstream at 60°C with a velocity of 10 m/s.

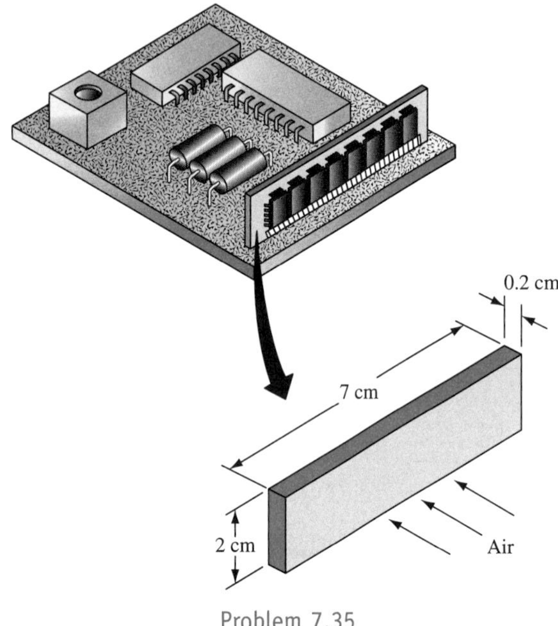

Problem 7.35

7.36 A long, half-round cylinder is placed in an airstream with its flat face downstream. An electrical resistance heater inside the cylinder maintains the cylinder surface temperature at 50°C. The cylinder diameter is 5 cm, the air velocity is 31.8 m/s, and the air temperature is 20°C. Determine the power input of the heater per unit length of cylinder. Neglect radiation heat transfer.

7.37 One method of storing solar energy for use during cloudy days or at night is to store it in the form of sensible heat in a rock bed, as shown in the sketch below. Suppose such a rock bed has been heated to 70°C and it is desired to heat a stream of air by blowing it through the bed. If the air inlet temperature is 10°C and the mass velocity of the air in the bed is 0.5 kg/s m^2, how long must the bed be in order for the initial outlet air temperature to be 65°C? Assume that the rocks are spherical, 2 cm in diameter, and that the bed void fraction is 0.5. (*Hint*: The surface area of the rocks per unit volume of the bed is $(6/D_p)(1 - \epsilon)$.)

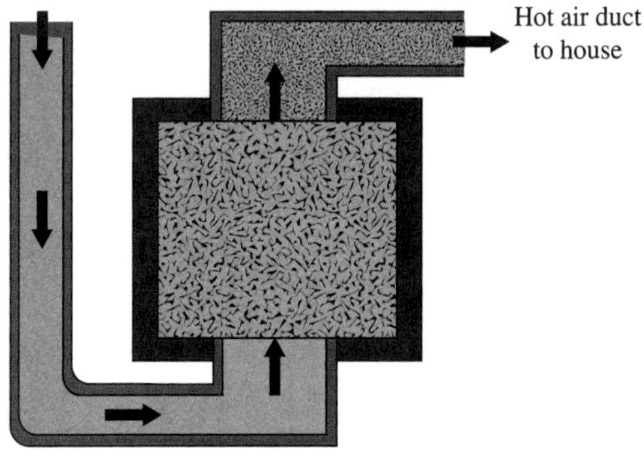

Problem 7.37

7.38 Suppose the rock bed in problem 7.37 has been completely discharged and the entire bed is at 10°C. Hot air at 90°C and 0.2 m/s is then used to recharge the bed. How long will it take until the first rocks are back up to 70°C, and what is the total heat transfer from the air to the bed?

7.39 An automotive catalytic convertor is a packed bed in which a platinum catalyst is coated on the surface of small alumina spheres. A metal container holds the catalyst pellets and allows engine exhaust gases to flow through the bed of pellets. The catalyst must be heated by the exhaust gases to 300°C before the catalyst can help oxidize unburned hydrocarbons in the gases. The time required to achieve this temperature is critical, because unburned hydrocarbons emitted by the vehicle during a cold start comprise a large fraction of the total emissions from the vehicle during an emission test. A fixed volume of catalyst is required, but the shape of the bed can be modified to increase the heat-up rate. Compare the heat-up time for a bed 5 cm in diameter and 20 cm long with one 10 cm in diameter and 5 cm long. The catalyst pellets are spherical, 5 mm in diameter, and have a density of 2 g/cm^3, a thermal conductivity of 12 W/m K, and a specific heat of 1100 J/kg K. The packed-bed void fraction is 0.5. Exhaust gas from the engine is at a temperature of 400°C, has a flow rate of 6.4 gm/s, and has the properties of air.

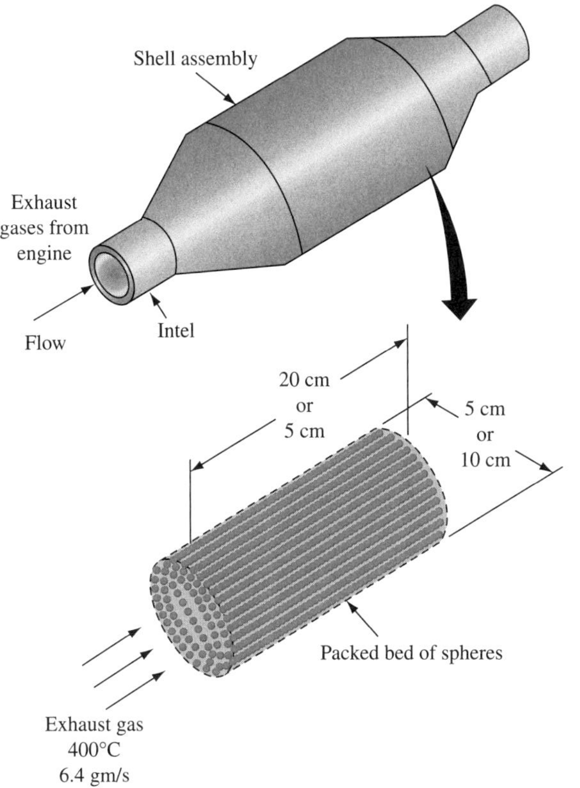

Problem 7.39

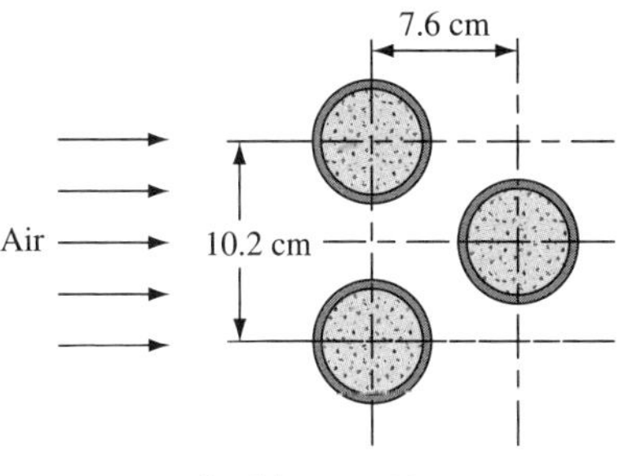

Problem 7.40

7.40 Determine the average heat transfer coefficient for air at 60°C flowing at a velocity of 1 m/s over a bank of 6-cm-OD tubes arranged as shown in the accompanying sketch. The tube-wall temperature is 117°C.

7.41 Repeat Problem 7.40 for a tube bank in which all of the tubes are spaced with their centerlines 7.5 cm apart.

7.42 Carbon dioxide gas at 1 atmosphere pressure is to be heated from 25°C to 75°C by pumping it through a tube bank at a velocity of 4 m/s. The tubes are heated by steam condensing within them at 200°C. The tubes have a 10-mm OD, are in an in-line arrangement, and have a longitudinal spacing of 15 mm and a transverse spacing of 17 mm. If 13 tube rows are required, what is the average heat transfer coefficient and what is the pressure drop of the carbon dioxide?

7.43 Estimate the heat transfer coefficient for liquid sodium at 1000°F flowing over a 10-row staggered-tube bank of 1-inch-diameter tubes arranged in an equilateral-triangular array with a 1.5 pitch-to-diameter ratio. The entering velocity is 2 ft/s, based on the area of the shell, and the tube surface temperature is 400°F. The outlet sodium temperature is 600°F.

7.44 Liquid mercury at a temperature of 315°C flows at a velocity of 10 cm/s over a staggered bank of 5/8-in. 16 BWG stainless steel tubes arranged in an equilateral-triangular array with a pitch-to-diameter ratio of 1.375. If water at 2 atm pressure is being evaporated inside the tubes, estimate the average rate of heat transfer to the water per meter length of the bank, if the bank is 10 rows deep and contains 60 tubes. The boiling heat transfer coefficient is 20,000 W/m^2 K.

7.45 Compare the rate of heat transfer and the pressure drop for an in-line and a staggered arrangement of a tube bank consisting of 300 tubes that are 6 ft long with a 1-in. OD. The tubes are to be arranged in 15 rows with longitudinal and transverse spacing of 2 in. The tube surface temperature is 200°F, and water at 100°F is flowing at a mass rate of 12,000 lb/s over the tubes.

7.46 Consider a heat exchanger consisting of 12.5-mm-OD copper tubes in a staggered arrangement with transverse spacing of 25 mm and longitudinal spacing of 30 mm with nine tubes in the longitudinal direction. Condensing steam at 150°C flows inside the tubes. The heat exchanger is used to heat a stream of air flowing at 5 m/s from 20°C to 32°C. What are the average heat transfer coefficient and pressure drop for the tube bank?

Design Problems

7.1 **Alternative Uses for the Alaskan Pipeline** (Chapter 7)
Recent studies have shown that the supply of crude oil from Alaska's North Slope will soon decline to subeconomic levels and that production will then cease. Alternatives are under consideration that would continue to make use of the Alaska pipeline and to generate revenues from the large natural gas resources in that region. The pipeline was designed to maintain crude oil at a sufficiently high temperature to allow it to be pumped while at the same time protecting the fragile Alaskan permafrost. From the standpoint of the existing thermal design of the pipeline, consider the feasibility of transporting the following alternatives: (i) natural gas, (ii) liquified natural gas, (iii) methanol, (iv) diesel fuel. Your considerations should include (a) temperature required to transport each candidate product, (b) insulating and heating capacity of the existing pipeline, (c) effect on the systems in place to protect permafrost, and (d) use of the existing crude oil pumping stations.

7.2 **Motorcycle Engine Cooling**
Motorcycle manufacturers offer engines with two methods of cooling: air cooling and liquid cooling. In air cooling, fins are applied to the outside of the cylinder and the cylinder is oriented to provide the best possible air flow. In liquid cooling, the engine cylinder is jacketed and a liquid coolant is circulated between the cylinder and the jacket. The coolant is then circulated to a heat exchanger where air flow is used to transfer heat from the coolant to the air. Discuss advantages and disadvantages of both arrangements and quantify your results with calculations. Considerations include: weight, cost, rider comfort, center of gravity, maintenance requirements, and compactness of design. As a baseline, consider a two-cylinder engine with cylinders of 3.30-in. diameter and 3.92-in. length producing a maximum of 80 hp at a thermal efficiency of 15%. Assume that the outer wall of the cylinder operates at a temperature of 200°C and that ambient air is at 40°C.

7.3 **Microprocessor Cooling** (Chapter 7)
Consider a microprocessor dissipating 50 W with dimensions 2-cm × 2-cm square and 0.5-cm high (see figure). In order to cool the microprocessor, it is necessary to mount it to a device called a heat sink, which serves two purposes. First, it distributes the heat from the relatively small microprocessor to a larger area; second, it provides extended heat transfer area in the form of fins. A small fan then can be used to provide forced-air cooling. The main constraints to the design of a heat sink are cost and size. For laptop computers, fan power is also an important consideration. Develop a heat-sink design that will maintain the microprocessor at 90°C or less and suggest ways to optimize the cooling system.

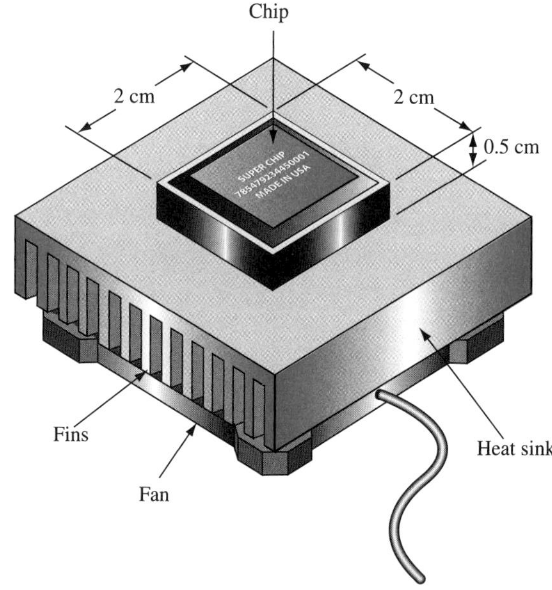

Design Problem 7.3

7.4 **Cooling Analysis of Aluminum Extrusion** (Chapters 3 and 7)
In Chapter 3, you were asked to determine the time required for an aluminum extrusion to cool to a maximum temperature of 40°C. Repeat these calculations, but determine the convection heat transfer coefficients over the extrusion, assuming that air is directed perpendicular to the right face of the extrusion at a velocity of

15 m/s. Conditions at the front resemble that of a jet impinging on a surface, whereas conditions on the upper and lower surfaces resemble those of flow over a plate; see accompanying sketch. The rear face presents a problem, and some estimates and constructive ideas about calculating the heat transfer coefficients will be left to the designer.

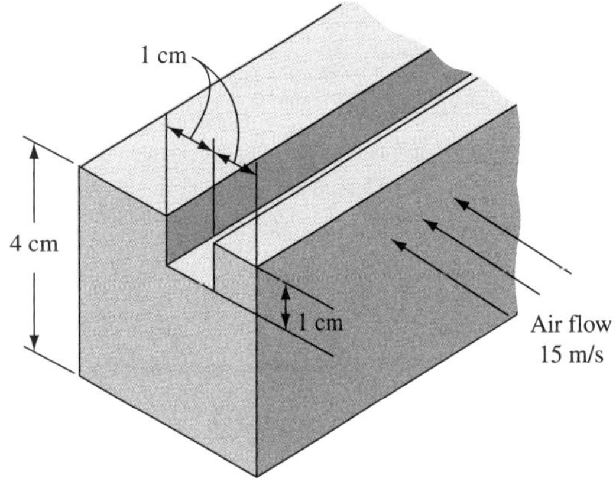

Design Problem 7.4

CHAPTER 8

Heat Exchangers

A front section of a typical automobile radiator, which is a tube-fin type compact heat exchanger, showing the inter-fin air-flow passages across flattened coolant-flow tubes.

Source: Courtesy of Philip Sayer/Alamy.

Concepts and Analyses to Be Learned

Heat exchangers are generally devices or systems in which heat is transferred from one flowing fluid to another. The fluids may be liquids or gases, and in some heat exchangers more than two fluids might flow. These devices may have a tubular structure, of which the double-pipe and shell-and-tube exchangers are perhaps the most prevalent, or a stacked-plate structure, which includes the plate-fin and plate-and-frame exchangers, among some other configurations. Perhaps the most conspicuous, and historically the oldest, applications can be found in a power plant. The steam generator or boiler, water-cooled steam condenser, boiler feed-water heater, and combustion air regenerator, as well as several other types of equipment are all heat exchangers. In most homes, common heat exchangers are the gas-fired hot water heater, and the evaporator and condenser coils of a central air-conditioning unit. All automobiles have a radiator and oil cooler, along with a few other heat exchangers. A study of this chapter will teach you:

- How to classify different types of heat exchangers and to characterize their structural and geometric features
- How to set up the thermal resistance network for the overall heat transfer coefficient
- How to calculate the log mean temperature difference (or LMTD) and to evaluate the thermal performance of a heat exchanger by the *F*-LMTD method

- How to determine heat exchanger effectiveness and to evaluate the thermal performance by the $\mathscr{E}$-NTU method
- How to model and evaluate the thermal and hydrodynamic performance of heat exchangers that employ heat transfer enhancement techniques, as well as microscale heat exchangers

8.1 Introduction

This chapter deals with the thermal analysis of various types of heat exchangers that transfer heat between two fluids. Two methods of predicting the performance of conventional industrial heat exchangers will be outlined, and techniques for estimating the required size and the most suitable type of heat exchanger to accomplish a specified task will be presented.

When a heat exchanger is placed into a thermal transfer system, a temperature drop is required to transfer the heat. The magnitude of this temperature drop can be decreased by utilizing a larger heat exchanger, but this will increase the cost of the heat exchanger. Economic considerations are important in engineering design, and in a complete engineering design of heat exchange equipment, not only the thermal performance characteristics but also the pumping power requirements and the economics of the system are important. The role of heat exchangers has taken on increasing importance recently as engineers have become energy conscious and want to optimize designs not only in terms of a thermal analysis and economic return on the investment but also in terms of the energy payback of a system. Thus economics, as well as such considerations as the availability and amount of energy and raw materials necessary to accomplish a given task, should be considered.

8.2 Basic Types of Heat Exchangers

A heat exchanger is a device in which heat is transferred between a warmer and a colder substance, usually fluids. There are three basic types of heat exchangers:

Recuperators. In this type of heat exchanger the hot and cold fluids are separated by a wall and heat is transferred by a combination of convection to and from the wall and conduction through the wall. The wall can include extended surfaces, such as fins (see Chapter 2), or other heat transfer enhancement devices.

Regenerators. In a regenerator the hot and cold fluids alternately occupy the same space in the exchanger core. The exchanger core or "matrix" serves as a heat storage device that is periodically heated by the warmer of the two fluids and then transfers heat to the colder fluid. In a *fixed matrix* configuration, the hot and cold fluids pass alternately through a stationary exchanger, and for continuous operation two or more matrices are necessary, as shown in Fig. 8.1(a) on the next page. One commonly

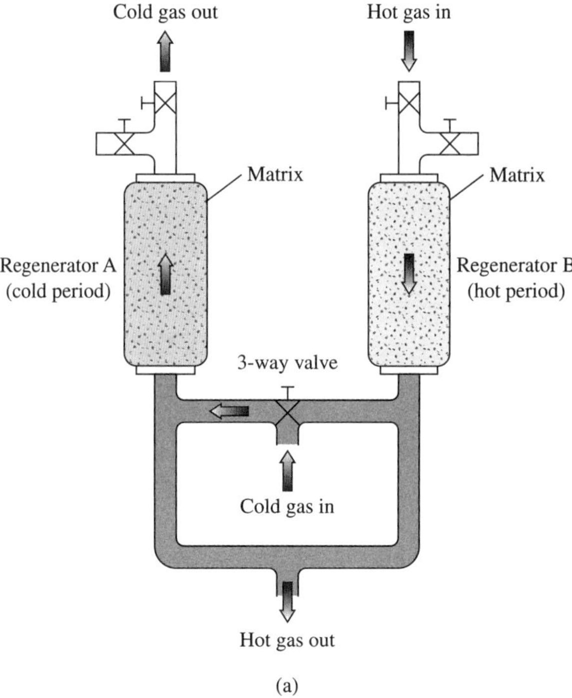

(a)

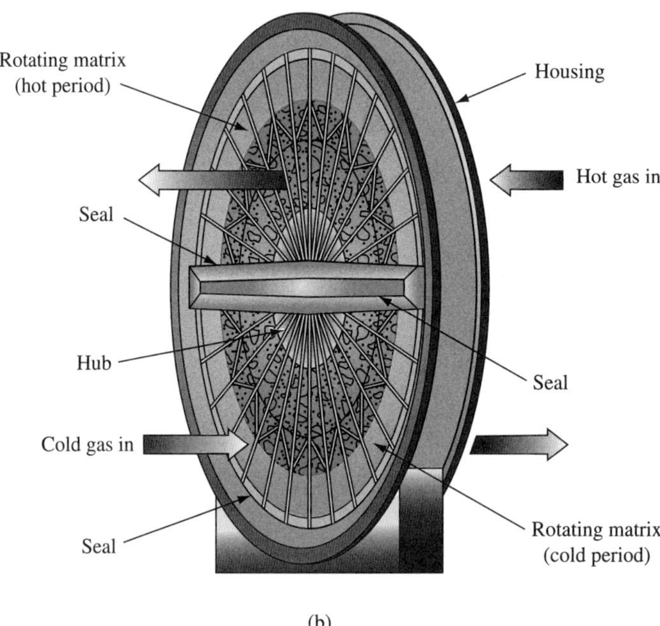

(b)

FIGURE 8.1 (a) Fixed dual-bed regenerator or system. (b) Rotary regenerator.

used arrangement for the matrix is the "packed bed" discussed in Chapter 7. Another approach is the *rotary regenerator* in which a circular matrix rotates and alternately exposes a portion of its surface to the hot and then to the cold fluid, as shown in Fig. 8.1(b). Hausen [1] gives a complete treatment of regenerator theory and practice.

Direct Contact Heat Exchangers. In this type of heat exchanger the hot and cold fluids contact each other directly. An example of such a device is a cooling tower in which a spray of water falling from the top of the tower is directly contacted and cooled by a stream of air flowing upward. Other direct contact systems use immiscible liquids or solid-to-gas exchange. An example of a direct contact heat exchanger used to transfer heat between molten salt and air is described in Bohn and Swanson [2]. The direct contact approach is still in the research and development stage, and the reader is referred to Kreith and Boehm [3] for further information.

This chapter deals mostly with the first type of heat exchanger and will emphasize the "shell-and-tube" design. The simplest arrangement of this type of heat exchanger consists of a tube within a tube, as shown in Fig. 8.2(a). Such an

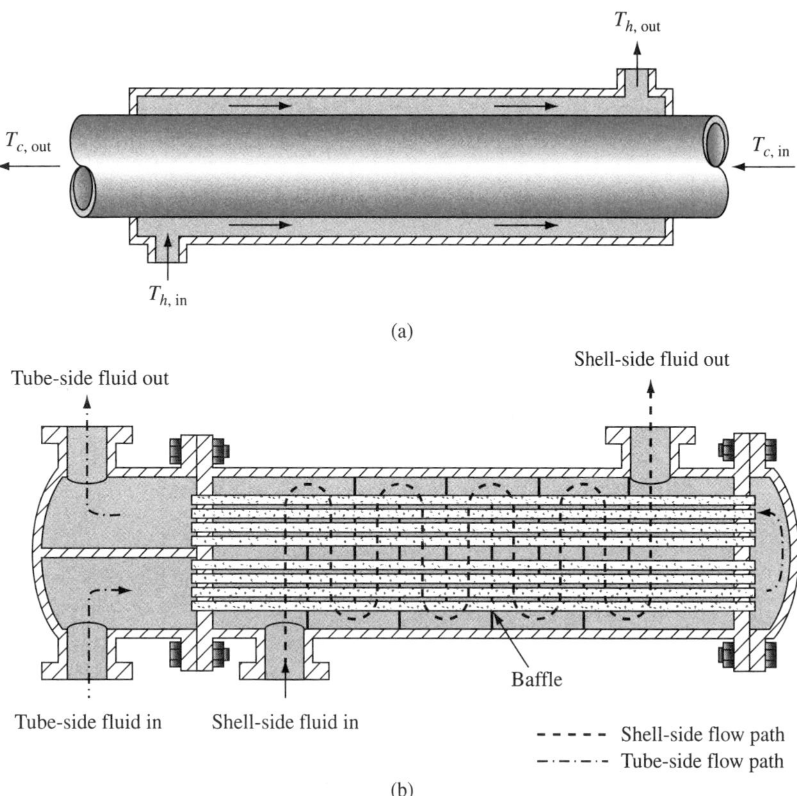

FIGURE 8.2 (a) Simple tube-within-a-tube counterflow heat exchanger. (b) Shell-and-tube heat exchanger with segmental baffles: two-tube passes, one-shell pass.

arrangement can be operated either in counterflow or in parallel flow, with either the hot or the cold fluid passing through the annular space and the other fluid passing through the inside of the inner pipe.

A more common type of heat exchanger that is widely used in the chemical and process industry is the shell-and-tube arrangement shown in Fig. 8.2(b). In this type of heat exchanger one fluid flows inside the tubes while the other fluid is forced through the shell and over the outside of the tubes. The fluid is forced to flow over the tubes rather than along the tubes because a higher heat transfer coefficient can be achieved in cross-flow than in flow parallel to the tubes. To achieve cross-flow on the shell side, baffles are placed inside the shell as shown in Fig. 8.2(b). These baffles ensure that the flow passes across the tubes in each section, flowing downward in the first, upward in the second, and so on. Depending on the header arrangements at the two ends of the heat exchanger, one or more tube passes can be achieved. For a two-tube-pass arrangement, the inlet header is split so that the fluid flowing into the tubes passes through half of the tubes in one direction, then turns around and returns through the other half of the tubes to where it started, as shown in Fig. 8.2(b). Three- and four-tube passes can be achieved by rearrangement of the header space. A variety of baffles have been used in industry (see Fig. 8.3), but the most common kind is the disk-and-doughnut baffle shown in Fig. 8.3(b).

In gas heating or cooling it is often convenient to use a cross-flow heat exchanger such as that shown in Fig. 8.4 on page 490. In such a heat exchanger, one of the fluids passes through the tubes while the gaseous fluid is forced across the tube bundle. The flow of the exterior fluid may be by forced or by natural convection. In this type of exchanger the gas flowing across the tube is considered to be *mixed*, whereas the fluid in the tube is considered to be *unmixed*. The exterior gas flow is mixed because it can move about freely between the tubes as it exchanges heat, whereas the fluid within the tubes is confined and cannot mix with any other stream during the heat exchange process. Mixed flow implies that all of the fluid in any given plane normal to the flow has the same temperature. Unmixed flow implies that although temperature differences within the fluid may exist in at least one direction normal to the flow, no heat transfer results from this gradient [4].

Another type of cross-flow heat exchanger that is widely used in the heating, ventilating, and air-conditioning industry is shown in Fig. 8.5 on page 490. In this arrangement gas flows across a finned tube bundle and is unmixed because it is confined to separate flow passages.

In the design of heat exchangers it is important to specify whether the fluids are mixed or unmixed, and which of the fluids is mixed. It is also important to balance the temperature drop by obtaining approximately equal heat transfer coefficients on the exterior and interior of the tubes. If this is not done, one of the thermal resistances may be unduly large and cause an unnecessarily high overall temperature drop for a given rate of heat transfer, which in turn demands larger equipment and results in poor economics.

The shell-and-tube heat exchanger illustrated in Fig. 8.2(b) has fixed *tube sheets* at each end, and the tubes are welded or expanded into the sheets. This type of construction has the lowest initial cost but can be used only for small temperature

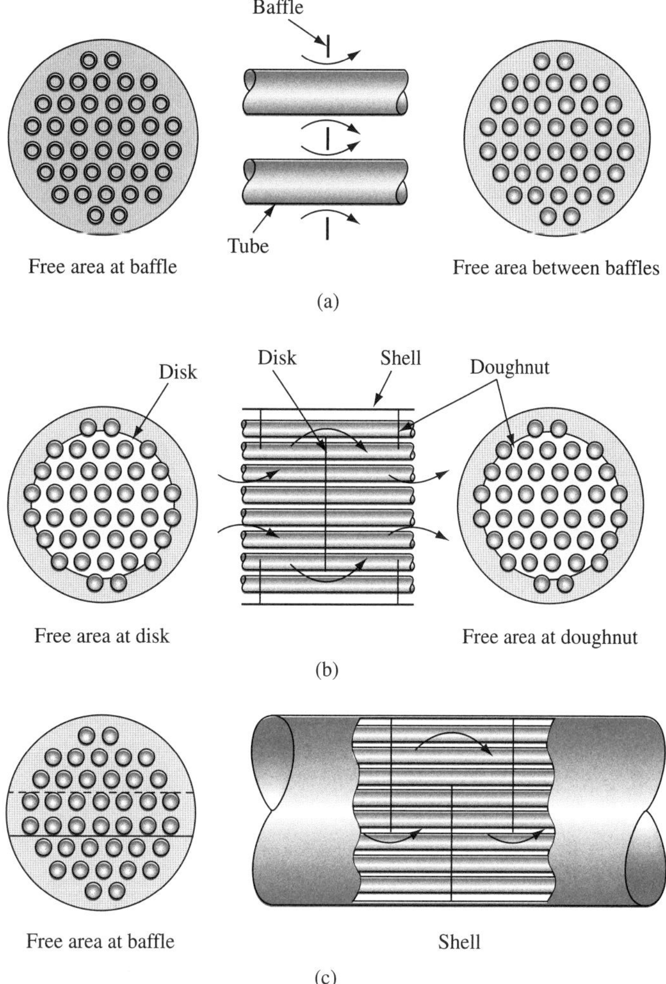

FIGURE 8.3 Three types of baffles used in shell-and-tube heat exchangers: (a) orifice baffle; (b) disk-and-doughnut baffle; (c) segmental baffle.

differences between the hot and the cold fluids because no provision is made to prevent thermal stresses due to the differential expansion between the tubes and the shell. Another disadvantage is that the tube bundle cannot be removed for cleaning. These drawbacks can be overcome by modification of the basic design, as shown in Fig. 8.6 on page 491. In this arrangement one tube sheet is fixed but the other is bolted to a floating-head cover that permits the tube bundle to move relative to the shell. The floating tube sheet is clamped between the floating head and a flange so that it is possible to remove the tube bundle for cleaning. The heat exchanger shown in Fig. 8.6 has one shell pass and two tube passes.

In the design and selection of a shell-and-tube heat exchanger, the power requirement and the initial cost of the unit must be considered. Results obtained

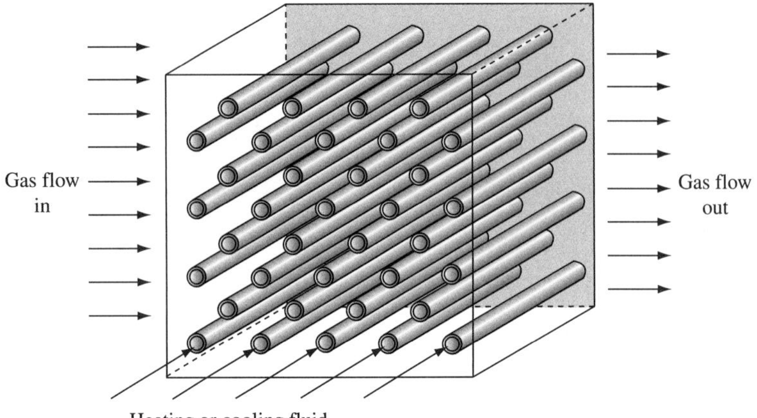

FIGURE 8.4 Cross-flow gas heater illustrating cross-flow with one fluid (gas) mixed, the other unmixed.

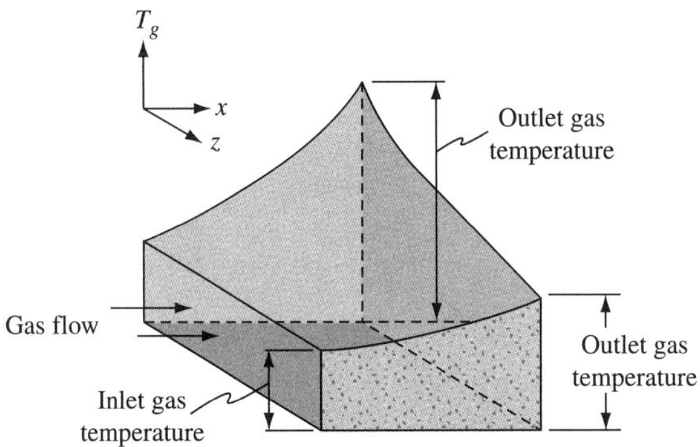

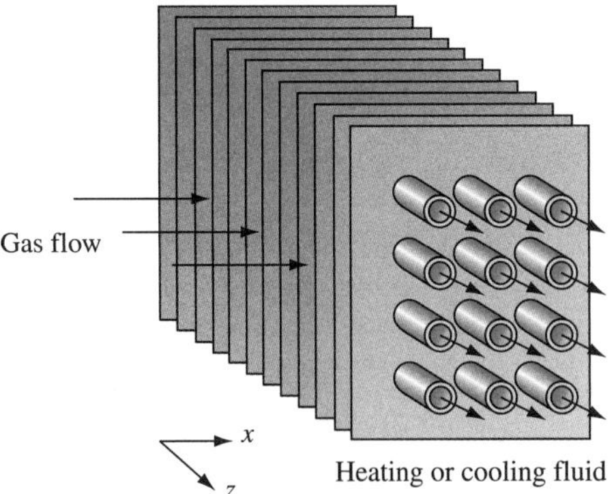

FIGURE 8.5 Cross-flow heat exchanger, widely used in the heating, ventilating, and air-conditioning industry. In this arrangement both fluids are unmixed.

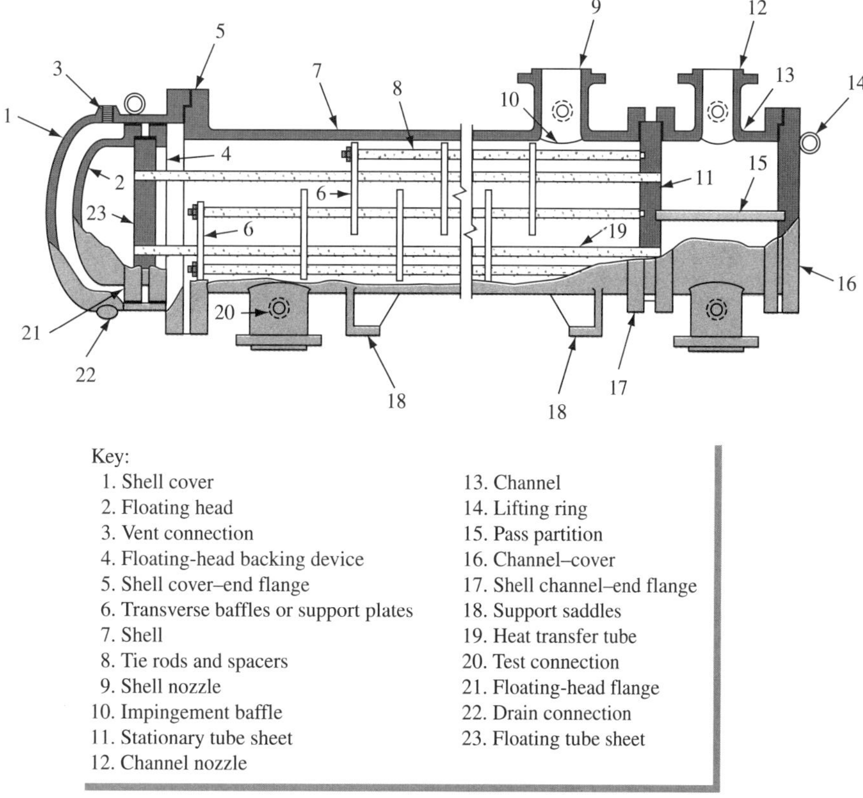

FIGURE 8.6 Shell-and-tube heat exchanger with floating head.
Source: Courtesy of the Tubular Exchanger Manufacturers Association.

by Pierson [5] show that the smallest possible pitch in each direction results in the lowest power requirement for a specified rate of heat transfer. Since smaller values of pitch also permit the use of a smaller shell, the cost of the unit is reduced when the tubes are closely packed. There is little difference in performance between inline and staggered arrangements, but the former are easier to clean. The Tubular Exchanger Manufacturers Association (TEMA) recommends that tubes be spaced with a minimum center-to-center distance of 1.25 times the outside diameter of the tube and, when tubes are on a square pitch, that a minimum clearance lane of 0.65 cm be provided.

Figure 8.7, on the next page is a photograph of a large baffled exchanger for vegetable-oil service. The flow of the shell-side fluid in baffled heat exchangers is partly perpendicular and partly parallel to the tubes. The heat transfer coefficient on the shell side in this type of unit depends not only on the size and spacing of the tubes and the velocity and physical properties of the fluid but also on the spacing and shape of the baffles. In addition, there is always leakage through the tube holes in the baffle and between the baffle and the inside of the shell, and there is bypassing between the tube bundle and the shell. Because of these complications, the heat transfer coefficient can be estimated only by approximate methods or from experience with

FIGURE 8.7 Heat exchanger tube bundle with baffles.
Source: Courtesy of the Aluminum Company of America.

similar units. According to one approximate method, which is widely used for design calculations [6], the average heat transfer coefficient calculated for the corresponding tube arrangement in simple cross-flow is multiplied by 0.6 to allow for leakage and other deviations from the simplified model. For additional information the reader is referred to Tinker [6], Short [7], Donohue [8], and Singh and Soler [9].

In some heat exchanger applications, the heat exchanger size and weight are of prime concern. This can be especially true for heat exchangers in which one or both fluids are gases, since the gas-side heat transfer coefficients are small and large heat transfer surface area requirements can result. *Compact heat exchangers* refer to heat exchanger designs in which large heat transfer surface areas are provided in as small a space as possible. Applications in which compact heat exchangers are required include (i) an automobile heater core in which engine coolant is circulated through tubes and the passenger compartment air is blown over the finned exterior surface of the tubes and (ii) refrigerator condensers in which the refrigerant is circulated inside tubes and cooled by room air circulated over the finned outside of the tubes.

Figure 8.8 shows another application, an automobile radiator. In Fig. 8.8 the engine coolant is pumped through the flattened, horizontal tubes while air from the engine fan is blown through the finned channels between the coolant tubes. The fins are brazed to the coolant tubes and help transfer heat from the exterior surfaces of the tube into the airstream. Experimental data are required to allow one to determine the gas-side heat transfer coefficient and pressure drop for compact heat exchanger cores like the one in Fig. 8.8. Fin design parameters that affect the heat transfer and pressure drop on the gas side include thickness, spacing, material, and length. Kays and London [10] have compiled heat transfer and pressure drop data for a large number of compact heat exchanger cores. For each core, the fin parameters listed above are given in addition to the hydraulic diameter on the gas side, the total heat

FIGURE 8.8 Vacuum brazed aluminum radiator.
Source: Courtesy of Ford Motor Company.

transfer surface area per unit volume, and the fraction of total heat transfer area that is fin area. Data in London [10] are presented in the form of the Stanton number and friction factor as a function of the gas-side Reynolds number. Given the heat exchanger requirements, the designer can estimate the performance of several candidate heat exchanger cores to determine the best design.

Given the large variety of applications and structural configurations of heat exchangers, as just discussed, it becomes important to provide a classification scheme to help in their selection process. Although several schemes have been proposed in the literature [11–13], somewhat reflecting the inherent difficulty in trying to categorize equipment that comes in different materials, shapes, and sizes for diverse usage, the following perhaps represent the simplest criteria [11] that can be adopted:

1. *The type of heat exchanger: (a) recuperator and (b) regenerator.* A recuperator, as discussed earlier, is the conventional heat exchanger in which heat is recovered or recouped by the cold fluid stream from the hot fluid stream. The two fluid streams flow simultaneously, possibly in a variety of flow arrangements, through the heat exchanger. In a regenerator, the hot and cold fluids alternately flow through the exchanger, which essentially acts as a transient energy storage and dissipation unit.
2. *The type of heat exchange process between the fluids: (a) indirect contact, or transmural, and (b) direct contact.* In a transmural heat exchanger, the hot and cold fluids are separated by a solid material, which is typically of either tubular or plate geometry. In direct contact heat exchanger, as the name suggests, both the hot and cold fluids flow into the same space without a partitioning wall.
3. *Thermodynamic phase or state of the fluids: (a) single phase, (b) evaporation or boiling, and (c) condensation.* This criterion refers to the state of phase

of the hot and cold fluids, and the three categories refer to cases where both fluids maintain single-phase flow and one of the two fluids undergoes flow evaporation or condensation.

4. *The type of construction or geometry: (a) tubular, (b) plate, and (c) extended or finned surface.* A typical example for each of the first two categories, respectively, is the shell-and-tube heat exchanger and the plate-and-frame [14] heat exchanger. An extended- or finned-surface exchanger could either have a tubular (tube-fin) or plate (plate-fin) geometry. It is often referred to as a compact heat exchanger, especially when it has a large surface area density, i.e., relatively large ratio of heat transfer surface area to volume.

Thus, based on this simple scheme, an automobile radiator, for example (see Fig. 8.8), would be classified as a transmural recuperator with single-phase fluid flows and a finned (tube-fin type construction) surface. This heat exchanger is often also characterized as a compact heat exchanger [10] because of its large area density. Likewise, a boiler feed-water heater, which is a shell-and-tube heat exchanger similar to that shown in Fig. 8.7, would be classified as a transmural recuperator of a tubular construction with condensation in one fluid (feed-water is heated by the condensation of steam extracted from a power turbine). Students should bear in mind, however, that classification schemes serve only as guidelines and that the actual design and selection of heat exchangers may involve several other factors [11–14].

8.3 Overall Heat Transfer Coefficient

The thermal analysis and design of a heat exchanger fundamentally requires the application of the first law of thermodynamics in conjunction with the principles of heat transfer. Students would recall from Chapter 1 the application of and differences between the thermodynamic and heat transfer models of a heat exchange device and/or system. This is illustrated in Fig. 8.9, where a simple representation of the two models is depicted for the case of a typical shell-and-tube heat exchanger. Here, for the overall heat exchanger, the thermodynamic model gives the overall or total energy transfer as

$$-q_{\text{loss}} + \sum \dot{E}_{\text{in}} - \sum \dot{E}_{\text{out}} = 0$$

This statement of the first law is not very useful in heat exchanger design. However, when restated by considering the hot and cold fluids separately along with their respective mass flow rate, inlet and outlet enthalpy (stated in terms of specific heat and temperature difference), it provides the model to determine heat transfer between the two fluid streams when $q_{\text{loss}} = 0$:

$$q = (\dot{m}c_p)_c(T_{c,\text{out}} - T_{c,\text{in}}) = (\dot{m}c_p)_h(T_{h,\text{in}} - T_{h,\text{out}}) \tag{8.1}$$

The heat transfer rate given by Eq. (8.1) can then be equated with the overall heat transfer coefficient, or the overall thermal resistance, and the true-mean temperature difference between the hot and cold fluids to complete the model.

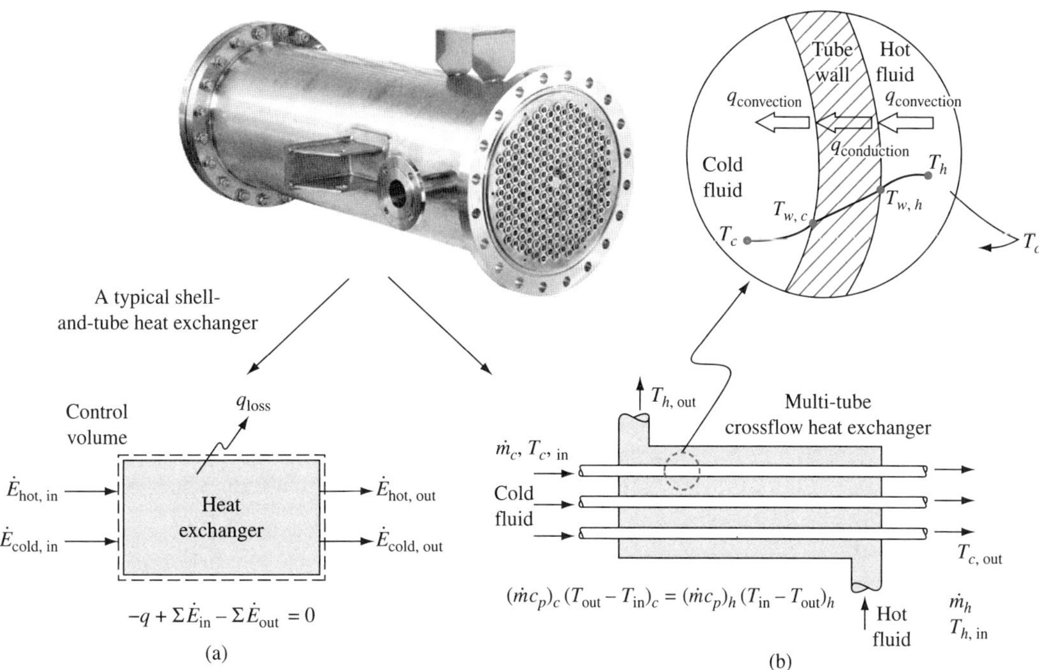

FIGURE 8.9 Application of and contrast between (a) a thermodynamic and (b) a heat transfer model for a typical shell-and-tube heat exchanger used in chemical processing.

Source: A typical shell- and tube heat exchanger courtesy of Sanjivani Phytopharma Pvt Ltd.

One of the first tasks in a thermal analysis of a heat exchanger is to evaluate the overall heat transfer coefficient between the two fluid streams. It was shown in Chapter 1 that the overall heat transfer coefficient between a hot fluid at temperature T_h and a cold fluid at temperature T_c separated by a solid plane wall is defined by

$$q = UA(T_h - T_c) \tag{8.2}$$

where $$UA = \frac{1}{\sum_{n=1}^{n=3} R_n} = \frac{1}{(1/h_1A_1) + (L/kA_k) + (1/h_2A_2)}$$

For a tube-within-a-tube heat exchanger, as shown in Fig. 8.2(a), the area at the inner heat transfer surface is $2\pi r_i L$ and the area at the outer surface is $2\pi r_o L$. Thus, if the overall heat transfer coefficient is *based on the outer area*, A_o,

$$U_o = \frac{1}{(A_o/A_ih_i) + [A_o \ln(r_o/r_i)/2\pi kL] + (1/h_o)} \tag{8.3}$$

while *on the basis of the inner area*, A_i, we get

$$U_i = \frac{1}{(1/h_i) + [A_i \ln(r_o/r_i)/2\pi kL] + (A_i/A_oh_o)} \tag{8.4}$$

If the tube is finned, Eqs. (8.3) and (8.4) should be modified as in Eq. (2.69). Although for a careful and precise design it is always necessary to calculate the individual heat transfer coefficients, for preliminary estimates it is often useful to have an approximate value of U that is typical of conditions encountered in practice. Table 8.1 lists a few typical values of U for various applications [15]. It should be noted that in many cases the value of U is almost completely determined by the thermal resistance at one of the fluid/solid interfaces, as when one of the fluids is a gas and the other a liquid or when one of the fluids is a boiling liquid with a very large heat transfer coefficient.

8.3.1 Fouling Factors

The overall heat transfer coefficient of a heat exchanger under some operating conditions, especially in the process industry, often cannot be predicted from thermal analysis alone. During operation with most liquids and some gases, a deposit gradually builds up on the heat transfer surface. The deposit may be rust, boiler scale, silt, coke, or any number of things. Its effect, which is referred to as *fouling*, is to

TABLE 8.1 Overall heat transfer coefficients for various applications (W/m^2 K)[a] (Multiply values in the table by 0.176 to get units of Btu/h ft^2°F.)

Heat Flow → to: ↓ **from:**	**Gas (stagnant)** $\bar{h}_c = 5 - 15$	**Gas (flowing)** $\bar{h}_c = 10 - 100$	**Liquid (stagnant)** $\bar{h}_c = 50 - 1{,}000$	**Liquid (flowing) Water** $\bar{h}_c = 1{,}000 - 3{,}000$ **Other Liquids** $\bar{h}_c = 500 - 2{,}000$	**Boiling Liquid Water** $\bar{h}_c = 3{,}500 - 60{,}000$ **Other Liquids** $\bar{h}_c = 1{,}000 - 20{,}000$
Gas (natural convection) $\bar{h}_c = 5 - 15$	Room/outside air through glass $U = 1–2$	Superheaters $U = 3–10$		Combustion chamber $U = 10–40$ + radiation	Steam boiler $U = 10–40$ + radiation
Gas (flowing) $\bar{h}_c = 10 - 100$		Heat exchangers for gases $U = 10–30$	Gas boiler $U = 10–50$		
Liquid (natural convection) $\bar{h}_c = 50 - 10{,}000$			Oil bath for heating $U = 25–500$	Cooling coil $U = 500–1{,}500$ with stirring	
Liquid (flowing) water $\bar{h}_c = 3{,}000 - 10{,}000$ other liquids $\bar{h}_c = 500 - 3{,}000$	Radiator central heating $U = 5–15$	Gas coolers $U = 10–50$	Heating coil in vessel water/water without stirring $U = 50–250$, with stirring $U = 500–2{,}000$	Heat exchanger water/water $U = 900–2{,}500$ water/other liquids $U = 200–1{,}000$	Evaporators of refrigerators $U = 300–1{,}000$
Condensing vapor water $\bar{h}_c = 5{,}000 - 30{,}000$ other liquids $\bar{h}_c = 1{,}000 - 4{,}000$	Steam radiators $U = 5–20$	Air heaters $U = 10–50$	Steam jackets around vessels with stirrers, water $U = 300–1{,}000$ other liquids $U = 150–500$	Condensers steam/water $U = 1{,}000–4{,}000$ other vapor/water $U = 300–1{,}000$	Evaporators steam/water $U = 1{,}500–6{,}000$ steam/other liquids $U = 300–2{,}000$

[a]Source: Adapted from Beek and Muttzall [15].

increase the thermal resistance. The manufacturer cannot usually predict the nature of the dirt deposit or the rate of fouling. Therefore, only the performance of clean exchangers can be guaranteed. The thermal resistance of the deposit can generally be obtained only from actual tests or from experience. If performance tests are made on a clean exchanger and repeated later after the unit has been in service for some time, the thermal resistance of the deposit (or *fouling factor*) R_d can be determined from the relation

$$R_d = \frac{1}{U_d} - \frac{1}{U} \tag{8.5a}$$

where U = overall heat transfer coefficient of clean exchanger
U_d = overall heat transfer coefficient after fouling has occurred
R_d = fouling factor (or unit thermal resistance) of deposit

A convenient working form of Eq. (8.5a) is

$$U_d = \frac{1}{R_d + 1/U} \tag{8.5b}$$

Fouling factors for various applications have been compiled by the Tubular Exchanger Manufacturers Association (TEMA) and are available in their publication [16]. A few examples are given in Table 8.2. The fouling factors should be applied as indicated in the following equation for the overall design heat transfer coefficient U_d of *unfinned* tubes with deposits:

$$U_d = \frac{1}{(1/\bar{h}_o) + R_o + R_k + (R_i A_o/A_i) + (A_o/\bar{h}_i A_i)} \tag{8.6}$$

where U_d = design overall coefficient of heat transfer, W/m^2 K, based on unit area of outside tube surface
$\bar{h}_o$ = average heat transfer coefficient of fluid on outside of tubing, W/m^2 K

TABLE 8.2 Typical fouling factors

Type of Fluid	Fouling Factor, R_d (m^2 K/W)
Seawater	
below 325 K	0.00009
above 325 K	0.0002
Treated boiler feedwater above 325 K	0.0002
Fuel oil	0.0009
Quenching oil	0.0007
Alcohol vapors	0.00009
Steam, non-oil-bearing	0.00009
Industrial air	0.0004
Refrigerating liquid	0.0002

Source: Courtesy of the Standards of Tubular Exchanger Manufacturers Association.

$\bar{h}_i$ = average heat transfer coefficient of fluid inside tubing, W/m^2 K
R_o = unit fouling resistance on outside of tubing, m^2 K/W
R_i = unit fouling resistance on inside of tubing, m^2 K/W
R_k = unit thermal resistance of tubing, m^2 K/W, based on outside tube surface area
$\frac{A_o}{A_i}$ = ratio of outside tube surface to inside tube surface area

8.4 Log Mean Temperature Difference

The temperatures of fluids in a heat exchanger are generally not constant but vary from point to point as heat flows from the hotter to the colder fluid. Even for a constant thermal resistance, the rate of heat flow will therefore vary along the path of the exchangers because its value depends on the temperature difference between the hot and the cold fluid in that section. Figures 8.10–8.13 illustrate the changes in temperature that may occur in either or both fluids in a simple shell-and-tube exchanger

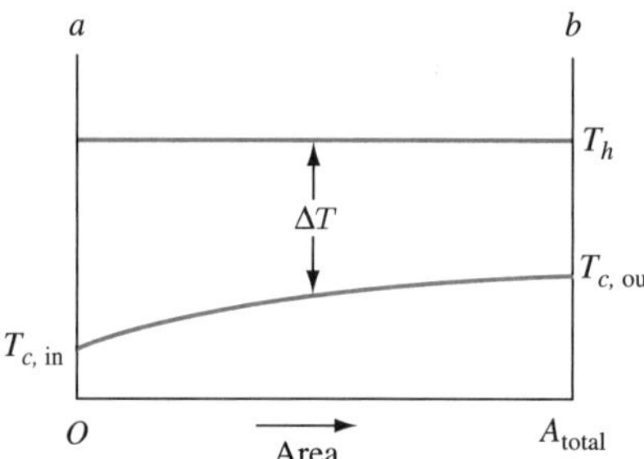

FIGURE 8.10 Temperature distribution in single-pass condenser.

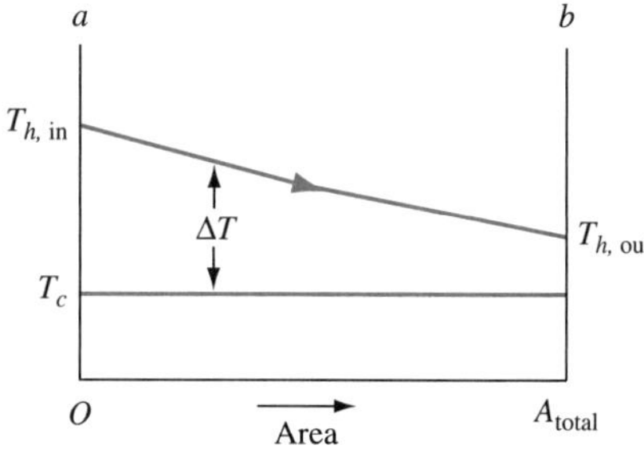

FIGURE 8.11 Temperature distribution in single-pass evaporator.

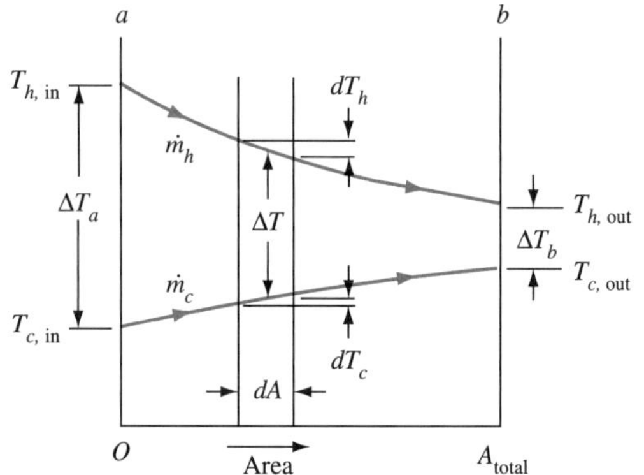

FIGURE 8.12 Temperature distribution in single-pass parallel-flow heat exchanger.

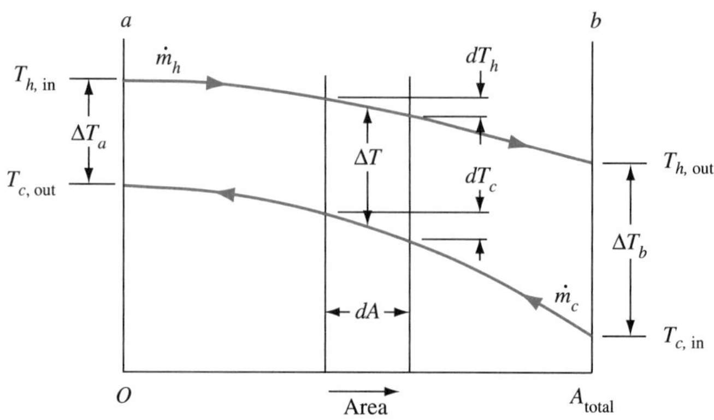

FIGURE 8.13 Temperature in single-pass counterflow heat exchanger.

[Fig. 8.2(a)]. The distances between the solid lines are proportional to the temperature differences ΔT between the two fluids.

Figure 8.10 illustrates the case in which a vapor is condensing at a constant temperature while the other fluid is being heated. Figure 8.11 represents a case where a liquid is evaporated at constant temperature while heat is flowing from a warmer fluid whose temperature decreases as it passes through the heat exchanger. For both of these cases the direction of flow of either fluid is immaterial, and the constant-temperature medium may also be at rest. Figure 8.12 represents conditions in a parallel-flow exchanger, and Fig. 8.13 applies to counterflow. No change of phase occurs in the latter two cases. Inspection of Fig. 8.12 shows that no matter how long the exchanger is, the final temperature of the colder fluid can never reach the exit temperature of the hotter fluid in parallel flow. For counterflow, on the other hand, the final temperature of the cooler fluid may exceed the outlet temperature of the hotter fluid, since a favorable temperature gradient exists all along the heat exchanger. An additional advantage of the counterflow arrangement is that for a given rate of heat flow, less surface area is required than in parallel flow. In fact, the counterflow arrangement is the most effective of all heat exchanger arrangements.

To determine the rate of heat transfer in any of the aforementioned cases, the equation

$$dq = U\,dA\,\Delta T \tag{8.7}$$

must be integrated over the heat transfer area A along the length of the exchanger. If the overall heat transfer coefficient U is constant, if changes in kinetic energy are neglected, and if the shell of the exchanger is perfectly insulated, Eq. (8.7) can be easily integrated analytically for parallel flow or counterflow. An energy balance over a differential area dA yields

$$dq = -\dot{m}_h c_{ph}\,dT_h = \pm\,\dot{m}_c c_{pc}\,dT_c = U\,dA(T_h - T_c) \tag{8.8}$$

where $\dot{m}$ is the mass rate of flow in kg/s, c_p is the specific heat at constant pressure in J/kg K, and T is the average bulk temperature of the fluid in K. The subscripts h and c refer to the hot and cold fluid, respectively; the plus sign in the third term applies to parallel flow and the minus sign to counterflow. If the specific heats of the fluids do not vary with temperature, we can write a heat balance from the inlet to an arbitrary cross section in the exchanger:

$$-C_h(T_h - T_h,\text{in}) = C_c(T_c - T_c,\text{in}) \tag{8.9}$$

where $C_h \equiv \dot{m}_h c_{ph}$, heat capacity rate of hotter fluid, W/K
$C_c \equiv \dot{m}_c c_{pc}$, heat capacity rate of colder fluid, W/K

Solving Eq. (8.9) for T_h gives

$$T_h = T_{h,\text{in}} - \frac{C_c}{C_h}(T_c - T_{c,\text{in}}) \tag{8.10}$$

from which we obtain

$$T_h - T_c = -\left(1 + \frac{C_c}{C_h}\right)T_c + \frac{C_c}{C_h}T_{c,\text{in}} + T_{h,\text{in}} \tag{8.11}$$

Substituting Eq. (8.11) for $T_h - T_c$ in Eq. (8.8) yields, after some rearrangement,

$$\frac{dT_c}{-[1 + (C_c/C_h)]T_c + (C_c/C_h)T_{c,\text{in}} + T_{h,\text{in}}} = \frac{U\,dA}{C_c} \tag{8.12}$$

Integrating Eq. (8.12) over the entire length of the exchanger (i.e., from $A = 0$ to $A = A_{\text{total}}$) yields

$$\ln\left\{\frac{-[1 + (C_c/C_h)]T_{c,\text{out}} + (C_c/C_h)T_{c,\text{in}} + T_{h,\text{in}}}{-[1 + (C_c/C_h)]T_{c,\text{in}} + (C_c/C_h)T_{c,\text{in}} + T_{h,\text{in}}}\right\} = -\left(\frac{1}{C_c} + \frac{1}{C_h}\right)UA$$

which can be simplified to

$$\ln\left[\frac{(1 + C_c/C_h)(T_{c,\text{in}} - T_{c,\text{out}}) + T_{h,\text{in}} - T_{c,\text{in}}}{T_{h,\text{in}} - T_{c,\text{in}}}\right] = -\left(\frac{1}{C_c} + \frac{1}{C_h}\right)UA \tag{8.13}$$

From Eq. (8.9) we obtain

$$\frac{C_c}{C_h} = \frac{T_{h,\text{out}} - T_{h,\text{in}}}{T_{c,\text{out}} - T_{c,\text{in}}} \tag{8.14}$$

which can be used to eliminate the heat capacity rates in Eq. (8.13). After some rearrangement we get

$$\ln\left(\frac{T_{h,\text{out}} - T_{c,\text{out}}}{T_{h,\text{in}} - T_{c,\text{in}}}\right) = [(T_{h,\text{out}} - T_{c,\text{out}}) - (T_{h,\text{in}} - T_{c,\text{in}})]\frac{UA}{q} \tag{8.15}$$

since

$$q = C_c\,(T_{c,\text{out}} - T_{c,\text{in}}) = C_h\,(T_{h,\text{in}} - T_{h,\text{out}})$$

Letting $T_h - T_c = \Delta T$, Eq. (8.15) can be rewritten as

$$q = UA\frac{\Delta T_a - \Delta T_b}{\ln(\Delta T_a/\Delta T_b)} \tag{8.16}$$

where the subscripts a and b refer to the respective ends of the exchanger and ΔT_a is the temperature difference between the hot and cold fluid streams at the inlet while ΔT_b is the temperature difference at the outlet end as shown in Figs. 8.12 and 8.13. In practice, it is convenient to use an average effective temperature difference $\overline{\Delta T}$ for the entire heat exchanger, defined by

$$q = UA\overline{\Delta T} \tag{8.17}$$

Comparing Eqs. (8.16) and (8.17), one finds that for parallel flow or counterflow,

$$\overline{\Delta T} = \frac{\Delta T_a - \Delta T_b}{\ln(\Delta T_a/\Delta T_b)} \tag{8.18}$$

The average temperature difference, $\overline{\Delta T}$, is called the *logarithmic mean temperature difference*, often designated by LMTD. The LMTD also applies when the temperature of one of the fluids is constant, as shown in Figs. 8.10 and 8.11. When $\dot{m}_h c_{ph} = \dot{m}_c c_{pc}$, the temperature difference is constant in counterflow and $\overline{\Delta T} = \Delta T_a = \Delta T_b$. If the temperature difference ΔT_a is not more than 50% greater than ΔT_b, the arithmetic mean temperature difference will be within 1% of the LMTD and may be used to simplify calculations.

The use of the logarithmic mean temperature is only an approximation in practice because U is generally neither uniform nor constant. In design work, however, the overall heat transfer coefficient is usually evaluated at a mean section halfway between the ends and treated as constant. If U varies considerably, numerical step-by-step integration of Eq. (8.7) may be necessary.

For more complex heat exchangers such as the shell-and-tube arrangements with several tube or shell passes and with cross-flow exchangers having mixed and unmixed flow, the mathematical derivation of an expression for the mean temperature difference becomes quite complex. The usual procedure is to modify the simple LMTD by correction factors, which have been published in chart form by Bowman et al. [17] and by TEMA [16]. Four of these graphs* are shown in Figs. 8.14–8.17 on page 502 through 504.

The ordinate of each is the correction factor F. To obtain the true mean temperature for any of these arrangements, the LMTD calculated for *counterflow* must be multiplied by the appropriate correction factor, that is,

$$\Delta T_{\text{mean}} = (\text{LMTD})(F) \tag{8.19}$$

*Correction factors for several other arrangements are presented in TEMA [16].

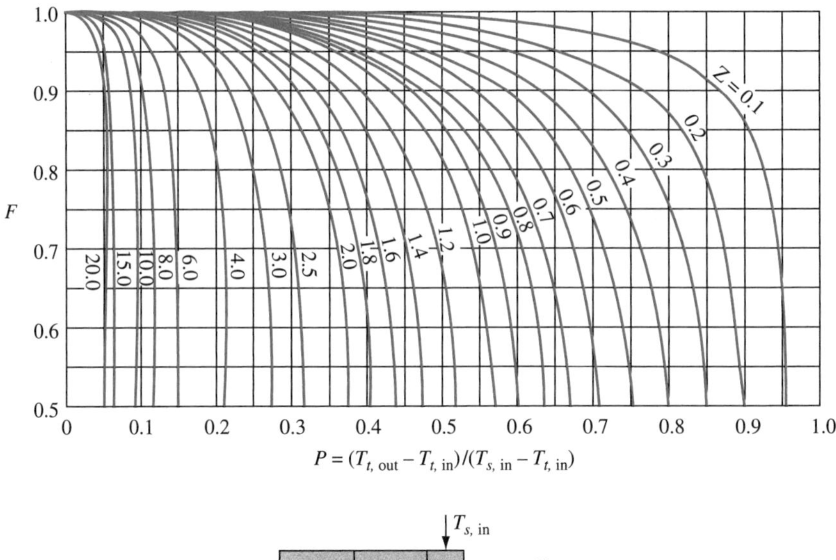

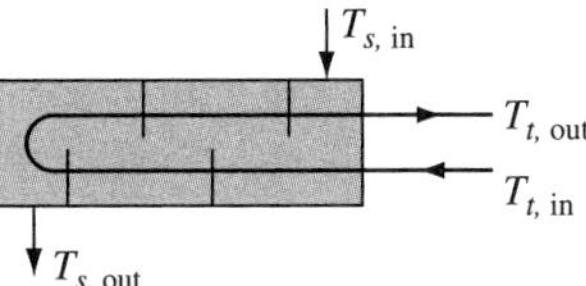

FIGURE 8.14 Correction factor to counterflow LMTD for heat exchanger with one shell pass and two (or a multiple of two) tube passes.

Source: Courtesy of the Tubular Exchanger Manufacturers Association.

The values shown on the abscissa are for the dimensionless temperature-difference ratio

$$P = \frac{T_{t.\text{out}} - T_{t.\text{in}}}{T_{s,\text{in}} - T_{t,\text{in}}} \tag{8.20}$$

where the subscripts t and s refer to the tube and shell fluid, respectively, and the subscripts "in" and "out" refer to the inlet and outlet conditions, respectively. The ratio P is an indication of the heating or cooling effectiveness and can vary from zero for a constant temperature of one of the fluids to unity for the case when the inlet temperature of the hotter fluid equals the outlet temperature of the cooler fluid. The parameter for each of the curves, Z, is equal to the ratio of the products of the mass flow rate times the heat capacity of the two fluids, $\dot{m}_t c_{pt}/\dot{m}_s c_{ps}$. This ratio is also equal to the temperature change of the shell fluid divided by the temperature change of the fluid in the tubes:

$$Z = \frac{\dot{m}_t c_{pt}}{\dot{m}_s c_{ps}} = \frac{T_{s,\text{in}} - T_{s,\text{out}}}{T_{t,\text{out}} - T_{t,\text{in}}} \tag{8.21}$$

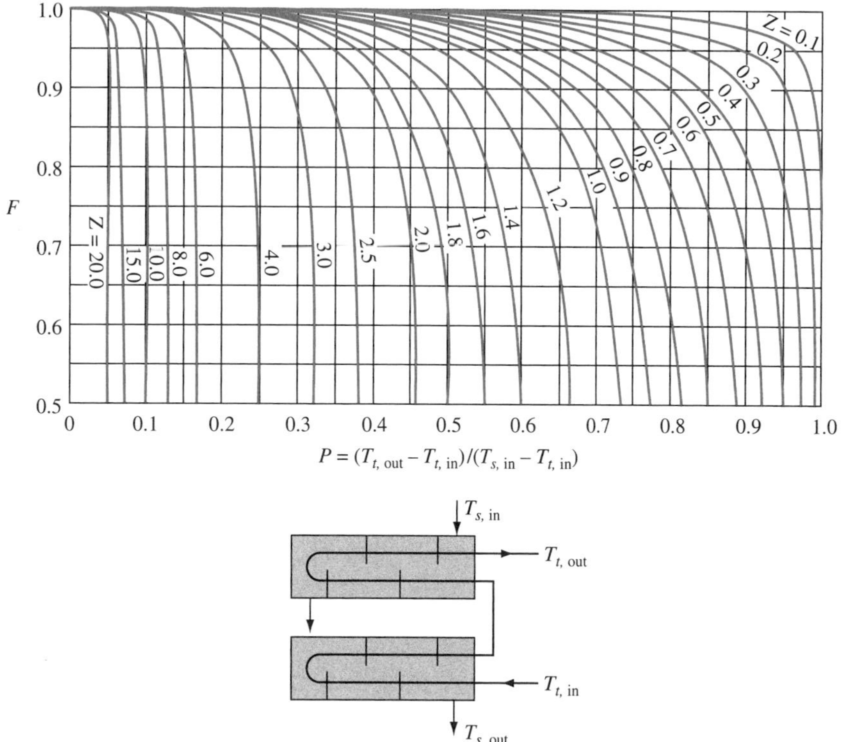

FIGURE 8.15 Correction factor to counterflow LMTD for heat exchanger with two shell passes and a multiple of two tube passes.

Source: Courtesy of the Tubular Exchanger Manufacturers Association.

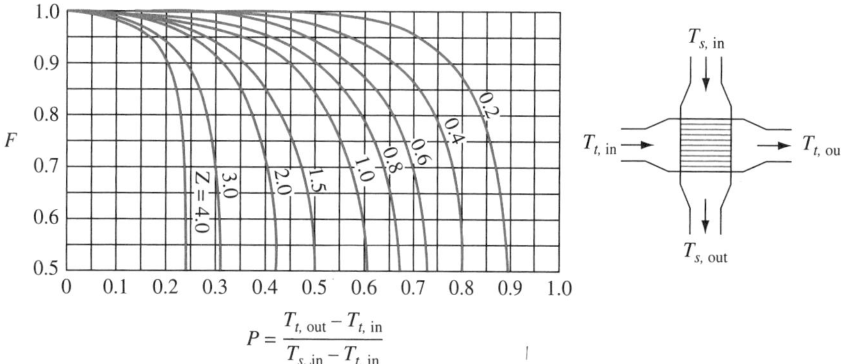

FIGURE 8.16 Correction factor to counterflow LMTD for cross-flow heat exchangers with the fluid on the shell side mixed, the other fluid unmixed, and one tube pass.

Source: Extracted from Bowman, Mueller, and Nagel [17], with permission of the publishers, the American Society of Mechanical Engineers.

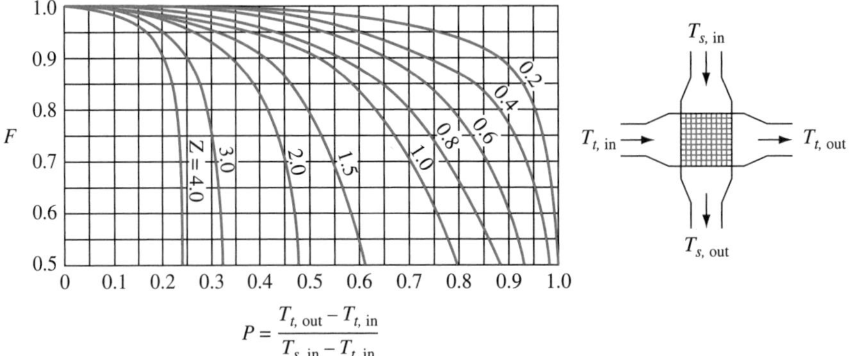

FIGURE 8.17 Correction factor to counterflow LMTD for a cross-flow heat exchanger with both fluids unmixed and one tube pass.

Source: Courtesy of R. A. Bowman, A. C. Mueller and W. M. Nagle, "Mean Temperature Difference in Design," Trans. ASME, vol. 62, pp. 283–294, 1940.

In applying the correction factors it is immaterial whether the warmer fluid flows through shell or tubes. If the temperature of either of the fluids remains constant, the direction of flow is also immaterial, since F equals 1 and the LMTD applies directly.

EXAMPLE 8.1 Determine the heat transfer surface area required for a heat exchanger constructed from a 0.0254-m-OD tube to cool 6.93 kg/s of a 95% ethyl alcohol solution (c_p = 3810 J/kg K) from 65.6°C to 39.4°C, using 6.30 kg/s of water available at 10°C. Assume that the overall coefficient of heat transfer based on the outer-tube area is 568 W/m^2 K and consider each of the following arrangements:

(a) Parallel-flow tube and shell
(b) Counterflow tube and shell
(c) Counterflow exchanger with 2 shell passes and 72 tube passes, the alcohol flowing through the shell and the water flowing through the tubes
(d) Cross-flow, with one tube pass and one shell pass, shell-side fluid mixed

SOLUTION The outlet temperature of the water for any of the four arrangements can be obtained from an overall energy balance, assuming that the heat loss to the atmosphere is negligible. Writing the energy balance as

$$\dot{m}_h c_{ph}(T_{h,\text{in}} - T_{h,\text{out}}) = \dot{m}_c c_{pc}(T_{c,\text{out}} - T_{c,\text{in}})$$

and substituting the data in this equation, we obtain

$$(6.93)(3810)(65.6 - 39.4) = (6.30)(4187)(T_{c,\text{out}} - 10)$$

from which the outlet temperature of the water is found to be 36.2°C. The rate of heat flow from the alcohol to the water is

$$q = \dot{m}_h c_{ph}(T_{h,\text{in}} - T_{h,\text{out}}) = (6.93 \text{ kg/s})(3810 \text{ J/kg K})(65.6 - 39.4)(\text{K}) \\ = 691{,}800 \text{ W}$$

(a) From Eq. (8.18) the LMTD for parallel flow is

$$\text{LMTD} = \frac{\Delta T_a - \Delta T_b}{\ln(\Delta T_a/\Delta T_b)} = \frac{55.6 - 3.2}{\ln(55.6/3.2)} = 18.4\,°\text{C}$$

From Eq. (8.16) the heat transfer surface area is

$$A = \frac{q}{(U)(\text{LMTD})} = \frac{(691{,}800 \text{ W})}{(568 \text{ W/m}^2 \text{ K})(18.4 \text{ K})} = 66.2 \text{ m}^2$$

The 830-m length of the exchanger for a 0.0254-m-OD tube would be too great to be practical.

(b) For the counterflow arrangement, the appropriate mean temperature difference is 65.6 − 36.2 = 29.4°C, because $\dot{m}_c c_{pc} = \dot{m}_h c_{ph}$. The required area is

$$A = \frac{q}{(U)(\text{LMTD})} = \frac{691{,}800}{(568)(29.4)} = 41.4 \text{ m}^2$$

which is about 40% less than the area necessary for parallel flow.

(c) For the two-shell-pass counterflow arrangement, we determine the appropriate mean temperature difference by applying the correction factor found from Fig. 8.15 to the mean temperature for counterflow:

$$P = \frac{T_{c,\text{out}} - T_{c,\text{in}}}{T_{h,\text{in}} - T_{c,\text{in}}} = \frac{36.2 - 10}{65.6 - 10} = 0.47$$

and the heat capacity rate ratio is

$$Z = \frac{\dot{m}_t c_{pt}}{\dot{m}_s c_{ps}} = 1$$

From the chart of Fig. 8.15, $F = 0.97$ and the heat transfer area is

$$A = \frac{41.4}{0.97} = 42.7 \text{ m}^2$$

The length of the exchanger for 72, 0.0254-m-OD tubes in parallel would be

$$L = \frac{A/72}{\pi D} = \frac{42.7/72}{\pi(0.0254)} = 7.4 \text{ m}$$

This length is not unreasonable, but if it is desirable to shorten the exchanger, more tubes could be used.

(d) For the cross-flow arrangement (Fig. 8.4), the correction factor is found from the chart of Fig. 8.16 to be 0.88. The required surface area is thus 47.0 m^2, about 10% larger than that for the exchanger in part (c).

8.5 Heat Exchanger Effectiveness

In the thermal analysis of the various types of heat exchangers presented in the preceding section, we used [Eq. (8.17)] expressed as

$$q = UA\,\Delta T_{\text{mean}}$$

This form is convenient when all the terminal temperatures necessary for the evaluation of the appropriate mean temperature are known, and Eq. (8.17) is widely employed in the design of heat exchangers to given specifications. There are, however, numerous occasions when the performance of a heat exchanger (i.e., U) is known or can at least be estimated but the temperatures of the fluids leaving the exchanger are not known. This type of problem is encountered in the selection of a heat exchanger or when the unit has been tested at one flow rate, but service conditions require different flow rates for one or both fluids. In heat exchanger design texts and handbooks, this type of problem is also referred to as a *rating problem*, where the outlet temperatures or the total heat load needs to be determined, given the size (A) and the convective performance (U) of the unit. The outlet temperatures and the rate of heat flow can be found only by a rather tedious trial-and-error procedure if the charts presented in the preceding section are used. In such cases it is desirable to circumvent entirely any reference to the logarithmic or any other mean temperature difference. A method that accomplishes this has been proposed by Nusselt [18] and Ten Broeck [19].

To obtain an equation for the rate of heat transfer that does not involve any of the outlet temperatures, we introduce the *heat exchanger effectiveness* $\mathscr{E}$. The heat exchanger effectiveness is defined as the ratio of the actual rate of heat transfer in a given heat exchanger to the maximum possible rate of heat exchange. The latter would be obtained in a counterflow heat exchanger of infinite heat transfer area. In this type of unit, if there are no external heat losses, the outlet temperature of the colder fluid equals the inlet temperature of the warmer fluid when $\dot{m}_c c_{pc} < \dot{m}_h c_{ph}$; when $\dot{m}_h c_{ph} < \dot{m}_c c_{pc}$, the outlet temperature of the warmer fluid equals the inlet temperature of the colder one. In other words, the effectiveness compares the actual heat transfer rate to the maximum rate whose only limit is the second law of thermodynamics. Depending on which of the heat capacity rates is smaller, the effectiveness is

$$\mathscr{E} = \frac{C_h(T_{h,\text{in}} - T_{h,\text{out}})}{C_{\min}(T_{h,\text{in}} - T_{c,\text{in}})} \tag{8.22a}$$

or

$$\mathscr{E} = \frac{C_c(T_{c,\text{out}} - T_{c,\text{in}})}{C_{\min}(T_{h,\text{in}} - T_{c,\text{in}})} \tag{8.22b}$$

where $C_{\min}$ is the smaller of the $\dot{m}_h c_{ph}$ and $\dot{m}_c c_{pc}$ magnitudes. It may be noted that the denominator in Eq. (8.22) is the thermodynamically maximum heat transfer possible between the hot and cold fluids flowing through the heat exchanger, given their respective inlet temperature and mass flow rate, or the maximum available energy. The numerator is the actual heat transfer accomplished in the unit, and hence the effectiveness $\mathscr{E}$ represents a thermodynamic performance of the heat exchanger.

Once the effectiveness of a heat exchanger is known, the rate of heat transfer can be determined directly from the equation

$$q = \mathscr{E}\, C_{\min}(T_{h,\text{in}} - T_{c,\text{in}}) \tag{8.23}$$

since

$$\mathscr{E}\, C_{\min}(T_{h,\text{in}} - T_{c,\text{in}}) = C_h(T_{h,\text{in}} - T_{h,\text{out}}) = C_c(T_{c,\text{out}} - T_{c,\text{in}})$$

Equation (8.23) is the basic relation in this analysis because it expresses the rate of heat transfer in terms of the effectiveness, the smaller heat capacity rate, and the difference between the inlet temperatures. It replaces Eq. (8.17) in the LMTD analysis but does not involve the outlet temperatures. Equation (8.23) is, of course, also suitable for design purposes and can be used instead of Eq. (8.17).

We shall illustrate the method of deriving an expression for the effectiveness of a heat exchanger by applying it to a parallel-flow arrangement. The effectiveness can be introduced into Eq. (8.13) by replacing $(T_{c,\text{in}} - T_{c,\text{out}})/(T_{h,\text{in}} - T_{c,\text{in}})$ by the effectiveness relation from Eq. (8.22b). We obtain

$$\ln\left[1 - \mathscr{E}\left(\frac{C_{\min}}{C_h} + \frac{C_{\min}}{C_c}\right)\right] = -\left(\frac{1}{C_c} + \frac{1}{C_h}\right)UA$$

or

$$1 - \mathscr{E}\left(\frac{C_{\min}}{C_h} + \frac{C_{\min}}{C_c}\right) = e^{-(1/C_c + 1/C_h)UA}$$

Solving for $\mathscr{E}$ yields

$$\mathscr{E} = \frac{1 - e^{-[1+(C_h/C_c)]UA/C_h}}{(C_{\min}/C_h) + (C_{\min}/C_c)} \tag{8.24}$$

When C_h is less than C_c, the effectiveness becomes

$$\mathscr{E} = \frac{1 - e^{-[1+(C_h/C_c)]UA/C_h}}{1 + (C_h/C_c)} \tag{8.25a}$$

and when $C_c < C_h$, we obtain

$$\mathscr{E} = \frac{1 - e^{-[1+(C_c/C_h)]UA/C_c}}{1 + (C_c/C_h)} \tag{8.25b}$$

The effectiveness for both cases can therefore be written in the form

$$\mathscr{E} = \frac{1 - e^{-[1+(C_{\min}/C_{\max})]UA/C_{\min}}}{1 + (C_{\min}/C_{\max})} \tag{8.26}$$

The foregoing derivation illustrates how the effectiveness for a given flow arrangement can be expressed in terms of two dimensionless parameters, the heat

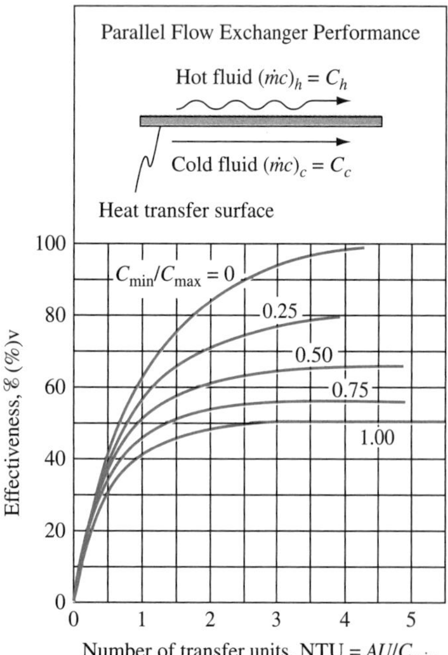

FIGURE 8.18 Heat exchanger effectiveness for parallel flow.

Source: With permission from Kays and London [10].

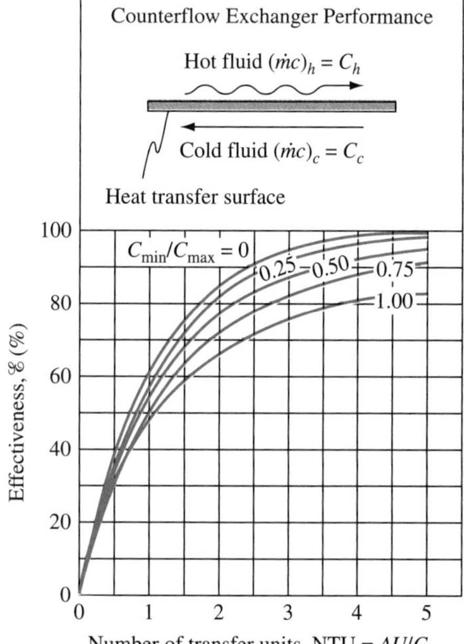

FIGURE 8.19 Heat exchanger effectiveness for counterflow.

Source: With permission from Kays and London [10].

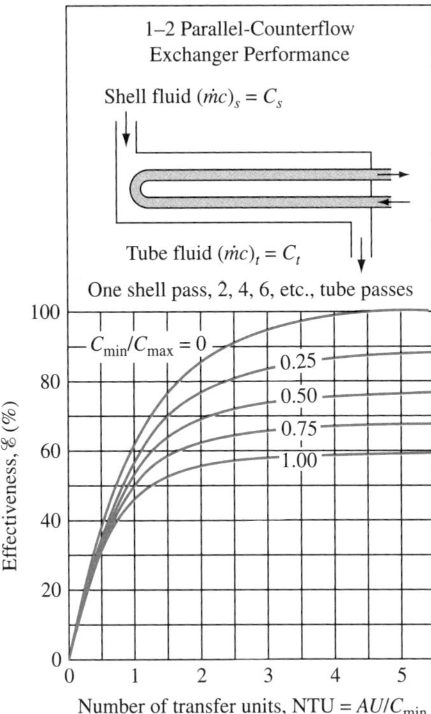

FIGURE 8.20 Heat exchanger effectiveness for shell-and-tube heat exchanger with one well-baffled shell pass and two (or a multiple of two) tube passes.

Source: With permission from Kays and London [10].

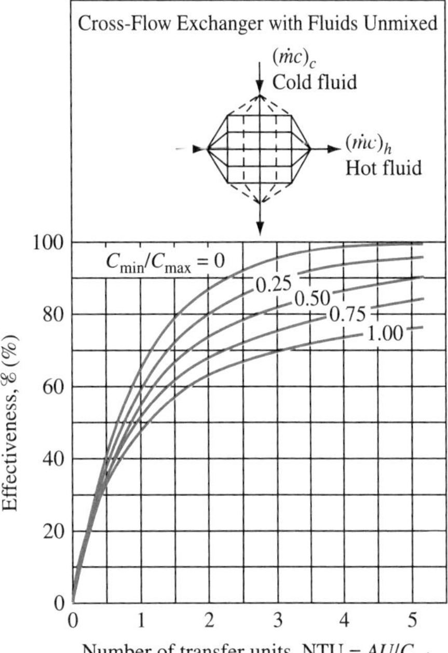

FIGURE 8.21 Heat exchanger effectiveness for cross-flow with both fluids unmixed.

Source: With permission from Kays and London [10].

capacity rate ratio C_{min}/C_{max} and the ratio of the overall conductance to the smaller heat capacity rate, UA/C_{min}. The latter of the two parameters is called the *number of heat transfer units* or NTU. The number of heat transfer units is a measure of the heat transfer size of the exchanger. The larger the value of NTU, the closer the heat exchanger approaches its thermodynamic limit. By analyses that, in principle, are similar to the one presented here for parallel flow, effectiveness can be evaluated for most flow arrangements of practical interest. The results have been put together by Kays and London [10] into convenient graphs from which the effectiveness can be determined for given values of NTU and C_{min}/C_{max}. The effectiveness curves for some common flow arrangements are shown in Figs. 8.18–8.22. The abscissas of these figures are the NTUs of the heat exchangers. The constant parameter for each curve is the heat capacity rate ratio C_{min}/C_{max}, and the effectiveness is read on the ordinate. Note that for an evaporator or condenser, $C_{min}/C_{max} = 0$, because if one fluid remains at constant temperature throughout the exchanger, its effective specific heat and thus its heat capacity rate are by definition equal to infinity.

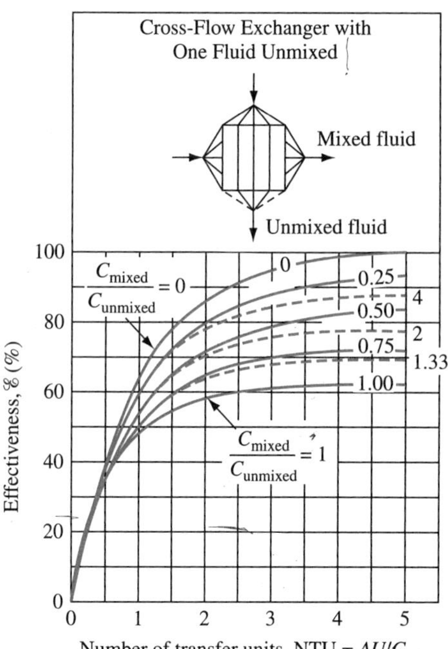

FIGURE 8.22 Heat exchanger effectiveness for cross-flow with one fluid mixed and the other unmixed. When $C_{mixed}/C_{unmixed} > 1$, NTU is based on $C_{unmixed}$.
Source: With permission from W. M. Kays and A. L. London [10].

EXAMPLE 8.2 From a performance test on a well-baffled single-shell, two-tube-pass heat exchanger, the following data are available: oil (c_p = 2100 J/kg K) in turbulent flow inside the tubes entered at 340 K at the rate of 1.00 kg/s and left at 310 K; water flowing on the shell side entered at 290 K and left at 300 K. A change in service conditions requires the cooling of a similar oil from an initial temperature of 370 K but at three-fourths of the flow rate used in the performance test. Estimate the outlet temperature of the oil for the same water flow rate and inlet temperature as before.

SOLUTION The test data can be used to determine the heat capacity rate of the water and the overall conductance of the exchanger. The heat capacity rate of the water is, from Eq. (8.14),

$$C_c = C_h \frac{T_{h,\text{in}} - T_{h,\text{out}}}{T_{c,\text{out}} - T_{c,\text{in}}} = (1.00 \text{ Kg/s})(2100 \text{ J/kg K}) \frac{340 - 310}{300 - 290}$$
$$= 6300 \text{ W/K}$$

and the temperature ratio P is, from Eq. (8.20),

$$P = \frac{T_{t,\text{out}} - T_{t,\text{in}}}{T_{s,\text{in}} - T_{t,\text{in}}} = \frac{340 - 310}{340 - 290} = 0.6$$

$$Z = \frac{300 - 290}{340 - 310} = 0.33$$

From Fig. 8.14, $F = 0.94$ and the mean temperature difference is

$$\Delta T_{\text{mean}} = (F)(\text{LMTD}) = (0.94)\frac{(340 - 300) - (310 - 290)}{\ln[(340 - 300)/(310 - 290)]} = 27.1 \text{ K}$$

From Eq. (8.17) the overall conductance is

$$UA = \frac{q}{\Delta T_{\text{mean}}} = \frac{(1.00 \text{ kg/s})(2100 \text{ J/kg K})(340 - 310)(\text{K})}{(27.1 \text{ K})} = 2325 \text{ W/K}$$

Since the thermal resistance on the oil side is controlling, a decrease in velocity to 75% of the original value will increase the thermal resistance by roughly the velocity ratio raised to the 0.8 power. This can be verified by reference to Eq. (6.62). Under the new conditions, the conductance, the NTU, and the heat capacity rate ratio will therefore be approximately

$$UA \simeq (2325)(0.75)^{0.8} = 1850 \text{ W/K}$$

$$\text{NTU} = \frac{UA}{C_{\text{oil}}} = \frac{(1850 \text{ W/K})}{(0.75)(1.00 \text{ kg/s})(2100 \text{ J/kg K})} = 1.17$$

and

$$\frac{C_{\text{oil}}}{C_{\text{water}}} = \frac{C_{\min}}{C_{\max}} = \frac{(0.75)(1.00 \text{ kg/s})(2100 \text{ J/kg K})}{(6300 \text{ W/K})} = 0.25$$

From Fig. 8.20 the effectiveness is equal to 0.61. Hence from the definition of $\mathscr{E}$ in Eq. (8.22a), the oil outlet temperature is

$$T_{\text{oil out}} = T_{\text{oil in}} - \mathscr{E}\,\Delta T_{\max} = 370 - [0.61(370 - 290)] = 321.2 \text{ K}$$

The next example illustrates a more complex problem.

EXAMPLE 8.3 A flat-plate-type heater (Fig. 8.23) is to be used to heat air with the hot exhaust gases from a turbine. The required airflow rate is 0.75 kg/s, entering at 290 K; the hot gases are available at a temperature of 1150 K and a mass flow rate of 0.60 kg/s.

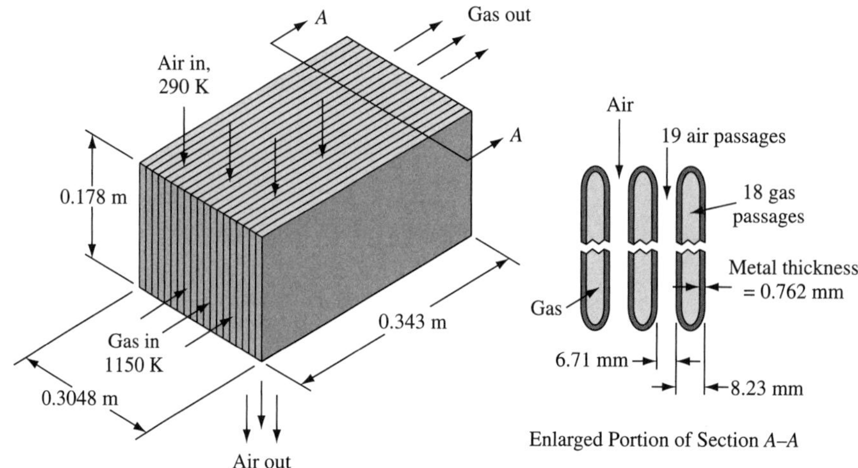

FIGURE 8.23 Flat-plate-type heater.

Determine the temperature of the air leaving the heat exchanger for the parameters listed below.

P_a = wetted perimeter on air side, 0.703 m
P_g = wetted perimeter on gas side, 0.416 m
A_g = cross-sectional area of gas passage (per passage), 1.6×10^{-3} m^2
A_a = cross-sectional area of air passage (per passage), 2.275×10^{-3} m^2
A = heat transfer surface area, 2.52 m^2

SOLUTION Inspection of Fig. 8.23 shows that the unit is of the cross-flow type, with both fluids unmixed. As a first approximation, the end effects will be neglected. The flow systems for the air and gas streams are similar to flow in straight ducts having the following dimensions:

L_a = length of air duct, 0.178 m

D_{Ha} = hydraulic diameter of air duct, $\dfrac{4A_a}{P_a} = 0.0129$ m

L_g = length of gas duct, 0.343 m

D_{Hg} = hydraulic diameter of gas duct, $\dfrac{4A_g}{P_g} = 0.0154$ m

A = heat transfer surface area, 2.52 m^2

The heat transfer coefficients can be evaluated from Eq. (6.63) for flow in ducts ($L_a/D_{Ha} = 13.8$, $L_g/D_{Hg} = 22.3$). A difficulty arises, however, because the temperatures of both fluids vary along the duct. It is therefore necessary to estimate an average bulk temperature and refine the calculations after the outlet and wall temperatures have been found. Selecting the average air-side bulk temperature to be

573 K and the average gas-side bulk temperature to be 973 K, the properties at those temperatures are, from Appendix 2, Table 28 (assuming that the properties of the gas can be approximated by those of air):

$$\begin{aligned} \mu_{\text{air}} &= 2.93 \times 10^{-5}\ \text{N s/m}^2 & \mu_{\text{gas}} &= 4.085 \times 10^{-5}\ \text{N s/m}^2 \\ \text{Pr}_{\text{air}} &= 0.71 & \text{Pr}_{\text{gas}} &= 0.73 \\ k_{\text{air}} &= 0.0429\ \text{W/m K} & k_{\text{gas}} &= 0.0623\ \text{W/m K} \\ c_{p_{\text{air}}} &= 1047\ \text{J/kg K} & c_{p_{\text{gas}}} &= 1101\ \text{J/kg K} \end{aligned}$$

The mass flow rates per unit area are

$$\left(\frac{\dot{m}}{A}\right)_{\text{air}} = \frac{(0.75\ \text{kg/s})}{(19)(2.275 \times 10^{-3}\ \text{m}^2)} = 17.35\ \text{kg/m}^2\ \text{s}$$

$$\left(\frac{\dot{m}}{A}\right)_{\text{gas}} = \frac{(0.60\ \text{kg/s})}{(18)(1.600 \times 10^{-3}\ \text{m}^2)} = 20.83\ \text{kg/m}^2\ \text{s}$$

The Reynolds numbers are

$$\text{Re}_{\text{air}} = \frac{(\dot{m}/A)_{\text{air}} D_{Ha}}{\mu_a} = \frac{(17.35\ \text{kg/m}^2\ \text{s})(0.0129\ \text{m})}{(2.93 \times 10^{-5}\ \text{kg/m s})} = 7640$$

$$\text{Re}_{\text{gas}} = \frac{(\dot{m}/A)_{\text{gas}} D_{Hg}}{\mu_g} = \frac{(20.83\ \text{kg/m}^2\ \text{s})(0.0154\ \text{m})}{(4.085 \times 10^{-5}\ \text{kg/m s})} = 7850$$

Using Eq. (6.63), the average heat transfer coefficients are

$$\begin{aligned} \bar{h}_{\text{air}} &= 0.023\, \frac{k_a}{D_{Ha}}\, \text{Re}_{\text{air}}^{0.8}\ \text{Pr}^{\,0.4} \\ &= 0.023\, \frac{0.0429}{0.0129}\, (7640)^{0.8} (0.71)^{0.4} \\ &= 85.2\ \text{W/m}^2\,\text{K} \end{aligned}$$

Since $L_a/D_{Ha} = 13.8$, we must correct this heat transfer coefficient for entrance effects, per Eq. (6.68). The correction factor is 1.377, so the corrected heat transfer coefficient is $(1.377)(85.2) = 117\ \text{W/m}^2\,\text{K} = \bar{h}_{\text{air}}$.

$$\begin{aligned} \bar{h}_{\text{gas}} &= (0.023)\, \frac{0.0623}{0.0154}\, (7850)^{0.8} (0.73)^{0.4} \\ &= 107.1\ \text{W/m}^2\,\text{K} \end{aligned}$$

Since $L_g/D_{Hg} = 22.3$, we must correct this heat transfer coefficient for entrance effects, per Eq. (6.69). The correction factor is $1 + 6(D_{Hg}/L_g) = 1.27$, so the corrected heat transfer coefficient is $(1.27)(107.1) = 136\ \text{W/m}^2\,\text{K} = \bar{h}_{\text{gas}}$.

The thermal resistance of the metal wall is negligible, therefore the overall conductance is

$$\begin{aligned} UA &= \frac{1}{\dfrac{1}{\bar{h}_a A} + \dfrac{1}{\bar{h}_g A}} = \frac{1}{\dfrac{1}{(117\ \text{W/m}^2\,\text{K})(2.52\ \text{m}^2)} + \dfrac{1}{(136\ \text{W/m}^2\,\text{K})(2.52\ \text{m}^2)}} \\ &= 158\ \text{W/K} \end{aligned}$$

The number of transfer units, based on the gas, which has the smaller heat capacity rate, is

$$\mathrm{NTU} = \frac{UA}{C_{\min}} = \frac{(158\ \mathrm{W/K})}{(0.60\ \mathrm{kg/s})(1101\ \mathrm{J/kg\ K})} = 0.239$$

The heat capacity-rate ratio is

$$\frac{C_g}{C_a} = \frac{(0.60)(1101)}{(0.75)(1047)} = 0.841$$

and from Fig. 8.21, the effectiveness is approximately 0.13. Finally, the average outlet temperatures of the gas and air are

$$\begin{aligned} T_{\text{gas out}} &= T_{\text{gas in}} - \mathscr{E}\,\Delta T_{\max} \\ &= 1150 - 0.13(1150 - 290) = 1038\ \mathrm{K} \end{aligned}$$

$$\begin{aligned} T_{\text{air out}} &= T_{\text{air in}} + \frac{C_g}{C_a}\mathscr{E}\,\Delta T_{\max} = 290 + (0.841)(0.13)(1150 - 290) \\ &= 384\ \mathrm{K} \end{aligned}$$

A check on the average air-side and gas-side bulk temperatures gives values of 337 K and 1094 K. Performing a second iteration with property values based on these temperatures yields values sufficiently close to the assumed values (573 K, 973 K) to make a third approximation unnecessary. To appreciate the usefulness of the approach based on the concept of heat exchanger effectiveness, it is suggested that this same problem be worked out by trial and error, using Eq. (8.17) and the chart in Fig. 8.17.

The effectiveness of the heat exchanger in Example 8.3 is very low (13%) because the heat transfer area is too small to utilize the available energy efficiently. The relative gain in heat transfer performance that can be achieved by increasing the heat transfer area is well represented on the effectiveness curves. A fivefold increase in area would raise the effectiveness to 60%. If, however, a particular design falls near or above the knee of these curves, increasing the surface area will not improve the performance appreciably but may cause an undue increase in the frictional pressure drop or heat exchanger cost.

EXAMPLE 8.4 A heat exchanger (condenser) using steam from the exhaust of a turbine at a pressure of 4.0-in. Hg abs. is to be used to heat 25,000 lb/h of seawater (c = 0.95 Btu/lb °F) from 60°F to 110°F. The exchanger is to be sized for one shell pass and four tube passes with 60 parallel tube circuits of 0.995-in.-ID and 1.125-in.-OD brass tubing (k = 60 Btu/h ft °F). For the clean exchanger the average heat transfer coefficients at the steam and water sides are estimated to be 600 and 300 Btu/h ft^2 °F, respectively. Calculate the tube length required for long-term service.

SOLUTION At 4.0-in. Hg abs. the temperature of condensing steam will be 125.4°F, so the required effectiveness of the exchanger is

$$\mathscr{E} = \frac{T_{c,\text{out}} - T_{c,\text{in}}}{T_{h,\text{in}} - T_{c,\text{in}}} = \frac{110 - 60}{125.4 - 60} = 0.765$$

For a condenser, $C_{\min}/C_{\max} = 0$, and from Fig. 8.20, NTU = 1.4. The fouling factors from Table 8.2 are 0.0005 h ft^2 °F/Btu for both sides of the tubes. The overall design heat-transfer coefficient per unit outside area of tube is, from Eq. (8.6),

$$U_d = \frac{1}{\dfrac{1}{600} + 0.0005 + \dfrac{1.125}{2 \times 12 \times 60}\ln\dfrac{1.125}{0.995} + \dfrac{0.0005 \times 1.125}{0.995} + \dfrac{1.125}{300 \times 0.995}}$$

$$= 152 \text{ Btu/h ft}^2\text{ °F}$$

The total area A_o is $20\pi D_o L$, and since $U_d A_o / C_{\min} = 1.4$, the length of the tube is

$$L = \frac{1.4 \times 25{,}000 \times 0.95 \times 12}{60 \times \pi \times 1.125 \times 152} = 12.3\text{ ft}$$

In practice, the flow through a cross-flow heat exchanger may not be strictly mixed or unmixed—the flow may be partially mixed. DiGiovanni and Webb [20] showed that the effectiveness of a heat exchanger in which one stream is unmixed and the other stream is partially mixed is

$$\mathscr{E}_{pm:u} = \mathscr{E}_{u:u} - y(\mathscr{E}_{u:u} - \mathscr{E}_{m:u}) \tag{8.27}$$

The subscripts on the effectiveness in Eq. (8.27) are *pm* for partially mixed, *m* for mixed, and *u* for unmixed, i.e., $\mathscr{E}_{m:u}$ is the effectiveness for a heat exchanger with one stream mixed and the other unmixed.

If one stream is mixed and the other is partially mixed

$$\mathscr{E}_{pm:m} = \mathscr{E}_{m:m} + y(\mathscr{E}_{u:m} - \mathscr{E}_{m:m}) \tag{8.28}$$

If both streams are partially mixed

$$\mathscr{E}_{pm:pm} = \mathscr{E}_{u:pm} - y(\mathscr{E}_{u:pm} - \mathscr{E}_{m:pm}) \tag{8.29}$$

In Eqs. (8.27) through (8.29) the parameter y is the fraction of mixing for the partially mixed stream. For an unmixed stream $y = 0$, and for a mixed stream $y = 1$. At the present time there is no general method for determining the fraction of mixing for a given heat exchanger. Since y is likely to be a strong function of heat exchanger geometry as well as the flow Reynolds number, experimental data are probably required for various heat exchanger geometries of interest to apply the degree-of-mixing correction. The uncertainty associated with the degree of mixing is greatest for high NTU designs.

8.6* Heat Transfer Enhancement

Heat transfer enhancement is the practice of modifying a heat transfer surface or the flow cross section to either increase the heat transfer coefficient between the surface and a fluid or the surface area so as to effectively sustain higher heat loads with a smaller temperature difference [21–22]. In previous chapters we have treated some practical examples of heat transfer enhancement, e.g., fins, surface roughness, twisted-tape insert, and coiled tube, which are generally referred to as passive techniques [21]. Heat transfer enhancement may also be achieved by surface or fluid vibration, electrostatic fields, or mechanical stirrers. These latter methods are often referred to as active techniques because they require the application of external power. Although active techniques have received attention in the research literature, their practical applications have been very limited. In this section, therefore, we shall focus on some specific examples of passive techniques, i.e., those based on modification of the heat transfer surface; a more complete and extended discussion of the full spectrum of enhancement techniques can be found in references Manglik [21] and Bergles [22].

Increases in heat transfer due to surface treatment can be brought about by increased turbulence, increased surface area, improved mixing, or flow swirl. These effects generally result in an increase in pressure drop along with the increase in heat transfer. However, with appropriate performance evaluation and concomitant optimization [21–22], significant heat transfer improvement relative to a smooth (untreated) heat transfer surface of the same nominal (base) heat transfer area can be achieved for a variety of applications. The increasing attractiveness of different heat transfer enhancement techniques are gaining industrial importance because heat exchangers offer the opportunity to: (1) reduce the heat transfer surface area required for a given application and thus reduce the heat exchanger size and cost, (2) increase the heat duty of the exchanger, and (3) permit closer approach temperatures. All of these can be visualized from the expression for heat duty for a heat exchanger, Eq. (8.17):

$$Q = UA\,\text{LMTD} \tag{8.17}$$

Any enhancement technique that increases the heat transfer coefficient also increases the overall conductance U. Therefore, in conventional and compact heat exchangers, one can reduce the heat transfer area A, increase the heat duty Q, or decrease the temperature difference LMTD, respectively, for fixed Q and LMTD, fixed A and LMTD, or fixed Q and A. Enhancement can also be used to prevent the overheating of heat transfer surfaces in systems with a fixed heat generation rate, such as in the cooling of electrical and electronic devices.

In any practical application, a complete analysis is required to determine the economic benefit of enhancement. Such an analysis must include a possible increased first-cost because of the enhancement, increased heat exchanger heat transfer performance, the effect on operating costs, and maintenance costs. Another concern in some industrial applications is the possibility of increased fouling of the heat exchange surface caused by the enhancement. Accelerated fouling can quickly eliminate any increase in the heat transfer coefficient achieved by enhancement of a clean surface. Nevertheless, in the present-day concerns of sustainable energy utilization and the need for conservation, the benefits of using enhancement techniques in most heat exchange systems cannot be overstated.

8.6.1 Applications

There is a very large, rapidly growing body of literature on the subject of heat transfer enhancement. Manglik and Bergles [23] have documented the latest cataloging of technical papers and reports on the subject and have discussed the status of recent advancements as well as the prospects of future developments in enhanced heat transfer technology. The taxonomy that has been developed [21–22] for the classification of the various enhancement techniques and their applications essentially considers the fluid flow condition (single-phase natural convection, single-phase forced convection, pool boiling, flow boiling, condensation, etc.) and the type of enhancement technique (rough surface, extended surface, displaced enhancement devices, swirl flow, fluid additives, vibration, etc.).

Table 8.3 shows how each enhancement technique applies to the different types of flow according to Bergles et al. [24]. Extended surfaces or fins are probably the most common heat transfer enhancement technique, and examples of different types of fins are shown in Fig. 8.24. The fin was discussed in Chapter 2 as an extended surface with primary application in gas-side heat transfer. The effectiveness of the fin in this application is based on the poor thermal conductivity of the gas relative to that of the fin material. Thus, while the temperature drop along the fin reduces its effectiveness somewhat, overall an increase in surface area and thus in heat transfer performance is realized. Several manufacturers have recently made available tubing with integral internal fins, and the prediction of the associated convective heat transfer coefficient has been highlighted in Chapter 6. Extended surfaces may also take the form of interrupted fins where the objective is to force the redevelopment of boundary layers. As discussed in Section 8.2, compact heat exchangers [10, 12] use extended surfaces to give a required heat transfer surface area in as small a volume as possible, and representative examples of such fins are shown in Fig. 8.24. This type of heat exchanger is important in applications such as

TABLE 8.3 Application of enhancement techniques to different types of flows[a]

	Single-Phase Natural Convection	Single-Phase Forced Convection	Pool Boiling	Flow Boiling	Condensation
Extended surfaces	c	c	c	o	c
Rough surfaces	o	c	o	c	c
Displaced enhancement devices	n	o	n	o	n
Swirl flow devices	n	c	n	c	o
Treated surfaces	n	c	c	o	c

[a]c = commonly practiced, o = occasionally practiced, n = not practiced.

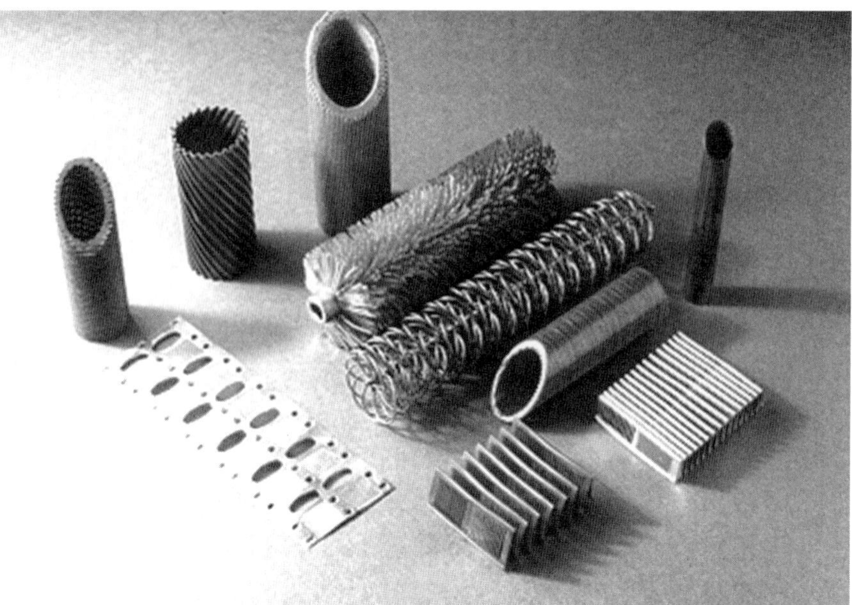

FIGURE 8.24 Examples of different types of finned tubes and plate fins used in tubular and compact tube-fin and plate-fin heat exchangers.
Source: Courtesy of Dr. Ralph Webb.

automobile radiators and gas turbine regenerators, where the overall size of the heat exchanger is of major concern.

Rough surfaces refer to small roughness elements approximately the height of the boundary layer thickness. In recent years, a variety of structured roughness elements of different geometries and surface distributions have been considered in the literature [21–22]. These roughness elements do not provide any significant increase in surface area; if there is an increase in area, then such surface modifications are classified as extended surfaces. Their effectiveness is based on promoting early transition to turbulent flow or promoting mixing between the bulk flow and the viscous sublayer in fully developed turbulent flow. The roughness elements may be randomly shaped, such as on a sand-grained surface, or regular, such as machined grooves or pyramids. Rough surfaces are primarily used to promote heat transfer in single-phase forced convection.

Displaced enhancement devices are inserted into the flow channel to improve mixing between the bulk flow and the heat transfer surface. A common example is the static mixer that is in the form of a series of corrugated sheets meant to promote bulk flow mixing. These devices are used most often in single-phase forced convection particularly in thermal processing of viscous media in the chemical industry so as to promote both fluid mixing and enhanced heat or mass transfer.

The most prominent and frequently used example of a swirl flow device is a twisted-tape insert, and its typical usage inside tubes of a shell-and-tube heat exchanger and the concomitant prediction of single-phase convective heat transfer coefficients have been considered in Chapter 6. Another example is an oval tube that

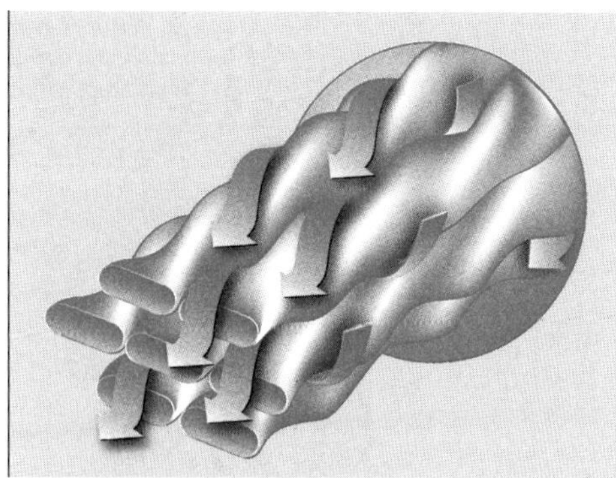

FIGURE 8.25 A schematic representation of a tube bundle of helically-twisted oval tubes and swirl flow of axial external flow; swirl flow is also generated inside the tubes.

is helically twisted about its axis, as shown in Fig. 8.25. Enhancement primarily arises due to secondary or helical swirl flows generated by the twisted flow geometry, and increased flow path length in the tube. Swirl flow devices are used for single-phase forced flow and in flow boiling [25].

Treated surfaces are used primarily in pool boiling and condensing applications. They consist of very small surface structures such as surface inclusions which promote nucleate boiling by providing bubble nucleation sites. Condensation can be enhanced by promoting the formation of droplets, rather than a film, on the condensing surface. This can be accomplished by coating the surface with a material that leaves the surface nonwetting. Boiling and condensation will be discussed in Chapter 10.

Figure 8.26 on the next page compares the performance of four enhancement techniques for single-phase forced convection in a tube with that for a smooth tube [26]. The basis of comparison is the heat transfer (Nusselt number) and pressure drop (friction factor) plotted as a function of the Reynolds number. One can see that at a given Reynolds number, all four enhancement techniques provide an increased Nusselt number relative to the smooth tube but at the expense of an even greater increase in the friction factor.

8.6.2 Analysis of Enhancement Techniques

We have previously noted the need for a comprehensive analysis of any candidate enhancement technique to determine its potential benefits. Since heat transfer enhancement can be used to accomplish several goals, no general procedure that would allow one to compare different enhancement techniques exists. A comparison such as that given in Fig. 8.26, which is limited to the thermal and hydraulic performance of the heat exchange surface, is often a useful starting point. Other factors that must be included in the analysis are the hydraulic diameter, the length of the

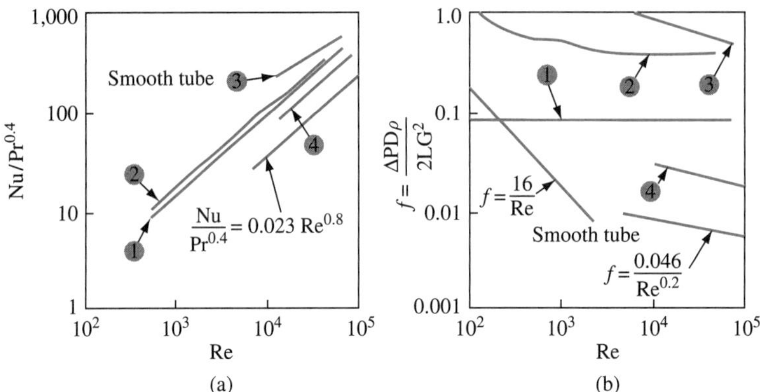

FIGURE 8.26 Typical data for turbulence promoters inserted inside tubes. (a) Heat transfer data, (b) friction data [26].

flow passages, and the flow arrangement (cross-flow, counterflow, etc.). In addition to these geometric variables, the flow rate per passage or Reynolds number and the LMTD can be varied or can be constrained for a given application. The factors that can be varied must be adjusted in the analysis to produce the desired goal, e.g., increased heat duty, minimum surface area, or reduced pressure drop. Table 8.4 lists the variables that should be considered in a complete analysis.

TABLE 8.4 Variables in the analysis of heat transfer enhancement

Symbol	Description	Comments
1. —	Type of enhancement technique	
2. $Nu(Re_{D_H})$	Thermal performance of the enhancement technique	Determined by choice of technique
3. $f(Re_{D_H})$	Hydraulic performance of the enhancement technique	Determined by choice of technique
4. Re_{D_H}	Flow Reynolds number	Probably an independent variable
5. D_H	Flow passage hydraulic diameter	May be determined by choice of technique
6. L	Flow passage length	Generally an independent variable with limits
7. —	Flow arrangement	May be determined by choice of technique
8. LMTD	Terminal flow temperatures	May be determined by the application
9. Q	Heat duty	Probably a dependent variable
10. A_s	Heat transfer surface area	Probably a dependent variable
11. Δp	Pressure drop	Probably a dependent variable

Fortunately, many applications constrain one or more of these variables, thereby simplifying the analysis. As an example, consider an existing shell-and-tube heat exchanger being used to condense a hydrocarbon vapor on the shell side with chilled water pumped through the tube side. It may be possible to increase the flow of vapor by increasing the water-side heat transfer since the vapor-side thermal resistance is probably negligible. Suppose the pressure drop on the water side is fixed due to pump constraints, and assume that it is necessary to keep the heat exchanger size and configuration the same to simplify installation costs. The water-side heat transfer could be increased by placing any of several devices such as swirl tapes or twisted-tape inserts inside the tubes, or wire-coil inserts to create structured [21–22] roughness on the tube inner surface. Assuming that thermal and hydraulic performance data are available for each enhancement technique to be considered, then items 1, 2, and 3 in Table 8.4, as well as 5, 6, 7, and 10, are known. We will adjust Re_{D_H}, which will affect the water outlet temperature or LMTD, Q, and Δp. Since the LMTD is not important (within reason), we can determine which surface provides the largest Q (and hence vapor flow) at a fixed Δp.

Several performance evaluation methods have been proposed in the literature [21–22], which are based on a variety of figures of merit that are applicable to different heat exchanger applications. Among these, Soland et al. [27] have outlined a useful performance ranking methodology that incorporates the thermal/hydraulic behavior of the heat transfer surface with the flow parameters and the geometric parameters for the heat exchanger. For each heat exchanger surface the method plots the fluid pumping power per unit volume of heat exchanger versus heat exchanger NTU per unit volume. These parameters are:

$$\frac{P_p}{V} = \frac{\text{pumping power}}{\text{volume}} \propto \frac{f\,\mathrm{Re}^3_{D_H}}{D^4_H} \tag{8.30}$$

$$\frac{\mathrm{NTU}}{V} = \frac{\mathrm{NTU}}{\text{volume}} \propto \frac{j\,\mathrm{Re}_{D_H}}{D^2_H} \tag{8.31}$$

Given the friction factor $f(\mathrm{Re})$, the heat transfer performance $\mathrm{Nu}(\mathrm{Re})$ or $j(\mathrm{Re})$ for the heat exchanger surface, and the flow passage hydraulic diameter D_H, one can easily construct a plot of the two parameters P/V and NTU/V.

In Eqs. (8.30) and (8.31) the Reynolds number is based on the flow area A_f, which ignores any enhancement:

$$\mathrm{Re}_{D_H} = \frac{GD_H}{\mu} \tag{8.32}$$

$$G = \frac{\dot{m}}{A_f}$$

where $\dot{m}$ is the mass flow rate in the flow passage of area A_f.

The friction factor is

$$f = \frac{\Delta p}{4(L/D_H)(G^2/2\rho g_c)} \tag{8.33}$$

where Δp is the frictional pressure drop in the core.

The j or Colburn factor is defined as

$$j = \frac{\bar{h}_c}{Gc_p} \mathrm{Pr}^{2/3} \tag{8.34}$$

where $\bar{h}_c$ is the heat transfer coefficient based on the bare (without enhancement) surface area A_b. The hydraulic diameter is defined as in Chapter 6 but can be written more conveniently in the form

$$D_H = \frac{4V}{A_b} \tag{8.35}$$

Using these definitions, a smooth tube of inside diameter D and a tube of inside diameter D with a twisted tape insert and with the same mass flow rate would have the same G, Re_D, A_b, and D but we would expect f and j to be larger for the latter tube.

Such a plot is useful for comparing two heat exchange surfaces because it allows a convenient comparison based on any of the following constraints:

1. Fixed heat exchanger volume and pumping power
2. Fixed pumping power and heat duty
3. Fixed volume and heat duty

These constraints can be visualized in Fig. 8.27, in which the $f\ \mathrm{Re}_D^3/D^4$ and $j\ \mathrm{Re}_D/D^2$ data are plotted for the two surfaces to be compared. From the baseline point labeled "o" in Fig. 8.27, comparisons based on the three constraints are labeled.

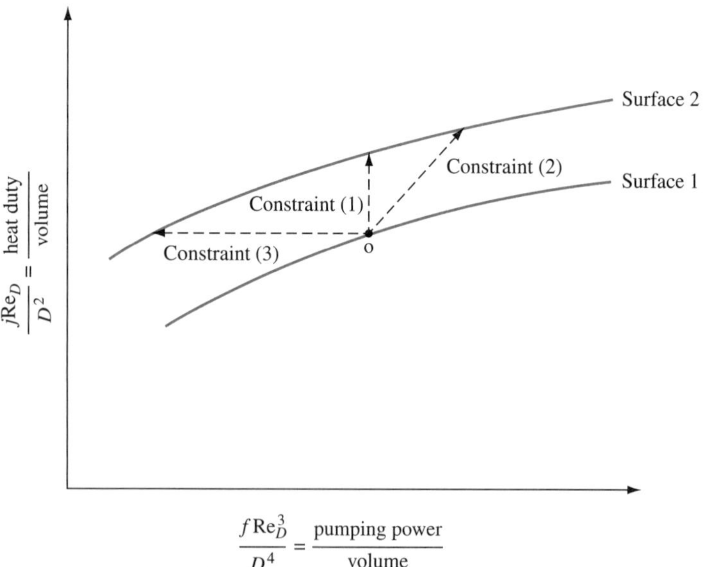

FIGURE 8.27 General comparison method of Soland et al. [27].

Source: Courtesy of T. Tinker, "Analysis of Fluid Flow Pattern in Shell-and-Tube Heat Exchangers and the Effect Distribution of the HeatnExchanger Performance," Inst. Mech. Eng., ASME Proc. General Discuss. Heat Transfer, pp. 89–115, September 1951.

A comparison based on constraint (1) can be made by constructing a vertical line through the baseline point. Comparing the two ordinate values where the vertical line intersects the curves allows one to compare the heat duty for each surface. The surface with the highest curve will transfer more heat. Constraint (2) can be visualized by constructing a line with slope +1. Comparing either the abscissa or ordinate where the line of slope +1 intersects the curves allows one to compare the heat exchanger volume required for each surface. The surface with the highest curve will require the least volume. Constraint (3) can be visualized by constructing a horizontal line. Comparing the abscissa where the line intersects the curves allows one to compare the pumping power for each surface. The surface with the highest curve will require the least pumping power.

EXAMPLE 8.5 Given the data in Fig. 8.26, compare the performance of wall protuberances and a twisted tape [surfaces (1) and (4) in Fig. 8.26] for a flow of air on the basis of fixed heat exchanger volume and pumping power. Assume that both surfaces are applied to the inside of a 1-cm-ID tube of circular cross section.

SOLUTION We must first construct the f (Re) and j (Re) curves for the two surfaces.

Curves (1) and (4) in Fig. 8.26(a) and (b) can be represented by straight lines with good accuracy. From the data in Fig. 8.26(a) and (b), these straight lines for the Nusselt numbers are

$$\mathrm{Nu}_1/\mathrm{Pr}^{0.4} = 0.054\ \mathrm{Re}_D^{0.805}$$
$$\mathrm{Nu}_4/\mathrm{Pr}^{0.4} = 0.057\ \mathrm{Re}_D^{0.772}$$

where the subscripts 1 and 4 denote surfaces 1 and 4.

Since $j = \mathrm{St\ Pr}^{2/3} = \mathrm{NuRe}_D^{-1}\mathrm{Pr}^{-1/3}$ we have

$$j_1 = 0.054\ \mathrm{Re}_D^{-0.195}\mathrm{Pr}^{1/15}$$

and

$$j_4 = 0.057\ \mathrm{Re}_D^{-0.228}\mathrm{Pr}^{1/15}$$

For the friction coefficient data we find

$$f_1 = 0.075\ \mathrm{Re}_D^{0.017}$$
$$f_4 = 0.222\ \mathrm{Re}_D^{-0.238}$$

In comparing the two surfaces we should restrict ourselves to the range

$$10^4 < \mathrm{Re}_D < 10^5$$

where the data for both surfaces are valid.

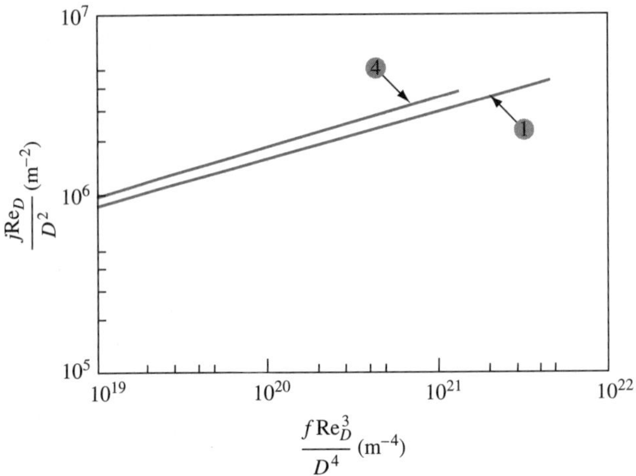

FIGURE 8.28 Comparison of wall protuberances and twisted tapes based on the method of Soland et al. [27].

Constructing the two comparison parameters, we have

$$\frac{f_1\,\text{Re}_D^3}{D_1^4} = \frac{0.075\,\text{Re}_D^{3.017}}{(0.01)^4} = 7.5 \times 10^6\,\text{Re}_D^{3.017} \quad \text{m}^{-4}$$

$$\frac{f_4\,\text{Re}_D^3}{D_4^4} = \frac{0.222\,\text{Re}_D^{2.76}}{(0.01)^4} = 2.22 \times 10^7\,\text{Re}_D^{2.76} \quad \text{m}^{-4}$$

$$\frac{j_1\,\text{Re}_D}{D_1^2} = \frac{0.054\,\text{Re}_D^{0.805}\,\text{Pr}^{1/15}}{(0.01)^2} = 527.8\,\text{Re}_D^{0.805} \quad \text{m}^{-2}$$

$$\frac{j_4\,\text{Re}_D}{D_4^2} = \frac{0.057\,\text{Re}_D^{0.772}\text{Pr}^{1/15}}{(0.01)^2} = 557.1\,\text{Re}_D^{0.772} \quad \text{m}^{-2}$$

These parameters are plotted in Fig. 8.28 for the Reynolds number range of interest. According to the specified constraint, a vertical line connecting the curves labeled (1) and (4) in Fig. 8.26 clearly demonstrates that surface 4, the twisted tape, is the better of the two surfaces. That is, for a fixed heat exchanger volume and at constant pumping power, the twisted tape enhancement will transfer more heat.

8.7* Microscale Heat Exchangers

With advancements in microelectronics and other high heat-flux dissipating devices, a variety of novel microscale heat exchangers have been developed to meet their cooling needs. Their structure usually incorporates microscale channels, which essentially exploit the benefits of high convection heat transfer coefficients in flows

(a)

(b)

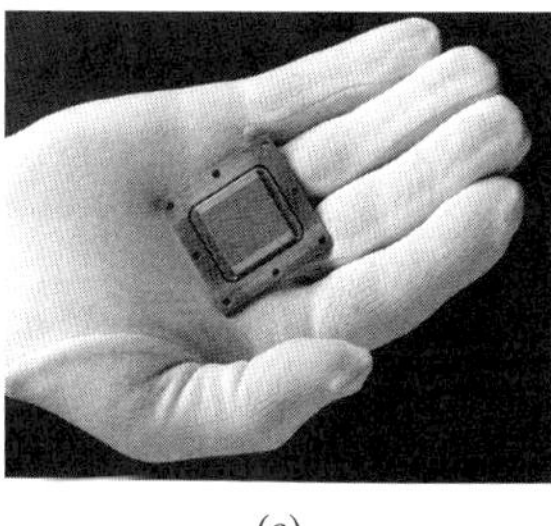

(c)

FIGURE 8.29 Typical microscale heat exchangers: (a) microchannels module manufactured by a laser sintering process; (b) details of the microstructure assembly of a typical micro heat exchanger; and (c) heat exchanger for microchip module cooling.

Source: (a) Courtesy of PennWell Corporation, (b) detail of the microstructure assembly of a cross-flow mhx made of stainless steel made by Institute for Micro Process Engineering, Karlsruhe Institute of Technology, Germany, (c) courtesy of Pacific Northwest National Laboratory.

through very small hydraulic-diameter ducts [28]. Applications of such heat exchangers include microchannel heat sinks, micro heat exchangers, and micro heat pipes, used in microelectronics, avionics, medical devices, space probes, and satellites, among others [28–30], and a few illustrative examples are depicted in Fig. 8.29.

To understand the implication of microchannels on convection heat transfer, consider laminar single-phase flows. Because of a very small hydraulic diameter D_h, which can range from a millimeter to a few microns in size, the flow tends to be fully developed and hence characterized by a constant Nusselt number. As a result, the heat transfer coefficient given by

$$h = \text{Nu}\left(\frac{k}{D_h}\right)$$

would increase substantially with decreasing hydraulic diameter. This was first explored by Tuckerman and Pease [30] for microelectronic cooling, and the exploitation of microchannels with both single- and two-phase flows continues to attract considerable research attention [28].

8.8 Closing Remarks

In this chapter we have studied the thermal design of heat exchangers in which two fluids at different temperatures flow in spaces separated by a wall and exchange heat by convection to and from and conduction through the wall. Such heat exchangers, sometimes called *recuperators*, are by far the most common and industrially important heat transfer devices. The most common configuration is the shell-and-tube heat exchanger, for which two methods of thermal analysis have been presented: the LMTD (log mean temperature difference) and the NTU or effectiveness method. The former is most convenient when all the terminal temperatures are specified and the heat exchanger area is to be determined, while the latter is preferred when the thermal performance or the area is known, specified, or can be estimated. Both of these methods are useful, but it is important to reemphasize the rather stringent assumptions on which they are based:

1. The overall heat transfer coefficient U is uniform over the entire heat exchanger surface.
2. The physical properties of the fluids do not vary with temperature.
3. Available correlations are satisfactory for predicting the individual heat transfer coefficients required to determine U.

Current design methodology is usually based on suitably chosen average values. When the spacial variation of U can be predicted, the appropriate value is an area average, $\bar{U}$, given by

$$\bar{U} = \frac{1}{A} \int_A U \, dA$$

The integration can be carried out numerically if necessary, but even this approach leaves the final result with a margin of error that is difficult to quantify. In the future, increased emphasis will probably be placed on computer-aided design (CAD), and the reader is encouraged to follow developments in this area. These tools will be particularly important in the design of condensers, and some preliminary information on this topic will be presented in Chapter 10.

In addition to recuperators, there are two other *generic* types of heat exchangers in use. In both of these types the hot and cold fluid streams occupy the same space, a channel with or without solid inserts. In one type, the *regenerator*, the hot and the cold fluid pass alternately over the same heat transfer surface. In the other type, exemplified by the *cooling tower*, the two fluids flow through the same passage simultaneously and contact each other directly. These types of exchangers are therefore often called *direct contact devices.* In many of the latter type the transfer of heat is accompanied by simultaneous transfer of mass.

Periodic flow regenerators have been used in practice only with gases. The regenerator consists of one or more flow passages that are partially filled either with solid pellets or with metal matrix inserts. During one part of the cycle, the inserts store internal energy as the warmer fluid flows over their surfaces. During the other part of the cycle, internal energy is released as the colder fluid passes through the regenerator and is heated. Thus, heat is transferred in a cyclic process. The principal advantage of the regenerator is a high heat-transfer effectiveness per unit weight and space. The major problem is to prevent leakage between the warmer and cooler fluids at elevated pressures. Regenerators have been used successfully as air pre-heaters in open-hearth and blast furnaces, in gas liquefication processes, and in gas turbines.

For preliminary estimates of shell-and-tube heat exchanger sizes and performance parameters, it is often sufficient to know the order of magnitude of the overall heat transfer coefficient under average service conditions. Typical values of overall heat transfer coefficients recommended for preliminary estimates are given in Table 8.5.

TABLE 8.5 Approximate overall heat transfer coefficients for preliminary estimates

Duty	Overall Coefficients, *U* (Btu/h ft^2 °F)	(W/m^2 K)
Steam to water		
instantaneous heater	400–600	2,270–3,400
storage-tank heater	175–300	990–1,700
Steam to oil		
heavy fuel	10–30	57–170
light fuel	30–60	170–340
light petroleum distillate	50–200	280–1,130
Steam to aqueous solutions	100–600	570–3,400
Steam to gases	5–50	28–280
Water to compressed air	10–30	57–170
Water to water, jacket water coolers	150–275	850–1,560
Water to lubricating oil	20–60	110–340
Water to condensing oil vapors	40–100	220–570
Water to condensing alcohol	45–120	255–680
Water to condensing Freon-12	80–150	450–850
Water to condensing ammonia	150–250	850–1,400
Water to organic solvents, alcohol	50–150	280–850
Water to boiling Freon-12	50–150	280–850
Water to gasoline	60–90	340–510
Water to gas oil or distillate	35–60	200–340
Water to brine	100–200	570–1,130
Light organics to light organics	40–75	220–425
Medium organics to medium organics	20–60	110–340
Heavy organics to heavy organics	10–40	57–200
Heavy organics to light organics	10–60	57–340
Crude oil to gas oil	30–55	170–310

Source: Adapted from Mueller [31].

For an up-to-date summary of specialized topics on the design and performance of heat exchangers, including evaporation and condensation, heat exchanger vibration, compact heat exchangers, fouling of heat exchangers, and heat exchange enhancement methods, the reader is referred to Shaw and Bell [32] and Hewitt [33].

References

1. H. Hausen, *Heat Transfer in Counterflow, Parallel Flow and Cross Flow*, McGraw-Hill, New York, 1983.
2. M. S. Bohn and L. W. Swanson, "A Comparison of Models and Experimental Data for Pressure Drop and Heat Transfer in Irrigated Packed Beds," *Int. J. Heat Mass Transfer*, vol. 34, pp. 2509–2519, 1991.
3. F. Kreith and R. F. Boehm, eds., *Direct Contact Heat Transfer*, Hemisphere, New York, 1987.
4. J. Taborek, "*F* and θ Charts for Cross-Flow Arrangements," Section 1.5.3 in *Handbook of Heat Exchanger Design*, vol. 1, E. U. Schlünder, ed., Hemisphere, Washington, D.C., 1983.

5. O. L. Pierson, "Experimental Investigation of Influence of Tube Arrangement on Convection Heat Transfer and Flow Resistance in Cross Flow of Gases over Tube Banks," *Trans. ASME*, vol. 59, pp. 563–572, 1937.
6. T. Tinker, "Analysis of the Fluid Flow Pattern in Shell-and-Tube Heat Exchangers and the Effect Distribution on the Heat Exchanger Performance," *Inst. Mech. Eng., ASME Proc. General Discuss. Heat Transfer*, pp. 89–115, September 1951.
7. B. E. Short, "Heat Transfer and Pressure Drop in Heat Exchangers," Bull. 3819, Univ. of Texas, 1938. (See also revision, Bull, 4324, June 1943.)
8. D. A. Donohue, "Heat Transfer and Pressure Drop in Heat Exchangers," *Ind. Eng. Chem.*, vol. 41, pp. 2499–2511, 1949.
9. K. P. Singh and A. I. Soler, *Mechanical Design of Heat Exchangers*, ARCTURUS Publishers, Inc., Cherry Hill, NJ., 1984.
10. W. M. Kays and A. L. London, *Compact Heat Exchangers*, 3rd ed., McGraw-Hill, New York, 1984.
11. G. F. Hewitt, G. L. Shires, and T. R. Bott, *Process Heat Transfer*, CRC Press, Boca Raton, FL, 1994.
12. R. K. Shah and D. P. Sekulic, *Fundamentals of Heat Exchanger Design*, Wiley, Hoboken, NJ, 2003.
13. A. P. Fraas, *Heat Exchanger Design*, 2nd ed., Wiley, Hoboken, NJ, 1989.
14. L. Wang, B. Sundén, and R. M. Manglik, *Plate Heat Exchangers: Design, Applications and Performance*, WIT Press, Southampton, UK, 2007.
15. W. J. Beek and K. M. K. Muttzall, *Transport Phenomena*, Wiley, New York, 1975.
16. TEMA, *Standards of the Tubular Exchanger Manufacturers Association*, 7th ed., Exchanger Manufacturers Association, New York, 1988.
17. R. A. Bowman, A. C. Mueller and W. M. Nagle, "Mean Temperature Difference in Design," *Trans. ASME*, vol. 62, pp. 283–294, 1940.
18. W. Nusselt, "A New Heat Transfer Formula for Cross-Flow," *Technische Mechanik und Thermodynamik*, vol. 12, 1930.
19. H. Ten Broeck, "Multipass Exchanger Calculations," *Ind. Eng. Chem.*, vol. 30, pp. 1041–1042, 1938.
20. M. A. DiGiovanni and R. L. Webb, "Uncertainty in Effectiveness-NTU Calculations for Crossflow Heat Exchangers," *Heat Transfer Engineering*, vol. 10, pp. 61–70, 1989.
21. R. M. Manglik, "Heat Transfer Enhancement," in *Heat Transfer Handbook*, A. Behan and A. D. Kraus, eds., Ch. 14, Wiley, Hoboken, NJ, 2003.
22. A. E. Bergles, "Techniques to Enhance Heat Transfer," in *Handbook of Heat Transfer*, 3rd ed., W. M. Rohsenow, J. P. Hartnett, and Y. I. Cho, eds., Ch. 11, McGraw-Hill, New York, 1998.
23. R. M. Manglik and A. E. Bergles, "Enhanced Heat and Mass Transfer in the New Millennium: A Review of the 2001 Literature," *Journal of Enhanced Heat Transfer*, vol. 11, no. 2, pp. 87–118, 2004.
24. A. E. Bergles, M. K. Jensen, and B. Shome, "Bibliography on Enhancement of Convective Heat and Mass Transfer," RPI Heat Transfer Laboratory, Rpt. HTL-23, 1995. See also A. E. Bergles, V. Nirmalan, G. H. Junkhan, and R. L. Webb, *Bibliography on Augmentation of Convective Heat and Mass Transfer-11*, Rept. HTL-31, ISU-ERI-Ames-84221, Iowa State University, Ames, Iowa, 1983.
25. R. M. Manglik and A. E. Bergles, "Swirl Flow Heat Transfer and Pressure Drop with Twisted-Tape Inserts," *Advances in Heat Transfer*, vol. 36, pp. 183–266, Academic Press, New York, 2002.
26. R. L. Webb and N. -K. Kim, *Principles of Enhanced Heat Transfer*, Taylor & Francis, Boca Raton, FL, 2005.
27. J. G. Soland, W. M. Mack, Jr., and W. M. Rohsenow, "Performance Ranking of Plate-Fin Heat Exchange Surfaces," *J. Heat Transfer*, vol. 100, pp. 514–519, 1978.
28. C. B. Sobhan and G. P. "Bud" Peterson, *Microscale and Nanoscale Heat Transfer: Fundamentals and Engineering Applications,* CRC Press, Boca Raton, FL, 2008.
29. R. Sadasivam, R. M. Manglik, and M. A. Jog, "Fully Developed Forced Convection Through Trapezoidal and Hexagonal Ducts," *International Journal of Heat and Mass Transfer*, vol. 42, no. 23, pp. 4321–4331, 1999.
30. D. B. Tuckerman and R. F. Pease, "High Performance Heat Sinking for VLSI," *IEEE Electron Device Letters*, vol. EDL-2, pp. 126–129, 1981.
31. A. C. Mueller, "Thermal Design of Shell-and-Tube-Heat Exchangers for Liquid-to-Liquid Heat Transfer," *Eng. Bull., Res. Ser. 121*, Purdue Univ. Eng. Exp. Stn., 1954.
32. R. K. Shaw and K. J. Bell, "Heat Exchangers," in F. Kreith, ed., *CRC Handbook of Thermal Engineering*, CRC Press, Boca Raton, FL, 2000.
33. G. F. Hewitt, ed., *Heat Exchanger Design Handbook*, Begell House, New York, 1998.

Problems

The problems for this chapter are organized as shown in the table below.

Topic	Problem Number
Finding the overall heat transfer coefficient	8.1–8.10
LMTD or effectiveness method, overall heat transfer coefficient given	8.11–8.34
LMTD or effectiveness method, overall heat transfer coefficient not given	8.35–8.52
Compact heat exchangers	8.53–8.55

8.1 In a heat exchanger, as shown in the accompanying figure, air flows over brass tubes of 1.8-cm ID and 2.1-cm OD containing steam. The convection heat transfer coefficients on the air and steam sides of the tubes are 70 W/m^2 K and 210 W/m^2 K, respectively. Calculate the overall heat transfer coefficient for the heat exchanger (a) based on the inner tube area and (b) based on the outer tube area.

8.2 Repeat Problem 8.1 but assume that a fouling factor of 0.00018 m^2 K/W has developed on the inside of the tube during operation.

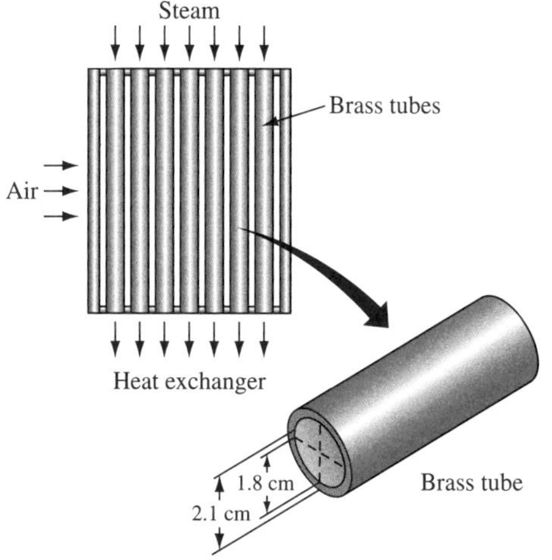

Problem 8.1

8.3 A light oil flows through a copper tube of 2.6-cm ID and 3.2-cm OD. Air flows perpendicular over the exterior of the tube as shown in the following sketch. The convection heat transfer coefficient for the oil is 120 W/m^2 K and for the air is 35 W/m^2 K. Calculate the overall heat transfer coefficient based on the outside area of the tube (a) considering the thermal resistance of the tube and (b) neglecting the resistance of the tube.

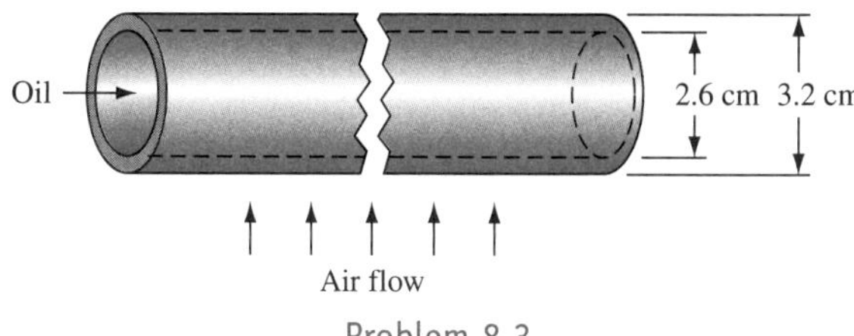

Problem 8.3

8.4 Repeat problem 8.3, but assume that fouling factors of 0.0009 m^2 K/W and 0.0004 m^2 K/W have developed on the inside and on the outside, respectively.

8.5 Water flowing in a long, aluminum tube is to be heated by air flowing perpendicular to the exterior of the tube. The ID of the tube is 1.85 cm, and its OD is 2.3 cm. The mass flow rate of the water through the tube is 0.65 kg/s, and the temperature of the water in the tube averages 30°C. The free-stream velocity and ambient temperature of the air are 10 m/s and 120°C, respectively. Estimate the overall heat transfer coefficient for the heat exchanger using appropriate correlations from previous chapters. State all your assumptions.

8.6 Hot water is used to heat air in a double-pipe heat exchanger as shown in the following sketch. If the heat transfer coefficients on the water side and on the air side are 100 Btu/h ft^2 °F and 10 Btu/h ft^2 °F, respectively, calculate the overall heat transfer coefficient based on the outer diameter. The heat exchanger pipe is 2-in., schedule 40 steel (k = 54 W/m K) with water inside. Express your answer in Btu/h ft^2 °F and W/m^2 °C.

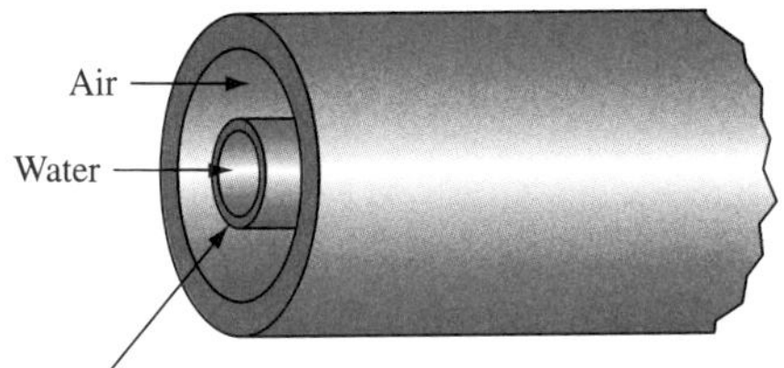

Problem 8.6

8.7 Repeat Problem 8.6, but assume that a fouling factor of 0.001 h ft^2/°F Btu based on the tube outside diameter has developed over time.

8.8 The heat transfer coefficient of a copper tube (1.9-cm ID and 2.3-cm OD) is 500 W/m^2 K on the inside and 120 W/m^2 K on the outside, but a deposit with a fouling factor of 0.009 m^2 K/W (based on the tube outside diameter) has built up over time. Estimate the percentage increase in the overall heat transfer coefficient if the deposit were removed.

8.9 In a shell-and-tube heat exchanger with $\bar{h}_i = \bar{h}_o = 5600$ W/m^2 K and negligible wall resistance, by what percent would the overall heat transfer coefficient (based on the outside area) change if the number of tubes were doubled? The tubes have an outside diameter of 2.5 cm and a tube wall thickness of 2 mm. Assume that the flow rates of the fluids are constant, the effect of temperature on fluid properties is negligible, and the total cross-sectional area of the tubes is small compared with the flow area of the shell.

8.10 Water at 80°F enters a No. 18 BWG 5/8-in. condenser tube made of nickel chromium steel (k = 15 Btu/h ft °F) at a rate of 5.43 gpm. The tube is 10 ft long, and its outside is heated by steam condensing at 120°F. Under these conditions the average heat transfer coefficient on the water side is 1750 Btu/h ft^2 °F. The heat transfer coefficient on the steam side can be taken as 2000 Btu/h ft^2 °F. On the interior of the tube, however, a scale with a thermal conductance equivalent to 1000 Btu/h ft^2 °F is forming. (a) Calculate the overall heat transfer coefficient U per square foot of exterior surface area after the scale has formed, and (b) calculate the exit temperature of the water.

8.11 Water is heated by hot air in a heat exchanger. The flow rate of the water is 12 kg/s and that of the air is 2 kg/s. The water enters at 40°C, and the air enters at 460°C. The overall heat transfer coefficient of the heat exchanger is 275 W/m^2 K based on a surface area of 14 m^2. Determine the effectiveness of the heat exchanger if it is (a) a parallel-flow type or (b) a cross-flow type (both fluids unmixed). Then calculate the heat transfer rate for the two types of heat exchangers described and the outlet temperatures of the hot and cold fluids for the conditions given.

8.12 Exhaust gases from a power plant are used to preheat air in a cross-flow heat exchanger. The exhaust gases enter the heat exchanger at 450°C and leave at 200°C. The air enters the heat exchanger at 70°C, leaves at 250°C, and has a mass flow rate of 10 kg/s. Assume the properties of the exhaust gases can be approximated by those of air. The overall heat transfer coefficient of the heat exchanger is 154 W/m^2 K. Calculate the heat exchanger surface area required if (a) the air is unmixed and the exhaust gases are mixed and (b) both fluids are unmixed.

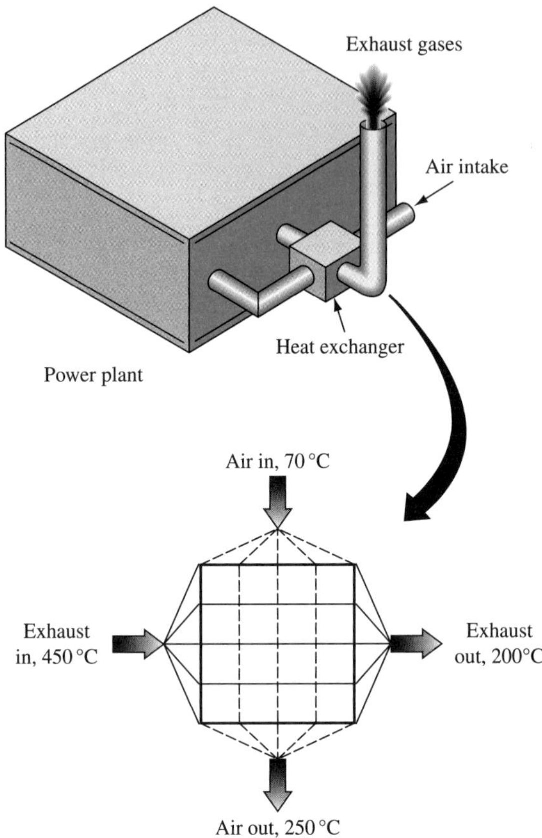

Problem 8.12

8.13 A shell-and-tube heat exchanger having one shell pass and four tube passes is shown schematically in the following sketch. The fluid in the tubes enters at 200°C and leaves at 100°C. The temperature of the fluid is 20°C entering the shell and 90°C leaving the shell. The overall heat transfer coefficient based on the surface area of 12 m^2 is 300 W/m^2 K. Calculate the heat transfer rate between the fluids.

8.14 Oil (c_p = 2.1 kJ/kg K) is used to heat water in a shell-and-tube heat exchanger with a single shell pass and two tube passes. The overall heat transfer coefficient is 525 W/m^2 K. The mass flow rates are 7 kg/s for the oil and 10 kg/s for the water. The oil and water enter the heat exchanger at 240°C and 20°C, respectively. The heat exchanger is to be designed so that the water leaves the heat exchanger with a minimum temperature of 80°C. Calculate the heat transfer surface area required to achieve this temperature.

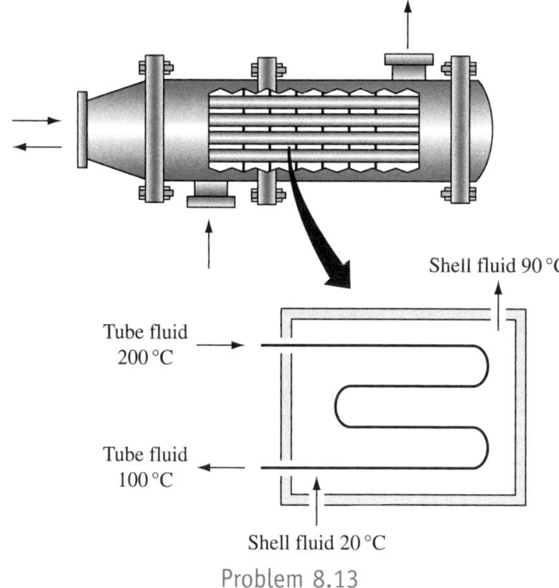

Problem 8.13

8.15 A shell-and-tube heat exchanger with two tube passes and a single shell pass is used to heat water by condensing steam in the shell. The flow rate of the water is 15 kg/s, and it is heated from 60°C to 80°C. The steam condenses at 140°C, and the overall heat transfer coefficient of the heat exchanger is 820 W/m^2 K. If there are 45 tubes with an OD of 2.75 cm, calculate the required tube length.

8.16 Benzene flowing at 12.5 kg/s is to be cooled continuously from 82°C to 54°C by 10 kg/s of water available at 15.5°C. Using Table 8.5, estimate the surface area required for (a) cross-flow with six tube passes and one shell pass, with neither of the fluids mixed, and (b) a counterflow exchanger with one shell pass and eight tube passes, with the colder fluid inside tubes.

8.17 Water entering a shell-and-tube heat exchanger at 35°C is to be heated to 75°C by an oil. The oil enters at 110°C and leaves at 75°C. The heat exchanger is arranged for counterflow with the water making one shell pass and the oil making two tube passes. If the water flow rate is 68 kg per minute and the overall heat transfer coefficient is estimated from Table 8.1 to be 320 W/m^2 K, calculate the required heat exchanger area.

8.18 Starting with a heat balance, show that the heat exchanger effectiveness for a counterflow arrangement is

$$\mathscr{E} = \frac{1 - \exp[-(1 - C_{min}/C_{max})\text{NTU}]}{1 - (C_{min}/C_{max})\exp[-(1 - C_{min}/C_{max})\text{NTU}]}$$

8.19 In the shell of a shell-and-tube heat exchanger with two shell passes and eight tube passes, 100,000 lb/h of water is heated from 180°F to 300°F. Hot exhaust gases having roughly the same physical properties as air enter the tubes at 650°F and leave at 350°F. The total surface area, based on the outer tube surface, is 10,000 ft^2. Determine (a) the log mean temperature difference if the heat exchanger is the simple counterflow type, (b) the correction factor F for the actual arrangement, (c) the effectiveness of the heat exchanger, and (d) the average overall heat transfer coefficient.

8.20 In gas turbine recuperators the exhaust gases are used to heat the incoming air and C_{min}/C_{max} is therefore approximately equal to unity. Show that for this case $\mathscr{E}$ = NTU/(1 + NTU) for counterflow and $\mathscr{E} = (1/2)(1 - e^{-2\text{NTU}})$ for parallel flow.

8.21 In a single-pass counterflow heat exchanger, 4536 kg/h of water enter at 15°C and cool 9071 kg/h of an oil having a specific heat of 2093 J/kg °C from 93°C to 65°C. If the overall heat transfer coefficient is 284 W/m^2 °C, determine the surface area required.

8.22 A steam-heated, single-pass tubular preheater is designed to raise 45,000 lb/h of air from 70°F to 170°F, using saturated steam at 375 psia. It is proposed to double the flow rate of air, and in order to be able to use the same heat exchanger and achieve the desired temperature rise, it is proposed to increase the steam pressure. Calculate the steam pressure necessary for the new conditions and comment on the design characteristics of the new arrangement.

8.23 For safety reasons, a heat exchanger performs as shown in (a) of the accompanying figure on the next page. An engineer suggests that it would be wise to double the heat transfer area so as to double the heat transfer rate. The suggestion is made to add a second, identical exchanger as shown in (b). Evaluate this suggestion, that is, show whether the heat transfer rate would double.

8.24 In a single-pass counterflow heat exchanger, 10,000 lb/h of water enters at 60°F and cools 20,000 lb/h of an oil having a specific heat of 0.50 Btu/lb °F from 200°F to 150°F. If the overall heat transfer coefficient is 50 Btu/h ft^2 °F, determine the surface area required.

8.25 Determine the outlet temperature of the oil in Problem 8.24 for the same initial fluid temperatures if the flow arrangement is one shell pass and two tube passes. The total area and average overall heat transfer coefficient are the same as those for the unit in Problem 8.24.

8.26 Carbon dioxide at 427°C is to be used to heat 12.6 kg/s of pressurized water from 37°C to 148°C while the gas temperature drops 204°C. For an overall heat transfer coefficient of 57 W/m^2 K, compute the required area of the exchanger in square feet for (a) parallel flow, (b) counterflow, (c) a 2–4 reversed current exchanger, and (d) cross-flow with the gas mixed.

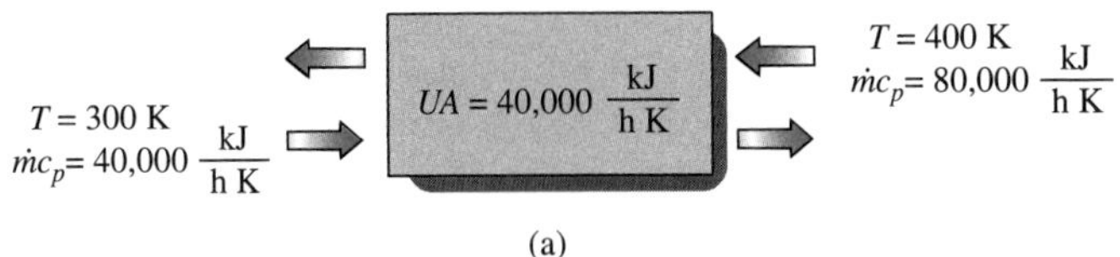

(a)

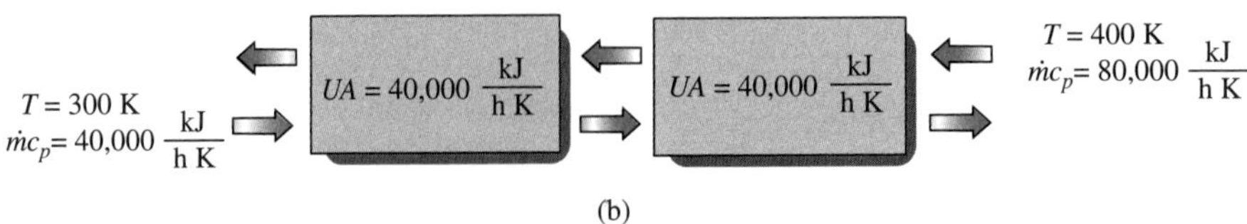

(b)

Problem 8.23

8.27 An economizer is to be purchased for a power plant. The unit is to be large enough to heat 7.5 kg/s of pressurized water from 71°C to 182°C. There are 26 kg/s of flue gases (c_p = 1000 J/kg K) available at 426°C. Estimate (a) the outlet temperature of the flue gases and (b) the heat transfer area required for a counterflow arrangement if the overall heat transfer coefficient is 57 W/m^2 K.

8.28 Water flowing through a pipe is heated by steam condensing on the outside of the pipe. (a) Assuming a uniform overall heat transfer coefficient along the pipe, derive an expression for the water temperature as a function of distance from the entrance. (b) For an overall heat transfer coefficient of 570 W/m^2 K based on the inside diameter of 5 cm, a steam temperature of 104°C, and a water flow rate of 0.063 kg/s, calculate the length required to raise the water temperature from 15.5°C to 65.5°C.

8.29 Water at a rate of 5.43 gpm and a temperature of 80°F enters a no. 18 BWG 5/8-in. condenser tube made of nickel chromium steel (k = 15 Btu/h ft °F). The tube is 10 ft long, and its outside is heated by steam condensing at 120°F. Under these conditions the average heat transfer coefficient on the water side is 1750 Btu/h ft^2 °F, and the heat transfer coefficient on the steam side can be taken as 2000 Btu/h ft^2 °F. On the interior of the tube, however, there is a scale having a thermal conductance equivalent to 1000 Btu/h ft^2 °F. (a) Calculate the overall heat transfer coefficient U per square foot of exterior surface area. (b) Calculate the exit temperature of the water.

8.30 It is proposed to preheat the water for a boiler using flue gases from the boiler stack. The flue gases are available at the rate of 0.25 kg/s at 150°C, with a specific heat of 1000 J/kg K. The water entering the exchanger at 15°C at the rate of 0.05 kg/s is to be heated to 90°C. The heat exchanger is to be of the reversed current type with one shell pass and four tube passes. The water flows inside the tubes, which are made of copper (2.5-cm ID, 3.0-cm OD). The heat transfer coefficient at the gas side is 115 W/m^2 K, while the heat transfer coefficient on the water side is 1150 W/m^2 K. A scale on the water side offers an additional thermal resistance of 0.002 m^2 K/W. (a) Determine the overall heat transfer coefficient based on the outer tube diameter. (b) Determine the appropriate mean temperature difference for the heat exchanger. (c) Estimate the required tube length. (d) What would be the outlet temperature and the effectiveness if the water flow rate is doubled, giving a heat transfer coefficient of 1820 W/m^2 K?

8.31 Hot water is to be heated from 10°C to 30°C at the rate of 300 kg/s by atmospheric pressure steam in a single-pass shell-and-tube heat exchanger consisting of 1-in. schedule 40 steel pipe. The surface coefficient on the steam side is estimated to be 11,350 W/m^2 K. An available pump can deliver the desired quantity of water provided the pressure drop through the pipes does not exceed 15 psi. Calculate the number of tubes in parallel and the length of each tube necessary to operate the heat exchanger with the available pump.

8.32 Water flowing at a rate of 12.6 kg/s is to be cooled from 90°C to 65°C by means of an equal flow rate of cold water entering at 40°C. The water velocity will be such that the overall coefficient of heat transfer U is 2300 W/m^2 K. Calculate the heat-exchanger surface area (in square meters) needed for each of the following arrangements: (a) parallel flow, (b) counterflow, (c) a multi-pass heat exchanger with the hot water making one pass through a well-balanced shell and the cold water making two passes through the tubes, and (d) a cross-flow heat exchanger with both sides unmixed.

8.33 Water flowing at a rate of 10 kg/s through a 50-tube double-pass shell-and-tube heat exchanger heats air that flows through the shell side. The length of the brass tubes is 6.7 m, and they have an outside diameter of 2.6 cm and an inside diameter of 2.3 cm. The heat transfer coefficients of the water and air are 470 W/m^2 K and 210 W/m^2 K, respectively. The air enters the shell at a temperature of 15°C and a flow rate of 16 kg/s. The temperature of the water as it enters the tubes is 75°C. Calculate (a) the heat exchanger effectiveness, (b) the heat transfer rate to the air, and (c) the outlet temperature of the air and water.

8.34 An air-cooled low-pressure steam condenser is shown in the following figure. The tube bank is four rows deep in the direction of air flow, and there are a total of 80 tubes. The tubes have a 2.2-cm ID and a 2.5-cm OD and are 9 m long with circular fins on the outside. The tube-plus-fin area is 16 times the bare tube area, i.e., the fin area is 15 times the bare tube area (neglect the tube surface covered by fins). The fin efficiency is 0.75. Air flows past the outside of the tubes. On a particular day the air enters at 22.8°C and leaves at 45.6°C. The air flow rate is 3.4×10^5 kg/h.

The steam temperature is 55°C and has a condensing coefficient of 10^4 W/m^2 K. The steam-side fouling coefficient is 10^4 W/m^2 K. The tube wall conductance per unit area is 10^5 W/m^2 K. The air-side fouling resistance is negligible. The air-side film heat transfer coefficient is 285 W/m^2 K (note that this value has been corrected for the number of transverse tube rows). (a) What is the log mean temperature difference between the two streams? (b) What is the rate of heat transfer? (c) What is the rate of steam condensation? (d) Estimate the rate of steam condensation if there were no fins.

8.35 Design (i.e., determine the overall area and a suitable arrangement of shell and tube passes) a tubular feed-water heater capable of heating 2300 kg/h of water from 21°C to 90°C. The following specifications are given: (a) saturated steam at 920 kPa absolute pressure is condensing on the outer tube surface, (b) the heat transfer coefficient on the steam side is 6800 W/m^2 K, (c) the tubes are made of copper, 2.5-cm OD, 2.3-cm ID, and 2.4 m long, and (d) the water velocity is 0.8 m/s.

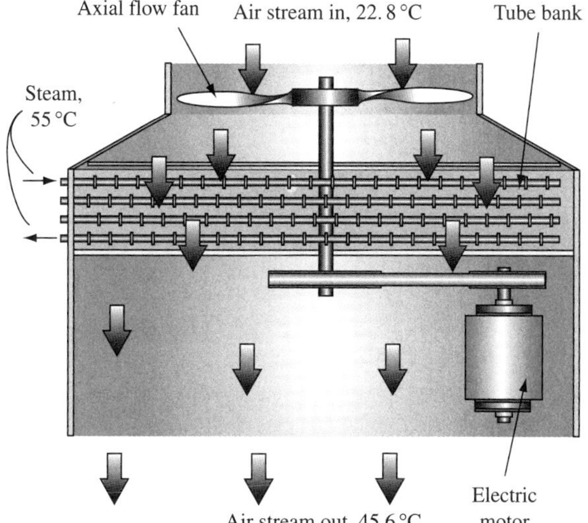

Problem 8.34

8.36 Two engineers are having an argument about the efficiency of a tube-side multipass heat exchanger compared to a similar exchanger with a single tube-side pass. Smith claims that for a given number of tubes and rate of heat transfer, more area is required in a two-pass exchanger than in a one pass, because the effective temperature difference is less. Jones, on the other hand, claims that because the tube-side velocity and hence the heat transfer coefficient are higher, less area is required in a two-pass exchanger.

With the conditions given below, which engineer is correct? Which case would you recommend, or what changes in the exchanger would you recommend?

Exchanger specifications:
200 tube passes total
1-inch OD copper tubes, 16 BWG

Tube-side fluid:
water entering at 16°C, leaving at 28°C, at a rate of 225,000 kg/h

Shell-side fluid:
Mobiltherm 600, entering at 50°C, leaving at 33°C
shell-side coefficient = 1700 W/m^2 K

8.37 A horizontal shell-and-tube heat exchanger is used to condense organic vapors. The organic vapors condense on the outside of the tubes, while water is used as the cooling medium on the inside of the tubes. The condenser tubes are 1.9-cm OD, 1.6-cm-ID copper tubes, 2.4 m in length. There are a total of 768 tubes. The water makes four passes through the exchanger.

Test data obtained when the unit was first placed into service are as follows:

water flow rate = 3700 liters/min
inlet water temperature = 29°C
outlet water temperature = 49°C
organic-vapor condensation temperature = 118°C

After three months of operation, another test made under the same conditions as the first (i.e., same water rate and inlet temperature and same condensation temperature) showed that the exit water temperature was 46°C. (a) What is the tube-side fluid (water) velocity? (b) What is the effectiveness, $\mathscr{E}$, of the exchanger at the times of the first and second test? (c) Assuming no changes in either the inside heat transfer coefficient or the condensing coefficient, negligible shell-side fouling, and no fouling at the time of the first test, estimate the tube-side fouling coefficient at the time of the second test.

8.38 A shell-and-tube heat exchanger is to be used to cool 200,000 lb/h (25.2 kg/s) of water from 100°F (38°C) to 90°F (32°C). The exchanger has one shell-side pass and two tube-side passes. The hot water flows through the tubes, and the cooling water flows through the shell. The cooling water enters at 75°F (24°C) and leaves at 90°F. The shell-side (outside) heat transfer coefficient is estimated to be 1000 Btu/h ft^2 °F (5678 W/m^2 K). Design specifications require that the pressure drop through the tubes be as close to 2 psi (13.8 kPa) as possible, that the tubes be 18 BWG copper tubing (1.24-mm wall-thickness), and that each pass be 16 feet (4.9 m) long. Assume that the pressure losses at the inlet and outlet are equal to one-and-one-half of a velocity head ($\rho U^2/2g_c$), respectively. For these specifications, what tube diameter and how many tubes are needed?

8.39 A shell-and-tube heat exchanger with the characteristics given below is to be used to heat 27,000 kg/h of water before it is sent to a reaction system. Saturated steam at 2.36 atm absolute pressure is available as the heating medium and will be condensed without subcooling on the outside of the tubes. From previous experience, the steam-side condensing coefficient can be assumed to be constant and equal to 11,300 W/m^2 K. If the water enters at 16°C, at what temperature will it leave the exchanger? Use reasonable estimates for fouling coefficients.

Heat exchanger specifications:

Tubes: 2.5-cm OD, 2.3-cm ID, horizontal copper tubes in six vertical rows
tube length = 2.4 m
total number of tubes = 52
number of tube-side passes = 2

8.40 Determine the appropriate size of a shell-and-tube heat exchanger with two tube passes and one shell pass to heat 70,000 lb/h (8.82 kg/s) of pure ethanol from 60°F to 140°F (15.6°C to 60°C). The heating medium is saturated steam at 22 psia (152 kPa) condensing on the outside of the tubes with a condensing coefficient of 15,000 W/m^2 K. Each pass of the exchanger has 50 copper tubes with an OD of 0.75 in. (1.91 cm) and a wall-thickness of 0.083 in. (0.211 cm). For the sizing, assume that the header cross-sectional area per pass is twice the total inside tube cross-sectional area. The ethanol is expected to foul the inside of the tubes with a fouling coefficient of 1000 Btu/h ft^2 °F (5678 W/m^2 K). After the size of the heat exchanger, i.e., the length of the tubes, is known, estimate the frictional pressure drop using the inlet loss coefficient of unity. Then estimate the pumping power required with a pump efficiency of 60% and the pumping cost per year at a rate of $0.10 per kWh.

8.41 A counterflow regenerator is used in a gas turbine power plant to preheat the air before it enters the combustor. The air leaves the compressor at a temperature of 350°C. Exhaust gas leaves the turbine at 700°C. The mass flow rates of air and gas are 5 kg/s. Take the c_p of air and gas to be equal to 1.05 kJ/kg K. Determine the required heat transfer area as a function of the regenerator effectiveness if the overall heat transfer coefficient is 75 W/m^2 K.

8.42 Determine the heat-transfer area requirements of Problem 8.41 if (a) a 1–2 shell-and-tube, (b) an unmixed cross-flow, and (c) a parallel flow heat exchanger are used.

8.43 A small space heater is constructed of 1/2-in., 18-gauge brass tubes that are 2 ft long. The tubes are arranged in equilateral, staggered triangles on $1\frac{1}{2}$-in. centers with four rows of 15 tubes each. A fan blows 2000 cfm of atmospheric pressure air at 70°F uniformly over the tubes (see the following sketch). Estimate (a) the heat transfer rate, (b) the exit temperature of the air, and (c) the rate of steam condensation, assuming that saturated steam at 2 psia inside the tubes is the heat source. State your assumptions. Work parts (a), (b), and (c) of this problem by two methods: first use the LMTD, which requires a trial-and-error or graphical solution, then use the effectiveness method.

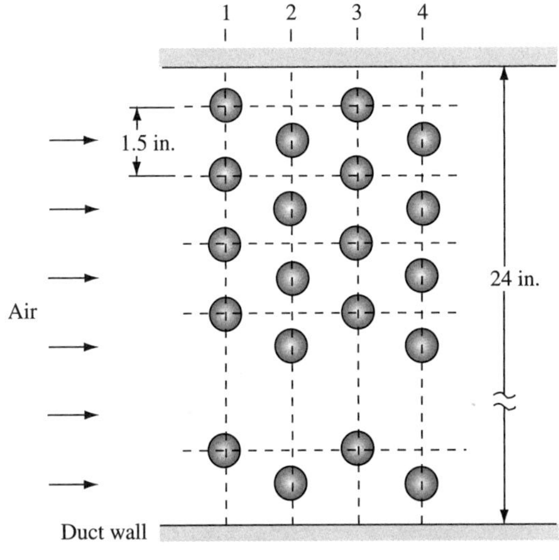

Problem 8.43

8.44 A one-tube pass cross-flow heat exchanger is being considered for recovering energy from the exhaust gases of a turbine-driven engine. The heat exchanger is constructed of flat plates forming an egg-crate pattern as shown in the following sketch. The velocities of the entering air (10°C) and exhaust gases (425°C) are both equal to 61 m/s. Assuming that the properties of the exhaust gases are the same as those of the air, estimate the overall heat transfer coefficient U for a path length of 1.2 m, neglecting the thermal resistance of the intermediate metal wall. Then determine the outlet temperature of the air, comment on the suitability of the proposed design, and if possible, suggest improvements. State your assumptions.

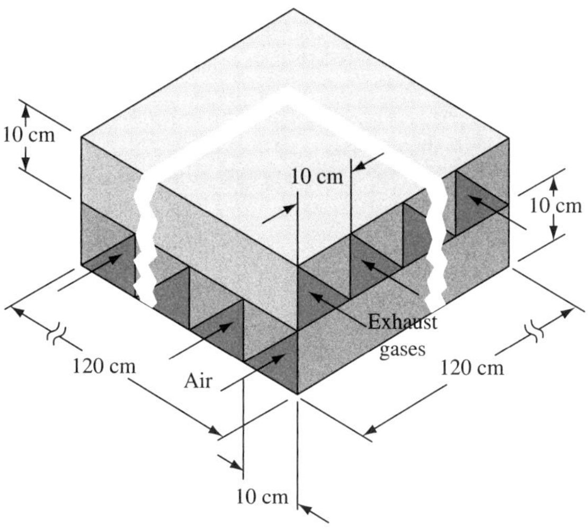

Problem 8.44

8.45 A shell-and-tube counterflow heat exchanger is to be designed for heating an oil from 80°F to 180°F. The heat exchanger has two tube passes and one shell pass. The oil is to pass through $1\frac{1}{2}$in. schedule 40 pipes at a velocity of 200 fpm, and steam is to condense at 215°F on the outside of the pipes. The specific heat of the oil is 0.43 Btu/lb °F, and its mass density is 58 lb/cu ft. The steam-side heat transfer coefficient is approximately 1800 Btu/h ft^2 °F, and the thermal conductivity of the metal of the tubes is 17 Btu/h ft °F. The results of previous experiments giving the oil-side heat transfer coefficients for the same pipe size at the same oil velocity as those to be used in the exchanger are:

ΔT (°F)	135	115	95	75	35	—
T_{oil} (°F)	80	100	120	140	160	180
h_{ci} (Btu/h ft^2 °F)	14	15	18	25	45	96

(a) Find the overall heat transfer coefficient U based on the outer surface area at the point where the oil is 100°F. (b) Find the temperature of the inside surface of the pipe when the oil temperature is 100°F. (c) Find the required length of the tube bundle.

8.46 A shell-and-tube heat exchanger in an ammonia plant is preheating 1132 m^3 of atmospheric pressure nitrogen per hour from 21°C to 65°C using steam condensing at 138,000 N/m^2. The tubes in the heat exchanger have an inside diameter of 2.5 cm. In order to change from ammonia synthesis to methanol synthesis, the same heater is to be used to preheat carbon monoxide from 21°C to 77°C, using steam condensing at 241,000 N/m^2. Calculate the flow rate that can be anticipated from this heat exchanger in kilograms of carbon monoxide per second.

8.47 In an industrial plant a shell-and-tube heat exchanger is heating pressurized dirty water at the rate of 38 kg/s from 60°C to 110°C by means of steam condensing at 115°C on the outside of the tubes. The heat exchanger has 500 steel tubes (ID = 1.6 cm, OD = 2.1 cm) in a tube bundle that is 9 m long. The water flows through the tubes while the steam condenses in the shell. If it can be assumed that the thermal resistance of the scale on the inside pipe wall is unaltered when the mass rate of flow is increased and that changes in water properties with temperature are negligible, estimate (a) the heat transfer coefficient on the water side and (b) the exit temperature of the dirty water if its mass rate of flow is doubled.

8.48 Liquid benzene (specific gravity = 0.86) is to be heated in a counterflow concentric-pipe heat exchanger from 30°C to 90°C. For a tentative design, the velocity of the benzene through the inside pipe (ID = 2.7 cm, OD = 3.3 cm) can be taken as 8 m/s. Saturated process steam at 1.38×10^6 N/m^2 is available for heating. Two methods of using this steam are proposed: (a) pass the process steam directly through the annulus of the exchanger—this would require that the latter be designed for the high pressure; (b) throttle the steam adiabatically to 138,000 N/m^2 before passing it through the heater. In both cases the operation would be controlled so that saturated vapor enters and saturated water leaves the heater. As an approximation, assume for both cases that the heat transfer coefficient for condensing steam remains constant at 12,800 W/m^2 K, that the thermal resistance of the pipe wall is negligible, and that the pressure drop for the steam is negligible. If the inside diameter of the outer pipe is 5 cm, calculate the mass rate of flow of steam (kg/s per pipe) and the length of heater required for each arrangement.

8.49 Calculate the overall heat transfer coefficient and the rate of heat flow from the hot gases to the cold air in the cross-flow tube bank of the heat exchanger shown in the accompanying illustration on the next page. The following operating conditions are given:

air flow rate = 3000 lb/h
hot gas flow rate = 5000 lb/h
temperature of hot gases entering exchanger = 1600°F
temperature of cold air entering exchanger = 100°F
Both gases are approximately at atmospheric pressure.

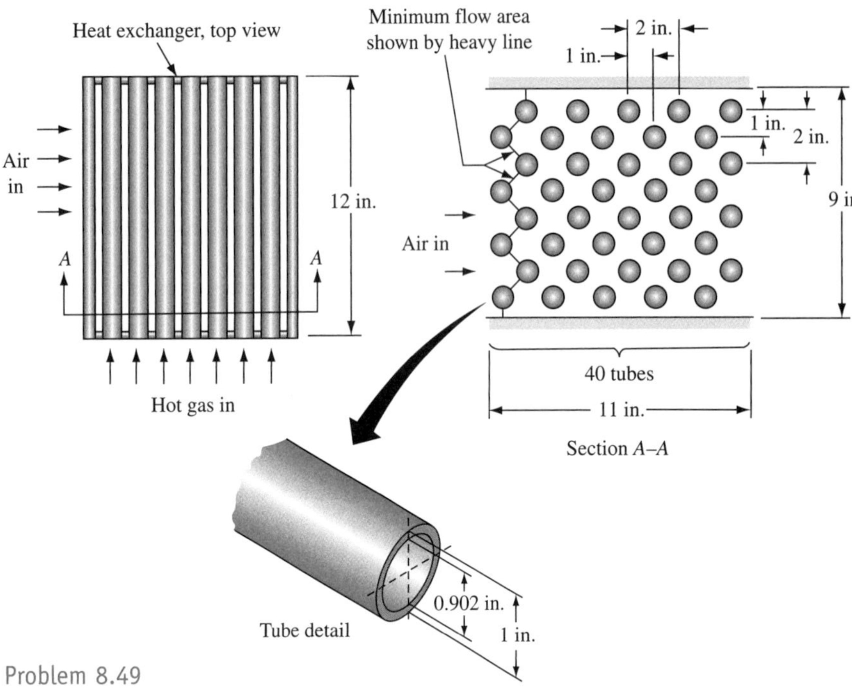

Problem 8.49

8.50 An oil having a specific heat of 2100 J/kg K enters an oil cooler at 82°C at the rate of 2.5 kg/s. The cooler is a counterflow unit with water as the coolant; the transfer area is 28 m^2, and the overall heat transfer coefficient is 570 W/m^2 K. The water enters the exchanger at 27°C. Determine the water rate required if the oil is to leave the cooler at 38°C.

8.51 While flowing at the rate of 1.25 kg/s in a simple counterflow heat exchanger, dry air is cooled from 65°C to 38°C by means of cold air that enters at 15°C and flows at a rate of 1.6 kg/s. It is planned to lengthen the heat exchanger so that 1.25 kg/s of air can be cooled from 65°C to 26°C with a counterflow current of air at 1.6 kg/s entering at 15°C. Assuming that the specific heat of the air is constant, calculate the ratio of the length of the new heat exchanger to the length of the original.

8.52 Saturated steam at 1.35 atm condenses on the outside of a 2.6-m length of copper tubing, heating 5 kg/h of water flowing in the tube. The water temperatures measured at 10 equally spaced stations along the tube length (see the sketch below) are:

Station	Temp °C
1	18
2	43
3	57
4	67
5	73
6	78
7	82
8	85
9	88
10	90
11	92

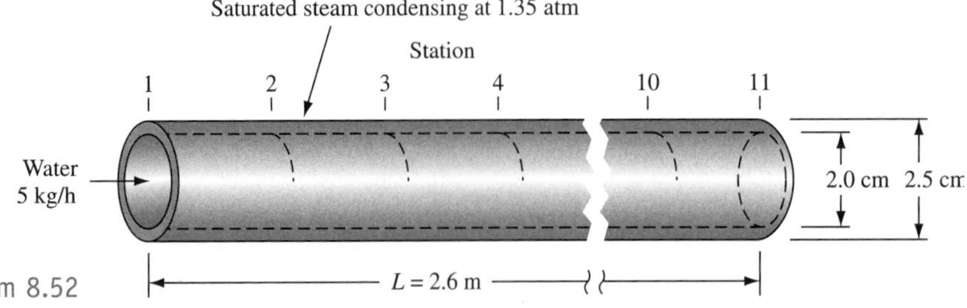

Problem 8.52

Calculate (a) the average overall heat transfer coefficient U_o based on the outside tube area, (b) the average water-side heat transfer coefficient h_w (assume the steam-side coefficient at h_s is 11,000 W/m^2 K), (c) the local overall coefficient U_x based on the outside tube area for each of the 10 sections between temperature stations, and (d) the local water-side coefficients h_{wx} for each of the 10 sections. Plot all items versus tube length. The tube dimensions are ID = 2 cm, OD = 2.5 cm. Temperature station 1 is at tube entrance and station 11 is at tube exit.

8.53 Calculate the water-side heat transfer coefficient and the coolant pressure drop per unit length of tube for the core of a compact air-to-water intercooler for a 5000-hp gas turbine plant. The water flows inside a flattened aluminum tube having the cross section shown below:

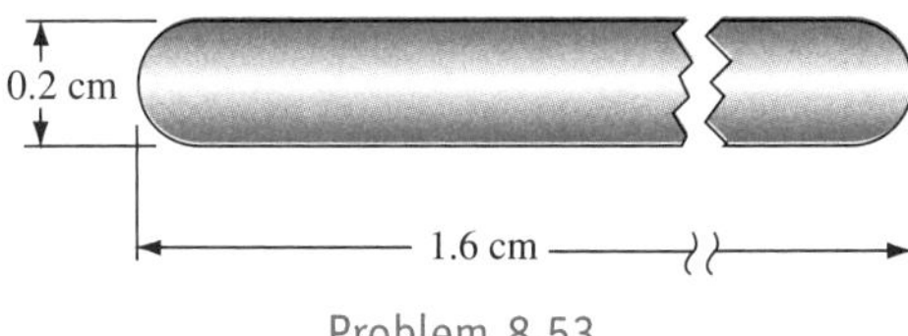

Problem 8.53

The inside diameter of the tube before it was flattened was 0.485 in. (1.23 cm) with a wall thickness (t) of 0.01 in. (0.025 cm). The water enters the tube at 60°F (15.6°C) and leaves at 80°F (26.7°C) at a velocity of 4.4 ft/s (1.34 m/s).

8.54 An air-to-water compact heat exchanger is to be designed to serve as an intercooler for a 5000-hp gas turbine plant. The exchanger is to meet the following heat transfer and pressure drop performance specifications:

Air-side operating conditions:		
Flow rate:	200,000 lb/h	(25.2 kg/s)
inlet temperature	720°R	(400 K)
outlet temperature	540°R	(300 K)
inlet pressure (p_1)	29.7 psia	(2.05×10^5 N/m^2)
pressure drop ratio	($\Delta p/p_1$)	7.6%
Water-side operating conditions:		
Flow rate:	400,000 lb/h	(50.4 kg/s)
inlet temperature	520°R	(289 K)

The exchanger is to have a cross-flow configuration with both fluids unmixed. The heat exchanger surface proposed for the exchanger consists of flattened tubes with continuous aluminum fins, specified as an 11.32-0.737-SR surface in Kays and London [10]. The heat exchanger is shown schematically:

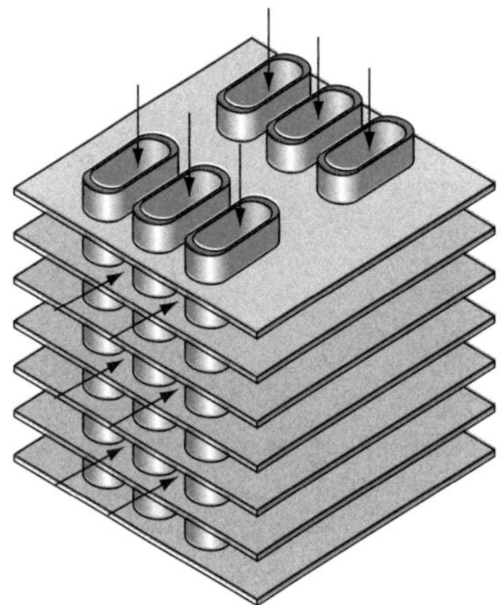

Problem 8.54

The measured heat transfer and friction characteristics for this exchanger surface are shown in the accompanying figure on the next page:

Geometric details for the proposed surface are:

Air side: flow passage hydraulic radius (r_h) = 0.00288 ft (0.0878 cm)
total transfer area/total volume (α_{air}) = 270 ft^2/ft^3 (886 m^2/m^3)
free-flow area/frontal area (σ) = 0.780
fin area/total area (A_f/A) = 0.845
fin metal thickness (t) = 0.00033 ft (0.0001 m)
fin length ($\frac{1}{2}$ distance between tubes, L_f) = 0.225 in. (0.00572 m)

Water side: tubes: specifications given in Problem 8.53
water-side transfer area/total volume (α_{H_2O}) = 42.1 ft^2/ft^3

The design should specify the core size, the air flow frontal area, and the flow length. The water velocity inside the tubes is 4.4 ft/s (1.34 m/s). See Problem 8.53 for the calculation of the water-side heat transfer coefficient.

Notes: (i) The free-flow area is defined such that the mass velocity, G, is the air mass flow rate per unit free-flow area; (ii) the core pressure drop is given by $\Delta p = fG^2L/2\rho r_h$ where L is the length of the core in the air flow direction; (iii) the fin length, L_f, is defined such that $L_f = 2A/P$ where A is the fin cross-sectional area for heat conduction and P is the effective fin perimeter.

8.55 Microchannel compact heat exchangers can be used to cool high heat flux microelectronic devices. The accompanying sketch on the next page shows a schematic view of a typical microchannel heat sink. Microfabrication

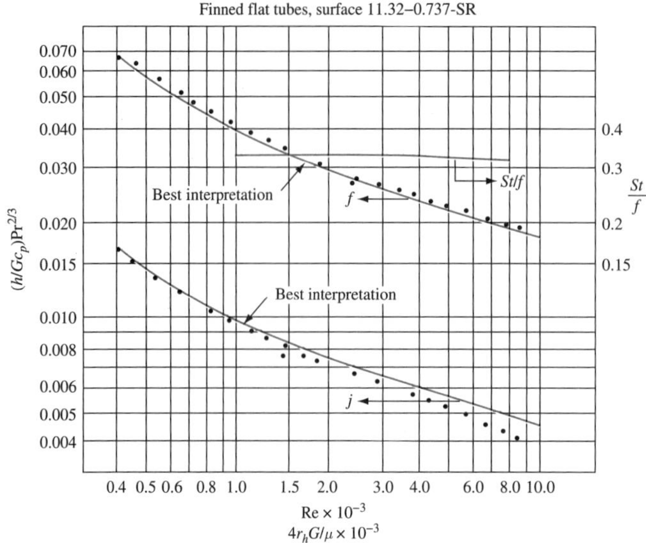

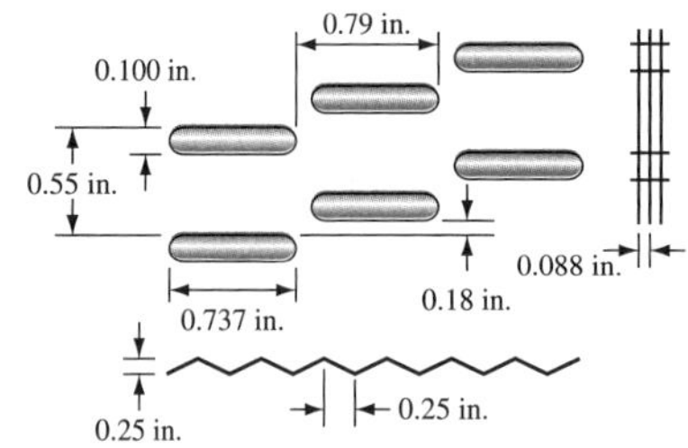

Problem 8.54

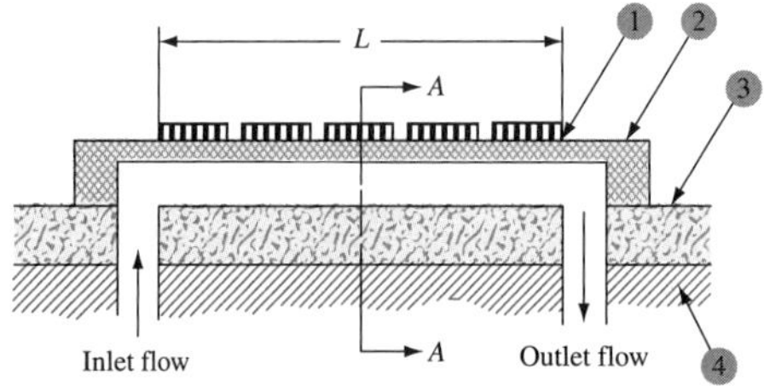

1 IC elements forming surface heat source
2 Microchannel heat sink
3 Cover plate
4 Manifold block

Problem 8.55

techniques can be used to mass produce aluminum channels and fins with the following dimensions:

$w_c = w_w = 50\ \mu m$
$b = 200\ \mu m$
$L = 1.0$ cm
$t = 100\ \mu m$

Assuming that there are a total of 100 fins and that water at 30°C is used as the cooling medium at a Reynolds number of 2000, estimate: (a) the water flow rate through all the channels, (b) the Nusselt number, (c) the heat transfer coefficient, (d) the effective thermal resistance between the IC elements forming the heat source and the cooling water, and (e) the rate of heat dissipation allowable if the temperature difference between the source and the water is not to exceed 100 K.

Design Problems

8.1 **Furnace Efficiency Improvement** (Chapter 8)
It is common practice in industry to recover thermal energy from the flue gas of a furnace. One method of using this thermal energy is to preheat the furnace combustion air with a heat exchanger that transfers heat from the flue gas to the combustion air stream. Design such a heat exchanger assuming that the furnace is fired with natural gas at a rate of 10 MW, uses combustion air at a rate of 90 standard cubic feet per second, and is 75% efficient before heat recovery is employed. Using the first law of thermodynamics, determine the temperature of the flue gas leaving the furnace before the heat exchanger is installed. Then determine the best design for the heat exchanger and calculate the outlet temperatures for both streams. The most important considerations will be capital cost of the heat exchanger, its maintenance costs, and the pressure drop on both the air side and the flue gas side.

8.2 **Condenser for a Steam Turbine** (Chapter 8)
Saturated steam vapor leaves a steam turbine at a mass-flow rate of 2 kg/s and a pressure of 0.5 atm, as shown in the following diagram. Design a heat exchanger to condense the vapor to the saturated liquid state using water at 10°C as the coolant. Use a condensing heat transfer coefficient in the middle range given in Table 1.5. In Chapter 10 you will calculate the condensing heat transfer coefficient.

8.3 **Waste-Heat Recovery** (Chapter 8)
Analyze the effectiveness of a heat exchanger intended to heat water with the flue gas from a combustion chamber as shown in the schematic diagram. The water is flowing through a finned tube, having dimensions shown in the schematic diagram, at a rate of 0.17 kg/s, while the flue gases are flowing through the annulus in the flow channels between the fins at a velocity of 10 m/s. The finned tubes may be constructed from carbon steel or copper. Determine the rate of heat transfer per unit length of tube from the gas to water at a water temperature of 200 K and a flue gas temperature of 700 K. Based on a cost-analysis comparing copper and steel, recommend the appropriate material to be used for this device.

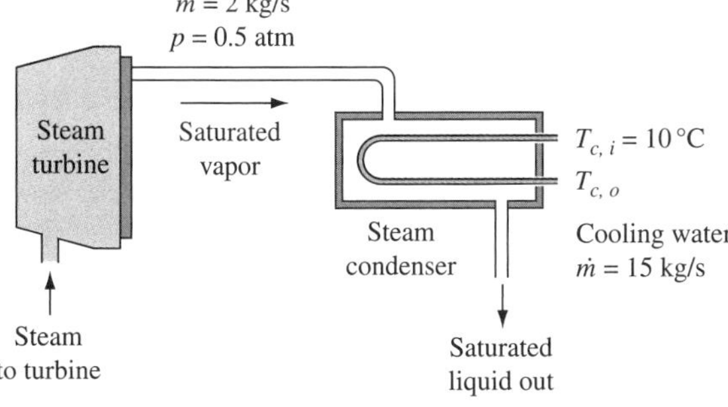

Design Problem 8.2

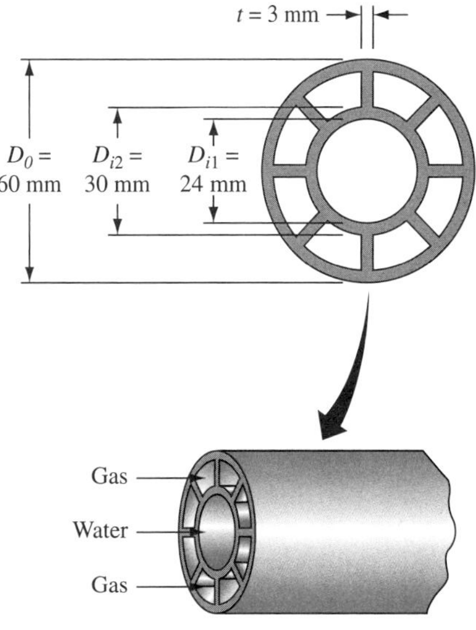

Design Problem 8.3

CHAPTER 9

Heat Transfer by Radiation

A satellite orbiting in space with its solar panels and heat rejecting radiators unfurled. The power generating system on the satellite receives solar energy by radiation and rejects waste heat by radiation on the dark side.

Source: Photo Courtesy of NASA.

Concepts and Analyses to Be Learned

Radiation heat transfer differs from that by convection and conduction because the driving potential is not the temperature, but the absolute temperature raised to the fourth power. Furthermore, heat can be transported by radiation without an intervening medium. Consequently, the integration of radiation heat transfer into an overall thermal analysis presents considerable challenges, including the need for carefully stated boundary conditions and assumptions necessary for the appropriate inclusion in the thermal circuit of a system. A study of this chapter will teach you:

- How to express the dependence of the monochromatic blackbody emissive power on wavelength and absolute temperature.
- How to express the relation between radiation intensity and emissive power.
- How to employ radiation properties such as emissivity, absorptivity, and transmissivity in heat transfer analysis, including their dependence on wavelength.
- How to define and use blackbody and graybody assumptions.
- How to evaluate a radiation shape factor for radiative heat transfer between different surfaces.
- How to set up an equivalent network for radiation in enclosures consisting of several surfaces.
- How to use MATLAB to solve radiation heat transfer problems.
- How to evaluate thermal problems when radiation is combined with convection and conduction.
- How to model the fundamentals of radiation in gaseous media.

9.1 Thermal Radiation

When a body is placed in an enclosure whose walls are at a temperature below that of the body, the temperature of the body will decrease even if the enclosure is evacuated. The process by which heat is transferred from a body by virtue of its temperature, without the aid of any intervening medium, is called *thermal radiation*. This chapter deals with the characteristics of thermal radiation and radiation exchange, that is, heat transfer by radiation.

The physical mechanism of radiation is not completely understood yet. Radiant energy is envisioned sometimes as transported by electromagnetic waves, at other times as transported by photons. Neither viewpoint completely describes the nature of all observed phenomena. It is known, however, that radiation travels with the speed of light c, equal to about 3×10^8 m/s in a vacuum. This speed is equal to the product of the frequency and the wavelength of the radiation, or

$$c = \lambda \nu$$

where λ = wavelength, m

ν = frequency, s^{-1}

The unit of wavelength is the meter, but it is usually more convenient to use the micrometer (μm), equal to 10^{-6} m [1 μm = 10^4Å (angstroms) or 3.94×10^{-5} in. (inches)]. In engineering literature, the micron (equal to a micrometer) is also used and is denoted by the symbol μ.

From the viewpoint of electromagnetic theory, the waves travel at the speed of light, while from the quantum point of view, energy is transported by photons that travel at that speed. Although all the photons have the same velocity, there is always a distribution of energy among them. The energy associated with a photon, e_p, is given by $e_p = h\nu$, where h is Planck's constant, equal to 6.625×10^{-34} J s, and ν is the frequency of the radiation in s^{-1}. The energy spectrum can also be described in terms of wavelength of radiation, λ, which is related to the propagation velocity and the frequency by $\lambda = c/\nu$.

Radiation phenomena are usually classified by their characteristic wavelength (Fig. 9.1). Electromagnetic phenomena encompasses many types of radiation, from short-wavelength gamma-rays and x-rays to long-wavelength radio waves. The wavelength of radiation depends on how the radiation is produced. For example, a metal bombarded by high-frequency electrons emits x-rays, while certain crystals can be excited to emit long-wavelength radio waves. *Thermal radiation* is defined as radiant energy emitted by a medium by virtue of its temperature. In other words, the emission of thermal radiation is governed by the temperature of the emitting body. The wavelength range encompassed by thermal radiation falls approximately between 0.1 and 100 μm. This range is usually subdivided into the ultraviolet, the visible, and the infrared, as shown in Fig. 9.1.

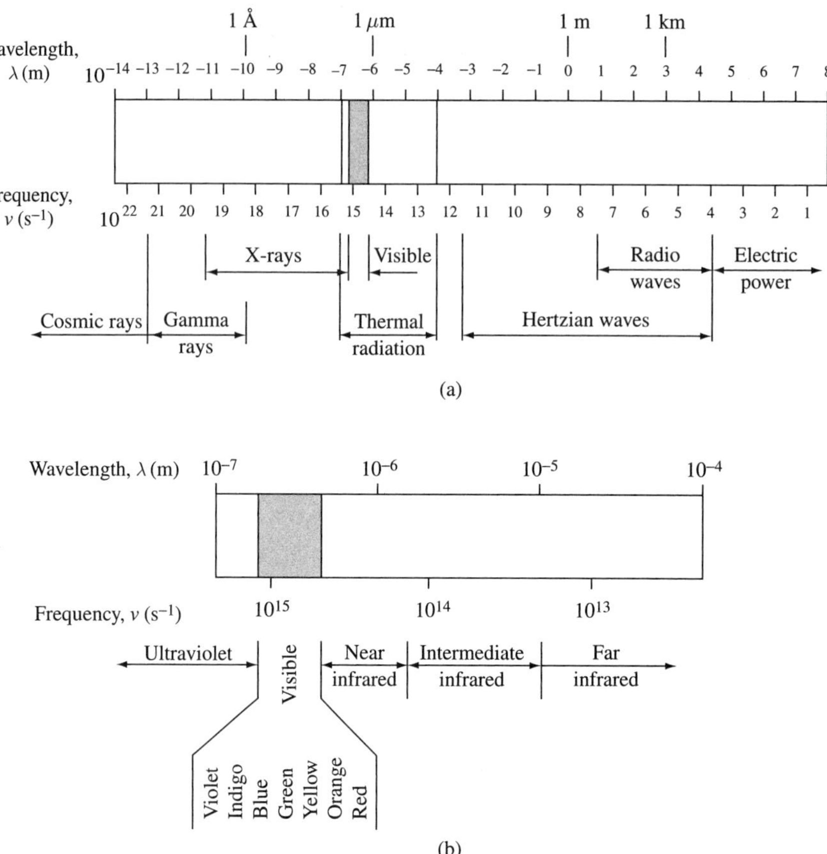

FIGURE 9.1 (a) Electromagnetic spectrum. (b) Thermal radiation portion of the electromagnetic spectrum.

Thermal radiation always encompasses a range of wavelengths. The amount of radiation emitted per unit wavelength is called *monochromatic radiation*; it varies with wavelength, and the word *spectral* is used to denote this dependence. The spectral distribution depends on the temperature and the surface characteristics of the emitting body. The sun, with an effective surface temperature of about 5800 K (10,400°R), emits most of its energy below 3 μm, whereas the earth, at a temperature of about 290 K (520°R), emits over 99% of its radiation at wavelengths longer than 3 μm. The difference in the spectral ranges warms a greenhouse inside even when the outside air is cool because glass permits radiation at the wavelength of the sun to pass, but it is almost opaque to radiation in the wavelength range emitted by the interior of the greenhouse. Thus, most of the solar energy that enters the greenhouse is trapped inside. In recent years, the combustion of fossil fuels has increased the amount of carbon dioxide in the atmosphere. Since carbon dioxide absorbs radiation in the solar spectrum, less energy escapes. This causes global warming, which is also called the "greenhouse effect."

9.2 Blackbody Radiation

A *blackbody*, or ideal radiator, is a body that emits and absorbs at any temperature the maximum possible amount of radiation at any given wavelength. The ideal radiator is a theoretical concept that sets an upper limit to the emission of radiation in accordance with the second law of thermodynamics. It is a standard with which the radiation characteristics of other media are compared.

For laboratory purposes, a blackbody can be approximated by a cavity, such as a hollow sphere, whose interior walls are maintained at a uniform temperature T. If there is a small hole in the wall, any radiation entering through it is partly absorbed and partly reflected at the interior surfaces. The reflected radiation, as shown schematically in Fig. 9.2, will not immediately escape from the cavity but will first repeatedly strike the interior surface. Each time it strikes, a part of it is absorbed; when the original radiation beam finally reaches the hole again and escapes, it has been so weakened by repeated reflection that the energy leaving the cavity is negligible. This is true regardless of the surface and composition of the wall of the cavity. Thus, a small hole in the walls surrounding a large cavity acts like a blackbody because practically all the radiation incident upon the hole is absorbed inside the cavity.

In a similar manner, the radiation emitted by the interior surface of a cavity is absorbed and reflected many times and eventually fills the cavity uniformly. If a blackbody at the same temperature as the interior surface is placed in the cavity, it receives radiation uniformly; that is, it is *irradiated isotropically*. The blackbody absorbs all of the incident radiation, and since the system consisting of the blackbody and the cavity is at a uniform temperature, the rate of emission of radiation by the body must equal its rate of irradiation (otherwise there would be a net transfer of energy as heat between two bodies at the same temperature in an isolated system, an obvious violation of the second law of thermodynamics). Denoting the rate at which radiant energy from the walls of the cavity is incident on the blackbody, that is, the *blackbody irradiation*, by G_b and the rate at which the blackbody emits energy by E_b, we thus obtain $G_b = E_b$. This means that the irradiation in a cavity whose walls

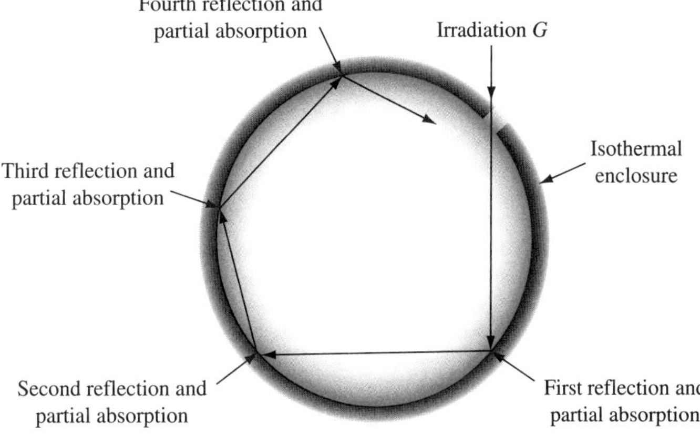

FIGURE 9.2 Schematic diagram of blackbody cavity.

are at a temperature T is equal to the emissive power of a blackbody at the same temperature. A small hole in the wall of a cavity will not disturb this condition appreciably, and the radiation escaping from it will therefore have blackbody characteristics. Since this radiation is independent of the nature of the surface, it follows that the *emissive power of a blackbody depends only on its temperature.*

9.2.1 Blackbody Laws

The spectral radiant energy emission per unit time and per unit area from a blackbody at wavelength λ in the wavelength range $d\lambda$ will be denoted by $E_{b\lambda}\,d\lambda$. The quantity $E_{b\lambda}$ is usually called the *monochromatic blackbody emissive power.* A relationship showing how the emissive power of a blackbody is distributed among the different wavelengths was derived by Max Planck in 1900 through his quantum theory. According to *Planck's law*, an ideal radiator at temperature T emits radiation according to the relation [1]

$$E_{b\lambda}(T) = \frac{C_1}{\lambda^5(e^{C_1/\lambda T} - 1)} \tag{9.1}$$

where $E_{b\lambda}$ = monochromatic emissive power of a blackbody at absolute temperature T, W/m^3(Btu/h ft$^2\mu$)

λ = wavelength, m (μ)

T = absolute temperature of the body, K (degrees °R = 460 + °F)

C_1 = first radiation constant

= 3.7415×10^{-16} W m^2 (1.1870×10^8 Btu/μ^4/h ft^2)

C_2 = second radiation constant
= 1.4388×10^{-2} m K ($2.5896 \times 10^4 \mu$ °R)

The monochromatic emissive power for a blackbody at various temperatures is plotted in Fig. 9.3 as a function of wavelength. Observe that at temperatures below 5800 K the emission of radiation energy is appreciable between 0.2 and about 50 μm. The wavelength at which the monochromatic emissive power is a maximum, $E_{b\lambda}(\lambda_{max}, T)$ decreases with increasing temperature.

The relationship between the wavelength λ_{max} at which $E_{b\lambda}$ is a maximum and the absolute temperature is called *Wien's displacement law* [1]. It can be derived from Planck's law by satisfying the condition for a maximum of $E_{b\lambda}$, or

$$\frac{dE_{b\lambda}}{d\lambda} = \frac{d}{d\lambda}\left[\frac{C_1}{\lambda^5(e^{C_2/\lambda T} - 1)}\right]_{T=\text{const}} = 0$$

The result of this operation is

$$\lambda_{max} T = 2.898 \times 10^{-3}\,\text{m K}\ (5216.4\,\mu\,°\text{R}) \tag{9.2}$$

The visible range of wavelengths, shown as a shaded band in Fig. 9.3, extends over a narrow region from about 0.4 to 0.7 μm. Only a very small amount of the total energy falls in this range of wavelengths at temperatures below 800 K. At higher temperatures, however, the amount of radiant energy within the visible

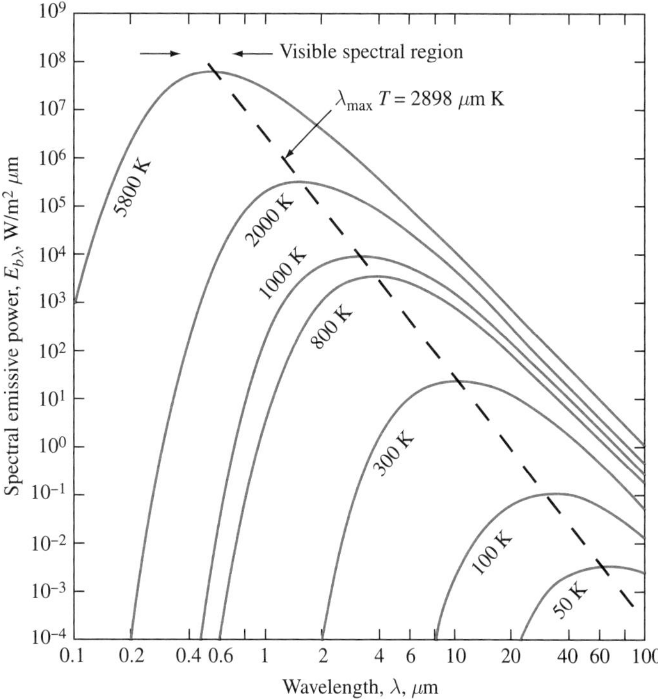

FIGURE 9.3 Monochromatic blackbody emissive power.

range increases and the human eye begins to detect the radiation. The sensation produced on the retina and transmitted to the optic nerve depends on the temperature, a phenomenon that is still used to estimate the temperatures of metals during heat treatment. At about 800 K, an amount of radiant energy sufficient to be observed is emitted at wavelengths between 0.6 and 0.7 μm, and an object at that temperature glows with a dull-red color. As the temperature is further increased, the color changes to bright red and yellow, becoming nearly white at about 1500 K. At the same time, the brightness also increases because more and more of the total radiation falls within the visible range.

Recall from Chapter 1 that the total emission of radiation per unit surface area, per unit time from a blackbody, is related to the fourth power of the absolute temperature according to the *Stefan-Boltzmann law*

$$E_b(T) = \frac{q_r}{A} = \sigma T^4 \tag{9.3}$$

where A = area of the blackbody emitting the radiation, m² (ft²)

T = absolute temperature of the area A in K (°R)

σ = Stefan-Boltzmann constant
$= 5.670 \times 10^{-8}\ \mathrm{W/m^2\ K^4} (0.1714 \times 10^{-8}\ \mathrm{Btu/h\ ft^2\ °R^4})$

The total emissive power given by Eq. (9.3) represents the total thermal radiation emitted over the entire wavelength spectrum. At a given temperature T, the area under a curve such as that shown in Fig. 9.3 is E_b. The total emissive power and the monochromatic emissive power are related by

$$\int_0^\infty E_{b\lambda}\, d\lambda = \sigma T^4 = E_b \tag{9.4}$$

Substituting Eq. (9.1) for $E_b\lambda$ and performing the integration indicated above shows that the Stefan-Boltzmann constant σ and the constants C_1 and C_2 in Planck's law are related by

$$\sigma = \left(\frac{\pi}{C_2}\right)^4 \frac{C_1}{15} = 5.670 \times 10^{-8}\ \text{W/m}^2\ \text{K}^4 \tag{9.5}$$

The Stefan-Boltzmann law shows that under most circumstances the effects of radiation are insignificant at low temperatures, owing to the small value for σ. At room temperature (~300 K) the total emissive power of a black surface is approximately 460 W/m^2. This value is only about one-tenth of the heat flux transferred from a surface to a fluid by convection, even when the convection heat transfer coefficient and temperature difference are reasonably low values of 100 W/m^2 K and 50 K, respectively. Therefore, at low temperatures we can often neglect radiation effects; however, we must include radiation effects at high temperatures because the emissive power increases with the fourth power of the absolute temperature.

9.2.2 Radiation Functions and Band Emission

For engineering calculations involving real surfaces it is often important to know the energy radiated at a specified wavelength or in a finite band between specific wavelengths λ_1 and λ_2, that is, $\int_{\lambda_1}^{\lambda_2} E_{b\lambda}(T)\, d\lambda$. Numerical calculations for such cases are facilitated by the use of the *radiation functions* [2]. The derivation of these functions and their application are illustrated below.

At any given temperature, the monochromatic emissive power is a maximum at the wavelength $\lambda_{\max} = 2.898 \times 10^{-3}/T$, according to Eq. (9.2). Substituting $\lambda_{\max}$ into Eq. (9.1) gives the maximum monochromatic emissive power at temperature T, $E_{b\lambda\max}(T)$, or

$$E_{b\lambda\max}(T) = \frac{C_1 T^5}{(0.002898)^5(e^{C_2/0.002898} - 1)} = 12.87 \times 10^{-6} T^5\ \text{W/m}^3 \tag{9.6}$$

If we divide the monochromatic emissive power of a blackbody, $E_{b\lambda}(T)$, by its maximum emissive power at the same temperature, $E_{b\lambda\max}(T)$, we obtain the dimensionless ratio

$$\frac{E_{b\lambda}(T)}{E_{b\lambda\max}(T)} = \left(\frac{2.898 \times 10^{-3}}{\lambda T}\right)^5 \left(\frac{e^{4.965} - 1}{e^{0.014388/\lambda T} - 1}\right) \tag{9.7}$$

where λ is in micrometers and T is in kelvin.

Observe that the right-hand side of Eq. (9.7) is a unique function of the product λT. To determine the monochromatic emissive power $E_{b\lambda}$ for a blackbody at given values of λ and T, evaluate $E_{b\lambda}/E_{b\lambda\max}$ from Eq. (9.7) and $E_{b\lambda\max}$ from Eq. (9.6) and multiply.

EXAMPLE 9.1 Determine (a) the wavelength at which the monochromatic emissive power of a tungsten filament at 1400 K is a maximum, (b) the monochromatic emissive power at that wavelength, and (c) the monochromatic emissive power at 5 μm.

SOLUTION From Eq. (9.2), the wavelength at which the emissive power is a maximum is

$$\lambda_{\max} = 2.898 \times 10^{-3}/1400 = 2.07 \times 10^{-6}\,\text{m}$$

From Eq. (9.6) at $T = 1400\,\text{K}$,

$$E_{b\lambda\max} = 12.87 \times 10^{-6} \times (1400)^5 = 6.92 \times 10^{10}\,\text{W/m}^3$$

At $\lambda = 5\,\mu\text{m}$, $\lambda T = 5 \times 1400 = 7.0 \times 10^3\,\text{mK}$; substituting this value into Eq. (9.7) we get

$$\frac{E_{b\lambda}(1400)}{E_{b\lambda\max}(1400)} = \left(\frac{2.898 \times 10^{-3}}{7.0 \times 10^{-3}}\right)^5\left(\frac{e^{4.965} - 1}{e^{0.014388/\lambda T} - 1}\right)$$

$$= (0.1216)\left(\frac{e^{4.965} - 1}{e^{2.055} - 1}\right) = 0.254$$

Thus, $E_{b\lambda}$ at 5 μm is 25.4% of the maximum value $E_{b\lambda\max}$, or $1.758 \times 10^{10}\,\text{W/m}^3$.

It is often necessary to determine the fraction of the total blackbody emission in a spectral band between wavelengths λ_1 and λ_2. To obtain the emission in a band, as shown in Fig. 9.4 by the shaded area, we must first calculate $E_b(0 - \lambda_1, T)$, the blackbody emission in the interval from 0 to λ_1 at T, or

$$\int_0^{\lambda_1} E_{b\lambda}(T)\,d\lambda = E_b(0 - \lambda_1, T) \tag{9.8}$$

This expression can be recast in a dimensionless form as a function of λT, the product of wavelength and temperature.

$$\frac{E_b(0 - \lambda_1 T)}{\sigma T^4} = \int_0^{\lambda_1 T} \frac{E_{b\lambda}}{\sigma T^5}\,d(\lambda T) \tag{9.9}$$

From Eqs. (9.6) and (9.7), the integrand in Eq. (9.9) is a function of λT only, and therefore Eq. (9.9) can be integrated between specified limits. The fraction of the

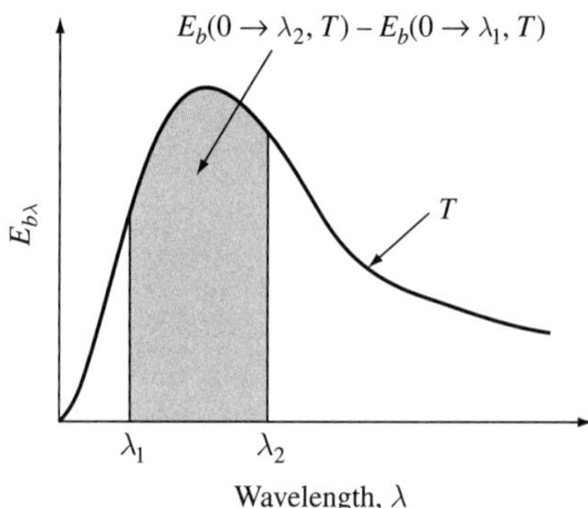

FIGURE 9.4 Radiation band and radiation function.

total blackbody emission between 0 and a given value of λ is presented in Fig. 9.5 and Table 9.1 as a universal function of λT.

To determine the amount of radiation emitted in the band between λ_1 and λ_2 for a black surface at temperature T, we evaluate the difference between the two integrals below

$$\int_0^{\lambda_2} E_{b\lambda}(T)\, d\lambda - \int_0^{\lambda_1} E_{b\lambda}(T)\, d\lambda = E_b(0 - \lambda_2 T) - E_b(0 - \lambda_1 T) \quad (9.10)$$

The procedure is illustrated in the following example.

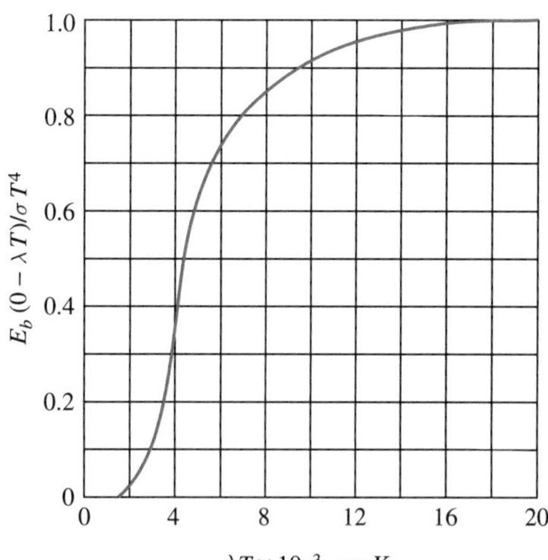

FIGURE 9.5 Ratio of blackbody emission between 0 and λ to the total emission, $E_b(0 - \lambda T)/\sigma T^4$ versus λT.

TABLE 9.1 Blackbody radiation functions

λT (mK $\times 10^3$)	$\dfrac{E_b(0-\lambda T)}{\sigma T^4}$	λT (mK $\times 10^3$)	$\dfrac{E_b(0-\lambda T)}{\sigma T^4}$
0.2	0.341796×10^{-26}	6.2	0.754187
0.4	0.186468×1^{-11}	6.4	0.769234
0.6	0.929299×10^{-7}	6.6	0.783248
0.8	0.164351×10^{-4}	6.8	0.796180
1.0	0.320780×10^{-3}	7.0	0.808160
1.2	0.213431×10^{-2}	7.2	0.819270
1.4	0.779084×10^{-2}	7.4	0.829580
1.6	0.197204×10^{-1}	7.6	0.839157
1.8	0.393449×10^{-1}	7.8	0.848060
2.0	0.667347×10^{-1}	8.0	0.856344
2.2	0.100897	8.5	0.874666
2.4	0.140268	9.0	0.890090
2.6	0.183135	9.5	0.903147
2.8	0.227908	10.0	0.914263
3.0	0.273252	10.5	0.923775
3.2	0.318124	11.0	0.931956
3.4	0.361760	11.5	0.939027
3.6	0.403633	12	0.945167
3.8	0.443411	13	0.955210
4.0	0.480907	14	0.962970
4.2	0.516046	15	0.969056
4.4	0.548830	16	0.973890
4.6	0.579316	18	0.980939
4.8	0.607597	20	0.985683
5.0	0.633786	25	0.992299
5.2	0.658011	30	0.995427
5.4	0.680402	40	0.998057
5.6	0.701090	50	0.999045
5.8	0.720203	75	0.999807
6.0	0.737864	100	1.000000

EXAMPLE 9.2 Silica glass transmits 92% of the incident radiation in the wavelength range between 0.35 and 2.7 μm and is opaque at longer and shorter wavelengths. Estimate the percentage of solar radiation that the glass will transmit. The sun can be assumed to radiate as a blackbody at 5800 K.

SOLUTION For the wavelength range within which the glass is transparent, $\lambda T = 2030\,\mu$m K at the lower limit and 15,660 μm K at the upper limit. From Table 9.1 we find

$$\frac{\int_0^{2030} E_{b\lambda}\, d\lambda}{\int_0^{\infty} E_{b\lambda}\, d\lambda} = 6.7\%$$

and

$$\frac{\int_0^{15,660} E_{b\lambda}\, d\lambda}{\int_0^{\infty} E_{b\lambda}\, d\lambda} = 97.0\%$$

Thus, 90.3% of the total radiant energy incident upon the glass from the sun is in the wavelength range between 0.35 and 2.7 μm, and 83.1% of the solar radiation is transmitted through the glass.

9.2.3 Intensity of Radiation

In our discussion so far, we have only considered the total amount of radiation leaving a surface, that is, the emissive power. The concept, however, is inadequate for a heat transfer analysis when the amount of radiation passing in a given direction and intercepted by some other body is sought. The amount of radiation passing in a given direction is described in terms of the intensity of radiation, I. Before defining the intensity of radiation, we must have measures of the direction and the space into which a body radiates. As shown in Fig. 9.6(a), a differential plane angle $d\alpha$ is defined as the ratio of an element of arc length dl on a circle to the radius r of that circle. Similarly, a *differential solid angle* $d\omega$, as defined in Fig. 9.6(b), is the ratio of the element of area dA_n on a sphere to the square of the radius of sphere, or

$$d\omega = \frac{dA_n}{r^2} \tag{9.11}$$

The unit of the solid angle is the *steradian* (sr).

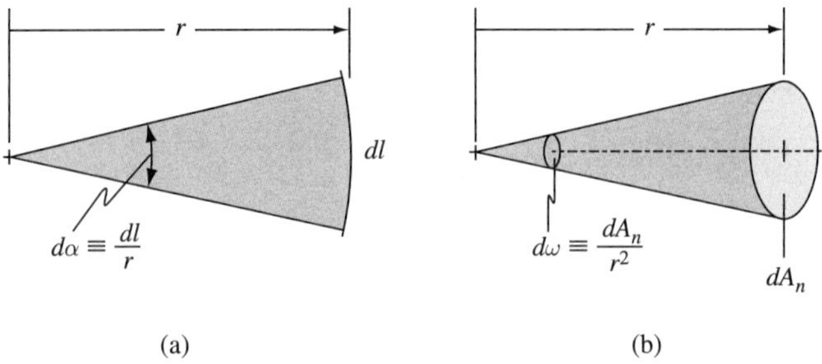

FIGURE 9.6 (a) Differential plane angle and (b) differential solid angle.

The rate of radiation heat flow per unit surface area emanating from a body and passing in a given direction can be measured by determining the radiation through an element on the surface of a hemisphere constructed around the radiating surface. If the radius of this hemisphere equals unity, the hemisphere has a surface area of 2π and subtends a solid angle of 2π steradians, or sr, about a point at the center of its base. The surface area on such a hemisphere with a radius of unity has the same numerical value as the so-called *solid angle* ω measured from the radiating surface element. The solid angle can be used to define simultaneously, the direction and the space into which radiation from a body propagates.

The *intensity of radiation* $I(\theta, \phi)$ is the energy emitted per unit area of emitting surface projected in the direction θ, ϕ per unit time into a solid angle $d\omega$ centered on a direction that can be defined in terms of the zenith angle θ, and the azimuthal angle ϕ in the spherical coordinate system of Fig. 9.7. The differential area dA_n in Fig. 9.7 is perpendicular to the (θ, ϕ) direction. But for a spherical surface, $dA_n = r\,d\theta\, r\ \sin\ d\phi$, and therefore

$$d\omega = \sin\theta\, d\theta\, d\phi \tag{9.12}$$

With the above definitions, the intensity of radiation $I(\theta, \phi)$ is the rate at which radiation is emitted in the direction (θ, ϕ) per unit area of the emitting surface normal to this direction, per unit solid angle centered about (θ, ϕ).

Since the projected area of emission from Fig. 9.7 is $dA_1 \cos\theta$, we obtain for the intensity of a black surface, $I_b(\theta, \phi)$,

$$I_b(\theta, \phi) = \frac{dq_r}{dA_1 \cos\theta\, d\omega}\ (\text{W/m}^2\ \text{sr}) \tag{9.13}$$

where dq_r is the rate at which radiation emitted from dA_1 passes through dA_n.

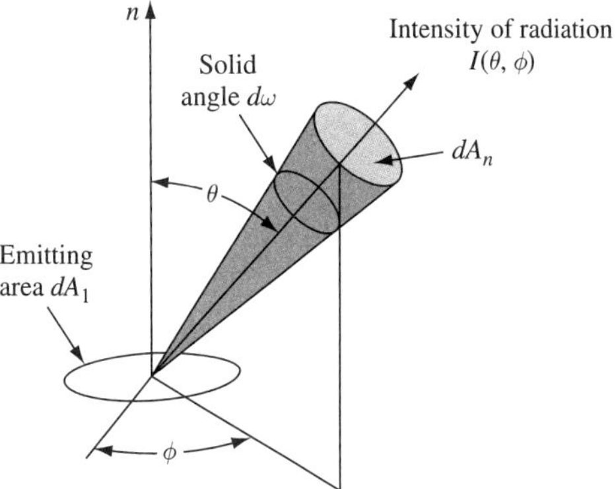

FIGURE 9.7 Schematic diagram illustrating intensity of radiation.

EXAMPLE 9.3 A flat, black surface of area $A_1 = 10\,\text{cm}^2$ emits 1000 W/m^2 sr in the normal direction. A small surface A_2 having the same area as A_1 is placed relative to A_1 as shown in Fig. 9.8, at a distance of 0.5 m. Determine the solid angle subtended by A_2 and the rate at which A_2 is irradiated by A_1.

SOLUTION Since A_1 is black, it is a diffuse emitter and its intensity I_b is independent of direction. Moreover, since both areas are quite small, they can be approximated as differential surface areas and the solid angle can be calculated from Eq. (9.11) or $d\omega_{2-1} = dA_{n,2}/r^2$.

The area $dA_{n,2}$ is the projection of A_2 in the direction *normal* to the incident radiation for dA_1, or $dA_{n,2} = dA_2 \cos\theta_2$, where θ_2 is the angle between the normal n_2 and the radiation ray connecting dA_1 and dA_2, that is, $\theta_2 = 30°$. Thus

$$d\omega_{2-1} = \frac{A_2 \cos\theta_2}{r^2} = \frac{10^{-3}\,\text{m}^2 \cos 30°}{(0.5\,\text{m})^2} = 0.00346\,\text{sr}$$

The irradiation of A_2 by A_1, $q_{r,1\to2}$, is

$$\begin{aligned} q_{r,1\to2} &= I_1 A_1 \cos\theta_1\, d\omega_{2-1} \\ &= \left(1000\,\frac{\text{W}}{\text{m}^2\,\text{sr}}\right)(10^{-3}\,\text{m}^2)(\cos 60°)(0.00346\,\text{sr}) = 0.00173\,\text{W} \end{aligned}$$

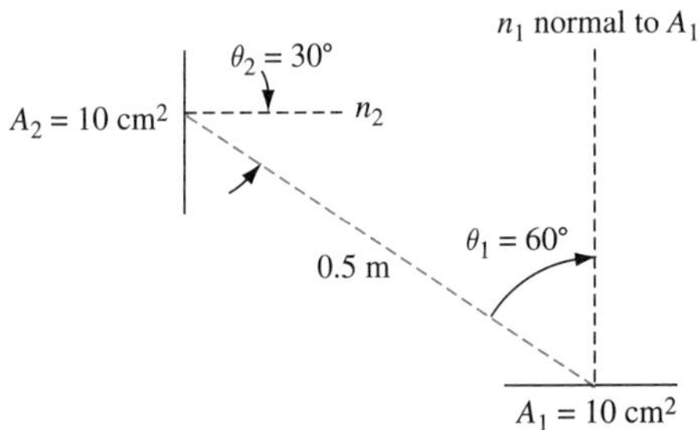

FIGURE 9.8 Sketch showing relation between A_1 and A_2 for Example 9.3.

9.2.4 Relation Between Intensity and Emissive Power

To relate the intensity of radiation to the emissive power, one simply determines the energy from a surface radiating into a hemispherical enclosure placed above it, as shown in Fig. 9.9. Since the hemisphere will intercept all the radiant rays emanating from the surface, the total amount of radiation passing through the hemispherical surface equals the emissive power. From Eq. (9.13), the rate of radiation emitted from dA_1 passing through dA_n is

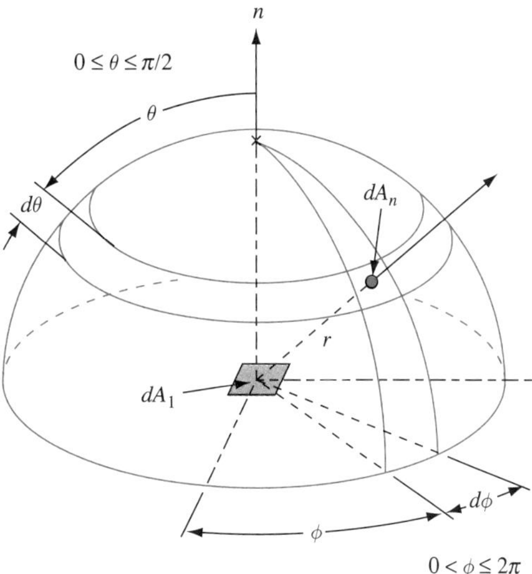

FIGURE 9.9 Radiation from a differential area dA_1 into surrounding hemisphere centered at dA_1.

$$\frac{dq_r}{dA_1} = I_b(\theta, \phi) \cos \theta \, d\omega \tag{9.14}$$

Substituting Eq. (9.12) for the solid angle $d\omega$ and integrating over the entire hemisphere yields the total rate of radiant emission per unit area, called the emissive power:

$$\left(\frac{q}{A}\right)r = \int_0^{2\pi} \int_0^{\pi/2} I_b(\theta, \phi) \cos \theta \sin \theta \, d\theta \, d\phi \tag{9.15}$$

In order to integrate Eq. (9.15), the variation of the intensity with θ and ϕ must be known. As will be discussed more fully in the next section, the intensity of real surfaces exhibits no appreciable variation with ϕ but does vary with θ. Although this variation can be taken into account, for most engineering calculations it can be assumed that the surface is diffuse and the intensity is uniform in all angular directions. Blackbody radiation is actually perfectly diffuse, and radiation from industrial rough surfaces approaches diffuse characteristics. If the intensity from a surface is independent of direction, it is said to conform to *Lambert's cosine law*. For a black surface, integration of Eq. (9.15) yields the *blackbody emissive power* E_b.

$$\left(\frac{q}{A}\right)_r = E_b = \pi I_b \tag{9.16}$$

Thus for a black surface, the emissive power equals π times the intensity. The same relation between emissive power and intensity holds for any surface that conforms to Lambert's cosine law.

The concept of intensity can be applied to the total radiation over the entire wavelength spectrum as well as to monochromatic radiation. The relation between the total and the monochromatic intensity I_λ is simply

$$I(\phi, \theta) = \int_0^\infty I_\lambda(\phi, \theta)\, d\lambda \tag{9.17}$$

If a surface radiates diffusely, it is also apparent that

$$E_\lambda = \pi I_\lambda \tag{9.18}$$

since I_λ is uniform in all directions.

9.2.5 Irradiation

To make a heat balance on a body, we need to know that not only is the radiation leaving but that the radiation incident is on the surface. This radiation orginates from emission and reflection occurring at other surfaces and will in general have a specific directional and spectral distribution. As shown in Fig. 9.10, the incident radiation can be characterized in terms of the incident spectral intensity, $I_{\lambda,i}$, defined as the rate at which radiant energy at wavelength λ impinges from direction (θ, ϕ) per unit area of the intercepting surface normal to this direction, per unit solid angle about the direction (θ, ϕ), per unit wavelength interval $d\lambda$ at λ. The term *irradiation* denotes the radiation incident from all directions on a surface. *Spectral irradiation*, G_λ(W/m^2 μm) is defined as the rate at which monochromatic radiation at wavelength λ is incident on a surface per unit area of that surface, or

$$G_\lambda = \int_0^{2\pi} \int_0^{\pi/2} I_{\lambda,i}(\lambda, \theta, \phi) \cos\theta \sin\theta \, d\theta \, d\phi \tag{9.19a}$$

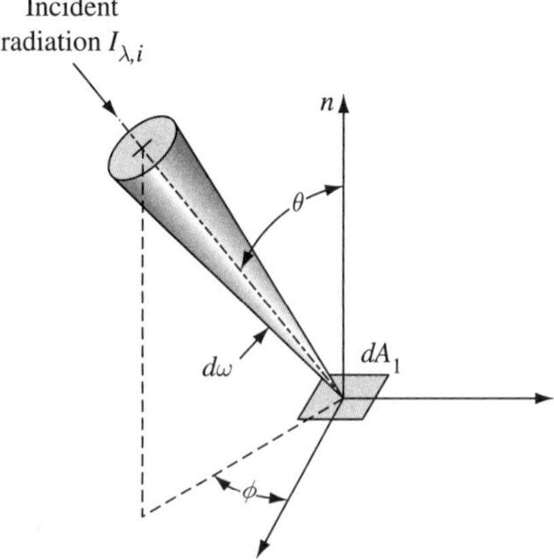

FIGURE 9.10 Radiation incident on a differential area dA_1 in a spherical coordinate system.

where $\sin\theta\, d\theta\, d\phi$ is the unit solid angle. Observe that the factor $\cos\theta$ originates from the fact that G_λ is a flux based on the actual surface area, whereas $I_{\lambda,i}$ is defined in terms of the projected area. The total irradiation represents the rate of radiation incident per unit area from all directions over all wavelengths and is given by

$$G = \int_0^\infty G_\lambda(\lambda)\, d\lambda = \int_0^\infty \int_0^{2\pi} \int_0^{\pi/2} I_{\lambda,i}(\lambda, \theta, \phi) \cos\theta \sin\theta\, d\theta\, d\phi\, d\lambda \quad (9.19b)$$

If the incident radiation is diffuse—that is, if the intercepting area is diffusely irradiated and $I_{\lambda,i}$ is independent of direction—it follows that

$$G = \pi I_i \quad (9.20)$$

9.3 Radiation Properties

Most surfaces encountered in engineering practice do not behave like blackbodies. To characterize the radiation properties of nonblack surfaces, dimensionless quantities such as the emissivity, absorptivity, and transmissivity are used to relate the emitting, absorbing, and transmitting capabilities of a real surface to those of a blackbody. Radiation properties of real surfaces are functions of wavelength, temperature, and direction. The properties that describe how a surface behaves as a function of wavelength are called monochromatic or spectral properties, and the properties that describe the distribution of radiation with angular direction are called directional properties. For precise heat transfer calculation, we must know the relative properties of the emitting surface as well as those of all other surfaces with which radiation exchange occurs.

Taking into account the spectral and directional properties of all surfaces, even if they are known, results in complex and involved analyses that can be carried out only by computer. However, engineering calculations of acceptable accuracy can usually be carried out by a simplified approach, using a single radiation property value averaged over the direction and wavelength range of interest. Radiation properties that are averaged over all wavelengths and directions are called total properties. Although we will almost exclusively use total radiation properties here, it is important to be aware of the spectral and directional characteristics of surfaces in order to account for them in problems in which these variations are significant. In this section, we will discuss radiation properties in order of increasing complexity, beginning with total properties, followed by spectral properties, and finally directional properties.

9.3.1 Radiation Properties

For most engineering calculations, total radiation properties as defined in this subsection are sufficiently accurate. The definition of total radiation properties is illustrated in Fig. 9.11. When radiation is incident on a surface at rate G, a portion of the total irradiation is absorbed in the material, a portion is reflected from the surface, and the remainder is transmitted through the body. The absorptivity, reflectivity, and transmissivity describe how the total irradiation is distributed.

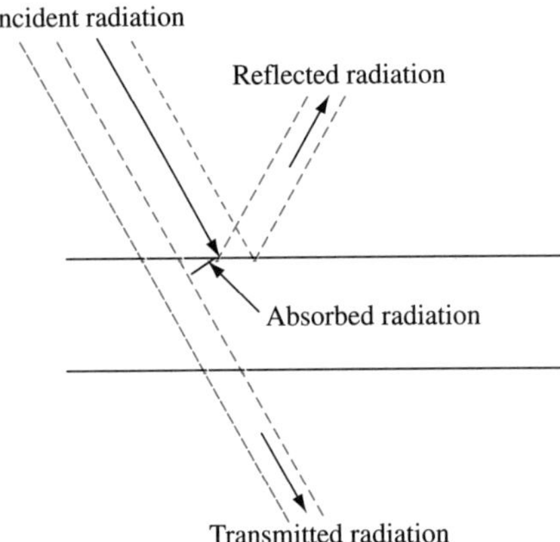

FIGURE 9.11 Schematic diagram illustrating incident, reflected, and absorbed radiation in terms of total radiation properties.

The *absorptivity* α of a surface is the fraction of the total irradiation absorbed by the body. The *reflectivity* ρ of a surface is defined as the fraction of the irradiation that is reflected from the surface. The *transmissivity* τ of a body is the fraction of the incident radiation that is transmitted. If an energy balance is made on a surface, as illustrated in Fig. 9.11, one obtains

$$\alpha G + \rho G + \tau G = G \tag{9.21}$$

From Eq. (9.21), it is apparent that the sum of the absorptivity, reflectivity, and transmissivity must equal unity:

$$\alpha + \rho + \tau = 1 \tag{9.22}$$

When a body is opaque it will not transmit any of the incident radiation, that is, $\tau = 0$. For an opaque body, Eq. (9.22) reduces to

$$\alpha + \rho = 1 \tag{9.23}$$

If a surface is also a perfect reflector, from which all irradiation is reflected, ρ is unity and the transmissivity as well as the absorptivity is zero. A good mirror approaches a reflectivity of 1. As mentioned previously, a blackbody absorbs all of the irradiation and therefore has an absorptivity equal to unity and a reflectivity equal to zero.

Another important total radiation property of real surfaces is the emissivity. The *emissivity* of a surface, ε, is defined as the total radiation emitted divided by the total radiation that would be emitted by a blackbody at the same temperature, or

$$\varepsilon = \frac{E(T)}{E_b(T)} = \frac{E(T)}{\sigma T^4} \tag{9.24}$$

Since a blackbody emits the maximum possible radiation at a given temperature, the emissivity of a surface is always between zero and unity. But when a surface is black, $E(T) = E_b(T)$ and $\varepsilon_b = \alpha_b = 1.0$.

9.3.2 Monochromatic Radiation Properties and Kirchhoff's Law

Total radiation properties can be obtained from monochromatic properties, which apply only at a single wavelength. Designating E_λ as the monochromatic emissive power of an arbitrary surface, the monochromatic hemispherical emissivity of the surface, ε_λ, is given by

$$\varepsilon_\lambda = \frac{E_\lambda(T)}{E_{b\lambda}(T)} \tag{9.25}$$

In other words, ε_λ is the fraction of the blackbody radiation emitted by the surface at wavelength λ. Similarly, the hemispherical monochromatic absorptivity of a surface, α_λ, is defined as the fraction of the total irradiation at wavelength λ that is absorbed by the surface,

$$\alpha_\lambda = \frac{G_{\lambda,\text{absorbed}}(T)}{G_\lambda(T)} \tag{9.26}$$

An energy balance on a monochromatic basis, similar to Eq. (9.22), yields

$$\alpha_\lambda + \rho_\lambda + \tau_\lambda = 1 \tag{9.27}$$

An important relation between ε_λ and α_λ can be obtained from *Kirchhoff's radiation law*, which states in essence that the monochromatic emissivity is equal to the monochromatic absorptivity for any surface. A rigorous derivation of this law has been presented by Planck [1], but the essential features can be illustrated more simply from the following consideration. Suppose we place a small body inside a black enclosure whose walls are fixed at temperature T (see Fig. 9.12). After thermal

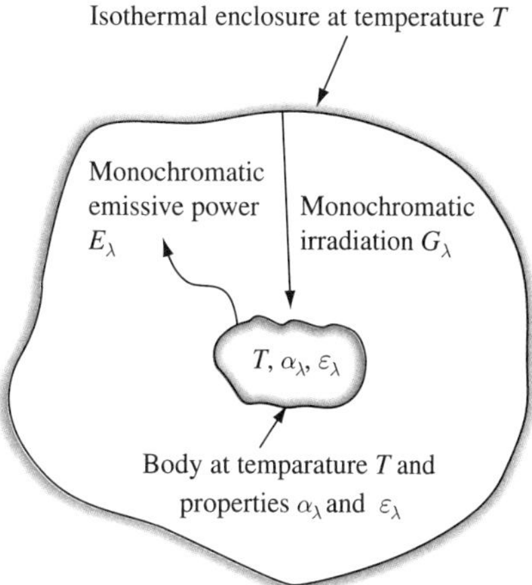

FIGURE 9.12 Radiation emitted and received at wavelength λ by a body in an isothermal enclosure at temperature T.

equilibrium is established, the body must attain the temperature of the walls. In accordance with the second law of thermodynamics, the body must, under these conditions, emit at every wavelength as much radiation as it absorbs. If the monochromatic radiation per unit time, per unit area incident on the body is $G_{b\lambda}$, the equilibrium condition is expressed by

$$E_\lambda = \alpha_\lambda G_{b\lambda} \tag{9.28}$$

or

$$\frac{E_\lambda}{\alpha_\lambda} = G_{b\lambda} \tag{9.29}$$

But since the incident radiation depends only on the temperature of the enclosure, it would be the same on any other body in thermal equilibrium with the enclosure, irrespective of the absorptivity of the body's surface. One can therefore conclude that the ratio of the monochromatic emissive power to the absorptivity at any given wavelength is the same for all bodies at thermal equilibrium. Since the absorptivity must always be less than unity and can be equal to one only for a perfect absorber, that is, a blackbody, Eq. (9.29) shows also that at any given temperature, the emissive power is a maximum for a blackbody. Thus, when $\alpha_\lambda = 1$, $E_\lambda = E_{b\lambda}$ and $G_{b\lambda} = E_{b\lambda}$ in Eq. (9.29). Replacing E_λ by $\varepsilon_\lambda E_{b\lambda}$ in Eq. (9.28) gives

$$\varepsilon_\lambda E_{b\lambda} = \alpha_\lambda G_{b\lambda} = \alpha_\lambda E_{b\lambda}$$

which shows that at any wavelength λ at temperature T,

$$\varepsilon_\lambda(\lambda, T) = \alpha_\lambda(\lambda, T) \tag{9.30}$$

as stated at the outset.

Although the above relation was derived under the condition that the body is in equilibrium with its surroundings, it is actually a general relation that applies under any conditions because both α_λ and ε_λ are surface properties that depend solely on the condition of the surface and its temperature. We can therefore conclude that unless changes in temperature cause physical alteration in the surface characteristics, the hemispherical monochromatic absorptivity equals the monochromatic emissivity of a surface.

The total hemispherical emissivity for a nonblack surface is obtained from Eqs. (9.4) and (9.25). Combining these two relations, we find that at a given temperature T the total hemispherical emittance is

$$\varepsilon(T) = \frac{E(T)}{E_b(T)} = \frac{\int_0^\infty \varepsilon_\lambda(\lambda) E_{b\lambda}(\lambda, T)\, d\lambda}{\int_0^\infty E_{b\lambda}(\lambda, T)\, d\lambda} \tag{9.31}$$

This relation shows that when the monochromatic emissivity of a surface is a function of wavelength, it will vary with the temperature of the surface, even though the monochromatic emissivity is solely a surface property. The reason for this variation is that the percentage of the total radiation that falls within a given wavelength band depends on the temperature of the emitting surface.

EXAMPLE 9.4 The hemispherical emissivity of an aluminum paint is approximately 0.4 at wavelengths below 3 μm and 0.8 at longer wavelengths, as shown in Fig. 9.13. Determine the total emissivity of this surface at a room temperature of 27°C and at a temperature of 527°C. Why are the two values different?

SOLUTION At room temperature the product λT at which the emissivity changes is equal to $3\,\mu\text{m} \times (27 + 273)\,\text{K} = 900\,\mu\text{m K}$, while at the elevated temperature $\lambda T = 2700\,\mu\text{m K}$. From Table 9.1 we obtain

$$\frac{E_b(0 \to \lambda T)}{\sigma T^4} \cong 0.0001 \quad \text{for} \quad \lambda T = 900\,\mu\text{m K}$$

$$\frac{E_b(0 \to \lambda T)}{\sigma T^4} = 0.140 \quad \text{for} \quad \lambda T = 2400\,\mu\text{m K}$$

Thus, the emissivity at 27°C is essentially equal to 0.8, while at 527°C Eq. (9.31) gives

$$\varepsilon = \frac{\int_0^{\lambda_1} \varepsilon_\lambda(\lambda) E_{b\lambda}(\lambda T)\, d\lambda + \int_{\lambda_i}^{\infty} \varepsilon_\lambda(\lambda) E_{b\lambda}(\lambda T)\, d\lambda}{\int_0^{\infty} E_{b\lambda}(\lambda T)\, d\lambda}$$

$$= (0.4)(0.14) + (0.8)(086) = 0.744$$

The reason for the difference in the total emissivity is that at the higher temperature, the percentage of the total emissive power in the low-emittance region of the paint is appreciable, while at the lower temperature practically all the radiation is emitted at wavelengths above 3 μm.

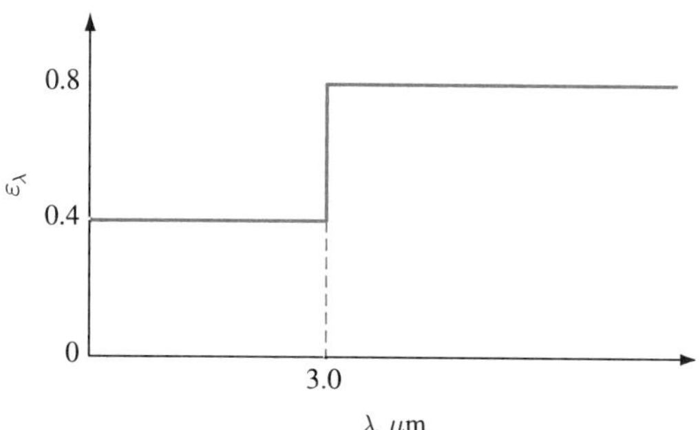

FIGURE 9.13 Spectral emissivity for paint in Example 9.4.

Similarly, the *total absorptivity* of a surface can be obtained from basic definitions. Consider a surface at temperature T subject to incident radiation from a source at T^* given by

$$G = \int_0^{\infty} G_\lambda(\lambda^*, T^*)\, d\lambda \tag{9.32}$$

where the asterisk is used to denote the conditions of the source. If the variation of the monochromatic absorptivity with wavelength of the receiving surface is given by $\alpha_\lambda(\lambda)$, the total absorptivity is

$$\alpha(\lambda^*, T^*) = \frac{\int_0^\infty \alpha_\lambda(\lambda) G_\lambda(\lambda^*, T^*)\, d\lambda}{\int_0^\infty G_\lambda(\lambda^*, T^*)\, d\lambda} \tag{9.33}$$

Note that the total absorptivity of a surface depends on the temperature and on the spectral characteristics of the incident radiation. Therefore, although the relation $\varepsilon_\lambda = \alpha_\lambda$ is always valid, the total values of absorptivity and emissivity are, in general, different for real surfaces.

9.3.3 Graybodies

Graybodies are surfaces with monochromatic emissivities and absorptivities whose values are independent of wavelength. Even though real surfaces do not meet this specification exactly, it is often possible to choose suitable average values for the emissivity and absorptivity, $\bar{\varepsilon}$ and $\bar{\alpha}$, to make the graybody assumption acceptable for engineering analysis. For a completely graybody, with the subscript g denoting gray,

$$\varepsilon_\lambda = \bar{\varepsilon} = \bar{\alpha} = \alpha_\lambda = \varepsilon_g = \alpha_g$$

The emissive power E_g is given by

$$E_g = \varepsilon_g \sigma T^4 \tag{9.34}$$

Thus, if the emissivity of a graybody is known at one wavelength, the total emissivity and the total absorptivity are also known. Moreover, the total values of absorptivity and emissivity are equal even if the body is not in thermal equilibrium with its surroundings. In practice, however, the choice of suitable average values should reflect the conditions of the source for the average absorptivity and the temperature of the surface of the body that receives and emits radiation for the choice of the average emissivity. A surface that is idealized as having uniform properties, but whose average emissivity is not equal to the average absorptivity, is called a selectively graybody.

EXAMPLE 9.5 The aluminum paint from Example 9.4 is used to cover the surface of a body that is maintained at 27°C. In one installation, this body is irradiated by the sun, in another by a source at 527°C. Calculate the effective absorptivity of the surface for both conditions, assuming the sun is a blackbody at 5800 K.

SOLUTION For the case of solar irradiation, we find from Table 9.1 for $\lambda T = 3\,\mu\text{m} \times 5800$ $K = 17{,}400\,\mu m K = 17.4 \times 10^{-3}$ mK that

$$\frac{E_b(0 \rightarrow \lambda T)}{\sigma T^4} = 0.98$$

This means that 98% of the solar radiation falls below 3 μm and the effective absorptivity is, from Eq. (9.33),

$$\alpha(\lambda_{\text{sun}}, T_{\text{sun}}) = \left(\int_0^{3\mu\text{m}} \alpha(\lambda) G_\lambda(\lambda_s, T_s)\, d\lambda + \int_{3\mu m}^{\infty} \alpha(\lambda) G_\lambda(\lambda_s, T_s)\, d\lambda\right) \Big/ \int_0^{\infty} G_\lambda(\lambda_s, T_s)\, d\lambda$$

$$= (0.4)(0.98) + (0.8)(0.02) = 0.408$$

For the second condition with the source at 527°C (800 K), the absorptivity can be calculated in a similar manner. However, the calculation is the same as for the emissivity at 800 K in Example 9.4 since $\varepsilon_\lambda = \alpha_\lambda$ and $\bar{\varepsilon} = \bar{\alpha}$ in equilibrium. Hence, $\bar{\alpha} = 0.744$ for a source at 800 K.

The preceding two examples illustrate the limits of graybody assumptions. Whereas it may be acceptable to treat the aluminum-painted surface as totally gray with an average $\bar{\alpha} = \bar{\varepsilon} = (0.8 + 0.744)/2 = 0.77$ for radiation exchange between it and a source at 800 K or less, for radiation exchange between the aluminum-painted surface and the sun such an approximation would lead to a serious error. The surface in the latter case would have to be treated as selectively gray with the averaged values for $\bar{\alpha}$ and $\bar{\varepsilon}$ equal to 0.408 and 0.80, respectively.

9.3.4 Real Surface Characteristics

Radiation from real surfaces differs in several aspects from blackbody or graybody radiation. Any real surface radiates less than a blackbody at the same temperature. Gray surfaces radiate a constant fraction ε_g of the monochromatic emissive power of a black surface at the same temperature T over the entire spectrum; real surfaces radiate a fraction ε_λ at any wavelength, but this fraction is not constant and varies with wavelength. Figure 9.14 on the next page, shows a comparison of spectral emission from black, gray, and real surfaces. Both gray and black surfaces radiate diffusely, and the shape of the spectroradiometric curve for a gray surface is similar to that for a black surface at the same temperature, with the height reduced proportionally by the numerical value of the emissivity.

The spectral emission from the real surface, shown by the wavy line in Fig. 9.14, differs in detail from the graybody spectral emission, but for the purpose of analysis the two may be sufficiently similar on the average to characterize the surface as approximately gray with $\varepsilon_g = 0.6$. The emissive power is given by Eq. (9.34):

$$E_{\text{real}} \cong \varepsilon_g \sigma T^4$$

Observe, however, that Fig. 9.14 compares the emissive power of the real surface with that of a gray surface with $\varepsilon_g = 0.6$ at a temperature of 2000 K. At wavelengths above 1.5 μm the fit is fairly good, but at wavelengths below 1.5 μm the emissivity of the real surface is only about 50% of that of the graybody. For temperatures below 2000 K, the difference will not introduce a serious error because most of the radiant emission occurs at wavelengths above 1.5 μm. At higher temperatures, however, it may be necessary to approximate the real surface with an emissivity value less than 0.6 for $\lambda < 1.5\,\mu$m. For the absorptivity to solar radiation, which falls largely below 2.0 μm, a value closer to 0.3 would be a good approximation.

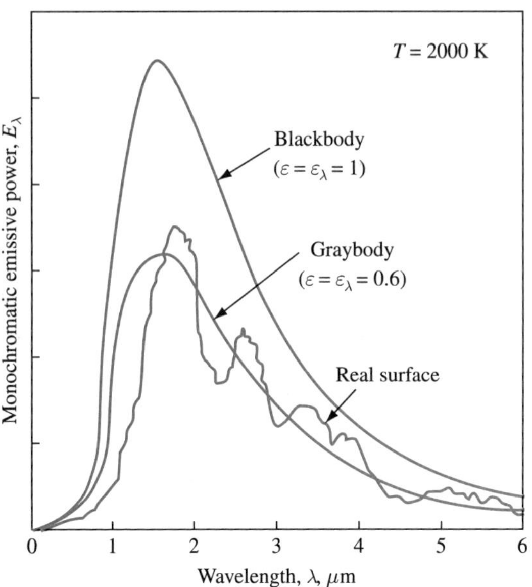

FIGURE 9.14 Comparison of hemispherical monochromatic emission for a black, a gray ($\varepsilon_g = 0.6$), and a real surface.

EXAMPLE 9.6 The spectral hemispherical emissivity of a painted surface is shown in Fig. 9.15. Using a selective gray approximation, calculate (a) the effective emissivity over the entire spectrum, (b) the emissive power at 1000 K, and (c) the percentage of solar radiation that this surface would absorb. Assume that solar radiation corresponds to a blackbody source at 5800 K.

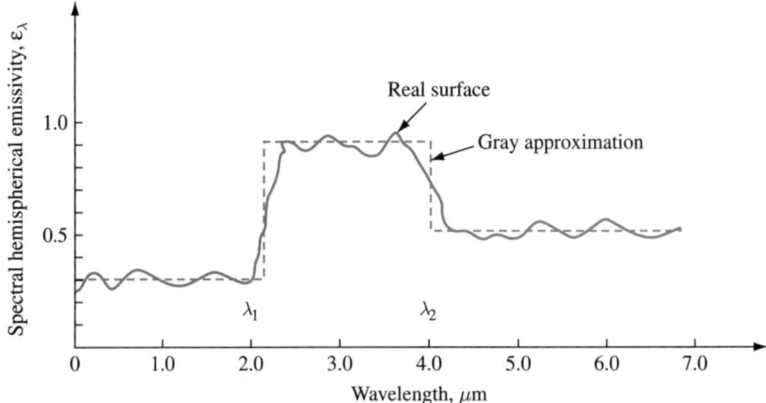

FIGURE 9.15 Hemispherical spectral emissivity of surface for Example 9.6.

SOLUTION We shall approximate the real surface characteristics by a three-band gray model. Below 2.0 μm the emissivity is 0.3, between 2.0 and 4.0 μm the emissivity is about 0.9, and above 4.0 μm the emissivity is about 0.5.

(a) The effective emissivity over the entire spectrum is

$$\bar{\varepsilon} = \frac{\int_0^\infty \varepsilon_\lambda E_{b\lambda}\, d\lambda}{\int_0^\infty E_{b\lambda}\, d\lambda}$$

$$= \varepsilon_1\left[\frac{E_b(0 \rightarrow \lambda_1 T)}{\sigma T^4}\right] + \varepsilon_2\left[\frac{E_b(0 \rightarrow \lambda_2 T) - E_b(0 \rightarrow \lambda_1 T)}{\sigma T^4}\right]$$

$$+ \varepsilon_3\left[\frac{E_b(0 \rightarrow \infty) - E_b(0 \rightarrow \lambda_2 T)}{\sigma T^4}\right]$$

From the data, $\lambda_1 T = 2 \times 10^{-3}$ mK and $\lambda_2 T = 4 \times 10^{-3}$ mK. Evaluating the blackbody emission in the three bands from Table 9.1 yields

$$\bar{\varepsilon} = (0.3)(0.0667) + 0.9(0.4809 - 0.0667) + 0.5(1.0 - 0.4809)$$

$$= 0.0200 + 0.373 + 0.255 = 0.6485$$

(b) The emissive power is then

$$E = \bar{\varepsilon}\sigma T^4 = (0.6485)(5.67 \times 10^{-8})(1000)^4$$

$$= 3.67 \times 10^4 \text{ W/m}^2$$

The emissive power of a black surface at 1000 K is, for comparison, 5.67×10^4 W/m^2.

(c) To calculate the average solar absorptivity we use Eq. (9.33):

$$\bar{\alpha}_s = \frac{\int_0^\infty \alpha_\lambda G_\lambda^*\, d\lambda}{\int_0^\infty G_\lambda^*\, d\lambda}$$

According to Kirchhoff's law, $\alpha_\lambda = \varepsilon_\lambda$ and therefore

$$\bar{\alpha}_s = \frac{\varepsilon_1 \int_0^{2\mu\text{m}} G_\lambda^*\, d\lambda}{\sigma T^4} + \frac{\varepsilon_2 \int_{2\mu\text{m}}^{4\mu\text{m}} G_\lambda^*\, d\lambda}{\sigma T^4} + \frac{\varepsilon_3 \int_{4\mu\text{m}}^{\infty} G_\lambda^*\, d\lambda}{\sigma T^4}$$

Assuming the sun radiates as a blackbody at 5800 K, we get from Table 9.1

$$\bar{\alpha}_s = (0.3)(0.941) + 0.9(0.990 - 0.94) + 0.5(1.0 - 0.99)$$

$$= 0.332$$

Thus, about 33% of the solar radiation would be absorbed. Note that the ratio of emissivity at 1000 K to absorptivity from a 5800 K source is almost 2.

For convenience, the total hemispherical emissivities of a selected group of industrially important surfaces at different temperatures are presented in Table 9.2. A more extensive tabulation of experimentally measured radiation properties for many surfaces has been prepared by Gubareff et al. [8]. Some general features and trends of their results are discussed below.

TABLE 9.2 Hemispherical emissivities of various surfaces

Material	Wavelength and Average Temperature				
	9.3 μm 310 K	5.4 μm 530 K	3.6 μm 800 K	1.8 μm 1700 K	0.6 μm Solar ~6000 K
Metals					
Aluminum					
polished	~0.04	0.05	0.08	~0.19	~0.3
oxidized	0.11	~0.12	0.18		
24-ST weathered	0.4	0.32	0.27		
surface roofing	0.22				
anodized (at 1000°F)	0.94	0.42	0.60	0.34	
Brass					
polished	0.10	0.10			
oxidized	0.61				
Chromium					
polished	~0.08	~0.17	0.26	~0.40	0.49
Copper					
polished	0.04	0.05	~0.18	~0.17	
oxidized	0.87	0.83	0.77		
Iron					
polished	0.06	0.08	0.13	0.25	0.45
cast, oxidized	0.63	0.66	0.76		
galvanized, new	0.23			0.42	0.66
galvanized, dirty	0.28			0.90	0.89
steel plate, rough	0.94	0.97	0.98		
oxide	0.96		0.85		0.74
molten				0.3–0.4	
Magnesium	0.07	0.13	0.18	0.24	0.30
Molybdenum filament			~0.09	~0.15	~0.2[b]
Silver					
polished	0.01	0.02	0.03		0.11
Stainless steel					
18–8, polished	0.15	0.18	0.22		
18–8, weathered	0.85	0.85	0.85		
Steel tube, oxidized		0.94			
Tungsten filament	0.03			~0.18	0.35[c]
Zinc					
polished	0.02	0.03	0.04	0.06	0.46
galvanized sheet	~0.25				
Building and Insulating Materials					
Asbestos paper	0.93	0.93			

(Continued)

TABLE 9.2 (*Continued*)

Asphalt	0.93		0.9		0.93
Brick					
red	0.93				0.7
fire clay	0.9		~0.7	~0.75	
silica	0.9		~0.75	0.84	
magnesite refractory	0.9			~0.4	
Enamel, white	0.9				
Marble, white	0.95		0.93		0.47
Paper, white	0.95		0.82	0.25	0.28
Plaster	0.91				
Roofing board	0.93				
Enameled steel, white				0.65	0.47
Asbestos cement, red				0.67	0.66
Paints					
Aluminized lacquer	0.65	0.65			
Cream paints	0.95	0.88	0.70	0.42	0.35
Lacquer, black	0.96	0.98			
Lampblack paint	0.96	0.97		0.97	0.97
Red paint	0.96				0.74
Yellow paint	0.95		0.5		0.30
Oil paints (all colors)	~0.94	~0.9			
White (ZnO)	0.95		0.91		0.18
Miscellaneous					
Ice	~0.97[d]				
Water	~0.96				
Carbon					
T-carbon, 0.9% ash	0.82	0.80	0.79		
filament	~0.72			0.53	
Wood	~0.93				
Glass	0.90				(Low)

[a]Since the emissivity at a given wavelength equals the absorptivity at that wavelength, the values in this table can be used to approximate the absorptivity to radiation from a source at the temperature listed. For example, polished aluminum will absorb 30% of incident solar radiation.
[b]At 3000 K.
[c]At 3600 K.
[d]At 273 K.
Source: Fischenden and Saunders [3], Hamilton and Morgan [4], Kreith and Black [5], Schmidt and Furthman [6], McAdams [7], and Gubareff et al. [8].

Figure 9.16 on the next page shows the measured monochromatic emissivity (or absorptivity) of some electrical conductors as a function of wavelength [9]. Polished surfaces of metals have low emissivities but, as shown in Fig. 9.17, the presence of an oxide layer may increase the emissivity appreciably. The monochromatic emissivity of an electrical conductor (e.g., see the curves for Al or Cu in Fig. 9.16) increases with decreasing wavelength. Consequently, in accordance with Eq. (9.31), the total emissivity of electrical conductors increases with increasing temperature, as illustrated in Fig. 9.18 on page 567 for several metals and one dielectric.

As a group, electrical nonconductors exhibit the opposite trend and have generally high values of infrared emissivity. Figure 9.19 on page 567 illustrates the variation of the monochromatic emissivity of several electrical nonconductors with wavelength.

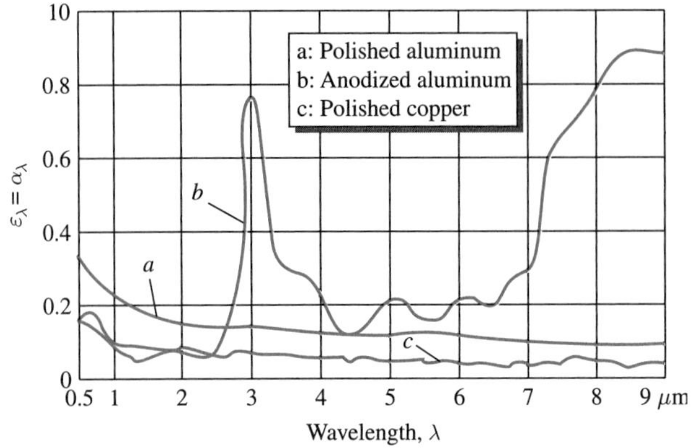

FIGURE 9.16 Variation of monochromatic absorptivity (or emissivity) with wavelength for three electrical conductors at room temperature.

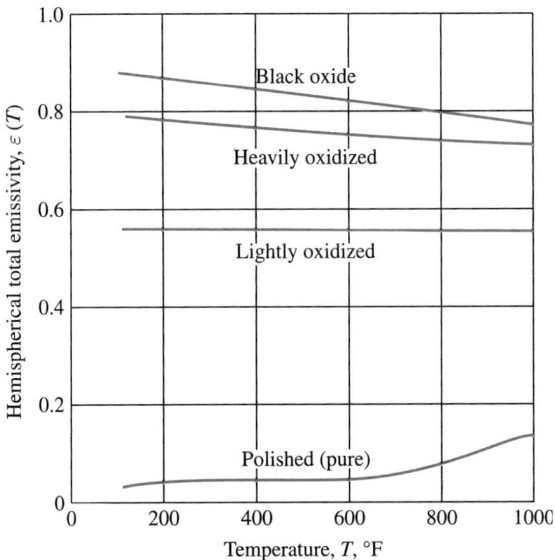

FIGURE 9.17 Effect of oxide coating on hemispherical total emissivity of copper.
Source: Data from Gubareff et al. [8].

For heat transfer calculations an average emissivity or absorptivity for the wavelength band in which the bulk of the radiation is emitted or absorbed is desired. The wavelength band of interest depends on the temperature of the body from which the radiation originates, as pointed out in Section 9.1. If the distribution of the monochromatic emissivity is known, the total emissivity can be evaluated from Eq. (9.31) and the total absorptivity can be calculated from Eq. (9.33) if the temperature and the spectral characteristics of the source are also specified. Sieber [9] evaluated the total

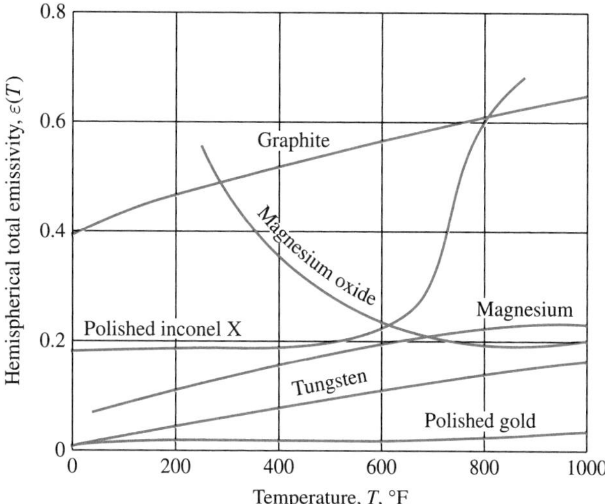

FIGURE 9.18 Effect of temperature on hemispherical total emissivity of several metals and one dielectric.

Source: Data from Gubareff et al. [8].

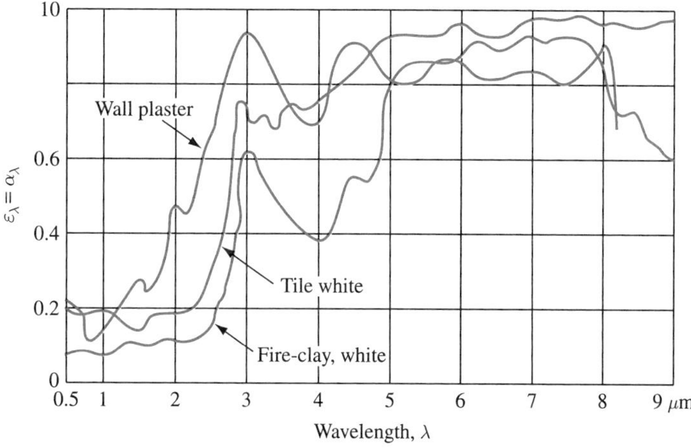

FIGURE 9.19 Variation of monochromatic absorptivity (or emissivity) with wavelength for three electrical nonconductors.

Source: According to Sieber [9].

absorptivity of the surfaces of several materials as a function of source temperature, with the receiving surfaces at room temperature and the emitter a blackbody. His results are shown in Fig. 9.20, where the ordinate is the total absorptivity for radiation normal to the surface and the abscissa is the source temperature. We observe that the absorptivity of aluminum, typical of good conductors, increases with increasing source temperature, whereas the absorptivity of nonconductors exhibits the opposite trend.

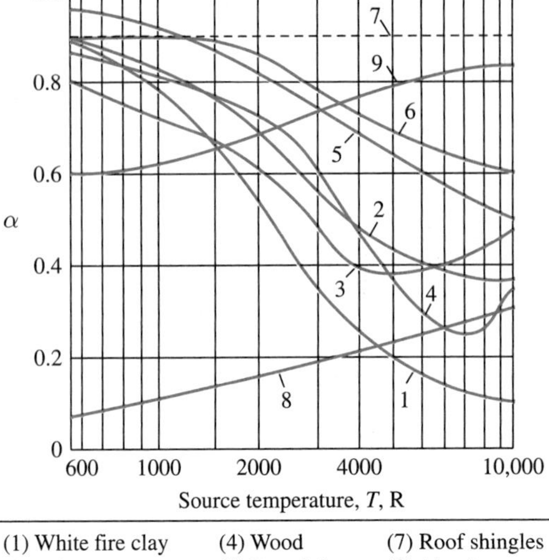

FIGURE 9.20 Variation of total absorptivity with source temperature for several materials at room temperature.

Source: According to Sieber [9].

Figure 9.21 illustrates that the emissivity of real surfaces is also a function of direction. The directional emissivity $\varepsilon(\theta, \phi)$ is defined as the intensity of radiation emitted from a surface in the direction θ, ϕ divided by the blackbody intensity:

$$\varepsilon(\theta, \phi) = \frac{I(\theta, \phi)}{I_b} \tag{9.35}$$

Referring to Eq. (9.25), the monochromatic hemispherical emissivity is defined by the relation

$$\varepsilon_\lambda = \frac{E_\lambda}{E_{b\lambda}} = \frac{\int_{\phi=0}^{2\pi}\int_{\theta=0}^{\pi/2} I_\lambda(\theta, \phi)\sin\theta\cos\theta\, d\theta\, d\phi}{\pi I_{b\lambda}} \tag{9.36}$$

but as mentioned previously, the variation of the emissivity with the azimuthal angle ϕ is usually negligible. If the emissivity is a function only of the elevation angle θ, Eq. (9.36) can be integrated over the angle ϕ and simplified to

$$\varepsilon_\lambda = \frac{2\pi\int_{\theta=0}^{\pi/2} I_\lambda(\theta)\sin\theta\cos\theta\, d\theta}{\pi I_b} \tag{9.37}$$

Substituting Eq. (9.35) for I_λ/I_b, we get

$$\varepsilon_\lambda = 2\int_{\theta=0}^{\pi/2} \varepsilon_\lambda(\theta)\sin\theta\cos\theta\, d\theta \tag{9.38}$$

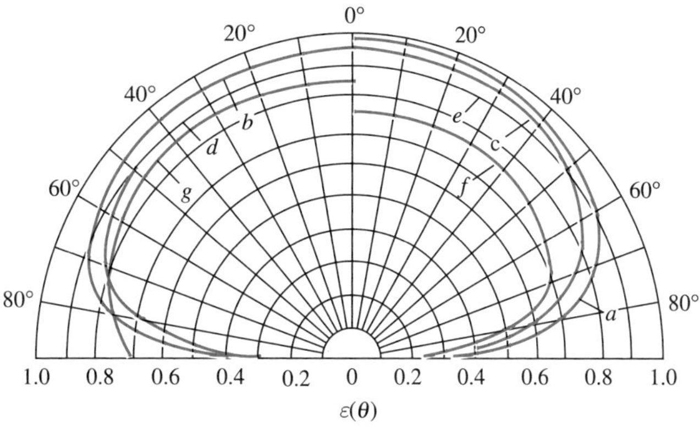

FIGURE 9.21 Variation of directional emissivity with elevation angle for several electrical nonconductors.

Source: From Schmidt and Eckert [10], with permission.

EXAMPLE 9.7 The directional emissivity of an oxidized surface at 800 K can be approximated by

$$\varepsilon(\theta) = 0.70 \cos \theta$$

Determine (a) the emissivity perpendicular to the surface, (b) the hemispherical emissivity, and (c) the radiant emissive power if the surface is 5 cm × 10 cm.

SOLUTION

(a) $\varepsilon(0)$, the emissivity for $\theta = 0°$ or $\cos\theta = 1$, is 0.70.

(b) The hemispherical emissivity is obtained by performing the integration indicated by Eq. (9.38):

$$\bar{\varepsilon} = 2\int_0^{\pi/2} 0.70 \cos^2\theta \sin\theta \, d\theta = -\left(\frac{1.4}{3}\right) \cos^3\theta \Big|_0^{\pi/2}$$

Substituting the above limits gives 0.467. Note that the ratio $\varepsilon(0)/\bar{\varepsilon} = 1.5$.

(c) The emissive power is

$$E = \bar{\varepsilon} A \sigma T^4 = (0.467)(5 \times 10^{-3}\,\text{m}^2)(5.67 \times 10^{-8}\,\text{W/m}^2\,\text{K}^4)(1800\,\text{K})^4$$
$$= 1390\,\text{W}$$

The polar plots in Fig. 9.21 and Fig. 9.22 on the next page illustrate the directional emissivity for some electrical nonconductors and conductors, respectively. In these plots θ is the angle between the normal to the surface and the direction of the radiant beam emitted from the surface. For surfaces whose radiation intensity follows Lambert's cosine law and depends only on the projected area, the emissivity curves

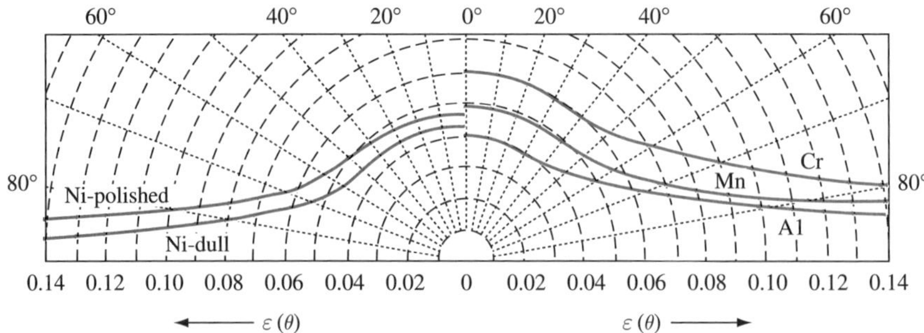

FIGURE 9.22 Variation of directional emissivity with elevation angle for several metals.

Source: From Schmidt and Eckert [10], with permission.

would be semicircles. Figure 9.21 shows that for nonconductors such as wood, paper, and oxide films, the emissivity decreases at large values of the emission angle θ, whereas for polished metals the opposite trend is observed (see Fig. 9.22). For example, the emissivity of polished chromium, which is widely used as a radiation shield, is as low as 0.06 in the normal direction but increases to 0.14 when viewed from an angle θ of 80°. Experimental data on the directional variation of emissivity is scant, and until more information becomes available, a satisfactory approximation for engineering calculations is to assume for polished metallic surfaces a mean value of $\bar{\varepsilon}/\varepsilon_n = 1.2$ and for nonmetallic surfaces $\bar{\varepsilon}/\varepsilon_n = 0.96$, where ε is the average emissivity through a hemispherical solid angle of 2π steradians and ε_n is the emissivity in the direction of the normal to the surface.

Reflectivity and Transmissivity When a surface does not absorb all of the incident radiation, the portion not absorbed will either be transmitted or reflected. Most solids are opaque and do not transmit radiation. The portion of the radiation that is not absorbed is therefore reflected back into hemispherical space. It can be characterized by the monochromatic hemispherical reflectivity ρ_λ defined as

$$\rho_\lambda = \frac{\text{radiant energy reflected per unit time-area-wavelength}}{G_\lambda} \tag{9.39}$$

or by the total reflectivity ρ, defined as

$$\rho = \frac{\text{radiant energy reflected per unit time-area}}{\int_0^\infty G_\lambda \, d\lambda} \tag{9.40}$$

For nontransmitting materials, the relations

$$\rho_\lambda = 1 - \alpha_\lambda \tag{9.41}$$

and

$$\rho = 1 - \alpha$$

must obviously hold at every wavelength and over the entire spectrum, respectively.

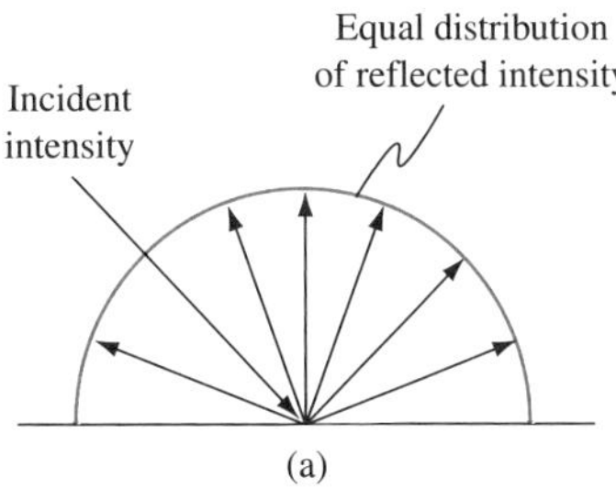

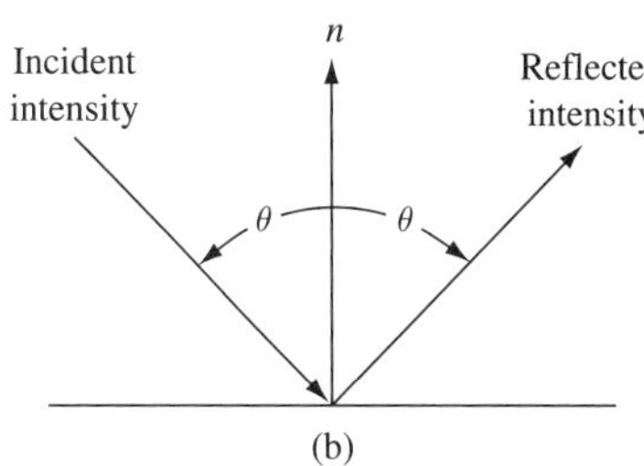

FIGURE 9.23 Schematic diagram illustrating (a) diffuse and (b) specular reflection.

For the most general case of a material that partly absorbs, partly reflects, and partly transmits radiation incident on its surface, we define τ_λ as the fraction transmitted at wavelength λ and τ as the fraction of the total incident radiation that is transmitted. Referring to Fig. 9.11, the monochromatic relation is

$$\rho_\lambda + \alpha_\lambda + \tau_\lambda = 1 \tag{9.42}$$

whereas the total relation between reflectivity, absorptivity, and transmissivity is given by Eq. (9.22). Glass, rock salt, and other inorganic crystals are examples of the few solids that, unless very thick, are to a certain degree transparent to radiation of certain wavelengths. Many liquids and all gases are also transparent.

There are two basic types of radiation reflections: *specular* and *diffuse*. If the angle of reflection is equal to the angle of incidence, the reflection is called specular. On the other hand, when an incident beam is reflected uniformly in all directions, the reflection is called diffuse. No real surface is either specular or diffuse. In general, reflection from highly polished and smooth surfaces approaches specular characteristics, while reflection from industrial "rough" surfaces approaches diffuse characteristics. An ordinary mirror reflects specularly in the visible wavelength range but not necessarily over the longer-wavelength range of thermal radiation.

Figure 9.23 illustrates, schematically, the behavior of diffuse and specular reflectors. For engineering calculations, industrially plated, machined, or painted surfaces can be treated as though they were diffuse, according to experiments by Schonhorst and Viskanta [11]. Methods for treating problems with surfaces that are partly specular and partly diffuse are presented in Sparrow and Cess [12], Siegel and Howe [13], and Hering and Smith [14].

9.4 The Radiation Shape Factor

In most practical problems involving radiation, the intensity of thermal radiation passing between surfaces is not appreciably affected by the presence of intervening media because, unless the temperature is so high as to cause ionization or dissociation, monatomic and most diatomic gases as well as air are transparent. Moreover, since most industrial surfaces can be treated as diffuse emitters and reflectors of radiation in a heat transfer analysis, a key problem in calculating radiation heat transfer between surfaces is to determine the fraction of the total diffuse radiation leaving one surface and being intercepted by another surface and vice versa. *The fraction of diffusely distributed radiation that leaves a surface A_i and reaches surface A_j is called the radiation shape factor F_{i-j}.* The first subscript appended to the radiation shape factor denotes the surface from which the radiation emanates, while the second subscript denotes the surface receiving the radiation. The shape factor is also often called the *configuration factor* or the *view factor*.

Consider two black surfaces A_1 and A_2, as shown in Fig. 9.24. The radiation leaving A_1 and arriving at A_2 is

$$q_{1\rightarrow 2} = E_{b1}A_1F_{1-2} \tag{9.43}$$

and the radiation leaving A_2 and arriving at A_1 is

$$q_{2\rightarrow 1} = E_{b2}A_2F_{2-1} \tag{9.44}$$

Since both surfaces are black, all the incident radiation will be absorbed and the net rate of energy exchange, $q_{1\rightleftharpoons 2}$, is

$$q_{1\rightleftharpoons 2} = E_{b1}A_1F_{1-2} - E_{b2}A_2F_{2-1} \tag{9.45}$$

If both surfaces are at the same temperature, $E_{b1} = E_{b2}$ then there can be no net heat flow between them. Therefore, $q_{1\rightleftharpoons 2} = 0$, and since neither areas nor shape factors are functions of temperature

$$A_1F_{1-2} = A_2F_{2-1} \tag{9.46}$$

Equation (9.46) is known as the *reciprocity theorem*. The net rate of transfer between any two black surfaces, A_1 and A_2, can thus be written in two forms

$$q_{1\rightleftharpoons 2} = A_1F_{1-2}(E_{b1} - E_{b2}) = A_2F_{2-1}(E_{b1} - E_{b2}) \tag{9.47}$$

Inspection of Eq. (9.47) reveals that the net rate of heat flow between two blackbodies can be determined by evaluating the radiation from either one of the surfaces to the other surface and replacing its emissive power by the difference between the

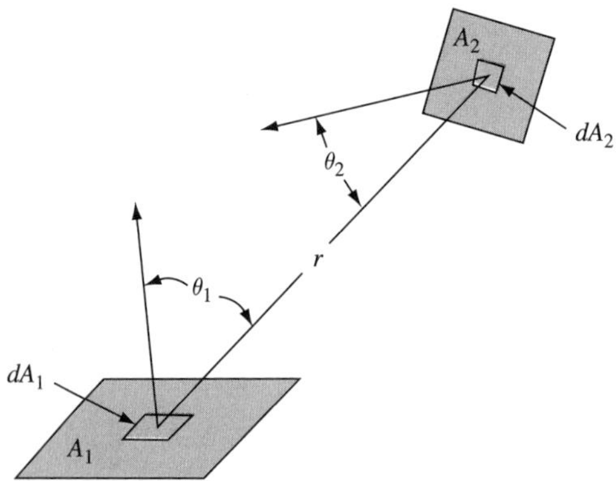

FIGURE 9.24 Nomenclature for geometric shape-factor derivation.

emissive powers of the two surfaces. Since the end result is independent of the choice of the emitting surface, one selects that surface whose shape factor can be determined more easily. For example, the shape factor F_{1-2} for any surface A_1 completely enclosed by another surface is unity. In general, however, the determination of a shape factor for any but the most simple geometric configuration is rather complex.

To determine the fraction of the energy leaving surface A_1 that strikes surface A_2, consider first the two differential surfaces dA_1 and dA_2. If the distance between them is r, then $dq_{1\to 2}$, the rate at which radiation from dA_1 is received by dA_2, is, from Eq. (9.13), given by

$$dq_{1\to 2} = I_1 \cos\theta_1 \, dA_1 \, d\omega_{1-2} \tag{9.48}$$

where I_1 = intensity of radiation from dA_1

$dA_1 \cos\theta_1$ = projection of area element dA_1 as seen from dA_2

$d\omega_{1-2}$ = solid angle subtended by receiving area dA_2 with respect to center point of dA_2

The subtended angle $d\omega_{1-2}$ is equal to the projected area of the receiving surface in the direction of the incident radiation divided by the square of the distance between dA_1 and dA_2, or, using the nomenclature of Fig. 9.24

$$d\omega_{1-2} = \cos\theta_2 \frac{dA_2}{r^2} \tag{9.49}$$

Substituting Eqs. (9.49) and (9.16) for $d\omega_{1-2}$ and I_1, respectively, into Eq. (9.48) yields

$$dq_{1\to 2} = E_{b1} \, dA_1 \left(\frac{\cos\theta_1 \cos\theta_2 \, dA_2}{\pi r^2} \right) \tag{9.50}$$

where the term in parentheses is equal to the fraction of the total radiation emitted from dA_1 that is intercepted by dA_2. By analogy, the fraction of the total radiation emitted from dA_2 that strikes dA_1 is

$$dq_{2\to 1} = E_{b2} \, dA_2 \left(\frac{\cos\theta_2 \cos\theta_1 \, dA_1}{\pi r^2} \right) \tag{9.51}$$

so that the net rate of radiant heat transfer between dA_1 and dA_2 is

$$dq_{1\rightleftharpoons 2} = (E_{b1} - E_{b2}) \frac{\cos\theta_1 \cos\theta_2 \, dA_1 \, dA_2}{\pi r^2} \tag{9.52}$$

To determine $q_{1\rightleftharpoons 2}$, the net rate of radiation between the entire surfaces A_1 and A_2, we can integrate the fraction in the preceding equation over both surfaces and obtain

$$q_{1\rightleftharpoons 2} = (E_{b1} - E_{b2}) \int_{A_1} \int_{A_2} \frac{\cos\theta_1 \cos\theta_2 \, dA_1 \, dA_2}{\pi r^2} \tag{9.53}$$

The double integral is conveniently written in shorthand notation as either $A_1 F_{1-2}$ or $A_2 F_{2-1}$, where F_{1-2} is called the shape factor evaluated on the basis of area A_1 and F_{2-1} is called the shape factor evaluated on the basis of A_2. The method of evaluation of the double integral is illustrated in the following example.

EXAMPLE 9.8 Determine the geometric shape factor for a very small disk A_1 and a large parallel disk A_2 located a distance L directly above the smaller one, as shown in Fig. 9.25.

SOLUTION From Eq. (9.53) the geometric shape factor is

$$A_1 F_{1-2} = \int_{A_1}\int_{A_2} \frac{\cos\theta_1 \cos\theta_2}{\pi r^2}\, dA_2\, dA_1$$

but since A_1 is very small, the shape factor is given by

$$A_1 F_{1-2} = \frac{A_1}{\pi}\int_{A_2} \frac{\cos\theta_1 \cos\theta_2}{r^2}\, dA_2$$

From Fig. 9.25, $\cos\theta_1 = \cos\theta_2 = L/r$, $r = \sqrt{\rho^2 + L^2}$, and $dA_2 = \rho\, d\phi\, d\rho$. Substituting these relations, we obtain

$$A_1 F_{1-2} = \frac{A_1}{\pi}\int_0^a \int_0^{2\pi} \frac{L^2}{(\rho^2 + L^2)^2}\, \rho\, d\rho\, d\phi$$

which can be integrated directly to yield

$$A_1 F_{1-2} = \frac{A_1 a^2}{a^2 + L^2} = A_2 F_{2-1}$$

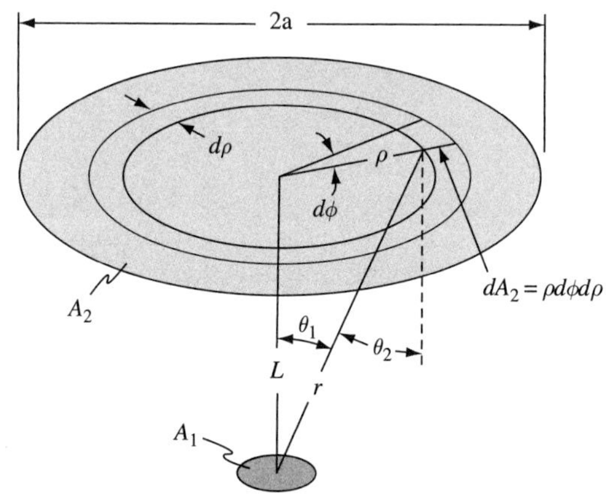

FIGURE 9.25 Nomenclature for evaluation of shape factor between two disks in Example 9.8.

Example 9.8 shows that the determination of a shape factor by evaluating the double integral of Eq. (9.53) is generally very tedious. Fortunately, the shape factors for a large number of geometric arrangements have been evaluated, and a majority of them can be found in references [3–7]. A selected group of practical interest is summarized in Table 9.3 and Figs. 9.26 to 9.30 found on pages 575–577.

TABLE 9.3 Geometric shape factors for use in Eqs. (9.47) and (9.55)

Surface Between Which Radiation Is Being Interchanged	Shape Factor, F_{1-2}
1. Infinite parallel planes.	1
2. Body A_1 completely enclosed by another body, A_2. Body A_1 cannot see any part of itself.	1
3. Surface element $dA(A_1)$ and rectangular surface (A_2) above and parallel to it, with on e corner of rectangle contained in normal to dA.	See Fig. 9.26
4. Element $dA(A_1)$ parallel circular disk (A_2) with its center directly above dA. (See Example 9.8.)	$\frac{a^2}{(a^2 + L^2)}$
5. Two parallel and equal squares, rectangles, or disks of width or diameter D, a distance L apart.	See Fig. 9.28 or Fig. 9.29
6. Two parallel disks of unequal diameter, distance L apart with centers on same normal to their planes, smaller disk A_1 of radius a, larger disk of radius b.	$\frac{1}{2a^2}[L^2 + a^2 + b^2 - \sqrt{(L^2 + a^2 + b^2)^2 - 4a^2b^2}]$
7. Two rectangles in perpendicular planes with a common side.	See Fig. 9.27
8. Radiation between an infinite plane A_1 and one or two rows of infinite parallel tubes in a parallel plane A_2 if the only other surface is a refractory surface behind the tubes.	See Fig. 9.30

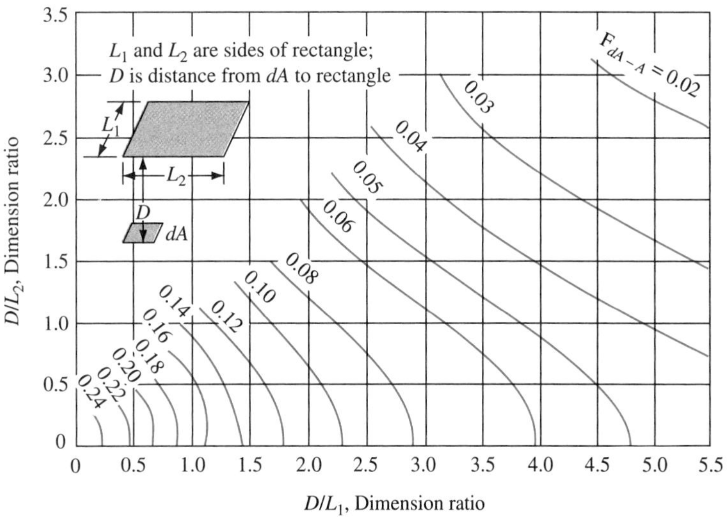

FIGURE 9.26 Shape factor for a surface element dA and a rectangular surface A parallel to it.

Source: From Hottel [15], with permission.

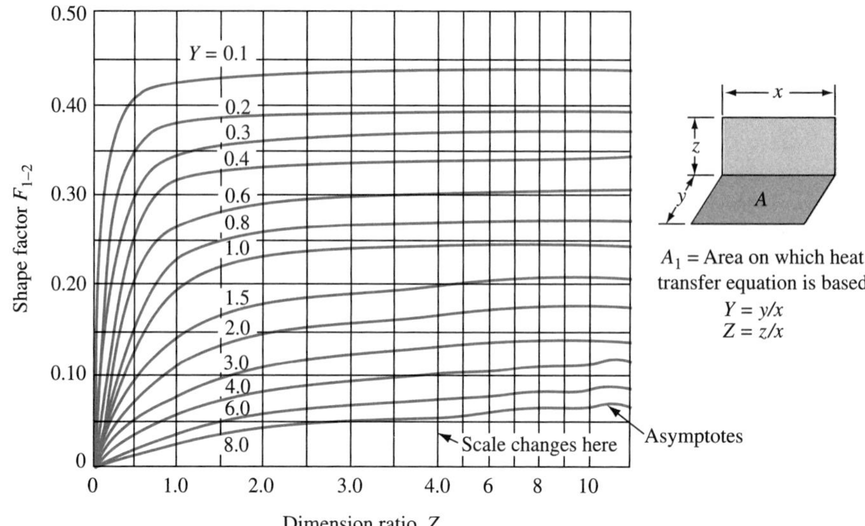

FIGURE 9.27 Shape factor for adjacent rectangles in perpendicular planes sharing a common edge.

Source: From Hottel [15], with permission.

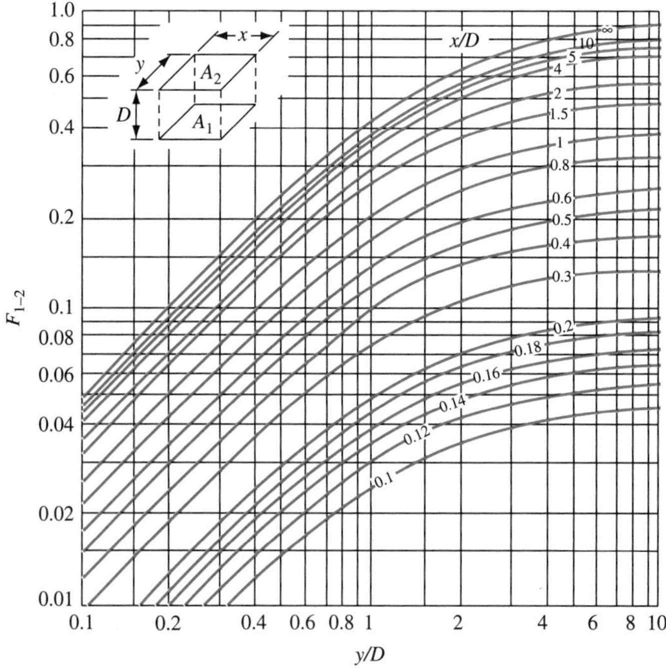

FIGURE 9.28 Shape factor for directly opposed rectangles.

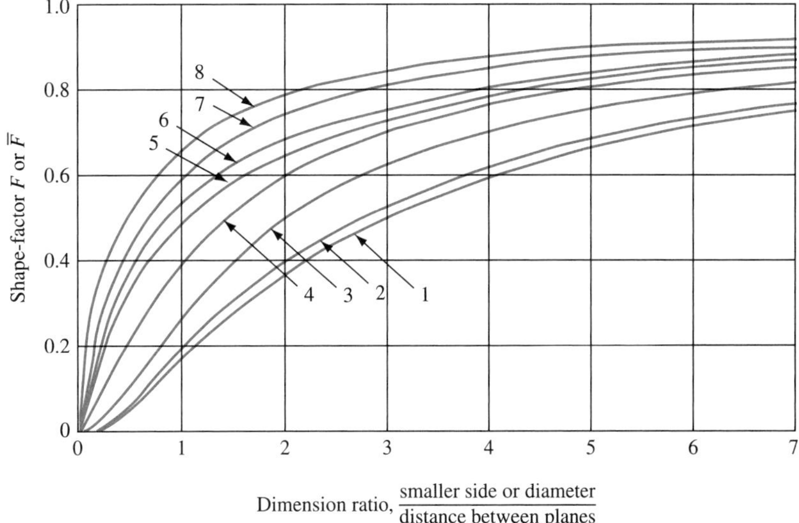

Radiation between parallel planes, directly opposed:
- 1, 2, 3, and 4: Direct radiation between the planes, F
- 5, 6, 7, and 8: Planes connected by nonconducting but reradiating walls, $\bar{F}$
- 1 and 5: Disks
- 2 and 6: Squares
- 3 and 7: 2:1 Rectangles
- 4 and 8: Long, narrow rectangles

FIGURE 9.29 Shape factors for equal and parallel squares, rectangles, and disks.

Source: From Hottel [15], with permission. See Eq. (9.65) for definition of $\bar{F}$.

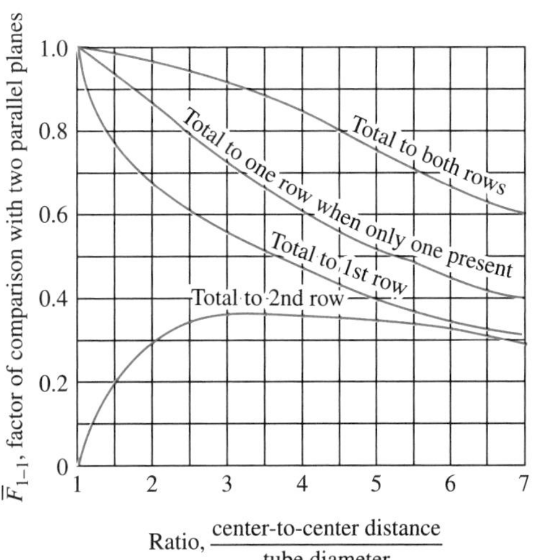

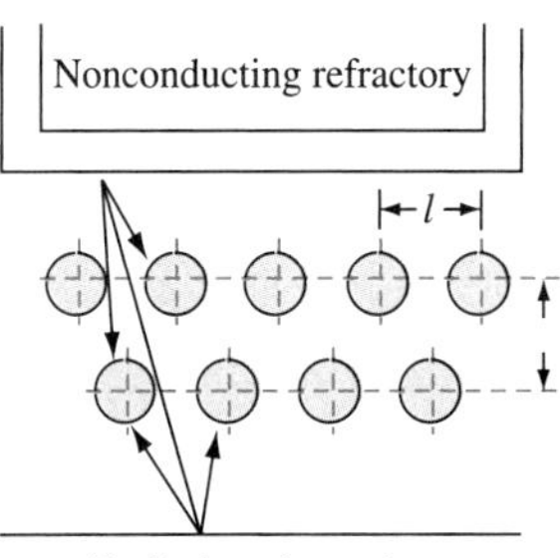

Ordinate is fraction of heat radiated from the plane A_1 to an infinite number of rows of tubes or to a plane replacing the tubes

FIGURE 9.30 Shape factor for a plane and one or two rows of tubes parallel to it.

Source: From Hottel [15], with permission.

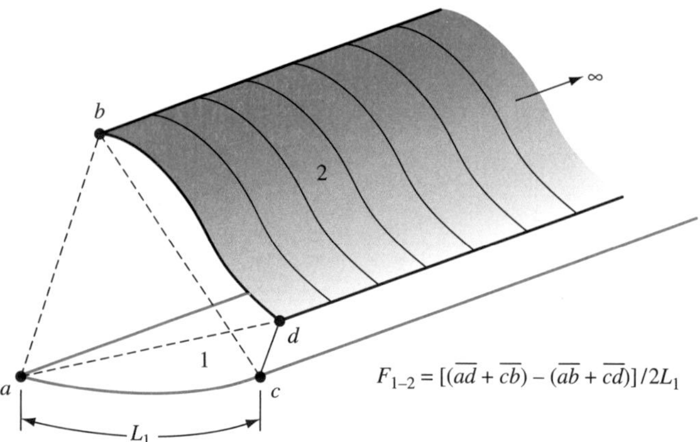

FIGURE 9.31 Schematic diagram illustrating the crossed-string method.

The shape factors for surfaces that are two-dimensional, infinitely long in one direction, and characterized by identical cross sections normal to the infinite direction can be determined by a simple procedure called the *crossed-string method.* Figure 9.31 shows two surfaces that satisfy the geometric restrictions for the crossed-string method. Hottel and Sarofim [16] have shown that the shape factor F_{1-2} is equal to the sum of the lengths of the crossed strings stretched between the ends of the two surfaces minus the sum of the lengths of the uncrossed strings divided by twice the length L_1. In the form of an equation,

$$F_{1-2} = \frac{(\overline{ad} + \overline{cb}) - (\overline{ad} + \overline{cd})}{2L_1} \tag{9.54}$$

EXAMPLE 9.9 A window arrangement consists of a long opening 1 m high and 5 m long. Under this window, as shown in Fig. 9.32, is a working table 2-m-wide. Determine the shape factor between the window and the table.

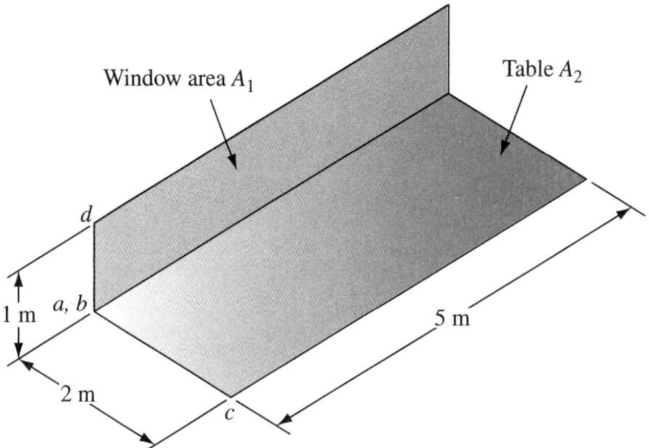

FIGURE 9.32 Window and table for Example 9.9.

SOLUTION Assume that the windows and the table are sufficiently long to be approximated as infinitely long surfaces. Then we can use the crossed-string method, and since for this case points a and b are the same, we have

$$ab = 0$$

$$cb = L_1 = 2\,\text{m}$$

$$ad = L_2 = 1\,\text{m}$$

$$cd = L_3 = \sqrt{5}\,\text{m}$$

and

$$F_{1-2} = \tfrac{1}{2}(1 + 2 - \sqrt{5}) = 0.382$$

Calculation of shape factors for arbitrary surfaces in three dimensions is quite complex and is therefore carried out numerically. In many problems of practical interest, there may be objects between two surfaces of interest that partially block the view from one of the surfaces to the other. This situation further complicates the calculation of the shape factors. Emery et al. [17] discuss and compare several numerical methods for calculating the shape factor between arbitrary surfaces.

9.4.1 Shape-Factor Algebra

The basic shape factors from the charts in Figs. 9.26 to 9.30 can be used to obtain shape factors for a larger class of geometries that can be built up from the elementary curves. This process is known as shape-factor algebra. Shape-factor algebra is based on the principle of conservation of energy. Suppose we want to determine the shape factor from surface A_1 to the combined areas $A_2 + A_3$ as shown in Fig. 9.33. We can write

$$F_{1\to(2+3)} = F_{1-2} + F_{1-3} \tag{9.55}$$

That is, the total shape factor is equal to the sum of its parts. Rewriting Eq. (9.55) as

$$A_1F_{1-2,3} = A_1F_{1-2} + A_1F_{1-3}$$

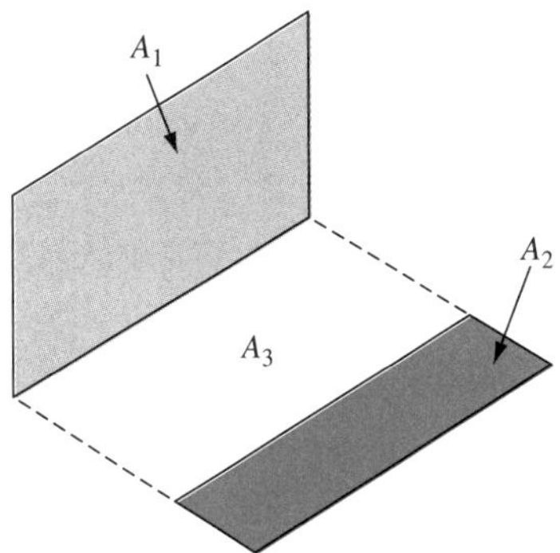

FIGURE 9.33 Schematic illustration of shape-factor algebra.

and using the reciprocity relations

$$A_1F_{1-2,3} = (A_2 + A_3)F_{2,3-1}$$
$$A_1F_{1-2} = A_2F_{2-1}$$
$$A_1F_{1-3} = A_3F_{3-1}$$

yields

$$(A_2 + A_3)F_{2,3-1} = A_2F_{2-1} + A_3F_{3-1} \tag{9.56}$$

This simple relation can be used to evaluate the shape factor F_{1-2} in terms of the shape factors for perpendicular rectangles with a common edge given in Fig. 9.27. Other combinations can be obtained in a similar manner. The following example illustrates the numerical evaluation procedure divided by 100.

EXAMPLE 9.10 Suppose an architect wants to evaluate the percentage of daylight entering through a store window A_1 that impinges on the floor area A_4 located relative to A_1 as shown in Fig. 9.34. Assuming that the light through the window is diffuse, evaluate the shape factor F_{1-4} that is equal to this percentage divided by 100.

SOLUTION Let $A_5 = A_1 + A_2$ and $A_6 = A_3 + A_4$. Using shape-factor algebra and applying Eq. 9.55 and Eq. 9.56 gives

$$A_5F_{5-6} = A_2F_{2-3} + A_2F_{2-4} + A_1F_{1-3} + A_1F_{1-4}$$
$$A_5F_{5-3} = A_2F_{2-3} + A_1F_{1-3}$$
$$F_{2-6} = F_{2-3} + F_{2-4}$$

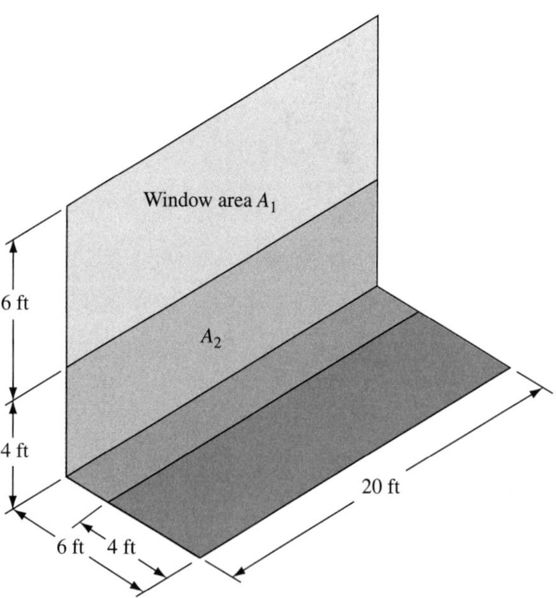

FIGURE 9.34 Sketch for Example 9.10.

Combining the three previous equations and solving for F_{1-4} gives

$$F_{1-4} = \frac{1}{A_1}(A_5F_{5-6} - A_2F_{2-6} - A_5F_{5-3} + A_2F_{2-3})$$

The shape factors for the right-hand side of this equation are plotted in Fig. 9.27. The values are:

$$F_{5-6} = 0.19$$
$$F_{2-6} = 0.32$$
$$F_{5-3} = 0.08$$
$$F_{2-3} = 0.19$$

Therefore,

$$F_{1-4} = \frac{1}{60}(100 \times 0.19 - 40 \times 0.32 - 100 \times 0.08 + 40 \times 0.19)$$
$$= 0.097$$

Thus, only about 10% of the light passing through the window will impinge on the floor area A_4.

9.5 Enclosures with Black Surfaces

To determine the net radiation heat transfer to or from a surface, it is necessary to account for radiation coming from all directions. This procedure is facilitated by figuratively constructing an enclosure around the surface and specifying the radiation characteristics of each surface. The surfaces comprising the enclosure for a given surface i are all the surfaces that can be seen by an observer standing on surface i in the surrounding space. The enclosure need not necessarily consist only of solid surfaces but may include open spaces denoted as "windows." Each such open window can be assigned an equivalent blackbody temperature corresponding to the entering radiation. If no radiation enters, a window acts like a blackbody at zero temperature, which absorbs all outgoing radiation and emits and reflects none.

The net rate of radiation loss from a typical surface A_i in an enclosure (see Fig. 9.35) consisting of N black surfaces is equal to the difference between the emitted radiation and the absorbed radiation, or

$$q_{i \rightleftharpoons \text{enclosure}} = A_i(E_{bi} - G_i) \tag{9.57}$$

where G_i is the radiation incident on surface i per unit time and unit area, called the irradiation.

The radiation incident on A_i comes from the other N surfaces in the enclosure. From a typical surface j, the radiation incident on i is $E_{bj}A_jF_{j-i}$. Summing the contributions from all N surfaces gives

$$A_iG_i = E_{b1}A_1F_{1-i} + E_{b2}A_2F_{2-i} + \cdots + E_{bN}A_NF_{N-i}$$

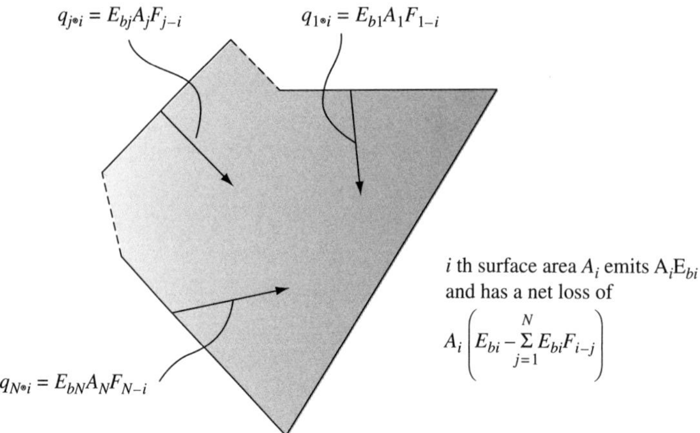

FIGURE 9.35 Schematic diagram of enclosure of N black surfaces with energy quantities incident upon and leaving surface i.

which can be written compactly in the form

$$A_iG_i = \sum_{j=1}^{N} E_{bj}A_jF_{j-i} \tag{9.58}$$

Using the reciprocity law, $A_iF_{i-j} = A_jF_{j-i}$, and substituting Eq. (9.58) for G_i in Eq. (9.57) yields for the net rate of radiation heat loss from *any* surface in an enclosure of black surfaces

$$q_{i \rightleftharpoons \text{enclosure}} = A_i\left(E_{bi} - \sum_{j=1}^{N} E_{bj}F_{i-j}\right) \tag{9.59}$$

An alternative approach to the problem is by extension of Eqs. (9.43) and (9.44). Since the radiant energy leaving any surface i must impinge on the N surfaces forming the enclosure,

$$\sum_{j=1}^{N} F_{i-j} = 1.0 \tag{9.60}$$

Equation (9.60) includes a term F_{i-i}, which is not zero when a surface is concave so that some radiation leaving surface i will be directly incident on it. The total emissive power of A_i is therefore distributed between the N surfaces according to

$$A_iE_{bi} = \sum_{j=1}^{N} E_{bi}A_iF_{i-j} \tag{9.61}$$

Introducing Eq. (9.61) for A_iE_{bi} in Eq. (9.59) gives the net rate of heat loss from surface i in the form

$$q_{i \rightleftharpoons \text{enclosure}} = \sum_{j=i}^{N} (E_{bi} - E_{bj})A_iF_{i-j} \tag{9.62}$$

Thus, the net heat loss can be calculated by summing the differences in emissive power and multiplying each by the appropriate area-shape factor.

An inspection of Eq. (9.62) shows that there is also an analogy between heat flow by radiation and the flow of electric current. If the blackbody emissive power E_b is considered to act as a potential and the area-shape factor A_iF_{i-j} as the conductance between two nodes at potentials E_{bi} and E_{bj}, then the resulting net flow of heat is analogous to the flow of electric current in an analogous network. Examples of networks for blackbody enclosures consisting of three and four heat transfer surfaces at given temperatures are shown in Figs. 9.36(a) and (b), respectively.

In engineering problems, there are situations when not the temperature but the heat flux is prescribed for one or more surfaces in an enclosure. In such cases, the temperatures of these surfaces are unknown. For the case when the net radiation heat transfer rate $q_{r,k}$ from one surface A_k is prescribed while the temperature is specified for all the other surfaces of the enclosure, Eq. (9.59) can be rearranged to solve for T_k. Since $E_{bk} = \sigma T_k^4$, one obtains

$$T_k = \left[\frac{\sum\limits_{j \neq k}^{N} \sigma T_j^4 F_{k-j} + (q_r/A)_k}{\sigma(1 - F_{k-k})} \right]^{1/4} \tag{9.63}$$

where $j = k$ is specifically excluded from the summation. Once T_k is known, the heat transfer rates at all other surfaces can be obtained from Eq. (9.62).

Of special interest is the case of a *no-flux* or *adiabatic surface*, which diffusely reflects and emits radiation at the same rate at which it receives it. Under steady-state conditions, the interior surfaces of refractory walls in industrial furnaces can be treated as adiabatic surfaces. The interior walls of these surfaces receive heat by convection as well as radiation and lose heat to the outside by conduction. In practice, however, the heat flow by radiation is so much larger than the difference between the heat flow by convection to and the heat flow by conduction from the surface that the walls act essentially as reradiators, that is, as no-flux surfaces.

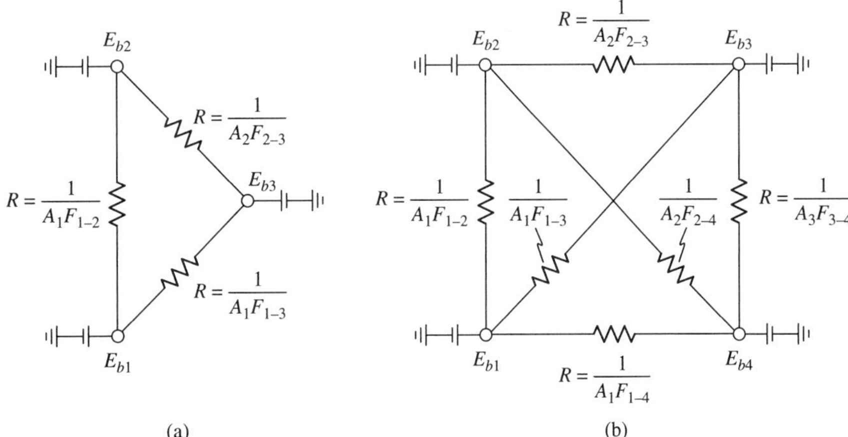

FIGURE 9.36 Equivalent networks for radiation in blackbody enclosures consisting of (a) three and (b) four surfaces.

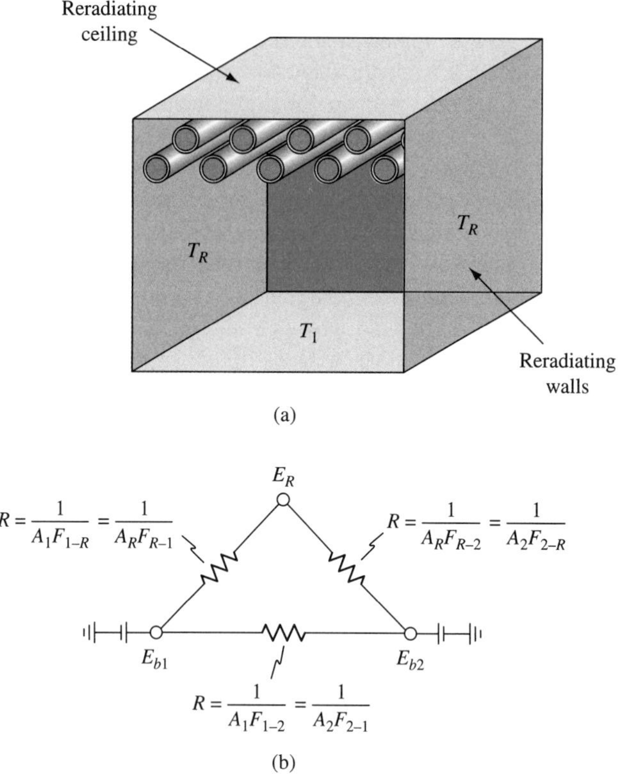

FIGURE 9.37 Simplified sketch of a furnace and equivalent network for radiation in an enclosure consisting of two black surfaces and an adiabatic surface.

A simplified sketch of a pulverized-fuel furnace is shown in Fig. 9.37(a). The floor is assumed to be at a uniform temperature T_1 radiating to a nest of oxidized-steel tubes at T_2 that fill the ceiling of the furnace. The side walls and the ceiling are assumed to act as reradiators at a *uniform temperature* T_R. If we let A_R denote the reradiating area and assume that the floor and the tubes are black, the equivalent network representing the radiation exchange between the floor and the tubes in the presence of the reradiating walls is that shown in Fig. 9.37(b). A part of the radiation emitted from A_1 goes directly to A_2, while the rest strikes A_R and is reflected from there. Of the reflected radiation, a part is returned to A_1, a part to A_2, and the rest to A_R for further reflection. However, since the refractory walls must get rid of all the incident radiation by either reflection or radiation, their emissive power will act in the steady state like a floating potential whose actual value. That is, its emissive power and temperature, depends only on the relative values of the conductances between E_R and E_{b1} and E_R and E_{b2}. Thus, the net effect of this rather complicated radiation pattern can be represented in the equivalent network by two parallel heat flow paths between A_1 and A_2, one having an effective conductance of A_1F_{1-2}, the other having an effective thermal conductance equal to

$$\frac{1}{1/A_1F_{1-R} + 1/A_2F_{2-R}}$$

The net heat flow by radiation between a black heat source and a black heat sink in such a simple furnace is then equal to

$$q_{1\rightleftarrows 2} = A_1(E_{b1} - E_{b2})\left(F_{1-2} + \frac{1}{1/F_{1-R} + A_1/A_2F_{2-R}}\right) \tag{9.64}$$

If neither of the surfaces can see any part of itself, F_{1-R} and F_{2-R} can be eliminated by using Eqs. (9.46) and (9.60). This yields, after some simplification,

$$q_{1\rightleftarrows 2} = A_1\sigma(T_1^4 - T_2^4)\frac{A_2 - A_1F_{1-2}^2}{A_1 + A_2 - 2A_1F_{1-2}} = A_1\bar{F}_{1-2}(E_{b1} - E_{b2}) \tag{9.65}$$

where $\bar{F}_{1-2}$ is the effective shape factor for the configuration shown in Fig. 9.37. The same result would, of course, be obtained from Eqs. (9.62) and (9.63). The details of this derivation are left as an exercise.

9.6 Enclosures with Gray Surfaces

In the preceding section, radiation between black surfaces was considered. The assumption that a surface is black simplifies heat transfer calculations because all of the incident radiation is absorbed. In practice, one can generally neglect reflections without introducing serious errors if the absorptivity of the radiating surfaces is larger than 0.9. There are, however, numerous problems involving surfaces of low absorptivity and emissivity, especially in installations where radiation is undesirable. For example, the inner walls of a thermos bottle are silvered in order to reduce the heat flow by radiation. Also, thermocouples for high-temperature work are frequently surrounded by radiation shields to reduce the difference between the indicated temperature and the temperature of the medium to be measured.

If the radiating surfaces are not black, the analysis becomes exceedingly difficult unless the surfaces are considered to be gray. The analysis in this section is limited to gray surfaces that follow Lambert's cosine law and also reflect diffusely. The radiation from such surfaces can be treated conveniently in terms of the *radiosity*, J, which is defined as the rate at which radiation leaves a given surface per unit area. The radiosity is the sum of radiation emitted, reflected, and transmitted. For opaque bodies that transmit no radiation, the radiosity from a typical surface i can be defined [18]

$$J_i = \rho_i G_i + \varepsilon_i E_{bi} \tag{9.66}$$

where J_i = radiosity, W/m^2

G_i = irradiation or radiation per unit time incident on a unit surface area, W/m^2

E_{bi} = blackbody emissive power, W/m^2

ρ_i = reflectivity

ε_i = emissivity

Consider the ith surface having area A_i in an enclosure consisting of N surfaces as shown in Fig. 9.35. To maintain surface i at temperature T_i, a certain amount of heat, q_i, must be supplied from some external source to make up for the net radiative loss in a steady-state condition. The net rate of heat transfer from a surface i by radiation is equal to the difference between the outgoing and the incoming radiation. Using the terminology of Eq. (9.66), the net rate of heat loss is the difference between the radiosity and the irradiation, or

$$q_i = A_i(J_i - G_i) \tag{9.67}$$

It should be noted that Eq. (9.67) is strictly valid only when the temperature as well as the irradiation over A_i is uniform. To satisfy both of these conditions simultaneously, it is sometimes necessary to subdivide a physical surface into smaller sections for the purpose of analysis.

If the surfaces exchanging radiation are gray, $\epsilon_i = \alpha_i$ and $\rho_i = (1 - \epsilon_i)$ for each of them. The irradiation G_i can then be eliminated from Eq. (9.67) by combining it with Eq. (9.66). This yields

$$q_i = \frac{A_i\varepsilon_i}{\rho_i}(E_{bi} - J_i) = \frac{A_i\varepsilon_i}{1 - \varepsilon_i}(E_{bi} - J_i) \tag{9.68}$$

Another relation for the *net rate of heat loss* by radiation from A_i can be obtained by evaluating the irradiation in terms of the radiosity of all the other surfaces that can be seen from it. The incident radiation G_i can be evaluated by the same approach used previously in a blackbody enclosure. The incident radiation consists of the portions of radiation from the other $N - 1$ surfaces that impinge on A_i. If the surface A_i can see a part of itself, a portion of the radiation emitted by A_i will also contribute to the irradiation. The shape factors for diffusely reflecting gray surfaces are obviously the same as for black surfaces, since they depend only on geometric relations defined by Eq. (9.53). We can, therefore, write in symbolic form

$$A_iG_i = J_1A_1F_{1-i} + J_2A_2F_{2-i} + \cdots + J_iA_iF_{i-1} + \cdots + J_jA_jF_{j-i} + \cdots + J_NA_NF_{N-i} \tag{9.69}$$

Using the reciprocity relations

$$A_1F_{1-i} = A_iF_{i-1}$$

$$A_2F_{2-1} = A_iF_{i-2}$$

$$A_NF_{N-i} = A_iF_{i-N}$$

Equation (9.69) can be written so that the only area appearing is A_i:

$$A_iG_i = J_1A_iF_{i-1} + J_2A_iF_{i-2} + \cdots + J_iA_iF_{i-i} + \cdots + J_jA_iF_{i-j} + \cdots + J_NA_iF_{i-N}$$

This can be expressed compactly as

$$G_i = \sum_{j=1}^{N} J_jF_{i-j} \tag{9.70}$$

Equation (9.70) is identical to Eq. (9.61) for a black enclosure, except that the blackbody emissive power has been replaced by the radiosity. Substituting the summation of Eq. (9.70) for G_i in Eq. (9.67) yields

$$q_i = A_i\left(J_i - \sum_{j=1}^{N} J_j F_{i-j}\right) \tag{9.71}$$

Equations (9.68) and (9.71) can be written for each of the N surfaces of the enclosure, giving $2N$ equations for $2N$ unknowns. There will always be N unknown J's, while the remaining unknowns will consist of q's or T's, depending on what boundary conditions are specified. The J's can always be eliminated, giving N equations relating the N unknown temperatures and net rates of radiation transfer.

In terms of an analogous electrical circuit, we could write Eq. (9.68) in the form

$$q_i = \frac{E_{bi} - J_i}{(1 - \varepsilon_i)/A_i\varepsilon_i} \tag{9.72}$$

and consider the rate of radiation heat transfer q_i as the current in a network between potentials E_{bi} and J_i with a resistance of $(1 - \varepsilon_i)/A_i\varepsilon_i$ between them. Since the effect of the system geometry on the net radiation between any two gray surfaces, A_i and A_k emitting radiation at the rates J_i and J_k, respectively, is the same as for geometrically similar black surfaces, it can be expressed in terms of the geometric shape factor defined by Eq. (9.53). The direct radiation exchange between any two opaque and diffuse surfaces A_i and A_j is given by

$$q_{i\rightleftharpoons j} = (J_i - J_j)A_iF_{i-j} = (J_i - J_j)A_jF_{j-i} \tag{9.73}$$

Equations (9.68) and (9.73) provide the basis for determining the net rate of radiant heat transfer between Graybodies in a gray enclosure by means of an equivalent network. The effect of the reflectivity and emissivity can be taken into account by connecting a *blackbody potential node* E_b to each of the nodal points in the network by means of a *finite resistance* $(1 - \varepsilon)/A\varepsilon$. In the case of a blackbody, this resistance is zero since $\varepsilon = 1$. In Fig. 9.38 the equivalent networks for radiation in an enclosure consisting of two and four graybodies are shown. For two-surface gray enclosures, such as two parallel and infinite plates, concentric cylinders of infinite height, and concentric spheres, the network reduces to a single line of resistances in series, as shown in Fig. 9.38(a).

To illustrate the procedure for calculating radiation heat transfer between gray surfaces, we will derive an expression for the rate of radiation heat transfer between two long concentric cylinders of areas A_1 and A_2 and temperatures T_1 and T_2, respectively, and compare the result with the network in Fig. 9.38(a).

Referring to Fig. 9.39, the shape factor for the smaller cylinder of area A_1 relative to the larger cylinder that encloses it, F_{1-2}, is 1.0. From Eq. (9.73), $A_1F_{1-2} = A_2F_{2-1}$ and $F_{2-1} = A_1/A_2$. Since surface 2 can partly view itself, from Eq. (9.60) we have also $F_{2-2} = 1 - (A_1/A_2)$. From Eqs. (9.68) and (9.71), the net rates of heat loss from A_1 and A_2 are

$$q_1 = \frac{A_1\varepsilon_1}{1 - \varepsilon_1}(E_{b1} - J_1) = A_1(J_1 - J_2)$$

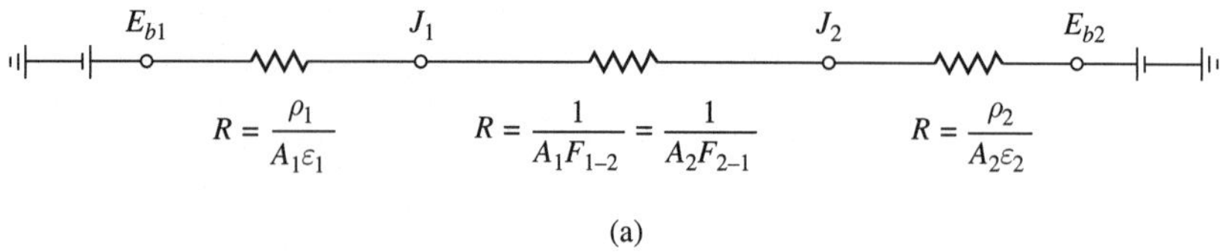

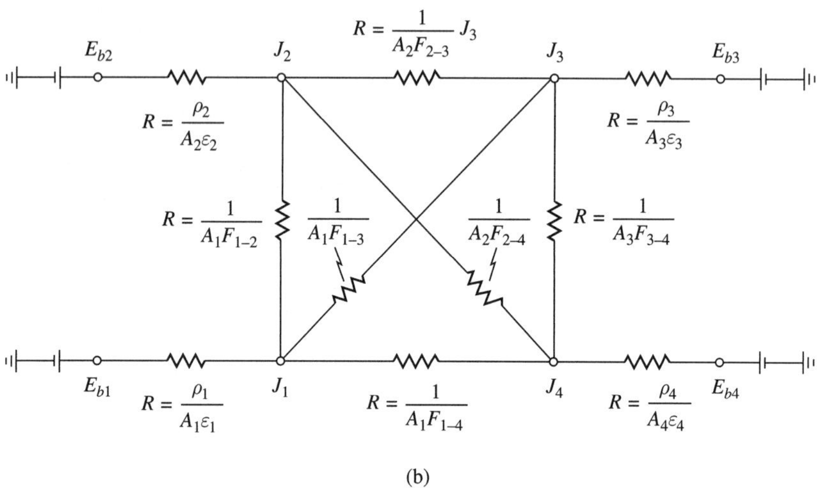

FIGURE 9.38 Equivalent networks for radiation in gray enclosures consisting of two and four surfaces: (a) two graybody surfaces and (b) four graybody surfaces.

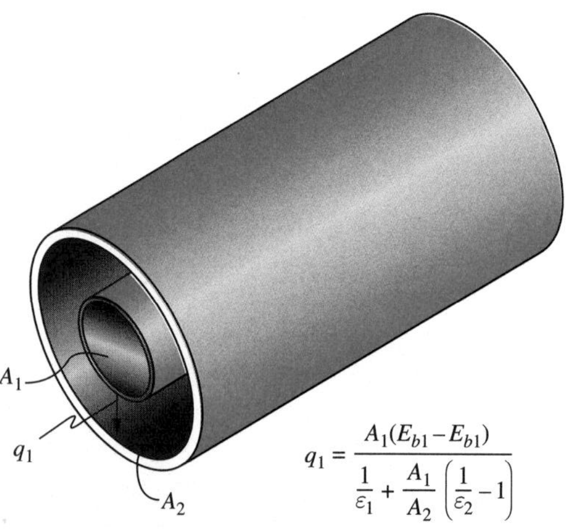

FIGURE 9.39 Radiation exchange between two gray-cylindrical surfaces.

and

$$q_2 = \frac{A_2\varepsilon_2}{1 - \varepsilon_2}(E_{b2} - J_2) = A_2(J_2 - J_1F_{2-1} - J_2F_{2-2})$$

Substituting the appropriate expressions for F_{2-1} and F_{2-2} yields the relation $q_2 = A_1(-J_1 + J_2) = -q_1$, as expected from an overall heat balance. Eliminating J_2 and substituting for J_1 in the heat loss equation for A_1 gives

$$q_1 = \frac{A_1(E_{b1} - E_{b2})}{1/\varepsilon_1 + (A_1/A_2)[(1 - \varepsilon_2)/\varepsilon_2]} \tag{9.74}$$

From the analogous network in Fig. 9.38(a), the sum of the three resistances is

$$\frac{1 - \varepsilon_1}{\varepsilon_1 A_1} + \frac{1}{A_1F_{1-2}} + \frac{1 - \varepsilon_2}{\varepsilon_2 A_2} = \frac{1}{A_1}\left[\frac{1}{\varepsilon_1} + \frac{A_1}{A_2}\left(\frac{1 - \varepsilon_2}{\varepsilon_2}\right)\right]$$

which gives the identical result for the net rate of heat loss from A_1, as expected.

The net rate of heat transfer in simple systems where radiation is transferred only between two gray surfaces can also be written in terms of an equivalent conductance $A_1\mathscr{F}_{1-2}$ in the form

$$q_{1\rightleftharpoons 2} = A_1\mathscr{F}_{1-2}(E_{b1} - E_{b2}) \tag{9.75}$$

In Eq. (9.75) A_1 is the smaller of the two surfaces and $\mathscr{F}_{1-2}$ is given below for some confiurations.

For two infinitely long concentric cylinders or two concentric spheres,

$$\mathscr{F}_{1-2} = \frac{1}{[(1 - \varepsilon_1)/\varepsilon_1] + 1 + [A_1(1 - \varepsilon_2)/A_2\varepsilon_2]} \tag{9.76}$$

For two equal parallel plates of the same emissivity ε spaced a finite distance apart,

$$\mathscr{F}_{1-2} = \frac{\varepsilon[1 + (1 - \varepsilon)F_{1-2}]}{1 + [(1 - \varepsilon)F_{1-2}]^2} \tag{9.77}$$

where the shape factor F_{1-2} can be obtained from Fig. 9.29. For two infinitely large parallel plates,

$$\mathscr{F}_{1-2} = \frac{1}{1/\varepsilon_1 + 1/\varepsilon_2 - 1} \tag{9.78}$$

For a small graybody of area A_1 inside a large enclosure of area $A_2 (A_1 \ll A_2)$,

$$\mathscr{F}_{1-2} = \varepsilon_1$$

In many real problems, radiation heat transfer will cause the internal energy and the temperature of a body to change. The heat transfer rate should then be interpreted as a quasi-steady-state result. Under those circumstances, the solution will require a transient analysis similar to that prescribed in Chapter 2, with the surface temperature of the body a function of time.

EXAMPLE 9.11 Liquified oxygen (boiling temperature, −297°F) is to be stored in a spherical container with 1 ft in diameter. The system is insulated by an evacuated space between the inner sphere and a surrounding 1.5-ft-ID concentric sphere as shown in Fig. 9.40. Both spheres are made of polished aluminum ($\varepsilon = 0.03$), and the temperature of the outer sphere is 30°F. Estimate the rate of heat flow by radiation to the oxygen in the container.

SOLUTION Although the internal energy of the oxygen will change, its temperature will remain constant since it is undergoing a change in phase. The absolute temperatures of the surfaces are

$$T_1 = 460 - 297 = 163\,\text{R}$$

$$T_2 = 460 + 30 = 490\,\text{R}$$

From Eq. (9.74) the rate of heat transfer from the inner sphere is

$$q_1 = \frac{A_1\sigma(T_1^4 - T_2^4)}{1/\varepsilon_1 + (A_1/A_2)[(1 - \varepsilon_2)/\varepsilon_2]} = \frac{\pi \times 0.1714(1.63^4 - 4.9^4)}{1/0.03 + (1/2.25)(0.97/0.03)}$$
$$= -6.5\,\text{Btu/h}$$

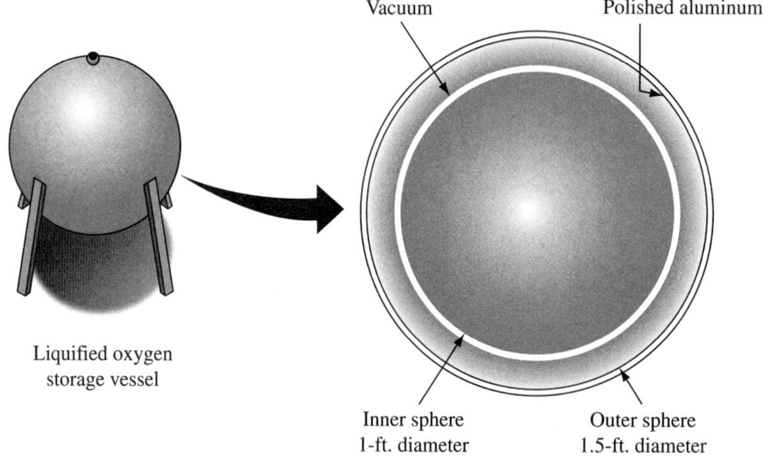

FIGURE 9.40 Sketch for Example 9.11.

Since the radiation heat transfer from A_1 is negative, the heat is actually transferred to the oxygen, as expected.

The radiant heat flow in an enclosure consisting of two gray surfaces connected by reradiating surfaces can also be solved without difficulty by means of the equivalent circuit. According to Eqs. (9.72) and (9.73), it is only necessary to replace E_{b1}

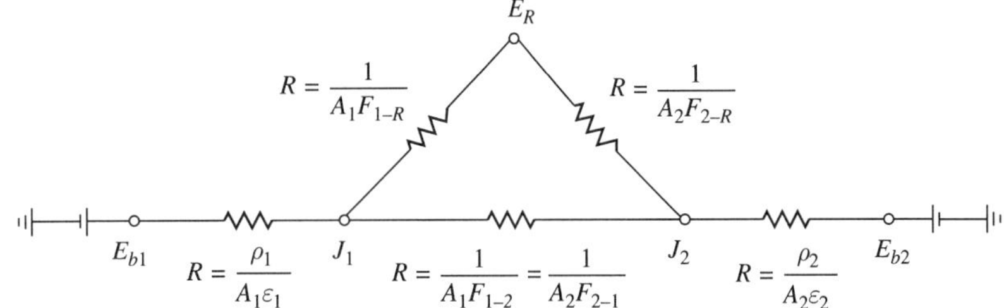

FIGURE 9.41 Analogous circuit for radiation in an enclosure consisting of two gray surfaces connected by a reradiating surface.

and E_{b2}, the potentials used in Section 9.5 for black surfaces, by J_1 and J_2 and connect the new potentials with the resistances $\rho_1/\varepsilon_1 A_1$ and $\rho_2/\varepsilon_2 A_2$ to their respective blackbody potentials E_{b1} and E_{b2}. The resulting network is shown in Fig. 9.41, and from it we see that the total conductance between E_1 and E_2 is now

$$A_1\mathcal{F}_{1-2} = \frac{1}{\dfrac{\rho_1}{\varepsilon_1 A_1} + \dfrac{\rho_2}{\varepsilon_2 A_2} + \dfrac{1}{A_1[F_{1-2} + 1/(1/F_{1-R} + A_1/A_2F_{2-R})]}}$$

where the last term of the denominator is the conductance for the blackbody network given by Eq. (9.64). The expression for the conductance can be recast into the more convenient form

$$A_1\mathcal{F}_{1-2} = \frac{1}{A_1\left(\dfrac{1}{\varepsilon_1} - 1\right) + \dfrac{1}{A_2}\left(\dfrac{1}{\varepsilon_2} - 1\right) + \dfrac{1}{A_1\bar{F}_{1-2}}} \tag{9.79}$$

where $A_1\bar{F}_{1-2}$ is the effective conductance for the blackbody network, equal to the inverse of the last term in the denominator of the original expression. The equation for the net radiant heat transfer per unit time between two gray surfaces at uniform temperatures in the presence of reradiating surfaces can then be written

$$q_{1\rightleftharpoons 2} = A_1\mathcal{F}_{1-2}\sigma(T_1^4 - T_2^4) \tag{9.80}$$

For enclosures consisting of several surfaces, the radiation heat transfer from any one of them can be calculated by drawing the analogous circuit and performing a circuit analysis. This analysis can be made by applying Kirchhoff's current law, which states that the sum of the currents entering a given node is zero. When a computer is available, the same result can be obtained by a matrix method outlined in Section 9.7.

9.7* Matrix Inversion

Matrix methods were used in Chapter 3 to solve conduction problems numerically. The matrix inversion method is also a powerful tool for solving radiation problems, although it requires certain assumptions and simplifications in practice. The method can be applied only when the radiation over each surface is uniform and each surface is isothermal. Any surface in the enclosure that does not meet these two requirements must be subdivided into smaller segments until the temperature and radiation flux over each are approximately uniform. However, with a computer, the addition of surfaces does not significantly increase the amount of work required to obtain a numerical solution [5, 13].

9.7.1 Enclosures with Gray Surfaces

The problem at hand is solving N linear algebraic equations in N unknowns. The equations are obtained by evaluating the emissivities of the surfaces and the shape factors between them and writing Eqs. (9.68) and (9.71) for each nodal point:

$$(q_i)''_{\text{net}} = \frac{\varepsilon_i}{\rho_i}(E_{bi} - J_i) = \frac{\varepsilon_i}{1 - \varepsilon_i}(E_{bi} - J_i) \tag{9.68}$$

and

$$(q_i)''_{\text{net}} = J_i - \sum_{j=1}^{j=N} J_j F_{i-j} \tag{9.71}$$

For a gray enclosure consisting of three surfaces at specified temperatures, this procedure yields

$$(q_1)''_{\text{net}} = \frac{\varepsilon_1}{1 - \varepsilon_1}(E_{b1} - J_1) = J_1 - J_1F_{1-1} - J_2F_{1-2} - J_3F_{1-3} \tag{9.81a}$$

$$(q_2)''_{\text{net}} = \frac{\varepsilon_2}{1 - \varepsilon_2}(E_{b2} - J_2) = J_2 - J_1F_{2-1} - J_2F_{2-2} - J_3F_{2-3} \tag{9.81b}$$

$$(q_3)''_{\text{net}} = \frac{\varepsilon_3}{1 - \varepsilon_3}(E_{b3} - J_3) = J_3 - J_1F_{3-1} - J_2F_{3-2} - J_3F_{3-3} \tag{9.81c}$$

In this set of equations, $N = 3$ and the three unknowns are the radiosities J_1, J_2, and J_3. The above set of equations can be recast into the more convenient form

$$\left(1 - F_{1-1} + \frac{\varepsilon_1}{1 - \varepsilon_1}\right)J_1 + (-F_{1-2})J_2 + (-F_{1-3})J_3 = \frac{\varepsilon_1}{1 - \varepsilon_1}E_{b1} \tag{9.82a}$$

$$(-F_{2-1})J_1 + \left(1 - F_{2-2} + \frac{\varepsilon_2}{1 - \varepsilon_2}\right)J_2 + (-F_{1-3})J_3 = \frac{\varepsilon_2}{1 - \varepsilon_2}E_{b2} \tag{9.82b}$$

$$(-F_{3-1})J_1 + (-F_{3-2})J_2 + \left(1 - F_{3-3} + \frac{\varepsilon_3}{1 - \varepsilon_3}\right)J_3 = \frac{\varepsilon_3}{1 - \varepsilon_3}E_{b3} \tag{9.82c}$$

Using matrix notation, we get

$$a_{11}J_1 + a_{12}J_2 + a_{13}J_3 = C_1 \tag{9.83a}$$

$$a_{21}J_1 + a_{22}J_2 + a_{23}J_3 = C_2 \tag{9.83b}$$

$$a_{31}J_1 + a_{32}J_2 + a_{33}J_3 = C_3 \tag{9.83c}$$

These equations can be written in the condensed matrix form presented in Chapter 3:

$$\mathbf{AJ} = \mathbf{C} \tag{9.84}$$

where $\mathbf{A}$ is the 3×3 matrix

$$\mathbf{A} = \begin{bmatrix} a_{11} & a_{12} & a_{13} \\ a_{21} & a_{22} & a_{23} \\ a_{31} & a_{32} & a_{33} \end{bmatrix} \tag{9.85}$$

and **J** and **C** are vectors consisting of three elements each:

$$\mathbf{J} = \begin{bmatrix} J_1 \\ J_2 \\ J_3 \end{bmatrix} \tag{9.86}$$

$$\mathbf{C} = \begin{bmatrix} \dfrac{\varepsilon_1}{1 - \varepsilon_1} E_{b1} \\ \dfrac{\varepsilon_2}{1 - \varepsilon_2} E_{b2} \\ \dfrac{\varepsilon_3}{1 - \varepsilon_3} E_{b3} \end{bmatrix} = \begin{bmatrix} C_1 \\ C_2 \\ C_3 \end{bmatrix} \tag{9.87}$$

For the general case of an enclosure with N surfaces the matrix will have the same form as Eq. (9.84), but

$$\mathbf{A} = \begin{bmatrix} a_{11} & a_{12} & \cdots & a_{1N} \\ a_{21} & a_{22} & \cdots & \\ a_{31} & & & \\ \vdots & & & \\ a_{N1} & a_{N2} & \cdots & a_{NN} \end{bmatrix}, \quad \mathbf{C} = \begin{bmatrix} C_1 \\ C_2 \\ \vdots \\ C_4 \end{bmatrix}, \quad \mathbf{J} = \begin{bmatrix} J_1 \\ J_2 \\ \vdots \\ J_N \end{bmatrix}$$

The off-diagonal elements of **A** are

$$a_{ij} = -F_{i-j} \qquad (i \neq j) \tag{9.88}$$

and the diagonal terms are

$$a_{ii} = \left(1 - F_{ii} + \frac{\varepsilon_i}{1 - \varepsilon_i}\right) \tag{9.89}$$

The elements of **C** are

$$C_i = \frac{\varepsilon_i}{1 - \varepsilon_i} E_{bi} \tag{9.90}$$

When a surface in the enclosure is black and its temperature T_i is specified, the radiosity J_i is equal to E_{bi}. Hence, it is no longer unknown and the terms in the matrix for a black element are

$$a_{ij} = 0 \quad (i \neq j) \tag{9.91}$$

$$a_{ii} = 1.0 \tag{9.92}$$

$$C_i = E_{bi} = \sigma T^4 \tag{9.93}$$

When the heat flux instead of the temperature is specified for a surface A_i, the off-diagonal elements of **A** remain the same as in Eq. (9.88). However, the diagonal elements, a_{ii}, become

$$a_{ii} = 1 - F_{ii} \tag{9.94}$$

and the elements in the **C** matrix are

$$C_i = (q_i)''_{\text{net}} \tag{9.95}$$

This can easily be verified for a three-surface enclosure by inspection of Eq. (9.81). For example, if the heat flux for surface 1 were specified, Eq. (9.82a) becomes, upon eliminating the unknown E_{b1},

$$(q_1)''_{\text{net}} = (1 - F_{1-1})J_1 + (-F_{1-2})J_2 + (-F_{1-3})J_3 \tag{9.96}$$

To obtain a numerical solution we must invert the matrix **A**. If $\mathbf{A}^{-1}$ denotes the inverse of **A**, the solution for the radiosities is given by

$$\mathbf{J} = \mathbf{A}^{-1}\mathbf{C} \tag{9.97}$$

where

$$\mathbf{A}^{-1} = \begin{bmatrix} b_{11} & b_{12} & \cdots & b_{1N} \\ b_{21} & \cdots & & \\ \vdots & & & \\ b_{N1} & b_{N2} & \cdots & b_{NN} \end{bmatrix} \tag{9.98}$$

The solution for each radiosity can then be written in the form of a series:

$$\begin{aligned} J_1 &= b_{11}C_1 + b_{12}C_2 + \cdots + b_{1N}C_N \\ J_2 &= b_{21}C_1 + b_{22}C_2 + \cdots + b_{2N}C_N \\ &\vdots \\ J_N &= b_{N1}C_1 + b_{N2}C_2 + \cdots + b_{NN}C_N \end{aligned} \tag{9.99}$$

In practical terms, the problem of solving the simultaneous linear algebraic equations for the radiosities reduces to the inversion of a matrix. Once the radiosities are known, the rate of heat flow can be obtained from Eq. (9.71) for each surface. When the heat flux is specified, Eq. (9.68) can be solved for the temperatures T_i,

$$T_i = \left[\frac{1 - \varepsilon_i}{\sigma\varepsilon_i}(q_i)''_{\text{net}} + J_i/\sigma\right]^{1/4} \tag{9.100}$$

The following examples illustrate this procedure.

EXAMPLE 9.12 The temperatures of the top and bottom surfaces of the frustum of the cone shown in Fig. 9.42 are maintained at 600 K and 1200 K, respectively, while side A_2 is perfectly insulated ($q_2 = 0$). If all surfaces are gray and diffuse, determine the net radiative exchange between the top and bottom surfaces, i.e., A_3 and A_1.

SOLUTION From Table 9.3 we find that $F_{31} = 0.333$ and from Eq. (9.60) we obtain $F_{32} = 1 - F_{31} = 0.667$.

According to the reciprocity theorem, $A_1F_{13} = A_3F_{31}$ and $A_2F_{23} = A_3F_{32}$. Therefore, $F_{13} = 0.147$ and $F_{23} = 0.130$. From Eq. (9.60) we get $F_{12} = 1 - F_{13} = 0.853$, and by reciprocity, $F_{21} = F_{12}A_1/A_2 = 0.372$. Finally, $F_{22} = 1 - F_{21} - F_{23} = 0.498$.

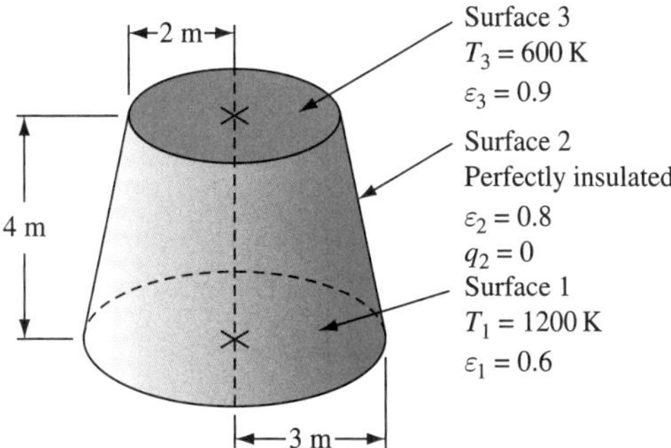

FIGURE 9.42 Schematic sketch of cone for Example 9.12.

According to the general relations given by Eqs. (9.68) and (9.71), the system of equations to be solved for this problem can be written:

$$E_{b1} \cdot \frac{\varepsilon_1}{1 - \varepsilon_1} = J_1\left(1 - F_{11} + \frac{\varepsilon_1}{1 - \varepsilon_1}\right) + J_2(-F_{12}) + J_3(-F_{13})$$

$$0 = J_1(-F_{21}) + J_2(1 - F_{22}) + J_3(-F_{23})$$

$$E_{b3} \cdot \frac{\varepsilon_3}{1 - \varepsilon_3} = J_1(-F_{31}) + J_2(-F_{32}) + J_3\left(1 - F_{33} + \frac{\varepsilon_3}{1 - \varepsilon_3}\right)$$

or in matrix notation $\mathbf{A} \cdot \mathbf{J} = \mathbf{C}$.

Linear algebraic systems of equation of the form $\mathbf{A} \cdot \mathbf{X} = \mathbf{B}$ can be readily solved to evaluate all the J's by either using MATLAB or writing a simple program in C++. The net rate of heat transfer between top and bottom, i.e., the value of $q_{3\rightleftharpoons 1}$, can then be determined from Eq. (9.73). Figure 9.43 gives the flow diagram

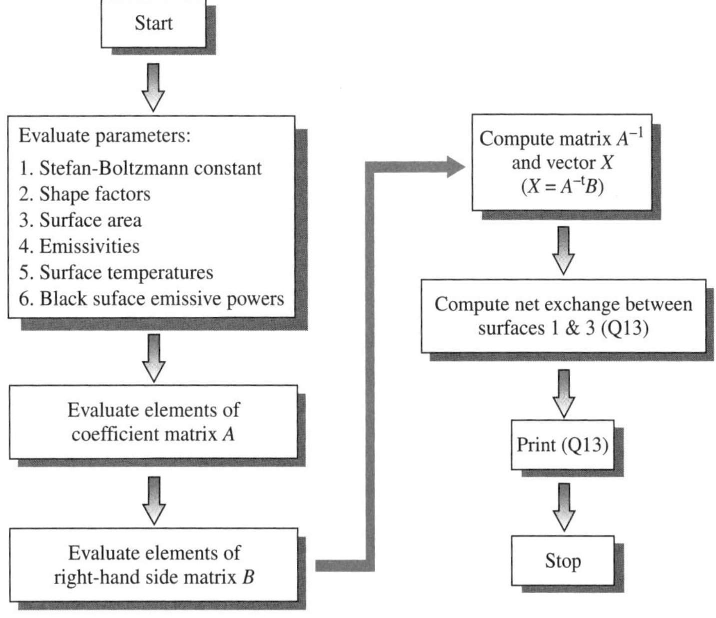

FIGURE 9.43 Flowchart for Example 9.12.

or algorithm for the computer operations to solve this problem. The MATLAB program and the solution are is presented in Table 9.4, and the symbols used in this program are defined in Table 9.5.

TABLE 9.4 MATLAB program for Example 9.12

```
% Provide all given inputs and constants of the problem
SIGMA=0.567E-07; % Stefan-Boltzmann constant (W/m^2/K^4)
AR(1)=9*pi; %Area(1)=R1^2*pi

% The physical parameters, e.g., shape factor and emissivity, are evaluated.

F(1,1)=0.0;
F(1,2)=0.853;
F(1,3)=0.147;
F(2,1)=0.372;
F(2,2)=0.498;
F(2,3)=0.130;
F(3,1)=0.333;
F(3,2)=0.667;
F(3,3)=0.0;
ESP(1)=0.6;
ESP(3)=0.9;
T(1)=1200;
T(3)=600;
EB(1)=SIGMA*T(1)^4;
EB(3)=SIGMA*T(3)^4;

% The values of the elements of the coefficient matrix A in the equation

%[A][X]=[B] are specified
A(1,1)=1-F(1,1)+ESP(1)/(1-ESP(1));
A(1,2)=-F(1,2);
A(1,3)=-F(1,3);
A(2,1)=-F(2,1);
A(2,2)=1-F(2,2);
A(2,3)=-F(2,3);
A(3,1)=-F(3,1);
A(3,2)=-F(3,2);
A(3,3)=1-F(3,3)+ESP(3)/(1-ESP(3));

% The values of the right-hand side vector B are specified.

B(1)=EB(1)*ESP(1)/(1-ESP(1));
B(2)=0;
B(3)=EB(3)*ESP(3)/(1-ESP(3));

% The inversion routine is used to solve for X

X=inv(A)*B' % solutions for J
```

TABLE 9.5 Symbols and function notations used in MATLAB program for Example 9.12

MATLAB Symbol	Heat Balance Equation Notation	Description	Units
A(I,J)	a_{ij}	coefficient of matrix elements	—
AR(1), AR(3)	A_1, A_3	lower and upper surface areas	m^2
B(I)	C_i	right-hand-side matrix elements	W/m^2
EB(1), EB(3)	E_{b1}, E_{b3}	blackbody emissive powers	W/m^2
ESP(1), etc.	ε_1, etc.	total hemispheric emissivity	—
F(1,1), F(1,2), etc.	F_{11}, F_{12}, etc.	shape factors	—
pi	π	3.1459 . . .	—
Q31	$q_{3\rightleftharpoons 1}$	net exchange between surfaces 3 and 1	W
SIGMA	σ	Stefan-Boltzmann constant (0.567×10^{-7})	W/m^2K^4
T(1), T(3)	T_1, T_3	surface temperatures	K
X(I)	J_i	radiosities (elements of solution vector)	W/m^3

EXAMPLE 9.13 Determine the temperature of surface 1 for the cone shown in Fig. 9.42 if $q_1 = 3 \times 10^5 \, W/m^2$ and $\varepsilon_3 = 1$. Assume that all other parameters are the same as in Example 9.12.

SOLUTION From Eqs. (9.94), (9.95), and (9.97), the following system of equations must be solved for J_1, J_2, and J_3.

$$
\begin{aligned}
q_1/A_1 &= J_1(1 - F_{11}) + J_2(-F_{12}) + J_3(-F_{13}) \\
0 &= J_1(-F_{21}) + J_2(1 - F_{22}) + J_3(-F_{23}) \\
E_{b3} &= J_3
\end{aligned}
$$

Once the J_1's are known, Eq. 9.100 gives T_1. The MATLAB program for the solution of this problem is shown in Table 9.6. Since it is very similar to the preceding program the flow diagram is essentially the same as that used in Example 9.12.

TABLE 9.6 MATLAB program for Example 9.13

```
% Provide all given inputs and constants of the problem
SIGMA=0.567E-07; % Stefan-Boltzmann constant (W/m^2/K^4)
F(1,1)=0.0; %F(I,J) shape factor
F(1,2)=0.853;
F(1,3)=0.147;
F(2,1)=0.372;
F(2,2)=0.498;
F(2,3)=0.130;
F(3,1)=0.333;
F(3,2)=0.667;
F(3,3)=0.0;
```

(Continued)

TABLE 9.6 *(Continued)*

```
AR(1)=9*pi; %Area(1)=R1^2*pi
ESP(1)=0.6; %ESP total hemispheric emissivity
ESP(3)=0.9;
Q1=300000;
T(3)=600;
EB(3)=SIGMA*T(3)^4; %EB blackbody emissive powers
% Evaluate elements of coefficient matrix
A(1,1)=1-F(1,1);
A(1,2)=-F(1,2);
A(1,3)=-F(1,3);
A(2,1)=-F(2,1);
A(2,2)=1-F(2,2);
A(2,3)=-F(2,3);
A(3,1)=0;
A(3,2)=0;
A(3,3)=1;
% Evaluate elements of right hand side matrix
B(1)=Q1/AR(1);
B(2)=0;
B(3)=EB(3);
% solve the system of equations for X
X=inv(A)*B';
T(1)=((X(1)+Q1*(1-ESP(1))/(AR(1)*ESP(1)))/SIGMA)^0.25
%solution for temperatures
T1=T(1) %Value for the required temperature in K
```

9.7.2 Enclosure with Nongray Surfaces

The method of approach used to calculate heat transfer in gray surface enclosures can easily be adapted to nongray surfaces. If the surface properties are functions of wavelength, they can be approximated by gray "bands" within which an average value of emissivity and absorptivity is used. Then, the same calculation method used previously for gray enclosures can be used to determine the radiation heat transfer within each band. The following example illustrates the procedure.

EXAMPLE 9.14 Determine the rate of heat transfer between two large parallel flat plates placed 2 in. apart, if one plate (A) is at 2040°F and the other (B) at 540°F. Plate A has an emissivity of 0.1 between 0 and 2.5 μm and an emissivity of 0.9 for wavelengths longer than 2.5 μm. The emissivity of plate B is 0.9 between 0 and 4.0 μm and 0.1 at longer wavelengths.

SOLUTION The shape factor F_{A-B} for two large parallel rectangular plates is 1.0 if end effects are negligible. The radiosity of A is given by

$$\int_0^\infty J_{\lambda A}\,d\lambda = \int_0^\infty \varepsilon_{\lambda A} E_{b\lambda A}\,d\lambda + \int_0^\infty \rho_{\lambda A} G_{\lambda A}\,d\lambda$$

and the radiosity of B by

$$\int_0^{\infty} J_{\lambda B}\, d\lambda = \int_0^{\infty} \varepsilon_{\lambda B} E_{b\lambda B}\, d\lambda + \int_0^{\infty} \rho_{\lambda B} G_{\lambda B}\, d\lambda$$

However, using spectral bands between 0 and 2.5 μm, 2.5 and 4.0 μm, and 4.0 μm or larger, the system obeys gray surface radiation laws within each band and the rate of heat transfer can be calculated from Eq. (9.75) in three bands as shown below:

Band 1:

$$q_{A \rightleftharpoons B}\Big|_{0}^{2.5\,\mu\text{m}} = \mathscr{F}_{A-B}(\varepsilon_A = 0.1, \varepsilon_B = 0.9) \times \left[\frac{E_{b,0-2.5}(T_A)}{E_{b,0-\infty}(T_A)}\sigma T_A^4 - \frac{E_{b,0-2.5}(T_B)}{E_{b,0-\infty}(T_B)}\sigma T_B^4\right]$$

Band 2:

$$q_{A \rightleftharpoons B}\Big|_{2.5\,\mu\text{m}}^{4.0\,\mu\text{m}} = \mathscr{F}_{A-B}(\varepsilon_A = 0.9, \varepsilon_B = 0.9) \times \left[\frac{E_{b,2.5-4.0}(T_A)}{E_{b,0-\infty}(T_A)}\sigma T_A^4 - \frac{E_{b,2.5-4.0}(T_B)}{E_{b,0-\infty}(T_B)}\sigma T_B^4\right]$$

Band 3:

$$q_{A \rightleftharpoons B}\Big|_{4.0\,\mu\text{m}}^{\infty} = \mathscr{F}_{A-B}(\varepsilon_A = 0.9, \varepsilon_B = 0.1) \times \left[\frac{E_{b,4.0-\infty}(T_A)}{E_{b,0-\infty}(T_A)}\sigma T_A^4 - \frac{E_{b,4.0-\infty}(T_B)}{E_{b,0-\infty}(T_B)}\sigma T_B^4\right]$$

where

$$\mathscr{F}_{A-B} = \frac{1}{1/\varepsilon_A + 1/\varepsilon_B - 1}$$

The percentage of the total radiation within a given band is obtained from Table 9.1. For example, $(E_{b,0-2.5}/E_{b,0-\infty})$ for a temperature of $T_A = 2500\,\text{R}$ is 0.375, and for a temperature of $T_B = 1000\,\text{R}$ it is about 0.004. Thus, for the first band,

$$q_{A \rightleftharpoons B1}\Big|_{0}^{2.5\,\mu\text{m}} = 0.10 \times 0.1714(0.375 \times 25^4 - 0.004 \times 10^4)$$
$$= 2530\,\text{Btu/h ft}^2$$

Similarly, for the second band,

$$q_{A \rightleftharpoons B2}\Big|_{2.5\,\mu\text{m}}^{4.0\,\mu\text{m}} = 23{,}000\,\text{Btu/h ft}^2$$

and for the third band,

$$q_{A \rightleftharpoons B3}\Big|_{4.0\,\mu\text{m}}^{\infty} = 1240\,\text{Btu/h ft}^2$$

Finally, summing over all three bands, the total rate of radiation heat transfer is

$$q_{A\rightleftharpoons B}\Big|_0^\infty = \sum_{N=1}^{N=3} q_{A\rightleftharpoons BN} = 2530 + 23{,}000 + 1240 = 26{,}770\,\text{Btu/h ft}^2$$

It should be noted that most of the radiation is transferred within the second band, where both surfaces are nearly black.

Enclosures consisting of several nongray surfaces can be treated in a similar manner by dividing the radiation spectrum into finite bands within which the radiation properties can be approximated by constant values. This procedure can become particularly useful when the enclosure is filled with a gas that absorbs and emits radiation only at certain wavelengths.

9.7.3* Enclosures with Absorbing and Transmitting Media

The method of analysis outlined in the preceding sections can be extended to solve problems in which heat is transferred by radiation in an enclosure containing a medium that is both absorbing and transmitting. Various glasses, plastics, and gases are examples of such media. To illustrate the method of approach, we will first consider radiation between two plates when the space between them is filled with a "gray" gas that does not reflect any of the incident radiation. The geometry is shown in Fig. 9.44(a). The two solid surfaces are at temperatures T_1 and T_2; the properties of the transmitting gas are denoted by the subscript m.

Kirchhoff's law applied to the transmitting gray gas requires that $\alpha_m = \varepsilon_m$, and since the reflectivity of the medium is zero,

$$\tau_m = 1 - \alpha_m = 1 - \varepsilon_m \tag{9.101}$$

We will derive the equations for the heat transfer rate between the surfaces by developing the thermal circuit for the problem. The portion of the total radiation leaving surface 1 that arrives at surface 2 after passing through the gas is

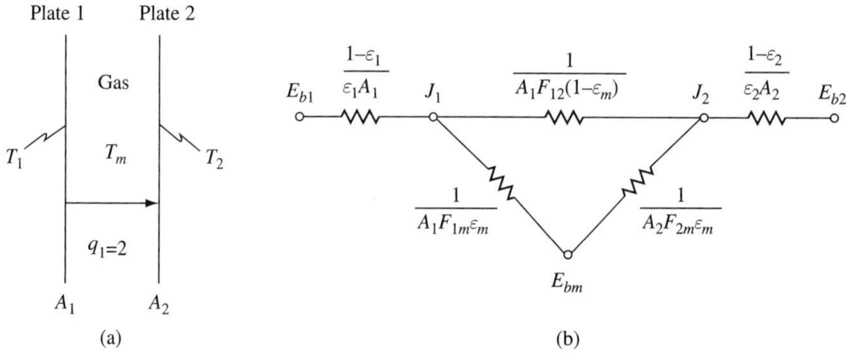

FIGURE 9.44 Electric analogy for radiation between finite plates separated by a gas.

$$J_1A_1F_{12}\tau_m$$

and the radiation from surface 2 that reaches 1 is

$$J_2A_2F_{12}\tau_m$$

The net rate of heat transfer between the two surfaces is therefore

$$q_{1\rightleftharpoons 2} = A_1F_{12}\tau_m(J_1 - J_2) = \frac{J_1 - J_2}{1/A_1F_{12}(1 - \varepsilon_m)} \tag{9.102}$$

Thus, for this case the equivalent resistance between nodal points J_1 and J_2 in a network will be $1/A_1F_{12}(1 - \varepsilon_m)$.

Radiation heat transfer also occurs between each of the surfaces and the gas. If the gas is at temperature T_m, it will emit radiation at a rate

$$J_m = \varepsilon_m E_{bm} \tag{9.103}$$

The fraction of the energy emitted by the gaseous medium that reaches surface 1 is

$$A_mF_{m-1}J_m = A_mF_{m-1}\varepsilon_m E_{bm} \tag{9.104}$$

Similarly, the fraction of the radiation leaving A_1 that is absorbed by the transparent medium is

$$J_1A_1F_{1m}\alpha_m = J_1A_1F_{1m}\varepsilon_m \tag{9.105}$$

The net rate of heat transfer by radiation between the gas and surface 1 is the difference between the radiation emitted by the gas toward A_1 and the radiation emanating from A_1 that is absorbed by the gas. Thus

$$q_{m\rightleftharpoons 1} = A_mF_{m1}\varepsilon_m E_{bm} - J_1A_1F_{1m}\varepsilon_m \tag{9.106}$$

Using the reciprocity theorem, $A_1F_{1m} = A_mF_{m1}$, the net exchange can be written in the form

$$q_{m=1} = \frac{E_{bm} - J_1}{1/A_1F_{1m}\varepsilon_m} \tag{9.107}$$

Similarly, the net exchange between the gas and A_2 is

$$q_{m=2} = \frac{E_{bm} - J_2}{1/A_2F_{2m}\varepsilon_m} \tag{9.108}$$

Using the above relations to construct an equivalent circuit, radiation between two surfaces at T_1 and T_2, respectively, separated by an absorbing medium at T_m, can be represented as shown in Fig. 9.44(b). If the gas is not maintained at a specified temperature but reaches an equilibrium temperature at which it emits radiation at the same rate at which it absorbs it, E_{bm} becomes a floating node in the network. For this case, the net rate of heat transfer between A_1 and A_2 is

$$q_{1=2} = \frac{\sigma(T_1^4 - T_2^4)}{\dfrac{1 - \varepsilon_1}{\varepsilon_1 A_1} + \dfrac{1 - \varepsilon_2}{\varepsilon_2 A_2} + \dfrac{1}{A_1[F_{1-2}\tau_m + 1/(F_{1-m}\varepsilon_m + A_1/A_2F_{2-m}\varepsilon_m)]}} \tag{9.109}$$

When A_1 and A_2 are so large that F_{1-2}, F_{1-m}, and F_{2-m} approach unity, the last factor in the denominator approaches $1/(A_1[\tau_m + 2/\varepsilon_m])$.

More complex enclosures with several surfaces can be treated by the matrix method once the appropriate thermal network has been drawn. Details for method of solution of such cases can be found in advanced radiation texts [12, 13].

9.8* Radiation Properties of Gases and Vapors

In this section, we shall consider some basic concepts of gaseous radiation. A comprehensive treatment of this subject is beyond the scope of this text, and the reader should consult references [13, 15, 19, 20–27] for details of the theoretical background and complete derivations of the calculation techniques.

Elementary gases such as O_2, N_2, H_2, and dry air have a symmetrical molecular structure and neither emit nor absorb radiation unless they are heated to such extremely high temperatures that they become ionized plasmas and electronic energy transformations occur. On the other hand, gases that have polar molecules with an electronic moment such as a dipole or quadrupole absorb and emit radiation in limited spectral ranges called bands. In practice, the most important of these gases are H_2O, CO_2, CO, SO_2, NH_3, and the hydrocarbons. These gases are asymmetric in one or more of their modes of vibration. During molecular collisions, rotation and vibrations of individual atoms in a molecule can be excited so that atoms that possess free electrical charges can emit electromagnetic waves. Similarly, when radiation of the appropriate wavelength impinges on such a gas, it can be absorbed in the process. We shall restrict our consideration here to the evaluation of the radiation properties of H_2O and CO_2. They are the most important gases in thermal radiation calculations and also illustrate the basic principles of gaseous radiation.

Typical changes in energy level due to changes in vibrational frequency or rotation manifest themselves in a strong peak at the wavelength corresponding to the vibrational transformation, with multiple rotational energy changes slightly above or below the peak. This process results in absorption or emission bands. The shape and width of these bands depend on the temperature and pressure of the gas, while the magnitude of the monochromatic absorptivity is primarily a function of the thickness of the gas layer. The absorption spectrum of steam shown in Fig. 9.45 illustrates the complexity of the process. The most important absorption bands for steam lie between 1.7 and 2.0 μm, 2.2 and 3.0 μm, 4.8 and 8.5 μm, and 11 and 25 μm.

Experimental measurements generally yield the absorptivity of a gas layer over a band width corresponding to the width of the spectrometer slit used. Thus, experimental data are usually presented in terms of the monochromatic absorptivity, as shown in Fig. 9.45. For most engineering calculations, however, the quantity of primary interest is the effective total absorptivity or emissivity. This quantity assumes that the gas is gray and, as shown on the next page, its value depends not only on the pressure, temperature, and composition, but also on the geometry of the radiating gas.

Whereas the emission and absorption of radiation are surface phenomena for opaque solids, in calculating the radiation emitted or absorbed by a gas layer, its thickness, pressure, and shape as well as its surface area must be taken into account. When monochromatic radiation at an intensity $I_{\lambda 0}$ passes through a gas layer of

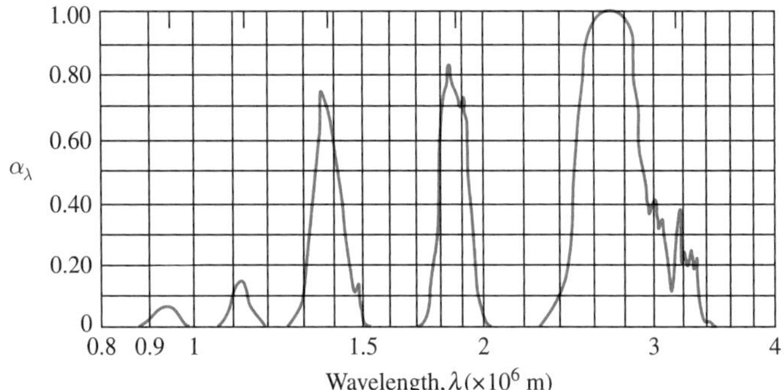

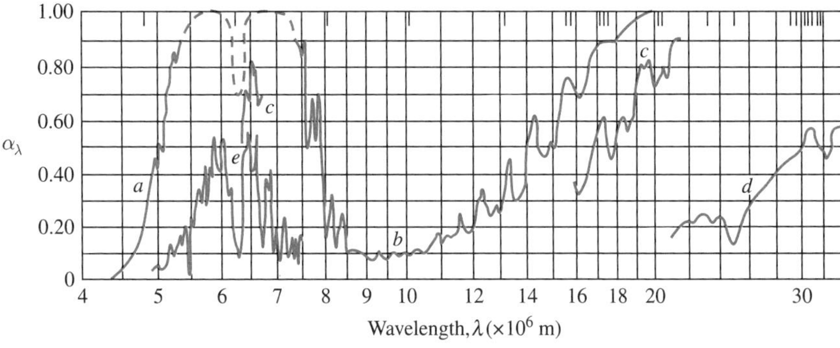

FIGURE 9.45 Monochromatic absorptivity of water vapor.

thickness L, the radiant-energy absorption in a differential distance dx is governed by the relation

$$dI_{\lambda x} = -k'_\lambda I_{\lambda x}\, dx \tag{9.110}$$

where $I_{\lambda x}$ = intensity at a distance x
k'_λ = monochromatic absorption coefficient, a proportionality constant whose value depends on the pressure and temperature of the gass

Integration between the limits $x = 0$ and $x = L$ yields

$$I_{\lambda L} = I_{\lambda 0} e^{-k'_\lambda L} \tag{9.111}$$

where $I_{\lambda L}$ is the intensity of radiation at L. The difference between the intensity of radiation entering the gas at $x = 0$ and the intensity of radiation leaving the gas layer at $x = L$ is the amount of energy absorbed by the gas

$$I_{\lambda 0} - I_{\lambda L} = I_{\lambda 0}(1 - e^{-k'_\lambda L}) = \alpha_{G\lambda} I_{\lambda 0} \tag{9.112}$$

The quantity in the parentheses represents the *monochromatic absorptivity of the gas*, $\alpha_G\lambda$, and according to Kirchhoff's law, it also represents the emissivity at the wavelength λ, $\varepsilon_{G\lambda}$. To obtain effective values of the emissivity or absorptivity, a summation over all of the radiation bands is necessary. We observe that, for large values of L, i.e., for thick layers, gas radiation approaches blackbody conditions within the wavelengths of its absorption bands.

For gas bodies of finite dimensions, however, the effective absorptivity or emissivity depends on the shape and the size of the gas body, since radiation is not confined to one direction. The precise method of calculating the effective absorptivity or emissivity is quite complex [15, 24–26], but for engineering calculations an approximate method developed by Hottel and Egbert [19, 28] yields results of satisfactory accuracy. Hottel evaluated the effective total emissivities of a number of gases at various temperatures and pressures, and presented his results in graphs similar to those shown in Figs. 9.46 and 9.47. The graphs apply strictly only to a system in which a hemispherical gas mass of radius L radiates to an element of surface located at the center of the base of a hemisphere. However, for shapes other than hemispheres, an effective beam length can be calculated. Table 9.7 on page 606 lists the constants by which the characteristic dimensions of several simple shapes are to be multiplied to obtain an equivalent mean hemispherical beam length L for use in Figs. 9.46 and 9.47. For

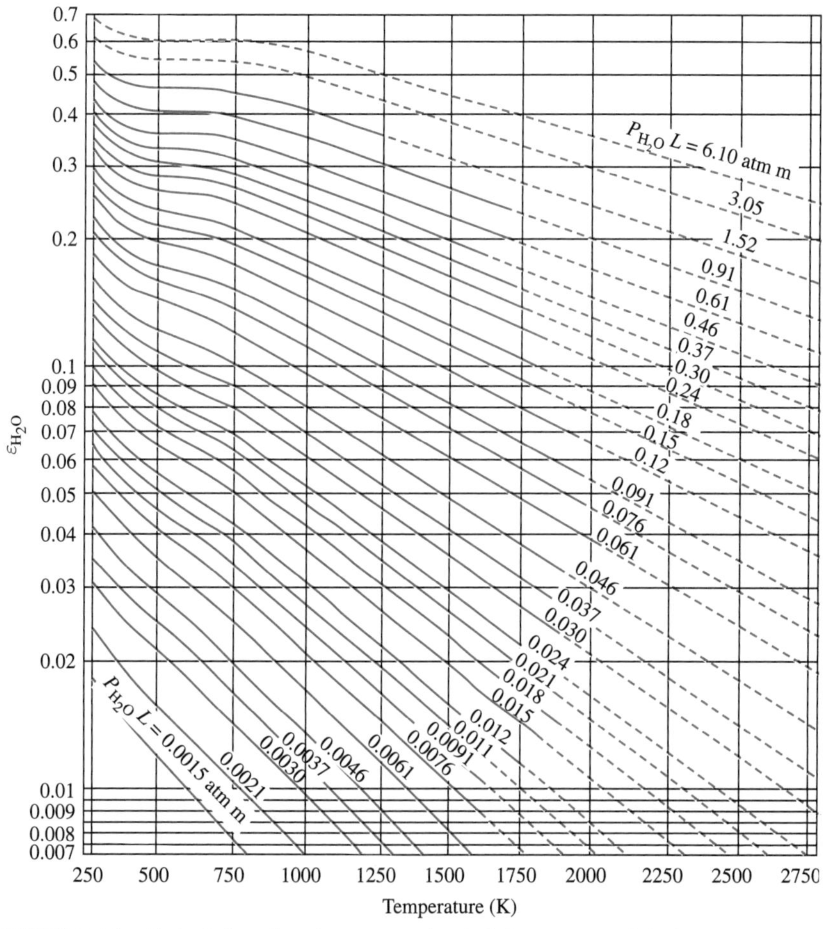

FIGURE 9.46 Emissivity of water vapor at a total pressure of 1 atm.

Source: HEAT TRANSMISSION by W. H. McAdams. Copyright 1954 by MCGRAW-HILL COMPANIES, INC. -BOOKS. Reproduced with permission of MCGRAW-HILL COMPANIES, INC. -BOOKS in the format Textbook via Copyright Clearance Center.

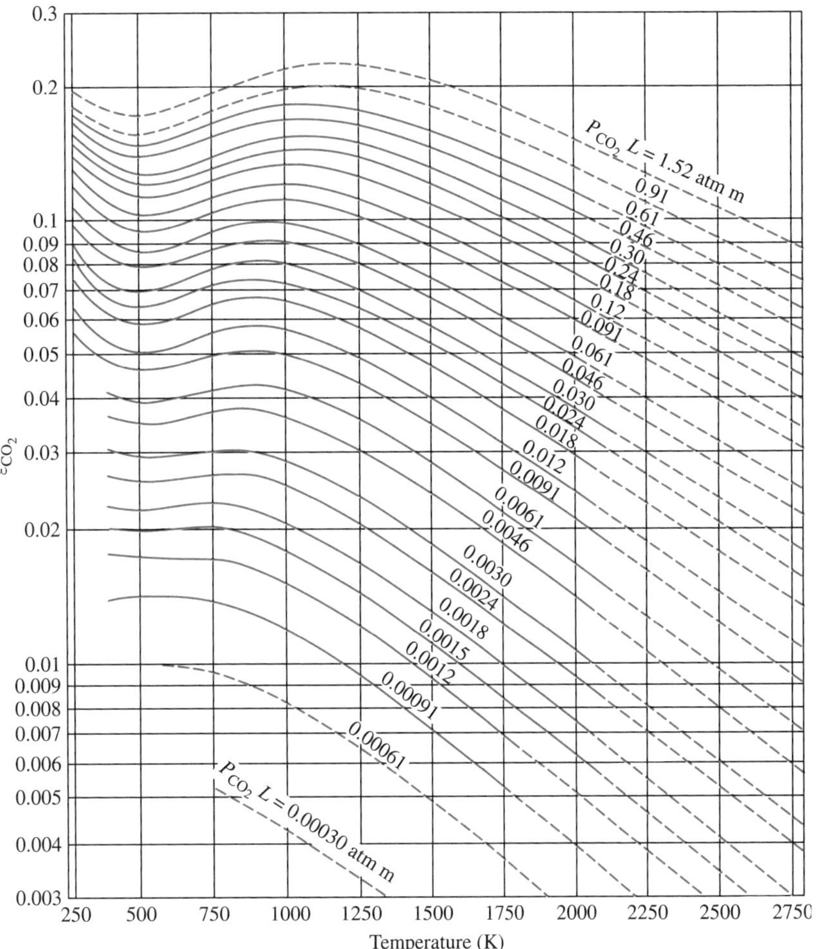

FIGURE 9.47 Emissivity of carbon dioxide at a total pressure of 1 atm.

Source: HEAT TRANSMISSION by W. H. McAdams. Copyright 1954 by MCGRAW-HILL COMPANIES, INC. -BOOKS. Reproduced with permission of MCGRAW-HILL COMPANIES, INC. -BOOKS in the format Textbook via Copyright Clearance Center.

approximate calculations and for shapes other than those listed in Table 9.7, L can be taken as 3.4 × volume/surface area.

In Figs. 9.46 and 9.47, the symbols P_{H_2O} and P_{CO_2} represent the partial pressures of the gases. The total pressure for both figures is 1 atm. When the total gas pressure differs from 1 atm, the values from Figs. 9.46 and 9.47 must be multiplied by a correction factor. The emissivities of H_2O and CO_2 at a total pressure P_T other than 1 atm are then given by the expressions [24]

$$(\varepsilon_{H_2O})P_T = C_{H_2O}(\varepsilon_{H_2O})P_T = 1 \tag{9.113a}$$

$$(\varepsilon_{CO_2})P_T = C_{CO_2}(\varepsilon_{CO_2})P_T = 1 \tag{9.113b}$$

and the correction factors C_{H_2O} and C_{CO_2} are plotted in Fig. 9.48 on the next page and Fig. 9.49 on page 607, respectively. When both H_2O and CO_2 exist in a mixture,

TABLE 9.7 Mean beam length of various gas shapes

Geometry	*L*
Sphere	2/3 (diameter)
Infinite cylinder	Diameter
Infinite parallel planes	2 (distance between planes)
Semi-infinite cylinder, radiating to center of base	Diameter
Right-circular cylinder, height equal to diameter:	
radiating to center of base	Diameter
radiating to whole surface	2/3 (diameter)
Infinite cylinder of half-circular cross section radiating to spot in middle of flat side	Radius
Rectangular parallelepipeds:	
cube	2/3 (edge)
1:1:4 radiating to 1 × 4 face	0.9 (shortest edge)
radiating to 1 × 1 face	0.86 (shortest edge)
radiating to all faces	0.891 (shortest edge)
Space outside infinite bank of tubes with centers on equilateral triangles:	
tube diameter = clearance	3.4 (clearance)
tube diameter = 1/2 (clearance)	4.44 (clearance)

Source: Rohsenow, Hartnett, and Ganic [29].

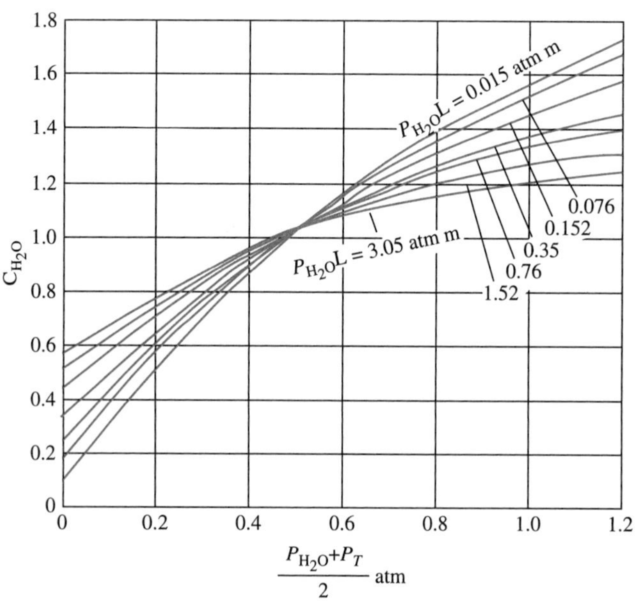

FIGURE 9.48 Correction factor for the emissivity of water vapor at pressures other than 1 atm.

Source: From Hottel and Egbert [19] and Egbert [25].

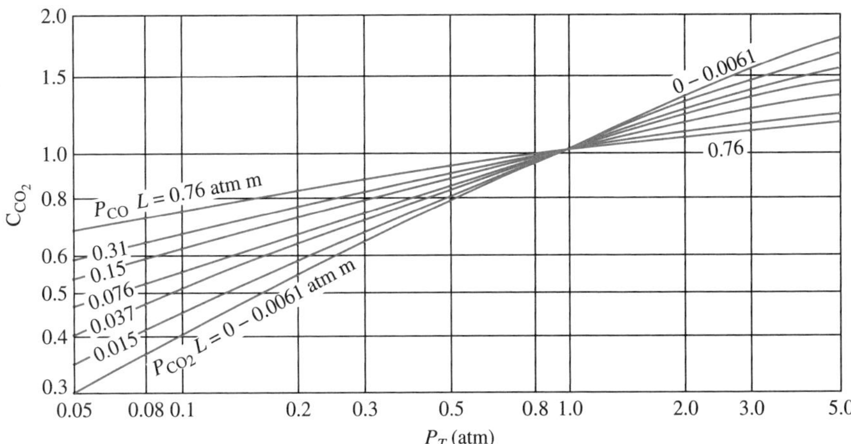

FIGURE 9.49 Correction factor for the emissivity of carbon dioxide at pressures other than 1 atm.

Source: From Hottel and Egbert [19].

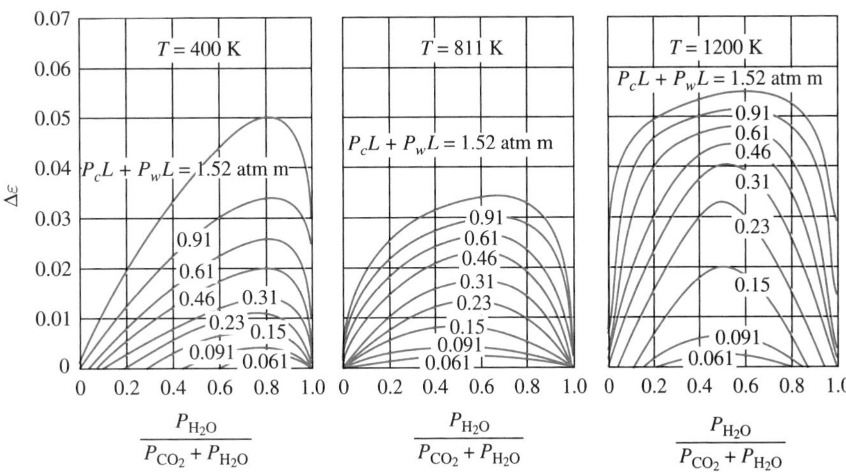

FIGURE 9.50 Factor $\Delta\varepsilon$ to correct the emissivity of a mixture of vapor and CO_2.

Source: From Hottel and Egbert [19].

the emissivity of the mixture can be calculated by adding the emissivity of the gases determined by assuming that each gas exists alone and then subtracting a factor $\Delta\varepsilon$, which accounts for emission in overlapping wavelength bands. The factor $\Delta\varepsilon$ for H_2O and CO_2 is plotted in Fig. 9.50. The emissivity of a mixture of H_2O and CO_2 is therefore given by the expression

$$\varepsilon_{mix} = C_{H_2O}(\varepsilon_{H_2O})P_T = 1 + C_{CO_2}(\varepsilon_{CO_2})P_T = 1 - \Delta\varepsilon \qquad (9.114)$$

EXAMPLE 9.15 Determine the emissivity of a gas mixture consisting of N_2, H_2O, and CO_2 at a temperature of 800 K. The gas mixture is in a sphere with diameter of 0.4 m, and the partial pressures of the gases are $P_{N_2} = 1\,\text{atm}$, $P_{H_2O} = 0.4\,\text{atm}$, and $P_{CO_2} = 0.6\,\text{atm}$.

SOLUTION The mean beam length for a spherical mass of gas is obtained from Table 9.7:

$$L = (2/3)D = 0.27\,\text{m}$$

The emissivities are given in Figs. 9.46 and 9.47, and appropriate values for the parameters to be used are

$$T = 800\,\text{K}$$
$$P_{H_2O}L = 0.107\,\text{atm m}$$
$$P_{CO_2}L = 0.160\,\text{atm m}$$

The emissivities for water vapor and carbon dioxide at 1 atm total pressure are, from Figs. 9.46 and 9.47, respectively,

$$(\varepsilon_{H_2O})P_{T=1} = 0.15$$
$$(\varepsilon_{CO_2})P_{T=1} = 0.125$$

N_2 does not radiate appreciably at 800 K, but since the total gas pressure is 2 atm, we must correct the 1-atm values for ε. From Figs. 9.48 and 9.49 the pressure correction factors are

$$C_{H_2O} = 1.62$$
$$C_{CO_2} = 1.12$$

The value for $\Delta\varepsilon$ used to correct for emission in overlapping wavelength bands is determined from Fig. 9.50:

$$\Delta\varepsilon = 0.014$$

Finally, the emissivity of the mixture can be obtained from Eq. (9.114):

$$\varepsilon_{mix} = 1.62 \times 0.15 + 1.12 \times 0.125 - 0.014 = 0.369$$

Gas absorptivity can be obtained from the emissivity charts shown previously by modifying the parameters in the charts. As an example, consider water vapor at a temperature of T_{H_2O} with incident radiation from a source surface at a temperature of T_s. The absorptivity of the H_2O vapor is approximately given by the relation

$$\alpha_{H_2O} = C_{H_2O}\varepsilon'_{H_2O}\left(\frac{T_{H_2O}}{T_s}\right)^{0.45} \tag{9.115}$$

if C_{H_2O} is taken from Fig. 9.48 and the value for the emissivity of water vapor ε'_{H_2O} from Fig. 9.46 is evaluated at temperature T_s and a pressure–mean beam length product equal to $P_{H_2O}L(T_s/T_{H_2O})$. Similarly, the absorptivity of CO_2 can be obtained from

$$\alpha_{CO_2} = C_{CO_2}\varepsilon'_{CO_2}\left(\frac{T_{CO_2}}{T_s}\right)^{0.65} \tag{9.116}$$

where the value for C_{CO_2} is taken from Fig. 9.49 and the value for ε'_{CO_2} is evaluated from Fig. 9.47 at a pressure–mean beam length product equal to $P_{CO_2}L(T_s/T_{CO_2})$.

EXAMPLE 9.16 Determine the absorptivity of a mixture of H_2O vapor and N_2 gas at a total pressure of 2.0 atm and a temperature of 500 K if the mean beam length is 0.75 m. Assume that the radiation passing through the gas is emitted by a source at 1000 K and the partial pressure of the water vapor is 0.4 atm.

SOLUTION Since nitrogen is transparent, the absorption in the mixture is due to the water vapor alone. From Eq. (9.115) the absorptivity of H_2O is

$$\alpha_{H_2O} = C_{H_2O}\varepsilon'_{H_2O}(T_{H_2O}/T_s)^{0.45}$$

The values of the parameters needed to evaluate the absorptivity of the gas are obtained from the data provided:

$$P_{H_2O} \cdot L = 0.4 \times 0.75 = 0.3 \text{ atm m}$$

$$\tfrac{1}{2}(P_T + P_{H_2O}) = \tfrac{1}{2}(2 + 0.4) = 1.2 \text{ atm}$$

From Figs. 9.46 and 9.48 we find

$$\varepsilon_{H_2O} = 0.29$$

$$C_{H_2O} = 1.40$$

Substituting the above values in Eq. (9.115) gives the absorptivity of the mixture

$$\alpha = 1.4 \times 0.29(500/1000)^{0.45} = 0.30$$

To calculate the rate of heat flow by radiation between a nonluminous gas at T_G and the walls of a blackbody container at T_s, the absorptivity α_G of the gas should be evaluated at the temperature T_s and the emissivity ε_G at the temperature T_G. The net rate of radiant heat flow is then equal to the difference between the emitted and absorbed radiation:

$$q_r = \sigma A_G(\varepsilon_G T_G^4 - \alpha_G T_s^4) \tag{9.117}$$

EXAMPLE 9.17 Flue gas at 2000°F containing 5% water vapor flows at atmospheric pressure through a 2-ft-square flue made of refractory brick. Estimate the rate of heat flow per foot length from the gas to the wall if the inner-wall surface temperature is 1850°F and the average convection heat transfer coefficient is 1 Btu/h ft^2 °F.

SOLUTION The rate of heat flow from the gas to the wall by convection per unit length is

$$q_c = \bar{h}_c A(T_{gas} - T_{surface})$$
$$= (1)(8)(150) = 1200 \text{ Btu/h ft length of flue}$$

To determine the rate of heat flow by radiation, we calculate first the effective beam length, or

$$L = \frac{3.4 \times \text{volume}}{\text{surface area}} = \frac{(3.4)(4)}{8} = 1.7 \text{ ft } (0.52 \text{ m})$$

The product of partial pressure and L is

$$pL = (0.05)(0.52) = 0.026\,\text{atm m}$$

From Fig. 9.46, for $pL = 0.026$ and $T_G = 1367\,\text{K}(2000°\text{F})$, we find $\epsilon_G = 0.035$. Similarly, we find $\alpha_G = 0.039$ at $T_s = 1283\,\text{K}(1850°\text{F})$. The pressure correction is negligible since $\bar{C}_p \cong 1$ according to Fig. 9.48. Assuming that the brick surface is black, the net rate of heat flow from the gas to the wall by radiation is, according to Eq. (9.117),

$$q_r = 0.171 \times 8[0.035(24.6)^4 - 0.039(23.1)^4] = 2340\,\text{Btu/h}$$

Therefore, the total heat flow from the gas to the duct is 3540 Btu/h. It is interesting to note that the small amount of moisture in the gas contributes about one-half of the total heat flow.

A recent review of radiation properties of gases showed that when the radiation properties of H_2O and CO_2 are evaluated from the graphs in Figs. 9.46–9.49, they can be used for industrial heat transfer calculations with satisfactory accuracy as long as the enclosure surface is not highly reflecting. But calculation of the radiant heat transfer in a gas-filled enclosure becomes considerably more complicated when the enclosure surfaces are not black and reflect a part of the incident radiation. When the emissivity of the enclosure is larger than 0.7, an approximate answer can be obtained by multiplying the rate of heat flow calculated from Eq. (9.117) by $(\varepsilon_s + 1)/2$, where ε_s is the emissivity of the enclosure surface. When the enclosure walls have smaller emissivities, the procedure outlined in Section 9.5 can be used, provided the assumption that all surfaces as well as the gases that are "gray" is acceptable. If one or more of the surfaces are not gray or if the gas cannot be treated as a graybody, a band approximation procedure similar to that in Example 9.14 must be used. Details for such refinement in the calculation procedures are presented in references [12, 13, 20, 29]. Scaling rules that extend the application of one-atmosphere total spectrum emissivity data to determine gas emissivities at higher and lower pressures are available [27].

9.9 Radiation Combined with Convection and Conduction

In the preceding sections of this chapter, we have considered radiation as an isolated phenomenon. Energy exchange by radiation is the predominant heat-flow mechanism at high temperatures because the rate of heat flow depends on the fourth power of the absolute temperature. In many practical problems, however, convection and conduction cannot be neglected, and in this section we shall consider problems that involve two or all three modes of heat flow simultaneously.

To include radiation in a thermal network involving convection and conduction is it often convenient to define a radiation heat transfer coefficient, $\bar{h}_r$, as

$$\bar{h}_r = \frac{q_r}{A_1(T_1 - T_2)} = \mathscr{F}_{1-2}\left[\frac{\sigma(T_1^4 - T_2^4)}{T_1 - T_2'}\right] \tag{9.118}$$

where A_1 = area upon which $\mathscr{F}_{1-2}$ is based, m^2

$T_1 - T_2'$ = a reference temperature difference, in K, in which T_2' may be chosen equal to T_2 or any other convenient temperature in the system

$\bar{h}_r$ = radiation heat transfer coefficient, $W/m^2\,K$

Once a radiation heat transfer coefficient has been calculated, it can be treated similarly to the convection heat transfer coefficient, because the rate of heat flow becomes linearly dependent on the temperature difference and radiation can be incorporated directly in a thermal network for which the temperature is the driving potential. Knowledge of the value of $\bar{h}_r$ is also essential in determining the overall conductance $\bar{h}$ for a surface to or from which heat flows by convection and radiation, since according to Chapter 1,

$$\bar{h} = \bar{h}_c + \bar{h}_r$$

If $T_2 = T_2'$, the bracketed expression in Eq. (9.118) is called the *temperature factor* F_T, and

$$\bar{h}_r = \mathscr{F}_{1-2} F_T \tag{9.119}$$

EXAMPLE 9.18

A butt-welded thermocouple (Fig. 9.51) having an emissivity of 0.8 is used to measure the temperature of a transparent gas flowing in a large duct whose walls are at a temperature of 440°F. The temperature indicated by the thermocouple is 940°F. If the convection heat transfer coefficient between the surface of the couple and the gas $\bar{h}_c$ is 25 Btu/h ft^2 °F, estimate the *true* gas temperature.

SOLUTION

The temperature of the thermocouple is below the gas temperature because the couple loses heat by radiation to the wall. Under steady-state conditions the rate of heat flow by radiation from the thermocouple junction to the wall equals the rate of heat flow by convection from the gas to the couple. We can write this heat balance as

$$q = \bar{h}_c A_T (T_G - T_T) = A_T \varepsilon \sigma (T_T^4 - T_{\text{wall}}^4)$$

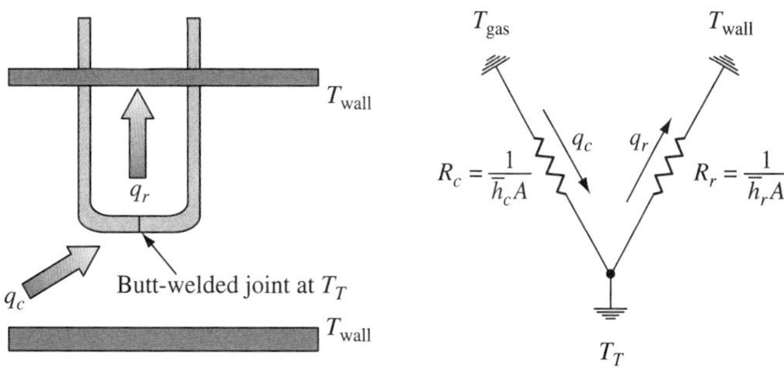

FIGURE 9.51 Physical system and thermal network for butt-welded thermocouple without radiation shield.

where A_T is the surface area, T_T the temperature of the thermocouple, and T_G the temperature of the gas. Substituting the data of the problem we obtain

$$\frac{q}{A_T} = 0.8 \times 0{,}1714\left[\left(\frac{1400}{100}\right)^4 - \left(\frac{900}{100}\right)^4\right] = 4410\,\text{Btu/h ft}^2$$

and the true gas temperature is

$$T_G = \frac{q}{\bar{h}_c A_T} + T_T = \frac{4410}{25} + 940 = 1116°\text{F}$$

In systems where heat is transferred simultaneously by convection and radiation, it is frequently not possible to determine the radiant heat transfer coefficient directly. Since the temperature factor F_T contains the temperatures of the radiation emitter and receiver, it can be evaluated only when both of these temperatures are known. If one of the temperatures depends on the rate of heat flow, that is, if one of the potentials in the network is "floating," one must assume a value for the floating potential and then determine whether that value will satisfy continuity of heat flow in the steady state. If the rate of heat flow to the potential node is not equal to the rate of heat flow from the node, another temperature must be assumed. The trial-and-error process is continued until the energy balance is satisfied. The general technique is illustrated in the next example.

EXAMPLE 9.19 Determine the correct gas temperature in Example 9.18 if the thermocouple is shielded by a thin, cylindrical radiation shield having an inside diameter four times as large as the outer diameter of the thermocouple. Assume that the convection heat transfer coefficient of the shield is 20 Btu/h ft^2 °F on both sides and that the emissivity of the shield, made of stainless steel, is 0.3 at 1000°F.

SOLUTION A sketch of the physical system is shown in Fig. 9.52. Heat flows by convection from the gas to the thermocouple and its shield. At the same time, heat flows by radiation from the thermocouple to the inside surface of the shield, is conducted through the shield, and flows by radiation from the outer surface of the shield to the walls of the duct. If we assume that the temperature of the shield is uniform (that is, if we neglect the thermal resistance of the conduction path because the shield is very thin), the thermal network is as shown in Fig. 9.52. The temperature of the duct wall T_w and the temperature of the thermocouple T_T are known, while the temperatures of the shield T_s and of the gas T_G must be determined. The latter two temperatures are floating potentials. A heat balance on the shield can be written as

$$\begin{matrix}\text{rate of heat flow from} \\ T_G \text{ and } T_T \text{ to } T_s\end{matrix} = \begin{matrix}\text{rate of heat flow} \\ \text{from } T_s \text{ to } T_w\end{matrix}$$

or

$$\bar{h}_{cs} 2A_s(T_G - T_s) + h_{rT}A_T(T_T - T_s) = \bar{h}_{rs}A_s(T_s - T_w)$$

A heat balance on the thermocouple yields

$$\bar{h}_{cT}A_T(T_G - T_T) = \bar{h}_{rT}A_T(T_T - T_s)$$

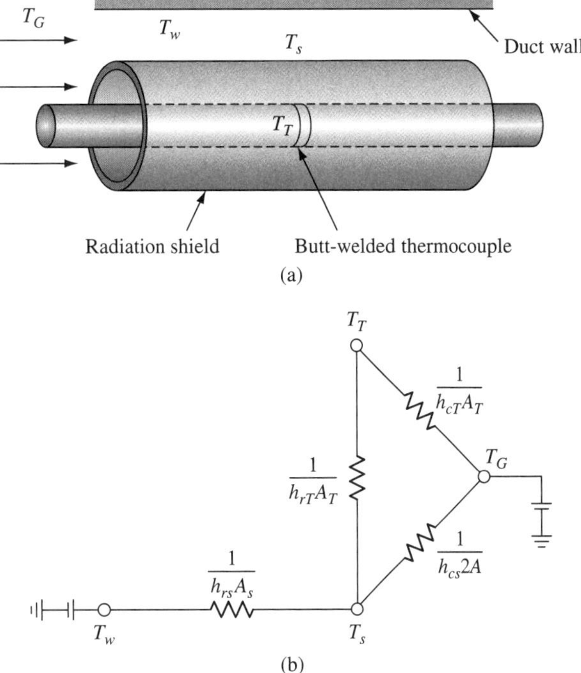

FIGURE 9.52 Physical system and thermal network for butt-welded thermocouple with radiation shield.

where the nomenclature is given in Fig. 9.51. Taking A_T as unity, A_s equals 4, and we obtain from Eq. (9.76)

$$A_T\mathscr{F}_{T-s} = \frac{1}{\dfrac{1-\varepsilon_T}{A_T\varepsilon_T} + \dfrac{1}{A_T} + \dfrac{1-\varepsilon_s}{A_s\varepsilon_s}} = \frac{1}{\dfrac{0.2}{0.8} + 1 + \dfrac{0.7}{(4)(0.3)}} = 0.547$$

and

$$A_s\mathscr{F}_{s-w} = A_s\varepsilon_s = (4)(0.3) = 1.2$$

Assuming a shield temperature of 900°F, we have, according to Eq. (9.118),

$$\bar{h}_{rT}A_T = A_T\mathscr{F}_{T-S}F_T = (0.547)(18.1) = 9.85$$

and

$$\bar{h}_{rs}A_s = A_s\mathscr{F}_{s-w}F_T = (1.2)(11.4) = 13.7$$

Substituting these values into the first heat balance permits the evaluation of the gas temperature, and we get

$$T_G = \frac{h_{rs}A_s(T_s - T_w) - h_{rT}A_T(T_T - T_s)}{(\bar{h}_{cs})(2A_s)} + T_s$$

$$= \frac{5750 - 581}{(20)(2)(4)} + 900 = 932°\text{F}$$

Since the temperature of the gas cannot be less than that of the thermocouple, the assumed shield temperature was too low. Repeating the calculations with a new shield temperature of 930°F yields $T_G = 970°\text{F}$. We now substitute this value to see if it satisfies the second heat balance and get:

$$\begin{matrix}\text{heat flow rate by convection} \\ \textit{to}\text{ thermocouple}\end{matrix} = 25A_T(970 - 940) = 750\,\text{Btu/h}$$

$$\begin{matrix}\text{net heat flow rate by radiation} \\ \textit{from}\text{ thermocouple}\end{matrix} = \bar{h}_{rT}A_T(T_T - T_s) = 203\,\text{Btu/h}$$

Since the rate of heat flow to the thermocouple exceeds the rate of heat flow from the thermocouple, our assumed shield temperature was too high. Repeating the calculations with an assumed shield temperature of 923°F yields a gas temperature of 966°F, which satisfies the heat balance on the thermocouple. The details of this calculation are left as an exercise.

A comparison of the results in Examples 9.18 and 9.19 shows that the indicated temperature of the unshielded thermocouple differs from the true gas temperature by 176°F, while the shielded couple reads only 26°F less than the true gas temperature. A double shield would reduce the temperature error to less than 10°F for the conditions specified in the example.

9.10 Closing Remarks

In this chapter, the basic characteristics of thermal radiation and methods for calculating heat exchange by radiation have been presented. Radiant energy emission is proportional to the absolute temperature raised to the fourth power, and radiation heat transfer therefore becomes increasingly important at higher temperatures. The ideal radiator, or "blackbody," is a concept convenient in the analysis of radiation heat transfer because it provides an upper limit to emission, absorption, and heat exchange by radiation. Blackbody radiation has geometric and spectral characteristics that can be treated analytically or numerically.

Real surfaces differ from black surfaces by virtue of their surface characteristics. Real surfaces always absorb and emit less radiation than black surfaces at the same temperature. Their surface characteristics can often be approximated by graybodies that emit and absorb a given fraction of blackbody radiation over the entire wavelength spectrum. Radiation heat transfer between real surfaces can be analyzed by assuming that the surfaces are gray or by using gray band approximations.

The geometric relation between bodies is characterized by the shape factor, which determines the amount of radiation leaving a given surface that is intercepted by another. Using the shape factors and surface characteristics, it is possible to construct equivalent networks for radiation between surfaces in an enclosure. These networks result in a series of linear relations that can be formulated as a matrix. The temperatures and rate of radiation heat transfer for each of the surfaces in an enclosure can be determined by a matrix inversion, which can be most conveniently performed with a computer. When radiation and convection occur simultaneously, the analysis requires solution of nonlinear equations, which can become complex, especially in systems with gaseous radiation. These types of problems usually require trial-and-error solutions.

References

1. M. Planck, *The Theory of Heat Radiation*, Dover, New York, 1959.
2. R. V. Dunkle, "Thermal-Radiation Tables and Applications," *Trans. ASME*, vol. 65, pp. 549–552, 1954.
3. M. Fischenden and O. A. Saunders, *The Calculation of Heat Transmission*. His Majesty's Stationery Office, London, 1932.
4. D. C. Hamilton and W. R. Morgan, "Radiant Interchange Configuration Factors," NACA TN2836, Washington, D.C., 1962.
5. F. Kreith and W. Z. Black, *Basic Heat Transfer*, Harper & Row, New York, 1980.
6. H. Schmidt and E. Furthman, "Über die Gesamtstrahlung fäster Körper," *Mitt. Kaiser-Wilhelm-Inst. Eisenforsch.*, Abh. 109, Dusseldorf, 1928.
7. W. H. McAdams, *Heat Transmission*, 3d ed., McGraw-Hill, New York, 1954.
8. G. G. Gubareff, J. E. Janssen, and R. H. Torborg, Thermal Radiation Properties Survey, Honeywell Research Center, Minneapolis, Minn., 1960.
9. W. Sieber, "Zusammensetzung der von Werk- und Baustoffen zurückgeworfene Wärmestrahlung," *Z. Tech. Phys.*, vol. 22, pp. 130–135, 1941.
10. E. Schmidt and E. Eckert, "Über die Richtungsverteilung der Wärmestrahlung von Oberflächen," *Forsch. Geb. Ingenieurwes.*, vol. 6, pp. 175–183, 1935.
11. J. R. Schonhorst and R. Viskanta, "An Experimental Examination of the Validity of the Commonly Used Methods of Radiant-Heat Transfer Analysis," *Trans. ASME, Ser. C., J. Heat Transfer*, vol. 90, pp. 429–436, 1968.
12. E. M. Sparrow and R. D. Cess, *Radiation Heat Transfer*, Hemisphere, New York, 1978.
13. R. Siegel and J. R. Howell, *Thermal Radiation Heat Transfer*, 3d ed., Hemisphere, New York, 1993.
14. R. G. Hering and T. F. Smith, "Surface Roughness Effects on Radiant Energy Interchange," *Trans. ASME, Ser. C., J. Heat Transfer*, vol. 93, pp. 88–96, 1971.
15. H. C. Hottel, "Radiant Heat Transmission," *Mech. Eng.*, vol. 52, pp. 699–704, 1930.
16. H. C. Hottel and A. F. Sarofim, *Radiative Heat Transfer*, pp. 31–39, McGraw-Hill, New York, 1967.
17. A. F. Emery, O. Johansson, M. Lobo, and A. Abrous, "A Comparative Study of Methods for Computing the Diffuse Radiation Viewfactors for Complex Structures," *J. Heat Transfer*, vol. 113, pp. 413–422, 1991.
18. A. K. Oppenheim, "The Network Method of Radiation Analysis," *Trans. ASME*, vol. 78, pp. 725–735, 1956.
19. H. C. Hottel and R. B. Egbert, "Radiant Heat Transmission from Water Vapor," *AIChE Trans.*, vol. 38, pp. 531–565, 1942.
20. C. L. Tien, "Thermal Radiation Properties of Gases," *Adv. Heat Transfer*, vol. 5, pp. 254–321, 1968.
21. R. Goldstein, "Measurements of Infrared Absorption by Water Vapor at Temperatures to 1000 K," *J. Quant. Spectrosc. Radiat. Transfer*, vol. 4, pp. 343–352, 1964.
22. D. K. Edwards and W. Sun, "Correlations for Absorption by the 9.4- and 10.4-micron CO_2 Bands," *Appl. Opt.*, vol. 3, p. 1501, 1964.
23. D. K. Edwards, B. J. Flornes, L. K. Glassen, and W. Sun, "Correlations of Absorption by Water Vapor at Temperatures from 300 to 1100 K," *Appl. Opt.*, vol. 4, pp. 715–722, 1965.
24. H. C. Hottel, in *Heat Transmission*, by W. C. McAdams, 3d ed., chapter 4, McGraw-Hill, New York, 1954.
25. R. B. Egbert, Sc.D. thesis, Massachussets Institute of Technology, 1941.
26. M. F. Modest, "Radiation," in F. Kreith, ed., *CRC Handbook of Thermal Engineering*, CRC Press, Boca Raton. FL, 2000.
27. D. K. Edwards and R. Matovosian, "Scaling Rules for Total Absorptivity and Emissivity of Gases," *J. Heat Transfer*, vol. 106, pp. 685–689, 1984.
28. H. C. Hottel, "Heat Transmission by Radiation from Nonluminous Gases," *AIChE Trans.*, vol. 19, pp. 173–205, 1927.
29. W. M. Rohsenow, J. P. Hartnett, and Y. I. Cho, eds., *Handbook of Heat Transfer*, McGraw-Hill, New York, 1998.

Problems

The problems for this chapter are organized by subject matter as shown below

Topic	Problem Number
Spectral characteristics of radiation	9.1–9.8
Shape factors and blackbody radiative exchange	9.9–9.14
Radiation in blackbody enclosures	9.15–9.22
Radiation in graybody enclosures	9.23–9.29
Gaseous radiation	9.30–9.32
Combined radiation and convection	9.33–9.53
Solar energy and global warming	9.54–9.60

9.1 For an ideal radiator (hohlraum) with a 10-cm-diameter opening, located in black surroundings at 16°C, calculate (a) the net radiant heat transfer rate for hohlraum temperatures of 100°C and 560°C, (b) the wavelength at which the emission is a maximum, (c) the monochromatic emission at λ_{max}, and (d) the wavelengths at which the monochromatic emission is 1% of the maximum value.

9.2 A tungsten filament is heated to 2700 K. At what wavelength is the maximum amount of radiation emitted? What fraction of the total energy is in the visible range (0.4 to 0.75 μm)? Assume that the filament radiates as a graybody.

9.3 Determine the total average hemispherical emissivity and the emissive power of a surface that has a spectral hemispherical emissivity of 0.8 at wavelengths less than 1.5 μm, 0.6 at wavelengths from 1.5 to 2.5 μm, and 0.4 at wavelengths longer than 2.5 μm. The surface temperature is 1111 K.

9.4 (a) Show that $E_{b\lambda}/T^5 = f(\lambda T)$ only. (b) For $\lambda T = 5000\,\mu$m K, calculate $E_{b\lambda}/T^5$.

9.5 Compute the average emissivity of anodized aluminum at 100°C and 650°C from the spectral curve in Fig. 9.16. Assume $\epsilon_\lambda = 0.8$ for $\lambda > 9\,\mu$m.

9.6 A large body of nonluminous gas at a temperature of 1100°C has emission bands between 2.5 and 3.5 μm and between 5 and 8 μm. At 1100°C the effective emissivity in the first band is 0.8 and in the second 0.6. Determine the emissive power of this gas in W/m^2.

9.7 A flat plate is in a solar orbit 150,000,000 km from the sun. It is always oriented normal to the rays of the sun, and both sides of the plate have a finish that has a spectral absorptivity of 0.95 at wavelengths shorter than 3 μm and 0.06 at wavelengths longer than 3 μm. Assuming that the sun is a 5550 K blackbody source with a diameter of 1,400,000 km, determine the equilibrium temperature of the plate.

9.8 By substituting Eq. 9.1 for $E_b\lambda(T)$ in Eq. (9.4) and performing the integration over the entire spectrum, derive a relationship between σ and constants C_1 and C_2 in Eq. (9.1).

9.9 Determine the ratio of the total hemispherical emissivity to the normal emissivity for a nondiffuse surface if the intensity of emission varies as the cosine of the angle measured from the normal.

9.10 Derive an expression for the geometric shape factor F_{1-2} for the rectangular surface A_1 shown below. A_1 is 1-m × 20-m and is placed parallel to and centered 5 m above a 20-m square surface A_2.

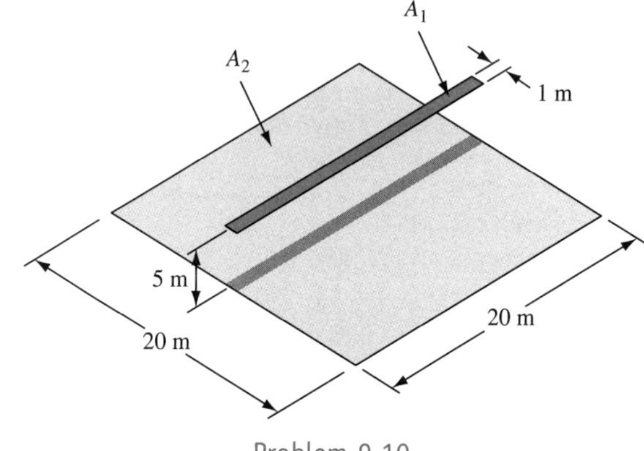

Problem 9.10

9.11 Determine the shape factor F_{1-4} for the geometrical configuration shown on the next page.

9.12 Determine the shape factor F_{1-2} for the geometrical configuration shown on the next page.

9.13 Using basic shape-factor definitions, estimate the equilibrium temperature of the planet Mars, which has a diameter of 4150 mi and revolves around the sun at a distance of 141×10^6 mi. The diameter of the sun is 865,000 mi. Assume that both the planet Mars and the sun act as blackbodies, with the sun having an equivalent blackbody temperature of 10,000 R. Then, repeat your calculations assuming that the albedo of Mars (the fraction of the incoming radiation returned to space) is 0.15.

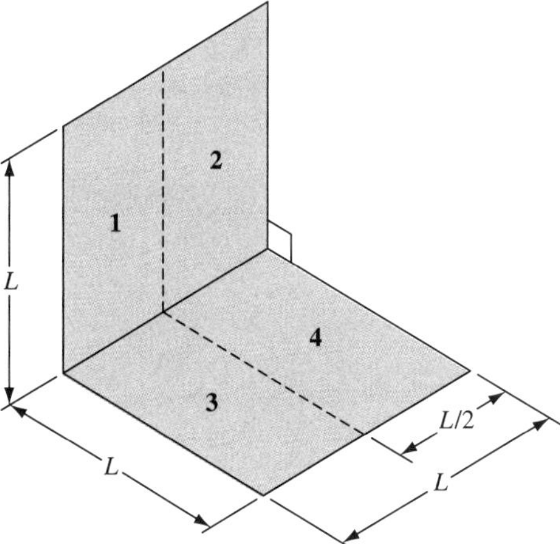

Problem 9.11

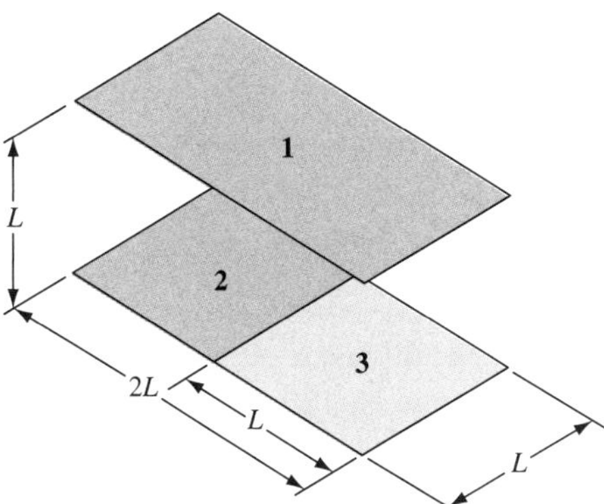

Problem 9.12

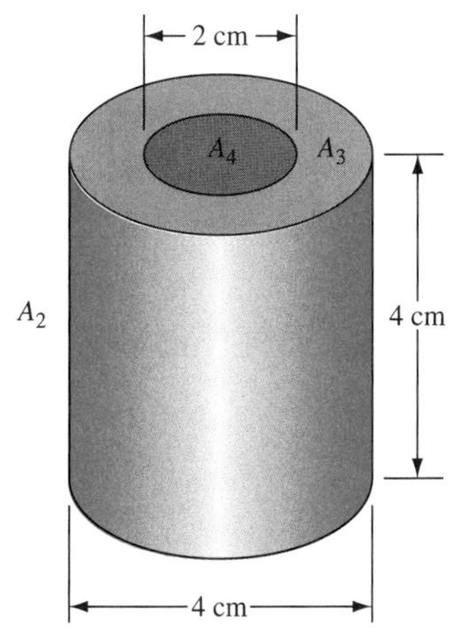

Problem 9.14

9.14 A 4-cm-diameter cylindrical enclosure with black surfaces, as shown in the accompanying sketch, has a 2-cm hole in the top cover. Assuming the walls of the enclosure are all at the same temperature, determine the percentage of the total radiation emitted from the walls that will escape through the hole in the cover.

9.15 Show that the temperature of the reradiating surface T_r in Fig. 9.37 is

$$T_R = \left(\frac{A_1 F_{1R} T_1^4 + A_2 F_{2R} T_2^4}{A_1 F_{1R} + A_2 F_{2R}} \right)^{1/4}$$

9.16 In the construction of a space platform, two structural members of equal size with surfaces that can be considered black are placed relative to each other as shown below. Assuming that the left member attached to the platform is at 500 K while the other is at 400 K and that the surroundings can be treated as though black at 0 K, calculate (a) the rate at which the warmer surface must be heated to maintain its temperature, (b) the rate of heat loss from the cooler surface to the surroundings, and (c) the net rate of heat loss to the surroundings for both members.

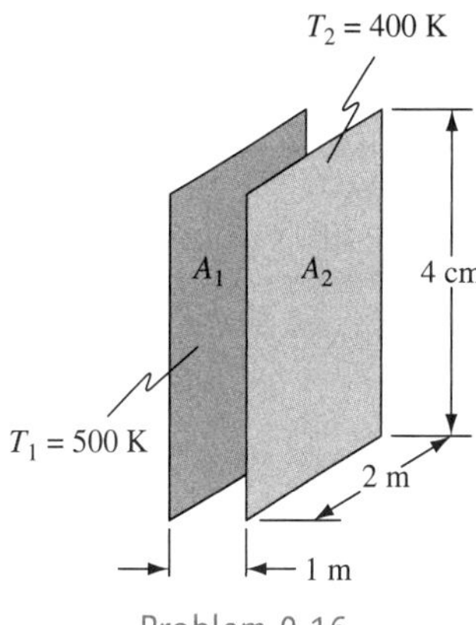

Problem 9.16

9.17 A radiation source is to be built as shown in the diagram for an experimental study of radiation. The base of the hemisphere is to be covered by a circular plate having a centered hole of radius $R/2$. The underside of the plate is to be held at 555 K by heaters embedded in its surface. The heater surface is black. The hemispherical surface is well-insulated on the outside. Assume gray diffuse processes and uniform distribution of radiation. (a) Find the ratio of the radiant intensity at the opening to the intensity of emission at the surface of the heated plate. (b) Find the radiant energy loss through the opening in watts for $R = 0.3$ m. (c) Find the temperature of the hemispherical surface.

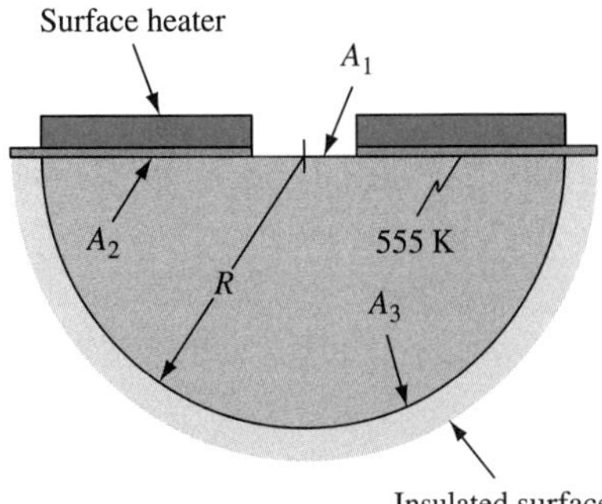

Problem 9.17

9.18 A large slab of steel 0.1-m thick contains a 0.1-m-diameter circular hole whose axis is normal to the surface. Considering the sides of the hole to be black, specify the rate of radiative heat loss from the hole in W and in Btu/h. The plate is at 811 K, and the surroundings are at 300 K.

9.19 A 15-cm-black disk is placed halfway between two black 3-m-diameter disks that are 7 m apart with all disk surfaces parallel to each other. If the surroundings are at 0 K, determine the temperature of the two larger disks required to maintain the smaller disk at 540°C.

9.20 Show that the effective conductance, $A_1\bar{F}_{1-2}$, for two black, parallel plates of equal area connected by reradiating walls at a constant temperature is

$$A_1\bar{F}_{1-2} = A_1\left(\frac{1 + F_{1-2}}{2}\right)$$

9.21 Calculate the net radiant-heat-transfer rate if the two surfaces in Problem 9.10 are black and are connected by a refractory surface with an area of 500 m^2. A_1 is at 555 K, and A_2 is at 278 K. What is the refractory surface temperature?

9.22 A black sphere (2.5 cm in diameter) is placed in a large infrared heating oven whose walls are maintained at 370°C. The temperature of the air in the oven is 90°C, and the heat transfer coefficient for convection between the surface of the sphere and the air is 30 W/m^2 K. Estimate the net rate of heat flow to the sphere when its surface temperature is 35°C.

9.23 The wedge-shaped cavity shown in the accompanying sketch consists of two long strips joined along one edge. Surface 1 is 1 m wide, has an emissivity of 0.4, and a temperature of 1000 K. The other wall has a temperature of 600 K. Assuming gray diffuse processes and uniform flux distribution, calculate the rate of energy loss from surfaces 1 and 2 per meter length.

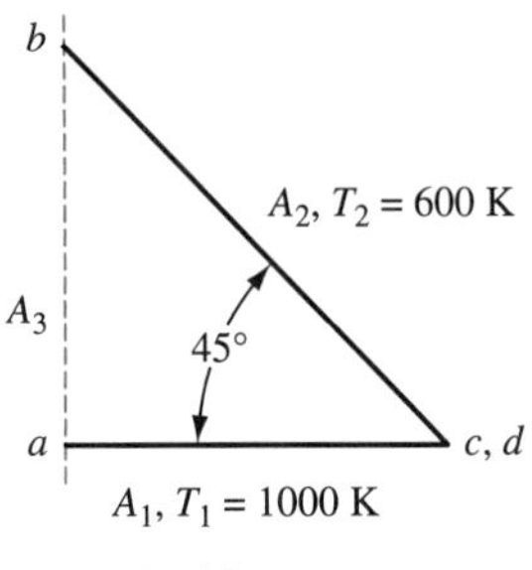

Problem 9.23

9.24 Derive an equation for the net rate of radiant heat transfer from surface 1 in the system shown in the accompanying sketch. Assume that each surface is at a uniform temperature and that the geometrical shape factor F_{1-2} is 0.1.

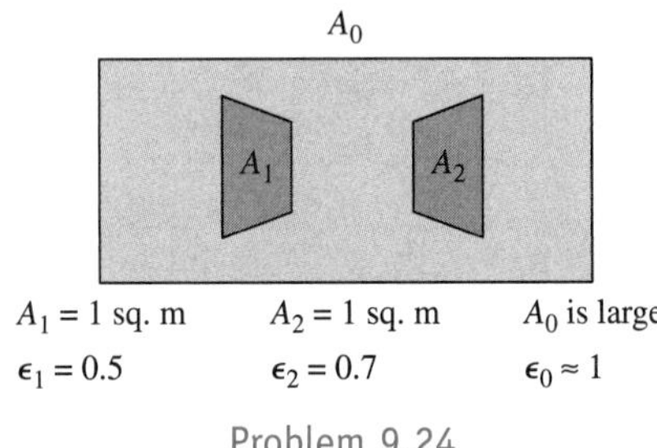

Problem 9.24

9.25 Two 5-ft-square, parallel flat plates are 1 ft apart. Plate A_1 is maintained at a temperature of 1540°F and A_2 at 460°F. The emissivities of the plates are 0.5 and 0.8, respectively. Considering the surroundings that are black at 0 R and including multiple interreflections, determine (a) the net radiant exchange between the plates, and (b) the heat input required by surface A_1 to maintain its temperature. The outer-facing surfaces of the plates are adiabatic.

9.26 Two concentric spheres 0.2 m and 0.3 m in diameter are to be used to store liquid air (133 K). The space between the spheres is evacuated. If the surfaces of the spheres have been flashed with aluminum and the liquid air has a latent heat of vaporization of 209 kJ/kg, determine the number of kilograms of liquid air evaporated per hour.

9.27 Determine the steady-state temperatures of two radiation shields placed in the evacuated space between two infinite planes at temperatures of 555 K and 278 K. The emissivity of all surfaces is 0.8.

9.28 Three thin sheets of polished aluminum are placed parallel to each other so that the distance between them is very small compared to the size of the sheets. If one of the outer sheets is at 280°C and the other outer sheet is at 60°C, calculate the temperature of the intermediate sheet and the net rate of heat flow by radiation. Convection can be ignored.

9.29 For each of the following situations, determine the rate of heat transfer between two 1-m × 1-m parallel flat plates placed 0.2-m apart and connected by reradiating walls. Assume that plate 1 is maintained at 1500 K and plate 2 at 500 K. (a) Plate 1 has an emissivity of 0.9 over the entire spectrum, and plate 2 has an emissivity of 0.1. (b) Plate 1 has an emissivity of 0.1 between 0 and 2.5 μm and an emissivity of 0.9 at wavelengths longer than 2.5 μm, while plate 2 has an emissivity of 0.1 over the entire spectrum. (c) The emissivity of plate 1 is the same as in part (b), and plate 2 has an emissivity of 0.1 between 0 and 4.0 μm and an emissivity of 0.9 at wavelengths larger than 4.0 μm.

9.30 A small sphere (1 in. in diameter) is placed in a heating oven. The oven cavity is a 1-ft cube filled with air at 14.7 psia; it contains 3% water vapor at 1000°F, and its walls are at 2000°F. The emissivity of the sphere is equal to 0.4–0.0001 T, where T is the surface temperature in °F. When the surface temperature of the sphere is 1000°F, determine (a) the total irradiation received by the walls of the oven from the sphere, (b) the net heat transfer by radiation between the sphere and the walls of the oven, and (c) the radiant heat transfer coefficient.

9.31 A 0.61-m-radius hemisphere (811 K in surface temperature) is filled with a gas mixture containing 6.67% CO_2 and water vapor at 0.5% relative humidity at 533 K and 2 atm pressure. Determine the emissivity and absorptivity of the gas and the net rate of radiant heat flow to the gas.

9.32 Two infinitely large, black, plane surfaces are 0.3 m apart, and the space between them is filled by an isothermal gas mixture at 811 K and atmospheric pressure. The gas mixture consists of 25% CO_2, 25% H_2O, and 50% N_2 by volume. If one of the surfaces is maintained at 278 K and the other at 1390 K, calculate (a) the effective emissivity of the gas at its temperature, (b) the effective absorptivity of the gas to radiation from the 1390 K surface, (c) the effective absorptivity of the gas to radiation from the 278 K surface, and (d) the net rate of heat transfer to the gas per square meter of surface area.

9.33 A manned spacecraft capsule has a shape of a cylinder 2.5 m in diameter and 9 m long (see the sketch below). The air inside the capsule is maintained at 20°C, and the convection heat transfer coefficient on the interior surface is 17 W/m^{2}K. Between the outer skin and the inner surface is a 15-cm layer of glass-wool insulation having a thermal conductivity of 0.017 W/mK. If the emissivity of the skin is 0.05 and there is no aerodynamic heating or irradiation from astronomical bodies, calculate the total heat transfer rate into space at 0 K.

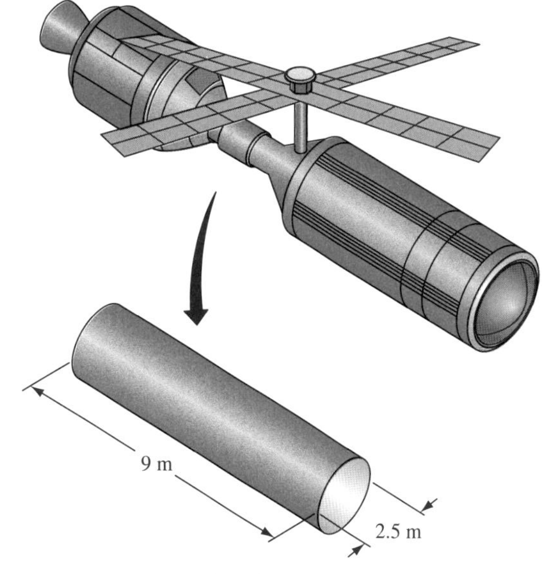

Problem 9.33

9.34 A 1-m × 1-m square solar collector is placed on the roof of a house. The collector receives a solar radiation flux of 800 W/m^2. Assuming that the surroundings act as a blackbody at an effective sky temperature of 30°C, calculate the equilibrium temperature of the collector (a) assuming its surface is black and that conduction and convection are negligible, and (b) assuming that the collector is horizontal and loses heat by natural convection.

9.35 A thin layer of water is placed in a pan 1 m in diameter in the desert. The upper surface is exposed to 300 K air, and the convection heat transfer coefficient between the upper surface of the water and the air is estimated to be 10 W/m^2 K. The effective sky temperature depends on atmospheric conditions and is often assumed to be 0 K for a clear night and 200 K for a cloudy night. Calculate the equilibrium temperature of the water on a clear night and a cloudy night.

9.36 Liquid nitrogen is stored in a dewar made of two concentric spheres with the space between them evacuated. The inner sphere has an outside diameter of 1 m, and the space between the two spheres is 0.1 m. The surfaces of both spheres are gray with an emissivity of 0.2. If the saturation temperature for nitrogen at atmospheric pressure is 78 K and its latent heat of vaporization is 2×10^5 J/kg, estimate its boil-off rate under the following conditions: (a) The outer sphere is at 300 K. (b) The outer surface of the surrounding sphere is black and loses heat by radiation to surroundings at 300 K. Assume convection is negligible. (c) Repeat part (b) but include the effect of heat loss by natural convection.

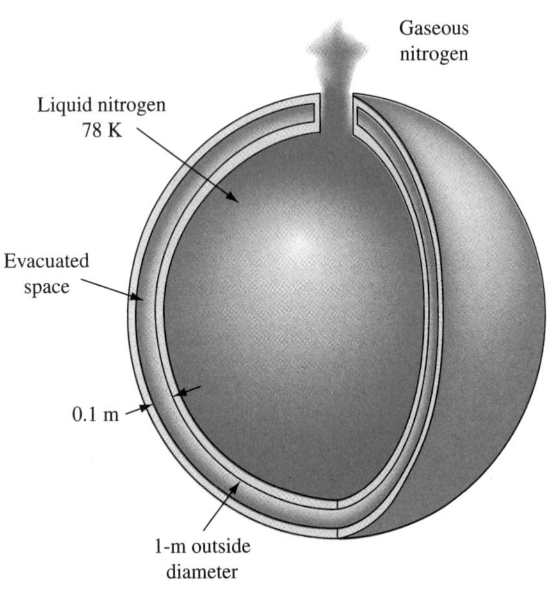

Problem 9.36

9.37 A package of electronic equipment is enclosed in a sheet-metal box that has a 0.3m-square base and is 0.15 m high. The equipment uses 1200 W of electrical power and is placed on the floor of a large room. The emissivity of the walls of the box is 0.80, and the surrounding temperature is 21°C. Assuming that the average temperature of the container wall is uniform, estimate that temperature.

9.38 An 0.2-m-OD oxidized steel pipe at a surface temperature of 756 K passes through a large room in which the air and the walls are at 38°C. If the heat transfer coefficient by convection from the surface of the pipe to the air in the room is 28 W/m^2 K, estimate the total heat loss per meter length of pipe.

9.39 A 6-mm-thick sheet of polished 304 stainless steel is suspended in a comparatively large vacuum-drying oven with black walls. The dimensions of the sheet are 30 cm $\times$ 30 cm, and its specific heat is 565 J/kg K. If the walls of the oven are uniformly at 150°C and the metal is to be heated from 10 to 120°C, estimate how long the sheet should be left in the oven if (a) heat transfer by convection can be neglected, and (b) the heat transfer coefficient is 3 W/m^{2}K.

9.40 Calculate the equilibrium temperature of a thermocouple in a large air duct if the air temperature is 1367 K, the duct-wall temperature is 533 K, the emissivity of the thermocouple is 0.5, and the convection heat transfer coefficient, $\bar{h}_c$, is 114 W/m^{2}K.

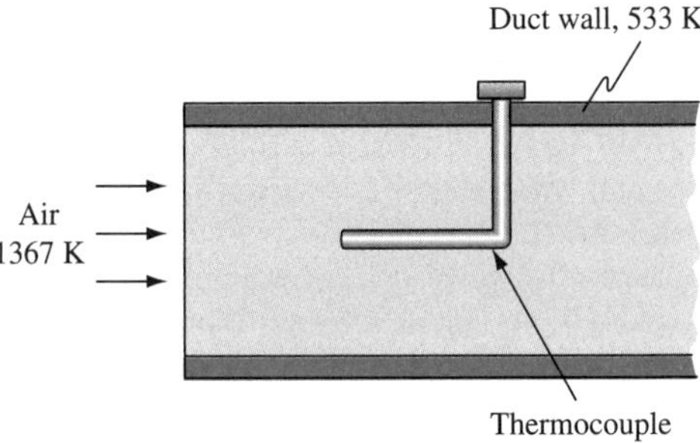

Problem 9.40

9.41 Repeat Problem 9.40 with the addition of a radiation shield with emissivity $\varepsilon_s = 0.1$.

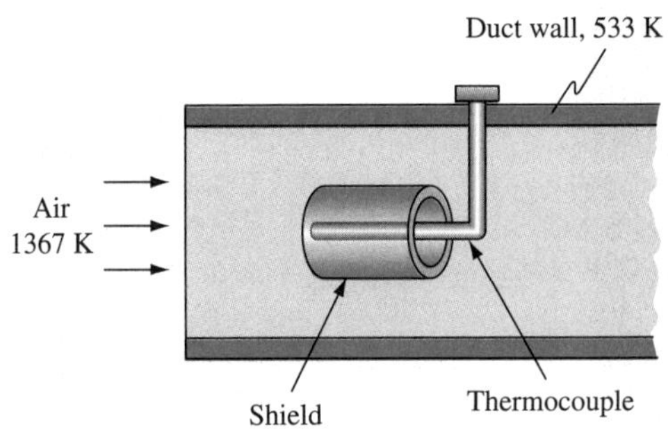

Problem 9.41

9.42 A thermocouple is used to measure the temperature of a flame in a combustion chamber. If the thermocouple temperature is 1033 K and the walls of the chamber are at 700 K, what is the error in the thermocouple reading due to radiation to the walls? Assume all surfaces are black and the convection coefficient is 568 W/m^2 K on the thermocouple.

9.43 A metal plate is placed in the sunlight. The incident radiant energy G is 780 W/m^2. The air and the surroundings are at 10°C. The heat transfer coefficient by natural convection from the upper surface of the plate is 17 W/m^2 K. The plate has an average emissivity of 0.9 at solar wavelengths and 0.1 at long wavelengths. Neglecting conduction losses on the lower surface, determine the equilibrium temperature of the plate.

9.44 A 2-ft-square section of panel heater is installed in the corner of the ceiling of a room having a 9-ft × 12-ft floor area with an 8-ft ceiling. If the surface of the heater, made from oxidized iron, is at 300°F and the walls and the air of the room are at 68°F in the steady state, determine (a) the rate of heat transfer to the room by radiation, (b) the rate of heat transfer to the room by convection ($h_c \approx 2$ Btu/h ft^2°F), and (c) the cost of heating the room per day if the cost of electricity is 7 cents per kW h.

9.45 In a manufacturing process a fluid is transported through a cellar maintained at a temperature of 300 K. The fluid is contained in a pipe having an external diameter of 0.4 m. The pipe surface has an emissivity of 0.5. To reduce heat losses, the pipe is surrounded by a thin shielding pipe having an ID of 0.5 m and an emissivity of 0.3. The space between the two pipes is effectively evacuated to minimize heat losses, and the inside pipe is at a temperature of 550 K. (a) Estimate the heat loss from the liquid per meter length. (b) If the fluid inside the pipe is an oil flowing at a velocity of 1 m/s, calculate the length of pipe for a temperature drop of 1 K.

9.46 One hundred pounds of carbon dioxide is stored in a high-pressure cylinder that is 10 in. in diameter (OD), 4 ft long, and 1/2 in. thick. The cylinder is fitted with a safety rupture diaphragm designed to fail at 2000 psig (with the specified charge, this pressure will be reached when the temperature increases to 120°F). During a fire, the cylinder is completely exposed to the irradiation from flames at 2000°F ($\varepsilon = 1.0$). For the specified conditions, $c_p = 0.60$ Btu/lb °F for CO_2. Neglecting the convection heat transfer, determine the length of time the cylinder can be exposed to this irradiation before the diaphragm fails if the initial temperature is 70°F and (a) the cylinder is bare oxidized steel ($\varepsilon = 0.79$), or (b) the cylinder is painted with aluminum paint ($\varepsilon = 0.30$).

9.47 A hydrogen bomb can be approximated by a fireball at a temperature of 7200 K, according to a report published in 1950 by the Atomic Energy Commission. (a) Calculate the total rate of radiant-energy emission in watts, assuming that the gas radiates as a blackbody and has a diameter of 1.5 km. (b) If the surrounding atmosphere absorbs radiation below 0.3 μm, determine the percent of the total radiation emitted by the bomb that is absorbed by the atmosphere. (c) Calculate the rate of irradiation on a 1-m^2 area of the wall of a house 40 km from the center of the blast if the blast occurs at an altitude of 16 km and the wall faces in the direction of the blast. (d) Estimate the total amount of radiation absorbed assuming that the blast lasts approximately 10 sec and that the wall is covered by a coat of red paint. (e) If the wall is made of oak whose flammability limit is estimated to be 650 K and has a thickness of 1 cm, determine whether the wood would catch on fire. Justify your answer by an engineering analysis carefully stating all assumptions.

9.48 An electric furnace is to be used for batch heating a certain material with a specific heat of 670 J/kg K from 20°C to 760°C. The material is placed on the furnace floor, which is 2 m × 4 m in area as shown in the accompanying sketch. The side walls of the furnace are made of a refractory material. A grid of round resistor rods is installed parallel to the plane of the roof but several inches below it. The resistors are 13 mm in diameter and are spaced 5 cm center to center. The resistor temperature is to be maintained at 1100°C; under these conditions the emissivity of the resistor surface is 0.6. If the top surface of the stock can be assumed to have an emissivity of 0.9, estimate the time required for heating a 6-metric-ton batch. External heat losses from the furnace can be neglected, the temperature gradient through the stock can be considered negligibly small, and steady-state conditions can be assumed.

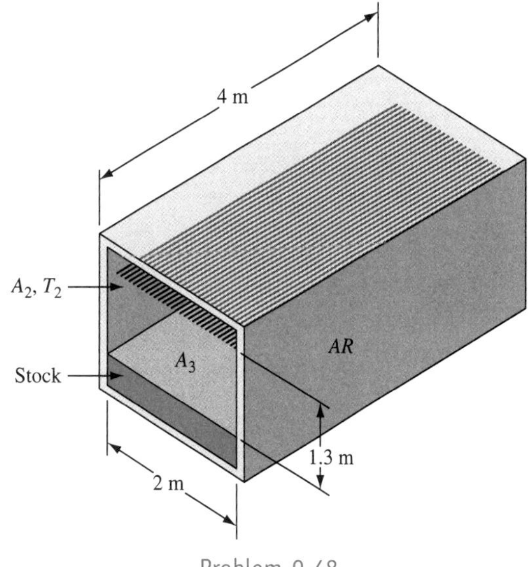

Problem 9.48

9.49 A rectangular, flat water tank is placed on the roof of a house with its lower portion perfectly insulated. A sheet of glass whose transmission characteristics are tabulated below is placed 1 cm above the water surface. Assuming that the average incident solar radiation is 630 W/m^2, calculate the equilibrium water temperature for a water depth of 12 cm if the heat transfer coefficient at the top of the glass is 8.5 W/m^2 K and the surrounding air temperature is 20°C. Disregard interreflections.

τ_λ of glass = 0 for wavelength form 0 to 0.35 μm
= 0.92 for wavelength from 0.35 to 2.7 μm
= 0 for wavelength larger than 2.7 μm

ρ_λ of glass = 0.08 for all wavelengths

9.50 Mercury is to be evaporated at 605°F in a furnace. The mercury flows through a 1-in. BWG no. 18 gauge 304 stainless-steel tube that is placed in the center of the furnace. The furnace cross section perpendicular to the tube axis is a square 8 in. × 8 in. The furnace is made of brick having an emissivity of 0.85, and its walls are maintained uniformly at 1800°F. If the convection heat transfer coefficient on the inside of the tube is 500 Btu/h ft^2 °F and the emissivity of the outer surface of the tube is 0.60, calculate the rate of heat transfer per foot of tube, neglecting convection within the furnace.

9.51 A 2.5-cm-diameter cylindrical-refractory crucible for melting lead is to be built for thermocouple calibration. An electrical heater immersed in the metal is shut off at some temperature above the melting point. The fusion-cooling curve is obtained by observing the thermocouple emf as a function of time. Neglecting heat losses through the wall of the crucible, estimate the cooling rate (W) for the molten lead surface (melting point 327.3°C, surface emissivity 0.8) if the crucible depth above the lead surface is (a) 2.5 cm and (b) 17 cm. Assume that the emissivity of the refractory surface is unity and the surroundings are at 21°C. (c) Noting that the crucible would hold about 0.09 kg of lead for which the heat of fusion is 23,260 J/kg, comment on the suitability of the crucible for the purpose intended.

9.52 A spherical satellite circling the sun is to be maintained at a temperature of 25°C. The satellite rotates continuously and is partly covered with solar cells having a gray surface with an absorptivity of 0.1. The rest of the sphere is to be covered by a special coating that has an absorptivity of 0.8 for solar radiation an an emissivity of 0.2 for the emitted radiation. Estimate the portion of the surface of the sphere that can be covered by solar cells. The solar irradiation can be assumed to be 1420 W/m^2 of surface perpendicular to the rays of the sun.

9.53 A 10-cm-square, electrically-heated plate is placed in a horizontal position 5 cm below a second plate of the same size, as shown schematically. The heating surface is gray (emissivity = 0.8), while the receiver has a black surface. The lower plate is heated uniformly over its surface with a power input of 300 W. Assuming that heat losses from the backs of the radiating surface and the receiver are negligible and that the surroundings are at a temperature of 27°C, calculate the following: (a) the temperature of the receiver, (b) the temperature of the heated plate, (c) the net radiation heat transfer between the two surfaces, and (d) the net radiation loss to the surroundings. (e) Estimate the effect of natural convection between the two surfaces on the rate of heat transfer.

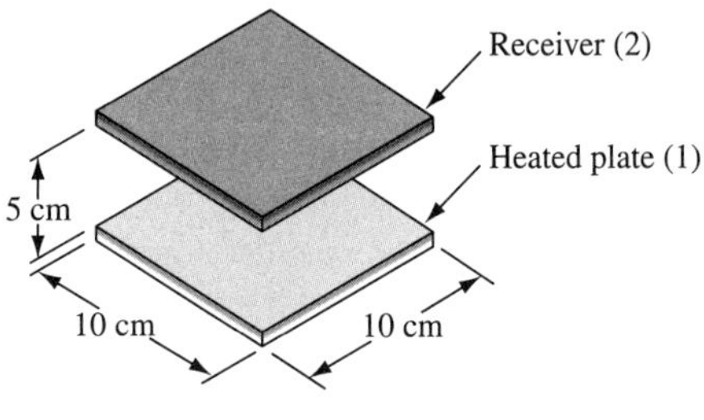

Problem 9.53

9.54 Estimate the temperature of the earth if there were no atmosphere to trap solar radiation. The diameter of the earth is approximately 1.27×10^7 m, and the distance between the sun and the earth is approximately 1.5×10^{11} m $\pm 1.7\%$. For your calculations, assume that the sun is a point source and that the earth moves in a circular motion around it. Furthermore, assume that the sun radiates as an equivalent blackbody at a temperature of 5760 K.

9.55 Repeat Problem 9.54 for the temperature of Mars. In this case, the student should carry out a simple library search to estimate the diameter of Mars and its approximate distance from the sun.

9.56 The diameter of the sun is 1.39×10^9 m. Estimate the percentage of the total radiation emitted by the sun, which approximates a blackbody at 5760 K, that is actually intercepted by the earth. Of the total radiation falling on the earth, about 70% falls on the ocean. Estimate the amount of radiation from the sun that falls on land, and then estimate the ratio of energy currently used worldwide and the amount of terrestrial solar energy that is available. Then, discuss why all of the energy cannot be harnessed.

9.57 As a result of the atmosphere surrounding the earth trapping some of the incoming solar radiation, the average temperature of the earth is approximately 15°C. Estimate the amount of radiation that is trapped by the atmosphere, including CO_2 and methane, which provides the shield to maintain the temperature at a level that can sustain living organisms. Then comment on the current concern about global warming as a result of an increasing percentage of CO_2 and methane in the atmosphere surrounding the earth.

9.58 A hypothetical PV solar cell in space can utilize solar radiation between 0.8 and 1.1 μm in wavelength. Estimate the maximum theoretical efficiency for this solar cell facing the sun using the ideal blackbody curve of the sun as the source. Supposing that all the radiation outside the spectral range utilized by the solar cell for generating electricity is dissipated into heat, estimate the rate at which a module of solar cells of 1.0 m^2 area would have to be cooled to maintain the temperature of the module below 90°F.

9.59 Repeat Problem 9.58 for a PV module in Phoenix on a sunny day at noon in an environment at 100°F. State your assumptions.

9.60 Estimate the rate at which heat needs to be supplied to an astronaut repairing the Hubble telescope in space. Assume that the emissivity of the spacesuit 0.5. Describe your model with a simple sketch and clearly state your assumptions.

Design Problems

9.1 **Energy-Efficient Electric Oven** (Chapter 9)
A large manufacturer of domestic electric ovens wants to explore more energy-efficient means for cooking with electricity. The baseline is the standard oven with electric heating elements having a volume sufficient to hold a 20-lb turkey. Investigate microwave heating, radiant elements, convection-assisted heating, or any other reasonable concepts or combinations of concepts that can improve the efficiency. You may also want to consider how the unit is insulated and how it is internally vented. Although the cost of the oven is important, energy consumption, reliability and speed of cooking are the primary concerns for this design.

9.2 **Advanced Insulation for a Hot-Water Heater** (Chapter 9)
Anticipating the next energy crisis, one forward-looking company would like to investigate advanced insulation systems for their line of hot-water heaters. They believe that a segment of the market will pay more for a hot-water heater that consumes less energy and is therefore less costly to operate. A possible additional benefit is that a thinner insulation package can allow for additional hot-water capacity and possibly faster recovery. Start with a baseline design for a commercially available hot-water heater. Investigate commercially available insulation systems, determine if any could provide these advantages, and quantify the cost and heat-transfer performance. You may also want to evaluate new insulation concepts that have appeared in the heat-transfer literature.

9.3 **Optical Pyrometer for Temperature Measurement** (Chapter 9)
An optical pyrometer is a device used to measure the temperature of high-temperature surfaces. In this instrument, an image of the hot surface is compared to the image of a heated filament whose temperature can be adjusted. When the color of the two images is the same, the unknown surface temperature equals that of the filament. Typically, a calibration table giving filament temperature versus filament-heating current is provided by manufacturers in their sales catalogs. Assuming the filament is contained in an evacuated optical chamber, design the filament and the power supply needed to achieve temperatures from 1000 K to 2500 K. Consider platinum and tungsten as candidate filament materials. Discuss the implication of the filament emissivity and the emissivity of the surface under analysis and how this will affect the accuracy of the measurement. Suggest methods that could be used to automate the device so that it could essentially become an on-line temperature-measurement device.

9.4 **Thermocouple Radiation Shield** (Continuation of Problem 1.2)
Design a radiation shield for the thermocouple described in Design Problem 1.2. Determine the accuracy of the thermocouple measurement with the shield as a function of air temperature and velocity. Suggest modifications to the shield that could be used to further improve the accuracy, e.g., painting or electroplating one or both surfaces. Are there any other modifications to the geometry of the thermocouple and shield that could provide further improvement?

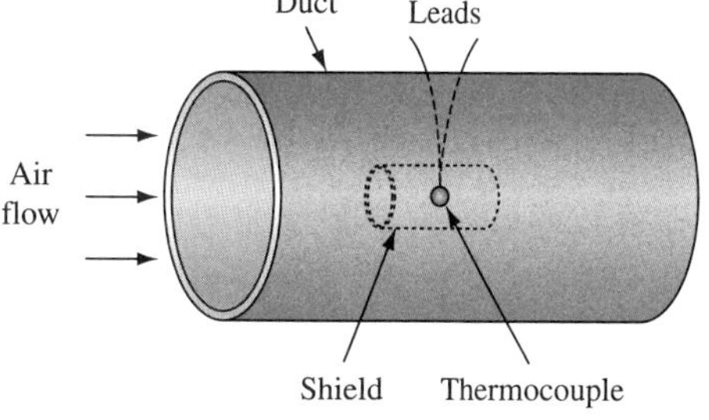

Design Problem 9.4

CHAPTER 10

Heat Transfer with Phase Change

Vapor bubble generation or ebullience behavior in pool boiling of water from a horizontal, electrically heated cylindrical heater a different heat flux levels: (a) in the partial nucleate boiling regime, and (b) in the fully developed nucleate boiling regime.

Source: Courtesy of Prof. Raj M. Manglik, Thermal-Fluids & Thermal Processing Laboratory, University of Cincinnati.

Concepts and Analyses to Be Learned

Heat transfer with phase change in a liquid-vapor medium (boiling or condensation) or a solid-liquid medium (melting or freezing) is very effective because the accommodation of latent heat ideally requires no temperature difference. The most common conventional applications are found in the boiler and condenser of a steam power plant, ice making, and metal casting in manufacturing. Some newer applications include immersion and microchannel cooling of microelectronics, evaporation and condensation in heat pipes, and crystal growth, among many others. The heat transfer processes in boiling, condensation, melting, and freezing are significantly more complex than those in single-phase conduction and convection. It is often difficult to model these processes mathematically, and therefore substantial experimentation is required to predict the energy exchange. A study of this chapter will teach you:

- How to characterize pool boiling behavior and its different regimes and to predict the corresponding heat transfer coefficients
- How to identify different flow regimes in forced convection boiling, calculate the heat transfer coefficient, and determine the critical heat flux at which burnout occurs
- How to model condensation heat transfer on a flat vertical plate as well as on the outside of a horizontal tube, determine the respective heat transfer coefficients, and apply them to condenser design
- How to evaluate and predict the performance of heat pipes
- How to model and analyze the heat transfer during melting and freezing

10.1 Introduction to Boiling

Heat transfer to boiling liquids is a convection process involving a change in phase from liquid to vapor. The phenomena of boiling heat transfer are considerably more complex than those of convection without phase change because in addition to all of the variables associated with convection, those associated with the phase change are also relevant. In liquid-phase convection, the geometry of the system, the viscosity, the density, the thermal conductivity, the expansion coefficient, and the specific heat of the fluid are sufficient to describe the process. In boiling heat transfer, however, the surface characteristics, the surface tension, the latent heat of vaporization, the pressure, the density, and possibly other properties of the vapor play an important part. Because of the large number of variables involved, neither general equations describing the boiling process nor general correlations of boiling heat transfer data are available. Considerable progress has been made, however, in gaining a physical understanding of the boiling mechanism [1–5]. By observing the boiling phenomena with the aid of high-speed photography, it has been found that there are distinct boiling regimes in which the heat transfer mechanisms differ radically. To correlate the experimental data it is therefore best to describe and analyze each of the boiling regimes separately.

10.2 Pool Boiling

10.2.1 Pool Boiling Regimes

To acquire a physical understanding of the characteristic phenomena of the various boiling regimes, we shall first consider a simple system consisting of a heating surface, such as a flat plate or a wire, submerged in a pool of water at saturation temperature without external agitation. Boiling in this situation is referred to as *pool boiling.* A familiar example of such a system is the boiling of water in a kettle on a stove. As long as the temperature of the surface does not exceed the boiling point of the liquid by more than a few degrees, heat is transferred to liquid near the heating surface by natural convection. The convection currents circulate the superheated liquid, and evaporation takes place at the free surface of the liquid. Although some evaporation occurs, the heat transfer mechanism in this process is simply natural convection, because only liquid is in contact with the heating surface.

As the temperature of the heating surface is increased, a point is reached at which vapor bubbles are formed and escape from the heated surface in certain places known as nucleation sites. Nucleation sites are very small inclusions in the surface that result from the process used to manufacture the surface. The inclusions are too small to admit liquid because of the liquid surface tension, and the resulting vapor pocket acts as a site for bubble growth and release. As a bubble is released, liquid flows over the inclusion, trapping vapor and thus providing a start for the next bubble. This process occurs simultaneously at a number of nucleation sites on the heating surface. At first the vapor bubbles are small and condense before reaching the surface, but as the temperature is raised further, they become more numerous and larger until they finally rise to the free surface. These phenomena can be observed when water boils in a kettle.

The various pool-boiling regimes are illustrated in Fig. 10.1 for a horizontal wire heated electrically in a pool of distilled water at atmospheric pressure with a corresponding saturation temperature of 100°C [6, 7]. In this curve the heat flux is plotted as a function of the temperature difference between the surface and the saturation temperature. This temperature difference, ΔT_x, is called the excess temperature above the boiling point, or the *excess temperature.* We observe that in regimes 2 and 3, the heat flux increases rapidly with increasing surface temperature. The process in these two regimes is called *nucleate boiling.* In the individual bubble regime, most of the heat is transferred from the heating surface to the surrounding liquid by a vapor-liquid exchange action [8]. As vapor bubbles form and grow on the heating surface, they push hot liquid from the vicinity of the surface into the colder bulk of the liquid. In addition, intense microconvection currents are set up as vapor bubbles are emitted and colder liquid from the bulk rushes toward the surface to fill the void. As the surface heat flux is raised and the number of bubbles increases to the point where they begin to coalesce, heat transfer by evaporation becomes more important and eventually predominates at very large heat fluxes in regime 3 [9].

If the excess temperature in a temperature-controlled system is raised to about 35°C, we observe that the heat flux reaches a maximum (about 10^6 W/m^2 in a pool of water), with further increases in the temperature causing a decrease in the rate of heat flow. This maximum heat flux, called the *critical heat flux,* is said to occur at the *critical excess temperature* (point *a* in Fig. 10.1).

The cause of the inflection point near *c* in the curve can be found by examining the heat transfer mechanism during boiling. At the onset of boiling, bubbles grow at nucleation sites on the surface until the buoyant force or currents of the surrounding liquid carry them away. But as the heat flux or the surface temperature is increased in nucleate boiling, the number of sites at which bubbles grow increases. The rate of growth of the bubbles increases simultaneously, and so does the frequency of formation. As the rate of bubble emission from a site increases, bubbles collide and coalesce with their predecessors [10]. This point marks the transition from regime 2 to regime 3 in Fig. 10.1. Eventually, successive bubbles merge into slugs and more-or-less continuous vapor columns [3, 5, 9].

As the maximum heat flux is approached, the number of vapor columns increases. But since each new column occupies space formerly occupied by liquid, there is a limit to the number of vapor columns that can be emitted from the surface. This limit is reached when the space between these columns is no longer sufficient to accommodate the streams of liquid that must move toward the hot surface to replace the liquid that evaporated to form the vapor columns.

If the surface temperature is raised further so that the ΔT_x at the maximum heat flux is exceeded, one of three situations can occur, depending on the method of heat control and the material of the heating surface [11]:

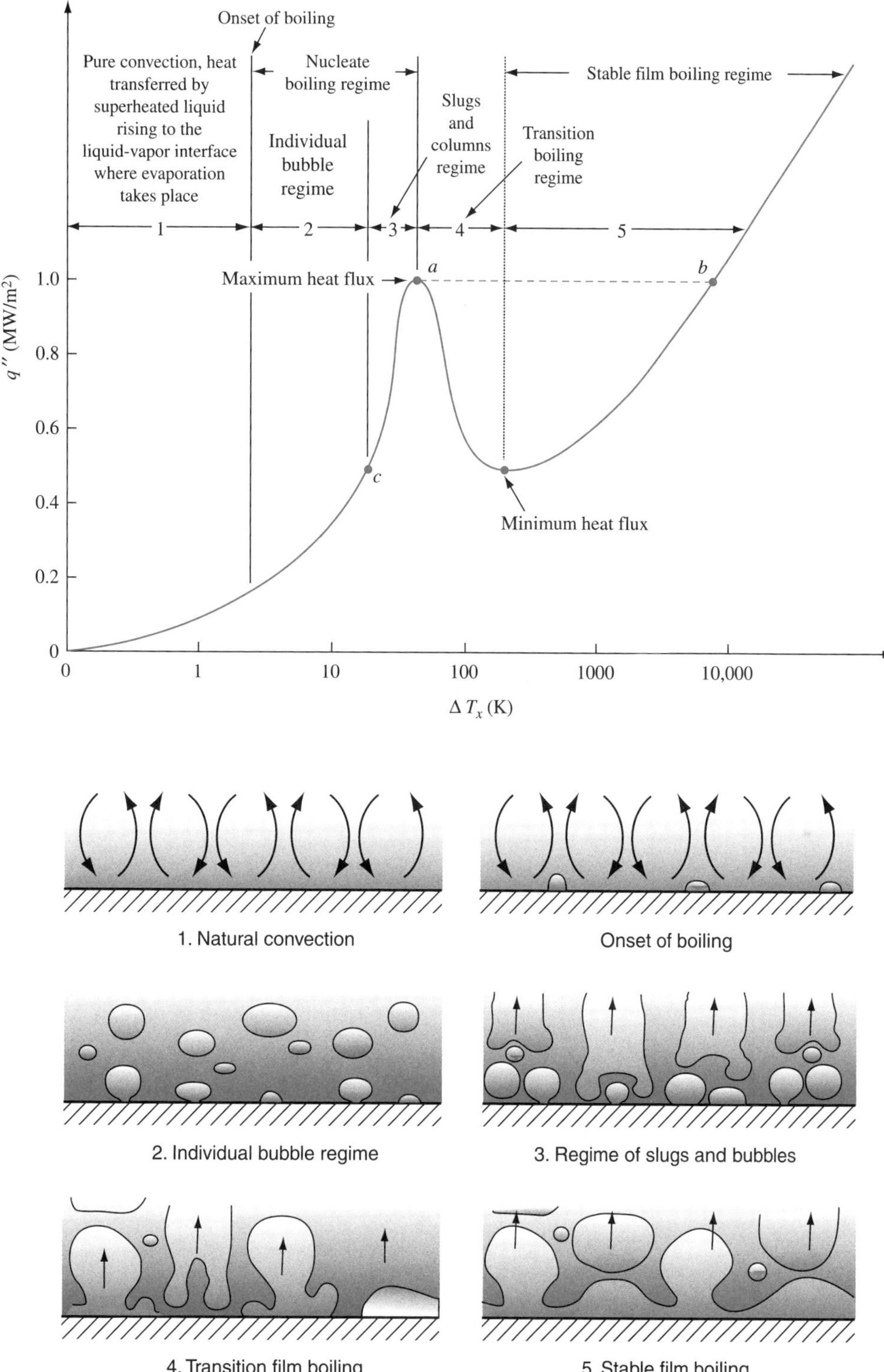

FIGURE 10.1 Typical boiling curves for a wire, tube, or horizontal surface in a pool of water at saturation temperature and atmospheric pressure with schematic representation of each boiling regime.

1. If the heater surface temperature is the independent variable and the heat flux is controlled by it, the mechanism will change to transition boiling and the heat flux will decrease. This corresponds to the operation in regime 4 in Fig. 10.1.
2. If the heat flux is controlled, as in an electrically heated wire, the surface temperature is dependent on it. Provided the melting point of the heater material is sufficiently high, a transition from nucleate to film boiling will take place, and the heater will operate at a very much higher temperature. This case corresponds to a transition from point *a* to point *b* in Fig. 10.1.
3. If the heat flux is independent but the heater material has a low melting point, burnout occurs. For a very short time, the heat supplied to the heater exceeds the amount of heat removed because when the peak heat flux is reached, an increase in heat generation is accompanied by a decrease in the rate of heat flow from the heater surface. Consequently, the temperature of the heater material will rise to the melting point, and the heater will burn out.

In the stable film-boiling regime, a vapor film blankets the entire heater surface, whereas in the transition film-boiling regime, nucleate and stable film boiling occurs alternately at a given location on the heater surface [12]. The photographs in Figs. 10.2 and 10.3 illustrate the nucleate- and film-boiling mechanisms on a wire submerged in water at atmospheric pressure. Note the film of vapor that completely covers the wire in Fig. 10.3. A phenomenon that closely resembles this condition can be observed when a drop of water falls on a red-hot stove. The drop does not evaporate immediately but dances on the stove because a steam film forms at the interface between the hot surface and the liquid and insulates the droplet from the surface.

10.2.2 Bubble Growth Mechanisms

When a fluid at its saturation temperature, T_{sat}, comes in contact with a heated surface at temperature $T_w > T_{sat}$, bubbles can form in the thermal boundary layer. The

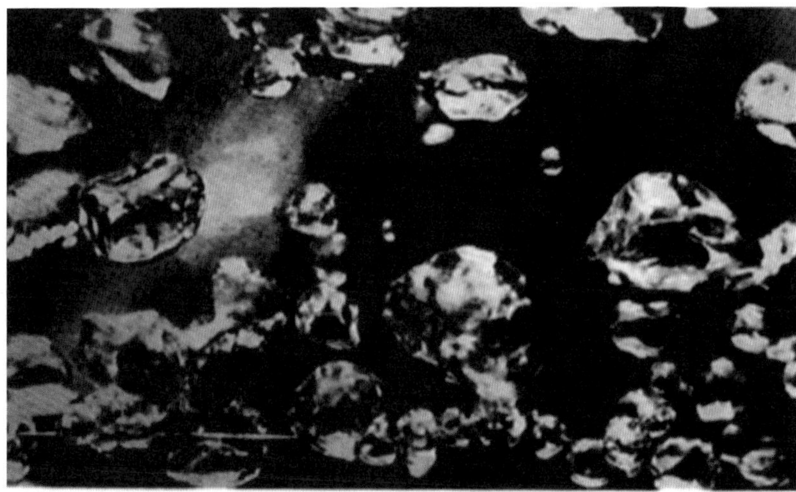

FIGURE 10.2 Photograph showing nucleate boiling on a wire in water.
Source: Courtesy of J. T. Castles.

FIGURE 10.3 Photograph showing film boiling on a wire in water.
Source: Courtesy of J. T. Castles.

bubble growth process is quite complex, but there are essentially two limiting conditions: inertia-controlled growth and heat-transfer-controlled growth. Carey has described these processes in detail [2]. In inertia-controlled growth, the heat transfer is very rapid and the growth of a bubble is limited by how fast it can push back the surrounding liquid. This condition exists during the initial growth stages, but in the later stages of growth when the bubble has become larger, the rate of heat transfer becomes the limiting factor and the interface motion is much slower.

The bubble growth process near a heated horizontal surface can be visualized as a sequence of stages shown schematically in Fig. 10.4 on the next page. After the departure of a bubble, liquid at the bulk fluid temperature rushes toward the hot surface. For a brief period of time heat from the surface is conducted into the liquid and superheats it, but bubble growth has not yet taken place. This time interval, t_w, is called the *waiting period.*

Once bubble growth begins, the thermal energy needed to vaporize liquid at the liquid-vapor interface comes, at least in part, from liquid adjacent to the bubble. Since the liquid immediately adjacent to the interface is highly superheated during the initial stages of bubble growth, transfer of heat to the interface is not a limiting factor. But as the embryonic bubble emerges from the nucleation site cavity, a rapid expansion is triggered as a result of the sudden increase in the radius of curvature of the bubble. The resulting rapid growth of the bubble is resisted primarily by the inertia of the liquid. For this inertia-controlled early stage of the bubble growth process, the bubble grows in a nearly hemispherical shape, as shown schematically in Fig. 10.4(c). In this stage a thin microlayer of liquid is left between the lower portion of the bubble interface and the heated surface as shown. This film, which is sometimes referred to as the *evaporation microlayer,* varies in thickness from nearly zero near the nucleation site cavity to a finite value at the edge of the hemispherical bubble. Heat is transferred across this film from the surface to the interface and vaporizes liquid at the surface directly. This film may evaporate completely near the cavity where nucleation

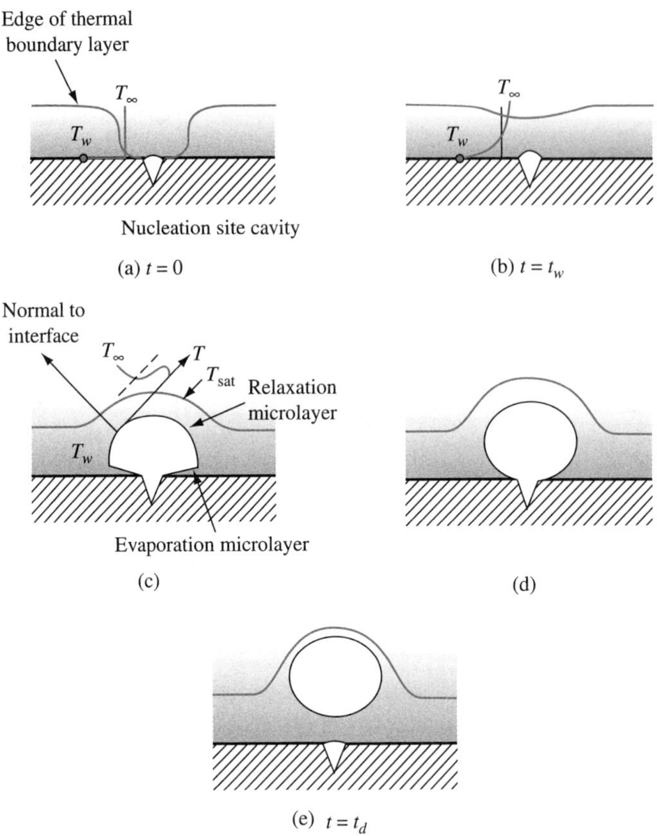

FIGURE 10.4 Stages in the bubble growth near a superheated surface in a fluid at saturation temperature.

began and thus elevate the surface temperature there significantly. When this occurs, the surface dries out and then rewets cyclically, and the surface temperature may fluctuate strongly with the repeated growth and release of bubbles.

The liquid region adjacent to the interface, which is sometimes referred to as the *relaxation microlayer,* is gradually depleted of its superheat as the bubble grows. The nature of the temperature profile in this region at an intermediate stage of the bubble growth process is indicated by the solid line in Fig. 10.4(c). The interface is at the saturation temperature corresponding to the ambient pressure in the liquid. The liquid temperature increases with increasing distance from the interface, reaches a peak, and then decreases toward the ambient temperature. As growth continues, heat transfer to the interface may become a limiting factor, and the bubble growth becomes heat-transfer controlled.

Once the bubble growth process becomes heat-transfer controlled, pressure and liquid inertia forces become relatively smaller and surface tension tends to pull the bubble into a more spherical shape. Thus, in undergoing the transition from inertia-controlled growth to heat-transfer–controlled growth, the bubble is transformed from a hemispherical shape to a more spherical configuration, as shown in Fig. 10.4(d).

Throughout the bubble growth process, interfacial tension acting along the contact line (where the interface meets the solid surface) tends to hold the bubble in place on the surface. Buoyancy, drag, lift, and/or inertial forces associated with motion of the surrounding fluid tend to pull the bubble away. These detaching forces become stronger as the bubble becomes larger [see Fig. 10.4(d)] and eventually is released at $t = t_d$ [see Fig. 10.4(e)].

The above description of the bubble growth process includes both inertia-controlled and heat-transfer–controlled growth regimes, but the occurrence or absence of either regime depends on the conditions under which bubble growth occurs. Very rapid, inertia-controlled growth is more likely to be observed under conditions that include high wall superheat, high imposed heat flux, a highly polished surface, low contact angle (highly wetting liquid), low latent heat of vaporization, and low system pressure (resulting in low vapor density). The first four items on this list result in the buildup of high superheat levels during the waiting period. The last two items result in very rapid volumetric growth of the bubble once the growth process begins. The first item and the last two imply that inertia-controlled growth is likely for large values of the product of the Jakob number (Ja) and the liquid-to-vapor density ratio, ρ_l/ρ_v. Ja is defined by

$$\mathrm{Ja}(\rho_l/\rho_v) = \frac{(T_\infty - T_{\mathrm{sat}})c_{pl}}{h_{fg}}\left(\frac{\rho_l}{\rho_v}\right)$$

The shape of the bubble is likely to be hemispherical when these conditions exist.

Conversely, heat-transfer–controlled growth of a bubble is more likely when conditions include low wall superheat, low imposed heat flux, a rough surface having many large and moderate-sized cavities, moderate contact angle (moderately wetting liquid), high latent heat of vaporization and moderate to high system pressure. All of these conditions result in slower bubble growth with smaller inertia effects or in a stronger dependence of bubble growth rate on heat transfer to the interface. The more of these conditions that are met, the greater is the likelihood that heat-transfer–controlled growth will result. Carey [2] has summarized the results of analyses for heat-transfer–controlled and inertia-controlled growths, leading to a description of the entire bubble cycle as well as of the heat transfer mechanism from a superheated wall to a saturated liquid in nucleate boiling. In recent years, Dhir [13] has provided results from mathematical and numerical simulations of the bubble dynamics process, in both nucleate pool boiling and film boiling regimes, and they provide additional insights into the associated heat transfer mechanisms. It may be noted that the theoretical and computational modeling of bubble dynamics in pool boiling is quite complex, which is beyond the scope of this textbook, and the interested student could pursue Dhir [13] and Stephan and Kern [14], among others.

When the surface temperature exceeds the saturation temperature, local boiling in the vicinity of the surface may take place even if the bulk temperature is below the boiling point. The boiling process in a liquid whose bulk temperature is below the saturation temperature but whose boundary layer is sufficiently superheated that bubbles form next to the heating surface is usually called *heat transfer to a subcooled liquid*, or *surface boiling*. The mechanisms of bubble formation and heat transfer are similar to those described for liquids at saturation temperature. However, the bubbles increase in number while their size and average lifetime decrease with decreasing bulk temperature at a given heat flux [15]. As a result of the increase in the bubble population, the agitation of the liquid caused by the motion of the bubbles is more intense in a subcooled liquid than in a pool of saturated liquid, and much larger heat fluxes can be attained before the critical temperature is achieved. The mechanism by which a typical bubble transfers heat in subcooled and degassed water is illustrated by the sketches in Fig. 10.5 [16]. The following lettered sequence of events corresponds to the lettered sketches in Fig. 10.5:

(a) The liquid next to the wall is superheated.
(b) A vapor nucleus of sufficient size to permit a bubble to grow has formed at a pit or scratch in the surface.
(c) The bubble grows and pushes the layer of superheated liquid above it away from the wall into the cooler liquid above. The resulting motion of the liquid is indicated by arrows.
(d) The top of the bubble surface extends into cooler liquid. The temperature in the bubble has dropped. The bubble continues to grow by virtue of the inertia of the liquid, but it grows at a slower rate than during stage (c) because it receives less heat per unit volume.
(e) The inertia of the liquid has caused the bubble to grow so large that its upper surface extends far into cooler liquid. It loses more heat by evaporation and convection than it receives by conduction from the heating surface.
(f) The inertial forces have been dissipated and the bubble begins to collapse. Cold liquid from above follows in its wake.
(g) The vapor phase has been condensed, the bubble has disappeared, and the hot wall is splashed by a stream of cold liquid at high velocity.
(h) The film of superheated liquid has settled and the cycle repeats.

The foregoing description of the life cycle of a typical bubble also applies qualitatively through stage (e) to liquids containing dissolved gases, to solutions of more than one liquid, and to saturated liquids. In these cases, however, the bubble does not

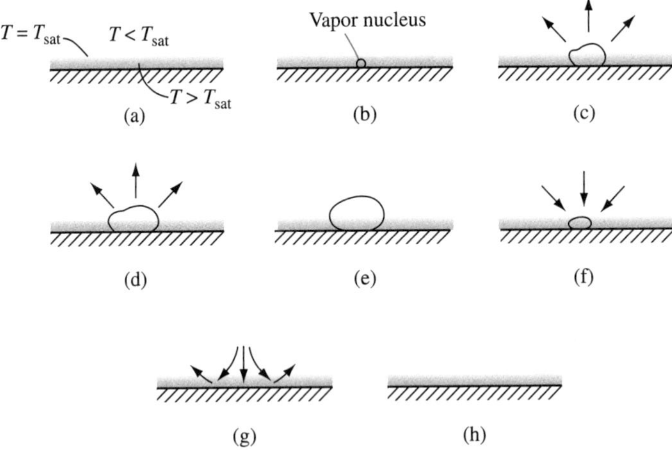

FIGURE 10.5 Flow pattern induced by a bubble in a subcooled boiling liquid.

collapse, but is carried away from the surface by buoyant forces or convection currents. In any case, a void is created and the surface is swept by cooler fluid rushing in from above. What eventually happens to the bubbles (whether they collapse on the surface or are swept away) has little influence on the heat transfer mechanism, which depends mostly on the pumping action and liquid agitation.

The primary variable controlling the bubble mechanism is the excess temperature. It should be noted, however, that in the nucleate boiling regime the total variation of the excess temperature irrespective of the fluid bulk temperature is relatively small for a very large range of heat flux. For design purposes the conventional heat transfer coefficient, which is based on the difference in temperature between the bulk of the fluid and the surface, is therefore only of secondary interest compared to the maximum heat flux attainable in nucleate boiling and the wall temperature at which boiling begins.

Generation of steam in the tubes of a boiler, vaporization of liquids such as gasoline in the chemical industry, and boiling of a refrigerant in the cooling coils of a refrigerator are processes that closely resemble those described above, except that in these industrial applications of boiling, the fluid generally flows past the heating surface. The heating surface is frequently the inside of a tube or a duct, and the fluid at the discharge end is a mixture of liquid and vapor. The foregoing descriptions of bubble formation and behavior also apply qualitatively to forced convection, but the heat transfer mechanism is further complicated by the motion of the bulk of the fluid. Boiling in forced convection will be discussed in Section 10.3.

10.2.3 Nucleate Pool Boiling

The dominant mechanism by which heat is transferred in single-phase forced convection is the turbulent mixing of hot and cold fluid particles. As shown in Chapter 4, experimental data for forced convection without boiling can be correlated by a relation of the type

$$\mathrm{Nu} = \phi(\mathrm{Re}, \mathrm{Pr})$$

where the Reynolds number, Re, is a measure of the turbulence and mixing motion associated with the flow. The increased heat transfer rates attained with nucleate boiling are the result of the intense agitation of the fluid produced by the motion of vapor bubbles. To correlate experimental data in the nucleate-boiling regime, the conventional Reynolds number in Eq. (4.20) is modified so that it is significant of the turbulence and mixing motion for the boiling process. A special type of Reynolds number, Re_b, which is a measure of the agitation of the liquid in nucleate-boiling heat transfer, is obtained by combining the average bubble diameter, D_b, the mass velocity of the bubbles per unit area, G_b, and the liquid viscosity, μ_l, to form the dimensionless modulus

$$\mathrm{Re}_b = \frac{D_b G_b}{\mu_l}$$

This parameter, often called the bubble Reynolds number, takes the place of the conventional Reynolds in nucleate boiling. If we use the bubble diameter D_b as the significant length in the Nusselt number, we have

$$\mathrm{Nu}_b = \frac{h_b D_b}{k_l} = \phi(\mathrm{Re}_b, \mathrm{Pr}_l) \tag{10.1}$$

where Pr_l is the Prandtl number of the saturated liquid and h_b is the *nucelate-boiling heat transfer coefficient*, defined as

$$h_b = \frac{q''}{\Delta T_x}$$

In nucleate boiling the excess temperature ΔT_x is the physically significant temperature potential. It replaces the temperature difference between the surface and the bulk of the fluid, ΔT, used in single-phase convection. Numerous experiments have shown the validity of this method, which obviates the need to know the exact temperature of the liquid and can therefore be applied to saturated as well as subcooled liquids.

Using experimental data on pool boiling as a guide, Rohsenow [17] modified Eq. (10.1) by means of simplifying assumptions. An equation found convenient for the reduction and correlation of experimental data [18] for many different fluids is

$$\frac{c_l \Delta T_x}{h_{fg}\mathrm{Pr}_l^n} = C_{sf}\left[\frac{q''}{\mu_l h_{fg}}\sqrt{\frac{g_c \sigma}{g(\rho_l - \rho_v)}}\right]^{0.33} \tag{10.2}$$

where c_l = specific heat of saturated liquid, J/kg K
q'' = heat flux, W/m^2
h_{fg} = latent heat of vaporization, J/kg
g = gravitational acceleration, m/s^2
ρ_l = density of the saturated liquid, kg/m^3
ρ_v = density of the saturated vapor, kg/m^3
σ = surface tension of the liquid-to-vapor interface, N/m
Pr_l = Prandtl number of the saturated liquid
μ_l = viscosity of the liquid, kg/ms
n = 1.0 for water, 1.7 for other fluids
C_{sf} = empirical constant that depends on the nature of the heating-surface fluid combination and whose numerical value varies from system to system

Use of Eq. (10.2) requires that property values be known very accurately. In particular, note the sensitivity of the Prandtl number effect on heat flux.

The most important variables affecting C_{sf} are the surface roughness of the heater, which determines the number of nucleation sites at a given temperature [12], and the angle of contact between the bubble and the heating surface, which is a measure of the wettability of a surface with a particular fluid. The sketches in Fig. 10.6 show that the contact angle θ decreases with greater wetting. A totally wetted surface has the smallest area covered by vapor at a given excess temperature and consequently represents the most favorable condition for efficient heat transfer. In the absence of quantitative information on the effect of wettability and surface conditions on the constant C_{sf}, its value must be determined empirically for each fluid-surface combination.

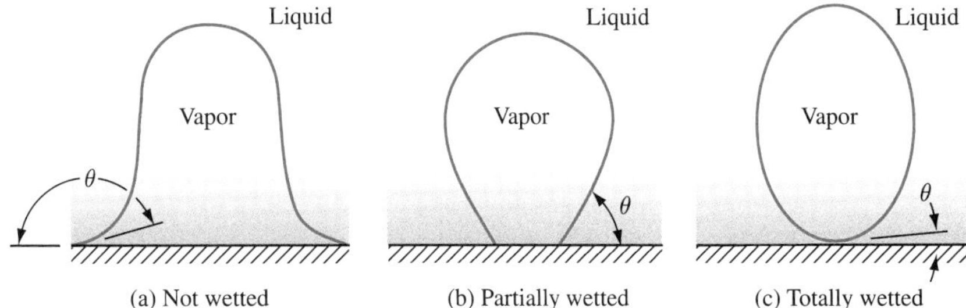

FIGURE 10.6 Effect of surface wettability on the bubble contact angle θ.

Figure 10.7 shows experimental data obtained by Addoms [19] for pool boiling of water on a 0.61-mm-diameter platinum wire at various saturation pressures. These data can be correlated using the dimensionless parameter

$$\frac{q''}{\mu_l h_{fg}}\sqrt{\frac{g_c\sigma}{g(\rho_l-\rho_v)}}$$

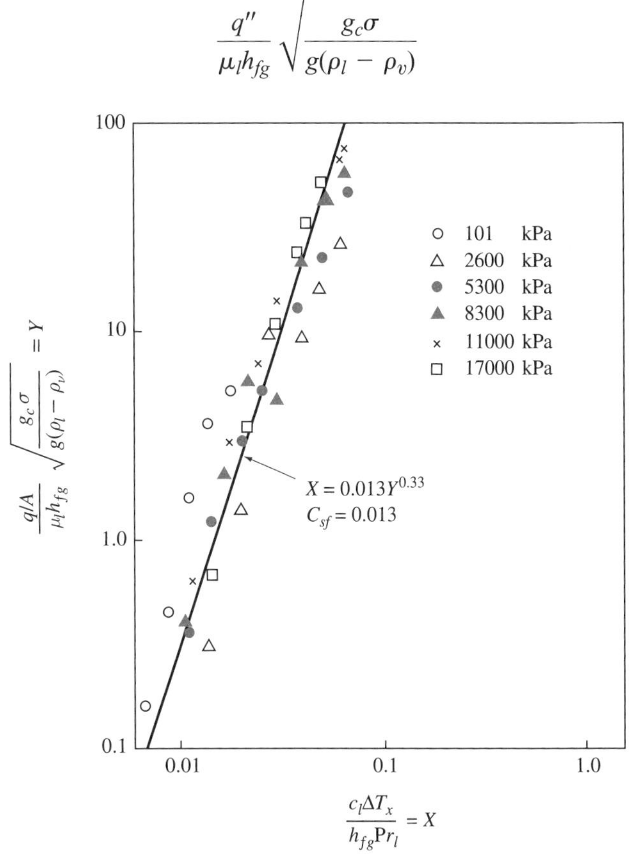

FIGURE 10.7 Correlation of pool-boiling heat transfer data for water by the method of Rohsenow.

Source: From Rohsenow [17], with permission of the publishers, the American Society of Mechanical Engineers; data from Addoms [19].

as the ordinate and $c_l\,\Delta T_x/h_{fg}\,\mathrm{Pr}_l$ as the abscissa. The slope of the best-fit straight line through the experimental points is 0.33; for water boiling on platinum, the value of C_{sf} is 0.013. For comparison, the experimental values of C_{sf} for a number of other fluid-surface combinations are listed in Table 10.1.

Selected values of the vapor-liquid surface tension for water at various temperatures are shown in Table 10.2 for use in Eq. (10.2).

The principal advantage of the Rohsenow correlation is that the performance of a particular fluid-surface combination in nucleate boiling at any pressure and heat flux can be predicted from a single test. One value of the heat flux q'' and the corresponding value of the excess temperature difference ΔT_x are all that are required to evaluate C_{sf} in Eq. (10.2). It should be noted, however, that Eq. (10.2) applies only to clean surfaces. For contaminated surfaces the exponent of Pr_l, n, has been found to vary between 0.8 and 2.0. Contamination also affects the other exponent in Eq. (10.2) and C_{sf}.

The geometric shape of the heating surface has no appreciable effect on the nucleate-boiling mechanism [20, 21]. This is not unexpected, since the influence of the bubble motion on the fluid conditions is limited to a region very near the surface. However, the size or diameter of a horizontal cylindrical heater has a

TABLE 10.1 Values of the coefficient C_{sf} in Eq. (10.2) for various liquid-surface combinations

Fluid-heating Surface Combination	C_{sf}
Water on scored copper [18][a]	0.0068
Water on emery-polished copper [18]	0.0128
Water-copper [25]	0.0130
Water on emery-polished, paraffin-treated copper [18]	0.0147
Water-brass [27]	0.0060
Water on Teflon coated stainless steel [18]	0.0058
Water on ground and polished stainless steel [18]	0.0080
Water on chemically etched stainless steel [18]	0.0133
Water on mechanically polished stainless steel [18]	0.0132
Water-platinum [19]	0.0130
n-Pentane on lapped copper [18]	0.0049
n-Pentane on emery-rubbed copper [18]	0.0074
n-Pentane on emery-polished copper [18]	0.0154
n-Pentane on emery-polished nickel [18]	0.0127
n-Pentane-chromium [26]	0.0150
Isopropyl alcohol-copper [25]	0.00225
n-Butyl alcohol-copper [25]	0.00305
Ethyl alcohol-chromium [26]	0.0027
Carbon tetrachloride on emery-polished copper [18]	0.0070
Carbon tetrachloride-copper [25]	0.0130
Benzene-chromium [26]	0.0100
50% K_2CO_3-copper [25]	0.00275
35% K_2CO_3-copper [25]	0.0054

[a]Numbers in brackets indicate references at the end of the chapter.

TABLE 10.2 Vapor-liquid surface tension for water

Surface Tension $\sigma(\times 10^3$ N/m)	Saturation Temperature °C
75.5	0
72.9	20
69.5	40
66.1	60
62.7	80
58.9	100
48.7	150
37.8	200
26.1	250
14.3	300
3.6	350

Source: N. B. Vargaftik. *Tables on the Thermophysical Properties of Liquids and Gases*, 2nd ed., Hemisphere. Washington. DC, 1975, p. 53.

significant influence on the nucleate boiling heat transfer [22, 23], and higher heat transfer coefficients are obtained by larger diameters in comparison with thin wires. This has been attributed to the formation of bubble boundary layer on the larger cylindrical surface accompanied by a more vigorous motion of large buoyancy-driven bubbles from the underside of the heater, which slide over the surface and "wipe" and detach other smaller growing bubbles in their path [22, 23].

For calculating the heat flux, Collier and Thome [24] recommend the following correlation equation as being simpler to use than Eq. (10.2).

$$q'' = 0.000481\,\Delta T_x^{3.33} p_{cr}^{2.3}\left[1.8\left(\frac{p}{p_{cr}}\right)^{3.17} + 4\left(\frac{p}{p_{cr}}\right)^{1.2} + 10\left(\frac{p}{p_{cr}}\right)^{10}\right]^{3.33} \qquad (10.3)$$

In Eq. (10.3) ΔT_x is the excess temperature in °C, p is the operating pressure in atm. p_{cr} is the critical pressure in atm, and q'' is in W/m^2.

10.2.4 Critical Heat Flux in Nucleate Pool Boiling

The Rohsenow method correlates data for all types of nucleate-boiling processes, including pool boiling of saturated or subcooled liquids and boiling of subcooled or saturated liquids flowing by forced or natural convection in tubes or ducts. Specifically, the correlation equation, Eq. (10.2), relates the boiling heat flux to the excess temperature, provided the relevant fluid properties and the pertinent coefficient C_{sf} are available. The correlation is restricted to nucleate boiling and does not reveal the excess temperature at which the heat flux reaches a maximum or what the value of this flux is when nucleate boiling breaks down and an insulating vapor film forms. As mentioned earlier, the maximum heat flux attainable with nucleate boiling is sometimes of greater interest to the designer than the exact surface temperature, because for efficient heat transfer [28] and operating safety [2, 29], particularly in high-performance constant-heat-input systems, operation in the film-boiling regime must be avoided.

Although no satisfactory theory exists for predicting boiling heat transfer coefficients, the maximum heat flux condition in nucleate pool boiling, that is, the critical heat flux, can be predicted with reasonable accuracy.

Close inspection of the nucleate-boiling regime (Fig. 10.1) shows [10] that it consists of at least two major subregimes. In the first region, which corresponds to low heat-flux densities, the bubbles behave as isolated entities and do not interfere with one another. But as the heat flux is increased the process of vapor removal from the heating surface changes from an intermittent to a continuous one, and as the frequency of bubble emission from the surface increases, the isolated bubbles merge into continuous vapor columns.

The stages of the transition process from isolated bubbles to continuous vapor columns are shown schematically in Fig. 10.8(a). The photographs in Figs. 10.8(b) and (c) show the two regimes for water boiling on a horizontal surface at atmospheric pressure [10]. At the transition from the region of isolated bubbles to that of vapor columns, only a small portion of the heating surface is covered by vapor. But as the heat flux is increased, the column diameter increases and additional vapor columns form. As the fraction of a cross-sectional area parallel to the heating surface occupied by vapor increases, neighboring vapor columns and the enclosed liquid begin to interact. Eventually a vapor generation rate is attained at which the close spacing between adjacent vapor columns leads to high relative velocities between the vapor moving away from the surface and the liquid streams flowing toward the surface to maintain continuity. The point of maximum heat flux occurs when the velocity of the liquid relative to the velocity of the vapor is so great that a further increase would either cause the vapor columns to drag the liquid away from the heating surface or cause the liquid streams to drag the vapor back toward the heating surface. Either case is obviously physically impossible without a decrease in the heat flux.

With this type of flow model as a guide, Zuber and Tribus [30] and Moissis and Berenson [10] derived analytical relations for the maximum heat flux from a horizontal surface. These relations are in essential agreement with an equation proposed earlier by Kutateladze [31] by empirical means. The Zuber equation [32] for the peak flux (in W/m^2) in saturated pool boiling is

$$q''_{\max.Z} = \frac{\pi}{24}\rho_v^{1/2} h_{fg}[\sigma g(\rho_l - \rho_v)g_c]^{1/4} \tag{10.4}$$

Lienhard and Dhir [33] recommend replacing the constant $\pi/24$ with 0.149.

Equation (10.4) predicts that water will sustain a larger peak heat flux than any of the common liquids because water has such a large heat of vaporization. Further inspection of Eq. (10.4) suggests ways to increase maximum heat flux. Pressure affects peak heat flux because it changes both the vapor density and the boiling point. Changes in the boiling point affect the heat of vaporization and the surface tension. For each liquid there exists, therefore, a certain pressure that

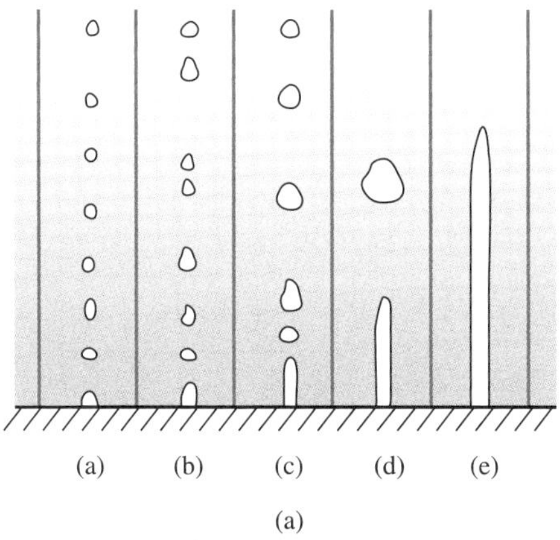

(a)

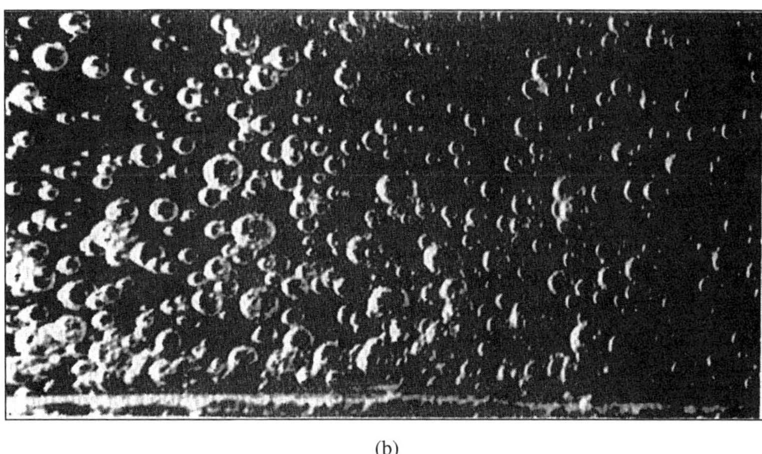

(b)

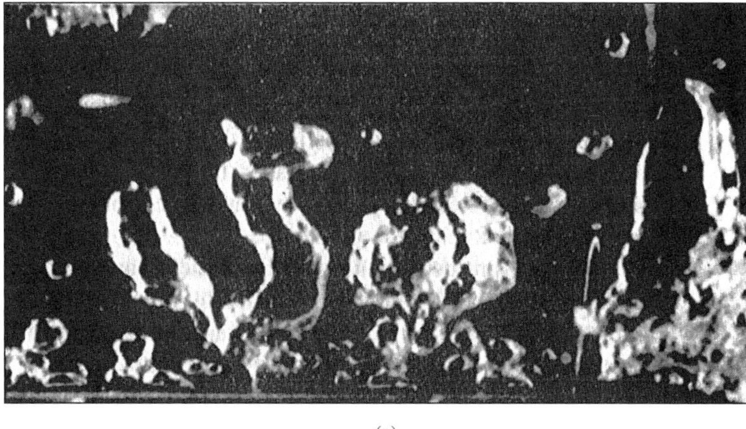

(c)

FIGURE 10.8 Transition from isolated-bubble regime to continuous-column regime in nucleate boiling. (a) Schematic sketch of transition. (b) Photograph of isolated-bubble regime for water at atmospheric pressure and a heat flux of 121,000 W/m^2. (c) Photograph of continuous-column regime for water at atmospheric pressure and a heat flux of 366,000 W/m^2.

Sources: (b) Courtesy of R. Moissis and P. J. Berenson, "On the Hydrodynamic Transitions in Nucleate Boiling," Trans. ASME Ser. C. J. Heat Transfer, Vol. 85, pp. 221–229, Aug. 1963, with permission of the publishers, the American Society of Mechanical Engineers. (c) Courtesy of R. Moissis and P. J. Berenson [9], with permission of the publishers, the American Society of Mechanical Engineers.

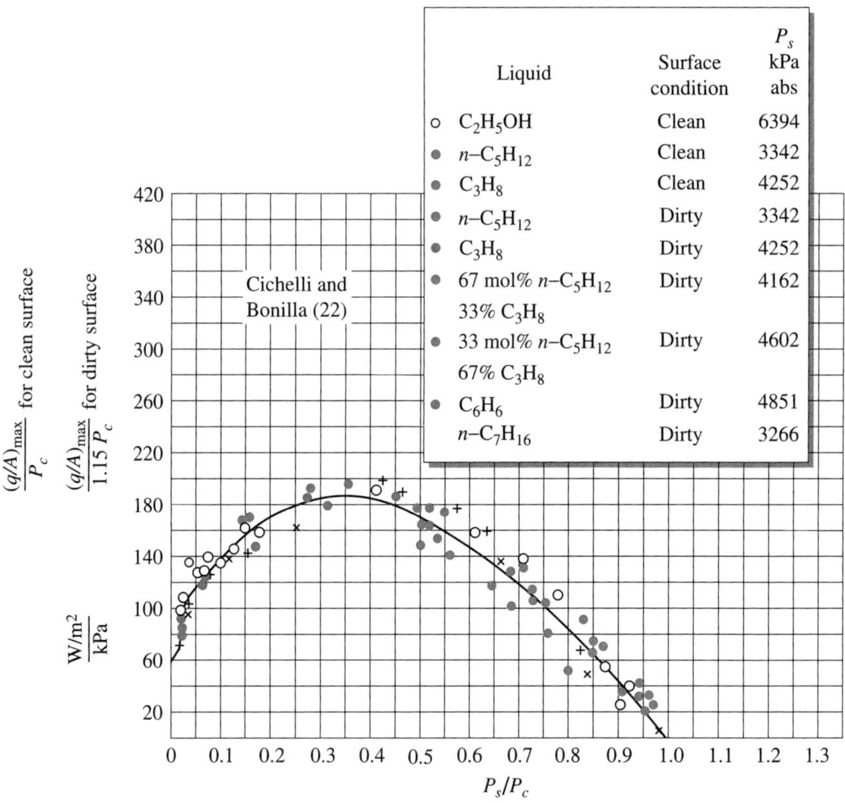

FIGURE 10.9 Peak heat flux in nucleate boiling at various pressures.
Source: Correlation of Cichelli and Bonilla [26], by permission.

yields the highest heat flux. This is illustrated in Fig. 10.9, where the peak nucleate-boiling heat flux is plotted as a function of the ratio of system pressure to critical pressure. For water the optimum pressure is about 10,300 kPa and the peak heat flux is about 3.8 MW/m^2. The quantity in brackets in Eq. (10.4) also shows that the gravitational field affects the peak heat flux. The reason for this behavior is that in a given field the liquid phase, by virtue of its higher density, is subjected to a larger force per unit volume than the vapor phase. Since this difference in forces acting on the two phases brings about a separation of the two phases, an increase in the field strength, as in a large centrifugal force field, increases the separating tendency and will also increase the peak flux. Conversely, experiments by Usiskin and Siegel [34] indicate that a reduced gravitational field decreases the peak heat flux in accordance with Eq. (10.4); in a field of zero gravity, vapor does not leave the heated solid and the critical heat flux tends toward zero.

In many practical applications the heater geometry is more complex than the infinite horizontal flat surface postulated by Zuber in the derivation of Eq. (10.4). However, this basic relationship can be applied to other geometries if a correction factor is applied. Lienhard and coworkers [35–38] have obtained experimental data for the critical heat flux in saturated pool boiling for square and round heated surfaces of finite size, cylinders, ribbons, and spheres. Since for each of these cases the heater has a finite size in at least one dimension, the scale characterizing the heater becomes an important parameter:

$$q''_{max} = q''_{max.Z} \cdot f(L/L_b)$$

where L_b is the bubble length scale defined by

$$L_b = \sqrt{\sigma/g(\rho_l - \rho_v)}$$

and $q''_{max.Z}$ is the maximum heat flux predicted by Zuber according to Eq. (10.4).

The ratio (L/L_b) characterizes the size of the heater relative to that of the vapor columns carrying vapor from the surface near the critical heat flux. Using this ratio and the Zuber flux from Eq. (10.4), Table 10.3 lists the ratios of the experimentally

TABLE 10.3 Correlations for the maximum pool boiling heat flux

Geometry	$\frac{q''_{max}}{q''_{max:Z}} =$	Range	Reference
Infinite heated flat plate	1.14	$\frac{L}{L_b} > 30$	[35]
Small heater of width or diameter L with vertical side walls	$\frac{135 L_b^2}{A_{heater}}$	$9 < \frac{L}{L_b} < 20$	[35]
Horizontal cylinder of radius R	$0.89 + 2.27 e^{-3.44\sqrt{R/L_b}}$	$\frac{R}{L_b} > 0.15$	[36]
Large horizontal cylinder of radius R	0.90	$\frac{R}{L_b} > 1.2$	[37]
Small horizontal cylinder of radius R	$0.94\left(\frac{R}{L_b}\right)^{-1/4}$	$0.15 < \frac{R}{L_b} < 1.2$	[37]
Large sphere of radius R	0.84	$4.26 < \frac{R}{L_b}$	[38]
Small sphere of radius R	$1.734\left(\frac{R}{L_b}\right)^{-1/2}$	$0.15 < \frac{R}{L_b} < 4.26$	[38]
Small horizontal ribbon oriented vertically with side height H-both sides heated	$1.18\left(\frac{H}{L_b}\right)^{-1/4}$	$0.15 < \frac{H}{L_b} < 2.96$	[37]
Small horizontal ribbon oriented vertically with side height H-back side insulated	$1.4\left(\frac{H}{L_b}\right)^{-1/4}$	$0.15 < \frac{H}{L_b} < 5.86$	[37]
Small slender horizontal cylindrical body of arbitrary cross section with transverse perimeter L_p	$1.4\left(\frac{L_p}{L_b}\right)^{-1/4}$	$0.15 < \frac{L_p}{L_b} < 5.86$	[37]
Small bluff body with characteristic dimension L	$C_0\left(\frac{L}{L_b}\right)^{-1/2}$	large $\frac{L}{L_b}$	[37]

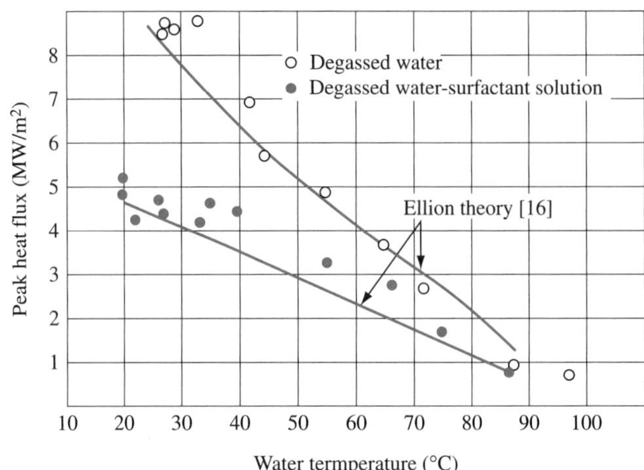

FIGURE 10.10 Effect of bulk temperature on peak heat flux in pool boiling.

Source: By permission from M. E. Ellion [16].

observed critical heat flux to the value predicted from Eq. (10.4) for various heater geometries. Also shown is the range of applicability. The improved $q''_{\max}$ correlation [33] for the infinitely large horizontal plate based on experimental measurements is also shown in the form $q''_{\max}/q''_{\max.Z} = 1.14$; this value corresponds to the recommendation to replace π/24 with 0.149 in Eq. (10.4). The accuracy of the correlation is about ± 20%.

When the liquid bulk is subcooled, the maximum heat flux can be estimated [32] from the equation

$$q''_{\max} = q''_{\max,\text{sat}}\left\{1 + \left[\frac{2k_l(T_{\text{sat}} - T_{\text{liquid}})}{\sqrt{\pi\alpha_l\tau}}\right]\frac{24}{\pi h_{fg}\rho_v}\left[\frac{\rho_v^2}{g_c\sigma g(\rho_l - \rho_v)}\right]^{1/4}\right\} \quad (10.5)$$

where

$$\tau = \frac{\pi}{3}\sqrt{2\pi}\left[\frac{g_c\sigma}{g(\rho_l - \rho_v)}\right]^{1/2}\left[\frac{\rho_v^2}{g_c\sigma g(\rho_l - \rho_v)}\right]^{1/4}$$

and $q''_{\max,\text{sat}}$ can be determined from Eq. (10.4). Figure 10.10 illustrates the influence of the bulk temperature on the peak heat flux for distilled water and a 1% aqueous solution of a surface-active agent boiling on a stainless steel heater. The addition of the surface-active agent decreased the surface tension of water from 72 to 34 dyne/cm, thus causing an appreciable decrease in the peak heat flux, an effect that is in agreement with Eq. (10.4) Noncondensable gases and nonwetting surfaces also reduce the peak heat flux at a given bulk temperature.

Westwater [11], Huber and Hoehne [39], and others found that certain additives (e.g., small amounts of Hyamine 1622) can increase the peak heat flux. Also, the presence of an ultrasonic or electrostatic field can increase the peak heat flux attainable in nucleate boiling.

EXAMPLE 10.1 Water at atmospheric pressure is boiling on a mechanically polished stainless steel surface that is heated electrically from below. Determine the heat flux from the surface to the water when the surface temperature is 106°C, and compare it with the critical heat flux for nucleate boiling. Repeat for the case of water boiling on a Teflon-coated stainless steel surface.

SOLUTION From Table 10.1, C_{sf} is 0.0132 for the mechanically polished surface. From Appendix 2, Table 13, $h_{fg} = 2250$ J/g, $\rho_l = 962$ kg/m^3, and $\rho_v = 0.60$ kg/m^3, $c_l = 4211$ J/kg °C $\text{Pr}_l = 1.75$, $\mu_l = 2.77 \times 10^{-4}$ kg/ms. From Table 10.2, the surface tension at 100°C is 58.8×10^{-3} N/m. Substituting these properties in Eq. (10.2) with $\Delta T_x = 106 - 100 = 6$°C gives

$$q'' = \left(\frac{c_l \Delta T_x}{C_{sf} h_{fg} \text{Pr}_l}\right)^3 \mu_l h_{fg} \sqrt{\frac{g(\rho_l - \rho_v)}{g_c \sigma}}$$

$$= \left[\frac{(4211 \text{ J/kg °C})(6\text{°C})}{(0.0132)(2.25 \times 10^6 \text{ J/kg})(1.75)}\right]^3 (2.77 \times 10^{-4} \text{ kg/m s})$$

$$\times\ (2.25 \times 10^6 \text{ J/kg}\left[\sqrt{\frac{962 \text{ kg/m}^3)(9.8 \text{ m/s}^2)}{58.8 \times 10^{-3} \text{ N/m}}}\right]$$

$$= 28{,}669 \text{ W/m}^2$$

Note that we have neglected the vapor density relative to the liquid density. To determine the critical heat flux, use Eq. (10.4):

$$q''_{\max.Z} = \frac{\pi}{24} \rho_v^{1/2} h_{fg} [\sigma g(\rho_l - \rho_v) g_c]^{1/4}$$

$$= \frac{\pi}{24} (0.60)^{1/2} (2.25 \times 10^6)[(58.8 \times 10^{-3})(9.8)(962)]^{0.25}$$

$$= 1.107 \times 10^6 \text{ W/m}^2$$

At 6°C excess temperature the heat flux is less than the critical value; therefore nucleate pool boiling exists. Had the calculated critical heat flux been less than the heat flux calculated from Eq. (10.2), film boiling would exist and the assumptions underlying the application of Eq. (10.2) would not be satisfied.

Since $q'' \sim C_{sf}^{-3}$, we have for the Teflon-coated stainless steel surface

$$q'' = 29{,}669\left(\frac{0.0132}{0.0058}\right)^3 = 349{,}700 \text{ W/m}^2$$

a remarkable increase in heat flux; however, it is still below the critical value.

In applying the theoretical equations for the critical heat flux in practice, a few words of caution are in order. Data have been presented in the literature indicating lower critical heat fluxes than those predicted from Eq. (10.4) or (10.5). Berenson [12] explains this as follows. Although boiling is a local phenomenon, in most experiments and industrial installations an average heat flux is measured or specified. Therefore, if different locations of a heating surface have different heat fluxes or different nucleate-boiling curves, the measured result will represent an average. But the largest local heat flux at a given temperature difference will always be higher than the average measured value, and if the heat flux is not uniform—for instance, if considerable differences in subcooling or in surface conditions exist or if gravitational variations occur (as around the periphery of a horizontal tube)—a burnout may occur locally even if the average value of heat flux is below the critical value.

Pool-boiling mechanisms can be enhanced by increasing the roughness of the surface and by specially shaped protrusions. Berenson [12] studied the effect of surface roughness on the pool boiling of pentane on a copper plate. He found that the heat flux increased and the excess temperature decreased appreciably with increasing surface roughness, which enhanced the available number of nucleation sites. The critical heat flux was only slightly increased, and the performance of the enhanced surface degraded to the smooth surface performance as trapped vapor leaked out of the cavities. However, as recently pointed out by Manglik and Jog [40], the size and shape of roughness, as well as its viability to produce active and stable nucleation sites, are rather difficult to characterize definitively. Different wetting behavior of various boiling liquids alters the performance of a rough heater. Premanufactured structured roughness, with specially shaped geometries, provides a better and more predictable enhanced boiling performance [40, 41].

Permanent enhancement can be achieved with specially shaped protrusions such as those shown in Fig. 10.11. According to Webb [42], who reviewed the performance of 29 special enhancement surfaces, there are two types: (1) very porous and (2) mechanically formed with deep cavities and small openings. In the latter type, surface tension at the narrow opening prevents degassing of the vapor trapped in the cavity. As can be seen from the data plotted in Fig. 10.11, some of these special surfaces achieve large increases in the nucleate-boiling heat flux compared to those of smooth surfaces. They can also operate under stable conditions and reach critical heat fluxes two or three times as high as those predicted by the Zuber-Kutateladze theory of Eq. (10.4). An extended review of different structured or specially produced surfaces, as well as several other techniques, and their enhanced boiling performance has been given by Manglik [41].

10.2.5 Pool Film Boiling

This boiling regime has less industrial significance because of the very high surface temperature encountered. As shown in Fig. 10.3, the surface is blanketed by a film of vapor. Heat transfer is by conduction across the vapor film and, at higher temperatures, by radiation from the surface to the liquid-vapor interface. Heat transfer to this interface produces the vapor bubbles seen in the photograph. The conduction heat transfer across the vapor film is relatively easy to analyze [43, 44].

For film boiling on tubes of diameter D, Bromley [43] recommends the following correlation equation for the heat transfer coefficient due to conduction alone:

$$\bar{h}_c = 0.62\left\{\frac{g(\rho_l - \rho_v)\rho_v k_v^3[h_{fg} + 0.68c_{pv}\Delta T_x]}{D\mu_v \Delta T_x}\right\}^{1/4} \tag{10.6}$$

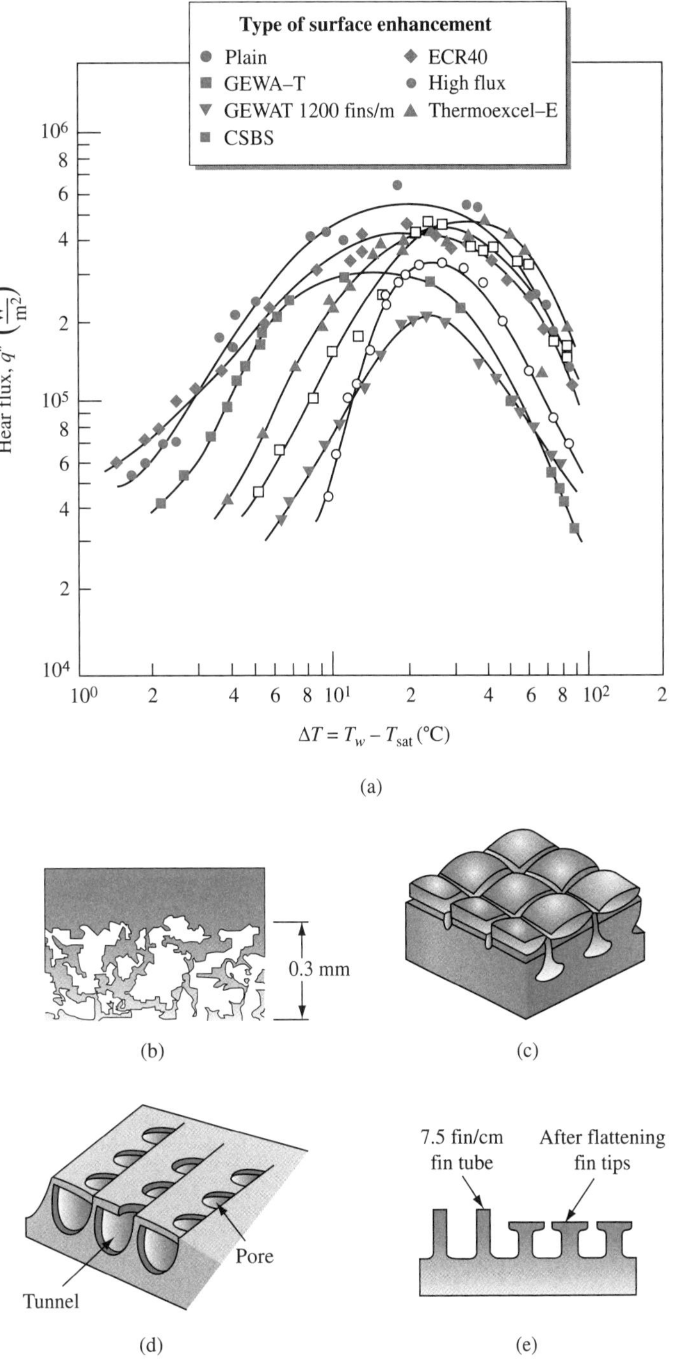

FIGURE 10.11 Nucleate boiling enhancement surfaces, after Webb (42); (a) comparison of single-tube pool boiling results for *p*-xylene at 1 atm; (b) high-flux surface; (c) ECR40; (d) Thermoexcel-E; (e) GEWA-T.

For very-large-diameter tubes and flat horizontal surfaces, Westwater and Breen [44] recommend

$$\bar{h}_c = \left(0.59 + 0.69\frac{\lambda}{D}\right)\left\{\frac{g(\rho_l - \rho_v)\rho_v k_v^3[h_{fg} + 0.68c_{pv}\Delta T_x]}{\lambda \mu_v \Delta T_x}\right\}^{1/4} \quad (10.7)$$

where

$$\lambda = 2\pi\left[\frac{g_c\sigma}{g(\rho_l - \rho_v)}\right]^{1/2}$$

To account for radiation from the surface, Bromley [43] suggests combining the two heat transfer coefficients in the form

$$\bar{h}_{\text{total}} = \bar{h}_c + 0.75\bar{h}_r \quad (10.8)$$

where $\bar{h}_c$ can be computed from Eq. (10.6) or (10.7). The radiation heat transfer coefficient $\bar{h}_r$ is calculated from Eq. (1.31) by assuming that the liquid-vapor interface and the solid are flat and parallel and that the interface has an emissivity of 1.0:

$$\bar{h}_r = \sigma\varepsilon_s\left(\frac{T_s^4 - F_{\text{sat}}^4}{T_s - T_{\text{sat}}}\right) \quad (10.9)$$

Here ε_s is the surface emissivity and T_s is the absolute surface temperature.

EXAMPLE 10.2 Repeat Example 10.1 using a surface temperature of 400°C for the mechanically polished stainless steel surface.

SOLUTION From Eq. (10.2) we note that $q'' \sim \Delta T_x^3$; therefore

$$q'' = 28{,}669 \times \left(\frac{300}{6}\right)^3 = 3.6 \times 10^9 \text{ W/m}^2$$

This exceeds the critical heat flux (1.107×10^6 W/m^2); therefore the system must be operating in the film boiling regime. From Appendix 2, Table 35, we find: k_c = 0.0249 W/m K, c_{pc} = 2034 J/kg K, $\mu_c = 12.1 \times 10^{-6}$ kg/m s. Using Eq. (10.7) for $D \rightarrow \infty$ we have

$$\lambda = 2\pi\left(\frac{58.8 \times 10^{-3}\,\text{N/m}}{(9.8\,\text{m/s}^2)(962\,\text{kg/m}^3)}\right)^{1/2} = 0.0157\,\text{m}$$

and

$$\bar{h}_c = (0.59)\left\{\frac{(9.8)(962)(0.60)(0.0249)^3[2250 + (0.68)(2034)(1000)(300)]}{(0.0157)(1.21 \times 10^{-6})(300)}\right\}^{1/4}$$

$$= 149.1\,\text{W/m}^2\,\text{K}$$

Since the surface is polished, $\varepsilon_s \approx 0.05$, and from Eq. (10.9) we see that $\bar{h}_r$ is negligible. The heat flux is therefore

$$q'' = (149.1\text{ W/m}^2\text{ K})(300\text{ K}) = 44{,}740\text{ W/m}^2$$

10.3 Boiling in Forced Convection

The heat transfer and pressure drop characteristics of forced-convection boiling play an important part in the design of boiling nuclear reactors, environmental control systems for spacecraft and space power plants, and other advanced power-production systems. Despite the large number of experimental and analytical investigations that have been conducted in the area of forced-convection boiling, it is not yet possible to predict all of the characteristics of this process quantitatively because of the great number of variables upon which the process depends and the complexity of the various two-phase flow patterns that occur as the quality of the vapor-liquid mixture (defined as the percentage of the total mass that is in the form of vapor at a given station) increases during vaporization. However, the forced-convection vaporization process has been photographed [45, 46], and it is possible to give a qualitative description of the process based on these photographic observations.

In most practical situations, a fluid at a temperature below its boiling point at the system pressure enters a duct in which it is heated so that progressive vaporization occurs. Figure 10.12 on the next page shows schematically what happens in a vertical duct in which a liquid is vaporized with low heat flux. Figure 10.12 includes a qualitative graph in which the heat transfer coefficient at a specific location is plotted as a function of the local quality. Because heat is continuously added to the fluid, the quality will increase with distance from the entrance.

The heat transfer coefficient at the inlet can be predicted from Eq. (6.63) with satisfactory accuracy. However, as the fluid bulk temperature increases toward its saturation point, which usually occurs only a short distance from the inlet in a system designed to vaporize the fluid, bubbles will begin to form at nucleation sites and will be carried into the main stream as in nucleate pool boiling. This regime, known as the *bubbly-flow regime*, is shown schematically in Fig. 10.12(a). Bubbly flow occurs with very low quality and consists of individual bubbles of vapor entrained in the main flow. In the very narrow quality range over which bubbly flow exists, the heat transfer coefficient can be predicted by superimposing liquid-forced-convection and nucleate-pool-boiling equations as long as the wall temperature is not so large as to produce film boiling (see Section 10.3.1).

As the vapor-volume fraction increases, the individual bubbles begin to agglomerate and form plugs or slugs of vapor as shown in Fig. 10.12(b). Although in this regime, known as the *slug-flow* regime, the mass fraction of vapor is generally much less than 1%, as much as 50% of the volume fraction may be vapor and the fluid velocity in the slug-flow regime may increase appreciably. The plugs of vapor are compressible volumes that also produce flow oscillations within the duct even if the entering flow is steady. Bubbles may continue to nucleate at the wall, and it is probable that the heat transfer mechanism in plug flow is the same as in the bubbly regime: a superposition of forced convection to a liquid and nucleate pool boiling. The heat transfer coefficient rises because of the increased liquid flow velocity, as can be seen in the graph in Fig. 10.12.

While both the bubbly and slug-flow regimes are interesting, it should be noted that for density ratios of importance in forced-convection evaporators, the quality in these two regimes is too low to produce appreciable vaporization. These regimes become important in practice only if the temperature difference is so large as to

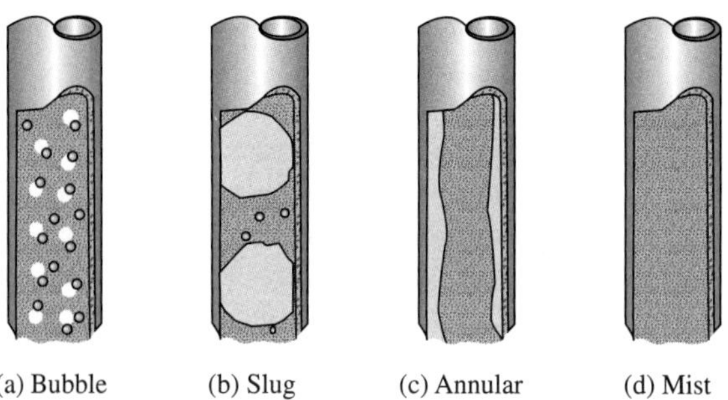

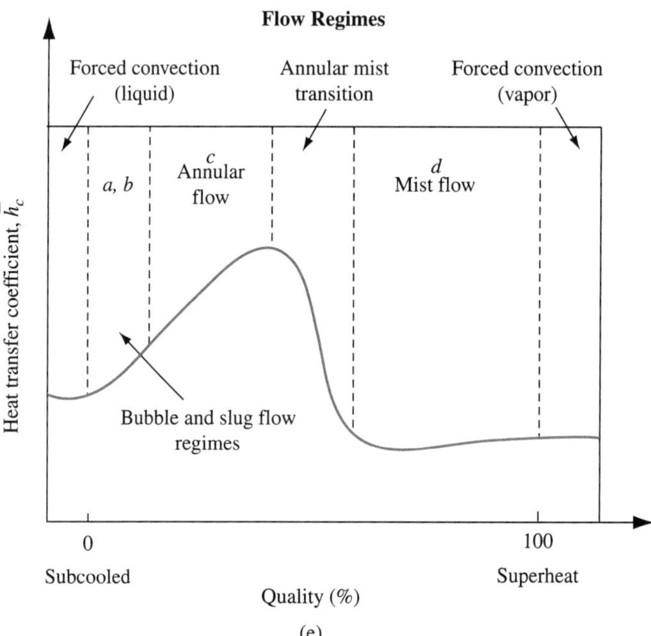

FIGURE 10.12 Characteristics of forced-convection vaporization in a vertical tube: heat transfer coefficient versus quality and type of flow regime.

cause film boiling or if the flow oscillations produced in the slug-flow regime cause instability in a system.

As the fluid flows farther along in the tube and the quality increases, a third flow regime, commonly known as the *annular-flow regime*, appears [see Fig. 10.12(c)]. In this regime the wall of the tube is covered by a thin film of liquid and heat is transferred through this liquid film. In the center of the tube, vapor is flowing at a higher velocity, and although there may be a number of active bubble nucleation sites at the wall, vapor is generated primarily by vaporization from the liquid-vapor interface inside the tube and not by the formation of bubbles inside the liquid annulus unless the heat flux is high. In addition to the liquid in the annulus at the wall, there may be a significant amount of liquid dispersed throughout the vapor core as droplets. The quality range for this type of flow is strongly affected by fluid properties and geometry. But it is generally believed that transition to the next flow regime, shown in Fig. 10.12(d), known as the *mist-flow regime*, occurs at qualities of about 25% or higher.

The transition from annular to mist flow is of great interest since this is presumably the point at which the heat transfer coefficient experiences a sharp decrease, as shown in the graph in Fig. 10.12. In systems with fixed heat flux a sharp increase in wall temperature results, while systems with fixed wall temperature will exhibit a sharp drop in heat flux. Generally, this point is referred to as the *critical heat flux*. Specifically, for low heat flux the condition is called *dryout* because the tube wall is no longer wetted by liquid. An important change takes place in the transition between annular and mist flow: in the former the wall is covered by a relatively high-conductivity liquid, whereas in the latter, due to complete evaporation of the liquid film, the wall is covered by a low-conductivity vapor. Berenson and Stone [45] observed that the wall-drying process occurs in the following manner: A small dry spot forms suddenly at the wall and grows in all directions as the liquid vaporizes because of conduction heat transfer through the liquid. The small strips of liquid remaining on the wall are almost stationary relative to the high-velocity vapor and the liquid droplets in the vapor core. The dominant heat transfer mechanism is conduction through the liquid film, and although nucleation may produce the initial dry spot on the wall, it has only a small effect on the heat transfer. It thus appears that the drying process in transition to mist flow is similar to the process that occurs with a thin film of liquid in a hot pan whose temperature is not great enough to cause nucleate boiling.

Most of the heat transfer in mist flow is from the hot wall to the vapor, and after the heat has been transferred into the vapor core, it is transferred to the liquid droplets there. Vaporization in mist flow actually takes place in the interior of the duct, not at the wall. For this reason the temperature of the vapor in the mist-flow regime can be greater than the saturation temperature, and thermal equilibrium may not exist in the duct. While the volume fraction of the liquid droplets is small, they account for a substantial mass fraction because of the high liquid-to-vapor density ratio.

These observations are consistent with Miles' theoretical stability analysis for a liquid film [47], which predicts that a liquid film is stable at sufficiently small Reynolds numbers irrespective of the vapor velocity. Since the Reynolds number of the liquid film in a forced-convection evaporator decreases as the quality increases, the liquid annulus will be stable at sufficiently high quality irrespective of the value of the vapor velocity.

The regimes of forced-convection boiling depend on the magnitude of the heat flux and can be visualized in Fig. 10.13. At high heat fluxes the annular-flow regime is not exhibited. The critical heat flux under these conditions occurs because of a transition from saturated nucleate boiling in the bubble/slug-flow regime to saturated film boiling in the mist-flow regime and is known as departure from nucleate boiling (DNB). At still higher heat fluxes, the critical heat flux results from a transition from subcooled boiling in the bubble-flow regime to subcooled film boiling in the mist-flow regime. This transition is also known as DNB. At the higher heat fluxes that produce DNB to subcooled film boiling, there are very large temperature increases and actual tube burnout may occur. At lower heat fluxes, where the transition is due to dryout, the temperature increase is much less and physical burnout is not likely.

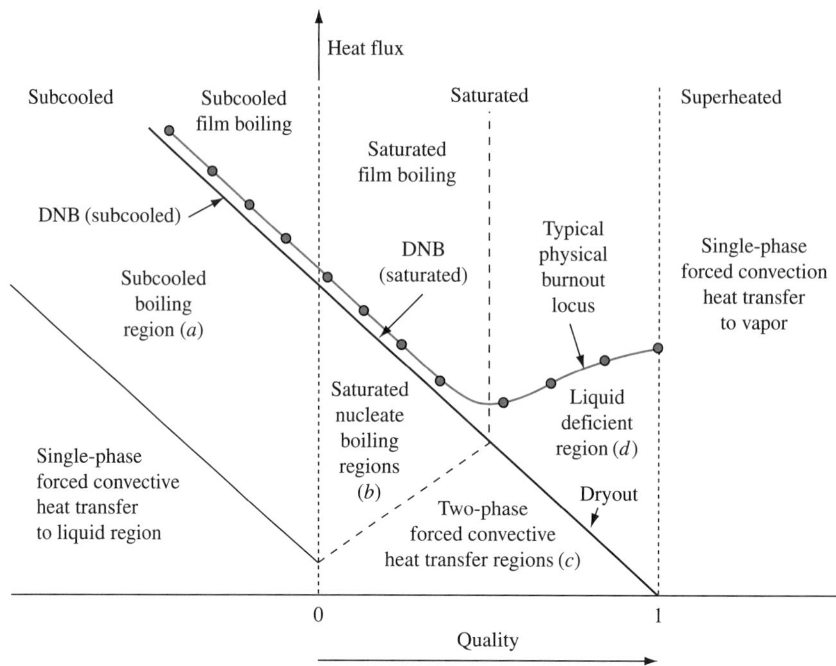

FIGURE 10.13 Regimes of two-phase forced-convection heat transfer as a function of quality with increasing heat flux as ordinate [24].

For horizontal tubes, the situation is more complex due to stratification of the vapor and liquid phases by gravity, especially at low-flow velocities. Much less data are available for the horizontal orientation than for the vertical orientation, but it is clear that the critical heat flux is strongly influenced. In addition, stratification may lead to overheating of the upper portions of the tube, where the vapor may become superheated before dryout occurs on the lower portion of the tube.

10.3.1 Nucleate Forced-Convection Boiling

The method of correlating data for nucleate pool boiling described in Section 10.2.2 has also been applied successfully to the boiling of fluids flowing inside tubes or ducts by forced convection [17] or natural [25] convection.

Figure 10.14 shows best-fit curves through boiling data, typical of subcooled forced convection in tubes or ducts [29, 48]. The system in which these data were obtained consisted of a vertical annulus containing an electrically heated stainless steel tube placed centrally in tubes of various diameters. The heater was cooled by degassed distilled water flowing upward at velocities from 0.3 to 3.7 m/s and pressures from 207 to 620 kPa. The scale of Fig. 10.14 is logarithmic. The ordinate is the heat flux q/A, and the abscissa is ΔT, the temperature difference between the heating surface and the bulk of the liquid. The dashed lines represent forced-convection conditions at various velocities and various degrees of subcooling. The solid lines indicate the deviation from forced convection caused by

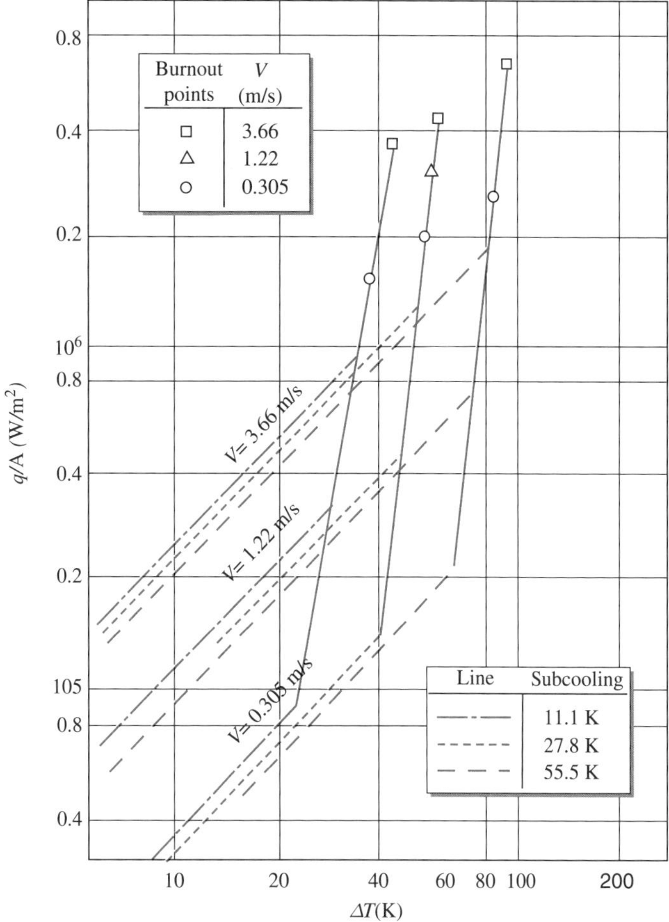

FIGURE 10.14 Typical boiling data for subcooled forced convection: heat flux versus temperature difference between surface and fluid bulk.

Source: By permission from McAdams et al. [48].

surface boiling. We note that the onset of boiling caused by increasing the heat flux depends on the velocity of the liquid and the degree of subcooling below its saturation temperature at the prevailing pressure. At lower pressures the boiling point at a given velocity is reached at lower heat fluxes. An increase in velocity increases the effectiveness of forced convection, decreases the surface temperature at a given heat flux, and thereby delays the onset of boiling. In the boiling region the curves are steep and the wall temperature is practically independent of the fluid velocity. This shows that the agitation caused by the bubbles is much more effective than turbulence in forced convection without boiling. The heat flux data with surface boiling are plotted separately in Fig. 10.15 against the excess temperature. The resulting curve is similar to that for nucleate boiling in a saturated pool shown in Fig. 10.1 and emphasizes the similarity of the boiling

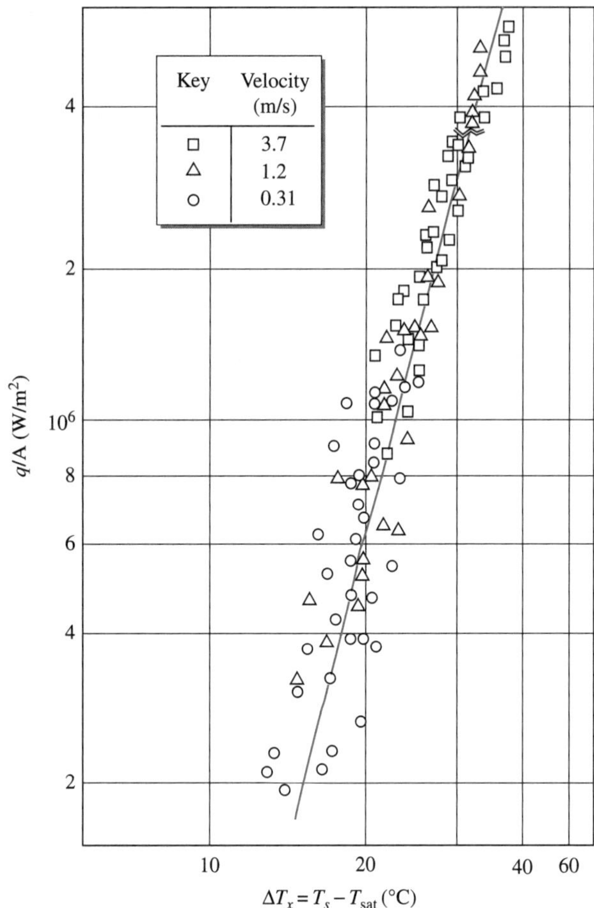

FIGURE 10.15 Approximate correlation of data for nucleate boiling with forced convection obtained by plotting heat flux versus excess temperature.

Source: By permission from McAdams et al. [48].

processes and their dependence on the excess temperature; in particular, the heat flux increases approximately with ΔT^3. However, there are not yet sufficient data to suggest that the fully developed boiling curve for forced convection will always follow pool-boiling data correlations.

To apply the pool-boiling correlation to forced-convection boiling, the total heat flux must be separated into two parts, a *boiling flux* q_b/A and a *convection flux* q_c/A, where

$$q_{\text{total}} = q_b + q_c$$

The boiling heat flux is determined by subtracting the heat flow rate accountable for by forced convection alone from the total flux:

$$q_b = q_{\text{total}} - A\bar{h}_c(T_s - T_b) \tag{10.10}$$

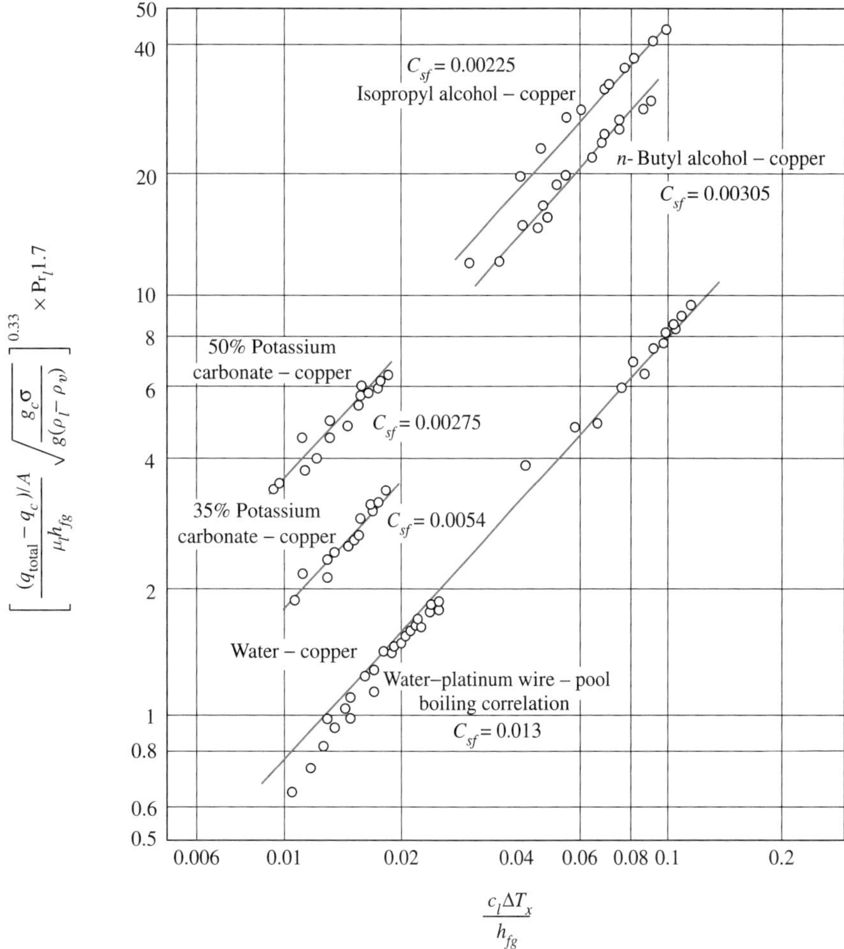

FIGURE 10.16 Correlation of data for convection boiling by the Rohsenow method.

Source: Adapted from Jens and Leppert [49], with permission of the publisher, the American Society of Naval Engineers.

where $\bar{h}_c$ is determined from Eq. (6.63)* using property values for the liquid phase. This value of q_b is to be determined by Eq. (10.2). The results of this method of correlating data for boiling superimposed on convection are shown in Fig. 10.16 for a number of fluid-surface combinations. Some of the data shown in Fig. 10.16 were obtained with subcooled liquids, others with saturated liquids containing various amounts of vapor.

---

*Rohsenow [17] recommends that the coefficient 0.023 in Eq. (6.63) be replaced by 0.019 in nucleate boiling.

10.3.2 Boiling with Net Vapor Production

Beyond the narrow range of quality in which bubble flow exists and Eq. (10.10) is valid, the bulk of the liquid will be at the saturation temperature. The heat transfer mechanism here is referred to as saturated nucleate boiling. Beyond this, in the annular regime, heat is transferred across the film of liquid on the wall. In these flow regimes, Chen [50] has proposed a correlation that assumes that the convection as well as the boiling heat transfer mechanisms play a role and that their effects are additive:

$$h = h_c + h_b$$

where

$$h_c = 0.023\left[\frac{G(1-x)D}{\mu_l}\right]^{0.8} \mathrm{Pr}_l^{0.4}\frac{k_l}{D}F \tag{10.11}$$

is the contribution of the annular region and

$$h_b = 0.00122\left(\frac{k_l^{0.79}c_l^{0.45}\rho_l^{0.49}g_c^{0.25}}{\sigma^{0.5}\mu_l^{0.29}h_{fg}^{0.24}\rho_v^{0.24}}\right)\Delta T_x^{0.24}\Delta p_{\mathrm{sat}}^{0.75}S \tag{10.12}$$

is the contribution of the nucleate boiling region. In Eqs. (10.11) and (10.12), SI units are used with Δp_{sat} (the change in vapor pressure corresponding to a temperature change ΔT_x) expressed in N/m^2. The parameter F can be calculated [51] from

$$F = 1.0 \qquad \text{when } \frac{1}{X_{tt}} < 0.1$$

$$F = 2.35\left(\frac{1}{X_{tt}} + 0.213\right)^{0.736} \qquad \text{when } \frac{1}{X_{tt}} > 0.1$$

where

$$\frac{1}{X_{tt}} = \left(\frac{x}{1-x}\right)^{0.9}\left(\frac{\rho_l}{\rho_v}\right)^{0.5}\left(\frac{\mu_v}{\mu_l}\right)^{0.1}$$

the parameter S is given by

$$S = (1 + 0.12\,\mathrm{Re}_{TP}^{1.14})^{-1} \quad \text{for} \quad \mathrm{Re}_{TP} < 32.5$$

$$S = (1 + 0.42\,\mathrm{Re}_{TP}^{0.78})^{-1} \quad \text{for} \quad 32.5 < \mathrm{Re}_{TP} < 70$$

$$S = 0.1 \quad \text{for} \quad \mathrm{Re}_{TP} > 70$$

with the Reynolds number Re_{TP} defined as

$$\mathrm{Re}_{TP} = \frac{G(1-x)D}{\mu_l}F^{1.25} \times 10^{-4}$$

This correlation has been tested against data for several systems (water, methanol, cyclohexane, pentane, heptane, and benzene) for pressures ranging from 0.5 to 35 atm and quality x ranging from 1 to 0.71 with an average deviation of 11%. Collier and Thome [24] describe how the Chen correlation can be extended to the subcooled boiling region.

EXAMPLE 10.3 Saturated liquid n-butyl alcohol, $C_4H_{10}O$, is flowing at 161 kg/h through a 1-cm-ID copper tube at atmospheric pressure. The tube wall temperature is held at 140°C by condensing steam at 361 kPa absolute pressure. Calculate the length of tube required to achieve a quality of 50%. The following property values can be used for the alcohol:

$\sigma = 0.0183$ N/m, surface tension

$h_{fg} = 591{,}500$ J/kg, heat of vaporization

$T_{sat} = 117.5$°C, atmospheric pressure boiling point

$P_{sat} = 2$ atm, saturation pressure corresponding to a saturation temperature of 140°C

$\rho_v = 2.3\ \text{kg/m}^3$, density of the vapor

$\mu_v = 0.0143 \times 10^{-3}$ kg/m s, viscosity of the vapor

SOLUTION The following property values are taken from Appendix 2, Table 19:

$$\rho_l = 737\ \text{kg/m}^3$$
$$\mu_l = 0.39 \times 10^{-3}\ \text{kg/m s}$$
$$c_l = 3429\ \text{J/kg K}$$
$$\text{Pr}_l = 8.2$$
$$k_l = 0.163\ \text{W/m K}$$
$$C_{sf} = 0.00305 \text{ from Table 10.1}$$

The mass velocity is

$$G = \frac{(161\,\text{kg/h})}{(3600\,\text{s/h})} \frac{4}{\pi(0.01\,\text{m})^2} = 569\,\text{kg/m}^2\,\text{s}$$

The Reynolds number for the liquid flow is

$$\text{Re}_D = \frac{GD}{\mu_l} = \frac{(569\,\text{kg/m}^2\,\text{s})(0.01\,\text{m})}{(0.39 \times 10^{-3}\,\text{kg/m s})} = 14{,}590$$

The contribution to the heat transfer coefficient due to the two-phase annular flow is

$$h_c = (0.023)(14{,}590)^{0.8}(8.2)^{0.4}\left(\frac{0.163\,\text{W/m k}}{0.01\,\text{m}}\right)(1 - x)^{0.8}F$$
$$= 1865(1 - x)^{0.8}F$$

Since the vapor pressure changes by 1 atm over the temperature range from T_{sat} to 140°C, we have $\Delta p_{sat} = 101{,}300\ \text{N/m}^2$. Therefore, the contribution to the heat transfer coefficient from nucleate boiling is

$$h_b = 0.00122\left[\frac{0.163^{0.79}3429^{0.45}737^{0.49}1^{0.25}}{0.0183^{0.5}(0.39 \times 10^{-3})^{0.29}591{,}300^{0.24}2.3^{0.24}}\right]$$
$$\times (140 - 117.5)^{0.24}(101{,}300)^{0.75}S$$

or $h_b = 8393S$.

The calculation for $1/X_{tt}$ becomes

$$\frac{1}{X_{tt}} = \left(\frac{x}{1-x}\right)^{0.9}\left(\frac{737}{2.3}\right)^{0.5}\left(\frac{0.0143}{0.39}\right)^{0.1} = 12.86\left(\frac{x}{1-x}\right)^{0.9}$$

Since the liquid is at saturation temperature, the heat flux over a length Δl can be related to an increase in quality by

$$\dot{m}h_{fg}\,\Delta x = q''\pi D\,\Delta l$$

Substituting the known quantities, we find

$$\Delta l = 842{,}031\,\frac{\Delta x}{q''}$$

where from Eq. (10.11) $h = h_c + h_b$ and $q'' = h\,\Delta T_x$.

We can now prepare a table showing stepwise calculations that track the increase in quality, from $x = 0$ to $\dot{x} = 0.50$, assuming that the steps Δx are small enough that the heat flux and other parameters are reasonably constant in that step.

x	Δx	$\frac{1}{X_{tt}}$	F	h_c (W/m² K)	Re_{TP}	S	h_b (W/m² K)	h (W/m² K)	q'' (W/m²)	Δl (m)	l (m)
0											0
0.01	0.01	0.206	1.24	2291	1.89	0.801	6728	9019	202927	0.041	0.041
0.05	0.04	0.909	2.56	4577	4.49	0.601	5045	9623	216509	0.156	0.197
0.10	0.05	1.78	3.90	6692	7.19	0.468	3922	10614	238820	0.176	0.373
0.20	0.10	3.69	6.41	9994	11.90	0.331	2780	12774	287419	0.293	0.666
0.30	0.10	6.00	9.01	12637	15.94	0.262	2197	14834	333755	0.252	0.919
0.40	0.10	8.93	11.98	14844	19.51	0.220	1846	16690	375523	0.224	1.143
0.50	0.10	12.86	15.59	16695	22.60	0.192	1616	18310	411984	0.204	1.347

The tube length required to reach 50% quality is 1.35 m.

Note the relative importance of the nucleate boiling contribution, h_b, and the two-phase flow contribution, h_c, along the tube.

10.3.3 Critical Heat Flux

Predictions of the critical heat flux for forced-convection systems are less accurate than those for pool boiling, primarily because of the number of variables involved and difficulties encountered in trying to perform controlled experiments to measure the critical heat flux or to determine its location.

A very large number of critical heat flux correlations have been proposed, primarily for boiling water in vertical round tubes with constant heat flux. An empirical critical heat flux correlation for forced convection has been developed by Griffith [52] and covers a wide range of conditions. Griffith correlated critical heat flux data for water, benzene, *n*-heptane, *n*-pentane, and ethanol at pressures varying from

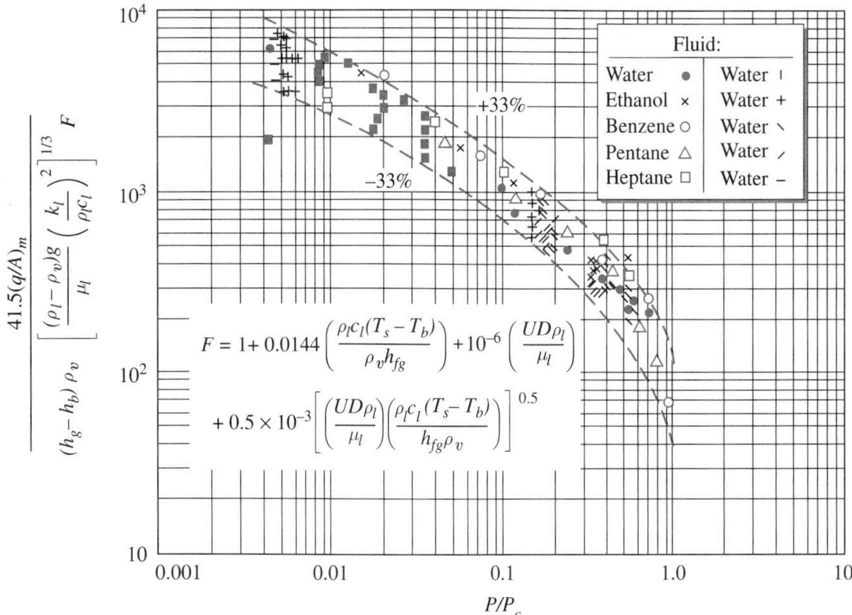

FIGURE 10.17 Peak heat flux correlation for forced-convection boiling and vaporization.

Source: Courtesy of Griffith [52] and the American Society of Mechanical Engineers.

0.5% to 96% of critical pressure, at velocities from 0 to 30 m/s, at subcooling from 0 to 138°C, and at qualities ranging from 0 up to 70%. The data used in this correlation were obtained in round tubes and rectangular channels. Figure 10.17 shows the correlated data, and an inspection of this figure suggests that the critical heat flux can apparently be predicted to within ±33% for the conditions used in this study. In Fig. 10.17, h_{fg} is the saturated vapor enthalpy and h_b is the bulk enthalpy of the fluid, which may be subcooled liquid, saturated liquid, or a two-phase flow mixture at some quality less than 70%.

The pressure drop in pipes and ducts with two-phase flow has been investigated by numerous authors. The problem is quite complex and no entirely satisfactory method of calculation is available. A very useful summary of the state of the art has been prepared by Griffith [53], who concludes, as do several others, that the best available method for predicting the pressure loss is that proposed by Lockhart and Martinelli [54]. The reader interested in this problem is referred to the detailed treatments by Tong [55] and Collier and Thome [24].

A very effective method for increasing the peak heat flux attainable in low-quality forced-convection boiling is to insert twisted tapes into a tube to produce a helical flow pattern that generates a centrifugal force field corresponding to many g's [56]. Gambill et al. [57] achieved a peak heat flux of 174 MW/m^2 in a swirl system with 61°C subcooled water at 5860 kPa flowing at a velocity of 30 m/s in a 0.5-cm-diameter tube; this is almost three times the energy flux emanating from the surface of the sun.

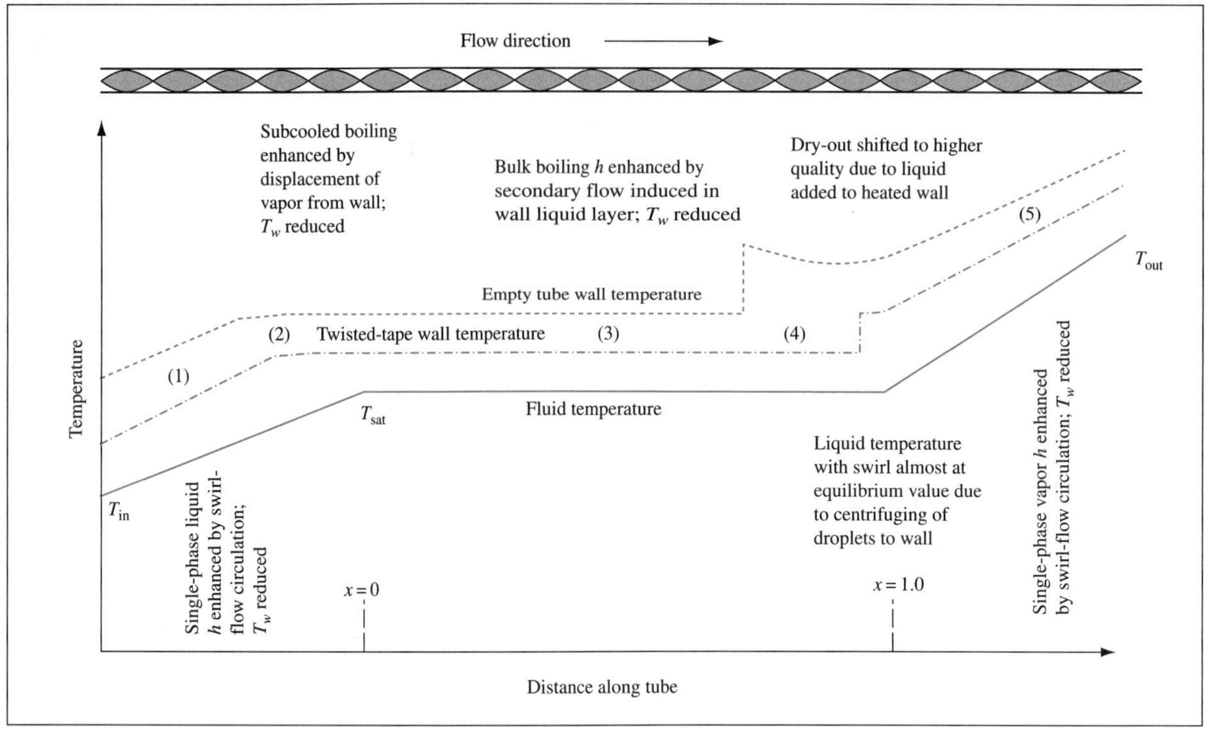

FIGURE 10.18 The influence of twisted-tape-induced swirl on the evolution of fluid bulk and tube-wall temperatures along the tube length in forced convection boiling with uniform heat flux, and fixed mass flux, pressure level, and inlet temperature [58].

Source: Courtesy of R. M. Manglik and A. E. Bergles, "Swirl Flow Heat Transfer and Pressure Drop with Twisted-Tape Inserts", Advances in Heat Transfer, Academic Press, New York, NY, Vol. 36, pp. 183–266, 2002.

A comprehensive review of the literature on twisted-tape inserts and their application in both single-phase flows and forced-convection boiling has been given by Manglik and Bergles [58]. To understand the effects of swirl generated by twisted-tape inserts in flow boiling, consider a uniformly heated, once-through boiler tube for generating superheated vapor; mass flux, pressure level, and inlet temperature are fixed. Thus, the intent of the twisted tape is to reduce the wall temperature, and Fig. 10.18 schematically depicts the progress of wall temperatures for both an empty tube and a tube fitted with a twisted tape [58]. Enhanced heat transfer coefficient in the single-phase region (1), followed by a rather small subcooled boiling region (2), results in a substantial reduction in the wall temperature. This is followed by bulk boiling (3) and dispersed-flow film boiling (4); when the liquid is finally evaporated, the single-phase vapor is heated up in region (5). With a twisted-tape insert, the wall temperature is reduced in all boiling regions, as shown in Fig. 10.18. In bulk boiling (3), the empty tube dries out (or reaches the critical heat flux condition) at an intermediate quality, with the wall temperature increasing sharply. Due to droplet cooling in the dispersed-flow film boiling region (4), the wall temperature decreases before increasing again as the vapor is superheated. After dry-out and extending into the quality region (5), the fluid is in a nonequilibrium state—i.e., the vapor is superheated—and there is more liquid in the form of droplets at saturation temperature.

With a twisted-tape insert in the bulk boiling region (3) of Fig. 10.18, the liquid is centrifuged to the wall so that a liquid film is maintained, and dry-out is thus delayed until a very high quality is achieved. The remaining droplets are again centrifuged to the wall, thereby reducing the temperature excursion. An equilibrium fluid condition is promoted, so that the wall temperature quickly settles down and tracks the fluid temperature (4). Beyond dry-out (5), because of the swirl-generated enhancement of the single-phase vapor heat transfer coefficient, the wall temperature is again reduced relative to that in an empty tube.

10.3.4 Heat Transfer Beyond the Critical Point

As suggested by Fig. 10.13, there are three critical transitions leading to a sudden increase in wall temperature for constant heat flux. Operation beyond the critical points involves (1) subcooled film boiling, (2) saturated film boiling, or (3) a liquid-deficient region (mist flow). For systems in which temperature is the independent variable, a fourth critical transition known as *transition boiling* exists.

Film Boiling In the film-boiling regime a central liquid core is surrounded by an irregular vapor film. As in pool film boiling, the presence of the vapor film simplifies analysis of this boiling regime. These analyses generally follow that for filmwise condensation as described in Section 10.4. For film boiling in vertical tubes, a correlation that agrees reasonably well with the analyses is the one recommended for pool boiling on the outside of horizontal tubes by Bromley [43], that is, Eq. (10.6).

Liquid-Deficient Region This regime results from a thinning of the annular liquid film on the heated surface, which ultimately results in wall dryout. Note from Fig. 10.13 that at higher heat flux the liquid-deficient region results from a transition from saturated film boiling, that is, DNB.

In the saturated film-boiling the flow pattern is inverted from that in the annular regime, Fig. 10.12. That is, a central liquid core is surrounded by a vapor film. As the thermodynamic quality is increased, the liquid core breaks up into droplets and the resulting liquid-deficient flow is similar to that resulting from transition from annular flow.

Liquid droplets periodically hit the wall, thereby producing significantly higher heat transfer coefficients than in the saturated film-boiling regime—thus physical burnout is unlikely. Heat transfer from the wall to the vapor and then from the vapor to the droplets allows a nonequilibrium state to exist since the vapor can become superheated in the presence of the droplets. Correlations developed for heat transfer in this regime are of two types: (1) purely empirical correlations and (2) empirical correlations that attempt to account for the nonequilibrium.

An empirical correlation developed by Groeneveld [59] is of the form of the Dittus-Boelter equation for single-phase, forced convection:

$$\frac{hD}{k_v} = a\left\{\text{Re}_v\left[x + \frac{\rho_v}{\rho_l}(1 - x)\right]\right\}^b \text{Pr}_v^c\left[1 - 0.1\left(\frac{\rho_l}{\rho_v} - 1\right)^{0.4}(1 - x)^{0.4}\right]^d \quad (10.13)$$

Table 10.4 on the next page gives values of *a, b, c,* and *d* for various geometries and the range of operating conditions over which the correlation is valid.

TABLE 10.4 Constants for Eq. (10.13)

Geometry	a	b	c	d	No. of Points	RMS Error, %
Tubes	1.09×10^{-3}	0.989	1.41	−1.15	438	11.5
Annuli	5.20×10^{-2}	0.668	1.26	−1.06	266	6.9
Tubes and annuli	3.27×10^{-3}	0.901	1.32	−1.50	704	12.4

Range of Data on Which Correlations are Based

Parameters and Units	Geometry: Tube	Geometry: Annulus
Flow direction	Vertical and horizontal	Vertical
Inside diameter, cm	0.25 to 2.5	0.15 to 0.63
Pressure, atm	68 to 215	34 to 100
G, kg/m^2 s	700 to 5300	800 to 4100
x, fraction by weight	0.10 to 0.90	0.10 to 0.90
Heat flux, kW/m^2	120 to 2100	450 to 2250
hD/k_v	95 to 1770	160 to 640
$\mathrm{Re}_v\left[x + \frac{\rho_v}{\rho_l}(1-x)\right]$	6.6×10^4 to 1.3×10^6	1.0×10^5 to 3.9×10^5
Pr	0.88 to 2.21	0.91 to 1.22
$1 - 0.1\left(\frac{\rho_l}{\rho_v} - 1\right)^{0.4}(1-x)^{0.4}$	0.706 to 0.976	0.610 to 0.963

Rohsenow [60] warns that all such purely empirical correlations should be used with caution. Collier and Thome [24] present additional correlation equations that account for nonequilibrium in the liquid-deficient regime.

Transition Boiling The transition-boiling regime is difficult to characterize in a quantitative manner [3]. Within the region the amount of vapor generated is not enough to support a stable vapor film but is too large to allow sufficient liquid to reach the surface to support nucleate boiling. Berenson [12] suggests, therefore, that nucleate and film boiling occur alternately at a given location. The process is unstable, and photographs show that liquid surges sometimes toward the heating surface and sometimes away from it. At times, this turbulent liquid becomes so highly superheated that it explodes into vapor [11]. From an industrial viewpoint, the transition-boiling regime is of little interest; equipment designed to operate in the nucleate-boiling region can be sized with more assurance and operated with more reproducible results. Tong and Young [61] have proposed a correlation for heat flux in this region.

10.4 Condensation

When a saturated vapor comes in contact with a surface at a lower temperature, condensation occurs. Under normal conditions, a continuous flow of liquid is formed over the surface and the condensate flows downward under the influence of gravity.

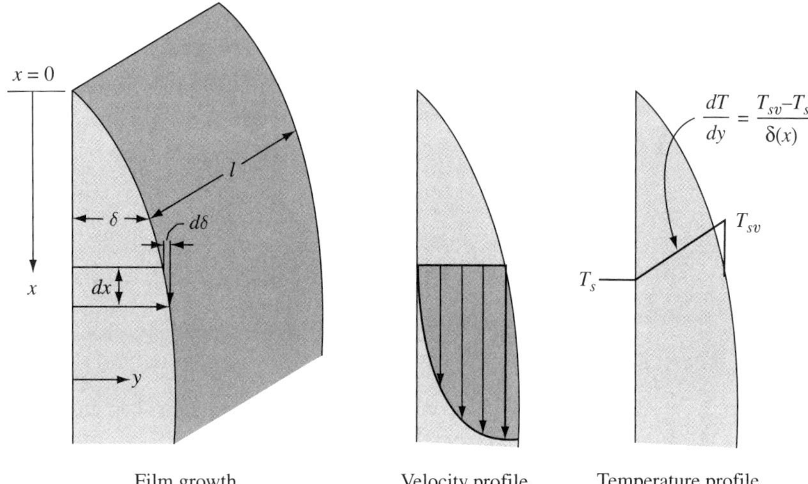

FIGURE 10.19 Filmwise condensation on a vertical surface: film growth, velocity profile, and temperature distribution.

Unless the velocity of the vapor is very high or the liquid film very thick, the motion of the condensate is laminar and heat is transferred from the vapor-liquid interface to the surface merely by conduction. The rate of heat flow therefore depends primarily on the thickness of the condensate film, which in turn depends on the rate at which vapor is condensed and the rate at which the condensate is removed. On a vertical surface the film thickness increases continuously from top to bottom, as shown in Fig. 10.19. As the plate is inclined from the vertical position, the drainage rate decreases and the liquid film becomes thicker. This, of course, causes a decrease in the rate of heat transfer.

10.4.1 Filmwise Condensation

Theoretical relations for calculating the heat transfer coefficients for filmwise condensation for pure vapors on tubes and plates were first obtained by Nusselt [62] in 1916. To illustrate the classical approach, we shall consider a plane vertical surface at constant temperature T_s on which a pure vapor at saturation temperature T_{sv} is condensing. As shown in Fig. 10.19, a continuous film of liquid flows downward under the action of gravity, and its thickness increases as more and more vapor condenses at the liquid-vapor interface. At a distance x from the top of the plate the thickness of the film is δ. If the flow of the liquid is laminar and is caused by gravity alone, we can estimate the velocity of the liquid by means of a force balance on the element $dx\ \delta l$. The downward force per unit depth l acting on the liquid at a distance greater than y from the surface is $(\delta - y)\ dx\ \rho_l g/g_c$. Assuming that the vapor outside the condensate layer is in hydrostatic balance ($dp/dx = \rho_v g/g_c$), a partially balancing force equal to $(\delta - y)\ dx\ \rho_v g/g_c$ will be present as a result of the pressure difference between the upper and lower faces of the element. The other force retarding the downward motion is the drag at the inner boundary of the element. Unless the vapor flows at a very high velocity, the shear at the free surface is quite small and can be neglected. The remaining force will then simply be the viscous shear (μ_l du/dy) dx at the vertical plane y. Under steady-state conditions the upward and downward forces are equal:

$$(\delta - y)(\rho_l - \rho_v)g = \mu_l \frac{du}{dy}$$

where the subscripts l and v denote liquid and vapor, respectively. The velocity u at y is obtained by separating the variables and integrating. This yields the expression

$$u(y) = \frac{(\rho_l - \rho_v)g}{\mu_l}\left(\delta y - \frac{1}{2}y^2\right) + \text{const}$$

The constant of integration is zero because the velocity u is zero at the surface, that is, $u = 0$ at $y = 0$.

The mass rate of flow of condensate per unit breadth Γ_c is obtained by integrating the local mass flow rate at the elevation x, $\rho u(y)$, between the limits $y = 0$ and $y = \delta$, or

$$\Gamma_c = \int_0^\delta \frac{\rho_l(\rho_l - \rho_v)g}{\mu_l}\left(\delta y - \frac{1}{2}y^2\right)dy = \frac{\rho_l(\rho_l - \rho_v)\delta^3}{3\mu_l}g \tag{10.14}$$

The change in condensate flow rate Γ_c with the thickness of the condensate layer δ is

$$\frac{d\Gamma_c}{d\delta} = \frac{g\rho_l(\rho_l - \rho_v)}{\mu_l}\delta^2 \tag{10.15}$$

Heat is transferred through the condensate layer solely by conduction. Assuming that the temperature gradient is linear, the average enthalpy change of the vapor in condensing to liquid and subcooling to the average liquid temperature of the condensate film is

$$h_{fg} + \frac{1}{\Gamma_c}\int_0^\delta \rho_l u c_{pl}(T_{sv} - T)\,dy = h_{fg} + \frac{3}{8}c_{pl}(T_{sv} - T_s)$$

and the rate of heat transfer to the wall is $(k/\delta)(T_{sv} - T_s)$, where k is the thermal conductivity of the condensate. In the steady state the rate of enthalpy change of the condensing vapor must equal the rate of heat flow to the wall:

$$\frac{q}{A} = k\frac{T_{sv} - T_s}{\delta} = \left[h_{fg} + \frac{3}{8}c_{pl}(T_{sv} - T_s)\right]\frac{d\Gamma_c}{dx} \tag{10.16}$$

Equating the expressions for $d\Gamma_c$ from Eqs. (10.15) and (10.16) gives

$$\delta^3 d\delta = \frac{k\mu_l(T_{sv} - T_s)}{g\rho_l(\rho_l - \rho_c)h'_{fg}}dx$$

where $h'_{fg} = h_{fg} + \frac{3}{8}c_{pl}(T_{sv} - T_s)$. Integrating between the limits $\delta = 0$ at $x = 0$ and $\delta = \delta$ at $x = x$ and solving for $\delta(x)$ yields

$$\delta = \left[\frac{4\mu_l kx(T_{sv} - T_s)}{g\rho_l(\rho_l - \rho_v)h'_{fg}}\right]^{1/4} \tag{10.17}$$

Since heat transfer across the condensate layer is by conduction, the local heat transfer coefficient h_{cx} is k/δ. Substituting the expression for δ from Eq. (10.17) gives the heat transfer coefficient as

$$h_{cx} = \left[\frac{\rho_l(\rho_l - \rho_v)gh'_{fg}k^3}{4\mu_l x(T_{sv} - T_s)}\right]^{1/4} \tag{10.18}$$

and the local Nusselt number at x is

$$\mathrm{Nu}_x = \frac{h_{cx}x}{k} = \left[\frac{\rho_l(\rho_l - \rho_v)gh'_{fg}x^3}{4\mu_l k(T_{sv} - T_s)}\right]^{1/4} \tag{10.19}$$

Inspection of Eq. (10.18) shows that the heat transfer coefficient for condensation decreases with increasing distance from the top as the film thickens. The thickening of the condensate film is similar to the growth of a boundary layer over a flat plate in convection. At the same time it is also interesting to observe that an increase in the temperature difference $(T_{sv} - T_s)$ causes a decrease in the heat transfer coefficient. This is caused by the increase in the film thickness as a result of the increased rate of condensation. No comparable phenomenon occurs in simple convection.

The average value of the heat transfer coefficient $\bar{h}_c$ for a vapor condensing on a plate of height L is obtained by integrating the local value h_{cx} over the plate and dividing by the area. For a vertical plate of unit width and height L we obtain by this operation the average heat transfer coefficient

$$\bar{h}_c = \frac{1}{L}\int_0^L h_{cx}\,dx = \frac{4}{3}h_{x=L} \tag{10.20}$$

or

$$\bar{h}_c = 0.943\left[\frac{\rho_l(\rho_l - \rho_v)gh'_{fg}k^3}{\mu_l L(T_{sv} - T_s)}\right]^{1/4} \tag{10.21}$$

It can easily be shown that for a surface inclined at an angle ψ with the horizontal, the average coefficient is

$$\bar{h}_c = 0.943\left[\frac{\rho_l(\rho_l - \rho_v)gh'_{fg}k^3 \sin\psi}{\mu_l L(T_{sv} - T_s)}\right]^{1/4} \tag{10.22a}$$

A modified integral analysis for this problem by Rohsenow [63], which is in better agreement with experimental data if $\mathrm{Pr} > 0.5$ and $c_{pl}(T_{sv} - T_s)/h'_{fg} < 1.0$, yields results identical to Eqs. (10.18) – (10.22a) except that h'_{fg} is replaced by $[h_{fg} + 0.68c_{pl}(T_{sv} - T_s)]$.

Chen [64] considered the interfacial shear and momentum effects and computed a correction factor to Eq. (10.22a):

$$\bar{h}'_c = \bar{h}_c\left(\frac{1 + 0.68A + 0.02AB}{1 + 0.85B - 0.15AB}\right)^{1/4} \tag{10.22b}$$

where $\bar{h}_c'$ is the corrected heat transfer coefficient, $\bar{h}_c$ is the heat transfer coefficient from Eq. (10.22a), and

$$A = \frac{c_l(T_{sv} - T_s)}{h_{fg}} < 2 \qquad \text{(upper limit of validity)}$$

$$B = \frac{k_l(T_{sv} - T_s)}{\mu_l h_{fg}} < 20 \quad \text{(upper limit of validity)}$$

and

$$0.05 < \text{Pr}_l < 1.0$$

Although the foregoing analysis was made specifically for a vertical flat plate, the development is also valid for the inside and outside surfaces of vertical tubes if the tubes are large in diameter compared with the film thickness. These results cannot be extended to inclined tubes, however. In such cases the film flow would not be parallel to the axis of the tube and the effective angle of inclination would vary with x.

The average heat transfer coefficient of a pure saturated vapor condensing on the outside of a sphere or a horizontal tube [see Fig. 10.20(a) and (b)] can be

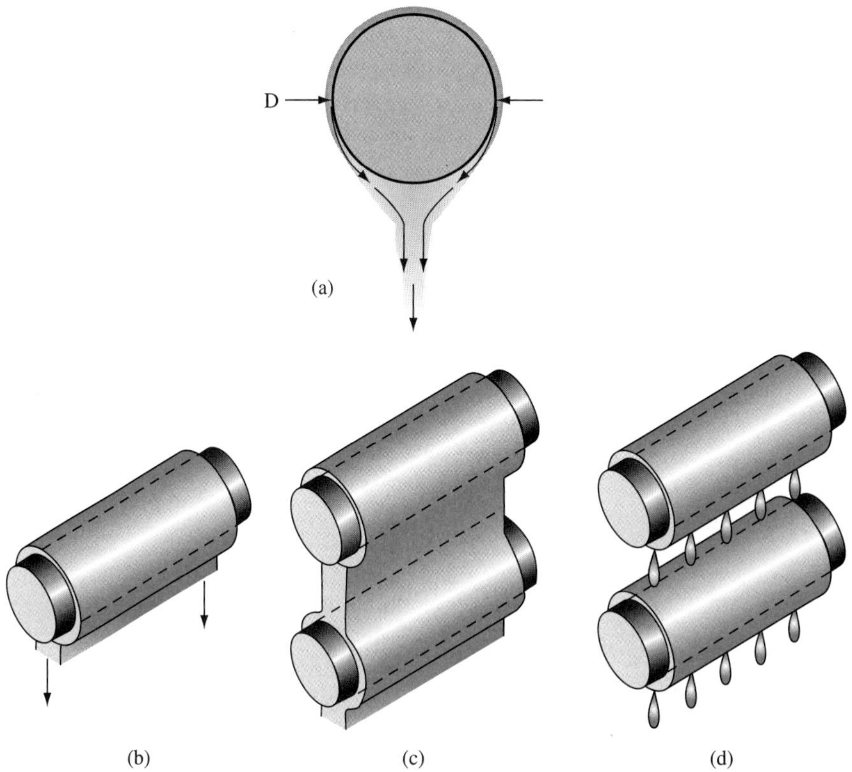

FIGURE 10.20 Film condensation on (a) a sphere, (b) a single horizontal tube, (c) a vertical tier of horizontal tubes with a continuous condensate sheet, and (d) a vertical tier of horizontal tubes with dripping condensate.

evaluated by the method used to obtain Eq. (10.21). For an outside diameter D it leads to the equation

$$\bar{h}_c = c\left[\frac{\rho_l(\rho_l - \rho_v)gh'_{fg}k^3}{D\mu_l(T_{sv} - T_s)}\right]^{1/4} \tag{10.23}$$

where $c = 0.815$ for a sphere and 0.725 for a tube.

If condensation occurs on N horizontal tubes so arranged that the condensate sheet from one tube flows directly onto the tube below [see Fig. 10.20(c)] the average heat transfer coefficient for the system can be estimated by replacing the tube diameter D in Eq. (10.23) by DN. This method will in general yield conservative results because condensate does not fall in smooth sheets from one row to another but drips from tube to tube as shown in Fig. 10.20(d).

Chen [64] suggested that since the liquid film is subcooled, additional condensation occurs on the liquid layer between tubes. Assuming that all the subcooling is used for additional condensation, Chen's analysis yields

$$\bar{h}_c = 0.728[1 + 0.2(N - 1)\mathrm{Ja}]\left[\frac{g\rho_l(\rho_l - \rho_v)k^3h'_{fg}}{ND\mu_l(T_{sv} - T_s)}\right]^{1/4} \tag{10.24}$$

where Ja was previously defined as $c_{pl}(T_{sv} - T_s)/h_{fg}$. Ja is called the Jakob number in honor of the German heat transfer researcher Max Jakob, who did pioneering work on phase change phenomena. Physically, Ja represents the ratio of the maximum sensible heat absorbed by the liquid to the latent heat of the liquid. When Ja is small, the latent heat absorption dominates and the correction factor can be neglected. Eq. (10.24) is in reasonably good agreement with experimental results, provided $[(N - 1)\mathrm{Ja}] < 2$.

In the preceding equations the heat transfer coefficient will be in W/m^2 °C if the other quantities are evaluated in the units listed below.

c_p = specific heat of vapor, J/kg °C
c_{pl} = specific heat of liquid, J/kg °C
k = thermal conductivity of liquid, W/m °C
ρ_l = density of liquid, kg/m^3
ρ_v = density of vapor, kg/m^3
g = acceleration of gravity, m/s^2
h_{fg} = latent heat of condensation or vaporization, J/kg
$h'_{fg} = h_{fg} + \frac{3}{8}c_{pl}(T_{sv} - T_s)$, J/kg
μ_l = viscosity of liquid, N s/m^2
D = tube diameter, m
L = length of plane surface, m
T_{sv} = temperature of saturated vapor, °C
T_s = wall surface temperature, °C

The physical properties of the liquid film in Eqs. (10.17) – (10.24) should be evaluated at an effective film temperature $T_{\text{film}} = T_s + 0.25(T_{sv} - T_s)$ [19]. When used in this manner, Nusselt's equations are satisfactory for estimating heat transfer coefficients for condensing vapors. Experimental data are in general agreement with Nusselt's theory when the physical conditions comply with the assumptions inherent in the analysis. Deviations from Nusselt's film theory occur when the condensate flow becomes turbulent, when the vapor velocity is very high [65], or when a special effort is made to render the surface nonwettable. All of these factors tend to increase the heat transfer coefficients and the Nusselt film theory will therefore always yield conservative results.

EXAMPLE 10.4 A 0.013-m-OD, 1.5-m-long tube is to be used to condense steam at 40,000 N/m^2, T_{sv} = 349 K. Estimate the heat transfer coefficients for this tube in (a) the horizontal position and (b) the vertical position. Assume that the average tube wall temperature is 325 K.

SOLUTION (a) At the average temperature of the condensate film [T_f = (349 + 325)/2 = 337 K], the physical property values pertinent to the problem are

$$k_l = 0.661 \text{ W/m K} \qquad \mu_l = 4.48 \times 10^{-4} \text{ N s/m}^2$$
$$\rho_l = 980.9 \text{ kg/m}^3 \qquad c_{pl} = 4184 \text{ J/kg K}$$
$$h_{fg} = 2.349 \times 10^6 \text{ J/kg} \quad \rho_v = 0.25 \text{ kg/m}^3$$

For the tube in the horizontal position, Eq. (10.23) applies and the heat transfer coefficient is

$$\bar{h}_c = 0.725\left[\frac{(980.9)(980.6)(9.81)(2.417 \times 10^6)(0.661)^3}{(0.013)(4.48 \times 10^{-4})(349 - 325)}\right]^{1/4}$$

$$= 10.680 \text{ W/m}^2\text{K}$$

(b) In the vertical position the tube can be treated as a vertical plate of area πDL and according to Eq. (10.21), the average heat transfer coefficient is

$$\bar{h}_c = 0.943\left[\frac{(980.9)(980.6)(9.81)(2.417 \times 10^6)(0.661)^3}{(4.48 \times 10^{-4})(349 - 325)}\right]^{1/4}$$

$$= 4239 \text{ W/m}^2\text{K}$$

Effect of Film Turbulence The preceding correlations show that for a given temperature difference, the average heat transfer coefficient is considerably larger when the tube is placed in a horizontal position, where the path of the condensate is shorter and the film thinner, than in a vertical position, where the path is longer and the film thicker. This conclusion is generally valid when the length of the vertical tube is more than 2.87 times the outer diameter, as can be seen by a comparison of Eqs. (10.21) and (10.23). However, these equations are based on the assumption that the flow of the condensate film is laminar, and consequently they do not apply when the flow of the condensate is turbulent. Turbulent flow is hardly ever reached on a horizontal tube but may be established over the lower portion of a vertical surface. When it does occur, the average heat transfer coefficient becomes larger as the

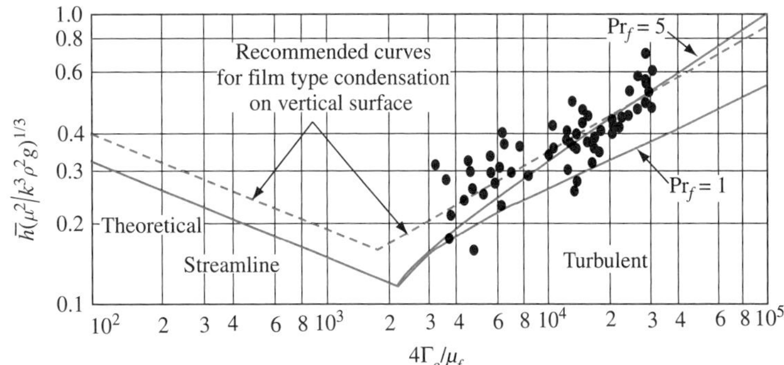

FIGURE 10.21 Effect of turbulence in film on heat transfer with condensation.

length of the condensing surface is increased because the condensate no longer offers as high a thermal resistance as it does in laminar flow. This phenomenon is somewhat analogous to the behavior of a boundary layer.

Just as a fluid flowing over a surface undergoes a transition from laminar to turbulent flow, so the motion of the condensate becomes turbulent when its Reynolds number exceeds a critical value of about 2000. The Reynolds number of the condensate film, Re_δ, when based on the hydraulic diameter [Eq. (6.2)], can be written as $\mathrm{Re}_\delta = (4A/P)\Gamma_c/\delta\mu_f$, where P is the wetted perimeter, equal to πD for a vertical tube, and A is the flow cross-sectional area, equal to $P\delta$. According to an analysis by Colburn [66], the local heat transfer coefficient for turbulent flow of the condensate can be evaluated from

$$h_{cx} = 0.056\left(\frac{4\Gamma_c}{\mu_f}\right)^{0.2}\left(\frac{k^3\rho^2 g}{\mu^2}\right)^{1/3}\mathrm{Pr}_f^{1/2} \tag{10.25}$$

To obtain average values of the heat transfer coefficient, integration of h_x over the surface using Eq. (10.18) for values of $4\Gamma_c/\mu_f$ less than 2000 and Eq. (10.25) for values larger than 2000 is necessary. The results of such calculations for two values of the Prandtl number are plotted as solid lines in Fig. 10.21, where some experimental data obtained with diphenyl in turbulent flow are also shown [67]. The heavy dashed line shown on the same graph is an empirical curve recommended by McAdams [21] for evaluating the average heat transfer coefficient of single vapors condensing on vertical surfaces.

EXAMPLE 10.5 Determine whether the flow of the condensate in Example 10.4 part (b) is laminar or turbulent at the lower end of the tube.

SOLUTION The Reynolds number of the condensate at the lower end of the tube can be written with the aid of Eq. (10.14) as

$$\mathrm{Re}_\delta = \frac{4\Gamma_c}{\mu_l} = \frac{4\rho_l^2 g\delta^3}{3\mu_l^2}$$

Substituting Eq. (10.17) for δ yields

$$\mathrm{Re}_\delta = \frac{4\rho_l^2 g}{3\mu_l^2}\left[\frac{4\mu_l k_l L(T_{sv} - T_s)}{g h_{fg}\rho_l^2}\right]^{3/4} = \frac{4}{3}\left[\frac{4k_l L(T_{sv} - T_s)\rho_l^{2/3} g^{1/3}}{\mu_l^{5/3} h'_{fg}}\right]^{3/4}$$

Inserting the numerical values for the problem into the expression above yields

$$\mathrm{Re}_\delta = \frac{4}{3}\left[\frac{4(0.661)(1.5)(349 - 325)(980.9)^{2/3}(9.81)^{1/3}}{(4.48 \times 10^{-4})^{5/3}(2.417 \times 10^6)}\right]^{3/4} = 564$$

Since the Reynolds number at the lower edge of the tube is below 2000, the flow of the condensate is laminar and the result obtained from Eq. (10.21) is valid.

Effect of High Vapor Velocity One of the approximations made in Nusselt's film theory is that the frictional drag between the condensate and the vapor is negligible. This approximation ceases to be valid when the velocity of the vapor is substantial compared with the velocity of the liquid at the vapor-condensate interface. When the vapor flows upward, it adds a retarding force to the viscous shear and causes the film thickness to increase. With downward flow of vapor, the film thickness decreases and heat transfer coefficients substantially larger than those predicted from Eq. (10.21) can be obtained. In addition, the transition from laminar to turbulent flow occurs at condensate Reynolds numbers of the order of 300 when the vapor velocity is high. Carpenter and Colburn [68] determined the heat transfer coefficients for condensation of pure vapors of steam and several hydrocarbons in a vertical tube 2.44-m long and 1.27-cm-ID, with inlet vapor velocities up to 152 m/s at the top. Their data are correlated reasonably well by the equation

$$\frac{\bar{h}_c}{c_{pl}G_m}\,\mathrm{Pr}_l^{0.50} = 0.046\sqrt{\frac{\rho_l}{\rho_v}f} \tag{10.26}$$

where Pr_l = Prandtl number of liquid
ρ_l = density of liquid, kg/m^3
ρ_v = density of vapor, kg/m^3
c_{pl} = specific heat of liquid, J/kg K
$\bar{h}_c$ = average heat transfer coefficient, W/m^2 K
f = pipe-friction coefficient evaluated at the average vapor velocity $= \tau_w/[G_m^2/2\rho_v]$
τ_w = wall shear stress, N/m^2
G_m = mean value of the mass velocity of the vapor, kg/s m^2

The value of G_m in Eq. (10.26) can be calculated from

$$G_m = \sqrt{\frac{G_l^2 + G_l G_2 + G_2^2}{3}}$$

where G_1 = mass velocity at top of tube
G_2 = mass velocity at bottom of tube

All physical properties of the liquid in Eq. (10.26) are to be evaluated at a reference temperature equal to $0.25T_{sv} + 0.75T_s$. These results can be used generally as an indication of the influence of vapor velocity on the heat transfer coefficient of condensing vapors when the vapor and the condensate flow in the same direction.

Soliman et al. [69] have modified the numerical coefficients in Eq. (10.26) on the basis of additional data:

$$\frac{\bar{h}_c}{c_{pl}G_m}\,\mathrm{Pr}_l^{0.35} = 0.036\sqrt{\frac{\rho_l}{\rho_v}}f \tag{10.27}$$

The effect of vapor velocity in a horizontal tube is complicated by the existence of several flow regimes created by the interaction of vapor and liquid within the tube. Collier and Thome [24] treat this problem in detail.

For condensation on the outside of the horizontal tube when the effect of vapor velocity cannot be ignored, Shekriladze and Gomelauri [70] developed the following correlation equation:

$$\bar{h}'_c = \left[\frac{1}{2}\bar{h}_s^2 + \left(\frac{1}{4}\bar{h}_s^4 + \bar{h}_c^4\right)^{1/2}\right]^{1/2} \tag{10.28}$$

where $\bar{h}'_c$ is the heat transfer coefficient corrected for the effect of vapor shear, $\bar{h}_c$ is the uncorrected heat transfer coefficient for condensation on horizontal tubes, Eq. (10.23), and $\bar{h}_s$, the contribution of the vapor shear to the heat transfer coefficient, is calculated from

$$\frac{\bar{h}_s D}{k_l} = 0.9\left(\frac{\rho_l U_\infty D}{\mu_l}\right)^{0.5} \quad \text{for} \quad \frac{\rho_l U_\infty D}{\mu_l} < 10^6 \tag{10.29}$$

$$\frac{\bar{h}_s D}{k_l} = 0.59\left(\frac{\rho_l U_\infty D}{\mu_l}\right)^{0.5} \quad \text{for} \quad \frac{\rho_l U_\infty D}{\mu_l} > 10^6 \tag{10.30}$$

where U_∞ is the vapor velocity approaching the tube.

Condensation of Superheated Vapor Although all of the preceding equations strictly apply only to saturated vapors, they can also be used with reasonable accuracy for condensation of superheated vapors. The rate of heat transfer from a superheated vapor to a wall at T_s will therefore be

$$q = A\bar{h}_c(T_{sv} - T_s) \tag{10.31}$$

where $\bar{h}_c$ = average value of the heat transfer coefficient determined from an equation appropriate to the geometric configuration with the same vapor at saturation conditions

T_{sv} = *saturation temperature* corresponding to the prevailing system pressure

10.4.2 Dropwise Condensation

When a condensing surface material prevents the condensate from wetting the surface, such as is the case for a metallic (nonoxide) coating, the vapor will condense in drops rather than as a continuous film [71]. This phenomenon is known as *dropwise condensation.* A large part of the surface is not covered by an insulating film under these conditions, and the heat transfer coefficients are four to eight times as high as in filmwise condensation. The ratio of condensate mass flux for dropwise condensation, $\dot{m}_D$, from the outside of a horizontal tube of diameter D to that for film condensation, $\dot{m}_f$, can be estimated [72] from

$$\frac{\dot{m}_D}{\dot{m}_f} = \left(\frac{\rho_l^2 D^2 g}{24.2\mu_l \dot{m}_f}\right)^{1/9} \tag{10.32}$$

For steam at atmospheric pressure and $\dot{m}_f = 0.014\,\text{kg/m}^2\,\text{s}$, Eq. (10.32) predicts a ratio of 6.5.

To calculate the heat transfer coefficient in practice, a conservative approach is to assume filmwise condensation because, even with steam, dropwise condensation can be expected only under carefully controlled conditions that cannot always be maintained in practice [73, 74]. Dropwide condensation of steam, however, can be a useful technique in experimental work when it is desirable to reduce the thermal resistance on one side of a surface to a negligible value.

10.5* Condenser Design

The evaluation of the heat transfer coefficients of condensing vapors, as can be seen from Eqs. (10.21) through (10.23), presupposes a knowledge of the temperature of the condensing surface. In practical problems, this temperature is generally not known because its value depends on the relative orders of magnitude of the thermal resistances in the entire system. The type of problem usually encountered in practice, whether it be a performance calculation for an existing piece of equipment or the design of equipment for a specific process, requires simultaneous evaluation of thermal resistances at the inner and outer surfaces of a tube or the wall of a duct. In most cases, the geometric configuration is either specified, as in the case of an existing piece of equipment, or assumed, as in the design of new equipment. When the desired rate of condensation is specified, the usual procedure is to estimate the total surface area required and then to select a suitable arrangement for a combination of size and number of tubes that meets the preliminary area specification. The performance calculation can then be made as though one were dealing with an existing piece of equipment, and the results can later be compared with the specifications. The flow rate of the coolant is usually determined by the allowable pressure drop or the allowable temperature rise. Once the flow rate is known, the thermal resistances of the coolant and the tube wall can be computed without difficulty. The heat transfer coefficient of the condensing fluid, however depends on the condensing-surface temperature, which can be computed only after the heat transfer coefficient is known. A trial-and-error solution is therefore necessary. One either assumes a surface temperature or, if more convenient, estimates the heat transfer coefficient on the condensing side and calculates the corresponding surface temperature. With this first approximation of the surface temperature, the heat transfer coefficient is then recalculated and compared with the assumed value. A second approximation is usually sufficient for satisfactory accuracy.

The orders of magnitude of the heat transfer coefficients for various vapors listed in Table 10.5 will aid in the initial estimates and reduce the amount of trial and error. We note that for steam the thermal resistance is very small, whereas for organic vapors it is of the same order of magnitude as the resistance offered to the flow of heat by water at a low turbulent Reynolds number. In the refrigeration industry and in some chemical processes, finned tubes have been used to reduce the thermal resistance on the condensing side. A method for dealing with condensation of finned tubes and tube banks is presented in [76]. When repeated calculations of the heat transfer coefficient for condensation of pure vapors are to be made, alignment charts devised by Chilton, Colburn, Genereaux, and Vernon and reproduced in McAdams [21] are convenient.

Mixtures of Vapors and Noncondensable Gases The analysis of a condensing system containing a mixture of vapors or a pure vapor mixed with noncondensable gas is considerably more complicated than the analysis of a pure-vapor system. The presence of appreciable quantities of a noncondensable gas will, in general, reduce the rate of heat transfer. If high rates of heat transfer are desired, it is considered good practice to vent the noncondensable gas, which otherwise will blanket the cooling surface and add considerably to the thermal resistance. Noncondensable gases also inhibit mass transfer by offering a diffusional resistance. A complete treatment of problems involving condensation of mixtures is beyond the scope of this text, and the reader is referred to McAdams [21] for a comprehensive summary of available information on this topic.

TABLE 10.5 Approximate values of heat transfer coefficients for condensation of pure vapors

Vapor	System	Approximate Range of $(T_{sv} - T_s)$ (K)	Approximate Range of Average Heat Transfer Coefficient (W/m^2 K)
Steam	Horizontal tubes, 2.5–7.5-cm-OD	3–22	11,400–22,800
Steam	Vertical surface 3.1 m high	3–22	5700–11,400
Ethanol	Vertical surface 0.15 m high	11–55	1100–1900
Benzene	Horizontal tube, 2.5-cm-OD	17–44	1400–2000
Ethanol	Horizontal tube, 5-cm-OD	6–22	1700–2600
Ammonia	Horizontal 5- to 7.5-cm annulus	1–4	1400–2600[a]

[a] Overall heat transfer coefficient U for water velocities between 1.2 and 24 m/s [75] inside the tube.

10.6* Heat Pipes

One of the main objectives of energy conversion systems is to transfer energy from a receiver to some other location where it can be used to heat a working fluid. The heat pipe is a novel device that can transfer large quantities of heat through small surface areas with small temperature differences. The method of operation of a heat pipe is shown schematically in Fig. 10.22. The device consists of a circular pipe with an annular layer of wicking material covering the inside. The core of the system is hollow in the center to permit the working fluid to pass freely from the heat addition end on the left to the heat rejection end on the right. The heat addition end is equivalent to an evaporator, and the heat rejection end corresponds to a condenser. The condenser and the evaporator are connected by an insulated section of length L. The liquid permeates the wicking material by capillary action, and when heat is added to the evaporator end of the heat pipe, liquid is vaporized in the wick and moves through the central core to the condenser end, where heat is removed. Then, the vapor condenses back into the wick and the cycle repeats.

A large variety of fluid and pipe material combinations have been used for heat pipes. Some typical working fluid and material combinations, as well as the temperature ranges over which they can operate, are presented in Table 10.6. The fourth and fifth columns of the table list measured axial heat fluxes and measured surface heat fluxes, and it is apparent that very high heat fluxes can be obtained [78, 79]

In order for a heat pipe to operate, the maximum capillary pumping head, $(\Delta p_c)_{\max}$, must be able to overcome the total pressure drop in the heat pipe. This pressure drop consists of three parts:

1. The pressure drop required to return the liquid from the condenser to the evaporator, Δp_e
2. The pressure drop required to move the vapor from the evaporator to the condenser, Δp_v
3. The potential head due to the difference in elevation between the evaporator and the condenser, Δp_g

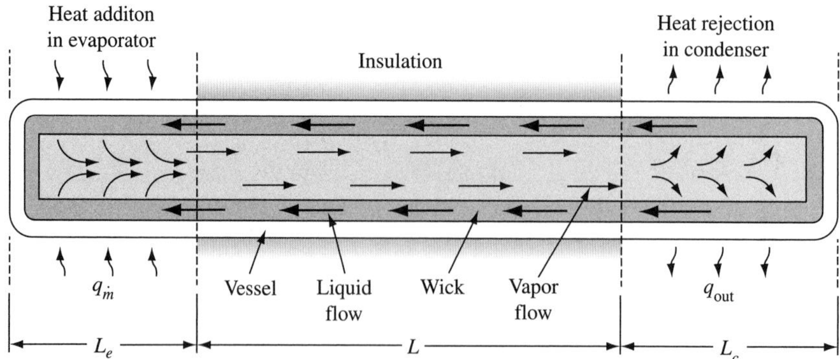

FIGURE 10.22 Schematic diagram of a heat pipe and the associated flow mechanisms.

TABLE 10.6 Some typical operating characteristics of heat pipes

Temperature Range (K)	Working Fluid	Vessel Material	Measured Axial Heat Flux[a] (W/cm^2)	Measured Surface Heat Flux[a] (W/cm^2)
230–400	Methanol[b]	Copper, nickel, stainless steel	0.45 at 373 K	75.5 at 373 K
280–500	Water	Copper, nickel	0.67 at 473 K	146 at 443 K
360–850	Mercury[c]	Stainless steel	25.1 at 533 K	181 at 533 K
673–1073	Potassium	Nickel, stainless steel	5.6 at 1023 K	181 at 1023 K
773–1173	Sodium	Nickel, stainless steel	9.3 at 1123 K	224 at 1033 K

[a]Varies with temperature.

[b]Using threaded artery wick.

[c]Based on sonic limit in heat pipe.

Source: Abstracted from Dutcher and Burke [77].

The condition for pressure equilibrium can thus be expressed in the form

$$(\Delta p_c)_{\max} \geq \Delta p_e + \Delta p_v + \Delta p_g \tag{10.33}$$

If this condition is not met, the wick will dry out in the evaporator region and the heat pipe will cease to operate.

The liquid pressure drop in flow through a homogeneous wick can be calculated from the empirical relation

$$\Delta p_e = \frac{\mu_l L_{\text{eff}} \dot{m}}{\rho_l K_w A_w} \tag{10.34}$$

where μ_l = liquid viscosity
$\dot{m}$ = mass flow rate
ρ_l = liquid density
A_w = wick cross-sectional area
K_w = wick permeability or wick factor
L_{eff} = effective length between the evaporator and condenser, given by

$$L_{\text{eff}} = L + \frac{L_e + L_c}{2} \tag{10.35}$$

where L_e = evaporator length
L_c = condenser length

The pressure drop through longitudinal, grooved wicks or composite wicks can be obtained by minor modifications of Eq. (10.34), as shown in [78].

The vapor pressure drop is generally small compared to the liquid pressure loss. As long as the velocity of the vapor is small compared to the velocity of sound, say less than 30%, one can neglect compressibility effects and calculate the viscous pressure loss Δp_v from incompressible flow relations. For steady-state laminar flow (see Chapter 6)

$$\Delta p_v = f \frac{L_{\text{eff}}}{D} \rho \bar{u}^2 = \frac{64 \mu_v \dot{m} L_{\text{eff}}}{\rho_v \pi D_v^4} \tag{10.36}$$

where D_v is the inside wick diameter in contact with the vapor and the subscript v denotes vapor properties.

In addition to the viscous drop, a pressure force is necessary to accelerate the vapor entering from the evaporator section to its axial velocity, but most of this loss is regained in the condenser, where the vapor stream is brought to rest. A more detailed treatment of the vapor pressure loss, including the pressure recovery in the evaporator, is presented in [78].

The pressure difference due to the hydrostatic or potential head of the liquid may be positive, negative, or zero, depending on the relative positions of the evaporator and condenser. The pressure difference Δp_g is given by

$$\Delta p_g = \rho_l g L \sin \phi \tag{10.37}$$

where ϕ is the angle between the axis of the heat pipe and the horizontal (positive when the evaporator is above the condenser).

The driving force in the wick is the result of surface tension. Surface tension is a force resulting from an imbalance of the natural attractions among a homogenous assembly of molecules. For example, a molecule near the surface of a liquid will experience a force directed inward due to the attraction of neighboring molecules below. One of the consequences of surface tension is that the pressure on a concave surface is less than that on a convex surface. The resulting pressure difference Δp is related to the surface tension σ_l and the radius of curvature r_c. For a hemispherical surface, the tension force action around the circumference is equal to $2\pi r_c \sigma_l$, and it is balanced by a pressure force over the surface equal to $\Delta p \pi r_c^2$. Hence

$$\Delta p = \frac{2\sigma_l}{r_c} \tag{10.38}$$

Another illustration of surface tension can be observed when a capillary tube is placed vertically in a wetting fluid; the fluid will rise in the tube due to capillary action, as shown in Fig. 10.23. A pressure balance then gives

$$\Delta p_c = \rho_l g h = \frac{2\sigma_l}{r_c} \cos \theta \tag{10.39}$$

where θ is the contact angle, which is between 0 and $\pi/2$ for wetting fluids. For a nonwetting fluid, θ is larger than $\pi/2$, and the liquid level in the capillary will be

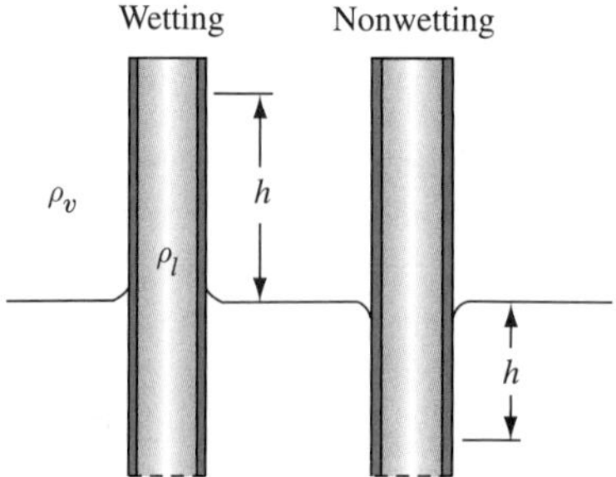

FIGURE 10.23 Capillary rise in a tube.

depressed below the surface. Hence, to obtain a capillary driving force only wetting fluids can be used in heat pipes.

Substituting Eqs. (10.34), (10.36), (10.37), and (10.39) for the pressure terms in the relation for the dynamic equilibrium, Eq. (10.33), yields one of the key design criteria for heat pipes:

$$\frac{2\sigma_l \cos\theta}{r_c} = \frac{\mu_l L_{\text{eff}} \dot{m}}{\rho_l K_w A_w} + \frac{64\mu_v \dot{m} L_{\text{eff}}}{\rho_r \pi D_v^4} + \rho_l g L_{\text{eff}} \sin\phi \tag{10.40}$$

If $(64\mu_v/\rho_v \pi D_v^4) \ll (\mu_l/\rho_l K_w A_w)$, the pressure drop of the vapor is negligible and the second term in Eq. (10.33) can be deleted in a preliminary design.

The maximum heat transport capability of a heat pipe due to wicking limitations is given by the relation

$$q_{\max} = \dot{m}_{\max} h_{fg} \tag{10.41}$$

where $\dot{m}_{\max}$ can be obtained from Eq. (10.40). Assuming $\cos\theta = 1$ and a negligible vapor flow pressure drop, one can solve for $\dot{m}_{\max}$ and combine the result with Eq. (10.41) to obtain the following expression for the maximum heat transport capability:

$$q_{\max} = \left(\frac{\rho_l \sigma_l h_{fg}}{\mu_l}\right)\left(\frac{A_w K_w}{L_{\text{eff}}}\right)\left(\frac{2}{r_c} - \frac{\rho_l g L_{\text{eff}} \sin\phi}{\sigma_l}\right) \tag{10.42}$$

In the above equation all the terms in the first parentheses $(\rho_l \sigma_l h_{fg}/\mu_l)$ are properties of the working fluid. This group is known as the figure of merit M:

$$M = \frac{\rho_l \sigma_l h_{fg}}{\mu_l} \tag{10.43}$$

and is plotted in Fig. 10.24 as a function of temperature for a number of heat pipe fluids.

The wick geometric properties are functions of A_w, K, and r_c. Table 39 in the Appendix presents data for pore size and permeability for a few wick materials and mesh sizes. They can be used for preliminary design as shown in Example 10.6 on page 682.

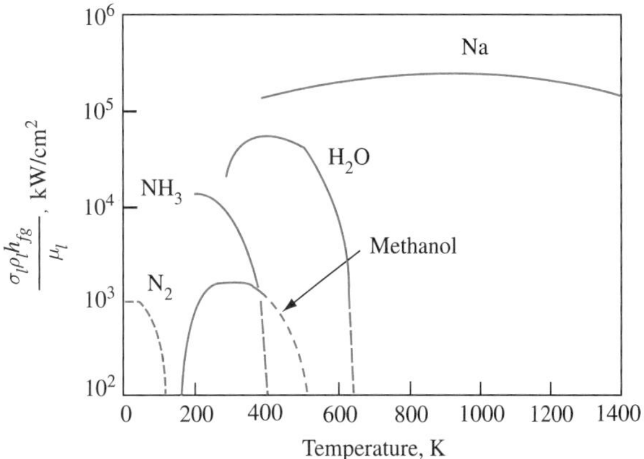

FIGURE 10.24 Figure of merit for several heat pipe working fluids as a function of temperature.

A widely used correlation between the maximum achievable heat transfer by a heat pipe and its dominant dimensions and operating parameters is

$$q_{\max} = \frac{A_w h_{fg} g \rho_l^2}{\mu_l} \left(\frac{l_w K_w}{L_{\text{eff}}} \right) \tag{10.44}$$

where A_w = wick cross-sectional area
g = gravitational acceleration
h_{fg} = heat of vaporization of liquid
ρ_l = liquid density
μ_l = liquid viscosity
l_w = wicking height of fluid in wick

The wicking height is given by

$$l_w = \frac{2\sigma_l}{r_c \rho_l g} \tag{10.45}$$

where σ_l = surface tension
r_c = effective pore radius

The maximum wicking height with sodium as the working fluid is about 38.5 cm, which is calculated by assuming an effective pore diameter of 8.6×10^{-3} cm. This is typical for a screen made with eight 4.1×10^{-3}-cm-diameter wires per square millimeter.

The most dominant parameters affecting the total power transfer capacity are the wick area, effective wicking height, and heat pipe length. For any effective wicking height, a wicking area can be selected to achieve the desired total power transfer if the operating temperature as well as the temperature drops at the evaporator section and the condenser section can be freely selected. However, when a limit to the upper operator temperature as well as to the temperature of the heat pipe at the condenser section exists, the wicking thickness might be determined by these temperature considerations. In general, the temperature drops and the operating temperature increases with increasing wick thickness. If the wick thickness is based on temperature and temperature-drop considerations, the maximum heat pipe length for a given power transfer is determined.

Although a heat pipe behaves like a structure of very high thermal conductance, it has heat transfer limitations that are governed by certain principles of fluid mechanics. The possible effects of these limitations on the capability of a heat pipe with a liquid-metal working fluid are shown in Fig. 10.25. Individual limitations indicated in the figure are discussed next.

10.6.1 Sonic Limitation

When heat is transferred from the evaporator section of a heat pipe to the condenser section, the rate of heat transfer q between the two sections is given by

$$q = \dot{m}_v h_{fg} \tag{10.46}$$

where $\dot{m}_v$ is the rate of mass flow of vapor at the evaporator exit and h_{fg} is the latent heat of the fluid. Because the latent energy of the working fluid is used instead of its

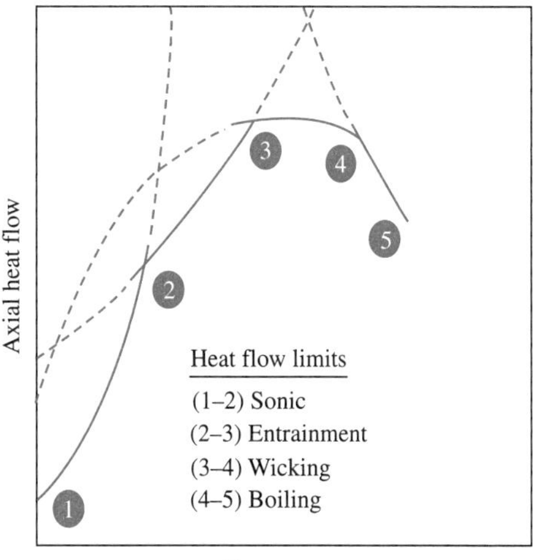

FIGURE 10.25 Limitations to heat transport in a heat pipe.

heat capacity, large heat transfer rates can be achieved with a relatively small mass flow. Furthermore, if the heat is transferred by high-density/low-velocity vapor, the transfer is nearly isothermal because only small pressure gradients are necessary to move the vapor.

To show the effect of vapor density and velocity on heat transfer, Eq. (10.46) can be modified by using the continuity equation

$$\dot{m}_v = \bar{\rho}_v \bar{u} A_v \tag{10.47}$$

where $\bar{\rho}_v$ is the radial average vapor density at the evaporator exit and A_v the cross-sectional area of vapor passage. By combining Eqs. (10.46) and (10.47) and rearranging, the result is

$$\frac{q}{A_v} = \bar{\rho}_v \bar{u} h_{fg} \tag{10.48}$$

where q/A_v is the axial heat flux based on the cross-sectional area of the vapor passage.

Equation (10.48) shows that the axial heat flux in a heat pipe can be held constant and the condenser environment adjusted to lower the pressure, temperature, and density of vapor until the flow at the evarporator exit becomes sonic. Once this occurs, pressure changes in the condenser will not be transmitted to the evaporator. This sonic limiting condition is represented in Fig. 10.25 by the solid curve between points 1 and 2. Some values for sonic heat flux limits as a function of evaporator exit temperature are given in Table 10.7 on the next page for Cs, K, Na, and Li.

Although heat pipes are normally not operated at sonic flow, such conditions have been encountered during startup with the working fluids listed in Table 10.7. Temperatures during such startups are always higher at the beginning of the heat pipe evaporator than at the evaporator exit.

TABLE 10.7 Sonic limitations of working heat-pipe fluids

Evaporator Exit Temperature (°C)	Heat Flux Limits (kW/cm^2)			
	Cs	K	Na	Li
400	1.0	0.5		
500	4.6	2.9	0.6	
600	14.9	12.1	3.5	
700	37.3	36.6	13.2	
800			38.9	1.0
900			94.2	3.9
1000				12.0
1100				31.1
1200				71.0
1300				143.8

10.6.2 Entrainment Limitation

Ordinarily, the sonic limitations just discussed do not cause dryout of the wick with attendant overheating of the evaporator. In fact, they often prevent the attainment of other limitations during startup. However, if the vapor density is allowed to increase without an accompanying decrease in velocity, some liquid from the wick-return system may be entrained. The onset of entrainment can be expressed in terms of a Weber number,

$$\frac{\rho_v \bar{u}^{-2} L_c}{2\pi\sigma_l} = 1 \tag{10.49}$$

where L_c is a characteristic length describing the pore size. Equation (10.49) simply expresses the ratio of vapor inertial forces to liquid surface tension forces. When this ratio exceeds unity, a condition develops that is very similar to that of a body of water agitated by high-velocity winds into waves that propagate until liquid is torn from their crests. Once entrainment begins in a heat pipe, fluid circulation increases until the liquid return path cannot accommodate the increased flow. This causes dry-out and overheating of the evaporator.

Because the wavelength of the perturbations at the liquid-vapor interface in a heat pipe is determined by the wick structure, the entrainment limit can be estimated by combining Eqs. (10.48) and (10.49) to give

$$\frac{q}{A_v} = h_{fg}\left(\frac{\lambda\pi\sigma_l\rho_v}{L_c}\right)^{1/2} \tag{10.50}$$

Equation (10.50) can then be used to obtain the type of curve represented by the solid line between points 2 and 3 in Fig. 10.25.

10.6.3 Wicking Limitation

Fluid circulation in a heat pipe is maintained by capillary forces that develop in the wick structure at the liquid-vapor interface. These forces balance the pressure losses due to the flow in the liquid and vapor phases; they manifest as many menisci that allow the pressure in the vapor to be higher than the pressure in the adjacent liquid in all parts of the system. When a typical meniscus is characterized by two principal radii of curvature (r_1 and r_2), the pressure drop ΔP_c across the liquid surface is given by

$$\Delta P_c = \sigma\left(\frac{1}{r_1} + \frac{1}{r_2}\right) \tag{10.51}$$

These radii, which are smallest at the evaporator end of the heat pipe, become even smaller as the heat transfer rate is increased. If the liquid wets the wick perfectly, the radii will be defined exactly by the pore size of the wick when a heat transfer limit is reached. Any further increase in heat transfer will cause the liquid to retreat into the wick, and drying and overheating will occur at the evaporator end of the system.

As indicated by Eq. (10.51), the capillary force in a heat pipe can be increased by decreasing the size of the wick pores that are exposed to vapor flow. However, if the pore size is also decreased in the remainder of the wick, the wicking limit might actually be reduced because of the increased pressure drop in the liquid phase. This is shown by *Poiseuille's equation* for the pressure drop through a capillary tube:

$$\Delta P_e = \frac{8\mu \dot{m}_l L}{\pi r^4 \rho} \tag{10.52}$$

where μ is the liquid viscosity, $\dot{m}_l$ is the rate of mass flow of liquid, r is the tube radius, ρ is the liquid density, and L is the tube length.

Equation (10.52) can be modified to obtain the liquid pressure drop at a particular heat transfer rate q for various wick structures. The equations in Fig. 10.26 on the next page give the pressure drop for the wick structures shown.

Although the artery wick system appears ideal, it requires an additional capillary network to distribute the liquid over surfaces that are used for heat addition and removal. Because of this complication, arteries are usually reserved for systems where boiling is likely to occur within the wick if the bulk of the liquid return network is located in the path of the incoming heat. The consequences of such boiling will be discussed later.

Equation (10.52b) is essentially the same as Eq. (10.52c), except that it involves a number of channels, N, and an effective channel radius r_e, which is obtained from the hydraulic diameter

$$\frac{D_H}{2} = r_e = 2\left(\frac{\text{flow area}}{\text{wetted perimeter}}\right)$$

Although open channels are subject to an interaction of vapor and liquid that causes waves but no liquid entrainment, the interaction can be suppressed by covering the channels with a layer of fine-mesh screen. Because the screen is located at the interface of liquid and vapor, the fine pores of the screen provide large capillary forces for fluid circulation, while the channels provide a less restrictive flow path for liquid return. This general type of structure is called a *composite wick*.

All screen-composite wicks can be made by wrapping a layer of a fine screen around a mandrel followed by a second layer of coarse screen. The assembly can be placed in a container tube, the diameter of which is then drawn down until the inner wall makes contact with the coarse screen. Next, the quantity $b/\varepsilon r_c^2$ in Eq. (10.52c)

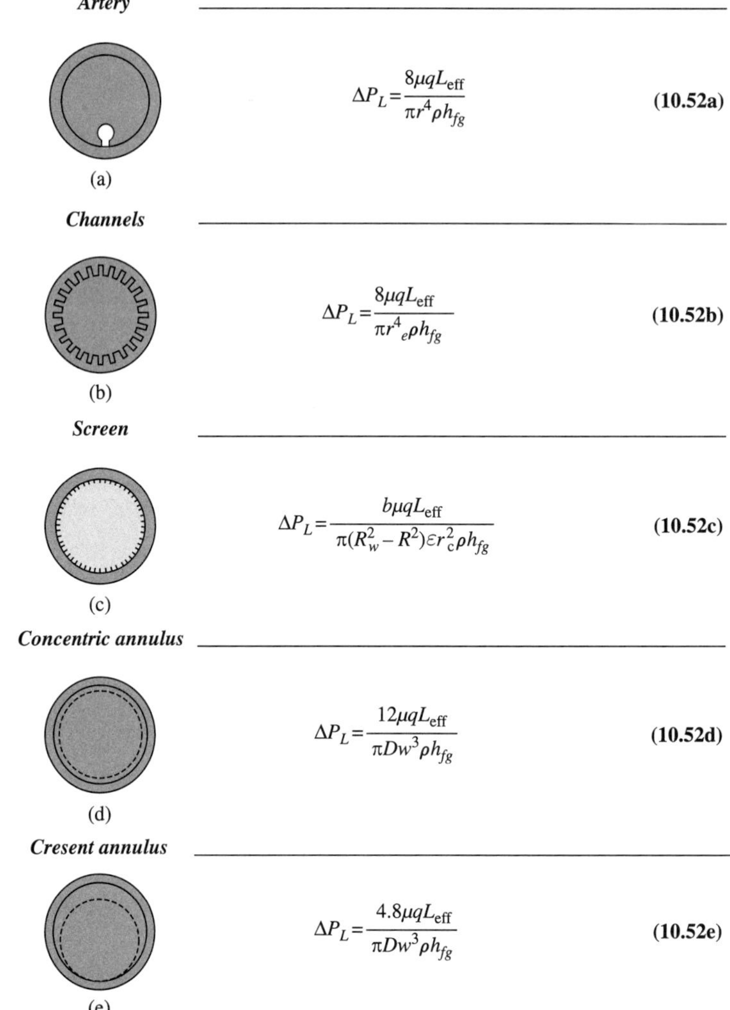

$$\Delta P_L = \frac{8\mu q L_{\text{eff}}}{\pi r^4 \rho h_{fg}} \tag{10.52a}$$

$$\Delta P_L = \frac{8\mu q L_{\text{eff}}}{\pi r^4{}_e \rho h_{fg}} \tag{10.52b}$$

$$\Delta P_L = \frac{b\mu q L_{\text{eff}}}{\pi (R_w^2 - R^2)\varepsilon r_c^2 \rho h_{fg}} \tag{10.52c}$$

$$\Delta P_L = \frac{12\mu q L_{\text{eff}}}{\pi D w^3 \rho h_{fg}} \tag{10.52d}$$

$$\Delta P_L = \frac{4.8\mu q L_{\text{eff}}}{\pi D w^3 \rho h_{fg}} \tag{10.52e}$$

FIGURE 10.26 Cross sections of various wick structures.

The quantities in the above equations are defined as follows:

L_{eff} = effective length of heat pipe	R = radius of vapor passage
r_e = effective channel radius	ε = screen void fraction
N = number of channels	r_c = effective radius of screen openings
b = screen tortuosity factor	D = mean diameter of annulus
R_w = outer radius of screen structure	w = width of annulus

can be determined by liquid-flow measurements through the screen before the mandrel is removed.

An ideal wick system for liquid-metal working fluids consists of an inner porous tube separated from an outer container tube by a gap that provides an unobstructed annulus for liquid return. The pressure drop in a concentric annulus is obtained by deriving Poiseuille's equation for flow between two parallel plates. Although not as precise as the equation for flow between concentric cylinders, it is easier to handle and is fairly accurate provided the width of the annulus is small compared to its mean diameter. Equation (10.52e) for a crescent annulus is obtained by assuming the displacement obeys a cosine function—the width of the annulus doubles at the top of the tube, becomes zero at the bottom, and remains unchanged on the sides.

In Fig. 10.25, the wicking limitation is represented by the solid line between points 3 and 4. Although this limitation is shown to occur at temperatures where essentially all the pressure drop is in the liquid phase, the effect of a significant vapor-pressure drop is indicated by the dashed extension line at lower temperatures.

10.6.4 Boiling Limitations

In most two-phase flow systems the formation of vapor bubbles in the liquid phase (boiling) enhances convection, which is required for heat transfer. Such boiling is often difficult to produce in liquid-metal systems because the liquid tends to fill the nucleation sites necessary for bubble formation. In a heat pipe, convection in the liquid is not required because heat enters the pipe by conduction through a thin, saturated wick. Furthermore, the formation of vapor bubbles is undesirable because they could cause hot spots and destroy the action of the wick. Therefore, heat pipes are usually heated isothermally before being used to allow the liquid to wet the inner heat pipe wall and to fill all but the smallest nucleation sites.

Boiling may occur at high-input heat fluxes and high operating temperatures. The curve between points 4 and 5 in Fig. 10.25 is based on the equations

$$p_i - p_l = \frac{2\sigma}{r} \tag{10.53}$$

$$\frac{q}{A} = \frac{k(T_w - T_v)}{t} \tag{10.54}$$

where p_i is the vapor pressure inside the bubble, p_l the pressure in adjacent liquid, r the radius of the largest nucleation site, A the heat input area, k the effective thermal conductivity of the saturated wick, T_w the temperature at the inside wall, T_v the temperature at the liquid-vapor interface, and t the thickness of the first layer in the wick [78].

Since the sizes of nucleation sites in a system are usually unknown, it is not possible to predict when boiling will occur. However, Eqs. (10.53) and (10.54) show how various factors influence boiling. For example, if nucleation sites are small, a large pressure difference will be required for bubbles to grow. For a given heat input flux, this pressure difference will depend on the thickness and thermal conductivity of the wick, on the saturation temperature of the vapor, and on the pressure drop in the vapor and liquid phases. This pressure drop is often overlooked because it is not a factor in the ordinary treatment of boiling.

Boiling is not a limitation with liquid metals, but when water is used as the working fluid, boiling may be a major heat transfer limitation because the thermal conductivity of the fluid is low and because it does not readily fill nucleation sites. Unfortunately, little experimental information is available concerning this limitation.

EXAMPLE 10.6 Determine the maximum heat transport capability and the liquid flow rate of a water heat pipe operating at 100°C and atmospheric pressure. The heat pipe is 30 cm long and has an inner diameter of 1 cm. The heat pipe is inclined at 30° with the evaporator above the condenser. The wick consists of four layers of phosphorus-bronze, 250-mesh wire screen (wire diameter of 0.045 mm) on the inner surface of the pipe as shown in Fig. 10.26(a).

SOLUTION The pressure balance relation to prevent dryout is

$$(\Delta p_c)_{\max} \geq \Delta p_l + \Delta p_v + \Delta p_g$$

As a first approximation in the analysis we will neglect the vapor pressure drop Δp_v. Substituting Eq. (10.39) for the capillary pumping head Δp_c and Eqs. (10.34) and (10.37) for the liquid pressure drop Δp_l and the gravitational heat Δp_g, respectively, gives

$$\frac{2\sigma_l \cos\theta}{r_c} = \frac{\mu_l q L_{\text{eff}}}{\rho_l h_{fg} A_w K_w} + \rho_l g L_{\text{eff}} \sin\phi$$

The area of the wick A_w is approximately

$$A_w = \pi D t = \pi(1\ \text{cm})(4)(0.0045\ \text{cm})$$
$$= 0.057\ \text{cm}^2$$

where t is the thickness of the four layers of wire mesh. The effective flow length L_{eff} is approximately 0.30 m. From Table 39 in Appendix 2 the pore radius r_c is 0.002 cm and the permeability K is $0.3 \times 10^{-10}\ \text{m}^2$. The water properties at 100°C are, from Table 13 in Appendix 2 and Table 10.2,

$$h_{fg} = 2.26 \times 10^6\ \text{J/kg}$$
$$\rho_l = 958\ \text{kg/m}^3$$
$$\mu_l = 279 \times 10^{-6}\ \text{N s/m}^2$$
$$\sigma_l = 58.9 \times 10^{-3}\ \text{N/m}$$

The maximum liquid flow rate through the wick can be obtained from the pressure balance equation. Assuming perfect wetting with $\theta = 0$, substituting $\dot{m}_{\max} h_{fg}$ for $q_{\max}$, and solving for $\dot{m}_{\max}$ yields

$$\dot{m}_{\max} = \left(\frac{2\sigma_l}{r_c} - \rho_l g L_{\text{eff}} \sin\phi\right) \frac{\rho_l h_{fg} A_w K}{\mu_l L_{\text{eff}} h_{fg}}$$

$$= \left[\frac{2 \times 58.9 \times 10^{-3}\ \text{N/m}}{0.002 \times 10^{-2}\ \text{m}} - (958\ \text{kg/m}^3)(9.81\ \text{m/s}^2)(0.30\ \text{m})(0.5)\right]$$

$$\times \left[\frac{(958\ \text{kg/m}^3)(0.057 \times 10^{-4}\ \text{m}^2)(0.3 \times 10^{-10}\ \text{m}^2)}{(279 \times 10^{-6}\ \text{N s/m}^2)(0.30\ \text{m})}\right]$$

$$= 9.0 \times 10^{-6}\ \text{kg/s}$$

The maximum heat transport capability is then, from Eq. (10.41),

$$\begin{aligned} q_{\max} &= \dot{m}_{\max} h_{fg} \\ &= (8.8 \times 10^{-6}\,\text{kg/s})(2.26 \times 10^{6}\,\text{J/kg}) \\ &= 19.8\,\text{W} \end{aligned}$$

Note that the heat transport capability could be increased significantly by adding two or three layers of 100-mesh screen.

For a more complete treatment of the heat-pipe theory and practice, the reader is referred to [78–81].

10.7* Freezing and Melting

Problems involving the solidification or melting of materials are of considerable importance in many technical fields. Typical examples in the field of engineering are the making of ice, the freezing of foods, and the solidification and melting of metals in casting processes. In geology, the solidification rate of the earth has been used to estimate the age of our planet. Whatever the field of application, the problem of central interest is the rate at which solidification or melting occurs.

We shall consider here only the problem of solidification, and it is left for the reader as an exercise to show that a solution to this problem is also a solution to the corresponding problem in melting. Figure 10.27 shows the temperature distribution in an ice layer on the surface of a liquid. The upper face is exposed to air at sub-freezing temperature. Ice formation occurs progressively at the solid-liquid interface as a result of heat transfer through the ice to the cold air. Heat flows by

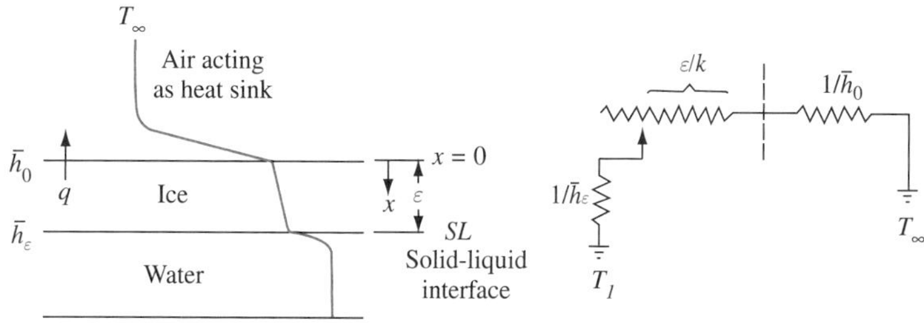

FIGURE 10.27 Temperature distribution for ice forming on water with air acting as heat sink, and simplified thermal circuit for the system with the heat capacity of the solid considered negligible.

convection from the water to the ice, by conduction through the ice, and by convection to the sink. The ice layer is subcooled except for the interface in contact with the liquid, which is at the freezing point. A portion of the heat transferred to the sink is used to cool the liquid at the interface *SL* to the freezing point and to remove its latent heat of solidification. The other portion serves to subcool the ice. Cylindrical or spherical systems can be described in a similar manner, but solidification can proceed either inward (as in the freezing of water inside a can) or outward (as in water freezing on the outside of a pipe).

The freezing of a slab can be formulated as a boundary-value problem in which the governing equation is the general conduction equation for the solid phase

$$\frac{\partial^2 T}{\partial x^2} = \frac{1}{\alpha}\frac{\partial T}{\partial t}$$

subject to the boundary condition that

$$-k\frac{\partial T}{\partial x} = \bar{h}_o(T_{x=0} - T_\infty) \qquad \text{at } x = 0$$

$$-k\frac{\partial T}{\partial x} = \rho L_f \frac{d\varepsilon}{dt} + \bar{h}_\varepsilon(T_l - T_{fr}) \quad \text{at } x = \varepsilon$$

where ε = distance to the solid-liquid interface, which is a function of time t
L_f = latent heat of fusion of the material
α = thermal diffusivity of the solid phase $(k/\rho c)$
ρ = density of the solid phase
T_l = temperature of the liquid
T_∞ = temperature of the heat sink
T_{fr} = freezing-point temperature
$\bar{h}_o$ = heat transfer coefficient at $x = 0$, the air-ice interface
$\bar{h}_\varepsilon$ = heat transfer coefficients at $x = \varepsilon$, the water-ice interface

The analytic solution of this problem is very difficult and has been obtained only for special cases. The reason for the difficulty is that the governing equation is a partial differential equation for which the particular solutions are unknown when physically realistic boundary conditions are imposed.

An approximate solution of practical value can, however, be obtained by considering the heat capacity of the subcooled solid phase as negligible relative to the latent heat of solidification. To simplify our analysis further, we shall assume that the physical properties of the ice, ρ, k, and c, are uniform, that the liquid is at the solidification temperature (i.e., $T_l = T_{fr}$ and $1/\bar{h}_\varepsilon = 0$), and that $\bar{h}_o$ and T_∞ are constant during the process.

The rate of heat flow per unit area through the resistance offered by the ice and the air, acting in series, as a result of the temperature potential $(T_{fr} - T_\infty)$ is

$$\frac{q}{A} = \frac{T_{fr} - T_\infty}{1/\bar{h}_o + \varepsilon/k} \tag{10.55}$$

This is the heat flow rate that removes the latent heat of fusion necessary for freezing at the surface $x = \varepsilon$, or

$$\frac{q}{A} = \rho L_f \frac{d\varepsilon}{dt} \tag{10.56}$$

where $d\varepsilon/dt$ is the volume rate of ice formation per unit area at the growing surface (m^3/hr m^2) and ρL_f is the latent heat per unit volume (J/m^3). Combining Eqs. (10.55) and (10.56) to eliminate the rate of heat flow yields

$$\frac{T_{fr} - T_\infty}{1/\bar{h}_o + \varepsilon/k} = \rho L_f \frac{d\varepsilon}{dt} \tag{10.57}$$

which relates the depth of ice to the freezing time. The variables ε and t can now be separated, and we get

$$d\varepsilon\left(\frac{1}{\bar{h}_o} + \frac{\varepsilon}{k}\right) = \frac{T_{fr} - T_\infty}{\rho L_f} dt \tag{10.58}$$

To make this equation dimensionless, let

$$\varepsilon^+ = \frac{\bar{h}_o \varepsilon}{k}$$

and

$$t^+ = t\bar{h}_o^2 \frac{T_{fr} - T_\infty}{\rho L_f k}$$

Substituting these dimensionless parameters into Eq. (10.58) yields

$$d\varepsilon^+(1 + \varepsilon^+) = dt^+ \tag{10.59}$$

If the freezing process starts at $t = t^+ = 0$ and continues for a time t, the solution of Eq. (10.59), obtained by integration between the specified limits, is

$$\varepsilon^+ + \frac{(\varepsilon^+)^2}{2} = t^+ \tag{10.60}$$

or

$$\varepsilon^+ = -1 + \sqrt{1 + 2t^+} \tag{10.61}$$

When the temperature of the liquid T_l is above the fusion temperature and the convection heat transfer coefficient at the liquid-solid interface is $\bar{h}_c$, the dimensionless equation corresponding to Eq. (10.59) in the foregoing simplified treatment becomes

$$\frac{(1 + \varepsilon^+)d\varepsilon^+}{1 + R^+T^+(1 + \varepsilon^+)} = dt^+ \tag{10.62}$$

where $R^+ = \dfrac{\bar{h}_\varepsilon}{\bar{h}_o}$

$$T^+ = \frac{T_l - T_{fr}}{T_{fr} - T_\infty}$$

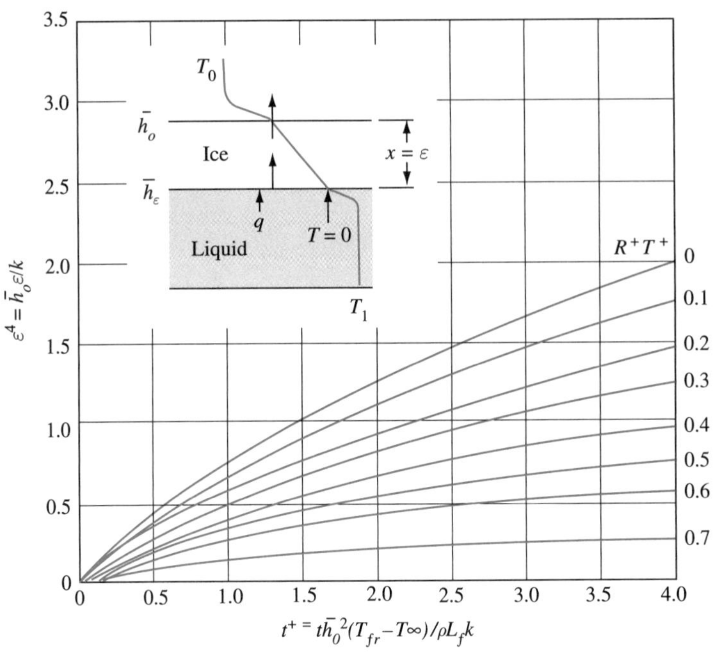

FIGURE 10.28 Solidification of slab: thickness versus time.

Source: From London and Seban [82], with permission of the publishers, the American Society of Mechanical Engineers.

and the other symbols represent the same dimensionless quantities used previously in Eq. (10.59).

For the boundary conditions that at $t^+ = 0$, $\varepsilon^+ = 0$, and at $t^+ = t^+$, $\varepsilon^+ = \varepsilon^+$, the solution of Eq. (10.62) becomes

$$t^+ = -\frac{1}{(R^+T^+)^2}\ln\left(1 + \frac{R^+T^+\varepsilon^+}{1 + R^+T^+}\right) + \frac{\varepsilon^+}{R^+T^+} \tag{10.63}$$

The results are shown graphically in Fig. 10.28, where the generalized thickness ε^+ is plotted against generalized time t^+, with the generalized *potential-resistance ratio* R^+T^+ as the parameter.

EXAMPLE 10.7 In the production of Flakice, ice forms in thin layers on a horizontal rotating drum that is partly submerged in water (see Fig. 10.29). The cylinder is internally refrigerated with a brine spray at −11°C. Ice formed on the exterior surface is peeled off as the revolving-drum surface emerges from the water.

For the operating conditions listed on the following page, estimate the time required to form an ice layer that is 0.25-cm thick.

water-liquid temperature = 4.4°C

liquid-surface conductance = 57 W/m^2 K

conductance between brine and ice (including metal wall) = 570 W/m^2 K

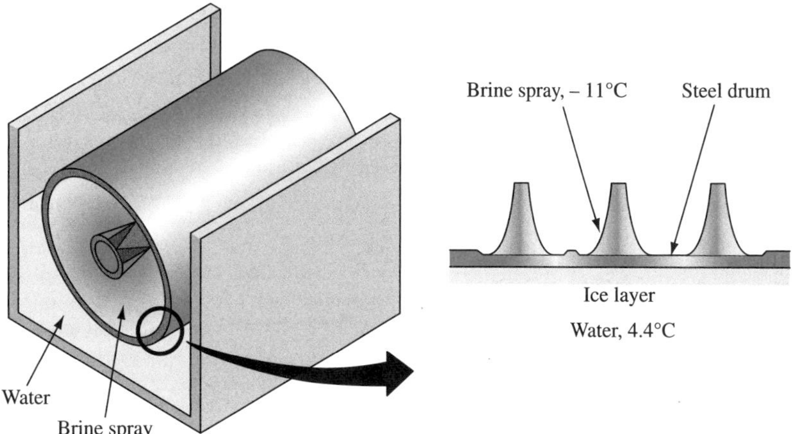

FIGURE 10.29 Schematic diagram for Example 10.7.

Use the following properties for ice: latent heat of fusion = 333.700 J/kg; thermal conductivity = 2.22 W/m K; density = 918 kg/m^3.

SOLUTION For the conditions stated above we have

$$R^+ = \frac{\bar{h}_\epsilon}{\bar{h}_o} = \frac{57}{570} = 0.1$$

$$T^+ = \frac{T_l - T_{fr}}{T_{fr} - T_\infty} = \frac{4.4 - 0}{0 - (-11)} = 0.4$$

$$\varepsilon^+ = \frac{\bar{h}_o \varepsilon}{k_{\text{ice}}} = \frac{(570\,\text{W/m}^2\,°\text{C})(0.0025\,\text{m})}{2.32\,\text{W/m}\,°\text{C}} = 0.614$$

We assume now that the ice is a sheet. This assumption is justified because the thickness of the ice is very small compared to the radius of curvature of the drum. The boundary conditions of this problem are then the same as those assumed in the solution of Eq. (10.63). Hence Eq. (10.63) is the solution to the problem at hand. Substituting numerical values for r^+, T^+, and ε^+ in Eq. (10.63) yields

$$t^+ = -\frac{1}{(0.04)^2} \ln\left(1 + \frac{0.0246}{1 + 0.04}\right) + \frac{0.614}{0.04} = 0.739$$

From the definition of t^+, the time t is

$$t = \frac{0.739 \rho L_f k}{\bar{h}_o^2 (T_{fr} - T_\infty)}$$

$$= \frac{(0.739)(918\,\text{kg/m}^3)(333{,}700\,\text{J/kg})(2.22\,\text{W/m K})}{(570\,\text{W/m}^2\,°\text{C})^2(11)(\text{K})} = 141\,\text{s}$$

An estimate of the error caused by neglecting the heat capacity of the solidified portion has been obtained by means of an electrical network simulating the freezing of a slab originally at the fusion temperature [83]. It was found that the error is not appreciable when $\varepsilon \bar{h}_o/k$ is less than 0.1 or when $L_f/(T_{fr} - T_\infty)c$ is larger than 1.5 [84]. In the intermediate range, the freezing rates predicted by the simplified analysis are too large. The solutions presented here are valid for ice and other substances that have large heats of fusion. An approximate method for predicting the freezing rate of steel and other metals, where $L_f/(T_{fr} - T_\infty)c$ may be less than 1.5, is presented in Cochran [84]. Numerical methods of solution for systems involving a change of phase are presented in Murray and Landis [85], and Lazaridis [86]. Melting and freezing in wedges and corners has been analyzed in Budhia and Kreith [87]. An overview of inelting and freezing is given in Lion [88].

References

1. A. E. Bergles, "Fundamentals of Boiling and Evaporation," in *Two-Phase Flow Heat Exchangers: Thermal-Hydraulic Fundamentals and Design*, S. Kakaç, A. E. Bergles, and E. O. Fernandes, eds., Kluwer Academic Publishers, Dordrecht, The Netherlands, pp. 159–200, 1988.
2. Van P. Carey. *Liquid-Vapor Phase-Change Phenovena*, Hemisphere Publ. Co., Washington. D.C., 1992, 2nd ed., Taylor & Francis, New York, NY, 2007.
3. D. P. Jordan. "Film and Transition Boiling." in *Advances in Heat Transfer*, vol. 5, T. F. Irvine, Jr., and J. P. Hartnett, eds., pp. 55–125, Academic Press. New York, 1968.
4. G. Leppert and C. C. Pitts, "Boiling." in *Advances in Heat Transfer*, vol. 1, T. F. Irvine, Jr., and J. P. Hartnett, eds., pp. 185–265. Academic Press. New York, 1968.
5. W. M. Rohsenow, "Boiling Heat Transfer." in *Developments in Heat Transfer*, W. M. Rohsenow, ed., pp. 169–260, MIT Press. Cambridge, Mass., 1964.
6. E. A. Farber and R. L. Scorah, "Heat Transfer to Water Boiling under Pressure." *Trans. ASME*, vol. 70, pp. 369–384, 1948.
7. S. Nukiyama, "Maximum and Minimum Values of Heat Transmitted from a Metal to Boiling Water under Atmospheric Pressure," *J. Soc. Mech. Eng. Jpn.*, vol. 37, no. 206, pp. 367–394, 1934.
8. K. Engelberg-Forster and R. Greif, "Heat Transfer to a Boiling Liquid—Mechanism and Correlations." *Trans. ASME. Ser. C. J. Heat Transfer*, vol. 81, pp. 43–53, Feb. 1959.
9. R. F. Gaertner, "Photographic Study of Nucleate Pool Boiling on a Horizontal Surface." ASME Paper 63-WA-76.
10. R. Moissis and P. J. Berenson, "On the Hydrodynamic Transitions in Nucleate Boiling," *Trans. ASME. Ser. C. J. Heat Transfer*, vol. 85, pp. 221–229, Aug. 1963.
11. J. W. Westwater, "Boiling Heat Transfer," *Am. Sci.*, vol. 47, no. 3, pp. 427–446, Sept. 1959.
12. P. J. Berenson, "Experiments on Pool-Boiling Heat Transfer," *Int. J. Heat Mass Transfer*, vol. 5, pp. 985–999, 1962.
13. V. K. Dhir, "Numerical Simulations of Pool-Boiling Heat Transfer," *AIChE Journal*, vol. 47, no. 4, pp. 813–834, 2001.
14. P. Stephan and J. Kern, "Evaluation of Heat and Mass Transfer Phenomena in Nucleate Boiling," *International Journal of Heat and Fluid Flow*, vol. 25, pp. 140–148, 2004.
15. F. C. Gunther, "Photographic Study of Surface Boiling Heat Transfer with Forced Convection," *Trans. ASME*, vol. 73, pp. 115–123, 1951.
16. M. E. Ellion, "A Study of the Mechanism of Boiling Heat Transfer," Memorandum 20–88. Jet Propulsion Laboratory. Pasadena, Calif., March 1954.
17. W. M. Rohsenow, "A Method of Correlating Heat-Transfer Data for Surface Boiling Liquids." *Trans. ASME*, vol. 74, pp. 969–975, 1952.
18. R. I. Vachon, G. H. Nix, and G. E. Tanger, "Evaluation of Constants for the Rohsenow Pool-Boiling Correlation," *Trans. ASME, Ser. C. J. Heat Transfer*, vol. 90, pp. 239–247, 1968.
19. J. N. Addoms, "Heat Transfer at High Rates to Water Boiling outside Cylinders," D.Sc. thesis, Dept. of Chemical Engineering, Massachusetts Institute of Technology, 1948.

20. W. H. McAdams et al., "Heat Transfer from Single Horizontal Wires to Boiling Water," *Chem. Eng. Prog.*, vol. 44, pp. 639–646, 1948.
21. W. H. McAdams, *Heat Transmission*, 3d ed., McGraw-Hill, New York, 1954.
22. K. Cornwell and S. D. Houston, "Nucleate Pool Boiling on Horizontal Tubes: A Convection-Based Correlation," *International Journal of Heat and Mass Transfer*, vol. 37, suppl. 1, pp. 303–309, 1994.
23. V. M. Wasekar and R. M. Manglik, "Pool Boiling Heat Transfer in Aqueous Solutions of an Anionic Surfactant," *Journal of Heat Transfer*, vol. 122, pp. 708–715, 2000.
24. J. G. Collier and J. R. Thome, *Convective Boiling and Condensation*, 3rd ed., Clarendon Press, Oxford, UK, 2001.
25. E. L. Piret and H. S. Isbin, "Natural Circulation Evaporation Two-Phase Heat Transfer," *Chem. Eng. Prog.*, vol. 50, p. 305, 1954.
26. M. T. Cichelli and C. F. Bonilla, "Heat Transfer to Liquids Boiling under Pressure," *Trans. AIChE*, vol. 41, pp. 755–787, 1945.
27. D. S. Cryder and A. C. Finalbargo, "Heat Transmission from Metal Surfaces to Boiling Liquids: Effect of Temperature of the Liquid on Film Coefficient," *Trans. AIChE*, vol. 33, pp. 346–362, 1937.
28. *Steam—Its Generation and Use*, Babcock & Wilcox Co., New York, 1955.
29. F. Kreith and M. J. Summerfield, "Heat Transfer to Water at High Flux Densities with and without Surface Boiling," *Trans. ASME*, vol. 71, pp. 805–815, 1949.
30. N. Zuber and M. Tribus, "Further Remarks on the Stability of Boiling Heat Transfer," Rept. 58–5, Dept. of Engineering, Univ. of Calif., Los Angeles, 1958.
31. S. S. Kutateladze, "A Hydrodynamic Theory of Changes in a Boiling Process under Free Convection," *Izv. Akad. Nauk SSSR Otd. Teckh. Nauk*, no. 4, p. 524, 1951.
32. N. Zuber, M. Tribus, and J. W. Westwater, "The Hydrodynamic Crisis in Pool Boiling of Saturated and Subcooled Liquids," in *Proceedings of the International Conference on Developments in Heat Transfer*, pp. 230–236, Am. Soc. of Mech. Eng. (ASME) New York, 1962.
33. J. H. Lienhard and V. K. Dhir, "Extended Hydrodynamic Theory of the Peak and Maximum Pool Boiling Heat Fluxes," NASA Contract. Rept. CR-2270, July 1973.
34. C. M. Usiskin and R. Siegel, "An Experimental Study of Boiling in Reduced and Zero Gravity Fields," *Trans. ASME, Ser. C*, vol. 83. pp. 243–253, 1961.
35. J. H. Lienhard, V. K. Dhir, and D. M. Riherd, "Peak Pool Boiling Heat Flux Measurements on Finite Horizontal Flat Plates." *ASME J. Heat Transfer*, vol. 95, pp. 477–482, 1973.
36. K. H. Son and J. H. Lienhard, "The Peak Pool Boiling Heat Flux on Horizontal Cylinders," *Int. J. Heat and Mass Transfer*, vol. 13, pp. 1425–1439, 1970.
37. J. H. Lienhard and V. K. Dhir, "Hydrodynamic Prediction of Peak Pool-Boiling Heat Fluxes from Finite Bodies," *ASME J. Heat Transfer*, vol. 95, pp. 152–158, 1973.
38. J. S. Ded and J. H. Lienhard, "The Peak Pool Boiling Heat Flux from a Sphere," *AIChE Journal*, vol. 18, pp. 337–342, 1972.
39. D. A. Huber and J. C. Hoehne, "Pool Boiling of Benzene, Diphenyl, and Benzene-Diphenyl Mixtures under Pressure," *Trans. ASME*, Ser. C, vol. 85, pp. 215–220, 1963.
40. R. M. Manglik and M. A. Jog, "Molecular-to-Large-Scale Heat Transfer with Multiphase Interfaces: Current Status and New Directions," *Journal of Heat Transfer*, vol. 131, no. 12, 2009.
41. R. M. Manglik, "Heat Transfer Enhancement," in *Heat Transfer Handbook*, A. Bejan and A. D. Kraus, eds., Chap. 14, Wiley, Hoboken, NJ, 2003.
42. R. L. Webb, "The Evolution of Enhanced Surface Geometries for Nucleate Boiling," *Heat Transfer Engineering*, vol. 2, pp. 46–69, 1981.
43. L. A. Bromley, "Heat Transfer in Stable Film Boiling," *Chem. Eng. Prog.*, vol. 46, pp. 221–227, 1950.
44. J. W. Westwater and B. P. Breen, "Effect of Diameter of Horizontal Tubes on Film Boiling Heat Transfer," *Chem. Eng. Prog.*, vol. 58, pp. 67–72, 1962.
45. P. J. Berenson and R. A. Stone. "A Photographic Study of the Mechanism of Forced-Convection Vaporization," AIChE Reprint No. 21, Symposium on Heat Transfer, San Juan, Puerto Rico, 1963.
46. K. Konmutsos, R. Moissis, and A. Spyridonos, "A Study of Bubble Departure in Forced Convection Boiling," *Trans. ASME, Ser. C., J. Heat Transfer*, vol. 90, pp. 223–230, 1968.
47. J. W. Miles, "The Hydrodynamic Stability of a Thin Film of Liquid in Uniform Shearing Motion," *J. Fluid Mech.*, vol. 8, pp. 592–610, 1961.
48. W. H. McAdams, W. E. Kennel, C. S. Minden, R. Carl, P. M. Picarnell, and J. E. Drew, "Heat Transfer at High Rates to Water with Surface Boiling," *Ind. Eng. Chem.*, vol. 41, pp. 1945–1953, 1949.
49. W. H. Jens G. Leppert, "Recent Developments in Boiling Research," and II. *J. Am. Eng.*, vol. 66, pp. 437–456, 1955; vol. 67, pp. 137–155, 1955.
50. J. C. Chen. "Correlation for Boiling Heat Transfer to Saturated Liquids in Convective Flow," *Ind. Eng. Chem. Proc. Des. Dev.*, vol. 5, p. 332, 1966.
51. J. G. Collier, "Forced Convective Boiling," *Two Phase Flow and Heat Transfer*, Chap. 8, pp. 247–248, Hemisphere, Washington, D.C., 1981.

52. P. Griffith, "Correlation of Nucleate-Boiling Burnout Data," ASME Paper 57-HT-21.
53. P. Griffith, "Two Phase Flow in Pipes." course notes. Massachusetts Institute of Technology, Cambridge, 1964.
54. R. W. Lockhart and R. C. Martinelli, "Proposed Correlation of Data for Isothermal Two-Component Flow in Pipes," *Chem. Eng. Prog.*, vol. 45, pp. 39–48, 1949.
55. L. S. Tong. *Boiling Heat Transfer and Two-Phase Flow*, Wiley, New York, 1965.
56. F. Kreith and M. Margolis. "Heat Transfer and Friction in Turbulent Vortex Flow," *Appl. Sci. Res.*, Sec. A, vol. 8, pp. 457–473, 1959.
57. W. R. Gambill, R. D. Bundy, and R. W. Wansbrough, "Heat Transfer, Burnout, and Pressure Drop for Water in Swirl Flow through Tubes with Internal Twisted Tapes," *Chem. Eng. Prog. Symp. Ser.*, no. 32, vol. 57, pp. 127–137, 1961.
58. R. M. Manglik and A. E. Bergles, "Swirl Flow Heat Transfer and Pressure Drop with Twisted-Tape Inserts," *Advances in Heat Transfer*, vol. 36, pp. 183–266, Academic Press, New York, 2002.
59. D. C. Groeneveld, "Post-Dryout Heat Transfer at Reactor Operating Correlations," Paper AECL-4513, National Topical Meeting on Water Reactor Safety, ANS, Salt Lake City, UT, 1973.
60. W. M. Rohsenow, J. P. Harnett, and Y. I. Cho, eds., *Handbook of Heat Transfer*, McGraw-Hill, New York, 1998.
61. L. S. Tong and J. D. Young, "A Phenomenological Transition and Film Boiling Heat Transfer Correlation," *Proc. 5th Int. Heat Transfer Conf.*, Tokyo, Sept. 1974.
62. W. Nusselt, "Die Oberflächenkondensation des Wasserdampfes," *Z. Ver: Dtsch. Ing.*, vol. 60, pp. 541 and 569, 1916.
63. W. M. Rohsenow, "Heat Transfer and Temperature Distribution in Laminar-Film Condensation," *Trans. ASME*, vol. 78, pp. 1645–1648, 1956.
64. M. M. Chen, "An Analytical Study of Laminar Film Condensation," part 1. "Flat Plates," and part 2, "Single and Multiple Horizontal Tubes," *Trans. ASME, Ser. C*, vol. 83, pp. 48–60, 1961.
65. W. M. Rohsenow, J. M. Weber, and A. T. Ling, "Effect of Vapor Velocity on Laminar and Turbulent Film Condensation," *Trans. ASME*, vol. 78, pp. 1637–1644, 1956.
66. A. P. Colburn, "The Calculation of Condensation Where a Portion of the Condensate Layer Is in Turbulent Flow," *Trans. AIChE.* vol. 30, p. 187, 1933.
67. C. G. Kirkbridge, "Heat Transfer by Condensing Vapors on Vertical Tubes," *Trans. AIChE*, vol. 30, p. 170, 1933.
68. E. F. Carpenter and A. P. Colburn, "The Effect of Vapor Velocity on Condensation Inside Tubes," in *Proceedings, General Discussion on Heat Transfer*, pp. 20–26, Inst. Mech. Eng. ASME, 1951.
69. M. Soliman, J. R. Schuster, and P. J. Berenson, "A General Heat Transfer Correlation for Annular Flow Condensation." *J. Heat Transfer*, vol. 90, pp. 267–276, 1968.
70. I. G. Shekriladze and V. I. Gomelauri, "Theoretical Study of Laminar Film Condensation of Steam." *Trans. Int. J. Heat Mass Transfer*, vol. 9, pp. 581–591, 1966.
71. T. B. Drew, W. M. Nagle, and W. Q. Smith, "The Conditions for Dropwise Condensation of Steam," *Trans. AIChE*, vol. 31, pp. 605–621, 1935.
72. R. S. Silver, "An Approach to a General Theory of Surface Condensers," *Proc. Inst. Mech. Eng.*, vol. 178, part 1, no. 14, pp. 339–376, 1964.
73. J. W. Rose, "On the Mechanism of Dropwise Condensation." *Int. J. Heat Mass Transfer*, vol. 10, pp. 755–762, 1967.
74. P. Griffith and M. S. Lee, "The Effect of Surface Thermal Properties and Finish on Dropwise Condensation," *Int. J. Heat Mass Transfer*, vol. 10, pp. 697–707, 1967.
75. A. P. Katz, H. J. Macintire, and R. E. Gould, "Heat Transfer in Ammonia Condensers," Bull. 209, Univ. Ill. Eng. Expt. Stn., 1930.
76. D. L. Katz, E. H. Young and G. Balekjian, "Condensing Vapors on Finned Tubes," *Petroleum Refiner*, pp. 175–178, Nov. 1954.
77. C. H. Dutcher and M. R. Burke, "Heat Pipes: A Cool Way to Cool Circuits," *Electronics*, pp. 93–100, February 16, 1970.
78. P. D. Dunn and D. A. Reay, *Heat Pipes*, 3d ed., Pergamon, New York, 1982.
79. S. W. Chi, *Heat Pipe Theory and Practice*, Hemisphere, Washington D.C., 1976.
80. R. Richter, "Solar Collector Thermal Power Systems," vol. 1, Rept. AFAPL-TR-74-89-1, Xerox Corp., Pasadena, Calif.: NTIS AD/A-000-940, National Technical Information Service, Springfield, Va., 1974.
81. L. Swanson, "Heat Pipes," in *CRC Handbook of Mechanical Engineering*, F. Kreith, ed., CRC Press, Boca Raton, FL, 1998.
82. A. L. London and R. A. Seban, "Rate of Ice Formation," *Trans. ASME*, vol. 65, pp. 771–778, 1943.
83. F. Kreith and F. E. Romie, "A Study of the Thermal Diffusion Equation with Boundary Conditions Corresponding to Freezing or Melting of Materials at the Fusion Temperature," *Proc. Phys. Soc.*, vol. 68, pp. 277–291, 1955.

84. D. L. Cochran, "Solidification Application and Extension of Theory," Tech. Rept. 24, Navy Contract N6-ONR-251, Stanford Univ., 1955.
85. W. D. Murray and F. Landis, "Numerical and Machine Solutions of Transient Heat Conduction Problems Involving Melting or Freezing," *Trans. ASME,* vol. 81, pp. 106–112, 1959.
86. A. Lazaridis, "A Numerical Solution of the Multi-Dimensional Solidification (or Melting) Problem," *Int. J. Heat Mass Transfer,* vol. 13, pp. 1459–1477, 1970.
87. H. Budhia and F. Kreith, "Heat Transfer with Melting or Freezing in a Wedge," *Int. J. Heat Mass Transfer,* vol. 16, 1973, pp. 195–211.
88. N. Lion, "Melting and Freezing," in *CRC Handbook of Thermal Engineering,* F. Kreith, ed., CRC Press, Boca Raton, Fl, 2000.
89. D. Y. Goswami, F. Kreith, and J. F. Kreider, *Principles of Solar Engineering,* 2nd ed., Taylor and Francis, Philadelphia, PA, Fig. 3.32, 2000.

Problems

The problems for this chapter are organized by subject matter as shown below.

Topic	Problem Number
Pool boiling	10.1–10.15
Film boiling	10.16–10.17
Convection boiling	10.18–10.19
Condensation	10.20–10.31
Freezing	10.32–10.36
Heat pipes	10.37–10.40

10.1 Water at atmospheric pressure is boiling in a pot with a flat-copper bottom on an electric range that maintains the surface temperature at 115°C. Calculate the boiling heat transfer coefficient.

10.2 Predict the nucleate-boiling heat transfer coefficient for water boiling at atmospheric pressure on the outside surface of a 1.5-cm-OD vertical-copper tube, 1.5 m long. Assume the tube surface temperature is constant at 10 K above the saturation temperature.

10.3 Estimate the maximum heat flux obtainable with nucleate pool boiling on a clean surface for (a) water at 1 atm on brass, (b) water at 10 atm on brass.

10.4 Determine the excess temperature at one-half of the maximum heat flux for the fluid-surface combinations in Problem 10.3.

10.5 In a pool boiling experiment in which water boiled on a large horizontal surface at atmospheric pressure, a heat flux of 4×10^5 W/m^2 was measured at an excess temperature of 14.5 K. What was the boiling surface material?

10.6 Compare the critical heat flux for a flat horizontal surface with that for a submerged horizontal wire of 3-mm diameter in water at saturation temperature and pressure.

10.7 For saturated pool boiling of water on a horizontal plate, calculate the peak heat flux at pressures of 10, 20, 40, 60, and 80% of the critical pressure p_c. Plot your results as q''_{max}/p_c versus p/p_c. The surface tension of water can be taken as $\sigma = 0.0743\ (1 - 0.0026\ T)$, where σ is in newtons per meter and T in degrees Centigrade. The critical pressure of water is 22.09 MPa.

10.8 A flat, stainless steel plate is 0.6 cm thick, 7.5 cm wide, and 0.3 m long is immersed horizontally at an initial temperature of 980°C in a large water bath at 100°C and at atmospheric pressure. Determine how long it will take this plate to cool to 540°C.

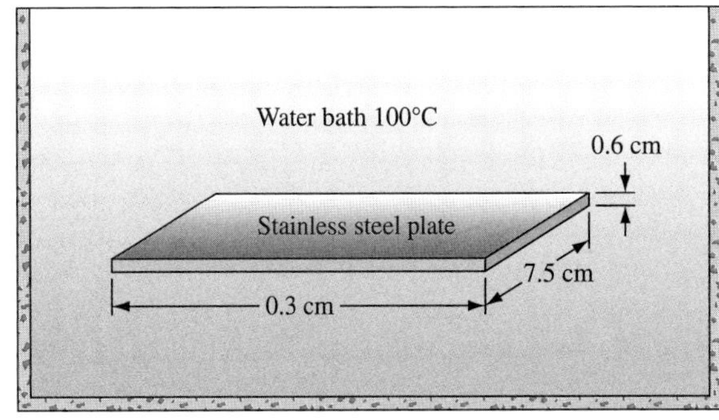

Problem 10.8

10.9 Calculate the maximum heat flux attainable in nucleate boiling with saturated water at 2 atm pressure in a gravitational field equivalent to one-tenth that of the earth.

10.10 Prepare a graph showing the effect of subcooling between 0°C and 50°C on the maximum heat flux calculated in Problem 10.9.

10.11 A thin-walled, horizontal copper tube of 0.5-cm OD is placed in a pool of water at atmospheric pressure and 100°C. Inside the tube an organic vapor is condensing, and the outside surface temperature of the tube is uniform at 232°C. Calculate the average heat transfer coefficient on the outside of the tube.

10.12 In boiling (and condensation) heat transfer, the convection heat transfer coefficient, h_c, is expected to depend on the difference between the surface and saturation temperatures, $\Delta T = (T_{\text{surface}} - T_{\text{saturation}})$, the body force arising from the density difference between liquid and vapor, $g(\rho_l - \rho_v)$, the latent heat, h_{fg}, the surface tension, σ, a characteristic length of the system, L, and the thermophysical properties of the liquid or vapor, ρ, c_p, k, and μ. Thus, we can write

$$h_c = h_c\{\Delta T, g(\rho_l - \rho_v), h_{fg}, \sigma, L, \rho, c_p, k, \mu\}$$

Determine (a) the number of dimensionless groups necessary to correlate experimental data, and (b) appropriate dimensionless groups that should include the Prandtl number, the Jakob number, and the Bond number $(g\,\Delta\rho L^2/\sigma)$.

10.13 Environmental concerns have recently motivated the search for replacements for chlorofluorocarbon refrigerants. An experiment has been devised to determine the applicability of one such replacement. A silicon chip is bonded to the bottom of a thin copper plate as shown in the sketch below. The chip is 0.2 cm thick and has a thermal conductivity of 125 W/m K. The copper plate is 0.1 cm thick, and there is no contact resistance between the chip and the copper plate. This assembly is to be cooled by boiling a saturated liquid refrigerant on the copper surface. The electronic circuit on the bottom of the chip generates heat uniformly at a flux of $q''_0 = 5 \times 10^4$ W/m^2. Assume that the sides and the bottom of the chip are insulated.

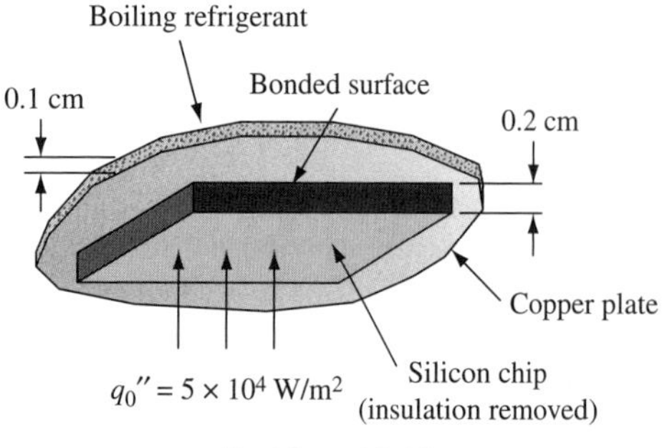

Problem 10.13

Calculate the steady-state temperature at the copper surface and the bottom of the chip, as well as the maximum heat flux in pool boiling, assuming that the boiling coefficient, C_{sf}, is the same as for n-pentane on lapped copper. The physical properties of this new coolant are $c_p = 1100$ J/kg K, $h_{fg} = 8.4 \times 10^4$ J/kg, $\rho_l = 1620$ kg/m^3, $\rho_v = 13.4$ kg/m^3, $\sigma = 0.081$ N/m, $\mu_l = 4.4 \times 10^{-4}$ k/m s, $T_{\text{sat}} = 60$°C and Pr$_l = 9.0$.

10.14 It has recently been proposed by Andraka et al. of Sandia National Laboratories Albuquerque, in *Sodium Reflux Pool-Boiler Solar Receiver On-Sun Test Results* (SAND89-2773, June 1992) that the heat flux from a parabolic dish solar concentrator could be delivered effectively to a Stirling engine by a liquid-metal pool boiler. The following sketch shows a cross-section of the pool boiler-receiver assembly. Solar flux is absorbed on the concave side of a hemispherical absorber dome, with boiling molten sodium metal on the convex side of the dome. The sodium vapor condenses on the engine heater tube as shown near the top of the figure. Condensing sodium transfers its latent heat to the engine working fluid that circulates inside the tube. Calculations indicate that a maximum heat flux of 75 W/cm^2 delivered by the solar concentrator to the absorber dome is to be expected.

After the receiver had been tested for about 50 hours, a small spot on the absorber dome suddenly melted and the receiver failed. Is it possible that the

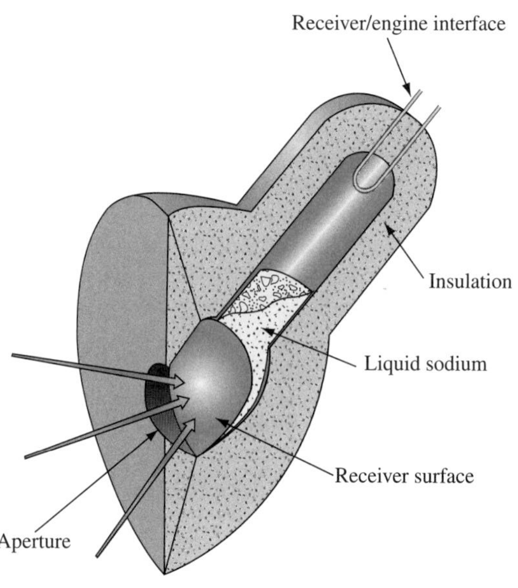

Problem 10.14

critical flux for the boiling sodium was exceeded? Use the following properties for the sodium: $\rho_v = 0.056$ kg/m^3, $\rho_l = 779$ kg/m^3, $h_{fg} = 4.039 \times 10^6$ J/kg, $\sigma_l = 0.138$ N/m, $\mu_l = 1.8 \times 10^{-4}$ kg/ms.

10.15 Calculate the peak heat flux for nucleate pool boiling of water at 3-atm pressure and 110°C on clean copper.

10.16 Calculate the heat transfer coefficient for film boiling of water on a 1.3-cm horizontal tube if the tube temperature is 550°C and the system is placed under a pressure of 0.5 atm.

10.17 A metal-clad electrical heating element of cylindrical shape, as shown in the sketch below, is immersed in water at atmospheric pressure. The element is 5-cm OD, and heat generation produces a surface temperature of 300°C. Estimate the heat flux under steady-state conditions and the rate of heat generation per unit length.

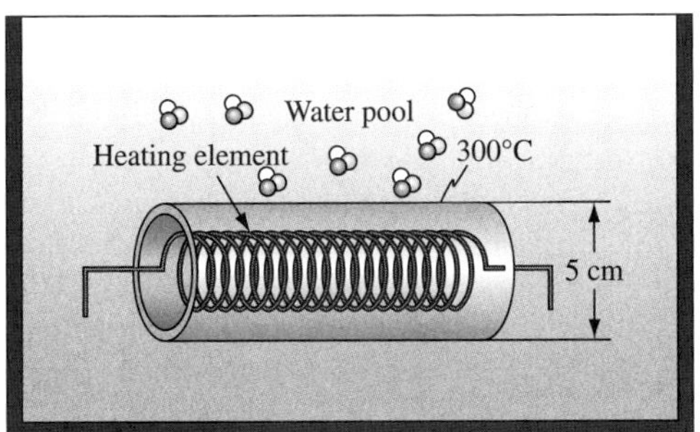

Problem 10.17

10.18 Calculate the maximum safe heat flux in the nucleate-boiling regime for water flowing at a velocity of 15 m/s through a 0.31-m-long, 1.2-cm-ID tube if the water enters at 1 atm and 100°C.

10.19 During the 1980s, solar thermal electric technology was commercialized with the installation of 350 MW electrical power-plant capacity in the California desert. The technology involved heating a heat transfer oil in receiver tubes placed at the focus of line-focus, parabolic trough solar concentrators. The heat transfer oil was then used to generate steam, which in turn powered a steam turbine/electrical generator. Since the transfer of heat from the oil to the steam creates a temperature drop and a resulting loss in thermal efficiency, alternatives are being considered for a future plant. In one alternative, steam would be generated directly inside the receiver tubes. Consider a situation in which a heat flux of 50,000 W/m^2 is absorbed on the outside surface of a 12.7-mm-ID, stainless steel, 316 tube with a wall thickness of 1.245 mm. Inside the tube, saturated liquid water at 300°C flows at a rate of 100 kg/h. Determine the tube wall temperature if the steam quality is to be increased to 0.5. Assume the viscosity of steam at the operating pressure is $\mu_v = 2.0 \times 10^{-5}$ kg/ms. Neglect heat losses from the outside of the receiver. See Goswami, Kreith, and Kreider [89] for a system description.

10.20 Calculate the average heat transfer coefficient for film-type condensation of water at pressures of 10 kPa and 101 kPa for (a) a vertical surface 1.5 m high, (b) the outside surface of a 1.5-cm-OD vertical tube 1.5 m long, (c) the outside surface of a 1.6-cm-OD horizontal tube 1.5 m long, and (d) a 10-tube vertical bank of 1.6-cm-OD horizontal tubes 1.5 m long. In all cases assume that the vapor velocity is negligible and that the surface temperatures are constant at 11°C below saturation temperature.

10.21 The inside surface of a 1-m-long, vertical, 5-cm-ID tube is maintained at 120°C. For saturated steam at 350 kPa condensing inside, estimate the average heat transfer coefficient and the condensation rate, assuming the steam velocity is small.

10.22 A horizontal, 2.5-cm-OD tube is maintained at a temperature of 27°C on its outer surface. Calculate the average heat transfer coefficient if saturated steam at 12 kPa is condensing on this tube.

10.23 Repeat Problem 10.22 for a tier of six horizontal 2.5-cm-OD tubes under similar thermal conditions.

10.24 Saturated steam at 34 kPa condenses on a 1-m-tall vertical plate whose surface temperature is uniform at 60°C. Compare the average heat transfer coefficient and the value of the coefficient 1/3m, 2/3m, and 1 m from the top. Also, find the maximum height for which the condensate film will remain laminar.

10.25 At a pressure of 490 kPa, the saturation temperature of sulfur dioxide (SO_2) is 32°C, the density is 1350 kg/m^3, the latent heat of vaporization is 343 kJ/kg, the absolute viscosity is 3.2×10^{-4} Ns/m^2, the specific heat is 1445 J/kg K and the thermal conductivity is 0.192 W/m K. If the SO_2 is to be condensed at 490 kPa on a 20-cm flat surface that is inclined at an angle of 45° and whose, temperature is maintained uniformly at 24°C, calculate (a) the thickness of the condensate film 1.3 cm from the bottom, (b) the average heat transfer coefficient, and (c) the rate of condensation in kilograms per hour.

10.26 Repeat Problem 10.25 (b) and (c) but assume that condensation occurs on a 5-cm-OD, horizontal tube.

10.27 Problem 10.12 indicated that the Nusselt number for condensation depends on the Prandtl number and four other dimensionless groups including the Jakob number (Ja), the Bond number (Bo), and a nameless group resembling the Grashof number $[\rho g(\rho_l - \rho_v)L^3/\mu^2]$.

Give a physical explanation for each of these three groups and explain when you expect Bo and Ja to exert a significant influence and when their influences are negligible.

10.28 Saturated methyl chloride at 62 psia condenses on a horizontal 10 × 10 bank of tubes. The 2-in.-OD tubes are equally spaced and are 4 in. apart center-to-center on rows and columns. If the surface temperature of the tubes is maintained at 45°F by pumping water through them, calculate the rate of condensation of methyl chloride in lb/h ft. The properties of saturated methyl chloride at 62 psia are shown in the list that follows:

saturation temperature = 60°F
heat of vaporization = 167 Btu/lb
liquid density = 58.8 lb/ft^3
liquid specific heat = 0.38 Btu/lb °F
liquid absolute viscosity = 1.344×10^{-4} lb/ft s
liquid thermal conductivity = 0.10 Btu/h ft °F

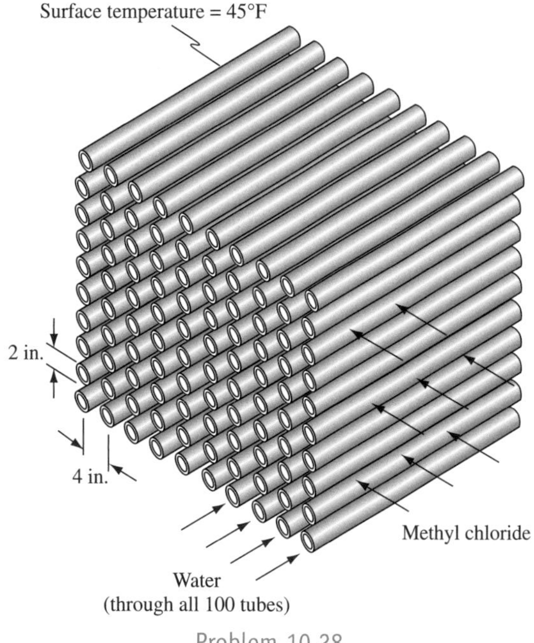

Problem 10.28

10.29 A vertical, rectangular water duct 1 m high and 0.10 m deep shown in the sketch is placed in an environment of saturated steam at atmospheric pressure. If the outer surface of the duct is about 50°C, estimate the rate of steam condensation per unit length.

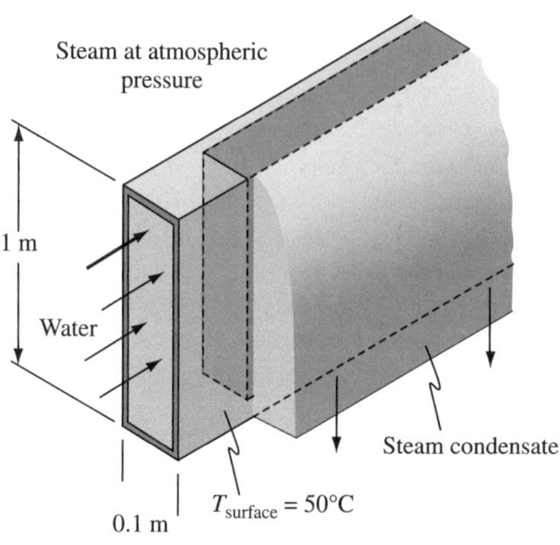

Problem 10.29

10.30 The 1-m-long, tube-within-a-tube, heat exchanger shown in the sketch is used to condense steam at 2 atm in the annulus. Water flows in the inner tube, entering at 90°C. The inner tube is made of copper with a 1.27-cm-OD and 1.0-cm-ID. (a) Estimate the water flow rate required to keep its outlet temperature below 100°C. (b) Estimate the pressure drop and the pumping power for the water in the heat exchanger, neglecting inlet and outlet losses.

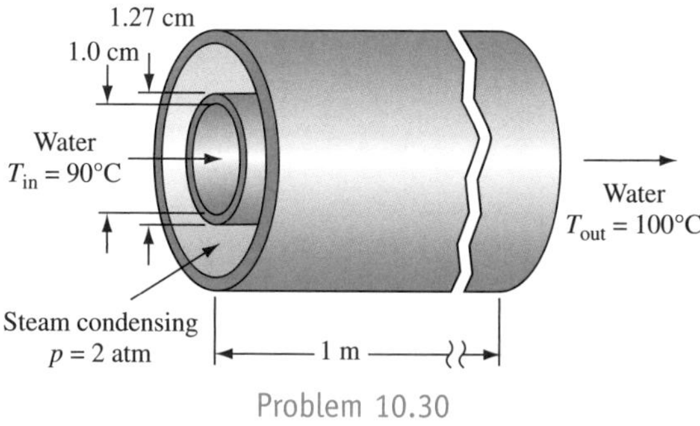

Problem 10.30

10.31 The one-pass condenser, heat exchanger shown in the sketch has 64 tubes arranged in a square array with 8 tubes per line. The tubes are 4 ft long and are made of copper with an outside diameter of 0.50 in. They are contained in a shell at atmospheric pressure. Water flows inside the tubes, whose outside wall temperature is 208.4°F. Calculate (a) the rate of steam condensation and (b) the temperature rise of the water if the flow rate per tube is 0.1 lb/s. Answer in SI and English units.

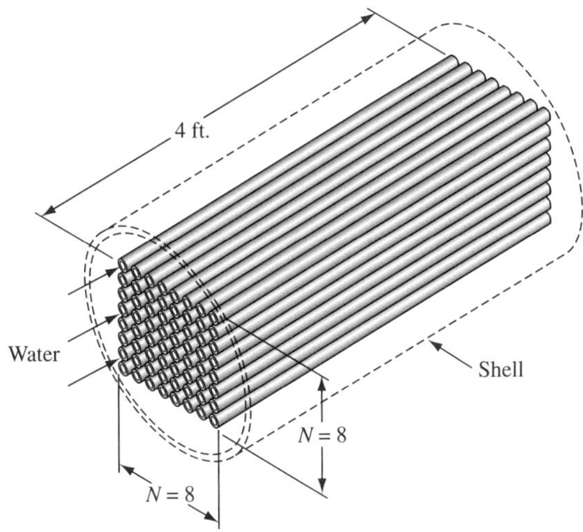

Problem 10.31

10.32 Show that the dimensionless equation for ice formation on the outside of a tube of radius r_0 is

$$t^* = \frac{r^{*2}}{2} \ln r^* + \left(\frac{1}{2R^*} - \frac{1}{4}\right)(r^{*2} - 1)$$

where

$$r^* = \frac{\varepsilon + r_0}{r_0} \qquad R^* = \frac{h_0 r_0}{k} \qquad t^* = \frac{(T_f - T_s)kt}{\rho L r_o^2}$$

Assume that the water is originally at the freezing temperature T_f, that the cooling medium inside the tube surface is just below the freezing temperature at a uniform temperature T_s, and that h_0 is the total heat transfer coefficient between the cooling medium and the pipe-ice interface. Also show the thermal circuit.

10.33 In the manufacture of can ice, cans having inside dimensions of 11 × 22 × 50 in. with 1 in. inside taper are filled with water and immersed in brine at a temperature of 10°F. For the purpose of a preliminary analysis, the actual ice can be considered as an equivalent cylinder having the same cross-sectional area as the can, and end effects can be neglected. The overall conductance between the brine and the inner surface of the can is 40 Btu/h ft^2 °F. Determine the time required to freeze the water, and compare it with the time necessary if the brine circulation rate is increased to reduce the thermal resistance of the surface to one-tenth of the value specified above. The latent heat of fusion of ice is 143.5 Btu/lb, its density is 57.3 lb/ft^3, and its thermal conductivity is 1.28 Btu/h ft °F.

10.34 Estimate the time required to freeze vegetables in thin, tin cylindrical containers 15 cm in diameter. Air at −12°C is blowing at 4 m/s over the cans, which are stacked to form one long cylinder. The physical properties of the vegetables before and after freezing can be taken as those of water and ice respectively.

10.35 Estimate the time required for nocturnal radiation to freeze a 3-cm thickness of water with ambient air and initial water temperatures at 4°C. Neglect evaporation effects.

10.36 The temperature of a 100-m-diameter cooling pond is 7°C on a winter day. If the air temperature suddenly drops to −7°C, calculate the thickness of ice formed after three hours.

10.37 On a rainy Monday afternoon, a wealthy banker calls Sherlock Holmes to arrange a breakfast appointment for the following day to discuss the collection of a loan from farmer Joe. When Holmes arrives at the home of the banker at 9 a.m. Tuesday, he finds the body of the banker in the kitchen. The farmer's house is located on the other side of a lake, approximately 10 km from the banker's home. Since there is no convenient road between the home of the farmer and that of the banker, Holmes phones the police to question the farmer.

The police arrive at the farmer's home within the hour and interrogate him about the death of the banker. The farmer claims to have been home all night. The tires on his truck are dry, and he explains that his boots were moist and soiled because he had been fishing at the lake early in the morning. The police then phone Holmes to eliminate farmer Joe as a murder suspect because he could not have been at the banker's home since Holmes spoke to him.

Holmes then calls the local weather bureau and learns that although the temperature had been between 2°C and 5°C for weeks, it had dropped to −30°C quite suddenly on Monday night. Remembering that a 3-cm layer of ice can support a man, Holmes takes out his slide rule and heat transfer text, lights his pipe, makes a few calculations, and then phones the police to arrest farmer Joe. Why?

10.38 Estimate the cross-sectional area required for a 30-cm-long methanol-nickel heat pipe to transport 100 Btu/h at atmospheric pressure.

10.39 Design a heat pipe cooling system for a spherical satellite that dissipates 5000 W/m^3, has a surface area of 5 m^2, and cannot exceed a temperature of 120°C. All the heat must be dissipated by radiation into space. State all your assumptions.

10.40 Compare the axial heat flux achievable by a heat pipe using water as the working fluid with that of a solid silver rod. Assume that both are 20 cm long, that the temperature difference for the rod from end to end is 100°C, and that the heat pipe operates at atmospheric pressure. State your other assumptions.

Design Problems

10.1 **Liquid Nitrogen Evaporator** (Chapter 10) Liquid nitrogen is typically delivered in large dewars by suppliers to a jobsite. In many applications nitrogen vapor is required, so it is necessary to provide a means for evaporating the liquid nitrogen. Design such an evaporator, capable of delivering nitrogen vapor at a rate of 125 g/min. Your design should consider equipment cost, required floor space and operating cost. You should also consider that most end-users do not want to be bothered with complex equipment. Hence, a very simple, yet effective and safe system is required.

10.2 **Electrical-Resistance Heater** (Chapters 2, 3, 6, and 10) In Chapters 2, 3, and 6 you determined the required heat transfer coefficients for water flowing over the outside surface of a heating element. Those solutions required an assumption that by limiting the heating element surface temperature to 100°C, surface boiling could be eliminated. Given the operating pressure of the system and your understanding of convective boiling heat transfer, determine whether the constant was too conservative. If it was, refine your hot water heater design.

10.3 **Condenser for a Steam Turbine** (Chapter 10) Repeat your design for Design Problem 8.2 in Chapter 8 but calculate the condensing heat transfer coefficient. Explain any differences in results.

10.4 **Laboratory Steam Generator** (Chapter 10) You are to design an electrically powered steam generator for use in a laboratory experiment. The boiler is to provide 1 g/s of dry saturated steam at 1.5 atm. The primary design considerations are cost, ease of use, and safety. For simplicity, you may consider a pool boiler arrangement, with an electric-resistance heating element similar to that used in a tea kettle. Discuss how the electrical heating element would be sized and attached, placed in the pool, water quality issues, and how you would control the device to ensure that it could produce steam continuously.

APPENDIXES

Appendix 1 The International System of Units A3

Appendix 2 Data Tables A6

APPENDIX 1

The International System of Units

The International System of Units (SI) has evolved from the MKS system, in which the meter is the unit of length, the kilogram is the unit of mass, and the second is the unit of time. The SI system is rapidly becoming the standard system of units throughout the industrialized world.

The SI system is based on seven units. Other derived units may be related to these seven base units through governing equations. The base units are listed in Table 1 along with the recommended symbols. Several defined units are listed in Table 2, while the derived units of interest in heat transfer and fluid flow are given in Table 3.

Standard prefixes can be used in the SI system to designate multiples of the basic units and thereby conserve space. The standard prefixes are listed in Table 4.

Table 5 contains an alphabetical listing of physical constants that are frequently used in heat transfer and fluid flow problems, along with their values in the SI system of units.

TABLE 1 SI base units

Quantity	Name of Unit	Symbol
Length	meter	m
Mass	kilogram	kg
Time	second	s
Electrical current	ampere	A
Thermodynamic temperature	kelvin	K
Luminous intensity	candela	cd
Amount of a substance	mole	mol

TABLE 2 SI defined units

Quantity	Unit	Defining Equation
Capacitance	farad, F	$1\ F = 1\ A\ s/V$
Electrical resistance	ohm, Ω	$1\ \Omega = 1\ V/A$
Force	newton, N	$1\ N = 1\ kg\ m/s^2$
Potential difference	volt, V	$1\ V = 1\ W/A$
Power	watt, W	$1\ W = 1\ J/s$
Pressure	pascal, Pa	$1\ Pa = 1\ N/m^2$
Temperature	kelvin, K	$K = °C + 273.15$
Work, heat, energy	joule, J	$1\ J = 1\ N\ m$

TABLE 3 SI derived units

Quantity	Name of Unit	Symbol
Acceleration	meter per second squared	m/s^2
Area	square meter	m^2
Density	kilogram per cubic meter	kg/m^3
Dynamic viscosity	newton-second per square meter	$N\ s/m^2$
Force	newton	N
Frequency	hertz	Hz
Kinematic viscosity	square meter per second	m^2/s
Plane angle	radian	rad
Power	watt	W
Radiant intensity	watt per steradian	W/sr
Solid angle	steradian	sr
Specific heat	joule per kilogram-kelvin	J/kg K
Thermal conductivity	watt per meter-kelvin	W/m K
Velocity	meter per second	m/s
Volume	cubic meter	m^3

TABLE 4 SI prefixes

Multiplier	Symbol	Prefix	Multiplier	Symbol	Prefix
10^{12}	T	tera	10^{-2}	c	centi
10^{9}	G	giga	10^{-3}	m	milli
10^{6}	M	mega	10^{-6}	μ	micro
10^{3}	k	kilo	10^{-9}	n	nano
10^{2}	h	hecto	10^{-12}	p	pico
10^{1}	da	deka	10^{-15}	f	femto
10^{-1}	d	deci	10^{-18}	a	atto

TABLE 5 Physical constants in SI units

Quantity	Symbol	Value
—	e	2.718281828
—	π	3.141592653
—	g_c	$1.00000\ kg\ m\ N^{-1}\ s^{-2}$
Avogadro constant	N_A	$6.022169 \times 10^{26}\ kmol^{-1}$
Boltzmann constant	k	$1.380622 \times 10^{23}\ J\ K^{-1}$
First radiation constant	$C_1 = 2\pi hc^2$	$3.741844 \times 10^{-16}\ W\ m^2$
Planck constant	h	$6.626196 \times 10^{-34}\ J\ s$
Second radiation constant	$C_2 = hc/k$	$1.438833 \times 10^{-2}\ m\ K$
Speed of light in a vacuum	c	$2.997925 \times 10^{8}\ m\ s^{-1}$
Stefan-Boltzmann constant	σ	$5.66961 \times 10^{-8}\ W\ m^{-2}\ K^{-4}$

TABLE 6 Conversion factors

Physical Quantity	**Symbol**	**Conversion Factor**
Area	A	1 ft^2 = 0.0920 m^2
		1 $in.^2$ = 6.452 × 10−4 m^2
Density	ρ	1 lb_m/ft^3 = 16.018 kg/m^3
		1 $slug/ft^3$ = 515.379 kg/m^3
Energy, heat	Q	1 Btu = 1055.1 J
		1 cal = 4.186 J
		1 (ft)(lb_f) = 1.3558 J
		1 (hp)(h) = 2.685 × 106 J
Force	F	1 lb_f = 4.448 N
Heat flow rate	q	1 Btu/h = 0.2931 W
		1 Btu/s = 1055.1 W
Heat flux	q''	1 Btu/(h)(ft^2) = 3.1525 W/m^2
Heat generation per unit volume	$\dot{q}_G$	1 Btu/(h)(ft^3) = 10.343 W/m^3
Heat transfer coefficient	h	1 Btu/(h)(ft^2)(°F) = 5.678 W/m^2 K
Length	L	1 ft = 0.3048 m
		1 in. = 2.54 cm = 0.0254 m
		1 mile = 1.6093 km = 1609.3 m
Mass	m	1 lb_m = 0.4536 kg
		1 slug = 14.594 kg
Mass flow rate	$\dot{m}$	1 lb_m/h = 0.000126 kg/s
		1 lb_m/s = 0.4536 kg/s
Power	P	1 hp = 745.7 W
		1 (ft)(lb_f)/s = 1.3558 W
		1 Btu/s = 1055.1 W
		1 Btu/h = 0.293 W
Pressure	p	1 $lb_f/in.^2$ = 6894.8 N/m^2 (Pa)
		1 lb_f/ft^2 = 47.88 N/m^2 (Pa)
		1 atm = 101,325 N/m^2 (Pa)
Specific heat capacity	c	1 Btu/(lb_m)(°F) = 4188 J/kg K
Temperature	T	T(°R) = (9/5)T(K)
		T(°F) = [T(°C)](9/5) + 32
		T(°F) = [T(K) − 273.15](9/5) + 32
Thermal conductivity	k	1 Btu/(h)(ft)(°F) = 1.731 W/m K
Thermal diffusivity	α	1 ft^2/s = 0.0929 m^2/s
		1 ft^2/h = 2.581 × 10^{-5} m^2/s
Thermal resistance	R_t	1 (h)(°F)/Btu = 1.8958 K/W
Velocity	U_∞	1 ft/s = 0.3048 m/s
		1 mph = 0.44703 m/s
Viscosity, dynamic	μ	1 lb_m/(ft)(s) = 1.488 N s/m^2
		1 centipoise = 0.00100 N s/m^2
Viscosity, kinematic	ν	1 ft^2/s = 0.0929 m^2/s
		1 ft^2/h = 2.581 × 10^{-5} m^2/s
Volume	V	1 ft^3 = 0.02832 m^3
		1 $in.^3$ = 1.6387 × 10^{-5} m^3
		1 gal (U.S. liq.) = 0.003785 m^3

APPENDIX 2

Data Tables

To facilitate conversion of property values from SI to English units, conversion factors have been incorporated into each table. For temperature-dependent data, temperature is listed in both systems of units. Property values are given in SI units (with the exception of Table 38); to obtain a property in English units, the property in SI units must be multiplied by the conversion factor at the top of the column. For example, suppose we wish to determine the absolute viscosity of water in English units, at 95°F. From Table 13 we have

$$\mu = \underset{\substack{\text{(SI value}\\ \text{from table)}}}{(719.8 \times 10^{-6})} \times \underset{\substack{\text{(conversion factor}\\ \text{from column head)}}}{(0.6720)} = 4.84 \times 10^{-4}\ \text{lb}_m/\text{ft s}$$

Tables 41 and 42 are expressed in English units only, because in the U.S. pipe and tubing sizes are generally specified in these dimensions.

Properties of Solids

TABLE 7 Normal emissivities of metals

Substance	State of Surface	Temperature (K)	Temperature (R)	Normal Emissivity ε_n[a]
Aluminum	polished plate	296	533	0.040
		498	896	0.039
	rolled, polished	443	797	0.039
	rough plate	298	536	0.070
Brass	oxidized	611	1100	0.22
	polished	292	526	0.05
		573	1031	0.032
	tarnished	329	592	0.202
Chromium	polished	423	761	0.058
Copper	black oxidized	293	527	0.780
	lightly tarnished	293	527	0.037
	polished	293	527	0.030
Gold	not polished	293	527	0.47
	polished	293	527	0.025
Iron	oxidized smooth	398	716	0.78
	ground bright	293	527	0.24
	polished	698	1256	0.144
Lead	gray oxidized	293	527	0.28
	polished	403	725	0.056
Molybdenum	filament	998	1796	0.096
Nickel	oxidized	373	671	0.41
	polished	373	671	0.045
Platinum	polished	498	896	0.054
		898	1616	0.104
Silver	polished	293	527	0.025
Steel	oxidized rough	313	563	0.94
	ground sheet	1213	2183	0.520
Tin	bright	293	527	0.070
Tungsten	filament	3300	5940	0.39
Zinc	tarnished	293	527	0.25
	polished	503	905	0.045

[a]Hemispherical emissivity values ε may be approximated by $\varepsilon = 1.2\varepsilon_n$ for bright metal surfaces, $\varepsilon = 0.95\varepsilon_n$ for other smooth surfaces, and $\varepsilon = 0.98\varepsilon_n$ for rough surfaces.

Source: K. Raznjevič, *Handbook of Thermodynamic Tables and Charts*, McGraw-Hill, New York, 1976.

TABLE 8 Normal emissivities of nonmetals

Substance	State of Surface	Temperature (K)	Temperature (R)	Normal Emissivity ε_n
Asbestos board		297	535	0.96
Brick	red, rough	293	527	0.93
Carbon filament		1313		0.53
Glass	smooth	293	527	0.93
Ice	smooth	273	491	0.966
	rough	273	491	0.985
Masonry	plastered	273	491	0.93
Paper		293	527	0.80
Plaster, lime	white, rough	293	527	0.93
Porcelain	glazed	293	527	0.93
Quartz	fused, rough	293	527	0.93
Rubber				
soft	gray	297	535	0.86
hard	black, rough	297	535	0.95
Wood				
beech	planed	343	617	0.935
oak	planed	294	529	0.885

Source: K. Raznjevič, *Handbook of Thermodynamic Tables and Charts*, McGraw-Hill, New York, 1976.

TABLE 9 Normal emissivities of paints and surface coatings

Substance	State of Surface	Temperature (K)	Temperature (R)	Normal Emissivity ε_n
Aluminum bronze		373	671	0.20–0.40
Aluminum enamel	rough	293	527	0.39
Aluminum paint	heated to 325°C	423–588	761–1058	0.35
Bakelite enamel		353	635	0.935
Enamel				
white	rough	293	527	0.90
black	bright	298	536	0.876
Oil paint		273–473	491–851	0.885
Red lead primer		293–373	527–671	0.93
Shellac, black	bright	294	529	0.82
	dull	348–418	626–752	0.91

Source: K. Raznjevič, *Handbook of Thermodynamic Tables and Charts*, McGraw-Hill, New York, 1976.

TABLE 10 Alloys

Metal	Composition (%)	Properties at 293 K (20°C, 68°F) ρ (kg/m^3) × 6.243 × 10^{-2} = (lb$_m$/ft^3)	c_p (J/kg K) × 2.388 × 10^{-4} = (Btu/lb$_m$ °F)	k (W/m K) × 0.5777 = (Btu/h ft °F)	$\alpha \times 10^5$ (m^2/s) × 3.874 × 10^4 = (ft^2/h)
Aluminum					
duralumin	94–96 Al, 3–5 Cu, trace Mg	2787	833	164	6.676
silumin	87 Al, 13 Si	2659	871	164	7.099
Copper					
aluminum bronze	95 Cu, 5 Al	8666	410	83	2.330
bronze	75 Cu, 25 Sn	8666	343	26	0.859
red brass	85 Cu, 9 Sn, 6 Zn	8714	385	61	1.804
brass	70 Cu, 30 Zn	8522	385	111	3.412
german silver	62 Cu, 15 Ni, 22 Zn	8618	394	24.9	0.733
constantan	60 Cu, 40 Ni	8922	410	22.7	0.612
Iron					
cast iron	~4 C	7272	420	52	1.702
wrought iron	0.5 CH	7849	460	59	1.626
Steel					
carbon steel	1 C	7801	473	43	1.172
	1.5 C	7753	486	36	0.970
chrome steel	1 Cr	7865	460	61	1.665
	5 Cr	7833	460	40	1.110
	10 Cr	7785	460	31	0.867
chrome nickel steel	15 Cr, 10 Ni	7865	460	19	0.526
	20 Cr, 15 Ni	7833	460	15.1	0.415
nickel steel	10 Ni	7945	460	26	0.720
	20 Ni	7993	460	19	0.526
	40 Ni	8169	460	10	0.279
	60 Ni	8378	460	19	0.493
nickel chrome steel	80 Ni, 15 C	8522	460	17	0.444
	40 Ni, 15 C	8073	460	11.6	0.305
manganese steel	1 Mn	7865	460	50	1.388
	5 Mn	7849	460	22	0.637
silicon steel	1 Si	7769	460	42	1.164
	5 Si	7417	460	19	0.555
stainless steel	type 304	7817	461	14.4	0.387
	type 347	7817	461	14.3	0.387
tungsten steel	1 W	7913	448	66	1.858
	5 W	8073	435	54	1.525

Source: E. R. G. Eckert and R. M. Drake, *Analysis of Heat and Mass Transfer,* McGraw-Hill, New York, 1972.

TABLE 11 Insulations and building materials

Material	Properties at 293 K (20°C, 68°F)			
	ρ (kg/m^3) × 6.243 × 10^{-2} = (lb$_m$/ft^3)	c_p (J/kg K) × 2.388 × 10^{-4} = (Btu/lb$_m$ °F)	k (W/m K) × 0.5777 = (Btu/h ft °F)	$\alpha \times 10^5$ (m^2/s) × 3.874 × 10^4 = (ft^2/h)
Asbestos	383	816	0.113	0.036
Asphalt	2120		0.698	
Bakelite	1270		0.233	
Brick				
common	1800	840	0.38–0.52	0.028–0.034
carborundum (50% SiC)	2200		5.82	
magnesite (50% MgO)	2000		2.68	
masonry	1700	837	0.658	0.046
silica (95% SiO_2)	1900		1.07	
zircon (62% ZrO_2)	3600		2.44	
Cardboard			0.14–0.35	
Cement, hard			1.047	
Clay (48.7% moisture)	1545	880	1.26	0.101
Coal, anthracite	1370	1260	0.238	0.013–0.015
Concrete, dry	2300	837	1.8	0.094
Cork, boards	150	1880	0.042	0.015–0.044
Cork, expanded	120		0.036	
Diatomaceous earth	466	879	0.126	0.031
Glass fiber	220		0.035	
Glass, window	2800	800	0.81	0.034
Glass, wool	50		0.037	
	100		0.036	
	200	670	0.040	0.028
Granite	2750		3.0	
Ice (0°C)	913	1830	2.22	0.124
Kapok	25		0.035	
Linoleum	535		0.081	
Mica	2900		0.523	
Pine bark	342		0.080	
Plaster	1800		0.814	
Plexiglas	1180		0.195	
Plywood	590		0.109	
Polystyrene	1050		0.157	
Rubber, Buna	1250		0.465	
hard (ebonite)	1150	2009	0.163	0.0062
spongy	224		0.055	
Sand, dry			0.582	
Sand, moist	1640		1.13	

(Continued)

TABLE 11 (*Continued*)

	Properties at 293 K (20°C, 68°F)			
	ρ (kg/m^3)	c_p (J/kg K)	k (W/m K)	$\alpha \times 10^5$ (m^2/s)
Material	$\times 6.243 \times 10^{-2}$ = (lb$_m$/ft^3)	$\times 2.388 \times 10^{-4}$ = (Btu/lb$_m$ °F)	$\times 0.5777$ = (Btu/h ft °F)	$\times 3.874 \times 10^4$ = (ft^2/h)
Sawdust	215		0.071	
Soil				
dry	1500	1842	~0.35	~0.0138
wet	1500		~2.60	0.0414
Wood				
oak	609–801	2390	0.17–0.21	0.0111–0.0121
pine, fir, spruce	416–421	2720	0.15	0.0124
Wood fiber sheets	200		0.047	
Wool	200		0.038	
85% magnesia			0.059	

Source: E. R. G. Eckert and R. M. Drake, *Analysis of Heat and Mass Transfer*, McGraw-Hill, New York, 1972; K. Raznjevič, *Handbook of Thermodynamic Tables and Charts*, McGraw-Hill, New York, 1976.

TABLE 12 Metallic elements[a]

Element	Thermal Conductivity k (W/m K)[b] 200 K −73°C	273 K 0°C 32°F	400 K 127°C 261°F	600 K 327°C 621°F	800 K 527°C 981°F	1000 K 727°C 1341°F	1200 K 927°C 1701°F	Properties at 293 K or 20°C or 68°F: ρ (kg/m^3)	c_p (J/kg K)	k (W/m K)	$\alpha \times 10^6$ (m^2/s)	Melting Temperature (K)
	× 0.5777 = (Btu/h ft °F)							$\times 6.243 \times 10^{-2}$ = (lb$_m$/ft^3)	$\times 2.388 \times 10^{-4}$ = (Btu/lb$_m$ °F)	× 0.5777 = (Btu/h ft °F)	3.874×10^4 = (ft^2/h)	
Aluminum	237	236	240	232	220			2,702	896	236	97.5	933
Antimony	30.2	25.5	21.2	18.2	16.8			6,684	208	24.6	17.7	904
Beryllium	301	218	161	126	107	89	73	1,850	1750	205	63.3	1550
Bismuth[c]	9.7	8.2						9,780	124	7.9	6.51	545
Boron[c]	52.5	31.7	18.7	11.3	8.1	6.3	5.2	2,500	1047	28.6	10.9	2573
Cadmium[c]	99.3	97.5	94.7					8,650	231	97	48.5	594
Cesium	36.8	36.1						1,873	230	36	83.6	302
Chromium	111	94.8	87.3	80.5	71.3	65.3	62.4	7,160	440	91.4	29.0	2118
Cobalt[c]	122	104	84.8					8,862	389	100	29.0	1765
Copper	413	401	392	383	371	357	342	8,933	383	399	116.6	1356
Germanium	96.8	66.7	43.2	27.3	19.8	17.4	17.4	5,360		61.6		1211
Gold	327	318	312	304	292	278	262	19,300	129	316	126.9	1336
Hafnium	24.4	23.3	22.3	21.3	20.8	20.7	20.9	13,280		23.1		2495
Indium	89.7	83.7	74.5					7,300		82.2		430
Iridium	153	148	144	138	132	126	120	22,500	134	147	48.8	2716
Iron	94	83.5	69.4	54.7	43.3	32.6	28.2	7,870	452	81.1	22.8	1810
Lead	36.6	35.5	33.8	31.2				11,340	129	35.3	24.1	601
Lithium	88.1	79.2	72.1					534	3391	77.4	42.7	454
Magnesium	159	157	153	149	146			1,740	1017	156	88.2	923
Manganese	7.17	7.68						7,290	486	7.78	2.2	1517

(*Continued*)

TABLE 12 (Continued)

Element	Thermal Conductivity k (W/m K)[b] 200 K −73°C	273 K 0°C 32°F	400 K 127°C 261°F	600 K 327°C 621°F	800 K 527°C 981°F	1000 K 727°C 1341°F	1200 K 927°C 1701°F	Properties at 293 K or 20°C or 68°F: ρ (kg/m³)	c_p (J/kg K)	k (W/m K)	$\alpha \times 10^6$ (m²/s)	Melting Temperature (K)
	× 0.5777 = (Btu/h ft°F)							$\times 6.243 \times 10^{-2}$ = (lb_m/ft^3)	$\times 2.388 \times 10^{-4}$ = (Btu/lb_m °F)	× 0.5777 = (Btu/h ft °F)	3.874×10^4 = (ft²/h)	
Mercury[c]	28.9							13,546				234
Molybdenum	143	139	134	126	118	112	105	10,240	251	138	53.7	2883
Nickel	106	94	80.1	65.5	67.4	71.8	76.1	8,900	446	91	22.9	1726
Niobium	52.6	53.3	55.2	58.2	61.3	64.4	67.5	8,570	270	53.6	23.2	2741
Palladium	75.5	75.5	75.5	75.5	75.5	75.5		12,020	247	75.5	25.4	1825
Platinum	72.4	71.5	71.6	73.0	75.5	78.6	82.6	21,450	133	71.4	25.0	2042
Potassium	104	104	52					860	741	103	161.6	337
Rhenium	51	48.6	46.1	44.2	44.1	44.6	45.7	21,100	137	48.1	16.6	3453
Rhodium	154	151	146	136	127	121	115	12,450	248	150	48.6	2233
Rubidium	58.9	58.3						1,530	348	58.2	109.3	312
Silicon	264	168	98.9	61.9	42.2	31.2	25.7	2,330	703	153	93.4	1685
Silver	403	428	420	405	389	374	358	10,500	234	427	173.8	1234
Sodium	138	135						971	1206	133	113.6	371
Tantalum	57.5	57.4	57.8	58.6	59.4	60.2	61	16,600	138	57.5	25.1	3269
Tin[c]	73.3	68.2	62.2					5,750	227	67.0	51.3	505
Titanium[c]	24.5	22.4	20.4	19.4	19.7	20.7	22	4,500	611	22.0	8.0	1953
Tungsten[c]	197	182	162	139	128	121	115	19,300	134	179	69.2	3653
Uranium[c]	25.1	27	29.6	34	38.8	43.9	49	19,070	113	27.4	12.7	1407
Vanadium	31.5	31.3	32.1	34.2	36.3	38.6	41.2	6,100	502	31.4	10.3	2192
Zinc	123	122	116	105				7,140	385	121	44.0	693
Zirconium[c]	25.2	23.2	21.6	20.7	21.6	23.7	25.7	6,570	272	22.8	12.8	2125

[a]Purity for all elements exceeds 99%.

[b]The expected percent errors in the thermal conductivity values are approximately within ±5% of the true values near room temperature and within about ±10% at other temperatures.

[c]For crystalline materials, the values are given for the polycrystalline materials.

Source: E. R. G. Eckert and R. M. Drake, *Analysis of Heat and Mass Transfer*, McGraw-Hill, New York, 1972; K. Raznjevič, *Handbook of Thermodynamic Tables and Charts*, 3rd ed., McGraw-Hill, New York, 1976; Y. S. Touloukian, 3rd ed., *Thermophysical Properties of Matter*, IFI/Plenum, New York, 1970.

Thermodyanamic Properties of Liquids

TABLE 13 Water at saturation pressure

Temperature, T			Density, ρ (kg/m^3)	Coefficient of Thermal Expansion, $\beta \times 10^4$ (1/K)	Specific Heat, c_p (J/kg K)	Thermal Conductivity, k (W/m K)	Thermal Diffusivity, $\alpha \times 10^6$ (m^2/s)	Absolute Viscosity, $\mu \times 10^6$ ($N\ s/m^6$)	Kinematic Viscosity, $\nu \times 10^6$ (m^2/s)	Prandtl Number, Pr	$\frac{g\beta}{\nu^2} \times 10^{-9}$ ($1/K\ m^3$)
°F	K	°C	$\times 6.243 \times 10^{-2}$ = (lb_m/ft^3)	$\times 0.5556$ = (1/R)	$\times 2.388 \times 10^{-4}$ = (Btu/lb_m °F)	$\times 0.5777$ = (Btu/h ft °F)	$\times 3.874 \times 10^4$ = (ft^2/h)	$\times 0.6720$ = ($lb_m/ft\ s$)	$\times 3.874 \times 10^4$ = (ft^2/h)		$\times 1.573 \times 10^{-2}$ = ($1/R\ ft^3$)
32	273	0	999.9	−0.7	4226	0.558	0.131	1794	1.789	13.7	—
41	278	5	1000	—	4206	0.568	0.135	1535	1.535	11.4	—
50	283	10	999.7	0.95	4195	0.577	0.137	1296	1.300	9.5	0.551
59	288	15	999.1	—	4187	0.585	0.141	1136	1.146	8.1	—
68	293	20	998.2	2.1	4182	0.597	0.143	993	1.006	7.0	2.035
77	298	25	997.1	—	4178	0.606	0.146	880.6	0.884	6.1	—
86	303	30	995.7	3.0	4176	0.615	0.149	792.4	0.805	5.4	4.540
95	308	35	994.1	—	4175	0.624	0.150	719.8	0.725	4.8	—
104	313	40	992.2	3.9	4175	0.633	0.151	658.0	0.658	4.3	8.833
113	318	45	990.2	—	4176	0.640	0.155	605.1	0.611	3.9	—
122	323	50	988.1	4.6	4178	0.647	0.157	555.1	0.556	3.55	14.59
167	348	75	974.9	—	4190	0.671	0.164	376.6	0.366	2.23	—
212	373	100	958.4	7.5	4211	0.682	0.169	277.5	0.294	1.75	85.09
248	393	120	943.5	8.5	4232	0.685	0.171	235.4	0.244	1.43	140.0
284	413	140	926.3	9.7	4257	0.684	0.172	201.0	0.212	1.23	211.7
320	433	160	907.6	10.8	4285	0.680	0.173	171.6	0.191	1.10	290.3
356	453	180	886.6	12.1	4396	0.673	0.172	152.0	0.173	1.01	396.5
392	473	200	862.8	13.5	4501	0.665	0.170	139.3	0.160	0.95	517.2
428	493	220	837.0	15.2	4605	0.652	0.167	124.5	0.149	0.90	671.4
464	513	240	809.0	17.2	4731	0.634	0.162	113.8	0.141	0.86	848.5
500	533	260	779.0	20.0	4982	0.613	0.156	104.9	0.135	0.86	1076
536	553	280	750.0	23.8	5234	0.588	0.147	98.07	0.131	0.89	1360
572	573	300	712.5	29.5	5694	0.564	0.132	92.18	0.128	0.98	1766

(Continued)

TABLE 13 (*Continued*)

Saturation Temperature T			Saturation Pressure $p \times 10^{-5}$ (N/m^2)	Specific Volume of Vapor v_g(m^3/kg)	Enthalpy h_f (kJ/kg)	Enthalpy h_g (kJ/kg)	Enthalpy h_{fg} (kJ/kg)
°F	K	°C	$\times 1.450 \times 10^{-4}$ = (psi)	$\times 16.02$ = (ft^3/lb$_m$)	$\times 0.430$ = (Btu/lb$_m$)	$\times 0.430$ = (Btu/lb$_m$)	$\times 0.430$ = (Btu/lb$_m$)
32	273	0	0.0061	206.3	−0.04	2501	2501
50	283	10	0.0122	106.4	41.99	2519	2477
68	293	20	0.0233	57.833	83.86	2537	2453
86	303	30	0.0424	32.929	125.66	2555	2430
104	313	40	0.0737	19.548	167.45	2574	2406
122	323	50	0.1233	12.048	209.26	2591	2382
140	333	60	0.1991	7.680	251.09	2609	2358
158	343	70	0.3116	5.047	292.97	2626	2333
176	353	80	0.4735	3.410	334.92	2643	2308
194	363	90	0.7010	2.362	376.94	2660	2283
212	373	100	1.0132	1.673	419.06	2676	2257
248	393	120	1.9854	0.892	503.7	2706	2202
284	413	140	3.6136	0.508	589.1	2734	2144
320	433	160	6.1804	0.306	675.5	2757	2082
356	453	180	10.027	0.193	763.1	2777	2014
392	473	200	15.551	0.127	852.4	2791	1939
428	493	220	23.201	0.0860	943.7	2799	1856
464	513	240	33.480	0.0596	1037.6	2801	1764
500	533	260	46.940	0.0421	1135.0	2795	1660
536	553	280	64.191	0.0301	1237.0	2778	1541
572	573	300	85.917	0.0216	1345.4	2748	1403

Source: K. Raznjevič, *Handbook of Thermodynamic Tables and Charts,* McGraw-Hill, New York, 1976.

TABLE 14 Freon 12 (CCL_2F_2), saturated liquid

Temperature, T			Density, ρ (kg/m³)	Coefficient of Thermal Expansion, $\beta \times 10^3$ (1/K)	Specific Heat, c_p (J/kg K)	Thermal Conductivity, k (W/m K)	Thermal Diffusivity, $\alpha \times 10^8$ (m²/s)	Absolute Viscosity, $\mu \times 10^4$ (N s/m²)	Kinematic Viscosity, $\nu \times 10^6$ (m²/s)	Prandtl Number, Pr	$\frac{g\beta}{\nu^2} \times 10^{-10}$ (1/K m³)
°F	K	°C	$\times 6.243 \times 10^{-2}$ = (lb_m/ft^3)	$\times 0.5556$ = (1/R)	$\times 2.388 \times 10^{-4}$ = (Btu/lb_m °F)	$\times 0.5777$ = (Btu/h ft °F)	$\times 3.874 \times 10^4$ = (ft²/h)	$\times 0.6720$ = (lb_m/ft s)	$\times 3.874 \times 10^4$ = (ft²/h)		$\times 1.573 \times 10^{-2}$ = (1/R ft³)
−58	223	−50	1547	2.63	875.0	0.067	5.01	4.796	0.310	6.2	26.84
−40	233	−40	1519		884.7	0.069	5.14	4.238	0.279	5.4	
−22	243	−30	1490		895.6	0.069	5.26	3.770	0.253	4.8	
−4	253	−20	1461		907.3	0.071	5.39	3.433	0.235	4.4	
14	263	−10	1429		920.3	0.073	5.50	3.158	0.221	4.0	
32	273	0	1397	3.10	934.5	0.073	5.57	2.990	0.214	3.8	6.68
50	283	10	1364		949.6	0.073	5.60	2.769	0.203	3.6	
68	293	20	1330		965.9	0.073	5.60	2.633	0.198	3.5	
86	303	30	1295		983.5	0.071	5.60	2.512	0.194	3.5	
104	313	40	1257		1001.9	0.069	5.55	2.401	0.191	3.5	
122	323	50	1216		1021.6	0.067	5.45	2.310	0.190	3.5	

Source: E. R. G. Eckert and R. M. Drake, *Analysis of Heat and Mass Transfer,* McGraw-Hill, New York, 1972.

TABLE 15 R-134a ($C_2H_2F_4$), saturated liquid

Temperature, T			Density, ρ (kg/m^3)	Coefficient of Thermal Expansion, $\beta \times 10^3$ $(1/K)$	Specific Heat, c_p (J/kg K)	Thermal Conductivity, k (W/m K)	Thermal Diffusivity, $\alpha \times 10^8$ (m^2/s)	Absolute Viscosity, $\mu \times 10^4$ $(N\ s/m^2)$	Kinematic Viscosity, $\nu \times 10^6$ (m^2/s)	Prandtl Number, Pr	$\frac{g\beta}{\nu^2} \times 10^{-10}$ $(1/K\ m^3)$
°F	K	°C	$\times 6.243 \times 10^{-2}$ $= (lb_m/ft^3)$	$\times 0.5556$ $= (1/R)$	$\times 2.388 \times 10^{-4}$ $= (Btu/lb_m\ °F)$	$\times 0.5777$ = (Btu/h ft °F)	$\times 3.874 \times 10^4$ $= (ft^2/h)$	$\times 0.6720$ $= (lb_m/ft\ s)$	$\times 3.874 \times 10^4$ $= (ft^2/h)$		$\times 1.573 \times 10^{-2}$ $= (1/R\ ft^3)$
−58	223	−50	1446	1.96	1238	0.116	6.46	5.551	0.384	5.9	13.03
−40	233	−40	1418	2.05	1255	0.111	6.21	4.722	0.333	5.4	18.11
−22	243	−30	1388	2.14	1273	0.106	5.99	4.064	0.293	4.9	24.45
−4	253	−20	1358	2.28	1293	0.101	5.76	3.53	0.260	4.5	33.03
14	263	−10	1327	2.43	1316	0.097	5.53	3.086	0.233	4.2	43.99
32	273	0	1295	2.59	1341	0.092	5.30	2.711	0.209	4.0	57.98
50	283	10	1261	2.81	1370	0.088	5.07	2.388	0.189	3.7	76.73
68	293	20	1225	3.08	1405	0.083	4.84	2.107	0.172	3.6	102.00
86	303	30	1188	3.43	1446	0.079	4.60	1.858	0.156	3.4	137.52
104	313	40	1147	3.91	1498	0.075	4.35	1.634	0.142	3.3	189.00
122	323	50	1102	4.59	1566	0.070	4.08	1.431	0.130	3.2	266.92

Source: ASHRAE Handbook, ASHRAE Inc., Atlanta, GA, 2007.

TABLE 16 Ammonia (NH_3), saturated liquid

Temperature, T			Density, ρ (kg/m^3)	Coefficient of Thermal Expansion, $\beta \times 10^3$ (1/K)	Specific Heat, c_p (J/kg K)	Thermal Conductivity, k (W/m K)	Thermal Diffusivity, $\alpha \times 10^8$ (m^2/s)	Absolute Viscosity, $\mu \times 10^4$ (N s/m^2)	Kinematic Viscosity, $\nu \times 10^6$ (m^2/s)	Prandtl Number, Pr	$\frac{g\beta}{\nu^2} \times 10^{-10}$ (1/K m^3)
°F	K	°C	$\times 6.243 \times 10^{-2}$ = (lb$_m$/ft^3)	$\times 0.5556$ = (1/R)	$\times 2.388 \times 10^{-4}$ = (Btu/lb$_m$ °F)	$\times 0.5777$ = (Btu/h ft °F)	$\times 3.874 \times 10^4$ = (ft^2/h)	$\times 0.6720$ = (lb$_m$/ft s)	$\times 3.874 \times 10^4$ = (ft^2/h)		$\times 1.573 \times 10^{-2}$ = (1/R ft^3)
−58	223	−50	703.7		4463	0.547	17.42	3.061	0.435	2.60	
−40	233	−40	691.7		4467	0.547	17.75	2.808	0.406	2.28	
−22	243	−30	679.3		4476	0.549	18.01	2.629	0.387	2.15	
−4	253	−20	666.7		4509	0.547	18.19	2.540	0.381	2.09	
14	263	−10	653.6		4564	0.543	18.25	2.471	0.378	2.07	
32	273	0	640.1	2.16	4635	0.540	18.19	2.388	0.373	2.05	1.51
50	283	10	626.2		4714	0.531	18.01	2.304	0.368	2.04	
68	293	20	611.8	2.45	4798	0.521	17.75	2.196	0.359	2.02	18.64
86	303	30	596.4		4890	0.507	17.42	2.081	0.349	2.01	
104	313	40	581.0		4999	0.493	17.01	1.975	0.340	2.00	
122	323	50	564.3		5116	0.476	16.54	1.862	0.330	1.99	

Source: E. R. G. Eckert and R. M. Drake, *Analysis of Heat and Mass Transfer*, McGraw-Hill, New York, 1972.

TABLE 17 Unused engine oil

Temperature, T			Density, ρ (kg/m^3)	Coefficient of Thermal Expansion, $\beta \times 10^3$ (1/K)	Specific Heat, c_p (J/kg K)	Thermal Conductivity, k (W/m K)	Thermal Diffusivity, $\alpha \times 10^{10}$ (m^2/s)	Absolute Viscosity, $\mu \times 10^3$ (N s/m^2)	Kinematic Viscosity, $\nu \times 10^6$ (m^2/s)	Prandtl Number, Pr	$\frac{g\beta}{\nu^2}$ (1/K m^3)
°F	K	°C	$\times 6.243 \times 10^{-2}$ = (lb$_m$/ft^3)	$\times 0.5556$ = (1/R)	$\times 2.388 \times 10^{-4}$ = (Btu/lb$_m$ °F)	$\times 0.5777$ = (Btu/h ft °F)	$\times 3.874 \times 10^4$ = (ft^2/h)	$\times 0.6720$ = (lb$_m$/ft s)	$\times 3.874 \times 10^4$ = (ft^2/h)		$\times 1.573 \times 10^{-2}$ = (1/R ft^3)
32	273	0	899.1		1796	0.147	911	3848	4280	471	
68	293	20	888.2	0.648	1880	0.145	872	799	900	104	7.85×10^3
104	313	40	876.1	0.691	1964	0.144	834	210	240	28.7	1.18×10^5
140	333	60	864.0	0.697	2047	0.140	800	72.5	83.9	10.5	9.72×10^5
176	353	80	852.0	0.704	2131	0.138	769	32.0	37.5	4.90	4.91×10^6
212	373	100	840.0	0.684	2219	0.137	738	17.1	20.3	2.76	1.63×10^7
248	393	120	829.0	0.697	2307	0.135	710	10.3	12.4	1.75	4.44×10^7
284	413	140	816.9	0.706	2395	0.133	686	6.54	8.0	1.16	1.08×10^8
320	433	160	805.9		2483	0.132	663	4.51	5.6	0.84	—

Source: E. R. G. Eckert and R. M. Drake, *Analysis of Heat and Mass Transfer*, McGraw-Hill, New York, 1972.

TABLE 18 Transformer oil (Standard 982-68)

Temperature, T			Density, ρ (kg/m^3)	Coefficient of Thermal Expansion, $\beta \times 10^3$ (1/K)	Specific Heat, c_p (J/kg K)	Thermal Conductivity, k (W/m K)	Thermal Diffusivity, $\alpha \times 10^{10}$ (m^2/s)	Absolute Viscosity, $\mu \times 10^3$ (N s/m^2)	Kinematic Viscosity, $\nu \times 10^6$ (m^2/s)	Prandtl Number, Pr $\times 10^{-2}$
°F	K	°C	$\times 6.243 \times 10^{-2}$ = (lb$_m$/ft^3)	$\times 0.5556$ = (1/R)	$\times 2.388 \times 10^{-4}$ = (Btu/lb$_m$ °F)	$\times 0.5777$ = (Btu/h ft °F)	$\times 3.874 \times 10^4$ = (ft^2/h)	$\times 0.6720$ = (lb$_m$/ft s)	$\times 3.874 \times 10^4$ = (ft^2/h)	
−58	223	−50	922		1700	0.116	742	29,320	31,800	4,286
−40	233	−40	916		1680	0.116	750	3,866	4,220	563
−22	243	−30	910		1650	0.115	764	1,183	1,300	170
−4	253	−20	904		1620	0.114	778	365.6	404	52
14	263	−10	898		1600	0.113	788	108.1	120	15.3
32	273	0	891		1620	0.112	778	55.24	67.5	8.67
50	283	10	885		1650	0.111	763	33.45	37.8	4.95
68	293	20	879		1710	0.111	736	21.10	24.0	3.26
86	303	30	873		1780	0.110	707	13.44	15.4	2.18
104	313	40	867		1830	0.109	688	9.364	10.8	1.57

Source: N. B. Vargaftik, *Tables on the Thermophysical Properties of Liquids and Gases*, 2nd ed., Hemisphere, Washington, DC, 1975.

TABLE 19 n-Butyl alcohol ($C_4H_{10}O$)

Temperature, T			Density, ρ (kg/m^3)	Coefficient of Thermal Expansion, $\beta \times 10^4$ (1/K)	Specific Heat, c_p (J/kg K)	Thermal Conductivity, k (W/m K)	Thermal Diffusivity, $\alpha \times 10^{10}$ (m^2/s)	Absolute Viscosity, $\mu \times 10^3$ (N s/m^2)	Kinematic Viscosity, $\nu \times 10^6$ (m^2/s)	Prandtl Number, Pr	$\frac{g\beta}{\nu^2} \times 10^{-6}$ (1/K m^3)
°F	K	°C	$\times 6.243 \times 10^{-2}$ = (lb$_m$/ft^3)	$\times 0.5556$ = (1/R)	$\times 2.388 \times 10^{-4}$ = (Btu/lb$_m$ °F)	$\times 0.5777$ = (Btu/h ft °F)	$\times 3.874 \times 10^4$ = (ft^2/h)	$\times 0.6720$ = (lb$_m$/ft s)	$\times 3.874 \times 10^4$ = (ft^2/h)		$\times 1.573 \times 10^{-2}$ = (1/R ft^3)
60	289	16	809		2258	0.168	901	3.36	4.16	45.2	
100	311	38	796	8.1	2542	0.166	816	1.92	2.41	29.4	1367
150	339	66	777	8.6	2852	0.164	743	1.00	1.29	17.4	5086
200	366	93	756		3166	0.163	666	0.57	0.76	11.1	
243.5	390.7	117.5	737		3429	0.163	769	0.39	0.53	8.2	
300	422	149						0.28			

TABLE 20 Commercial aniline

Temperature, T			Density, ρ (kg/m^3)	Coefficient of Thermal Expansion, $\beta \times 10^4$ (1/K)	Specific Heat, c_p (J/kg K)	Thermal Conductivity, k (W/m K)	Thermal Diffusivity, $\alpha \times 10^{10}$ (m^2/s)	Absolute Viscosity, $\mu \times 10^3$ (N s/m^2)	Kinematic Viscosity, $\nu \times 10^6$ (m^2/s)	Prandtl Number, Pr	$\frac{g\beta}{\nu^2} \times 10^{-6}$ (1/K m^3)
°F	K	°C	$\times 6.243 \times 10^{-2}$ = (lb$_m$/ft^3)	$\times 0.5556$ = (1/R)	$\times 2.388 \times 10^{-4}$ = (Btu/lb$_m$ °F)	$\times 0.5777$ = (Btu/h ft °F)	$\times 3.874 \times 10^4$ = (ft^2/h)	$\times 0.6720$ = (lb$_m$/ft s)	$\times 3.874 \times 10^4$ = (ft^2/h)		$\times 1.573 \times 10^{-2}$ = (1/R ft^3)
60	289	16	1025		2011	0.173	839	4.84	4.72	56.0	
100	311	38	1009	8.82	2052	0.173	837	2.53	2.51	30.0	1373
150	339	66	985	8.86	2115	0.170	816	1.44	1.46	18.0	4100
200	366	93	961		2157	0.166	803	0.91	0.947	11.8	
300	422	149	921		2261	0.161	775	0.48	0.521	6.8	

TABLE 21 Benzene (C_6H_6)

Temperature, T °F	K	°C	Density, ρ (kg/m³)	Coefficient of Thermal Expansion, $\beta \times 10^3$ (1/K)	Specific Heat, c_p (J/kg K)	Thermal Conductivity, k (W/m K)	Thermal Diffusivity, $\alpha \times 10^{10}$ (m²/s)	Absolute Viscosity, $\mu \times 10^3$ (N s/m²)	Kinematic Viscosity, $\nu \times 10^6$ (m²/s)	Prandtl Number, Pr	$\frac{g\beta}{\nu^2} \times 10^{-6}$ (1/K m³)
			$\times 6.243 \times 10^{-2}$ = (lb$_m$/ft³)	$\times 0.5556$ = (1/R)	$\times 2.388 \times 10^{-4}$ = (Btu/lb$_m$ °F)	$\times 0.5777$ = (Btu/h ft °F)	$\times 3.874 \times 10^4$ = (ft²/h)	$\times 0.6720$ = (lb$_m$/ft s)	$\times 3.874 \times 10^4$ = (ft²/h)		$\times 1.573 \times 10^{-2}$ = (1/R ft³)
60	239	16	883	1.08	1675	0.161	1089	0.685	0.776	7.2	19,072
80	300	27	875		1759	0.159	1035	0.589	0.673	6.5	
100	311	38	865		1843	0.151	911	0.522	0.604	5.1	
150	339	66	857		1926			0.387	0.452	4.5	
200	366	93						0.302		4.0	

TABLE 22 Organic compounds at 20°C, 68° F

Liquid	Chemical Formula	Density, ρ (kg/m³)	Coefficient of Thermal Expansion, $\beta \times 10^4$ (1/K)	Specific Heat, c_p (J/kg K)	Thermal Conductivity, k (W/m K)	Thermal Diffusivity, $\alpha \times 10^9$ (m²/s)	Absolute Viscosity, $\mu \times 10^4$ (N s/m²)	Kinematic Viscosity, $\nu \times 10^6$ (m²/s)	Prandtl Number, Pr	$\frac{g\beta}{\nu^2} \times 10^{-8}$ (1/K m³)
		$\times 6.243 \times 10^{-2}$ = (lb$_m$/ft³)	$\times 0.5556$ = (1/R)	$\times 2.388 \times 10^{-4}$ = (Btu/lb$_m$ °F)	$\times 0.5777$ = (Btu/h ft °F)	$\times 3.874 \times 10^4$ = (ft²/h)	$\times 0.6720$ = (lb$_m$/ft s)	$\times 3.874 \times 10^4$ = (ft²/h)		$\times 1.573 \times 10^{-2}$ = (1/R ft³)
Acetic acid	$C_2H_4O_2$	1049	10.7	2031	0.193	90.6				
Acetone	C_3H_6O	791	14.3	2160	0.180	105.4	3.31	0.418	3.97	802.6
Chloroform	$CHCl_3$	1489	12.8	967	0.129	89.6	5.8	0.390	4.35	825.3
Ethyl acetate	$C_4H_8O_2$	900	13.8	2010	0.137	75.7	4.49	0.499	6.59	543.5
Ethyl alcohol	C_2H_6O	790	11.0	2470	0.182	93.3	12.0	1.52	16.29	46.7
Ethylene glycol	$C_2H_6O_2$	1115		2382	0.258	97.1	199	17.8	183.7	
Glycerol	$C_3H_8O_3$	1260	5.0	2428	0.285	93.2	14,800	1175	12,609	0.0000355
n-Heptane	C_7H_{16}	684	12.4	2219	0.140	92.2	4.09	0.598	6.48	340.1
n-Hexane	C_6H_{14}	660	13.5	1884	0.137	11.02	3.20	0.485	4.40	562.8
Isobutyl alcohol	$C_4H_{10}O$	804	9.4	2303	0.134	72.4	39.5	4.91	67.89	3.82
Methyl alcohol	CH_4O	792	11.9	2470	0.212	108.4	5.84	0.737	6.80	214.9
n-Octane	C_8H_{18}	720	11.4	2177	0.147	93.8	5.4	0.750	8.00	198.8
n-Pentane	C_5H_{12}	626	16.0	2177	0.136	99.8	2.29	0.366	3.67	1171
Toluene	C_7H_8	866	10.8	1675	0.151	104.1	5.86	0.677	6.50	231.1
Turpentine	$C_{10}H_{16}$	855	9.7	1800	0.128	83.2	14.87	1.74	20.91	31.4

Source: K. Raznjevič, *Handbook of Thermodynamic Tables and Charts*, McGraw-Hill, New York, 1976.

Heat Transfer Fluids

TABLE 23 Mobiltherm 600

Temperature, T			Density, ρ (kg/m^3)	Coefficient of Thermal Expansion, $\beta \times 10^3$ (1/K)	Specific Heat, c_p (J/kg K)	Thermal Conductivity, k (W/m K)	Thermal Diffusivity, $\alpha \times 10^{10}$ (m^2/s)	Absolute Viscosity, $\mu \times 10^3$ ($N\ s/m^2$)	Kinematic Viscosity, $\nu \times 10^6$ (m^2/s)	Prandtl Number, Pr	$\frac{g\beta}{\nu^2} \times 10^{-6}$ ($1/K\ m^3$)
°F	K	°C	$\times 6.243 \times 10^{-2}$ = (lb_m/ft^3)	$\times 0.5556$ = (1/R)	$\times 2.388 \times 10^{-4}$ = (Btu/lb_m °F)	$\times 0.5777$ = (Btu/h ft °F)	$\times 3.874 \times 10^4$ = (ft^2/h)	$\times 0.6720$ = ($lb_m/ft\ s$)	$\times 3.874 \times 10^4$ = (ft^2/h)		$\times 1.573 \times 10^{-2}$ = ($1/R\ ft^3$)
50	283	10	953	0.621	1549	0.123	833				
122	323	50	929	0.637	1680	0.120	769	30.28	32.60	424	5.9
212	373	100	899	0.658	1859	0.116	694	5.48	6.10	87.9	173
302	423	150	870	0.680	2031	0.113	640	2.04	2.34	36.6	1218
392	473	200	839	0.705	2209	0.110	594	1.05	1.25	21.0	4425
482	523	250	810	0.730	2386	0.106	545	0.64	0.790	14.5	11,470

Source: P. L. Geiringer, *Handbook of Heat Transfer Media*, Krieger, New York, 1977.

TABLE 24 Molten nitrate salt (60% $NaNO_3$, 40% KNO_3, by weight)

Temperature, T		Density, ρ (kg/m^3)	Coefficient of Thermal Expansion, $\beta \times 10^4$ (1/K)	Specific Heat, c_p (J/kg K)	Thermal Conductivity, k (W/m K)	Thermal Diffusivity, $\alpha \times 10^7$ (m^2/s)	Absolute Viscosity, $\mu \times 10^3$ (N s/m^2)	Kinematic Viscosity, $\nu \times 10^6$ (m^2/s)	Prandtl Number, Pr	$\frac{g\beta}{\nu^2} \times 10^{-9}$ (1/K m^3)
°F	°C	$\times 6.243 \times 10^{-2}$ = (lb_m/ft^3)	× 0.5556 = (1/R)	$\times 2.388 \times 10^{-4}$ = (Btu/lb_m °F)	× 0.5777 = (Btu/h ft °F)	$\times 3.874 \times 10^4$ = (ft^2/h)	×0.6720 = (lb_m/ft s)	$\times 3.874 \times 10^4$ = (ft^2/h)		$\times 1.573 \times 10^{-2}$ = (1/R ft^3)
572	300	1899	3.370	1495	0.500	1.761	3.26	1.717	9.747	1.122
662	350	1867	3.321	1503	0.510	1.817	2.34	1.253	6.896	2.074
752	400	1836	3.486	1512	0.519	1.870	1.78	0.969	5.186	3.638
842	450	1804	3.548	1520	0.529	1.929	1.47	0.815	4.224	5.241
932	500	1772	3.612	1529	0.538	1.986	1.31	0.739	3.723	6.483
1022	550	1740	3.678	1538	0.548	2.048	1.19	0.684	3.340	7.714
1112	600	1708		1546	0.557	2.109	0.99	0.580	2.748	

Source: Sandia National Laboratories, SAND87-8005, "A Review of the Chemical and Physical Properties of Molten Alkali Nitrate Salts and Their Effect on Materials Used for Solar Central Receivers," 1987.

Liquid Metals

TABLE 25 Bismuth

Temperature, T			Density, ρ (kg/m^3)	Coefficient of Thermal Expansion, $\beta \times 10^3$ (1/K)	Specific Heat, c_p (J/kg K)	Thermal Conductivity, k (W/m K)	Thermal Diffusivity, $\alpha \times 10^5$ (m^2/s)	Absolute Viscosity, $\mu \times 10^4$ (N s/m^2)	Kinematic Viscosity, $\nu \times 10^7$ (m^2/s)	Prandtl Number, Pr	$\frac{g\beta}{\nu^2} \times 10^{-9}$ (1/K m^3)
°F	K	°C	$\times 6.243 \times 10^{-2}$ = (lb_m/ft^3)	× 0.5556 = (1/R)	$\times 2.388 \times 10^{-4}$ = (Btu/lb_m °F)	× 0.5777 = (Btu/h ft °F)	$\times 3.874 \times 10^4$ = (ft^2/h)	×0.6720 = (lb_m/ft s)	$\times 3.874 \times 10^4$ = (ft^2/h)		$\times 1.573 \times 10^{-2}$ = (1/R ft^3)
600	589	316	10,011	0.117	144.5	16.44	1.14	16.22	1.57	0.014	46.5
800	700	427	9,867	0.122	149.5	15.58	1.06	13.39	1.35	0.013	65.6
1000	811	538	9,739	0.126	154.5	15.58	1.03	11.01	1.08	0.011	106
1200	922	649	9,611		159.5	15.58	1.01	9.23	0.903	0.009	
1400	1033	760	9,467		164.5	15.58	1.01	7.89	0.813	0.008	

TABLE 26 Mercury

Temperature, T °F	K	°C	Density, ρ (kg/m^3) × 6.243 × 10^{-2} = (lb$_m$/ft^3)	Coefficient of Thermal Expansion, $\beta \times 10^4$ (1/K) × 0.5556 = (1/R)	Specific Heat, c_p (J/kg K) × 2.388 × 10^{-4} = (Btu/lb$_m$ °F)	Thermal Conductivity, k (W/m K) × 0.5777 = (Btu/h ft °F)	Thermal Diffusivity, $\alpha \times 10^{10}$ (m^2/s) × 3.874 × 10^4 = (ft^2/h)	Absolute Viscosity, $\mu \times 10^4$ (N s/m^2) ×0.6720 = (lb$_m$/ft s)	Kinematic Viscosity, $\nu \times 10^6$ (m^2/s) × 3.874 × 10^4 = (ft^2/h)	Prandtl Number, Pr	$\frac{g\beta}{\nu^2} \times 10^{-10}$ (1/K m^3) × 1.573 × 10^{-2} = (1/R ft^3)
32	273	0	13,628		140.3	8.20	42.99	16.90	0.124	0.0288	
68	293	20	13,579	1.82	139.4	8.69	46.06	15.48	0.114	0.0249	13.73
122	323	50	13,506		138.6	9.40	50.22	14.05	0.104	0.0207	
212	373	100	13,385		137.3	10.51	57.16	12.42	0.0928	0.0162	
302	423	150	13,264		136.5	11.49	63.54	11.31	0.0853	0.0134	
392	473	200	13,145		157.0	12.34	69.08	10.54	0.0802	0.0134	
482	523	250	13,026		135.7	13.07	74.06	9.96	0.0765	0.0103	
600	588.7	315.5	12,847		134.0	14.02	81.50	8.65	0.0673	0.0083	

Source: E. R. G. Eckert and R. M. Drake, *Analysis of Heat and Mass Transfer*, McGraw-Hill, New York, 1972.

TABLE 27 Sodium

Temperature, T °F	K	°C	Density, ρ (kg/m^3) × 6.243 × 10^{-2} = (lb$_m$/ft^3)	Coefficient of Thermal Expansion, $\beta \times 10^3$ (1/K) × 0.5556 = (1/R)	Specific Heat, c_p (J/kg K) × 2.388 × 10^{-4} = (Btu/lb$_m$ °F)	Thermal Conductivity, k (W/m K) × 0.5777 = (Btu/h ft °F)	Thermal Diffusivity, $\alpha \times 10^5$ (m^2/s) × 3.874 × 10^4 = (ft^2/h)	Absolute Viscosity, $\mu \times 10^4$ (N s/m^2) ×0.6720 = (lb$_m$/ft s)	Kinematic Viscosity, $\nu \times 10^7$ (m^2/s) × 3.874 × 10^4 = (ft^2/h)	Prandtl Number, Pr	$\frac{g\beta}{\nu^2} \times 10^{-9}$ (1/K m^3) × 1.573 × 10^{-2} = (1/R ft^3)
200	367	94	929	0.27	1382	86.2	6.71	6.99	7.31	0.0110	4.96
400	478	205	902	0.36	1340	80.3	6.71	4.32	4.60	0.0072	16.7
700	644	371	860		1298	72.4	6.45	2.83	3.16	0.0051	
1000	811	538	820		1256	65.4	6.19	2.08	2.44	0.0040	
1300	978	705	778		1256	59.7	6.19	1.79	2.26	0.0038	

Thermodynamic Properties of Gases

TABLE 28 Dry air at atmospheric pressure

Temperature, T			Density, ρ (kg/m^3)	Coefficient of Thermal Expansion, $\beta \times 10^3$ (1/K)	Specific Heat, c_p (J/kg K)	Thermal Conductivity, k (W/m K)	Thermal Diffusivity, $\alpha \times 10^6$ (m^2/s)	Absolute Viscosity, $\mu \times 10^6$ (N s/m^2)	Kinematic Viscosity, $\nu \times 10^6$ (m^2/s)	Prandtl Number, Pr	$\frac{g\beta}{\nu^2} \times 10^{-8}$ (1/K m^3)
°F	K	°C	$\times 6.243 \times 10^{-2}$ = (lb$_m$/ft^3)	$\times 0.5556$ = (1/R)	$\times 2.388 \times 10^{-4}$ = (Btu/lb$_m$ °F)	$\times 0.5777$ = (Btu/h ft °F)	$\times 3.874 \times 10^4$ = (ft^2/h)	$\times 0.6720$ = (lb$_m$/ft s)	$\times 3.874 \times 10^4$ = (ft^2/h)		$\times 1.573 \times 10^{-2}$ = (1/R ft^3)
32	273	0	1.252	3.66	1011	0.0237	19.2	17.456	13.9	0.71	1.85
68	293	20	1.164	3.41	1012	0.0251	22.0	18.240	15.7	0.71	1.36
104	313	40	1.092	3.19	1014	0.0265	24.8	19.123	17.6	0.71	1.01
140	333	60	1.025	3.00	1017	0.0279	27.6	19.907	19.4	0.71	0.782
176	353	80	0.968	2.83	1019	0.0293	30.6	20.790	21.5	0.71	0.600
212	373	100	0.916	2.68	1022	0.0307	33.6	21.673	23.6	0.71	0.472
392	473	200	0.723	2.11	1035	0.0370	49.7	25.693	35.5	0.71	0.164
572	573	300	0.596	1.75	1047	0.0429	68.9	29.322	49.2	0.71	0.0709
752	673	400	0.508	1.49	1059	0.0485	89.4	32.754	64.6	0.72	0.0350
932	773	500	0.442	1.29	1076	0.0540	113.2	35.794	81.0	0.72	0.0193
1832	1273	1000	0.268	0.79	1139	0.0762	240	48.445	181	0.74	0.00236

Source: K. Raznjevič, *Handbook of Thermodynamic Tables and Charts*, McGraw-Hill, New York, 1976.

TABLE 29 Carbon dioxide at atmospheric pressure

Temperature, T			Density, ρ (kg/m^3)	Coefficient of Thermal Expansion, $\beta \times 10^3$ (1/K)	Specific Heat, c_p (J/kg K)	Thermal Conductivity, k (W/m K)	Thermal Diffusivity, $\alpha \times 10^4$ (m^2/s)	Absolute Viscosity, $\mu \times 10^6$ $(N\ s/m^2)$	Kinematic Viscosity, $\nu \times 10^6$ (m^2/s)	Prandtl Number, Pr	$\frac{g\beta}{\nu^2} \times 10^{-6}$ $(1/K\ m^3)$
°F	K	°C	$\times 6.243 \times 10^{-2} = (lb_m/ft^3)$	$\times 0.5556 = (1/R)$	$\times 2.388 \times 10^{-4} = (Btu/lb_m\ °F)$	$\times 0.5777 = (Btu/h\ ft\ °F)$	$\times 3.874 \times 10^4 = (ft^2/h)$	$\times 0.6720 = (lb_m/ft\ s)$	$\times 3.874 \times 10^4 = (ft^2/h)$		$\times 1.573 \times 10^{-2} = (1/R\ ft^3)$
−63	220	−53	2.4733		783	0.01080	0.0592	11.105	4.490	0.818	
−9	250	−23	2.1657		804	0.01288	0.0740	12.590	5.813	0.793	
81	300	27	1.7973	3.33	871	0.01657	0.1058	14.958	8.321	0.770	472
171	350	77	1.5362	2.86	900	0.02047	0.1480	17.205	11.19	0.755	224
261	400	127	1.3424	2.50	942	0.02461	0.1946	19.32	14.39	0.738	118
351	450	177	1.1918	2.22	980	0.02897	0.2481	21.34	17.90	0.721	67.9
441	500	227	1.0732	2.00	1013	0.03352	0.3084	23.26	21.67	0.702	41.8
531	550	277	0.9739	1.82	1047	0.03821	0.3750	25.08	25.74	0.685	26.9
621	600	327	0.8938	1.67	1076	0.04311	0.4483	26.83	30.02	0.668	18.2

Source: E. R. G. Eckert and R. M. Drake, *Analysis of Heat and Mass Transfer*, McGraw-Hill, New York, 1972.

TABLE 30 Carbon monoxide at atmospheric pressure

Temperature, T			Density, ρ (kg/m^3)	Coefficient of Thermal Expansion, $\beta \times 10^3$ (1/K)	Specific Heat, c_p (J/kg K)	Thermal Conductivity, k (W/m K)	Thermal Diffusivity, $\alpha \times 10^4$ (m^2/s)	Absolute Viscosity, $\mu \times 10^6$ (N s/m^2)	Kinematic Viscosity, $\nu \times 10^6$ (m^2/s)	Prandtl Number, Pr	$\frac{g\beta}{\nu^2} \times 10^{-6}$ (1/K m^2)
°F	K	°C	$\times$ 6.243 $\times 10^{-2}$ = (lb$_m$/ft^3)	$\times$ 0.5556 = (1/R)	$\times$ 2.388 $\times 10^{-4}$ = (Btu/lb$_m$ °F)	$\times$ 0.5777 = (Btu/h ft °F)	$\times$ 3.874 $\times 10^4$ = (ft^2/h)	$\times$ 0.6720 = (lb$_m$/ft s)	$\times$ 3.874 $\times 10^4$ = (ft^2/h)		$\times$ 1.573 $\times 10^{-2}$ = (1/R ft^3)
−63	220	−53	2.4733		783	0.01080	0.0592	11.105	4.490	0.818	
−63	220	−53	1.554		1043	0.01906	0.1176	13.88	8.90	0.758	
−9	250	−23	0.841		1043	0.02144	0.1506	15.40	11.28	0.750	
81	300	27	1.139	3.33	1042	0.02525	0.2128	17.84	15.67	0.737	133
171	350	77	0.974	2.86	1043	0.02883	0.2836	20.09	20.62	0.728	65.9
261	400	127	0.854	2.50	1048	0.03226	0.3605	22.19	25.99	0.722	36.3
351	450	177	0.758	2.22	1055	0.04360	0.4439	24.18	31.88	0.718	21.4
441	500	227	0.682	2.00	1064	0.03863	0.5324	26.06	38.19	0.718	13.4
531	550	277	0.620	1.82	1076	0.04162	0.6240	27.89	44.97	0.721	8.83
621	600	327	0.569	1.67	1088	0.04446	0.7190	29.60	52.06	0.724	6.04

Source: E. R. G. Eckert and R. M. Drake, *Analysis of Heat and Mass Transfer*, McGraw-Hill, New York, 1972.

TABLE 31 Helium at atmospheric pressure

Temperature, T			Density, ρ (kg/m^3)	Coefficient of Thermal Expansion, $\beta \times 10^3$ (1/K)	Specific Heat, c_p (J/kg K)	Thermal Conductivity, k (W/m K)	Thermal Diffusivity, $\alpha \times 10^4$ (m^2/s)	Absolute Viscosity, $\mu \times 10^6$ (N s/m^2)	Kinematic Viscosity, $\nu \times 10^6$ (m^2/s)	Prandtl Number, Pr	$\frac{g\beta}{\nu^2} \times 10^{-6}$ (1/K m^3)
°F	K	°C	$\times 6.243 \times 10^{-2}$ = (lb$_m$/ft^3)	$\times 0.5556$ = (1/R)	$\times 2.388 \times 10^{-4}$ = (Btu/lb$_m$ °F)	$\times 0.5777$ = (Btu/h ft °F)	$\times 3.874 \times 10^4$ = (ft^2/h)	$\times 0.6720$ = (lb$_m$/ft s)	$\times 3.874 \times 10^4$ = (ft^2/h)		$\times 1.573 \times 10^{-2}$ = (1/R ft^3)
−454	3	−270			5200	0.0106		0.842			
−400	33	−240	1.466		5200	0.0353	0.04625	5.02	3.42	0.74	
−200	144	−129	3.380	6.94	5200	0.0928	0.5275	12.55	37.11	0.70	49.4
−100	200	−73	0.2435	5.00	5200	0.1177	0.9288	15.66	64.38	0.694	11.8
0	255	−18	0.1906	3.92	5200	0.1357	1.3675	18.17	95.50	0.70	4.22
200	366	93	0.1328	2.73	5200	0.1691	2.449	23.05	173.6	0.71	0.888
400	477	204	0.1020	2.10	5200	0.197	3.716	27.50	269.3	0.72	0.284
600	589	316	0.08282	1.70	5200	0.225	5.215	31.13	375.8	0.72	0.118
800	700	427	0.07032	1.43	5200	0.251	6.661	34.75	494.2	0.72	0.0574
981	800	527	0.06023	1.25	5200	0.275	8.774	38.17	634.1	0.72	0.0305
1161	900	627	0.05286	1.11	5200	0.298	10.834	41.36	781.3	0.72	0.0178

Source: E. R. G. Eckert and R. M. Drake, *Analysis of Heat and Mass Transfer*, McGraw-Hill, New York, 1972.

TABLE 32 Hydrogen at atmospheric pressure

Temperature, T			Density, ρ (kg/m^3)	Coefficient of Thermal Expansion, $\beta \times 10^3$ (1/K)	Specific Heat, c_p (J/kg K)	Thermal Conductivity, k (W/m K)	Thermal Diffusivity, $\alpha \times 10^4$ (m^2/s)	Absolute Viscosity, $\mu \times 10^6$ (N s/m^2)	Kinematic Viscosity, $\nu \times 10^6$ (m^2/s)	Prandtl Number, Pr	$\frac{g\beta}{\nu^2} \times 10^{-6}$ (1/K m^3)
°F	K	°C	$\times 6.243 \times 10^{-2}$ = (lb$_m$/ft^3)	$\times 0.5556$ = (1/R)	$\times 2.388 \times 10^{-4}$ = (Btu/lb$_m$ °F)	$\times 0.5777$ = (Btu/h ft °F)	$\times 3.874 \times 10^4$ = (ft^2/h)	$\times 0.6720$ = (lb$_m$/ft s)	$\times 3.874 \times 10^4$ = (ft^2/h)		$\times 1.573 \times 10^{-2}$ = (1/R ft^3)
−369	50	−223	0.50955		10,501	0.0362	0.0676	2.516	4.880	0.721	
−279	100	−173	0.24572	10.0	11,229	0.0665	0.2408	4.212	17.14	0.712	333.8
−189	150	−123	0.16371	6.67	12,602	0.0981	0.475	5.595	34.18	0.718	55.99
−100	200	−73	0.12270	5.00	13,540	0.1282	0.772	6.813	55.53	0.719	15.90
−9	250	−23	0.09819	4.00	14,059	0.1561	1.130	7.919	80.64	0.713	6.03
81	300	27	0.08185	3.33	14,314	0.182	1.554	8.963	109.5	0.706	2.72
171	350	77	0.07016	2.86	14,436	0.206	2.031	9.954	141.9	0.697	1.39
261	400	127	0.06135	2.50	14,491	0.228	2.568	10.864	177.1	0.690	0.782
351	450	177	0.05462	2.22	14,499	0.251	3.164	11.779	215.6	0.682	0.468
441	500	227	0.04918	2.00	14,507	0.272	3.817	12.636	257.0	0.675	0.297
621	600	327	0.04085	1.67	14,537	0.315	5.306	14.285	349.7	0.664	0.134
800	700	427	0.03492	1.43	14,574	0.351	6.903	15.89	455.1	0.659	0.0677
981	800	527	0.03060	1.25	14,675	0.384	8.563	17.40	569	0.664	0.0379
1341	1000	727	0.02451	1.00	14,968	0.440	11.997	20.16	822	0.686	0.0145
2192	1200	927	0.02050	0.833	15,366	0.488	15.484	22.75	1107	0.715	0.00667

Source: E. R. G. Eckert and R. M. Drake, *Analysis of Heat and Mass Transfer,* McGraw-Hill, New York, 1972.

TABLE 33 Nitrogen at atmospheric pressure

Temperature, T			Density, ρ (kg/m³)	Coefficient of Thermal Expansion, $\beta \times 10^3$ (1/K)	Specific Heat, c_p (J/kg K)	Thermal Conductivity, k (W/m K)	Thermal Diffusivity, $\alpha \times 10^4$ (m²/s)	Absolute Viscosity, $\mu \times 10^6$ (N s/m²)	Kinematic Viscosity, $\nu \times 10^6$ (m²/s)	Prandtl Number, Pr	$\frac{g\beta}{\nu^2} \times 10^{-6}$ (1/K m³)
°F	K	°C	$\times 6.243 \times 10^{-2}$ = (lb$_m$/ft³)	$\times 0.5556$ = (1/R)	$\times 2.388 \times 10^{-4}$ = (Btu/lb$_m$ °F)	$\times 0.5777$ = (Btu/h ft °F)	$\times 3.874 \times 10^4$ = (ft²/h)	$\times 0.6720$ = (lb$_m$/ft s)	$\times 3.874 \times 10^4$ = (ft²/h)		$\times 1.573 \times 10^{-2}$ = (1/R ft³)
−279	100	−173	3.4808		1072	0.00945	0.0253	6.86	1.97	0.786	
−100	200	−73	1.7108	5.00	1043	0.01824	0.1022	12.95	7.57	0.747	855.6
81	300	27	1.1421	3.33	1041	0.02620	0.2204	17.84	15.63	0.713	133.7
261	400	127	0.8538	2.50	1046	0.03335	0.3734	21.98	25.74	0.691	37.00
441	500	227	0.6824	2.00	1056	0.03984	0.5530	25.70	37.66	0.684	13.83
621	600	327	0.5687	1.67	1076	0.04580	0.7486	29.11	51.19	0.686	6.25
800	700	427	0.4934	1.43	1097	0.05123	0.9466	32.13	65.13	0.691	3.31
981	800	527	0.4277	1.25	1123	0.05609	1.1685	34.84	81.46	0.700	1.85
1161	900	627	0.3796	1.11	1146	0.06070	1.3946	37.49	91.06	0.711	1.31
1341	1000	727	0.3412	1.00	1168	0.06475	1.6250	40.00	117.2	0.724	0.714
1521	1100	827	0.3108	0.909	1186	0.06850	1.8591	42.28	136.0	0.736	0.482
	1200	927	0.2851	0.833	1204	0.07184	2.0932	44.50	156.1	0.748	0.335

Source: E. R. G. Eckert and R. M. Drake, *Analysis of Heat and Mass Transfer,* McGraw-Hill, New York, 1972.

TABLE 34 Oxygen at atmospheric pressure

Temperature, T			Density, ρ (kg/m^3)	Coefficient of Thermal Expansion, $\beta \times 10^3$ (1/K)	Specific Heat, c_p (J/kg K)	Thermal Conductivity, k (W/m K)	Thermal Diffusivity, $\alpha \times 10^4$ (m^2/s)	Absolute Viscosity, $\mu \times 10^6$ (N s/m^2)	Kinematic Viscosity, $\nu \times 10^6$ (m^2/s)	Prandtl Number, Pr	$\frac{g\beta}{\nu^2} \times 10^{-6}$ (1/K m^3)
°F	K	°C	$\times 6.243 \times 10^{-2}$ = (lb$_m$/ft^3)	$\times 0.5556$ = (1/R)	$\times 2.388 \times 10^{-4}$ = (Btu/lb$_m$ °F)	$\times 0.5777$ = (Btu/h ft °F)	$\times 3.874 \times 10^4$ = (ft^2/h)	$\times 0.6720$ = (lb$_m$/ft s)	$\times 3.874 \times 10^4$ = (ft^2/h)		$\times 1.573 \times 10^{-2}$ = (1/R ft^3)
−279	100	−173	3.992		948	0.00903	0.0239	7.768	1.946	0.815	
−189	150	−123	2.619	6.67	918	0.01367	0.0569	11.49	4.387	0.773	3398
−100	200	−73	1.956	5.00	913	0.01824	0.1021	14.85	7.593	0.745	850.5
−9	250	−23	1.562	4.00	916	0.02259	0.1579	17.87	11.45	0.725	299.2
80	300	27	1.301	3.33	920	0.02676	0.2235	20.63	15.86	0.709	129.8
171	350	77	1.113	2.86	929	0.03070	0.2968	23.16	20.80	0.702	64.8
261	400	127	0.9755	2.50	942	0.03461	0.3768	25.54	26.18	0.695	35.8
351	450	177	0.8682	2.22	957	0.03828	0.4609	27.77	31.99	0.694	21.3
441	500	227	0.7801	2.00	972	0.04173	0.5502	29.91	38.34	0.697	13.3
531	550	277	0.7096	1.82	988	0.04517	0.6441	31.97	45.05	0.700	8.79
621	600	327	0.6504	1.67	1004	0.04832	0.7399	33.92	52.15	0.704	6.02

Source: E. R. G. Eckert and R. M. Drake, *Analysis of Heat and Mass Transfer*, McGraw-Hill, New York, 1972.

TABLE 35 Steam (H_2O) at atmospheric pressure

Temperature, T			Density, ρ (kg/m^3)	Coefficient of Thermal Expansion, $\beta \times 10^3$ (1/K)	Specific Heat, c_p (J/kg K)	Thermal Conductivity, k (W/m K)	Thermal Diffusivity, $\alpha \times 10^4$ (m^2/s)	Absolute Viscosity, $\mu \times 10^6$ (N s/m^2)	Kinematic Viscosity, $\nu \times 10^6$ (m^2/s)	Prandtl Number, Pr	$\frac{g\beta}{\nu^2} \times 10^{-6}$ (1/K m^3)
°F	K	°C	$\times 6.243 \times 10^{-2}$ = (lb$_m$/ft^3)	$\times 0.5556$ = (1/R)	$\times 2.388 \times 10^{-4}$ = (Btu/lb$_m$ °F)	$\times 0.5777$ = (Btu/h ft °F)	$\times 3.874 \times 10^4$ = (ft^2/h)	$\times 0.6720$ = (lb$_m$/ft s)	$\times 3.874 \times 10^4$ = (ft^2/h)		$\times 1.573 \times 10^{-2}$ = (1/R ft^3)
212	373	100	0.5977		2034	0.0249	0.204	12.10	20.2	0.987	
225	380	107	0.5863		2060	0.0246	0.204	12.71	21.6	1.060	
261	400	127	0.5542	2.50	2014	0.0261	0.234	13.44	24.2	1.040	41.86
351	450	177	0.4902	2.22	1980	0.0299	0.307	15.25	31.1	1.010	22.51
441	500	227	0.4405	2.00	1985	0.0339	0.387	17.04	38.6	0.996	13.16
531	550	277	0.4005	1.82	1997	0.0379	0.475	18.84	47.0	0.991	8.08
621	600	327	0.3652	1.67	2026	0.0422	0.573	20.67	56.6	0.986	5.11
711	650	377	0.3380	1.54	2056	0.0464	0.666	22.47	66.4	0.995	3.43
800	700	427	0.3140	1.43	2085	0.0505	0.772	24.26	77.2	1.000	2.35
891	750	477	0.2931	1.33	2119	0.0549	0.883	26.04	88.8	1.005	1.65
981	800	527	0.2739	1.25	2152	0.0592	1.001	27.86	102.0	1.010	1.18
1071	850	577	0.2579	1.18	2186	0.0637	1.130	29.69	115.2	1.019	0.872

Source: E. R. G. Eckert and R. M. Drake, *Analysis of Heat and Mass Transfer*, McGraw-Hill, New York, 1972.

TABLE 36 Methane at atmospheric pressure

Temperature, T			Density, ρ (kg/m^3)	Coefficient of Thermal Expansion, $\beta \times 10^3$ (1/K)	Specific Heat, c_p (J/kg K)	Thermal Conductivity, k (W/m K)	Thermal Diffusivity, $\alpha \times 10^4$ (m^2/s)	Absolute Viscosity, $\mu \times 10^6$ (N s/m^2)	Kinematic Viscosity, $\nu \times 10^6$ (m^2/s)	Prandtl Number, Pr	$\frac{g\beta}{\nu^2} \times 10^{-6}$ (1/K m^3)
°F	K	°C	$\times 6.243 \times 10^{-2}$ = (lb$_m$/ft^3)	$\times 0.5556$ = (1/R)	$\times 2.388 \times 10^{-4}$ = (Btu/lb$_m$ °F)	$\times 0.5777$ = (Btu/h ft °F)	$\times 3.874 \times 10^4$ = (ft^2/h)	$\times 0.6720$ = (lb$_m$/ft s)	$\times 3.874 \times 10^4$ = (ft^2/h)		$\times 1.573 \times 10^{-2}$ = (1/R ft^3)
−112	193	−80	1.014	5.18		0.0207		7.4	7.30		954
−76	213	−60	0.9187	4.69		0.0230		8.1	8.82		592
−40	233	−40	0.8399	4.29		0.0260		8.8	10.48		383
−4	253	−20	0.7735	3.95		0.0278		9.5	12.28		257
32	273	0	0.7168	3.66	2165	0.0302	0.195	10.35	14.43	0.74	174
68	293	20	0.6679	3.41	2222	0.0332	0.224	10.87	16.27	0.73	126
122	323	50	0.6058	3.10	2307	0.0372	0.266	11.80	19.48	0.73	80.1
212	373	100	0.5246	2.68	2448			13.31	25.37		40.8
302	423	150	0.4626	2.36	2628			14.71	31.80		22.9
392	473	200	0.4137	2.11	2807			16.05	38.80		13.8
482	523	250	0.3742	1.91	2991			17.25	46.10		8.8
572	573	300	0.3415	1.75	3175			18.60	54.47		5.8

Source: N. B. Vargaftik, *Tables on the Thermophysical Properties of Liquids and Gases*, 2nd ed., Hemisphere, Washington, DC, 1975.

TABLE 37 Ethane at atmospheric pressure

Temperature, T			Density, ρ (kg/m^3)	Coefficient of Thermal Expansion, $\beta \times 10^3$ (1/K)	Specific Heat, c_p (J/kg K)	Thermal Conductivity, k (W/m K)	Thermal Diffusivity, $\alpha \times 10^4$ (m^2/s)	Absolute Viscosity, $\mu \times 10^6$ (N s/m^2)	Kinematic Viscosity, $\nu \times 10^6$ (m^2/s)	Prandtl Number, Pr	$\frac{g\beta}{\nu^2} \times 10^{-6}$ (1/K m^3)
°F	K	°C	$\times 6.243 \times 10^{-2}$ = (lb$_m$/ft^3)	$\times 0.5556$ = (1/R)	$\times 2.388 \times 10^{-4}$ = (Btu/lb$_m$ °F)	$\times 0.5777$ = (Btu/h ft °F)	$\times 3.874 \times 10^4$ = (ft^2/h)	$\times 0.6720$ = (lb$_m$/ft s)	$\times 3.874 \times 10^4$ = (ft^2/h)		$\times 1.573 \times 10^{-2}$ = (1/R ft^3)
−103	198	−75	1.870	5.05		0.0114		6.52	3.49		4066
32	273	0	1.356	3.66	1647	0.0183	0.0819	8.55	6.31	0.77	901
68	293	20	1.263	3.41	1731	0.0207	0.0947	9.29	7.36	0.78	617
104	313	40	1.183	3.19	1815	0.0235	0.109	9.86	8.33	0.76	451
140	333	60	1.112	3.00	1899	0.0265	0.126	10.50	9.44	0.75	330
176	353	80	1.049	2.83	1983	0.0296	0.142	11.11	10.66	0.75	244
212	373	100	0.992	2.68	2067	0.0328	0.160	11.67	11.76	0.74	190
248	393	120	0.942	2.54	2152			12.30	13.06		146
302	423	150	0.875	2.36	2279			12.78	14.61		108
392	473	200	0.783	2.11	2490			14.09	17.99		63.9
482	523	250	0.708	1.91	2680			15.26	21.55		40.3

Source: N. B. Vargaftik, *Tables on the Thermophysical Properties of Liquids and Gases*, 2nd ed., Hemisphere, Washington, DC, 1975.

TABLE 38 The atmosphere[a]

Altitude, (ft)	Altitude, (m)	Absolute Temperature (R) $\times \frac{5}{9}$ = (K)	Absolute Pressure (lb_f/ft^2) × 47.88 = (N/m^2)	Pressure Ratio	Density (lb/ft^3) × 16.02 = (Kg/m^3)	Density Ratio	Speed of Sound (ft/s) × 0.3048 = (m/s)
0	0	518	2,116	1.00	7.65×10^{-2}	1.00	1,120
5,000	1524	500	1,758	8.32×10^{-1}	6.60×10^{-2}	8.61×10^{-1}	1,100
10,000	3048	483	1,456	6.87×10^{-1}	5.66×10^{-2}	7.38×10^{-1}	1,080
20,000	6096	447	972	4.59×10^{-1}	4.08×10^{-2}	5.33×10^{-1}	1,040
30,000	9144	411	628	2.97×10^{-1}	2.88×10^{-2}	3.76×10^{-1}	997
40,000	12,192	392	392	1.85×10^{-1}	1.88×10^{-2}	2.45×10^{-1}	973
50,000	15,240	392	243	1.15×10^{-1}	1.16×10^{-2}	1.52×10^{-1}	973
60,000	18,288	392	151	7.13×10^{-2}	7.32×10^{-3}	9.45×10^{-2}	973
70,000	21,336	392	94.5	4.47×10^{-2}	4.51×10^{-3}	5.90×10^{-2}	974
80,000	24,384	392	58.8	2.78×10^{-2}	2.80×10^{-3}	3.67×10^{-2}	974
90,000	27,432	392	36.6	1.73×10^{-2}	1.67×10^{-3}	2.28×10^{-2}	974
100,000	30,480	392	22.8	1.08×10^{-3}	1.1×10^{-3}	1.4×10^{-2}	975
150,000	45,720	575	3.2	1.5×10^{-3}	9.7×10^{-4}	1.3×10^{-3}	1,190
200,000	60,960	623	0.73	3.6×10^{-4}	2.2×10^{-5}	2.9×10^{-4}	1,240
300,000	91,440	487	0.017	9.0×10^{-6}	6.9×10^{-7}	9.0×10^{-6}	1,110
400,000	121,920	695	0.0011	5.2×10^{-7}	2.7×10^{-8}	3.5×10^{-7}	1,430
500,000	152,400	910	1.2×10^{-4}	8.5×10^{-8}	3.1×10^{-9}	4.1×10^{-8}	
600,000	182,880	1,130	4.1×10^{-5}	1.9×10^{-8}	5.7×10^{-10}	7.5×10^{-9}	
700,000	213,360	1,350	1.3×10^{-5}	6.2×10^{-9}	1.5×10^{-10}	1.9×10^{-9}	
800,000	243,840	1,570	4.6×10^{-6}	2.2×10^{-9}	4.6×10^{-11}	6.0×10^{-10}	
900,000	274,320	1,800	1.9×10^{-6}	9.0×10^{-10}	1.7×10^{-11}	2.2×10^{-10}	

[a]Sources of atmospheric property data: C. N. Warfield, "Tentative Tables for the Properties of the Upper Atmosphere," *NACATN* 1200, 1947; H. A. Johnson, M. W. Rubsein, F. M. Sauer, E. G. Slack, and L. Fossner, "The Thermal Characteristics of High Speed Aircraft," AAF, AMC, Wright Field, TR 5632, 1947; J. P. Sutton, *Rocket Propulsion Elements*, 2nd ed., McGraw-Hill, New York, 1957.

Miscellaneous Properties and Error Function

TABLE 39 Heat pipe wick pore size and permeability data[a]

Materials and Mesh Size	Capillary Height[b] (cm)	Pore Radius (cm)	Permeability (m^2)	Porosity (%)
Glass fiber	25.4		0.061×10^{-11}	
Monel beads				
30–40	14.6	0.052[c]	4.15×10^{-10}	40
70–80	39.5	0.019[c]	0.78×10^{-10}	40
100–140	64.6	0.013[c]	0.33×10^{-10}	40
140–200	75.0	0.009	0.11×10^{-10}	40
Felt metal				
FM1006	10.0	0.004	1.55×10^{-10}	
FM1205		0.008	2.54×10^{-10}	
Nickel powder				
200 μm	24.6	0.038	0.027×10^{-10}	
500 μm	>40.0	0.004	0.081×10^{-11}	
Nickel fiber				
0.01 mm diameter	>40.0	0.001	0.015×10^{-11}	68.9
Nickel felt		0.017	6.0×10^{-10}	89
Nickel foam		0.023	3.8×10^{-9}	96
Copper foam		0.021	1.9×10^{-9}	91
Copper powder (sintered)	156.8	0.0009	1.74×10^{-12}	52
45–56 μm		0.0009		28.7
100–125 μm		0.0021		30.5
150–200 μm		0.0037		35
Nickel 50	4.8	0.0305	6.635×10^{-10}	62.5
Copper 60	3.0		8.4×10^{-10}	
Nickel				
100 (3.23)		0.0131	1.523×10^{-10}	
120 (3.20)	5.4		6.00×10^{-10}	
120[d] (3.20)	7.9	0.019	3.50×10^{-10}	
2[e] × 120 (3.25)			1.35×10^{-10}	
Nickel				
200	23.4	0.004	0.62×10^{-10}	
2 × 200			0.81×10^{-10}	
Nickel[d]				
2 × 250		0.002		
4[e] × 250		0.002		
325		0.0032		
Phosphorus/bronze		0.0021	0.296×10^{-10}	

[a]Abstracted from P. D. Dunn and D. A. Reay, *Heat Pipes*, 3rd ed., Pergamon, New York, 1982.
[b]Obtained with water, unless stated otherwise.
[c]Particle diameter.
[d]Oxidized.
[e]Denotes number of layers.

TABLE 40 Solar absorptivities (α_s) and total hemispherical emissivities (ε_h) of selected building elements

Material	Color	Surface Treatment/ Surface Condition	Solar Absorptivity (α_s)	Total Hemispherical Thermal Emissivity (ε_h)
Aluminum	matt-silver	as received	0.28 ± 0.02	0.07 ± 0.01
	bright silver	mirror-finish	0.24 ± 0.03	0.04 ± 0.01
Aluminum paint	bright silver	hand-coated	0.35 ± 0.02	0.56 ± 0.01
Anodized aluminum	light green	anodized in 2–4% oxalic acid for 20 min at a current density of 2.20 amp/dm^2 at 5–12 V	0.55 ± 0.02	0.29 ± 0.01
Asbestos	gray	dry surface	0.73 ± 0.02	0.89 ± 0.02
		wet surface	0.92 ± 0.02	0.92 ± 0.02
Austenitic stainless steel	matt silver	unpolished	0.42 ± 0.02	0.23 ± 0.01
AISI 321	silver-gray	mirror-finish	0.38 ± 0.01	0.15 ± 0.01
	light blue	mirror-polished and chemically oxidized for 12 min in 0.6M aqueous solution of chromic and sulfuric acid at 90°C.	0.85 ± 0.01	0.18 ± 0.01
	light blue	thermally oxidized for 10 min at 1043 K under normal atmospheric conditions	0.85 ± 0.03	0.14 ± 0.01
Bricks	bright red	thinned and smoothed; dry surface	0.65 ± 0.02	0.85 ± 0.02
		wet surface	0.88 ± 0.02	0.91 ± 0.02
Cement	light gray	a thin coat dried on a mirror-finish aluminum plate having ε_h of 0.04	0.67 ± 0.02	0.88 ± 0.02
Clay	dark gray	a thin coat dried on a mirror-finish aluminum plate having ε_h of 0.04	0.76 ± 0.02	0.92 ± 0.02
Concrete	light pink	nonreflective smooth surface	0.65 ± 0.02	0.87 ± 0.02
Copper	light red	mirror-finish	0.27 ± 0.03	0.03 ± 0.01
Enamels	white	hand coated on a mirror-finish aluminum plate having ε_h of 0.04	0.28 ± 0.02	0.90 ± 0.01
	black		0.93 ± 0.02	0.90 ± 0.01
	blue		0.68 ± 0.02	0.87 ± 0.01
	red		0.65 ± 0.02	0.87 ± 0.01
	yellow		0.46 ± 0.02	0.88 ± 0.01
	green		0.78 ± 0.02	0.90 ± 0.01
Galvanized iron	silver-gray	bright-finish	0.39 ± 0.03	0.05 ± 0.01
	dark brown	heavily weathered and rusted	0.90 ± 0.02	0.90 ± 0.02
Lacquer	colorless and transparent	hand-coated film on a mirror-finish aluminum plate having ε_h of 0.04	transparent	0.88 ± 0.01
"Makrolon"	colorless and transparent	commercially available plastic	transparent (τ_s = 0.88 ± 0.02)	0.88 ± 0.02

(Continued)

TABLE 40 (*Continued*)

Material	Color	Surface Treatment/ Surface Condition	Solar Absorptivity (α_s)	Total Hemispherical Thermal Emissivity (ε_h)
Marble	slightly off-white	nonreflective	0.40 ± 0.03	0.88 ± 0.02
Mosaic tiles	chocolate	nonreflective	0.82 ± 0.02	0.82 ± 0.02
Paper	white	—	0.27 ± 0.03	0.83 ± 0.03
Plywood	dark brown	as received	0.67 ± 0.03	0.80 ± 0.02
Porcelain tiles	white	reflective glazed surface	0.26 ± 0.03	0.85 ± 0.02
Roofing tiles	bright red	as received; dry surface	0.65 ± 0.02	0.85 ± 0.02
		wet surface	0.88 ± 0.02	0.91 ± 0.02
Sand	off-white	dry	0.52 ± 0.02	0.82 ± 0.03
	dull red	dry	0.73 ± 0.02	0.86 ± 0.03
Steel	bright gray	mirror-finish	0.41 ± 0.03	0.05 ± 0.01
	dark brown	weathered and heavily rusted	0.89 ± 0.02	0.92 ± 0.02
Stone	light pink	nonreflective smooth surface	0.65 ± 0.02	0.87 ± 0.02
"Sun-lite" fiberglass	colorless and transparent	as received from Kalwall, U.S.A.	transparent (τ_s = 0.88 ± 0.02)	0.87 ± 0.02
Tin	silver-bright	mirror-finish	0.30 ± 0.03	0.04 ± 0.01
Varnish	colorless and transparent	hand-coated film on a mirror-finish aluminum plate having ε_h of 0.04	transparent	0.90 ± 0.02
Window glass	colorless and transparent	no treatment given	transparent (τ_s = 0.88 ± 0.02)	0.86 ± 0.02
Whitewash	white	a thick layer of whitewash deposited on a mirror-finish aluminum plate having ε_h of 0.04	0.19 ± 0.02	0.80 ± 0.02
Wood	light brown	planed and thinned	0.59 ± 0.03	0.90 ± 0.02

Source: V. C. Sharma and A. Sharma, "Solar Properties of Some Building Elements," *Energy*, vol. 14, pp. 805–810, 1989.

TABLE 41 Steel pipe dimensions[a]

Nominal Pipe Size (in.)	Outside Diameter (in.)	Schedule No.	Wall Thickness (in.)	Inside Diameter (in.)	Cross-Sectional Area Metal (in.^{2})	Inside Cross-Sectional Area (ft^2)
$\frac{1}{8}$	0.405	40[b]	0.068	0.269	0.072	0.00040
		80[c]	0.095	0.215	0.093	0.00025
$\frac{1}{4}$	0.540	40[b]	0.088	0.364	0.125	0.00072
		80[c]	0.119	0.302	0.157	0.00050
$\frac{3}{8}$	0.675	40[b]	0.091	0.493	0.167	0.00133
		80[c]	0.126	0.423	0.217	0.00098
$\frac{1}{2}$	0.840	40[b]	0.109	0.622	0.250	0.00211
		80[c]	0.147	0.546	0.320	0.00163
		160	0.187	0.466	0.384	0.00118
$\frac{3}{4}$	1.050	40[b]	0.113	0.824	0.333	0.00371
		80[c]	0.154	0.742	0.433	0.00300
		160	0.218	0.614	0.570	0.00206
1	1.315	40[b]	0.133	1.049	0.494	0.00600
		80[c]	0.179	0.957	0.639	0.00499
		160	0.250	0.815	0.837	0.00362
$1\frac{1}{4}$	1.660	40[b]	0.140	1.380	0.699	0.01040
		80[c]	0.191	1.278	0.881	0.00891
		160	0.250	1.160	1.107	0.00734
$1\frac{1}{2}$	1.900	40[b]	0.145	1.610	0.799	0.01414
		80[c]	0.200	1.500	1.068	0.01225
		160	0.281	1.338	1.429	0.00976
2	2.375	40[b]	0.154	2.067	1.075	0.02330
		80[c]	0.218	1.939	1.477	0.02050
		160	0.343	1.689	2.190	0.01556
$2\frac{1}{2}$	2.875	40[b]	0.203	2.469	1.704	0.03322
		80[c]	0.276	2.323	2.254	0.02942
		160	0.375	2.125	2.945	0.02463
3	3.500	40[b]	0.216	3.068	2.228	0.05130
		80[c]	0.300	2.900	3.016	0.04587
		160	0.437	2.626	4.205	0.03761
$3\frac{1}{2}$	4.000	40[b]	0.226	3.548	2.680	0.06870
		80[c]	0.318	3.364	3.678	0.06170
4	4.500	40[b]	0.237	4.026	3.173	0.08840
		80[c]	0.337	3.826	4.407	0.07986
		120	0.437	3.626	5.578	0.07170
		160	0.531	3.438	6.621	0.06447
5	5.563	40[b]	0.258	5.047	4.304	0.1390
		80[c]	0.375	4.813	6.112	0.1263
		120	0.500	4.563	7.953	0.1136
		160	0.625	4.313	9.696	0.1015

(Continued)

TABLE 41 (*Continued*)

Nominal Pipe Size (in.)	Outside Diameter (in.)	Schedule No.	Wall Thickness (in.)	Inside Diameter (in.)	Cross-Sectional Area Metal (in.^{2})	Inside Cross-Sectional Area (ft^2)
6	6.625	40[b]	0.280	6.065	5.584	0.2006
		80[c]	0.432	5.761	8.405	0.1810
		120	0.562	5.501	10.71	0.1650
		160	0.718	5.189	13.32	0.1469
8	8.625	20	0.250	8.125	6.570	0.3601
		30[b]	0.277	8.071	7.260	0.3553
		40[b]	0.322	7.981	8.396	0.3474
		60	0.406	7.813	10.48	0.3329
		80[c]	0.500	7.625	12.76	0.3171
		100	0.593	7.439	14.96	0.3018
		120	0.718	7.189	17.84	0.2819
		140	0.812	7.001	19.93	0.2673
		160	0.906	6.813	21.97	0.2532
10	10.75	20	0.250	10.250	8.24	0.5731
		30[b]	0.307	10.136	10.07	0.5603
		40[b]	0.365	10.020	11.90	0.5475
		60[c]	0.500	9.750	16.10	0.5185
		80	0.593	9.564	18.92	0.4989
		100	0.718	9.314	22.63	0.4732
		120	0.843	9.064	26.24	0.4481
		140	1.000	8.750	30.63	0.4176
		160	1.125	8.500	34.02	0.3941
12	12.75	20	0.250	12.250	9.82	0.8185
		30[b]	0.330	12.090	12.87	0.7972
		40	0.406	11.938	15.77	0.7773
		60	0.562	11.626	21.52	0.7372
		80	0.687	11.376	26.03	0.7058
		100	0.843	11.064	31.53	0.6677
		120	1.000	10.750	36.91	0.6303
		140	1.125	10.500	41.08	0.6013
		160	1.312	10.126	47.14	0.5592
14	14.0	10	0.250	13.500	10.80	0.9940
		20	0.312	13.376	13.42	0.9750
		30	0.375	13.250	16.05	0.9575
		40	0.437	13.126	18.61	0.9397
		60	0.593	12.814	24.98	0.8956
		80	0.750	12.500	31.22	0.8522
		100	0.937	12.126	38.45	0.8020
		120	1.062	11.876	43.17	0.7693
		140	1.250	11.500	50.07	0.7213
		160	1.406	11.188	55.63	0.6827

[a] Based on A.S.A. Standards B36.10.
[b] Designates former "standard" sizes.
[c] Former "extra strong."

TABLE 42 Average properties of tubes

Diameter		Thickness		External			Internal				
External (in.)	Internal (in.)	BWG Gage	Nom Wall (in.)	Circumference (in.)	Surface per Lineal Foot (ft^2)	Lineal Feet of Tube per Square Foot of Surface	Transverse Area ($in.^2$)	Volume or Capacity per Lineal Foot ($in.^3$)	Volume or Capacity per Lineal Foot (ft^3)	Volume or Capacity per Lineal Foot U.S. Gal.	Length of Tube Containing 1 ft^3 (ft)
$\frac{5}{8}$	0.527	18	.049	1.9635	0.1636	6.1115	0.218	2.616	0.0015	0.011	661
	0.495	16	.065	↓	↓	↓	0.193	2.316	0.0013	0.010	746
	0.459	14	.083				0.166	1.992	0.0011	0.009	867
$\frac{3}{4}$	0.652	18	.049	2.3562	0.1963	5.0930	0.334	4.008	0.0023	0.017	431
	0.620	16	.065	↓	↓	↓	0.302	3.624	0.0021	0.016	477
	0.584	14	.083				0.268	3.216	0.0019	0.014	537
	0.560	13	.095				0.246	2.952	0.0017	0.013	585
1	0.902	18	.049	3.1416	0.2618	3.8197	0.639	7.668	0.0044	0.033	225
	0.870	16	.065	↓	↓	↓	0.595	7.140	0.0041	0.031	242
	0.834	14	.083				0.546	6.552	0.0038	0.028	264
	0.810	13	.095				0.515	6.180	0.0036	0.027	280
$1\frac{1}{4}$	1.152	18	.049	3.9270	.3272	3.0558	1.075	12.90	0.0075	0.056	134
	1.120	16	.065	↓	↓	↓	0.985	11.82	0.0068	0.051	146
	1.084	14	.083				0.923	11.08	0.0064	0.048	156
	1.060	13	.095				0.882	10.58	0.0061	0.046	163
	1.032	12	.109				0.836	10.03	0.0058	0.043	172

(Continued)

TABLE 42 (*Continued*)

Diameter		Thickness		External			Internal				
								Volume or Capacity per Lineal Foot			
External (in.)	Internal (in.)	BWG Gage	Nom Wall (in.)	Circumference (in.)	Surface per Lineal Foot (ft^2)	Lineal Feet of Tube per Square Foot of Surface	Transverse Area ($in.^2$)	($in.^3$)	(ft^3)	U.S. Gal.	Length of Tube Containing 1 ft^3 (ft)
	1.402	18	.049	4.7124	.3927	2.5465	1.544	18.53	0.0107	0.080	93
	1.370	16	.065	↓	↓	↓	1.474	17.69	0.0102	0.076	98
$1\frac{1}{2}$	1.334	14	.083	↓	↓	↓	1.398	16.78	0.0097	0.073	103
	1.310	13	.095	↓	↓	↓	1.343	16.12	0.0093	0.070	107
	1.282	12	.109	↓	↓	↓	1.292	15.50	0.0090	0.067	111
	1.620	16	.065	5.4978	.4581	2.1827	2.061	24.73	0.0143	0.107	70
	1.584	14	.083	↓	↓	↓	1.971	23.65	0.0137	0.102	73
$1\frac{3}{4}$	1.560	13	.095	↓	↓	↓	1.911	22.94	0.0133	0.099	75
	1.532	12	.109	↓	↓	↓	1.843	22.12	0.0128	0.096	78
	1.490	11	.120	↓	↓	↓	1.744	20.92	0.0121	0.090	83
	1.870	16	.065	6.2832	.5236	1.9099	2.746	32.96	0.0191	0.143	52
	1.834	14	.083	↓	↓	↓	2.642	31.70	0.0183	0.137	55
2	1.810	13	.095	↓	↓	↓	2.573	30.88	0.0179	0.134	56
	1.782	12	.109	↓	↓	↓	2.489	29.87	0.0173	0.129	58
	1.760	11	.120	↓	↓	↓	2.433	29.20	0.0169	0.126	59

TABLE 43 The error function

x	*erf(x)*	*x*	*erf(x)*	*x*	*erf(x)*
0.00	0.00000	0.76	0.71754	1.52	0.96841
0.02	0.02256	0.78	0.73001	1.54	0.97059
0.04	0.04511	0.80	0.74210	1.56	0.97263
0.06	0.06762	0.82	0.75381	1.58	0.97455
0.08	0.09008	0.84	0.76514	1.60	0.97635
0.10	0.11246	0.86	0.77610	1.62	0.97804
0.12	0.13476	0.88	0.78669	1.64	0.97962
0.14	0.15695	0.90	0.79691	1.66	0.98110
0.16	0.17901	0.92	0.80677	1.68	0.98249
0.18	0.20094	0.94	0.81627	1.70	0.98379
0.20	0.22270	0.96	0.82542	1.72	0.98500
0.22	0.24430	0.98	0.83423	1.74	0.98613
0.24	0.26570	1.00	0.84270	1.76	0.98719
0.26	0.28690	1.02	0.85084	1.78	0.98817
0.28	0.30788	1.04	0.85865	1.80	0.98909
0.30	0.32863	1.06	0.86614	1.82	0.98994
0.32	0.34913	1.08	0.87333	1.84	0.99074
0.34	0.36936	1.10	0.88020	1.86	0.99147
0.36	0.38933	1.12	0.88679	1.88	0.99216
0.38	0.40901	1.14	0.89308	1.90	0.99279
0.40	0.42839	1.16	0.89910	1.92	0.99338
0.42	0.44749	1.18	0.90484	1.94	0.99392
0.44	0.46622	1.20	0.91031	1.96	0.99443
0.46	0.48466	1.22	0.91553	1.98	0.99489
0.48	0.50275	1.24	0.92050	2.00	0.99532
0.50	0.52050	1.26	0.92524	2.10	0.997020
0.52	0.53790	1.28	0.92973	2.20	0.998137
0.54	0.55494	1.30	0.93401	2.30	0.998857
0.56	0.57162	1.32	0.93806	2.40	0.999311
0.58	0.58792	1.34	0.94191	2.50	0.999593
0.60	0.60386	1.36	0.94556	2.60	0.999764
0.62	0.61941	1.38	0.94902	2.70	0.999866
0.64	0.63459	1.40	0.95228	2.80	0.999925
0.66	0.64938	1.42	0.95538	2.90	0.999959
0.68	0.66378	1.44	0.95830	3.00	0.999978
0.70	0.67780	1.46	0.96105	3.20	0.999994
0.72	0.69143	1.48	0.96365	3.40	0.999998
0.74	0.70468	1.50	0.96610	3.60	1.000000

Correlation Equations for the Physical Properties

The source for these tables is C. L. Yaws, *Physical Properties—A Guide to the Physical, Thermodynamic and Transport Property Data of Industrially Important Chemical Compounds*, McGraw-Hill, New York, 1977. A more recent edition of this book (C. L. Yaws, *Chemical Properties Handbook: Physical, Thermodynamic, Environmental, Transport, Safety, and Health Related Properties for Organic and Inorganic Chemicals*, McGraw-Hill, New York, 1999) provides equations with additional terms in the polynomial equations. For engineering calculations, however, the simpler versions in the following tables are sufficient.

TABLE 44 Heat capacities of ideal gases

	$c_p = A + BT + CT^2 + DT^3$, cal/(g-mol K) for T in K[a]					
Compound	**A**	**$B \times 10^3$**	**$C \times 10^6$**	**$D \times 10^9$**	**c_p at 298 K, cal/(g-mol)(K)**	**Range, K**
Carbon dioxide, CO_2	5.14	15.4	−9.94	2.42	8.91	298–1,500
Carbon monoxide, CO	6.92	−0.65	2.80	−1.14	6.94	298–1,500
Helium, He	4.97	—	—	—	4.97	298–1,500
Hydrogen, H_2	6.88	−0.022	0.21	0.13	6.90	298–1,500
Nitrogen, N_2	7.07	−1.32	3.31	−1.26	6.94	298–1,500
Oxygen, O_2	6.22	2.71	−0.37	−0.22	6.99	298–1,500
Water, H_2O	8.10	−0.72	3.63	−1.16	8.18	298–1,500
Methane, CH_4	5.04	9.32	8.87	−5.37	8.53	298–1,500
Ethane, C_2H_6	2.46	36.1	−7.0	−0.46	12.57	298–1,500
Propane, C_3H_8	−0.58	69.9	−32.9	6.54	17.50	298–1,500
Nitrogen dioxide, NO_2	5.53	13.2	−7.96	1.71	8.80	298–1,500
Ammonia, NH_3	6.07	8.23	−0.16	−0.66	8.49	298–1,500

[a] $\dfrac{\text{cal}}{\text{g-mol K}} \times \dfrac{4186}{\mathcal{M}} = \dfrac{J}{\text{kg K}}$ where $\mathcal{M}$ is the molecular weight.

TABLE 45 Viscosities of gases at low pressure

	$\mu_G = A + BT + CT^2$, micropoise for T in K				
Compound	**A**	**$B \times 10^2$**	**$C \times 10^6$**	**μ_G at 25°C, micropoise[a]**	**Range, °C**
Carbon dioxide, CO_2	25.45	45.49	−86.49	153.4	−100 to 1,400
Carbon monoxide, CO	32.28	47.47	−96.48	165.2	−200 to 1,400
Helium, He	54.16	50.14	−89.47	195.7	−160 to 1,200
Hydrogen, H_2	21.87	22.2	−37.51	84.7	−160 to 1,200
Nitrogen, N_2	30.43	49.89	−109.3	169.5	−160 to 1,200
Oxygen, O_2	18.11	66.32	−187.9	199.2	−160 to 1,000
Water, H_2O	−31.89	41.45	−8.272	90.14	0 to 1,000
Methane, CH_4	15.96	34.39	−81.40	111.9	−80 to 1,000
Ethane, C_2H_6	5.576	30.64	−53.07	92.2	−80 to 1,000
Propane, C_3H_8	4.912	27.12	−38.06	82.4	−80 to 1,000
Nitrogen dioxide, NO_2			Equation not applicable		
Ammonia, NH_3	−9.372	38.99	−44.05	103	−200 to 1,200

[a] micropoise $\times 10^{-7}$ = kg/m s

TABLE 46 Thermal conductivities of gases at ~1 atm

	$k_G = A + BT + CT^2 + DT^3$, micro cal/(cm s K) for T in K					
Compound	A	$B \times 10^2$	$C \times 10^4$	$D \times 10^8$	k_G at 25°C micro cal/(s)(cm)(K)[a]	Range, °C
Carbon dioxide, CO_2	−17.23	19.14	0.1308	−2.514	40.3	−90 to 1,400
Carbon monoxide, CO	1.21	21.79	−0.8416	1.958	59.3	−160 to 1,400
Helium, He	88.89	93.04	−1.79	3.09	351.20	−160 to 800
Hydrogen, H_2	19.34	159.74	−9.93	37.29	417.22	−160 to 1,200
Nitrogen, N_2	0.9359	23.44	−1.21	3.591	61.02	−160 to 1,200
Oxygen, O_2	−0.7816	23.8	−0.8939	2.324	62.8	−160 to 1,200
Water, H_2O	17.53	−2.42	4.3	−21.73	42.8	0 to 800
Methane, CH_4	−4.463	20.84	2.815	−8.631	80.4	0 to 1,000
Ethane, C_2H_6	−75.8	52.57	−4.593	39.74	51.1	0 to 750
Propane, C_3H_8	4.438	−1.122	5.198	−20.08	42	0 to 1,000
Nitrogen dioxide, NO_2	−33.52	26.46	−0.755	1.071	38.9	25 to 1,400
Ammonia, NH_3	0.91	12.87	2.93	−8.68	63.03	0 to 1,400

[a] $\frac{\text{micro cal}}{\text{cm s K}} \times 4.186 \times 10^{-4} = \text{W/m K}$

TABLE 47 Heat capacities of saturated liquids

	$c_p = A + BT + CT^2 + DT^3$, cal/g K for T in K					
Compound	A	$B \times 10^3$	$C \times 10^6$	$D \times 10^9$	c_p, cal/(g)(K)[a]	Range, °C
Nitrogen dioxide, NO_2	−1.625	18.99	−61.72	68.77	0.37 @ 21.2°C	−11.2 to 140
Carbon monoxide, CO	0.5645	4.798	−143.7	911.95	0.515 @ −191.5°C	−205 to −150
Carbon dioxide, CO_2	−19.30	254.6	−1,095.5	1,573.3	0.46 @ −30°C	−56.5 to 20
Methanol, CH_3OH	0.8382	−3.231	8.296	−0.1689	0.608 @ 25°C	−97.6 to 220
Ethanol, C_2H_5OH	−0.3499	9.559	−37.86	54.59	0.58 @ 25°C	−114.1 to 180
n-Propanol, C_3H_7OH	−0.2761	8.573	−34.2	49.85	0.57 @ 25°C	−126.2 to 200
n-Butanol, C_4H_9OH	−0.7587	12.97	−46.12	58.59	0.56 @ 25°C	−89.3 to 200
Ammonia, NH_3	−1.923	31.1	−110.9	137.6	1.05 @ −33.43°C	−77.4 to 100
Water, H_2O	0.6741	2.825	−8.371	8.601	1.0 @ 25°C	0 to 350
Hydrogen, H_2	3.79	−329.8	12,170.9	−2,434.8	2.1 @ −252.8°C	−259.4 to −245
Nitrogen, N_2	−1.064	59.47	−768.7	3,357.3	0.49 @ −195.8°C	−209.9 to −160
Oxygen, O_2	−0.4587	32.34	−395.1	1,575.7	0.405 @ −183.0°C	−218.4 to −130
Helium, He	−1.733	1,386.0	−293,133	27,280,000	0.96 @ −268.9°C	−270 to −268.5
Methane, CH_4	1.23	−10.33	72.0	−107.3	0.824 @ −161.5°C	−182.6 to −110
Ethane, C_2H_6	0.1388	8.481	−56.54	126.1	0.583 @ −88.2°C	−183.2 to 20
Propane, C_3H_8	0.3326	2.332	−13.36	30.16	0.532 @ −42.1°C	−187.7 to 80

[a] $\frac{\text{cal}}{\text{g K}} \times 4186 = \frac{J}{\text{kg K}}$

TABLE 48 Viscosities of saturated liquids

$$\log \mu_L = A + \frac{B}{T} + CT + DT^2, \text{ centipoise for } T \text{ in K}$$

Compound	A	B	$C \times 10^2$	$D \times 10^6$	μ_L, centipoise[a]	Range, °C
Nitrogen dioxide, NO_2	−8.431	932.6	2.759	−37.54	0.39 @ 25°C	−11.2 to 158.0
Carbon monoxide, CO	−2.346	105.2	0.4613	−19.64	0.21 @ −200°C	−205.0 to 140.1
Carbon dioxide, CO_2	−1.345	21.22	1.034	−34.05	0.06 @ 25°C	−56.5 to 31.1
Methanol, CH_3OH	−99.73	7,317	46.81	−745.3	0.53 @ 25°C	−97.6 to −40.0
	−17.09	2,096	4.738	−48.93		−40.0 to 239.4
Ethanol, C_2H_5OH	−2.697	700.9	0.2682	−4.917	1.04 @ 25°C	−105.0 to 243.1
n-Propanol, C_3H_7OH	−5.333	1,158	0.8722	−9.699	1.94 @ 25°C	−72.0 to 263.6
n-Butanol, C_4H_9OH	−4.222	1,130	0.4137	−4.328	2.61 @ 25°C	−60.0 to 289.8
Water, H_2O	−10.73	1,828	1.966	−14.66	0.90 @ 25°C	0.0 to 374.2
Hydrogen, H_2	−4.857	25.13	14.09	−2,773	0.016 @ −256.0°C	−259.4 to −240.2
Nitrogen, N_2	−12.14	376.1	12.00	−470.9	0.18 @ −200.0°C	−209.9 to −195.8
Oxygen, O_2	−2.072	93.22	0.6031	−27.21	0.47 @ −210°C	−218.4 to −118.5
Helium, He	4.732	−2.990	−586.0	1,417,000	0.0034 @ −270.0°C	−272.0 to −271.6
	−3.442	1.002	32.22	−35,650		−270.5 to −268.0
Methane, CH_4	−11.67	499.3	8.125	−226.3	0.14 @ −170.0°C	−182.6 to −82.6
Ethane, C_2H_6	−4.444	290.1	1.905	−41.64	0.032 @ 25°C	−183.2 to 32.3
Propane, C_3H_8	−3.372	313.5	1.034	−20.26	0.091 @ 25°C	−187.7 to 96.7

[a]centipoise × 10^{-3} = kg/(m s)

TABLE 49 Thermal conductivities of liquids

	$k_L = A + BT + CT^2$, micro cal/(cm s K) for T in K				
Compound	A	$B \times 10^2$	$C \times 10^4$	k_L, (micro cal)/(s)(cm)(K)[a]	**Range, °C**
Nitrogen dioxide, NO_2	519.74	6.22	−25.73	317 @ 25°C	−11 to 142
Carbon monoxide, CO	475.48	3.31	−214.26	360 @ −200°C	−205 to −145
Carbon dioxide, CO_2	972.06	−201.53	−22.99	184 @ 25°C	−56 to 26
Methanol, CH_3OH	770.13	−114.28	2.79	459.2 @ 25°C	−97.6 to 210.0
Ethanol, C_2H_5OH	628.0	−91.88	5.28	404 @ 25°C	−114.1 to 190
n-Propanol, C_3H_7OH	442.74	−8.04	−5.29	368 @ 25°C	−126.2 to 220
n-Butanol, C_4H_9OH	546.51	−64.42	0.316	361 @ 25°C	−89.3 to 230.0
Water, H_2O	−916.62	1,254.73	−152.12	1,452 @ 25°C	0 to 350
Hydrogen, H_2	−20.41	2,473.70	−5,347.26	268 @ −250°C	−259 to −241
Nitrogen, N_2	627.99	−368.91	−22.57	275 @ −182.5°C	−209 to −152
Oxygen, O_2	583.79	−210.49	−48.31	355 @ −183°C	−218 to −135
Helium, He	−954.21	1.55×10^5	-5.0×10^6	200 @ −271.3°C	−271.3 to −271.0
	98.35	−4,376.85	9.05×10^4	50 @ −270.0°C	−271.0 to −268.3
Methane, CH_4	722.72	−144.42	−76.36	325 @ −120°C	−182.6 to −90.0
Ethane, C_2H_6	699.31	−165.88	−4.87	170 @ 25°C	−183.2 to 20
Propane, C_3H_8	623.51	−126.79	−2.12	234 @ 25°C	−187.7 to 80.0

[a] $\frac{\text{micro cal}}{\text{cm s K}} \times 4.186 \times 10^{-4} = \frac{\text{W}}{\text{m K}}$

TABLE 50 Densities of saturated liquids

$$\rho = AB^{-(1-Tr)^{2/T}},\ \mathrm{g/cm^3},\ Tr = T(\mathrm{K})/(T_c + 273.15)$$

[T(K) = temperature of the liquid in kelvin]

Compound	*A*	*B*	T_c, °C	ρ, g/cm³	Range, °C
Nitrogen dioxide, NO_2	0.5859	0.2830	158.0	1.43 @ 25°C	−11.2 to 158.00
Carbon monoxide, CO	0.2931	0.2706	−140.1	0.79 @ −191.52°C	−205.0 to −140.1
Carbon dioxide, CO_2	0.4576	0.2590	31.1	0.71 @ 25°C	−56.5 to 31.1
Methanol, CH_3OH	0.2928	0.2760	239.4	0.79 @ 25°C	−97.6 to 239.4
Ethanol, C_2H_5OH	0.2903	0.2765	243.1	0.79 @ 25°C	−114.1 to 243.1
n-Propanol, C_3H_7OH	0.2915	0.2758	263.6	0.80 @ 25°C	−126.2 to 263.6
n-Butanol, C_4H_9OH	0.2633	0.2477	289.8	0.80 @ 25°C	−89.3 to 289.8
Ammonia, NH_3	0.2312	0.2471	132.4	0.60 @ 25°C	−77.74 to 132.4
Water, H_2O	0.3471	0.2740	374.2	1.00 @ 25°C	0.0 to 374.2
Hydrogen, H_2	0.0315	0.3473	−240.2	0.07 @ −252.78°C	−259.4 to −240.2
Nitrogen, N_2	0.3026	0.2763	−146.8	0.81 @ −195.81°C	−209.9 to −146.8
Oxygen, O_2	0.4227	0.2797	−118.5	1.14 @ −183.16°C	−218.4 to −118.5
Helium, He	0.0747	0.4406	−268.0	0.12 @ −268.9°C	−271 to −268.0
Methane, CH_4	0.1611	0.2877	−82.6	0.42 @ −161.5°C	−182.6 to −82.6
Ethane, C_2H_6	0.2202	0.3041	32.3	0.33 @ 25°C	−183.2 to 32.3
Propane, C_3H_8	0.2204	0.2753	96.7	0.49 @ 25°C	−187.7 to 96.7

APPENDIX 3

Tridiagonal Matrix Computer Programs

Solution of a Tridiagonal System of Equations

Computer programs that demonstrate a commonly used algorithm for solving a system of equations, which can be written in the form of a tridiagonal matrix, are given below. The derivation of the algorithm is given by S. V. Patankar in *Numerical Heat Transfer and Fluid Flow* (Hemisphere Publishing Corporation, Washington, DC, 1980). These programs consider a sample ten-element matrix and are written for (a) MATLAB and in (b) C++ and (c) FORTRAN programming languages. Of the two languages, C++ is the currently used scientific programming language, and the program in the much older FORTRAN is included because several commercial and open-source codes that are used today were written in this language.

(a) Computer Program for MATLAB

```
% Matlab program that demonstrates the solution of a Tridiagonal
  Matrix with
% a User Defined Function
% TRIDIAG with N = 10

clc;
clear all;

% Declare Variables
N = 10;          % N is the Dimension of the Square Matrix
I = 1:N;         % I is a For Loop Variable
A = [1 0.9 0.8 1.1 .95 .85 1.15 .7 .75 1.2];
% A is the Vector of Diagonal Elements

B = [-0.6 -0.5 -0.4 -0.7 -0.6 -0.4 -0.6 -0.4 -0.8 0];
% B is the vector of the Super-Diagonal Elements

C = [0 -0.3 -0.2 -0.7 -0.5 -0.1 -0.3 -0.2 -0.1 -0.5];
% C is the vector of the Sub-Diagonal Elements

D = [0.1666 0.2022 0.2177 0.5155 0.5906 0.5489 1.075 0.8755 1.4728
    1.6056];
% D is the Right Hand side Vector
```

```
% Call the user Defined Function "tridiag"

tridiag (N, A, B, C, D);

% End of Program.
% User Defined Function "tridiag"

function z = tridiag (N, A, B, C, D)

%User Defined Function "tridiag" solves a TRIDIAGONAL System:

% | A(1)  -B(1)                                          | |T(1)   |   |D(1)   |
% |-C(2)   A(2)  -B(2)                                   | |T(2)   |   |D(2)   |
% |.        .      .      .       .                      | |  .    |   | .     |
% |.        .    -C(i)   A(i)    -B(i)                   | |T(i)   | = |D(i)   |
% |.        .      .      .       .                      | |  .    |   | .     |
% |.        .      .   -C(N-1)  A(N-1)  -B(N-1)          | |T(N-1)|    |D(N-1)|
% |.        .      .      .     -C(N)    A(N)            | |T(N)   |   |D(N)   |

% where N is the size of the system

%% Forming Tridiagonal Matrix for Coefficients A, B and C
I = 1:N;
A1 = diag (A); % A1 is square matrix of order N with Vector A on
     diagonal.

B1 = diag (B, 1);
B1 (:, N+1) = [];
B1 (N+1, :) = []; % B1 is a square Matrix of order N with vector B
              on Super diagonal.

C1 = diag (C, -1);
C1 (1, :) = [];
C1 (:, 1) = []; % C1 is a square Matrix of order N with vector C on
            Sub-Diagonal.

P = A1 + (-B1) + (-C1) ; % Setting Negative Values of the Sub-Diagonal
    and Super-Diagonal Matrix and adding to the Diagonal matrix.
% P is the required Coefficient Tridiagonal Matrix.

T = inv (P)*D' ; % T is the solution matrix which is obtained by
    matrix inversion method.

%% Program Output print commands

fprintf ('I\t  A\t  B\t  C\t  D\t\t  T  \n');
fprintf ('-------------------------------------------------- \n');

Y = [I;A;B;C;D;T'];
fprintf ('%2.0i \t %2.2f \t %2.2f \t %2.2f \t %2.4f \t %2.4f \n', Y);

% End of function "tridiag".
```

Output:

I	A	B	C	D	T
1	1.00	−0.60	0.00	0.1666	0.0999
2	0.90	−0.50	−0.30	0.2022	0.1112
3	0.80	−0.40	−0.20	0.2177	0.1444
4	1.10	−0.70	−0.70	0.5155	0.1999
5	0.95	−0.60	−0.50	0.5906	0.2779
6	0.85	−0.40	−0.10	0.5489	0.3778
7	1.15	−0.60	−0.30	1.0750	0.5001
8	0.70	−0.40	−0.20	0.8755	0.6443
9	0.75	−0.80	−0.10	1.4728	0.8111
10	1.20	0.00	−0.50	1.6056	1.0000

(b) Computer Program in C++

This program first defines and sets up the matrix coefficients, then calls a subroutine “tridiag” to perform the actual matrix inversion, or solution. The tridiag subroutine can be incorporated into any computer simulation program written in C++ that requires the solution of a tridiagonal system of equations.

```
/*C++ program for solving a given Tridiagonal Matrix using the
  Thomas algorithm*/
/*The size of the Tridiagonal Matrix in this example is taken
  to be 10*/
/*Including the necessary header files*/
#include<stdio.h>
#include<iostream>
#include<conio.h>
#include<math.h>
#include<fstream>
using namespace std;

/*Defining a function which takes the diagonal, super-diagonal and
  sub-diagonal elements of the Tridiagonal Matrix along with the right
  hand array elements and the size of the matrix to solve the matrix*/
/*The Tridiagonal Matrix is of the general form*/
/*

 | A(1)   -B(1)                                          | |T(1)   |     |D(1)   |
 |-C(2)    A(2)  -B(2)                                   | |T(2)   |     |D(2)   |
 |  .       .      .       .        .                    | | .     |     | .     |
 |  .       .    -C(i)    A(i)     -B(i)                 | |T(i)   |  =  |D(i)   |
 |  .       .      .       .        .                    | | .     |     | .     |
 |  .       .      .    -C(N-1)   A(N-1)  -B(N-1)        | |T(N-1) |     |D(N-1) |
 |  .       .      .       .      -C(N)     A(N)         | |T(N)   |     |D(N)   |
*/
```

```
/*N is the size of the matrix*/
void tridiag(int m, double W[10], double X[10], double Y[10], double
  Z[10])
{
   /*W, X, Y and Z are the diagonal, super-diagonal, sub-diagonal
and right hand array elements*/
      /*m is the size of the Tridiagonal Matrix*/
      double P[10]={0};
      double Q[10]={0};
      double T[10]={0};
      /*P and Q are the Recursion Variables*/
      /*T is the temperature variable or the Solution Array*/
      /* Calculate the initial values of the Recursion Variables*/
      P[0]=X[0]/W[0];
      Q[0]=Z[0]/W[0];
      /*Calculate the subsequent values of the Recursion Variables*/
      for(int i=1;i<m;i++)
      {
           P[i]=X[i]/(W[i]-(Y[i]*P[i-1]));
           Q[i]=(Z[i]+(Y[i]*Q[i-1]))/(W[i]-(Y[i]*P[i-1]));
      }
      /*Back Calculate for T*/
      T[m-1]=Q[m-1];
      for(int j=m-2;j>>=0;j--)
      {
           T[j]=(P[j]*T[j+1])+Q[j];
      }
      /*Display the Solution Array*/
      for(int i=0;i<m;i++)
       {
     cout<<"\n";
     cout<<T[i];
       }
}
/*End of the Tridiagonal Solving Function*/
int main()
{
      ofstream outdata;
      /*Declare the size of the Tridiagonal Matrix*/
       int n=10;
      /*Declare the Diagonal Elements*/
       double A[] = {1,.9,.8,1.1,.95,.85,1.15,.7,.75,1.2};
      /*Set the negative values of the Super-diagonal Elements*/
       double B[] = {-.6,-.5,-.4,-.7,-.6,-.4,-.6,-.4,-.8,0};
      /*Set the negative values of the Sub-diagonal Elements*/
       double C[] = {0,-.3,-.2,-.7,-.5,-.1,-.3,-.2,-.1,-.5};
      /*Set the right hand side array elements*/
       double D[] =
{.1666,.2022,.2177,.5155,.5906,.5489,1.075,.8755,1.4278,1.6056};
      /*Call the solving subroutine defined*/
      tridiag(n,A,B,C,D);
}
/*End of the Program*/
```

The output of this program with the solution of the matrix is the same as given in the previous MATLAB example.

(c) Computer Program in FORTRAN

As in the previous case, this program first defines and sets up the matrix coefficients, then calls a subroutine TRIDIAG to perform the actual matrix inversion, or solution. Again, the subroutine TRIDIAG can be incorporated into any other computer program that is written in FORTRAN and that requires the solution of a tridiagonal system of equations.

```
C
C    ##### PROGRAM PATANKAR. FOR #####
C
C    AN EXAMPLE FORTRAN PROGRAM THAT DEMONSTRATES
C    THE TRIDIAGONAL MATRIX SOLVING SUBROUTINE
C    TRIDIAG WITH N = 10.

C    DECLARE VARIABLES
     INTEGER I, N
     PARAMETER (N = 10)
     REAL*8 A(N), B(N), C(N), D(N), P(N), Q(N), T(N)

C    I IS A DO LOOP VARIABLE
C    N IS THE DIMENSION OF THE SQUARE MATRIX
C    A IS THE VECTOR OF DIAGONAL ELEMENTS
C    B IS THE VECTOR OF THE SUPER-DIAGONAL ELEMENTS
C    C IS THE VECTOR OF THE SUB-DIAGONAL ELEMENTS
C    D IS THE RIGHT HAND SIDE VECTOR
C    P IS A RECURSION VARIABLE
C    Q IS A RECURSION VARIABLE
C    T IS THE SOLUTION VECTOR
C    SET THE DIAGONAL ARRAY ELEMENTS
     DATA A/ 1, .9, .8, 1.1, .95, .85, 1.15, .7, .75, 1.2/
C    SET THE NEGATIVE VALUES OF THE SUPER-DIAGONAL ARRAY ELEMENTS
     DATA B/ -.6, -.5, -.4, -.7, -.6, -.4, -.6, -.4, -.8, 0/
C    SET THE NEGATIVE VALUES OF THE SUB-DIAGONAL ARRAY ELEMENTS
     DATA C/ 0, -.3, -.2, -.7, -.5, -.1, -.3, -.2, -.1, -.5/

C    SET THE RIGHT HAND SIDE ARRAY ELEMENTS
     DATA D/ .1666, .2022, .2177, .5155, .5906, .5489, 1.075,
     &        .8755, 1.4728, 1.6056/

C    CALL THE SOLVING SUBROUTINE
     CALL TRIDIAG (N, A, B, C, D, P, Q, T)

C    PRINT THE INUT DATA AND RESULTS TO THE SCREEN
     WRITE (6, 100)
     WRITE (6, *)
     DO 20 I=1, N, 1
           WRITE (6, 110) I, A(I), B(I), C(I), D(I), T(I)
20 CONTINUE

100 FORMAT (3X, 'I', 8X, 'A', 7X, 'B', 6X, 'C', 7X, 'D', 9X, 'T')
110 FORMAT (2X, I2, 5X, F5.2, 4X, F3.1, 4X, F3.1, 4X, F7.4, 3X, F7.4)

C    END OF THE EXAMPLE PROGRAM
     END
```

```
C
C    ***** SUBROUTINE TRIDIAG *****
C

C          SUBROUTINE TRIDIAG Solves a Tridiagonal System :
C
C | A(1)  -B(1)                                        | |T(1)  |    |D(1)  |
C |-C(2)   A(2)  -B(2)                                 | |T(2)  |    |D(2)  |
C |  .      .      .      .       .       .           | |  .   |    | .    |
C |         .    -C(i)   A(i)     -B(i)     .           | |T(i)  | =  |D(i)  |
C |  .      .      .      .       .       .           | |  .   |    | .    |
C |                    -C(N-1)  A(N-1)  -B(N-1)         | |T(N-1)|    |D(N-1)|
C |                             -C(N)    A(N)           | |T(N)  |    |D(N)  |
C
C      where N is the size of the system

   SUBROUTINE TRIDIAG (N, A, B, C, D, P, Q, T)

C   DECLARE VARIABLES
    INTEGER N, I
    REAL*8 A(N), B(N), C(N), D(N), P(N), Q(N), T(N)

C   CALCULATE RECURSION VARIABLES
    P(1) = B(1)/A(1)
    Q(1) = D(1)/A(1)
    DO 10 I = 2, N, 1
        P(I) = B(I)/(A(I) -C(I) * P(I-1))
        Q(I) = (D(I)+C(I) * Q(I-1))/(A(I)-C(I) * P(I-1))
10   CONTINUE
C   BACK SUBSTITUTE FOR T(I)
    T(N) = Q(N)
    DO 20 I=N-1, 1, -1
          T(I) = P(I) *T(I+1)+Q(I)
20   CONTINUE

C   END OF SUBROUTINE TRIDIAG
    RETURN
    END
```

PROGRAM OUTPUT

I	A	B	C	D	T
1	1.00	-.6	.0	.1666	.0999
2	.90	-.5	-.3	.2022	.1112
3	.80	-.4	-.2	.2177	.1444
4	1.10	-.7	-.7	.5155	.1999
5	.95	-.6	-.5	.5906	.2779
6	.85	-.4	-.1	.5489	.3778
7	1.15	-.6	-.3	1.0750	.5001
8	.70	-.4	-.2	.8755	.6443
9	.75	-.8	-.1	1.4728	.8111
10	1.20	.0	-.5	1.6056	1.0000

APPENDIX 4

Computer Codes for Heat Transfer

A brief and representative listing of some popular computer codes and software packages that are commercially available is given below with their respective Web site URLs. These software packages are often used to solve different heat transfer problems by both industrial practitioners and academic researchers. These codes are generally intended to solve very complex problems that may involve complicated and unusual geometry as well as various modes of heat transfer, including conduction, convection, radiation, boiling, and condensation. In the cases of boiling and condensation, or two-phase flows, additional modeling of the two-phase interfaces is sometimes required. Note that this listing is neither comprehensive nor an endorsement of any software, and many other codes and software packages may be available commercially.

Name of Code	Web Site URL
ADINA-FSI	http://www.adina.com/index.shtml
ANSWER™	http://www.acricfd.com/
Ansys CFX	http://www.ansys.com/products/fluid-dynamics/cfx/
Autodesk® Algor® Simulation	http://usa.autodesk.com/
CFD2000	http://www.adaptive-research.com/
COMSOL Multiphysics®	http://www.comsol.com/
FLUENT	http://www.fluent.com/
InThermal	http://cae-net.com/v2/?page id=21
MSC Nastran	http://www.mscsoftware.com/Contents/Products/
OpenFOAM®: open source CFD	http://www.openfoam.com/
PHOENICS	http://www.cham.co.uk/
STAR-CD	http://www.cd-adapco.com/

APPENDIX 5

The Heat Transfer Literature

A limitation of any textbook is the depth to which the material can be covered. Textbooks can only provide the background needed to understand principles and prepare the student to handle more complex "real-world" problems.

We have made an effort to present up-to-date information in this book, but before starting to solve a real-life heat transfer problem, one should become familiar with work in the area performed by other experts. A few hours spent in a library can save many hours "reinventing the wheel." In addition to specialized textbooks and handbooks, several periodicals are devoted to heat transfer and provide the most current literature available. Conference proceedings are also a valuable source of information. Although the articles in these sources have been reviewed by specialists in the field before publication, it is important to critically evaluate each paper and not assume that the work is infallible.

The following list includes the most important English-language heat transfer journals, with information on the publisher, and frequency of publication:

Journal of Heat Transfer, published monthly by the American Society of Mechanical Engineers (ASME International).

International Journal of Heat and Mass Transfer, published in 26 issues in one volume each year by Elsevier.

International Journal of Heat and Fluid Flow, published bimonthly by Elsevier.

Numerical Heat Transfer, Part A: Applications, published in two volumes each year, with 12 issues in each volume by Taylor & Francis.

AIChE Journal, published monthly by Wiley InterScience (for the American Institute of Chemical Engineers).

Journal of Fluid Mechanics, published biweekly by Cambridge University Press.

Advances in Heat Transfer, published annually and/or half-yearly by Elsevier (Academic Press).

Advances in Chemical Engineering, published annually and/or half-yearly by Elsevier (Academic Press).

Journal of Enhanced Heat Transfer, published quarterly by Begell House.

Heat and Mass Transfer, published monthly by Springer.

Experimental Thermal and Fluid Science, published in eight issues in one volume each year by Elsevier.

International Journal of Multiphase Flow, published monthly by Elsevier.

International Journal of Transport Phenomena, published quarterly by Old City Publishing (in association with the Pacific Center of Thermal-Fluids Engineering).

Heat Transfer Engineering, published in 14 issues each year by Taylor & Francis.

Heat Transfer—Asian Research, published in eight issues in one volume each year by Wiley InterScience.

INDEX

U

V

W

Z